D1251890

D

Autonomous Mobile Robots: Perception, Mapping, and Navigation

WITHDRAWN

1951-1991
40 YEARS OF SERVICE
IEEE COMPUTER SOCIETY
A member society of the
Institute of Electrical and Electronics Engineers, Inc.

AUTONOMOUS MOBILE ROBOTS

S. Sitharama Iyengar
Alberto Elfes

1951-1991
IEEE Computer Society Press

The Institute of Electrical and
Electronics Engineers, Inc.

Autonomous Mobile Robots: Perception, Mapping, and Navigation

S.S. Iyengar and Alberto Elfes

1951-1991

IEEE Computer Society Press
Los Alamitos, California

Washington • Brussels • Tokyo

IEEE COMPUTER SOCIETY PRESS TUTORIAL

Tennessee Tech Library
Cookeville, TN

Library of Congress Cataloging-in-Publication Data

Autonomous mobile robots / edited by S.S. Iyengar and Alberto Elfes.
p. cm.
Includes bibliographical references.
Contents: v. 1. Perception, mapping, and navigation -- v. 2. Control, planning, and architecture.
ISBN 0-8186-2018-8 (pbk. : v. 1). -- ISBN 0-8186-9018-6 (case : v. 1). -- ISBN 0-8186-2116-8 (pbk. : v. 2). -- ISBN 0-8186-9116-6 (case : v. 2)
1. Mobile robots. I. Iyengar, S. Sitharama. II. Elfes, Alberto. TJ211.415.A87 1991
629.8'92--dc20 CIP 91-13280

Published by the
IEEE Computer Society Press
10662 Los Vaqueros Circle
PO Box 3014
Los Alamitos, CA 90720-1264

© 1991 by the Institute of Electrical and Electronics Engineers, Inc. All rights reserved.

Copyright and Reprint Permissions: Abstracting is permitted with credit to the source. Libraries are permitted to photocopy beyond the limits of US copyright law, for private use of patrons, those articles in this volume that carry a code at the bottom of the first page, provided that the per-copy fee indicated in the code is paid through the Copyright Clearance Center, 27 Congress Street, Salem, MA 01970. Instructors are permitted to photocopy isolated articles, without fee, for non-commercial classroom use. For other copying, reprint, or republication permission, write to IEEE Copyrights Manager, IEEE Service Center, 445 Hoes Lane, PO Box 1331, Piscataway, NJ 08855-1331.

IEEE Computer Society Press Order Number 2018
Library of Congress Number 91-13280
IEEE Catalog Number 91EH0341-8
ISBN 0-8186-6018-X (microfiche)
ISBN 0-8186-9018-6 (case)

Additional copies can be ordered from

IEEE Computer Society Press
Customer Service Center
10662 Los Vaqueros Circle
PO Box 3014
Los Alamitos, CA 90720-1264

IEEE Service Center
445 Hoes Lane
PO Box 1331
Piscataway, NJ 08855-1331

IEEE Computer Society
13, avenue de l'Aquilon
B-1200 Brussels
BELGIUM

IEEE Computer Society
Ooshima Building
2-19-1 Minami-Aoyama
Minato-ku, Tokyo 107
JAPAN

Technical Editor: Krishna Kavi
Production Editor: Robert Werner
Copy editing by Henry Ayling
Printed in the United States of America by McNaughton & Gunn, Inc.

The Institute of Electrical and Electronics Engineers, Inc.

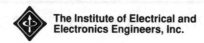

Dedication

To the memory of my brother, S.G. Rajan, who taught me everything I have learned.

S.S. Iyengar

To the memory of my father, Albert Elfes, a true scientist and a Christian humanitarian.

Alberto Elfes

TABLE OF CONTENTS

Chapter 1: Introduction

Building intelligent robotic systems that can reason while functioning in unstructured environments is a challenging task. Due to the absence of dynamic human interaction, such autonomous systems possess unique and exacting computational requirements. Intelligent, self-sufficient inference systems are essential if robots are to operate continuously in unpredictable environments. These computational requirements are even more stringent for autonomous mobile robots designed to operate in hazardous terrains. In addition to an on-board knowledge base large enough to handle various operating environments, these systems also require the ability to monitor and react to dynamic, unexpected events in real time. Moreover, they must guarantee intelligent responses to events while making optimal use of limited, on-board resources.

Despite the importance of "intelligence" to future robotic systems, issues regarding the design of intelligent machines have never been fairly and completely explored in a single publication. It is fallacious to believe that enough work has already been done on intelligent autonomous systems. For these reasons, we believe that a tutorial is needed to promote fruitful research in this important field.

Motivation

For the past few years, interest has grown in potential contributions towards intelligent autonomous machines within universities and federal agencies – including the DoE, DoD, NSF, and NASA – with respect to American productivity and security. Such intelligent machines must be capable of autonomous decision making and action. To minimize risk to humans, while assuring reliable performance in working environments, applications for intelligent autonomous machines typically involve work in hazardous, tedious, or physically uncomfortable situations. The robotics and automation community is being swept by broad, pervasive technological demands. Successful deployment of industrial tele-operated and reprogrammable robots is expanding this technology to more exacting scientific applications in remote, unstructured, and hazardous environments. That expansion is driven by techno-economic objectives (enhanced productivity, profitability, and quality) and by a desire to replace humans in the performance of hazardous, strenuous, or repetitive tasks. The sensing, navigation, and decision-making capabilities required for autonomous machines to operate within unstructured environments (in combat, or in space) present many technical challenges that we can test experimentally. Research areas include sensor-based knowledge acquisition, problem solving under uncertainty, path planning and navigation, multisensor interpretation and integration, robot design and control, advanced parallel computing, and machine learning. While theory and implementation continue to advance rapidly, research is scattered within numerous sources.

Guidelines

As is frequently the case in preparing research-paper collections, we were faced with numerous difficult choices when selecting material for this tutorial. Our problems were increased in the mobile-robot domain because mobile robotics is fundamentally multidisciplinary, incorporating technologies from mechanical and electrical engineering, control theory, computer vision, robotics, estimation theory, artificial intelligence, operations research, programming languages, and physics. Consequently, a vast amount of material is directly or indirectly relevant to mobile-robot research.

Our original goal was to develop a tutorial on mobile robots, rather than an overview of recent research. This led us to prepare a collection broader in scope than a research review would have been. Instead of focusing strictly on recent mobile-robot papers, we decided to include papers discussing relevant approaches and technologies in areas that contribute to mobile robotics. This approach makes our tutorial more useful to more researchers by providing a wider perspective of mobile robotics, and by helping to bridge the gap between practitioners of diverse backgrounds.

The material compiled herein should be useful to several reader categories. In particular, we have endeavored to provide (1) a panoramic overview of mobile robotics to readers unfamiliar with the domain, (2) a starting point in terms of relevant literature for researchers beginning to engage in active mobile robotics research, and (3) a convenient compilation of relevant papers providing a useful reference base for researchers already engaged in mobile robotics applications.

We expect this tutorial to be useful as a ready resource for undergraduate or graduate-level seminars and courses on robotics, computer vision, or mobile robotics. We have selected papers based on the relevance, quality, and significance of material presented, and on their potential or actual applicability to mobile robotics. On the other hand, rather than concentrating on a specific component technology (range-based sensing or road-following, for example), we have provided a cross section of mobile robotics research. Moreover, we have included material from different research groups and schools of thought. This collection includes

- **Papers of historical value,**
- **Papers providing relevant theoretical and practical insights,**
- **Papers describing a specific system that has been tested experimentally,**
- **Papers focusing on important application domains,** and
- **Papers overviewing a supporting technology.**

One strategic decision we made early in the selection and compilation process concerned whether this collection should include any papers on legged mobile robots. Legged locomotion has progressed rapidly in the last 10 years. When we compare legged mobile robots to wheeled mobile robots, we find a substantial overlap in sensing, mapping, world modeling, and task-level planning. On the other hand, we find shared areas for which specific requirements differ; for example, legged mobile robots must carefully sense specific positions at which foot placement will take place, thereby placing special requirements on perception and navigation modules. A more extreme case concerns the study of vehicular kinematics, dynamics, and control, where issues addressed for statically or dynamically stable legged locomotion differ significantly from those arising for current wheeled mobile robots. Overall, we decided not to cover legged mobile robots in depth herein; however, we also decided to provide interested readers with a "bonus" by including two survey papers on legged locomotion.

Outline

This tutorial is organized into four chapters:

(1) **Introduction;**
(2) **Robotic sensing and perception;**
(3) **Mapping and real-world modeling;** and
(4) **Path planning and navigation.**

Our companion volume — *Autonomous Mobile Robots: Control, Planning, and Architecture* — is organized into five chapters:

(1) **Introduction;**
(2) **Mobile robot modeling and control;**
(3) **Task-level planning, decision making, and control;**
(4) **Systems and applications;** and
(5) **An introduction to legged locomotion.**

We classified and arranged papers under those topics chiefly for organizational and didactic reasons. Due to the multidisciplinary nature of mobile robotics, this classification can be somewhat arbitrary. To alleviate this problem, some cross-referencing guidelines follow.

• **Mobile robot sensing and perception** — Topics include sensing modalities for mobile robots, proximity sensors, sonar laser range finders, stereo vision, sensor calibration, sensor interpretation, spatial perception, sensor integration and fusion, robotic perception using sensor- and model-based approaches, and sensor control.

- **Mapping and real-world modeling** — Topics include spatial representation and reasoning models, probabilistic models, geometric models, topological models, mapping strategies, real-world modeling from sensor data, incorporation of prior models, object recognition, and landmark tracking.
- **Path planning and navigation** — Topics include path planning and obstacle avoidance, multi-objective and multiconstraint path planning, road following, terrain-acquisition strategies, motion solving, and position estimation.
- **Mobile robot modeling and control** — Topics include wheeled mobile robot modeling and control, kinematic and dynamic modeling, and control issues.
- **Task-level planning, decision making, and control** — Topics include task-level planning, prior planning for precompiled plans, reactive planning, dynamic planning, integration of prior plans and local information, system-level decision making and control, optimal decision making under uncertainty, architectures for planning and control, blackboard-based and distributed architectures, subsumption, and massively parallel architectures.
- **Systems and applications** — Topics include descriptions of successful systems in various domains, and recent systems currently under development (including planetary exploration rovers), autonomous underwater vehicles, vehicles for hazardous environment, road following, and aids for the handicapped.
- **An Introduction to legged locomotion** — Topics include a survey of issues in legged locomotion.

Further Reading

Collections such as this should always be complemented by current material, given the speed with which a dynamic field like mobile robotics progresses. We draw the reader's attention to the following relevant conferences and journals: Annual conferences include the IEEE International Conference on Robotics and Automation, the IEEE International Conference on Computer Vision, the IEEE International Conference on Computer Vision and Pattern Recognition, the IEEE Conference on Decision and Control, the International Symposium on Robotics Research, the International Joint Conference on Artificial Intelligence, the Conference of the American Association for Artificial Intelligence, and the DARPA Image Understanding Workshops. Relevant journals include the *IEEE Transactions on Robotics and Automation*; the *IEEE Transactions on Pattern Analysis and Machine Intelligence*; the *IEEE Transactions on Automatic Control*; the *IEEE Transactions on Systems, Man, and Cybernetics*; the *Artificial Intelligence Journal*; the *Journal of Computer Vision, Graphics, and Image Processing*; the *Journal of the Optical Society of America*; the *International Journal of Computer Vision*; the *International Journal of Robotics Research*; the *Robotica* journal; and the *Journal of Machine Vision and Applications*.

Chapter 2: Robot Sensing and Perception

Robotic perception has traditionally been perceived of as passive. More recently, researchers have advocated that perception is an active process, and have argued for explicitly planning and controlling a robot system's sensory activities. Although sensor interpretation and world modeling are fundamental for robots to operate in the real world, robotic perception is still one of the weakest components of current robotic systems. One reason for this is the influence of AI-based approaches to computer vision, which to a large extent have (until recently) dominated image and sensor understanding, as can be seen in reviews of relevant literature. These approaches emphasized the use of high-level knowledge-based reasoning and problem-solving techniques and heuristics to handle the fundamental underconstraints of sensor data, and sought to develop task-independent and general-purpose image-understanding systems. Unfortunately, resulting AI-based computer vision systems have frequently shown themselves to be narrow in scope, brittle in real applications, ponderous in size, baroque in structure, and inefficient in processing.

In recent years, some researchers have emphasized the development of appropriate physical models for sensors and their interaction with the real world, and have explored the application of well-established estimation-theoretic models to sensor interpretation problems. This change of emphasis has spurred significant advances in the potential capabilities of robotic perception systems, and is reflected to some extent in the choice of papers presented here.

It is not our purpose to provide a tutorial on computer vision as a whole. Other books and selections of papers providing introductory material, and covering a broader set of issues in computer vision, can be found in the existing literature.[1-5] Rather, this chapter provides a selection of papers in sensor interpretation, sensor integration, and the recovery of world models particularly relevant to a robot's spatial perception and reasoning in real-world scenarios. Topics in this chapter are intimately related to and complemented by material in the next chapter, which covers topics in robotic mapping and real-world representation, and in Chapter 4 of our companion volume.

Some papers in this chapter are tutorial in nature. Jarvis overviews different kinds of range sensors and interpretation techniques used in depth sensing. Dhond and Aggarwal contribute their recent tutorial on stereo depth and shape perception. Luo and Kay describe a taxonomy for research in multisensor integration and fusion.

Traditional approaches to robotic perception have emphasized the use of geometric representations. Ayache and Faugeras discuss in depth the handling of parameter uncertainty in geometric representations. In his paper on occupancy grids, Elfes radically departs from geometric models, developing a stochastic grid-based representation for spatial perception. The use of recursive estimation techniques for depth recovery from stereo, stereo and motion, and laser range finder data is discussed by Matthies et al., Grosso et al., and Hebert et al., respectively (as well as by Elfes, and Ayache and Faugeras) and represents a significant breakthrough in robotic perception. Moravec discusses his milestone effort in developing a real-world stereo mapping and navigation system.

Position or motion estimation is an important problem for autonomous mobile robots. Rigaud and Marcé discuss absolute localization methods for underwater robotic vehicles using acoustic sensors. Calibration is also an important (but often neglected) component of robotic perception. Tsai overviews techniques in camera calibration.

Sensor integration has been a long-standing topic of interest in robotic perception, since it is generally expected that combining information from multiple sensors will potentially lead to better world models. While a body of theoretical work exists in this area, relatively little has been done experimentally in the context of mobile robots. Matthies and Elfes discuss the integration of sonar and stereo for mobile robot mapping and navigation. Hebert et al. discuss the combination of range and intensity data. Kriegman et al. and Thorpe et al. also discuss multisensor systems in Chapter 4 of our companion volume.

References

1. *Robotics Science,* M. Brady, ed., MIT Press, Cambridge, Mass., 1989.
2. T. Kanade, "Computer Vision," in *The Handbook of Artificial Intelligence,* Vol. 3, William Kaufmann, Los Altos, Calif., 1982.
3. *Readings in Computer Vision,* M.A. Fischler and O. Firschein, eds., William Kaufmann, Los Altos, Calif., 1987.
4. D. Ballard and C. Brown, *Computer Vision,* Prentice-Hall, Englewood Cliffs, N.J., 1982.
5. B.K.P. Horn, *Robot Vision,* MIT Press, Cambridge, Mass., 1986.

A Perspective on Range Finding Techniques for Computer Vision

R. A. JARVIS

Reprinted from *IEEE Transactions on Pattern Analysis and Machine Intelligence*, Vol. PAMI-5, No. 2, March 1983, pages 122-139. Copyright © 1983 by The Institute of Electrical and Electronics Engineers, Inc. All rights reserved.

Abstract—**In recent times a great deal of interest has been shown, amongst the computer vision and robotics research community, in the acquisition of range data for supporting scene analysis leading to remote (noncontact) determination of configurations and space filling extents of three-dimensional object assemblages. This paper surveys a variety of approaches to generalized range finding and presents a perspective on their applicability and shortcomings in the context of computer vision studies.**

Index Terms—**Computer vision, range finding.**

I. Introduction

It IS well documented in the psychological literature [1]-[3] that humans use a great variety of vision-based depth cues, combinations from this repertoire often serving as confirming strategies with various weighting factors depending upon the visual circumstances. These cues include texture gradient, size perspective (diminution of size with distance), binocular perspective (inward turning eye muscle feedback and stereo disparity), motion parallax, aerial perspective (haziness, etc. associated with distance), relative upward locations in the visual field, occlusion effects, outline continuity (complete objects look closer), and surface shading variations. In difficult circumstances, various of these cues provide evidence of feasible interpretations; resolving ambiguity when it exists depends not only on the sensory information immediately available but also on previously formed precepts and consequent expectation factors. It is interesting to note that binocular convergence (related to muscle driven inward turning of the eyes) adjusts the scale of the stereo disparity system. Of the human vision depth cues, from a geometric point of view, convergence and disparity are unambiguously related to distance whereas, without movement, perspective depth cues are intrinsically ambiguous as can be easily demonstrated by the Ames room illusions [2]. Thus, convergence and disparity are good candidates for depth estimation in computer vision studies. However, the apparatus available to support depth estimation in machine vision can extend beyond anthropomorphically based cues; these will be detailed later.

Categorization of the various types of range finding techniques is useful in providing a structure for detailed discussion. Direct and active range finding includes ultrasonic and light time-of-flight estimation and triangulation systems. All involve a controlled energy beam and reflected energy detection.

Manuscript received April 10, 1981; revised September 2, 1982.
The author is with the Department of Computer Science, Australian National University, Canberra, Australia.

Most other range finding methods can be generally classified as image based, but further refinement is helpful. Passive (or relatively so) monocular image-based range finding includes texture gradient analysis, photometric methods (surface normals from reflectance), occlusion effects, size constancy, and focusing methods. Contrived lighting approaches include striped and grid lighting, patterned lighting, and Moiré fringe analysis. This leaves methods based either on motion or multiple relative positions of camera or scene; these include reconstruction from multiple views, stereo disparity, retinal flow, and other motion related techniques. Most of these are really geometric triangulation systems of one kind or another. In fact, almost every circumstance that includes at least two views of a scene is potentially exploitable as a range finding mechanism of a triangulation kind; equally true is the drawback that all triangulation methods potentially suffer from the problem of "missing parts" not included in more than one view. In contrast, a coaxial source/detector arrangement for a time-of-flight laser range finder is not subject to this malady.

In general, passive methods have a wider range of applicability since no artificial source of energy is involved and natural outdoor scenes (lit by the sun) fall within this category. However, for complex scenes, the degree of ambiguity in need of resolution is likely to be higher if intrusive methods such as using ultrasonics, laser-beams, and striped lighting are not applicable. On the other hand, ranging methods using structured light sources or time of flight measuring devices, although perhaps contributing little to our understanding of human vision, are certainly acceptable in indoor factory environments where these active approaches are consistent with other instrumentation methodologies.

It will be assumed that the type of range finding required is that which results in a "rangepic," an array of distance estimates from a known point or plane with adjacency constraints corresponding to those of two-dimensional intensity imagery. Not only does this allow a direct correspondence to be made with intensity imagery, but also indicates the amount of information associated with the results and puts some time constraints on range data acquisition if robotic manipulation is to be carried out in a reasonable time span. That a "rangepic" is in fact an "image" is more easily argued in this context; however, a purist might argue that the use of directly acquired range data puts this analysis outside the scope of legitimate computer vision. That both intensity and range data can be remotely acquired to plan manipulation trajectories, is, however, most valuable. The combination of range and intensity data to this end is worthy of careful study because of the high potential of resolving scene interpretation ambiguities in this way, without heavy dependence on semantically derived guidance which might severely restrict the breadth of applicability.

The paper first deals with two contrived lighting ranging methods. Then follows coverage of the monocular passive techniques of relative range from occlusion cues, range from texture gradient, range from focusing, and surface orientation from brightness. The multiple camera position techniques of stereo disparity and camera motion into the scene are then addressed. This is followed by a section on Moiré fringe range contouring, which, although in the category of contrived lighting methods, is presented later in the paper both because more specialized instrumentation is involved and because a photographic intermediate step makes it unsuitable for real-time range analysis. Then follows the artificial beam energy source methods which range to one point at a time: simple triangulation active ranging, ultrasonic and laser time-of-flight active methods, and a streak camera approach which provides an interesting and fast method for measuring light transit times with great accuracy.

Although some of the methods presented may seem to have considerable drawbacks which could throw doubt on why they are included, it was felt that a wide representative spread of ranging approaches should be described to provoke thought and development in this important, relatively new field, bearing in mind that new technologies could change the feasibility status of various methods for particular applications.

II. Contrived Lighting Range Finding

In many laboratory situations where experiments in computer vision are intended to have applications in the component handling, inspection and assembly industry, special lighting effects to both reduce the computational complexity, and improve the reliability of 3-D object analysis is entirely acceptable. That similar methods are not applicable in generalized scene analysis, particularly out of doors, is of no great concern. This class of range finding method involves illuminating the scene with controlled lighting and interpreting the pattern of the projection in terms of the surface geometry of the objects.

A. Striped Lighting (See Fig. 1)

Here the scene is lit by a sheet of light usually produced with a laser beam light source and a cylindrical lens, but projecting a slit using a standard slide projector is also feasible. This sheet of light is scanned across the scene, producing a single light stripe for each position. When the light source is displaced from a viewing TV camera, the camera view of the stripe shows displacements along a stripe which are proportional to depth; a kink indicates a change of plane and a discontinuity a physical gap between surfaces. The proportionality constant between the beam displacement and depth is dependent upon the displacement of the source from the camera so that more accurate depth measurements can be made with larger displacements; however, larger parts of the scene viewable from the source position (lightable) are not seen from the TV camera—the depth of such portions cannot be measured in this way. High source energies permit operation in normal ambient lighting conditions.

It is simpler to analyze one stripe at a time in the image since line identification for tracing purposes becomes difficult with multiple lines, particularly when discontinuities occur. Using specially identifiable adjacent stripes using dashes or color coding could reduce the number of images requiring analysis. In [4] Shirai and Suwa use a simple stripe lighting scheme with a rotating slit projector and TV camera to recognize polyhedral shapes. The fact that the projected lines are equispaced, parallel, and straight on planar surfaces are taken advantage of in reducing the computational complexity; only the endpoints of each straight line segment are found and line grouping procedures are used to identify distinct planar surfaces. The geometry of the relationship between depth and displacement of a point on a stripe is not confounded by non-planar surfaced objects but point by point depth analysis involving one binary image for each stripe is expensive computationally. Furthermore, if a rangepic on a uniform grid is required, some extrapolation and interpolation calculations will be required.

In [5] Agin and Binford describe a laser ranging system capable of moving a sheet of light of controllable orientation across the scene. A helium neon laser emitting 35 mW of red light at a wavelength of 6328 Å was thought adequate for use with the vidicon camera used. Their aim was to derive descriptions of curved objects based on the generalized cylinder model [6]. The stripe line from an image at each of a number of positions of a rotating mirror were analyzed, the orientation of the sheet of light rotated by 90°, and a second mirror scan sweep taken. This process resulted in data for an overlapping grid of laser lines covering the scene. Some 5-10 min were involved in the process. Range data are derivable from these data since the relative position of the laser beam with respect to the camera can be determined by a calibration process. The disadvantages of the apparatus as cited by the authors include slowness of data collection and low-level processing, the monochromaticity of the laser source in restricting the hue of the objects to the scanned, and the hazards present during the use of lasers in an uncontrolled environment.

The work reported by Popplestone *et al.* [7] differs from Shirai's in that it deals with both cylindrical and plane surfaces and from both Shirai's and Agin's work in the development of body models specifically suited to solving juxtaposition problems in automatic assembly. The stripe analysis hardware takes advantage of the facts that 1) a nearly vertical stripe intersects each horizontal scan of a TV camera only once and that 2) since each TV line scan is at constant speed, the time from the start of that line at which the video signal indicates an intersection "blip" is proportional to the distance of the stripe from the left edge of the image. The stripe finder electronics returns the relevant timing data to the controlling minicomputer which is fast enough to collect the data for one complete stripe in one TV frame time (1/50 s). Hardware details are given in [8].

It is evident that without special video signal timing electronics, many frames, each with little relevant information,

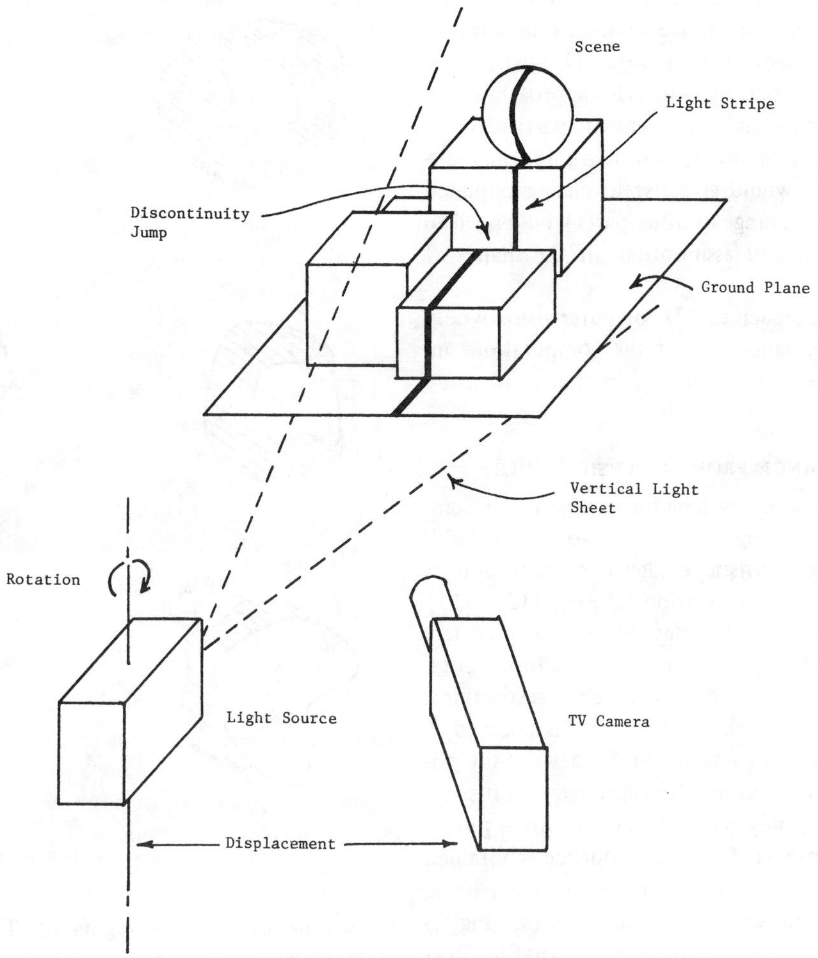

Fig. 1. Striped lighting apparatus.

have to be stored and analyzed, leading to a time-consuming range extraction process. Röcker and Kiessling [9] discuss this problem along with difficulties associated with other ranging techniques. In particular, they point out that, if more is to be extracted from a single image frame by using parallel grid illuminations, the strike identification problem causes a number of restrictions related to the following.

1) The image should contain parts of the supporting plane surface.

2) Shadows cause line interruptions.

3) Top surface lines should be distinguishable from ground plane lines.

4) Scenes with more than one object should not have hidden object planes.

This last point would appear to affect nearly all types of ranging and all image based scene analysis, for that matter.

B. Grid Coding

Will and Pennington [10] describe a method by which the locations and orientations of planar areas of polyhedral solids are extracted through linear frequency domain filtering applied to images of scenes illuminated by a high contrast rectangular grid of lines. Edges are defined through the intersections of the extracted planes. Fast Fourier 2-D transforms are used for rapid computation and segmentation in the Fourier domain used to identify the planes in the grid coded image. Once again the TV camera is offset from the illumination source. The transformation matrix which would restore the individual distorted squares back to their original form in the projected grid contains the local surface normal directions; but here Fourier domain analysis is used instead. The grid coded planar areas map into a 2-D Fourier transform which is a crossed set of harmonically related delta functions in the spatial frequency domain. Separation of the delta functions identifying the planes is equivalent to bandpass filtering; the inverse transform is a reconstruction of the isolated planes in the image domain. Higher level processing can then deduce the object structures in terms of the identified planes. A filter consisting of a 1° sector of a circle with radial direction all-pass response was applied to the 2-D Fourier spectrum to produce an energy versus angle function, the peaks of which are associated with individual planes; a set of filters was designed to straddle each peak; each filter then passed those parts of the 2-D Fourier spectrum corresponding to the individual planes; inverse transforms reconstruct the planes in the image domain.

Towards the end of the paper, an interesting alternative approach is mentioned. If a photographic camera is moved transversely to the dominant direction to the scene, each image point gives rise to a streak of length inversely proportional to range. These streaks are coded by shuttering the camera at constant intervals; each is encoded as an array of points whose period is proportional to range. This periodicity can be detected in the Fourier domain. This approach would seem to be time-consuming and inconvenient, particularly if photographic processing is involved. Furthermore, only distinct points in the scene would give rise to clear modulated streaks on the image plane; range to other parts would need to be calculated on the basis of assumptions of the shapes of surfaces and other clues.

The Fourier analysis approach to 3-D computer vision would be feasible for robot guidance only if the computations involved could be completed quickly enough, perhaps with array processing support.

III. Relative Range from Occlusion Cues

Rosenburg *et al.* [11] have developed a technique for computing the relative relationships of "in-front-of," "behind," and "equidistant" using heuristic evidence of occlusion in monocular color imagery. A relaxation labeling [12], [13] process is used to produce a depth map which is used to test the consistency of a depth graph derived from occlusion cues. The scheme functions without domain-specific restrictions. A segmented image is used as input—each region is assumed to be distinguished from adjacent regions on the basis of the primary features of color and texture. It is assumed that if some of these regions represent only parts of objects, this is purely the result of occlusion effects. Occlusion evidence is obtained by examining clusters of adjacent regions, each cluster being evaluated in terms of six different occlusion cases (see Fig. 2) ordered in decreasing evidence of occlusion. In Fig. 2(a) region A is totally contained within region B. This represents the strongest evidence of occlusion, there being no breaks in the occluded region. In Fig. 2(b) the occluded object is broken somewhat. Region A is surrounded by region B on at least 75 percent of its perimeter and the fragment region C has the same primary feature properties as region B. It is likely that B and C are parts of the same object. Removing region C gives rise to Fig. 2(c) which is ranked below (b). In Fig. 2(d), the reduction of the occluded region B weakens the clue further—A is surrounded by B on at least 50 percent of its perimeter. Again the additional area C with feature properties similar to B strengthens the occlusion hypothesis a little. In Fig. 2(e), the lack of region C weakens the hypothesis. In Fig. 2(f) the extent of the adjacency of region B is reduced, A is surrounded by B on at least 25 percent of its perimeter; once again C can help.

Where there are distinct occlusion related groupings with no occlusion clues between the groupings, relative depth relationships between the groups cannot be determined in this way. Probabilistic relaxation labeling [12] is used both to resolve/reduce possible contradictions in local occlusion data based on the six classes given above and to establish the "equidistant" relationship. Labels are attached to each region indicating hypotheses about depth level of the underlying physical objects. If N depths are used (depth level 1 is foreground and depth level N is background) and λ_α, $\alpha = 1, 2, \cdots, N$ are the labels used, $p_i(\lambda_\alpha)$ is the probability that λ_α is the correct depth level for region i. The total number of regions in occlusion clusters is taken as an upper bound on N. Initially all $p_i(\lambda)$ are set to $1/N$.

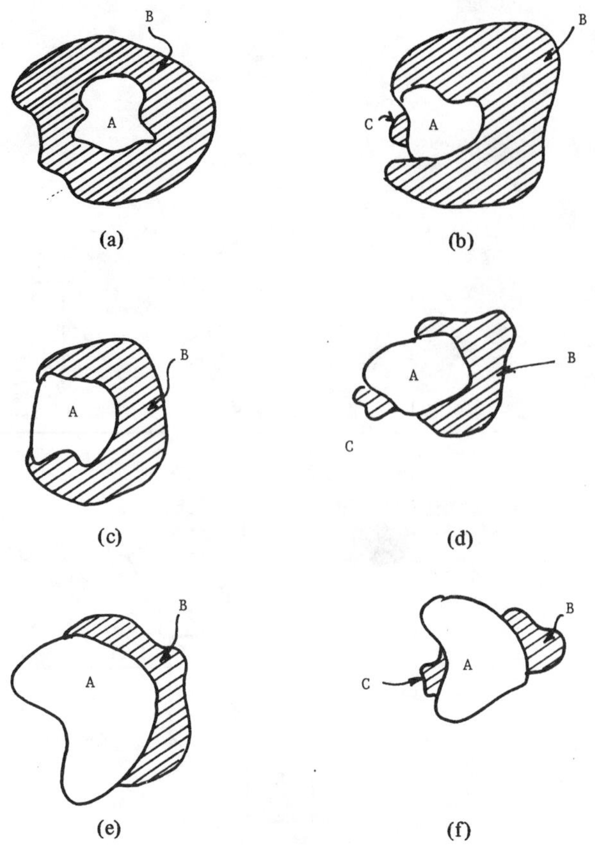

Fig. 2. Occlusion cases.

These probabilities are updated using

$$p_i^{k+1}(\lambda) = \frac{p_i^k(\lambda)\,[1 + q_i^k(\lambda)]^\gamma}{\sum_\lambda p_i^k(\lambda)\,[1 + q_i^k(\lambda)]^\gamma}$$

maintaining

$$\sum_{\alpha=1}^{N} p_i^{k+1}(\lambda_\alpha) = 1 \qquad \text{for each } i$$

where we have the following.

1) $q_i^k(\lambda) = \sum_{j \in \mathrm{NEIGH}(i)} c_{ij} \sum_{\alpha=1}^{N} r_{ij}(\lambda, \lambda'_\alpha)\, p_j^k(\lambda'_\alpha)$.

2) k is the iteration number and $\gamma > 1$ an accelerating factor. NEIGH(i) is the adjacency neighborhood set of i.

3) $r_{ij}(\lambda, \lambda')$ is the compatibility (or consistency) between label λ on region i and label λ' on region j. In this example,

$$r_{ij}(\lambda, \lambda') = -1 \quad \text{if } \lambda \geqslant \lambda'$$
$$= +1 \quad \text{if } \lambda < \lambda'$$

if i is an occluding region and j the corresponding occluded region.

4) c_{ij} is relative certainty of inferences attached to each of the six occlusion cases for region pair i, j. The assignment used

in this paper is

$$c_{ij} = \begin{cases} 0.6 & \text{for Fig. 2(a)} \\ 0.5 & \text{for Fig. 2(b)} \\ 0.4 & \text{for Fig. 2(c)} \\ 0.3 & \text{for Fig. 2(d)} \\ 0.2 & \text{for Fig. 2(e)} \\ 0.1 & \text{for Fig. 2(f).} \end{cases}$$

Convergence usually occurs after applying the update function a medium number of times (in paper, 22 iterations used for example presented). γ can be changed at any iteration if it seems helpful.

The final step is the determination of depth relationships of each region with each other region; this process is not presented here.

The most glaring weakness in this approach is the restraint of correct segmentation of the scene in the first place. In many practical situations one would wish that range information would help resolve segmentation ambiguities, not range information itself to depend upon the lack of these. Total reliance on occlusion cues is also a weakness in that object groups not linked by occlusion relationships cannot be relatively placed in the depth map. However, the approach is most ingeneous and deserving of attention, particularly because of its monocular application.

IV. Depth from Texture Gradient

Texture gradient refers to the increasing fineness of visual texture with depth observed when viewing a 2-D image of a 3-D scene containing approximately uniformly textured planes or objects. Gibson [14] placed considerable emphasis on this effect in terms of human depth cues, particularly when associated with the ground plane. It is intrinsically a monocluar phenomenon particularly useful in range analysis on natural outdoor scenes where uniform visual texture is a dominant manifestation.

Bajcsy and Lieberman [15] have developed a method of measuring texture gradients in the domain of natural outdoor scenes based on Fourier descriptors which are claimed to vary in a manner consistent with surface geometries in three dimensions. The Gibson [14] point of view is supported—surfaces are the primary objects of the visual world; these reflect light, some of which is projected on the retina. The basic surface classes are longitudinal (parallel to line of sight) and frontal (transverse to line of sight); longitudinal surfaces are associated with distance perception.

The Bajcsy and Lieberman texture operator is developed as follows.

The 2-D discrete Fourier transform of a digitized image window considered as a real function $g(x, y)$ of two spatial integer variables x, y is

$$F(n,m) = \frac{1}{p^2} \sum_{x=0}^{p-1} \sum_{y=0}^{p-1} g(x,y) \exp\left[-2\pi i(xn + ym)/p\right]$$

where p is the dimension of the square image window array ($0 \leqslant x, y \leqslant p$, all integers).

The power spectrum is

$$P(n,m) = [F_{R_e}^2(n,m) + F_{I_m}^2(n,m)]^{1/2}$$

where $F_{R_e}(n,m)$ and $F_{I_m}(n,m)$ are the real and complex parts, respectively, of $F(n,m)$. The phase spectrum is

$$\psi(n,m) = \arctan\left[F_{I_m}(n,m)/F_{R_e}(n,m)\right].$$

The power spectrum, being invarient to translation but not rotation preserves the visual pattern directionality of the image; the phase spectrum contains position information in the image and is not relevant for texture cues. Transforming the power spectrum from Cartesian (n, m) to polar (r, ϕ) coordinates aids extraction of directionality information. In each direction ϕ, $P(r, \phi)$ is a one-dimensional function $P_\phi(r)$; for each frequency, r, $P(r, \phi)$ is a one-dimensional function, $P_r(\phi)$. The paper is not clear in distinguishing $P_\phi(r)$ functions for each ϕ from a single function $P(r)$ formed by integrating over ϕ nor in distinguishing between $P_r(\phi)$ for each r and a single function $P(\phi)$ formed by integrating over r.

In [15] in fact, an example is given where $P(\phi)$ is calculated by summing the energy spectrum $P(n, m)$ in equiangular sectors and then finding $P(r)$ for each direction indicated by peaks in $P(\phi)$ by summing the energy spectrum $P(n, m)$ within the associated ϕ sector and within rectangular annuluses of radius r. The nomenclature used is not consistent.

Peaks in $P(\phi)$ indicate texture directionality—a few distinct peaks indicate strong directionality properties, a uniform function indicates nondirectionality of texture. A significant peak is taken as one higher than the mean by 1.5 times the standard deviation. In the nondirectional texture case a uniform $P(r)$ indicates a noisy texture and a peaky $P(r)$ a blob-like texture. A lack of texture (smooth image) gives rise to a large $P(r)$ at $r = 0$ (zero frequency component).

The quantitative components used for the texture descriptor include:

1) average gray value (zero frequency constant),

2) the number and angle values of prime textural directionalities,

3) maximum power corresponding to each prime directionality,

4) the r corresponding to the maximum power in each prime direction, and

5) their corresponding spatial frequencies and wavelength.

Qualitative components include:

1) texture class (bloblike, monodirectional, noisy, homogeneous, etc.),

2) contrast (sharp, medium, weak),

3) brightness (bright, dark),

4) granulation (large, medium, small).

The depths derivable from texture gradient are only relative unless the actual size of the texture element is known as a basis of calibration.

Using the geometric model of Fig. 3, l_i is the texture element size as projected on the image plane, Y_i the center of the window in which l_i is found. The ground plane texture elements are all the same size (say = t). Y_A, Y_B, Y_C are Y value image projections of points A, B, C which are on the ground plane. The relative distances of A, B, and C to the image plane are to be determined.

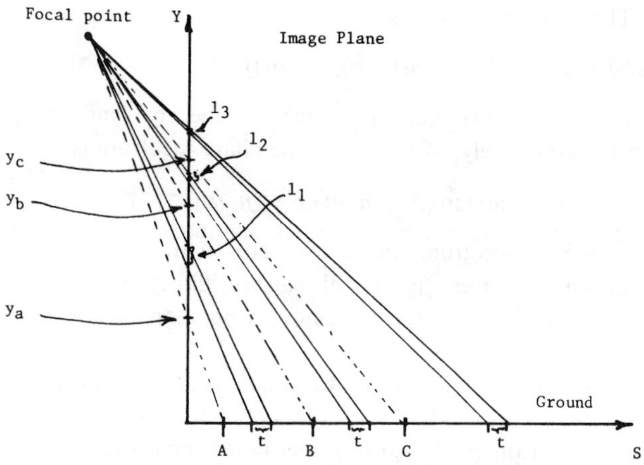

Fig. 3. Texture gradient geometric model.

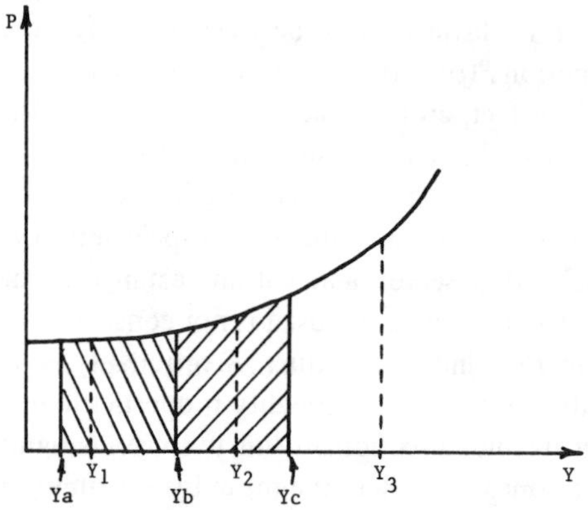

Fig. 4. Projection function.

$P = (1/k_1)(ds/dy)$ is a projection function which indicates how distance on the ground plane is related to distance on the image plane where k_1 is a proportionality constant depending on geometric system parameters. Distance in 3-D space is given by

$$S = k_1 \int P\, dy.$$

The texture descriptors extracted earlier can be used to produce a form of P,

$$P_i(Y_i) = k_2 t / l_i(Y_i)$$

where Y_i identifies the associated windows in the image plane and k_2 is another constant

$$\frac{\text{distance } AC}{\text{distance } AB} = \frac{\int_{YA}^{YC} P^*\, dy}{\int_{YA}^{YB} P^*\, dy}$$

where P^* is a curve fit approximation to P.

This is shown in Fig. 4. The l_i's are the texture wavelengths or elements sizes measured in adjacent windows along the vertical direction of the image. Thus relative distances can be found in terms of texture gradient without needing to know focal length, height above ground, etc.

Window size requires careful consideration since relativity to texture coarseness is needed to give reliable texture features.

This method of relative range measurement would seem to have several drawbacks. Firstly, the regions of the image over which the texture features are to be extracted must be uniformly textured in the 3-D sense. Prior segmentation is required. Secondly, application is restricted to highly textured scenes. Thirdly, computational cost, despite use of fast FFT algorithms, would be high.

Other texture coarseness measures [16], [17] might be substituted in the above method to derive relative depths.

V. Range from Focusing

Knowledge of the focal length and focal plane to image plane distances permits evaluation of focal plane to object distance (range) for components of the 3-D scene in sharp focus. The sharpness of focus needs to be measured on windows on the image over a range of lens positions to determine the range of the corresponding components of the scene. Prior segmentation is not required, but sufficient visual "business" is required to enable sharp focus. Large lens apertures shorten the depth of focus and enhance focus position discrimination of objects at different ranges. Horn [18] provides some technical details on focusing relationships and Jarvis [19] suggests some simple computational formulas for sharpness of focus evaluation. The method is essentially simple and a direct calibration procedure can be used to associate lens positions for various ranges in focus, thus obviating the need for mathematical derivation of this function. The method becomes increasingly inaccurate with range (see Fig. 5).

Jarvis [19] suggests the following focus sharpness measures, chosen on the basis of computational simplicity, effectiveness, consistency and possible direct hardware implementation:

1) entropy (comentropy) $= E = -\sum_x P(x) \ln P(x)$.

2) variance $= V = \frac{1}{N}\sum_{i=1}^{N}(x_i - \bar{x})^2$; $\bar{x} = \frac{1}{N}\sum_{i=1}^{N} x_i$

3) sum modulus difference $= \text{SMD} = \sum_{i=2}^{N} |x_i - x_{i-1}|$.

For each window on the image one need only find the lens position (and thus range) which maximizes these functions; no texture details are required and no absolute values are important.

Instead of considering each of a set of rectangular grid based windows in the image over a set of lens positions, it is also possible, for any one complete image, to identify those portions which are in focus and thus derive the range of the corresponding objects.

Once again, as with the texture gradient approach, the range to visually homogeneous regions of the image cannot be determined directly. However, only one camera position is involved and no special apparatus (except perhaps a computer controlled motor to adjust the lens position) is required in addition to standard digitization equipment if the focus sharpness calculations are to be computed rather than determined with

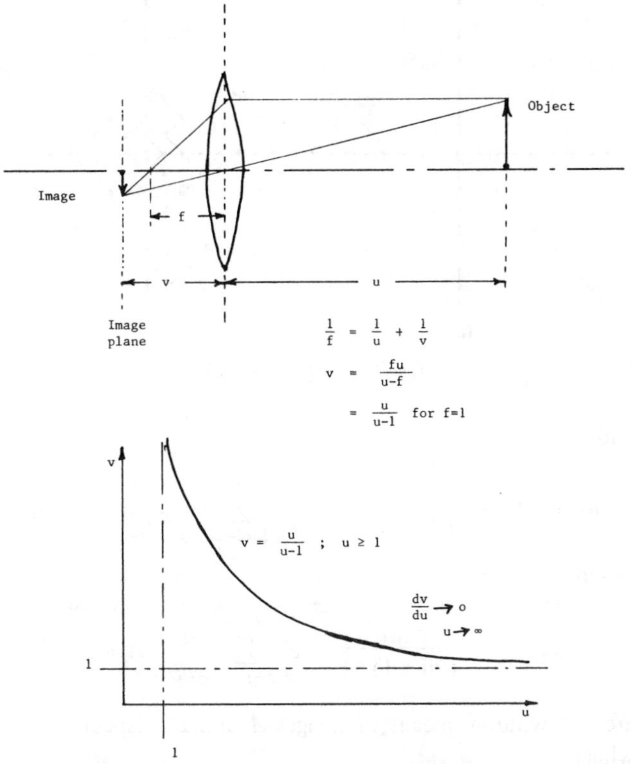

Fig. 5. Depth from focus principle.

specialized analog electronic hardware. In the former case, calculations, although straightforward, could be lengthy.

VI. Surface Orientation from Image Brightness

As early as 1970 Horn [20] raised issues concerning the recovery of surface shape from shading information in the image. In more recent times there has been a renewal of interest in this approach to scene analysis as it represents a generalized analysis strategy independent of domain specific restrictions. Only the recent work by Ikeuchi and Horn [21] will be briefly presented here as representative of this approach to surface orientation recovery. As mentioned earlier, surface orientation permits relative range information over parts of the scene to be calculated by integration; discontinuities frustrate absolute range determinations over the entire scene. Central to the method is the concept of a reflectance map which captures the relationship between image intensity (shading) and surface orientation.

Denoting the slope components of a surface patch as p, q

$$p = \partial z/\partial x$$

$$q = \partial z/\partial y$$

where z is the depth coordinate and the brightness distribution on the surface gradient space, $R(p, q)$, is called the "reflectance map."

The dominant image brightness relationship to reflectance map is

$$E_p(x, y) = R(p, q, p_s, q_s)$$

where

$E_p(x, y)$	is the image irradiance in the image plane at (x, y) (orthographic projection assumed)
(p_s, q_s)	are the direction components of the light source
$R(p, q, p_s, q_s)$	is the reflectance map function defined over surface orientation and light source position.

The central mechanism of surface normal recovery is to calculate the $R(p, q, p_s, q_s)$ map off-line for the surface material of the scene and to determine p, q for each x, y image point from solving a set of $E_p(x, y) = R(p, q, p_s, q_s)$ equations with different light source positions but with camera and scene stationary. The paper deals with surfaces with high specular reflectance properties which the authors suggest are typical of industrial objects, but the general approach is not restricted to these types of surfaces. A considerable amount of off-line computation of the reflectance map function is required but on-line scene analysis is largely by table look-up which is rapid. Surfaces with indirect illuminations from adjacent components cannot be analyzed reliably in this way and the method would be restricted to objects of the one type of surface for which the reflectance map has been calculated off-line. The accuracy with which image intensity can be evaluated would also seem critical to the method. In an industrial hand/eye coordination system these restrictions may not be prohibitive.

VII. Range from Stereo Disparity

Stereo disparity refers to the phenomenon by which the image of a 3-D object point shifts as the camera is moved laterally to the depth coordinate axis. For two such camera positions, simple geometry indicates that the image displacement (disparity) is inversely proportional to depth as measured from the camera (see Fig. 6). The image of a point at an infinite distance along the optical axis can be used as a reference position in both images. Disparity relative to a line through this reference point on the image at right angles to the camera shift direction is inversely proportional to depth. (In the limit, the image of the infinitely distant point does not shift at all.)

It is necessary to establish correspondence or matching of points between the two images to derive the depth relationship. If this correspondence is to be determined from the image data there must be sufficient visual information at the matching points to establish a unique pairing relationship. Two basic problems arise in relation to this requirement. The first arises at parts of the image where uniformity of intensity or color makes matching impossible, the second when the image of some part of the scene appears in only one view of a stereo pair because of occlusion effects (the missing parts problem) or because of the limited field of view captured on the images. The further apart the two camera positions, the potentially more accurate the disparity depth calculation–but the more prevalent the missing parts problem and smaller the field of view overlap.

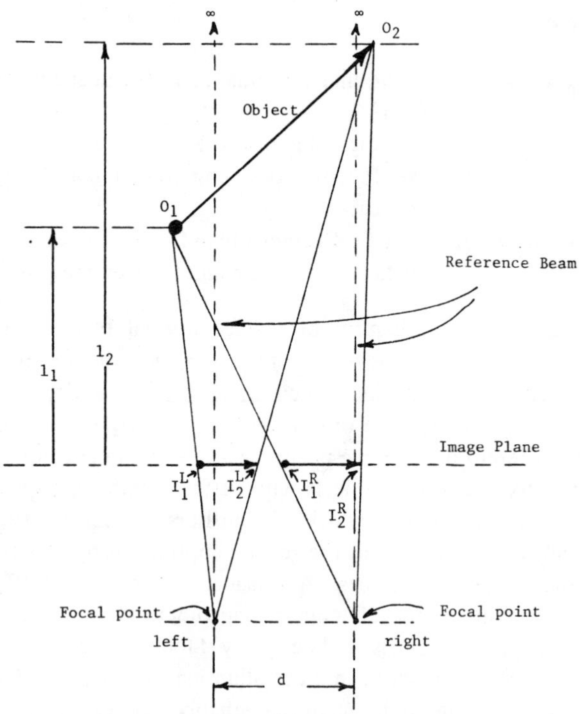

Fig. 6. Stereo disparity geometry.

If the correspondence problem is to be tackled using correlation maximization over windows of the image pair, the correlation shift need only be in the direction of the camera movement axis if this is known.

There is quite a lot of literature on the stereo disparity range finding problem and [22], [24]-[26] are only a sample from this field.

The solution of the correspondence may be effectively sought over the entire overlap areas of the stereo pair of images if the scene is visually "busy" over most of its imaged extent. When large areas of the image are relatively featureless, correlation window matching attempts would prove futile and it is more reliable and expedient to preselect portions for matching on the basis of scene "busyness" measure of a textural or line structure sensitive type. Hopefully, these preselected areas are strategically sufficiently well placed to allow extrapolation and interpolation based depth estimates to be reliably made for unrepresented portions of the scene. Certainly in man-made environments with planar faced solids, this approach is likely to be fruitful. The "busyness" measures that are likely to be suitable include many of the same measures that relate to texture quantification [15]-[17] and also those suggested earlier which have proven useful [19] as focus sharpness measures.

Levine *et al.* [22] apply the following correlation measure over $(2u+1) \times (2v+1)$ windows of the stereo image pair $A(i,j)$, $B(i,j)$; the windows are centered off some point (i,j):

$$\phi(p) = \frac{1}{(2u+1)(2v+1)} \sum_{\xi=i-u}^{i+u} \sum_{\eta=j-v}^{j+v} \cdot \frac{\{[A(\xi,\eta)\, B(\xi,\eta+p)] - \mu_A(i,j)\, \mu_B(i,j+p)\}}{\sigma_A(i,j)\, \sigma_B(i,j+p)}$$

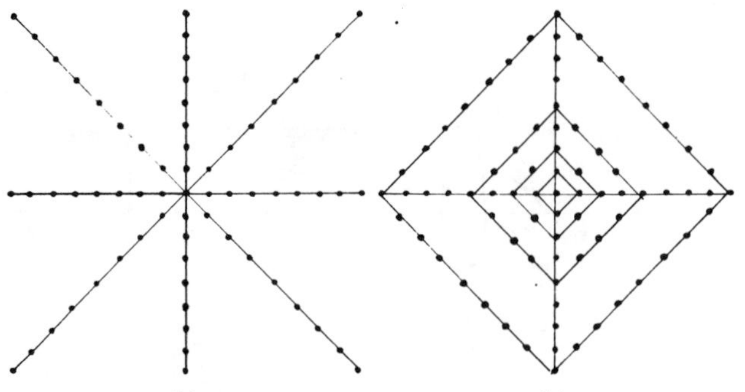

Fig. 7. Correlation masks.

and

$$\mu_B(i,j+p) = \frac{1}{(2u+1)(2v+1)} \sum_{\xi=i-u}^{i+u} \sum_{\eta=j+p-v}^{j+p+v} B(\xi,\eta)$$

where

$$\mu_A(i,j) = \frac{1}{(2u+1)(2v+1)} \sum_{\xi=i-u}^{i=u} \sum_{\eta=j-v}^{j=v} A(\xi,\eta)$$

are the window means of images A and B, respectively, and where

$$\sigma_A^2(i,j) = \frac{1}{(2u+1)(2v+1)} \sum_{\xi=i-u}^{i+u} \sum_{\eta=j-v}^{j+v} \cdot \{[A(\xi,\eta)]^2 - [\mu_A(i,j)]^2\}$$

and

$$\sigma_B^2(i,j+p) = \frac{1}{(2u+1)(2v+1)} \sum_{\xi=i-u}^{i+u} \sum_{\eta=j+p-v}^{j+d+v} \cdot \{[B(\xi,\eta)]^2 - [\mu_B(i,j)]^2\}$$

are the corresponding window variances.

The correlation shift p is along the axis in the direction of camera displacement. The maximum $\phi(p)$ occurs at the sought disparity p^*, linking the corresponding pair of points of the two images, one in each.

The window size can be adjusted and is an important factor in the analysis. If too small a window is used, false matches can occur through random effects, the sample size being inadequate for reliable peak finding over the $\phi(p)$ function. Too large a correlation window leads to poor spatial discrimination of picture cells corresponding to different depths in the scene. In general $\phi(p)$ is a multimodal function which needs to be searched over using global optimization techniques [23], the simplest but most expensive of which is exhaustive search. Hierarchical searches including coarse and fine components can reduce the search cost considerably.

Yakimovsky and Cunningham [24] present details of a camera model and calibration system to support stereo disparity range finding. They continue with the description of stereo correlation algorithms in which specially configured masks are used instead of the more usual rectangular image window (see Fig. 7). This first type of mask [Fig. 7(b)] is a set of concentric diamonds $D_0, D_1, \cdots, D_k$ where $D_0 = T_1$, the reference image point (I_1, J_1) and

$$D_i = \{I, J: |I_1 - I| + |J_1 - J| = d_i\}, \qquad i = 1, \cdots, k.$$

Typical values of d_i are

$$d_1 = 1, \; d_2 = 2, \; d_3 = 4, \; d_4 = 8$$

giving an $N = 61$ point mask.

The second type of mask [Fig. 7(a)] consists of four line segments defined by integer k:

1) horizontal $(I_1 - k, J_1)$ to $(I_1 + k, J_1)$
2) vertical $(I, J_1 - k)$ to $(I, J_1 + k)$
3) 45° $(I_1 - k, J_1 + k)$ to $(I_1 + k, J_1 + k)$
4) -45° $(I_1 - k, J_1 + k)$ to $(I_1 + k, J_1 - k)$.

k is typically =8 producing an $N = 65$ point mask.

In both cases the mask definitions can be interpreted as sequences of N displacement pairs $(\Delta I_i, \Delta J_i)$, $i = 1, \cdots, N$, each point, m_i, on the mask being defined as $(I + \Delta I_i, J + \Delta J_i)$ with the mask centered on (I, J). To find T_2 in the second image of a stereo pair which represents the same scene point P which appears at T_1 in the reference image a mask correlation search is carried out along a line segment S. The search is indexed by k in the X direction.

The image intensity value X_i at each point m_k of the mask centered at T_1 on the reference image is stored in an N element array.

With the mask centered on $T_k = (I_k, J_k)$ in the second image, the intensity value Y_i at each point $m_i = (I_k + \Delta I_i, J_k + \Delta J_i)$ is sampled.

The correlation function

$$C_k = \sum_{i=1}^{N} (X_i - \overline{X})(Y_i - \overline{Y}) \Big/ \left(\sum_{i=1}^{N} (X_i - \overline{X})^2 \sum_{i=1}^{N} (Y_i - \overline{Y})^2 \right)^{1/2}$$

$$-1 < C_k \leqslant 1$$

where

$$\overline{X} = \sum_{i=1}^{N} X_i/N \quad \text{and} \quad \overline{Y} = \sum_{i=1}^{N} Y_i/N$$

is transformed to

$$N \cdot \sum_{i=1}^{N} X_i Y_i - \left(\sum_{i=1}^{N} X_i \sum_{i=1}^{N} Y_i \right) \Big/ \left\{ \left[N \sum_{i=1}^{N} X_i^2 - \left(\sum_{i=1}^{N} X_i \right)^2 \right] \cdot \left[N \sum_{i=1}^{N} Y_i^2 - \left(\sum_{i=1}^{N} Y_i \right)^2 \right] \right\}^{1/2}$$

to minimize computation. Since the X_i's remain fixed while T_1 is fixed during the search for the corresponding point in the second image T_2, the maximization of the following suffices to determine the value of k which maximizes C_k:

$$(D \cdot |D|) \Big/ \left[N \cdot \sum_{i=1}^{N} Y_i^2 - \left(\sum_{i=1}^{N} Y_i \right)^2 \right]$$

where

$$D = \sum_{i=1}^{N} Z_i Y_i - S_k \sum_{i=1}^{N} Y_i$$

$$S_k = \sum_{i=1}^{N} X_i$$

$$Z_i = N X_i \qquad \text{for each } T_k.$$

Since S_k and Z_i are constant for the search over k, the main computational effort is in computing

$$\sum_{i=1}^{N} Y_i, \quad \sum_{i=1}^{N} Y_i^2, \quad \text{and} \quad \sum_{i=1}^{N} Z_i Y_i.$$

Two tests are applied to determine whether a particular mask at T_1 will produce reliable correlation peaks corresponding to a proper match. If the intensity variance of the mask points is less than 0.3 of the camera noise variance, the mask is not considered suitable for correlation matching, there being insufficient visual information present. If the autocorrelation function of T_1 centered mask points and mask point sets of a sequence of shifts along the X direction on the same image shows a sharp drop from 1 away from the reference position, the correlation with the other image proceeds. These measures are consistent with the "busyness" measures suggested earlier in this section.

Moravec [25] used a TV camera on a horizontal rack to gather a sequence of nine images at equal intervals of camera displacement. The high degree of redundancy this sequence provides was exploited to improve the accuracy of range estimation based on disparity. An experiment begins with a camera calibration phase with a visual chart which automatically establishes focal length and image distortion parameters. Then with the cameras pointing at the scene, localized features which can unambiguously be detected from different views are selected. Image regions with high contrasts in orthogonal directions (corners) are ideal. An "interest operator" subroutine attempts to select a scattering of such regions so that each object might be represented a few times. Sums of squares of adjacent pixel intensity difference in each of the directions, horizontal, vertical, left diagonal, and right diagonal are calculated over small square windows and the minimum of these four directional variance measures used as that window's "interest" measure. Points of interest valuable for disparity locally maximal interest points with other images by searching a whole image area or a specified rectangular subimage.

A hierarchical search which begins with a coarse strategy applied to a reduced resolution image and proceeds by refining the search into finer and finer resolution images guided by higher level results is used. The fifth image of the nine camera images sequence is used as the reference both for the interest operator phase and the correlation match phase. The correlator attempts to match selected high interest features from that image with each of the other eight images. Since the camera shift is horizontal, the search is restricted to narrow horizontal bands. After the correlation search, each feature's position in each nine images is known. The 36 image pairings are used for stereo disparity analysis for each feature. The range estimations for each pair are considered to be the means of normal distributions with standard deviations inversely proportional to the relative camera shift, the distribution area being scaled by a confidence measure based on the correlation measures (the product of the two best match correlation figures with respect to the reference image using the value 1 for one factor when the central image is involved amongst the 36 pairs) and by the projection of the feature shift on the X axis. For each feature, the peak of the summed 36 distribution functions provides the overall range estimation with considerable reliability. False matches tend to produce distributions which fail to gain reinforcement from the others. In all, this method of combining the 36 estimations is statistically sound and most intelligent in refining the solution. Note that only eight correlation match searches for each feature of the reference image are involved.

Baker [26] describes a stereo pair range analysis technique based on edge data in the images. The use of edge data fulfills the basic requirement of visual "busyness" (at least in the direction across the edge) for reliable correlation matching, at the same time reducing the computational cost. Camera shift is in the direction of horizontal scan lines and only edges with a vertical component in their slope are used in the correlation process (i.e., edges are associated with sharp differences in an intensity plot along a horizontal scan line). The correspondence problem is attacked one horizontal line at a time using edge correlation procedures for finding the best association of first and second image edges; the information used is strictly local for this phase. At a more global level, edge continuity constraints are used to confirm or reject these edge pairings. This second phase is termed "cooperative continuity enforcement." Inconsistent pairings involving those edges where nearest image space connections (as seen in either image) are with edges other than given by the correlation based link, are removed. Those that remain hopefully provide reliable range data from the corresponding disparity values. This approach is an excellent example of filtering local information through a global constraints function to preserve consistency and thus improve reliability. This method should be particularly useful for colinear edged planar surfaced objects under edge enhancing lighting conditions.

The work by Marr and Poggio [27] has excited a considerable amount of interest among computer vision and psychophysicists alike, particularly as their proposals for solving the stereo disparity correspondence problem by the use of cooperative computational processes contains clues of human neurophysiological function in this same domain. Julesz's [28] findings regarding the human interpretation of random dot stereograms when viewed binocularly to yield patterns separated in depth suggests a mechanism of local processing which inspired Marr and Poggio in their computer vision work on stereo disparity analysis. A cooperative algorithm is one which operates in parallel upon a large array of inputs to yield a global (consistent) organization through local interacting constraints. In this case the constraints are derived from the physical 3-D world of solid objects where

1) a point on a surface has unique position in space at any time instant, and

2) the surfaces of objects are smooth compared to their range from the viewer and matter, divided into objects, is cohesive.

Only identifiable features on surfaces are suitable for matching stereoscopically; lines, edges, shadows, other markings, etc. in the images usually have a physical existence in the 3-D scene.

The above two constraints can be mapped into rules for combining descriptions (including positions) of identifiable features in the left and right images of a stereo pair.

1) *Uniqueness:* Each item (feature) can be assigned at most, only one disparity value.

2) *Continuity:* Disparity varies smoothly almost everywhere, i.e., discontinuities corresponding to depth change occur only relatively infrequently in the image when compared with the total area.

Computational cells for each x, y position in the image pair and for each possible disparity value d, evaluate the state for triples in x, y, d to represent actual disparity match points by using iterative processes with the local neighborhood constraint conditions inhibiting and supporting candidature at each step. The stable states for the cells represent a disparity solution. The computational cost, when the algorithm is processed on a conventional serial machine would probably be large, but specialized array processing would be most effective in reducing this cost.

The form of the iterative equations is

$$C_{xyd}^{(n+1)} = \sigma\left\{\left(\sum_{x'y'd \in S(x,y,d)} C_{x'y'd'}^{(n)}\right) - \xi\left(\sum_{x'y'd' \in O(x,y,d)} C_{x'y'd'}^{(n)}\right) + C_{xyd}^{(0)}\right\}$$

where

1) $C_{xyd}^{(n)}$ is the state of the cell at position (x, y) with disparity d at iteration n;

2) S and O identify supportive neighbors and inhibitive neighbors, respectively, in the vicinity of x, y, d;

3) σ is a sigmoid function ("S" shaped curve) with range $[0, 1]$;

4) ξ is the inhibition constant.

From the paper, it would seem that when σ is a simple threshold function, the process converges for a wide range of parameter values. A number of difficulties are encountered in regarding this process as a theory of human vision stereo. These concern the human tolerance for the defocusing of one image, the movements of the eyes as a stereo pair of images come into fusion and the hysterisis effect by which there is a delay in matching but fusion remains for subsequent separation of the images in the pair beyond the distance for which fusion was initially impossible.

More recently, Marr *et al.* [29]-[31] have proposed an alternative theory of stereo vision computation which has strong links with low-level biological visual mechanisms. It is based on initially extracting edges with mask operators of various sizes convolved over both left and right images and extracting the zero crossings for each. The stereo correspondence problem is then solved by using the disparity matches of the gross line structures for the results of using the large masks to guide the matchings at finer, higher resolution.

Neurophysiological studies carried out on cats and monkeys indicate a lateral inhibition local operator which can be modeled mathematically as the difference of two Gaussian distributions:

$$G_1(x,y) - G_2(x,y) = \frac{1}{2\Pi\sigma_1} e^{2\sigma_1^2/-r^2} - \frac{1}{2\Pi\sigma_2} e^{2\sigma_2^2/-r^2}$$

where r is radius from the center at the point of reference (x, y) and σ_1, σ_2 are standard deviations which correspond to scale factors for excitatory (G_1) and inhibitory (G_2) distributions, respectively. This is approximately equivalent to the application of a Gaussian smoothing operator followed by the application of the Laplacian operator

$$\nabla^2 = \frac{\partial^2}{\partial x^2} + \frac{\partial^2}{\partial y^2}$$

which is a nonoriented second derivative.

Since convolving the image I with G, the Gaussian smoothing mask, and then applying ∇^2 is equivalent to applying, in one pass, the convolution of the product mask $\nabla^2 G$, the computational cost of convolutions over finely quantized images is reduced:

$$\nabla^2(G * I) = (\nabla^2 G) * I$$

where $*$ is the convolution operator.

The shape of the $\nabla^2 G$ mask is given by

$$\left(2 - \frac{r^2}{\sigma^2}\right) e^{-r^2/2\sigma^2}.$$

The various scales of edges are extracted by applying the discretized form of this function, with a geometric progression of sizes (equivalent to adjusting σ) and extracting the zero crossings (which correspond to extrema of the first derivatives of the smoothed images).

Intermediate between the extraction of zero crossings for a sequence of scaled $\nabla^2 G$ mask convolutions and application of stereo correspondence algorithms, a representation called the "raw primal sketch" [32] is created by segmenting collections of zero-crossing contours into sequences of short line segments and evaluating, for each, its position, orientation, length, and rate at which $\nabla^2 G$ changes across the segment. This representation aids the left/right image matching process which, as mentioned earlier, is applied at the crudest scale level first, these results then being used to guide finer matches at higher resolution.

It is given in [33] that the first implementation of the Marr-Hildreth theory took in the order of 3 h to compute the coarse level zero crossings of a 512 X 512 pixel image and a prototype hardware implementation some 30 min. In [34] is a report of a hardware implementation which can complete the zero crossings in under 0.25 s. Note, however, that the smallest operator used on something like a 512 X 512 pixel image (N^2) is 35 X 35 pixels (M^2) and a convolution pass involves some M^2N^2 multiplications and slightly fewer additions.

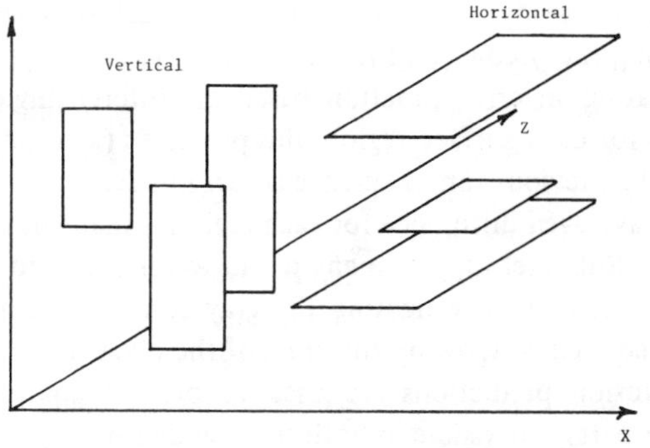

Fig. 8. Horizontal and vertical surface assumption.

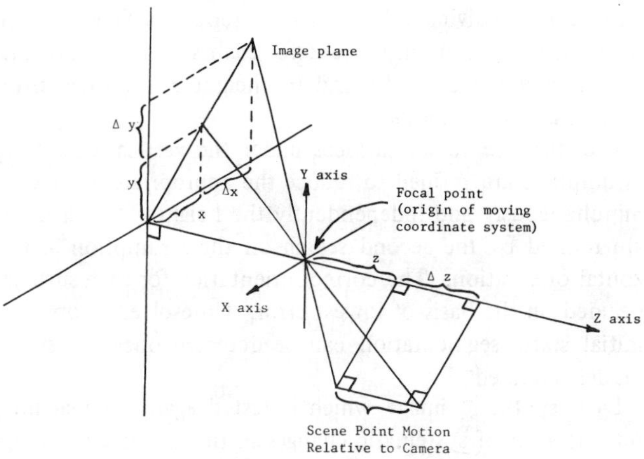

Fig. 9. Relationship between distance in the scene and image displacement.

VIII. Range from Camera Motion

In this case camera motion is not restricted to a limited lateral displacement as for stereo disparity evaluation. Two approaches in this class are represented by Williams [35] and Prazdny [36].

Williams makes the simplifying assumption that all surfaces are planar and orientated in one of only two directions, vertical and horizontal (see Fig. 8). The relationship between the camera relative movement of a point in the scene and the corresponding displacement in the image is illustrated in Fig. 9, where from similar triangles

$$\Delta x = \frac{x\Delta z}{z} \quad \text{and} \quad \Delta y = \frac{y\Delta z}{z}$$

where z is the distance to the scene point at time t_1, Δz is the camera movement since t_0, (x, y) are the coordinates of the corresponding image point at t_0, and $(\Delta x, \Delta y)$ are its displacement components at t_1. The relative distance between camera and scene has diminished between t_0 and t_1. All points in the image move radially outward from a point on the image plane called the focus of expansion. It is assumed that the position of this point is known and that there is no movement in the scene itself. An initial static segmentation is used in defining the extent of the planar regions.

A simple model of planar surfaces (either vertical or horizontal), used to express a 3-D scene interpretation, is used to predict image dynamics which are tested against image data by applying the prediction (in reverse time) to an image at t_1 and calculating an error function based on differencing (pixel by pixel) for each surface region, this predicted (synthetic) image with the previous time sequence actual image at t_0.

The averaged difference for each region is that surface's error value. Subpixel displacement resolution is achieved by interpolation based on weighting the gray levels of each pixel in the predicted window by the areas of the pixel cells involved.

Occlusion predictions are used to prevent companions of region parts not visible in both time sequence images of a pair. A search process to reduce the error measure refines the scene interpretation model. The search is split into two independent parts, each involving only one parameter per surface. The first search is to find the distance Z for each surface assuming it is vertical and the other to find the height Y for each surface assuming it is horizontal.

The distance to all surfaces under the vertical orientation assumption are refined to reduce the appropriate error value. Simultaneously and independently the heights of each surface are refined by the second search on the assumption of horizontal orientation. The correct orientation for each surface is decided on the basis of lowest error. Unresolved errors in the initial static segmentation can be detected once the surface model is refined.

Each synthetic image which is tested against a real image reflects a set of systematic changes in the distance and height of every surface—these are simply increments and decrements of Z and Y for each hypothesized surface. The global minimum error for each surface is sought using a hill climbing algorithm (see [23] for survey on global search methods) with fractional perturbations on the best values of Zs and Ys found so far. The fractional perturbations are diminished as extrema are approached and a number of simple stopping criteria are applied for each surface independently.

The overall approach is a conventional optimization strategy applied independently for each surface. Since the error evaluations are based on average pixel value differences over segmentation regions for each of the Z and Y values for the corresponding surface the process is computationally expensive, particularly if a large number of surfaces are involved. Again, accuracy would diminish for distant objects as corresponding image displacements with camera motion would be small. Overall, this method represents an interesting approach worthy of further investigation but would seem to be both tedious and not particularly reliable, especially since the initial static segmentation is carried out without the aid of range information. Also, the sharpness with which a synthetic to real image segment segment match can be achieved would depend upon the visual "busyness" in those regions. Furthermore, hill climbing techniques are intrinsically only able to find local extrema which are obviously not guaranteed to be the global ones sought if multimodal search performance index functions are involved.

In [35] Prazdny presents a rather elegant method of recovering instantaneous egomotion (observer motion) parameters and a surface normal map (from which relative range information can be derived) starting with optical (or retinal) flow data in the form of the instantaneous positional velocity field (planar retina based) which is regarded as being provided

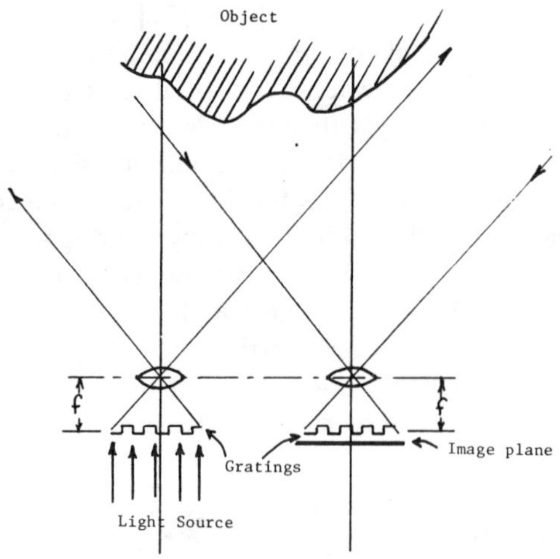

Fig. 10. Moiré fringe range contour apparatus.

by some procedure not yet defined. The vector geometry is complex but elegant and the method involves an iterative solution of a set of three third degree equations in three unknowns for retinal point sets. The author admits that the greatest weakness in the approach is the assumption concerning the provision of the instantaneous positional velocity field. Apart from the assumption concerning the provision of this field defined on the observer's retinal plane, the only requirements are the smoothness of the observer trajectory and the rigidity of the objects in the scene. However, absolute distance to an object is not recoverable and must be seen as another weakness of the method. The computational complexity would also be considerable.

IX. Moiré Fringe Range Contours

A Moiré fringe interference pattern formed by illuminating a scene with shadow patterns through an equispaced optical grating and viewing the scene through an identical grating in a camera displaced laterally from the light projection system (see Fig. 10) represents contours of equal range, but the sign information indicating increasing or decreasing range between adjacent contour lines is missing. Completing two experiments with a known movement of the scene objects between observations or using a phase shifted second grating does allow sign recovery, but contour correspondence problems make range recovery difficult.

Idesawa *et al.* [37], [38] describe an ingeneous method whereby automatic range recovery is possible by modifying the standard Moiré fringe method through use of a high spatial resolution image tube. In this method, the second grating is replaced by a "virtual grating" formed by a set of equispaced scan lines of the image tube system. Sampling along these lines is equivalent to the superposition associated with the standard configuration of Fig. 10. Suppression of unwanted lines not associated with the intensity peaks and valleys which form the range contours is carried out to clarify the contour patterns.

Contour lines for different range levels can be produced simply by changing the phase or the pitch (spacing) of the scanning lines, while using just one grating shadow lit image.

The required contour change sign information is recoverable in this process (using only phase shift will suffice). The image tube spatial accuracy and reproducibility was an order of

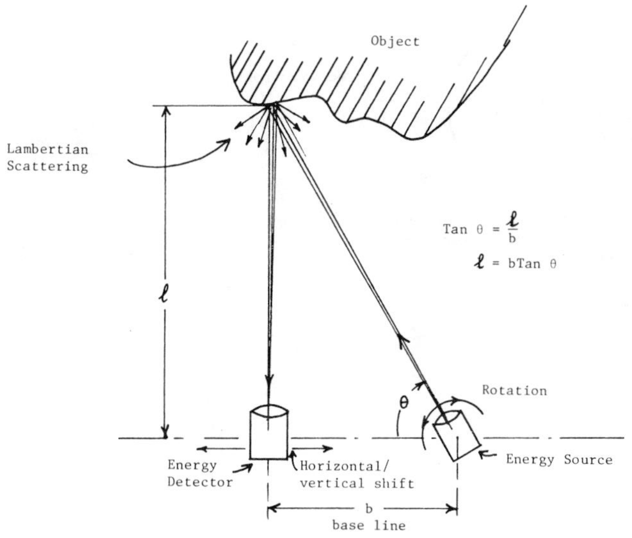

Fig. 11. Simple triangulation range finding geometry.

magnitude higher than for an ordinary cathode ray tube. Unfortunately, the experiment required the photographic recording of the grating lit image and this was scanned in a flying spot scanner mode using the high resolution image tube as the spot light source. If high resolution solid-state or vidicon TV cameras could be used directly it would be a great advantage for real time range analysis of scenes which may be required to be robotically navigated through or manipulated. The basic concept is certainly worth exploring further as the potentials are high. However, once again, as for many of the ranging methods presented earlier, although relative ranges over contiguous surfaces can be measured this way, the absolute range to a partially occluded surface cannot be recovered if there is no range contour continuum to that surface.

X. Simple Triangulation Range Finder

Perhaps the most obvious method of absolute range finding is to use simple one spot at a time triangulation. In a sense, this is a one-dimensional version of stripe light ranging and no image analysis is required. The image of a small portion of the scene is focused upon a light detector. A narrow beam of light from a source laterally displaced from the detector is swept over the scene. The known directions associated with source and detector orientation at the instant the detector "sees" the light spot on the scene are sufficient to recover range if the displacement between the source and detector is known. It is sensible, of course, to sweep the light beam only in the plane defined by the line from the scene to the detector and the line from the light source to the detector. If the detecting system is made to "look" at a raster sequence of scene points, sweeping the source beam in the suitable plane for each position and recording the relevant angles when a "strike" is detected, a reliable rangepic can be easily constructed (see Fig. 11). Sometimes no strike will be detected because of occlusion or surface absorbance. The larger the base line distance between detector and source, the more accurate the ranging but more prevalent the "missing parts" problem caused by directional occlusion. Also, closer ranges can be more accurately measured.

Fig. 12(b) shows a 64 X 64 rangepic of the scene of Fig. 12(a) obtained by using the infrared range detection components from an inexpensive Cannon AF 35 mm camera and mounting them on the pen carriage of an *XY* plotter, which was driven in a raster sequence. The experimental setup is shown in Fig. 12(c). The scan time was in the vicinity of 50 min but there is no inherent reason why the whole process could not be speeded up by an order of magnitude or two with the design of suitable apparatus. Range accuracy for the above example was not great but this could easily be improved upon also. This range scanner is also discussed in [39]. The use of an infrared source permitted range finding in normal lighting conditions (or in the dark).

(a)

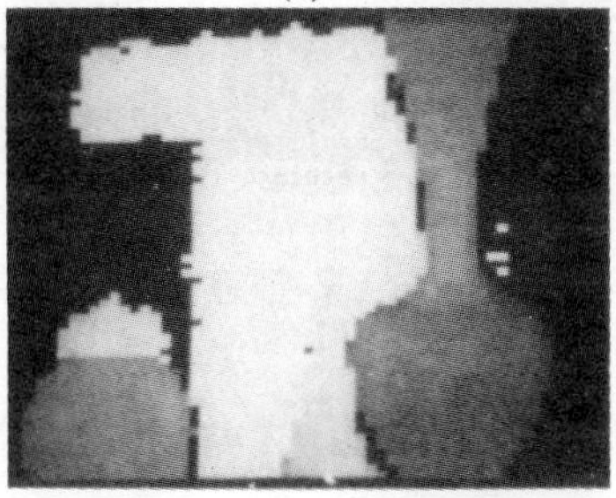

(b)

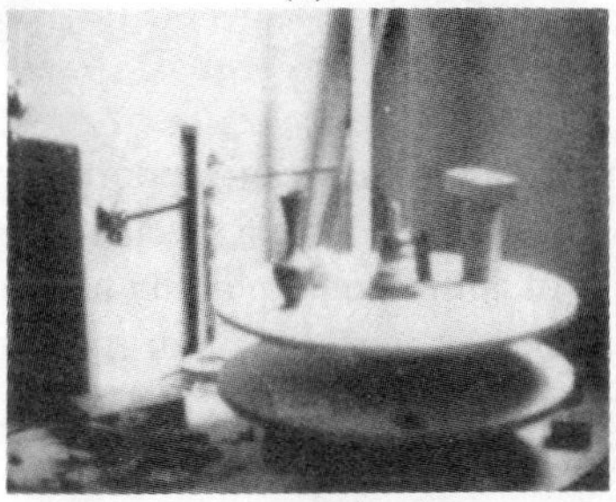

(c)

Fig. 12. (a) Color image. (b) Pseudocolor rangepic of scene depicted in (a). (c) Infrared range scanner experimental setup.

XI. Time-of-Flight Range Finders

A distinct turning away from triangulation based range finding with its inherent "missing parts" problems and diminishing accuracy with range is exemplified in the existence of time-of-flight ranging apparatus where energy source and detection windows can be coaxial and range accuracy maintained over depth up to the point where reliable signal detection is no longer possible. The main two representatives in this category are ultrasonic range finders and laser range finders, the speed of sound and the speed of light, respectively, being the most relevant parameters. No image analysis is involved, nor are assumptions concerning the planar properties or otherwise of the objects in the scene relevant. Furthermore, absolute range is directly available and rangepic registration with imagery easily achieved. Since the range measurements are image independent, they are a legitimate source of com-

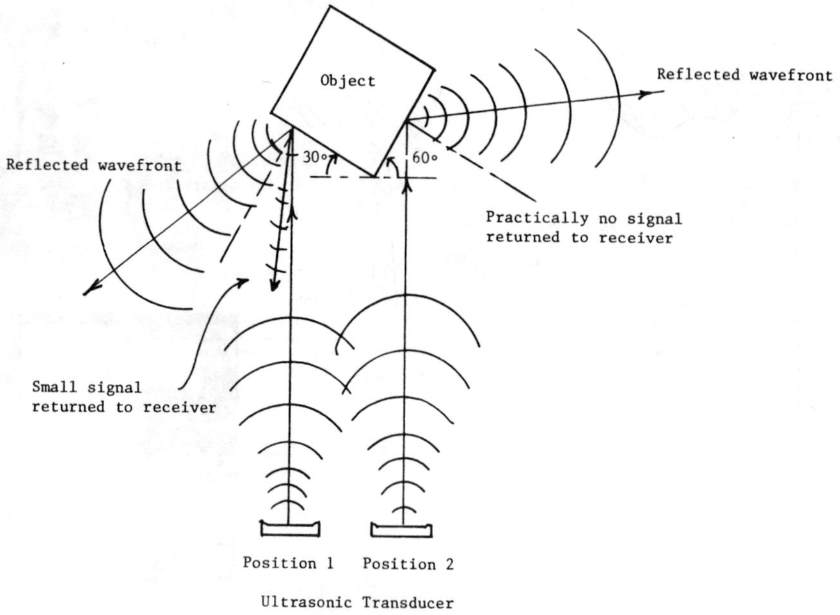

Fig. 13. Ultrasonic signal reflection.

plementary information that can support reliable scene segmentation, unrestricted by domain specificity.

A whole range of anthropomorphically phrased questions are entirely sidestepped as direct time-of-flight ranging has no analog in human depth perception, although, of course, it is relevant to bat navigation.

A. Ultrasonic Range Finding

Polaroid makes available an ultrasonic ranging system kit based on the transducer and electronics of their ultrasonic range finder cameras. This kit, which includes a test board and range read-out display and a fairly comprehensive manual, is ideal for exploring the advantages and disadvantages of ultrasonic rangefinding for any particular application.

A 1 ms ultrasonic "chirp" consisting of 56 pulses at four frequencies, 60, 57, 53, and 50 kHz, is transmitted by a simple electrostatic transducer of about $1\frac{1}{2}$ in diameter and exhibiting a beam pattern with a major forward lobe of about 30° solid angle. The signal reflected off an object within range is detected by the same transducer and processed by an amplifier whose gain and bandwidth are adjusted during delay between "chirp" and "bounce back" so as to improve the signal noise ratio and the reliability of signal detection. The mixed frequency "chirp" is used to lower the probability of signal cancellation for certain target topographies. Ranges from 0.9 to 35 ft are recovered with an accuracy of about 1 in. Two basic problems are encountered in using such a device in an attempt to derive a rangepic of acceptable spatial resolution for computer vision studies perhaps involving robotic manipulation. The first is that the 30° solid angle of the main lobe of the transducers beam pattern does not allow better than about 4 X 4 resolutions of a 90° solid angle field. Special acoustic focusing devices could improve this resolution as could using arrays to sharpen the directionality; a simple sound absorbing plastic foam tube is also effective in narrowing the directionality to about a 10° solid angle but even this only gives at best a 10 X 10 resolution over a 90° solid angle field [40]. The second problem is a more fundamental one and concerns the intrinsic properties of acoustic waves and reflecting surfaces. If the transducer disk is pointed at more than about 40° to the normal of a large hard surface there is a tendency for the acoustic wave to bounce off mostly concentrated in a direction where the angle of incidence and the angle of reflection are equal (see Fig. 13); consequently, little energy is reflected directly back to the detector directly from this surface. Sometimes other objects in the path of the reflected beam may return signals back to the sensor via reflection off the plane, thus producing a false reading. No return energy is a better result since it would indicate an invalid situation for the range finder. This reflection effect is explained in terms of Huygen's principal and the undulations of the surface material in relation to the wavelength of the energy. A simple particle theory analogy is the way in which many elastic balls whose sizes (diameters, say) correspond to energy wavelength would bounce off a relatively smooth surface (whose undulations are small in comparison with the ball's diameter) in a highly predictable way with a small probability of returning towards the incident direction. When the surface is relatively undulating in comparison with the ball size a stream of balls striking a small portion of this surface would bounce off in all directions with equal probability; some energy is detectable along the incident direction. For light sources the surface needs to be almost mirror smooth before this specular reflectance effect is noticeable since the wavelengths involved are much smaller than in the ultrasonic range. Thus for a large number of commonly encountered surface materials the ultrasonic device cannot measure range at incidence angles more than about 40°. The scattering of reflected energy with equal probability in all directions in a hemisphere on the surface is known as lambertian scattering in the theory of light.

In summary, range finding using a system like the Polaroid range finder kit is not suitable for producing medium to high resolution rangepics over scenes containing hard objects with surfaces whose normals are in arbitrary directions. However, for crude navigation purposes, the device would be most useful as an obstacle detector.

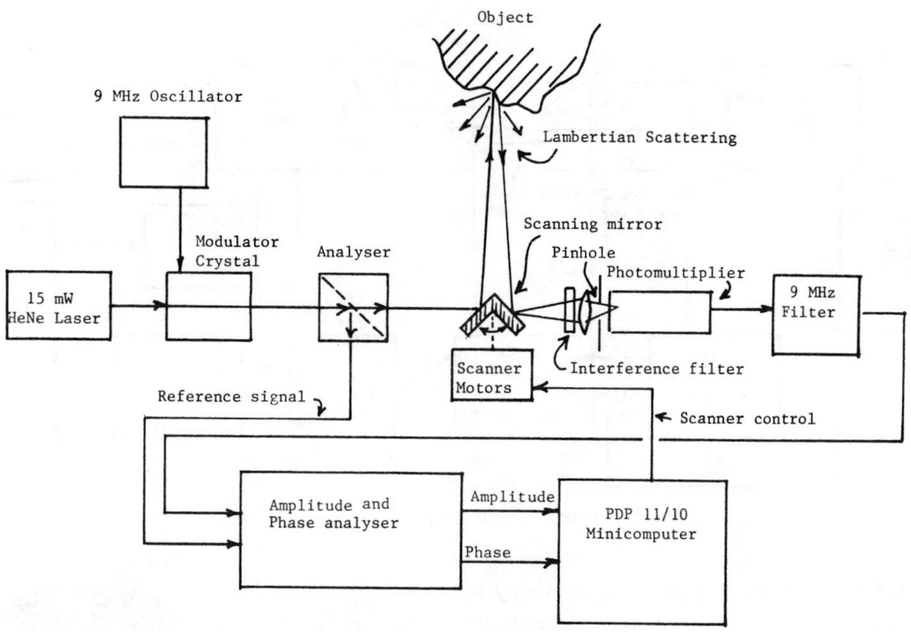

Fig. 14. Phase detection laser range finder block diagram.

B. Laser Range Finders

There are two basic laser range finder designs dependent upon time of flight to and from an object point whose range is sought. The first kind measures phase shift in a continuous wave modulated laser beam between leaving the source and returning to the detector coaxially. The second measures the time a laser pulse takes to go from the source, bounce off a target point (approximately lambertian surface assumed) and return coaxially to a detector. As light travels at approximately 1 ft/ns, the supporting instrumentation must be capable of 50 ps time resolution for range accuracy in the vicinity of $\pm\frac{1}{4}$ in. For both approaches an intrinsically large dynamic range of return energy is involved both because of the inverse fourth power range law involved and because of the variable reflectance properties of the target surfaces.

The modulated beam phase shift measuring version is represented by the instrument built at the Stanford Research Institute, reported by Duda and Nitzan [41] and detailed by Nitzan *et al.* in [42]. A simplified block diagram of this instrument is shown in Fig. 14. A scanning mirror unit points the modulated continuous laser beam at a raster scan of positions in a scene and captures a coaxial portion of the lambertian scattered beam for a receiver chain consisting of an interference filter, photomultiplier, logarithmic amplifier, and phase detector, the last using a sample of the direct source beam as a phase reference. Range is recovered through phase shift measurements and reflected intensity by energy measurements. The ratio of returned energy over source energy, when corrected for range gives the intrinsic surface property of the target known as albedo which is independent of both the surface orientation and the illumination. The combination of intensity and range information is a powerful complementary source of information for supporting scene segmentation and other scene analysis problems. The coaxial paths of source and reflected beams ensure not only that no shadows are cast on the scene by any one object or edge on another surface but also that there are no parts of the scene which can be illuminated by the source but not "seen" by the detector. The "missing parts" problem, which is prevalent in all triangulation based ranging including stereo disparity, is entirely absent. A 15 mW He-Ne laser ($\lambda = 632.8$ nm) was used and a 9 mHz modulation applied. The wide dynamic range ($\approx$100 dB) of the reflected energy and low energy laser used made it necessary to integrate over many measurements for each position to reduce the uncertainty of measurement to an acceptable degree. This proved time consuming as a typical 128 X 128 rangepic of 7-8 bits accuracy in the 1-5 m range took 2 h. If further developments in technique and instrumentation can reduce this time by three orders of magnitude, a device most useful for near real-time robotic hand/eye coordination tasks would result. Adding color discrimination would also be valuable.

The direct time of flight pulse laser range finder alternative is represented by the instrument developed at CALTECH's Jet Propulsion Laboratory and reported by Lewis and Johnston [43]. The block diagram is shown in Fig. 15.

A solid-state gallium arsenide pulse laser emitting at wavelength $\lambda = 840$ nm was used as the energy source and a photomultiplier with a gallium arsenide photosurface with spectral sensitivity to match as the detector. Again, as for the phase shift type system, a mirror scanning system was used to deflect the beam and to collect a coaxial component of the reflected beam. The output from the detector is passed onto a chain of sensitive instrumentation normally associated with nuclear physics experiments. A constant fraction discriminator produces a time pulse corresponding to a point on the input signal at a constant fraction of the peak; this ensures stable pulse arrival timing independent of intensity which has a large dynamic range. The time between a reference time pulse produced at the moment of laser firing and the output of the constant fraction discriminator is converted into a relatively wide (2 μs) pulse whose height is proportional to the required time interval (time to pulse height converter). This height is averaged over many pulses and digitized for trans-

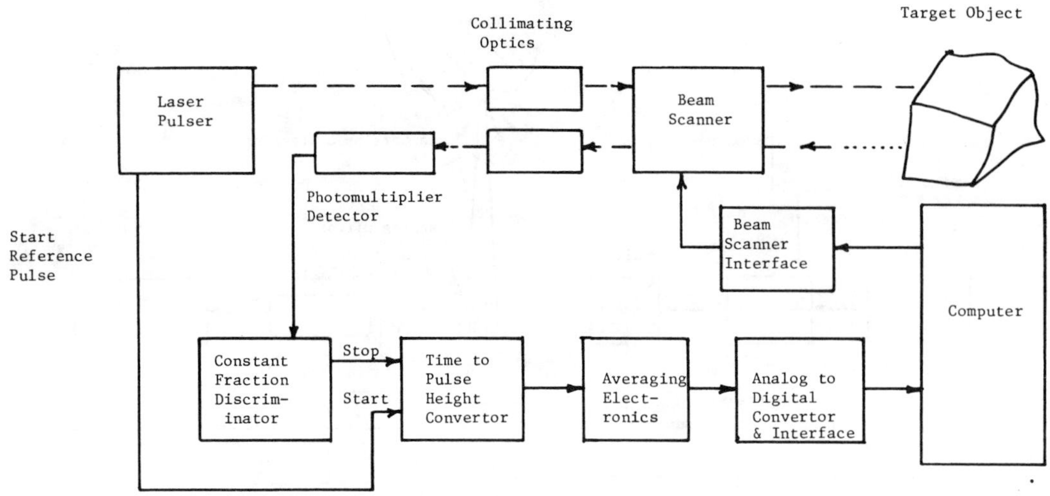

Fig. 15. Time-of-flight laser range finder schematic.

mission to the controlling computer. Its ranging accuracy is about 2 cm and once again the repetition requirements to increase signal/noise to acceptable levels constrained the speed with which the rangepic could be produced. The authors state that reliable range collection at beyond the rate of 100 points/s would be prohibited by the basic noise in the system. A range accuracy approaching 2 cm in the 1-3 m range was achieved. A 128 X 128 rangepic would, under ideal conditions take about 3 min to collect, still a long time in terms of a convenient vision-robotic manipulation or navigation cycle.

More recently Jarvis [44] has constructed a laser rangefinder using the same configuration as the Lewis and Johnston [43] instrument. Using a low powered infrared laser (820 nm) with a 100 ps pulse repeatable at 10 kHz, this instrument is capable of acquiring a noisy 64 X 64 rangepic in 4 s; a rather better result is achieved in 40 s with a resolution in the range 1-4 m of about $\frac{1}{2}$ cm. Example rangepics are shown in Fig. 16. Better accuracy in a shorter time can be achieved by increasing the laser power.

The laser ranging instruments described tend to be expensive to construct ($10 000-25 000) but with speed improvements they represent an effective direct attack upon the ranging problem with a wide variety of applications both on the laboratory bench and out of doors. Computational cost is minimal, all the complexity being delegated to a specialized piece of optoelectronic hardware.

C. Streak Camera Range Finders

Streak cameras (temporal dispersers) [45] accelerate photoelectrons from a photocathode (upon which incident light falls through a slit aperture) towards a positively charged mesh, and on passing through it, they are deflected in one direction (transverse to the acceleration) by an electrostatic field swept at variable speeds, the sweep being triggered by some timing reference (see Fig. 17). The intensity variation across this "streak" in the direction of the deflecting sweep gives the temporal intensity profile of the incident light at a time scale given by the sweep velocity. These cameras are capable of 10 ps resolution and are therefore suitable for discriminating light transit time variations corresponding to differences of light path distances of the order of 0.1 in (light travels at ≈1 ft/ns). Thus light energy arrival time variation due to reflection of a short duration laser pulse from objects at various distances from the temporal disperser camera can be used for time of flight range estimation. If a cylindrical lens is used to illuminate a scene with a line of pulsed laser light, it should be possible to obtain range information for one such line at a time. The light "sheet" from the laser can be deflected transversely by a scanning mirror system to give full coverage of the scene.

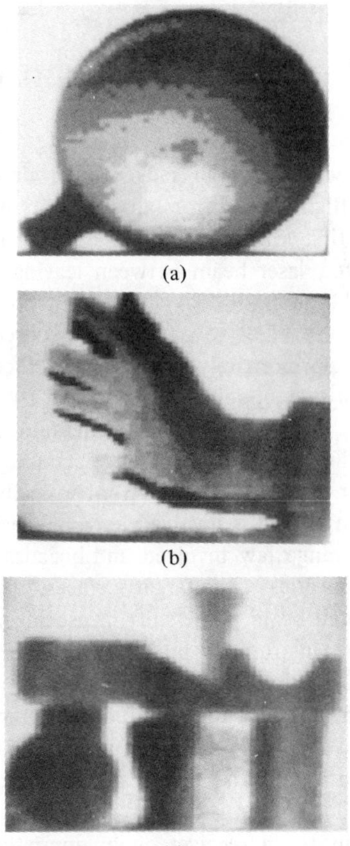

(a)

(b)

(c)

Fig. 16. Examples of 64 × 64 rangepics from a laser time-of-flight range scanner (darker is closer). (a) Six inch diameter plastic funnel. (b) Human hand. (c) Block scheme.

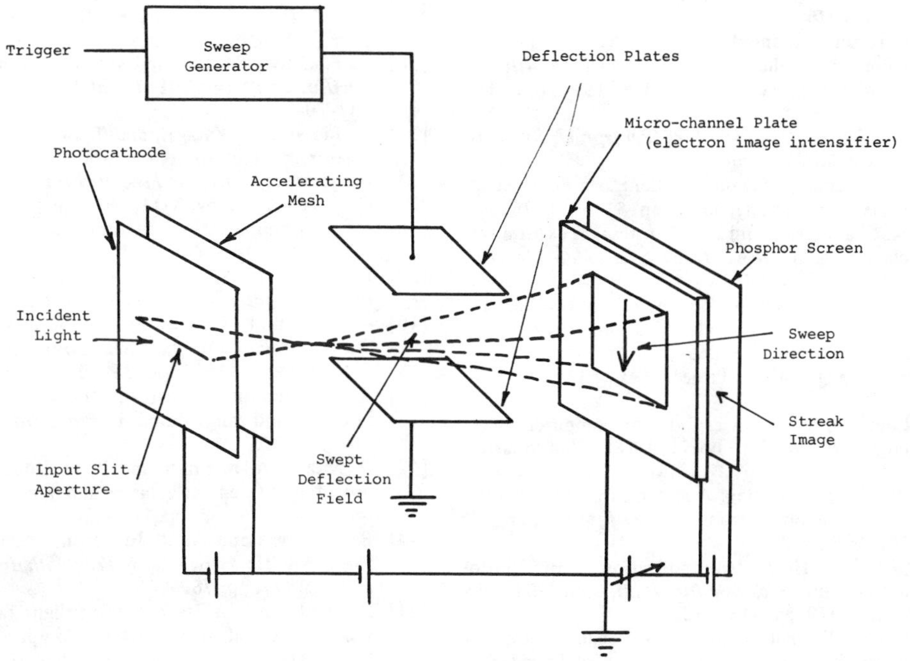

Fig. 17. Streak camera.

XII. Conclusion

Of the various approaches to range finding covered in this paper no one method would seem clearly superior to the rest. All appear to have drawbacks which fall into one or more of the following categories:

1) missing parts problem,
2) computational complexity,
3) time-consuming in improvement of signal/noise ratio,
4) limited to indoor application,
5) limited to highly textured or line structured scenes,
6) limited surface orientation,
7) limited spatial resolution.

From the viewpoint of sheer simplicity, it is hard to improve upon triangulation schemes involving one point at a time. In terms of potential, it would seem that direct laser time-of-flight range finders could, in theory, eliminate all the above problems provided an intense enough energy source could be provided—this would be at considerable expense and could also create a hazardous environment for humans. In terms of anthropomorphically posed questions, stereo disparity, occlusion, photometric and texture gradient methods would prove more interesting. From the practical standpoint of vision driven robotics, however, these approaches would hardly seem worthwhile as they are all, to some extent, indirect.

What is quite clear is that capturing the third dimension through nonimage-based range finding is of great utility in 3-D scene analysis since many of the ambiguities of interpretation arising from occasional lack of correspondence between object boundaries and inhomogeneities of intensity, texture, and color can be thus trivially resolved [46], [47].

For example, two objects of the same color and texture but at different ranges, which are in visual juxtaposition from the camera viewpoint may be difficult to separate through image analysis alone but can easily be detected as separate through rangepic analysis. An interesting approach to 3-D scene analysis using registered range and intensity data is given in [48]; in this case range analysis dominates the method and intensity data are used only when necessary. Collecting range data independently of intensity imagery analysis strengthens its ambiguity resolving potential. In both a technical and literal way, range data of this sort is orthogonal to the other data sources.

References

[1] R. N. Haber and M. Hershenson, *The Psychology of Visual Perception.* Holt, Rinehart and Winston, 1973.
[2] R. L. Gregory, *The Intelligent Eye.* New York: McGraw-Hill, 1970.
[3] M. L. Braunstein, *Depth Perception Through Motion.* New York: Academic, 1976.
[4] Y. Shirai and M. Suwa, "Recognition of polyhedrons with a range finder," in *Proc. 2nd Int. Joint Conf. Artificial Intell.*, London, Sept. 1971, pp. 80-87.
[5] G. J. Agin and T. O. Binford, "Computer description of curved objects," in *Proc. Int. Joint Conf. Artificial Intell.*, Stanford Univ., Aug. 20-23, 1973, pp. 629-640.
[6] T. O. Binford, "Visual perception by computer," in *Proc. IEEE Conf. Syst. Contr.*, Miami, FL, Dec. 1971.
[7] R. J. Popplestone, C. M. Brown, A. P. Ambler, and G. F. Crawford, "Forming models of plane-and-cylinder faceted bodies from light stripes," in *Proc. 4th Int. Joint Conf. Artificial Intell.*, 1975, pp. 664-668.
[8] G. F. Crawford, "The stripe finder hardware," Dep. Artificial Intell., Univ. Edinburgh, 1974.
[9] F. Röcker and A. Kiessling, "Methods for analysing three dimensional scenes," in *Proc. 4th Int. Joint Conf. Artificial Intell.*, 1975, pp. 669-673.
[10] P. M. Will and K. S. Pennington, "Grid coding: A preprocessing technique for robot and machine vision," in *Proc. 2nd Int. Joint Conf. Artificial Intell.*, Sept. 1971, pp. 66-68.
[11] D. Rosenberg, M. D. Levine, and S. W. Zucker, "Computing relative depth relationships from occlusion cues," in *Proc. 4th Int. Joint Conf. Pattern Recognition*, Kyoto, Japan, Nov. 7-10, 1978, pp. 765-769.
[12] A. Rosenfeld, R. A. Hummel, and S. W. Zucker, "Scene labeling

by relaxation operations," *IEEE Trans. Syst., Man, Cybern.*, vol. SMC-6, pp. 420–443, 1976.
[13] S. W. Zucker, A. Rosenfeld, and L. S. Davis, "General purpose models: Expectations about the unexpected," in *Proc. 4th Int. Joint Conf. Artificial Intell.*, Tbilisi, Sept. 3-8, 1975, pp. 716-721.
[14] J. J. Gibson, *The Senses Considered as Perceptual Systems.* Boston, MA: Houghton-Mifflin, 1966.
[15] R. Bajcsy and L. Lieberman, "Texture gradient as a depth cue," *Comput. Graphics Image Processing*, vol. 5, pp. 52–67, 1976.
[16] R. M. Haralick, K. Shanmugan, and I. H. Dinstein, "Textural features for image classification," *IEEE Trans. Syst., Man, Cybern.*, vol. SMC-3, pp. 610–621, Nov. 1973.
[17] J. S. Weszka, C. R. Dyer, and A. Rosenfeld, "A comparative study of texture measures for terrain classification," *IEEE Trans. Syst., Man, Cybern.*, vol. SMC-6, pp. 269–285, Apr. 1976.
[18] B. K. P. Horn, "Focussing," M.I.T., Project MAC, AI Memo. 160, May 1968.
[19] R. A. Jarvis, "Focus optimisation criteria for computer image processing," *Microscope*, vol. 24, pp. 163–180, 2nd quarter, 1976.
[20] B. K. P. Horn, "Shape from shading: A method for obtaining the shape of a smooth opaque object from one view," M.I.T., Project MAC, MAC TR-79, Nov. 1970.
[21] K. Ikeuchi and B. K. P. Horn, "An application of the photometric stereo method," in *Proc. 6th Int. Joint Conf. Artificial Intell.*, Tokyo, Japan, 1979, pp. 413–415.
[22] M. D. Levine, D. A. O'Handley, and G. M. Yagi, "Computer determination of depth maps," *Comput. Graphics Image Processing*, vol. 2, pp. 134–150, 1973.
[23] R. A. Jarvis, "Optimisation in adaptive control: A selective survey," *IEEE Trans. Syst., Man, Cybern.*, vol. SMC-5, pp. 83–94, Jan. 1975.
[24] Y. Yakimovsky and R. Cunningham, "A system for extracting three-dimensional measurements from a stereo pair of TV cameras," *Comput. Graphics Image Processing*, vol. 7, pp. 195–210, 1978.
[25] H. P. Moravec, "Visual mapping by a robot rover," in *Proc. 6th Int. Joint Conf. Artificial Intell.*, 1979, pp. 598–620.
[26] H. H. Baker, "Edge based stereo correlation," in *Proc. ARPA Image Understanding Workshop*, Univ. Maryland, Apr. 1980.
[27] D. Marr and T. Poggio, "Cooperative computation of stereo disparity," M.I.T., A.I. Lab., Memo. 364, June 1976.
[28] B. Julesz, "Binocular depth perception without familiarity cues," *Science*, vol. 145, pp. 356–362, 1964.
[29] D. Marr and T. Poggio, "Computational approaches to image understanding," M.I.T., A.I. Lab., see also *Proc. R. Soc. London B*, vol. 204, pp. 301–328, 1979.
[30] D. Marr and E. C. Hildreth, "Theory of edge detection," in *Proc. R. Soc. London B*, vol. 207, pp. 187–217, 1980.
[31] E. C. Hildreth, "Edge detection in man and machine," *Robotics Age*, pp. 8–14, Sept./Oct. 1981.
[32] D. Marr, "Early processing of visual information," *Phil. Trans. R. Soc. London B*, vol. 275, pp. 483–524, 1980.
[33] M. Brady, Computational approaches to image understanding," M.I.T., A.I. Lab., A.I. Memo. 653, Oct. 1981.
[34] H. K. Nishihara and N. C. Larson, "Toward a real time implementation of the Marr-Poggio stereo matcher," in *Proc. Image Understanding Workshop*, Lee Bauman, Ed., 1981.
[35] T. D. Williams, "Depth from camera motion in a real world scene," *IEEE Trans. Pattern Anal. Machine Intell.*, vol. PAMI-2, pp. 511–516, Nov. 1980.
[36] K. Prazdny, "Motion and structure from optical flow," in *Proc. 6th Int. Joint Conf. Artificial Intell.*, Tokyo, Japan, 1979, pp. 702–704.
[37] M. Idesawa, T. Yatagai, and T. Soma, "A method for automatic measurement of three-dimensional shape by new type of Moiré fringe topography," in *Proc. 3rd Int. Joint Conf. Artificial Intell.*, Coronada, CA, Nov. 8–11, 1976, pp. 708–712.
[38] —, "Scanning Moiré method and automatic measurement of 3D shapes," *Appl. Opt.*, vol. 16, pp. 2152–2162, Aug. 1977.
[39] R. A. Jarvis, "A computer vision and robotics laboratory," *IEEE Computer*, pp. 8–24, June 1982.
[40] —, "A mobile robot for computer vision research," in *Proc. 3rd Australian Comput. Sci. Conf.*, A.N.U., Canberra, A.C.T., Jan. 31–Feb. 1, 1980, pp. 39–51.
[41] R. O. Duda and D. Nitzan, "Low-level processing of registered intensity and range data," in *Proc. 3rd Int. Joint Conf. Artificial Intell.*, 1976.
[42] D. Nitzan, A. E. Brain, and R. O. Duda, "The measurement and use of registered reflectance and range data in scene analysis," *Proc. IEEE*, vol. 65, pp. 206–220, Feb. 1977.
[43] R. A. Lewis and A. R. Johnston, "A scanning laser rangefinder for a robotic vehicle," in *Proc. 5th Int. Joint Conf. Artificial Intell.*, 1977, pp. 762–768.
[44] R. A. Jarvis, "A laser time-of-flight range scanner for robotic vision," Australian Nat. Univ., Comput. Sci. Tech. Rep. TR-CS-81-10; also in preparation for publication in *IEEE Trans. Pattern Anal. Machine Intell.*
[45] Y. Tsuchiya, E. Inuzuka, Y. Suzui, and W. Yu, "Ultrafast streak camera," in *Proc. 13th Int. Congr. High Speed Photography and Photonics*, Tokyo, Japan, Aug. 20–25, 1978.
[46] R. A. Jarvis, "Expedient 3D robot colour vision," Australian Nat. Univ., Comput. Sci. Tech. Rep., 1982.
[47] —, "Vision driven robotics in a partially structured environment," Australian Nat. Univ., Comput. Sci. Tech. Rep. TR-CS-82-03, 1982.
[48] R. O. Duda, D. Nitzan, and P. Barrett, "Use of range and reflectance data to find planar surface regions," *IEEE Trans. Pattern Anal. Machine Intell.*, vol. PAMI-1, pp. 259–271, July 1979.

R. A. Jarvis received the Ph.D. degree in electrical engineering from the University of Western Australia in 1968.

He is currently a reader in computer science at the Australian National University, Canberra, Australia, where he was Head of the Department of Computer Science from 1976 to 1979. He spent 1969, 1970, and 1977 as a Visiting Professor in Electrical Engineering at Purdue University, West Lafayette, IN. His current research interests include digital computing technology, pattern recognition, image processing, computer vision, and robotics.

Structure from Stereo—A Review

UMESH R. DHOND, STUDENT MEMBER, IEEE, AND J. K. AGGARWAL, FELLOW, IEEE

Reprinted from *IEEE Transactions on Systems, Man, and Cybernetics*, Vol. 19, No. 6, November/December 1989, pages 1489-1510. Copyright © by The Institute of Electrical and Electronics Engineers, Inc. All rights reserved.

***Abstract*—Major recent developments in establishing stereo correspondence for the extraction of the 3-D structure of a scene are reviewed. Broad categories of stereo algorithms are identified based upon differences in image geometry, matching primitives, and the computational structure used. Performance of these stereo techniques on various classes of test images is reviewed and the possible direction of future research is indicated.**

I. Introduction

A major portion of the research efforts of the computer vision community has been directed towards the study of the three-dimensional (3-D) structure of objects using machine analysis of images. Analysis of video images in stereo has emerged as an important passive method for extracting the 3-D structure of a scene. Earlier, Barnard and Fischler [6] presented a review covering the major steps involved in stereo analysis, the evaluation criteria for stereo algorithms, and a survey of the different approaches to computational stereo developed starting from the mid-70's up to 1981. In this paper we review the computational structure of the major schemes that have evolved in the past decade for recovering depth using stereo.

The basic principle involved in the recovery of depth using passive imaging is triangulation. Many active range sensing techniques are also based upon the triangulation principle. However, in active ranging techniques that use triangulation, the nature of the problem is different in that the triangle for recovering depth is predefined by three points—the light source, the illuminated spot in the scene, and its image point. Thus, in active methods that use triangulation, the correspondence problem has already been solved by using an artificial source of illumination.

In stereopsis, which is a passive technique, the triangulation needs to be achieved with the help of only the existing ambient illumination. Hence a correspondence needs to be established between features from two images that correspond to some physical feature in space. Then, provided the position of centers of projection, the effective focal length, the orientation of the optical axis, and the sampling interval of each camera are known, the depth can be reconstructed using triangulation. Based upon this basic correspondence problem, a particular matching paradigm can be constructed depending upon the specific matching features used, the number of cameras used, the positioning of the cameras, and the scene domain.

The problem of passive range sensing is important where there are overriding circumstantial constraints that prevent the use of artificial illumination or other active sources of radiation. Applications of stereo-based depth measurement include automated cartography, aircraft navigation, autonomous land rovers, robotics, industrial automation and stereomicroscopy.

In the following sections, we identify broad categories of matching algorithms depending upon various factors like the imaging geometry, the matching primitives, as well as the matching strategy used. Within each category, the implementation details of the contemporary approaches will be highlighted. Section II gives an overview of the major steps involved in the process of stereopsis, namely, preprocessing, stereo matching, and depth reconstruction. Section III examines various computational theories of stereopsis that have been motivated by the human visual system. Sections IV through X describe the major computational techniques that have been successfully tested in the past decade for solving the stereo correspondence problem. In Section IV we review area-based correlation schemes. Relaxation labeling processes have been used by many researchers to iteratively impose global consistency constraints on multiple matches for the purpose of disambiguation, which we describe in Section V. Many stereo algorithms use edge segments obtained from fitting piecewise linear curves to connected edges. Section VI describes two approaches for using edge segments in stereo matching. Stereo algorithms that utilize hierarchical computational structures are described in Section VII. In Section VIII, we examine the use of dynamic programming methods for stereo matching. The trinocular camera setup and the resulting matching paradigm, with both point- and segment-based matching algorithms, are reviewed in Section IX. Section X briefly describes the formulation of the correspondence problem using structural descriptions. Section XI deals with the aspect of performance evaluation of stereo algorithms and the various classes of test data used. Section XII contains concluding remarks.

II. The Process of Stereopsis

The major steps involved in the process of stereopsis are preprocessing, establishing correspondence, and recovering depth. In this section we shall briefly examine each of them.

Manuscript received October 17, 1988; revised March 30, 1989. This work was supported in part by a grant from the Army Research Office under Contract DAAL03-87-K-0089 and in part by a grant from the National Science Foundation under Contract NSF/ECS-8513123.

The authors are with the Computer and Vision Research Center, College of Engineering, ENS Building, University of Texas, Austin, TX 78712.

IEEE Log Number 8930349.

EH0341-8/91/0000/0025$01.00 © 1989 IEEE

A. Preprocessing

Preprocessing of images is an important component of stereopsis. During this stage image locations satisfying certain well-defined feature characteristics are identified in each image. They have to be chosen carefully because the subsequent matching strategy shall make extensive use of these feature characteristics.

Some of the earlier stereo algorithms used area-based matching schemes in which area patches from two images were matched [18], [48]. Points of interest were located in one image using certain *interest operators*. Moravec [48] proposed one such interest operator that computed the local maxima of a directional variance measure over a 4×4 (or 8×8) window around a point. The sums of squares of differences of adjacent pixels were computed along all four directions (horizontal, vertical, and two diagonal), and the minimum sum was chosen as the value returned by the operator. The site of the local maximum of the values returned by the interest operator was chosen as a feature point whose stereo counterpart was to be found.

By and large most of the contemporary stereo algorithms match features directly rather than areas (in Section II-B we shall examine the issues regarding area-based and feature-based matching). Hence, the importance of good feature detectors has increased. Since physical discontinuities in a scene usually project as local changes in gray-level intensity in an image, edges have been increasingly used as matching primitives. A large number of edge operators have been proposed that compute the direction of orientation as well as the strength of an edge. Most of the edge operators currently in use can be classified [4] into three main categories:

1) Operators that approximate certain mathematical derivative operators (such as the Laplacian operator);
2) Operators that involve convolution of the image with a set of templates tuned to different orientations; and
3) Operators that fit local gray-level intensity values surrounding a point with (edge) surface models and extract edge parameters from the model.

The Marr–Hildreth edge operator [39] has been used by many algorithms for locating edge points during the feature extraction process. The operator convolves a mask approximating the Laplacian of Gaussian ($\nabla^2 G$) function (see Section III-B) over the entire image and labels the zero-crossings of the convolution output as edge points. The edge orientation on a zero-crossing contour is given by the direction of the gradient of the convolution output. The edge strength is proportional to the magnitude of the gradient of the convolution output. Recently, Torre, and Poggio [73] have also studied the problem of using differential operators for edge detection.

Grimson [19]–[21], Mayhew and Frisby [44], [59], Kim and Aggarwal [35], and Ayache and Faverjon [1], Ayache and Lustman [2], among others, use the Marr–Hildreth operator (or a simplified version thereof) for feature point extraction. The edge detectors proposed by Canny [11], and Deriche [13] are also used fairly widely for low-level feature extraction. Baker and Binford [3] and Ohta and Kanade [54] locate peaks of the magnitude of the first derivative of the intensity profile along a scan line as feature points for matching. Some of the other popular gradient edge detectors are the Roberts, the Sobel, and the Prewitt operators [4]. Haralick [27] has proposed a step-edge detector based upon the second directional derivative, and compared its performance with the Marr–Hildreth zero-crossing detector and the Prewitt gradient operator. [22] and [26] contain interesting discussions about the comparison of the Marr–Hildreth and the Haralick edge operators.

Medioni and Nevatia [46] used a set of oriented step-edge masks (Type II) spaced at 30° intervals to extract edge points. Ito and Ishii [30], Harwood and Peitikäinen [57], and others have used Type (II) edge operators consisting of eight template masks tuned to the eight directions of the compass. The mask giving the maximum output decides the orientation and magnitude of the edge. The edges obtained using this type of operators need further processing in the form of edge thinning and edge linking.

Raju, Binford, and Shekhar [62] have used an operator of Type (III) described in [50] to detect an *edgel* (an edge element). The edgel operator fits a directional *tanh*-surface to a window in the image. Edgels are characterized by their position and orientation. Ballard and Brown [4], and Rosenfeld and Kak [68] contain a more in-depth treatment of edge detectors.

Linear edge segments have also been used as matching primitives for stereo by Medioni and Nevatia [46], Ayache and Faverjon [1], Ayache, and Lustman [2], Hansen, Ayache, and Lustman [25] and others. In the segment-based stereo algorithm of Medioni and Nevatia [46] edge points were extracted using Type (II) edge operators and fitted with piecewise-linear edge segments using the Nevatia–Babu algorithm [52]. Each edge segment description consisted of the coordinates of its endpoints, its orientation, and the average contrast (absolute value) in gray-level intensity along a direction normal to its orientation. Ayache and Faverjon [1] obtained edge points by using two methods. The first involved locating the zero-crossings in the output of the convolution of the intensity image with a difference-of-averages filter, and connecting them to obtain a chain of edge points. Then, the magnitude of intensity gradient along each chain was computed by the Sobel operator and portions of chains having connected points with the magnitude of intensity gradient below a certain threshold were discarded. The second method used a modified version of the Canny edge detector [11]. In both cases, the intermediate results were chains of connected edges. Each chain of connected edges was then approximated by a set of linear edge segments using a polynomial approximation algorithm [56]. Each of the resulting edge segments was described using the coordinates of its midpoint, its length, and its orientation.

Thus, two of the major types of features extracted from images are edge points and line segments.

B. Matching

Matching is perhaps the most important stage in stereo computation. Given two (or more) views of a scene, correspondence needs to be established among *homologous* features, that is, features that are projections of the same physical identity in each view. Matching strategies can be differentiated in the broadest sense according to the primitives used for matching as well as the imaging geometry. Differences in the matching primitives separate area-based matching from feature-based matching. Imaging geometry creates distinctions that separate parallel-axis stereo from nonparallel axis stereo, and binocular stereo from trinocular (and other multiocular) stereo paradigms. Local search procedures for possible matches are governed by the projection geometry of the imaging system, and are expressed in terms of the epipolar constraints. Various local properties of the features to be matched are used in order to achieve a reasonable amount of success in the *local matching* process. The *global consistency* of the local matches is then tested by figural continuity[1] or other similar constraints.

Area-based stereo techniques use correlation among brightness (intensity) patterns in the local neighborhood of a pixel in one image with brightness patterns in a corresponding neighborhood of a pixel in the other image. First, a point of interest is chosen in one image. A cross-correlation measure is then used to search for a point with a matching neighborhood in the other image. The area-based techniques have a disadvantage in that they use intensity values at each pixel directly, and are hence sensitive to distortions as a result of changes in viewing position (perspective) as well as changes in absolute intensity, contrast, and illumination. Also, the presence of occluding boundaries in the correlation window tends to confuse the correlation-based matcher, often giving an erroneous depth estimate.

Feature-based stereo techniques use symbolic features derived from intensity images rather than image intensities themselves. Hence, these systems are more stable towards changes in contrast and ambient lighting. The features used most commonly are either edge points or edge segments (derived from connected edge points) that may be located with subpixel precision. Also feature-based methods allow for simple comparisons between attributes of the features being matched, and are hence faster than correlation-based area matching methods.

Stereo matching paradigms are also characterized by the particular imaging geometry being used. Factors that could be changed include, but are not limited to, the mutual orientation of the optical axes of the cameras (either parallel or nonparallel) and the number of cameras used

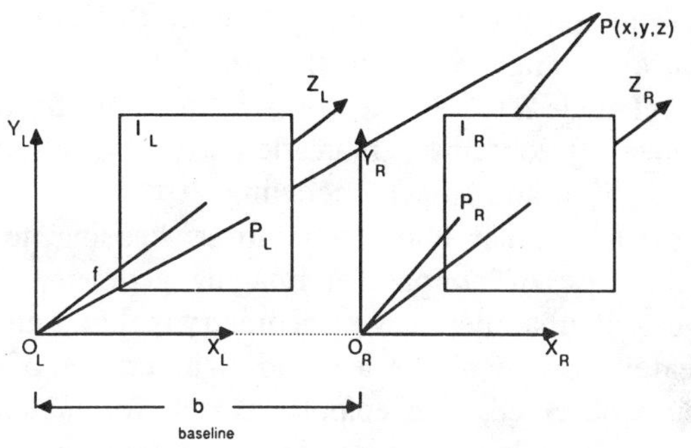

Fig. 1. Parallel axis stereo geometry.

(either two or more than two). The imaging geometry of a conventional stereo imaging system involves a pair of cameras with their optical axes mutually parallel and separated by a horizontal distance denoted as the stereo baseline. The cameras have their optical axes perpendicular to the stereo baseline, and their image scanlines parallel to the baseline (horizontal). Since the displacement between the optical centers of the two cameras is purely horizontal, the position of corresponding points in the two images can differ only in the horizontal component. Fig. 1 shows the imaging geometry of a stereo pair of cameras. The two cameras are represented by their equivalent pinhole approximation models with their image planes, I_L and I_R, reflected about their centers of projections, O_L and O_R, respectively. The origin of the world coordinate system is at O_L, the effective focal length of each camera is f, and the stereo baseline is b. The world coordinate axes X_W, Y_W, and Z_W coincide with the coordinate axes of the left camera, X_L, Y_L, and Z_L, respectively. Let $P_L(x_L, y_L, z_L)$ and $P_R(x_R, y_R, z_R)$ be the projections of the 3-D scene point $P(x, y, z)$. The rays of projection $\overline{PO_L}$ and $\overline{PO_R}$ define the plane of projection of the 3-D scene point called the epipolar plane. For a given point P_L in the left image, its corresponding match point P_R in the right image must lie on the line of intersection of the epipolar plane and the image plane that is called the epipolar line. The epipolar line in the right image corresponding to a point P_L in the left image defines the search space within which the corresponding matchpoint P_R should lie in the right image. Thus the epipolar constraint is obtained as a result of the imaging geometry of the stereo camera system and helps limit the search space in the correspondence problem for stereo analysis. In the conventional parallel-axis geometry, all epipolar planes intersect the image planes along horizontal lines.

However if the optical axis of any one of the cameras were not parallel to the world z-direction, then the epipolar lines in the image would appear inclined to the horizontal. Fig. 2 depicts a special case when the coordinate axes (Z_R and X_R) of only the right camera have been rotated by a pan angle ϕ (about the y-axis, Y_R) to Z_R' and X_R', respectively. Then the epipolar lines in the right image L_1, L_2, and L_3, corresponding to points P_{1L}, P_{2L}, and P_{3L} intersect at point E_R known as the epipole center of the right image. In general, the coordinate system of each

[1]The concept of figure continuity constraint and its various interpretations are discussed in Section III.

camera could have a pan angle ϕ (about the world y-direction), a tilt angle θ (about the world x-direction) as well as a roll angle α (about the world z-direction). Barnard and Fischler [4] contains a more detailed description of image acquisition and camera modeling.

Thus, extra epipolar line computations become necessary in the case of nonparallel imaging geometry. The advantage of nonparallel imaging geometry is that it allows for a greater overlap of the left and right images of the scene being observed. The epipolar search for matching edge points is usually aided by certain geometric similarity constraints like similarity of edge orientation or edge strength. This matching process is also referred to as local matching.

The match points obtained as a result of imposing the epipolar constraint on the local matching search could result in two or more candidate matches being judged as having almost equal possibility for getting matched. Or worse, an incorrect match point might satisfy the local matching constraints (epipolar constraint and geometric property constraint) and get chosen as a good match. The disparity obtained by computing the relative displacement of the matching feature points in the two images is used to extract the 3-D depth of the scene point that projects on the two matched points. Thus, if certain assumptions can be made regarding the nature of surfaces in the 3-D scene being observed, they could be used to determine the consistency of the disparities obtained as a result of the local matching, or guide the epipolar search so as to avoid inconsistent/false matching. An inherent assumption that is usually made about objects is that their surfaces are predominantly smooth. The smoothness in depth is expected to result in the smoothness of disparities obtained as a result of the matching process. This is formulated in the form of a regional disparity continuity constraint. Also the contours on the scene surface project on each image as continuous (or piecewise continuous) curves, which is the motivation behind the figural continuity constraint. Hence physical features on objects that satisfy the surface smoothness assumption and project on the stereo pair of images as image features would satisfy some form of the disparity continuity and figural continuity constraints. This is otherwise referred to as global matching. Thus, local matching and global matching can be regarded as two phases of the stereo matching process.

C. 3-D Structure Determination

The conventional parallel, axis stereo geometry provides a disparity value d for each matched pair of points $P_L(x_L, y_L)$ and $P_R(x_R, y_R)$ (see Fig. 1) as, $d = x_L - x_R$. By considering similar triangles, the world coordinates of the scene point $P(x, y, z)$ can be easily obtained as

$$x = \frac{bx_L}{d}, \qquad y = \frac{by_L}{d}, \quad \text{and} \quad z = \frac{bf}{d}$$

where b is the stereo baseline and f is the effective focal length of the camera.

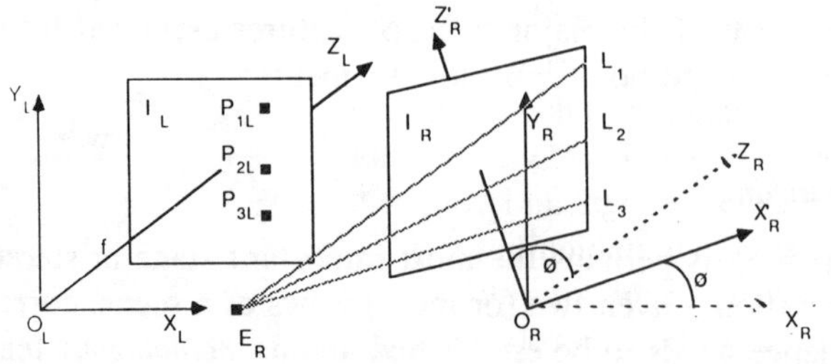

Fig. 2. Nonparallel axis stereo geometry.

The 3-D reconstruction process of nonparallel stereo systems ([1], [2], [30], [76]) requires a more general approach, since closed form solutions may not exist for many cases. The lines joining the center of projection and the image point in each of the stereo images are projected backwards into space. Then the point in space that minimizes the sum of its distance from each of the back-projected lines is chosen as the estimated 3-D position of the matched point. For nonparallel imaging systems using edge segments as matching features ([1], [2]), the end points of the matched edge segments are projected backwards in space and the 3-D position of the segment is determined using a similar minimization criterion.

III. Computational Theory of Stereopsis

Marr and Poggio [41] proposed a feature-point based computational model of human stereopis. Grimson [19], [21] developed a computer implementation of their algorithm and demonstrated the effectiveness of this model on standard psychological test images (random dot stereograms) as well as on natural images. A number of additional psychophysical predictions of the Marr–Poggio model have been tested and several modifications have been proposed [17], [44], [49], [59]. After extensive testing, Grimson [20] embodied these modifications in a newer version of his implementation. We shall briefly review the implementation of Marr–Poggio theory [19], examine the problems associated with it, and review the modifications which appear in Grimson's new implementation [20].

A. Marr-Poggio Theory

Marr and Poggio [41] based their computational structure of the stereo fusion problem upon biological evidence. Some of the early works by neurophysiologists and psychologists on subjects like existence of independent spatial-frequency-tuned channels [17], [31], [33], [45], cooperative processes [31], [32], [42], and vergence eye movements [63]–[65], [74], [75] in the human and other biological vision systems were used to formulate the outline of the theory.

The Marr–Poggio theory [41] proposed that the human visual processor solved the stereo matching problem in five main steps. 1) The left and right images are filtered at twelve different orientation-specific masks each approximated by the difference of two Gaussian functions with space-constants in the ratio 1:1.75. 2) Zero-crossings in the filtered images are found by scanning them along lines perpendicular to the orientation of the mask. 3) For each mask size, matching takes place between the zero-crossing segments extracted from each filtered image output that are of the same sign and roughly the same orientation. Local matching ambiguities are resolved by considering the disparity sign of nearby unambiguous matches. 4) Matches obtained from wider masks control vergence movements aiding matches among output of smaller masks; 5) The correspondence results are stored in a dynamic buffer called the 2.5-D sketch.

Marr and Poggio [41] formulate two basic rules for matching left- and right-image descriptions. Each item in an image can be assigned to one and only one disparity value (uniqueness). Secondly, matter is cohesive. Hence disparity varies smoothly almost everywhere, except where depth discontinuities occur at surface boundaries (continuity).

B. Grimson's Implementation

Grimson [19] implemented the computational theory of Marr and Poggio [41] and addressed certain implementation details that were not covered earlier by the Marr–Poggio theory.

1) Feature Extraction: Marr and Hildreth [39] have shown theoretically that, provided two simple conditions on the image intensity function in the neighborhood of an edge are satisfied, intensity changes occurring at a particular scale may be detected by locating the zero-crossings in the output of the $\nabla^2 G$ (Laplacian of Gaussian) filter. Instead of convolving each image with 12 directional *DoG* operators, each of which yield an approximation to the second directional derivative, Grimson [19] used the Laplacian of Gaussian ($\nabla^2 G$) operator and grouped the zero-crossing points in 12 directional bins. The precise form of the operator is given in polar coordinates (r, θ) by

$$\nabla^2 G(r,\theta) = \left[\frac{r^2 - 2\sigma^2}{\sigma^4}\right] \exp\left[\frac{-r^2}{2\sigma^2}\right] \tag{1}$$

where σ is the Gaussian space-constant. This is a rotationally symmetric function shaped like an inverted Mexican hat (Fig. 3). The width of the central negative region is given by $w_{2-D} = 2\sqrt{2}\,\sigma$. Grimson used three [20] or four [19] different sizes of filters for his images.

2) Matching: The algorithm begins with images filtered by the largest filters because the reduced density of zero-crossings makes matching easier. The overall matching strategy of Grimson [19] uses a coarse-to-fine iterative approach with disparities found at coarser resolutions used to guide match-point search at finer resolutions. Marr [38], [41] studied the probability distribution of the interval between adjacent zero-crossings of the same sign obtained from the convolution of random dot stereograms with the Laplacian of Gaussian filter. The results indicated that if the disparity between the images is less than $\pm(\omega/2)$, a search for matches within the range $\pm(\omega/2)$ will yield only the correct match with probability 0.95. However the alternate strategy of using a search space with range $\pm\,\omega$ is used by Grimson [19] since it allows one to search for matches over a larger disparity range and yet get unambiguous and correct matches with probability 0.5. In Grimson's implementation [19] for each zero crossing $P_L(x, y)$ in the left image, possible candidate matches $P_R'(x', y)$ are searched for along the epipolar line in the right image such that,

$$x + d_i - \omega \leqslant x' \leqslant x + d_i + \omega \tag{2}$$

as shown in Fig. 4(a), where d_i is the estimated disparity and ω ($= 2\sqrt{2}\,\sigma$) is the width of the *LoG* filter. Zero-crossings in the left and right images having the same contrast sign and approximately the same orientation (within $\pm 30°$) are matched. If only one match is found within the $\pm\,\omega$ region, then that match is accepted as unambiguous, and the disparity is recorded.

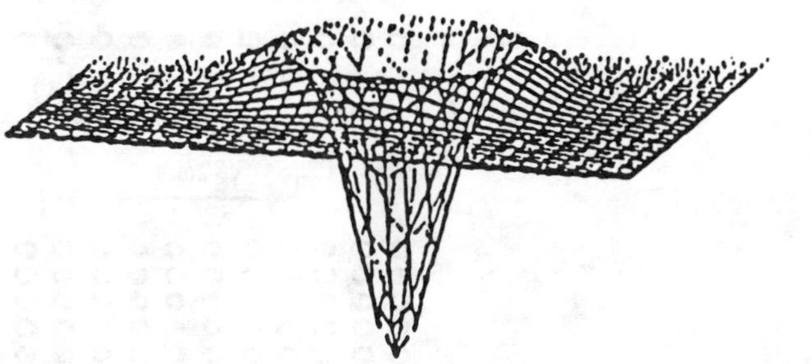

Fig. 3. 2-D Laplacian of Gaussian.

3) Disambiguation of multiple matches: If more than one match is found within the $\pm\,\omega$ region, then the one having disparity of the same type (convergent, divergent, or zero) as the dominant disparity in the neighborhood is accepted. Otherwise the match at that point is left ambiguous. This can be regarded as the pulling effect which is described in the psychophysical experiments of Julesz and Chang [32]. Each 2-D array of matched results is scanned and if the percentage of matched points is < 0.7 then all matches in that region are discarded.

C. Grimson's Modified Implementation of Marr–Poggio Theory

Grimson's earlier implementation [19] of the Marr–Poggio theory [41] imposes a regional continuity check on disparity. Later, Grimson [20] highlights some of the problems associated with the earlier implementation of the Marr–Poggio theory and presents a modified implementation.

1) Figural Continuity: Grimson's implementation [19] of the Marr–Poggio theory [41] used a regional continuity check on disparity in order to validate the matches. Grimson [20] observed that this caused difficulties in propagation of disparity at occluding boundaries between objects and along thin elongated surfaces. Elsewhere the matched feature points tended to form extended contours. Hence the figural continuity constraint of Mayhew and Frisby [44] that required continuity of disparity along contours was deemed more appropriate.

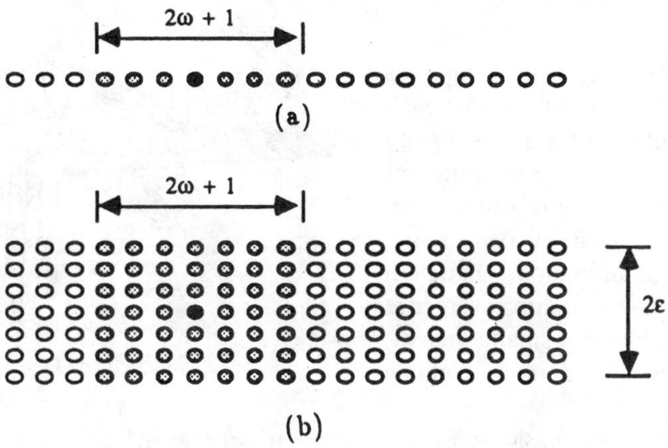

Fig. 4. Search space in Grimson's algorithm. $P_R'(x', y)$ is shown as black dot. Search space is shown as shaded dots around P_R'. (a) Original implementation. (b) Modified implementation.

Vertical Disparity: There is psychophysical evidence [15], [16], [53] to suggest that the human vision system does resort to eye movements in order to correct gross vertical misalignments in the images. Accordingly, the Marr–Poggio algorithm [41] uses a strict epipolar matching strategy (see Section III-B) after aligning the images in the vertical direction. However, local distortions due to perspective effects, noise in early processing, and discretization effects cause deterioration in matching performance at finer resolutions [20]. In the modified stereo algorithm, for a zero-crossing at a point $P_L(x, y)$ in the left image, Grimson [20] searches for the corresponding zero-crossing match points $P_R'(x', y')$ in the region

$$\{(x', y') \mid x + d_i - \omega \leqslant x' \leqslant x + d_i + \omega; \quad y - \epsilon \leqslant y' \leqslant y + \epsilon\} \quad (3)$$

where ω and d_i denote quantities described in (2), and $(2\epsilon + 1)$ is the height of the search space in the vertical direction (see Fig. 4(b)).

D. Mayhew–Frisby Theory of Disparity Gradient

1) Figural Continuity: Mayhew and Frisby [44] propose a new interpretation of the surface continuity constraint, to include figural continuity. They extend the Marr–Poggio concept [41] of continuity to imply that edges of surfaces and surface markings would also be continuous resulting in continuity of disparity along figural contours. Baker and Binford [3], and Ohta and Kanade [54] also used a similar figural continuity constraint along with the added restriction of left-to-right ordering of edges for stereo matching.

2) Cross-Channel Activity: Mayhew and Frisby [44] postulate the existence of interaction between several spatial-frequency-tuned channels in parallel, as against the sequential coarse-to-fine process proposed by Marr and Poggio [41]. In simple terms, the rule for cross-channel correspondence requires that any feature attribute or pattern at a disparity location should be supported by a similar feature attribute or pattern in other spatial frequency channels within a certain disparity range, and that dissimilar cross-channel activity patterns should be rejected as figurally rivalrous.

In one stereo algorithm implemented by Mayhew and Frisby [44], the contrast-signed zero-crossings and peaks of the convolution of each image with the $\nabla^2 G$ operator are encoded at each location for three spatial-frequency tuned channels , as a triplet. Fig. 5 shows a schematic representation of using cross-channel activity according to Mayhew and Frisby [44]. The top row of Fig. 5 shows a triplet of measurement primitives found at a location in one image (say, left). Primitive values are marked +, − and · (nil) to signify positive, negative, or nonexistent zero-crossings, respectively. Beneath them are triplets at candidate match-points in the other image (say, right). The bottom row shows the result of the binocular cross-channel correspondence. Correct matches are marked *M* and incorrect (rivalrous) matches are marked *R*. If only one image has a primitive at a particular channel, the entry is marked *U*; and nil entries that match are ignored (marked ·).

3) Disparity Gradient Limit: Burt and Julesz [10] provide evidence supporting the claim that, for binocular fusion of random dot stereograms by the human visual system, the disparity gradient must not exceed 1. Pollard, Mayhew, and Frisby [59] suggest that for most natural scene surfaces, including jagged ones, the disparity gradients between correct matches is usually <1, whereas it is very rare among incorrect matches obtained for the same set of images.

Consider the binocular parallel imaging geometry as shown in Fig. 6 with image centers $O_L(x_{OL}, y_O)$ and $O_R(x_{OR}, y_O)$, separated by baseline b. Let $A_L(x_{AL}, y_A)$, $B_L(x_{BL}, y_B)$ in the left image and $A_R(x_{AR}, y_A)$, $B_R(x_{BR}, y_B)$ in the right image be the projections of the world points A_P and B_P, respectively. Then a cyclopean space is defined such that the origin $OC(x_{OC}, y_{OC})$ is defined as,

$$x_{OC} = \frac{x_{OL} + x_{OR}}{2} \quad \text{and} \quad y_{OC} = y_O. \quad (4)$$

Let A_P and B_P have disparities $d_A = x_{AL} - x_{AR}$ and $d_B = x_{BL} - x_{BR}$, respectively, and cyclopean images A_C and B_C, respectively. The disparity gradient D_g would then be

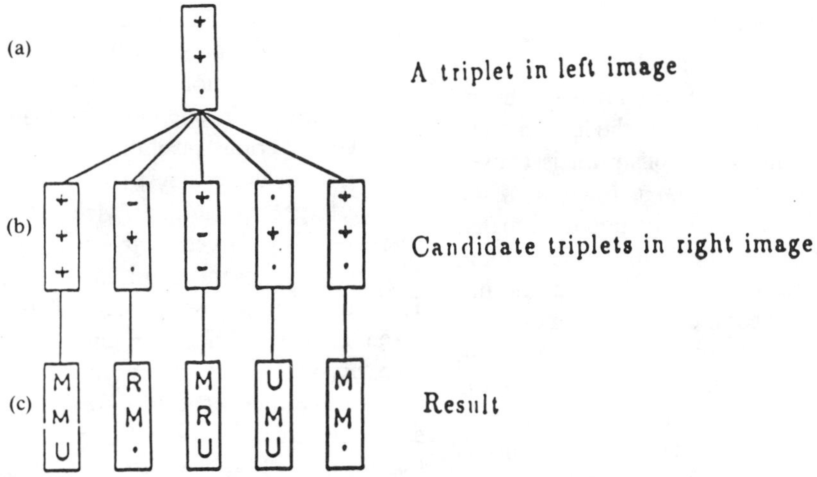

Fig. 5. Cross-channel correspondence.

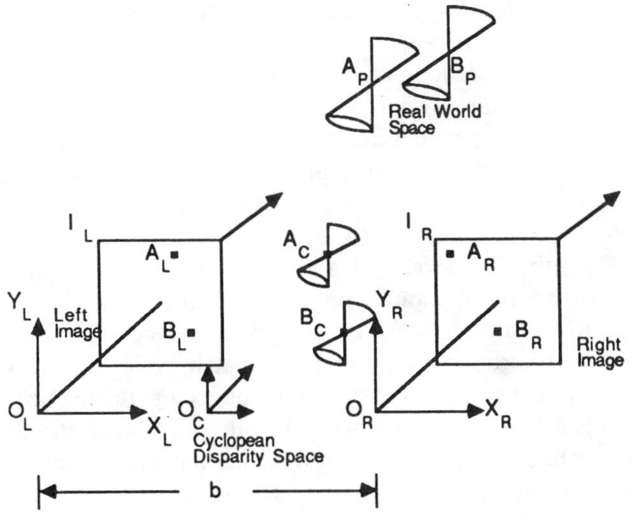

Fig. 6. Imaging geometry for PMF algorithm.

$D_g = (d_A - d_B)/d(A_C B_C)$, where $d(A_C B_C)$ is the cyclopean separation between A_C and B_C. A disparity gradient limit of 1 defines a cone-shaped forbidden zone for the point A_C in the cyclopean space, that is, any point within this forbidden zone violates the criterion for disparity gradient limit of 1.

Pollard, Mayhew, and Frisby [59] propose a new PMF algorithm that imposes a disparity gradient limiting constraint among correct matches. During the matching process, the matching strength of each potential match is evaluated as a sum of the support it receives from all potential matches in the neighborhood that satisfy the disparity limit criterion. In a match-pair of feature points (a_i, b_j), the support for the candidate match b_j is computed as a weighted sum of the number of potential matches in the neighborhood of b_j that have a disparity gradient less than 1, and vice-versa for the support of a_i. The PMF algorithm is interested only in the positive support to a potential match offered by surrounding matches that satisfy the within-disparity-gradient-limit criterion, and is unaffected by surrounding matches that exceed the disparity gradient limit. The uniqueness constraint is propagated using a discrete relaxation scheme such that if two image primitives a_i and b_j have the highest matching strength among their respective lists of candidate matches, then the match-pair (a_i, b_j) is considered as correct. Pollard, Mayhew, and Frisby [59] also show that since a disparity gradient limit in cyclopean space translates to a gradient limit in the real-world space it is possible even for planar surfaces to violate the disparity gradient limit criterion provided they have a sufficiently steep slope.

E. The Coherence Principle

Prazdny [61] has suggested the coherence principle to encompass the cohesiveness of matter [41] as well as the disparity continuity principles, which hold for opaque surface only. It recognizes the case of transparent objects. It allows the occurrence of a discontinuous disparity field if it is a result of several interlaced continuous disparity fields, each corresponding to a piecewise smooth surface. Two disparities facilitate each other if they possibly contain information about the same surface. When they do not interact at all they possibly contain information about different surfaces.

Prazdny [61] suggests one similarity function to quantify similarity between neighboring disparities. A Gaussian similarity function $s(i, j)$ is defined as

$$s(i, j) = \frac{1}{c|i-j|\sqrt{2\pi}} \exp\left[\frac{-|d_i - d_j|^2}{2c^2|i-j|^2}\right]. \qquad (5)$$

This algorithm uses the quantity $|d_i - d_j|/|i-j|$ in the exponent of the Gaussian that is the disparity gradient used in Julesz [9]. However, in the Burt and Julesz algorithm increase in disparity difference results in inhibition of support, whereas in the coherence principle of Prazdny [61], there is no inhibition. The disparity gradient used by Prazdny [61] is also similar to that in the PMF algorithm [59].

F. Multifeature-Based Matching

Kass [34] proposes the use of matching coefficients obtained from a large number of uncorrelated (independent) measurements to contribute towards the local matching constraint. If the local matching constraint is chosen appropriately it is postulated that a large fraction of the points in the first image will have only one potential match in the second image making it unnecessary to use global consistency measures. Kass [34] has used a stochastic image model to substantiate this computational framework.

The local matching constraint proposed by Kass [34] relies on a representation of local intensity variation in the form of functionals $f_i(p, I)$, $1 < i < n$ for each point p in image I. No single image measurement (functional) is expected to contain all the information about the correspondence of a pair of image points. The functionals have been chosen to be orthogonal (low cross-correlation), linear, shift-invariant operators. At each point p, a vector $F(p, I) = (f_1(p, I), f_2(p, I), \cdots, f_n(p, I))$ is formed. Since each functional $f_i(p, I)$ in the representation defines a similarity measure for correspondence, $F(p_L, I_L) - F(p_R, I_R)$ is expected to be very small in each component, if p_L and p_R are truly matched. If p_L and p_R do not correspond, $F(p_L, I_L) - F(p_R, I_R)$ will most probably have at least one large component.

A predicate $matchp(p_L, p_R)$ is defined such that $matchp(p_L, p_R)$ is true if and only if,

$$|f_i(p_L, I_L) - f_i(p_R, I_R)| < k_i \sigma(f_i(p, I_L)), \quad \forall\, i \in \{1, 2, \cdots, n\}$$

where $\sigma(x)$ denotes the square root of the expected value of x^2 and k_i are appropriate scaling constants. (p_L, p_R) is considered a correct match if $matchp(p_L, p_R)$ evaluates to true. In formulating matchp, Kass suggests one set of functionals $\mathscr{F}^* = \mathscr{F}_\sigma \cup \mathscr{F}_{\sigma s} \cup \mathscr{F}_{\sigma s^2}$ as the set of first and second partial derivatives of the Gaussian-smoothed images, with the sizes of the space constants being σ, σs, and σs^2, respectively. Each $\mathscr{F}_\sigma$ is the set of four partial derivatives with space constant σ as given by

$$\mathscr{F}_\sigma = \{\partial f_\sigma/\partial x, \partial f_\sigma/\partial y, \partial^2 f_\sigma/\partial x^2, \partial^2 f_\sigma/\partial y^2\}$$

f_σ being the Gaussian smoothing mask

$$f_\sigma = \frac{1}{s\pi\sigma^2}\exp\left[\frac{-(x^2+y^2)}{2\sigma^2}\right].$$

It is also shown that for synthetic stereo images derived from stationary Gaussian white noise if $s \geqslant 2.5$, the 12 functionals will have sufficiently low cross-correlations so that they can be regarded as approximately independent.

IV. Area-Based Stereo

Much of the earlier work done in stereo matching involves the use of correlation measures to match neighborhoods of points in the given images. Moravec [48] has used area-based correlation with a coarse-to-fine strategy to find corresponding match points. Initially feature points are identified in each image by the Moravec interest operator [48] that measures directional variance of image intensities in four directions surrounding a given pixel. Given a feature point P in one (source) image, the target image is searched at various resolutions ($\times 16$, $\times 8$, $\times 4$, and so on) starting from the coarsest. At each resolution the position in the target image that yields the highest correlation coefficient is enlarged to the next finer level of resolution. The process continues till the $\times 1$ resolution is reached. The same correlation process is applied to nine images taken two at a time to give 36 (${}^9C_2$) possible stereo disparity values for each point of interest. The disparities and correlation coefficients are combined into a histogram, and a confidence measure is defined based upon the histogram peak. Matches with a confidence measure above a certain threshold are accepted.

Gennery [18] developed a high-resolution correlator that used the matches provided by the previous correlation matcher and produced an improved estimate of the matching point based upon the statistics of noise in the image intensities. This high-resolution correlator not only provided improved match points but also gave an estimate of the accuracy of the match in the form of variances and covariance of the (x, y) coordinates of the match in the second image.

Hannah [23] developed a correlation-based stereo system for an autonomous aerial vehicle. A modified Moravec operator is used to select control points. Autocorrelation in the neighborhood of a candidate match point is used to evaluate the goodness of a match. Subpixel matching accuracy is achieved through parabolic interpolation of correlation values. In Stereosys [24], Hannah has implemented a hierarchical correlation-based stereo system. Images of lower resolution (say, $n \times n$) are obtained by smoothing $2n \times 2n$ images by a Gaussian window and resampling. The points for which matches are to be searched for are picked by an interest operator as in [23]. A hill-climbing procedure is used to search for a match-point whose neighborhood results in a maximum in normalized cross-correlation with that of the original interesting point. Matches are propagated over the finer resolution images in the hierarchy. Matches found at the finest level are checked by reversing the role of left and right images, and repeating the hierarchical search starting from the just-found matching point. These initial matches are used to guide the match point search of neighboring points using the disparity continuity constraint. Finally the sparse density map is interpolated to construct a dense disparity map.

V. Relaxation Process in Stereo

Relaxation labeling is a fairly general model proposed earlier by Rosenfeld, Hummel, and Zucker [67] for scene labeling. In the paradigm of matching a stereo pair of images using relaxation labeling, a set of feature points (nodes) are identified in each image, and the problem involves assigning unique labels (or matches) to each node

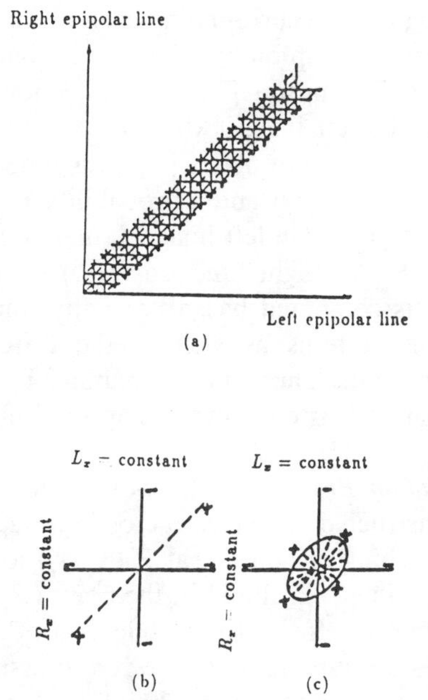

Fig. 7. Excitatory and inhibitory neighborhoods for Marr–Poggio cooperative algorithm. (a) Network of nodes for one scanline pair. (b) Linear excitatory neighborhood. (c) Disc-shaped excitatory neighborhood.

out of a discrete feature space (list of possible matches). For each candidate pair of matches, a matching probability is updated iteratively depending upon the matching probabilities of neighboring nodes so that stronger neighboring matches improve the chances of weaker matches in a globally consistent manner. This interaction between neighboring matches is motivated by the existence of cooperative processes in the biological vision systems postulated by Julesz [31], Julesz and Chang [32], Marr and Poggio [42], and others.

A. Marr–Poggio Cooperative Algorithm

Marr and Poggio [42] and Marr, Palm, and Poggio [40] have used the neighborhood information of matchable primitives in a simple iterative scheme. For each scanline pair in the stereo images (Fig. 7), a two-dimensional interconnected network of nodes (or cells) is set up. The horizontal and vertical connections are described as inhibitory, meaning all cells along each horizontal (or vertical) line inhibit each other, so that finally only one match remains on each horizontal (or vertical) line (uniqueness constraint). The diagonal connections are termed excitatory, meaning they favor diagonally adjacent matches to have the same disparity (disparity continuity). Fig. 7(b) shows the local disposition of the excitatory (+) and inhibitory (−) linkages in the neighborhood of a cell in the network. The bold lines (L_x = constant and R_x = constant) denote inhibitory interactions, and the dotted lines (diagonal, with slope = 1) denote excitatory interactions. A two-dimensional disparity continuity constraint can be effected by considering a disc-shaped excitatory neighborhood (Fig. 7(c)). In the cooperative process, let $C^t_{x,y;d}$ denote the state of a cell at time t corresponding to the coordinate (x, y) in

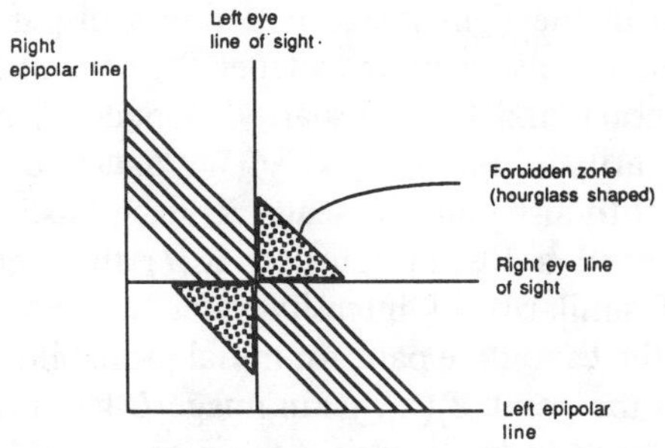

Fig. 8. Neighborhoods for Drumheller's implementation.

the left-image matching $(x + d, y)$ in the right image. Initially the nodes which represent possible stereo matchpoints are loaded with 1's and all others are loaded with 0's. Thus, when the iterations begin, each cell adds the states of the neighboring excitatory potential matches in $S(x, y, d)$ (the excitatory neighborhood) to the previous state, and subtracts from it a weighted sum of the states of the neighboring inhibitory potential matches in $O(x, y, d)$ (the inhibitory neighborhood). This iterative updating can be represented by the relation

$$C^{t+1}_{x,y;d} = \sigma\Big\{ \sum_{x',y';d' \in S(x,y,d)} C^t_{x',y';d'} - \epsilon \sum_{x',y';d' \in O(x,y,d)} C^t_{x',y';d'} + C^0_{x,y;d} \Big\} \quad (6)$$

where, ϵ is the weighting factor for the inhibitor effect, and σ is the threshold function. This algorithm was shown to obtain stereo fusion for random dot stereograms successfully. It represents a very simple mechanism for the propagation of the uniqueness and the continuity constraints among neighboring match-points for disambiguation of multiple stereo matches in an iterative manner.

Drumheller and Poggio [14] mapped the previous cooperative stereopsis model of Marr, Palm and Poggio [40] on the Connection Machine [28], using the north–east–west–south (NEWS) mechanism for near-neighbor communication. The uniqueness constraint in Drumheller and Poggio's implementation [14] imposed an hour-glass shaped forbidden zone (see Fig. 8) and did not allow more than one match in the entire forbidden zone, unless the scene contained transparent or narrow occluding objects. Other variations of the cooperative model are proposed by Prazdny [61], Pollard, Mayhew, Porrill, and Frisby [60] and Marroquin [43]. Barnard and Thompson [7] and Kim and Aggarwal [35] have used the principle of cooperative processing to formulate the relaxation-based algorithms which incorporate more complicated disambiguating constraints.

B. Barnard–Thompson Algorithm

1) Computation of Feature Attributes: Barnard and Thompson [7] extract feature points (nodes) from each image using the Moravec interest operator. Each node a_i, at position $Z_l(x_i, y_i)$ in the left image L is assigned a set of labels L_i that represent the possible candidate matches

$Z_R(x_j, y_j)$ in the right image R within a disparity range. Every label set also contains a label l^* in the initial stage, which denotes undefined disparity. A node a_i has undefined disparity if point $Z_l(x_i, y_i)$ in image L does not correspond to any point in image R. Each label l of node a_i is assigned a weight function $w_i(l)$ that reflects the degree of similarity of intensity values in the neighborhoods of the candidate pair. An initial probability estimate $p_i^0(l)$ that the point $Z_l(x_i, y_i)$ in image L has a disparity l is then derived from the weight function $w_i(l)$.

2) Relaxation Process: The initial probabilities $p_i^k(l)$ computed from similarity in gray-level intensity values surrounding the match points are now updated iteratively to impose global consistency. That is, the probability $p_i^k(l)$ is increased if the neighbors of a_i have high probability values for disparities close to l. In particular, at the kth iteration, for a node a_i at (x_i, y_i) having neighbors a_j at (x_j, y_j), a quantity $q_i^k(l)$ is defined as

$$q_i^k(l) = \sum_{\|l - l'\| \leqslant 1} p_j^k(l')$$

where only those neighbors a_j are considered whose disparity label l' differs from l by $\leqslant 1$ in both $x-$ and $y-$ directions. The $q_i^k(l)$ serves as a measure of consistency of disparity in the neighborhood because it increases if more neighbors of $a(i)$ have disparities closer to l. The probabilities $p_i^k(l)$ are updated at the kth iteration as

$$\hat{p}_i^{k+1}(l) = p_i^k(l) * \left(a + b * q_i^k(l)\right), \; l \neq l^*$$

and

$$\hat{p}_i^{k+1}(l^*) = \hat{p}_i^k(l^*) \tag{7}$$

where a is the rate constant to delay the suppression of unlikely labels (prevents $\hat{p}_i^{k+1}(l)$ from going to 0 if $q_i^k(l) = 0$), and b controls the speed of convergence.

The iterative procedure is continued either until the probabilities reach a steady state or a predetermined number of iterations are completed. This relaxation-based algorithm essentially imposes a disparity continuity constraint in the neighborhood of each matchpoint, favoring labels (disparities) consistent with the strong labels (disparities) occurring in the immediate neighborhood. This constraint is similar to the disparity continuity constraint over a region proposed by Marr–Poggio [41].

C. Kim–Aggarwal Algorithm

Kim and Aggarwal [35] propose a relaxation scheme that combines three disambiguating constraints, namely, continuity of disparity, figural continuity, and smoothness of probability (certainty) of matching. A conventional parallel-axis binocular setup is considered. Edge points are extracted by convolving each image with the *LoG* operator and locating the zero-crossings in the output.

1) Matching Primitives: A novel set of matching primitives is used. Depending upon the connectivity of the surrounding zero-crossings, 16 zero-crossing patterns are identified (see Fig. 9). A similarity measure is defined between two zero-crossing points depending upon the zero-crossing pattern surrounding each of them. The relaxation process is set up on lines similar to that explained in Barnard and Thompson [7]. The collection of all zero-crossings in the left image, which do not have horizontal patterns, form the set of nodes $\{a_i\}$. Each node is assigned a set of labels $L_i = \{l_j\}$ and a probability $p_i(l_j)$ that node a_i at point $Z_l(x_i, y_i)$ in left image L matches $Z_r(x_j, y_j)$ in right image R. A weight function $w_i(l_j)$ for node a_i with disparity l_j is computed based upon the similarity of the zero-crossing patterns as well as the difference in the intensity gradients. The initial probability $p_i^0(l_j)$ that node a_i has disparity l_j is computed using the weight functions $w_i(l_j)$.

2) Relaxation Process: A three-dimensional probability array is constructed on the zero-crossing map. The probability of matching of node a_i at $Z_l(x_i, y_i)$ to a point in R at disparity value l_j is stored in the point $(Z_l(x_i, y_i), l_j)$ in the 3-D array. In effect, it is a collection of 2-D arrays of probabilities corresponding to the zero-crossing map, with each 2-D array representing probabilities for one disparity value. $Z_l(x_f, y_f)$ and $Z_l(x_s, y_s)$ are the first and second neighboring zero-crossing points of $Z_l(x_i, y_i)$ (among the total 8 neighborhood points). The $p_i^k(l_j)$, $p_f^k(l_j)$, and $p_s^k(l_j)$ represent the entries in the 3-D array at positions $(Z_l(x_i, y_i), l_j)$, $(Z_l(x_f, y_f), l_j)$ and $(Z_l(x_s, y_s), l_j)$, respectively. The procedure for updating the matching probabilities is given by

$$p_i^{k+1}(l_j) = p_i^k(l_j) + c * F\left(p_i^k(l_j)\right) * \left(p_S^k\right) - d * p_i^k(l_j) * I(P_{FS}) \tag{8}$$

where,

$$p_F^k = \max\left[p_f^k(l_j - 1), p_f^k(l_j), p_f^k(l_j + 1)\right]$$

$$p_S^k = \max\left[p_s^k(l_j - 1), p_s^k(l_j), p_s^k(l_j + 1)\right]$$

$$F\left(p_i^k(l_j)\right) = \begin{cases} \left[p_i^k(l_j)\right]^2 & 0 \leqslant p_i^k(l_j) \leqslant 0.5 \\ p_i^k(l_j) * \left(1 - p_i^k(l_j)\right) & 0.5 < p_i^k(l_j) < 1 \end{cases}$$

$$I(P_{FS}) = \begin{cases} 0 & p_F^k + p_S^k \neq 0 \\ 1 & p_F^k + p_S^k = 0. \end{cases}$$

The foregoing formula combines three constraints—disparity continuity, figural continuity, and smoothness of probability of matching. The function $F(p_i^k(l_j))$ controls the rate of convergence in two ways: 1) It reduces the tendency to converge fast to the most probable disparity value so that less-probable values may still have chances to compete; and 2) If all other conditions are the same, the magnitudes of increases of higher probabilities are higher. In other words, the third term in (8) implements the figural continuity criterion proposed by Mayhew and Frisby [44]. The p_F^k and p_S^k check the existence of nonzero probabilities for a match with disparity l_j in the connected neighborhood of $Z_l(x_i, y_i)$. If the connected zero-crossings do not have disparities within the disparity gradient limit ± 1 of l_j, $I(P_{FS})$ is set and the probability $\hat{p}_i^{k+1}(l_j)$ is decre-

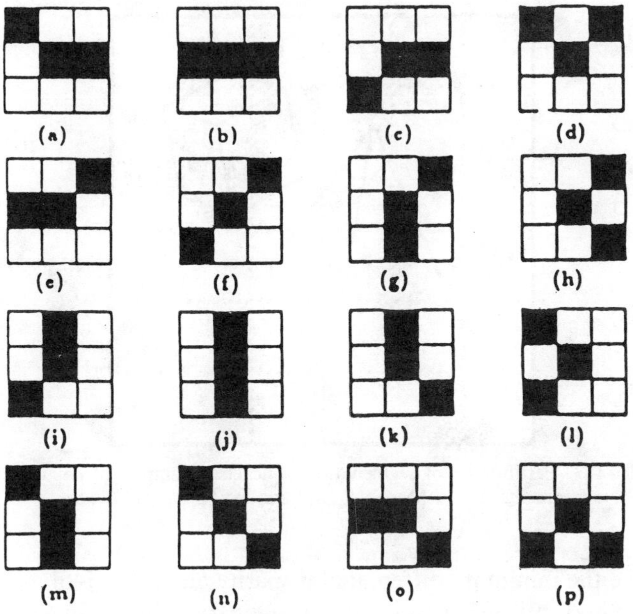

Fig. 9. Zero-crossing patterns.

mented. The second term in (8) reinforces the probability at $(Z_l(x_i, y_i), l_j)$ if the points in the neighborhood have a nonzero probability for disparities close to l_j by ± 1. This is similar to the area-based disparity continuity constraint which checks for surface smoothness and is in common with the implementation of Barnard and Thompson [7]. However the control of the rate of convergence using $F(p_i^k(l_j))$ is more flexible rather than a set of constants as used by Barnard and Thompson [7].

VI. Stereo Matching Using Edge Segments

The use of piecewise-linear approximations to connected edge points as matching primitives has been shown to be a viable alternative to matching of individual edge points ([1], [2], [25], [46]). Linear edge segments have certain advantages over single-edge points in the matching process. Firstly when edge points are grouped into a piecewise-linear segment, positional error at an isolated point has little effect on the position and orientation of the edge segment and most of the remaining edge points lie very close to the best fit. Secondly the edge connectivity constraint, which states that connected edge points in one image must match to connected edge points in the other image, must be imposed as an explicit disambiguating constraint while matching point-like features as against while matching line segments. On the other hand, due to possible fragmentation of edge segments during preprocessing, allowance has to be made for matching a single segment in one image with two or more segments in the other image, and vice versa.

A. Minimum Differential Disparity Algorithm

Medioni and Nevatia [46] describe a segment-based matching algorithm that uses a disparity continuity constraint called the minimal differential disparity criterion applied over neighboring edge segments.

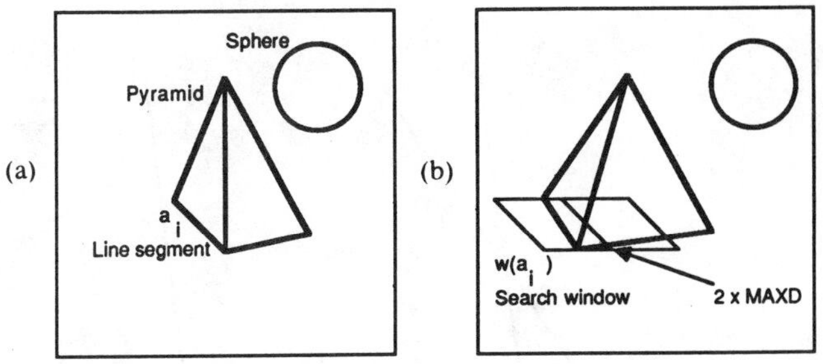

Fig. 10. Parallelogram-shaped search window. (a) Left image. (b) Right image.

1) Feature Extraction: The two stereo images, L and R, are brought into vertical alignment and a parallel-axis imaging geometry is assumed such that the epipolar lines run along horizontal scan lines. The feature-extraction stage described by Nevatia and Babu [52] is used to extract linear edge segments. Each edge segment is described by the coordinates of its end points, its orientation, and the average contrast in gray-level intensity (absolute value) along a direction normal to its orientation.

2) Matching Algorithm: Let $\mathscr{A} = \{a_i\}$ be the set of line segments in L, and $\mathscr{B} = \{b_j\}$ be the set of line segments in R. For each segment a_i in L, the search space is defined by a parallelogram-shaped window $w(a_i)$ in R whose one side is a_i and the other side is a horizontal vector of length $2 \times \text{MAX}\,D$, where $\text{MAX}\,D$ is the upper limit on the expected disparity (see Fig. 10). Similarly for each segment b_j in R, a window $w(b_j)$ is defined in L. Thus for a match (a_i, b_j), a_i lies in $w(b_j)$ and b_j lies in $w(a_i)$. Two segments x and y are said to be overlapping if by sliding either one of them along a direction parallel to the epipolar line, they can be made to intersect. Segments a_i in L and b_j in R can match only if a_i and b_j overlap, they have similar contrast in gray-levels, and have similar orientations. $\mathscr{S}_p(a_i) \subseteq w(a_i)$ denotes the set of all possible matches for a_i of L. A segment a_i in one image can be matched to two (or more) segments $b_{i1}, b_{i2}, \cdots, b_{in}$ in the other image provided none of the candidates $b_{i1}, \cdots, b_{in}$ overlap with each other.

An evaluation function $v^t(i, j)$ is computed iteratively to determine the merit of each match (a_i, b_j) as

$$\begin{aligned} v^{t+1}&(i, j) \\ &= \sum_{a_h \in w(b_j)} \min_{b_k \text{ verifies } C_1(a_h)} \lambda_{ijhk} |d_{hk} - d_{ij}| / \text{card}(b_j) \\ &\quad + \sum_{b_k \in w(a_i)} \min_{a_h \text{ verifies } C_2(b_k)} \lambda_{ijhk} |d_{hk} - d_{ij}| / \text{card}(a_i) \qquad (9) \end{aligned}$$

where λ_{ijhk} denotes the smaller of the overlap lengths for the match-pairs (a_i, b_j) and (a_h, b_k). The card (a_i) and card (b_j) are the number of segments in $w(a_i)$ and $w(b_j)$, respectively. Condition $C_1(a_h)$ allows for a_i and a_h to be matched to the same segment $b_j (= b_k)$ only if a_i and a_h do not overlap, and vice versa for condition $C_2(b_k)$. This allows for the possibility that if a_i and a_h are parts of a fragmented segment, they can get mapped to a single (unfragmented) segment b_j. The evaluation function $v(i, j)$ is updated during each iteration depending upon the dis-

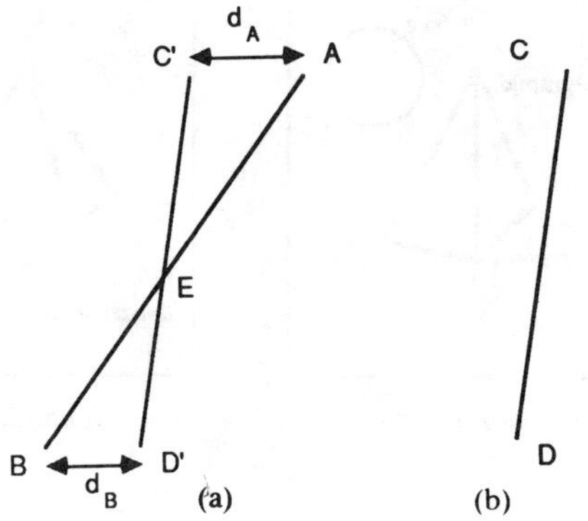

Fig. 11. Disparity change across linear segments. (a) Left image. (b) Right image.

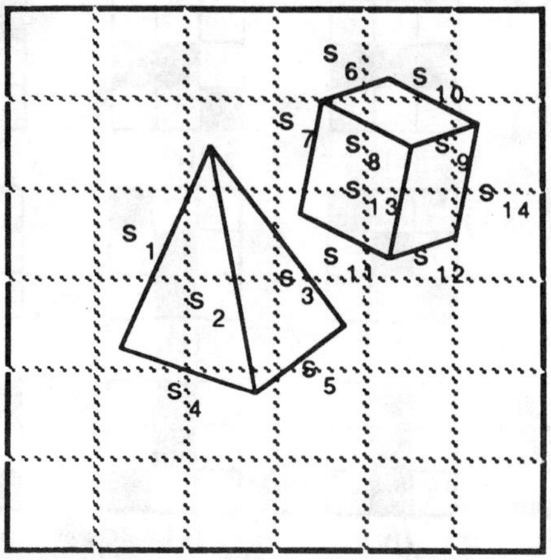
Fig. 12. Partitioning of segments into buckets. Examples of neighbors of segments are S_2: $\{S_1, S_3, S_4, S_5\}$ and S_3: $\{S_1, S_2, S_5, S_7, S_{11}\}$.

parities between the segments neighboring a_i and b_j, and their respective preferred matches. For each segment a_h in the window $w(b_j)$ (recall, $w(b_j)$ defines a neighborhood of a_i) a preferred match b_k is found such that, $|d_{hk} - d_{ij}|$ is minimized. During the first iteration, the selection of b_k for each a_h is done from among the complete set $\mathcal{S}_p(a_h)$ since the set of preferred matches is empty. For each a_i, the match which yields the lowest $v(i, j)$ is chosen as the preferred match.

Since this matching algorithm minimizes the disparity difference among matched line segments in a neighborhood it is termed as the minimum differential disparity algorithm. This, in effect, imposes a condition that the matched line segment pairs, when reconstructed in space, form 3-D contours of surfaces that are smooth almost everywhere. Thus this matching algorithm has implemented the surface continuity constraint proposed by Marr and Poggio [41] for a paradigm of stereo matching that uses line segments as matching primitives.

Recently, Mohan, Medioni, and Nevatia [47] have proposed a scheme to detect and correct local segment-matching errors based upon disparity variation across linear segments. Let AB and CD (Fig. 11) be matching linear segments (or linear approximations to segments), and let $C'D'$ be the position of CD when superposed such that pixels with zero disparity coincide. Then it can be shown that disparity varies linearly along the length of the matched segments, and

$$\frac{d_A}{|AE|} = \frac{d_B}{|EB|} = \text{constant} \tag{10}$$

where d_A and $-d_B$ are the disparities associated with the points A and B, respectively.

Next we examine the matching algorithm of Ayache and Faverjon [1] that implements the disparity gradient limit approach for imposing a surface smoothness constraint on the reconstructed scene, and for disambiguation of false matches within the framework of segment-based matching.

B. Ayache–Faverjon Algorithm

Ayache and Faverjon [1] use descriptions of edge segments with the coordinates of the midpoint, the length of the segment, and its orientation for stereo matching. Unlike the minimum differential disparity algorithm [46], this method utilizes a generalized nonparallel axis imaging geometry and uses disparity between midpoints of matching line segments rather than average disparity between corresponding points that lie on matching line segments. A neighborhood graph is used to store the information regarding the adjacency of line segments in each image and a disparity gradient limit criterion (defined for line segments) is used to guide the global correspondence search. A neighborhood graph is constructed for each image using nodes to represent edge segments, and links to connect the nodes satisfying certain neighborhood relationships. Thus, each segment s_j has a list of neighbors that is obtained as a union of buckets of segments $\{b_k\}$ attached to windows $\{w_k\}$ that it intersects (see Fig. 12). The global matching stage uses a specialized representation of potential matches called the disparity graph. The idea is to use the disparity graph to propagate these matches within their neighborhoods to recover subsets of 3-D segments lying on a smooth surface patch.

1) Local Matching Constraints: A pair of line segments a_i and b_j in the left and right images, respectively, constitutes a pair of potential matches if they satisfy the geometrical similarity constraint for line segments and their midpoints satisfy the epipolar constraint. A pair of edge segments whose length ratio and orientation difference lies below a preset threshold satisfies the geometrical similarity constraint. For the midpoint I_L of an edge segment a_i, a corresponding point I_R is searched for along the corresponding epipolar line near an expected disparity value. Ayache and Faverjon [1] compute disparity in the case of a pair of edge segments as follows (Fig. 13). If $\overline{P_R'Q_R'}$ (part of segment b_j) with midpoint I_L be a candidate match-segment for $\overline{P_LQ_L}$ (part of segment a_i) with center I_L, then the disparity d_{ij} between a_i and b_j is defined by

$$d_{ij} = E_RI_R - E_LI_L \tag{11}$$

where, E_L and E_R are the epipole centers in the left and right images, respectively.

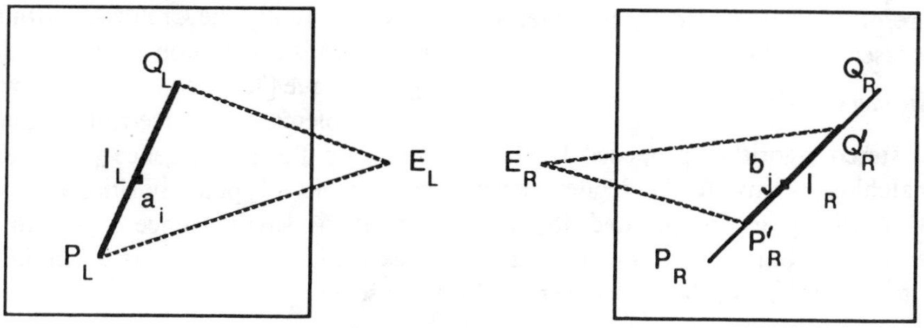

Fig. 13. Disparity for matched segment pair.

2) Global Matching Constraints: The global matching scheme of the Ayache–Faverjon algorithm [1] consists of a prediction and recursive propagation process. A disparity graph is constructed with nodes as pairs of potential matches (a_i, b_j) between the left and right images, and edges that connect pairs of nodes $(a_i, b_j),(a'_i, b'_j)$ that are adjacent segments in their respective neighborhood graphs. The allowable difference in disparity among neighboring nodes of matched pairs in the disparity graph is called the disparity gradient limit and corresponds to an ϵ variation in depth. For each node of the disparity graph (a_i, b_j), the neighborhood graphs of (a_i) and (b_j) are recursively explored for potential matched pairs that have disparities within the allowable disparity interval. Out of the potential matches the one with disparity closest to the predicted disparity is chosen. This favors those matches of line segments that make the 3-D scene maximally smooth in the sense of surface continuity as proposed by Marr and Poggio [41].

VII. Hierarchical Approaches to Stereo Matching

In this section we consider algorithms that utilize a hierarchical computational structure for stereo matching. The hierarchical structure of the algorithms allows matching information to be interchanged amongst various levels of matching computations, thus imposing global consistency in the disparity map. Apart from the Marr–Poggio–Grimson algorithm [19], [41] considered in Section III, the computational models of Terzopoulos [71], Hoff and Ahuja [29], and Lim and Binford [37] are some examples of hierarchical approaches to stereo matching.

A. Concurrent Multilevel Relaxation

Terzopoulos [71], [72] has developed an efficient multilevel relaxation computational model for low-level visual processing in concurrent mode. Conventional multigrid schemes employ recursive coordination of computations and flow of intermediate results starting from the coarsest level and proceeding successively to the finest level. Results at any level are used as approximations for the next level. With the advent of massively parallel architectures, such sequential algorithms result in inefficient use of hardware because most of the time is spent performing relaxations on only a single level, while processors at other levels (if configured in a multilevel architecture) remain idle. The concurrent strategy of Terzopoulos [71] maintains processors on all levels busy performing simultaneous relaxation operations. The concurrent strategy seeks to optimize a multilevel objective functional, with each term having three components: 1) A discrete version of the given functional at each level of a multigrid hierarchy; 2) an additive functional coupling each level (except the finest) to its next finer level; and 3) an additive functional coupling each level (except the coarsest) to the next coarser level. A concurrent multigrid algorithm for the problem of computing visible surface representations as formulated in [70] has been implemented.

B. Surfaces from Stereo: An Integrated Approach

Hoff and Ahuja [29] have argued in favor of integrating the steps of stereo matching and surface interpolation. Objects have faces that have a smooth variation in surface normals. Object surface meet on ridges that are smooth (or piecewise smooth) curves in 3-D space. They propose an integration of the matching and surface fitting processes in a way that the correctness of the choice of matches could be judged by the type of surface it produces.

Consider a stereo pair of $4n \times 4n$ images. Edge points are extracted using the Laplacian of Gaussian ($\nabla^2 G$) operator at three resolutions $- n \times n$, $2n \times 2n$, and $4n \times 4n$. Initial matching is performed in both left-to-right and right-to-left directions. For each feature point P_i in, say, the left image a set of candidate match points $\{Q_i\}$ is selected from the right image according to similarity of local (or geometric) properties of feature points. A set of parameterized functions, planar and quadratic, are fitted to circular image regions centered at each grid point (x, y) in sequence. First, up to two planar patches are chosen at each grid point (x_i, y_i) that give the best least-squares fit-rating with the observed disparity z_i. Secondly, quadratic patches are fitted at each grid point to the above combinations of matches. The quadratic surface containing the most points is kept as the best fit for that grid point. Next, depth and orientation contours are detected by fitting bipartite planar patches and detecting discontinuities between the two halves. The bipartite planar patches are actually circular patches divided into two halves by a diameter with a given 3-D orientation. Finally a smooth surface is interpolated away from contours to yield a piecewise-smooth surface map at each resolution. Matching of edges at finer resolutions is guided by the interpolated surface at the coarser resolution.

C. From Objects to Surfaces to Edgels

In the hierarchical stereo algorithm proposed by Lim and Binford [37], matching begins at the highest level (objects). Results of matching are propagated to each successive lower level (surface boundaries, junctions, and edgels) and are used to guide the matching of lower-level features.

Edgels are detected using the Nalwa operator [50] that fits a tanh surface to each window in the image. Edges are linked into connected edges and curves (straight lines or conic sections) are fitted using best-fit criteria of Nalwa and Pauchon [51]. Surfaces are identified by tracing the boundaries of connected curves using both left-wall following as well as right-wall following strategies. Bodies are identified as groups of surfaces that share edges. The ordering information of surfaces in a body in the left-to-right as well as the top-to-down directions is saved to be used later as a matching constraint.

Matching of bodies is attempted at the highest level. Bodies that lie within the limits of corresponding upper and lower epipolar lines (i.e., having the same extent) are candidate matches. Multiple candidate matches are disambiguated using the left-to-right ordering of bodies along epipolar lines, the number of surfaces in the bodies, and the ordering of surfaces in the bodies. Next the system attempts to match surfaces that have the same extent and so on, down to edge segments and edgels. The advantage of this hierarchical stereo system is that the depth map obtained is already segmented and ready for surface interpolation.

D. Hierarchical Stochastic Optimization

Barnard [5] has implemented a solution to the stereo matching problem using a stochastic optimization technique called microcanonical annealing. Poggio, Torre, and Koch [58] have posed the stereomatching problem as imposing a regularization criterion on the stereo images,

$$\min: \epsilon = \int\int \left\{ \left[\nabla^2 G \circ \left(I_L(x, y) - I_R(x + D(x, y), y) \right) \right]^2 + \lambda (\nabla D)^2 \right\} dx\, dy \quad (12)$$

where $I_L(x, y)$ and $I_R(x, y)$ are continuous intensity functions in the left and right images, respectively, $\nabla^2 G$ is the *LoG* operator, ∇D is the gradient of disparity, and λ is a constant. The first term of the integrand in (12) can be understood as a measure of the difference in image brightness values of corresponding points, and the second term as a measure of disparity gradient.

In the discrete version, (12) can be represented as minimizing the total potential energy $E = \Sigma E(i, j)$. Finding a disparity map $D(x, y)$ that results in the minimal energy constitutes a solution to the stereo correspondence problem. A stochastic optimization technique called microcanonical annealing using the Creutz algorithm [12] is used to control the combinatorial explosion of the search involved. Actually the Creutz algorithm [12] (microcanonical annealing) is a variation of the standard simulated annealing technique [36] used for solving combinatorial optimization problems. A coarse-to-fine method of computation speeds up the convergence process. At the coarser level, the number of pixel positions as well as the range of disparity is small. Hence a ground state can be reached quickly which can serve as an initial estimate for the next finer scale.

VIII. Stereo Matching by Dynamic Programming

Baker and Binford [3] use the Viterbi algorithm, a dynamic programming technique, to partition the stereo matching problem recursively based upon the constraint that a left-to-right ordering of edges is preserved along a scanline in a stereo image pair. In this edge-based technique, each edge is treated as a doublet, with a left half-edge and a right half-edge. The dynamic programming procedure is repeatedly applied for matching edge points on each scanline pair. The first and second passes of the Viterbi algorithm (preliminary edge correlation) match half-edges in the left image to those in the right image, and vice-versa. Next, a cooperative procedure uses an edge connectivity constraint to identify surface contours that are not continuous in disparity. That is, a connected sequence of edges in one image should match a connected sequence of edges in the other (both L-to-R and R-to-L). Finally, an intensity-based Viterbi correlation performed between intensity pixels from scanline intervals lying between the paired edges in the two images yields a denser depth map.

Ohta and Kanade [54] have also used pixel intensities of scanline intervals (delimited by edge points) to guide the *intrascanline* matching search by dynamic programming. This intrascanline search is formulated as a path-finding problem in a 2-D search space in which vertical and horizontal axes are the right and left scanlines, respectively. This is achieved by defining a cost function associated with each partial path based upon variances of gray-level intensities of the scanline intervals being matched. The edges are numbered from left to right on each scanline, with two ends of each scanline being also treated as nodes. If there are M nodes in the left scanline and N nodes in the right scanline, the solution to the intra-scanline search could be represented as a path comprised of a sequence of straight lines form node (0,0) to node (M, N) with the optimum cost. The cost of the optimal path from node (0,0) to node m is denoted by $D(m)$, and is the sum of the costs of its primitive paths. A primitive path between nodes k and m is a partial path that contains no vertices as in Fig. 14. The cost of the optimal path $D(m)$ is obtained by recursively adding the cost of each newly added primitive path to the already existing partial optimal path. The results of this intrascanline search are used to establish global consistency among matches achieved in neighboring scanlines using an *interscanline* search. The interscanline search is aimed at imposing a consistency

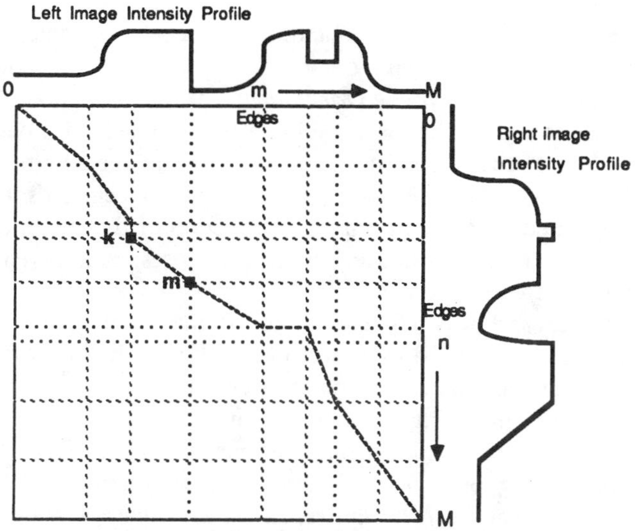

Fig. 14. 2-D search plane for intra-scanline search.

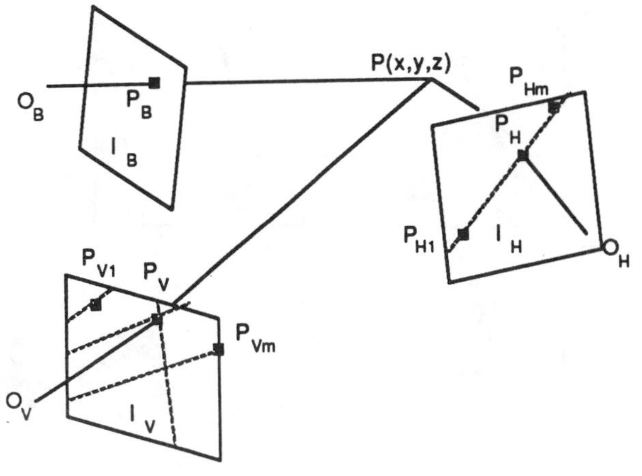

Fig. 15. Trinocular imaging geometry.

constraint among matches obtained at each scanline using edge connectivity. The problem is posed as that of finding the least-cost path between 3-D nodes in a 3-D search space. Each 3-D node is formed as a collection of the 2-D nodes connected across scanlines. The optimal path in the 3-D search space is obtained by recursively adding an optimal 3-D primitive path to the existing optimal partial path.

The approaches of Baker and Binford [3], and Ohta and Kanade [54] are based on the assumption that the ordering of edges remains unchanged for a stereo pair. The ordering of corresponding edges does not remain intact in scenes having large differences in depth, especially if these features were derived from thin, ribbon-like overlapping objects. Such a system is also liable to get confused in case of scenes with repetitive features especially if some of the features are missing in one of the images.

IX. Trinocular Stereo

The trinocular approach to the stereo problem has been proposed recently as an alternative means to conduct the correspondence search. The basic advantage of the third camera has been the extra epipolar geometry constraints offered by the three cameras. Provided the centers of projection of the three cameras are noncolinear, the true match points in the three images satisfy the condition that they must lie on the conjugate epipolar lines of the other two cameras. This allows for disambiguation of the multiple candidate matches that are found during local binocular-type correspondence search.

A. Edge-Based Trinocular Stereo

Yachida, Kitamura, and Kimachi [76] use an edge-based trinocular algorithm to obtain 3-D information about objects. Consider three cameras with centers of projection O_B, O_H, and O_V at known positions and with their optical axes having known orientations (Fig. 15). The *trinocular epipolar constraint* works as follows: For any point P_B in the image plane I_B, let there be multiple match point candidates $\{P_{H_1}, P_{H_2}, \cdots, P_{H_m}\}$ along the epipolar line l_{BH}. A set of epipolar lines $\{l_{HV_i}\}$ is constructed in I_V for each candidate $P_{H_i} \in \{P_{H_1}, P_{H_2}, \cdots, P_{H_m}\}$. At the intersection P_{V_i} of each l_{HV_i} and l_{BV}, the presence of an edge point P_{V_i} is tested. Each triplet (P_B, P_{H_i}, P_{V_i}) is tested for local similarity of feature attributes, and the best match is considered. In case some matching ambiguities still persist, the matchpoint candidate that yields a disparity closest to that of the points in the surrounding neighborhood is considered the best match. Ito and Ishii [30] have proposed a trinocular algorithm that uses a similar epipolar search procedure and a matching coefficient based upon the difference in gray-level intensity values in a 5×5 neighborhood of the candidate points. If any of the edge points do not get matched in the first pass due to occlusion, special one-sided matching coefficients are used in a second pass to match occluded points.

Ohta, Watanabe, and Ikeda [55] use a third camera and a relaxation procedure to improve the depth map obtained from binocular stereo. The camera geometry involves a left (L), a right (R), and an upper (U) camera, all having axes parallel to each other. The two image pairs $L-U$ and $L-R$ are processed independently using binocular stereo [54] to give two separate depth maps, H-depth and V-depth, respectively. The H-depth and V-depth values thus obtained are then combined into one depth image using a relaxation process. In this scheme, the trinocular geometry is used only to provide additional depth values that would be available from using two simultaneous binocular matching processes operating on mutually orthogonal epipolar lines.

Peitikäinen and Harwood [57] have used a three-view system with a parallel-axis geometry. The camera geometry involves a base camera (B) and two other cameras, H and V, displaced in the horizontal and vertical directions, respectively. Local features of the edge points like edge orientations and intensity contrast are used as local similarity attributes. In addition to the trinocular epipolar constraints, a postprocessing algorithm using connectivity of contours is also used to disambiguate multiple matches.

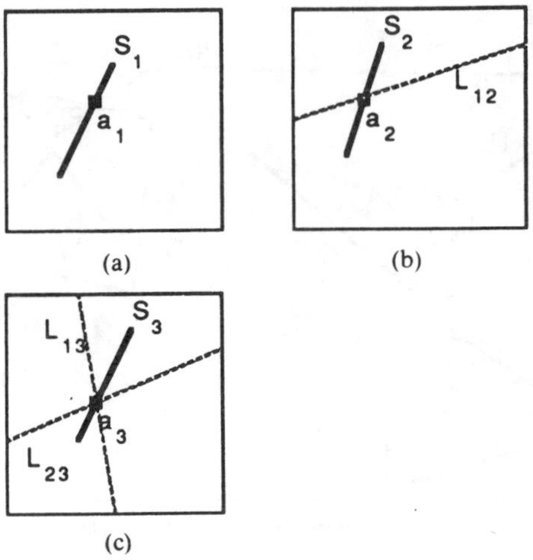

Fig. 16. Trinocular segment matching. (a) Image 1. (b) Image 2. (c) Image 3.

B. Segment-Based Trinocular Stereo

Ayache and Lustman [2], and Hansen, Ayache, and Lustman [25] have applied the segment-based (binocular) matching technique of Ayache and Faverjon [1] to three views.

In [2], Ayache and Lustman employ a prediction and verification scheme using neighborhood graphs of linear segments in three images that is an extension of the earlier binocular algorithm [1]. For any segment S_1 in image 1, if a triplet (S_1, S_2, S_3) can be found to satisfy the trinocular epipolar constraint of lines L_{12}, L_{23}, and L_{13} (see Fig. 16), and have sufficient similarity in local geometric properties then it is retained as a potential triplet.

Subsequently Hansen, Ayache, and Lustman [25] made further developments in the trinocular matching system using image rectification in the preprocessing stage. The original images are reprojected, as shown in Fig. 17 (for a binocular system), on a new image plane that is parallel to the plane containing the centers of projection of the cameras. As a result, the conjugate epipolar lines become parallel to each other in an image, and align themselves with the image coordinate frame. This reduces the search for matches to the horizontal and vertical lines, thus speeding up the matching process. A problem occurs when, due to some noise in preprocessing, a single segment (say, S_1) in the image 1 gets broken up into two (or more) segments, say S_2 and S_2' in image 2 (see Fig. 18). Hansen, Ayache, and Lustman [25] handle the problem by allowing flexibility in the order in which images are traversed for hypothesis generation. The problem of broken segments was also mentioned earlier by Peitikäinen and Harwood [57].

X. Structural Stereopsis

Boyer and Kak [8] have proposed the use of *structural descriptions* of image primitives and certain *information theoretic measures* defined on the basis of the structural descriptions to formulate the stereo matching problem.

The structural description of each image is derived from a skeletal or *stick-figure* representation of objects in the scene. The edges of the skeletons form a set of primitives over which the following binary scalar parametric relations are defined. *1*) *Pairwise Orientation*: The mean orientation of the straight line segment joining the centroids of a pair of skeletal edges. *2*) *Distance*: A function of the length of the straight line distance joining the two centroids of skeletal edge. *3*) *End-distance*: A function of the length of the straight line distance joining the closest pair of end points between two edges.

Boyer and Kak [8] have modified the exact matching approach developed by Shapiro and Haralick [69] in favor of an information theoretic approach for achieving inexact structural matching by defining interprimitive distance measures and relational inconsistency measures. The stereomatching problem is framed as a consistent labeling problem. The set of primitives in the left image $\boldsymbol{P} = \{p_i\}$ form the object set and the set of primitives in the right image $\boldsymbol{Q} = \{q_j\}$ form the label set. The labeling process utilizes two kinds of information: knowledge about the attributes of each label ($\mathscr{L}$) and knowledge about the relationship between labels ($\mathscr{R}$). $\mathscr{L}$ captures the information regarding the primitive distortion process (due to perspective effects as well as noise effects) and consists of a set of conditional probabilities of an attribute taking on a specific value in the right image, given its value in the left image. $\mathscr{R}$ captures the changes in the values of relational constraint parameters. It consists of a set of conditional probabilities, each item in the set being the probability that a relational parameter would take on a particular value in the right image having known its value in the left image. An event $p_i^{q_j}$ is defined as the left image primitive p_i being assigned to the right image primitive q_j in the stereo mapping h. The solution to the consistent labeling problem is considered optimal if the prob$[p_1^{k_1}, p_2^{k_2}, \cdots, p_n^{k_n}]$ is maximum, given the information in $\mathscr{L}$ and $\mathscr{R}$. That is

$$\max \text{OPM}: \text{prob}\left[p_1^{k_1}, p_2^{k_2}, \cdots, p_n^{k_n} | \mathscr{L}, \mathscr{R}\right] = \text{prob}\left[h | \mathscr{L}, \mathscr{R}\right]. \quad (13)$$

where, OPM is the optimal probability measure. The following basic assumptions are made in this probabilistic model: 1) Information in $\mathscr{L}$ is independent of the information in $\mathscr{R}$. This is based upon the idea that relational information is perceived by higher-level cognition processes that may be independent of lower-level processes required for the perception of primitive attributes. 2) The *a priori* probabilities of any particular event $p_i^{k_i}$ is constant, which translates to the fact that no advance information is available about the correct mapping function. Based upon these assumptions (13) becomes

$$\text{OPM} = \left(\text{prob}\left[\bigcap_i p_i^{k_i} \varepsilon \mathscr{L}\right]\right) \cdot \left(\text{prob}\left[\bigcap_i p_i^{k_i} \varepsilon \mathscr{R}\right]\right). \quad (14)$$

The two terms in the right hand side expression are referred to as the $\mathscr{L}$-term and $\mathscr{R}$-term. An information-theoretic interprimitive distance measure $\text{DIST}_h(\boldsymbol{P}, \boldsymbol{Q})$ is formulated to represent the dissimilarity between the sets of primitives $\boldsymbol{P}$ and $\boldsymbol{Q}$ under a specific mapping h: $\boldsymbol{P} \rightarrow \boldsymbol{Q}$,

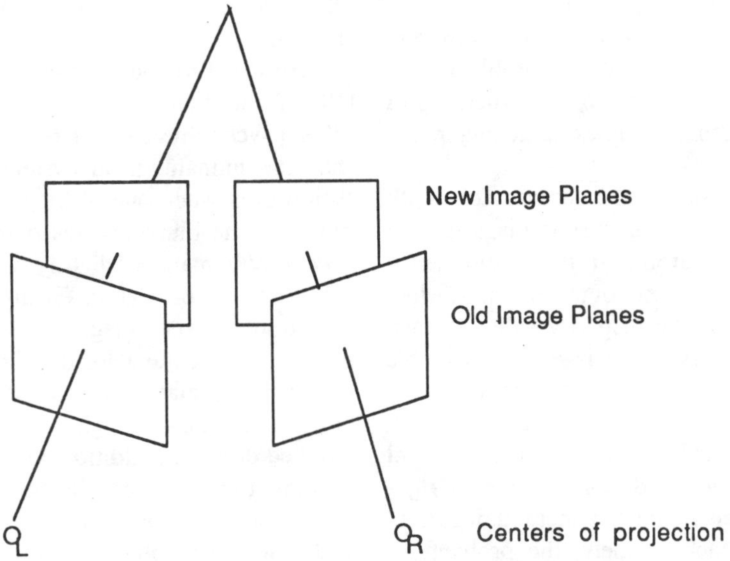

Fig. 17. Rectification of two images by reprojection.

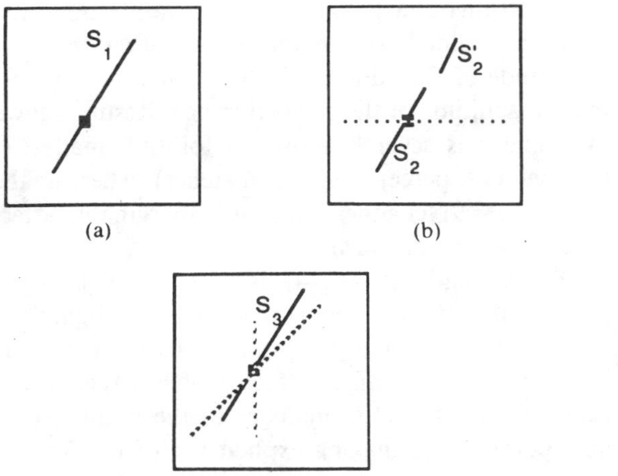

Fig. 18. Matching broken segments. (a) Image 1. (b) Image 2. (c) Image 3.

given the information $\mathscr{L}$ about the primitive attribute distortion process between the two images. Also, a relational inconsistency measure $\mathrm{INC}_h(\boldsymbol{R}, \boldsymbol{S})$ is formulated to measure the distortion of relational parameters between the sets of primitives ($\boldsymbol{R}$ and $\boldsymbol{S}$ represent parametric relations between elements of the sets of primitives $\boldsymbol{P}$ and $\boldsymbol{Q}$, respectively). The DIST_h and INC_h are defined to be $\mathrm{DIST}_h(\boldsymbol{P}, \boldsymbol{Q}) = -\log[\mathscr{L}$ term] and $\mathrm{INC}_h(\boldsymbol{R}, \boldsymbol{S}) = -\log[\mathscr{R}$ term]. Then (14) becomes

$$\text{Min: } \mathrm{DIST}_h(\boldsymbol{P}, \boldsymbol{Q}) + \mathrm{INC}_h(\boldsymbol{R}, \boldsymbol{S}). \qquad (15)$$

The matching of the structural descriptions of the two images is performed in the following steps. For each primitive p_i in the left image, a match pool of potential primitives $\{q_j\}$ is obtained. This is achieved by accessing a look-up table of attributes and computing the distance between the two primitives. Any right primitive whose distance from a left primitive lies within a certain threshold is included in the match pool. The match pool for each left primitive is then stored in a best-first order. A *nilmap entry* is added as the last entry of any match pool if the cost associated with the best-fitting primitive in that pool were to exceed a certain threshold value, which signifies that the particular primitive in the left image may not have a matching primitive in the right image. Finally the consistent labeling problem is solved using a backtracking tree search. Out of the resulting list of possible mappings between the two primitive sets, the one that has the lowest value for $\mathrm{INC}_h(\boldsymbol{R}, \boldsymbol{S}) + \mathrm{DIST}_h(\boldsymbol{P}, \boldsymbol{Q})$ is chosen as the solution to the consistent labeling formulation of the stereo matching problem.

XI. Results and Discussion

In this section we shall review the experimental results of some of the stereo matching algorithms and the characteristics of the test images used therein. Testing of stereo algorithms has not been standardized as yet in the research community. Different algorithms have each been tested on different sets of stereo images. Without standardized test procedures, it is difficult to comment on the relative merits of stereo algorithms. However, one can identify the classes of images that have been used to test the algorithms, examine their performance, and know more about the domain of applicability of the algorithms.

The scene domains used for testing stereo algorithms have ranged from simple blocks world images to outdoor/aerial scenes. The block world images ([8], [30], [46], [54], [55], [76]) are typically scenes depicting an assortment of objects with polyhedral, cylindrical, conical, or spherical surfaces characterized by sharp physical boundaries and/or surface markings, all laid against a sharply contrasting background. Since the features being matched are few and most of them correspond to object boundaries, these images serve well as test images. Indoor (laboratory) scenes ([1], [2], [25], [57]) represent a higher degree of complexity in that the background of objects is no longer controlled. This makes the matching task more complicated. Also, many straight line edges (provided by doors, windows, and furniture) with their repetitive structure add to the complexity of the correspondence problem. The outdoor/aerial scenes are by far the most unstructured of scene domains and pose more complex matching problems.

Secondly, the task of computing the accuracy of depth estimates and the correctness of matches is plagued with the problem of lack of reliable ground truth measurements. For example, in the case of random dot stereograms, exact knowledge is available about the disparity value at each pixel in the stereo pair. Such exactness is seldom available in natural outdoor scenes or indoor laboratory scenes. Hence accuracy of the depth estimates is, at best, determined at a few selected points by actual measurement and compared with the results obtained from the stereo algorithm. Finally, stereopsis being a passive method, it suffers from the additional drawbacks namely, the problem of false matches and the sparseness of the resulting depth maps.

The test images used can be broadly classified into the following categories:

1) Psychophysical test patterns,
2) Indoor scenes,
3) Synthetic scenes,
4) Outdoor/aerial scenes.

We shall discuss, in brief, the experimental results obtained in each category.

A. Psychophysical Test Patterns

Grimson [19] used random dot stereograms to compare the performance of the computer implementation of the Marr–Poggio theory [41] of stereo fusion with the results of numerous psychophysical experiments conducted on the human vision system. Random dot stereograms are pairs of images, each consisting of two or more planar patches of random dots such that when a stereo pair is fused by humans a 3-D structure can be perceived. Since the disparity value is known at each pixel position for random dot stereograms, they can be used to test the correctness of the algorithm's performance. Typical 3-D structures used by Grimson [19], [21] include a square block rising out of a planar background, a series of square planes arranged on top of each other like a wedding cake, and a rectangular staircase pattern. For most random dot stereograms, a 50 percent dot density was used. Each stereogram was analyzed at four spatial channels with $w = 4$, 9, 17 and 35 pixels. Disparities obtained at coarser channels were used to guide the fusion at finer channels. In case of the pattern with a central square separated in depth from a second plane, out of 11 847 zero-crossing points only three (roughly 0.03 percent) were wrongly matched. Similar patterns with dot densities 25 percent, 10 percent, and 5 percent gave percent mismatch errors of 0.07 percent, 0.04 percent, and 0.06 percent, respectively. The wedding cake pattern (at 50 percent dot density) gave 0.06 percent mismatches. Almost all of the mismatches occurred at the boundary between the planes.

Grimson also found that the computer implementation [19] of the Marr–Poggio theory was in agreement with other psychophysical test results. Julesz [31] found earlier that the human vision system could perform binocular fusion even when one of the images of the stereo pair was blurred. The blurring caused the flat surfaces to be perceived by humans as slightly warped; nevertheless the 3-D structure was preserved. Grimson used Gaussian smoothing to blurr one image of the random dot stereogram before running the algorithm on the computer. The resulting disparity map was consistent with the 3-D structure, but it had a slightly higher number of errors in the reconstructed depth. In addition, Grimson studied [19] the effect of adding low-frequency as well as high-frequency noise to the random dot patterns. The results were in agreement with the psychophysical evidence found by Julesz and Miller [33] that stereo fusion is possible for noisy stereograms if the spectrum of the noise is sufficiently far from the spectrum of the pattern. In one example, high-frequency noise was added to one image such that the maximum magnitude of the added noise was twice that of the maximum magnitude of the original image. Results showed that matching was severely impaired for the smallest ($w = 4$) channel (17 percent wrong matches) whereas the next larger ($w = 9$) channel was only marginally affected (6 percent wrong matches).

Mayhew and Frisby [44] used stereograms of textured patterns in order to support the role of the figural continuity constraint in their computational theory of stereopsis. They report a significant (factor of 35) reduction in the ratio of potential false matches to the number of matchable points, after making explicit use of the figural continuity constraint.

Pollard, Mayhew, and Frisby [59] have tested the disparity gradient limit approach for imposing global consistency among matches using random dot stereograms. They report 98 percent correct matches for random dot stereograms that have disparity gradients up to 1.0. The matching performance degrades to about 50 percent correct matches for a disparity gradient of 1.8.

B. Indoor Scenes

The simplest of the indoor scenes are composed of a few objects scattered against a featureless (usually dark) background. Several stereo algorithms ([20], [30], [35], [46], [54], [55], [57], and [76]) have been tested on blocks world images. Grimson [20] presents results of matching stereo images of dark blocks placed against a bright background. In a typical blocks world scene, out of 2703 zero-crossing points as many as 1780 (65.9 percent) are reported to have been matched. The difficulty of matching blocks world images increases in the presence of occluding objects and repetitive features on the object surfaces. Medioni and Nevatia [46] and Ohta and Kanade [54] each report a matching example of a blocks scene (containing the Rubik cube) that has the aforementioned characteristics. Ohta and Kanade [54] compared the number of mismatches (or inconsistencies in matching) before and after the interscanline search for the blocks world scene. The global constraint imposed by the interscanline search was shown to reduce the number of mismatches by more than a factor 4. Ayache and Faverjon [1] and Medioni and Nevatia [46] have reported the matching result of a stereo image pair depicting an industrial part. The objects being viewed are essentially similar and provide a means for comparing the performance of stereo algorithms on common ground.

Indoor scenes of real-life laboratory environments have also been used for verifying stereo matching algorithms. Moravec's [48] stereo algorithm was used by a mobile cart to navigate its way around obstacles. The stereo system identified world position, the height of each obstacle, and the associated positional error caused by the pixel resolution of the camera and built an internal map of its immediate surroundings. The cart made successive short runs punctuated by halts during which the internal map was updated. The cart made successful runs in both indoor and outdoor environments. Kim and Aggarwal [35] tested their algorithm on indoor room scenes. Estimated depth was checked against actual (measured) depth at a few selected points. Percent error in depth varied between 0.17 percent and 3.7 percent. Percentage of false matches was as low as 2 percent for an optimum choice of parameters in the relaxation process. Ayache and Faverjon [1], and Hansen, Ayache, and Lustman [25] used indoor scenes for segment-based matching. The results report a maximum of 2 percent mismatches after applying global consistency validation.

C. Synthetic Scenes

Barnard and Thompson [7], Medioni and Nevatia [46], and Ohta and Kanade [54] have used a synthetic image (obtained from Control Data Corporation) for testing their algorithms. The correctness of matches was checked manually. Ito and Ishii [30] have tested their trinocular matching algorithm using a synthetic image of a pyramid-shaped block. Accuracy of depth estimates was found at selected points and compared with the actual depth. The process was then repeated with an actual block placed under similar conditions. In the experimental results with synthetic as well as real blocks world images of Ito and Ishii [30], the estimated maximum percent positional error of the selected 3-D points was within ± 0.5 percent with the maximum percent measured error.

D. Outdoor Scenes

Grimson [20] tested his implementation of the Marr–Poggio theory on a number of aerial terrain images ("Phoenix" and "Ft. Sill" images). An interesting case depicts a highway interchange scene ("Boeing" image) that consists of a number of thin, elongated, and closely-spaced contours, each at different depths. The difficulty caused by such "spaghetti" contour scenes is evident by comparing the percentage matching errors of the highway aerial scene with that of the urban aerial scene (obtained from the Univ. of British Columbia, Vancouver) consisting mostly of buildings and a few roads. The same matching algorithm that resulted in 0.07 percent matching errors for the urban scene gave as high as 2.53 percent errors in the highway interchange scene. Ohta and Kanade [54] have also tested their algorithm on aerial scenes of the Washington D.C. area ("Pentagon" and "White House" images).

E. Discussion

One of the major differences among the different stereo algorithms discussed in this paper is the way they handle the global consistency of the matches obtained. As was mentioned earlier in Section II-B, a stereo algorithm can detect false positive matches obtained as a result of the local matching procedure by looking for other matches in the neighborhood that are consistent in disparity with that particular match. The disparity value for a given match is easily translated to a depth estimate by inverting the perspective projection equations. The prime motivation for imposing some sort of continuity constraint on the disparity values is that a mismatch would result in a disparity value that would translate into a strikingly discontinuous depth estimate as compared to the other neighboring points.

Marr and Poggio [41] have proposed a coarse-to-fine approach for propagation of the disparity continuity in the neighborhood of the matches. A purely region-based approach, as in [41], for imposing disparity continuity does not work very well when the scene is composed of a large number of thin, ribbon-like overlapping objects at various depths that partially overlap each other. This was recognized by many researchers like Grimson [20], Mayhew and Frisby [44], Kim and Aggarwal [35], Baker and Binford [3], and Barnard and Thompson [7], who among others have also included a figural continuity constraint in their stereo implementations. In the figural continuity constraint, a potential match (i, j) is favored if all their connected, neighboring matches (h, k) also have similar disparities. Figural continuity is used in segment-based matching in an implicit manner by which connected edge points are grouped together into segments and are matched as a group. Medioni and Nevatia [46] apply the minimal-differential-disparity rule for edge segments (a_i, b_j) by taking into account the disparity of each of the edge points in the segments and then taking an average disparity d_{ij}. In the segment-based matching of Ayache and Faverjon [1], for a match pair (a_i, b_j), the disparity d_{ij} does not explicitly take into account the disparities of the individual points in the edge segments but is computed using the positions of midpoint of a_i (I_L) and its corresponding potential match in segment b_j (I_R). The edge segments are treated as one unit, and the disparity of neighboring edge segment matches is constrained by a disparity gradient limit (defined specially for edge segments).

Kim and Aggarwal [35] have used a relaxation (cooperative) algorithm that uses a smoothness constraint on the probability of matching, in addition to the aforementioned constraints of regional continuity of disparity and figural continuity. Also, rather than using individual edge points or edge segments, they use 16 distinct edge (zero-crossing) patterns as matching primitives.

A disparity gradient limit approach is proposed in the PMF algorithm by Pollard, Mayhew, and Frisby [59] as an alternative to the figural continuity criterion, and it allows for matching of smooth as well as jagged surfaces.

The left-to-right (L-to-R) ordering of edges has also been used as a global matching constraint to disambiguate multiple matches and identify false positive matches. Baker and Binford [3], and Ohta and Kanade [54] have used the L-to-R ordering constraint in their dynamic programming algorithm to do the intrascanline search. In the Ayache–Faverjon algorithm [1], an L–R ordering relationship is used in building the disparity graph of edge segments from the left and right images. The disparity graph guides the prediction of matching hypotheses and thus controls the matching search. However, it must be noted that an L–R ordering constraint is not universally valid for guiding a binocular search. In the presence of transparent objects and/or thin, ribbon-like objects (also called spaghetti contours), differences in depth could result in the reversal of the L–R ordering of feature primitives on a scanline.

Apart from the factors discussed previously, the performance of various stereo algorithms can be dependent upon a lot of implementation details like the choice of threshold factors and the rate constants used to control the convergence of iterative algorithms. Also the behavior of a stereo algorithm in widely different scene domains needs to be understood carefully before choosing any one algorithm to be used in an application.

In this paper the authors have presented a broad review of the major recent developments in stereo algorithms. Experimental results of a variety of computational techniques have been grouped according to the scene domains on which the tests were conducted, and comparisons are made wherever possible. However it must be noted that the matching statistics like percentage error in depth, and percentage of mismatches mentioned in this paper appear as they were quoted in the respective technical publications of the said author(s) and were not observed under strictly identical conditions. Hence, if the reader is interested in building an application of a stereovision depth finder for a specific scene domain, a certain degree of caution needs to be exercised in the interpretation of the performance statistics and in understanding the trade-offs between various approaches.

XII. Conclusion

In this paper we have presented a review of the major techniques developed in the recent past for recovering the 3-D structure of a scene from analysis of stereo images. We have outlined the three main stages of stereo analysis, namely, preprocessing, establishing correspondence, and recovering depth. Based upon the differences in matching primitives as well as the imaging geometry being used, distinctions were made between area-based and feature-based matching, between parallel-axis and nonparallel axis stereo, between point-based and segment-based matching, and between binocular and trinocular matching.

We described the computational theory of stereopsis formulated by Marr and Poggio [41], which is motivated by a model of the human stereo vision system, and that formulates the basic constraints of uniqueness and regional continuity. Mayhew and Frisby developed further upon the figural continuity [44] and disparity gradient limit [59] criteria that impose global consistency constraints in order to disambiguate false matches. In the successive sections, we describe the different approaches developed for solving the stereo correspondence problem: area-based matching [18], [48], relaxation labeling [7], [35], dynamic programming [54], hierarchical approaches [5], [29], [37], [71], segment-based matching [1], [46], trinocular matching (edge-based [30], [55], [57], [76], as well as segment-based [2], [25]), and structural matching [8]. The performance of various approaches was discussed for different classes of test images and the difficulties involved in the evaluation of stereo algorithms were addressed.

The major issue involved in the stereo analysis of images is the correspondence problem. Algorithms need to be improved to give a lower percentage of false matches as well as better accuracy of depth estimates. Performance of algorithms needs to be evaluated over a broad range of image types in order to test their robustness. Most of the stereo work done so far has been limited to developing basic stereo matching capabilities for working with simplistic images. A great deal of research in stereo is needed in order to not only overcome the abovementioned difficulties but also to apply stereo techniques to solve more real-world problems.

Acknowledgment

The authors wish to thank E. Grimson and various anonymous reviewers for several insightful suggestions that enhanced the usefulness of this paper. They are grateful to J. J. Rodríguez and F. Arman for their helpful comments and for careful proofreading of this paper.

References

[1] N. Ayache and B. Faverjon, "Efficient registration of stereo images by matching graph descriptions of edge segments," *Int. J. Comput. Vision*, pp. 107–131, 1987.

[2] N. Ayache and F. Lustman, "Fast and reliable trinocular stereovision," in *Proc. 1st Int. Conf. Comput. Vision*, June 8–11, 1987, pp. 422–427.

[3] H. H. Baker and T. O. Binford, "Depth from edge and intensity based stereo," in *Proc. 7th Int. Joint Conf. Artificial Intell.*, Vancouver, Canada, Aug. 1981, pp. 631–636.

[4] D. H. Ballard and C. M. Brown, *Computer Vision*. Englewood Cliffs, NJ: Prentice-Hall, 1982.
[5] S. T. Barnard, "Stochastic stereo matching over scale," in *Proc. DARPA Image Understanding Workshop*, Cambridge, MA, pp. 769–778, Apr. 6–8, 1988.
[6] S. T. Barnard and M. A. Fischler, "Computational stereo," *Comput. Surveys*, vol. 14, no. 4, pp. 553–572, Dec. 1982.
[7] S. T. Barnard and W. B. Thompson, "Disparity analysis of images," *IEEE Trans. Pattern Anal. Machine Intell.*, vol. PAMI-2, no. 4, pp. 333–340, July 1980.
[8] K. L. Boyer and A. C. Kak, "Structural stereopsis for 3-D vision," *IEEE Trans. Pattern Anal. Machine Intell.*, vol. PAMI-10, no. 2, pp. 144–166, Mar. 1988.
[9] P. Burt and B. Julesz, "A disparity gradient limit for binocular fusion," *Science*, vol. 208, pp. 615–617, 1980.
[10] ___, "Modifications of the classical notion of Panum's fusional area," *Perception*, vol. 9, pp. 671–682, 1980.
[11] J. F. Canny, "A computational approach to edge detection," *IEEE Trans. Pattern Anal. Machine Intell.*, vol. PAMI-8, no. 6, pp. 679–698, Jan. 1985.
[12] M. Creutz, "Microcanonical Monte Carlo simulation," *Physical Rev. Lett.*, vol. 50, pp. 1141–1414, 1983.
[13] R. Deriche, "Using Canny's criteria to derive a recursively implemented optimal edge detector," *Int. J. Computer Vision*, vol. 1, no. 2, May 1987.
[14] M. Drumheller and T. Poggio, "On parallel stereo," in *Proc. IEEE Int. Conf. Robotics and Automation*, Apr. 7–10, 1986, pp. 1439–1448.
[15] A. L. Duwaer and G. van den Brink, "Diplopia thresholds and the initiation of vergence eye movements," *Vision Res.*, vol. 21, pp. 1727–1737, 1981.
[16] ___, "What is the diplopia threshold?," *Perception Psychophys.*, vol. 29, pp. 295–309, 1981.
[17] J. P. Frisby and J. E. W. Mayhew, "The role of spatial frequency tuned channels in vergence control," *Vision Res.*, vol. 20, pp. 727–732, 1981.
[18] D. B. Gennery, "Object detection and measurement using stereo vision," in *Proc. ARPA Image Understanding Workshop*, College Park, MD, Apr. 1980, pp. 161–167.
[19] W. E. L. Grimson, "A computer implementation of a theory of human stereo vision," *Phil. Trans. Royal Soc. London*, vol. B292, pp. 217–253, 1981.
[20] ___, "Computational experiments with a feature-based stereo algorithm," *IEEE Trans. Pattern Anal. Machine Intell.*, vol. PAMI-7, no. 1, pp. 17–34, Jan. 1985.
[21] ___, *From Images to Surfaces: A Computational Study of the Human Early Visual System*. Cambridge, MA: M.I.T. Press, 1981.
[22] W. E. L. Grimson and E. C. Hildreth, "Comments on digital step edges from zero-crossings of second directional derivatives," *IEEE Trans. Pattern Anal. Machine Intell.*, vol. PAMI-7, no. 1, pp. 121–126, Jan. 1985.
[23] M. J. Hannah, "Bootstrap stereo," in *Proc. ARPA Image Understanding Workshop*, College Park, MD, Apr. 1980, pp. 201–208.
[24] ___, "SRI's baseline stereo system," in *Proc. DARPA Image Understanding Workshop*, Miami Beach, FL, Dec. 1985, pp. 149–155.
[25] C. Hansen, N. Ayache, and F. Lustman, "High-speed trinocular stereo for mobile-robot navigation," in *Proc. NATO Adv. Res. Workshop Highly Redundant Sensor Systems*, Il Chiocco, Italy, May 16–20, 1988.
[26] R. M. Haralick, "Author's reply," *IEEE Trans. Pattern Anal. Machine Intell.*, vol. PAMI-7, no. 1, pp. 126–128, Jan. 1985.
[27] ___, "Digital step edges from zero crossing of second directional derivatives," *IEEE Trans. Pattern Anal. Machine Intell.*, vol. PAMI-6, no. 1, pp. 58–68, Jan. 1984.
[28] D. Hillis, "The connection machine," Ph.D. dissertation, Dept. Elect. Eng. Comput. Sci., M.I.T., Cambridge, MA, 1985.
[29] W. Hoff and N. Ahuja, "Surfaces from stereo: Integrating feature matching, disparity estimation, and contour detection," *IEEE Trans. Pattern Anal. Machine Intell.*, vol. PAMI-11, no. 2, pp. 121–136, Feb. 1989.
[30] M. Ito and A. Ishii, "Three view stereo analysis," *IEEE Trans. Pattern Anal. Machine Intell.*, vol. PAMI-8, no. 4, pp. 524–532, July 1986.
[31] B. Julesz, *Foundations of Cyclopean Perception*. Chicago, IL: Univ. of Chicago Press, 1971.
[32] B. Julesz and J. J. Chang, "Interaction between pools of binocular disparity detectors tuned to different disparities," *Biol. Cybern.*, vol. 22, pp. 107–120, 1976.
[33] B. Julesz and J. E. Miller, "Independent spatial frequency tuned channels in binocular fusion and rivalry," *Perception*, vol. 4, pp. 125–143, 1975.
[34] M. Kass, "Computing visual correspondence," in *Proc. DARPA Image Understanding Workshop*, Arlington, VA, June 1983, pp. 54–60.
[35] Y. C. Kim and J. K. Aggarwal, "Positioning 3-D objects using stereo images," *IEEE J. Robotics and Automation*, vol. RA-3, no. 4, pp. 361–373, Aug. 1987.
[36] P. J. M. van Laarhoven and E. H. L. Aarts, *Simulated Annealing: Theory and Applications*. Dordrecht, Holland: D. Riedel Publishing Co., 1987.
[37] H. S. Lim and T. O. Binford, "Stereo correspondence: A hierarchical approach," in *Proc. DARPA Image Understanding Workshop*, Los Angeles, CA, pp. 234–241, Feb. 1987.
[38] D. Marr, *Vision*. San Francisco, CA: Freeman, 1982.
[39] D. Marr and E. Hildreth, "Theory of edge detection," *Proc. Royal Soc. London*, vol. B207, pp. 187–217, 1980.
[40] D. Marr, G. Palm, and T. Poggio, "Analysis of a cooperative stereo algorithm," *Biol. Cybern.*, vol. 28, pp. 223–229, 1978.
[41] D. Marr and T. Poggio, "A computational theory of human stereo vision," *Proc. Royal Soc. London*, vol. B204, pp. 301–328, 1979.
[42] ___, "Cooperative computation of stereo disparity," *Science*, vol. 194, pp. 283–287, 1976.
[43] J. L. Marroquin, "Design of cooperative networks," AI Lab, Mass. Inst. Technol., Cambridge, MA, working paper 253, 1983.
[44] J. E. W. Mayhew and J. P. Frisby, "Psychophysical and computational studies towards a theory of human stereopsis," *Artificial Intell.*, vol. 17, pp. 349–385, 1981.
[45] ___, "Rivalrous texture stereograms," *Nature*, vol. 264, pp. 53–56, 1976.
[46] G. Medioni and R. Nevatia, "Segment-based stereo matching," *Comput. Vision, Graphics, Image Processing*, vol. 31, pp. 2–18, 1985.
[47] R. Mohan, G. Medioni, and R. Nevatia, "Stereo error detection, correction, and evaluation," *IEEE Trans. Pattern Anal. Machine Intell.*, vol. PAMI-11, no. 2, pp. 113–120, Feb. 1989.
[48] H. P. Moravec, "Towards automatic visual obstacle avoidance," in *Proc. 5th Int. Joint Conf. Artificial Intell.*, 1977, p. 584.
[49] P. Mowforth, J. E. W. Mayhew and J. P. Frisby, "Vergence eye movements made in response to spatial frequency filtered random dot stereograms," *Perception*, vol. 10, pp. 299–304, 1981.
[50] V. S. Nalwa and T. O. Binford, "On detecting edges," *IEEE Trans. Pattern Anal. Machine Intell.*, vol. PAMI-8, no. 6, pp. 699–714, Nov. 1986.
[51] V. S. Nalwa and E. Pauchon, "Algorithms for edgel aggregation and edge description," in *Proc. DARPA Image Understanding Workshop*, Miami Beach, FL, Dec. 1985, pp. 176–185.
[52] R. Nevatia and K. Babu, "Linear feature extraction and description," *Comput. Graphics, Image Processing*, vol. 13, pp. 257–269, 1980.
[53] K. R. K. Nielsen and T. Poggio, "Vertical image registration in human stereopsis," AI Lab, Mass. Inst. Technol., Cambridge, MA, Memo 743, 1983.
[54] Y. Ohta and T. Kanade, "Stereo by intra- and inter-scanline search," *IEEE Trans. Pattern Anal. Machine Intell.*, vol. PAMI-7, no. 2, pp. 139–154, Mar. 1985.
[55] Y. Ohta, and M. Watanabe, and K. Ikeda, "Improving depth map by trinocular stereo," in *Proc. 8th Int. Conf. Pattern Recognition*, Paris, France, Oct. 27–31, 1986, pp. 519–521.
[56] T. Pavlidis, *Structural Pattern Recognition*. New York: Springer-Verlag, 1977.
[57] M. Peitikäinen and D. Harwood, "Depth from three camera stereo," in *Proc. IEEE CS Conf. Pattern Recognition*, Miami Beach, FL, June 22–26, 1986, pp. 2–8.
[58] T. Poggio, V. Torre, and C. Koch, "Computational vision and regularization theory," *Nature*, vol. 317, pp. 314–319, 1985.
[59] S. B. Pollard, J. E. W. Mayhew, and J. P. Frisby, "PMF: A stereo correspondence algorithm using a disparity gradient limit," *Perception*, vol. 14, pp. 449–470, 1981.
[60] S. B. Pollard, J. E. W. Mayhew, J. Porrill, and J. P. Frisby, "Disparity gradient, Lipschitz continuity, and computing binocular correspondences," U. of Sheffield, Artificial Intelligence Vision Research Unit, Tech. Rep. 010, 1985.

[61] K. Prazdny, "Detection of binocular disparities," *Biol. Cybernetics*, vol. 52, pp. 93–99, 1985.

[62] G. V. S. Raju, T. O. Binford, and S. Shekhar, "Stereo matching using Viterbi algorithm," in *Proc. DARPA Image Understanding Workshop*, Los Angeles, CA, Feb. 23–25, 1987, pp. 766–776.

[63] C. Rashbass and G. Westheimer, "Disjunctive eye movements," *J. Physiology*, vol. 159, pp. 339–360, 1961.

[64] C. Rashbass and G. Westheimer, "Independence of conjunctive and disjunctive eye movements," *J. Physiology*, vol. 159, pp. 361–364, 1961.

[65] L. A. Riggs and E. W. Niehl, "Eye movements recorded during convergence and divergence," *J. Opt. Soc. Amer.*, vol. 50, pp. 913–920, 1960.

[66] G. Robinson, "Edge detection by compass gradient mask," *Comput. Graphics Image Processing*, vol. 6, pp. 492–572, 1977.

[67] A. Rosenfeld, R. A. Hummel, and S. W. Zucker, "Scene labeling by relaxation operation," *IEEE Trans. Syst. Man Cybern.*, vol. SMC-6, pp. 420–423, June 1976.

[68] A. Rosenfeld and A. C. Kak, *Digital Picture Processing.* New York: Academic Press, 1976.

[69] L. G. Shapiro and R. M. Haralick, "Structural descriptions and inexact matching," *IEEE Trans. Pattern Anal. Machine Intell.*, vol. PAMI-3, no. 5, pp. 504–519, Sept. 1981.

[70] D. Terzopoulos, "Computing visible-surface representations," AI Lab, Mass. Inst. Technol., Cambridge, MA, Memo 800, 1985.

[71] ——, "Concurrent multilevel relaxation," in *Proc. DARPA Image Understanding Workshop*, Miami Beach, FL, Dec. 1985, pp. 156–161.

[72] ——, "Multilevel computational processes for visual surface reconstruction," *Comput. Vision Graphics Image Processing*, vol. 24, pp. 52–96, 1983.

[73] V. Torre and T. A. Poggio, "On edge detection," *IEEE Trans. Pattern Anal. Machine Intell.*, vol. PAMI-8, no. 2, pp. 147–163, Mar. 1986.

[74] G. Westheimer and D. E. Mitchell, "The sensory stimulus for disjunctive eye movements," *Vision Res.*, vol. 9, pp. 749–755, 1969.

[75] R. H. Williams and D. H. Fender, "The synchrony of binocular saccadic eye movements," *Vision Res.*, vol. 17, pp. 303–306, 1969.

[76] M. Yachida, Y. Kitamura, and M. Kimachi, "Trinocular vision: New approach for correspondence problem," in *Proc. 8th Int. Conf. Pattern Recognition*, Paris, France, Oct. 27–31, 1986, pp. 1041–1044.

Umesh R. Dhond (SM'89) was born in Bombay, India, on March 13, 1964. He received the B.Tech. degree in electrical engineering from Indian Institute of Technology, Bombay, in 1985, and the M.S. degree in electrical and computer engineering from Louisiana State University, Baton Rouge, in 1987.

He is currently a Research Assistant at Computer and Vision Research Center, The University of Texas at Austin, and is working towards the Ph.D. degree. His research interests include computer vision, image processing, and artificial intelligence.

J. K. Aggarwal (S'62–M'65–SM'74–F'76) received the B.S. degree in mathematics and physics from the University of Bombay, India, in 1956, the B.Eng. degree from the University of Liverpool, England, in 1960, and the M.S. and Ph.D. degrees from the University of Illinois, Urbana, in 1961 and 1964, respectively.

He joined the University of Texas in 1964 as an Assistant Professor and has since held positions as Associate Professor (1968) and Professor (1972). Currently he is the John J. McKetta Energy Professor of Electrical and Computer Engineering and Computer Sciences at the University of Texas, Austin. Further he was a Visiting Assistant Professor at Brown University, Providence, RI (1968), and a Visiting Associate Professor at the University of California, Berkeley, during 1969–70. He has published numerous technical papers and several books, *Notes on Nonlinear Systems* (1972), *Nonlinear Systems: Stability Analysis* (1977), *Computer Methods in Image Analysis* (1977), *Digital Signal Processing* (1979), and *Deconvolution of Seismic Data* (1982). His current research interests are image processing and computer vision.

Dr. Aggarwal is an active member of IEEE Computer Society, ACM, AAAI, the International Society for Optical Engineering, the Pattern Recognition Society, and Eta Kappa Nu. He was Co-Editor of the Special Issue on Digital Filtering and Image Processing of the IEEE TRANSACTIONS ON CIRCUITS AND SYSTEMS, March 1975, and on Motion and Time Varying Imagery, IEEE TRANSACTIONS PATTERN ANALYSIS AND MACHINE INTELLIGENCE, November 1980, and Editor of the two-volume Special Issue on Motion of *Computer Vision, Graphics and Image Processing*, January and February 1983. He was the General Chairman for the IEEE Computer Society Conference and Pattern Recognition and Image Processing, Dallas, TX, 1981, and was the Program Chairman for the First Conference on Artificial Intelligence Applications sponsored by the IEEE Computer Society and AAAI, Denver, CO, 1984. Currently he is an Associate Editor of the journals *Pattern Recognition*, *Image and Vision Computing*, and *Computer Vision, Graphics and Image Processing*. Further, he is a member of the IEEE Transnational Relations Committee, member of the Editorial Board of *IEEE Press*, and the Chairman of the IEEE Computer Society Technical Committee on PAMI.

The Stanford Cart and the CMU Rover

HANS P. MORAVEC, MEMBER, IEEE

Invited Paper

Reprinted from *Proceedings of the IEEE*, Vol. 71, No. 7, July 1983, pages 872-884. Copyright © 1983 by The Institute of Electrical and Electronics Engineers, Inc. All rights reserved.

Abstract—The Stanford Cart was a remotely controlled TV-equipped mobile robot. A computer program was written which drove the Cart through cluttered spaces, gaining its knowledge of the world entirely from images broadcast by an on-board TV system. The CMU Rover is a more capable, and nearly operational, robot being built to develop and extend the Stanford work and to explore new directions.

The Cart used several kinds of stereopsis to locate objects around it in three dimensions and to deduce its own motion. It planned an obstacle-avoiding path to a desired destination on the basis of a model built with this information. The plan changed as the Cart perceived new obstacles on its journey.

The system was reliable for short runs, but slow. The Cart moved 1 m every 10 to 15 min, in lurches. After rolling a meter it stopped, took some pictures, and thought about them for a long time. Then it planned a new path, executed a little of it, and paused again. It successfully drove the Cart through several 20-m courses (each taking about 5 h) complex enough to necessitate three or four avoiding swerves; it failed in other trials in revealing ways.

The Rover system has been designed with maximum mechanical and control system flexibility to support a wide range of research in perception and control. It features an omnidirectional steering system, a dozen on-board processors for essential real-time tasks, and a large remote computer to be helped by a high-speed digitizing/data playback unit and a high-performance array processor. Distributed high-level control software similar in organization to the Hearsay II speech-understanding system and the beginnings of a vision library are being readied.

By analogy with the evolution of natural intelligence, we believe that incrementally solving the control and perception problems of an autonomous mobile mechanism is one of the best ways of arriving at general artificial intelligence.

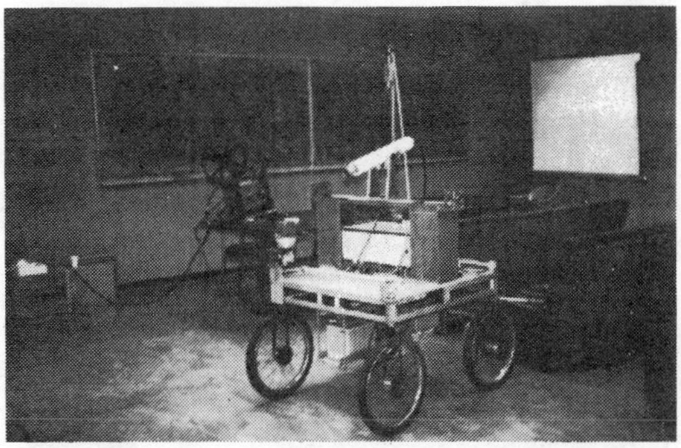

Fig. 1. The Stanford Cart.

Fig. 2. The Cart on an obstacle course.

INTRODUCTION

EXPERIENCE with the Stanford Cart [8], [9], [11], a minimal computer-controlled mobile camera platform, suggested to me that, while maintaining such a complex piece of hardware was a demanding task, the effort could be worthwhile from the point of view of artificial intelligence and computer vision research. A roving robot is a source of copious and varying visual and other sensory data which force the development of general techniques if the controlling programs are to be even minimally successful. By contrast, the (also important) work with disembodied data and fixed robot systems often focuses on relatively restricted stimuli and small image sets, and improvements tend to be in the direction of specialization. Drawing an analogy with the natural world, I believe it is no mere coincidence that in all cases imaging eyes and large brains evolved in animals that first developed high mobility.

Manuscript received December 17, 1982; revised February 23, 1983. The Stanford Cart work conducted at the Stanford University Artificial Intelligence Laboratory, was supported over the years 1973–1980 by the Defense Advanced Research Projects Agency, the National Science Foundation, and the National Aeronautics and Space Administration. The CMU Rover has been supported at the Carnegie-Mellon University Robotics Institute since 1981 by the Office of Naval Research under Contract N00014-81-0503.

The author is with the Robotics Institute, Carnegie-Mellon University, Pittsburgh, PA 15213.

THE STANFORD CART

The Cart [10] was a minimal remotely controlled TV-equipped mobile robot (Fig. 1) which lived at the Stanford Artificial Intelligence Laboratory (SAIL). A computer program was written which drove the Cart through cluttered spaces, gaining its knowledge of the world entirely from images broadcast by the on-board TV system (Fig. 2).

The Cart used several kinds of stereo vision to locate objects around it in three dimensions (3D) and to deduce its own motion. It planned an obstacle-avoiding path to a desired destination on the basis of a model built with this information. The plan changed as the Cart perceived new obstacles on its journey.

The system was reliable for short runs, but slow. The Cart moved 1 m every 10 to 15 min, in lurches. After rolling a meter it stopped, took some pictures, and thought about them for a long time. Then it planned a new path, executed a little of it, and paused again.

It successfully drove the Cart through several 20-m courses (each taking about 5 h) complex enough to necessitate three or four avoiding swerves. Some weaknesses and possible im-

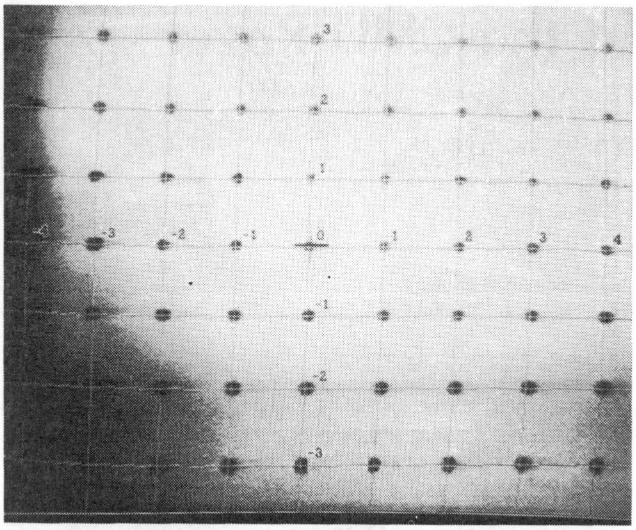

Fig. 3. A Cart's eye view of the calibration grid. The Cart camera's focal length and distortions were determined by parking the Cart a precise distance in front of, and aiming its camera at, a carefully painted array of spots pattern, and running a calibration program. The program located the spots in the image and fitted a 2D, third-degree polynomial which converted actual positions in the image to coordinates in an ideal unity focal length camera. The picture presented here was obtained by running a corresponding grid in the unity focal length frame through the inverse function of the polynomial so obtained and superimposing it on the raw spot image.

provements were suggested by these and other, less successful, runs.

A Cart Run

A run began with a calibration of the Cart's camera. The Cart was parked in a standard position in front of a wall of carefully painted spots. A calibration program noted the disparity in position of the spots in the image seen by the camera with their position predicted from an idealized model of the situation. It calculated a distortion correction polynomial which related these positions, and which was used in subsequent ranging calculations (Fig. 3).

The Cart was then manually driven to its obstacle course (littered with large and small debris) and the obstacle-avoiding program was started. It began by asking for the Cart's destination, relative to its current position and heading. After being told, say, 50 m forward and 20 to the right, it began its maneuvers. It activated a mechanism which moved the TV camera, and digitized nine pictures as the camera slid in precise steps from one side to the other along a 50-cm track.

A subroutine called the *Interest Operator* was applied to one of these pictures. It picked out 30 or so particularly distinctive regions (features) in this picture. Another routine called the *Correlator* looked for these same regions in the other frames (Fig. 4). A program called the *Camera Solver* determined the 3D position of the features with respect to the Cart from their apparent movement from image to image (Fig. 5).

The *Navigator* planned a path to the destination which avoided all the perceived features by a large safety margin. The program then sent steering and drive commands to the Cart to move it about a meter along the planned path. The Cart's response to such commands was not very precise. The camera was then operated as before, and nine new images were acquired. The control program used a version of the Correlator to find as many of the features from the previous location as possible in the new pictures, and applied the camera solver. The program

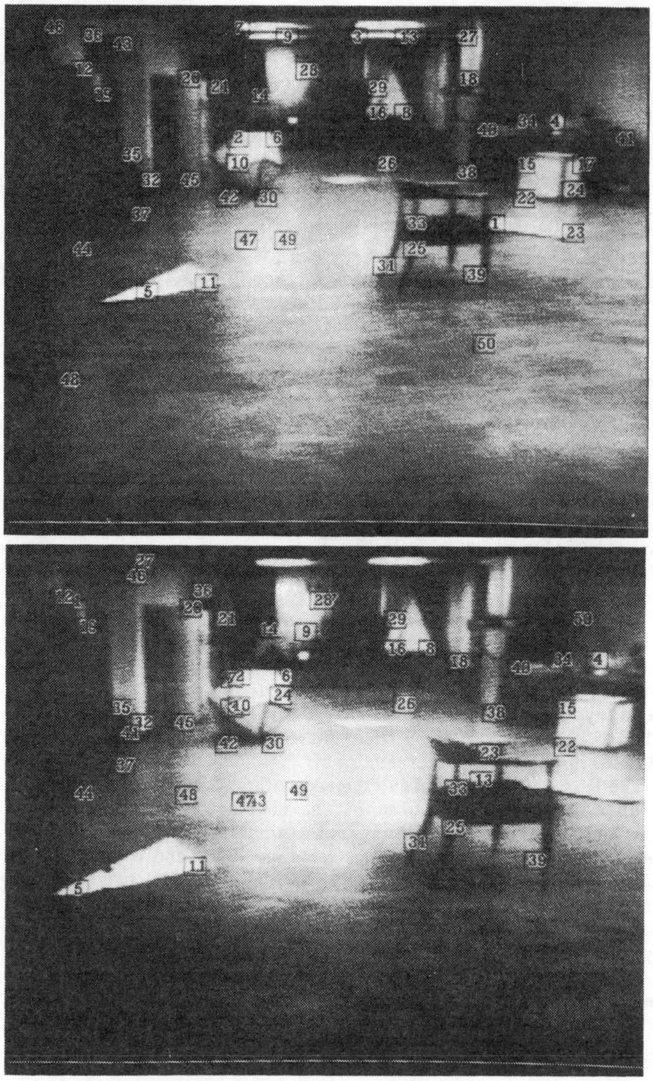

Fig. 4. Interest Operator and Correlator results. The upper picture shows points picked out by an application of the Interest Operator. The lower picture shows the Correlator's attempt to find the same points in an image of the same scene taken from a different point of view.

then deduced the Cart's actual motion during the step from the apparent 3D shift of these features. Some of the features were pruned during this process, and the Interest Operator was invoked to add new ones.

This repeated until the Cart arrived at its destination or until some disaster terminated the program. Figs. 6 and 7 document the Cart's internal world model at two points during a sample run.

Some Details

The Cart's vision code made extensive use of a reductions of each acquired image. Every digitized image was stored as the original picture accompanied by a pyramid of smaller versions of the image reduced in linear size by powers of two, each successive reduction obtained from the last by averaging four pixels into one.

Camera Calibration

The camera's focal length and geometric distortion were determined by parking the Cart a precise distance in front of a wall of many spots and one cross. A program digitized an

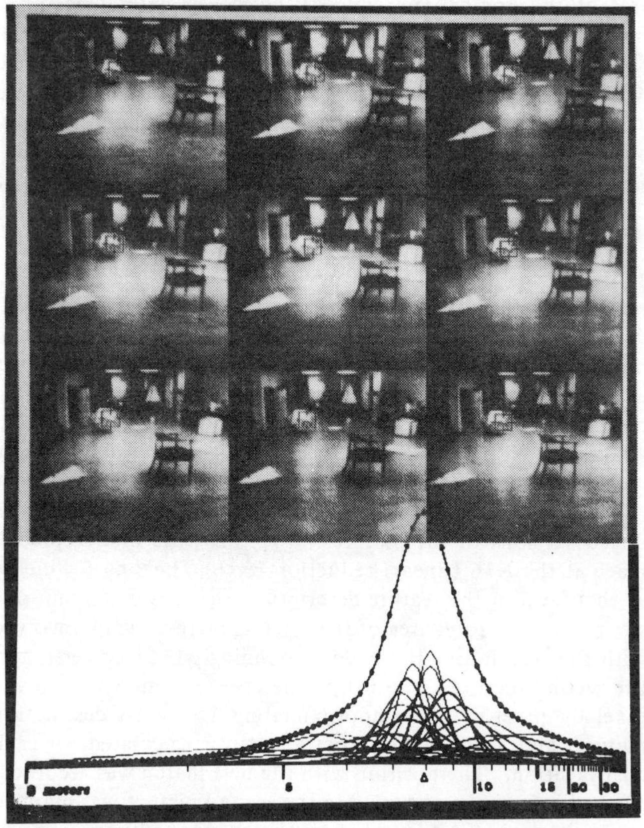

Fig. 5. Slider stereo. A typical ranging. The nine pictures are from a slider scan. The Interest Operator chose the marked feature in the central image, and the Correlator found it in the other eight. The small curves at bottom are distance measurements of the feature made from pairs of the images. The large beaded curve is the sum of the measurements over all 36 pairings. The horizontal scale is linear in inverse distance.

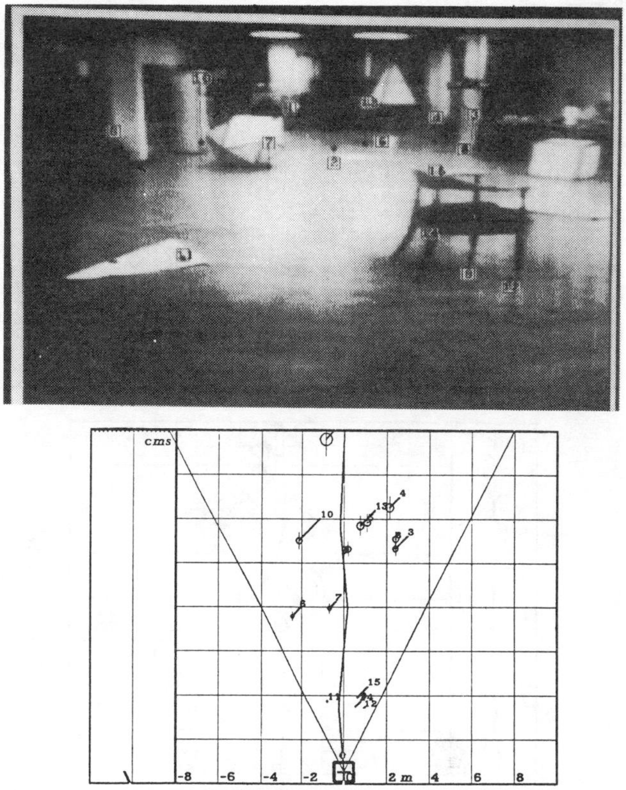

Fig. 6. A Cart obstacle run. This and the following diagram are plan views of the Cart's internal world model during a run of the obstacle-avoiding program. The grid cells are 2 m^2, conceptually on the floor. The Cart's own position is indicated by the small heavy square, and by the graph, indicating height, calibrated in centimeters, to the left of the grid. Since the Cart never actually leaves or penetrates the floor, this graph provides an indication of the overall accuracy. The irregular, tick marked, line behind the Cart's position is the past itinerary of the Cart as deduced by the program. Each tick mark represents a stopping place. The picture at top of the diagrams is the view seen by the TV camera. The two rays projecting forward from the Cart position show the horizontal boundaries of the camera's field of view (as deduced by the camera calibration program). The numbered circles in the plan view are features located and tracked by the program. The centers of the circles are the vertical projections of the feature positions onto the ground. The size of each circle is the uncertainty (caused by finite camera resolution) in the feature's position. The length of the 45° line projecting to the upper right, and terminated by an identifying number, is the height of the feature above the ground, to the same scale as the floor grid. The features are also marked in the camera view, as numbered boxes. The thin line projecting from each box to a lower blob is a stalk which just reaches the ground, in the spirit of the 45° lines in the plan view. The irregular line radiating forwards from the Cart is the planned future path. This changes from stop to stop, as the Cart fails to obey instructions properly, and as new obstacles are detected. The small ellipse a short distance ahead of the Cart along the planned path is the planned position of the next stop.

image of the spot array, located the spots and the cross, and constructed a two-dimensional (2D) polynomial that related the position of the spots in the image to their position in an ideal unity focal length camera, and another polynomial that converted points from the ideal camera to points in the image. These polynomials were used to correct the positions of perceived objects in later scenes (Fig. 3).

The algorithm began by determining the array's approximate spacing and orientation. It reduced by averaging and trimmed the picture to 64 by 64, calculated the Fourier transform of the reduced image, and took its power spectrum, arriving at a 2D transform symmetric about the origin, and having strong peaks at frequencies corresponding to the horizontal and vertical as well as half-diagonal spacings, with weaker peaks at the harmonics. It multiplied each point $[i, j]$ in this transform by point $[-j, i]$ and points $[j - i, j + i]$ and $[i + j, j - i]$, effectively folding the primary peaks onto one another. The strongest peak in the 90° wedge around the y axis gave the spacing and orientation information needed by the next part of the process.

The Interest Operator described later was applied to roughly locate a spot near the center of the image. A special operator examined a window surrounding this position, generated a histogram of intensity values within the window, decided a threshold for separating the black spot from the white background, and calculated the centroid and first and second moment of the spot. This operator was again applied at a displacement from the first centroid indicated by the orientation and spacing of the grid, and so on, the region of found spots growing outward from the seed.

A binary template for the expected appearance of the cross in the middle of the array was constructed from the orientation/spacing data from the Fourier transform. The area around each of the found spots was thresholded on the basis of the expected cross area, and the resulting two-valued pattern was convolved with the cross template. The closest match in the central portion of the picture was declared to be the origin.

Two least squares polynomials (one for X and one for Y) of third (or sometimes fourth) degree in two variables, relating the actual positions of the spots to the ideal positions in a unity focal length camera, were then generated and written into a file. The polynomials were used in the obstacle avoider to cor-

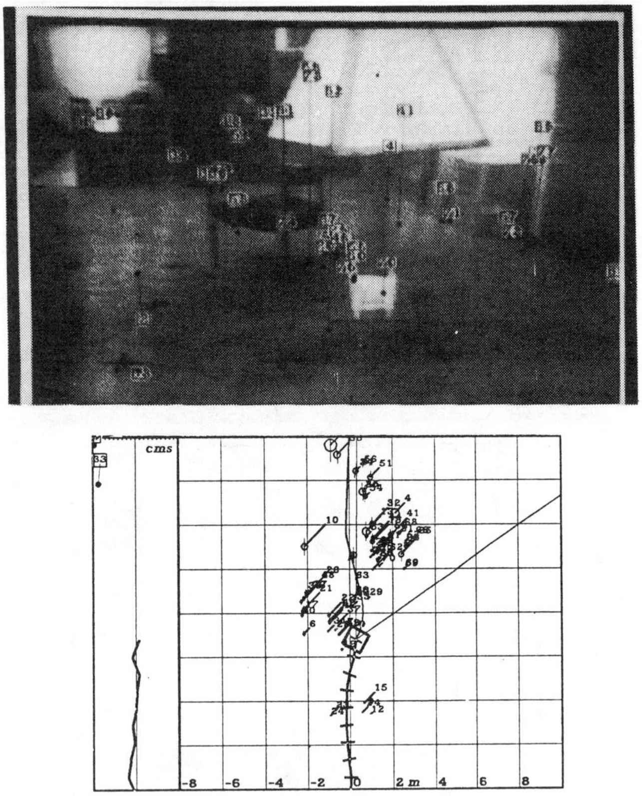

Fig. 7. After the eleventh lurch the Cart has rounded the chair, the icosahedron, and is working on the cardboard tree. The world model has suffered some accumulated drift error, and the oldest acquired features are considerably misplaced.

rect for camera roll, tilt, lens focal length, and long-term variations in the vidicon geometry.

Interest Operator

The Cart vision code dealt with localized image patches called features. A feature is conceptually a point in the 3D world, but it was found by examining localities larger than points in pictures. A feature was good if it could be located unambiguously in different views of a scene. A uniformly colored region or a simple edge is not good because its parts are indistinguishable. Regions, such as corners, with high contrast in orthogonal directions are best.

New features in images were picked by a subroutine called the *Interest Operator*, which returned regions that were local maxima of a directional variance measure, defined below. The idea was to select a relatively uniform scattering of good features over the image, so that a few would likely be picked on every visible object while textureless areas and simple edges were avoided.

Directional variance was measured over small square windows. Sums of squares of differences of pixels adjacent in each of four directions (horizontal, vertical, and two diagonals) over each window were calculated, and the window's interest measure was the minimum of these four sums. Features were chosen where the interest measure had local maxima. The chosen features were stored in an array, sorted in order of decreasing interest measure (Fig. 4–top).

Once a feature was chosen, its appearance was recorded as series of excerpts from the reduced image sequence. A 6 by 6 window was excised around the feature's location from each of the variously reduced pictures. Only a tiny fraction of the area of the original (unreduced) image was extracted. Four times as much area (but the same number of pixels) of the X2 reduced image was stored, sixteen times as much of the X4 reduction, and so on, until at some level we had the whole image. The final result was a series of 6 by 6 pictures, beginning with a very blurry rendition of the whole picture, gradually zooming in linear expansions of two, to a sharp closeup of the feature.

Correlation

Deducing the 3D location of features from their projections in 2D images requires that we know their position in two or more such images. The *Correlator* was a subroutine that, given a feature description produced by the interest operator from one image, found the best match in a different, but similar, image. Its search area could be the entire new picture, or a rectangular subwindow.

The search used a coarse to fine strategy that began in reduced versions of the pictures. Typically, the first step took place at the X16 (linear) reduction level. The 6 by 6 window at that level in the feature description, that covered about one seventh of the total area of the original picture, was convolved with the search area in the correspondingly reduced version of the second picture. The 6 by 6 description patch was moved pixel by pixel over the approximately 15 by 16 destination picture, and a correlation coefficient was calculated for each trial position. The position with the best match was recorded. The 6 by 6 area it occupied in the second picture was mapped to the X8 reduction level, where the corresponding region was 12 pixels by 12. The 6 by 6 window in the X8 reduced level of the feature description was then convolved with this 12 by 12 area, and the position of best match was recorded and used as a search area for the X4 level. The process continued, matching smaller and smaller, but more and more detailed windows, until a 6 by 6 area was selected in the unreduced picture (Fig. 4–bottom).

This "divide and conquer" strategy was, in general, able to search an entire picture for a match more quickly (because most of the searching was done at high reduction levels) and more reliably (because context up to and including the entire picture guided the search) than a straightforward convolution over even a very restricted search area.

Slider Stereo

At each pause on its computer-controlled itinerary, the Cart slid its camera from left to right on a 52-cm track, taking nine pictures at precise 6.5-cm intervals. Points were chosen in the fifth (middle) of these nine images, either by the Correlator to match features from previous positions, or by the Interest Operator. The camera slid parallel to the horizontal axis of the (distortion corrected) camera coordinate system, so the parallax-induced displacement of features in the nine pictures was purely horizontal.

The Correlator was applied eight times to look for the points chosen in the central image in each of other eight pictures. The search was restricted to a narrow horizontal band. This had little effect on the computation time, but it reduced the probability of incorrect matches. In the case of correct matches, the distance to the feature was inversely proportional to its displacement from one image to another. The uncertainty in such a measurement is the difference in distance a shift one pixel in the image would make. The uncertainty varies inversely with the physical separation of the camera positions where the pictures were taken (the stereo baseline). Long baselines give more accurate distance measurements.

After the correlation step the program knew a feature's position in nine images. It considered each of the 36 (9 values taken 2 at a time) possible image pairings as a stereo baseline, and recorded the estimated (inverse) distance of the feature in a histogram. Each measurement added a little normal curve to the histogram, with mean at the estimated distance, and standard deviation inversely proportional to the baseline, reflecting the uncertainty. The area under each curve was made proportional to the product of the goodness of the matches in the two images (in the central image this quantity is taken as unity), reflecting the confidence that the correlations were correct. The distance to the feature was indicated by the largest peak in the resulting histogram, if this peak was above a certain threshold. If below, the feature was forgotten (Fig. 5).

The Correlator sometimes matched features incorrectly. The distance measurements from incorrect matches in different pictures were not consistent. When the normal curves from 36 pictures pairs are added up, the correct matches agree with each other, and build up a large peak in the histogram, while incorrect matches spread themselves more thinly. Two or three correct correlations out of the eight usually built a peak sufficient to offset a larger number of errors. In this way, eight applications of a mildly reliable Operator interacted to make a very reliable distance measurement.

Motion Stereo

After having determined the 3D location of objects at one position, the computer drove the Cart about a meter forward. At the new position, it slid the camera and took nine pictures. The Correlator was applied in an attempt to find all the features successfully located at the previous position. Feature descriptions extracted from the central image at the last position were searched for in the central image at the new stopping place.

Slider Stereo then determined the distance of the features so found from the Cart's new position. The program now knew the 3D position of the features relative to its camera at the old and the new locations. Its own movement was deduced from 3D coordinate transform that related the two.

The program first eliminated mismatches in the correlations between the central images at the two positions. Although it did not yet have the coordinate transform between the old and new camera systems, the program knew the distance between pairs of feature positions should be the same in both. It made a matrix in which element $[i, j]$ is the absolute value of the difference in distances between points i and j in the first and second coordinate systems divided by the expected error (based on the one pixel uncertainty of the ranging). Each row of this matrix was summed, giving an indication of how much each point disagreed with the other points. The idea is that while points in error disagree with virtually all points, correct positions agree with all the other correct ones, and disagree only with the bad ones. The worst point was deleted, and its effect removed from the remaining points in the row sums. This pruning was repeated until the worst error was within the error expected from the ranging uncertainty.

After the pruning, the program had a number of points, typically 10 to 20, whose position error was small and pretty well known. The program trusted these, and recorded them in its world model, unless it had already done so at a previous position. The pruned points were forgotten forevermore.

The 3D rotation and translation that related the old and new Cart position was then calculated by a Newton's method iteration that minimized the sum of the squares of the distances between the transformed first coordinates and the raw coordinates of the corresponding points at the second position, with each term divided by the square of the expected uncertainty in the 3D position of the points involved.

Path Planning

The Cart vision system modeled objects as simple clouds of features. If enough features are found on each nearby object, this model is adequate for planning a noncolliding path to a destination. The features in the Cart's 3D world model can be thought of as fuzzy ellipsoids, whose dimensions reflect the program's uncertainty of their position. Repeated applications of the Interest Operator as the Cart moves caused virtually all visible objects to be become modeled as clusters of overlapping ellipsoids.

To simplify the problem, the ellipsoids were approximated by spheres. Those spheres sufficiently above the floor and below the Cart's maximum height were projected on the floor as circles. The 1-m^2 Cart itself was modeled as a 3-m circle. The path-finding problem then became one of maneuvering the Cart's 3-m circle between the (usually smaller) circles of the potential obstacles to a desired location. It is convenient (and equivalent) to conceptually shrink the Cart to a point, and add its radius to each and every obstacle. An optimum path in this environment will consist of either a straight run between start and finish, or a series of tangential segments between the circles and contacting arcs (imagine loosely laying a string from start to finish between the circles, then pulling it tight).

The program converted the problem to a shortest path in graph search. There are four possible paths between each pair of obstacles because each tangent can approach clockwise or counterclockwise. Each tangent point became a vertex in the graph, and the distance matrix of the graph (which had an entry for each vertex pair) contained sums of tangential and arc paths, with infinities for blocked or impossible routes. The shortest distance in this space can be found with an algorithm whose running time is $O(n^3)$ in the number of vertices, and the Cart program was occasionally run using this exact procedure. It was run more often with a faster approximation that made each obstacle into only two vertices (one for each direction of circumnavigation).

A few other considerations were essential in path planning. The charted routes consisted of straight lines connected by tangent arcs, and were thus plausible paths for the Cart, which steered like an automobile. This plausibility was not necessarily true of the start of the planned route, which, as presented thus far, did not take the initial heading of the Cart into account. The plan could, for instance, include an initial segment going off 90° from the direction in which the Cart pointed, and thus be impossible to execute. This was handled by including a pair of "phantom" obstacles along with the real perceived ones. The phantom obstacles had a radius equal to the Cart's minimum steering radius, and were placed, in the planning process, on either side of the Cart at such a distance that after their radius was augmented by the Cart's radius (as happened for all the obstacles), they just touched the Cart's centroid, and each other, with their common tangents being parallel to the direction of the Cart's heading. They effectively blocked the area made inaccessible to the Cart by its maneuverability limitations (Fig. 6).

Lozano-Pérez and Wesley [7] describe an independently developed, but very similar, approach to finding paths around polygonal obstacles.

Path Execution

After the path to the destination had been chosen, a portion of it had to be implemented as steering and motor commands and transmitted to the Cart. The control system was primitive. The drive motor and steering motors could be turned on and off at any time, but there existed no means to accurately determine just how fast or how far they had gone. The program made the best of this bad situation by incorporating a model of the Cart that mimicked, as accurately as possible, the Cart's actual behavior. Under good conditions, as accurately as possible means about 20 percent; the Cart was not very repeatable, and was affected by ground slope and texture, battery voltage, and other less obvious externals.

The path executing routine began by excising the first 0.75 m of the planned path. This distance was chosen as a compromise between average Cart velocity, and continuity between picture sets. If the Cart moved too far between picture-digitizing sessions, the picture would change too much for reliable correlations. This is especially true if the Cart turns (steers) as it moves. The image seen by the camera then pans across the field of view. The Cart had a wide angle lens that covers 60° horizontally. The 0.75 m, combined with the turning radius limit (5 m) of the Cart resulted in a maximum shift in the field of view of 15°, one quarter of the entire image.

The program examined the Cart's position and orientation at the end of the desired 0.75-m lurch, relative to the starting position and orientation. The displacement was characterized by three parameters; displacement forward, displacement to the right, and change in heading. In closed form, the program computed a path that accomplished this movement in two arcs of equal radius, but different lengths. The resulting trajectory had a general "S" shape. Rough motor timings were derived from these parameters. The program then used a simulation that took into account steering and drive motor response to iteratively refine the solution.

Cart Experiments

The system described above only incompletely fulfills some of the hopes I had when the work began many years ago.

One of the most serious limitations was the excruciating slowness of the program. In spite of my best efforts, and many compromises in the interest of speed, it took 10 to 15 min of real time to acquire and consider the images at each meter long lurch, on a lightly loaded DEC KL-10. This translated to an effective Cart velocity of 3 to 5 m an hour. Interesting obstacle courses (two or three major obstacles, spaced far enough apart to permit passage within the limits of the Cart's size and maneuverability) were about 20 m long, so interesting Cart runs took 5 h.

The reliability of individual moves was high, as it had to be for a 20-lurch sequence to have any chance of succeeding, but the demanding nature of each full run and the limited amount of time available for testing (discussion of which is beyond the scope of this paper) after the bulk of the program was debugged, ensured that many potential improvements were left untried. Three full (about 20-m) runs were digitally recorded and filmed, two indoors and one outdoors. Two indoor false starts, aborted by failure of the program to perceive an obstacle, were also recorded. The two long indoor runs were nearly perfect.

In the first long indoor run, the Cart successfully slalomed its way around a chair, a large cardboard icosahedron, and a cardboard tree then, at a distance of about 16 m, encountered a cluttered wall and backed up several times trying to find a way around it (Figs. 6 and 7 are snapshots from this run).

The second long indoor run involved a more complicated set of obstacles, arranged primarily into two overlapping rows blocking the goal. I had set up the course hoping the Cart would take a long, picturesque (the runs were being filmed) "S" shaped path around the ends of the rows. To my chagrin, it instead tried for a tricky shortcut. The Cart backed up twice to negotiate the tight turn required to go around the first row, then executed several tedious steer forward/backup moves, lining itself up to go through a gap barely wide enough in the second row. This run had to be terminated, sadly, before the Cart had gone through the gap because of declining battery charge and increasing system load.

The outdoor run was less successful. It began well; in the first few moves the program correctly perceived a chair directly in front of the camera, and a number of more distant cardboard obstacles and sundry debris. Unfortunately, the program's idea of the Cart's own position became increasingly wrong. At almost every lurch, the position solver deduced a Cart motion considerably smaller than the actual move. By the time the Cart had rounded the foreground chair, its position model was so far off that the distant obstacles were replicated in different positions in the Cart's confused world model, because they had been seen early in the run and again later, to the point where the program thought an actually existing distant clear path was blocked. I restarted the program to clear out the world model when the planned path became too silly. At that time the Cart was 4 m in front of a cardboard icosahedron, and its planned path lead straight through it. The newly reincarnated program failed to notice the obstacle, and the Cart collided with it. I manually moved the icosahedron out of the way, and allowed the run to continue. It did so uneventfully, though there were continued occasional slight errors in the self-position deductions. The Cart encountered a large cardboard tree towards the end of this journey and detected a portion of it only just in time to squeak by without colliding (Fig. 2 was a photograph taken during this run).

The two short abortive indoor runs involved setups nearly identical to the two-row successful long run described one paragraph ago. The first row, about 3 m in front of the Cart's starting position contained a chair, a real tree (a small cypress in a planting pot), and a polygonal cardboard tree. The Cart saw the chair instantly and the real tree after the second move, but failed to see the cardboard tree ever. Its planned path around the two obstacles it did see put it on a collision course with the unseen one. Placing a chair just ahead of the cardboard tree fixed the problem, and resulted in a successful run. The finished program never had trouble with chairs.

Problems

These tests revealed some weaknesses in the program. The system did not see simple polygonal (bland and featureless) objects reliably, and its visual navigation was fragile under certain conditions. Examination of the program's internal workings suggested some causes and possible solutions.

The program sometimes failed to see obstacles lacking sufficient high contrast detail within their outlines. In this regard, the polygonal tree and rock obstacles I whimsically constructed to match diagrams from a 3D drawing program, were a terrible mistake. In none of the test runs did the programs ever fail to see a chair placed in front of the Cart, but half the time they did fail to see a pyramidal tree or an icosahedral rock made of clean white cardboard. These contrived obstacles were picked up reliably at a distance of 10 to 15 m, silhouetted against a relatively unmoving (over slider travel and Cart lurches) background, but were only rarely and sparsely seen at closer range, when their outlines were confused by a rapidly shifting background, and their bland interiors provided no purchase for the interest operator or correlator. Even when the artificial obstacles were correctly perceived, it was by virtue of only two to four features. In contrast, the program usually tracked five to ten features on nearby chairs.

In the brightly sunlit outdoor run, the artificial obstacles had another problem. Their white coloration turned out to be much brighter than any "naturally" occurring extended object. These super bright, glaring, surfaces severely taxed the very limited dynamic range of the Cart's vidicon/digitizer combination. When the icosahedron occupied 10 percent of the camera's field of view, the automatic target voltage circuit in the electronics turned down the gain to a point where the background behind the icosahedron appeared nearly solid black.

The second major problem exposed by the runs was glitches in the Cart's self-position model. This model was updated after a lurch by finding the 3D translation and rotation that best related the 3D position of the set of tracked features before and after the lurch. In spite of the extensive pruning that preceded this step (and partly because of it, as is discussed later), small errors in the measured feature positions sometimes caused the solver to converge to the wrong transform, giving a position error beyond the expected uncertainty. Features placed into the world model before and after such a glitch were not in the correct relative positions. Often an object seen before was seen again after, now displaced, with the combination of old and new positions combining to block a path that was in actuality open.

This problem showed up mainly in the outdoor run. I had observed it indoors in the past, in simple mapping runs, before the entire obstacle avoider was assembled. There appear to be two major causes for it, and a wide range of supporting factors.

Poor seeing, resulting in too few correct correlations between the pictures before and after a lurch, was one culprit. The highly redundant nine-eyed stereo ranging was very reliable, and caused few problems, but the nonredundant correlation necessary to relate the position of features before and after a lurch was error prone. Sometimes the mutual-distance invariance pruning that followed was overly aggressive, and left too few points for a stable least squares coordinate fit.

The outdoor runs encountered another problem. The program ran so slowly that shadows moved significantly (up to a half meter) between lurches. Their high contrast boundaries were favorite points for tracking, enhancing the program's confusion.

Quick Fixes

Though elaborate (and thus far untried in our context) methods such as edge matching may greatly improve the quality of automatic vision in the future, subsequent experiments with the program revealed some modest incremental improvements that would have solved most of the problems in the test runs.

The issue of unseen cardboard obstacles turns out to be partly one of overconservatism on the program's part. In all cases where the Cart collided with an obstacle it had correctly ranged a few features on the obstacle in the prior nine-eyed scan. The problem was that the much more fragile correlation between vehicle forward moves failed, and the points were rejected in the mutual distance test. Overall, the nine-eyed stereo produced very few errors. If the path planning stage had used the prepruning features (still without incorporating them permanently into the world model) the runs would have proceeded much more smoothly. All of the most vexing false negatives, in which the program failed to spot a real obstacle, would have been eliminated. There would have been a very few false positives, in which nonexistent ghost obstacles would have been perceived. One or two of these might have caused an unnecessary swerve or backup, but such ghosts would not pass the pruning stage, and the run would have recovered after the initial, noncatastrophic, glitch.

The self-position confusion problem is related, and in retrospect may be considered a trivial bug. When the Path Planner computed a route for the Cart, another subroutine took a portion of this plan and implemented it as a sequence of commands to be transmitted to the Cart's steering and drive motors. During this process, it ran a simulation that modeled the Cart acceleration, rate of turning, and so on, and which provided a prediction of the Cart's position after the move. With the old hardware, the accuracy of this prediction was not great, but it nevertheless provided much *a priori* information about the Cart's new position. This information was used, appropriately weighted, in the least squares coordinate system solver that deduced the Cart's movement from the apparent motion in 3D of tracked features. It was not used, however, in the mutual distance pruning step that preceeded this solving. When the majority of features had been correctly tracked, failure to use this information did not hurt the pruning. But when the seeing was poor, it could make the difference between choosing a spuriously agreeing set of mistracked features and the small correctly matched set. Incorporating the prediction into pruning, by means of a heavily weighted point that the program treats like another tracked feature, removed almost all the positioning glitches when the program was fed the pictures from the outdoor run.

More detail on all these areas can be found in [10].

The CMU Rover

The major impediments to serious extensions of the Cart work were limits to available computation, resulting in debilitatingly long experimental times, and the very minimal nature of the robot hardware, which precluded inexpensive solutions for even most basic functions (like "roll a meter forward").

We are addressing these problems at CMU in an ambitious new effort centered around a new, small but sophisticated mobile robot dubbed the CMU Rover. The project so far has been focused on developing a smoothly functional and highly capable vehicle and associated support system which will serve a wide variety of future research.

The shape, size, steering arrangements, and on-board as well as external processing capabilities of the Rover system were chosen to maximize the flexibility of the system (naturally limited by present-day techniques).

The robot is cylindrical, about a meter tall and 55 cm in diameter (Fig. 8) and has three individually steerable wheel assemblies which give it a full three degrees of freedom of mobility in the plane (Figs. 9 and 10). Initially it will carry a

Fig. 8. The CMU Rover.

Fig. 9. The Rover wheelbase. The steering angle and drive of each wheel pair is individually controlled. The rover's trajectory will be an arc about any point in the floor plane if lines through the axles of all three wheels intersect at the center of that arc.

Fig. 10. The Rover wheel assembly. The steering motor is shown attached to the wheel assembly, part of the drive motor is shown detached.

TV camera on a pan/tilt/slide mount, several short-range infrared and long-range sonar proximity detectors, and contact switches. Our design calls for about a dozen on-board processors (at least half of them powerful 16-bit MC68000's) for high-speed local decision making, servo control, and communication (Fig. 13).

Serious processing power, primarily for vision, is to be provided at the other end of a remote-control link by a combination of a host computer VAX 11/780 an ST-100 array processor (a new machine from a new company, Star Technologies Inc., which provides 100 million floating-point operations per second) and a specially designed high-performance analog data acquisition and generation device. The Stanford Cart used 15 min of computer time to move a meter. With this new CMU hardware, and some improved algorithms, we hope to duplicate (and improve on) this performance in a system that runs at least ten times as fast, leaving room for future extensions.

The 15 min for each meter-long move in the Cart's obstacle-avoiding program came in three approximately equal chunks. The first 5 min were devoted to digitizing the nine pictures from a slider scan. Though the SAIL computer had a flash digitizer which sent its data through a disk channel into main memory at high speed, it was limited by a poor sync detector. Often the stored picture was missing scanlines, or had lost vertical sync, i.e., had rolled vertically. In addition, the image was quantized to only 4 bits per sample; 16 grey levels. To make one good picture the program digitized 30 raw images in rapid succession, intercompared them to find the largest subset of nearly alike pictures (on the theory that the nonspoiled ones would be similar, but the spoiled ones would differ even from each other) and averaged this "good" set to obtain a less noisy image with 6 bits per pixel. The new digitizing hardware, which can sample a raw analog waveform, and depends on software in the array processor to do sync detection, should cut the total time to under 1 s per picture. The next 5 min was spent doing the low-level vision; reducing the images, sometimes filtering them, applying the interest operator, especially the correlator, and statistical pruning of the results. The array processor should be able to do all this nearly 100 times faster. The last 5 min were devoted to higher level tasks; maintaining the world model, path planning, and generating graphical documentation of the program's thinking. Some steps in this section may be suitable for the array processor, but in any case we have found faster algorithms for much of it; for instance, a shortest path in graph algorithm which makes maximum use of the sparsity of the distance matrix produced during the path planning.

We hope eventually to provide a manipulator on the Rover's topside, but there is no active work on this now. We chose the high steering flexibility of the current design partly to ease the requirements on a future arm. The weight and power needed can be reduced by using the mobility of the Rover to substitute for the shoulder joint of the arm. Such a strategy works best if the Rover body is given a full three degrees of freedom (X,

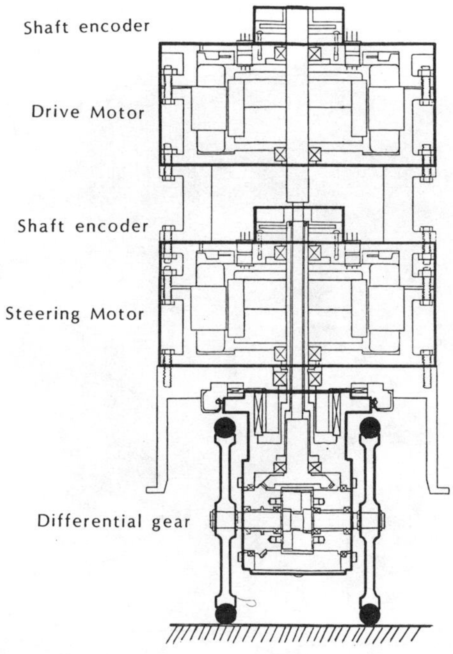

Fig. 11. Diagram of the Rover wheel assembly.

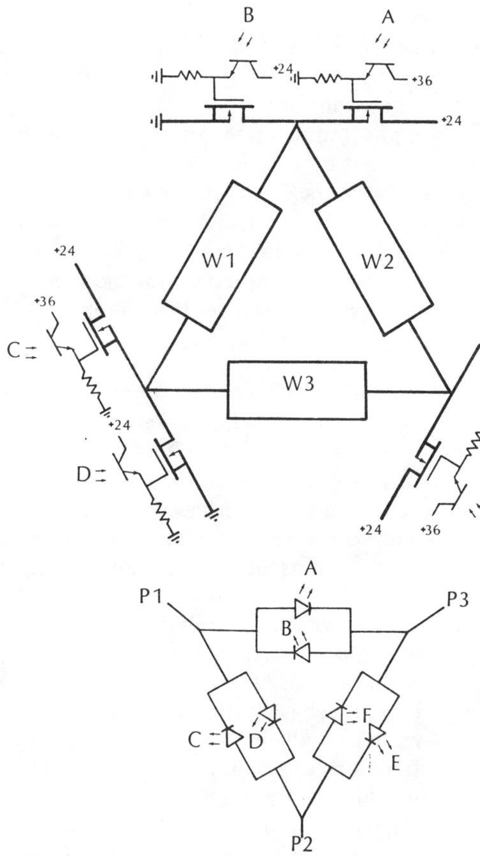

Fig. 12. Rover brushless motor drive circuitry. Each samarium–cobalt brushless motor in the wheel assemblies is sequenced and servoed by its own microprocessor. A processor generates three bipolar logic signals *P1*, *P2*, and *P3* which control the currents through the three phase motor windings *W1*, *W2*, and *W3* through the IR light-emitting diode (LED) phototransistor optical links *A* through *F* shown. Each phototransistor controls a power field-effect transistor which switches power to the windings. The circuitry in the upper, optically isolated, portion of the diagram is contained within its motor's housing, on an annular circuit board, using the housing as heat sink. The LED's poke through holes in the cover. Motor direction is controlled by sequencing order; torque is adjusted by pulsewidth modulation.

Y, and angle) in the plane of the floor. Conventional steering arrangements, as in cars, give only two degrees at any instant.

Rover Details

Three degrees of freedom of mobility are achieved by mounting the chassis on three independently steerable wheel assemblies (Figs. 9–12). The control algorithm for this arrangement at every instant orients the wheels so that lines through their axles meet at a common point. Properly orchestrated, this design permits unconstrained motion in any (2D) direction, and simultaneous independent control of the robot's rotation about its own vertical axis. An unexpected benefit of this agility is the availability of a "reducing gear" effect. By turning about the vertical axis while moving forward the robot derives a mechanical advantage for its motors. For a given motor speed, the faster the Rover spins, the slower it travels forward, and the steeper the slope it can climb. (Visualization of this effect is left as an exercise for the reader.)

To permit low-friction steering while the robot is stationary, each assembly has two parallel wheels connected by a differential gear. The drive shaft of the differential goes straight up into the body of the robot. A concentric hollow shaft around this one connects to the housing of the differential (Fig. 11). Turning the inner shaft causes the wheels to roll forwards or backwards, turning the outer one steers the assembly, with the two wheels rolling in a little circle. The assemblies were manufactured for us by Summit Gear Corp.

Each shaft is connected to a motor and a 4000-count/revolution optical shaft encoder (Datametrics K3). The two motors and two encoders are stacked pancake fashion on the wheel assembly, speared by the shafts. There are no gears except for the ones in the differential. (Fig. 11 shows a schematic cross section of a complete motor/wheel-assembly structure, Fig. 10 a partially assembled stack in the flesh.)

The motors are brushless with samarium–cobalt permanent-magnet rotors and three-phase windings (Inland Motors BM-3201). With the high-energy magnet material, this design has better performance when the coils are properly sequenced than a conventional rotating coil motor. The coils for each are energized by six power MOSFET's (Motorola MTP1224) mounted in the motor casing and switched by six optoisolators (to protect the controlling computers from switching noise) whose LED's are connected in bidirectional pairs in a delta configuration, and lit by three logic signals connected to the vertices of the delta (Fig. 12).

The motor sequencing signals come directly from on-board microprocessors, one for each motor. These are CMOS (Motorola MC146805 with Hitachi HM6116 RAM's) to keep power consumption low. Each processor pulsewidth modulates and phases its motor's windings, and observes its shaft encoder, to servo the motor to a desired motion (supplied by yet another processor, a Motorola 68000, the *Conductor* as a time parameterized function). Though the servo loop works in its present form, several approximations were necessary in this real-time task because of the limited arithmetic capability of the 6805. We will be replacing the 6805 with the forthcoming MC68008, a compact 8-bit bus version of the 68000.

The shaft encoder outputs and the torques from all the motors, as estimated by the motor processors, are monitored by another processor, the *Simulator*, a Motorola MC68000 (with all CMOS support circuitry the power requirement for our

32K 68000 is under 1 W. The new high-performance 74HC series CMOS allows operation at full 10-MHz speed.), which maintains a dead-reckoned model of the robot's position from instant to instant. The results of this simulation (which represents the robot's best position estimate) are compared with the desired position, produced by another 68000, the *Controller*, in the previously introduced the *Conductor*, which orchestrates the individual motor processors. The Conductor adjusts the rates and positions of the individual motors in an attempt to bring the Simulator in line with requests from the Controller, in what amounts to a highly nonlinear feedback loop.

Other on-board processors are as follows:

Communication — A 68000 which maintains an error corrected and checked packet infrared link with a large controlling computer (a VAX 11/780 helped out by an ST-100 array processor and a custom high-speed digitizer) which will do the heavy thinking. Programs run in the Controller are obtained over this link.

Sonar — A 6805 which controls a number of Polaroid sonar ranging devices around the body of the Rover. These will be used to maintain a rough navigation and bump-avoidance model. All measurements and control functions of this processor and the following ones are available (on request over a serial link) to the Controller.

Camera — A 6805 which controls the pan, tilt, and slide motors of the onboard TV camera. The compact camera broadcasts its image on a small UHF or microwave transmitter. The signal is received remotely and the video signal captured by a high-bandwidth digitizer system and then read by the remote VAX. There are tentative plans for a minimal vision system using a 68000 with about 256K of extra memory on-board the Rover, for small vision tasks when the Rover is out of communication with the base system.

Proximity — A 6805 which monitors several short-range modulated infrared proximity detectors which serve as a last line of defense against collision, and which sense any drop off in the floor, and contact switches.

Utility — A 6805 which senses conditions such as battery voltage and motor temperature, and which controls the power to nonessential but power-hungry systems like the TV camera and transmitter.

Communication between processors is serial, via Harris CMOS UART's, at a maximum speed of 256 kBd. The Conductor talks with the motor processors on a shared serial line and the Controller communicates with the Sonar, Camera, Proximity, Utility, and any other peripheral processors by a similar method.

The processors live in a rack on the second storey of the robot structure (Figs. 8 and 13), between the motor and battery assembly (first floor) and the camera plane (penthouse). Fig. 13 shows the initial interconnection.

The Rover is powered by six sealed lead-acid batteries (Globe gel-cell 12230) with a total capacity of 60 A · h at 24 V. The motors are powered directly from these, the rest of the circuitry derives its power indirectly from them through switching dc/dc converters (Kepco RMD-24-A-24 and Semiconductor Circuits U717262). Each 6805 processor draws about one eighth of a watt, each 68000 board only 1 W.

Physically, the robot is a meter tall and 55 cm in diameter. It weighs 90 kg. The maximum acceleration is one quarter *g*, and the top speed is 10 km/h. With appropriate on-board programming the motions should be very smooth. The great steering flexibility will permit simulation of other steering systems such as those of cars, tanks, and boats and other robots by changes in programming.

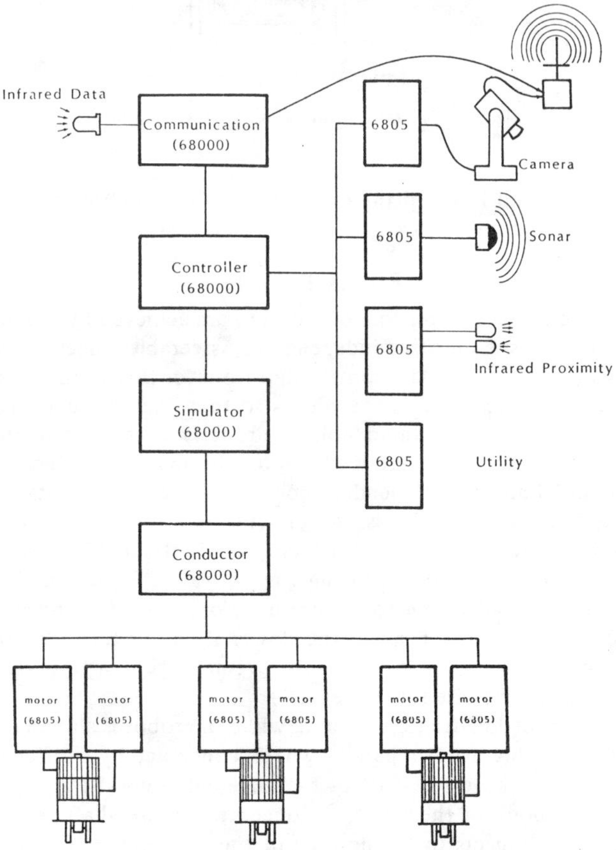

Fig. 13. The Rover's on-board processors.

Progress

As of this writing (January 1983) the robot's major mechanical and electronic structures are complete, mostly tested, and being assembled. An instability in the servo control algorithms which had held us up for several months has finally been solved, and we expect baby's first "steps" early in 1983. The digitizer unit which will receive and generate video and other data for the robot is under construction. Its specifications include four channels each with 2 Mbytes of memory and two ports able to transfer data at 100 Mbytes/s.

Promises

The high-level control system has become very interesting. Initially, we had imagined a language in which to write scripts for the on-board Controller similar to the **AL** manipulator language developed at Stanford [6], from which the commercial languages **VAL** at Unimation [15] and the more sophisticated **AML** [14] at IBM were derived. Paper attempts at defining the structures and primitives required for the mobile application revealed that the essentially linear control structure of these state-of-the-art arm languages was inadequate for a rover. The essential difference is that a rover, in its wanderings, is regularly "surprised" by events it cannot anticipate, but with which it must deal. This requires that routines for responding to various situations can be activated in arbitrary order, and run concurrently.

We briefly examined a production system, as used in many "expert systems," as the framework for the requisite real-time concurrency, but have now settled on a structure similar to that developed for the CMU **Hearsay II** speech understanding project [5]. Independent processes will communicate via messages posted on a commonly accessible data structure we call a *Blackboard*. The individual processes, some of which will run under control of a spare real-time operating system on one or more of the onboard 68000's, others of which will exist at the other end of a radio link on the VAX, change their relative priority as a consequence of relevant messages on the blackboard. For instance, a note from several touch sensors signaling a collision is taken as a cue by the avoidance routine to increase its running rate, and to post messages which trigger the motor coordinating routines to begin evasive actions. We plan to implement the multiple processes required for this task on each of several of the on-board 68000's with the aid of a compact (4K), efficient real-time operating system kernel called VRTX available from Hunter & Ready. A more detailed description of the state of this work may be found in [4].

Other interesting preliminary thinking has resulted in a scheme by which a very simple arm with only three actuators will enable the robot, making heavy use of its great steering flexibility, to enter and leave through a closed standard office door (Fig. 14).

Stepping into a more speculative realm, we are considering approaches to model-based vision [2] which would permit recognition of certain classes of objects seen by the robot. Discussions with the distributed sensor net crew here at CMU [13] has raised the possibility of equipping the robot with ears, so it could passively localize sound, and thus perhaps respond correctly, both semantically and geometrically, to a spoken command like "Come here!" (using, in addition, speech understanding technology also developed at CMU [16]).

We are also toying with the idea of a phased array sonar with about 100 transducers operating at 50 kHz which, in conjunction with the high-speed analog conversion device mentioned above and the array processor, would be able to produce a modest resolution depth map (and additional information) of a full hemisphere in about 1 s, by sending out a single spherical pulse, then digitally combining the returned echoes from the individual transducers with different delay patterns to synthesize narrow receiving beams.

Philosophy

It is my view that developing a responsive mobile entity is the surest way to approach the problem of general intelligence in machines.

Though computers have been programmed to do creditable jobs in many *intellectual* domains, competent performance in *instinctive* domains like perception and common sense reasoning is still elusive. I think this is because the instinctive skills are fundamentally much harder. While human beings learned most of the intellectual skills over a few thousand years, the instinctive skills were genetically honed for hundreds of millions of years, and are associated with large, apparently efficiently organized, fractions of our brain; vision, for example, is done by a specialized 10 percent of our neurons. Many animals share our instinctive skills, and their evolutionary record provides clues about the conditions that foster development of such skills. A universal feature that most impresses me in this context is that *all animals that evolved perceptual and behavioral competence comparable to that of humans first adopted a mobile way of life.*

This is perhaps a moot point in the case of the vertebrates, which share so much of human evolutionary history, but it is dramatically confirmed among the invertebrates. Most molluscs are sessile shellfish whose behavior is governed by a nervous system of a few hundred neurons. Octopus and squid are molluscs that abandoned life in the shell for one of mobility; as a consequence, they developed imaging eyes, a large (annular!) brain, dexterous manipulators, and an unmatched million-channel color display on their surfaces. By contrast, no sessile animal nor any plant shows any evidence of being even remotely this near to the human behavioral competence.

My conclusion is that *solving the day to day problems of developing a mobile organism steers one in the direction of general intelligence, while working on the problems of a fixed entity is more likely to result in very specialized solutions.*

I believe our experience with the control language for the Rover vis a vis the languages adequate for a fixed arm, is a case in point. My experiences with computer vision during the Cart work reinforce this opinion; constantly testing a program against fresh real-world data is nothing at all like optimizing a program to work well with a limited set of stored data. The variable and unpredictable world encountered by a rover applies much more selection pressure for generality and robustness than the much narrower and more repetitive stimuli experienced by a fixed machine. Mobile robotics may or may not be the fastest way to arrive at general human competence in machines, but I believe it is one of the surest roads.

Related Work

Other groups have come to similar conclusions, and have done sophisticated mobile robot work in past [12], [17]. The robotics group at Stanford has acquired a new, experimental,

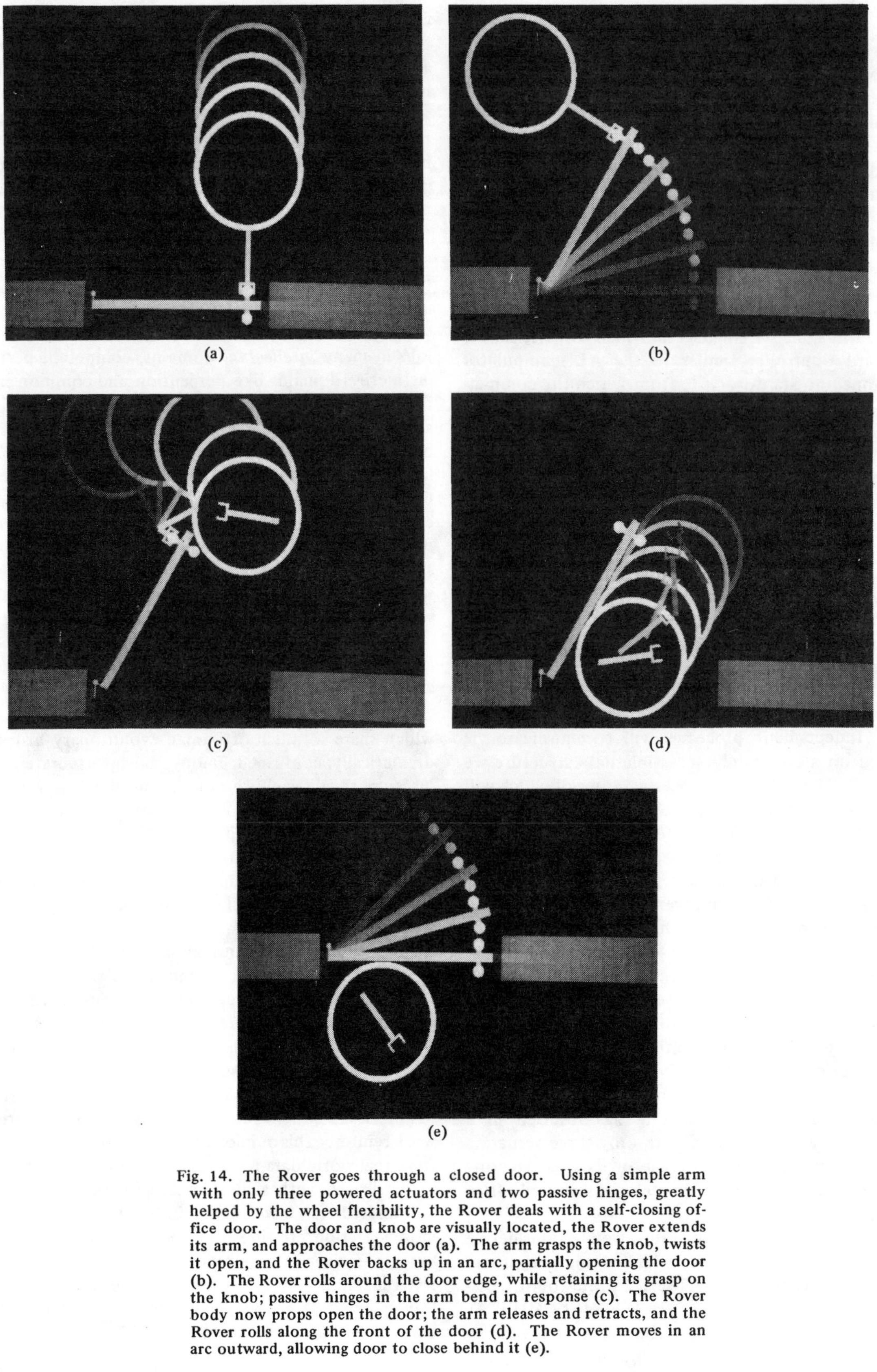

Fig. 14. The Rover goes through a closed door. Using a simple arm with only three powered actuators and two passive hinges, greatly helped by the wheel flexibility, the Rover deals with a self-closing office door. The door and knob are visually located, the Rover extends its arm, and approaches the door (a). The arm grasps the knob, twists it open, and the Rover backs up in an arc, partially opening the door (b). The Rover rolls around the door edge, while retaining its grasp on the knob; passive hinges in the arm bend in response (c). The Rover body now props open the door; the arm releases and retracts, and the Rover rolls along the front of the door (d). The Rover moves in an arc outward, allowing door to close behind it (e).

mobile robot from Unimation Inc., and plans research similar to ours [1]. This new Unimation rover [3] is very similar in size, shape, and mechanical capabilities to the machine we are building. It achieves a full three degrees of freedom of floor-plane mobility by use of three novel "omnidirectional" wheels which, by virtue of rollers in place of tires, can freely move in a direction broadside to the wheel plane as well as performing the usual wheel motion under motor control.

References

[1] T. O. Binford, "The Stanford mobile robot," personal communication, Stanford Univ. Comput. Sci. Dept., Stanford, CA, Oct. 1982.

[2] R. A. Brooks, "Symbolic reasoning among 3-D models and 2-D images," Ph.D. dissertation, Stanford Univ., Stanford, CA, June 1981.

[3] B. Carlisle and B. Shimano, "The Unimation mobile robot," personal communication, Unimation Inc., Mountain View, CA, Aug. 1981.

[4] A. Elfes and S. N. Talukdar, "A distributed control system for the CMU Rover," 8th Int. Joint Conf. on Artificial Intelligence, Karlsruhe, West Germany (IJCAI), Aug. 1983.

[5] L. D. Erman and V. R. Lesser, "The HEARSAY-II speech-understanding system: Integrating knowledge to resolve uncertainty," *Commun. ACM*, vol. 23, no. 6, June 1980.

[6] R. Goldman and S. Mujtaba, *AL User's Manual.* 3rd ed., Computer Science STAN-CS-81-889 (Rep. AIM-344), Stanford Univ., Dec. 1981.

[7] T. Lozano-Pérez and M. A. Wesley, "An algorithm for planning collision-free paths among polyhedral obstacles," *Commun. ACM*, vol. 22, no. 10, pp. 560–570, Oct. 1979.

[8] H. P. Moravec, "Towards automatic visual obstacle avoidance," in *Proc. 5th Int. Joint Conf. on Artificial Intelligence* (MIT, Cambridge, MA), pp. 584 (IJCAI), Aug. 1977.

[9] ——, "Visual mapping by a robot rover," in *Proc. 6th Int. Joint Conf. on Artificial Intelligence* (Tokyo, Japan), pp. 599–601 (IJCAI), Aug. 1979.

[10] ——, "Obstacle avoidance and navigation in the real world by a seeing robot rover," Ph.D. dissertation, Stanford Univ., Sept. 1980 (published as *Robot Rover Visual Navigation.* Ann Arbor, MI: UMI Research Press, 1981.

[11] ——, "Rover visual obstacle avoidance," in *Proc. 7th Int. Joint Conf. on Artificial Intelligence* (Vancouver, B.C., Canada), pp. 785–790 (IJCAI), Aug. 1981.

[12] B. Raphael, *The Thinking Computer.* San Francisco, CA: W. H. Freeman and Company, 1976.

[13] R. F. Rashid and G. G. Robertson, "Accent, a communication oriented operating system kernel," Carnegie-Mellon Univ., Pittsburgh, PA, Tech. Rep. CS-81-123, Oct. 1981.

[14] R. H. Taylor, P. D. Summers, and J. M. Meyer, "AML: A manufacturing language," IBM Thomas J. Watson Research Center, Yorktown Heights, NY, Res. Rep. RC-9389, Apr. 1982.

[15] Unimation, Inc., "User's guide to VAL, a robot programming and control system, version II," Unimation Inc., Tech. Rep., Feb. 1979.

[16] A. Waibel and B. Yegnanarayana, "Comparative study of nonlinear time warping techniques for speech understanding," Carnegie-Mellon Univ., Comput. Sci. Dep., Pittsburgh, PA, Tech. Rep. 125, 1981.

[17] Y. Yakimovsky and R. Cunningham, "A system for extracting three-dimensional measurements from a stereo pair of TV cameras," *Comput. Graphics Image Proces.*, vol. 7, pp. 195–210, 1978.

Occupancy Grids: A Stochastic Spatial Representation for Active Robot Perception

Alberto Elfes

Autonomous Robotics Laboratory
Department of Computer Sciences
IBM T. J. Watson Research Center
Yorktown Heights, NY 10598
Phone: (914)784-7944
E-mail (Internet): ELFES@IBM.COM

Abstract

In this paper we provide an overview of a new framework for robot perception, real-world modelling, and navigation that uses a stochastic tesselated representation of spatial information called the Occupancy Grid. *The Occupancy Grid is a multi-dimensional random field model that maintains probabilistic estimates of the occupancy state of each cell in a spatial lattice. Bayesian estimation mechanisms employing stochastic sensor models allow incremental updating of the Occupancy Grid using multi-view, multi-sensor data, composition of multiple maps, decision-making, and incorporation of robot and sensor position uncertainty. We present the underlying stochastic formulation of the Occupancy Grid framework, and discuss its application to a variety of robotic tasks. These include range-based mapping, multi-sensor integration, path-planning and obstacle avoidance, handling of robot position uncertainty, incorporation of pre-compiled maps, recovery of geometric representations, and other related problems. The experimental results show that the Occupancy Grid approach generates dense world models, is robust under sensor uncertainty and errors, and allows explicit handling of uncertainty. It supports the development of robust and agile sensor interpretation methods, incremental discovery procedures, and composition of information from multiple sources. Furthermore, the results illustrate that robotic tasks can be addressed through operations performed directly on the Occupancy Grid, and that these operations have strong parallels to operations performed in the image processing domain.*

1 Introduction

Autonomous robot systems require the ability to recover robust spatial models of the surrounding world from sensory information and to efficiently utilize these models in robot planning and control tasks. These capabilities enable the robot to interact coherently with its environment, both by adequately interpreting the available sensor data so as to reach appropriate conclusions about the real world for short-term decisions, and by acquiring and manipulating a rich and substantially complete world model for long-term planning and decision-making.

Traditional approaches to robot perception have emphasized the use of geometric sensor models and heuristic assumptions to constrain the sensor interpretation process, and the use of geometric world models as the basis for planning robotic tasks [10, 5, 2]. "Low-level" sensing procedures extract geometric features such as line segments or surface patches from the sensor data, while "high-level" sensor interpretation modules use prior geometric models and heuristic assumptions about the environment to constrain the sensor interpretation process. The resulting deterministic geometric descriptions of the environment of the robot are subsequently used as the basis for other robotic activities, such as obstacle avoidance, path-planning and navigation, or planning of grasping and assembly operations. These approaches, which incorporate what we have characterized as the *Geometric Paradigm* in robot perception, have several shortcomings [5]. Generally speaking, the Geometric Paradigm leads to sparse and brittle world models; it requires early decisions in the interpretation of the sensor data for the instantiation of specific model primitives; it does not provide appropriate mechanisms for handling the uncertainty and errors intrinsic to the sensory information; and it relies heavily on the accurateness and adequacy of the prior world models and of the heuristic assumptions used. Overall, these approaches are of limited use in more complex scenarios, such as those encountered by mobile robots. Autonomous or semi-autonomous vehicles for planetary exploration, operation in hazardous environments, submarine exploration and servicing, mining and industrial applications have to explore and operate in unknown and unstructured environments, handle unforeseen events, and perform in real time.

In this paper, we review a new approach to robot perception and world modelling that uses a probabilistic tesselated

"Occupancy Grids: A Stochastic Spatial Representation for Active Robot Perception" by A. Elfes. Reprinted from the *Proceedings of the Sixth Conference on Uncertainty in AI*, July 1990. Copyright © 1990 American Association for Artificial Intelligence. Copies of this and other AAAI Proceedings are available from Morgan Kaufmann Publishers, Inc., 2929 Campus Drive, San Mateo, CA 94403.

representation of spatial information called the *Occupancy Grid* [8, 5, 10, 11]. The Occupancy Grid is a multi-dimensional random field that maintains stochastic estimates of the occupancy state of each cell in a spatial lattice. The cell estimates are obtained by interpreting sensor range data using probabilistic models that capture the uncertainty in the spatial information provided by the sensors. Bayesian estimation procedures allow the incremental updating of the Occupancy Grid using readings taken from several sensors and from multiple points of view. As a result, the disambiguation of sensor data is performed not through heuristics or prior models, but by additional sensing and through the use of adequate sensing strategies.

In subsequent sections, we provide an overview of the Occupancy Grid formulation and discuss how the Occupancy Grid framework provides a unified approach to a number of tasks in mobile robot perception and navigation. These tasks include range-based mapping, multiple sensor integration, path-planning and obstacle avoidance, handling of robot position uncertainty and other related problems. We show that a number of robotic problem-solving activities can be performed directly on the Occupancy Grid representation, precluding the need for the recovery of deterministic geometric descriptions. We also draw some parallels between operations on Occupancy Grids and related image processing operations.

2 The Occupancy Grid Framework

In this section, we provide a brief outline of the Occupancy Grid formulation, while in the succeeding sections we discuss several applications of the Occupancy Grid framework to mobile robot mapping and navigation. The scenarios under consideration in this paper involve a mobile robot operating in unknown and unstructured environments, and carrying a complement of sensors that provide range information directly (sonar, scanning laser rangefinders) or indirectly (stereo systems). A qualitative overview of some parts of this work is found in [11]; preliminary experimental results have been reported in [4, 8, 12], while a more detailed discussion is available in [10, 5]. More recently, we are applying the Occupancy Grid framework to the active control of robot perception [9] and as part of a multi-level performance-oriented mobile robot architecture [6].

2.1 The Occupancy Grid Representation

The Occupancy Grid is a multi-dimensional (typically 2D or 3D) tesselation of space into cells, where each cell stores a probabilistic estimate of its state. Formally, an *Occupancy Field* $O(\mathbf{x})$ can be defined as a discrete-state stochastic process defined over a set of continuous spatial coordinates $\mathbf{x} = (x_1, x_2, \ldots)$, while the *Occupancy Grid* is defined over a discrete spatial lattice. Consequently, the Occupancy Grid corresponds to a discrete-state (binary) random field [22]. A *realization* of the Occupancy Grid is obtained by estimating the state of each cell from sensor data.

More generally, the cell state can be used to encode a number of properties, represented using a random vector associated with each lattice point of the random field, and estimated accordingly. Properties of interest for robot planning could include occupancy, observability, reachability, connectedness, danger, reflectance, etc. We refer to such general world models, which are again instances of random fields, as *Inference Grids* [5]. In this paper, we are mainly interested in *spatial* models for robot perception, and will restrict ourselves to the estimation of a single property, the *occupancy state* of each cell.

2.2 Estimating the Occupancy Grid

In the Occupancy Grid, the state variable $s(C)$ associated with a cell C is defined as a discrete random variable with two states, *occupied* and *empty*, denoted OCC and EMP. Since the states are exclusive and exhaustive, $P[s(C) = \text{OCC}] + P[s(C) = \text{EMP}] = 1$. Each cell has, therefore, an associated probability mass function that is estimated by the sensing process.

To construct a map of the robot's environment, two processing stages are involved. First, a sensor range measurement r is interpreted using a stochastic sensor model. This model is defined by a probability density function (p.d.f.) of the form $p(r \mid z)$, where z is the actual distance to the object being detected. Secondly, the sensor reading is used in the updating of the cell state estimates of the Occupancy Grid. For simplicity, we will derive the interpretation and updating steps for an Occupancy Grid defined over a single spatial coordinate, and outline the generalization to more dimensions.

In the continuous case, the random field $O(x)$ is described by a probability mass function defined for every x and is written as $O(x) = P[s(x) = \text{OCC}](x)$, the probability of the state of x being *occupied*. The probability of x being *empty* is obviously given by $P[s(x) = \text{EMP}](x) = 1 - P[s(x) = \text{OCC}](x)$. The conditional probability of the state of x being occupied given a sensor reading r will be written as $O(x \mid r) = P[s(x) = \text{OCC} \mid r](x)$. For the discrete case, the Occupancy Grid corresponds to a sampling of the random field over a spatial lattice. We will represent the probability of a cell C_i being occupied as $O(C_i) = P[s(C_i) = \text{OCC}](C_i)$, and the conditional probability given a sensor reading r as $O(C_i \mid r) = P[s(C_i) = \text{OCC} \mid r](C_i)$. When only a single cell C_i is being referenced, we will use the more succinct notation $P[s(C_i) = \text{OCC}]$.

We now consider a range sensor characterized by a sensor model defined by the p.d.f. $p(r \mid z)$, which relates the reading r to the true parameter space range value z. Determining an optimal estimate $\hat{z}$ for the parameter z is a straightforward estimation step, and can be done using Bayes' formula and MAP estimates [3, 21]. Recovering a model of the environment as a whole, however, leads to a more complex estimation problem. In general, obtaining an optimal estimate of the occupancy grid $O(C_i \mid r)$ would require determining the conditional probabilities of all possible world configurations. For the two-dimensional case of a map with $m \times m$ cells, a total of 2^{m^2} alternatives are possible, leading to a non-trivial estimation problem. To avoid this combinatorial explosion of grid configurations, the cell states are estimated as *independent* random variables. As a result, the Occupancy Grid corresponds to a Markov Random Field (MRF) of order 0 [22]. The independence assumption can be justified conceptually by the fact that in general there are no causal relationships between the occupancy states of different cells, and can be justified from an engineering point of view, because the resulting models are adequate for the range of tasks in which they are being applied. Finally, the computational simplicity intrinsic in the use of zero-order MRFs allows the development of very agile perception systems. On the other hand, there are applications, such as precise shape recovery, where more complex Occupancy Grid estimation models using higher-order MRFs are required [13, 15].

To determine how a sensor reading is used in estimating the state of the cells of the Occupancy Grid, we start by applying Bayes' theorem to a single cell C_i:

$$P[s(C_i) = \text{OCC} \mid r] = \frac{p[r \mid s(C_i) = \text{OCC}]\, P[s(C_i) = \text{OCC}]}{\sum_{s(C_i)} p[r \mid s(C_i)]\, P[s(C_i)]} \quad (1)$$

Notice that the $p[r \mid s(C_i)]$ terms that are required in this equation do not correspond directly to the sensor model $p(r \mid z)$, since the latter implicitly relates the range reading to the detection of a single object surface. In other words, the sensor model can be rewritten as:

$$p(r \mid z) = p[r \mid s(C_i) = \text{OCC} \wedge s(C_k) = \text{EMP}, k < i] \quad (2)$$

To derive the distributions for $p[r \mid s(C_i)]$, it is necessary to perform an estimation step over all possible world configurations. This can be done using Kolmogoroff's theorem [18]:

$$p[r \mid s(C_i) = \text{OCC}] = \sum_{\{G_{s(C_i)}\}} \left(p[r \mid s(C_i) = \text{OCC}, G_{s(C_i)}] \times P[G_{s(C_i)} \mid s(C_i) = \text{OCC}]\right) \quad (3)$$

where $G_{s(C_i)} = (s(C_1) = s_1, \cdots, s(C_{i-1}) = s_{i-1}, s(C_{i+1}) = s_{i+1}, \cdots, s(C_n) = s_n)$ stands for a specific grid configuration with $s(C_i) = \text{OCC}$, and $\{G_{s(C_i)}\}$ represents all possible grid configurations under that constraint. In the same manner, $p[r \mid s(C_i) = \text{EMP}]$ can be computed as:

$$p[r \mid s(C_i) = \text{EMP}] = \sum_{\{G_{s(C_i)}\}} \left(p[r \mid s(C_i) = \text{EMP}, G_{s(C_i)}] \times P[G_{s(C_i)} \mid s(C_i) = \text{EMP}]\right) \quad (4)$$

The configuration probabilities $P[G_{s(C_i)} \mid s(C_i)]$ are determined from the individual prior cell state probabilities. These, in turn, can be obtained from experimental measurements for the areas of interest, or derived from other considerations about likelihoods of cell states. We have opted for the use of non-informative or maximum entropy priors [1], which in this case assign equal probability values to the two possible states:

$$P[s(C_i) = \text{OCC}] = P[s(C_i) = \text{EMP}] = 1/2 \quad (5)$$

Using the cell independence assumption, these priors are used to determine the configuration probabilities $P[G_{s(C_i)} \mid s(C_i)]$, needed in Eqs. 3 and 4. Finally, Eq. 2 is used in the computation of the distributions $p[r \mid s(C_i)]$. The full derivation of these terms is found in [5]; we only remark that because there are subsets of configurations that are *indistinguishable* under a single sensor observation r, it is possible to derive closed form solutions of these equations for certain sensor models, and to compute numerical solutions in other cases.

To illustrate the approach, consider the case of an ideal sensor, characterized by the p.d.f. $p(r \mid z) = \delta(r - z)$, where δ is the Kronecker delta. For this case, the following closed form solution of Eq. 1 results (Fig. 1):

$$P[s(C_i) = \text{OCC} \mid r] = \begin{cases} 0 & \text{for } x < r, x \in C_i \\ 1 & \text{for } x, r \in C_i \\ 1/2 & \text{for } x > r, x \in C_i \end{cases} \quad (6)$$

which is an intuitively appealing result: if an ideal sensor measures a range value r, the corresponding cell has occupancy probability 1; the preceding cells are empty and have occupancy probability 0; and the subsequent cells have not been observed and are therefore unknown, having occupancy probability 1/2.

As another example, consider a range sensor whose measurements are corrupted by Gaussian noise of zero mean and variance σ^2. The corresponding sensor p.d.f. is given by:

$$p(r \mid z) = \frac{1}{\sqrt{2\pi}\sigma} \exp\left(\frac{-(r-z)^2}{2\sigma^2}\right) \quad (7)$$

This equation can be used in the numerical evaluation of Eqs. 3 and 4. A plot of a typical cell occupancy profile obtained for this sensor from Eq. 1 is shown in Fig. 2.

To extend the derivation to two spatial dimensions, consider the example of a range sensor characterized by Gaussian uncertainty in both range and angle, given by the variances σ_r^2 and σ_θ^2. In this case, the sensor p.d.f. can be represented in polar coordinates as:

$$p(r \mid z, \theta) = \frac{1}{2\pi\sigma_r\sigma_\theta} \exp\left[-\frac{1}{2}\left(\frac{(r-z)^2}{\sigma_r^2} + \frac{\theta^2}{\sigma_\theta^2}\right)\right] \quad (8)$$

In this formula, the dependency of the random variable r on z and θ is decoupled, a reasonable assumption for a first-order

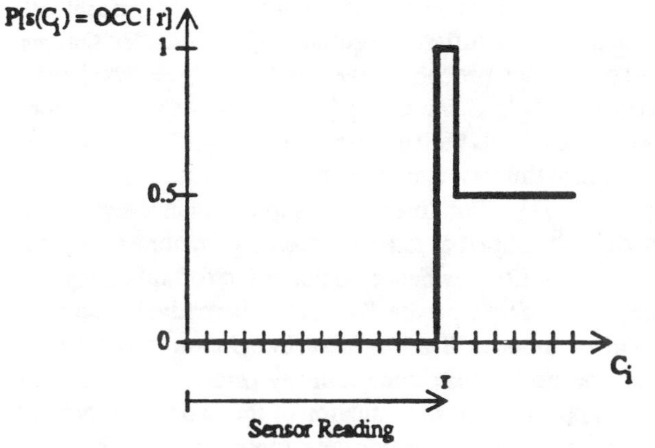

Figure 1: Occupancy Probability Profile for an ideal sensor, given a range measurement r.

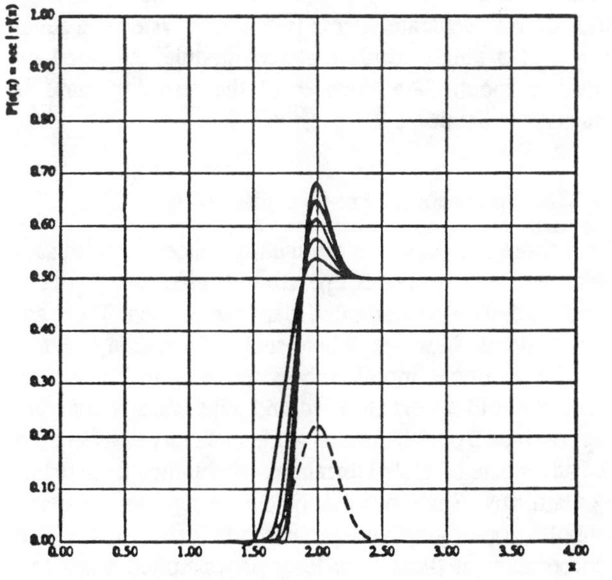

Figure 2: Occupancy Probability Profiles obtained from a sensor with Gaussian distribution. The sensor model $p(r \mid z)$ is shown superimposed (dashed line). Several successive updates of the cell occupancy probabilities are plotted, with the sensor positioned at $x = 0.0$ and with $r = 2.0$. The grid was initialized with $P[s(x) = \text{OCC}](x) = 0.5$. The profiles show that the Occupancy Grid converges towards the behaviour of the ideal sensor.

model of certain kinds of range sensors. Consequently, the estimation of the two-dimensional Occupancy Grid can be performed conveniently in polar coordinates (ρ, φ), using fundamentally the same formulation as above (Eqs. 3 and 4) and applying Eq. 8 to recover the distributions $p[r \mid s(C_{\rho_i \varphi_j})]$. These in turn are used to obtain the polar Occupancy Grid $P[s(C_{\rho_i \varphi_j}) \mid r]$. To generate the corresponding two-dimensional cartesian Occupancy Grid, the polar grid can be scanned and resampled. A 2D cartesian Occupancy Grid is shown in Fig. 3, obtained from a single sonar reading. Similar derivations can be performed for 3D Occupancy Grids.

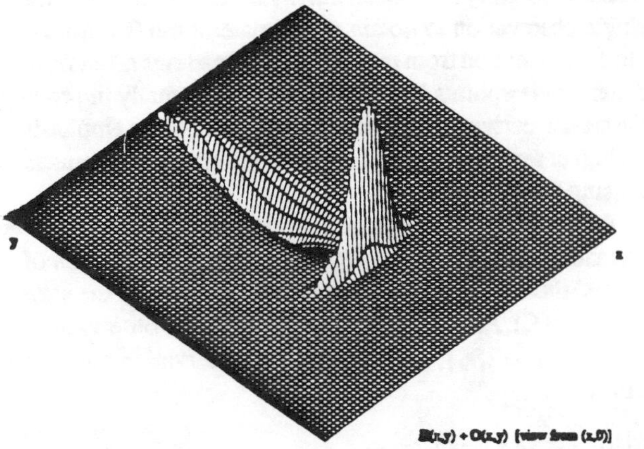

Figure 3: Two-Dimensional Sonar Occupancy Grid. The occupancy profile shown corresponds to a range measurement taken by a sonar sensor positioned at the upper left, pointing to the lower right. The plane corresponds to the UNKNOWN (1/2) level.

The sensor model actually used in most of our experiments, and that can be tailored to a large class of range sensors, is expressed as

$$p(\mathbf{r} \mid \mathbf{z}) = D(\mathbf{z})G(\mathbf{r}, \mathbf{z}, \Sigma(\mathbf{z})) \tag{9}$$

where, in addition to the multivariate Gaussian range measurement noise encoded in G, we also take into account the detection probability of an object at distance $\mathbf{z}$, $D(\mathbf{z})$, and the dependency of the range covariance on the object distance, expressed as $\Sigma(\mathbf{z})$. These terms were obtained through calibration experiments performed in our laboratory.

2.3 Updating the Occupancy Grid

Due to the intrinsic limitations of sensor systems, recovering a description of the world from sensory information is fundamentally an underconstrained problem. As mentioned previously, this has historically been addressed by the heavy use of prior models and simplifying heuristic assumptions about the robot's environment, leading to slow and brittle systems. Within the Occupancy Grid framework, the underconstrainedness of the sensor data is handled instead by the use of active perception strategies to resolve sensor ambiguity and uncertainty. Rather than relying on a single observation to obtain an estimate of the Occupancy Grid, information from multiple sensor readings taken from different viewpoints is composed to incrementally improve the sensor-derived map. This leads naturally to an emphasis on higher sensing rates and on the development of adequate sensing strategies.

To allow the incremental composition of sensory information, we use the sequential updating formulation of Bayes' theorem [5]. Given the current estimate of the state of a cell $s(C)$, $P[s(C_i) = \text{OCC} \mid \{r\}_t]$, based on observations $\{r\}_t = \{r_1, \cdots, r_t\}$, and given a new observation r_{t+1}, we can write:

$$P[s(C_i) = \text{OCC} \mid \{r\}_{t+1}] = \\ = \frac{p[r_{t+1} \mid s(C_i) = \text{OCC}]\, P[s(C_i) = \text{OCC} \mid \{r\}_t]}{\sum_{s(C_i)} p[r_{t+1} \mid s(C_i)]\, P[s(C_i) \mid \{r\}_t]} \quad (10)$$

In this formula, the previous estimate of the cell state, $P[s(C_i) = \text{OCC} \mid \{r\}_t]$, serves as the prior and is obtained directly from the Occupancy Grid. Tables for the sensor model-derived terms, $p[r_{t+1} \mid s(C_i)]$, can be computed offline for use in the recursive estimation procedure, allowing fast map updating. The new cell state estimate $P[s(C_i) = \text{OCC} \mid \{r\}_{t+1}]$ is subsequently stored again in the map. An example of this Bayesian updating procedure is shown in Fig. 2.

2.4 Sensor Integration

To increase the capabilities and the performance of robotic systems in general, a variety of sensing devices are necessary to support the different kinds of tasks to be performed. This is particularly important for mobile robots, where multiple sensor systems can provide higher levels of fault-tolerance and safety. Additionally, qualitatively different sensors have different operational characteristics and failure modes, and can therefore complement each other.

Within the Occupancy Grid framework, sensor integration can be performed using a formula similar to Eq. 10 for the combination of estimates provided by different sensors [5]. This allows the updating of the *same* Occupancy Grid by multiple sensors operating independently. Consider two independent sensors S_1 and S_2, characterized by sensor models $p_1(r \mid z)$ and $p_2(r \mid z)$. In this case, the integration of readings r_{S_1} and r_{S_2}, measured by sensors S_1 and S_2, respectively, can be done using:

$$P[s(C_i) = \text{OCC} \mid r_{S_1}, r_{S_2}] = \\ = \frac{p[r_{S_2} \mid s(C_i) = \text{OCC}]\, P[s(C_i) = \text{OCC} \mid r_{S_1}]}{\sum_{s(C_i)} p[r_{S_2} \mid s(C_i)]\, P[s(C_i) \mid r_{S_1}]} \quad (11)$$

A different estimation problem occurs when separate Occupancy Grids are maintained for each sensor system, and integration of these sensor maps is performed at a later stage by composing the corresponding cell probability estimates. This requires the combination of probabilistic evidence from different sources [1]. Consider the two cell occupancy probabilities $P_1 = P_{S_1}[s(C_i) = \text{OCC} \mid \{r\}_{t_1}]$ and $P_2 = P_{S_2}[s(C_i) = \text{OCC} \mid \{r\}_{t_2}]$, obtained from separate Occupancy Grids built using sensors S_1 and S_2. The general solution to this problem involves the use of a *Superbayesian* approach [1]. For linear sensor performance evaluation models, the Superbayesian estimation procedure is reduced to a probabilistic evidence combination formula known as the *Independent Opinion Pool* [1]. Alternatively, the same result is obtained if a Bayesian integration is performed, with the use of maximum entropy priors. This method, when applied to the combination of the two sensor-derived estimates, P_1 and P_2, yields the simple formula [5]:

$$P[s(C_i) = \text{OCC} \mid P_1, P_2] = \frac{P_1\, P_2}{P_1\, P_2 + (1 - P_1)\,(1 - P_2)} \quad (12)$$

In previous work, described in [12, 16], the Independent Opinion Pool method was used to integrate Occupancy Grids derived separately from two sensor systems, a sonar array and a single-scanline stereo module, mounted on a mobile robot. An example of the resulting maps is presented in Section 3.2.

2.5 Incorporation of Pre-Compiled Maps

Throughout this paper we are mainly concerned with scenarios where the robot is operating in unknown environments, so that no pre-compiled maps can be used. There are other contexts, however, where such information *is* available. For example, mobile robots operating inside nuclear facilities could access detailed and substantially accurate maps derived from blueprints, while planetary rovers could take advantage of global terrain maps obtained from orbiting platforms. Such information can be represented using symbolic, topological and geometric models [14, 5]. The incorporation of these high-level pre-compiled maps can be done within the Occupancy Grid framework using the same methodology outlined in the previous sections. To provide a common representation, the geometric models are scan-converted into an Occupancy Grid, with occupied and empty areas being assigned the corresponding probabilities. These pre-compiled maps can subsequently be used as priors for sensor maps, or can simply be treated as another source of information to be integrated with sensor-derived maps [5].

2.6 Decision-Making

For certain applications, it may be necessary to make discrete choices concerning the state of a cell C. The *optimal estimate* is provided by the *maximum a posteriori* (MAP) decision rule [3], which can be written in terms of occupancy probabilities as:

$$\begin{cases} C \text{ is OCCUPIED} & \text{if } P(s(C) = \text{OCC}) > P(s(C) = \text{EMP}) \\ C \text{ is EMPTY} & \text{if } P(s(C) = \text{OCC}) < P(s(C) = \text{EMP}) \\ C \text{ is UNKNOWN} & \text{if } P(s(C) = \text{OCC}) = P(s(C) = \text{EMP}) \end{cases} \quad (13)$$

Additional factors, such as the cost involved in making different choices, can be taken into account by using other decision criteria, such as minimum-cost or minimum-risk estimates [21]. Depending on the specific application, it may also be of interest to define an UNKNOWN band, as opposed to a single thresholding value. As shown in [5], however, many robotic tasks can be performed directly on the Occupancy Grid, precluding the need to make discrete choices concerning the state of individual cells. In path-planning, for example, we define the cost of a path in terms of a risk factor directly related to the corresponding cell probabilities [8].

3 Using Occupancy Grids for Mobile Robot Mapping

We now proceed to illustrate the Occupancy Grid approach by discussing some applications of Occupancy Grids to autonomous mobile robots. In this section, we summarize the use of Occupancy Grids in sensor-based mobile robot *Mapping*, while in Section 4 we provide an overview of the use of Occupancy Grids in mobile robot *Navigation*. The experimental results shown here have been mostly obtained in operating environments that can be adequately described by two-dimensional maps. We have recently started to extend our work to the generation and manipulation of 3D Occupancy Grids [5].

One possible flow of processing for sensor-based robot mapping applications is outlined below and summarized in Fig. 4. As the mobile robot explores and maps its environment, the incoming sensor readings are interpreted using the corresponding probabilistic sensor models.The map of the world that the robot acquires from a single sensor reading is called a *Sensor View*. Various Sensor Views taken from a single robot position can be composed into *Local Sensor Maps*, which can be maintained separately for each sensor type. A composite description of the robot's surroundings is obtained through sensor integration of separate Local Sensor Maps into a *Robot View* (as mentioned previously, Robot Views can be generated directly from the integration of different sensors). As a result, the Robot View encapsulates the information recovered at a single mapping location. As the robot explores its surroundings, Robot Views taken from multiple data-gathering positions are composed into a *Global Map* of the environment. This requires relative registration of the Robot Views, an issue that is addressed in Section 4.

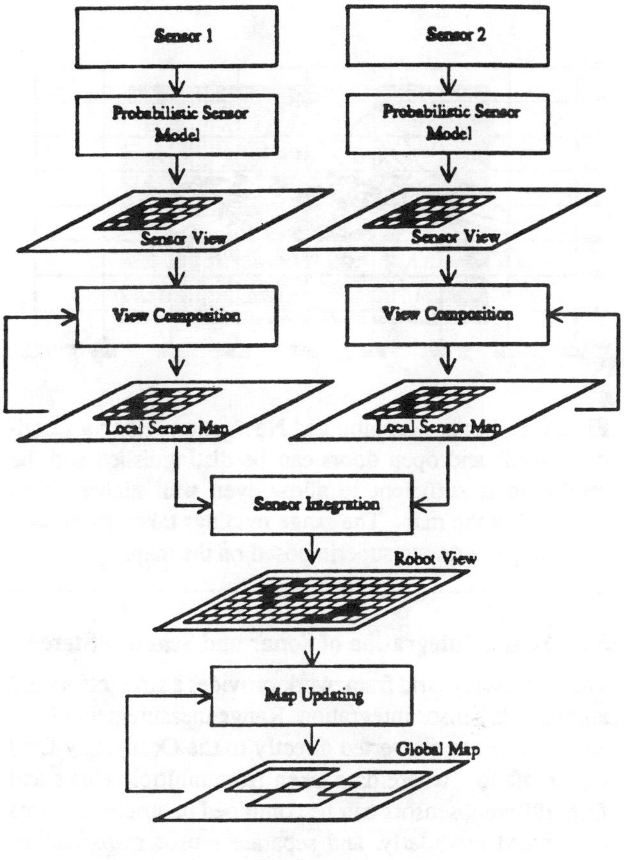

Figure 4: A Framework for Occupancy Grid Based Robot Mapping.

3.1 Sonar-Based Mapping

The Occupancy Grid representation was first developed in the context of sonar-based mapping experiments [4, 7, 8]. The functional limitations of sonar sensors and the need to recover robust and dense maps of the robot's environment precluded the use of simple geometric interpretation methods [8] and led to the investigation of tesselated probabilistic representations. Initial results using a heuristic approach called Certainty Grids [17, 4, 7, 8] were encouraging, and led to the development of the Occupancy Grid framework. To test the framework, we implemented an experimental system for sonar-based mapping and navigation for autonomous mobile robots called *Dolphin* [7, 8]. A number of indoor and outdoor experiments were performed (see, for example, [8, 5]). Fig. 5 presents a sonar map obtained during navigation down a corridor. The experimental work has shown that the cell updating mechanisms are computationally fast, allowing a high sensing to computation ratio, and that the framework can be equally well applied to other kinds of sensors [5, 12, 10].

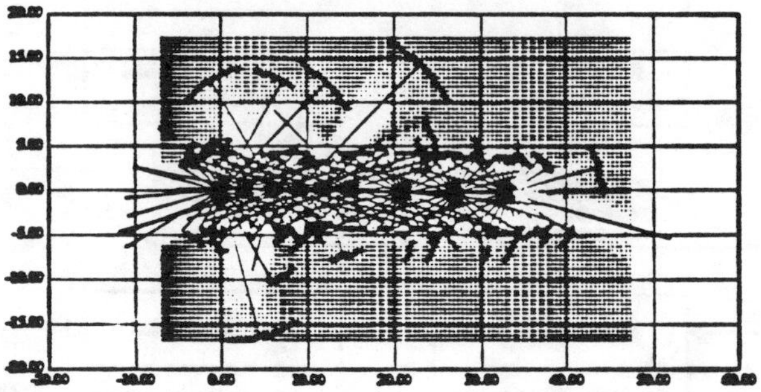

Figure 5: Sonar Mapping and Navigation Along a Corridor. Walls and open doors can be distinguished and the resolution is sufficient to allow even wall niches to be observed in the map. The range readings taken from each robot stop are drawn superimposed on the map.

3.2 Sensor Integration of Sonar and Scanline Stereo

The Occupancy Grid framework provides a straightforward approach to sensor integration. Range measurements from each sensor are converted directly to the Occupancy Grid representation, where data taken from multiple views and from different sensors can be combined naturally. Sensors are treated modularly, and separate sensor maps can be maintained concomitantly with integrated maps, allowing independent or joint sensor operation. In joint work with Larry Matthies, we have performed experiments in the integration of data from two sensor systems: a *sonar sensor array* and a *single-scanline stereo module* that provides horizontal depth profiles, both mounted on a mobile robot. This allows the generation of improved maps that take advantage of the complementarity of the sensors [12, 16]. A typical set of maps is shown in Fig. 6.

4 Using Occupancy Grids for Robot Navigation

For autonomous robot navigation, a number of concerns have to be addressed. In this section, we briefly outline the use of Occupancy Grids in path-planning and obstacle avoidance, estimating and updating the robot position, and incorporating the positional uncertainty of the robot into the mapping process (Fig. 7). A detailed discussion is found in [5].

4.1 Path-Planning and Obstacle Avoidance

In the *Dolphin* system, path-planning and obstacle avoidance are performed using potential functions and an A* search algorithm that operates directly on the Occupancy Grid. The path-planning operation minimizes a multi-objective cost function $f(\mathbf{P})$, defined over the path $\mathbf{P} = \{(x_0, y_0, \theta_0), \cdots, (x_n, y_n, \theta_n)\}$, that takes into account

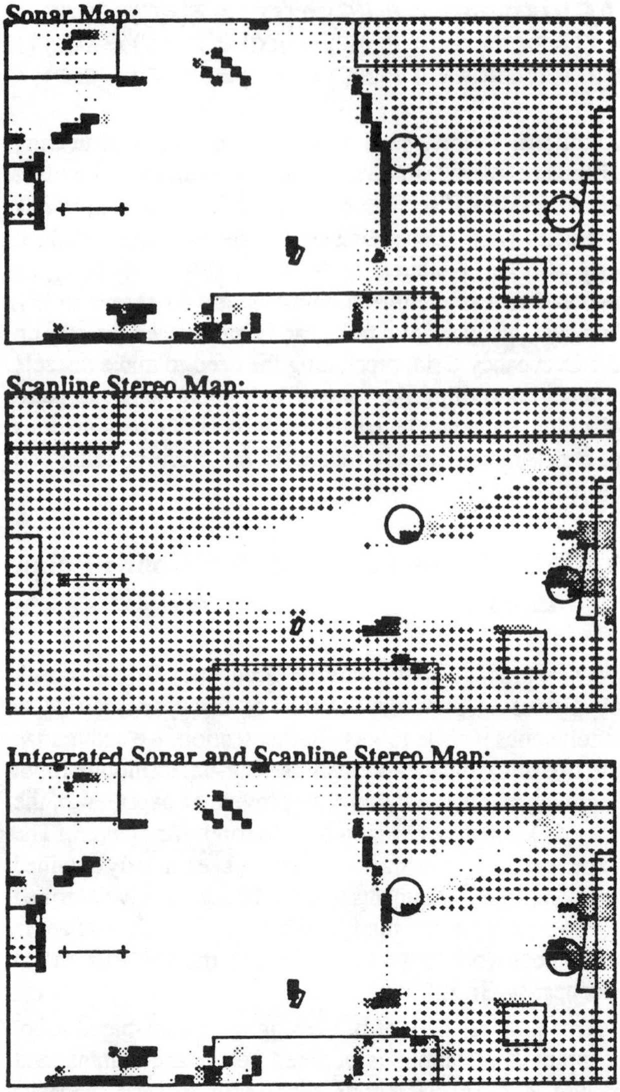

Figure 6: Sensor Integration of Sonar and Scanline Stereo. Occupancy Grids generated separately for sonar and scanline stereo, and jointly through sensor integration are shown. *Occupied* regions are marked by shaded squares, *empty* areas by dots fading to white space, and *unknown* spaces by + signs.

both the occupancy probabilities of the cells being traversed and the total distance to the robot's destination:

$$f(\mathbf{P}) = \tau_c \sum_{\forall C \in \mathbf{P}} \Gamma(C) + \tau_d \, \text{length}(\mathbf{P}) \qquad (14)$$

where τ_c and τ_d weigh the component costs that are associated with the cell occupancy probabilities and with the distance to the goal, respectively. The function $\Gamma(C)$ expresses the cost of traversing a single cell, and is de-

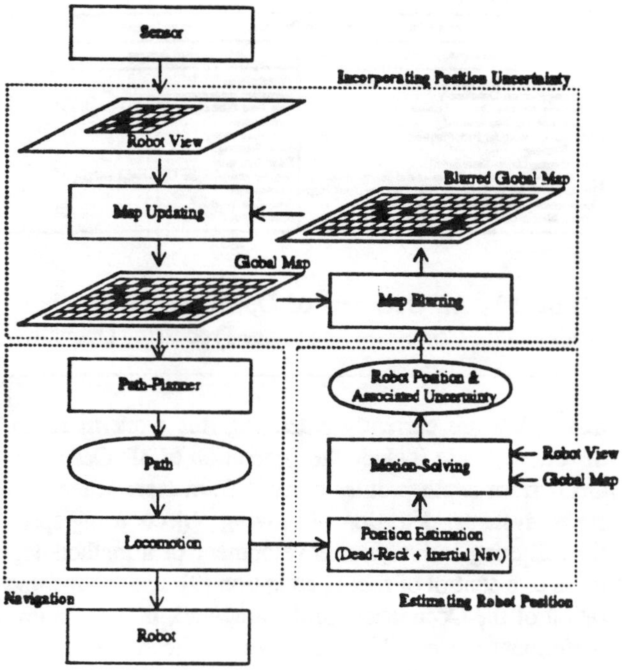

Figure 7: A Framework for Occupancy Grid-Based Robot Navigation.

fined directly as a non-linear function of the occupancy probability $P[s(C) = \mathrm{OCC}]$ (see [5]).

4.2 Handling Robot Position Uncertainty

To desambiguate sensor information and recover accurate and complete descriptions of the environment of operation of a robot, it is necessary to integrate sensor data acquired from multiple viewing positions. To allow the composition of these multiple views into a coherent model of the world, accurate information concerning the relative transformations between data-gathering positions is necessary to allow precise registration of the views for subsequent integration. For mobile robots that move around in unstructured environments, recovering precise position information poses a major problem. Over longer distances, dead-reckoning estimates are not sufficiently reliable; consequently, motion-solving methods that use landmark tracking or map matching approaches are usually applied to reduce the registration imprecision due to motion. Furthermore, the positional error is compounded over sequences of movements as the robot traverses its environment. This leads to the need for explicitly handling positional uncertainty and taking it into account when composing sensor information.

To represent and estimate the robot position as the vehicle explores its environment, we use the *Approximate Transformation* (AT) framework [19]. A robot motion M, defined with respect to some coordinate frame, is represented as $\tilde{M} = <\hat{M}, \Sigma_M>$, where $\hat{M}$ is the estimated (nominal) position, and Σ_M is the associated covariance matrix that captures the positional uncertainty. The parameters of the robot motion are determined from dead-reckoning and inertial navigation estimates, which can be composed using the AT *merging* operation, while the updating of the robot position uncertainty over several moves is done using the AT *composition* operation [19].

4.3 Motion-Solving

For more precise position estimation, a multi-resolution correlation-based motion-solving procedure is employed. It searches for an optimal registration between the new Robot View and the current Global Map, by matching the corresponding Occupancy Grids before map composition [17].

4.4 Incorporating Positional Uncertainty into the Mapping Process

After estimating the registration between the new Robot View and the Global Map, the associated uncertainty is incorporated into the map updating process as a blurring or convolution operation performed on the Occupancy Grid. We distinguish between *World-Based Mapping* and *Robot-Based Mapping* [5, 11].

In *World-Based Mapping*, the motion of the robot is related to the observer or world coordinate frame, and the current Robot View is blurred by the robot's positional uncertainty prior to composition with the Global Map. If we represent the Global Map by M_G, the current Robot View by V_R, the robot position by the AT $\tilde{R} = <\hat{R}, \Sigma_R>$, the blurring operation by the symbol $\tilde{\otimes}$ and the composition of maps by the symbol $\tilde{\oplus}$, we can express the world-based mapping procedure as:

$$M_G \leftarrow M_G \tilde{\oplus} (V_R \tilde{\otimes} \tilde{R}) \quad (15)$$

Since the global robot position uncertainty increases with every move, the effect of this updating procedure is that the new Views become progressively more blurred, adding less and less useful information to the Global Map. Observations seen at the beginning of the exploration are "sharp", while recent observations are "fuzzy". From the point of view of the inertial observer, the robot eventually "dissolves" in a cloud of probabilistic smoke.

For *Robot-Based Mapping* (Fig. 7), the registration uncertainty of the Global Map due to the recent movement of the robot is estimated, and the Global Map is blurred by this uncertainty prior to composition with the current Robot View. This mapping procedure can be expressed as:

$$M_G \leftarrow V_R \tilde{\oplus} (M_G \tilde{\otimes} \tilde{R}) \quad (16)$$

A consequence of this method is that observations performed in the remote past become increasingly uncertain, while recent observations have suffered little blurring.

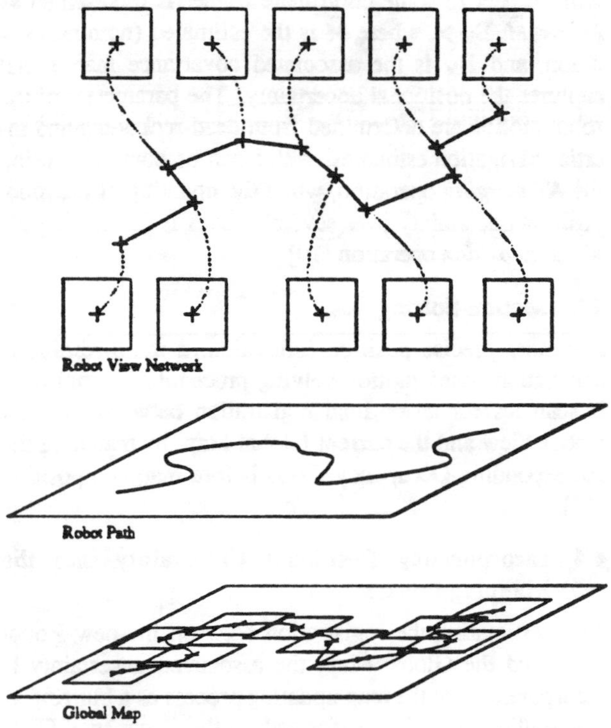

Figure 9: Maintaining a Dual Representation. A stochastic graph with the individual Robot Views is maintained in conjunction with the Global Map.

From the point of view of the robot, the immediate surroundings (which are of relevance to its current navigational tasks) are "sharp". The robot is leaving, so to speak, an expanding "probabilistic trail" of weakening observations behind it (see Fig. 8).

It should be noted, however, that the local spatial relationships observed within a Robot View still hold. So as not to lose this information, we use a two-level spatial representation, incorporating Occupancy Grids and Approximate Transformations. On one level, the individual Views are stored attached to the nodes of an AT graph (a *stochastic map* [20]) that describes the movements of the robot. Coupled to this, a Global Map is maintained that represents the robot's current overall knowledge of the world (Fig. 9).

5 Other Applications

In the previous sections, we have seen that Occupancy Grids provide a unified approach to a number of issues in Robotics and Computer Vision. Additional tasks that we have addressed include the recovery of geometric descriptions from Occupancy Grids [7, 8], incorporation of pre-compiled maps [5], landmark recognition, prediction of sensor readings from Occupancy Grids, and related problems. We are currently extending this work in several directions; these include the generation of 3D Occupancy Grids from depth profiles derived from laser scanners or stereo systems, detection of moving objects using space-time filtering techniques, development of a methodology for active control of robot perception [9], and the incorporation of the Occupancy Grid framework in a multi-level performance-oriented mobile robot architecture [6].

Comparison of Operations on Occupancy Grids and on Images	
Occupancy Grids	**Images**
Labelling cells as Occupied, Empty or Unknown	Thresholding
Handling Position Uncertainty	Blurring/Convolution
Removing Spurious Spatial Readings	Low-Pass Filtering
Motion Solving/Map Matching	Correlation
Obstacle Growing for Path-Planning	Region Growing
Path-Planning	Edge Tracking
Determining Object Boundaries	Edge Detection
Extracting and Labelling Occupied and Empty Areas	Segmentation/Region Colouring/Labelling
Prediction of Sensor Readings from User-Provided Maps	Convolution
Incorporating User-Provided Maps	Scan-Conversion
Object Motion Detection over Map Sequences	Space-Time Filtering

Figure 10: An Overview of Operations on Occupancy Grids and the Corresponding Image Processing Operations.

It should be noted that many robotic tasks can be performed on Occupancy Grids using operations that are similar or equivalent to computations performed in the image processing domain. Table 10 provides a qualitative overview and comparison of some of these operations.

We finalize our remarks with a note concerning low-level versus high-level representations. It is interesting to observe that in Robotics and Computer Vision there has been historically a slow move from very high-level (stylized) representations of blocks-world objects to the recovery of simple spatial features in very constrained real images; from there to the recovery of surface patches; and recently towards "dense", tesselated representations of spatial information such as the Occupancy Grid. A parallel evolution from sparse, high-level or exact descriptions to dense, lower-level and sometimes approximate descriptions can be seen in some other computational fields, such as Computer Graphics and Finite Element Analysis.

6 Conclusions

We have reviewed in this paper the Occupancy Grid framework and presented results from its application to mobile robot mapping and navigation tasks in unknown and unstructured environments. The Occupancy Grid approach supports agile and robust sensor interpretation methods, incremental discovery procedures, composition of information from multiple sensors and over multiple positions of the robot, and explicit handling of uncertainty. Furthermore, the world models recovered from sensor data can be used efficiently in robotic planning and problem-solving activities. The results lead us to suggest that the Occupancy Grid framework provides a novel approach to robot perception and spatial reasoning that has the characteristics of

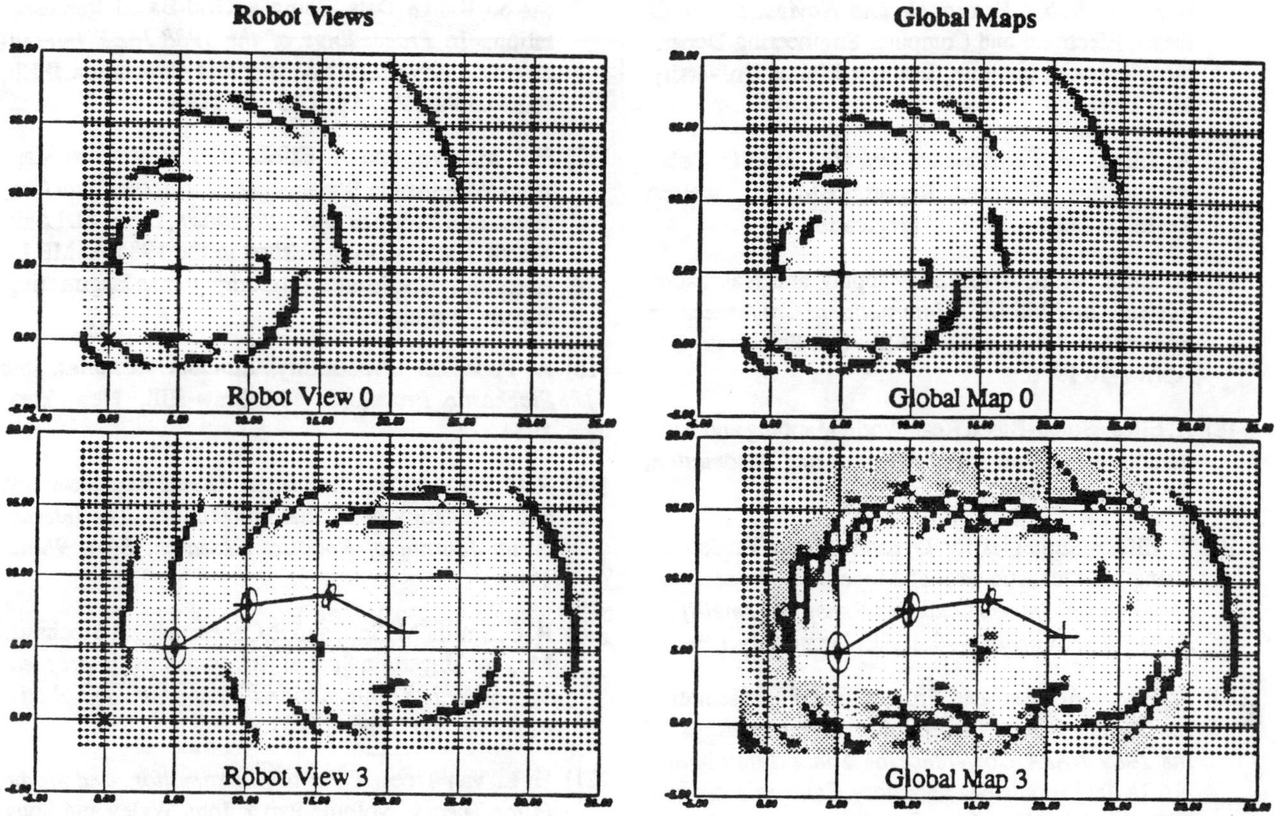

Figure 8: Incorporating Motion Uncertainty into the Mapping Process. For robot-centered mapping, the Global Map is blurred by the robot position uncertainty (shown using the corresponding covariance ellipses) prior to composition with the Robot View. Two stages of the process are shown.

robustness and generality necessary for real-world robotic applications.

Acknowledgments

The author wishes to thank Hans Moravec, Sarosh Talukdar, Peter Cheeseman, Radu Jasinschi, Larry Matthies, José Moura, Michael Meyer and Larry Wasserman for their comments and suggestions concerning some of the issues discussed in this paper.

Most of the research discussed in this paper was performed when the author was with the Mobile Robot Lab, Robotics Institute, Carnegie-Mellon University, and with the Engineering Design Research Center, CMU. It was supported in part by the Office of Naval Research under Contract N00014-81-K-0503. The author was supported in part by a graduate fellowship from the Conselho Nacional de Desenvolvimento Científico e Tecnológico - CNPq, Brazil, under Grant 200.986-80, in part by the Mobile Robot Lab, CMU, and in part by the Engineering Design Research Center, CMU.

The views and conclusions contained in this document are those of the author and should not be interpreted as representing the official policies, either expressed or implied, of the funding agencies.

References

[1] J. O. Berger. *Statistical Decision Theory and Bayesian Analysis*. Springer-Verlag, Berlin, 1985. Second Edition.

[2] P. J. Besl and R. C. Jain. Three-Dimensional Object Recognition. *ACM Computing Surveys*, 17(1), March 1985.

[3] A. E. Bryson and Y. C. Ho. *Applied Optimal Control*. Blaisdell Publishing Co., Waltham, MA, 1969.

[4] A. Elfes. Multiple Levels of Representation and Problem-Solving Using Maps From Sonar Data. In *Proceedings of the DOE/CESAR Workshop on Planning and Sensing for Autonomous Navigation*, Oak Ridge National Laboratory, UCLA, Los Angeles, August 18-19 1985.

[5] A. Elfes. *Occupancy Grids: A Probabilistic Framework for Robot Perception and Navigation*. PhD thesis, Electrical and Computer Engineering Department/Robotics Institute, Carnegie-Mellon University, May 1989.

[6] A. Elfes. *A Performance-Oriented Mobile Robot Architecture*. Research Report, IBM T. J. Watson Research Center, 1990. In preparation.

[7] A. Elfes. A Sonar-Based Mapping and Navigation System. In *1986 IEEE International Conference on Robotics and Automation*, IEEE, San Francisco, CA, April 7-10 1986.

[8] A. Elfes. Sonar-Based Real-World Mapping and Navigation. *IEEE Journal of Robotics and Automation*, RA-3(3), June 1987.

[9] A. Elfes. Strategies for Dynamic Robot Perception Using a Stochastic Spatial Model. In *Proceedings of the IEEE International Workshop on Intelligent Robots and Systems 1990*, IEEE, Japan, July 1990.

[10] A. Elfes. A Tesselated Probabilistic Representation for Spatial Robot Perception. In *Proceedings of the 1989 NASA Conference on Space Telerobotics*, NASA/Jet Propulsion Laboratory, California Institute of Technology, Pasadena, CA, January 31 - February 2 1989.

[11] A. Elfes. Using Occupancy Grids for Mobile Robot Perception and Navigation. *IEEE Computer Magazine, Special Issue on Autonomous Intelligent Machines*, June 1989.

[12] A. Elfes and L. Matthies. Sensor Integration for Robot Navigation: Combining Sonar and Stereo Range Data in a Grid-Based Representation. In *Proceedings of the 26th IEEE Conference on Decision and Control*, IEEE, Los Angeles, CA, December 9 - 11 1987.

[13] S. Geman and D. Geman. Stochastic Relaxation, Gibbs Distributions, and the Bayesian Restoration of Images. *IEEE Transactions on Pattern Analysis and Machine Intelligence*, PAMI-6(6), November 1984.

[14] D. J. Kriegman, E. Triendl, and T. O. Binford. A Mobile Robot: Sensing, Planning and Locomotion. In *Proceedings of the 1987 IEEE International Conference on Robotics and Automation*, IEEE, Raleigh, NC, April 1987.

[15] J. L. Marroquin. *Probabilistic Solution of Inverse Problems*. PhD thesis, Artificial Intelligence Lab, Computer Science Department, Massachusetts Institute of Technology, September 1985.

[16] L. Matthies and A. Elfes. Integration of Sonar and Stereo Range Data Using a Grid-Based Representation. In *Proceedings of the 1988 IEEE International Conference on Robotics and Automation*, IEEE, Philadelphia, PA, April 25 - 29 1988.

[17] H.P. Moravec and A. Elfes. High-Resolution Maps from Wide-Angle Sonar. In *IEEE International Conference on Robotics and Automation*, IEEE, St. Louis, March 1985. Also published in the 1985 ASME International Conference on Computers in Engineering, Boston, August 1985.

[18] A. Papoulis. *Probability, Random Variables, and Stochastic Processes*. McGraw-Hill, New York, 1984.

[19] R. C. Smith and P. Cheeseman. On the Representation and Estimation of Spatial Uncertainty. *The International Journal of Robotics Research*, 5(4), Winter 1986.

[20] R. C. Smith, M. Self, and P. Cheeseman. A Stochastic Map for Uncertain Spatial Relationships. In *Proceedings of the 1987 International Symposium on Robotics Research*, MIT Press, 1987.

[21] H. L. Van Trees. *Detection, Estimation, and Modulation Theory*. Volume Part I, John Wyley and Sons, New York, N.Y., 1968.

[22] E. Vanmarcke. *Random Fields: Analysis and Synthesis*. MIT Press, Cambridge, MA, 1983.

Maintaining Representations of the Environment of a Mobile Robot

NICHOLAS AYACHE AND OLIVIER D. FAUGERAS

Abstract—**In this paper we describe our current ideas related to the problem of building and updating 3-D representation of the environment of a mobile robot that uses passive Vision as its main sensory modality. Our basic tenet is that we want to represent both geometry and uncertainty. We first motivate our approach by defining the problems we are trying to solve and give some simple didactic examples. We then present the tool that we think is extremely well-adapted to solving most of these problems: the extended Kalman filter (EKF). We discuss the notions of minimal geometric representations for 3-D lines, planes, and rigid motions. We show how the EKF and the representations can be combined to provide solutions for some of the problems listed at the beginning of the paper, and give a number of experimental results on real data.**

I. Introduction

In the last few years, Computer Vision has gone extensively into the area of three-dimensional (3-D) analysis from a variety of sensing modalities such as stereo, motion, range finders, and sonars. A book that brings together some of this recent work is [24].

Most of these sensing modalities start from pixels which are then converted into 3-D structures. A characteristic of this work as compared to previous work (like in image restoration, for example) where images were the starting and the ending point is that noise in the measurements is, of course, still present but, contrary to what has happened in the past, it has to be taken into account all the way from pixels to 3-D geometry.

Another aspect of the work on 3-D follows from the observation that if noise is present, it has to be evaluated, i.e., we need models of sensor noise (sensor being taken here in the broad sense of sensory modality), and reduced. This reduction can be obtained in many ways. The most important ones are as follows:

- First, the case of one sensor in a fixed position: it can repeat its measurements and thus maybe obtain better estimations.
- Second, the case of a sensor that can be moved around: given its measurements in a given position, what is the best way to move in order to reduce the uncertainty and increase the knowledge of the environment in a way that is compatible with the task at hand.
- Third, is the case of several different sensors that have to combine their measurements in a meaningful fashion.

Interesting work related to those issues has already emerged which is not reported in [24]. In the area of robust estimation procedures and models of sensors noise, Hager and Mintz [22] and McKendall and Mintz [27] have started to pave the ground. Bolle and Cooper [12] have developed maximum likelihood techniques to combine range data to estimate object positions. Darmon [16] applies the Kalman filter formalism to the detection of moving objects in sequences of images. Durrant-Whyte [18], in his Ph.D. dissertation has conducted a thorough investigation of the problems posed by multi-sensory systems. Applications to the navigation of a mobile robot have been discussed by Crowley [15], Smith and Cheeseman [32], and Matthies and Shafer [28]. The problem of combining stereo views has been attacked by Ayache and Faugeras [3], [4], [19], Porril *et al.* [30], and Kriegman [25]. It also appears that the linearization paradigm extensively used in this paper has been already used in the photogrammetry field [26].

Several problems related to these preliminary studies need more attention. Modeling sensor noise in general and more specifically visual sensor noise appears to us an area where considerable progress can be achieved; relating sensor noise to geometric uncertainty and the corresponding problem of representing geometric information with an eye toward describing not only the geometry but also the uncertainty on this geometry are key problems to be investigated further as is the problem of combining uncertain geometric information produced by different sensors.

II. What are the Problems that we are Trying to Solve

We have been focusing on a number of problems arising in connection with a robot moving in an indoor environment and using passive vision and proprioceptive sensory modalities such as odometry. Our mid-term goals are to incrementally build on the robot an increasing set of sensing and reasoning capabilities such as:

- build local 3-D descriptions of the environment,
- use the descriptions to update or compute motion descriptions where the motion is either the robot's motion or others,
- fuse the local descriptions of neighboring places into more global, coherent, and accurate ones,
- "discover" interesting geometric relations in these descriptions,
- "discover" semantic entities and exhibit "intelligent" behavior.

We describe how we understand each of these capabilities and what are the underlying difficulties.

Manuscript received April 5, 1988; revised May 2, 1989. This work was partially supported by the Esprit Project P940.

The authors are with INRIA—Rocquencourt, Domaine de Voluceau, Rocquencourt, B. P. 105—78153, Le Chesney Cedex, France.

IEEE Log Number 8930460

Reprinted from *IEEE Transactions on Robotics and Automation*, Vol. 5, No. 6, December 1989, pages 804-819. Copyright © 1989 by The Institute of Electrical and Electronics Engineers, Inc. All rights reserved.

EH0341-8/91/0000/0071$01.00 © 1989 IEEE

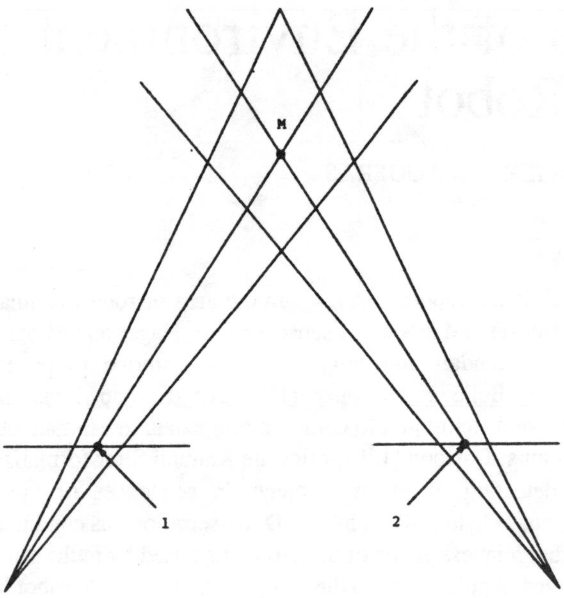

Fig. 1. Effect of pixel noise on 3-D reconstruction.

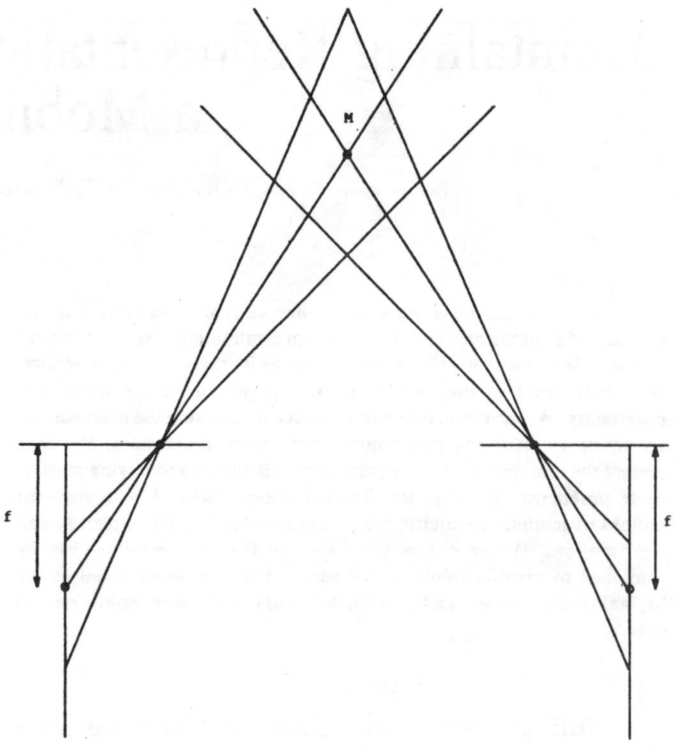

Fig. 2. Effect of calibration errors on 3-D reconstruction.

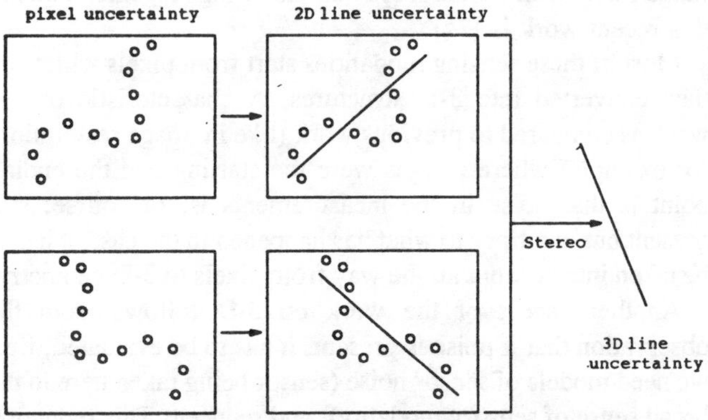

Fig. 3. From pixel uncertainty to 3-D line uncertainty.

A. Build Local 3-D Descriptions of the Environment

Up until now, our main source of 3-D information has been Stereo [5], [9] even though we have made considerable progress toward the use of structure from motion as well [21]. In any case, the problems are very similar for both sensing modalities and we concentrate on Stereo. As announced in the Introduction, our main concern is to track uncertainty all the way from pixel noise to geometric descriptions. Fig. 1 shows, for example, that in a Stereo system, if pixels positions are imperfectly known, then the corresponding 3-D point varies in an area with a quite anisotropic diamond shape. This is a clear example of a relation between pixel uncertainty and geometric (the position of point *M*) uncertainty. Another source of uncertainty in Stereo is the calibration uncertainty. In a stereo rig, intrinsic parameters of the cameras such as focal length, and extrinsic parameters such as relative position and orientation of the cameras have to be calculated. Fig. 2 shows the effect on the reconstruction of a point *M* of an uncertainty on the focal lengths of the two cameras. Again, *M* varies in a diamond-like shape. Of course, this source of uncertainty adds to the previous pixel uncertainty.

Another example of the propagation of uncertainty is given in Fig. 3 where pixels in left and right images are grouped into line segments: pixel uncertainty is converted into 2-D line uncertainty. Line segments are then matched and used to reconstruct 3-D line segments: 2-D line uncertainty and calibration uncertainty are converted into 3-D uncertainty.

Yet another set of examples of this kind of propagation is shown in Fig. 4 where coplanar and cocylindrical line segments are grouped together; again, the question is, what is the uncertainty on the plane or on the cylinder? (the uncertainty on the position of the lines, plane, and cylinder is represented on the picture in a symbolic manner by ellipses).

From these examples, we see that the main problem that needs to be solved in order to build local 3-D descriptions of the environment is how geometric uncertainty propagates when we build more complex primitives from simpler ones. This, in turn, generates two questions:

1) How do we represent geometric primitives?
2) How do we represent uncertainty on these primitives?

B. Update Position and Motion Information

Fig. 5 shows a measurement of a physical point made in two positions 1 and 2 of a mobile vehicle. In position 1, it "sees" *M* with some uncertainty represented by the ellipse around it. In position 2, it "sees" *P* with another uncertainty. Assuming that the displacement between 1 and 2 is exactly known, it is possible to express *P* and *M* in the same coordinate system. If the displacement estimate is wrong, as it is in Fig. 5, the two zones of uncertainty do not intersect and it is very unlikely that the observer will realize that the points *M* and *P* are instances of the same physical point. If we now take into account the uncertainty on the displacement (assuming that we can estimate it) we have Fig. 6 where the combination of

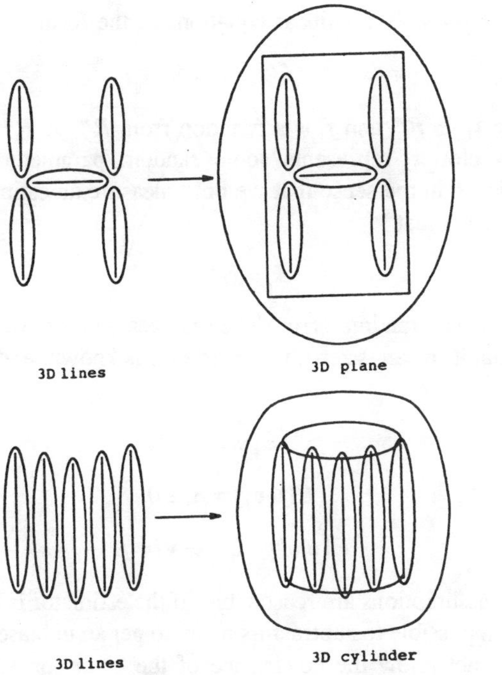

Fig. 4. From 3-D line uncertainty to 3-D surface uncertainty.

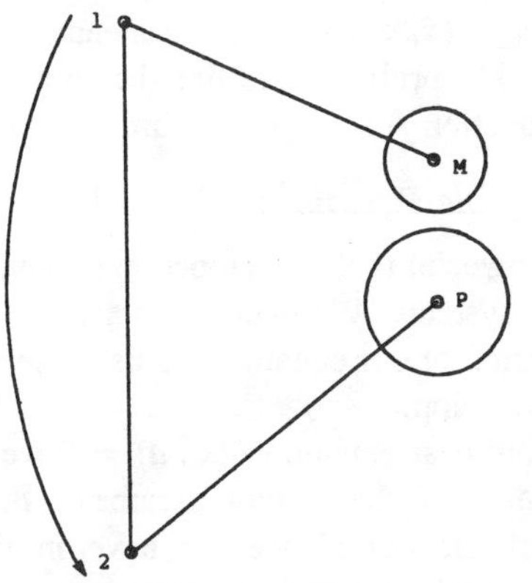

Fig. 5. Measuring a point in two positions (wrong displacement estimation).

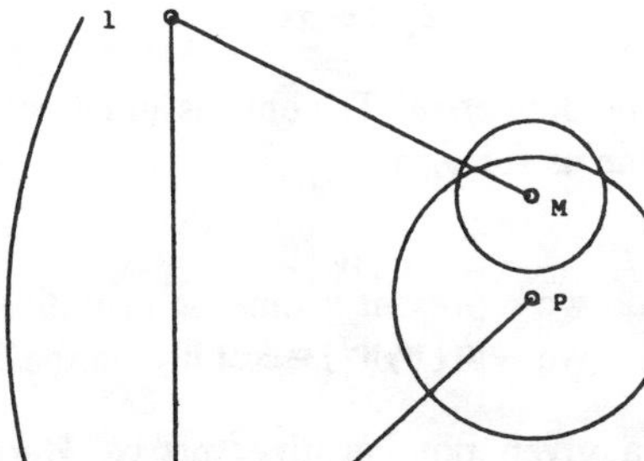

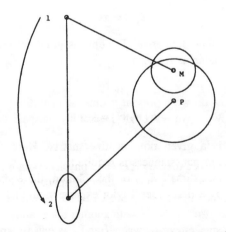

Fig. 6. Measuring a point in two positions (displacement and uncertainty estimation).

displacement uncertainty and measurement uncertainty produces a larger ellipse around P which intersects the one around M: the observer can now infer that the probability of M and P being the same physical point is quite high and use the two measurements to obtain a better estimate of the displacement

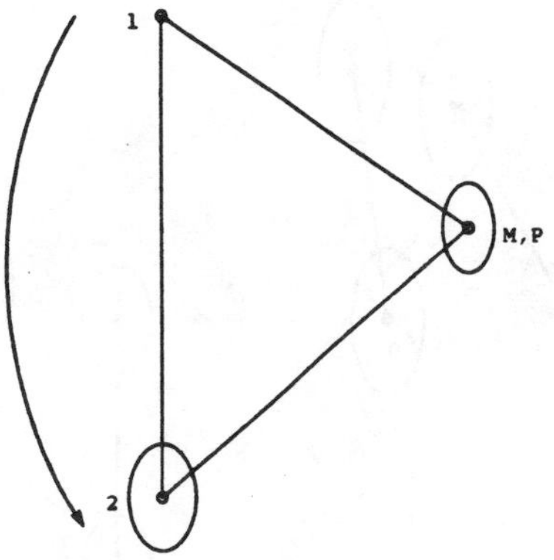

Fig. 7. Improving the estimation of the points position.

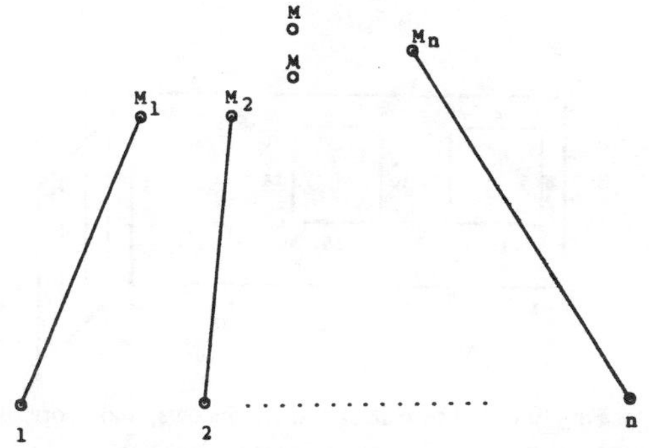

Fig. 8. Fusing n points measured from different positions.

and reduce its uncertainty. We explain how to do this in Section V. The measurements can also be used to produce better estimates of the positions (Fig. 7). This is related to what we call geometric fusion.

C. Fusing Geometric Entities

Fig. 8 shows a slightly more general case than what is depicted in Fig. 7. The mobile vehicle has measured the physical point M in n positions numbered from 1 and n. Each measurement yields a point M_i, $i = 1, \cdots, n$ and some uncertainty in the coordinate system attached to the robot. Displacement uncertainty is also available. Using the ideas described in Section V, we can improve the estimates of the displacements and reduce their uncertainty by discovering that points $M_1, \cdots, M_n$ are all instanciations of the same point. We can also use this observation to reduce the uncertainty on, let us say M_1, by combining the n measurements and produce a point $\mathcal{M}$, fusion of $M_1, \cdots, M_n$, as well as its related uncertainty. The points $M_1, \cdots, M_n$ can then be erased from the representation of the environment, they can be forgotten. What remains is the point $\mathcal{M}$ expressed in the coordinate system attached to position 1, for example, and the displacement from 1 to 2, 2 to 3, etc...., which allows us to express $\mathcal{M}$ in the other coordinate systems.

Fusing geometric entities is therefore the key to "intelligent" forgetting which, in turn, prevents the representation of the environment from growing too large.

D. Discovering "Interesting" Geometric Relations

Using this approach also allows us to characterize the likelihood that a given geometric relation exists between a

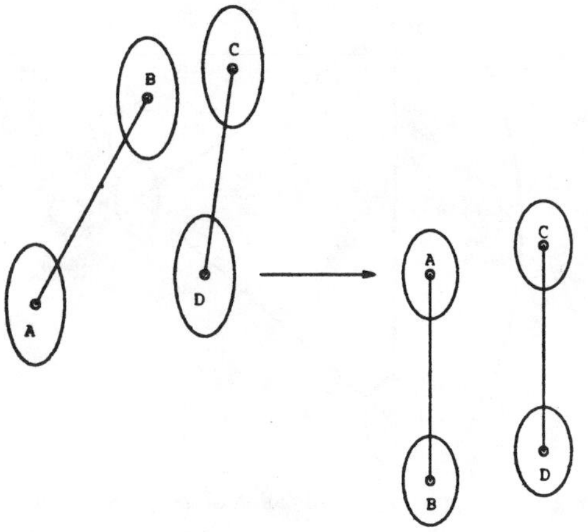

Fig. 9. Discovering that AB and CD are parallel.

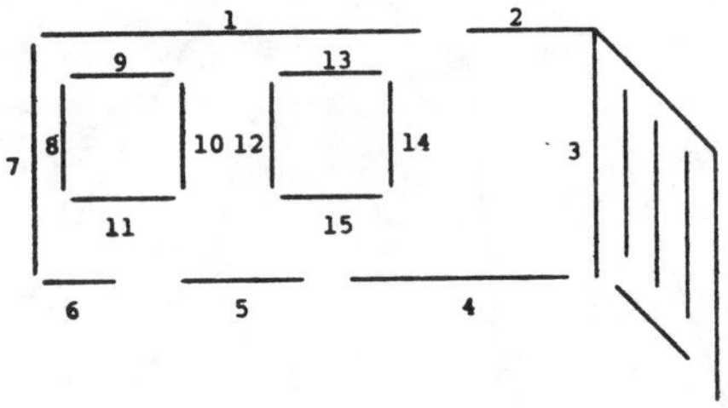

Fig. 10. Hypothesizing walls, windows, and doors.

number of geometric entities and to use this information to obtain better estimates of these entities and reduce their uncertainty. For example, as shown in Fig. 9, segments AB and CD which have uncertainty attached to their endpoints have a high likelihood to be parallel. Assuming that they are, we can update their position (they become more parallel) and reduce the uncertainty of their endpoints. The same reasoning can be used, for the relation "to be perpendicular."

E. Discovering Semantic Entities

Fig. 10 shows the kind of "semantic" grouping that is of interest to us in the context of a mobile robot moving indoors, to combine geometry and some *a priori* description of the environment. The line segments numbered from 1 to 15 are found, using the ideas described in Section II-D, to be coplanar with a high probability; the corresponding plane is found to be vertical with a very high probability which can be deduced from the geometric uncertainty of the line segments. This observation can then be used to infer that the plane has a high probability to be a wall. If we also observe that segments 8 to 11 and 12 to 15 form approximately two rectangles this can be used to infer that they have a high probability to be parts of a window or a door.

III. What is the Tool that we are Using

In this section, we introduce the Extended Kalman Filter (EKF) formalism which is applied in Sections IV and V to solve the problems we have just listed in Section II.

A. Unifying the Problems

In all of these previously listed problems, we are confronted with the estimation of an unknown parameter $\mathbf{a} \in R^n$ given a set of k possibly nonlinear equations of the form

$$f_i(\mathbf{x}_i, \mathbf{a}) = 0 \tag{1}$$

where $\mathbf{x}_i \in R^m$ and f_i is a function from $R^m \times R^n$ into R^p. The vector $\mathbf{x}_i$ represents some random parameters of the function f_i in the sense that we only measure an estimate $\hat{\mathbf{x}}_i$ of them, such that

$$\hat{\mathbf{x}}_i = \mathbf{x}_i + \mathbf{v}_i \tag{2}$$

where $\mathbf{v}_i$ is a random error. The only assumption we make on $\mathbf{v}_i$ is that its mean is zero, its covariance is known, and that it is a white noise

$$E[\mathbf{v}_i] = 0$$

$$E[\mathbf{v}_i\mathbf{v}_i^t] = \mathbf{\Lambda}_i \geq 0$$

$$E[\mathbf{v}_i\mathbf{v}_j^t] = 0 \qquad \forall i \neq j.$$

These assumptions are reasonable. If the estimator is biased, it is often possible to subtract its mean to get an unbiased one. If we do not know the covariance of the error (or at least an upper bound of it), the estimator is meaningless. If two measurements $\hat{\mathbf{x}}_i$ and $\hat{\mathbf{x}}_j$ are correlated, we take the concatenation of them $\hat{\mathbf{x}}_k = (\hat{\mathbf{x}}_i, \hat{\mathbf{x}}_j)$ and the concatenated vector function $f_k = [f_i^t, f_j^t]^t$. The problem is to find the optimal estimate $\hat{\mathbf{a}}$ of $\mathbf{a}$ given the function f_i and the measurements $\hat{\mathbf{x}}_i$.

B. Linearizing the Equations

The most powerful tools developed in parameter estimation are for linear systems. We decided to apply these tools to a linearized version of our equations. This is the EKF approach that we now develop.

For each nonlinear equation $f_i(\mathbf{x}_i, \mathbf{a}) = \mathbf{0}$ we need to know an estimate $\hat{\mathbf{a}}_{i-1}$ of the sougth parameter $\mathbf{a}$, and again a measure S_i of the confidence we have in this estimate.[1] Actually, we model probalistically the current estimate $\hat{\mathbf{a}}_{i-1}$ of $\mathbf{a}$ by assuming that

$$\hat{\mathbf{a}}_{i-1} = \mathbf{a} + \mathbf{w}_i \tag{3}$$

where $\mathbf{w}_i$ is a random error. The only assumptions we make on $\mathbf{w}_i$ are the same as for $\mathbf{v}_i$, i.e.,

$$E[\mathbf{w}_i] = 0$$

$$E[\mathbf{w}_i\mathbf{w}_i^t] = S_i \geq \mathbf{0}$$

where S_i is a given non-negative matrix. Here again, no assumption of gaussianness is required.

Having an estimate $\hat{\mathbf{a}}_{i-1}$ of the solution, the equations are linearized by a first-order Taylor expansion around $(\hat{\mathbf{x}}_i, \hat{\mathbf{a}}_{i-1})$

$$f_i(\mathbf{x}_i, \mathbf{a}) = \mathbf{0} \approx f_i(\hat{\mathbf{x}}_i, \hat{\mathbf{a}}_{i-1}) + \frac{\widehat{\partial f_i}}{\partial \mathbf{x}}(\mathbf{x}_i - \hat{\mathbf{x}}_i) + \frac{\widehat{\partial f_i}}{\partial \mathbf{a}}(\mathbf{a} - \hat{\mathbf{a}}_{i-1}) \tag{4}$$

where the derivatives $\widehat{\partial f_i}/\partial \mathbf{x}$ and $\widehat{\partial f_i}/\partial \mathbf{a}$ are estimated at $(\hat{\mathbf{x}}_i, \hat{\mathbf{a}}_{i-1})$.

[1] In practice, we shall see that only an initial estimate $(\hat{\mathbf{a}}_0, S_0)$ of $\mathbf{a}$ is required prior to the first measurement $\hat{\mathbf{x}}_1$, while the next ones $(\hat{\mathbf{a}}_i, S_i)$ are provided automatically by the Kalman filter itself.

Equation (4) can be rewritten as

$$y_i = M_i a + u_i \tag{5}$$

where

$$y_i = -f_i(\hat{x}_i, \hat{a}_{i-1}) + \frac{\widehat{\partial f_i}}{\partial a}\hat{a}_{i-1}$$

$$M_i = \frac{\widehat{\partial f_i}}{\partial a}$$

$$u_i = \frac{\widehat{\partial f_i}}{\partial x}(x_i - \hat{x}_i).$$

Equation (5) is now a linear measurement equation, where y_i is the new measurement, M_i is the linear transformation, u_i is the random measurement error. Both y_i and M_i are readily computed from the actual measurement $\hat{x}_i$, the estimate $\hat{a}_{i-1}$ of a, the function f_i, and its first derivative. The second-order statistics of u_i are derived easily from those of v_i

$$E[u_i] = 0$$

$$W_i \triangleq E[u_i u_i^t] = \frac{\widehat{\partial f_i}}{\partial x}\Lambda_i \frac{\widehat{\partial f_i^t}}{\partial x}.$$

C. Recursive Kalman Filter

When no gaussianness is assumed on the previous random errors u_i, v_i, and w_i, the Kalman filter equations provide the best (minimum variance) linear unbiased estimate of a. This means that among the estimators which seek a_k as a linear combination of the measurements $\{y_i\}$, it is the one which minimizes the expected error norm squared

$$E[(\hat{a}_k - a)^t(\hat{a}_k - a)]$$

while verifying

$$E[\hat{a}_k] = a.$$

The recursive equations of the Kalman filter which provide a new estimate $(\hat{a}_i, S_i)$ of a from $(\hat{a}_{i-1}, S_{i-1})$ are as follows [23]:

$$\hat{a}_i = \hat{a}_{i-1} + K_i(y_i - M_i\hat{a}_{i-1}) \tag{6}$$

$$K_i = S_{i-1}M_i^t(W_i + M_i S_{i-1} M_i^t)^{-1} \tag{7}$$

$$S_i = (I - K_i M_i)S_{i-1} \tag{8}$$

or equivalently

$$S_i^{-1} = S_{i-1}^{-1} + M_i^t W_i^{-1} M_i. \tag{9}$$

One can see that the previously estimated parameter $\hat{a}_{i-1}$ is corrected by an amount proportional to the current error $y_i - M_i\hat{a}_{i-1}$ called the innovation. The proportionality factor K_i is called the Kalman gain. At the end of the process, $\hat{a}_k$ is the final estimate and S_k represents the covariance of the estimation error

$$S_k = E[(\hat{a}_k - a)(\hat{a}_k - a)^t].$$

The recursive process is initialized by $\hat{a}_0$, an initial estimate of a, and S_0, its error covariance matrix. Actually, the criterion minimized by the final estimate $\hat{a}_k$ is

$$C = (a - \hat{a}_0)^t S_0^{-1}(a - \hat{a}_0) + \sum_{i=1}^{k}(y_i - M_i a)^t W_i^{-1}(y_i - M_i a). \tag{10}$$

It is interesting to note that the first term of (10) measures the squared distance of a from an initial estimate, weighted by its covariance matrix, while the second term is nothing else but the classical least square criterion, i.e., the sum of the squared measurement errors weighted by the covariance matrices. Indeed, initializing the process with an arbitrary $\hat{a}_0$ and $S_0^{-1} = 0$, criterion (10) provides the classical least square estimate $\hat{a}_k$ obtained from the measurements only, while the initial estimate does not play any role.

The enormous advantage of such a recursive solution is that if we decide, after a set of k measurements $\{\hat{x}_i\}$, to stop the measures, we only have to keep $\hat{a}_k$ and $\hat{S}_k$ as the whole memory of the measurement process. If we decide later to take into account additional measurements, we simple have to initialize $\hat{a}_0 \sim \hat{a}_k$ and $S_0 \sim S_k$ and to process the new measurements to obtain exactly the same solution as if we had processed all the measurements together.

D. Gaussian Assumption

Up to now, we did not introduce any Gaussian assumptions on the random measurement errors $v_i = x_i - \hat{x}_i$ of (2) and on the prior estimate error $w_0 = a - \hat{a}_0$ of (3). However, in practice, these errors usually come from a sum of independent random processes, which tend toward a Gaussian process (Central Limit theorem). If we actually identify v_i and w_0 with Gaussian processes, i.e.,

$$v_i \sim N(0, \Lambda_i)$$

$$w_0 \sim N(0, S_0)$$

then, it follows that the noise u_i in (5) is also Gaussian, i.e.,

$$u_i \sim N(0, W_i)$$

and that all the successive estimates provided by the recursive Kalman filter are also Gaussian, with mean a and covariance S_k

$$\hat{a}_k \sim N(a, S_k).$$

Moreover, in this case, the Kalman filter provides the best (minimum variance) unbiased estimate $\hat{a}_k$ among all, even nonlinear, filters. This estimate $\hat{a}_k$ is also the maximum likelihood estimator of a. This comes from the fact that in the Gaussian case, the solution is the conditional mean $\hat{a}_k = E[a/y_1, \cdots, y_k]$ which both minimizes the variance and maximizes the likelihood while being expressed as a linear combination of the measurements y_i. Therefore, in this case, the minimum variance and minimum variance linear estimates are the same; namely, the estimate $\hat{a}_k$ provided by the Kalman filter [23].

In conclusion, in the Gaussian case, the Kalman filter provides the best estimate with the advantage of preserving gaussianness of all the implied random variables, which means that no information on the probability density functions of the parameters is lost while keeping only their mean and covariance matrix.

E. Rejecting Outlier Measurements

At iteration i, we have an estimate $\hat{a}_{i-1}$ and an attached covariance matrix S_{i-1} for parameter a. We also have a noisy measurement $(\hat{x}_i, \Lambda_i)$ of x_i and we want to test the plausibility of this measurement with respect to the equation $f_i(x_i, a) = 0$.

If we consider again a first-order expansion of $f_i(x_i, a)$ around $(\hat{x}_i, \hat{a}_{i-1})$ (4), considering that $(\hat{x}_i - x_i)$ and $(\hat{a}_{i-1} - a)$ are independent centered Gaussian processes, we see that $f_i(\hat{x}_i, \hat{a}_{i-1})$ is also (up to a linear approximation) a centered Gaussian process whose mean and covariance are given by

$$E[f_i(\hat{x}_i, \hat{a}_{i-1})] = 0$$

$$Q_i = E[f_i(\hat{x}_i, \hat{a}_{i-1}) f_i(\hat{x}_i, \hat{a}_{i-1})^t] = \widehat{\frac{\partial f_i}{\partial x}} \Lambda_i \widehat{\frac{\partial f_i}{\partial x}}^t + \widehat{\frac{\partial f_i}{\partial a}} S_{i-1} \widehat{\frac{\partial f_i}{\partial a}}^t .$$

Therefore, if the rank of Q_i is q, the generalized Mahalanobis distance

$$d(\hat{x}_i, \hat{a}_{i-1}) = [f_i(\hat{x}_i, \hat{a}_{i-1})]^t Q_i^{-1} [f_i(\hat{x}_i, \hat{a}_{i-1})] \qquad (11)$$

has a χ^2 distribution with q degrees of freedom.[2]

Looking at a χ^2 distribution table, it is therefore possible to reject an outlier measurement $\hat{x}_i$ at a 95-percent confidence rate by setting an appropriate threshold ϵ on the Mahalanobis distance, and by keeping only those measurements $\hat{x}_i$ which verify

$$d(\hat{x}_i, \hat{a}_{i-1}) < \epsilon. \qquad (12)$$

We shall see in the experimental section at the end of this paper how this formalism can be used in practice, and how well it fits with reality.

IV. Geometric Representations

In this section, we give the details of the geometric representations that we have found useful at various stages of our work. It is first important to note that we have been dealing so far only with points, lines, and planes, i.e., with affine geometric entities. This may appear to be quite a restriction on the type of environments that we can cope with. This is indeed the case but there are a number of reasons why we think that our approach is quite reasonable.

1) The obvious one is that for the kind of environment that our mobile robot moves into, these primitives are very likely to cover most of the geometric features of importance.

2) A second reason is that more complicated curved features can be first approximated with affine primitives which are then grouped into more complicated nonaffine primitives.

3) A third reason is that we believe that the techniques we have developed for representing and combining uncertainty of affine primitives are generic and directly applicable to nonaffine primitives.

[2] If $q < p$ = the size of the measurement vector f_i, Q_i^{-1} is the pseudo-inverse of Q_i.

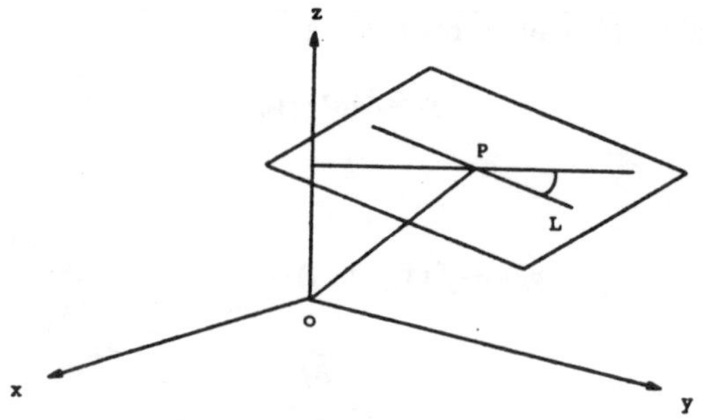

Fig. 11. A possible 3-D line representation.

Let us now discuss specifically lines, planes, and rigid displacements.

A. Line Segments

The 3-D segments that we deal with are usually constructed from stereo [4], [9]. Their endpoints may be quite unreliable, even though they can be of some use from time to time, and we largely depend on the infinite lines supporting those line segments.

We concentrate here on how to represent 3-D lines. The obvious representation we mention here only for pedagogical reasons, is by two points; this representation is six-dimensional and, as we will see next, not minimal. Another way to represent a line is to choose a point on it (three parameters), and a unit vector defining its direction (two parameters). The corresponding representation is five-dimensional and, again, not minimal. In fact, the set of affine 3-D lines is a manifold of dimension 4 for which we will exhibit later an atlas of class C^∞.[3] This implies that a minimal representation of a straight line has four parameters.

One such representation can be obtained by considering the normal to the line from the origin (if the line goes through the origin it is the same as a vector line and can be defined by two parameters only). The point of intersection between the normal and the line is represented by three parameters. If we now consider (see Fig. 11) the plane normal at P to OP, the line is in that plane and can be defined by one more parameter, its angle with an arbitrary direction, for example, the line defined by P, and one of the axis of coordinates (in Fig. 11, the z axis). Of course, when the line is parallel to the xy plane this direction is not defined and we must use either the x or the y axis. This brings up an interesting point, namely, that a global minimal representation for affine lines, i.e., one which can be used for all such lines, does not exist. We must choose the representation as a function of the line orientation. Mathematically, this means that the manifold of the affine straight lines cannot be defined with only one map. This is quite common and is also true for affine planes and rotations of R^3, as will be shown next.

The previous representation for a line is not in fact the one

[3] Grossly speaking, a manifold of dimension d is a set that can be defined locally by d parameters. When the functions that transform one set of parameters into another are p times differentiable, the manifold is said to be of class C^p. For more details, see [14].

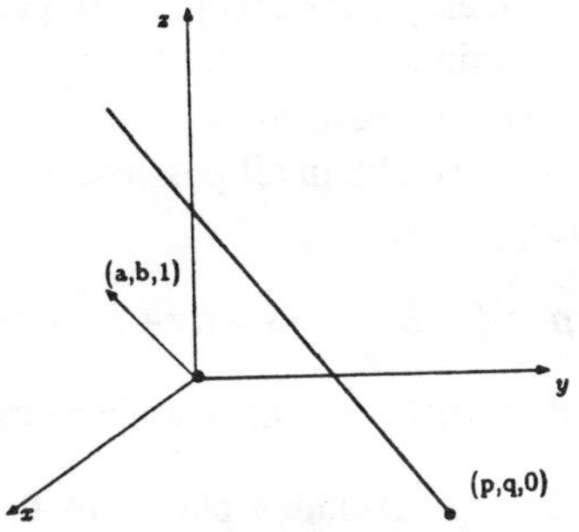

Fig. 12. A better 3-D line representation.

we have been using. In effect, the parameters involved in the previous representation are usually combined in a highly nonlinear manner in the measurement equations expressing geometric relationships between geometric entities (cf. next section), which is not good for the extended Kalman filtering approach. Also, the angular parameter must be assigned some fixed bounds (for instance]0, π[), which might cause some problems during a recursive evaluation with the Kalman filter. This latter constraint also appears in the representation recently proposed by Roberts [31].

Therefore, we prefer the following representation in which the retained parameters are usually combined linearly in the measurement equations, and are not constrained to any bounded interval. This representation considers a line (not perpendicular to the z axis) as the intersection of a plane parallel to the y axis, and a plane parallel to the x axis

$$\begin{cases} x = az + p \\ y = bz + q. \end{cases} \tag{13}$$

The intersection is represented by the four-dimensional vector $L = [a, b, p, q]^T$ which has the following geometric interpretation (see Fig. 12): The direction of the line is that of the vector $[a, b, 1]^T$, and the point of intersection of the line with the xy plane has coordinates p and q. Since the last coordinate of the direction vector is equal to 1, the line cannot be perpendicular to the z axis or parallel to the xy plane. If we want, and we do in practice, represent such lines, we must choose another representation, for example

$$\begin{cases} y = ax + p \\ z = bx + q \end{cases} \tag{14}$$

which cannot represent lines parallel to the yz plane, or perpendicular to the x axis, or

$$\begin{cases} z = ay + p \\ x = by + q \end{cases} \tag{15}$$

which excludes lines parallel to the zx plane.

Each representation defines a one-to-one mapping between R^4 and a subset (in fact an open subset) of the set of affine 3-D lines and it can be shown that these three mappings define on this set a structure of C^∞ manifold for which they form an atlas. In practice, this means the representation is not exactly four-dimensional, but is made of the four numbers a, b, p, and q and an integer i taking the values 1, 2, and 3 to indicate which map 13, 14, or 15 we are currently using.

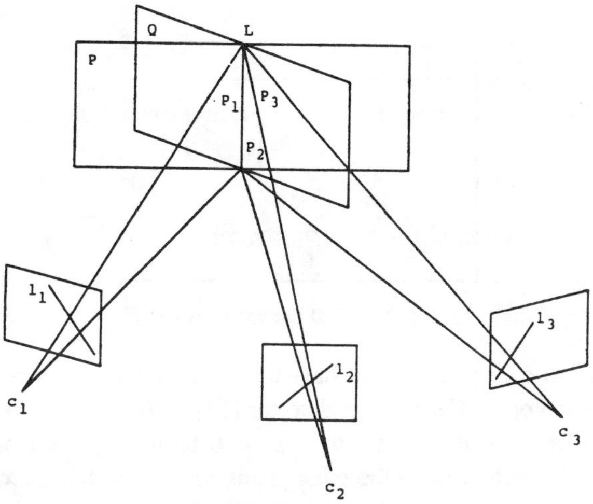

Fig. 13. Reconstruction of 3-D lines.

The fact that the set of affine 3-D lines has been given a structure of C^∞ manifold implies that the a', b', p', q' of a given representation are C^∞ functions of the a, b, p, q of another representation for all lines for which the two representations are well defined (for example, all lines not parallel to the xy and yz planes). The representation of a line also includes a 4×4 covariance matrix Δ_L on the vector L.

It is interesting at this stage to trace the computation of this covariance matrix all the way from pixel to 3-D. In order to do this, we must briefly explain how 3-D lines are computed in our current Stereo system [9]. We use three cameras as indicated in Fig. 13. In theory, the three planes defined by the 2-D lines l_1, l_2, and l_3 and the optical centers C_1, C_2, and C_3 belong to the same pencil and intersect along the 3-D line L. In practice they do not because of noise, and we have to find the "best" line satisfying the measurements, i.e., l_1, l_2, and l_3. This can be done by using the idea of pencil of planes, described more fully in [21]. We assume that in the coordinate system attached to camera 1, for example, the equation of the ith plane P_i, $i = 1, 2, 3$, is given by

$$u_i x + v_i y + w_i z + r_i = 0$$

where the three four-vectors $P_i = [u_i, v_i, w_i, r_i]^T$ are known, as well as their covariance matrix Λ_{P_i} (we show later how to compute them). If we use representation (13) for the 3-D line, it is represented as the intersection of the two planes P of the equation $x = az + p$ and Q of the equation $y = bz + q$. Writing that the five planes P, Q, and P_i, $i = 1, 2, 3$, form a pencil allows us to write six equations

$$\begin{cases} w_i + au_i + bv_i = 0, \\ r_i + pu_i + qu_i = 0, \end{cases} \quad i = 1, 2, 3$$

in the four unknowns a, b, p, and q.

We can apply directly the Kalman formalism to these measurement equations and choose $\boldsymbol{a} = [a, b, p, q]^T$, and x_i as the four-vector P_i. We can therefore simply compute an estimate $\hat{a}$ of $\boldsymbol{a}$ and its covariance matrix $\Lambda_{\hat{a}}$ from the P_i's and Λ_{P_i}'s.

Let us now show how we can compute the P_i's and Λ_{P_i}'s. Each line l_i, $i = 1, 2, 3$, is obtained by fitting a straight line to

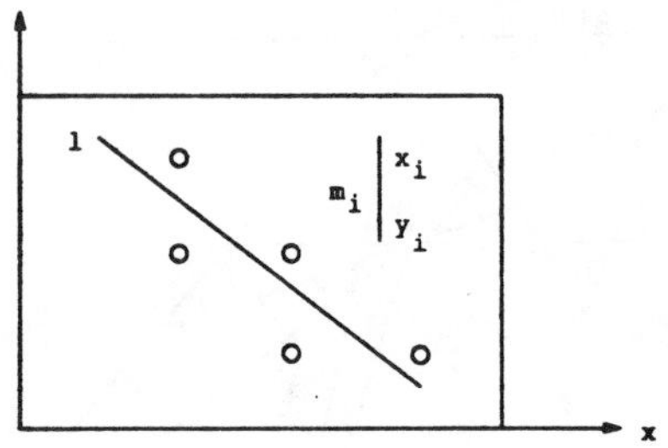

Fig. 14. 2-D line approximation.

a set of edge pixels which have been detected using a modified version of the Canny edge detector [13], [17]. Looking at Fig. 14, let $x \cos \theta + y \sin \theta - p = 0$ be the equation of the line l which is fit to the edge pixels m_i of coordinates x_i, y_i, $(0 \leq \theta < 2\pi, \rho \geq 0)$. We assume that the measured edge pixels are independent and corrupted by a gaussian isotropic noise and take the parameter $\boldsymbol{a}$ equal to $[\theta, \rho]^T$ and the measurement $\boldsymbol{x}$ as the vector $[x, y]^T$. The measurement equation is therefore

$$f(\boldsymbol{x}, \boldsymbol{a}) = x \cos \theta + y \sin \theta - \rho.$$

Applying the EKF formalism to the n-edge pixels forming the line provides the best estimate $\hat{\boldsymbol{a}}$ of the line parameters and its covariance matrix. Having done this for all three cameras, it is easy to deduce the equations of the three planes P_i and the covariance matrices on their coefficients.

B. Planes

Planes can receive pretty much the same treatment as lines. A plane is defined by three parameters, and this is minimal. A possible representation is the representation by the normal $\hat{n}$ (a unit norm vector), and the distance d to the origin. The problem with this representation is that it is not unique since $(-\boldsymbol{n}, -d)$ represents the same plane. It is possible to fix that problem by assuming that one component of $\boldsymbol{n}$, say n_z, is positive, i.e., we consider planes not parallel to the z axis. For these planes we must choose another convention, for example, that n_z is positive. Again, this works well for planes not parallel to the x axis. The third possible representation is to assume n_y positive which excludes planes parallel to the y axis.

So, we have three one-to-one mappings of open subsets of the product $S_2 \times R$, where S_2 is the usual Gaussian sphere into open subsets of the set of planes

$$(\boldsymbol{n}, d),\ n_z > 0 \rightarrow \text{planes not parallel to } Oz$$

$$(\boldsymbol{n}, d),\ n_x > 0 \rightarrow \text{planes not parallel to } Ox$$

$$(\boldsymbol{n}, d),\ n_y > 0 \rightarrow \text{planes not parallel to } Oy.$$

It is easy to show that these three mappings define on the set of 3-D planes a structure of C^∞ manifold of dimension 3.

One practical disadvantage of the previous representations is that the normal $\boldsymbol{n}$ is constrained to lie on the unit sphere S_2, i.e., it must satisfy the constraint $\|\boldsymbol{n}\| = 1$. A possibly simpler representation is obtained by considering the mapping from R^3 to the set of 3-D planes defined by

$$p_1 : (a, b, c) \rightarrow ax + by + z + c = 0. \tag{16}$$

This can represent all planes except those parallel of Oz and it is a one-to-one continuous mapping from R^3 to the open subset of the set of 3-D planes constituted of the planes not parallel to the z axis. In order to obtain all possible planes, we must also consider the mappings

$$p_2 : (a, b, c) \rightarrow x + ay + bz + c = 0 \tag{17}$$

$$p_3 : (a, b, c) \rightarrow bx + y + az + c = 0. \tag{18}$$

p_2 (respectively, p_3) excludes planes to the x axis (respectively, the y axis). It is easy to show that p_1, p_2, p_3 also define on the set of 3-D planes a structure of C^∞ manifold of dimension 3.

C. Rigid Displacements

In a previous paper [4], [6] we have proposed the use of the exponential representation of rotations. This is the same as saying that a rotation is defined by its axis $\boldsymbol{u}$ (a unit vector) and its angle θ. The vector $\boldsymbol{r} = \theta\boldsymbol{u}$ can be used to represent the rotation and we have

$$R = e^H$$

where H is an antisymmetric matrix representing the cross product with the vector $\boldsymbol{r}$ (i.e., $Hx = r \times x$, for all x). In this case, the rotation is represented by the three coordinates of $\boldsymbol{r}$, i.e., by three independent numbers. There are several other possible representations for rotations, the most widely known being the one with orthogonal matrices or quaternions. Their main disadvantage is that an orthogonal matrix is defined by nine numbers subject to six quadratic constraints, whereas a quaternion is defined by four numbers subject to one quadratic constraint. These constraints are not easy to deal with in the EKF formalism and, moreover, these two representations are more costly than the exponential one.

Let us see how we can define a structure of manifold on the set of rotations using this representation. If we allow θ to vary over the semi-open interval $[0, 2\pi[$, the vector $\boldsymbol{r}$ can vary in the open ball $B(0, 2\pi)$ of R^3 of radius 2π. But the mapping f: $B(0, 2\pi)$ into the set of rotations is not one to one because $(\boldsymbol{u}, \pi)$ and $(-\boldsymbol{u}, \pi)$ represent the same rotation. To enforce uniqueness we can assume that one of the coordinates, for example u_z, of the rotation axis $\boldsymbol{u}$ is positive. We can then represent uniquely the open subset of the set of rotations for which the axis is not perpendicular to the z axis, and has a positive component along the axis, and the mapping is continuous. If we consider the open set of rotations defined by $(\boldsymbol{u}, \theta)$, $u_z < 0$, we have another one-to-one continuous mapping. With these two mappings, we cannot represent rotations with an axis perpendicular to the z axis. In order to obtain all possible rotations, we have to introduce the other four mappings defined by $(\boldsymbol{u}, \theta)$ and $u_x > 0$ (respectively, $u_x < 0$, $u_y > 0$, $u_y < 0$) which represent rotations with an axis not perpendicular to the x axis (respectively, the y axis). We are sill missing the null vector, i.e., we have no representation for the null rotation, the identity matrix. In order to include it, we have to add a seventh map by considering for example the rotations defined by the "small" open ball $B(0, \epsilon)$ where ϵ

must be smaller than π. These seven mappings define on the set of rotations a structure of C^∞ manifold of dimension 3.[4]

It is interesting that in all three cases (3-D lines, planes, and rotations), unique global representation does not exist and that we must deal with at least three local mappings.

It is now instructive to study how the group of rigid displacements operates on the representations for lines and planes.

1) Applying Rigid Displacement to Lines: The easiest way to derive how representation (13) changes under rotation and translation is by considering that the line is defined by two points M_1 and M_2 of coordinates (x_1, y_1, z_1) and (x_2, y_2, z_2). It is then easy to verify that

$$a=\frac{x_2-x_1}{z_2-z_1} \quad b=\frac{y_2-y_1}{z_2-z_1}$$

$$p=\frac{z_2x_1-z_1x_2}{z_2-z_1} \quad q=\frac{z_2y_1-z_1y_2}{z_2-z_1}.$$

Introducing the vector $M_1M_2 = [A, B, C]^T$, we have $a = A/C$, and $b = B/C$. a and b are therefore only sensitive to rotation

$$M_1M_2 \rightarrow R\, M_1M_2$$

$$\begin{bmatrix} A \\ B \\ C \end{bmatrix} \rightarrow \begin{bmatrix} A' \\ B' \\ C' \end{bmatrix} = R \begin{bmatrix} A \\ B \\ C \end{bmatrix}$$

$$\begin{bmatrix} a \\ b \end{bmatrix} \rightarrow \begin{bmatrix} a' \\ b' \end{bmatrix} = \begin{bmatrix} A'/C' \\ B'/C' \end{bmatrix}.$$

This yields

$$a'=\frac{r_1\cdot m}{r_3\cdot m} \quad b'=\frac{r_2\cdot m}{r_3\cdot m}$$

where $m = [a, b, 1]^T$, and the r_i's are the row vectors of matrix R. This is true only if $r_3 \cdot m \neq 0$; if $r_3 \cdot m = 0$, the transformed line is perpendicular to the z axis and representation (14) or (15) must be used.

To treat the case of p and q, let us introduce $P = p(z_2 - z_1) = pC$ and $Q = q(z_2 - z_1) = qC$. It is easy to show that

$$\begin{bmatrix} P \\ Q \end{bmatrix} = \begin{bmatrix} 0 & -1 & 0 \\ 1 & 0 & 0 \end{bmatrix} OM_1 \times OM_2 = H(OM_1 \times OM_2).$$

This allows us to study how P and Q change under rotation and translation

$$OM_1 \rightarrow R\, OM_1 + t \qquad OM_2 \rightarrow R\, OM_2 + t.$$

Therefore

$$\begin{bmatrix} P \\ Q \end{bmatrix} \rightarrow \begin{bmatrix} P' \\ Q' \end{bmatrix} = H(R(OM_1 \times OM_2) + t \times R\, M_1M_2).$$

Using the previous notations, $M_1M_2 = [A, B, C]^t$, and $OM_1 \times OM_2 = [Q, -P, X]^t$ where X is unknown. But noticing that $M_1M_2 \cdot (OM_1 \times OM_2) = 0$ we have

$$AQ - BP + CX = 0$$

and therefore

$$X = \frac{BP - AQ}{C} = bP - aQ.$$

C is not equal to 0 since by definition, the line is not perpendicular to the z axis. Putting everything together

$$\begin{bmatrix} P' \\ Q' \end{bmatrix} = CH\left(R\begin{pmatrix} q \\ -p \\ bp-aq \end{pmatrix} + t \times Rm\right).$$

Finally

$$\begin{bmatrix} p' \\ q' \end{bmatrix} = \begin{bmatrix} P'/C' \\ Q'/C' \end{bmatrix} = \frac{C}{C'} H\left(R\begin{pmatrix} q \\ -p \\ bp-aq \end{pmatrix} + t \times Rm\right)$$

and we know from the previous derivation that $C/C' = 1/r_3 \cdot M$, therefore

$$\begin{bmatrix} p' \\ q' \end{bmatrix} = \frac{1}{r_3\cdot m} H(Rp + t \times Rm)$$

where we have taken $p = [q, -p, bp - aq]^t$.

2) Applying Rigid Displacements to Planes: Given a plane represented by its normal n and its distance to the origin d, if we apply to it a rotation along an axis going through the origin represented by a matrix R followed by a translation represented by a vector t, the new plane is represented by Rn and $d - t \cdot Rn$ [20].

This allows us to compute how the representation (16), for example, is transformed by the rigid displacement. From the previous observation:

$$\begin{pmatrix} a \\ b \\ 1 \end{pmatrix} \rightarrow R \begin{pmatrix} a \\ b \\ 1 \end{pmatrix} \quad \text{and} \quad c \rightarrow c - t \cdot R \begin{pmatrix} a \\ b \\ 1 \end{pmatrix}.$$

Introducing the three row vectors r_1, r_2, r_3 of matrix R, we have, assuming that $r_3 \cdot m \neq 0$.

$$a'=\frac{r_1\cdot m}{r_3\cdot m} \quad b'=\frac{r_2\cdot m}{r_3\cdot m} \quad c'=\frac{c-t\cdot Rm}{r_3\cdot m}$$

if $r_3 \cdot m = 0$, this means that we cannot use the same representation for the transformed plane since it is parallel to the z axis, therefore, we must choose the representation (17) or (18).

V. Registration, Motion, and Fusion of Visual Maps

In this section we show how to solve the problems listed in Section II within the formalism and the representations detailed in Sections III and IV.

A. Initial Assumptions

We are given two visual maps $\mathcal{V}$ and $\mathcal{V}'$, each of them attached to a coordinate reference frame $\mathfrak{F}$ and $\mathfrak{F}'$ (see Fig. 15).

[4] In [10] and [11] one can find an atlas of rotations with only four maps.

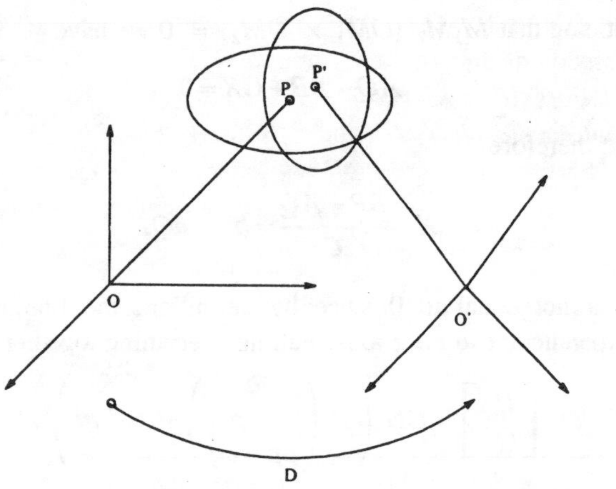

Fig. 15. The general registration motion fusion problem.

TABLE I
RELATIONS BETWEEN THE PRIMITIVES

Relations	Points	Lines	Planes
Points	$\equiv$	$\subset$	$\subset$
Lines		$\equiv \parallel \perp$	$\subset \parallel \perp$
Planes			$\equiv \parallel \perp$

Each visual map $\mathcal{V}$ is composed of primitives $\mathcal{P}$, described by a parameter vector $\boldsymbol{P}$. We have an estimate $\hat{\boldsymbol{P}}_0$ of $\boldsymbol{P}$ and an error covariance matrix $\boldsymbol{W}_{P_0}$.

The coordinate frames $\mathfrak{F}$ and $\mathfrak{F}'$ are related by a rigid displacement $\mathfrak{D}$ such that each point M' of $\mathfrak{F}'$ is related to a point M of $\mathfrak{F}$ by the relation

$$O'M' = R\, OM + t$$

where $\boldsymbol{R}$ is the rotation matrix and $\boldsymbol{t}$ the translation vector of the displacement $\mathfrak{D}$. We also have an estimate $\hat{D}_0$ of D, with an error covariance matrix $\boldsymbol{W}_{D_0}$.

B. Defining Geometric Relations

We define a set of geometric relations between the primitive $\mathcal{P}$ and $\mathcal{P}'$ of two visual maps $\mathcal{V}$ and $\mathcal{V}'$. These relations are given in Table I.

The list of relations/primitives is not exhaustive but only demonstrative. The relation "identical" expresses the fact that the primitives $\mathcal{P}$ and $\mathcal{P}'$ represented in $\mathcal{V}$ and $\mathcal{V}'$ actually describe the same physical primitive. The relation "included" expresses that $\mathcal{P}$ describes a physical primitive which is part of the physical primitive described by $\mathcal{P}'$. The relations "parallel" and "orthogonal" are interpreted in a similar fashion.

Each geometric relation can be expressed by a vector equation of the form

$$f_i(P, P', D) = \mathbf{0}. \tag{19}$$

C. Expressing Geometric Relations

We rewrite (19) for the geometric relations of Table I. We denote by $\bar{P}$ the parameters of the primitive $\bar{\mathcal{P}} = D(\mathcal{P})$, the image of $\mathcal{P}$ by the rigid displacement D. The computation of $\bar{P}$ from P is, in the case of points, $\overline{OM} = R\, OM + t$. The case of lines and planes was detailed in the previous section. The measurement equations are as follows:

Point-Point:

$$\text{relation } \equiv\text{: } O'M' - \overline{OM} = \mathbf{0}.$$

Point-Line: assuming the line is not orthogonal to the z axis:

$$\text{relation } \subset\text{: } \begin{cases} \bar{x} - a'\bar{z} - p' = 0 \\ \bar{y} - b'\bar{z} - q' = 0. \end{cases}$$

Point-Plane: assuming the plane is not parallel to the z axis:

$$\text{relation } \subset\text{: } a'\bar{x} + b'\bar{y} + \bar{z} + c' = 0.$$

Line-Line: assuming the two lines are not orthogonal to the z axis:

$$\text{relation } \equiv\text{: } (a', b', c', d')^t - (\bar{a}, \bar{b}, \bar{c}, \bar{d})^t = \mathbf{0}$$

$$\text{relation } \parallel\text{: } (a', b')^t - (\bar{a}, \bar{b})^t = \mathbf{0}$$

$$\text{relation } \perp\text{: } a'\bar{a} + b'\bar{b} + 1 = 0.$$

Line-Plane: assuming the line is not orthogonal and the plane not parallel to the z axis:

$$\text{relation } \subset\text{: } \begin{cases} a'\bar{a} + b'\bar{b} + 1 = 0 \\ a'\bar{p} + b'\bar{q} + c' = 0 \end{cases}$$

$$\text{relation } \parallel\text{: } a'\bar{a} + b'\bar{b} + 1 = 0$$

$$\text{relation } \perp\text{: } (a', b')^t - (\bar{a}, \bar{b})^t = \mathbf{0}.$$

Plane-Plane: assuming the plane is not parallel to the z axis:

$$\text{relation } \equiv\text{: } (a', b', c')^t - (\bar{a}, \bar{b}, \bar{c})^t = \mathbf{0}$$

$$\text{relation } \parallel\text{: } (a', b')^t - (\bar{a}, \bar{b})^t = \mathbf{0}$$

$$\text{relation } \perp\text{: } a'\bar{a} + b'\bar{b} + 1 = 0.$$

This approach should be compared to that of [29].

D. Registration

1) Principle: The registration (or matching) of two primitives $\mathcal{P}$ and $\mathcal{P}'$ consists in detecting that their parameters $\boldsymbol{P}$ and $\boldsymbol{P}'$ verify (19) for one of the above listed geometric relations, with respect to the current noisy estimates ($\hat{P}_0$, W_{P_0}), ($\hat{P}'_0$, $W_{P'_0}$), and ($\hat{D}_0$, W_{D_0}) of $\boldsymbol{P}$, $\boldsymbol{P}'$, and $\boldsymbol{D}$.

This "detection" is done by computing between each pair of primitive the generalized Mahalanohis distance given by (11), and by matching a pair of primitives each time the χ^2 acceptance test given by inequality (12) is verified, i.e., when

$$d(\hat{P}_0, W_{P_0}, \hat{P}'_0, W_{P'_0}, \hat{D}_0, W_{D_0}) < \epsilon. \tag{20}$$

2) Reliability: The above-described registration procedure detects what would be called *plausible* matches between geometric primitives. When the uncertainty attached to the primitives parameters is large, it may happen that a plausible match is false. In order to improve the reliability of the procedure, one can use a strategy (inspired by [16]) which starts by registrating primitives whose parameters have a small covariance matrix, or primitives which can be matched unambiguously. Such a strategy is exemplified in the experimental results section of this paper and also in another paper [8].

3) Efficiency: In order to avoid a $O(n^2)$ complexity algorithm, it is of course possible to use additional control structures to select a subset of candidate primitives for each test. For instance, to test the relation "$\equiv$" between points or lines, bucketing techniques can be used with efficiency (see for instance [1], [7]).

E. Motion

Having registered two primitives $\mathcal{P}$ and $\mathcal{P}'$, the motion problem consists in reducing the uncertainty on the *motion parameters* $\boldsymbol{D}$ while taking into account the uncertainty on the parameters $\boldsymbol{P}$, $\boldsymbol{P}'$, and $\boldsymbol{D}$.

This is done by setting $\boldsymbol{a} = \boldsymbol{D}$ and $\boldsymbol{x} = (\boldsymbol{P}, \boldsymbol{P}')^t$, and by using the relation equation (19) as a measurement equation (1)

$$f_i(\boldsymbol{x}, \boldsymbol{a}) \equiv f_i((\boldsymbol{P}, \boldsymbol{P}'), \boldsymbol{D}) = 0.$$

Starting from the initial estimate $\hat{\boldsymbol{a}}_0 = \hat{\boldsymbol{D}}_0$, $S_0 = W_{D_0}$, and using the measurement $\hat{\boldsymbol{x}}_1 = (\hat{P}_0, P_0')^t)$ with

$$W_1 = \begin{pmatrix} W_{P_0} & 0 \\ 0 & W_{P_0'} \end{pmatrix}$$

one applies the EKF formalism to obtain a new estimate $\hat{\boldsymbol{a}}_1$ of the motion with a reduced covariance matrix $S_1 < S_0$. (In the sense $S_0 - S_1$ is nonnegative).

This process is recursively repeated: at iteration i, if a new pair of primitives can be registered with the new motion estimate $(\hat{\boldsymbol{a}}_{i-1}, S_{i-1})$, the additional measurement equations they bring lead to a new better estimate $\hat{\boldsymbol{a}}_i$ of the motion with a still reduced covariance matrix S_i. This process ends after the matching of k primitives with a final estimate $(\hat{\boldsymbol{a}}_k, S_k)$ of the motion parameter $\boldsymbol{D}$.

F. Fusion

1) General Fusion: The fusion problem is exactly the dual of the motion problem, as it consists, after the registration of two primitives, in reducing the uncertainty on the *primitive parameters* $\boldsymbol{P}$ and $\boldsymbol{P}'$ while taking into account the uncertainty on the parameters $\boldsymbol{P}$, $\boldsymbol{P}'$, and $\boldsymbol{D}$.

This is done by "switching the attention," i.e., by choosing $\boldsymbol{a} = (\boldsymbol{P}, \boldsymbol{P}')^t$ and $\boldsymbol{x} = \boldsymbol{D}$ while using again the relation equation (19) as a measurement equation (1)

$$f_i(\boldsymbol{x}, \boldsymbol{a}) \equiv f_i(\boldsymbol{D}, (\boldsymbol{P}, \boldsymbol{P}')) = 0.$$

The initial estimate is taken as $\hat{\boldsymbol{a}}_0 = (\hat{P}_0, \hat{P}_0')^t$ and

$$S_0 = \begin{pmatrix} W_{P_0} & 0 \\ 0 & W_{P_0'} \end{pmatrix}$$

and one uses the measurement $\hat{\boldsymbol{x}}_1 = \hat{\boldsymbol{D}}_0$ with $W_1 = W_{D_0}$ to apply the EKF formalism and obtain a new estimate $\hat{\boldsymbol{a}}_1$ of the primitive parameters with a reduced covariance matrix $S_1 < S_0$.

If additional relations hold between these primitives and other ones, the same treatment allows for a further reduction in their parameters uncertainty, and therefore a more accurate estimation of the primitive parameters.

2) Forgetting Primitives: After the treatment of a constraint, the parameters $\boldsymbol{P}_1$ and $\boldsymbol{P}_1'$ of the primitives are usually correlated, which means that the covariance matrix

$$\text{cov}(\hat{P}_1, \hat{P}_1') = \begin{pmatrix} W_{P_1} & W_{P_1 P_1'} \\ W_{P_1' P_1} & W_{P_1} \end{pmatrix}$$

contains $W_{P_1' P_1} = W^t_{P_1 P_1'} \neq 0$.

Therefore, it is no longer possible to treat independently $\mathcal{P}$ and $\mathcal{P}'$ in successive measurement equations. One has to consider them as a new primitive, either by keeping only one of them, or the union of them.

For instance if one updates the parameters of $\mathcal{P}'$ with those of an "identical" primitive $\mathcal{P}$ observed in a previous visual map, one keeps only the updated parameters of $\mathcal{P}'$ in the new map, with their covariance matrix $W_{p'}$, forgetting the previous parameters $\mathcal{P}$ after having used them.

On the other hand, if one updates the parameters of two lines by detecting that they are orthogonal, one keeps the new primitive formed by the union of the updated two lines, with the corresponding covariance matrix. One must use this kind of relation carefully, in order to control the size of the state parameter $\boldsymbol{a}$.

3) Autofusion: In the special case where $\mathcal{V} \equiv \mathcal{V}'$, all primitives come from the same visual map, and the motion parameters vanish as they correspond to the identity transform and are perfectly known.

Nevertheless, one can still detect the previous geometric relations between pairs of primitives $\mathcal{P}$ and $\mathcal{P}'$, and use them to reduce the uncertainty on the primitives parameters.

VI. Experimental Results

The basic principles presented in this paper were tested on a variety of synthetic and real data. The interested reader can find registration and motion results with real points and lines in [3], registration and fusion results with synthetic and real points and lines in [2], and results on the building of global 3-D maps from passive stereovision in [9]. In this paper we only present results of the motion estimation from two 3-D maps from passive stereovision in [9]. In this paper we only present results of the motion estimation from two 3-D maps, the fusion of several inaccurate 3-D maps, and the detection of colinearity within a single 3-D map (what we called "autofusion"). In each of these examples, the 3-D map is made of 3-D lines.

A. Registration and Motion

Fig. 16 shows the edges of a triplet of images taken by the mobile robot in a first position. From these edges, the trinocular stereovision system computes a set of 3-D segments. Each 3-D segment is represented by the parameters (a, b, p, q) of the 3-D line supporting it and by the error covariance computed—as explained in Section IV-B—from the uncertainty on the edge points in the three images (we took an isotropic Gaussian density function of covariance 1 pixel around each edge point). Each 3-D line is bounded by two endpoints obtained from the endpoints measured in the three images which are projected on the reconstructed 3-D line.

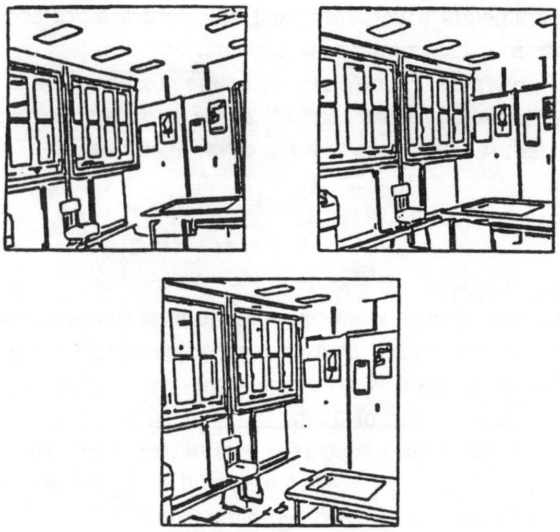

Fig. 16. Triplet of images taken in position 1.

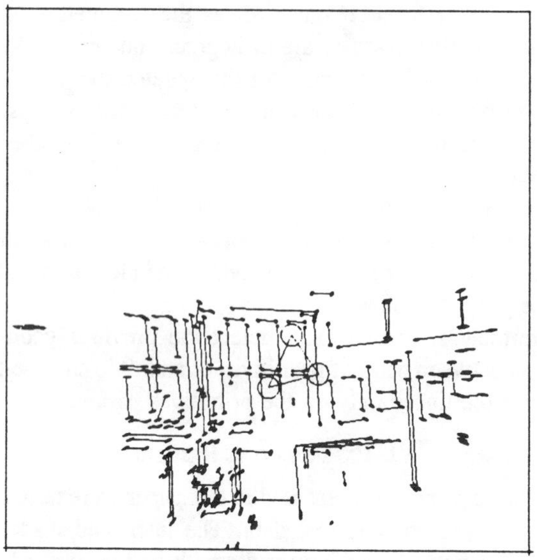

Fig. 17. Front view of reconstructed 3-D lines.

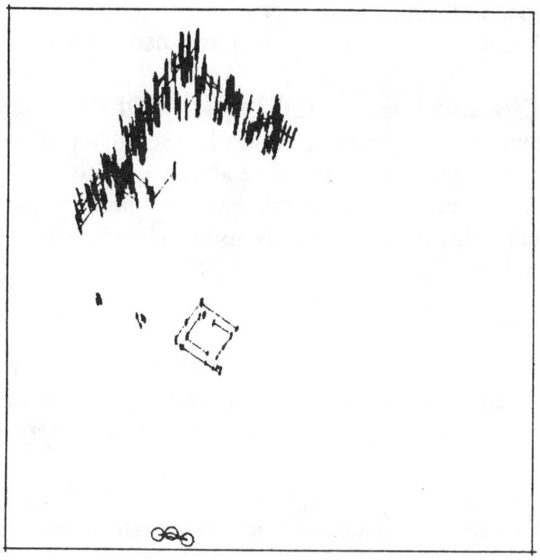

Fig. 18. Top view of reconstructed 3-D lines.

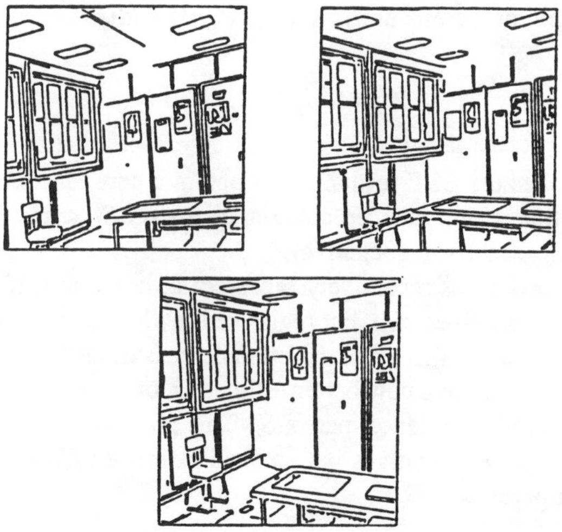

Fig. 19. Triplet of images taken in position 2.

We show in Figs. 17 and 18, respectively, the horizontal and vertical projections of the reconstructed 3-D segments. We also show the uncertainty attached to the reconstructed 3-D lines by showing the uncertainty it produces on the coordinates of their endpoints. The 95-percent confidence regions of the endpoints positions are ellipsoids whose projections are the ellipses shown in Figs. 17 and 18. One can see the anisotropic distribution of the uncertainty on the three coordinates of the points and its variation as a function of their position relative to the cameras (the projections of the three optical centers of the cameras correspond to the vertices of the triangle located grossly in the middle of the front view and at the bottom of the top view. Also, the circles around these vertices have been given an arbitrary radius of 20 cm to allow the reader to estimate the uncertainty attached to the other primitives).

The robot now moves a little, a new triplet of images is taken (Fig. 19) and another set of 3-D lines is computed. Initially, the robot is given a very crude estimate of its motion between the two views. Applying this crude estimate to the 3-D lines obtained in position 1, and projecting them in one of the images obtained in position 2 (the image of camera 3), one obtains the crude superimposition observed in Fig. 20. Solid lines are the transformed 3-D segments computed in position 1, while the dotted lines are the 2-D segments observed in position 2.

We now ask the system to discover the relation "≡" between the 3-D lines (see Section V-C) reconstructed in position 1 and 2, given the initial crude motion estimate and its uncertainty. The program takes each 3-D line in position 1, applies the noisy current motion estimate to place it in the 3-D map obtained in position 2 with a new covariance matrix (combining the initial uncertainty with the motion uncertainty), and computes its Mahalanobis distance (11) to all the other lines of position 2 (see Section V-D).

The program detects a match each time a pair of lines passes the χ^2 test of (12). If a line in position 1 can be matched to several lines in position 2, this is an *ambiguous* match, and nothing is done. On the other hand, each time an *unambiguous* match is found, the parameters of the motion are updated

Fig. 20. Superimposition of 3-D segments of position 1 with 2-D edges of position 2 (crude initial motion estimate).

Fig. 21. Superimposition of 3-D segments of position 1 with 2-D edges of position 2 (final motion estimate).

as it is explained in Section V-E. As the uncertainty on motion decreases after each new match, some previously ambiguous matches can now become unambiguous. Therefore, the entire matching process is repeated until no more lines can be matched (three iterations in this example). The final estimate of the motion is very accurate as can be seen in Fig. 21 where the obtained superimposition is now almost perfect.

Applying exactly the same technique to a set of six triplets of views taken during the motion of the robot (Figs. 22–27), the system was able to build a global 3-D map of the room shown in Fig. 28 where rotating segments at the bottom right are the computed successive robot positions. Fig. 29 gives a hand sketched semantic interpretation of this global map.

B. Registration, Motion, and Fusion

In this experiment, the robot is looking from four different positions at a regular pattern (Figs. 30 and 31) formed by vertical lines floating in front of horizontal lines, and builds in each position a local 3-D map. Exactly the same technique as in the previous example was used to register each successive

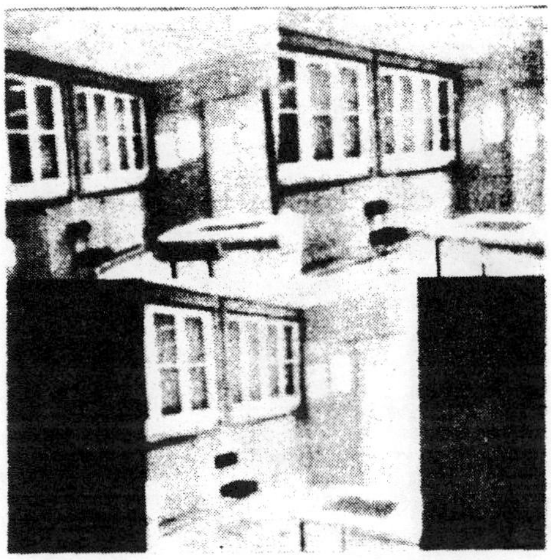

Fig. 22. First triplet of laboratory images.

Fig. 23. Second triplet of laboratory images.

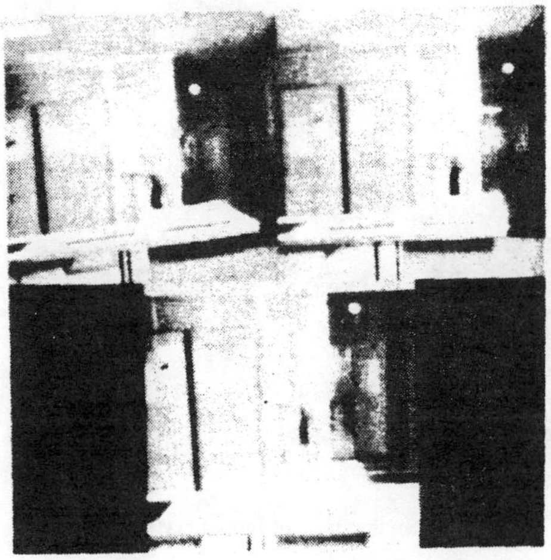

Fig. 24. Third triplet of laboratory images.

local 3-D map, and put all of them in a single absolute reference frame. Fig. 32 shows the resulting 3-D map before fusion. Fusion is achieved by discovering the relation "≡" computed between lines in the global 3-D map, and taking into account the uncertainty on the 3-D lines due to their reconstruction and to the successive motion estimations. Fusion yields a reduction from 1808 to 650 segments and improves accuracy, as can be seen by looking at the front and

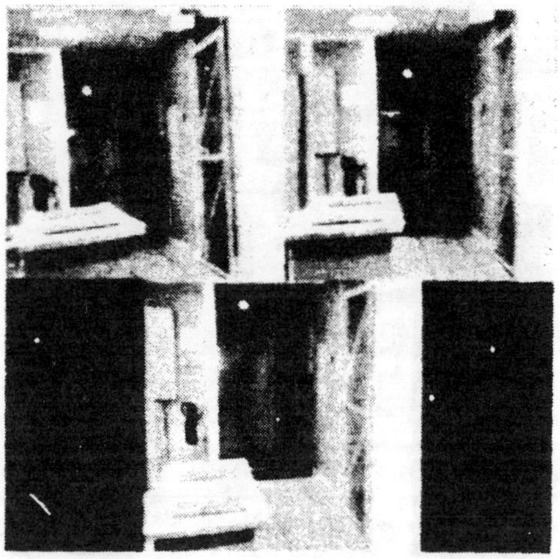

Fig. 25. Fourth triplet of laboratory images.

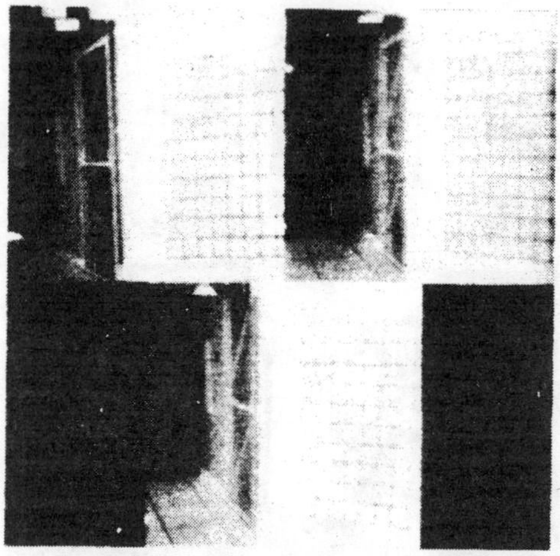

Fig. 26. Fifth triplet of laboratory images.

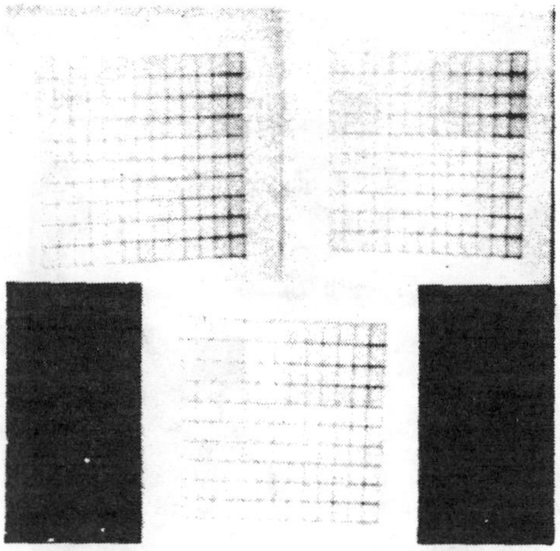

Fig. 27. Sixth triplet of laboratory images.

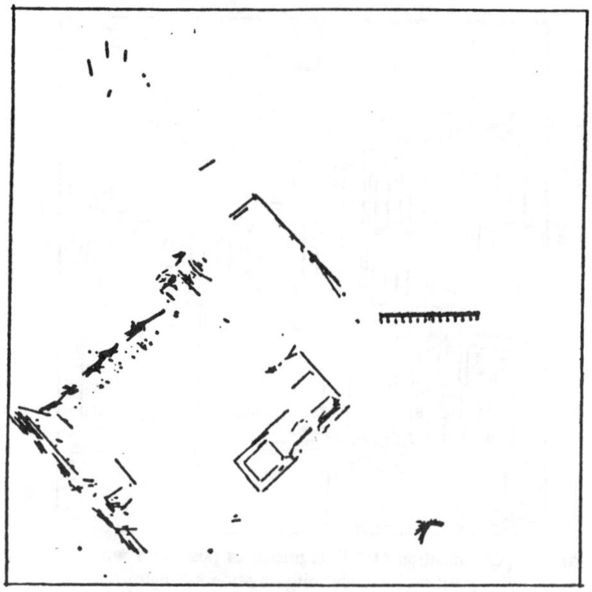

Fig. 28. Top view of a global 3-D map of the room computed from six local 3-D maps.

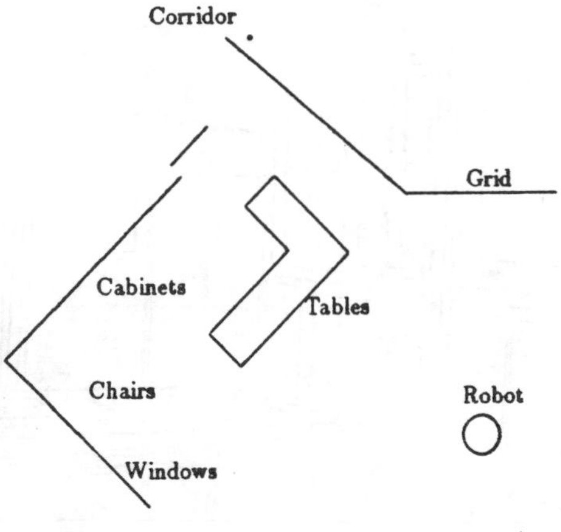

Fig. 29. Semantic interpretation of the previous global 3-D map.

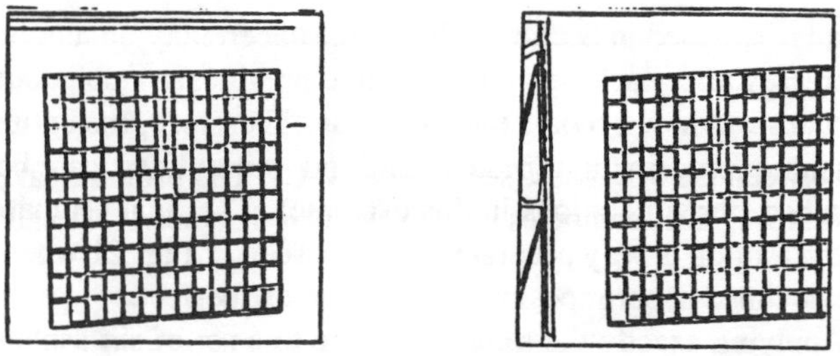

Fig. 30. A regular grid observed from position 1 and 2.

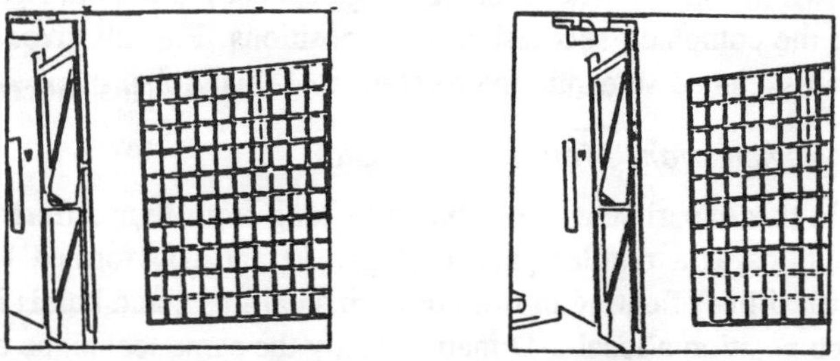

Fig. 31. A regular grid observed from positions 3 and 4.

top view of the reconstructed 3-D pattern after fusion (Fig. 33).

C. Detecting Colinearity in Space

In this experiment, the robot is looking at the regular pattern only once. We show in Fig. 34 the vertical and horizontal projections of the initially reconstructed 3-D segments. We also show in Fig. 35 the uncertainty attached to reconstructed 3-D lines by showing the uncertainty it produces on the

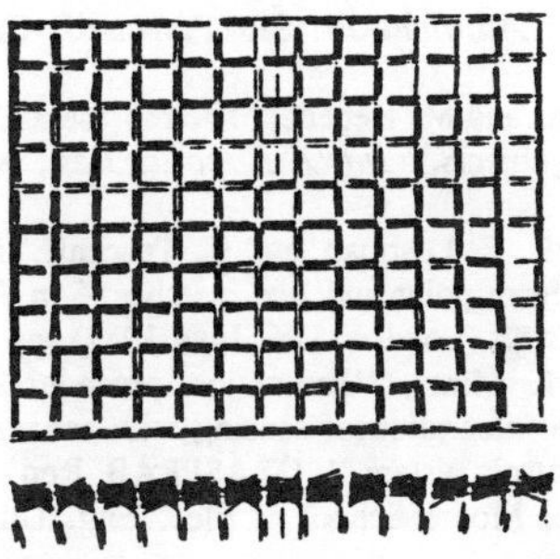

Fig. 32. Front an top view of reconstructed 3-D lines before identical lines are detected.

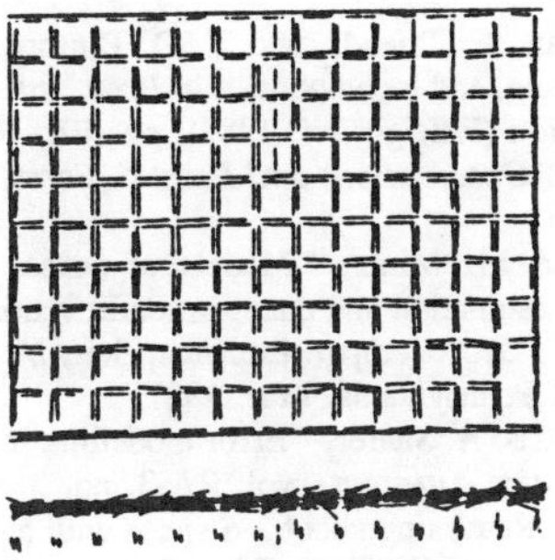

Fig. 33. Front and top view of 3-D lines after the fusion of identical lines.

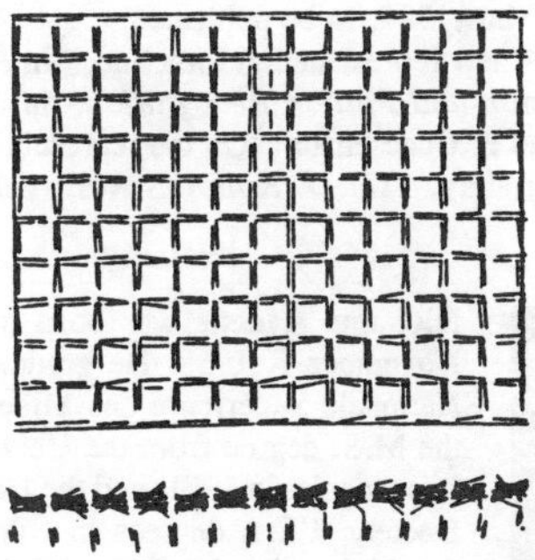

Fig. 34. Front and top view of reconstructed 3-D lines, before colinearity is detected.

coordinates of their endpoints (in the same way as in the first experiment).

We now ask the system to discover the relation "≡" between the 3-D lines (see Section V-C). The program takes a first 3-D line, computes its Mahalanobis distance (11) to all the other lines of the scene, and accepts the first line which passes the χ^2 test of (12) (see Section V-D). The two lines are fused using the technique of Section V-F and one keeps only the parameters of the optimal line representing both of them with an updated covariance matrix. The remaining lines are now compared to this new virtual line still with the Mahalanobis distance of (11) but with the new updated covariance matrix, while the χ^2 test of (12) remains unchanged. This process is repeated until no more lines can be matched with the first one, and then repeated with all the remaining unmatched lines.

The result is a reduced set of virtual lines on which the endpoints of the original segments have been projected, as shown in Fig. 36. The uncertainty on the line parameters has been greatly reduced: Fig. 37 shows the resulting uncertainty on the lines endpoints, which agrees very well with the reality.

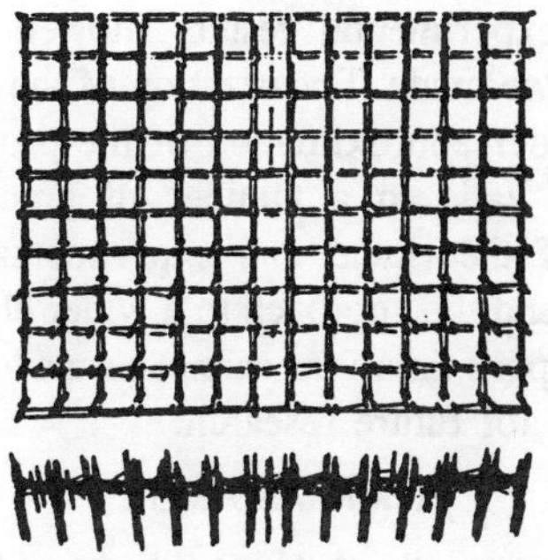

Fig. 35. Initial uncertainty attached to 3-D lines endpoints.

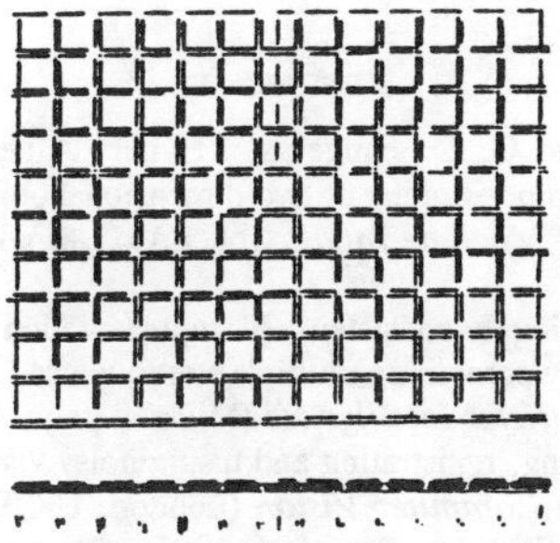

Fig. 36. Front and top view of 3-D lines when colinearity is discovered and enforced.

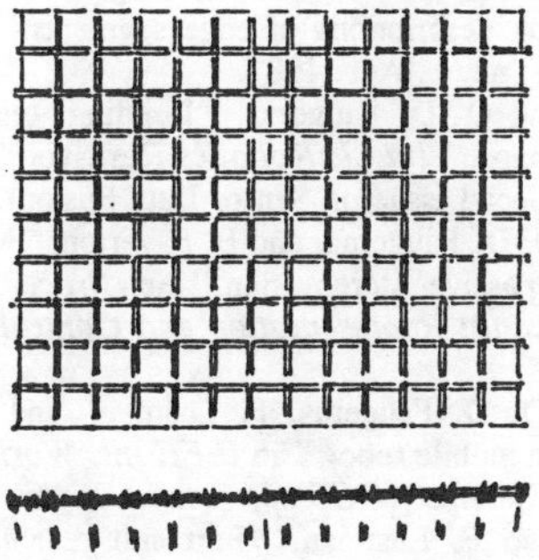

Fig. 37. Uncertainty attached to 3-D lines endpoints after the fusion of colinear segments.

VII. Conclusion

In this paper we have proposed a methodology for building and maintaining a geometric representation of the environment of a mobile robot. This methodology has the following salient features:

Representation: 1) We use geometric primitives to describe the environment and rigid displacements to describe the motion. These entities are described with a minimal number of parameters. 2) Uncertainty is modeled by a probability density function of these parameters. 3) Relationships between geometric entities are represented by algebraic equations on their parameters.

Algorithms: detecting geometric relationships, computing or updating the parameters of geometric entities (both for primitives and displacements) is done by recursive prediction-and-verification algorithms including the Extended Kalman Filter. These algorithms, better detailed in [8], [10], and [11], take into account prior knowledge to compute and propagate uncertainties.

Finally, the experimental results showed that the major approximations we made (linearization of the algebraic equations, second-order approximation of the probability density functions) were valid in a number of practical cases. Of course, a lot of theoritical and experimental work is still necessary to extend the approach to a wider class of problems for which such approximations cannot be made. This might be a good direction for future research.

Acknowledgment

The authors want to thank N. Gaudechoux for her precious help in the preparation of this paper and the reviewers for their helpful comments.

References

[1] N. Ayache and O. D. Faugeras, "Hyper: A new approach for the recognition and positioning of two-dimensional objects," *IEEE Trans. Pattern Anal. Machine Intell.*, vol. PAM1-8, no. 1, pp. 44–54, Jan. 1986.

[2] ——, "Building a consistent 3d representation of a mobile robot environment by combining multiple stereo views," in *Proc. Int. Joint Conf. on Artificial Intelligence* (Milano, Italy, Aug. 1987).

[3] ——, "Building, registrating and fusing noisy visual maps," in *Proc. Int. Conf. on Computer Vision* (London, UK, June 1987), pp. 73–82. Also an INRIA Int. Rep. 596, 1986.

[4] ——, "Maintaining representations of the environment of a mobile robot," in *Int. Symp. on Robotics Research* (Santa Cruz, CA, Aug. 1987), pp. 337–350.

[5] N. Ayache and B. Faverjon, "Efficient registration of stereo images by matching graph descriptions of edge segments," *Int. J. Computer Vision*, vol. 1, no. 2, Apr. 1987.

[6] N. Ayache and O. D. Faugeras, "Building, registrating and fusing noisy visual maps," *Int. J. Robotics Res.*, vol. 7, no. 6, pp. 45–65, Dec. 1988 (Special Issue on Sensor Data Fusion).

[7] N. Ayache, O. D. Faugeras, and B. Faverjon, "Matching depth maps obtained by passive stereovision," in *Proc. 3rd Workshop on Computer Vision: Representation and Control*, pp. 197–204, Oct. 1985.

[8] N. Ayache, O. D. Faugeras, F. Lustman, and Z. Zhang, "Visual navigation of a mobile robot," in *IEEE Int. Workshop on Intelligent Robots and Systems (IROS'88)* (Tokyo, Japan, Oct. 1988).

[9] N. Ayache and F. Lustman, "Fast and reliable passive trinocular stereovision," in *Proc. Int. Conf. on Computer Vision* (London, UK, June 1987), pp. 422–427.

[10] N. Ayache, "Construction et fusion de représentations visuelles tridimensionnelles; applications à la robotique mobile," Thèse d'Etat, Université de Paris-Sud, Orsay, May 1988. INRIA Int. Rep.

[11] ——, *Vision Stéréoscopique et Perception Multisensorielle—Application à la Robotique Mobile.* Inter-Editions, 1989. English translation will be available from MIT Press in 1990.

[12] R. M. Bolle and D. B. Cooper, "On optimally combining pieces of information, with application to estimating 3D complex-object position from range data," *IEEE Trans. Pattern Anal. Machine Intell.*, vol. PAMI-8, pp. 619–638, 1986.

[13] J. Canny, "A computational approach to edge detection," *IEEE Trans. Pattern Anal. Machine Intell.*, vol. PAMI-8, no. 6, pp. 679–698, 1986.

[14] M. P. do Carmo, *Differential Geometry of Curves and Surfaces.* Englewood Cliffs, NJ: Prentice Hall, 1976.

[15] J. L. Crowley, "Representation and maintenance of a composite surface model," in *Proc. Int. Conf. on Robotics and Automation* (San Francisco, CA, Apr. 1986), pp. 1455–1462.

[16] C. Darmon, "A new recursive method to detect moving objects in a sequence of images," in *Proc. IEEE Conf. on Pattern Recognition and Image Processing*, pp. 259–261, June 1982.

[17] R. Deriche, "Using Canny's criteria to derive an optimal edge detector recursively implemented," *Int. J. Computer Vision*, vol. 2, Apr. 1987.

[18] H. F. Durrant-Whyte, "Consistent integration and propagation of disparate sensor observations," in *Proc. Int. Conf. on Robotics and Automation* (San Francisco, CA, Apr. 1986), pp. 1464–1469.

[19] O. D. Faugeras, N. Ayache, and B. Faverjon, "Building visual maps by combining noisy stereo measurements," in *Proc. Int. Conf. on Robotics and Automation* (San Francisco, CA, Apr. 1986), pp. 1433–1438.

[20] O. D. Faugeras and M. Hébert, "The representation, recognition, and locating of 3d objects," *Int. J. Robotics Res.*, vol. 5, no. 3, pp. 27–52, 1986.

[21] O. D. Faugeras, F. Lustman, and G. Toscani, "Motion and structure from motion from point and line matches," in *Proc. Int. Conf. on Computer Vision* (London, UK, June 1987), pp. 25–34.

[22] Hager and M. Mintz, "Estimation procedures for robust sensor control, in the integration of sensing with actuation to form a robust intelligent control system," GRASPLAB Rep. 97, Dep. Comput. Informat. Sci., Moore School of Elec. Eng. Univ. of Pennsylvania, Mar. 1987.

[23] A. M. Jazwinsky, *Stochastic Processes and Filtering Theory.* New York, NY: Academic Press, 1970.

[24] T. Kanade, *Three-Dimensional Machine Vision.* New York, NY: 1987.

[25] D. J. Kriegman, E. Triendl, and T. O. Binford, "A mobile robot: Sensing, planning, and locomotion," in *Proc. Int. Conf. on Robotics and Automation* (Raleigh, NC 1987), pp. 402–408.

[26] E. M. Mikhail, *Observations and Least Squares.* University Press of America, 1976.

[27] McKendall and M. Mintz, "Models of sensor noise and optimal algorithms for estimation and quantization in vision systems," GRASPLAB Rep. 97, Dep. Comput. Informat. Sci., Moore School of Elec. Eng., Univ. of Pennsylvania, Mar. 1987.

[28] L. Matthies and S. A. Shafer, "Error modelling in stereo navigation," *IEEE J. Robotics Automat.*, vol. RA-3, no. 3, pp. 239–248, 1987.

[29] J. L. Mundy, "Reasoning about 3-d space with algebraic deduction," in O. D. Faugeras and G. Giralt, Eds., *Robotics Research, The Third International Symposium.* Cambridge, MA: MIT Press, 1986, pp. 117–124.

[30] J. Porrill *et al.*, "Optimal combination and constraints for geometrical sensor data," Mar. 1987, to be published.

[31] K. Roberts, "A new representation for a line," in *Proc. Int. Conf. on Computer Vision and Pattern Recognition*, pp. 635–640, 1988.

[32] R. C. Smith and P. Cheeseman, "On the representation and estimation of spatial uncertainty," *Int. J. Robotics Res.*, vol. 5, no. 4, pp. 56–68, 1987.

Nicholas Ayache was born in Paris, France, on November 1, 1958. He graduated from the Ecole Nationale Supérieure des Mines in 1980, received the M.S. degree from the University of California, Los Angeles in 1981, and the Docteur Ingénieur and Docteur d'Etat degrees in computer science from the University of Paris XI in 1983 and 1988, respectively.

He is currently the Research Director of the Medical Images and Robotics Project *Epidaure* at INRIA (Institut National de Recherche en Informatique et en Automatique, Le Chesnay, France), and an Associate Professor at the University of Paris XI and Ecole Centrale. His research interests are in Computer Vision, Robotics, Artificial Intelligence, and Medical Applications. He is the author of the book *Stereovision and Multisensor Perception.*

Olivier D. Faugeras is Director of Research at INRIA (National Institute for Research in Computer Science and Control Theory) where he leads the Robotics and Computer Vision Project. His research interest include Computer Vision, Robotics, Shapes Representation, Computational Geometry, and Architectures for Vision. He is also lecturer in Applied Mathematics at the Ecole Polytechnique in Palaiseau where he teaches Computer Vision and Computational Geometry. He also teaches Computer Vision and Robotics in the "Magistère de Mathématique et Informatique" at the Ecole Normale Supérieure de la rue d'Ulm and at l'Ecole Polytechnique Féderérale of Lausanne in Switzerland. He is Associated Editor of several international scientific journals including: IEEE TRANSACTIONS ON PATTERN ANALYSIS AND MACHINE INTELLIGENCE, *International Journal of Computer Vision, Internation Journal of Robotics Research, Pattern Recognition Letters, Signal Processing, and Robotics and Autonomous Systems.* In April 1989, he received from the French Science Academy the "Institut de France—Fondation Fiat" prize for his work in Vision and Robotics.

Kalman Filter-based Algorithms for Estimating Depth from Image Sequences

LARRY MATTHIES AND TAKEO KANADE
Department of Computer Science, Carnegie Mellon University, Pittsburgh, PA 15213;
Schlumberger Palo Alto Research, 3340 Hillview Ave., Palo Alto, CA 94304
RICHARD SZELISKI
Digital Equipment Corporation, 1 Kendall Square, Building 700, Cambridge, MA 02139

Abstract
Using known camera motion to estimate depth from image sequences is an important problem in robot vision. Many applications of depth-from-motion, including navigation and manipulation, require algorithms that can estimate depth in an on-line, incremental fashion. This requires a representation that records the uncertainty in depth estimates and a mechanism that integrates new measurements with existing depth estimates to reduce the uncertainty over time. Kalman filtering provides this mechanism. Previous applications of Kalman filtering to depth-from-motion have been limited to estimating depth at the location of a sparse set of features. In this paper, we introduce a new, pixel-based (*iconic*) algorithm that estimates depth and depth uncertainty at each pixel and incrementally refines these estimates over time. We describe the algorithm and contrast its formulation and performance to that of a feature-based Kalman filtering algorithm. We compare the performance of the two approaches by analyzing their theoretical convergence rates, by conducting quantitative experiments with images of a flat poster, and by conducting qualitative experiments with images of a realistic outdoor-scene model. The results show that the new method is an effective way to extract depth from lateral camera translations. This approach can be extended to incorporate general motion and to integrate other sources of information, such as stereo. The algorithms we have developed, which combine Kalman filtering with iconic descriptions of depth, therefore can serve as a useful and general framework for low-level dynamic vision.

1 Introduction

Using known camera motion to estimate depth from image sequences is important in many applications of computer vision to robot navigation and manipulation. In these applications, depth-from-motion can be used by itself, as part of a multimodal sensing strategy, or as a way to guide stereo matching. Many applications require a depth estimation algorithm that operates in an on-line, incremental fashion. To develop such an algorithm, we require a depth representation that includes not only the current depth estimate, but also an estimate of the uncertainty in the current depth estimate.

Previous work [3, 5, 9, 10, 16, 17, 25] has identified Kalman filtering as a viable framework for this problem, because it incorporates representations of uncertainty and provides a mechanism for incrementally reducing uncertainty over time. To date, applications of this framework have largely been restricted to estimating the positions of a sparse set of trackable features, such as points or line segments. While this is adequate for many robotics applications, it requires reliable feature extraction and it fails to describe large areas of the image. Another line of work has addressed the problem of extracting dense displacement or depth estimates from image sequences. However, these previous approaches have either been restricted to two-frame analysis [1] or have used batch processing of the image sequence, for example via spatiotemporal filtering [11].

In this paper we introduce a new, pixed-based (*iconic*) approach to incremental depth estimation and compare it mathematically and experimentally to a feature-based approach we developed previously [16]. The new approach represents depth and depth variance at every pixel and uses Kalman filtering to extrapolate and update the pixel-based depth representation. The algorithm uses correlation to measure the optical flow and to estimate the variance in the flow, then uses the known camera motion to convert the flow field into a depth map. It then uses the Kalman filter to generate an updated depth map from a weighted combination of the new measurements and the prior depth estimates. Regularization is employed to smooth the depth map

"Kalman Filter-Based Algorithms for Estimating Depth from Image Sequences" by L. Matthies, T. Kanade, and R. Szeliski. Reprinted from *International Journal of Computer Vision*, 3, 1989, pages 209-236. Copyright © 1989 by Kluwer Academic Publishers, reprinted with permission.

and to fill in the underconstrained areas. The resulting algorithm is parallel, uniform, and can take advantage of mesh-connected or multiresolution (pyramidal) processing architectures.

The remainder of this paper is structured as follows. In the next section, we give a brief review of Kalman filtering and introduce our overall approach to Kalman filtering of depth. Next, we review the equations of motion, present a simple camera model, and examine the potential accuracy of the method by analyzing its sensitivity to the direction of camera motion. We then describe our new, pixel-based depth-from-motion algorithm and review the formulation of the feature-based algorithm. Next, we analyze the theoretical accuracy of both methods, compare them both to the theoretical accuracy of stereo matching, and verify this analysis experimentally using images of a flat scene. We then show the performance of both methods on images of realistic outdoor scene models. In the final section, we discuss the promise and the problems involved in extending the method to arbitrary motion. We also conclude that the ideas and results presented apply directly to the much broader problem of integrating depth information from multiple sources.

2 Estimation Framework

The depth-from-motion algorithms described in this paper use image sequences with small frame-to-frame camera motion [4]. Small motion minimizes the correspondence problem between successive images, but sacrifices depth resolution because of the small baseline between consecutive image pairs. This problem can be overcome by integrating information over the course of the image sequence. For many applications, it is desirable to process the images incrementally by generating updated depth estimates after each new image is acquired, instead of processing many images together in a batch. The incremental approach offers real-time operation and requires less storage, since only the current estimates of depth and depth uncertainty need to be stored.

The Kalman filter is a powerful technique for doing incremental, real-time estimation in dynamic systems. It allows for the integration of information over time and is robust with respect to both system and sensor noise. In this section, we first present the notation and the equations of the Kalman filter, along with a simple example. We then sketch the application of this framework to motion-sequence processing and discuss those parts of the framework that are common to both the iconic and the feature-based algorithms. The details of these algorithms are given in sections 4 and 5, respectively.

2.1. *Kalman Filter*

The Kalman filter is a Bayesian estimation technique used to track stochastic dynamic systems being observed with noisy sensors. The filter is based on three separate probabilistic models, as shown in table 1. The first model, the ***system model***, describes the evolution over time of the current state vector u_t. The transition between states is characterized by the known transition matrix Φ_t and the addition of Gaussian noise with a covariance Q_t. The second model, the ***measurement*** (or ***sensor***) ***model***, relates the measurement vector d_t to the current state through a measurement matrix H_t and the addition of Gaussian noise with a covariance R_t. The third model, the ***prior model***, describes the knowledge about the system state $\hat{u}_0$ and its covariance P_0 before the first measurement is taken. The sensor and process noise are assumed to be uncorrelated.

Table 1. **Kalman filter equations.**

Models	system model	$u_t = \Phi_{t-1} u_{t-1} + \eta_t, \ \eta_t \sim N(0, Q_t)$
	measurement model	$d_t = H_t u_t + \xi_t, \ \xi_t \sim N(0, R_t)$
	prior model	$E[u_0] = \hat{u}_0, \ \mathrm{cov}[u_0] = P_0$
	(other assumptions)	$E[\eta_t \xi_j^T] = 0$
Prediction phase	state estimate extrapolation	$\hat{u}_t^- = \Phi_{t-1} \hat{u}_{t-1}^+$
	state covariance extrapolation	$P_t^- = \Phi_{t-1} P_{t-1}^+ \Phi_{t-1}^T + Q_{t-1}$
Update phase	state estimate update	$\hat{u}_t^+ = \hat{u}_t^- + K_t[d_t - H_t \hat{u}_t^-]$
	state coviarance update	$P_t^+ = [I - K_t H_t] P_t^-$
	Kalman gain matrix	$K_t = P_t^- H_t^T [H_t P_t^- H_t^T R_t]^{-1}$

To illustrate the equations of table 1, we will use the example of a ping-pong-playing robot that tracks a moving ball. In this example, the state consists of the ball position and velocity, $u = [x\ y\ z\ \dot{x}\ \dot{y}\ \dot{z}\ 1]^T$, where x and y lie parallel to the image plane (y is up), and z is parallel to the optical axis. The state transition matrix models the ball dynamics, for example by the matrix

$$\Phi_t = \begin{bmatrix} 1 & 0 & 0 & \Delta t & 0 & 0 & 0 \\ 0 & 1 & 0 & 0 & \Delta t & 0 & 0 \\ 0 & 0 & 1 & 0 & 0 & \Delta t & 0 \\ 0 & 0 & 0 & -\beta & 0 & 0 & 0 \\ 0 & 0 & 0 & 0 & -\beta & 0 & -g\Delta t \\ 0 & 0 & 0 & 0 & 0 & -\beta & 0 \\ 0 & 0 & 0 & 0 & 0 & 0 & 1 \end{bmatrix}$$

where Δt is the time step, β is the coefficient of friction and g is gravitational acceleration. The process noise matrix Q_t models the random disturbances that influence the trajectory. If we assume that the camera uses orthographic projection and uses a simple algorithm to find the "center of mass" (x,y) of the ball, then the sensor can then be modeled by the matrix

$$H_t = \begin{bmatrix} 1 & 0 & 0 & 0 & 0 & 0 & 0 \\ 0 & 1 & 0 & 0 & 0 & 0 & 0 \end{bmatrix}$$

which maps the state u to the measurement d. The uncertainty in the sensed ball position can be modeled by a 2×2 covariance matrix R_t.

Once the system, measurement, and prior models have been specified (i.e., the upper third of table 1), the Kalman filter algorithm follows from the formulation in the lower two thirds of table 1. The algorithm operates in two phases: extrapolation (prediction) and update (correction). At time t, the previous state and covariance estimates, $\hat{u}^+_{t-1}$ and P^+_{t-1}, are extrapolated to predict the current state $\hat{u}^-_t$ and covariance P^-_t. The predicted covariance is used to compute the new Kalman gain matrix K_t and the updated covariance matrix P^+_t. Finally, the measurement residual $d_t - H_t\hat{u}^-_t$ is weighted by the gain matrix K_t and added to the predicted state u^-_t to yield the updated state u^+_t. A block diagram for the Kalman filter is given in figure 1.

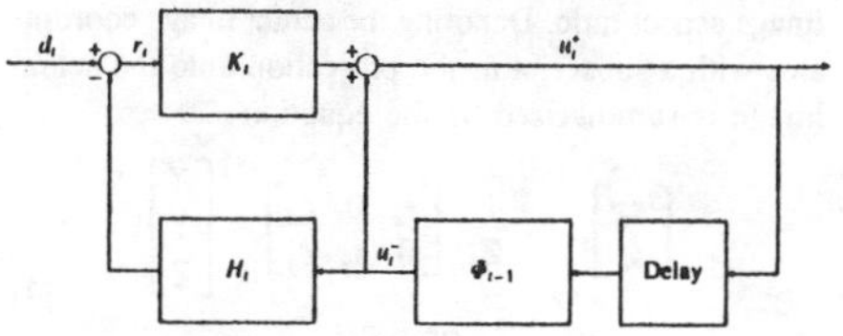

Fig. 1. Kalman filter block diagram.

2.2. *Application to Depth from Motion*

To apply the Kalman filter estimation framework to the depth-from-motion problem, we specialize each of the three models (system, measurement, and prior) and define the implementations of the extrapolation and update stages. This section briefly previews how these components are chosen for the two depth-from-motion algorithms described in this paper. The details of the implementation are left to sections 4 and 5.

The first step in designing a Kalman filter is to specify the elements of the state vector. The iconic depth-from-motion algorithm estimates the depth at each pixel in the current image, so the state vector in this case is the entire depth map.[1] Thus, the diagonal elements of the state covariance matrix P_t are the variances of the depth estimates at each pixel. As discussed shortly, we implicitly use off-diagonal elements of the inverse covariance matrix P_0^{-1} as part of the update stage of the filter, but do not explicitly model them anywhere in the algorithm because of the large size of the matrix. For the feature-based approach, which tracks edge elements through the image sequence, the state consists of a 3D position vector for each feature. We model the full covariance matrix of each individual feature, but treat separate features as independent.

The system model in both approaches is based on the same motion equations (section 3.1), but the implementations of the extrapolation and update stages differ because of the differences in the underlying representations. For the iconic method, the extrapolation stage uses the depth map estimated for the current frame, together with knowledge of the camera motion, to predict the depth and depth variance at each pixel in the next frame. Similarly, the update stage uses measurements of depth at each pixel to update the depth and variance estimates at each pixel. For the feature-based method, the extrapolation stage predicts the position vector and covariance matrix of each feature for the next image, then uses measurements of the image coordinates of the feature to update the position vector and the covariance matrix. Details of the measurement models for each algorithm will be discussed later.

Finally, the prior model can be used to embed prior knowledge about the scene. For the iconic method, for example, smoothness constraints requiring nearby image points to have similar disparity can be modeled easily by off-diagonal elements of the inverse of the prior covariance matrix P_0 [29]. Our algorithm incorporates

[1]Our actual implementation uses inverse depth (called "disparity") See section 4.

this knowledge as part of a smoothing operation that follows the state update stage. Similar concepts may be applicable to modeling figural continuity [20,24] in the edge-tracking approach, that is, the constraint that connected edges must match connected edges; however, we have not pursued this possibility.

3 Motion Equations and Camera Model

Our system and measurement models are based on the equations relating scene depth and camera motion to the induced image flow. In this section, we review these equations for an idealized camera (focal length = 1) and show how to use a simple calibration model to relate the idealized equations to real cameras. We also derive an expression for the relative uncertainty in depth estimates obtained from lateral versus forward camera translation. This expression shows concretely the effects of camera motion on depth uncertainty and reinforces the need for modeling the uncertainty in computed depth.

3.1. Equations of Motion

If the inter-frame camera motion is sufficiently small, the resulting optical flow can be expressed to a good approximation in terms of the instantaneous camera velocity [6, 13, 33]. We will specify this in terms of a translational velocity **T** and an angular velocity **R**. In the camera coordinate frame (figure 2), the motion of a 3D point **P** is described by the equation

$$\frac{d\mathbf{P}}{dt} = -\mathbf{T} - \mathbf{R} \times \mathbf{P}$$

Expanding this into components yields

$$\begin{aligned} dX/dt &= -T_x - R_y Z + R_z Y \\ dY/dt &= -T_y - R_z X + R_x Z \\ dZ/dt &= -T_z - R_x Y + R_y X \end{aligned} \qquad [1]$$

Now, projecting (X, Y, Z) onto an ideal, unit focal length image,

$$x = \frac{X}{Z} \qquad y = \frac{Y}{Z} \; ,$$

taking the derivatives of (x, y) with respect to time, and substituting in from equation (1) leads to the familiar equations of optical flow [33]:

$$\begin{bmatrix} \Delta x \\ \Delta y \end{bmatrix} = \frac{1}{Z} \begin{bmatrix} -1 & 0 & x \\ 0 & -1 & y \end{bmatrix} \begin{bmatrix} T_x \\ T_y \\ T_z \end{bmatrix} + \begin{bmatrix} xy & -(1+x^2) & y \\ (1+y^2) & -xy & -x \end{bmatrix} \begin{bmatrix} R_x \\ R_y \\ R_z \end{bmatrix} \qquad [2]$$

These equations relate the depth Z of the point to the camera motion **T**, **R** and the induced image displacements or optical flow $[\Delta x \;\; \Delta y]^T$. We will use these equations to measure depth, given the camera motion and optical flow, and to predict the change in the depth map between frames. Note that parameterizing (2) in terms of the inverse depth $d = 1/Z$ makes the equations linear in the "depth" variable. Since this leads to a simpler estimation formulation, we will use this parameterization in the balance of the paper.

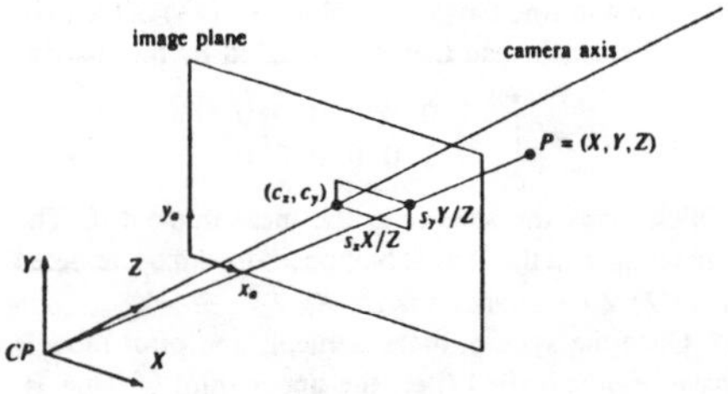

Fig. 2. Camera model. *CP* is the center of projection.

3.2. Camera Model

Relating the ideal flow equations to real measurements requires a camera model. If optical distortions are not severe, a pin-hole camera model will suffice. In this paper we adopt a model similar to that originated by Sobel [27] (figure 2). This model specifies the origin (c_x, c_y) of the image coordinate system and a pair of scale factors (s_x, s_y) that combine the focal length and image aspect ratio. Denoting the actual image coordinates with a subscript a, the projection onto the actual image is summarized by the equation

$$\begin{aligned} \begin{bmatrix} x_a \\ y_a \end{bmatrix} &= \frac{1}{Z} \begin{bmatrix} s_x & 0 & c_x \\ 0 & s_y & c_y \end{bmatrix} \begin{bmatrix} X \\ Y \\ Z \end{bmatrix} \\ &= \frac{1}{Z} \, CP \end{aligned} \qquad [3]$$

C is known as the *collimation matrix*. Thus, the ideal image coordinates (x, y) are related to the actual image coordinates by

$$x_a = s_x x + c_x \qquad y_a = s_y y + c_y$$

Equations in the balance of the paper will primarily use ideal image coordinates for clarity. These equations can be re-expressed in terms of actual coordinates using the transformations above.

3.3. Sensitivity Analysis

Before describing our Kalman filter algorithms, we will analyze the effect of different camera motions on the uncertainty in depth estimates. Given specific descriptions of real cameras and scenes, we can obtain bounds on the estimation accuracy of depth-from-motion algorithms using perturbation or covariance analysis techniques based on first-order Taylor expansions [8]. For example, if we solve the motion equations for the inverse depth d in terms of the optical flow, camera motion, and camera model,

$$d = F(\Delta x, \Delta y, T, R, c_x, c_y, s_x, s_y) \qquad [4]$$

then the uncertainty in depth arising from uncertainty in flow, motion, and calibration can be expressed by

$$\delta d = J_f \delta f + J_m \delta m + J_c \delta c \qquad [5]$$

where J_f, J_m, and J_c are the Jacobians of (4) with respect to the flow, motion, and calibration parameters, respectively, and δf, δm, and δc are perturbations of the respective parameters. We will use this methodology to draw some conclusions about the relative accuracy of depth estimates obtained from different classes of motion.

It is well known that camera rotation provides no depth information. Furthermore, for a translating camera, the accuracy of depth estimates increases with increasing distance of image features from the *focus of expansion* (FOE), the point in the image where the translation vector (**T**) pierces the image. This implies that the 'best' translations are parallel to the image plane and that the 'worst' are forward along the camera axis. We will give a short derivation that demonstrates the relative accuracy obtainable from forward and lateral camera translation. The effects of measurement uncertainty on depth-from-motion calculations is also examined in [26].

For clarity, we consider only one-dimensional flow induced by translation along the X or Z axes. For an ideal camera, lateral motion induces the flow

$$\Delta x_l = \frac{-T_x}{Z} \qquad [6]$$

whereas forward motion induces the flow

$$\Delta x_f = \frac{xT_z}{Z} \qquad [7]$$

The inverse depth (or disparity) in each case is

$$d_l = \frac{1}{Z} = \frac{-\Delta x_l}{T_x}$$

$$d_f = \frac{\Delta x_f}{xT_z}$$

Therefore, perturbations of δx_l and δx_f in the flow measurements Δx_l and Δx_f yield the following perturbations in the disparity estimates:

$$\delta d_l = \frac{\delta x_l}{|T_x|}$$

$$\delta d_f = \frac{\delta x_f}{|xT_z|}$$

These equations give the error in the inverse depth as a function of the error in the measured image displacement, the amount of camera motion, and position of the feature in the field of view. Since we are interested in comparing forward and lateral motions, a good way to visualize these equations is to plot the relative depth uncertainty, $\delta d_f / \delta d_l$. Assuming that the flow perturbations δx_l and δx_f are equal, the relative uncertainty is

$$\frac{\delta d_f}{\delta d_l} = \frac{\delta x_f / |xT_z|}{\delta x_l / |T_x|} = \frac{|T_x|}{|xT_z|}$$

The image coordinate x indicates where the object appears in the field of view. Figure 3 shows that x equals the tangent of the angle θ between the object and the camera axis. The formula for the relative uncertainty is thus

$$\frac{\delta d_f}{\delta d_l} = \frac{|T_x|}{|T_z \tan\theta|} \qquad [9]$$

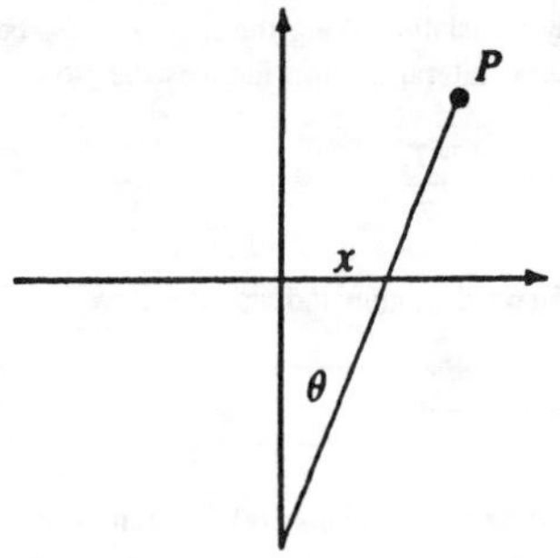

Fig. 3. Angle between object and camera axis is θ.

This relationship is plotted in figure 4 for $T_x = T_z$. At 45° from the camera axis, depth uncertainty is equal for forward and lateral translations. As this angle approaches zero, the ratio of uncertainty grows, first slowly then increasingly rapidly. As a concrete example, for the experiments in section 6.2 the field of view was approximately 36°, so the edges of the images were 18° from the camera axis. At this angle, the ratio of uncertainties is 3.1; halfway from the center to the edge of the image, at 9°, the ratio is 6.3. In general, for practical fields of view, the accuracy of depth extracted from forward motion will be effectively unusable for a large part of the image.

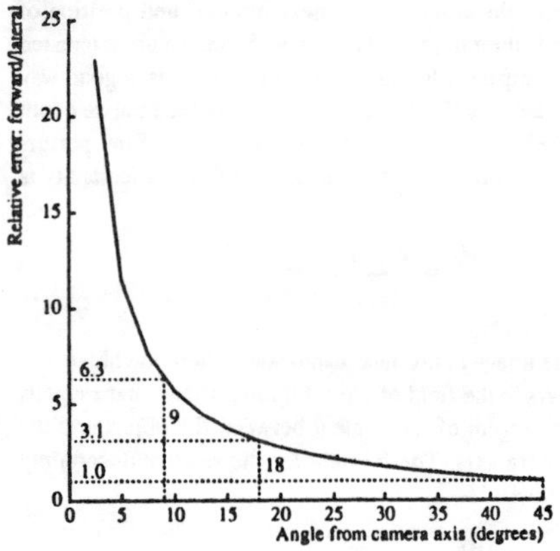

Fig. 4. Relative depth uncertainty for forward vs. lateral translation.

By setting $\delta d_f/\delta d_l = 1$, equation (9) also expresses the relative distances the camera must move forward and laterally to obtain equally precise depth estimates. An alternate interpretation for figure 4 is that it expresses the relative precision of stereo and depth-from-motion in a dynamic, binocular stereo system.

We draw several conclusions from this analysis. First, it underscores the value of representing depth uncertainty as we describe in the following sections. Second, for practical depth estimation, forward motion is effectively unusable compared with lateral motion. Finally, we can relate these results to dynamic, binocular stereo by noting that depth from forward motion will be relatively ineffective for constraining or confirming binocular correspondence.

4 Iconic Depth Estimation

This section describes the incremental, iconic depth-estimation algorithm we have developed. The algorithm processes each new image as it arrives, extracting optical flow at each pixel using the current and previous intensity images, then integrates this new information with the existing depth estimates.

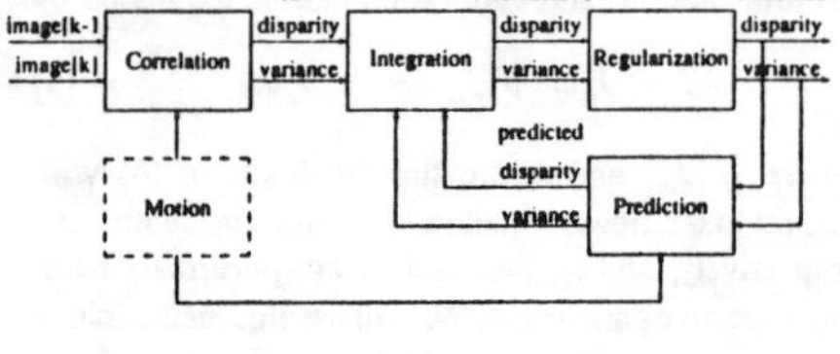

Fig. 5. Iconic depth-estimation block diagram.

The algorithm consists of four main stages (figure 5). The first stage uses correlation to compute estimates of optical flow vectors and their associated covariance matrixes. These are converted into disparity (inverse-depth) measurements using the known camera motion. The second stage integrates this information with the disparity map predicted from the previous time step. The third stage uses regularization-based smoothing to reduce measurement noise and to fill in areas of unknown disparity. The last stage uses the known camera motion to predict the disparity field that will be seen in the next frame and resamples the field to keep it iconic (pixel-based).

4.1. Measuring Disparity

The first stage of the Kalman filter computes measurements of disparity from the difference in intensity between the current image and the previous image. This computation proceeds in two parts. First, a two-dimensional optical flow vector is computed at each point using a correlation-based algorithm. The uncertainty in this vector is characterized by a bivariate Gaussian distribution. Second, these vectors are converted into disparity measurements using the known camera motion and the motion equations developed in section 3.1.

This two-part formulation is desirable for several reasons. First, it allows probabilistic characterizations of uncertainty in flow to be translated into probabilistic characterizations of uncertainty in disparity. This is especially valuable if the camera motion is also uncertain, since the equations relating flow to disparity can be extended to model this as well [25]. Second, by characterizing the level of uncertainty in the flow, it allows us to evaluate the potential accuracy of the algorithm independent of how flow is obtained. Finally, bivariate Gaussian distributions can capture the distinctions between knowing zero, one, or both components of flow [1, 11, 22], and therefore subsume the notion of the aperture problem.

The problem of optical flow estimation has been studied extensively. Early approaches used the ratio of the spatial and temporal image derivatives [12], while more recent approaches have used correlation between images [1] or spatiotemporal filtering [11]. In this paper we use a simple version of correlation-based matching. This technique, which has been called the *sum of squared differences* (SSD) method [1], integrates the squared intensity difference between two shifted images over a small area to obtain an error measure

$$e_t(\Delta x, \Delta y; x, y) = \iint [f_t(x - \Delta x + \lambda, y - \Delta y + \eta) - f_{t-1}(x + \lambda, y + \eta)]^2 \, d\lambda \, d\eta$$

where f_t and f_{t-1} are the two intensity images, and $w(\lambda, \eta)$ is a weighting function. The SSD measure is computed at each pixel for a number of possible flow values. In [1], a coarse-to-fine technique is used to limit the range of possible flow values. In our images the possible range of values is small (since we are using small-motion sequences), so a single-resolution algorithm suffices.[2] The resulting error surface $e_t(\Delta x, \Delta y; x, y)$ is approximately parabolic in shape. The lowest point of this surface defines the flow measurement and the shape of the surface defines the covariance matrix of the measurement.

To convert the displacement vector $[\Delta x \ \Delta y]^T$ into a disparity measurement, we assume that the camera motion $(\mathbf{T}, \mathbf{R})$ is given. The optical flow equation (2) can then be used to estimate depth as follows. First we abbreviate (2) to

$$\begin{bmatrix} \Delta x \\ \Delta y \end{bmatrix} = d \begin{bmatrix} t_x \\ t_y \end{bmatrix} + \begin{bmatrix} r_x \\ r_y \end{bmatrix} + \xi \qquad [10]$$

where d is the inverse depth and ξ is an error vector representing noise in the flow measurement. The noise ξ is assumed to be a bivariate Gaussian random vector with a zero mean and a covariance matrix P_m computed by the flow estimation algorithm. Equation (10) can be re-expressed in the following standard form for linear estimation problems:

$$\Delta \mathbf{x} = \begin{bmatrix} \Delta x \\ \Delta y \end{bmatrix} - \begin{bmatrix} r_x \\ r_y \end{bmatrix} = d \begin{bmatrix} t_x \\ t_y \end{bmatrix} + \xi = \mathbf{H}d + \xi \qquad [11]$$

The optimal estimate of the disparity d is then [19]

$$d = (H^T P_m^{-1} H)^{-1} H^T P_m^{-1} \Delta \mathbf{x} \qquad [12]$$

and the variance of this disparity measurement is

$$\sigma_d^2 = (H^T P_m^{-1} H)^{-1} \qquad [13]$$

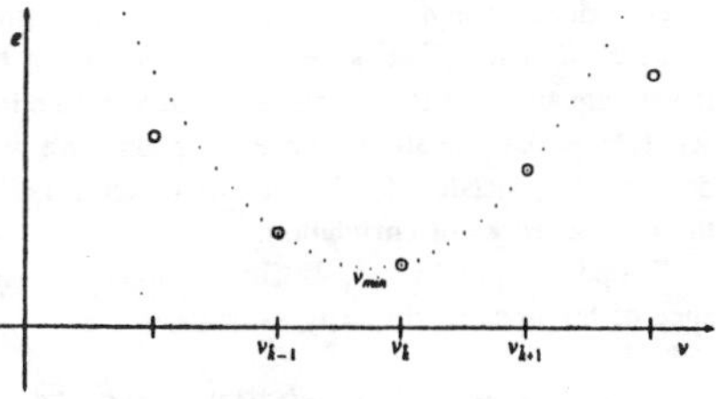

Fig. 6. Parabolic to fit to SSD error surface.

[2] It may be necessary to use a larger search range at first, but once the estimator has "latched on" to a good disparity map, the predicted disparity and disparity variance can be used to limit the search by computing confidence intervals.

This measurement process has been implemented in a simplified form, under the assumption that the flow is parallel to the image raster. To improve precision, each scan line of two successive images is magnified by a factor of four by cubic interpolation. The SSD measure e_k is computed at each interpolated subpixel displacement v_k, using a 5×5-pixel window. The minimum error $(v_{\hat{k}}, e_{\hat{k}})$ is found and a parabola

$$e(v) = av^2 + bv + c$$

is fit to this point and its two neighbors $(v_{\hat{k}-1}, e_{\hat{k}-1})$ and $(v_{\hat{k}+1}, e_{\hat{k}+1})$ (figure 6). The minimum of this parabola establishes the flow estimate (to sub-sub-pixel precision). Appendix A shows that the variance of the flow measurement is

$$\mathrm{var}(e) = \frac{2\sigma_n^2}{a}$$

where σ_n^2 is the variance of the image noise process. The appendix also shows that adjacent flow estimates are correlated over both space and time. The significance of this fact is considered in the following two sections and in section 6.1.

4.2. Updating the Disparity Map

The next stage in the iconic depth estimator is the integration of the new disparity measurements with the predicted disparity map (this step is omitted for the first pair of images). If each value in the measured and the predicted disparity maps is not correlated with its neighbors, then the map updating can be done at each pixel independently. In this case, the covariance matrices R_t and P_t^- of table 1 are diagonal, so the matrix equations of the update phase decompose into separate scalar equations for each pixel. We will describe the procedure for this case first, then consider the consequences of correlation.

To update a pixel value, we first compute the variance of the updated disparity estimate

$$p_t^+ = [(p_t^-)^{-1} + (\sigma_d^2)^{-1}]^{-1} = \frac{p_t^- \sigma_d^2}{p_t^- + \sigma_d^2}$$

and the Kalman filter gain K

$$K = \frac{p_t^+}{\sigma_d^2} = \frac{p_t^-}{p_t^- + \sigma_m^2}$$

We then update the disparity value by using the Kalman filter update equation

$$u_t^+ = u_t^- + K(d - u_t^-)$$

where u_t^- and u_t^+ are the predicted and updated disparity estimates and d is the new disparity measurement. This update equation can also be written as

$$u_t^+ = p_t^+ \left(\frac{u_t^-}{p_t^-} + \frac{d}{\sigma_d^2} \right)$$

This shows that the updated disparity estimate is a linear combination of the predicted and measured values, inversely weighted by their respective variances.

As noted in the previous section, the depth measurements d are actually correlated over both space and time. This induces correlations in the updated depth estimates u_t^+ and implies that the measurement covariance matrix R_t and the updated state covariance matrix P_t^+ will not be diagonal in a complete stochastic model for this problem. We currently do not model these correlations because of the large expense involved in computing and storing the entire covariance matrices. Finding more concise models of the correlation is a subject for future research.

4.3. Smoothing the Map

The raw depth or disparity values obtained from optical flow measurements can be very noisy, especially in areas of uniform intensity. We employ smoothness constraints to reduce the noise and to "fill in" underconstrained areas. The earliest example of this approach is that of Horn and Schunck [12]. They smoothed the optical flow field (u, v) by jointly minimizing the error in the flow equation

$$\mathcal{E}_b = E_x u + E_y v + E_t$$

(E is image intensity) and the departure from smoothness

$$\mathcal{E}_c^2 = |\nabla u|^2 + |\nabla v|^2$$

The smoothed flow was that which minimized the total error

$$\mathcal{E}^2 = \iint (\mathcal{E}_b^2 + \alpha^2 \mathcal{E}_c^2)\, dx\, dy$$

where α is a blending constant. More recently, this approach has been formalized using the theory of regularization [31] and extended to use two-dimensional confidence measures equivalent to local covariance estimates [1, 22].

For our application, smoothing is done on the disparity field, using the inverse variance of the disparity estimate as the confidence in each measurement. The smoother we use is the generalized piecewise continuous spline under tension [32], which uses finite element relaxation to compute the smoothed field. The algorithm is implemented with a three-level coarse-to-fine strategy to speed convergence and is amenable to implementation on a parallel computer.

Surface smoothness assumptions are violated where discontinuities exist in the true depth function, in particular at object boundaries. To reduce blurring of the depth map across such boundaries, we incorporate a discontinuity detection procedure in the smoother. After several iterations of smoothing have been performed, depth discontinuities are detected by thresholding the angle between the view vector and the local surface normal (appendix B) and doing nonmaximum suppression. This is superior to applying edge detection directly to the disparity image, because it properly takes into account the 3D geometry and perspective projection. Once discontinuities have been detected, they are incorporated into the piecewise continuous smoothing algorithm and a few more smoothing iterations are performed. Our approach to discontinuity detection, which interleaves smoothing and boundary detection, is similar to Terzopoulos' continuation method [32]. The alternative of trying to estimate the boundaries in conjunction with the smoothing [14] has not been tried, but could be implemented within our framework. An interesting issue we have not explored is the propagation of detected discontinuities between frames.

The smoothing stage can be viewed as the part of the Kalman filtering algorithm that incorporates prior knowledge about the smoothness of the disparity map. As shown in [29], a regularization-based smoother is equivalent to a prior model with a correlation function defined by the degree of the stabilizing spline (e.g., membrane or thin plate). In terms of table 1, this means that the prior covariance matrix P_0 is nondiagonal. The resulting posterior covariance matrix of the disparity map contains off-diagonal elements modeling the covariance of neighboring pixels. Note that this reflects the surface smoothness model and is distinct from the measurement-induced correlation discussed in the previous section. An optimal implementation of the Kalman filter would require transforming the prior model covariance during the prediction stage and would significantly complicate the algorithm. Our choice to explicitly model only the variance at each pixel, with covariance information implicitly modeled in a fixed regularization stage, has worked well in practice.

4.4. Predicting the Next Disparity Map

The extrapolation stage of the Kalman filter must predict both the depth and the depth uncertainty for each pixel in the next image. We will describe the disparity extrapolation first, then consider the uncertainty extrapolation.

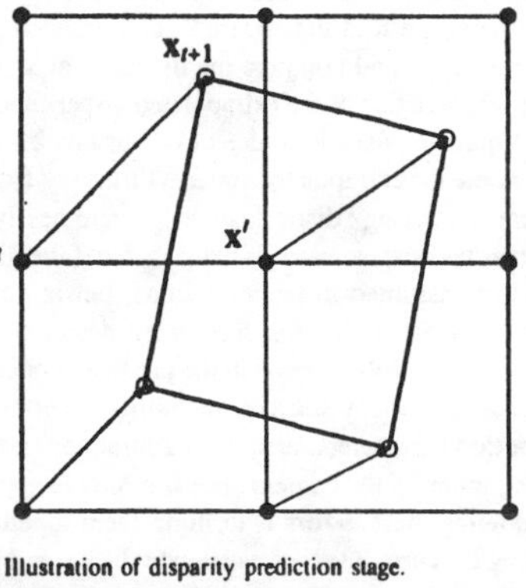

Fig. 7. Illustration of disparity prediction stage.

Our approach is illustrated in figure 7. At time t, the current disparity map and motion estimate are used to predict the optical flow between images t and $t + 1$, which in turn indicates where the pixels in frame t will 'move to' in the next frame:

$$\begin{aligned} x_{t+1} &= x_t + \Delta x_t \\ y_{t+1} &= y_t + \Delta y_t \end{aligned}$$

The flow estimates are computed with equation (2), assuming that Z, **T**, and **R** are known.[3] Next we predict what the new depth of this point will be using the equations of motion. From (2) we have

$$\begin{aligned} \Delta Z_t &= -T_z - R_x Y_t + R_y X_t \\ &= -T_z - R_x y_t Z_t + R_y x_t Z_t \end{aligned}$$

so that the predicted depth at x_{t+1}, y_{t+1} is

$$\begin{aligned} Z_{t+1} &= Z_t + \Delta Z_t \\ &= (1 - R_x y_t + R_y x_t) Z_t - T_z \\ &= \alpha Z_t - T_z \end{aligned}$$

An estimate of the inverse depth can be obtained by inverting this equation, yielding

$$u_{t+1}^- = \frac{u_t^+}{\alpha - T_z u_t^+} \qquad [14]$$

[3]There will be uncertainty in x_{t+1} and y_{t+1} due to uncertainty in the motion and disparity estimates. We ignore this for now.

This equation is nonlinear in the state variable, so it deviates from the form of linear system model illustrated in table 1. Nonlinear models are discussed in a number of references, such as [19].

In general, this prediction process will yield estimates of disparity between pixels in the new image (figure 7), so we need to resample to obtain predicted disparity at pixel locations. For a given pixel $\mathbf{x}'$ in the new image, we find the square of extrapolated pixels that overlap $\mathbf{x}'$ and compute the disparity at $\mathbf{x}'$ by bilinear interpolation of the extrapolated disparities. Note that it may be possible to detect occlusions by recording where the extrapolated squares turn away from the camera. Detecting "disocclusions," where newly visible areas become exposed, is not possible if the disparity field is assumed to be continuous, but is possible if disparity discontinuities have been detected.

Uncertainty will increase in the prediction phase due to errors from many sources, including uncertainty in the motion parameters, errors in calibration, and inaccurate models of the camera optics. A simple approach to modeling these errors is to lump them together by inflating the current variance estimates by a small multiplicative factor in the prediction stage. Thus, the variance prediction associated with the disparity prediction of equation (14) is

$$p_{t+1}^{-} = (1 + \epsilon)p_t^{+} \quad [15]$$

In the Kalman filtering literature this is known as exponential age-weighting of measurements [19], because it decreases the weight given to previous measurements by an exponential function of time. This is the approach used in our implementation. We first inflate the variance in the current disparity map using equation (15), then warp and interpolate the variance map in the same way as the disparity map. A more exact approach is to attempt to model the individual sources of error and to propagate their effects through the prediction equations. Appendix C examines this for uncertain camera motion.

5 Feature-Based Depth Estimation

The dense, iconic depth-estimation algorithm described in the previous section can be compared with existing depth-estimation methods based on sparse feature tracking. Such methods [2, 5, 10, 16] typically define the state vector to be the parameters of the 3D object being tracked, which is usually a point or straight-line segment. The 3D motion of the object between frames defines the system model of the filter and the perspective projection of the object onto each image defines the measurement model. This implies that the measurement equations (the perspective projection) are nonlinear functions of the state variables (e.g., the 3D position vector); this requires linearization in the update equations and implies that the error distribution of the 3D coordinates will *not* be Gaussian. In the case of arbitrary camera motion, a further complication is that it is difficult to reliably track features between frames. In this section, we will describe in detail an approach to feature-based Kalman filtering for lateral camera translation that tracks edgels along each scan line and avoids nonlinear measurement equations. The restriction to lateral motion simplifies the comparison of the iconic and feature-based algorithms performed in the following section; it also has valuable practical applications in the context of manipulator-mounted cameras and in bootstrapping binocular stereo correspondence. Extensions to arbitrary motion can be based on the method presented here.

5.1. Kalman Filter Formulation for Lateral Motion

Lateral camera translation considerably simplifies the feature tracking problem, since in this case features flow along scan lines. Moreover, the position of a feature on a scan line is a linear function of the distance moved by the camera, since

$$\Delta x = -T_x d \Leftrightarrow x_t = x_0 - tT_x d$$

where x_0 is the position of the feature in the first frame and d is the inverse depth of the feature. The *epipolar-plane-image* method [4] exploits these characteristics by extracting lines in "space-time" (epipolar plane) images formed by concatenating scan lines from an entire image sequence. However, sequential estimation techniques like Kalman filtering are a more practical approach to this problem because they allow images to be processing on-line by incrementally refining the depth model.

Taking x_0 and d as the state variables defining the location of the feature, instead of the 3D coordinates X and Z, keeps the entire estimation problem linear. This is advantageous because it avoids the approximations needed for error estimation with nonlinear equations. For point features, if the position of the feature in each image is given by the sequence of measurements $\tilde{\mathbf{x}} = [\tilde{x}_0, \tilde{x}_1, \ldots, \tilde{x}_n]^T$, knowledge of the camera position for each image allows the feature location to be determined by fitting a line to the measurement vector $\tilde{\mathbf{x}}$:

$$\tilde{\mathbf{x}} = \mathbf{H} \begin{bmatrix} x_0 \\ d \end{bmatrix} \qquad [16]$$

where $\mathbf{H}$ is a $(n + 1) \times 2$ matrix whose first column contains all 1's and whose second column is defined by the camera position for each frame, relative to the initial camera position. This fit can be computed sequentially by accumulating the terms of the normal equation solution for x_0 and d. The covariance matrix Σ of x_0 and d can be determined from the covariance matrix of the measurement vector $\tilde{\mathbf{x}}$.

The approach outlined above uses the position of the feature in the first frame x_0 as one of the two state variables. We can reformulate this in terms of the current frame by taking x_t and d to be the state variables. Assuming that the camera motion is exact and that measured feature positions have normally distributed uncertainty with variance σ_e^2, the initial state vector and covariance matrix are expressed in terms of ideal image coordinates as

$$x_1 = \tilde{x}_1$$

$$d = \frac{\tilde{x}_0 - \tilde{x}_1}{T_1}$$

$$P_0^+ = \sigma_e^2 \begin{bmatrix} 1 & -1/T_1 \\ -1/T_1 & 2/T_1^2 \end{bmatrix}$$

where T_1 is the camera translation between the first and second frame. The covariance matrix comes from applying standard linear error propagation methods to the equations for x_1 and d [19].

After initialization, if T_t is the translation between frames $t - 1$ and t, the motion equations that transform the state vector and covariance matrix to the current frame are

$$u_t^- = \begin{bmatrix} x_t^- \\ d_t^- \end{bmatrix} = \begin{bmatrix} 1 & -T_t \\ 0 & 1 \end{bmatrix} \begin{bmatrix} x_{t-1}^+ \\ d_{t-1}^+ \end{bmatrix} = \Phi_t u_{t-1}^+ \qquad [17]$$

$$P_t^- = \Phi_t P_{t-1}^+ \Phi_t^T \qquad [18]$$

The superscript minuses indicate that these estimates do not incorporate the measured edge position at time t. The newly measured edge position $\tilde{x}_t$ is incorporated by computing the updated covariance matrix P_t^+, a gain matrix K, and the updated parameter vector u_t^+:

$$P_t^+ = \{(P_t^-)^{-1} + S\}^{-1}$$

$$\text{where} \quad S = \frac{1}{\sigma_e^2} \begin{bmatrix} 0 & 0 \\ 0 & 1 \end{bmatrix}$$

$$K = \frac{1}{\sigma_t^2} P_t^+ \begin{bmatrix} 0 \\ 1 \end{bmatrix}$$

$$u_t^+ = u_t^- + K [\tilde{x}_t - x_t^-]$$

Since these equations are linear, we can see how uncertainty decreases as the number of measurements increases by computing the sequence of covariance matrices P_t, given only the measurement uncertainty σ_e^2 and the sequence of camera motions T_t. This is addressed in section 6.1

Note that the equations above can be generalized to arbitrary, uncertain camera motion using either the x, y, d image-based parameterization of point locations or an X, Y, Z three-dimensional parameterization. The choice of parameterization may affect the conditioning of general depth-from-motion algorithms, but we have addressed this question to date.

5.2. *Feature Extraction and Matching*

To implement the feature-based depth estimator, we must specify how to extract feature positions, how to estimate the noise level in these positions, and how to track features from frame to frame. For lateral motion, with image flow parallel to the scanlines, tracking edgels on each scanline is a natural implementation. Therefore, in this section we will describe how we extract edges to subpixel precision, how we estimate the variance of the edge positions, and how we track edges from frame to frame.

For one-dimensional signals, estimating the variance of edge positions has been addressed in [7]. We will review this analysis before considering the general case. In one dimension, edge extraction amounts to finding the zero crossings in the second derivative of the Gaussian-smoothed signal, which is equivalent to finding zero-crossings after convolving the image with a second derivative of Gaussian operator,

$$F(x) = \frac{d^2G(x)}{dx^2} * I(x)$$

We assume that the image I is corrupted by white noise with variance σ_n^2. Splitting the response of the operator into that due to the signal F_s, and that due to noise, F_n, edges are marked where

$$F_s(x) + F_n(x) = 0 \qquad [19]$$

An expression for the edge variance is obtained by taking a first-order Taylor expansion of the deterministic part of the response in the vicinity of the zero crossing, then taking mean square values. Thus, if the zero crossing occurs at x_0 in the noise free signal and $x_0 + \delta x$ in the noisy signal, we have

$$F(x_0 + \delta x) \approx F_s(x_0) + F_s'(x_0)\,\delta x + F_n(x_0 + \delta x) = 0 \qquad [20]$$

so that

$$\delta x = \frac{-(F_n(x_0 + \delta x) + F_s(x_0))}{F_s'(x_0)} \qquad [21]$$

The presence of a zero crossing implies that $F_s(x_0) = 0$ and the assumption of zero mean noise implies that $E[F_n(x_0)] = 0$. Therefore, the variance of the edge position is

$$E[\delta x^2] = \sigma_e^2 = \frac{\sigma_n^2 E[(F_n(x_o))^2]}{(F_s'(x_0))^2} \qquad [22]$$

In a discrete implementation, $E[(F_n(x_0))^2]$ is the sum of the squares of the coefficients in the convolution mask. $F_s'(x_0)$ is the slope of the zero crossing and is approximated by fitting a local curve to the filtered image. The zero crossing of this curve gives the estimate of the subpixel edge position.

For two-dimensional images, an analogous edge operator is a directional derivative filter with a derivative of Gaussian profile in one direction and a Gaussian profile in the orthogonal direction. Assuming that the operator is oriented to take the derivative in the direction of the gradient, the analysis above will give the variance of the edge position in the direction of the gradient (see [23] for an alternate approach). However, for edge tracking along scanlines, we require the variance of the edge position in the scanline direction, not the gradient direction. This is straightforward to compute for the difference of Gaussian *(DOG)* edge operator; the required variance estimate comes directly from equations (19)–(22), replacing F with the *DOG* and F' with the partial derivative $\partial/\partial x$. Details of the discrete implementation in this case are similar to those described above. Experimentally, the cameras and digitizing hardware we use provide 8-bit images with intensity variance $\sigma_n^2 \approx 4$.

It is worth emphasizing that estimating the variance of edge positions is more than a mathematical nicety; it is valuable in practice. The uncertainty in the position of an edge is affected by the contrast of the edge, the amount of noise in the image, and, in matching applications such as this one, by the edge orientation. For example, in tracking edges under lateral motion, edges that are close to horizontal provide much less precise depth estimates than edges that are vertical. Estimating variance quantifies these differences in precision. Such quantification is important in predictive tracking, fitting surface models, and applications of depth-from-motion to constraining stereo. These remarks of course apply to image features in general, not just to edges.

Tracking features from frame to frame is very simple if either the camera motion is very small or the feature depth is already known quite accurately. In the former case, a search window is defined that limits the feature displacement to a small number of pixels from the position in the previous image. For the experiments described in section 6, tracking was implemented this way, with a window width of two pixels. Alternatively, when the depth of a feature is already known fairly accurately, the position of the feature in a new image can be predicted from equation (17) to be

$$x_t^- = x_{t-1}^+ - T_t d_{t-1}^+$$

the variance of the prediction can be determined from equation (18), and a search window can be defined as a confidence interval estimated from this variance. This allows tight search windows to be defined for existing features even when the camera motion is not small. A simplified version of this procedure is used in our implementation to ensure that candidate edge matches are consistent with the existing depth model. The predefined search window is scanned for possible matches, and these are accepted only if they lie within some distance of the predicted edge location. Additional acceptance criteria require the candidate match to have properties similar to those of the feature in the previous image; for edges, these properties are edge orientation and edge strength (gradient magnitude or zero-crossing slope). Given knowledge of the noise level in the image, this comparison function can be defined probabilistically as well, but we have not pursued this direction.

Finally, if the noise level in the image is unknown it can be estimated from the residuals of the observations after x and d have been determined. Such methods are discussed in [21] for batch-oriented techniques analogous to equation (16) and in [18] for Kalman filtering.

6 Evaluation

In this section, we compare the performance of the iconic and feature-based depth estimation algorithms in three ways. First, we perform a mathematical analysis of the reduction in depth variance as a function of time. Second, we use a sequence of images of a flat scene to determine the quantitative performance of the two approaches and to check the validity of our analysis. Third, we test our algorithms on images of realistic scenes with complicated variations in depth.

6.1. *Mathematical Analysis*

We wish to compare the theoretical variance of the depth estimates obtained by the iconic method of section 4 to those obtained by the feature-based method of section 5. We will also compare the accuracy of both methods to the accuracy of stereo matching with the first and last frames of the image sequence. To do this, we will derive expressions for the depth variance as a function of the number of frames processed, assuming a constant noise level in the images and constant camera motion between frames. For clarity, we will assume this motion is $T_x = 1$.

6.1.1. Iconic Approach. For the iconic method, we will ignore process noise in the system model and assume that the variance of successive flow measurements is constant. For lateral motion, the equations developed in section 2 can be simplified to show that the Kalman filter simply computes the average flow [30]. Therefore, a sequence of flow measurements $\Delta x_1, \Delta x_2, \ldots, \Delta x_t$ is equivalent to the following batch measurement equation

$$\Delta \mathbf{x} = \begin{bmatrix} \Delta x_1 \\ \Delta x_2 \\ \vdots \\ \Delta x_t \end{bmatrix} = \begin{bmatrix} 1 \\ 1 \\ \vdots \\ 1 \end{bmatrix} d = \mathbf{H}d$$

Estimating d by averaging the flow measurements implies that

$$d = \frac{1}{t} H^T \Delta \mathbf{x} = \frac{1}{t} \sum_{i=1}^{t} \Delta x_i \qquad [23]$$

If the flow measurements were independent with variance $2\sigma_n^2/a$, where σ_n^2 is the noise level in the image (appendix A), the resulting variance of the disparity estimate would be

$$\frac{2\sigma_n^2}{ta} \qquad [24]$$

However, the flow measurements are not actually independent. Because noise is present in every image, flow measurements between frames $i - 1$ and i will be correlated with measurements for frames i and $i + 1$. Appendix A shows that a sequence of correlation-based flow measurements that track the same point in the image sequence will have the following covariance matrix:

$$P_m = \frac{\sigma_n^2}{a} \begin{bmatrix} 2 & -1 & & & & \\ -1 & 2 & -1 & & & \\ & -1 & \cdot & & & \\ & & & \cdot & & \\ & & & & \cdot & \\ & & & & 2 & -1 \\ & & & & -1 & 2 \end{bmatrix}$$

where σ_n^2 is the level of noise in the image and a reflects the local slope of the intensity surface. With this covariance matrix, averaging the flow measurements actually yields the following variance for the estimated flow:

$$\sigma_{\bar{x}}^2(t) = \frac{1}{t^2} H^T P_m H = \frac{2\sigma_n^2}{t^2 a} \qquad [25]$$

This is interesting and rather surprising. Comparing equations (24) and (25), the correlation structure that exists in the measurements means that the algorithm converges faster than we first expected.

With correlated measurements, averaging the flow measurements in fact is a suboptimal estimator for d. The optimal estimator is obtained by substituting the expressions for H and P_m into equation (12) and (13). This estimator does not give equal weight to all flow measurements; instead, measurements near the center of the sequence receive more weight than those near the end. The variance of the depth estimate is

$$\sigma_{\bar{x}^*}^2(t) = \frac{12\sigma_n^2}{t(t + 1)(t + 2)a}$$

The optimal convergence is cubic, whereas the convergence of the averaging method we implemented is quadratic. Developing an incremental version of the optimal estimator requires extending our Kalman filter formulation to model the correlated nature of the measurements. This extension is currently being investigated.

6.1.2. Feature-based approach. For the feature-based approach, the desired variance estimates come from computing the sequence of covariance matrices P_t, as mentioned at the end of section 5.1. A closed form expression for this matrix is easier to obtain from the batch method suggested by equation (16) than from the Kalman filter formulation and yields an equivalent result. Taking the constant camera translation to be $T_x = 1$ for simplicity, equation (16) expands to

$$\tilde{\mathbf{x}} = \begin{bmatrix} \tilde{x}_0 \\ \tilde{x}_1 \\ \cdot \\ \cdot \\ \cdot \\ \tilde{x}_t \end{bmatrix} = \begin{bmatrix} 1 & 0 \\ 1 & -1 \\ \cdot & \cdot \\ \cdot & \cdot \\ \cdot & \cdot \\ 1 & -t \end{bmatrix} \begin{bmatrix} x_0 \\ d \end{bmatrix} = H\mathbf{u} \quad [26]$$

Recall that $\tilde{x}_i$ are the edge positions in each frame, x_0 is the best fit edge position in the first frame, and d is the best fit displacement or flow between frames. Since we assume that the measured edge positions $\tilde{x}_i$ are independent with equal variance σ_e^2, we find that

$$P_F = \begin{bmatrix} \sigma_x^2 & \sigma_{xd} \\ \sigma_{xd} & \sigma_d^2 \end{bmatrix} = \sigma_e^2 \begin{bmatrix} \sum_{i=0}^{t} 1 & -\sum_{i=0}^{t} i \\ -\sum_{i=0}^{t} i & \sum_{i=0}^{t} i^2 \end{bmatrix}^{-1} \quad [27]$$

The summations can be expressed in closed form, leading to the conclusion that

$$\sigma_F^2(t) = \frac{12\sigma_e^2}{t(t+1)(t+2)} \quad [28]$$

The variance of the displacement or flow estimate d thus decreases as the cube of the number of images. This expression is identical in structure to the optimal estimate for the iconic approach, the only difference being the replacement of the variable of the SSD minimum by the variance of the edge position. Thus, if our estimators incorporate appropriate models of measurement noise, the iconic and feature-based methods theoretically achieve the same rate of convergence. This is surprising, given that the basic Kalman filter for the iconic method maintains only one state parameter *(d)* for each pixel, whereas the feature-based method maintains two per feature *(x_0 and d)*. We suspect that an incremental version of the optimal iconic estimator will require the same amount of state as the feature-based method.

6.1.3. Comparison with Stereo. To compare these methods to stereo matching on the first and last frames of the image sequence, we must scale the stereo disparity and its uncertainty to be commensurate with the flow between frames. This implies dividing the stereo disparity by t and the uncertainty by t^2. For the iconic method, we assume that the uncertainty in a stereo measurement will be the same as that for an individual flow measurement. Thus, the scaled uncertainty is

$$\sigma_{IS}^2(t) = \frac{2\sigma_n^2}{t^2 a}$$

This is the same as is achieved with our incremental algorithm which processes all of the intermediate frames. Therefore, processing the intermediate frames (while ignoring the temporal correlation of the measurements) may improve the reliability of the matching but in this case it does not improve precision.

For the feature-based approach, the uncertainty in stereo disparity is twice the uncertainty σ_e^2 in the feature position; the scaled uncertainty is therefore

$$\sigma_{FS}^2(t) = \frac{2\sigma_e^2}{t^2}$$

In this case using the intermediate frames helps, since

$$\frac{\sigma_F(t)}{\sigma_{FS}(t)} = \frac{1}{O(\sqrt{t})}$$

Thus, extracting depth from a small-motion image sequence has several advantages over stereo matching between the first and last frames. The ease of matching is increased, reducing the number of correspondence errors. Occlusion is less of a problem, since it can be predicted from early measurements. Finally, better accuracy is available by using the feature-based method or the optimal version of the iconic method.

6.2. *Quantitative Experiments: Flat Images*

The goals of our quantitative evaluation were to examine the actual convergence rates of the depth estimators, to assess the validity of the noise models, and to compare the performance of the iconic and feature-based algorithms. To obtain ground truth depth data, we used the facilities of the Calibrated Imaging Laboratory at CMU to digitize a sequence of images of a flat-mounted poster. We used a Sony XC-37 CCD camera with a 16-mm lens, which gave a field of view of 36 degrees. The poster was set about 20 inches (51 cm) from the camera. The camera motion between frames was 0.04

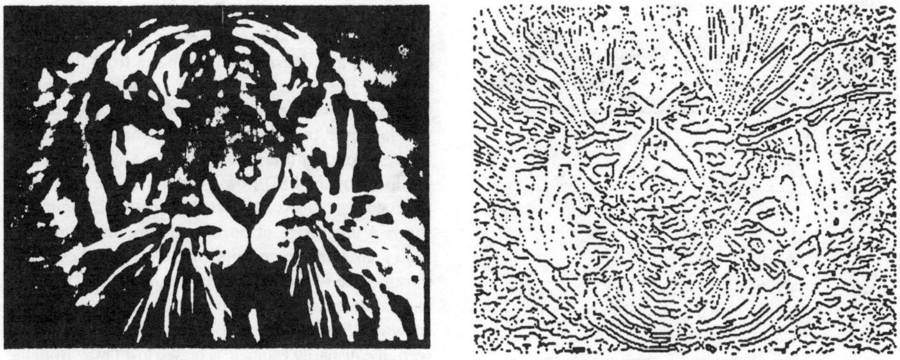

Fig. 8. Tiger image and edges.

inches (1 mm), which gave an actual flow of approximately two pixels per frame in 480×512 images. For convenience, our experiments were run on images reduced to 240×256 by Gaussian convolution and subsampling. The image sequence we will discuss here was taken with vertical camera motion. This proved to give somewhat better results than horizontal motion; we attribute this to jitter in the scanline clock, which induces more noise in horizontal flow than in vertical flow.

Figure 8 shows the poster and the edges extracted from it. For both the iconic and the feature-based algorithms, a ground truth value for the depth was determined by fitting a plane to the measured values. The level of measurement noise was then estimated by computing the RMS deviation of the measurements from the plane fit. Optical aberrations made the flow measurements consistently smaller near the periphery of the image than the center, so the RMS calculation was performed over only the center quarter of the image. Note that all experiments described in this section did *not* use regularization to smooth the depth estimates, so the results show only the effect of the Kalman filtering algorithm.

To determine the reliability of the flow variance estimates, we grouped flow measurements produced by the SSD algorithm according to their estimated variances, took sample variances over each group, and plotted the SSD variance estimates against the sample variances (figure 9). The strong linear relationship indicates fairly reliable variance estimates. The deviation of the slope of the line from the ideal value of one is due to an inaccurate estimate of the image noise (σ_n^2).

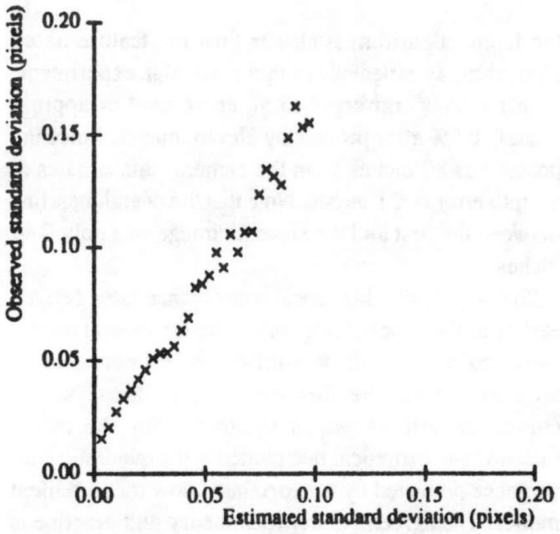

Fig. 9. Scatter plot.

To examine the convergence of the Kalman filter, the RMS depth error was computed for the iconic and the feature-based algorithms after processing each image in the sequence. We computed two sets of statistics, one for "sparse" depth and one for "dense" depth. The sparse statistic computes the RMS error for only those pixels where both algorithms gave depth estimates (that is, where edges were found), whereas the dense statistic computes the RMS error of the iconic algorithm over the full image. Figure 10 plots the relative RMS errors as a function of the number of images processed. Comparing the sparse error curves, the convergence rate of

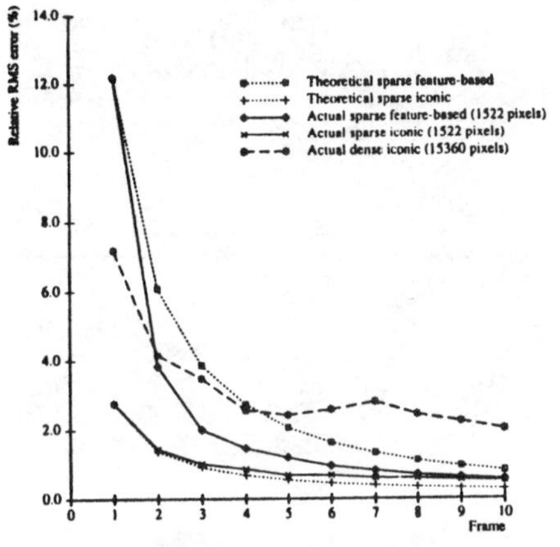

Fig. 10. RMS error in depth estimate.

the iconic algorithm is slower than the feature-based algorithm, as expected. In this particular experiment, both methods converged to an error level of approximately 0.5% after processing eleven images. Since the poster was 20 inches from the camera, this equates to a depth error of 0.1 inches. Note that the overall baseline between the first and the eleventh image was only 0.44 inches.

To compare the theoretical convergence rates derived earlier to the experimental rates, the theoretical curves were scaled to coincide with the experimental error after processing the first two frames. These scaled curves are also shown in figure 10. For the iconic method, the theoretical rate plotted is the quadratic convergence predicted by the correlated flow measurement model. The agreement between theory and practice is quite good for the first three frames. Thereafter, the experimental RMS error decreases more slowly; this is probably due to the effects of unmodeled sources of noise. For the feature-based method, the experimental error initially decreases faster than predicted because the implementation required new edge matches to be consistent with the prior depth estimate. When this requirement was dropped, the results agreed very closely with the expected convergence rate.

Note that the comparison between theoretical and experimental results also allows us to estimate the precision of the subpixel edge extractor. The variance of a disparity estimate is twice the variance of the edge positions. Since the frame-to-frame displacement in this image sequence was one pixel and the relative RMS error was 12% for the first disparity estimate, the RMS error in edge localization was $0.12/\sqrt{2} \approx 0.09$ pixels.

Finally, Figure 10 also compares the RMS error for the sparse and dense depth estimates from the iconic method. The dense flow field is considerably noisier than the flow estimates that coincide with edges, though still just over two percent error by the end of eleven frames. Some of this error is due to a systematic bias produced by the SSD flow estimator in the vicinity of ramp edges. This is illustrated in figure 11. Figure 11a shows a test image of horizontal bars and figure 11b shows an intensity profile of a vertical slice taken through one of the light-to-dark transitions. Disparity and variance estimates computed along this profile are shown in figures 11c and 11d, respectively. As can be seen, the disparity estimate is biased low (away from the "true" value in the central flat part) on one side of the discontinuity, and biased "high" on the other. This bias can also be confirmed by using an analytic model of a ramp edge. Fortunately, the variance estimates reflect this large error, so regularization-based smoothing can compensate for this systematic error. We conclude that the dense depth estimates do provide fairly good depth information.

6.3. *Qualitative Experiments: Real Scenes*

We have also tested the iconic and edge-based algorithms on complicated, realistic scenes obtained from the Calibrated Imaging Laboratory. Two sequences of ten images were taken with camera motion of 0.05 inches (1.27 mm) between frames; one sequence moved the camera vertically, the other horizontally. The overall range of motion was therefore 0.5 inches (1.27 cm); this compares with distances to objects in the scene of 20 to 40 inches (51 to 102 cm).

Figure 12 shows one of the images. Figures 13a–d show a reduced version of the image, the edges extracted from it with an oriented Canny operator [7], and depth maps produced by applying the iconic algorithm to the horizontal and vertical image sequences, respectively. Lighter areas in the depth maps are nearer. The main structure of the scene is recovered quite well in both cases, though the results with the horizontal sequence are considerably more noisy. This is most likely due to scanline jitter, as mentioned earlier. Edges oriented parallel to the direction of flow cause some scene structure to be observable in one sequence but not the other. This is most noticeable near the center of the scene, where a thin vertical object appears in

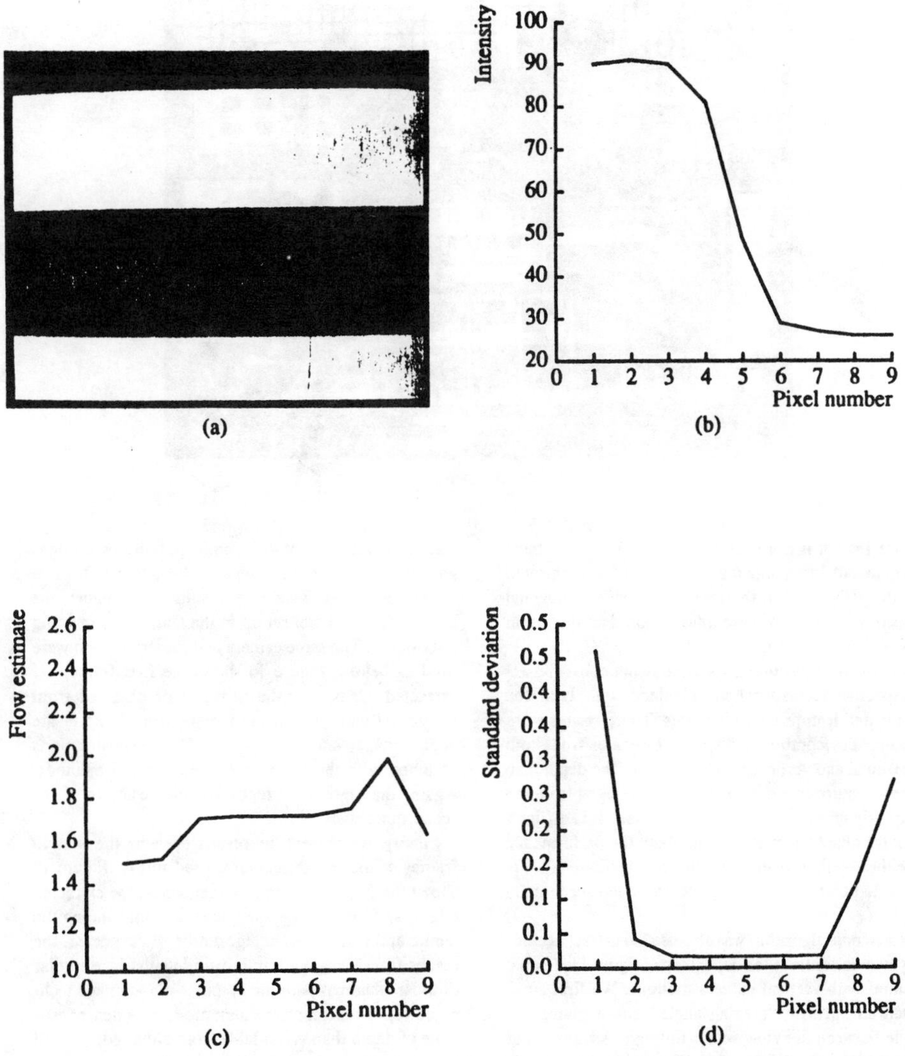

Fig. 11. Bias of subpixel correlations: (a) test image, (b) intensity profile acorss a light-to-dark transition, (c) estimated flow along the intensity profile, (d) estimated variance along the profile.

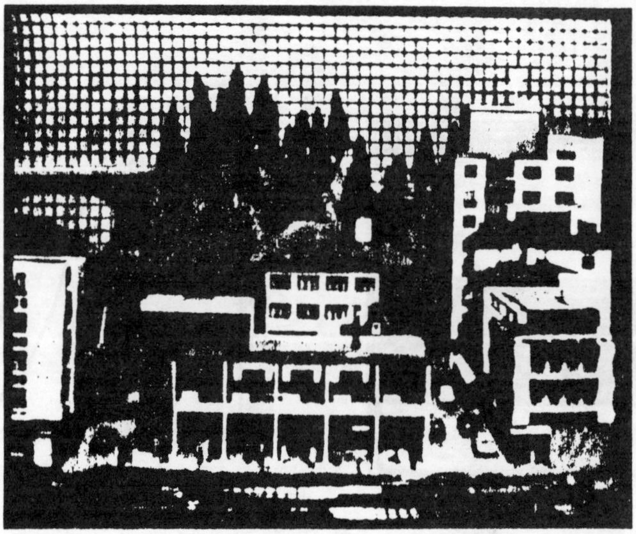

Fig. 12. CIL image.

figure 13c but is not visible in figure 13d. This object corresponds to an antenna on the top of a foreground building (figure 13a). In general, motion in orthogonal directions will yield more information than motion in any single direction.

Figure 14 shows intensity-coded depth maps and 3D perspective reconstructions obtained with both the iconic and feature-based methods. These results were produced by combining disparity estimates from both horizontal and vertical camera motion. The depth map for the feature-based approach was produced from the sparse depth estimates by regularization. It is difficult to make quantitative statements about the performance of either method from this data, but qualitatively it is clear that both recover the structure of the scene quite well.

The iconic algorithm was also used to extract occluding boundaries from the depth map of figure 13c (iconic method with vertical camera motion). We first computed an intrinsic "grazing angle" image giving the angle between the view vector through each pixel and the normal vector of the local 3D surface. Edge detection and thresholding were applied to this image to find pixels where the view vector and the surface normal were nearly perpendicular. The resulting boundaries are shown along with the depth map in figure 15. The method found most of the prominent building outlines and the outline of the bridge in the upper left.

Figures 16 and 17 show the results of our algorithms on a different model set up in the Calibrated Imaging Laboratory. The same camera and camera motion were used as before. Figure 16 shows the first frame, the extracted edges, and the depth maps obtained from horizontal and vertical motion. Figure 17 shows the depth maps and the perspective reconstructions obtained with the iconic and feature-based methods. Again, the algorithms recovered the structure of the scene quite well.

Finally, we present the results of using the first 10 frames of the image sequence used in [4]. Figure 18 shows the first frame from the sequence, the extracted edges, and the depth maps obtained from running the iconic and feature-based algorithms. As expected, the results from using the feature-based method are similar to those obtained with the epipolar-plane image technique [4]. The iconic algorithm produces a denser estimate of depth than is available from either edge-based technique. These results show that the sparse (edge-based) batch-processing algorithm for small motion sequences introduced in [4] can be extended to use dense depth maps and incremental processing.

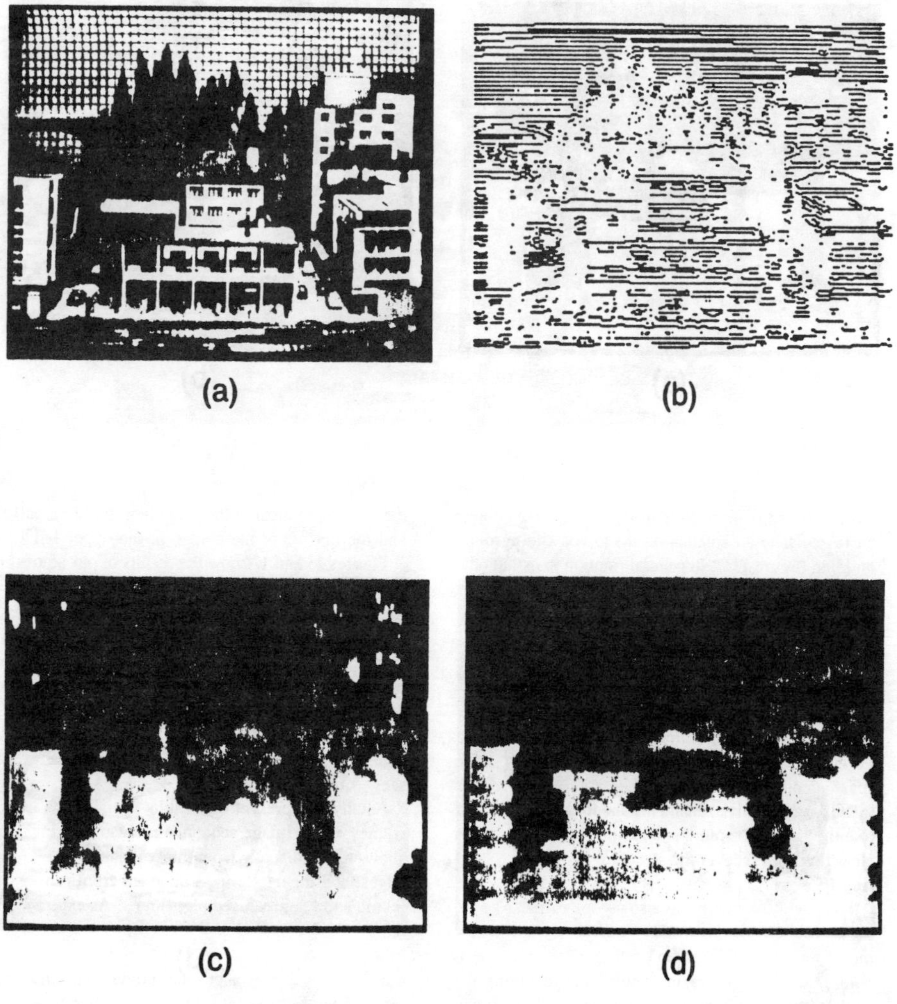

Fig. 13. CIL depth maps: (a) first frame, (b) edges, (c) horizontal-motion depth map, (d) vertical-motion depth map.

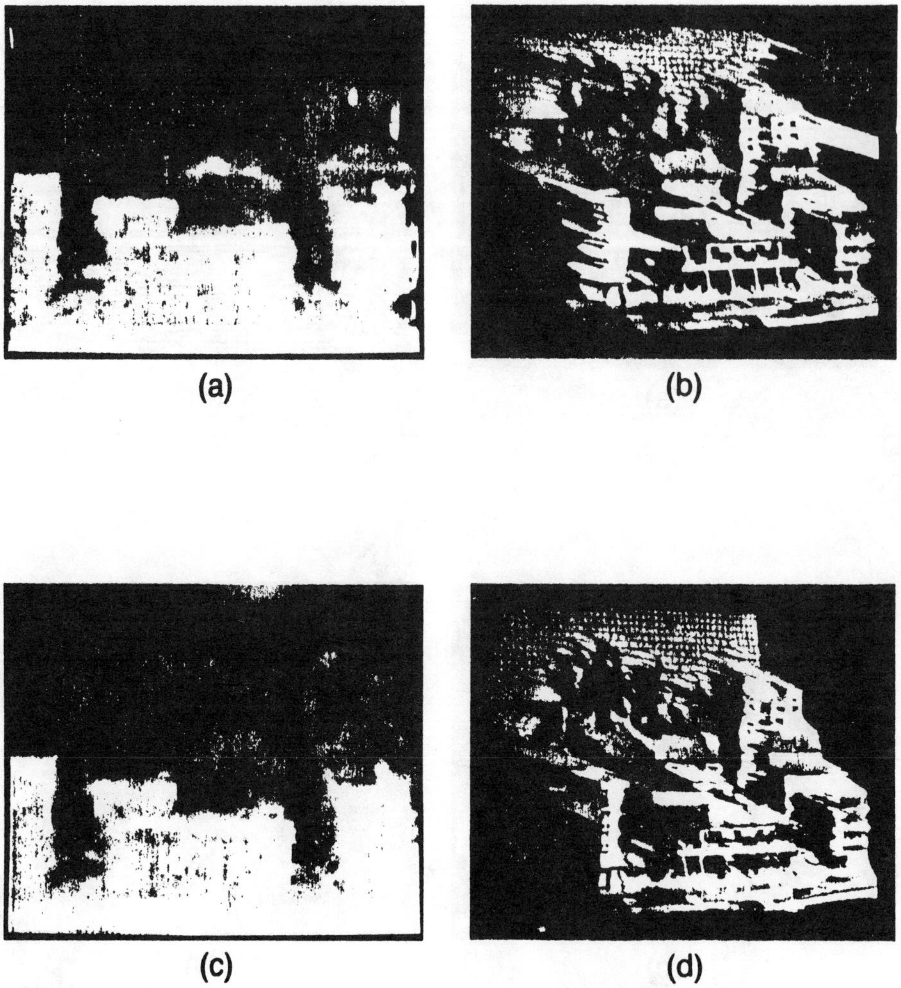

Fig. 14. CIL orthogonal motion results: (a) iconic method depth map, (b) perspective view, (c) feature-based method depth map, (d) perspective view.

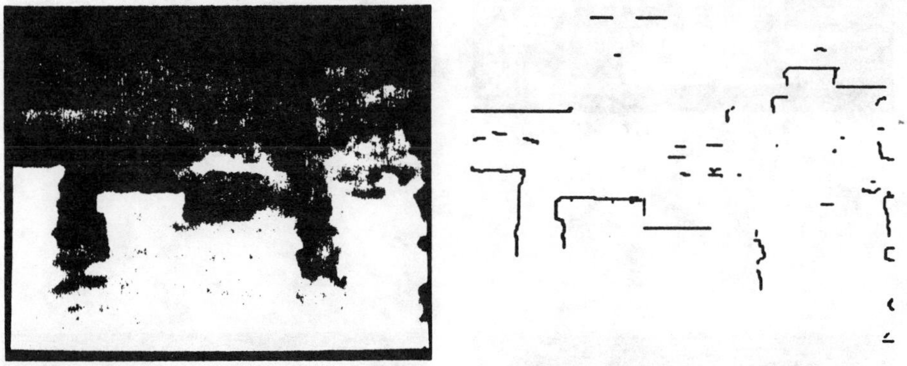

(a) (b)

Fig. 15. Occluding boundaries: (a) vertical-motion depth map, (b) occluding boundaries.

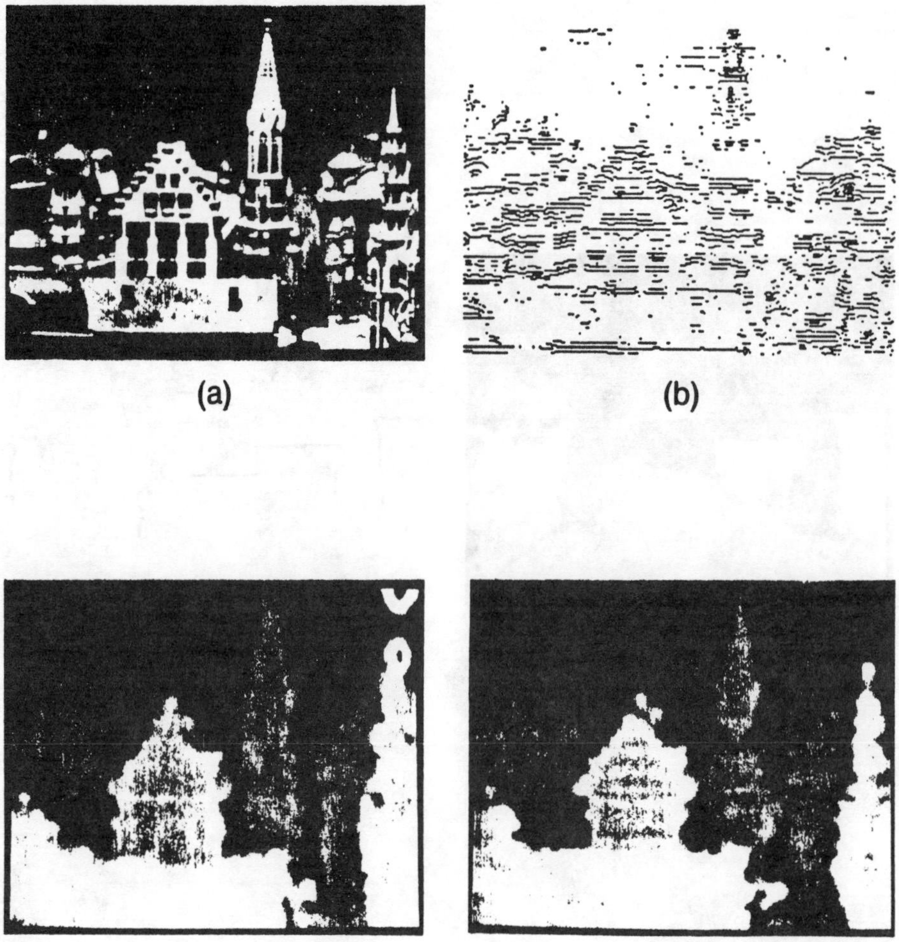

(a) (b)

(c) (d)

Fig. 16 CIL-2depth maps: (a) first frame, (b) edges, (c) horizontal-motion depth map, (d) vertical-motion depth map.

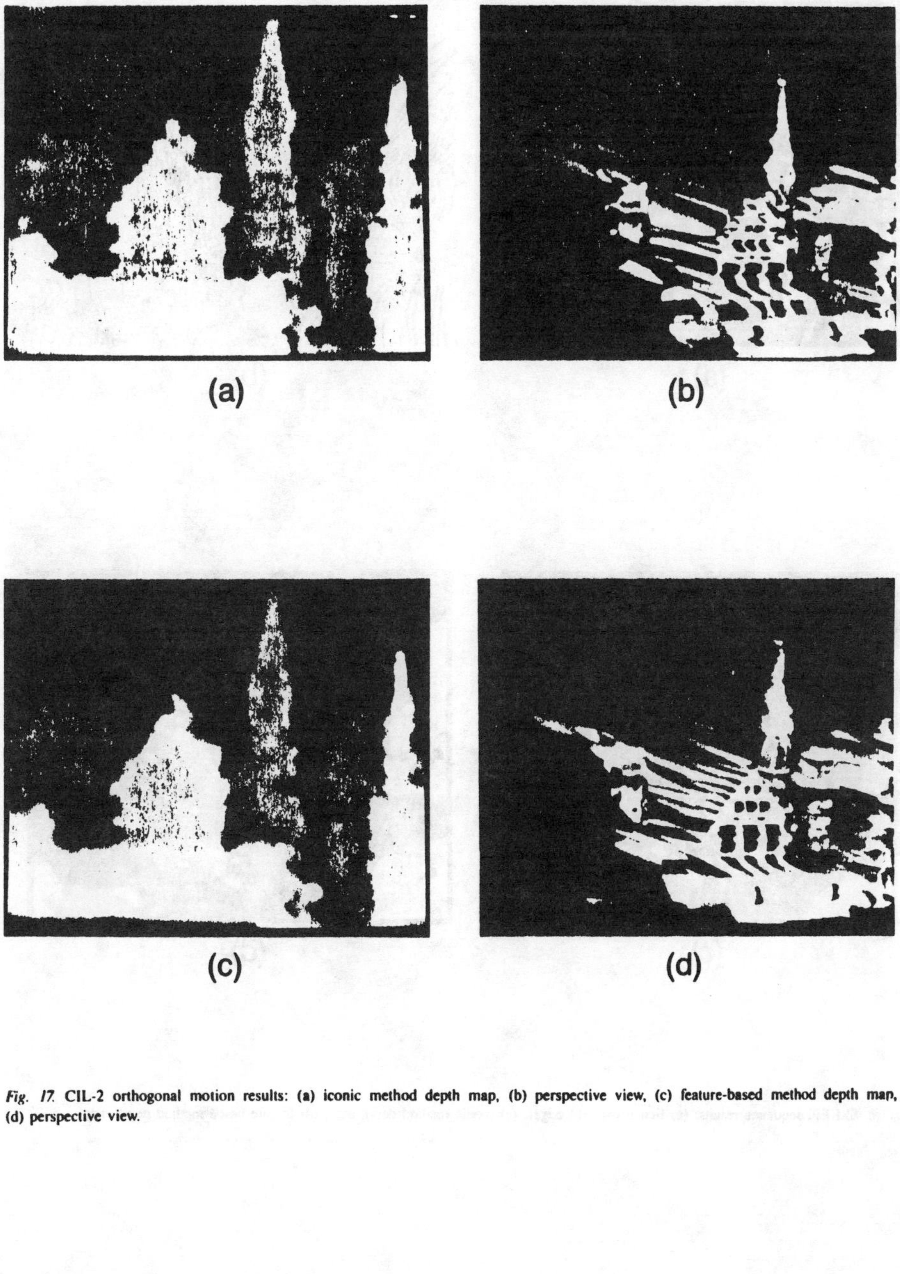

Fig. 17. CIL-2 orthogonal motion results: (a) iconic method depth map, (b) perspective view, (c) feature-based method depth map, (d) perspective view.

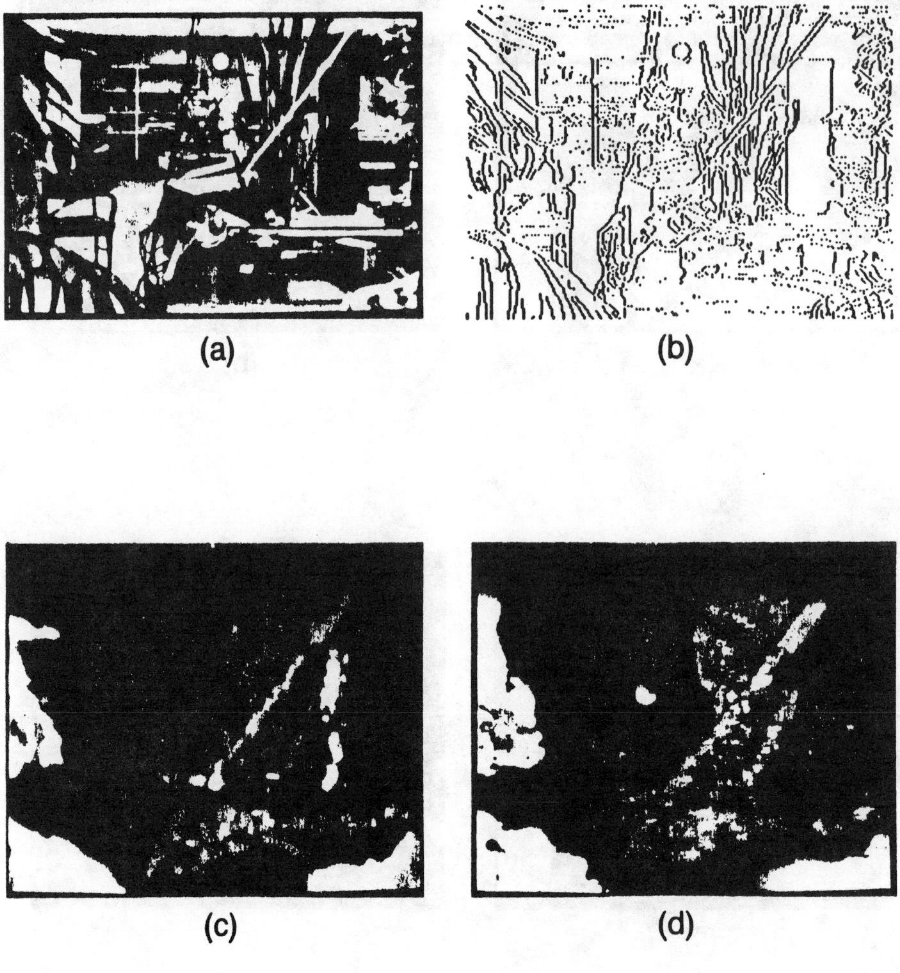

Fig. 18. SRI EPI sequence results: (a) first frame, (b) edges, (c) iconic method depth map, (d) feature-based method depth map.

7 Conclusions

This paper has presented a new algorithm for extracting depth from known motion. The algorithm processes an image sequence taken with small interframe displacements and produces an on-line estimate of depth that is refined over time. The algorithm produces a dense, iconic depth map and is suitable for implementation on parallel architectures.

The on-line depth estimator is based on Kalman filtering. A correlation-based flow algorithm measures both the local displacement at each pixel and the confidence (or variance) of the displacement. These two "measurement images" are integrated with predicted depth and variance maps using a weighted least-squares technique derived from the Kalman filter. Regularization-based smoothing is used to fill in areas of unknown disparity and to reduce the noise in the flow estimates. The current maps are extrapolated to the next frame by image warping, using knowledge of the camera motion, and are resampled to keep the maps iconic.

The algorithm has been implemented for lateral camera translation, evaluated mathematically and experimentally, and compared with a feature-based algorithm that uses Kalman filtering to estimate the depth of edges. The mathematical analysis shows that the iconic approach will have a slower convergence rate because it only keeps one element of state per pixel (the disparity), while the feature-based approach keeps both the disparity and the subpixel position of the feature. However, an optimal implementation of the iconic method (which takes into account temporal correlations in the measurements) has the potential to equal the convergence rate and accuracy of the symbolic method. Experiments with images of a flat poster have confirmed this analysis and given quantitative measures of the performance of both algorithms. Finally, experiments with images of a realistic outdoor-scene model have shown that the new algorithm performs well on images with large variations in depth and that occluding boundaries can be extracted from the resulting depth maps.

The lateral motion implementations developed in this paper have several potential robotics applications. For example, manipulators with cameras mounted near the end-effector can use small translations to acquire shape information about a workpiece. In addition, binocular stereo systems can take great advantage of such motions by using depth from motion to constrain binocular correspondence. The degrees of freedom necessary for this are already available in manipulator-mounted systems and can be designed into vehicle-mounted stereo systems.

Extensions

The algorithms described in this paper can be extended in several ways. The most straightforward extension is to the case of nonlateral motion. As sketched in section 4, this can be accomplished by designing a correlation-based flow estimator that produces two-dimensional flow vectors and an associated covariance matrix estimate [1]. This approach can also be used when the camera motion is uncertain, or when the camera motion is variable (e.g., for widening baseline stereo [34]). The alternative of searching only along epipolar lines during the correlation phase may be easier to implement, but is less general.

More research is required into the behavior of the correlation based flow and confidence estimator. In particular, we have observed that our current estimator produces biased estimates in the vicinity of intensity step edges. The correlation between spatially adjacent flow estimates, which is currently ignored, should be integrated into the Kalman filter framework. More sophisticated representations for the intensity and depth fields are also being investigated [28].

Finally, as noted above, the incremental depth from motion algorithms can be used to initiate stereo fusion. Work is currently in progress investigating the integration of depth-from-motion and stereo [15]. We believe that the framework presented in this paper will prove to be useful for integrating information from multiple visual sources and for tracking such information in a dynamic environment.

Acknowledgement

This research was sponsored in part by DARPA, monitored by the Air Force Avionics Lab under contract F33615-87-C-1499 and in part by a postgraduate fellowship from the FMC Corporation. Data for this research was partially provided by the Calibrated Imaging Laboratory at CMU. The views and conclusions contained in this document are those of the authors and should not be interpreted as representing the official policies, either expressed or implied, of the funding agencies.

Appendix A: Optic Flow Computation

In this appendix, we will analyze the performance of a simple correlation-based flow estimator, the sum-of-squared-differences (SSD) estimator [1]. This estimator selects at each pixel the disparity that minimizes the SSD measure

$$e(\bar{d};x) = \int w(\lambda)[f_1(x + \bar{d} + \lambda) - f_0(x + \lambda)]^2 \, d\lambda$$

where $f_0(x)$ and $f_1(x)$ are the two successive image frames, and $w(\lambda)$ is a symmetric, non-negative weighting function. To analyze its performance, we will assume that the two image frames are generated from an underlying true intensity image, $f(x)$, to which uncorrelated (white) Gaussian noise with variance σ_n^2 has been added:

$$f_0(x) = f(x) + n_0(x),$$
$$f_1(x) = f(x - d) + n_1(x)$$

Using this model, we can rewrite the error measure as*

$$e(\bar{d};x) = \int w(\lambda)[f(x + \bar{d} - d + \lambda) - f(x + \lambda) + n_1(x + \lambda) - n_0(x + \lambda)]^2 \, d\lambda$$

If $\bar{d} \approx d$, we can use a Taylor series expansion to obtain

$$\begin{aligned} e(\bar{d};x) &= \int w(\lambda)[f'(x + \lambda)](\bar{d} - d)^2 \\ &\quad + 2w(\lambda)f'(x + \lambda) \\ &\quad \times [n_1(x + \lambda) - n_0(x + \lambda)] \\ &\quad \times (\bar{d} - d) + w(\lambda) \\ &\quad \times [n_1(x + \lambda) - n_0(x + \lambda)]^2 \, d\lambda \\ &= a(x)(\bar{d} - d)^2 + 2[b_1(x) - b_0(x)] \\ &\quad \times (\bar{d} - d) + c(x) \end{aligned}$$

where

$$a(x) = \int w(\lambda)[f'(x + \lambda)]^2 \, d\lambda$$

$$b_i(x) = \int w(\lambda)f'(x + \lambda)n_i(x + \lambda) \, d\lambda$$

$$c(x) = \int w(\lambda)[n_1(x + \lambda) - n_0(x + \lambda)]^2 \, d\lambda$$

*This equation is actually incorrect, since it should contain $n_1(x + \bar{d} - d + \lambda)$ instead of $n_1(x + \lambda)$. The effect of including the correct term is to add small random terms involving integrals of $w(\lambda)$, $w'(\lambda)$, $f'(x + \lambda)$, $f''(x + \lambda)$, and $n_1(x)$ to the quadratic coefficients $a(x)$, $b_1(x)$, and $c(x)$ that are derived below. This intentional omission has been made to simplify the presentation.

The four coefficients $a(x)$, $b_0(x)$, $b_1(x)$, and $c(x)$ define the shape of the error surface $e(\bar{d};x)$. The first coefficient, $a(x)$, is related to the average "roughness" or "slope" of the intensity surface, and determines the confidence given to the disparity estimate (see below). The second and third coefficients, $b_0(x)$ and $b_1(x)$, are independent, zero-mean Gaussian random variables that determine the difference between $\bar{d}$ and d, i.e., the error in flow estimator. The fourth coefficient, $c(x)$, is a chi-squared-distributed random variable with mean $(2\sigma_n^2 \int w(\lambda) \, d\lambda)$, and defines the computed error at $\bar{d} = d$.

To estimate the disparity at point x given the error surface $e(\bar{d};x)$, we find the $\hat{d}$ such that

$$e(\hat{d};x) = \min_{\bar{d}}(\bar{d};x)$$

From the above quadratic equation,‡ we can compute $\hat{d}(x)$ as

$$\hat{d}(x) = d + \frac{b_0(x) - b_1(x)}{a(x)}$$

To calculate the variance in this estimate, we must first calculate the variance in $b_i(x)$,

$$\text{var}\,(b_i(x)) = \sigma_n^2 \int w^2(\lambda)[f'(x + \lambda)]^2 \, d\lambda$$

If we set $w(x) = 1$ on some finite interval, and zero elsewhere, this variance reduces to $\sigma_n^2 a(x)$, and we obtain

$$\text{var}\,(\hat{d}) = \frac{2\sigma_n^2}{a(x)}$$

In addition to calculating the disparity-estimate variance, we can compute its covariance with other estimates either in the same frame or in a subsequent frame. As described in section 6.1, knowing the correlation between adjacent or successive measurements is important in obtaining good overall uncertainty estimates.

To determine the correlation between two adjacent disparity estimates, $\hat{d}(x)$ and $\hat{d}(x + \Delta x)$, we must first determine the correlation between $b_i(x)$ and $b_i(x + \Delta x)$,

$$\begin{aligned} &\langle b_i(x)b_i(x + \Delta x)\rangle \\ &= \iint w(\lambda)w(\eta)f'(x + \lambda)f'(x + \Delta x + \eta) \\ &\quad \times \langle n_i(x + \lambda)n_i(x + \Delta x + \eta)\rangle \, d\lambda \, d\eta \end{aligned}$$

‡The true equation (when higher order Taylor series terms are included) is a polynomial series in $(\bar{d}-d)$ with random coefficients of decreasing variance. This explains the "rough" nature of the $e(\bar{d};x)$ observed in practice.

$$= \iint w(\lambda)w(\eta)f'(x + \lambda)f'(x + \Delta x + \eta) \times \delta(\lambda - \Delta x - \eta)\sigma_n^2 \, d\lambda \, d\eta$$

$$= \sigma_n^2 \int w(\lambda)w(\lambda - \Delta x)[f'(x + \lambda)]^2 \, d\lambda$$

For a slowly varying gradient $f'(x)$, this correlation is proportional to the autocorrelation of the weighting function,

$$R_w(\Delta x) = \int w(\lambda)w(\lambda + \Delta x) \, d\lambda$$

For the simple case of $w(\lambda) = 1$ on $[-s, s]$, we obtain

$$R_d(x, x + \Delta x) = \frac{2\sigma_n^2}{a(x)} \left(1 - \frac{|x|}{2s}\right) \quad \text{for } |x| \leq 2s$$

The correlation between two successive measurements in time is easier to compute. Since

$$f_2(x + 2d) = f(x) + n_2(x)$$

we can show that the flow estimate obtained from the second pair of frames is

$$\hat{d}_2(x) = d + \frac{b_2(x) - b_1(x)}{a(x)}$$

The covariance between $\hat{d}_1(x)$ and $\hat{d}_2(x)$ is

$$\text{cov}(\hat{d}_1(x), \hat{d}_2(x)) = \langle(\hat{d}_1(x) - d)(\hat{d}_2(x) - d)\rangle = -\frac{\sigma_n^2}{a(x)}$$

and the covariance matrix of the sequence of measurements $\hat{d}_i$ is

$$P_m = \frac{\sigma^2}{a} \begin{bmatrix} 2 & -1 & & & & \\ -1 & 2 & -1 & & & \\ & -1 & \cdot & & & \\ & & & \cdot & & \\ & & & & \cdot & \\ & & & & 2 & -1 \\ & & & & -1 & 2 \end{bmatrix}$$

This structure is used in section 6.1 to estimate the theoretical accuracy and convergence rate of the iconic depth from motion algorithm.

Appendix B: Three-Dimensional Discontinuity Detection

To calculate a discontinuity in the depth map, we compute the angle between the local normal **N** and the view vector **V**. The surface normal at pixel value (r, c) is computed by using the 3D locations of the three points:

$$\mathbf{P}_0 = (X_0, Y_0, Z_0) = (x_0, y_0, 1)\frac{T_x}{d_0}$$

where

$$x_0 = \frac{c - c_x}{s_x}, y_0 = -\frac{r - c_y}{s_y}$$

$$\mathbf{P}_1 = (X_1, Y_1, Z_1) = (x_1, y_1, 1)\frac{T_x}{d_1}$$

where

$$x_1 = \frac{c + 1 - c_x}{s_x} = x_0 + \frac{1}{s_x},$$

$$y_1 = -\frac{r - c_y}{s_y} = y_0$$

$$\mathbf{P}_2 = (X_2, Y_2, Z_2) = (x_2, y_2, 1)\frac{T_x}{d_2}$$

where

$$x_2 = \frac{c - c_x}{s_x} = x_0,$$

$$y_2 = -\frac{r + 1 - c_y}{s_y} = y_0 - \frac{1}{s_y}$$

We can obtain the normal from the cross product of the two vectors

$$\mathbf{Q}_1 = \mathbf{P}_1 - \mathbf{P}_0 = T_x\left[\frac{1}{s_x d_1} + x_0\left(\frac{1}{d_1} - \frac{1}{d_0}\right),\ y_0\left(\frac{1}{d_1} - \frac{1}{d_0}\right),\ \left(\frac{1}{d_1} - \frac{1}{d_0}\right)\right]$$

$$= \frac{T_x}{d_0 d_1}\left[\frac{d_0}{s_x} - x_0(d_1 - d_0),\ -y_0(d_1 - d_0),\ -(d_1 - d_0)\right]$$

$$\mathbf{Q}_2 = \mathbf{P}_2 - \mathbf{P}_0 = T_x\left[x_0\left(\frac{1}{d_2} - \frac{1}{d_0}\right),\ -\frac{1}{s_y d_2} + y_0\left(\frac{1}{d_2} - \frac{1}{d_0}\right),\ \left(\frac{1}{d_2} - \frac{1}{d_0}\right)\right]$$

$$= \frac{T_x}{d_0 d_2}\left[-x_0(d_2 - d_0),\ -\frac{d_0}{s_y} - y_0(d_2 - d_0),\ -(d_2 - d_0)\right]$$

$$\mathbf{Q}_1 \times \mathbf{Q}_2 \propto \left[-\frac{d_0(d_1 - d_0)}{s_y},\ \frac{d_0(d_2 - d_0)}{s_x},\ -\frac{d_0^2}{s_x s_y} + \frac{x_0 d_0(d_1 - d_0)}{s_y} - \frac{y_0 d_0(d_2 - d_0)}{s_x}\right]$$

Simplifying we obtain

$$\mathbf{N} = (-s_x\Delta_1, s_y\Delta_2, -d_0 + x_0 s_x \Delta_1 - y_0 s_y \Delta_2)$$

$$\mathbf{V} = (x_0, y_0, 1)$$

$$\mathbf{N}\cdot\mathbf{V} = -d_0$$

$$\cos\theta = \frac{\mathbf{N}\cdot\mathbf{V}}{|\mathbf{N}||\mathbf{V}|}$$

where $\Delta_1 = d_0(d_1 - d_0)$ and $\Delta_2 = d_0(d_2 - d_0)$. To implement the edge detector, we require that

$$\cos\theta < \cos\theta_t$$

or

$$[s_x^2\Delta_1^2 + s_y^2\Delta_2^2 + (-d_0 + x_0 s_x \Delta_1 - y_0 s_y \Delta_2)^2] \times (x_0^2 + y_0^2 + 1) > d_0^2 \sec^2\theta_t$$

If the field of view of the camera is small, we have near orthographic projection, and the above equations simplify to

$$\mathbf{N} = \left[-\frac{s_x\Delta_1}{d_0}, \frac{s_y\Delta_2}{d_0}, -1\right] = (p, q, -1)$$

$$\mathbf{V} = (x_0, y_0, 1)$$

and this reduces to the familiar gradient-based threshold

$$p^2 + q^2 > \tan^2\theta_t$$

Appendix C: Prediction Equations

To predict the new disparity map and variance map from the current maps, we will first map each pixel to its new location and value, and then use interpolation to resample the map. For simplicity, the development given here shows only the one-dimensional case, i.e., disparity d as a function of x. The extension to two dimensions is straightforward.

The motion equations for a point in the pixel map (x, d) are

$$x' = x + t_x d + r_x$$

$$d' = d + t_z$$

We will assume that the points which define the patch under consideration have the same t_x, r_x, and t_z values. These three parameters are actually stochastic variables, due to the uncertainty in camera motion. For the lateral-motion case, we assume that the mean of t_x is known and nonzero, while the means of r_x and t_z are zero.

We can write the vector equations for the motion of the points in a patch as

$$\mathbf{x}' = \mathbf{x} + t_x\mathbf{d} + r_x\mathbf{e}$$

$$\mathbf{d}' = \mathbf{d} + t_z\mathbf{e}$$

where

$$\mathbf{x} \sim N(\hat{\mathbf{x}}, \Sigma_x),\quad t_x \sim N(\hat{t}_x, \sigma_{t_x}^2),$$

$$r_x \sim N(0, \sigma_{r_x}^2)$$

$$\mathbf{d} \sim N(\hat{\mathbf{d}}, \Sigma_d),\quad r_z \sim N(0, \sigma_{t_z}^2),$$

$$\mathbf{e} = [1\ldots1]^T$$

The Jacobian of this vector equation is

$$\frac{\partial(\mathbf{x}', \mathbf{d}')}{\partial(\mathbf{x}, \mathbf{d}, t_x, r_x, t_z)} = \begin{bmatrix} \mathbf{I} & t_x\mathbf{I} & \mathbf{d} & \mathbf{e} & 0 \\ 0 & \mathbf{I} & 0 & 0 & \mathbf{e} \end{bmatrix}^T$$

and the variance of the predicted points is

$$\mathrm{var}\,(\mathbf{x}', \mathbf{d}')$$

$$= \begin{bmatrix} \Sigma_x + t_x^2\Sigma_d + \mathbf{d}\mathbf{d}^T\sigma_{t_x}^2 + \mathbf{e}\mathbf{e}^T\sigma_{r_x}^2 & t_x\Sigma_d \\ t_x\Sigma_d & \Sigma_d + \mathbf{e}\mathbf{e}^T\sigma_{t_z}^2 \end{bmatrix}$$

To obtain the new depth and variance at a point x, we must define an interpolation function for the patch surrounding this point. For a linear interpolant, the equation is

$$d = d_i \frac{(x_{i+1} - x)}{(x_{i+1} - x_i)} + d_{i+1} \frac{(x - x_i)}{(x_{i+1} - x_i)}$$

$$= (1 - \lambda) d_i + \lambda d_{i+1}, \quad \text{where } \lambda = \frac{(x - x_i)}{(x_{i+1} - x_i)}$$

$$\frac{\partial d}{\partial d_i} = \frac{(x_{i+1} - x)}{(x_{i+1} - x_i)} = (1 - \lambda)$$

$$\frac{\partial d}{\partial x_i} = \frac{-(d_{i+1} - d_i)(x_{i+1} - x_i)}{(x_{i+1} - x_i)^2}$$

$$= -m(1 - \lambda) \quad \text{where } m = \frac{(d_{i+1} - d_i)}{(x_{i+1} - x_i)}$$

and the associated Jacobian is

$$\frac{\partial(d)}{\partial(x_i, x_{i+1}, d_i, d_{i+1})} = \begin{bmatrix} -m(1 - \lambda) & -m\lambda & (1 - \lambda) & \lambda \end{bmatrix}$$

The variance of the new depth estimate is thus

$$\begin{aligned} \text{var}(d) = {} & m^2[(1 - \lambda)^2 \sigma^2_{x_i} + \lambda^2 \sigma^2_{x_{i+1}}] \\ & + (1 - t_x m)^2[(1 - \lambda)^2 \sigma^2_{d_i} + \lambda^2 \sigma^2_{d_{i+1}}] \\ & + m^2[d^2 \sigma^2_{t_x} + \sigma^2_{t_x}] + \sigma^2_{t_z} \end{aligned}$$

Each of the above four terms can be analyzed separately. The first term in the above equation, which involves $\sigma^2_{x_i}$, depends on the positional uncertainty of the points in the old map. It can either be ignored (if each disparity element represents the disparity at its *center*), or $\sigma^2_{x_i}$ can be set to 1/2. The second term is a blend of the variances at the two endpoints of the interpolated interval. Note that for $\lambda = 1/2$, the variance is actually reduced by half (the average of two uncertain measurements is more certain). It may be desirable to use a pure blend $[(1 - \lambda)\sigma^2_{d_i} + \lambda\sigma^2_{d_{i+1}}]$ to eliminate this bias. The second term also encodes the interaction between the disparity uncertainty and the disparity gradient m. The third term encodes the interaction between the disparity gradient and the camera translation and pan uncertainty. The final term is the uncertainty in camera forward motion, which should in practice be negligible.

References

1. P. Anandan, "Computing dense displacement fields with confidence measures in scenes containing occlusion," *Proc. DARPA Image Understanding Workshop*, pp. 236–246, 1984.
2. N. Ayache and O.D. Faugeras, "Maintaining representations of the environment of a mobile robot," *Proc. 4th Intern. Symp. Robotics Res.* 1987.
3. H.H. Baker, "Multiple-image computer vision," *Proc. 41st Photogrammetric Week*, Stuttgart, West Germany, pp. 7-19, 1987.
4. R.C. Bolles, H.H. Baker, and D.H. Marimont, "Epipolar-plane image analysis: An approach to determining structure from motion," *Intern. J. Computer Vision* 1:7–55, 1987.
5. T.J. Broida and R. Chellappa, "Kinematics and structure of a rigid object from a sequence of noisy images," *Proc. Workshop on Motion: Representation and Analysis*, pp. 95–100, 1986.
6. A.R. Bruss and B.K.P. Horn, "Passive navigation," *Comput. Vision, Graphics, and Image Process.* 21:3–20, 1983.
7. J. Canny, "A computational approach to edge detection,," *IEEE Trans. PAMI* 8:679–698, 1986.
8. J.R. Wertz (ed.), *Spacecraft Attitude Determination and Control*. D. Reidel: Dordrecht, 1978.
9. O.D. Faugeras, N. Ayache, B. Faverjon, and F. Lustman, "Building visual maps by combining noisy stereo measurements," *Proc. IEEE Intern. Conf. Robotics and Automation*, San Francisco, California pp. 1433–1438; 1986.
10. J. Hallam, "Resolving observer motion by object tracking," *Proc. 8th Intern. Joint Conf. Artif. Intelli.* Karlsruhe, 1983.
11. D.J. Heeger, "Optical flow from spatiotemporal filters," *Proc. 1st Intern. Conf. Computer Vision*, London, pp. 181–190, 1987.
12. B.K.P. Horn and B.G. Schunck, "Determining optical flow," *Artificial Intelligence*, 17:185–203, 1981.
13. H.C. Longuet-Higgins and K. Prazdny, "The interpretation of a moving retinal image," *Proc. Roy. Soc. London* B **208**:385–397, 1980.
14. J. Marroquin, S. Mitter, and T. Poggio, "Probabilistic solution of ill-posed problems in computational vision," *J. Am. Stat. Assoc.* **82**:76–89, 1987.
15. L.H. Matthies, "Dynamic stereo." Ph.D. thesis, Carnegie Mellon University, 1989.
16. L.H. Matthies and T. Kanade, "The cycle of uncertainty and constraint in robot perception," *Proc. Intern. Symp. Robotics Research*, 1987.
17. L.H. Matthies and S.A. Shafer, "Error modeling in stereo navigation," *IEEE J. Robotics and Automation*, pp. 239–248, 1987.
18. P.S. Maybeck, *Stochastic Models, Estimation, and Control*, vol. 2. Academic Press: New York, 1982.
19. P.S. Maybeck, *Stochastic Models, Estimation, and Control*, vol. 1. Academic Press, New York, 1979.
20. J.E.W. Mayhew and J.P. Frisby, "Psychophysical and computational studies towards a theory of human stereopsis," *Artificial Intelligence*, 17:349–408, 1981.
21. E.M. Mikhail, *Observations and Least Squares*. University Press of America: Lanham, MD, 1976.
22. H.-H. Nagel and W. Enkelmann, "An investigation of smoothness constraints for the estimation of displacement vector fields from image sequences," *IEEE Trans. PAMI* 8:565–593, 1986.
23. V. Nalwa, "On detecting edges," *IEEE Trans. PAMI* 8:699–714, 1986.

3-D Object Reconstruction Using Stereo and Motion

ENRICO GROSSO, GIULIO SANDINI AND MASSIMO TISTARELLI

Reprinted from *IEEE Transactions on Systems, Man, and Cybernetics*, Vol. 19, No. 6, November/December 1989.
Copyright © 1989 by The Institute of Electrical and Electronics Engineers, Inc. All rights reserved.

Abstract **—The extraction of reliable range data from images is investigated, considering, as a possible solution, the integration of different sensor modalities. Two different algorithms are used to obtain independent estimates of depth from a sequence of stereo images. The results are integrated on the basis of the uncertainty of each measure. The stereo algorithm uses a coarse-to-fine control strategy to compute disparity. An algorithm for depth-from-motion is used exploiting the constraint imposed by active motion of the cameras. To obtain a three-dimensional (3-D) description of the objects, the motion of the cameras is purposively controlled, as to move around the objects in view, while the direction of gaze is kept still toward a fixed point in space. This egomotion strategy, which is similar to that adopted by the human visuomotor system, allows a better exploration of partially occluded objects and simplifies the motion equations. The algorithm has been tested on real scenes, demonstrating a low sensitivity to image noise, mainly due to the integration of independent measures. An experiment, performed on a real scene containing several objects, is presented.**

I. Introduction

ONE OF THE PRIMARY goals of early vision is that of extracting volumetric measures about the observed objects in a scene from a continuous flow of visual information. In humans this task is accomplished using many different sources of information coming from multiple sensor modalities.

So far many methods have been proposed to acquire information about the three-dimensional (3-D) structure of the world, with the aim of building a feasible and handy representation of the environment [1]–[9].

It is our opinion that, at present, all computationally reasonable algorithms for range estimation suffer from errors and uncertainities peculiar to each method. For example, the illumination condition is a weak point in deriving shape from shading [10], [11] and the computation of stereo disparity fails when the matching is performed on edges parallel to the epipolar lines [3], [12]. A possible solution is to select and to integrate, according to a reliability measure, the results obtained from different information sources to obtain a unique representation of the environment [13], [14].

In this paper we present an example of the integration of range data computed from stereo matching and optical flow, ending with a 3-D (volumetric) representation of the solids in view. The experimental setup is based on a pair of cameras, with a coplanar optical axis directed toward a common fixation point, moving around an object and tracking an environmental point [15], [16] (i.e., the movement is performed keeping the fixation point still).

Along with the measure of depth, an explicit estimation of uncertainty is carried out. This uncertainty value, transformed into a *reliability* map and associated with the corresponding *depth* map, is used, along with the instantaneous position in space and the geometry of the stereo pair (i.e., *proprioceptive information*), to update the volumetric representation of the environment continuously [14], [17].

In fact, in spite of the great deal of information, obtained from the different viewpoints, it is clear that from a *bas relief* (i.e., a depth image) only a partial description of the 3-D shape can be derived. To complete and refine this information it is necessary to "move around," exploring the environment actively [18]. This is true not only because occluded objects can become evident from different viewpoints, but also because during the motion depth information can be derived from motion parallax. The tracking strategy, adopted to drive the egomotion, allows the active inspection of the environment and of the objects in the scene from different viewpoints.

In principle, the continuous flow of information, represented as continuously changing depth images derived from stereo measures and motion parallax, needs to be cast into an incremental representation. This casting process acts like an accumulator where only the "new" information changes the current description whereas the redundant (or duplicate) information does not affect it.

In our approach the casting process makes use of a geometric description of the world in terms of a 3-D array of voxels. The visual *bas relief* computed from each viewpoint is used to update this volumetric description.

During the integration process, the reliability map is used to weight the depth measure with respect to the

Manuscript received September 20, 1988; revised March 30, 1989. This work was supported in part by grants from ESPRIT (Project P419) and from the Italian National Council of Research and in part by an ELSAG SpA fellowship.

The authors are with the Department of Communication, Computer and Systems Science, University of Genoa, Via Opera Pia 11a, 1-16145 Genova, Italy.

IEEE Log Number 8930369.

accumulated one. Each voxel of the volumetric accumulator stores, at each instant of time, the measure of probability of "empty space." It is worth noting that this analogic geometric representation of the environment is certainly not sufficient for high-level processing (for example, recognition); on the other hand its primary use is to help the accumulation of depth information. More "complete" geometric description, including also surface information, can be derived from the voxel representation and logically linked to it.

II. Acquisition of Range Data

The outline of the experiment described in this paper is presented in Fig. 1. The images were acquired from four positions in space (east, west, south, and north), moving a stereo pair around a set of objects while tracking a fixed point on the surface of the central object. The distance of the cameras from the fixation point was kept constant during the movement (a circular trajectory) and equal to 103 cm.

From each position in space eleven stereo pairs were acquired. The displacement between successive position is 4° along the circular trajectory, around a vertical axis centered on the fixation point. The total angular span, between the first and last image of each sequence, is, consequently, 40°.

On the eleven stereo pairs, the third was used to compute depth from stereo, while all the remaining images were used in the motion algorithm to compute the depth map, also relative to the third image.

All the images were acquired at 256×256 pixels with 8 bits of resolution in intensity, with two CCD cameras (COHU 4713) and a VDS 7001 Eidobrain image processing system.

A. Disparity Extraction and Edge Matching

The first part of the algorithm is based on the computation of the cross correlation between corresponding square patches of the stereo pair; the images over which the cross correlation is performed are obtained by convolving the originals with a Laplacian of Gaussian operator and representing only the sign of the filtered images [19], [3]. The estimation of correspondence is performed, using a coarse-to-fine approach, in three successive steps. At each step the correlation is computed at a different spatial frequency band (i.e., filtering the images with a $\nabla^2 G$ mask of different size) going from low to high spatial frequencies; as a consequence also the size of the correlation patches decreases at each step (it is directly proportional to the size of the mask). At the end of each step, a measure of disparity is obtained which is successively refined during the following steps.

Finally, the maximum precision in disparity is achieved, performing an explicit edge matching between the zero crossings of the right and left image, extracted at the highest resolution scale. This process is performed using the disparity value computed by the three-step cross-correlation procedure. In particular, starting from an edge point in the left image, the corresponding contour point in the right image is searched in a neighborhood of the disparity computed during the previous phase. The amplitude of the search space is determined by the amplitude of the $\nabla^2 G$ mask used to filter the image at the higher resolution scale.

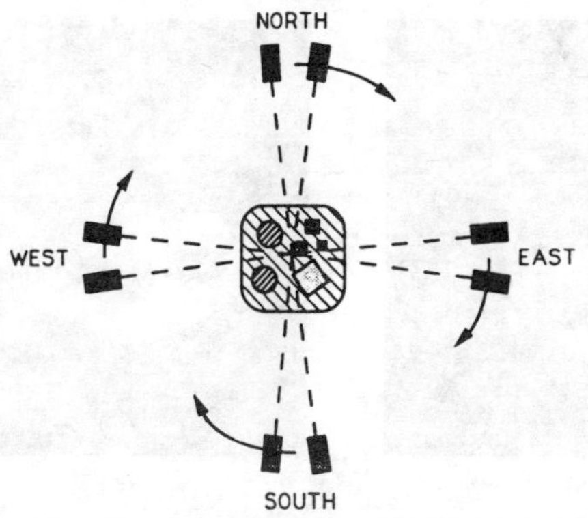

Fig. 1. Schematic representation of experimental setup. Square in middle represents tray with objects on top; initial positions of stereo cameras, with respect to objects, are indicated as north, south, east, and west. Arrows indicate direction of movement of cameras acquiring sequences.

The correlation values computed in a region with uniform shade (i.e., lacking in significant edges) have a low reliability and can produce many errors on the final results. For this reason the correlation is performed only over the regions of higher contrast, which correspond to the image areas whose energy, measured on the left image convolved with a Laplacian or Gaussian operator, is greatest.

Overlap constraints, with a threshold on the minimum energy value, limits the overall number of patches used to perform the correlation. In practice, the patches can be positioned directly on the edges, while the slope of the zero crossings (extracted from the convolved images) is used as the sorting value for the selection of the best image patches.

The correlation measure is weighted using measures of slope and spatial orientation of the gradient computed over the filtered images. As a consequence the disparity, identified by the peak of the correlation function, is also a function of local orientation. The value of correlation is used as a reliability factor [20], [21].

In Fig. 2, the four stereo pairs used in the experiment are presented. In Fig. 3 the result of the correlation process is shown, performed on the first stereo pair filtered with a $\nabla^2 G$ mask with standard deviation σ equal to 8, 4, and 2 pixels; the grey level is proportional to the disparity of the patches. The map presented in Fig. 3(a) represents the final disparity obtained from the regional part of the algorithm.

A planar model is used to determine depth from stereo (see Fig. 4). In this case the depth of a world point with respect to the left camera is a function of six independent

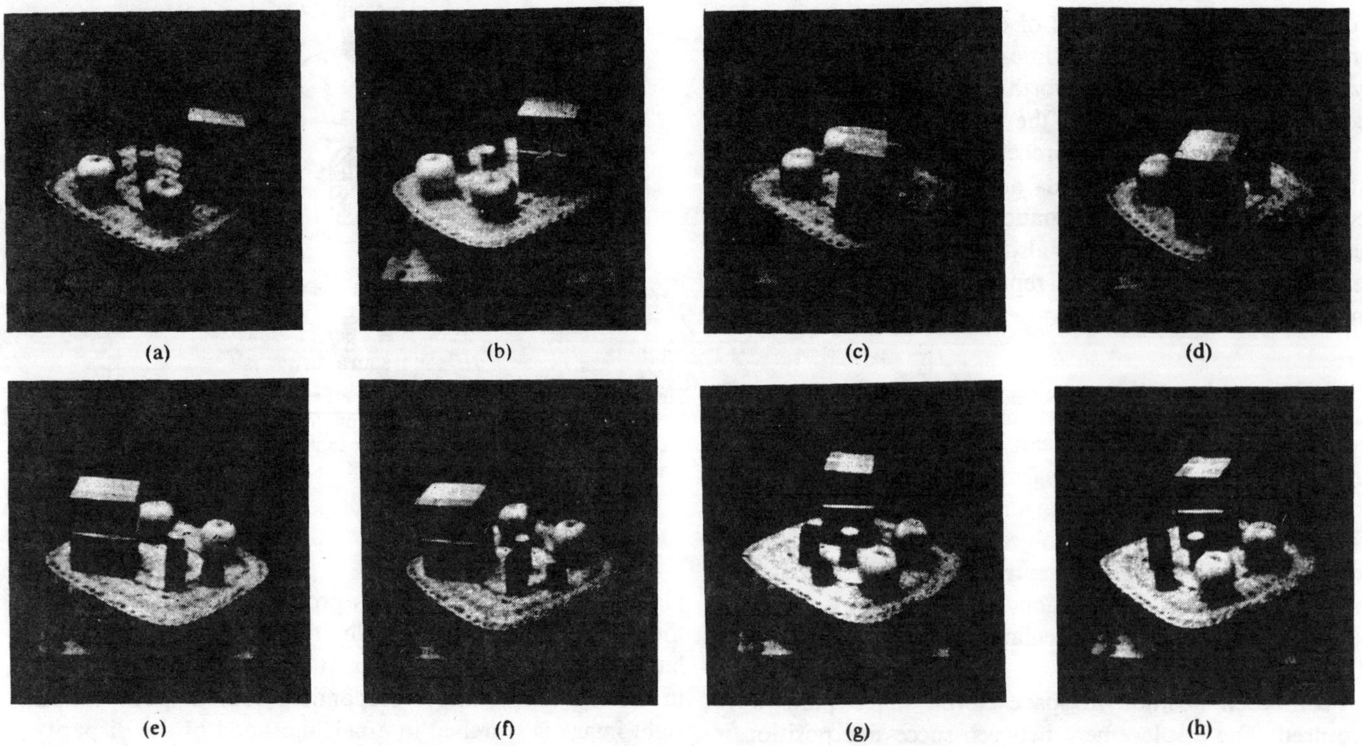

Fig. 2. Stereo pairs used in experiment. Resolution of images is 256×256 pixels.

parameters:

$$Z_s = Z_s(x_0, x_1, \theta, F, l, m) = \sqrt{F^2 + x_0^2}\,\frac{l[x_1 \sin\theta + F\cos\theta] + [F+m][F\sin\theta - x_1\cos\theta] + Fx_1}{[x_1x_0 + F^2]\sin\theta + F[x_0 - x_1]\cos\theta}. \tag{1}$$

The four parameters θ, l, m, F (refer to the scheme depicted in Fig. 4) depend upon a calibration procedure, while x_0 and x_1 are two corresponding points in the left and right image, respectively (i.e., $x_1 = x_0 + d$, where d is image disparity).

The depth map obtained from the first stereo pair is presented in Fig. 3(b) The depth values are computed at the edge points obtained from the convolved image at the highest resolution; the final depth map is computed, for all the image points, with a linear interpolation. Along with depth also the associated uncertainty measure is presented in Fig. 3(c); this measure reflects the reliability of the computed depth (see Section III-A).

B. Estimation of Depth from Motion

The estimation of the optic flow from an image sequence is based on a gradient technique in which proprioceptive knowledge of the egomotion parameters is used to constrain and solve, in closed form, the motion equations. The measurement is performed at the contour points obtained by filtering the images with a $\nabla^2 G$ operator and extracting the zero crossings [22]. The procedure for the computation of the optic flow is divided into the following steps:

- computation of the velocity component perpendicular to the local orientation of the contour; for each contour point the component $V^{\perp}$ is computed as the ratio between the time derivative and the local edge slope;
- computation of the true direction of motion.
- The computation of the direction of motion requires, in the case of general motion of the camera in a steady environment, the knowledge of at least seven variables related to the egomotion (six for displacements and rotations and the focal length). A reduction of the parameters required is achieved constraining the movement of the observer.

In this approach the motion of the camera was constrained as to keep the fixation point still during the motion around the objects in view. As a consequence the egomotion parameters that need to be measured are the distances of the camera from the fixation point (D_1 and D_2) measured at successive time instants, the rotation angles θ, ϕ, and ψ (measured for each sampled frame) and the focal length of the camera F (for explanation of the symbols refer to Fig. 5). From these parameters the direction of the flow due to the translation of the sensor $\vec{V}_t$ and the vector $\vec{V}_r$, due to the rotational part of motion, are computed.

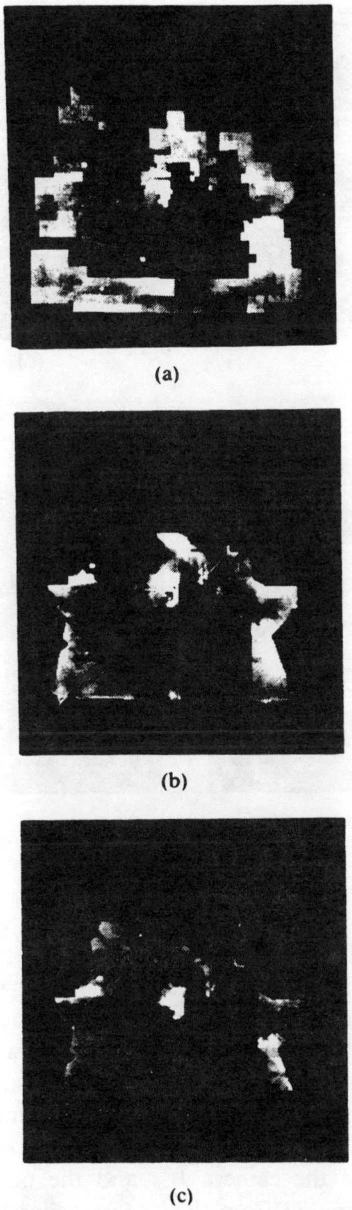

Fig. 3. Results of stereo algorithm relative to single view (topmost pair of Fig. 2). (a) Results of regional correlation step; grey level codes image disparity (b) Depth map obtained by linear interpolation of contour values, obtained after edge matching refinement (depth is proportional to gray level). (c) Associated uncertainty (uncertainty is proportional to gray level).

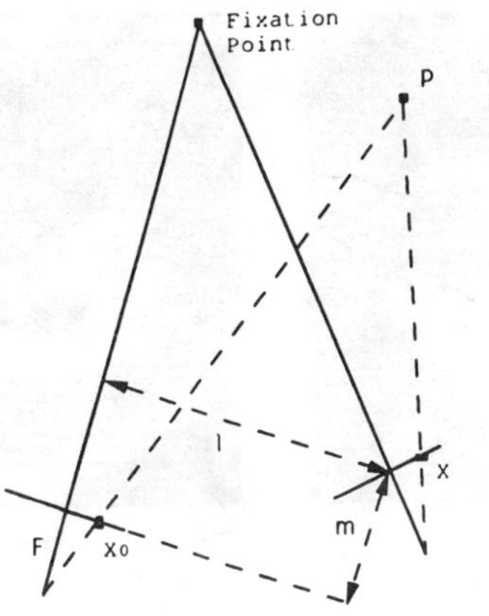

Fig. 4. Geometry of stereo setup.

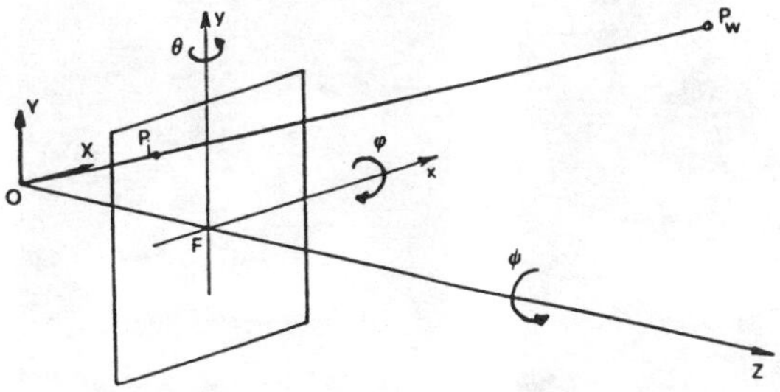

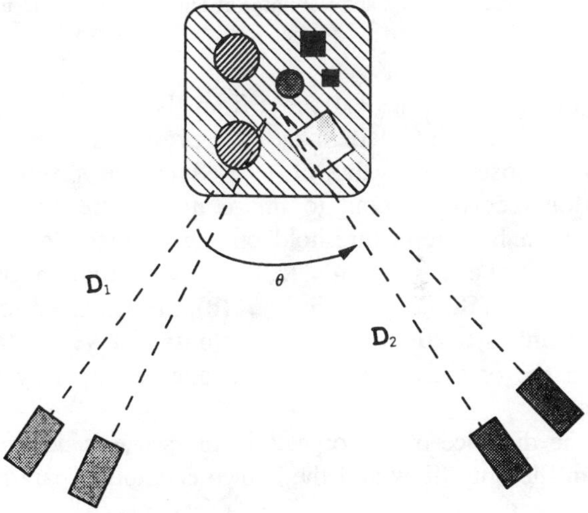

Fig. 5. Schematic representation of camera coordinate system with egomotion parameters used in motion algorithm.

The direction of the flow field is obtained solving, in closed form, a set of non-linear equations, which incorporates the known egomotion parameters and the motion constraints:

- matching of corresponding contours of successive image pairs (instantaneous optic flow).

The optic flow computed at the previous steps is refined searching for the first zero crossing in the successive image along the computed direction. This kind of search is motivated by the fact that, capturing the images with a high sampling frequency and by the smoothing operated by the Gaussian filtering, it is unlikely to find more than one contour between corresponding edges.

A confidence measure is associated with the matched points which reflects the likelihood of the match to be correct. This measure is obtained comparing the edge orientation and slope at corresponding contour points (this topic is further discussed in Section III-A).

To achieve a sufficient range of velocity for distance computation, the instantaneous optic flows, resulting after the edge matching, are joined together, giving *a global optic flow*, relative to a part of the sequence.

The images in Fig. 6 are the first and last left images of one of the stereo sequences acquired for the experiment (refer to Section II. and Fig. 1 for more explanation of the acquisition procedure).

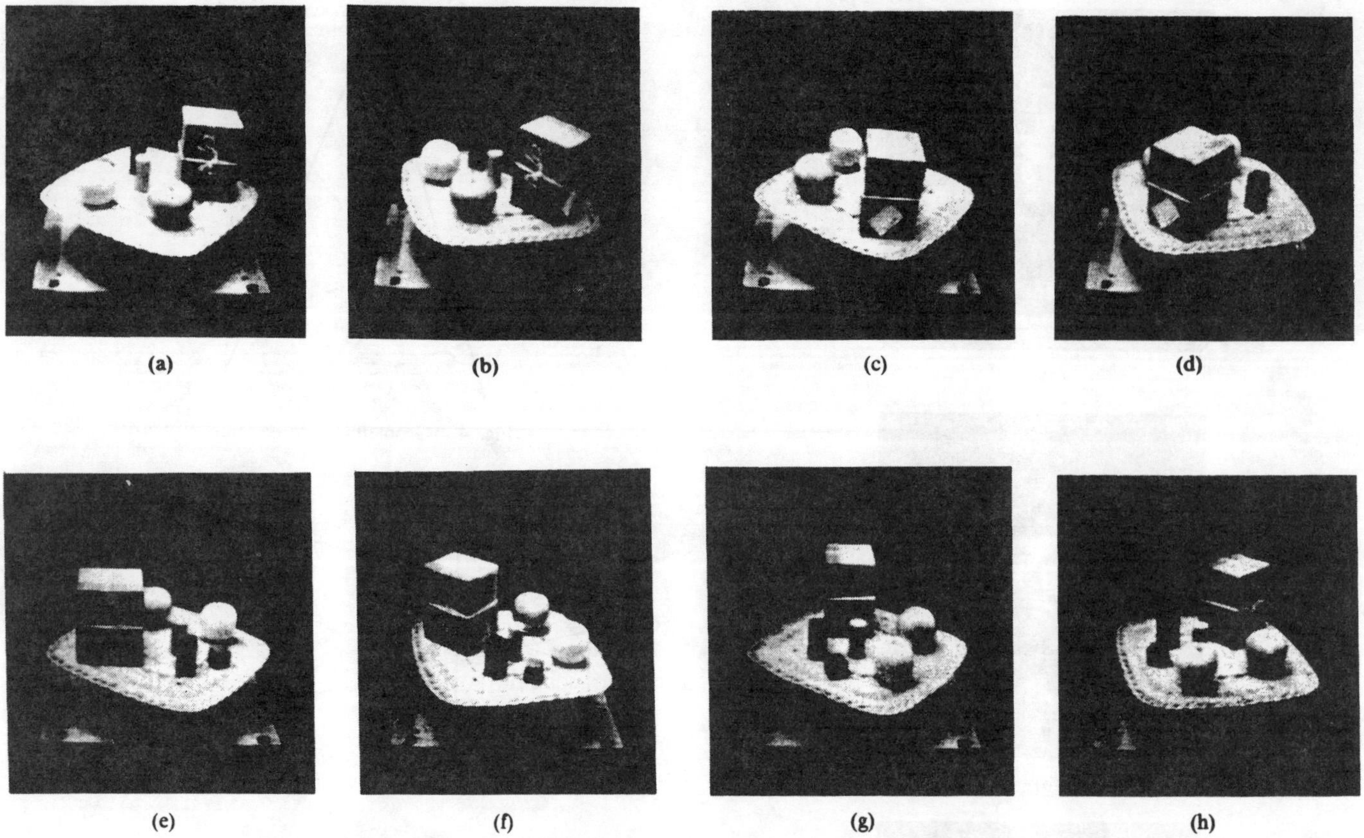

Fig. 6. First and last image of four sequences of 11 images used in experiment. Fixation point was kept still during motion of camera. Resolution of images is 256 × 256 pixels.

The zero crossings extracted from the first, third, fifth, seventh, and ninth left images are presented in Fig. 7(a) superimposed (the value of σ used is four pixels). The contours corresponding to image noise were eliminated using an hysteresis threshold on the average slope measured at the edge points [23]. A representation of the global optic flow is given in Fig. 7(b), the displayed vectors are evenly spaced along image contours and were obtained from the original flow by taking one vector every other vector.

The distance of the objects in the scene is determined from the optic flow and the known egomotion parameters

$$|Z_m + W_Z| = \frac{D_f W_Z}{|\vec{V}_t|}. \tag{2}$$

$\vec{V}_t$ is the component of the image velocity vector due to camera translation: it is computed by subtracting the rotational component $\vec{V}_r$ from the whole velocity $\vec{V}$; D_f is the displacement of the considered contour point from the focus of expansion (FOE) or the focus of contraction (FOC)[1]; W_Z is the velocity of the camera along the optic axis; Z_m is the distance of the world point from the camera along the direction of the optic axis.

The information acquired at the contour points is not a depth map; in the experiments we compute a dense depth map by interpolating the values at the contour points in a linear manner. The procedure described in this section has been applied to the optic flow of the analyzed sequence. The velocity of the camera W_Z and the position of the FOE were computed from the known egomotion parameters D_1, D_2, and the rotation angles θ and ψ (the camera did not rotate about the X axis).

In Fig. 7(c) the depth map of the analyzed scene is presented. It was obtained, for all image points, by a linear interpolation of the depth values computed at contour points. The gray level is proportional to depth. The uncertainty relative to the depth map shown in Fig. 7(c) is presented in Fig. 7(d); the intensity is proportional to the uncertainty of the measured depth.

III. Depth Uncertainty and Data Integration

Numerous sources of noise must be considered in processing images. Geometric distortions of the sensors and aliasing, in addition to the inevitable discretization of the image plane, are among the most important causes of errors. Also, stereo geometry and the egomotion parameters are among the potential sources of error. The reliability of these parameters is directly related to the accuracy of the measurement device.

[1]A *focus* in the optic flow represents the intersection of an imaginary straight line, along the direction of translation and passing through the *convergence point* of the optical system, with the image plane. In particular, in the case of egomotion, a FOE is produced if the camera moves towards the scene, then the velocity vectors are radiating from the focus; if the camera is moving away we have a FOC, and the velocity vectors collapse on it.

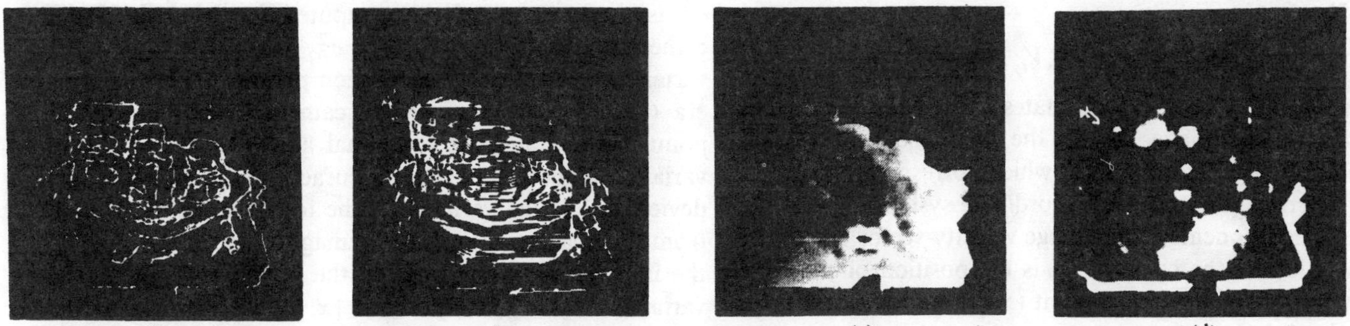

(a) (b) (c) (d)

Fig. 7. (a) Zero crossings of four images of sequence. (b) Optical flow. (c) Depth map obtained by linear interpolation of depth values computed from optical flow. (depth is proportional to gray level). (d) Associated uncertainty (uncertainty is proportional to gray level).

These errors should be taken into account estimating range from images. The problem can be faced by determining, for each parameter, an uncertainty region or a suitable probability distribution (Gaussian normal, triangular, etc.) [24]–[26].

Stochastic models of visual processes have already been proposed in the past [25], [27]–[29]. In particular, concerning 3-D vision, research has been mainly devoted to the refinement of depth through uncertainty measurements. Many researchers have identified Kalman filtering as a viable solution for this problem, as it explicitly incorporates a representation of uncertainty and allows incremental refinement of the measurement over time. Ayache and Faugeras [30] developed an elegant formalism (extended Kalman filter) to build and refine a 3-D representation of an observed scene. While it is very powerful, their system, which is based on the matching of sparse object features like points and lines, is not suitable to represent volumetric objects and their spatial occupancy, though it is most appropriate for path planning in robotic navigation. Matthies and Kanade [31], [26] applied the Kalman formalism for motion estimation from stereo images and for incremental depth measurement from known camera motion. Only two special cases of camera motion are considered, which, on the other hand, seem to maximize the accuracy in depth estimation (relative to other translational trajectories). However, the system computes only a depth map of the scene; a 3-D description of objects is not provided nor is an explicit integration between stereo- and motion-derived information performed.

Poggio and his associates at MIT [32], [33] followed another line of research, investigating random Markov fields (RMF's) as a tool for merging visual modalities, detecting discontinuities in image features, and incrementally refining a representation of the observed scene. This approach, first proposed by Geman and Geman [29], is general, as it can be applied to heterogeneous image measurements, and provides a nice representation for both dense feature maps and their discontinuities. Up to now the system has been limited to integration of visual maps without an explicit representation of volumetric objects.

In this paper we present a method for the integration of range data computed from stereo matching and optical flow, based on the incremental accumulation of dense depth maps, with their uncertainty, into a 3-D (volumetric) representation of the observed scene. The 3-D model of the objects is continuously updated according to a simplified version of Kalman filter.

In our scheme some quantities are directly measured (position and intensity of pixels, stereo and egomotion parameters) to compute the depth value. Starting from the uncertainty of the measured quantities, the goal is then to obtain an uncertainty measure of depth. A further purpose is to strengthen the estimate of the values of the uncertain quantities, for example, using repeated measures of the same object or integrating different sensor modalities. In the same way, the correct position of edge points depends upon the procedure of contour extraction, while the uncertainty in the disparity and velocity measures are determined according to the computational model.

A. Uncertainty Analysis

In Section II-B we addressed the computation of visual motion, aimed at determining depth maps from dynamic views of a static scene. The proposed approach is subject to errors due to the finite accuracy of the measured parameters and, particularly, to the computational scheme. To evaluate, and eventually reduce, the amount of errors in the computed parameters (optical flow and depth), a statistical analysis of the computational process has been performed. The basic idea is that of considering the measurement process as stochastic, where the state variables are Gaussian with known or measurable variance and mean values corresponding to their actual values.

A generic function $h(\cdot)$ of $s(i, j)$ variables

$$Z = h[s(1,1), \cdots, s(M-1, N-1)]$$

produces, by linear approximation, a Gaussian output statistic with mean and variance defined as follows:

$$\bar{Z} = h[\bar{s}(1,1), \cdots, \bar{s}(M-1, N-1)]$$
$$\sigma_Z^2 = \boldsymbol{J}\boldsymbol{V}\boldsymbol{J}^T \tag{3}$$

where $\boldsymbol{V}$ is the covariance matrix for the sequence $s(i, j)$ and $\boldsymbol{J}$ is the Jacobian of the function $h(\cdot)$.

To estimate depth from motion, the unknown depth is expressed as a function of the known parameters, as stated by (2):

$$Z_m = Z_m(x, y, V_{tx}, V_{ty}, W_Z, D_f, F) \tag{4}$$

where x and y are the coordinates of the considered point on the image plane, W_z is the component of camera velocity along the optic axis, which corresponds to the Z axis referred to the camera coordinate system, V_{tx} and V_{ty} are the components of the image velocity vector due to the translation of the camera, D_f is the position of the FOE with respect to the image point (x, y), and F is the focal length of the camera.

Considering (2), the distance D_f of a pixel P_i from the FOE can be expressed in the following way:

$$D_f = |\mathrm{FOE}_x - x, \mathrm{FOE}_y - y| = \frac{|FW_X - xW_Z, FW_Y - yW_Z|}{|W_Z|}$$

assuming that

$$N = \sqrt{[FW_X - xW_Z]^2 + [FW_Y - yW_Z]^2}\,; \tag{5}$$

then (2) can be rewritten as

$$Z_m = \frac{N}{|\vec{V}_t|} + W_z = \frac{M}{|\vec{V}_t|}$$

$$M = N + |\vec{V}_t| W_Z. \tag{6}$$

Moreover, assumed the tracking egomotion strategy, the camera translation is computed from the parameters D_1, D_2, ϕ, and θ (refer to Fig. 5):

$$W_X = D_2 \cos\phi \sin\theta$$
$$W_Y = D_2 \sin\phi$$
$$W_Z = D_1 - D_2 \cos\phi \cos\theta$$

The translational component of the image velocity is computed as difference between the optical flow and the rotational component of image velocity, which is computed from the rotation angles of the camera ϕ, θ, ψ. Hence the depth function Z_m results:

$$Z_m = Z_m(x, y, V_x, V_y, D_1, D_2, \phi, \theta, \psi, F).$$

Considering all the state variables as Gaussian and uncorrelated, the mean value of depth is assumed equal to Z_m, while its variance is expressed, using a linear approximation, as

$$\sigma_Z^2 = N^2 \left[\frac{\partial}{\partial Vx}\frac{1}{|\vec{V}_t|}\right]^2 \sigma_{Vx}^2 + N^2 \left[\frac{\partial}{\partial Vy}\frac{1}{|\vec{V}_t|}\right]^2 \sigma_{Vy}^2$$
$$+ \left[\frac{\partial}{\partial x}\frac{N}{|\vec{V}_t|}\right]^2 \sigma_x^2 + \left[\frac{\partial}{\partial y}\frac{N}{|\vec{V}_t|}\right]^2 \sigma_y^2 + \left[\frac{\partial}{\partial F}\frac{N}{|\vec{V}_t|}\right]^2 \sigma_F^2$$
$$+ \frac{1}{|\vec{V}_t|^2}\left[\frac{\partial}{\partial D_1} M\right]^2 \sigma_{D1}^2 + \frac{1}{|\vec{V}_t|^2}\left[\frac{\partial}{\partial D_2} M\right]^2 \sigma_{D2}^2$$
$$+ \left[\frac{\partial}{\partial \theta} Z_m\right]^2 \sigma_\theta^2 + \left[\frac{\partial}{\partial \phi} Z_m\right]^2 \sigma_\phi^2 + N^2\left[\frac{\partial}{\partial \psi}\frac{1}{|\vec{V}_t|}\right]^2 \sigma_\psi^2 \tag{7}$$

where σ_Z^2 represents the variance of depth, σ_x^2, σ_y^2 represent the errors in the localization of the contour point, and σ_F^2 is the variance of the computed focal length of the camera expressed in pixels. σ_{D1}^2, σ_{D2}^2 and σ_θ^2, σ_ϕ^2, σ_ψ^2 are the variances of the known egomotion parameters of the camera (i.e., the distances of the camera from the fixation point D_1, D_2 and the rotational angles ϕ, θ, ψ). These variances depend upon the accuracy of the measurement devices, while the variance of the focal length is obtained from the characteristics of the imaging sensor (position of the image center, deviation of the optical axis, etc.). The variance of the pixel position $[x, y]$ corresponds to the error in the localization of the contour, due to the $\nabla^2 G$ filtering (which is assumed to be approximately equal to half the standard deviation σ of the mask [34]). A better approximation can be obtained computing the statistic of the image (see, for example, [35]).

The right side on the first line of (7) constitutes the part that makes explicit the dependency of depth uncertainty from the variance of the optical flow. This is the last factor to be determined to estimate the uncertainty of Z_m.

The model underlying the estimation of visual motion is quite complex, as it involves an initial differentiation to recover the component $V^\perp$ of velocity, followed by a computation of the direction of velocity from proprioceptive data (the egomotion parameters) and a final matching procedure to refine the estimation. The final flow of a long sequence is obtained accumulating the partial flow fields relative to image pairs.

An accurate analysis should take into account the propagation of the errors due to the matching and the accumulation processes [36]. As to the present paper the variance of the flow field is determined from these observations.

- Corresponding contour points should exhibit the same characteristics (such as edge slope and local orientation).
- The velocity of a straight edge segment cannot be determined if it is moving along a direction parallel to its orientation.
- The accuracy in the estimation of the component $V^\perp$ (which is used to compute the optical flow) is inversely proportional to the edge slope. In fact,

$$V^\perp = -\frac{\partial I^g/\partial t}{|\nabla I^g|} \Rightarrow E_V = \frac{\partial I^g/\partial t}{|\nabla I^g|^2} E_{\nabla I} - \frac{1}{|\nabla I^g|} E_{It}$$

where I^g represents a zero crossing point of the image I filtered with the $\nabla^2 G$ operator, $|\nabla I^g|$ is the edge slope, E_V, $E_{\nabla I}$, and E_{It} are the measurement errors of the component $V^\perp$ of velocity, the edge slope, and the image time derivative $I_t = \partial I^g/\partial t$, respectively.

Taking these considerations into account, the variances of the x and y components of the flow field are determined in the following way:

$$\sigma_{Vx}^2 = V_x^2 \left[\frac{1}{|\nabla I^g|^2} + \Delta\eta^2 + \frac{1}{\Delta\omega^2}\right]\frac{1}{k}$$

$$\sigma_{Vy}^2 = V_y^2 \left[\frac{1}{|\nabla I^g|^2} + \Delta\eta^2 + \frac{1}{\Delta\omega^2}\right]\frac{1}{k}. \tag{8}$$

$\Delta\eta$ is the difference between the local orientation of the contour and the direction of the velocity vector $\vec{V}$, $\Delta\omega$ is the difference between the orientation of the corresponding contour points in the first and last frame of the sequence, and k is a normalizing factor bounding $[(1/|\nabla I^g|^2)+\Delta\eta^2+(1/\Delta\omega^2)](1/k)$ between zero and one.

Processing stereo images to estimate environmental depth involves the comparison of the gray-scale maps (to compute image disparity) and the use of external parameters relative to the acquisition sensors (to compute depth).

As in the case of depth-from-motion, the stereo algorithm provides an uncertain measure of distance:

$$Z_s = Z_s(l, m, \theta, F, x_0, d).$$

According to (3), the variance of depth is computed using a linear approximation:

$$\sigma_z^2 = \left[\frac{\partial Z_s}{\partial m}\right]^2 \sigma_m^2 + \left[\frac{\partial Z_s}{\partial l}\right]^2 \sigma_l^2 + \left[\frac{\partial Z_s}{\partial \theta}\right]^2 \sigma_\theta^2 + \left[\frac{\partial Z_s}{\partial f}\right]^2 \sigma_f^2 + \left[\frac{\partial Z_s}{\partial x_0}\right]^2 \sigma_{x0}^2 + \left[\frac{\partial Z_s}{\partial d}\right]^2 \sigma_d^2. \quad (9)$$

All parameters are considered stochastic Gaussian variables with known variance: among them the stereo pair parameters (θ, lm) are dynamically computed during image acquisition, whereas F, which is more properly a camera parameter, can be derived from camera–lens technical specifications. The uncertainty of these quantities depends on the accuracy of the calibration phase, and it is partially affected by external factors or noise (measurement errors). The position of the point x_0 and its uncertainty, depend upon the procedure of contour extraction.

Analyzing the case of egomotion, we have investigated the influence of the errors in the measured parameters (as well as of visual motion) in the estimation of depth. For the stereo algorithm we introduce some different considerations to determine the uncertainty of image disparity. An uncertainty estimate is possible, in this case, only in the presence of some precise simplifications and assumptions, as follows.

- An image is a spatially nonstationary random process: we can still use a simple description such as
$$f(i,j) = \bar{f}(i,j) + s(i,j), \qquad 0 \div M-1, 0 \div N-1.$$
- $s(i,j)$ is a high-frequency component which has stationary statistics and a mean value equal to zero. In more detail $s(i,j)$ are uncorrelated Gaussian variables in the range $(0 \div M-1, 0 \div N-1)$.
$$p(s) = k \exp\left[-\frac{1}{2} s^T \mathbf{R}_s^{-1} s\right].$$
- The correlation function, which depends on disparity d, is modeled in exponential form [3]. If a value of disparity d_0 corresponds to the peak of the correlation function, we can write
$$R(d) = A^2 \exp[-\alpha |(d-d_0)|]. \quad (10)$$

A is the maximum intensity value of the images, while α determines the peak amplitude. The value of α can be easily stated considering that $R(d)$ should not have central amplitude greater than the amplitude of the convolution mask used for the last step of the pyramidal correlation process.

Under these assumptions it is possible to evaluate the disparity variance in relation to the statistical parameters used in the stereo algorithm. The correlation between two images I_1 and I_2, performed on a patch with dimension $H \times K$ shifted horizontally by τ can be expressed as

$$\begin{aligned} R(\tau, i, j) &= \frac{1}{HK} \sum_{m=0}^{H-1} \sum_{n=0}^{K-1} I_1(i+m, j+n) I_2(i+m+\tau, j+n) \\ &= \frac{1}{HK} \sum_{m=0}^{H-1} \sum_{n=0}^{K-1} \bar{I}_1(i+m, j+n) \bar{I}_2(i+m+\tau, j+n) \\ &\quad + S_1(i+m, j+n) S_2(i+m+\tau, j+n) \\ &\quad + \bar{I}_1(i+m, j+n) S_2(i+m+\tau, j+n) \\ &\quad + S_1(i+m, j+n) \bar{I}_2(i+m+\tau, j+n) \end{aligned} \quad (11)$$

where

$$E[R(\tau, i, j)] = \frac{1}{HK} \sum_{m=0}^{H-1} \sum_{n=0}^{K-1} \bar{I}_1(i+m, j+n) \cdot \bar{I}_2(i+m+\tau, j+n).$$

$R(\tau, i, j)$ is a function of $S_1(i+m, j+n)$ and $S_2(i+m+\tau, j+n)$ statistical variables for each m and n. The variance of $R(\tau, i, j)$ is obtained differentiating (11) with respect to S_1 and S_2:

$$\sigma_R^2(\tau, i, j) = \sum_{m=0}^{H-1} \sum_{n=0}^{K-1} \left[\frac{\bar{I}_1(i+m, j+n)}{HK}\right]^2 \cdot \operatorname{var} S_2(i+m+\tau, j+n) + \left[\frac{\bar{I}_2(i+m+\tau, j+n)}{HK}\right]^2 \cdot \operatorname{var} S_1(i+m, j+n)$$

where norm is a normalization factor for I_1 and I_2.

It is worth noting that considering images filtered with a $\nabla^2 G$ operator, as in the algorithm described in Section II-A, the variance of the correlation does not increase. In fact, as the $\nabla^2 G$ operator preserves the range of intensity values of the image, the filtering does not increase the variance of the image.

Posing $\operatorname{var} S_1(i+m, j+n) = \operatorname{var} S_2(i+m+\tau, j+n) = \sigma^2$ and bounding the range of $\bar{I}_1(i,j)$ and $\bar{I}_2(i,j)$ with A we obtain

$$\sigma_R^2(\tau, i, j) \leq \sum_{m=0}^{H-1} \sum_{n=0}^{K-1} \frac{2\sigma^2 A^2}{H^2 K^2} = \frac{2\sigma^2 A^2}{HK}.$$

In our case the side of the patch is $H = K = 4w$, hence

$$\sigma_R^2(i, j) < \frac{A^2\sigma^2}{8w^2} \tag{12}$$

which approximates the variance of the correlation between two images.

The variance of disparity is now computed inverting (10) and differentiating $d(R)$ with respect to R:

$$d = d_0 \pm \frac{\ln R - \ln A^2}{\alpha}$$

$$\sigma_d^2 = \left[\frac{\partial d}{\partial R}\right]^2 \sigma_R^2 = \frac{1}{\alpha^2 R^2}\sigma_R^2. \tag{13}$$

Substituting (12) in (13), we obtain

$$\sigma_d^2 < \frac{A^2\sigma^2}{R^2\alpha^2 8w^2} \tag{14}$$

where α is the width of the peak of the correlation function, R is the estimated maximum of the correlation function, A is the maximum intensity value of the image, w is the amplitude of the central lobe of the $\nabla^2 G$ mask used for the filtering of the stereo pair, and σ is the variance of the $s(i, j)$ process (which corresponds to the image noise).

B. Volumetric Integration

Both the stereo and motion algorithm described in the previous paragraphs provide a depth map of the scene from the same view point. The peculiar errors of each algorithm are coded in the uncertainty measure associated with the computed depth.

As explained in Section II, four depth maps and the associated uncertainty maps were computed for both stereo and motion. The integration of stereo and motion is performed by projecting into a 3-D voxel representation of space the stereo and motion *bas reliefs*, weighted with the relative uncertainty measures.

In other words, each partial representation embedded in the visual *bas reliefs* can be used to generate and/or update a full 3-D representation of the solids in view. As in the present paper, the 3-D integration is performed using the *bas reliefs* obtained from different viewpoints to *carve* a 3-D array of voxels representing the viewed space.

If the depth of an image point $\bar{p}$ has a variance σ_p, then we can suppose that along the line of sight crossing that image point the space is entirely empty for distances less than $(\bar{p} - \sigma_p)$ and vice versa: the space is full for distances greater than $(\bar{p} + \sigma_p)$. All the intermediate values represent uncertain values. In this way a sort of rind is associated with the depth map. The thickness of this rind is proportional to σ_p.

From a procedural point of view, the accumulation of incoming depth images is performed in the following way:

- a 3-D matrix of voxels is defined representing the "work space";
- the position of the observer with respect to the work space is computed (or known);
- a line is traced from the position of the observer through the depth image, and all the voxels crossed by the line are modified according to probability of empty space.

TABLE I

x_i Current Value Stored in the Matrix	y_{i+1} New Value of Occupation Probability	x_{i+1} Final Value of Occupation Probability
U	U	x_i
O	U	x_i
S	U	x_i
U	O	y_{i+1}
O	O	x_i
S	O	x_i
U	S	y_{i+1}
O	S	y_{i+1}
S	S	update probability according to (15)

U Unknown.
O Occluded.
S Seen.

To perform the actual accumulation, we must consider that a single view cannot carry information on the space which is not seen; this occurs for the objects outside the visual field and for occluded objects. In fact, we subdivide the space in three parts.

- *Seen space* (S): This portion of the work space lies inside the visual field and, at the same time, belongs to the rind $(\bar{p} - \sigma_p) < p < (\bar{p} + \sigma_p)$. This measure is different from zero in the zones where disparity information is present.
- *Occluded space* (O): It belongs to vision cone but lies behind the rind (i.e., at $p > (\bar{p} + \sigma_p)$.
- *Unknown space* (U): This is the space external to the vision cone.

Every time a new depth map is obtained, the voxel work space is updated in accordance with Table I (initially the array is set to unknown). The update is performed, voxel by voxel, projecting the new depth map into the existing volumetric representation. If a new depth map does not add any information (*unknown* probability), the voxel is not updated. Also, if the new value if *occluded*, the voxel is not updated, unless it was already unknown (fourth row of Table I); in this case we label the voxel as occluded.

It is worth noting the operational equivalence between the occluded and unknown space. On the other hand, the distinction is not redundant because the two situations have a very different meaning: the observer has a strong interest in the occluded space, which must be reduced as much as possible (the knowledge of occluded space can be used, for example, to drive exploratory strategies). On the contrary, unknown space represents "all the universe" and must be considered in particular situations only.

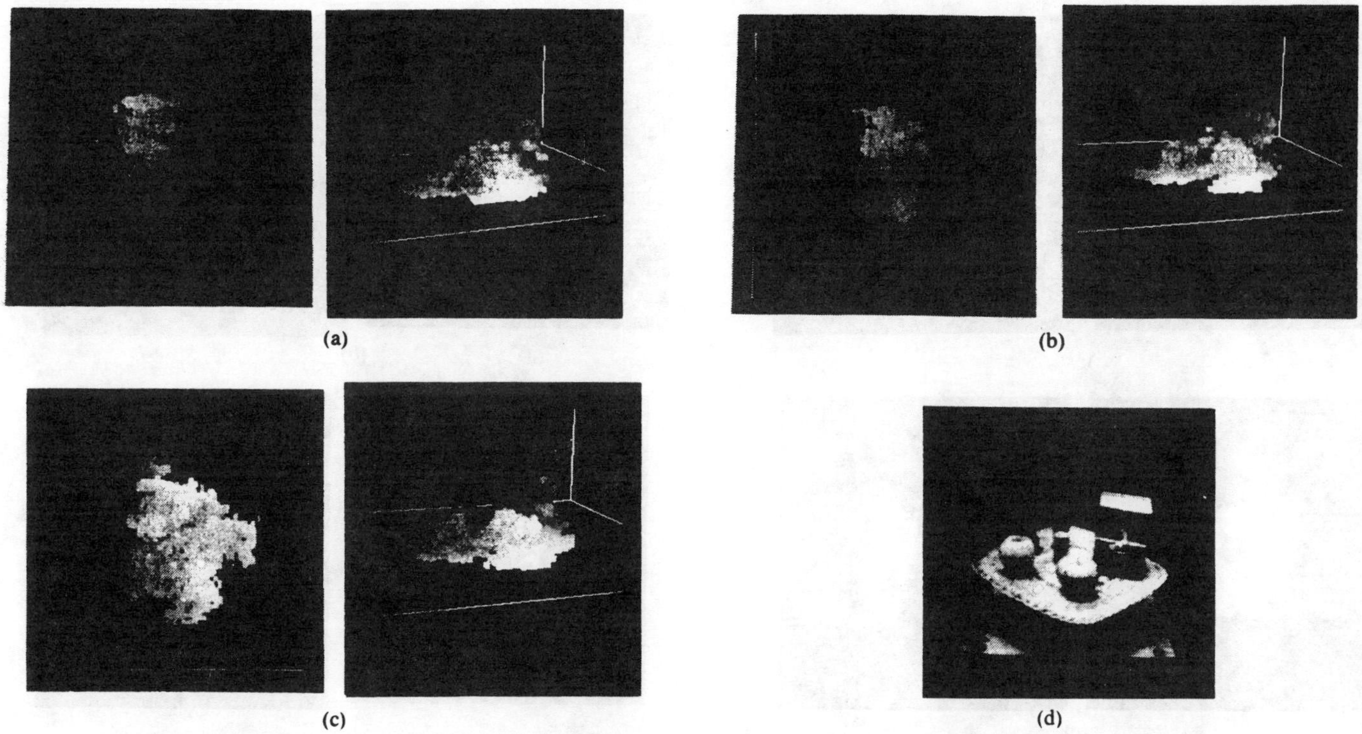

Fig. 8. 3-D volumetric integration of depth maps. Perspective representation of voxel-based accumulator. Gray-level codes distance (brighter means closer). (a) Integration of four depth maps derived from four stereo pairs of Fig. 2, (b) Integration of four depth maps derived from optical flow, (c) Integration of both stereo and motion-derived information. Left column: objects seen from above. Right column: object seen approximately from point of view of Fig. 8(d). Tall package is visible (brighter areas on top views) as are two apples. For this picture any information coming from below tray have been clipped (as consequence square base is not visible). (d) Original picture of scene observed from same point of view of perspective representation in (a), (b), and (c).

The most interesting case occurs when the new visual *bas relief* carries a new occupation probability for a voxel which labels it as *seen*: if it was unknown or occluded, then it is set to seen with the new probability value; if it was already seen, then the probability value is updated according to the new information. In the latter case, a simplified version of Kalman filter is used to update the voxel probability:

$$x_{i+1} = x_i + k_i[y_{i+1} - x_i]$$
$$k_{i+1}^{-1} = k_i^{-1} + 1$$
$$k_0^{-1} = 0 \qquad x_0 = \text{unknown}. \tag{15}$$

K_i is the Kalman gain factor, x_i is the current probability of the considered voxel, x_{i+1} is the updated voxel probability, and y_{i+1} is the probability (inverse normalized uncertainty) of newly computed depth map. This expression achieves the arithmetic average among all considered probability values, updating the uncertainty associated to full space. However, it is required to store the k_i coefficient for each voxel, doubling the memory necessary to maintain a comprehensive description of the work space.

An example of integration is shown in Fig. 8; in Fig. 8(a) the integration of the four depth maps derived from stereo is shown. In Fig. 8(b) the result of the same procedure is shown for the motion-derived depth maps. In Fig. 8(c), finally, the integration of both stereo and motion-derived depth images is shown. The probability of occupation is, for the three images of Fig. 8, set to 0.35. As it can be noticed, the approximation is better in Fig. 8(c) than in both Fig. 8(a) and (b). In Fig. 9 the result obtained raising the value of probability of filled space is shown: raising the voxel probability, the volume of the solid is reduced, until the achievement of a nucleus is certainly full. Notice that the filled space increases if the certainty measure is lowered.

IV. Conclusion

The fusion of different sensor modalities constitutes one of the hot topics in today's computer vision. The main advantage of the integration of multiple sensorial outputs is the robustness of the overall measurement process; moreover, the precision of the measures can be considerably improved. This is especially important when dealing with 3-D vision and its applications in robotics, like obstacle avoidance or automatic vehicle guidance, etc. In those and other cases, it is dangerous to produce erroneous measures, while it is necessary to know their reliability.

In this paper we have presented two algorithms for recovering the 3-D structure of objects from stereo matching and motion parallax, as examples of visual processes that can independently compute environmental depth. The stereo algorithm is characterized by robustness, due to regional correlation, and precision achieved with the final edge matching. The structure-from-motion algorithm takes

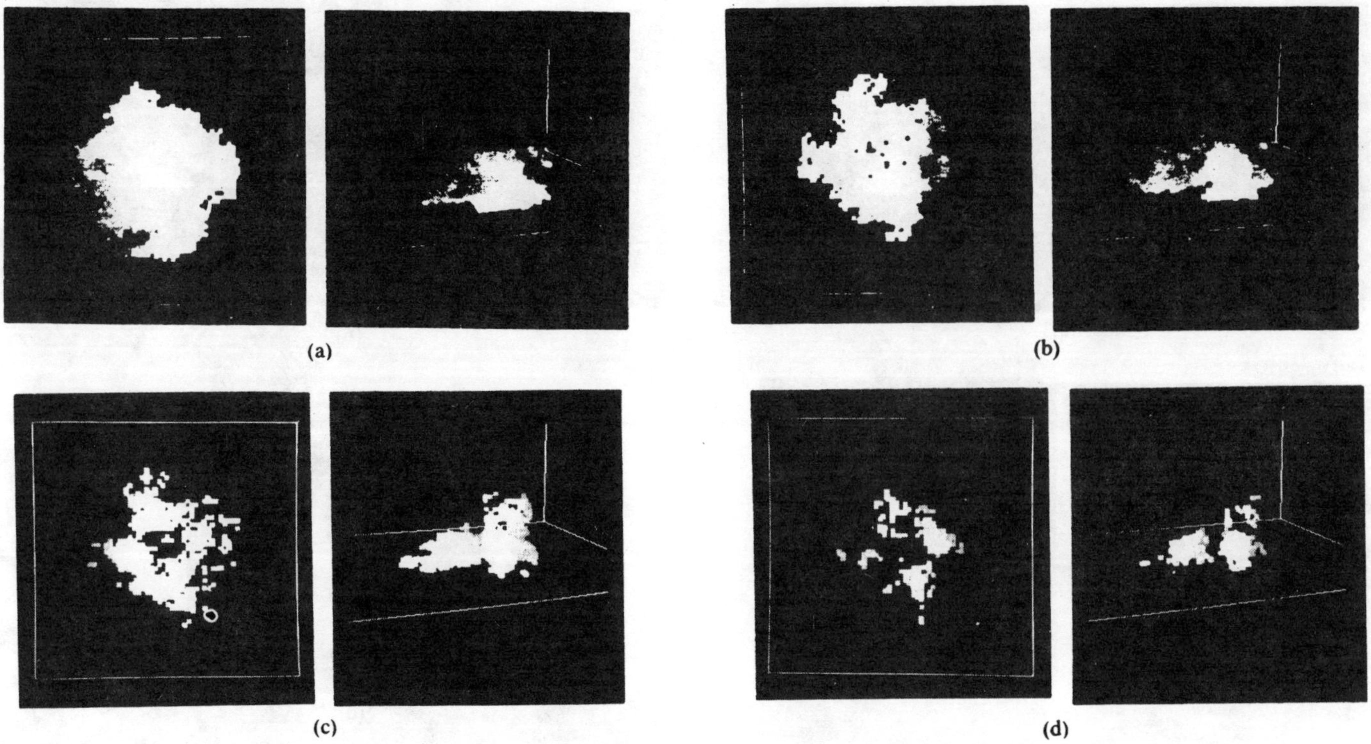

Fig. 9. Same representation as in Fig. 8. Results obtained by raising probability of filled space. From top to bottom probability changes from 0.15 to 0.75. For higher levels of probability tray disappears and only some voxels around package and two apples (i.e., largest objects) remains. Left column: objects seen from above. Right column: object seen approximately from point of view of Fig. 8(d).

advantage of a particular motion strategy of the observer, which allows the simplification of the motion constraint equations. The optic flow of a long sequence (obtained by matching successive image pairs) is used to compute depth. The depth is computed, in both cases, at the contour points; a dense depth map is obtained performing a linear interpolation of the depth values.

Both the stereo and motion algorithms provide a visual *bas relief* (a depth map) from several viewpoints, while the camera is moving around. Different measures are combined, continuously updating a volumetric description of the scene. We used a volumetric representation to model the environment: a cube of voxel, in which the *bas reliefs* are projected (with a ray-casting procedure) from the different viewpoints, carving the shape of the objects.

Each measure is characterized by an uncertainty value: the *bas reliefs* are projected into the work space weighted with their probability. To compute the uncertainty relative to each visual modality, a stochastic model of the processes has been developed in which all the parameters involved are assumed to be uncorrelated Gaussian variables with known variance and mean values corresponding to the measured values. The depth maps are integrated into the working space using Kalman filtering to update the volumetric representation.

In the experiment presented, the great advantage of multisensor integration is evident, as the final representation is more precise than that obtained using each single measurement process.

In the present paper we used only two visual processes to compute environmental depth, but the outputs of any other shape_from... algorithm or of ranging devices like ultrasound or laser range finders, could be added into the volumetric description of the world. For the future, we are developing a technique to segment the volumetric representation, isolating single objects, and to transform the voxel-based description to one based on the superficial characteristics and other geometric features of the observed scene.

References

[1] Y. F. Wang, M. J. Magee, and J. K. Aggarwal, "Matching three-dimensional objects using silhouettes," *IEEE Trans. Pattern Anal. Machine Intell.*, vol. PAMI-6, no. 4, pp. 513–518, 1984.

[2] L. Massone, P. Morasso, and R. Zaccaria, "Shape from occluding contours," presented at the SPIE Symp. Intelligent Robots and Computer Vision, Cambridge, MA, November 4–8, 1984.

[3] H. K. Nishihara, "PRISM: A practical real-time imaging stereo matcher," *Opt. Eng.*,vol. 23 no. 5 pp. 536–545, 1984.

[4] D. Marr, *Vision*. San Francisco CA: Freeman, 1982.

[5] K. Prazdny, "Egomotion and relative depth map from optical flow," *Biol. Cybern.*, vol. 36, pp. 87–102, 1980.

[6] D. T. Lawton, "Processing translational motion sequences," *CVGIP*, vol. 22, pp. 116–144, 1983.

[7] T. D. Williams, "Depth from camera motion in a real world scene," *IEEE Trans. Pattern Anal. Machine Intell.*, vol. PAMI-2, no. 6, pp. 511–516, 1980.

[8] T. M. Strat and M. A. Fischler, "One eyed stereo: A general approach to modeling 3-D scene geometry," *IEEE Trans. Pattern Anal. Machine Intell.*, vol. PAMI-8, no. 6, pp. 730–741, 1986.

[9] M. Brady, "Artificial intelligence and robotics," MIT A. I. Laboratory, A. I. Memo 756, Cambridge, MA, Feb. 1984.

[10] B. K. P. Horn, "Understanding image intensities," *Artificial Intell.*, vol. 8, 1977.

[11] K. Ikeuchi and B. K. P. Horn, "Numerical shape from shading and occluding boundaries," *Artificial Intell.*, vol. 17, 1981.

[12] G. Sandini, M. Straforini, and V. Torrre, "3-D reconstruction of silhouettes," in *Proc. 4th Intl. ROVISEC*, London, UK, 1984, pp. 173–182.
[13] G. Sandini and M. Tistarelli, "Recovery of depth information: Camera motion as an integration to stereo," in *Proc. Workshop on Motion: Representation and Analysis*, Kiawah Island Resort, May 7–9, 1986. pp.39–43.
[14] P. Morasso, G. Sandini, and M. Tistarelli, "Active vision: Integration of fixed and mobile cameras," in *NATO ARW on Sensors and Sensory Systems for Advanced Robots*. Berlin, Germany: Springer-Verlag, 1986, pp. 449–462.
[15] A. Bandopadhay, B. Chandra, and D. H. Ballard, "Active navigation: Tracking an environmental point considered beneficial," in *Proc. Workshop on Motion: Representation and Analysis*, Kiawah Island Resort, May 7–9, 1986, pp. 23–29.
[16] G. Sandini, V. Tagliasco, and M. Tistarelli, "Analysis of object motion and camera motion in real scenes," in *Proc. IEEE Intl. Conf. Robotics & Automation*, San Francisco, CA, Apr. 7–10, 1986, pp. 627–633.
[17] G. Sandini, P. Morasso, and M. Tistarelli, "Motor and spatial aspects in artificial vision," in *Proc. 4th Intl. Symp. Robotics Research*, Aug. 1987 Cambridge, MA: MIT Press, pp. 351–358.
[18] A. Bandopadhay, J. Y. Aloimonos, and I. Weiss, "Active vision," *Int Comput. Vision*, vol. 1, no. 4, pp. 333–356, Jan. 1988.
[19] K. Ikeuchi, H. Nishihara, B. K. P. Horn, P. Sobalvarro, and S. Nagata, "Determining grasp points using photometric stereo and the PRISM binocular stereo system," *Int. J. Robotics Res.* vol. 5, no. 1, pp. 46–65, 1986.
[20] C. Frigato, E. Grosso, and G. Sandini, "Integration of edge stereo information," DIST-Univ. of Genoa, Esprit P419 Tech. Rep. TKW1-WP1-DI3, 1987.
[21] ____, "Extraction of 3-D information and volumetric uncertainty from multiple stereo images," *Proc. ECAI*, Mũnchen, Germany, Aug. 1988, pp. 683–688.
[22] D. Marr and E. Hildreth, "Theory of edge detection," *Proc. Roy. Soc. London*, Ser. B, no. 207, pp. 187–217, 1980.
[23] G. Sandini and V. Torre, "Thresholding techniques for zero crossings," in *Proc. Winter 85 Topical Meeting Machine Vision*, Incline Village, NV. 1985, pp. ThD5-1–ThD5-4.
[24] M. A. Snyder, "Uncertainty analysis of image measurements," in *Proc. DARPA Image Understanding Workshop*, 1987.
[25] L. Matthies and S. A. Shafer, "Error modeling in stereo navigation," *IEEE Robot. Automation.*, vol. RA-3. no. 3, pp. 239–248, June 1987.
[26] L. Matthies and T. Kanade, "Using uncertainty models in visual motion and depth estimation," in *Proc. 4th Int. Symp. Robotics Research*, Santa Cruz, CA, Aug. 1987, pp. 120–138.
[27] J. Marroquin, S. Mitter, and T. Poggio, "Probabilistic solution of ill-posed problems in computational vision," *J. Amer. Statis. Assoc.*, vol. 82, no. 397, pp. 76–89, Mar. 1987.
[28] J. L. Marroquin, "Deterministic Bayesian estimation of Markov random fields with applications in computer vision," in *Proc. Int. Conf. Computer Vision*, Washington, DC, 1987.
[29] S. Geman and D. Geman, "Stochastic relaxation, Gibbs distributions and the Bayesian restoration of images," *IEEE Trans. Pattern Anal. Machine Intell.*, vol. PAMI-6, 1984.
[30] N. Ayache and O. D. Faugeras, "Maintaining representations of the environment of a mobile robot," *Proc. 4th Int. Symp. Robotics Research*, August 1987. Cambridge, MA: MIT Press, pp. 337–350.
[31] L. Matthies and T. Kanade, "The cycle of uncertainty and constraint in robot perception," in *Proc. 4th Int. Symp. Robotics Research*, Aug. 1987. Cambridge, MA: MIT Press, pp. 327–336.
[32] T. Poggio, "The MIT vision machine," in *Proc. DARPA Image Understanding Workshop*, Cambridge, MA, Apr. 1988, pp. 177–198.
[33] E. Gamble and T. Poggio, "Integration of intensity edges with stereo and motion," MIT A.I. Laboratory, Boston, MA, A.I. Memo 970, Feb. 1984.
[34] A. Huertas and G. Medioni, "Detection of intensity changes with subpixel accuracy using Laplacian–Gaussian masks," *IEEE Trans. Pattern Anal. Machine Intell.*, vol. PAMI-8, no. 5, pp. 651–664, Sept. 1986.
[35] L. Matthies, R. Szeliski, and T. Kanade, "Kalman filter-based algorithms for estimating depth from image sequences," Carnegie–Mellon Univ., Pittsburgh, PA CMU-RI-TR-88-1, 1988.
[36] M. Tistarelli and G. Sandini, "Uncertainty analysis in visual motion and depth estimation from active egomotion," in *Proc. IEEE/SPIE Intl. Conf. Applications of Artificial Intelligence VII*, Orlando, FL, Mar. 28–30, 1989.

Enrico Grosso was born on November 29, 1963, in Serravalle, Italy. He received the degree in electrical engineering in 1987 from the University of Genoa. He is a Ph.D. student at the Dipartimento di Ingegneria Elettrica of the University of Palermo.

Since 1985 he has been working at the "Dipartmento di Informatica, Sistemistica e Telematica" of the University of Genoa on the topic of artificial vision with particular emphasis on stereo analysis and 3-D reconstruction from multiple views.

Giulio Sandini was born on September 7, 1950 in Correggio, Italy. He received a degree in electrical engineering in 1976 from the University of Genoa, Italy.

Since 1976 he has worked on models of the visual system and on electrophysiology of the cat visual cortex at the Laboratorio di Neurofisiologia del CNR di Pisa. In 1978 and 1979 he was a Visiting Scientist at the Harvard Medical School in Boston, developing a system for topographic analysis of brain electrical activity. Since 1980 he has worked on Image Processing and Computer Vision particularly in the areas of low-level vision and feature extraction, at the Department of Communication, Computer and Systems Science of the University of Genoa, where he is currently a Associate Professor.

Massimo Tistarelli was born on November 11, 1962 in Genoa, Italy. He received a degree in electronic engineering from the University of Genoa, Italy. He is currently a Ph.D student.

Since 1984 he has worked on Image Processing and Computer Vision at the Laboratory of Robotic of the Department of Communication, Computer and Systems Science of the University of Genoa.

In 1986 he was a Research Assistant at the Department of Computer Science of Trinity College, in Dublin, developing a system for the analysis of image data, mainly aimed at the investigation of low-level visual processes.

In 1989 he was a visiting scientist at Thinking Machines Co., developing parallel algorithms for dynamic image processing on the Connection Machine. His research interests include Robotics, Artificial Intelligence, Image Processing and Computer Vision particularly in the area of three-dimensional and dynamic scene analysis.

3-D Vision Techniques for Autonomous Vehicles

Martial Hebert
Takeo Kanade
InSo Kweon [1]

"3-D Vision Technique for Autonomous Vehicles" by M. Hebert, T. Kanade, and I. Kweon, from *NSF Range Image Understanding Workshop*, 1988, pages 273-337. Reprinted from *Analysis and Interpretation of Range Images*, R.C. Jain and A.K. Jain, eds. Copyright © 1988 Springer-Verlag New York, Inc., reprinted with permission.

7.1 Introduction

A mobile robot is a vehicle that navigates autonomously through an unknown or partially known environment. Research in the field of mobile robots has received considerable attention in the past decade due to its wide range of potential applications, from surveillance to planetary exploration, and the research opportunities it provides, including virtually the whole spectrum of robotics research from vehicle control to symbolic planning (see for example [Har88b] for an analysis of the research issues in mobile robots). In this paper we present our investigation of some the issues in one of the components of mobile robots: perception. The role of perception in mobile robots is to transform data from sensors into representations that can be used by the decision-making components of the system. The simplest example is the detection of potentially dangerous regions in the environment (*i.e.* obstacles) that can be used by a path planner whose role is to generate safe trajectories for the vehicle. An example of a more complex situation is a mission that requires the recognition of specific landmarks, in which case the perception components must produce complex descriptions of the sensed environment and relate them to stored models of the landmarks.

There are many sensing strategies for perception for mobile robots, including single camera systems, sonars, passive stereo, and laser range finders. In this report, we focus on perception algorithms for range sensors that provide 3-D data directly by active sensing. Using such sensors has the advantage of eliminating the calibration problems and computational costs inherent in passive techniques such as stereo. We describe the range sensor that we used in this work in Section 7.2. Even though we tested our algorithm on one specific range sensor, we believe that the sensor characteristics of Section 7.2 are fairly typical of a wide range of sensors [Bes88a].

Research in perception for mobile robots is not only sensor-dependent but it is also dependent on the environment. A considerable part of the global research effort has concentrated on the problem of perception for mobile robot navigation in indoor environments, and our work in natural outdoor environments through the Autonomous Land Vehicle and Planetary Exploration projects is an important development. This report describes some of the techniques we have developed in this area of research. The aim of our work is to produce models of the environment, which we call the *terrain*, for path planning and object recognition.

---

[1]The Robotics Institute, Carnegie Mellon University, 5000 Forbes Avenue, Pittsburgh PA 15213. This research was sponsored in part by the Defense Advanced Research Projects Agency, DoD, through ARPA Order 5351, monitored by the US Army Engineer Topographic Laboratories under contract DACA76-85-C-0003, by the National Science Foundation contract DCR-8604199, by the Digital Equipment Corporation External Research Program, and by NASA grant NAGW-1175. The views and conclusions contained in this document are those of the authors and should not be interpreted as representing the official policies, either expressed or implied, of the funding agencies.

The algorithms for building a terrain representation from a single sensor frame are discussed in Section 7.3 in which we introduce the concept of dividing the terrain representation algorithms into three levels depending on the sophistication of the path planner that would use the representation, and on the anticipated difficulty of the terrain. Since a mobile robot is by definition a dynamic system, it must process not one, but many observations along the course of its trajectory. The 3-D vision algorithms must therefore be able to reason about representations that are built from sensory data taken from different locations. We investigate this type of algorithms in Section 7.4 in which we propose algorithms for matching and merging multiple terrain representations.

Finally, the 3-D vision algorithms that we propose are not meant to be used in isolation, they have to be eventually integrated in a system that include other sensors. A typical example is the case of road following in which color cameras can track the road, while a range sensor can detect unexpected obstacles. Another example is a mission in which a scene must be interpreted in order to identify specific objects, in which case all the available sensors must contribute to the final scene analysis. We propose some algorithms for fusing 3-D representations with representations obtained from a color camera in Section 7.5. We also describe the application of this sensor fusion to a simple natural scene analysis program.

Perception techniques for mobile robots have to be eventually validated by using real robots in real environments. We have implemented the 3-D vision techniques presented in this report on three mobile robots developed by the Field Robotics Center: the Terregator, the Navlab, and the Ambler. The Terregator (Figure 7.1) is a six-wheeled vehicle designed for rugged terrain. It does not have any onboard computing units except for the low-level control of the actuators. All the processing was done on Sun workstations through a radio connection.

We used this machine in early experiments with range data, most notably the sensor fusion experiments of Section 7.5.

The Navlab [SW88] (Figure 7.2) is a converted Chevy van designed for navigation on roads or on mild terrains. The Navlab is a self-contained robot in that all the computing equipment is on board. The results presented in Sections 7.3.3 and 7.3.4 come from the 3-D vision module that we integrated in the Navlab system [THKS88]. The Ambler [BW88] is an hexapod designed for the exploration of Mars (Figure 7.3). This vehicle is designed for navigation on very rugged terrain including high slopes, rocks, and wide gullies. This entirely new design prompted us to investigate alternative 3-D vision algorithms that are reported in Section 7.3.5. Even though the hardware for the Ambler does not exist at this time, we have evaluated the algorithms through simulation and careful analysis of the planetary exploration missions.

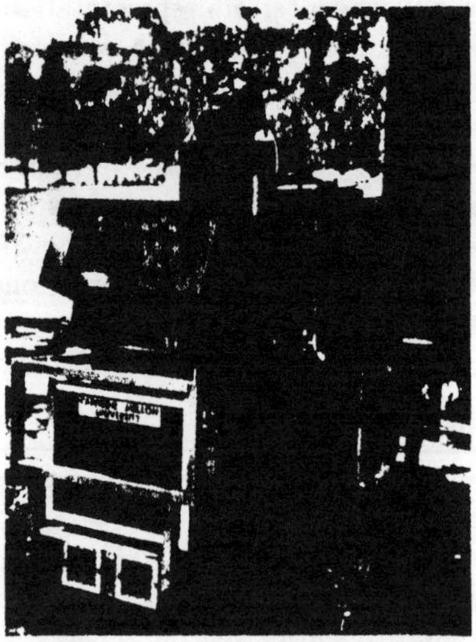

FIGURE 7.1. The Terregator

7.2 Active range and reflectance sensing

The basic principle of active sensing techniques is to observe the reflection of a reference signal (sonar, laser, radar..etc.) produced by an object in the environment in order to compute the distance between the sensor and that object. In addition to the distance, the sensor may report the intensity of the reflected signal which is related to physical surface properties of the

FIGURE 7.2. The Navlab

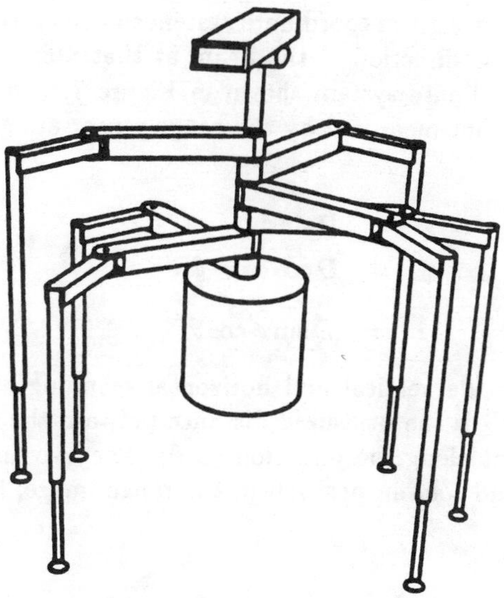

FIGURE 7.3. The Ambler

object. In accordance with tradition, we will refer to this type of intensity data as "reflectance" data even though the quantity measured is not the actual reflectance coefficient of the surface.

Active sensors are attractive to mobile robots researchers for two main reasons: first, they provide range data without the computation overhead associated with conventional passive techniques such as stereo vision, which is important in time critical applications such as obstacle detection. Second, it is largely insensitive to outside illumination conditions, simplifying considerably the image analysis problem. This is especially important for images of outdoor scenes in which illumination cannot be controlled or predicted. For example, the active reflectance images of outside scenes do not contain any shadows from the sun. In addition, active range finding technology has developed to the extent that makes it realistic to consider it as part of practical mobile robot implementations in the short term [Bes88a].

The range sensor we used is a time-of-flight laser range finder developed by the Environmental Research Institute of Michigan (ERIM). The basic principle of the sensor is to measure the difference of phase between a laser beam and its reflection from the scene [ZPFL85]. A two-mirror scanning system allows the beam to be directed anywhere within a 30° × 80° field of view. The data produced by the ERIM sensor is a 64 × 256 range image, the range is coded on eight bits from zero to 64 feet, which corresponds to a range resolution of three inches. All measurements are all relative since the sensor measures differences of phase. That is, a range value is known *modulo* 64 feet. We have adjusted the sensor so that the range value 0 corresponds to the mirrors for all the images presented in this report. In addition to range images, the sensor also produces active reflectance images of the same format (64 × 256 × 8 bits), the reflectance at each pixel encodes the energy of the reflected laser beam at each point. Figure 7.5 shows a pair of range and reflectance images of an outdoor scene. The next two Sections describe the range and reflectance data in more details.

7.2.1 From range pixels to points in space

The position of a point in a given coordinate system can be derived from the measured range and the direction of the beam at that point. We usually use the Cartesian coordinate system shown in Figure 7.4, in which case the coordinates of a point measured by the range sensor are given by the equations[2]:

$$\begin{aligned} x &= D\sin\theta \\ y &= D\cos\phi\cos\theta \\ z &= D\sin\phi\cos\theta \end{aligned} \qquad (7.1)$$

where ϕ and θ are the vertical and horizontal scanning angles of the beam direction, and D is the measured distance between the scanner and the closest scene point along the direction (ϕ, θ). The two angles are derived from the row and column position in the range image, (r, c), by the equations:

$$\begin{aligned} \theta &= \theta_0 + c \times \Delta\theta \\ \phi &= \phi_0 + r \times \Delta\phi \end{aligned} \qquad (7.2)$$

where θ_0 (respectively ϕ_0) is the starting horizontal (respectively vertical) scanning angle, and $\Delta\theta$ (respectively $\Delta\phi$) is the angular step between two consecutive columns (respectively rows). Figure 7.6 shows an overhead view of the scene of Figure 7.5, the coordinates of the points are computed using Eq. (7.3).

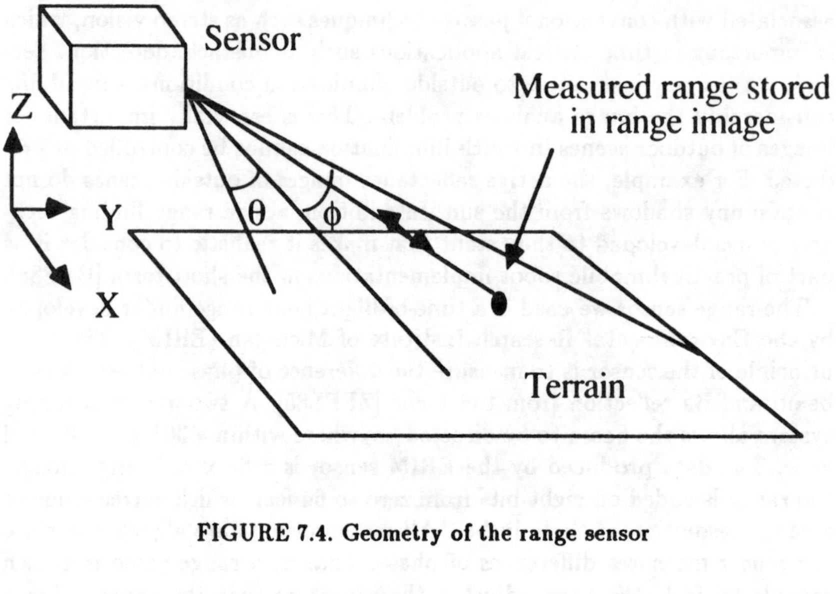

FIGURE 7.4. Geometry of the range sensor

[2]Note that the reference coordinate system is not the same as in [HK85] for consistency reasons

7.2.2 REFLECTANCE IMAGES

A reflectance image from the ERIM sensor is an image of the energy reflected by a laser beam. Unlike conventional intensity images, this data provides us with information which is to a large extent independent of the environmental illumination. In particular, the reflectance images contain no shadows from outside illumination. The measured energy does depend,

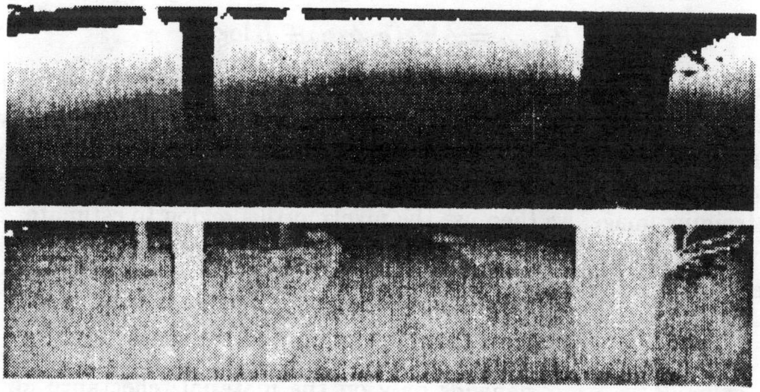

FIGURE 7.5. Range and reflectance images

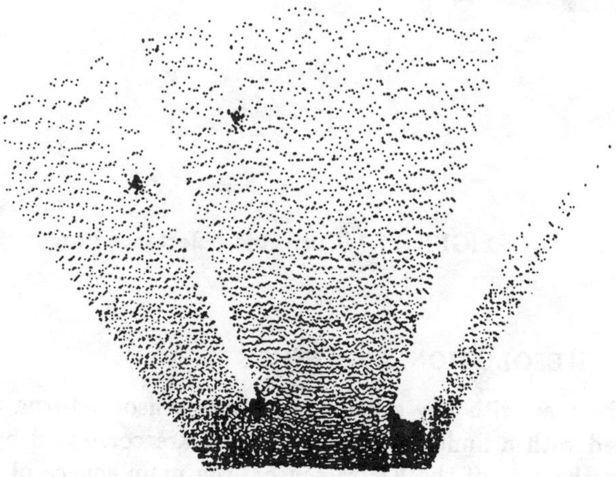

FIGURE 7.6. Overhead view

however, on the shape of the surface and its distance to the sensor. We correct the image so that the pixel values are functions only of the material reflectance. The measured energy, P_{return}, depends on the specific material reflectance, ρ, the range, D, and the angle of incidence, γ:

$$P_{return} = \frac{K\rho\cos\gamma}{D^2} \tag{7.3}$$

Due to the wide range of P_{return}, the value actually reported in the reflectance image is compressed by using a log transform. That is, the digitized value, P_{image} is of the form [WPZ87]:

$$P_{image} = A\log(\rho\cos\gamma) + B\log D \tag{7.4}$$

where A and B are constants that depend only on the characteristics of the laser, the circuitry used for the digitization, and the physical properties of the ambiant atmosphere. Since A and B cannot be computed directly, we use a calibration procedure in which a homogeneous flat region is selected in a training image; we then use the pixels in this region to estimate A and B by least-squares fitting Eq. (7.4) to the actual reflectance/range data. Given A and B, we correct subsequent images by:

$$P_{new-image} = (P_{image} - B\log D)/A \tag{7.5}$$

The value $P_{new-image}$ depends only on the material reflectance and the angle of incidence. This is a sufficient approximation for our purposes since for smooth surfaces such as smooth terrain, the $\cos\gamma$ factor does not vary widely. For efficiency purposes, the right-hand side of (7.5) is precomputed for all possible combinations (P_{image}, D) and stored in a lookup table. Figure 7.5 shows an example of an ERIM image, and Figure 7.7 shows the resulting corrected image.

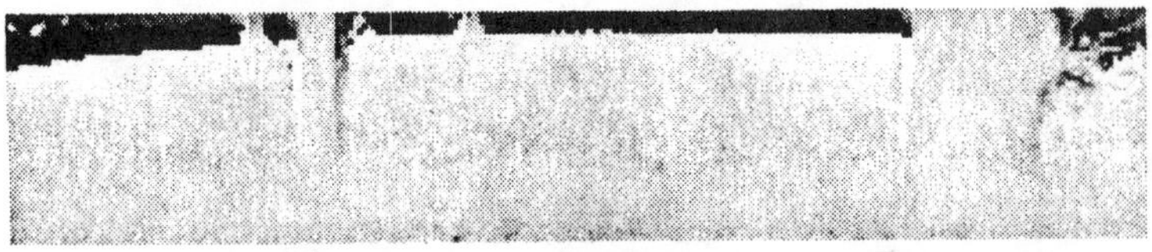

FIGURE 7.7. Corrected reflectance image

7.2.3 RESOLUTION AND NOISE

As is the case with any sensor, the range sensor returns values that are measured with a limited resolution which are corrupted by measurement noise. In the case of the ERIM sensor, the main source of noise is due to the fact that the laser beam is not a line in space but rather a cone whose opening is a 0.5° solid angle (the instantaneous field of view). The value returned at each pixel is actually the average of the range of values over a 2-D area, the *footprint*, which is the intersection of the cone with the target surface (Figure 7.8). Simple geometry shows that the area of the footprint is proportional to the square of the range at its center. The size of the footprint also depends on the angle θ between the surface normal and the beam as shown in Figure 7.8. The size of the footprint is roughly inversely proportional to $\cos\theta$ if we assume that the footprint is small enough and that θ is almost constant. Therefore, a first order approximation of the standard deviation of the range noise, σ is given by:

$$\sigma \propto \frac{D^2}{\cos\theta} \tag{7.6}$$

The proportionality factor in this equation depends on the characteristics of the laser transmitter, the outside illumination, and the reflectance ρ of the surface which is assumed constant across the footprint in this first order approximation. We validated the model of Equation 7.6 by estimating the RMS error of the range values on a sequence of images. Figure 7.9 shows the standard deviation with respect to the measured range. The Figure shows that σ follows roughly the D^2 behavior predicted by the first order model. The footprint affects all pixels in the image.

There are other effects that produce distortions only at specific locations in the image. The main effect is known as the "mixed point" problem and is illustrated in Figure 7.8 in which the laser footprint crosses the edge between two objects that are far from each other. In that case, the returned range value is some combination of the range of the two objects but does not have any physical meaning. This problem makes the accurate detection of occluding edges more difficult. Another effect is due to the reflectance properties of the observed surface; if the surface is highly specular then no laser reflection can be observed. In that case the ERIM sensor returns a value of 255. This effect is most noticeable on man-made objects that contain a lot of polished metallic surfaces. It should be mentioned, however, that the noise characteristics of the ERIM sensor are fairly typical of the behavior of active range sensors [BJ85c].

7.3 Terrain representations

The main task of 3-D vision in a mobile robot system is to provide sufficient information to the path planner so that the vehicle can be safely steered through its environment. In the case of outdoor navigation, the task is to convert a range image into a representation of the terrain. We use the word "terrain" in a very loose sense in that we mean both the ground surface

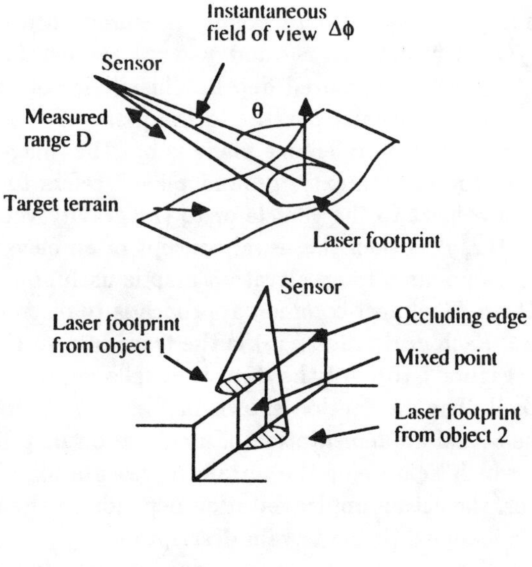

FIGURE 7.8. Sources of noise in range data

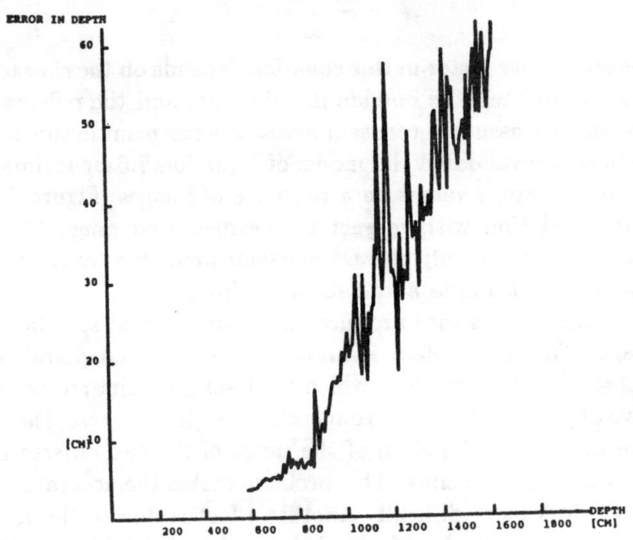

FIGURE 7.9. Noise in range data

and the objects that may appear in natural environments (*e.g.* rocks or trees). In this Section we discuss the techniques that we have implemented for the Navlab and Mars Rover systems. We first introduce the concept of the elevation map as a basis for terrain representations and its relationship with different path planning techniques. The last four Sections spell out the technical details of the terrain representation algorithms.

7.3.1 THE ELEVATION MAP AS THE DATA STRUCTURE FOR TERRAIN REPRESENTATION

Even though the format of the range data is an image, this may not be the most suitable structuring of the data for extracting information. For example , a standard representation in 3-D vision for manipulation is to view a range image as a set of data points measured on a surface of the equation $z = f(x, y)$ where the $x-$ and $y-$axes are parallel to the axis of the image and z is the measured depth. This choice of axis is natural since the image plane is usually parallel to the plane of the scene. In our case, however, the "natural" reference plane is not the image plane but is the ground plane. In this context, "ground plane" refers to a plane that is horizontal with respect to the vehicle or to the gravity vector. The representation $z = f(x, y)$ is then the usual concept of an elevation map. To transform the data points into an elevation map is useful only if one has a way to access them. The most common approach is to discretize the (x, y) plane into a grid. Each grid cell (x_i, y_i) is the trace of a vertical column in space, its *field* (Figure 7.10). All the data that falls within a cell's field is stored in that cell. The description shown in Figure 7.10 does not necessarily reflect the actual implementation of an elevation map but is more of a framework in which we develop the terrain representation algorithms. As we shall see later, the actual implementation depends on the level of detail that needs to be included in the terrain description.

Although the elevation map is a natural concept for terrain representations, it exhibits a number of problems due to the conversion of a regularly sampled image to a different reference plane [KHK88a]. Although we propose solutions to these problems in Section 7.3.5, it is important to keep them in mind while we investigate other terrain representations. The first problem is the sampling problem illustrated in Figure 7.11. Since we perform some kind of image warping, the distribution of data points in the elevation map is not uniform, and as a result conventional image processing algorithms cannot be applied directly to the map. There are two ways to get around the sampling problem: We can either use a base structure that is not a regularly spaced grid, such as a Delaunay triangulation of the data points [OR87], or we can interpolate between data points to build a dense elevation map. The former solution is not very practical because of the complex algorithms required to access data points and their

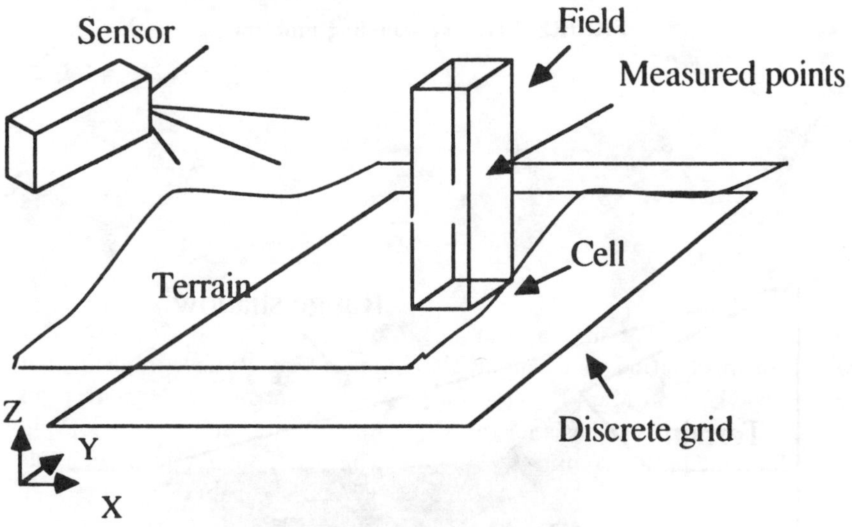

FIGURE 7.10. Structure of an elevation map

neighborhoods. We describe an implementation of the latter approach in Section 7.3.5. A second problem with elevation maps is the representation of the range shadows created by some objects (Figure 7.12). Since no information is available within the shadowed regions of the map, we must represent them separately so that no interpolation takes place across them and no "phantom" features are reported to the path planner. Finally, we have to convert the noise on the original measurements into a measure of uncertainty on the z value at each grid point (x, y). This conversion is difficult due to the fact that the sensor's uncertainty is most naturally represented with respect to the direction of measurement (Figure 7.13) and therefore spreads across a whole region in the elevation map.

7.3.2 TERRAIN REPRESENTATIONS AND PATH PLANNERS

The choice of a terrain representation depends on the path planner used for actually driving the vehicle. For example, the family of planners derived from the Lozano-Perez's A^* approach [LP79] uses discrete obstacles represented by 2-D polygons. By contrast, planners that compare a vehicle model with the local terrain [DHR88,Ste88] use some intermediate representation of the raw elevation map. Furthermore, the choice of a terrain representation and a path planner in turn depend on the environment in which the vehicle has to navigate. For example, representing only a small number of discrete upright objects may be appropriate if it is known in advance that the terrain is mostly flat, (*e.g.* a road) with a few obstacles (*e.g.* trees) while cross-country navigation requires a more detailed description of the eleva-

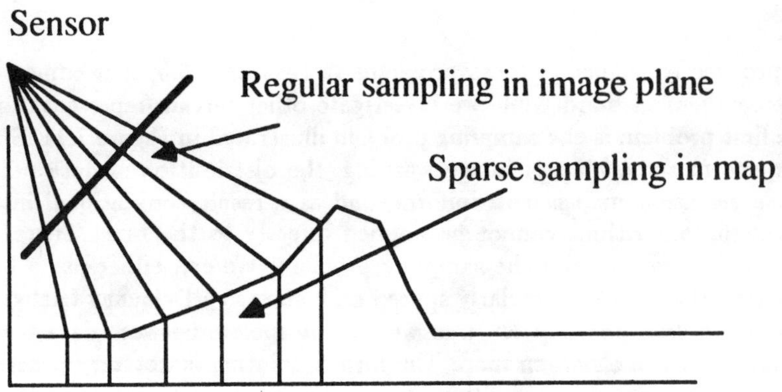

FIGURE 7.11. The sampling problem

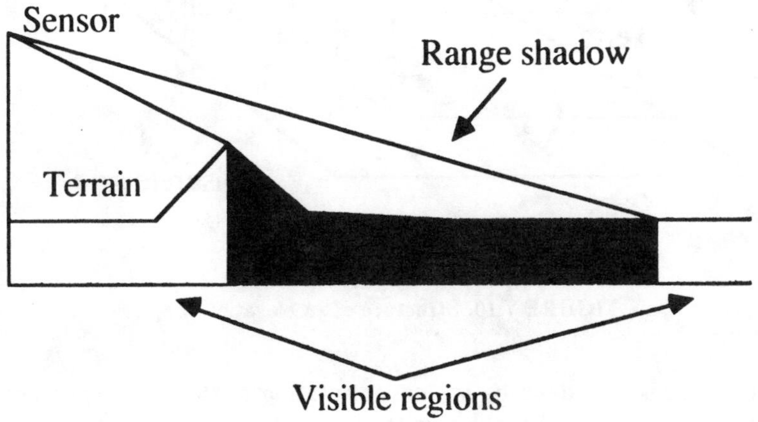

FIGURE 7.12. An example of a range shadow

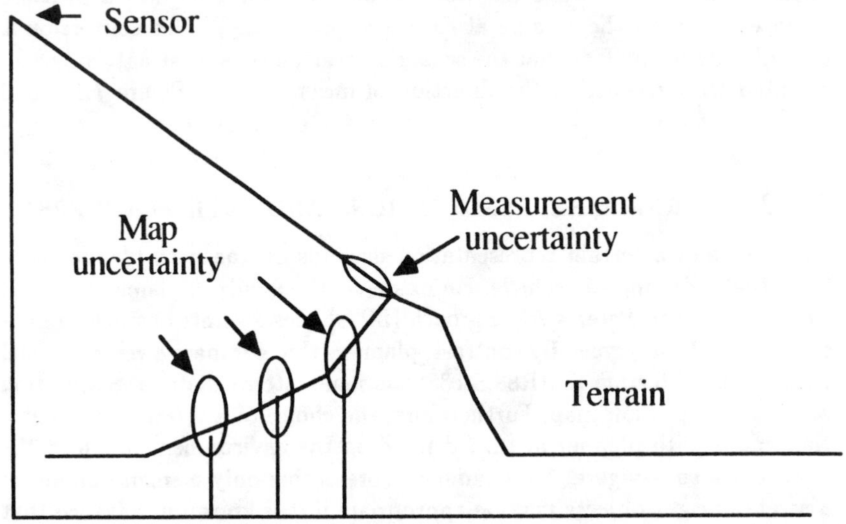

FIGURE 7.13. Representing uncertainty

tion map. Generating the most detailed description and then extracting the relevant information is not an acceptable solution since it would significantly degrade the performance of the system in simple environments. Therefore, we need several levels of terrain representation corresponding to different resolutions at which the terrain is described (Figure 7.14). At the low resolution level we describe only discrete obstacles without explicitly describing the local shape of the terrain. At the medium level, we include a description of the terrain through surface patches that correspond to significant terrain features. At that level, the resolution is the resolution of the operator used to detect these features. Finally, the description with the highest resolution is a dense elevation map whose resolution is limited only by the sensor. In order to keep the computations involved under control, the resolution is typically related to the size of the vehicle's parts that enter in contact with the terrain. For example, the size of one foot is used to compute the terrain resolution in the case of a legged vehicle.

7.3.3 Low resolution: Obstacle map

The lowest resolution terrain representation is an obstacle map which contains a small number of obstacles represented by their trace on the ground plane. Several techniques have been proposed for obstacle detection. The Martin-Marietta ALV [DM86,Dun88,TMGM88] detects obstacles by computing the difference between the observed range image and pre-computed images of ideal ground at several different slope angles. Points that are far

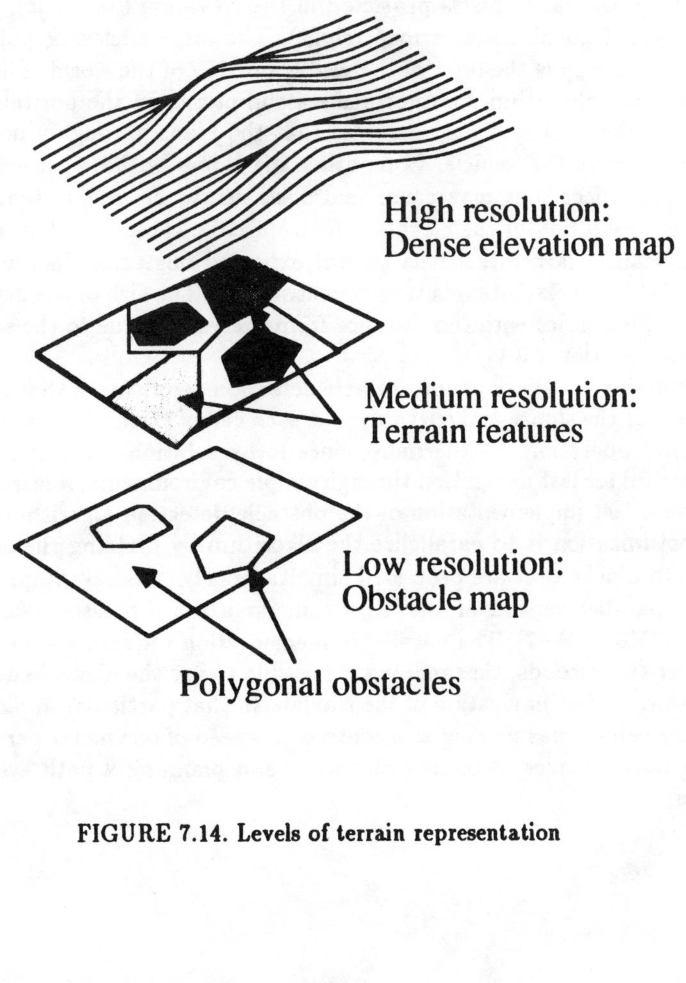

FIGURE 7.14. Levels of terrain representation

from the ideal ground planes are grouped into regions that are reported as obstacles to a path planner. A very fast implementation of this technique is possible since it requires only image differences and region grouping. It makes, however, very strong assumptions on the shape of the terrain. It also takes into account only the positions of the potential obstacle point, and as a result a very high slope ridge that is not deep enough would not be detected.

Another approach proposed by Hughes AI group [DHR87] is to detect the obstacles by thresholding the normalized range gradient, $\Delta D/D$, and by thresholding the radial slope, $D\Delta\phi/\Delta D$. The first test detects the discontinuities in range, while the second test detects the portion of the terrain with high slope. This approach has the advantage of taking a vehicle model into account when deciding whether a point is part of an obstacle. We used the terrain map paradigm to detect obstacles for the Navlab. Each cell of the terrain contains the set of data points that fall within its field (Figure 7.10). We can then estimate surface normal and curvatures at each elevation map cell by fitting a reference surface to the corresponding set of data points. Cells that have a high curvature or a surface normal far from the vehicle's idea of the vertical direction are reported as part of the projection of an obstacle. Obstacle cells are then grouped into regions corresponding to individual obstacles. The final product of the obstacle detection algorithm is a set of 2-D polygonal approximations of the boundaries of the detected obstacles that is sent to an A^*-type path planner (Figure 7.15). In addition, we can roughly classify the obstacles into holes or bumps according to the shape of the surfaces inside the polygons.

Figure 7.16 shows the result of applying the obstacle detection algorithm to a sequence of ERIM images. The Figure shows the original range images (top), the range pixels projected in the elevation map (left), and the resulting polygonal obstacle map (right). The large enclosing polygon in the obstacle map is the limit of the visible portion of the world. The obstacle detection algorithm does not make assumptions on the position of the ground plane in that it only assumes that the plane is roughly horizontal with respect to the vehicle. Computing the slopes within each cell has a smoothing effect that may cause real obstacles to be undetected. Therefore, the resolution of the elevation map must be chosen so that each cell is significantly larger than the typical expected obstacles. In the case of Figure 7.16, the resolution is twenty centimeters. The size of the detectable obstacle also varies with the distance from the vehicle due to the sampling problem (Section 7.3.1).

One major drawback of our obstacle detection algorithm is that the computation of the slopes and curvatures at each cell of the elevation map is an expensive operation. Furthermore, since low-resolution obstacle maps are most useful for fast navigation through simple environments, it is important to have a fast implementation of the obstacle detection algorithm. A natural optimization is to parallelize the algorithm by dividing the elevation map into blocks that are processed simultaneously. We have implemented such a parallel version of the algorithm on a ten-processor Warp computer [WK86,KW87]. The parallel implementation reduced the cycle time to under two seconds, thus making it possible to use the obstacle detection algorithm for fast navigation of the Navlab. In that particular implementation, the vehicle was moving at a continuous speed of one meter per second, taking range images, detecting obstacles, and planning a path every four meters.

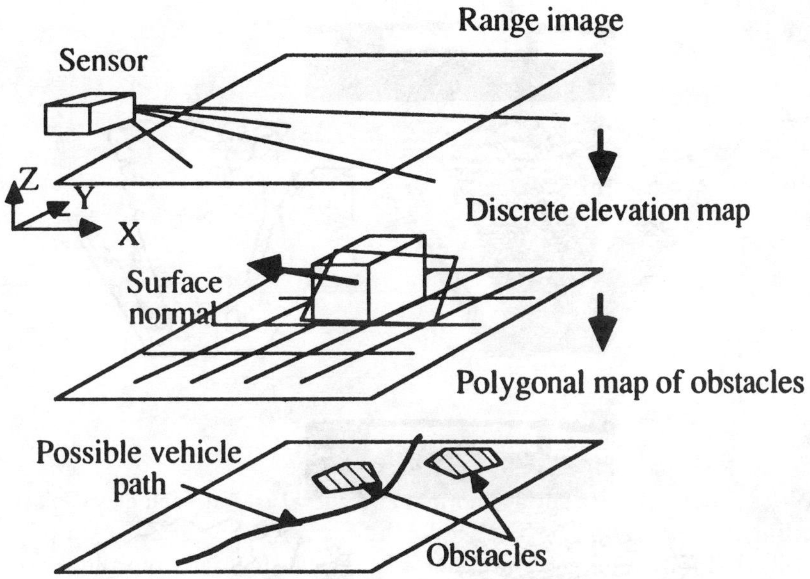

FIGURE 7.15. Building the obstacle map

7.3.4 Medium resolution: Polygonal terrain map

Obstacle detection is sufficient for navigation in flat terrain with discrete obstacles, such as following a road bordered by trees. We need a more detailed description when the terrain is uneven as in the case of cross-country navigation. For that purpose, an elevation map could be used directly [DHR88] by a path planner. This approach is costly because of the amount of data to be handled by the planner which does not need such a high resolution description to do the job in many cases (although we will investigate some applications in which a high resolution representation is required in Section 7.3.5). An alternative is to group smooth portions of the terrain into regions and edges that are the basic units manipulated by

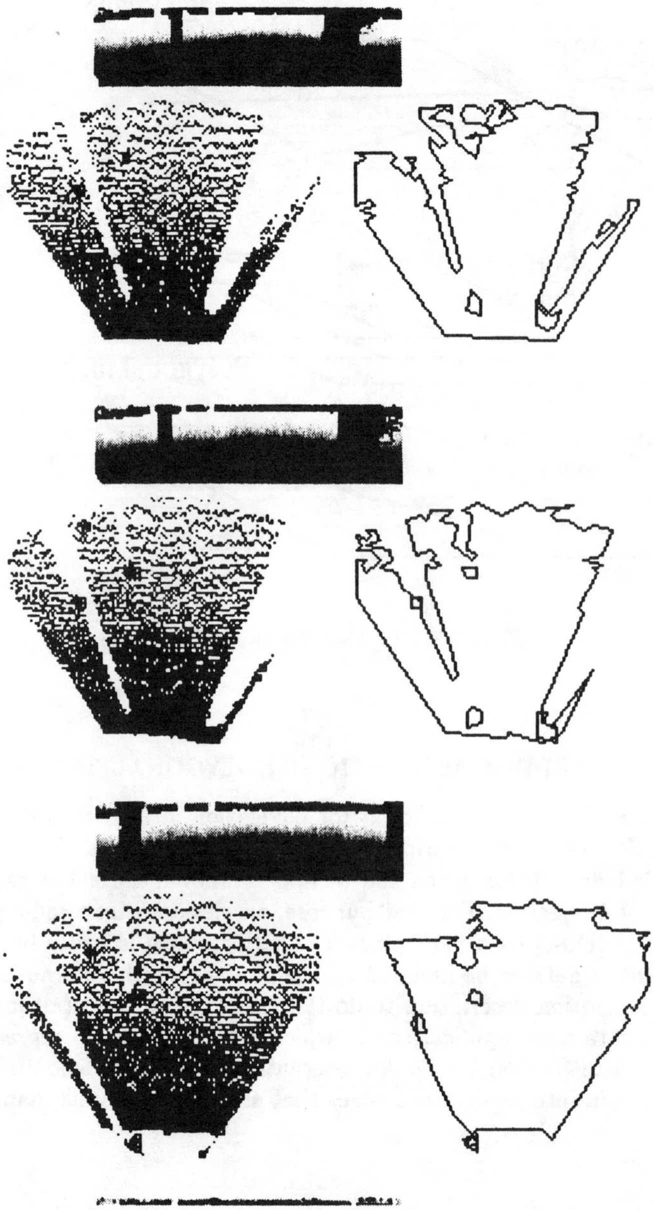

FIGURE 7.16. Obstacle detection on a sequence of images

the planner. This set of features provides a compact representation of the terrain thus allowing for more efficient planning [Ste88].

The features used are of two types: smooth regions, and sharp terrain discontinuities. The terrain discontinuities are either discontinuities of the elevation of the terrain, as in the case of a hole, or discontinuities of the surface normals, as in the case of the shoulder of a road [BC86]. We detect both types of discontinuities by using an edge detector over the elevation map and the surface normals map. The edges correspond to small regions on the terrain surface. Once we have detected the discontinuities, we segment the terrain into smooth regions. The segmentation uses a region growing algorithm that first identifies the smoothest locations in the terrain based on the behavior of the surface normals, and then grows regions around those locations. The result of the processing is a covering of the terrain by regions corresponding either to smooth portions or to edges.

The final representation depends on the planner that uses it. In our case, the terrain representation is embedded in the Navlab system using the path planner described in [Ste88]. The basic geometric object used by the system is the three-dimensional polygon. We therefore approximate the boundary of each region by a polygon. The approximation is done in a way that ensures consistency between regions in that the polygonal boundaries of neighboring regions share common edges and vertices. This guarantees that no "gaps" exist in the resulting polygonal mesh. This is important from the point of view of the path planner since such gaps would be interpreted as unknown portions of the terrain. Each region is approximated by a planar surface that is used by the planner to determine the traversability of the regions. Since the regions are not planar in reality, the standard deviation of the parameters of the plane is associated with each region.

Figure 7.18 shows the polygonal boundaries of the regions extracted from the image of Figure 7.17. In this implementation, the resolution of the elevation map is twenty centimeters. Since we need a dense map in order to extract edges, we interpolated linearly between the sparse points of the elevation map. Figure 7.17 shows the interpolated elevation map. This implementation of a medium resolution terrain representation is integrated in the Navlab system and will be part of the standard core system for our future mobile robot systems.

7.3.5 High resolution: Elevation maps for rough terrain

The elevation map derived directly from the sensor is sparse and noisy, especially at greater distances from the sensor. Many applications, however, need a dense and accurate high resolution map. One way to derive such a map is to interpolate between the data points using some mathematical approximation of the surface between data points. The models that

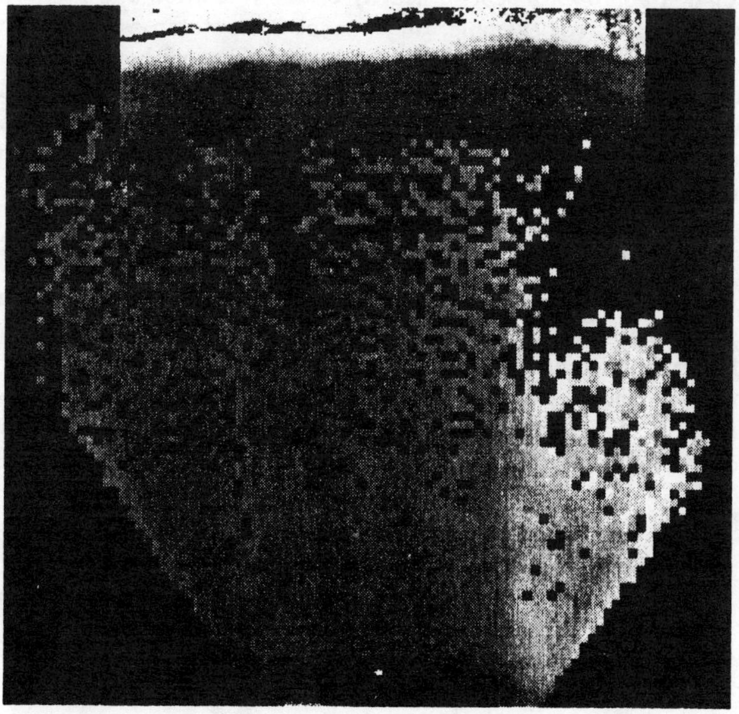

FIGURE 7.17. Range image and elevation map

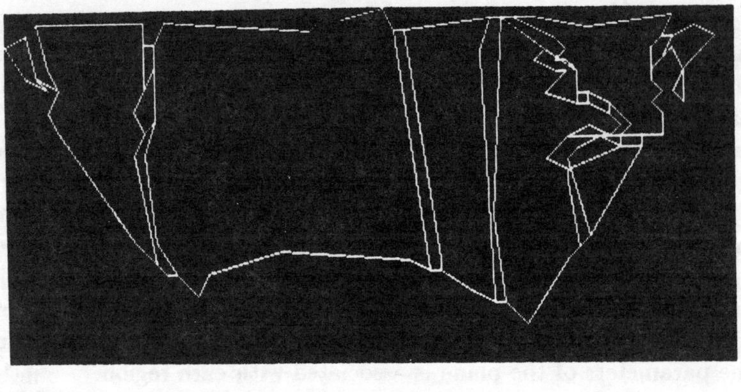

FIGURE 7.18. Polygonal boundaries of terrain regions

can be used include linear, quadratic, or bicubic surfaces [OR87]. Another approach is to fit a surface globally under some smoothness assumptions. This approach includes the family of regularization algorithms [BZ87b] in which a criterion of the form:

$$\int \|h_{data} - h_{interpolation}\|^2 + \lambda \int f(h_{interpolation}) \qquad (7.7)$$

is minimized, where f is a regularization function that reflects the smoothness model (*e.g.* thin plate). Two problems arise with both interpolation approaches: They make apriori assumptions on the local shape of the terrain which may not be valid (*e.g.* in the case of very rough terrain), and they do not take into account the image formation process since they are generic techniques independent of the origin of the data. In addition, the interpolation approaches depend heavily on the resolution and position of the reference grid. For example, they cannot compute an estimate of the elevation at an (x, y) position that is not a grid point without resampling the grid. We propose an alternative, the *locus* algorithm [KHK88a], that uses a model of the sensor and provides interpolation at arbitrary resolution without making any assumptions on the terrain shape other than the continuity of the surface.

The locus algorithm for the optimal interpolation of terrain maps

The problem of finding the elevation z of a point (x, y) is trivially equivalent to computing the intersection of the surface observed by the sensor and the vertical line passing through (x, y). The basic idea of the locus algorithm is to convert the latter formulation into a problem in image space (Figure 7.19). A vertical line is a curve in image space, the *locus*, whose equation as a function of ϕ is:

$$D = D_l(\phi) = \sqrt{\frac{y^2}{\cos^2\phi} + x^2} \qquad (7.8)$$

$$(7.9)$$

$$\theta = \theta_l(\phi) = \arctan\frac{x\cos\phi}{y}$$

where ϕ, θ, and D are defined as in Section 7.2. Equation (7.9) was derived by inverting Equation (7.2), and assuming x and y constant. Similarly, the range image can be viewed as a surface $D = I(\phi, \theta)$ in ϕ, θ, D space. The problem is then to find the intersection, if it exists, between a curve parametrized by ϕ and a discrete surface. Since the surface is known only from a sample of data, the intersection cannot be computed analytically. Instead, we have to search along the curve for the intersection point. The search proceeds in two stages: We first locate the two scanlines of the range image, ϕ_1 and ϕ_2, between which the intersection must be located, that is the two consecutive scanlines such that, $Diff(\phi_1) = D_l(\phi_1) - I(\phi_1, \hat{\theta}_l(\phi_1))$ and $Diff(\phi_2) = D_l(\phi_1) - I(\phi_2, \hat{\theta}_l(\phi_2))$ have opposite signs, where $\hat{\theta}_l(\phi)$ is the image column that is the closest to $\theta_l(\phi)$. We then apply a binary search between ϕ_1 and ϕ_2. The search stops when the difference between the two angles ϕ_n and ϕ_{n+1}, where $Diff(\phi_n)$ and $Diff(\phi_{n+1})$ have opposite signs, is lower than a threshold ϵ. Since there are no pixels between ϕ_1 and ϕ_2, we have to perform a local quadratic interpolation of the image in order to compute $\theta_l(\phi)$ and $D_l(\phi)$ for $\phi_1 < \phi < \phi_2$. The control points for the interpolation are the four pixels that surround the intersection point (Figure 7.20). The final result is a value ϕ that is converted to an elevation value by applying Equation (7.2) to $\phi, \theta_l(\phi), D_l(\phi)$. The resolution of the elevation is controlled by the choice of the parameter ϵ.

The locus algorithm enables us to evaluate the elevation at any point since we do not assume the existence of a grid. Figure 7.21 shows the result of applying the locus algorithm on range images of uneven terrain, in this case a construction site. The Figure shows the original range images and the map displayed as an isoplot surface. The centers of the grid cells are ten centimeters apart in the (x, y) plane.

We can generalize the locus algorithm from the case of a vertical line to the case of a general line in space. This generalization allows us to build maps using any reference plane instead of being restricted to the (x, y) plane. This is important when, for example, the sensor's (x, y) plane is not orthogonal to the gravity vector. A line in space is defined by a point $u = [u_x, u_y, u_z]^t$, and a unit vector $v = [v_x, v_y, v_z]^t$. Such a line is parametrized in λ by the relation $p = u + \lambda v$ if p is a point on the line. A general line is still a curve in image space that can be parametrized in ϕ by eliminating λ. We can then compute the intersection between the curve and the image surface by using the same algorithm as before except with the new equation of the locus curve.

The representation of the line by the pair (u, v) is not optimal since it uses six parameters while only four parameters are needed to represent a line in space. For example, this can be troublesome if we want to compute the Jacobian of the intersection point with respect to the parameters of the line. A better alternative [KM63] is to represent the line by its slopes in x and y and by its intersection with the plane $z = 0$ (See [Rob88] for a complete survey of 3-D line representations). The equation of the line then becomes:

$$\begin{aligned} x &= az + p \\ y &= bz + q \end{aligned} \tag{7.10}$$

We can still use the same technique to compute the locus because we can switch between the (a, b, p, q) and (u, v) representations.

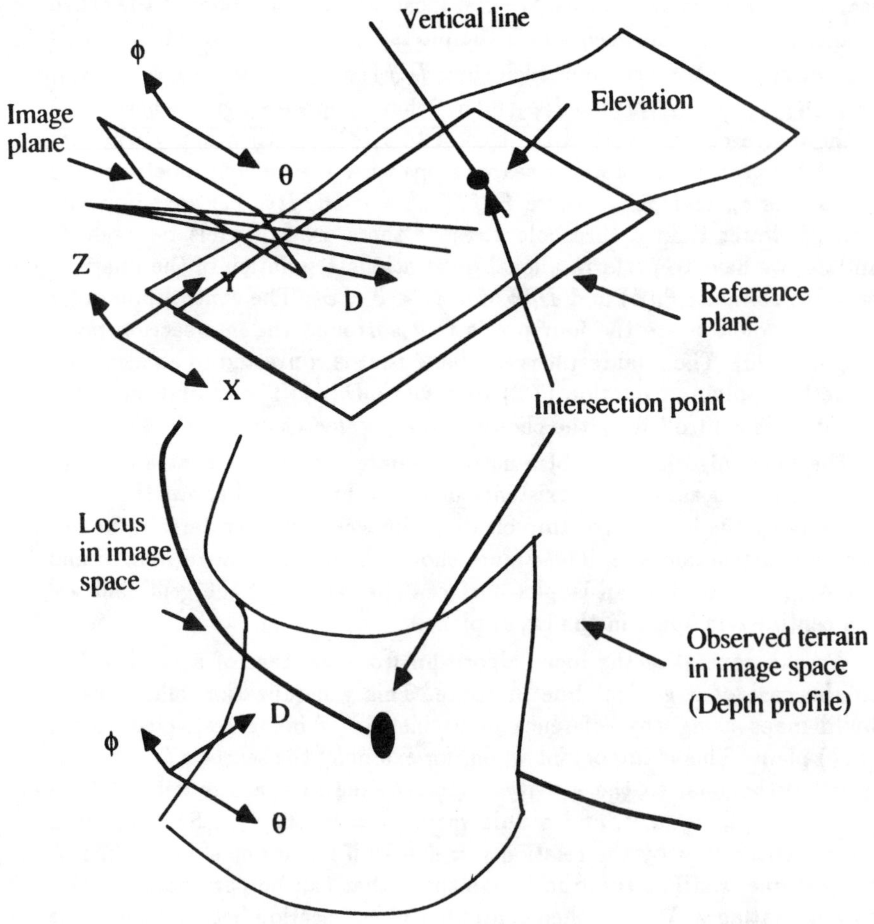

FIGURE 7.19. The locus algorithm for elevation maps

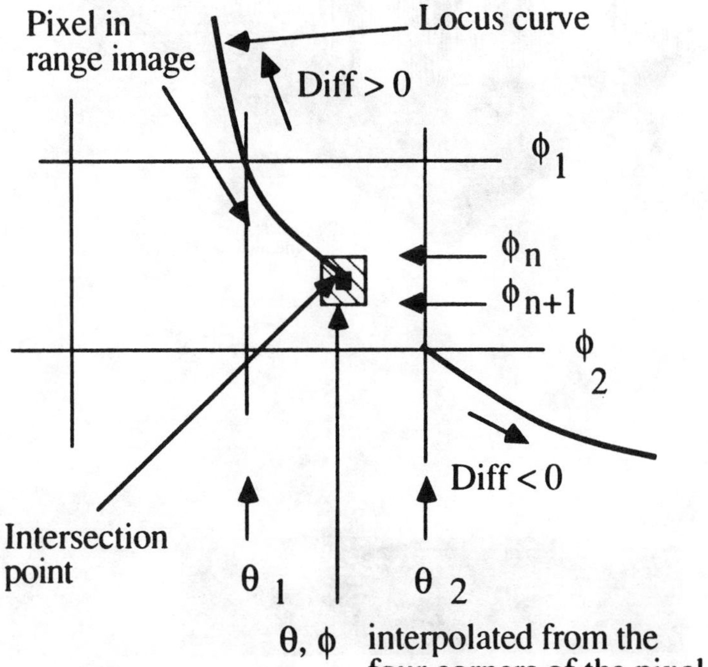

FIGURE 7.20. Image interpolation around the intersection point

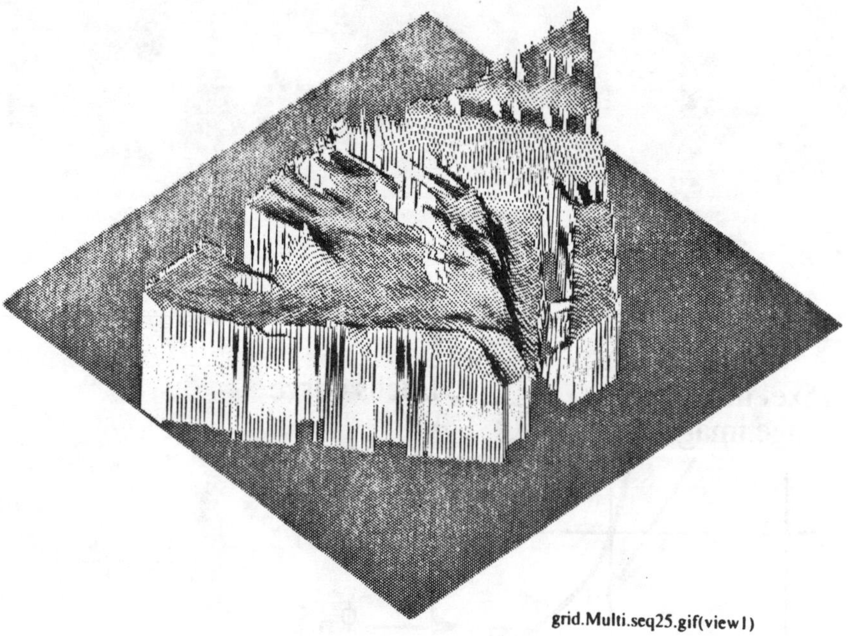

FIGURE 7.21. The locus algorithm on range images

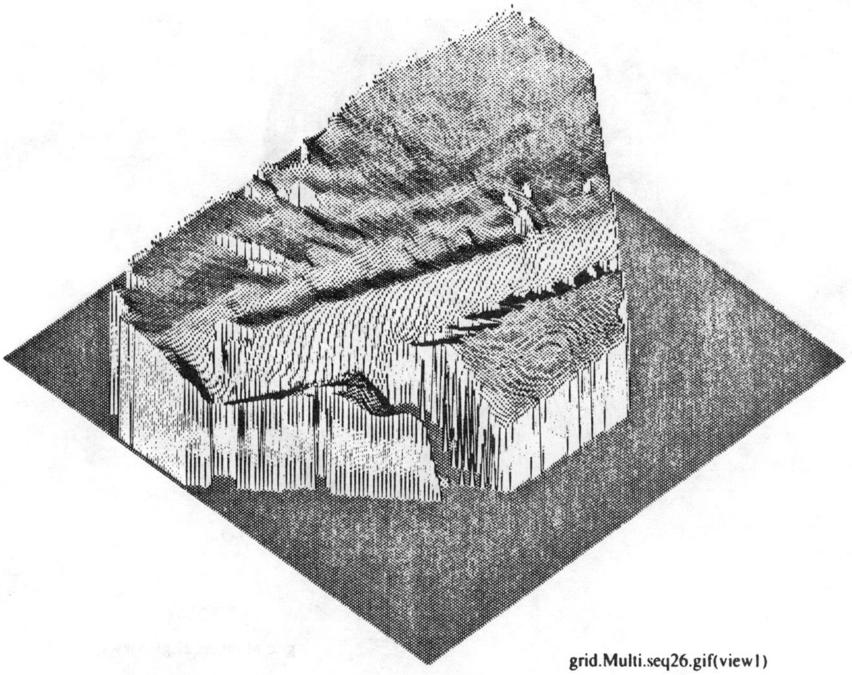

FIGURE 7.21. The locus algorithm on range images (Continued)

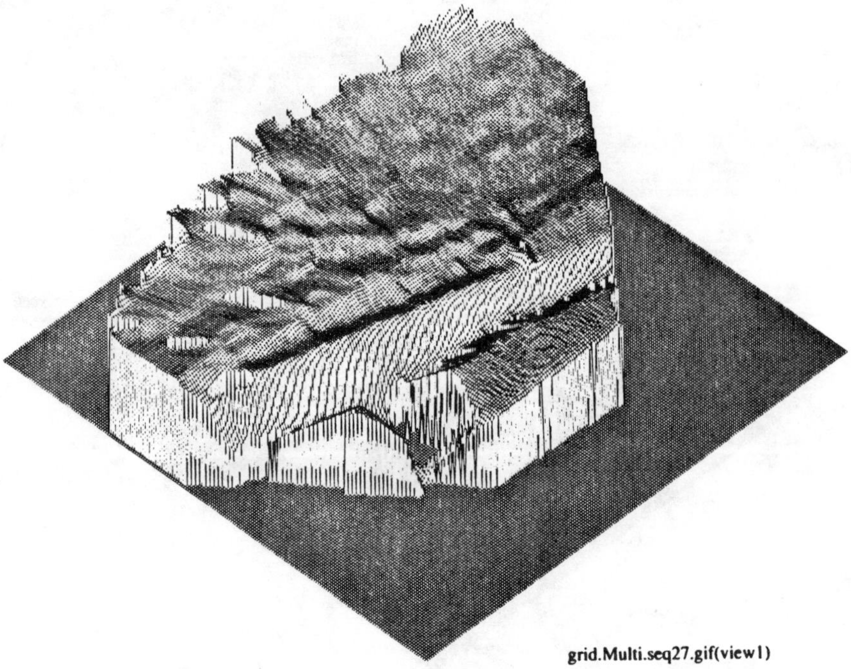

FIGURE 7.21. The locus algorithm on range images (Continued)

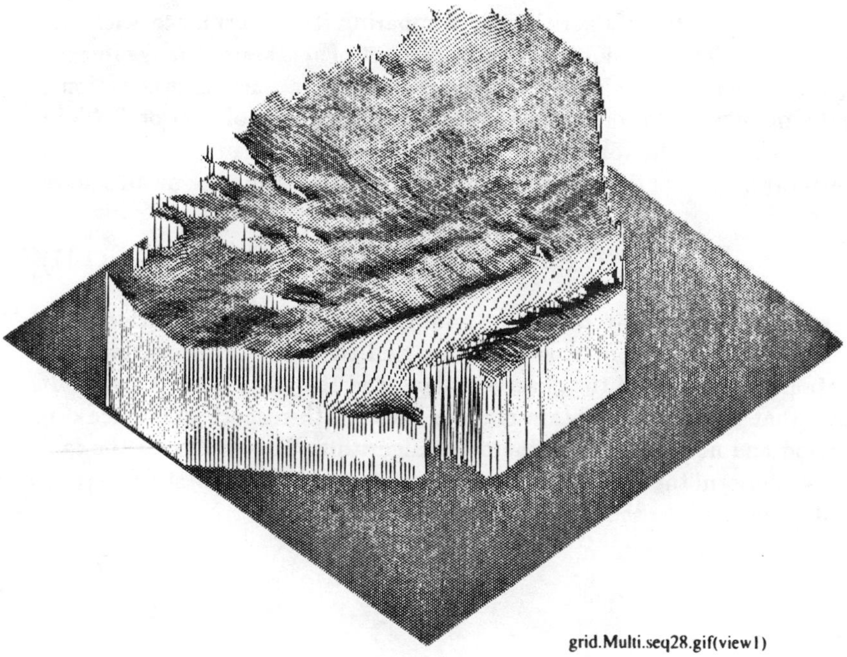

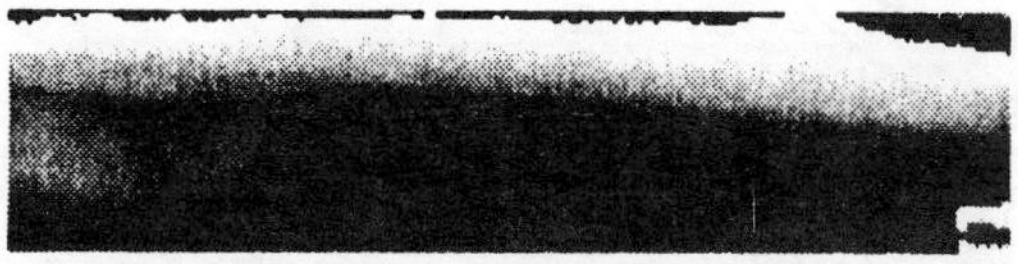

FIGURE 7.21. The locus algorithm on range images (Continued)

Evaluating the locus algorithm

We evaluate the locus algorithm by comparing its performance with the other "naive" interpolation algorithms on a set of synthesized range images of simple scenes. The simplest scenes are planes at various orientations. Furthermore, we add some range noise using the model of Section 7.2.3 in order to evaluate the robustness of the approach in the presence of noise. The performances of the algorithms are evaluated by using the mean square error:

$$E = \frac{\sum_{i=1}^{N}(h_i - \tilde{h}_i)^2}{N} \tag{7.11}$$

where h_i is the true elevation value and $\tilde{h}_i$ is the estimated elevation. Figure 7.22 plots E for the locus algorithm and the naive interpolation as a function of the slope of the observed plane and the noise level. This result shows that the locus algorithm is more stable with respect to surface orientation and noise level than the other algorithm. This is due to the fact that we perform the interpolation in image space instead of first converting the data points into the elevation map.

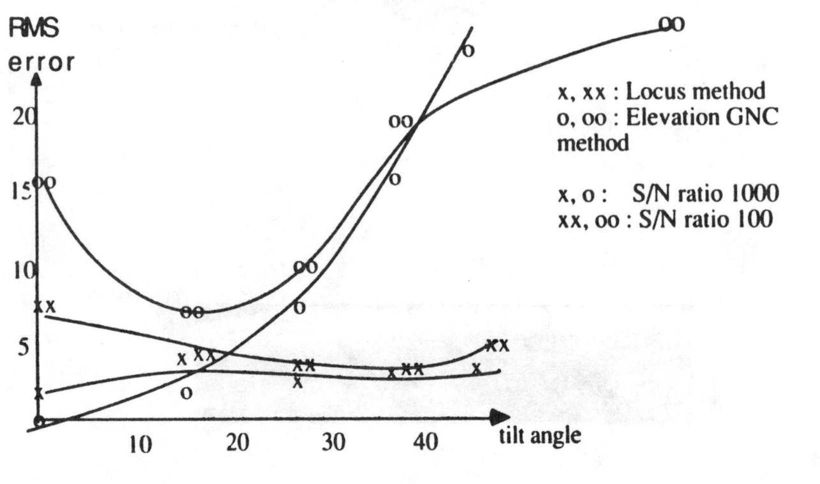

FIGURE 7.22. Evaluation of the locus algorithm on synthesized images

Representing the uncertainty

We have presented in Section 7.2.3 a model of the sensor noise that is a Gaussian distribution along the direction of measurement. We need to transform this model into a model of the noise, or uncertainty, on the elevation values returned by the locus algorithm. The difficulty here is that the uncertainty in a given range value spreads to many points in the elevation map, no matter how the map is oriented with respect to the image plane (Figure 7.13). We cannot therefore assume that the standard deviation of an elevation is the same as the one of the corresponding pixel in the range image. Instead, we propose to use the nature of the locus algorithm itself to derive a meaningful value for the elevation uncertainty. To facilitate the explanation, we consider only the case of the basic locus algorithm of Section 7.3.5 in which we compute an elevation z from the intersection of the locus of a vertical line with a depth profile from a range image. Figure 7.23 shows the principle of the uncertainty computation by considering a locus curve that corresponds to a line in space and the depth profile from the range image in the neighborhood of the intersection point, each point on the depth profile has an uncertainty whose density can be represented by a Gaussian distribution as computed in Section 7.2.3. The problem is to define a distribution of uncertainty along the line. The value of the uncertainty reflects how likely is the given point to be on the actual surface given the measurements.

Let us consider an elevation h along the vertical line. This elevation corresponds to a measurement direction $\phi(h)$ and a measured range $d'(h)$. If $d(h)$ is the distance between the origin and the elevation h, we assign to h the confidence [Sze88a]:

$$l(h) = \frac{1}{\sqrt{2\pi}\sigma(d'(h))} e^{-\frac{(d(h)-d'(h))^2}{2\sigma(d'(h))^2}} \tag{7.12}$$

where $\sigma(d'(h)$ is the variance of the measurement at the range $d'(h)$. Equation 7.12 does not tell anything about the shape of the uncertainty distribution $l(h)$ along the h axis except that it is maximum at the elevation h_o at which $d(h) = d'(h)$, that is the elevation returned by the locus algorithm. In order to determine the shape of $l(h)$, we approximate $l(h)$ around h_o by replacing the surface by its tangent plane at h_o. If α is the slope of the plane, and H is the elevation of the intersection of the plane with the z axis, we have:

$$\sigma(d'(h)) = K\frac{H^2(a^2+h^2)}{(a\tan\alpha + h)^2} \tag{7.13}$$

$$\frac{(d(h)-d'(h))^2}{2\sigma(d'(h))^2} = \frac{(h-h_o)^2(a\tan\alpha+h)^2}{K^2H^4(a^2+h^2)} \tag{7.14}$$

where a is the distance between the line and the origin in the $x-y$ plane and K is defined in Section 7.2.3 by $\sigma(d) \approx Kd^2$. By assuming that h is close to h_o, that is $h = h_o + \epsilon$ with $\epsilon \ll h_o$, and by using the fact that $H = h_o + a\tan\alpha$, we have the approximations:

$$\sigma(d'(h)) \approx K(a^2+h_o^2) \tag{7.15}$$

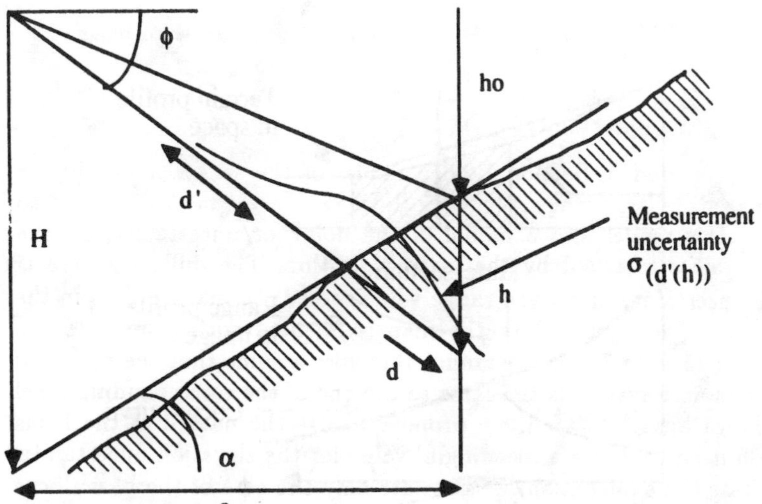

FIGURE 7.23. Computing the uncertainty from the locus algorithm

$$\frac{(d(h) - d'(h))^2}{2\sigma(d'(h))^2} \approx \frac{(h - h_o)^2}{2K^2H^2(a^2 + h_o^2)} \tag{7.16}$$

In the neighborhood of h_o, Equation 7.16 shows that $(d(h) - d'(h))^2/2\sigma(d'(h))^2$ is quadratic in $h - h_o$, and that $\sigma(d'(h))$ is constant. Therefore, $l(h)$ can be approximated by a Gaussian distribution of variance:

$$\sigma_h^2 = K^2H^2(a^2 + h_o^2) = K^2H^2d_o^2 \tag{7.17}$$

Equation 7.17 provides us with a first order model of the uncertainty of h derived by the locus algorithm. In practice, the distance $D(h) = (d(h) - d'(h))^2/2\sigma(d'(h))^2$ is computed for several values of h close to h_o, the variance σ_h is computed by fitting the function $(h - h_o)^2/2\sigma_h^2$ to the values of $D(h)$. This is a first order model of the uncertainty in the sense that it takes into account the uncertainty on the sensor measurements, but it does not include the uncertainty due to the locus algorithm itself, in particular the errors introduced by the interpolation.

Detecting the range shadows

As we pointed out in Section 7.3.1, the terrain may exhibit range shadows in the elevation map. It is important to identify the shadow regions because the terrain may have any shape within the boundaries of the shadows, whereas the surface would be smoothly interpolated if we applied the locus algorithm directly in those areas. This may result in dangerous situations for the robot if a path crosses one of the range shadows. A simple idea would be to detect empty regions in the raw elevation map, which are the projection of images in the map without any interpolation. This approach does not work because the size of the shadow regions may be on the order of the average distance between data points. This is especially true for shadows that are at some distance from the sensor in which case the distribution of data points is very sparse. It is possible to modify the standard locus algorithm so that it takes into account the shadow areas. The basic idea is that a range shadow corresponds to a strong occluding edge in the image (Figure 7.12). An (x, y) location in the map is in a shadow area if its locus intersects the image at a pixel that lies on such an edge (Figure 7.24).

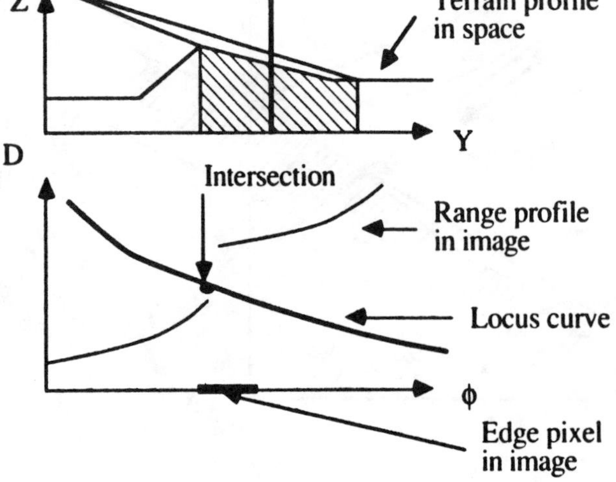

FIGURE 7.24. Detecting range shadows

We implement this algorithm by first detecting the edges in the range image by using a standard technique, the GNC algorithm [BZ87b]. We chose this algorithm because it allows us to vary the sensitivity of the edge detector across the image, and because it performs some smoothing of the image as a side effect. When we apply the locus algorithm we can then record the fact that the locus of a given location intersects the image at an edge pixel. Such map locations are grouped into regions that are the reported range shadows. Figure 7.25 shows an overhead view of an elevation map computed by the locus algorithm, the white points are the shadow points, the gray level of the other points is proportional to their uncertainty as computed in the previous Section.

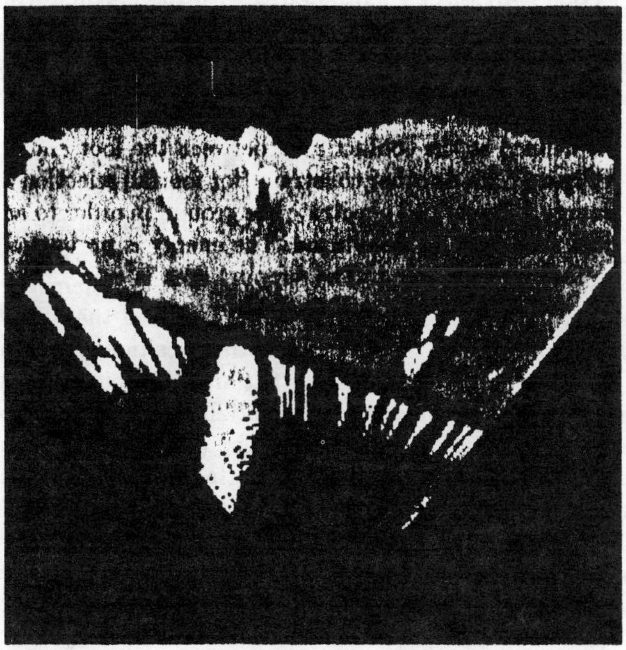

FIGURE 7.25. Shadow regions in an elevation map

An application: footfall selection for a legged vehicle

The purpose of using the locus algorithm for building terrain is to provide high resolution elevation data. As an example of an application in which such a resolution is needed, we briefly describe in this Section the problem of perception for a legged vehicle [Kwe88]. One of the main responsibilities of perception for a legged vehicle is to provide a terrain description that enables the system to determine whether a given foot placement, or *footfall*, is safe. In addition, we consider the case of locomotion on very rugged terrain such as the surface of Mars.

A foot is modeled by a flat disk of diameter 30 cms. The basic criterion

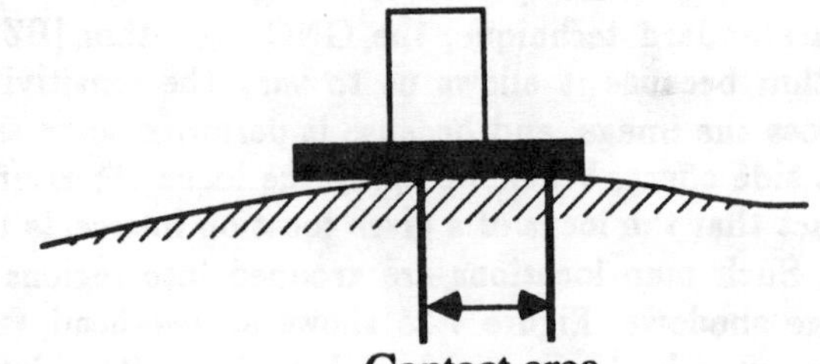

FIGURE 7.26. Footfall support area

for footfall selection is to select a footfall area with the maximum support area which is defined as the contact area between the foot and the terrain as shown in Figure 7.26. Another constraint for footfall selection is that the amount of energy necessary to penetrate the ground in order to achieve sufficient support area must be minimized. The energy is proportional to the depth of the foot in the ground. The support area is estimated by counting the number of map points within the circumference of the disk that are above the plane of the foot. This is where the resolution requirement originates because the computation of the support area makes sense only if the resolution of the map is significantly smaller than the diameter of the foot. Given a minimum allowed support area, S_{min}, and the high resolution terrain map, we can find the optimal footfall position within a given terrain area: First, we want to find possible flat areas by computing surface normals for each footfall area in a specified footfall selection area. Footfalls with a high surface normal are eliminated. The surface normal analysis, however, will not be sufficient for optimal footfall selection. Second, the support area is computed for the remaining positions. The optimal footfall position is the one for which the maximum elevation, h_{opt} that realizes the minimum support area S_{min} is the maximum across the set of possible footfall positions. Figure 7.27 shows a plot of the surface area with respect to the elevation from which h_{min} can be computed.

Extracting local features from an elevation map

The high resolution map enables us to extract very local features, such as points of high surface curvature, as opposed to the larger terrain patches of Section 7.3.4. The local features that we extract are based on the magnitude of the two principal curvatures of the terrain surface. The curvatures are computed as in [PB85] by first smoothing the map, and then computing the derivatives of the surface for solving the first fundamental form. Figure 7.28 shows the curvature images computed from an elevation map using the locus algorithm. The resolution of the map is ten centimeters. Points of high curvature correspond to edges of the terrain, such as the edges of

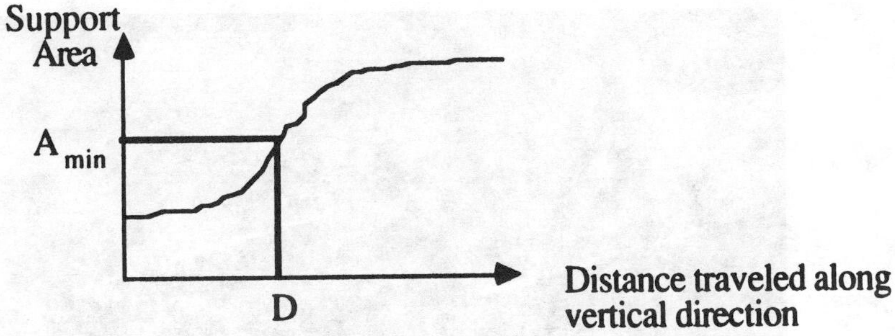

FIGURE 7.27. Support area versus elevation

a valley, or to sharp terrain features such as hills, or holes. In any case, the high curvature points are viewpoint-independent features that can be used for matching. We extract the high curvature points from both images of principal curvature. We group the extracted points into regions, then classify each region as point feature, line, or region, according to its size, elongation, and curvature distribution. Figure 7.28 shows the high curvature points extracted from an elevation map. The two images correspond to the two principal curvatures. Figure 7.29 shows the three types of local features detected on the map of Figure 7.28 superimposed in black over the original elevation map. The Figure shows that while some features correspond merely to local extrema of the surface, some such as the edges of the deep gully are characteristic features of the scene. This type of feature extraction plays an important role in Section 7.4 for combining multiple maps computed by the locus algorithm.

7.4 Combining multiple terrain maps

We have so far addressed the problem of building a representation of the environment from sensor data collected at one fixed location. In the case of mobile robots, however, we have to deal with a stream of images taken along the vehicle's path. We could ignore this fact and process data from each viewpoint as if it were an entirely new view of the world, thus forgetting whatever information we may have extracted at past locations. It has been observed that this approach is not appropriate for mobile robot navigation, and that there is a need for combining the representations computed

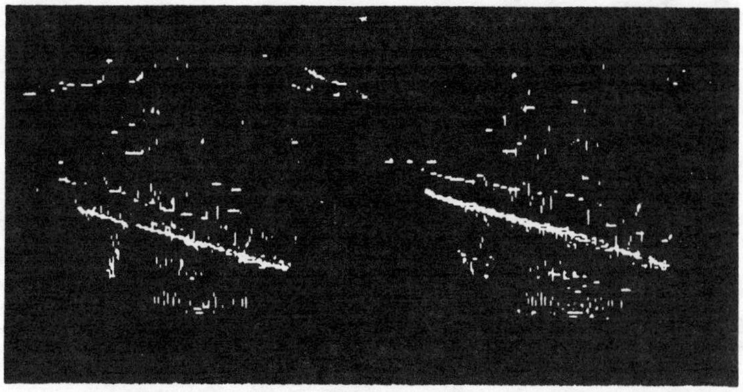

FIGURE 7.28. The high curvature points of an elevation map

FIGURE 7.29. Local features from a high resolution elevation map

from different vantage points into a coherent map. Although this has been observed first in the context of indoor mobile robots [FAF86,GCV84], the reasoning behind it holds true in our case. First of all, merging representations from successive viewpoints will produce a map with more information and better resolution than any of the individual maps. For example, a tall object observed by a range sensor creates an unknown area behind it, the range shadow, where no useful information can be extracted (Section 7.3.1). The shape and position of the range shadow changes as we move to another location; merging images from several locations will therefore reduce the size of the shadow, thus providing a more complete description to the path planner (Figure 7.30). Another reason why merging maps increases the resolution of the resulting representation concerns the fact that the resolution of an elevation map is significantly better at close range. By merging maps, we can increase the resolution of the parts of the elevation map that were originally measured at a distance from the vehicle.

The second motivation for merging maps is that the position of the vehicle at any given time is uncertain. Even when using expensive positioning systems, we have to assume that the robot's idea of its position in the world will degrade in the course of a long mission. One way to solve this problem is to compute the position with respect to features observed in the world instead of a fixed coordinate system [SC86,MS87]. That requires the identification and fusion of common features between successive observations in order to estimate the displacement of the vehicle (Figure 7.31). Finally, combining maps is a mission requirement in the case of an exploration mission in which the robot is sent into an unknown territory to compile a map of the observed terrain.

Many new problems arise when combining maps: representation of uncertainty, data structures for combined maps, predictions from one observation to the next etc. We shall focus on the terrain matching problem, that is the problem of finding common features or common parts between terrain maps so that we can compute the displacement of the vehicle between the two corresponding locations and then merge the corresponding portions of the terrain maps. We always make the reasonable assumption that a rough estimate of the displacement is available since an estimate can always be computed either from dead reckoning or from past terrain matchings.

7.4.1 THE TERRAIN MATCHING PROBLEM: ICONIC *vs.* FEATURE-BASED

In the terrain matching problem, as in any problem in which correspondences between two sets of data must be found, we can choose one of two approaches: feature-based or iconic matching. In feature-based matching, we first have to extract two sets of features (F_i^1) and (F_j^2) from the two views to be matched, and to find correspondences between features,

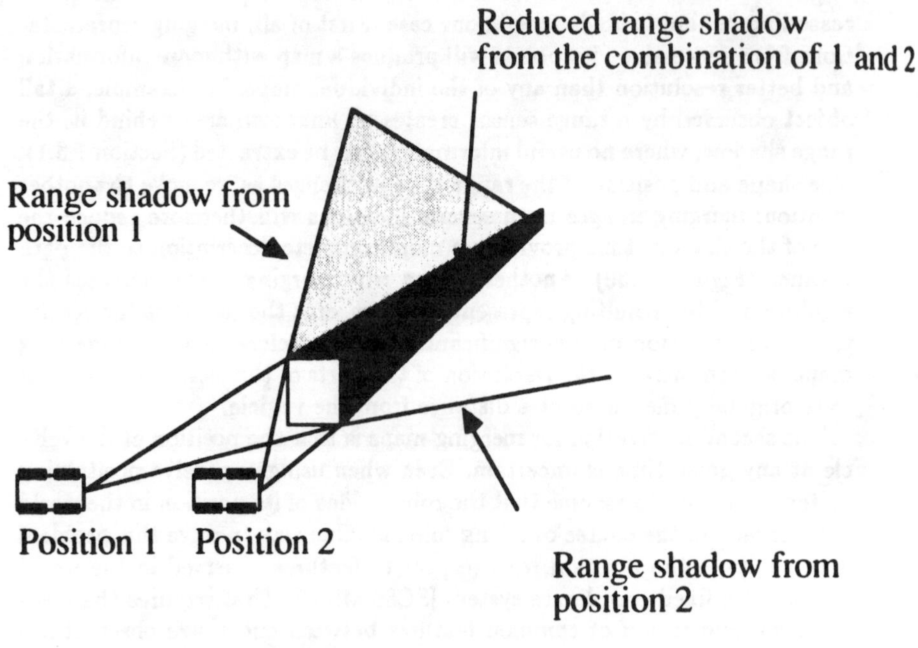

FIGURE 7.30. Reducing the range shadow

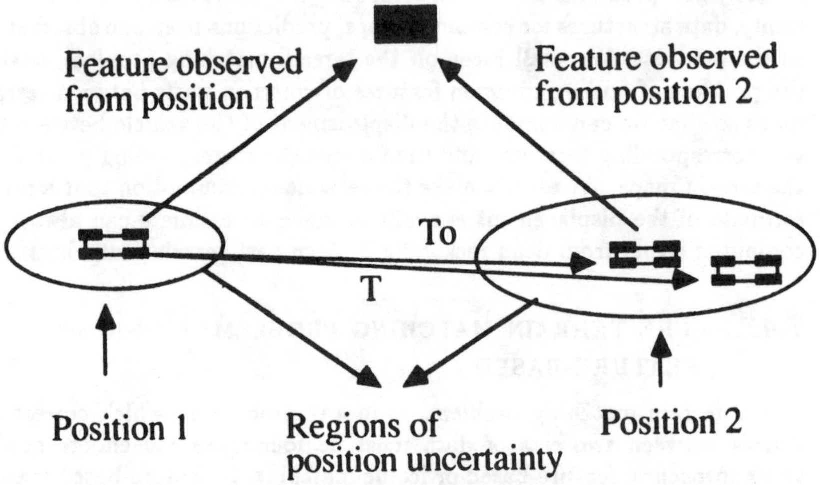

FIGURE 7.31. Matching maps for position estimation

$(F^1_{i_k}, F^2_{j_k})$ that are globally consistent. We can then compute the displacement between the two views from the parameters of the features, and finally merge them into one common map. Although this is the standard approach to object recognition problems [BJ85c], it has also been widely used for map matching for mobile robots [FAF86,KTB87,MS87,Bro85,Asa88,Tho84]. In contrast, iconic approaches work directly on the two sets of data points, P^1 and P^2 by minimizing a cost function of the form $F(T(P^2), P^1)$ where $T(P^2)$ is the set of points from view 2 transformed by a displacement T. The cost is designed so that its minimum corresponds to a "best" estimate of T in some sense. The minimization of F leads to an iterative gradient-like algorithm. Although less popular, iconic techniques have been successfully applied to incremental depth estimation [MS87,Luc85] and map matching [Sze88b,Elf87].

The proponents of each approach have valid arguments. The feature-based approach requires a search in the space of possible matches which may lead to a combinatorial explosion of the matching program. On the other hand, iconic approaches are entirely predictable in terms of computational requirements but are usually quite expensive since the size of the points sets P^i is typically on the order of several thousands. As for the accuracy of the resulting displacement T, the accuracy of iconic techniques can be better than the resolution of the sensors if we iterate the minimization of F long enough, while any feature extraction algorithm loses some of the original sensor accuracy. Furthermore, feature matching could in theory be used even if no a-priori knowledge of T, T_0, is available while iconic approaches require T_0 to be close to the actual displacement because of the iterative nature of the minimization of F.

Keeping these tenets in mind, we propose to combine both approaches into one terrain matching algorithm. The basic idea is to use the feature matching to compute a first estimate $\hat{T}$ given a rough initial value T_0, and then to use an iconic technique to compute an accurate estimate $\hat{\hat{T}}$. This has the advantage of retaining the level of accuracy of iconic techniques while keeping the computation time of the iconic stage under control because the feature matching provides an estimate close enough to the true value. We describe in detail the feature-based and iconic stages in the next three sections.

7.4.2 FEATURE-BASED MATCHING

Let F^1_i and F^2_j be two sets of features extracted from two images of an outdoor scene, I_1 and I_2. We want to find a transformation $\hat{T}$ and a set of pairs $C_k = (F^1_{i_k}, F^2_{j_k})$ such that $F^2_{j_k} \approx \hat{T}(F^1_{i_k})$, where $T(F)$ denotes the transformed by T of a feature F. The features can be any of those discussed in the previous Sections: points or lines from the local feature extractor, obstacles represented by a ground polygon, or terrain patches represented by their surface equation and their polygonal boundaries. We first investigate the feature matching algorithm independently of any particular feature type so that we can then apply it to any level of terrain representation.

For each feature F^1_i, we can first compute the set of features F^2_{ij} that could correspond to F^1_i given an initial estimate T_0 of the displacement. The F^2_{ij}'s should lie in a prediction region centered at $T_0(F^1_i)$. The size of the prediction region depends on the confidence we have in T_0 and in the feature extractors. For example, the centers of the polygonal obstacles of Section 7.3.4 are not known accurately, while the curvature points from Section 7.3.5. can be accurately located. The confidence on the displacement T is represented by the maximum distance δ between a point in image 1 and the transformed of its homologue in image 2, $||Tp^2 - p^1||$, and by the maximum angle ϵ, between a vector in image 2 and the transformed of its homologue in image 1 by the rotation part of T. The prediction is then defined as the set of features that are at a Cartesian distance lower than δ, and at an angular distance lower than ϵ from $T_0(F_i^2)$. The parameters used to determine if a feature belongs to a prediction region depend on the type of that feature. For example, we use the direction of a line for the test on the angular distance, while the center of an obstacle is used for the test on the Cartesian distance. Some features may be tested only for orientation, such as lines, or only for position, such as point features. The features in each prediction region are sorted according to some feature distance $d(F_i^1, T_0(F_{ij}^2))$ that reflects how well the features are matched. The feature distance depends also on the type of the feature: for points we use the usual distance, for lines we use the angles between the directions, and for polygonal patches (obstacles or terrain patches) we use a linear combination of the distance between the centers, the difference between the areas, the angle between the surface orientations, and the number of neighboring patches. The features in image 1 are also sorted according to an "importance" measure that reflects how important the features are for the matching. Such importance measures include the length of the lines, the strength of the point features (*i.e.* the curvature value) , and the size of the patches. The importance measure also includes the type of the features because some features such as obstacles are more reliably detected than others, such as point features.

Once we have built the prediction regions, we can search for matches between the two images. The search proceeds by matching the features F_i^1 to the features F_{ij}^2 that are in their prediction region starting at the most important feature. We have to control the search in order to avoid a combinatorial explosion by taking advantage of the fact that each time a new match is added both the displacement and the future matches are further constrained. The displacement is constrained by combining the current estimate T with the displacement computed from a new match (F_i^1, F_{ij}^2). Even though the displacement is described by six components, the number of components of the displacement that can be computed from one single match depends on the type of features involved: point matches provide only three components, line matches provide four components (two rotations and two translations), and region matches provide three components. We therefore combine the components of T with those components of the new match that can be computed. A given match prunes the search by constraining the future potential matches in two ways: if connectivity relations between features are available, as in the case of terrain patches, then a match (F_i^1, F_{ij}^2) constrains the possible matches for the neighbors of F_i^1) in that they have to be adjacent to F_{ij}^2. In the case of points or patches, an additional constraint is induced by the relative placement of the features in the scene: two matches, (F_i^1, F_{ij}^2) and $(F_{i'}^1, F_{i'j'}^2)$, are compatible only if the angle between the vectors $w^1 = \overline{F_{i'}^1 F_i^1}$ and $w^2 = \overline{F_{i'j'}^2 F_{ij}^2}$ is lower than π, provided the rotation part of T is no greater than π which is the case in realistic situations. This constraint means that the relative placement of the features remains the same from image to image which is similar to the classical ordering constraint used in stereo matching.

The result of the search is a set of possible matchings, each of which is a set of pairs $S = (F_{i_k}^1, F_{j_k}^2)_k$ between the two sets of features. Since we evaluated T simply by combining components in the course of the search, we have to evaluate T for each S in order to get an accurate estimate. T is estimated by minimizing an error function of the form:

$$E = \sum_k d(F_{i_k}^1 - T(F_{i_k}^2)) \tag{7.18}$$

The distance $d(.)$ used in Equation (7.18) depends on the type of the features involved: For point features, it is the usual distance between two points; for lines it is the weighted sum of the angle between the two lines and the distance between the distance vectors of the two lines; for regions it is the weighted sum of the distance between the unit direction vectors and the distance between the two direction vectors. All the components of T can be estimated in general by minimizing E. We have to carefully identify, however, the cases in which insufficient features are present in the scene to fully constrain the transformation. The matching S that realizes the minimum E is reported as the final match between the two maps while the corresponding displacement $\hat{T}$ is reported as the best estimate of the displacement between the two maps. The error $E(\hat{T})$ can then be used to represent the uncertainty in T.

This approach to feature based matching is quite general so that we can apply it to many different types of features, provided that we can define the distance $d(.)$ in Equation (7.18), the importance measure, and the feature measure. The approach is also fairly efficient as long as δ and ϵ do not become too large, in which case the search space becomes itself large. We describe two implementations of the feature matching algorithm in the next two Sections.

Example: Matching polygonal representations

We have implemented the feature-based matching algorithm on the polygonal descriptions of Section 7.3.4 and 7.3.3. The features are in this case:

- The polygons describing the terrain parametrized by their areas, the equation of the underlying surface, and the center of the region
- The polygons describing the trace of the major obstacles detected (if any).
- The road edges found in the reflectance images if the road detection is reliable enough. The reliability is measured by how much a pair of road edges deviates from the pair found in the previous image.

The obstacle polygons have a higher weight in the search itself because their detection is more reliable than the terrain segmentation, while the terrain regions and the road edges contribute more to the final estimate of the displacement since their localization is better. Once a set of matches and a displacement T are computed, the obstacles and terrain patches that are common between the current map and a new image are combined into new polygons, the new features are added to the map while updating the connectivity between features.

This application of the feature matching has been integrated with the rest of the Navlab system. In the actual system, the estimates of the displacement T_0 are taken from the central database that keeps track of the vehicle's position. The size of prediction region is fixed with $\delta =$ one meter, and $\epsilon = 20^\circ$. This implementation of the feature matching has performed successfully over the course of runs of several hundred meters. The final product of the matching is a map that combines all the observations made during the run, and a list of updated obstacle descriptions that are sent to a map module at regular intervals. Since errors in determining position tend to accumulate during such long runs, we always keep the map centered around the current vehicle position. As a result, the map representation is always accurate close to the current vehicle position. As an example, Figure 7.34 shows the result of the matching on five consecutive images separated by about one meter. The scene in this case is a road bordered by a few trees. Figure 7.32 shows the original sequence of raw range and reflectance images, Figure 7.33 shows perspective views of the corresponding individual maps, and Figure 7.34 is a rendition of the combined maps using the displacement and matches computed from the feature matching algorithm. This last display is a view of the map rotated by 45° about the x axis and shaded by the values from the reflectance image.

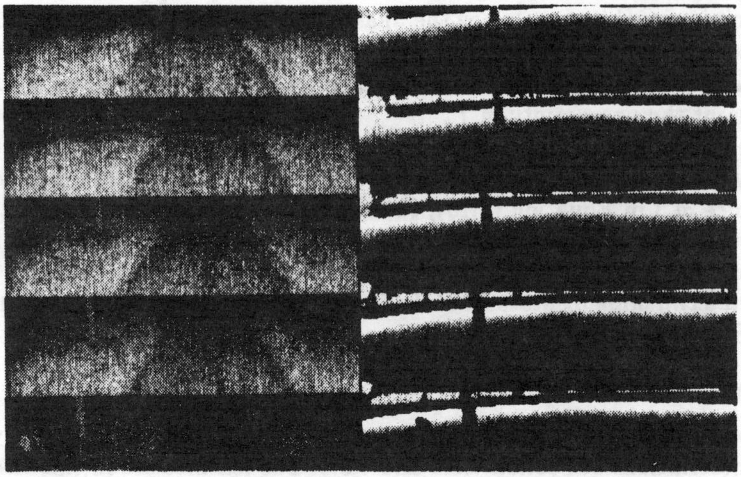

FIGURE 7.32. A sequence of range and reflectance images

Example: Matching local features from high resolution maps

Matching local features from high resolution maps provides the displacement estimate for the iconic matching of high resolution maps. The primitives used for the matching are the high curvature points and lines described in Section 7.3.5. The initial matches are based on the similarity of the length of the lines and the similarity of the curvature strength of the points. The search among candidate matches proceeds as described in Section 7.4.2. Since we have dense elevation at our disposal in this case, we can evaluate a candidate displacement over the entire map by summing up the squared differences between points in one map and points in the transformed map. Figure 7.35 shows the result of the feature matching on a pair of maps. The top image shows the superimposition of the contours and features of the two maps using the estimated displacement (about one meter translation and 4° rotation), while the bottom image shows the correspondences between the point and line features in the two maps. The lower map is transformed by T with respect to the lower right map. Figure 7.36 shows the result of the feature matching in a case in which the maps are separated by a very large displacement. The lower left display shows the area that is common between the two maps after the displacement. Even though the resulting displacement is not accurate enough to reliably merge the maps, it is close enough to the optimum to be used as the starting point of a minimization algorithm.

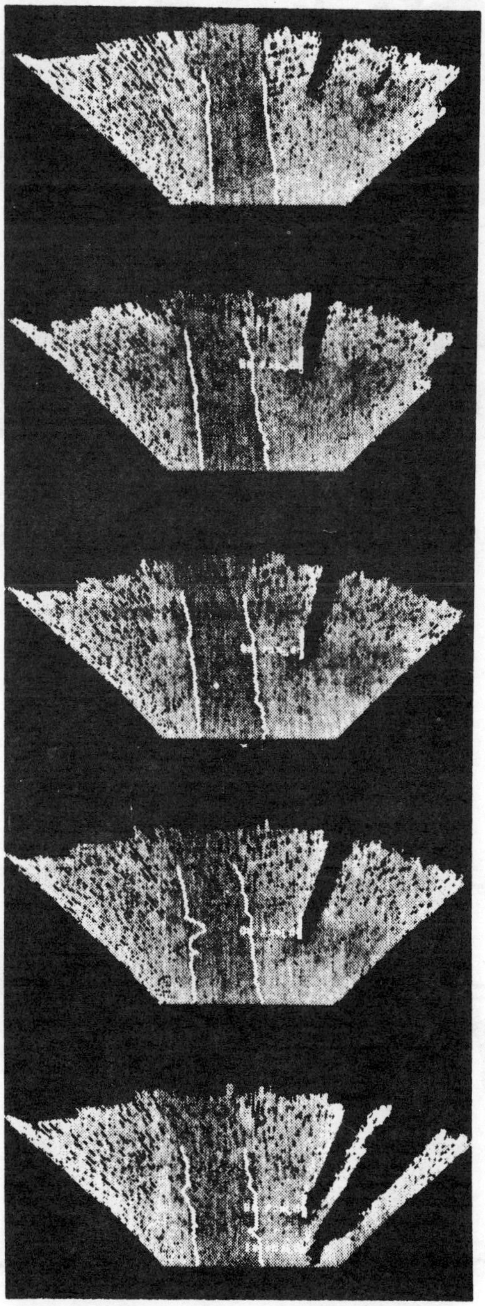

FIGURE 7.33. Individual maps

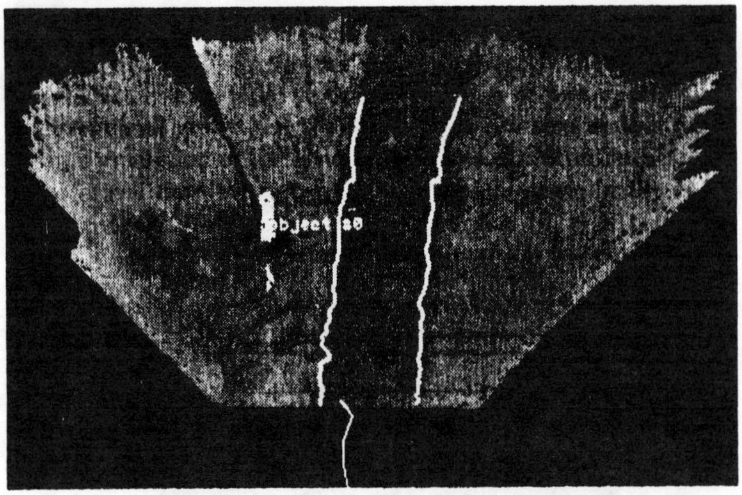

FIGURE 7.34. Perspective view of the combined map

FIGURE 7.35. Matching maps using local features

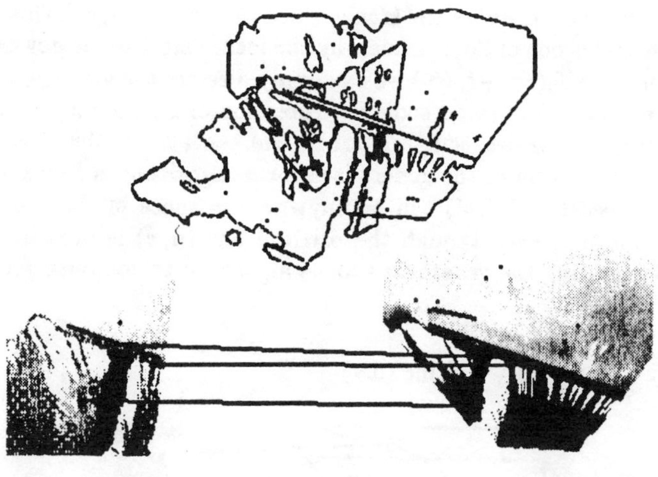

FIGURE 7.36. Matching maps using local features (large rotation component)

7.4.3 ICONIC MATCHING FROM ELEVATION MAPS

The general idea of the iconic matching algorithm is to find the displacement T between two elevation maps from two different range images that minimizes an error function computed over the entire combined elevation map. The error function E measures how well the first map and the transformed of the second map by T do agree. The easiest formulation for E is the sum of the squared differences between the elevation at a location in the first map and the elevation at the same location computed from the second map using T. To be consistent with the earlier formulation of the locus algorithm, the elevation at any point of the first map is actually the intersection of a line containing this point with the range image. We need some additional notations to formally define E: R and t denote the rotation and translation parts of T respectively, $f_i(u,v)$ is the function that maps a line in space described by a point and a unit vector to a point in by the generalized locus algorithm of Section 7.3.5 applied to image i. We have then:

$$E = \sum \|f_1(u,v) - g(u,v,T)\|^2 \tag{7.19}$$

where $g(u,v,T)$ is the intersection of the transformed of the line (u,v) by T with image 2 expressed in the coordinate system of image 1 (Figure 7.37). The summation in Equation (7.19) is taken over all the locations (u,v) in the first map where both $f_1(u,v)$ and $g(u,v,T)$ are defined. The lines (u,v) in the first map are parallel to the z-axis. In other words:

$$g(u,v,T) = T^{-1}(f_2(u',v')) = R'f_2(u',v') + t' \tag{7.20}$$

where $T^{-1} = (R', t') = (R^{-1}, -R^{-1}t)$ is the inverse transformation of T, and $(u', v') = (Ru+t, Rv)$ is the transformed of the line (u, v). This Equation demonstrates one of the reasons why the locus algorithm is powerful: in order to compute $f_2(Ru+t, Rv)$ we can apply directly the locus algorithm, whereas we would have to do some interpolation or resampling if we were using conventional grid-based techniques. We can also at this point fully justify the formulation of the generalized locus algorithm in Section 7.3.5: The transformed line (u', v') can be anywhere in space in the coordinate system of image 2, even though the original line (u, v) is parallel to the z-axis, necessitating the generalized locus algorithm to compute $f_2(u', v')$.

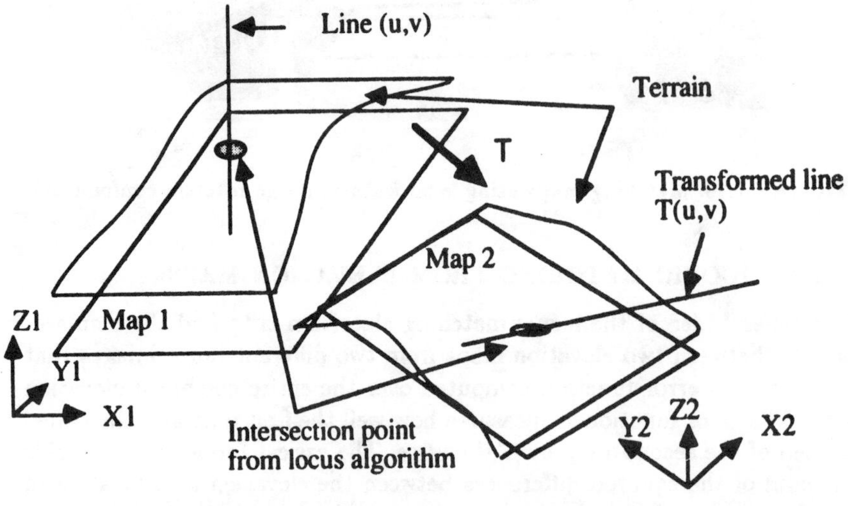

FIGURE 7.37. Principle of the iconic matching algorithm

We now have to find the displacement T for which E is minimum. If $\nu = [\alpha, \beta, \gamma, t_x, t_y, t_z]^t$ is the 6-vector of parameters of T, where the first three components are the rotation angles and the last three are the components of the translation vector, then E reaches a minimum when:

$$\frac{\partial E}{\partial \nu} = 0 \qquad (7.21)$$

Assuming an initial estimate T_0, such a minimum can be found by an iterative gradient descent of the form:

$$\nu^{i+1} = \nu^i + k\frac{\partial E}{\partial \nu}(\nu^i) \qquad (7.22)$$

where ν^i is the estimate of ν at iteration i. From Equation (7.19), the derivative of E can be computed by:

$$\frac{\partial E}{\partial \nu} = -2\sum(f_1(u,v) - g(u,v,T))\frac{\partial g}{\partial \nu}(u,v,T) \tag{7.23}$$

From Equation (7.20), we get the derivative of g:

$$\frac{\partial g}{\partial \nu}(u,v,T) = R'\frac{\partial f_2}{\partial \nu}(u',v') + \frac{\partial R'}{\partial \nu}f_2(u',v') + \frac{\partial t'}{\partial \nu} \tag{7.24}$$

The derivatives appearing in the last two components in Equation (7.24) are the derivatives of the transformation with respect to its parameters which can be computed analytically. The last step to compute the derivative of $g(u,v,T)$ is therefore to compute the derivative of $f_2(u',v')$ with respect to ν. We could write the derivative with respect to each component ν_i of ν by applying the chain rule directly:

$$\frac{\partial f_2}{\partial \nu_i}(u',v') = \frac{\partial f_2}{\partial u}\frac{\partial u'}{\partial \nu_i} + \frac{\partial f_2}{\partial v}\frac{\partial v'}{\partial \nu_i} \tag{7.25}$$

Equation (7.25) leads however to unstabilities in the gradient algorithm because, as we pointed out in Section 7.3.5, the (u,v) representation is an ambiguous representation of lines in space. We need to use a non ambiguous representation in order to correctly compute the derivative. According to equation (7.11), we can use interchangeably the (u,v) representation and the unambiguous (a,b,p,q) representation. Therefore by considering f_2 as a function of the transform by T, $l' = (a',b',p',q')$, of a line $l = (a,b,p,q)$ in image 1, we can transform Equation (7.25) to:

$$\frac{\partial f_2}{\partial \nu_i}(l') = \frac{\partial f_2}{\partial l'}\frac{\partial l'}{\partial \nu_i} \tag{7.26}$$

Since the derivative $\partial f_2/\partial l'$ depends only on the data in image 2, we cannot compute it analytically and have to estimate it from the image data. We approximate the derivatives of f_2 with respect to a, b, p, and q by differences of the type:

$$\frac{\partial f_2}{\partial a} = \frac{f(a+\Delta a, b, p, q) - f(a,b,p,q)}{\Delta a} \tag{7.27}$$

Approximations such as Equation (7.27) work well because the combination of the locus algorithm and the GNC image smoothing produces smooth variations of the intersection points.

The last derivatives that we have to compute to complete the evaluation of $\partial E/\partial \nu$ are the derivatives of l' with respect to each motion parameter ν_i. We start by observing that if $X = [x,y,z]^t$ is a point on the line of parameter l, and $X' = [x',y',z']^t$ is the transformed of X by T that lies on a line of parameter l', then we have the following relations from Equation (7.11):

$$\begin{aligned} x = az + p, x' = a'z' + p' \\ y = bz + q, y' = b'z' + q' \end{aligned} \tag{7.28}$$

By eliminating X and X' between Equation (7.28) and the relation $X' = RX + t$, we have the relation between l and l':

$$a' = \frac{R_x.V}{R_z.V}, \quad p' = R_x.U + t_x - a'(R_z.U + t_z) \tag{7.29}$$
$$b' = \frac{R_y.V}{R_z.V}, \quad q' = R_y.U + t_y - b'(R_z.U + t_z)$$

where R_x, R_y, R_z are the row vectors of the rotation matrix R, $A = [a, b, 1]^t$, $B = [p, q, 0]^t$. We now have l' as a function of l and T, making it easy to compute the derivatives with respect to ν_i from Equation (7.29).

In the actual implementation of the matching algorithm, the points at which the elevation is computed in the first map are distributed on a square grid of ten centimeters resolution. The lines (u, v) are therefore vertical and pass through the centers of the grid cells. E is normalized by the number of points since the since of the overlap region between the two maps is not known in advance. We first compute the $f_1(u, v)$ for the entire grid for image 1, and then apply directly the gradient descent algorithm described above. The iterations stop either when the variation of error ΔE is small enough, or when E itself is small enough. Since the matching is computationally expensive, we compute E over an eight by eight meter window in the first image. The last test ensures that we do not keep iterating if the error is smaller than what can be reasonably achieved given the characteristics of the sensor. Figure 7.38 shows the result of combining three high resolution elevation maps. The displacements between maps are computed using the iconic matching algorithm. The maps are actually combined by replacing the elevation $f_1(u, v)$ by the combination:

$$\frac{\sigma_1 f_1 + \sigma_2 f_2}{\sigma_1 + \sigma_2} \tag{7.30}$$

where σ_1 and σ_2 are the uncertainty values computed as in Section 7.3.5. Equation (7.30) is derived by considering the two elevation values as Gaussian distributions. The resulting mean error in elevation is lower than ten centimeters. We computed the initial T_0 by using the local feature matching of Section 7.4.2. This estimate is sufficient to ensure the convergence to the true value. This is important because the gradient descent algorithm converges towards a local minimum, and it is therefore important to show that T_0 is close to the minimum. Figure 7.39 plots the value of the ν_i's with respect to the number of iterations. These curves show that E converges in a smooth fashion. The coefficient k that controls the rate of convergence is very conservative in this case in order to avoid oscillations about the minimum.

Several variations of the core iconic matching algorithm are possible. First of all, we assumed implicitly that E is a smooth function of ν; this not true in general because the summation in Equation (7.19) is taken only

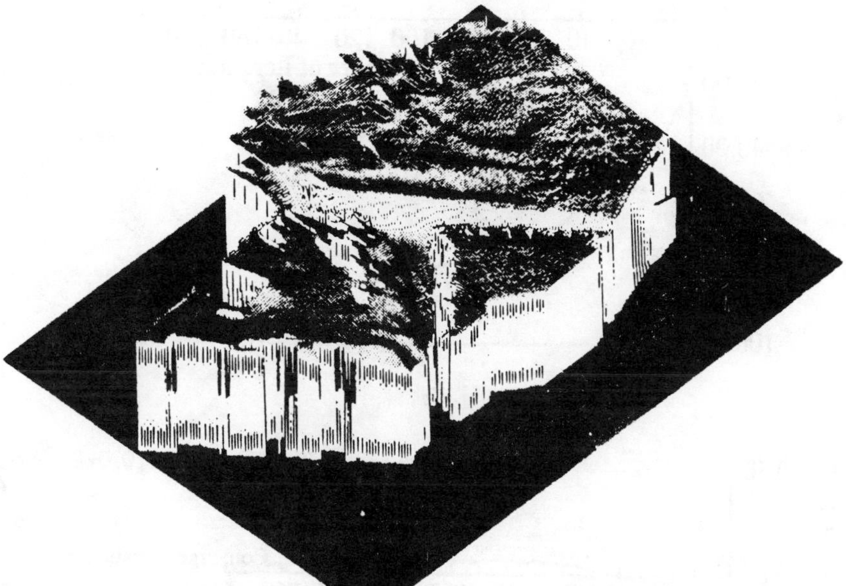

FIGURE 7.38. Combining four maps by the iconic matching algorithm

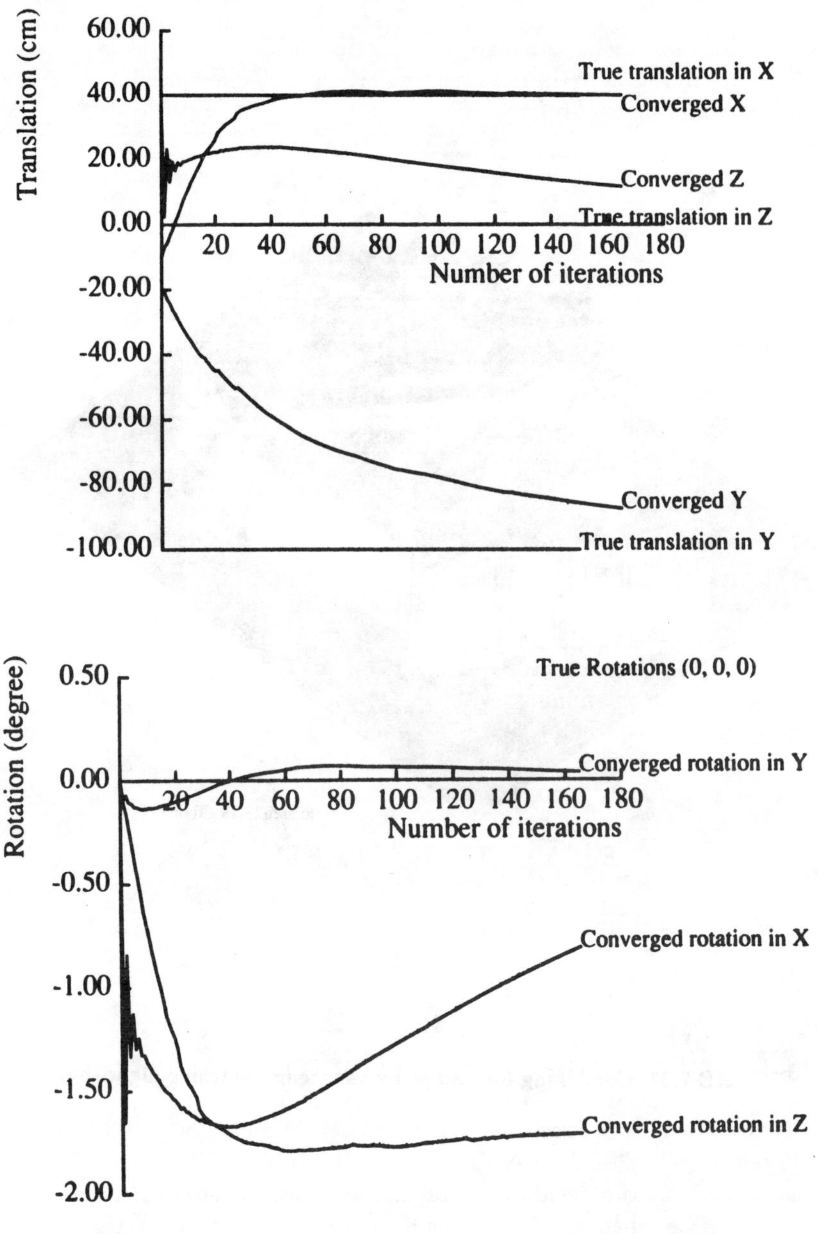

FIGURE 7.39. Convergence rate of the matching algorithm

over the regions in which both f_1 and g are defined, that is the intersection of the regions of map 1 and 2 that is neither range shadows nor outside of the field of view. Such a summation implicitly involves the use of a non-differentiable function that is 1 inside the acceptable region and 0 outside. This does not affect the algorithm significantly because the changes in ν from one iteration to the next are small enough. A differentiable formulation for E would be of the form:

$$E = \sum \mu_1(u,v)\mu_2(T(u,v))\|f_1(u,v) - g(u,v,T)\|^2 \tag{7.31}$$

where $\mu_i(u,v)$ is a function that is at most 1 when the point is inside a region where $f_i(u,v)$ is defined and vanishes as the point approaches a forbidden region, that is a range shadow or a region outside of the field of view. The summation in Eq. 7.31 is taken over the entire map. In order to avoid a situation in which the minimum is attained when the two maps do not overlap ($E = 0$), we must also normalize E by the number of points in the overlap region. For E to be still smooth, we should therefore normalize by:

$$\sum \mu_1(u,v)\mu_2(u,v) \tag{7.32}$$

In addition to E being smooth, we also assumed that matching the two maps entirely determines the six parameters of T. This assumption may not be true in all cases. A trivial example is one in which we match two images of a flat plane, where only the vertical translation can be computed from the matching. The gradient algorithm does not converge in those degenerate cases because the minimum $T(\nu)$ may have arbitrarily large values within a surface in parameter space. A modification of the matching algorithm that would ensure that the algorithm does converge to some infinite value changes Equation (7.19) to:

$$E = \sum \|f_1(u,v) - g(u,v,T)\|^2 + \sum_i \lambda_i \nu_i^2 \tag{7.33}$$

The effect of the weights λ_i is to include the constraint that the ν_i's do not increase to infinity in the minimization algorithm.

7.5 Combining range and intensity data

In the previous Section we have concentrated on the use of 3-D vision as it relates solely to the navigation capabilities of mobile robots. Geometric accuracy was the deciding factor in the choice of representations. This is appropriate when all we need is sufficient information for the vehicle to navigate through its environment. In many cases, however, we need more than geometric descriptions. The most important case is the landmark recognition problem in which we have to identify stored object models in a scene. In that case, geometric information may not be sufficient to unambiguously identify the sought object. Other types of information, such as surface markings, may also be needed. In general, tasks that require some kind of semantic interpretation of an observed scene involve the use of other types of information in addition to the geometric descriptions. Even though this has received relatively little attention in the field of mobile robotics, we anticipate a increasing need for research in this direction as the navigation issues become better understood.

As a first step, we address in this Section the problem of combining 3-D data with data from other sensors. The most interesting problem is the combination of 3-D data with color images since these are the two most common sensors for outdoor robots. Since the sensors have different fields of view and positions, we first present an algorithm for transforming the images into a common frame. As an example of the use of combined range/color images, we describe a simple scene analysis program in Section 7.5.3.

7.5.1 THE GEOMETRY OF VIDEO CAMERAS

The video camera is a standard color vidicon camera equipped with wide-angle lenses. The color images are 480 rows by 512 columns, and each band is coded on eight bits. The wide-angle lens induces a significant geometric distortion in that the relation between a point in space and its projection on the image plane does not obey the laws of the standard perspective transformation. We alleviate this problem by first transforming the actual image into an "ideal" image: if (R, C) is the position in the real image, then the position (r, c) in the ideal image is given by:

$$r = f_r(R, C), c = f_c(R, C) \tag{7.34}$$

where f_r and f_c are third order polynomials. This correction is cheap since the right-hand side of (7.34) can be put in lookup tables. The actual computation of the polynomial is described in [Mor80] The geometry of the ideal image obeys the laws of the perspective projection in that if $P = [x, y, z]^t$ is a point in space, and (r, c) is its projection in the ideal image plane, then:

$$r = fx/z, c = fy/z \tag{7.35}$$

where f is the focal length. In the rest of the paper, row and column positions will always refer to the positions in the ideal image, so that perspective geometry is always assumed.

7.5.2 THE REGISTRATION PROBLEM

Range sensor and video cameras have different fields of view, orientations, and positions. In order to be able to merge data from both sensors, we

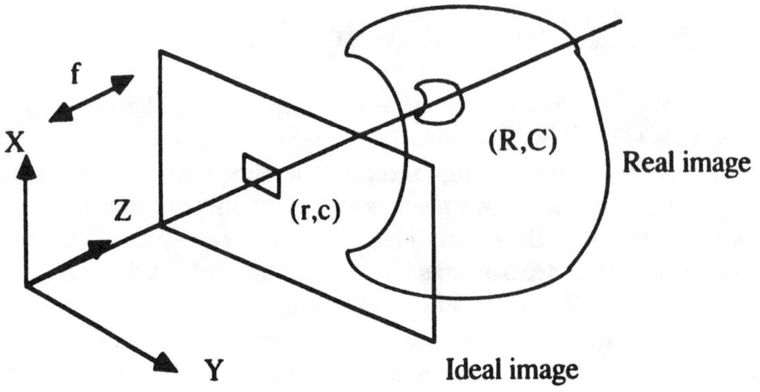

FIGURE 7.40. Geometry of the video camera

first have to estimate their relative positions, known as the calibration or registration problem (Figure 7.41). We approach the problem as a minimization problem in which pairs of pixels are selected in the range and video images. The pairs are selected so that each pair is the image of a single point in space as viewed from the two sensors. The problem is then to find the best calibration parameters given these pairs of points and is further divided into two steps: we first use a simple linear least-squares approach to find a rough initial estimate of the parameters, and then apply a non-linear minimization algorithm to compute an optimal estimate of the parameters.

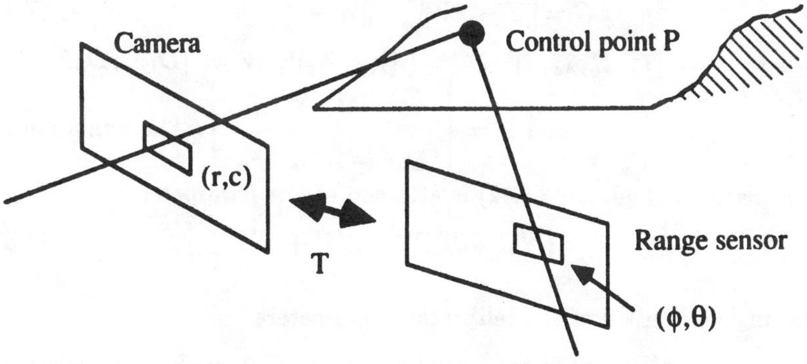

FIGURE 7.41. Geometry of the calibration problem

The calibration problem as a minimization problem

Let P_i be a point in space, with coordinates P_i^e with respect to the range sensor, and coordinates P_i^c with respect to the video camera. The relationship between the two coordinates is:

$$P_i^c = RP_i^e - T \tag{7.36}$$

where R is a rotation matrix, and T is a translation vector. R is a non-linear function of the orientation angles of the camera: pan (α), tilt (β), and rotation (γ). P_i^e can be computed from a pixel location in the range image. P_i^c is not completely known, it is related to the pixel position in the video image by the perspective transformation:

$$z_i^c r_i = f x_i^c \tag{7.37}$$

$$z_i^c c_i = f y_i^c \tag{7.38}$$

where f is the focal length. Substituting (7.36) into (7.37) and (7.38) we get:

$$R_z P_i^e r_i - T_z r_i - f R_x P_i^e + T_x' = 0 \tag{7.39}$$

$$R_z P_i^e c_i - T_z c_i - f R_y P_i^e + T_y' = 0 \tag{7.40}$$

where R_x, R_y, and R_z are the row vectors of the rotation matrix R, and $T_y' = fT_y$, $T_x' = fT_x$.

We are now ready to reduce the calibration problem to a least-squares minimization problem. Given n points P_i, we want to find the transformation (R, T) that minimizes the left-hand sides of equations (7.39) and (7.40). We first estimate T by a linear least-squares algorithm, and then compute the optimal estimate of all the parameters.

Initial estimation of camera position

Assuming that we have an estimate of the orientation R, we want to estimate the corresponding T. The initial value of R can be obtained by physical measurements using inclinometers. Under these conditions, the criterion to be minimized is:

$$C = \sum_{i=1}^{n} [(A_i - T_z B_i - fC_i + T'_x)^2 + (D_i - T_z E_i - fF_i + T'_y)^2] \quad (7.41)$$

where $A_i = R_z P_i^e r_i$, $B_i = r_i$, $C_i = R_x P_i^e$, $D_i = R_z P_i^e c_i$, $E_i = c_i$, and $F_i = R_y P_i^e$ are known and T_z, T'_x, T'_y, f are the unknowns.

Equation (7.41) can be put in matrix form:

$$C = \|U - AV\|^2 - \|W - BV\|^2 \quad (7.42)$$

where $V = [T'_x, T'_y, T_z, f]^t$, $U = [A_1, .., A_n]^t$, $W = [D_1, .., D_n]^t$, $A = \begin{bmatrix} B_i & 0 & -1 & C_i \\ . & . & . & \\ B_n & 0 & -1 & C_n \end{bmatrix}$, and $B = \begin{bmatrix} E_i & -1 & 0 & F_i \\ . & . & . & \\ E_n & -1 & 0 & F_n \end{bmatrix}$. The minimum for the criterion of Equation (7.42) is attained at the parameter vector:

$$V = (A^t A + B^t B)^{-1}(A^t U + B^t W) \quad (7.43)$$

Optimal estimation of the calibration parameters

Once we have computed the initial estimate of V, we have to compute a more accurate estimate of (R, T). Since R is a function of (α, β, γ), we can transform the criterion from equation (7.41) into the form:

$$C = \sum_{i=1}^{n} \|I_i - H_i(S)\|^2 \quad (7.44)$$

where I_i is the 2-vector representing the pixel position in the video image, $I_i = [r_i, c_i]^t$, and S is the full vector of parameters, $S = [T'_x, T'_y, T_z, f, \alpha, \beta, \gamma]^t$. We cannot directly compute C_{min} since the functions H_i are non-linear; instead we linearize C by using the first order approximation of H_i [Low80]:

$$C \approx \sum_{i=1}^{n} \|I_i - H_i(S_0) - J_i \Delta S\|^2 \quad (7.45)$$

where J_i is the Jacobian of H_i with respect to S, S_0 is the current estimate of the parameter vector, and $\Delta S = S - S_0$. The right-hand side of (7.45) is minimized when its derivative with respect to ΔS vanishes, that is:

$$\sum_{i=1}^{n} J_i^t J_i \Delta S + J_i^t \Delta C_i = 0 \quad (7.46)$$

where $\Delta C_i = I_i - H_i(S_0)$. Therefore, the best parameter vector for the linearized criterion is:

$$\Delta S = -\sum_{i=1}^{n} (J_i^t J_i)^{-1} J_i^t \Delta C_i \quad (7.47)$$

Equation (7.47) is iterated until there is no change in S. At each iteration, the estimate S_0 is updated by: $S_0 \leftarrow S_0 + \Delta S$.

Implementation and performance

The implementation of the calibration procedure follows the steps described above. Pairs of corresponding points are selected in a sequence of video and range images. We typically use twenty pairs of points carefully selected at interesting locations in the image (*e.g.* corners). An initial estimate of the camera orientation is $(0, \beta, 0)$, where β is physically measured using an inclinometer. The final estimate of S is usually obtained after less than ten iterations. This calibration procedure has to be applied only once, as long as the sensors are not displaced.

Once we have computed the calibration parameters, we can merge range and video images into a colored-range image. Instead of having one single fusion program, we implemented this as a library of fusion functions that can be divided in two categories:

1. Range → video: This set of functions takes a pixel or a set of pixels (r^e, c^e) in the range image and computes the location (r^c, c^c) in the video image. This is implemented by directly applying Equations (7.39) and (7.40).

2. Video → range: This set of functions takes a pixel or a set of pixels (r^c, c^c) in the video image and computes the location (r^e, c^e) in the range image. The computed location can be used in turn to compute the location of a intensity pixel in 3-D space by directly applying Equation (7.3). The algorithm for this second set of functions is more involved because a pixel in the video image corresponds to a line in space (Figure 7.40) so that Equations (7.39) and (7.40) cannot be applied directly. More precisely, a pixel (r^c, c^c) corresponds, after transformation by (R, T), to a curve C in the range image. C intersects the image at locations (r^e, c^e), where the algorithm reports the location (r^e, c^e) that is the minimum among all the range image pixels that lie on C of the distance between (r^c, c^c) and the projection of (r^e, c^e) in the video image (using the first set of functions). The algorithm is summarized on Figure 7.42.

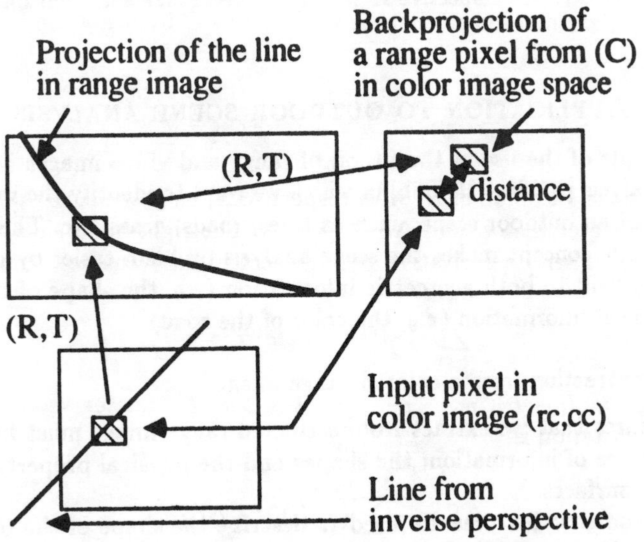

FIGURE 7.42. Geometry of the "video → range" transformation

Figure 7.43 shows the colored-range image of a scene of stairs and sidewalks, the image is obtained by mapping the intensity values from the color image onto the range image. Figure 7.44 shows a perspective view of the colored-range image. In this example [GMKO86], we first compute the location of each range pixel (r^e, c^e) in the video image, and then assign the color value to the 64 x 256 colored-range image. The final display is obtained by rotating the range pixels, the coordinates of which are computed using Equation (7.3).

FIGURE 7.43. Colored-range image of stairs

FIGURE 7.44. Perspective view of registered range and color images

7.5.3 Application to outdoor scene analysis

An example of the use of the fusion of range and video images is outdoor scene analysis [HK85,KHK88b] in which we want to identify the main components of an outdoor scene, such as trees, roads, grass, etc. The colored-range image concept makes the scene analysis problem easier by providing data pertinent to both geometric information (*e.g.* the shape of the trees) and physical information (*e.g.* the color of the road).

Feature extraction from a colored-range image

The features that we extract from a colored-range image must be related to two types of information: the shapes and the physical properties of the observed surfaces.

The geometric features are used to describe the shape of the objects in the scene. We propose to use two types of features: regions that correspond to smooth patches of surface, and edges that correspond either to transitions between regions, or to transitions between objects (occluding edges). Furthermore, we must be able to describe the features in a compact way. One common approach is to describe the regions as quadric patches, and the edges as sets of tri-dimensional line segments. More sophisticated descriptions are possible [BJ85c], such as bicubic patches or curvature descriptors. We use simpler descriptors since the range data is relatively low resolution, and we do not have the type of accurate geometric model that is suited for using higher order geometric descriptors. The descriptors attached to each geometric feature are:

- The parameters describing the shape of the surface patches. That is the parameters of the quadric surface that approximate each surface patch.

- The shape parameters of the surface patches such as center, area, and elongations.

- The 3-D polygonal description of the edges.

- The 3-D edge types: convex, concave, or occluding.

The surface patches are extracted by fitting a quadric of equation $X^t AX + B^t X + C = 0$ to the observed surfaces, where X is the Cartesian coordinate vector computed from a pixel in the range image. The fitting error,

$$E(A,B,C) = \sum_{X_i \in patch} [X_i^t AX_i + B^t X_i + C]^2 \tag{7.48}$$

is used to control the growing of regions over the observed surfaces. The parameters A, B, C are computed by minimizing $E(A,B,C)$ as in [FH86].

The features related to physical properties are regions of homogeneous color in the video image, that is regions within which the color values vary smoothly. The choice of these features is motivated by the fact that an homogeneous region is presumably part of a single scene component, although the converse is not true as in the case of the shadows cast by an object on an homogeneous patch on the ground. The color homogeneity criterion we use is the distance $(X-m)^t \Sigma^{-1} (X-m)$ where m is the average mean value on the region, Σ is the covariance matrix of the color distribution over the region, and X is the color value of the current pixel in $(red, green, blue)$ space. This is a standard approach to color image segmentation and pattern recognition. The descriptive parameters that are retained for each region are:

- The color statistics (m, Σ).

- The polygonal representation of the region border.

- Shape parameters such as center or moments.

The range and color features may overlap or disagree. For example, the shadow cast by an object on a flat patch of ground would divide one surface patch into two color regions. It is therefore necessary to have a cross-referencing mechanism between the two groups of features. This mechanism provides a two-way direct access to the geometric features that intersect color features. Extracting the relations between geometric and physical features is straightforward since all the features are registered in the colored-range image.

An additional piece of knowledge that is important for scene interpretation is the spatial relationships between features. For example, the fact that a vertical object is connected to a large flat plane through a concave edge may add evidence to the hypothesis that this object is a tree. As in this example, we use three types of relational data:

- The list of features connected to each geometric or color feature.

- The type of connection between two features (convex / concave / occluding) extracted from the range data.

- The length and strength of the connection. This last item is added to avoid situations in which two very close regions become accidentally connected along a small edge.

Scene interpretation from the colored-range image

Interpreting a scene requires the recognition of the main components of the scene such as trees or roads. Since we are dealing with natural scenes, we cannot use the type of geometric matching that is used in the context of industrial parts recognition [BJ85c]. For example, we cannot assume that a given object has specific quadric parameters. Instead, we have to rely on "fuzzier" evidence such as the verticality of some objects or the flatness of others. We therefore implemented the object models as sets of properties that translate into constraints on the surfaces, edges, and regions found in the image. For example, the description encodes four such properties:

- $P1$: The color of the trunk lies within a specific range $\Longrightarrow$ constraint on the statistics (m, Σ) of a color region.
- $P2$: The shape of the trunk is roughly cylindrical $\Longrightarrow$ constraint on the distribution of the principal values of the matrix A of the quadric approximation.
- $P3$: The trunk is connected to a flat region by a concave edge $\Longrightarrow$ constraint on the neighbors of the surface, and the type of the connecting edge.
- $P4$: The tree has two parallel vertical occluding edges $\Longrightarrow$ constraint on the 3-D edges description.

Other objects such as roads or grass areas have similar descriptions. The properties P_{ij} of the known object models M_j are evaluated on all the features F_k extracted from the colored-range image. The result of the evaluation is a score S_{ijk} for each pair (P_{ij}, F_k). We cannot rely on individual scores since some may not be satisfied because of other objects, or because of segmentation problems. In the tree trunk example, one of the lateral occluding edges may itself be occluded by some other object, in which case the score for $P4$ would be low while the score for the other properties would still be high. In order to circumvent this problem, we first sort the possible interpretations M_j for a given feature F_k according to all the scores $(S_{ij})_i$. In doing this, we ensure that all the properties contribute to the final interpretation and that no interpretations are discarded at this stage while identifying the most plausible interpretations.

We have so far extracted plausible interpretations only for individual scene features F_k. The final stage in the scene interpretation is to find the interpretations (M_{j_k}, F_k) that are globally consistent. For example, property $P3$ for the tree implies a constraint on a neighboring region, namely that this has to be a flat ground region. Formally, a set of consistency constraints C_{mn} is associated with each pair of objects (M_m, M_n). The C_{mn} constraints are propagated through the individual interpretations (M_{j_k}, F_k) by using the connectivity information stored in the colored-range feature description. The propagation is simple considering the small number of features remaining at this stage.

The final result is a consistent set of interpretations of the scene features, and a grouping of the features into sets that correspond to the same object. The last result is a by-product of the consistency check and the use of connectivity data. Figure 7.45 shows the color and range images of a scene which contains a road, a couple of trees, and a garbage can. Figure 7.46 shows a display of the corresponding colored-range image in which the white pixels are the points in the range image that have been mapped into the video image. This set of points is actually sparse because of the difference in resolutions between the two sensors, and some interpolation was performed to produce the dense regions of Figure 7.46.

Only a portion of the image is registered due to the difference in field of view between the two sensors (60° for the camera versus 30° in the vertical direction for the range sensor). Figure 7.47 shows a portion of the image in which the edge points from the range image are projected on the color image. The edges are interpreted as the side edges of the tree and the connection between the ground and the tree. Figure 7.48 shows the final scene interpretation. The white dots are the main edges found in the range image. The power of the colored-range image approach is demonstrated by the way the road is extracted. The road in this image is separated into many pieces by strong shadows. Even though the shadows do not satisfy the color constraint on road region, they do perform well on the shape criterion (flatness), and on the consistency criteria (both with the other road regions, and with the trees). The shadows are therefore interpreted as road regions and merge with the other regions into one road region. This type of reasoning is in general difficult to apply when only video data is used unless one uses stronger models of the objects such as an explicit model of a shadowed road region. Using the colored-range image also makes the consistency propagation a much easier task than in purely color-based scene interpretation programs [Oht84].

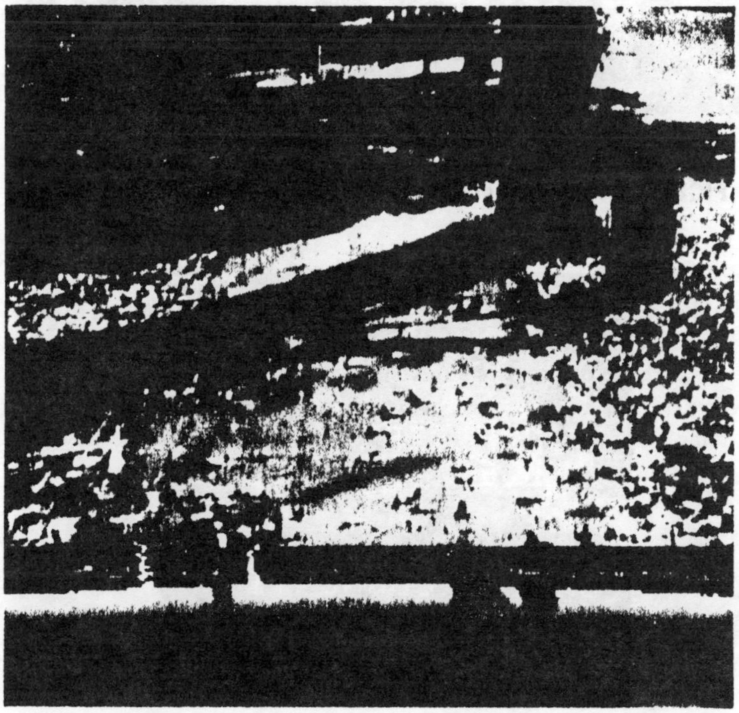

FIGURE 7.45. Color and range images of an outdoor scene

7.6 Conclusion

We have described techniques for building and manipulating 3-D terrain representations from range images. We have demonstrated these techniques on real images of outdoor scenes. Some of them (Sections 7.3.3, 7.3.4, and 7.4.2) were integrated in a large mobile robot system that was suc-

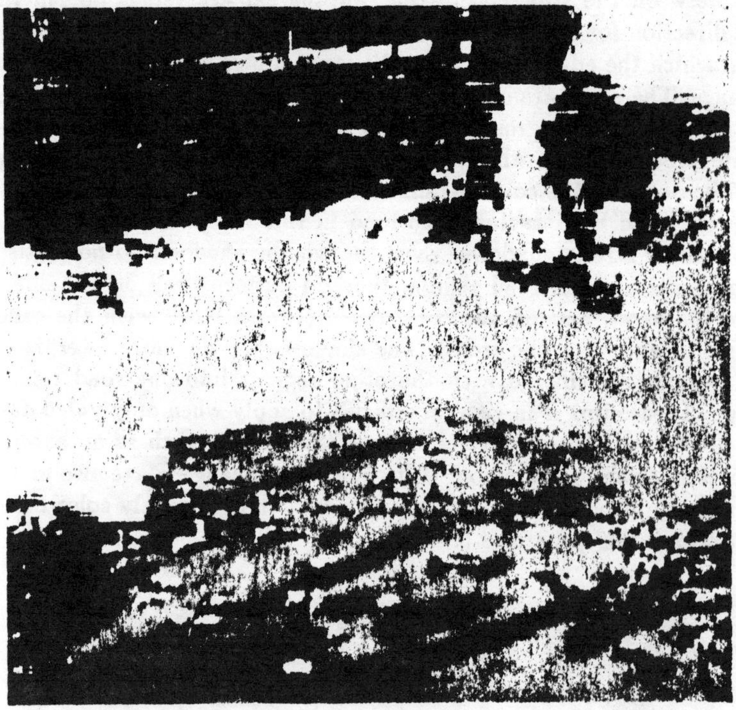

FIGURE 7.46. A view of the corresponding colored-range image

FIGURE 7.47. Edge features from the colored-range image

FIGURE 7.48. Final scene interpretation

cessfully tested in the field. We expect that the module that manipulates and creates these terrain representations will become part of the standard core system of our outdoor mobile robots, just as a local path planner or a low-level vehicle controller are standard modules of a mobile robot system independent of its application. This work will begin by combining the polygonal terrain representation of Section 7.3.4 with the path planner of [Ste88] in order to generate the basic capabilities for an off-road vehicle.

Many issues still remain to be investigated. First of all, we must define a uniform way of representing and combining the uncertainties in the terrain maps. Currently, the uncertainty models depend heavily on the type of sensor used and on the level at which the terrain is represented. Furthermore, the displacements between terrain maps are known only up to a certain level of uncertainty. This level of uncertainty must be evaluated and updated through the matching of maps, whether iconic or feature-based. Regarding the combination of the 3-D representations with representations from other sensors, we need to define an algorithm for sensor registration that is general enough for application to a variety of situations. The algorithms presented in Section 7.5 are still very dependent on the sensors that we used, and on the intended application. Registration schemes such as [GTK88] would enable us to have a more uniform approach to the problem. An added effect of using such a registration algorithm is that we could explicitly represent errors caused by the combination of the sensors, which we did not do in Section 7.5. Another issue concerns our presentation of the three levels of terrain representation, the matching algorithms, and the sensor combination algorithms as separate problems. We should define a common perceptual architecture to integrate these algorithms in a common representation that can be part of the core system of a mobile robot. Finally, we have tackled the terrain representation problems mainly from a geometrical point of view. Except in Section 7.5, we did not attempt to extract semantic interpretations from the representations. A natural extension of this work is to use the 3-D terrain representations to identify known objects in the scene. Another application along these lines is to use the terrain maps to identify objects of interest, such as terrain regions for sampling tasks for a planetary explorer [Kwe88]. Although we have performed some preliminary experiments in that respect [HK88,BW88], extracting semantic information from terrain representations remains a major research area for outdoor mobile robots.

Absolute location of underwater robotic vehicles by acoustic data fusion

Rigaud V. , IFREMER-IRISA* and **Marcé L.**, LIB-UBO †

IRISA Campus de Beaulieu, Rennes 35042 France
e.mail : rigaud@irisa.fr or marce@irisa.fr

Abstract

This paper presents a study on computer assistance for the manned guidance of ROV. In our approach, it is possible to estimate an optimal location by fusing data deriving from log and gyroscope or a non optimal location by composition of compass, depthmeter and log data. But this relative reckoning is achieved taking into account a hypothesis on the value of the undercurrent. Therefore, we propound a second model of reckoning module to improve the ROV position close to an offshore jacket, fusing data given by an echo-sounders belt and taking into account an a priori CAD map of the surroundings. This new estimate is independent of the undercurrent, and it is possible with the help of the operator to match this absolute model with the relative one in an asynchronous way. This matching gives a new hypothesis on the value of the undercurrent and resets the dead-reckoning module.

Introduction

Manned guidance of an autonomous submarine is a very difficult activity which requires from the pilots to be skilled to integrate data coming from different, partial and highly noisy knowledge sources. The pilot builds his own internal model of the surroundings from the integration of various sensorial informations[11].
But the characteristics shared by the surroundings constraints (undercurrent, turbidity, loss of transmission, unforeseen obstacles) , the inherent inaccuracy in sensing , the large beamdwidth of sensors and the ambiguities in sensor interpretation assigns a heavy intellectual load to the pilot.
In this context, we propound to help operators in the tasks of location and piloting close to an offshore jacket, with synthetic visualisation and fusion data tools.
The emergence of multisensor fusion techniques in environment modelling [2] or sensor data interpretation [3], shows that fusion allows to expand the dimension of the observed state space in spatial or temporal resolution. Data fusion can be static if we use synchronous measurements at a given time, or dynamic if we accumulate data on a given time period.
Anyway fusion is an optimisation method on the informations given by an accumulation process on redundant data (description of a state variable by several measurements). Our approach consists of building two fused maps. The first is produced by a classical dead reckoning module using "non cognitive sensors" like depthmeter, compass, log and possibly a gyroscope. It allows us to automatically compute the relative position of the ROV taking into account a hypothesis on the value of the undercurrent, fusing eventual redundant data and propagating the non redundant ones. An absolute reckoning module using the fusion of data given by the echo-sounders belt, taking into account an a priori map of the surroundings, allows us to build the second map of the possible positions of ROV, independent of the undercurrent . We will show that the operator can use it in a synchronous way to match the absolute model with the relative one.Though we include the "man" in the command loop, to compare, validate or weaken hypothesis on external conditions (value of the undercurrent...). This matching gives a new hypothesis on the value of the undercurrent and resets the dead-reckoning module.

*French institute for sea exploitation
†University of Brest

Localization modulo a hypothesis on undercurrent

In this chapter we will describe a classical incremental dead-reckoning module, using "non cognitive sensors": log, depthmeter, compass and possibly a gyroscope. Depending on whether we dispose of a gyroscope or not, it is possible to fuse (optimal estimation) or only to propagate relative data.

Synchronous optimal filter

More and more often underwater vehicles are fitted with cheaper inertial units . Then measurements of attitude and vertical position are available with accuracy. In this case, if we know an estimate of the starting location and the initial covariance matrix, it is possible to compute an optimal location of ROV. We estimate this optimal location by fusing data deriving from log and gyroscope, integrating measurements of acceleration and mixing them to the speed relative measurement values (relative to the undercurrent) with the help of a synchronous Kalman filter[5]. But this optimal estimate of the speed vector is computed "modulo" an hypothesis on the value of the undercurrent relative to the bottom.
The Kalman filter dynamic fusion allows to compute the estimated location, the variance of the a posteriori error and the gain for each iteration . However the interest of this estimation is limited because of the inertial sensors drift and the absence of estimation process on the hypothetic undercurrent value .

Propagation of relative measurements

If we don't have inertial data, nevertheless it is possible to supply the pilots with a non optimal estimate of ROV location, propagating non cognitive measurements from the compass, the log and the depthmeter.
For a good attitude servoing (horizontal) it is possible to reduce the problem of location to a more general 2D location process in an isodepth plane determined by dephtmeter.

Reprinted from the *Proceedings of the IEEE International Conference on Robotics and Automation*, 1990, pages 1310-1315.
Copyright © 1990 by The Institute of Electrical and Electronics Engineers, Inc. All right reserved.

Log gives at regular intervals of time kT the speed value V_{m_k} of the robot with respect to the undercurrent $V_{c_k} = (V_{c_x}, V_{c_y})_k$ in the ROV frame. Let $||V_{m_k}||$ be the module and α_k the orientation of this speed. Compass gives the ROV attitude θ_k with respect to the magnetic North. Let V_k be the speed with respect to the sea-bottom (cf. fig.1)

Fig. 1: *Global and Local frame*

The pilot gives an hypothesis $V_{c_{hypo_k}}$ on the value of undercurrent V_{c_k}, from his own experience or a priori charts.

$$V_k = V_{m_k} + \boxed{V_{c_{hypo_k}}} \tag{1}$$

It is possible to define the variables V_{m_k}, α_k, θ_k as random variables whose distributions are supposed to be normal, independent and with variance respectively equal to $\sigma_v, \sigma_\alpha, \sigma_\theta$. At time kT, (x_k, y_k) is deduced from the values of the three variables V_{m_k}, α_k et θ_k up to time $(k-1)T$, and from the initial position (x_0, y_0). The estimated location is:

$$\begin{aligned} \hat{x}_k &= \hat{x}_{k-1} + ||V_{m_{k-1}}|| * cos(\alpha_{k-1} + \theta_{k-1}) + V_{c_{xhypo_{k-1}}} \\ \hat{y}_k &= \hat{y}_{k-1} + ||V_{m_{k-1}}|| * sin(\alpha_{k-1} + \theta_{k-1}) + V_{c_{yhypo_{k-1}}} \end{aligned} \tag{2}$$

The best would be to give the joint probability density arising from this transformation, but it is a complex calculation requiring to compute the associated Jacobian. We propound a more "robotics approach" which consists to compute iteratively, in real time the two first moments [9] of the compound random variables x and y. If the propagated distribution of the estimated position of the ROV is normal, then it is possible to show the ellipse representing an equiprobability profile of the associated normal law.

It is possible to validate the assumption on the normality of the propagated distribution, for a great number of agregations of independent measurements with the help of the *stable estimation principle*:

For a great number of iterative transformations we have:

$$\begin{aligned} \hat{x}_n &= \hat{x}_0 + \textstyle\sum_{i=0}^{n-1} ||V_{m_i}|| cos(\theta_i + \alpha_i) + nV_{c_{xhypo_0}} \\ \hat{y}_n &= \hat{y}_0 + \textstyle\sum_{i=0}^{n-1} ||V_{m_i}|| sin(\theta_i + \alpha_i) + nV_{c_{xhypo_0}} \end{aligned} \tag{3}$$

From the stable estimation theorem, if x_i (id. y_i) are independent and continuous variables, then the mean and variance of the new variable, $\sum_{i=0}^{n-1} x_i$ (id.$\sum_{i=0}^{n-1} y_i$) are $\mu = \sum \mu_i$ and $\sigma^2 = \sum \sigma_i^2$ and the associated distribution is normal. An expand in a Taylor expansion of the equations (2) around the respective mean values of $||V_{m_{k-1}}||$, α_{k-1} et θ_{k-1},that is to say $\bar{V}_{k-1}$, $\bar{\alpha}_{k-1}$, $\bar{\theta}_{k-1}$, gives for the x components:

$$\begin{aligned} \hat{x}_k - \hat{x}_{k-1} \simeq\ & \bar{V}_{k-1} * cos(\bar{\alpha}_{k-1} + \bar{\theta}_{k-1}) \\ & + \Delta v_{k-1} * cos(\bar{\alpha}_{k-1} + \bar{\theta}_{k-1}) \\ & - \bar{V}_{k-1} * sin(\bar{\alpha}_{k-1} + \bar{\theta}_{k-1}) * (\Delta\alpha_{k-1} + \Delta\theta_{k-1}) \end{aligned} \tag{4}$$

with $\Delta\zeta_{k-1} = ||\zeta_{mk-1}|| - \bar{\zeta}_{k-1}$, $\zeta \in \{V, \alpha, \theta\}$. Then the value of variance is :

$$\Delta\hat{x}(k) = (\hat{x}(k) - \hat{x}(k-1)) - (\bar{\hat{x}}(k) - \bar{\hat{x}}(k-1)) \tag{5}$$

$$\overline{\Delta\hat{x}_k^2} = cos^2(\bar{\alpha}_{k-1} + \bar{\theta}_{k-1}) * \sigma_V^2 + sin^2(\bar{\alpha}_{k-1} + \bar{\theta}_{k-1}) * (\sigma_\theta^2 + \sigma_\alpha^2) \tag{6}$$

Though it is possible to compute in a synchronous way the moments of the iterative non linear projections, to give an idea of the precision on the estimation of the ROV location to the pilot (cf. fig.4).

But in any case the relative location of ROV built in time, depends on the value of the undercurrent or follows the drift of the inertial sensors. Then we propound an absolute reckoning module, independent of undercurrent, using the result of the fusion of acoustic range data, given by an horizontal echo-sounders belt.

Absolute location by acoustic data fusion

In this section we show the methodological tools for a discrete absolute reckoning module, built by accumulation and Bayesian fusion on acoustic data taking into account an a priori map of the surroundings. We will also present the rules to model sensory and a priori knowledges with discrete representation, and we will give a way of finding optimal sampling steps for these models.

"The robot seen from the world"

In robotics, numerous works propound to build an incremental map of the perceived world, with the help of range finder sensors (acoustic sounders, laser telemeters...) [7][6]. But the dynamic generation of these maps implies that motions of the robot are well known.

With incremental methods, if the synthetized world model is dense (numerous perceived obstacles, a typical land robotic view), it is possible to build and to refine a map of "the world seen from the robot". It is also possible to reset the position of the robot with correlative methods (matching of the sensorial map and an a priori map of the surroundings). But the error on displacement estimation must be small enough to keep spatial consistency from one observation to the other.

In underwater surroundings these classical automatic mapping methods, and the associate hypothesis (high density on a priori and sensorial informations) are not effective because of the intrinsic conditions of the world, the poor performances of the sensors, the fuzziness and the dispersal of data.

Therefore we propound to compute an absolute location of "the ROV seen from the world" with an accumulation/fusion method on acoustic range finding data, taking directly into account the a priori knowledge on the surroundings.

The accumulation process

The knowledge sources we consider for the absolute model building are, sensors measurements and sensors characteristics, a sampled a priori map of the surroundings, attitude, aiming direction of acoustic transducers and depth.

If we suppose for simplification that the vertical servoing is good, it is possible to reduce the reckoning to a plane. A regular grid representation is used to model the location of the vehicle in that plane [7][6][4]. Each cell is associated with an accumulator, the value of which describes the probability of ROV presence .

Each echo-sounder is modeled by a spatial distribution function which gives the probability of presence for every echo measurement. Every sample of this model is potentially the one which physically generates the echo. Then we have as many hypothesis as potential samples, which can be considered as a priori knowledge sources.

Each measurement cycle gives a set of robocentric data noted $M \equiv \{M_i | M_i = (\rho_i, \theta_i)_{i=1}^{n}\}$ (n is the number of sounders on the belt).

If we know the attitude of the ROV(then sounders aiming directions), it is possible to find the position of the ROV with triangulation on the a priori data base. Accumulation consists in regarding all the acoustically perceptible obstacles as potential targets: it is a vote method described by Marcé and al.[6] .

Let n be the dimension of robocentric data set, and m the dimension of the sampled a priori reference map $R \equiv \{Rj | Rj = (X_j, Y_j)_{j=1}^{m}\}$. Let $P = (X, Y)$ be the position of ROV. The possible equations are:

$$\left.\begin{array}{l} X - Xj = \rho_i * cos(\theta_i) = X_{cij} \\ Y - Yj = \rho_i * sin(\theta_i) = Y_{cij} \end{array}\right\}_{[i=1,j=1]}^{[i=n,j=m]} \tag{7}$$

The world is divided in cells with step Δ, of wich addresses are :

$$N_{xij} = int(\frac{X_{cij} + X_j}{\Delta}), \quad N_{yij} = int(\frac{Y_{cij} + Y_j}{\Delta}) \tag{8}$$

For each sensory robocentric data it is possible to increment the associated cellular accumulator for all potential a priori obstacles. The accumulation is a parameter estimation method: P is the parameter to estimate with a set of n*m parametric equations :

$$M_i = f(R_j, P) + e_i \;\Big\}_{i=1,j=1}^{i=n,j=m} \tag{9}$$

with e_i the noise on the set of measurements M.

In practice we build a grid in the parameter space, that is to say the space of all ROV possible locations, and we compute with all potential estimators given by equations (7) the best estimate of P, in a least square sense.

Optimal steps

We build three discrete models, for which we describe the best steps to solve the absolute location problem in an optimal way.

• The first is the ROV possible locations world model.

It is possible to show that the accumulation method is a method building maximum likelihood estimators for an uniform noise on data[12]. The quality of this estimation depends on the size of the cellular accumulators.

We compute the optimal step of the regular accumulation grid with the help of the CRAMER and RAO results, on maximum likelihood estimation techniques. The inversion of the information matrix (called Fisher matrix) I leads to the minimal variance matrix, and then it is possible to compute the minimal step of the cells for an "optimal" estimation. The Fisher matrix is :

$$I = ||E(-(\frac{d^2 lnL(\rho_i\theta_i X_j Y_j)}{d(\rho_i\theta_i X_j Y_j)}))|| \tag{10}$$

with L the likelihood function of the system (10).

$$L(\rho_i\theta_i X_j Y_j)_{(1,1)}^{(n,m)} = \prod_{(1,1)}^{(n,m)} f(\rho_i\theta_i X_j Y_j) \tag{11}$$

We compute quickly an optimal step of the accumulation space , considering that for each real data measurement there is only one candidate sample which produces the echo in the a priori obstacle data base. It seems to be a good hypothesis because the accumulation method could not be more accurate than the estimation if the positions of the reactive obstacles were known. Then the steps we are computing in that way are perhaps too large but in an other way we have to take steps large enough for real time application on a sequential computer. If there is no noise on the a priori data we have:

$$\begin{array}{lll} \theta_i = \phi_i + \delta_{\theta_i} & with & \delta_{\theta_i} \leadsto \mathcal{N}(0, \epsilon_\theta) \\ \rho_i = \varphi_i + \delta_{\rho_i} & with & \delta_{\rho_i} \leadsto \mathcal{N}(0, \epsilon_\rho) \end{array} \tag{12}$$

ρ_i, θ_i are the real values of φ_i and ϕ_i with respective errors δ_{ρ_i} et δ_{θ_i}.

Let $x_i = X - X_i$, $y_i = Y - Y_i$ and

$$\varphi_i = ((x_i)^2 + (y_i)^2)^{1/2}, \phi_i = Atan((y_i)/(x_i)) \tag{13}$$

Obviously, we have:

$$\begin{array}{l} E(Atan((y_i)/(x_i)) = E(\phi_i) = \theta_i \\ E(((x_i)^2 + (y_i)^2)^{1/2}) = E(\varphi_i) = \rho_i \end{array} \tag{14}$$

and the Fisher matrix I is :

$$\begin{pmatrix} \begin{array}{l} 1/\epsilon_\theta^2 \sum_1^n \frac{y_i^2}{(x_i^2+y_i^2)^2} \\ +1/\epsilon_{\rho^2} \sum_1^n \frac{x_i^2}{(x_i^2+y_i^2)} \end{array} & \begin{array}{l} -1/\epsilon_{\theta^2} \sum_1^n \frac{y_i * x_i}{(x_i^2+y_i^2)^2} \\ +1/\epsilon_{\rho^2} \sum_1^n \frac{x_i * y_i}{(x_i^2+y_i^2)} \end{array} \\ \\ \begin{array}{l} -1/\epsilon_{\rho^2} \sum_1^n \frac{x_i * y_i}{(x_i^2+y_i^2)} \\ +1/\epsilon_{\theta^2} \sum_1^n \frac{y_i * x_i}{(x_i^2+y_i^2)^2} \end{array} & \begin{array}{l} +1/\epsilon_{\theta^2} \sum_1^n \frac{x_i^2}{(x_i^2+y_i^2)^2} \\ +1/\epsilon_{\rho^2} \sum_1^n \frac{y_i^2}{(x_i^2+y_i^2)} \end{array} \end{pmatrix} \tag{15}$$

The inversion of this matrix gives the minimal variances:

$$\begin{array}{l} varminx = \frac{1/\epsilon_{\theta^2} \sum_1^n (x_i^2/(x_i^2+y_i^2)^2) + 1/\epsilon_{\rho^2} \sum_1^n (y_i^2/(x_i^2+y_i^2))}{DetI} \\ varminy = \frac{1/\epsilon_{\theta^2} \sum_1^n (y_i^2/(x_i^2+y_i^2)^2) + 1/\epsilon_{\rho^2} \sum_1^n (x_i^2/(x_i^2+y_i^2))}{DetI} \end{array} \tag{16}$$

Then the sampling steps are:

$$stg_x \geq (varminx)^{1/2}/k_x \quad and \quad stg_y \geq (varminy)^{1/2}/k_y \tag{17}$$

with k_x and $k_y \geq 1$. For a regular grid $stg = sup(stg_x, stg_y)$. The values of k_x and k_y depend on the distribution shape of potential ROV location in the grid, and the step depends on the value of measurements(ρ_i, θ_i).

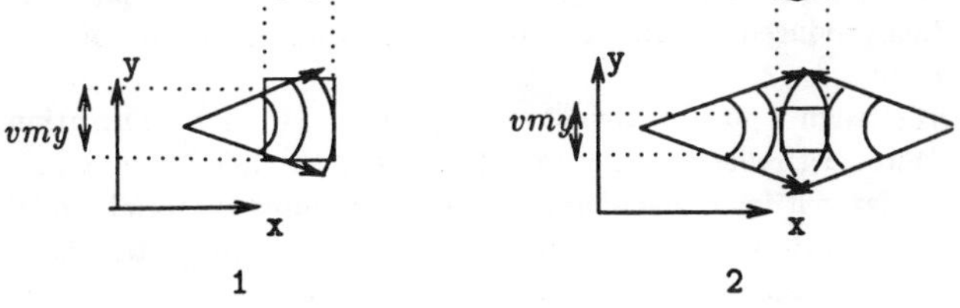

Fig. 2: *Evolution of optimal step with measurements*
1:M $= (\rho_1, \theta_1) : stg_1 = min(\epsilon_\rho^2, \epsilon_\theta^2 * \rho_1^2)$
2:M $= (\rho_1, \theta_1), (\rho_1, \theta_1 + \pi) stg_2 = stg_1/2$

The accumulation grid sizes are computed for each set of measurement at regular time. If we are far from the jacket or if there is a little echoes number the step is large. Fig. 2 shows that the better the triangulation is (a large echoes number) , the less the step is large that is to say the better the estimation is .

• The step of the accumulation grid allows to build the step of a discrete sounders model.

A sounder is not a perfect telemeter. The angular resolution is very poor. To express it, Moravec and Elfes [1] described in their certainty grid model, a discrete model of sounders. It is a two state stochastic model, reduced to a one state model by Stewart [11]. But Moravec, Elfes and Stewart tried to build an incremental map of the surroundings rather than to use a priori information on the world to locate the robot.

In our method, these models are too heavy for computation. We adopt a simplest weighted vectorial model. For each measurement (ρ_i, θ_i) we build a set of vectors $(\rho_i, \theta_i{}^{+}_{-}\epsilon_{\theta_i})$ weighted by a statistic weight w_i which expresses the confidence in angular resolution . We compute the angular sampling step ϵ_{θ_i} with respect to the size of the step of the accumulation grid *stg* and to the range measurement ρ_m with: $\epsilon_{\theta_i} = Arcos(1 - (stg)^2/4 * (\rho_m))$

• The a priori obstacles are modelised by sampled polyhedrons. The value of the sampling step is less or equal to the grid step to avoid holes in the representation of the obstacles in the accumulation grid as shown in fig. 3 .

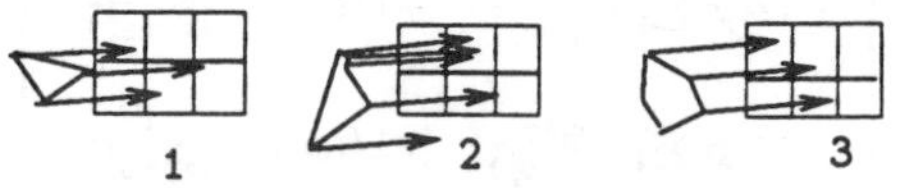

Fig. 3: *1:obstacle step too large 2:not regular ,3:good step = grid step*

However a too small step leads to a useless increase of the computing time.

These discrete models built to optimize the accumulation process are used in the process of data fusion to combine independent knowledges sources i.e. all the data coming from the individual sensors.

The accumulation/fusion method

For each sounder it is possible to build a local accumulation grid which describes all the possible locations of the ROV for a given measurement. But a measurement is modelised by a set of weighted robocentric data. We don't know which sample of the sounder model generates the echo!

• Berger[8] propounds a method named the "linear opinion pool" in the Bayesian team to process the different sources of information produced , in our case, by the individual sample of sounder model.

We assign a positive weight w_i ($\sum_{i=1}^{m} w_i$=1) to each information source π_i, reflecting the probability that the ROV is in a particular cell for a given element of the weighted vectorial model of one acoustic measurement. Then the total distribution for a given measurement built with the accumulation process is

$$\pi(P|_M) = \sum_{i=1}^{m} w_i * \pi_i(P|_M) \quad (18)$$

For a given sounder k, we compute a set of weighted accumulation grid for each vectorial sample of

$$Modele_k \equiv \{(\rho_k, \theta_{ki} \pm \epsilon_{\theta_i}, w_i)\}_{(w_i<1, \sum w_i=1)}.$$

Then we fuse them with the linear opinion pool formula as shown in [10].

• We fuse the sensorial map corresponding to several sensor measurements in a fused grid using the "independent opinion pool" used to fuse independent information in the Bayesian analysis[8]:

$$\pi(P|_M) = k[\prod_{i=1}^{m} \pi_i(P|_M)] \quad (19)$$

with k a normalization constant.

From two independent sensorial grids we obtain a global fused grid which expresses the distribution of the fusion of ROV location.

The fused grid for a set of measurements, gives the distribution of absolute possible ROV locations, independently of undercurrent value.

But this "spatial" data fusion runs on simultaneous measurements and assumes that the ROV does not move during data acquisition.

Data acquisition on an interval of time

While the robot is not moving, it is possible to validate all the acoustic range data, before running the accumulation/fusion processing. If the robot moves it is possible to accumulate data on a computed time period. These validation intervals are determined by the comparison of relative estimation module and absolute estimation module accuracies. Variances on relative estimation, given by $(\Delta\hat{x}_k^2, \Delta\hat{y}_k^2)$ grow independently of undercurrent value in time. When these values are less than the computed optimal step value *maxi(varminx , varminy)* of the grid , and when the ROV's displacements are less than dimension of the accumulation cells, it is possible to accumulate data.

Nevertheless, if the relative hypothesis on the value of undercurrent seems too bad, the relative estimation of location would be very dangerous to use in the validation process, because the drift between real and estimated relative location would quickly excede the size *stg* of the accumulation cells without control. To avoid it, the choice of a very large hypothesis on the undercurrent value allows to compute a suboptimal but safe validation interval.

The validation process was the last one for the absolute modelization of ROV location. Now we will describe the utilisation of this model by the pilot.

Matching of relative and absolute models

Matching of the absolute and the relative modules allows us to get a new hypothesis on the undercurrent value,and to reinit the relative estimation (new estimated location and covariance resetting).

This matching is automatic or manual. For each valid measurements set, the accumulation/fusion process gives a distribution of possible ROV location in a plane. We present this discrete distribution to the pilot, with grey level image or 3D profile. When the discrete step of accumulation/fusion grid is small enough, it is possible to automatically look for the cells of which the counter has the maximal value. For a small set of information it is also possible to have more than one cell at maximum value. In that case the pilot decides to select one of the hypothetical cells from

his own experience or decides to invalidate the matching of the two models, waiting for a best absolute estimation.
In the next section we will show results for the relative and the absolute data.

Results on real data

Relative module

Here the dead-reckoning interface is shown. The display shows the evolution of relative covariance in a plane and a 3D synthetic view of "what the robot observes from the relative estimated location":

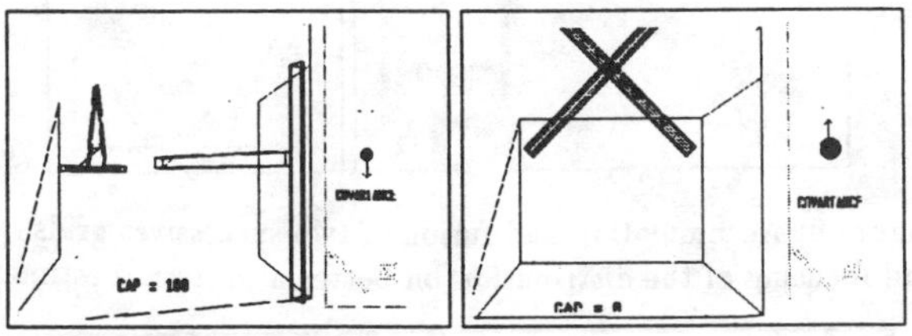

Figure 4: Relative module and covariance evolution

Accumulation process test on sonar image

We tested the accumulation algorithm on sonar images acquired in a pool with known obstacles (in size and position). Up to a certain level, each pixel is ranked as a given measurement (ρ_i, θ_i) with respect to the known simulated ROV position ROV_{simu}. Then we run the accumulation algorithm for all the measurements on the a priori data base. Results are shown with grey level overwritten on the original sensorial image (with a step of 5 pixels for reproduction facilities). The great number of data leads to an accumulation step lower than 1 pixel, but the screen physical resolution constrains us to compute a suboptimal absolute location(fig. 5. and 6.). The computed location $RO\hat{V}_{accu}$ (cell with the biggest counter value) is very precise (at least 1 pixel) and for 10^3 sensorial pixels (echoes) the algorithm takes 1mn on SUN3.

Accumulation/Fusion process on real data

We tested the accumulation-fusion process on real data acquired with an experimental equipment fitted out with 8 Mesotech '980' sounders with a 30 degrees conical beamwidth, equally arranged on a belt, and with a Mesotech '971' sonar in the center of the belt.
On fig. 7 , we model all the obstacles in the a priori data base shown in fig. 4, that is to say, the walls of the pool and three obstacles. Then we represent the ROV possible locations for one cycle with 8 sounders, as the robot is static. The first step of the algorithm is to clip all the possible obstacles in the world (no a priori information on ROV location), then for a set of measurements we compute the optimal accumulation step, we regularize the a priori concurrent obstacles and we run the accumulation process. fig. 7 shows also the estimated location (cells with higher counter) after a high pass filtering.
For more than one measurement cycle for short ROV displacement (in the validation interval) we run the validation process, and fuse successive grids: fig. 8 shows how best is the concentration of the distribution of the fused grids.
We have also tested the method with an approximate and imprecise data base.
If obstacles are submodelized then distribution is more diffuse and we have often lots of disconnected maxima as shown on fig. 9. Then it is difficult to make decision on the real location. The pilot gives new hypothesis, or uses a classical incremental modelling process [7] to validate new targets in the a priori data base. If obstacles are overmodelized, or if we model non existent obstacles, the inherent overmodelization leads to ghost areas with counters yet lower than the good estimate as shown on fig. 10.

Conclusion

The accumulation/fusion is a method which allows to integrate a priori data, in the ROV location problem. But this algorithm needs lots of computing capacity, in memory and computing time. Our optimization method on discrete steps allows to optimize the calculation, but often the constraining criteria on the method is the calculation time, which leads to get bigger steps, and consequently suboptimal estimation. Development of a dedicated computer architecture, will be the future for all these discrete methods.
In an other way , sensibility of these methods to noises and mistakes on the a priori data base are under evaluation. Lack of representation leads to a dispersion and overmodelization leads to ghosts (with counters yet lower than the good estimate) in the accumulation distribution. We are now studying the possibility to modify a priori data bases and to refine them on line.

References

[1] Elfes A. A tesselated probabilistic representation for spatial robot perception and navigation. *JPL Nasa Space Telerobotics Worshop*, February 1989.Pasadena.

[2] Kak A. and Chen S., editors. *Worshop on Spatial Reasoning and Multisensor-Fusion*, AAAI, Morgan Kauffman, October 5-7 1987.

[3] Weaver C., editor. *Sensor Fusion*, SPIE, 4-6 April 1988.

[4] Zhao C.J. *Localisation absolue d'un robot mobile autonome par télémétrie laser.* PhD thesis, INSA. Rennes I, April 1984.

[5] Chui C.K and Chen G. *Kalman filtering with real time applications.* Springer Verlag, 1987.

[6] Marcé L. Place H. and Zhao C.J. Comparison of methods for absolute location of a mobile robot in non polyhedral environments. *SPIE*, November 1984.

[7] Moravec H. Sensor fusion in certainty grids for mobile robots. *AI magazine*, vol. 2, Summer 1988.

[8] Berger J.O. *Statistical decision theory and Bayesian analysis. Second edition in Springer series in Statistics*, Springer Verlag, 1988.

[9] Cheeseman P. and Smith R.C. On the representation and estimation of spatial uncertainty. *Int. Journal of Robotic Research*, vol. 5:pp. 56–68, 1987.

[10] Rigaud V. and Marcé L. Acoustic data fusion for an absolute submarine navigation. In LAAS CNRS, editor, *First*

IARP Workshop on multisensor fusion and environment modelling, IARP, Toulouse France, October 1989.

[11] Stewart W.K. A non deterministic approach to 3D modelling underwater. *Symposium on Unmanned Unthetered Submersible Technology*, vol. 1:pp. 283–309, June 1987.

[12] Linnick Y.V. *Méthode des moindres carrés*. Dunod, 1963.

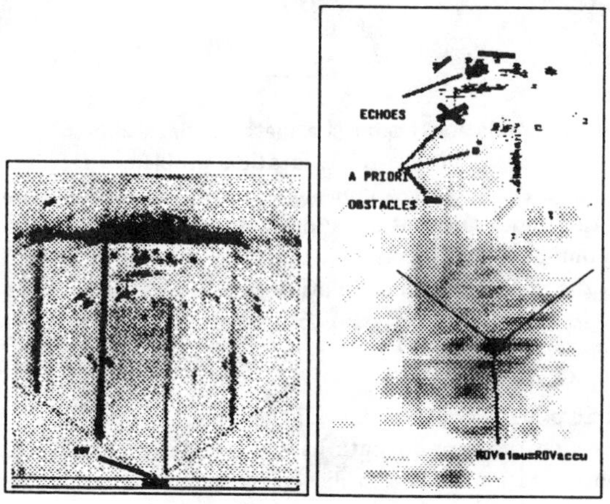

Figure 5: Accumulation on a sonar image test: 1-image data, 2-Resulting distribution

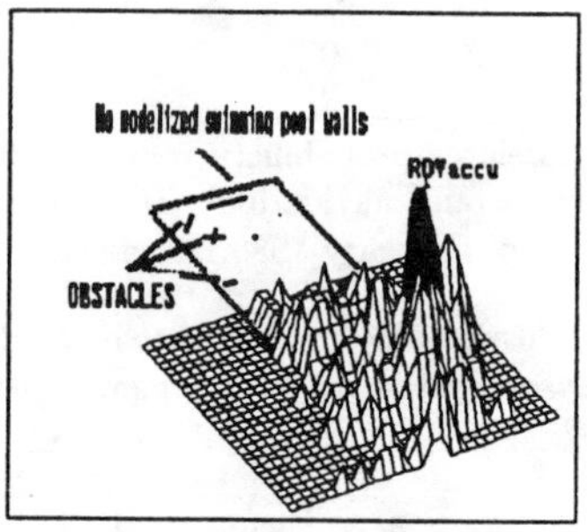

Figure 6: 3D view of the accumulation grid

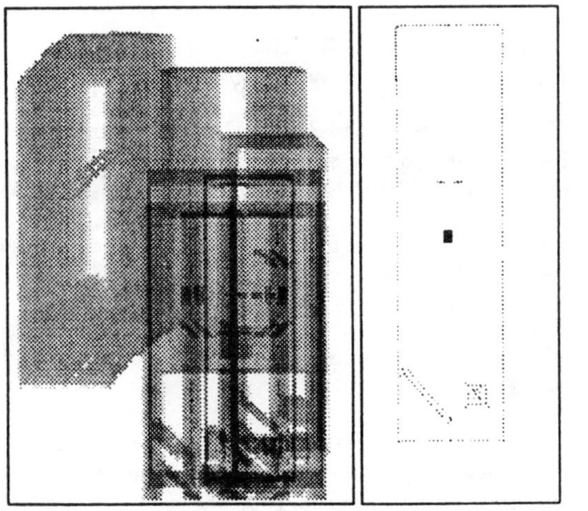

Figure 7: Accumulation of 8 sounders measurements in a pool

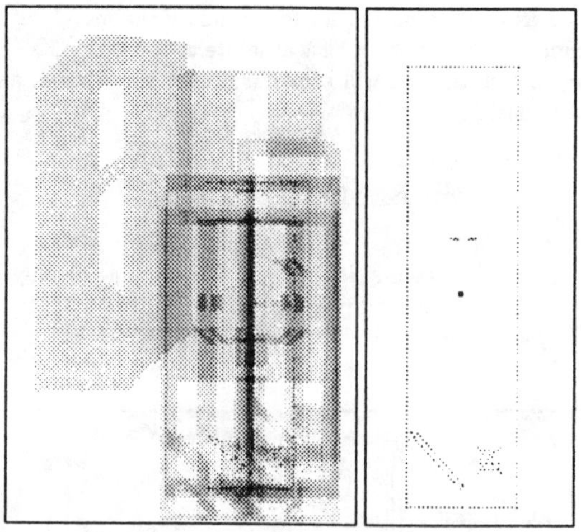

Figure 8: Accumulation and fusion of two successives grids and focusing of the distribution on better a posteriori estimate

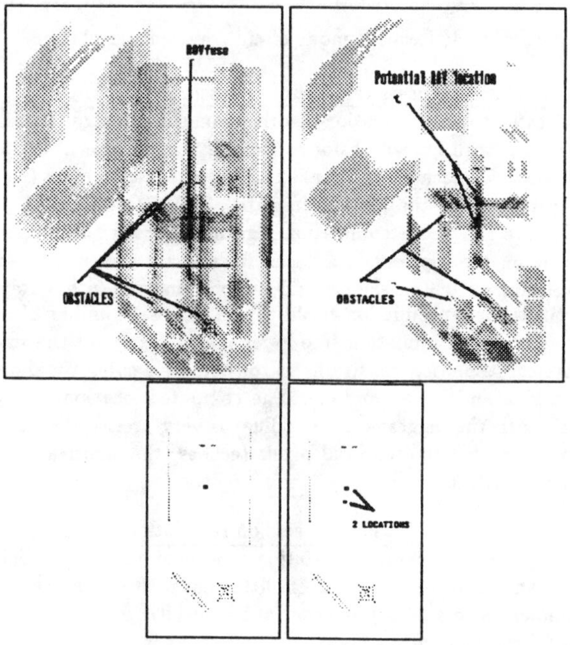

Figure 9: Submodelisation: Two possible locations

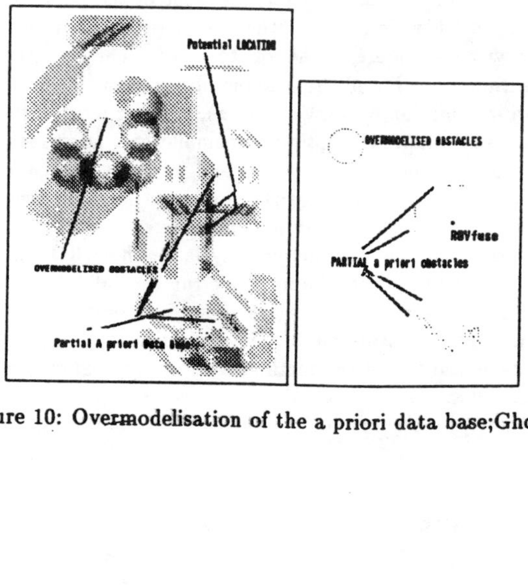

Figure 10: Overmodelisation of the a priori data base;Ghosts

Synopsis of Recent Progress on Camera Calibration for 3D Machine Vision

Roger Y. Tsai

IBM T. J. Watson Research Center
Yorktown Heights, NY 10598

1.1 Introduction

Camera calibration is the problem of determining the elements that govern the relationship or transformation between the 2D image that a camera perceives and the 3D information of the imaged object. Simply speaking, there are four main reasons why camera calibration is important:

- Only when the camera is properly calibrated can its 2-dimensional image coordinates be translated into real-world locations or constraints of locations for the objects in its field of view.
- In model based 3D Vision, it is important to model and predict the performance or accuracy capability of any vision algorithm, in order to plan the proper strategy for sensing. Without a solid understanding of camera calibration, the chain is broken and sensing strategy cannot be formed or realized.
- Part of the task of camera calibration is camera location determination relative to the calibration plate. This task can be applied to the general task of model based object location determination or 3D object tracking.
- Camera calibration is critical for vision-based robot calibration, robot hand-to-hand calibration and other vision related geometric calibration (see Tsai and Lenz, 1987B, 1988,and Lenz and Tsai, 1988).

There are two kinds of parameters that define this 2D/3D relationship:

- Intrinsic
  These are the parameters that characterize the inherent properties of the camera and optics:
  - Image Center (C_x, C_y)
  - Image X and Y scale factors
  - Lens principal distance ("effective focal length")
  - Lens distortion coefficients

Note that focal length and the image scales cannot be uniquely determined at the same time, although the aspect ratio is unique. You can increase the focal length and decrease the image scale without influencing the transformation between 2D and 3D. It is inherent in any imaging system but will not influence the goodness of fit between the model and the world. However, for solid state, discrete array camera (DAC), since the pixel spacing in the sensor plane can be known very accurately from camera manufacturers' data sheet, this ambiguity is not present. In spite of this, the x scale factor still needs to be estimated in any case since raster scan converts image signal into analog waveform and then sampled by the frame grabber. The conversion always entails uncertainty and needs to be calibrated.

Reprinted from *Robotics Review 1*, 1989, pages 147-159, "Synopsis of Recent Progress on Camera Calibration for 3-D Machine Vision" by R.Y. Tsai, by permission of The MIT Press, Cambridge, Massachusetts.

- Extrinsic
  These are the parameters indicating the position and orientation of the camera with respect to the world coordinate system:
  - Translation (T_x, T_y, T_z)
  - Rotation about X, Y, Z axes

In this synopsis, we will first make a brief survey of all the camera calibration techniques that came into existence before 1985. Then, we will make some critical comparison among four post-85 techniques. The reasons for a break at 85 are twofold. First is that, simply speaking, prior to 1985, most of the camera calibration techniques are polarized between approaches closely related to classical Photogrammetry approach where accuracy is emphasized, and the approaches geared for automation, where speed and autonomy are emphasized. The former will be called category I and the latter category II. After 1985, the techniques are more versatile and tend to combine strengths from both categories. It is natural to discuss them separately. The second reason is that we are making some critical comparisons among selected recent advancement, and there must be a break for distinguishing "recent" from the older ones. We choose 1985 for this break.

We use three criteria for comparison of the post-85 techniques. They are speed, accuracy and autonomy. Some people might wonder why should speed be important for calibration. Well, if it can be done fast with less cost or effort, why not. Actually, the more important reason is that for automation environment, the camera may be mounted on the robot hand, which may bounce around wildly. The calibration parameters may change frequently and require fast recalibration. Also, in automation environment, the camera parameters might have to be changed purposely to meet changing tasks requirements. Furthermore, part of the calibration results are position and orientation of the world coordinate system relative to the camera, which can be applied to model based object location determination. Obviously, it pays very well to do it fast, even in real time.

Summary of Pre-85 Calibration Techniques

For analysis purposes, it is convenient to classify the techniques into the following three groups:

Category I: Techniques involving full scale nonlinear optimization

The transformation between the 3D object coordinates and 2D image coordinate is a nonlinear function of all the calibration parameters. Category I is the classical approach and it attempts to do nonlinear optimization to obtain the best estimate of the calibration parameters by minimizing the residual error of the above nonlinear equations. The main advantage is that no approximation is involved and the camera model can be quite elaborate. The problem is the same as that associated with any full scale nonlinear optimization, which is that it is very computation intensive, and needs a good initial guess. The latter of course makes automation difficult. Also, it needs a good nonlinear optimization software package, which may be cumbersome to setup for a mini or personal computer. The DLT (Direct Linear Transformation) is also included in this category if lens distortion is considered (see Abdel-Aziz and Karara, 1971, 1974; Karara,1979). DLT normally should belong to category II below. But when lens distortion is included, full scale nonlinear optimization again needs to be done. Typical examples for category I are: Faig, 1975; Abdel-Aziz and Karara, 1971, 1974; Karara, 1979; Brown, 1971; Sobel, 1974; Gennery, 1979; Malhotra, 1971; Wong, 1975; Okamoto, 1981, 1984.

Category II: Techniques involving computing perspective transformation matrix first using linear equation solving

Although the equations characterizing the transformation from 3D world coordinate to 2D image coordinate are nonlinear function of the camera internal parameters (intrinsic) and external orientation and position parameters (extrinsic), it is possible to linearize the problem by ignoring lens distortion and treating the coefficients of the 3×4 perspective transformation matrix as unknowns (see Duda and Hart, 1973, for a definition of perspective transformation matrix). These coefficients are functions of the camera parameters. Given the 3D world coordinates of a number of points and the corresponding 2D image coordinates, the coefficients in the perspective transformation matrix can be solved by least square solution of an overdetermined systems of linear equations. Given the perspective transformation matrix, the camera model parameters can then be computed if needed. The advantage obviously is that nonlinear search is avoided and automation is possible since no initial guess needs to be made. But there are several problems:

- Many investigators have found that ignoring lens distortion is unacceptable when doing 3D measurement (e.g., Itoh, Miyanchi and Ozwa, 1984, Luh and Klassen, 1985; Faugeras and Toscani, 1987).
- The resultant rotation matrix is usually not orthonormal. Recently, Grosky and Tamburino tried to correct this problem, and is reviewed after this pre-85 analysis.
- Although the equations are linearized, the number of unknowns for the linear equations is usually greater than necessary. For example, the two stage technique (Tsai, 1986; Tsai and Lenz, 1988) requires solving linear equation of five unknowns is stage 1 and linear equations in three unknowns in stage 2 only, while category II techniques generally need to solve linear equations with more than 11 unknowns.
- The calibration points cannot be coplanar. Allowing coplanar calibration points greatly facilitates the process of producing highly accurate calibration points as well as making the illumination and feature extraction problem much easier. Due to the artificial linearization of the unknowns in category II techniques, coplanar points cause a singularity problem. Recently, Grosky and Tamburino tried to resolve this issue, and will be reviewed below.

A partial list of publications related to category II techniques are: Yakimovsky and Cunningham, 1978; Sutherland, 1974; Strat, 1984; Ganapathy, 1984; Abdel-Aziz and Karara, 1971, 1974; Karara, 1979; Hall, Tio, McPherson and Sadjadi, 1982.

Category III: Two Plane Method

The two-plane methods model the transformation from the image coordinates to the 3D coordinates on several calibration planes to be linear, and once each individual linear transformation (one to each plane) is estimated, the rest of the 3D points within these calibration planes are interpolated. The advantage is the same as that for category II techniques, i.e., the computation is linear. But all the problems associated with category II come in here too. Particular, the number of unknowns is at least 24 (12 for each plane), much larger than the degrees of freedom, and larger than necessary for accurate and fast computation. The two-plane method developed by Martins, Birk and Kelly (1981) theoretically can be applied in general without having any restrictions on the extrinsic camera parameters. However, for the experimental results they reported, the relative orientation between the camera coordinate system and the object world coordinate system was assumed to be known (no relative rotation). In such case, the average error is about 4 mil with a distance of 25 inches. This is comparable to the accuracy obtained using Category II and the Two-Stage Calibration Technique (Tsai, 1986). A general calibration using the two-plane technique was proposed by Isaguirre, Pu and Summers (1985). Full scale nonlinear optimization is needed. No experimental results were reported. A partial list of publications related to category II techniques are: Martins, Birk and Kelley, 1981; Isaguirre, Pu and Summers, 1985.

Critical Comparison of Several Selective post-85 Techniques

Four methods are selected for comparisons. Brief introductions are first given, followed by critical comparisons. Then, in the conclusion, there will be some final recommendations.

Brief Introduction of four selected techniques

Method A: Faugeras and Toscani, 1987

This method basically falls into category II for the pre-85 survey, so all the introduction there apply here too. Since category II methods inherently ignore lens distortion, so does method I, except that after the linear coefficients are estimated, the lens distortion is corrected at a separate step. Furthermore, in order to estimate a measure of uncertainty of the calibration results, Kalman filtering is applied, assuming zero mean independent Gaussian noise. Also, in order to make the homogeneous transformation matrix separable into a product of two matrices, one for intrinsic, and one for extrinsic, they propose an alternative means which uses the actual intrinsic and extrinsic camera parameters as unknowns and minimize a nonlinear function in the formulation of extended Kalman filtering. The residual error is minimized by iterative linear approximation.

Method B: Grosky and Tamburino, 1987

This method again falls into Category II in the pre-85 survey, so all descriptions applies to this well, except that during the transformation from the homogenous linear coefficients to the actual intrinsic and extrinsic parameters, the orthonormality of the rotation matrix is preserved. Also, with a slightly different formulation, the calibration points can be coplanar. These are the two main features of this method. Depending on how many camera parameter are unknown, the formulation for the final solution is different. Each case needs to be worked out analytically and separately. Usually, in addition to solving linear equations with 11 unknowns, a polynomial equation of single unknown and of third or fourth order needs to be solved. As characteristic of the category II methods, lens distortion is not considered.

Method C: Tsai, 1987; Tsai and Lenz, 1987A, 1988, Lenz and Tsai, 1987, Tsai, 1988

This method avoids large scale nonlinear optimization while at the same time maintaining exact modelling without making any approximations, by decoupling the calibration parameters into two groups; each group can be solved easily and rapidly. A physical constraint is sought for that is only dependent by a subset of the calibration parameters and independent of the others. If such constraint can be found, then equations can be setup that is only a function of the former subset of parameters, not the latter. Furthermore, this equation should be easy to solve. It turns out that the best candidate for such physical constraint is the Radial Alignment Constraint (see Tsai, 1987). It is quite a simple one, yet it is capable of decoupling the parameters. The resulting equations can be used to solve for most of the extrinsic parameters by computing solutions of a linear matrix equation with five unknowns (if calibration points are coplanar) or seven unknowns (if calibration points are noncoplanar), plus some tricks. The solution of the linear equation is not exactly camera parameters, but there is a one-to-one correspondence between the two. The nice thing is, there exists a trick to convert one to the other efficiently. The rest of the parameters, mostly comprising intrinsic parameters, can be solved trivially. One other goal is to allow the calibration points to reside on a single plane. In turns out that the above mentioned approach allows this to happen.

Method D: Goshtasby, 1987

This method is for calibrating lens distortion. Lens distortion is formulated in terms of Bezier patches. The x distortion and y distortion are treated separately. A square grid is put in front of the camera. Assuming that the plane holding the grid is parallel to the image plane, and the x and y axes of the grid is aligned with the x and y axes of the image plane,and the patches are used as a transformation functions that map the deformed image into the ideal image. Iterative approach is employed to obtain the control vertices of the Bezier patch that best fit the transformation that carries the distorted image sample points to the ideal image sample points.

Critical Comparisons:

- Speed

  We discuss the speed of two separate tasks involved in a normal camera calibration:

  - Speed of image feature extraction

    Although image feature extraction is only a preprocessing step for camera calibration, it is influenced in part by certain generic nature of the calibration itself. This generic nature is primarily whether the calibration points can be coplanar or not. If they can be coplanar, then the calibration plate can be a glass plate with many circles or disks printed on it (using photographic process, for instance), and the illumination can be back lighting. It is hard to overemphasize how much contribution this single factor has on the speed, robustness and ease of feature extraction. Methods B and C require only single plane calibration points, and it is reported in the publication for Method C that one is able to extract image features of 36 points with an accuracy of 1/30 pixel within 65 milliseconds, including the frame grabbing time, using minicomputer and off-the-shelf general purpose image processing hardware. Although the same feature extraction algorithm can be implemented for Method A with equal speed, Method A requires multiple plane calibration, making it a necessity to move the calibration target or the camera to a different location with highly accurate moving mechanism, and perform another feature extraction.[1] One of course will start with a static set of calibration target points that are not coplanar. But this would make it vastly more difficult to produce a calibration target with more high accuracy while at the same time easy for the camera to extract image features. Illumination may easily cause problems frequently. Another issue is about calibration point patterns (note that this has nothing to do with the calibration algorithm itself). In the paper for Method A, it is mentioned that rectangular grids are preferred rather than circles since they think the perspective distortion would cause a mismatch between the 3D circle center and the centroid of the distorted ellipse. Through some analysis, we have found that such deviation is extremely small for a normal calibration setup (easily within 1/30 pixel). Using the circles makes the feature extraction a lot easier and faster.

- Speed of camera calibration itself

  If the basic approach in Method A (rather than extended Kalman filtering in Method A) is considered, then Method A and B should be similar in speed, since they both solve linear least square problems with about 11 unknowns. Method C however, only requires solving linear equations of five and two unknowns, and can be done in 25 milliseconds using a normal minicomputer. Note that this includes both intrinsic and extrinsic calibration. If extended Kalman filtering of Method A is used, then it is equivalent to full scale nonlinear optimization, and should be more time consuming. Also, for Method A, the distortion correction requires solving for linear affine transformation coefficients for about a hundred regions (eight coefficients for each region). It is not clear how efficient it would be. So far as distortion estimation is concerned, Method D seems to be similar to Method A and less efficient than Method C.

- Accuracy

  Since Method B does not calibrate distortion, its accuracy is naturally lower in general comparing with Method A and C. Also, experimental results on the accuracy for Method B is not yet available. In principle, Method A and C should be similar in accuracy unless tangential lens distortion is big. In all the lenses we have tested, tangential lens distortion is less than 1/30 of a pixel. This is also supported by the Manual of Photogrammetry, 1980, which states that the tangential distortion is only of historical importance, and is much less than the radial distortion. The publication for Method A listed above however indicated that for a set of experiments being performed, Method A is better than Method C (still within the same order of magnitude). But they indicated that it is probably due to the experiment setup since calibration plane was almost parallel to the image plane when using Method C. When using single plane calibration in Method C, it is necessary to tilt the calibration plane at least 30 degrees to make it work. Another reason could be that during those tests, the new center and scale calibration schemes developed by Lenz and Tsai, 1987, were not done (these were developed after the original paper for Method C were published). Still another reason could be that the tangential distortion is too big for the lens used during the tests and Method C suffers some error. There is one reason why we think Method A "might" suffer some error too due to the separation of lens distortion correction from the rest of the calibration. Method A first assumes there is no distortion. Based on that, the whole calibration is done. Then, the distortion is to be estimated. When distortion is significant, this will create a bias on the solution for the rest of the parameters, which are estimated before the correction of the distortion. We do not know how great this bias is, so this concern may not be important, at least not for all cases.

[1] Furthermore, if the calibration glass plate is sitting on top of the backlighting illuminator, which in turn has to sit on the translation device, the parallelism between the top and bottom surface of the illuminator will influence the accuracy significantly.

One attractiveness of Method A is that Kalman filtering is employed, and therefore, if one is interested in obtaining an statistical measure of the uncertainty as the calibration point comes in one by one, Method A suits the purpose well.

When multiplane calibration is used for Method C, a certain set of the parameters characterizing the motion of the xyz translation stage carrying the calibration plate can be modelled into the equation to be solved for so that the xyz stage need not be very accurate to give highly accurate results.

For correcting distortion, Method D requires precise alignment of the calibration plate and the image plane. Unless that alignment is achieved, the accuracy should be less than Method A and C.

- Autonomy

  There are a few factors to be discussed. The first is the dependence on *a priori* knowledge or information. If the basic approach of Method A (as opposed to extended Kalman filtering of Method A) is considered, which does not allow the independence of intrinsic parameters with the world coordinate system, then both Method A and C do not need a *a priori* information. Although Method C makes use of the manufacturer supplied information on the pixel distance in y direction, it does not harm automation in any sense for the reason explained in the paragraph following the definition of intrinsic parameters in the Introduction. Furthermore, the manufacturer's information on the the pixel distance on the actual DAC sensor plane is highly accurate, due to the manufacturing process of the DAC sensor.

Conclusion

The importance of camera calibration may be self evident by observing the wealth of work poured into this area in the past decade. The importance of camera calibration is not just for making accurate 3D measurement, but also for helping 3D model based vision system to model the performance or capability of any particular sensing strategy. The purpose of this synopsis is to review the key features of the existing techniques, and make critical comparisons, so that the community may benefit from it when comes to choosing a particular calibration approach. From the analysis above, the following is a brief summary:

From a user's point of view,

1. if you are most concerned with the simplicity of the setup and the speed and accuracy of feature extraction, use a technique that allows single plane calibration. This eliminates the need for a very expensive and accurate xyz translation stage for moving the calibration plate or camera for creating non-coplanar calibration points. Note that although one can create non-coplanar calibration target in a static setup, it makes the construction of the setup, image feature extraction, as well as illumination a lot more difficult. Use Method B or C if single plane calibration is desired.
2. if you are most concerned about accuracy, then use Method A or C. If you have reasons to believe that "tangential" lens distortion is abnormally large, use Method A or D. We think that such occasion rarely occurs. Even if it does occur, some of the tangential distortion people experienced may not be as significant as it appears to be since they estimate the distortion after the rest of the calibration is done, and some bias is already introduced.

3. if you are most concerned with speed, then use Method C. It only takes 25 milliseconds on a minicomputer (not including image feature extraction). Comparing with all calibration methods that does not ignore lens distortion, this is at least two or three (even four) orders of magnitude faster. Even considering those methods that ignore lens distortion, Method C is still many times faster.
4. if you are most concerned with autonomy, use Method A, B or C.
5. if you have an application where the calibration points must come in sequentially, and you are interested in how each calibration point influences the statistical uncertainty of the results as the calibration points come in one by one, use the Kalman filtering approach in Method A.
6. if you are in a situation where a large portion of the calibration parameters are known already, then Method B is intended for it, although Method A and C suit the purpose equally well.

Each method has its own unique merit, and the above comparisons and recommendation should help the user make proper decision depending on his or her own needs.

References

1. Abdel-Aziz, Y.I. and Karara, H.M., 1971, Direct Linear Transformation into Object Space Coordinates in Close-Range Photogrammetry, *Symposium on Close-Range Photogrammetry,* University of Illinois at Urbana-Champaign, Urbana, Illinois, January 26-29, 1-18.
2. Abdel-Aziz, Y.I. and Karara, H.M., 1974, *Photogrammetric Potential of Non-Metric Cameras,* Civil Engineering Studies, Photogrammetry Series No. 36, University of Illinois at Urbana-Champaign, Urbana, Illinois, March..
3. Brown, Duane C., 1971, Close-Range Camera Calibration, *Photogrammetric Engineering,* Vol. 37, No. 8, 855-866.
4. Cohen, R.R. and Feigenbaum, E.A., editors, 1982, *The Handbook of Artificial Intelligence,* Vol. III, Heuris Tech Press, William Kaufmann, Inc.
5. Dainis, A. and Juberts, M., 1985, Accurate Remote Measurement of Robot Trajectory Motion, *Proceedings of Int. Conf. on Robotics and Automation,* 92-99.
6. Duda, R.O. and Hart, P.E., 1973, *Pattern Recognition and Scene Analysis,* New York, Wiley.
7. Faig, W., 1975, Calibration of Close-Range Photogrammetry Systems: Mathematical Formulation, *Photogrammetric Engineering and Remote Sensing,* Vol. 41, No. 12, 1479-1486.
8. Faugeras, O. and Toscani, G., 1987, Camera Calibration for 3D Computer Vision, *Proceedings of the International Workshop on Industrial Applications of Machine Vision and Machine Intelligence,* Seiken Symposium, Tokyo, Japan, February 2-5.
9. Ganapaphy, S., 1984, Decomposition of Transformation Matrices for Robot Vision, *Proceedings of Int. Conf. on Robotics and Automation,* 130-139.
10. Gennery, D.B., 1979, Stereo-Camera Calibration, *Proceedings Image Understanding Workshop,* November, 101-108.
11. Goshtasby, A., 1987, Correction of Image Deformation from Lens Distortion, Technical Report, Department of Computer Science, University of Kentucky, Lexington, Kentucky 40506-0027.
12. Grosky, W. and Tamburino, L., 1987, A Unified Approach to the Linear Camera Calibration Problem, Technical Report, Department of EE and CS, University of Michigan, Ann Arbor, Michigan.

13. Hall, E.L., Tio, M.B.K., McPherson, C.A. and Sadjadi, F.A.,1982, Curved Surface Measurement and Recognition for Robot Vision, Conference Record, *IEEE Workshop on Industrial Applications of Machine Vision,* May 3-5.

14. Itoh, H., Miyauchi, A. and Ozawa, S., 1984, Distance Measuring Method using only Simple Vision Constructed for Moving Robots, *7th Int. Conf. on Pattern Recognition,* Montreal, Canada, July 30-August 2, Vol. 1, p. 192.

15. Isaguirre, A., Pu, P. and Summers, J., 1985, A New Development in Camera Calibration: Calibrating A Pair of Mobile Cameras, *Proceedings of Int. Conf. on Robotics and Automation,* 74-79.

16. Karara, H.M., editor, 1979, *Handbook of Non-Topographic Photogrammetry,* American Society of Photogrammetry.

17. Lenz, R. and Tsai, R., 1987, Techniques for Calibration of the Scale Factor and Image Center for High Accuracy 3D Machine Vision Metrology, *Proceedings of IEEE International Conference on Robotics and Automation,* Raleigh, NC, Also to appear in IEEE Trans. on PAMI.

18. Lenz, R. and Tsai, R., 1988, Calibrating a Cartesian Robot with Eye-on-Hand Configuration Independent of Eye-to-Hand Relationship, *Proceedings of IEEE International Conference on CVPR,* Ann Arbor, MI. Also to appear in IEEE Trans. on PAMI.

19. Lowe, D.G., Solving for the Parameters of Object Models from Image Descriptions, *Proceedings Image Understanding Workshop,* April 1980, 121-127.

20. Luh, J.Y. and Klaasen, J.A., 1985, A Three-Dimensional Vision by Off-Shelf System with Multi-Cameras, *IEEE Transactions on Pattern Analysis and Machine Intelligence,* Vol. PAMI-7, No. 1, January, 35-45.

21. Malhotra, 1971, A Computer Program for the Calibration of Close-Range Cameras, *Proceedings of Symposium on Cloase Range Photogrammetric Systems,* Urbana, Ill.

22. *Manual of Photogrammetry,* 1980, fourth edition, American Society of Photogrammetry.

23. Martins, H.A., Birk, J.R. and Kelley, R.B., 1981, Camera Models Based on Data from Two Calibration Planes, *Computer Graphics and Image Processing,* 17, 173-180.

24. Moravec, H., 1981, *Robot Rover Visual Navigation,* UMI Research Press.

25. Okamoto, A., 1981, Orientation and Construction of Models, Part I: The Orientation Problem in Close-Range Photogrammetry, *Photogrammetric Engineering and Remote Sensing,* Vol. 47, No. 10, 1437-1454.

26. Okamoto, A., 1984, The Model Construction Problem Using the Collinearity Condition, *Photogrammetric Engineering and Remote Sensing,* Vol. L, No. 6, 705-711.

27. Sobel, I., 1974, On Calibrating Computer Controlled Cameras for Perceiving 3-D Scenes, *Artificial Intelligence,* 5, 185-198.

28. Strat, T.M., 1984, Recovering the Camera Parameters from a Transformation Matrix, *Proceedings: DARPA Image Understanding Workshop,* Oct., pp. 264-271.

29. Sutherland, I., 1974, Three-Dimensional Data Input by Tablet, *Proceedings of the IEEE,* Vol. 62, No. 4, April, 453-461.

30. Tsai, R., 1987, A Versatile Camera Calibration Technique for High Accuracy 3D Machine Vision Metrology using Off-the-Shelf TV Cameras and Lenses, IEEE Journal of Robotics and Automation, Vol. RA-3, No. 4, August. A preliminary version appeared in 1986 IEEE International Conference on Computer Vision and Pattern Recognition, Miami, Florida, June 22-26.

31. Tsai, R., 1988, Review of RAC-based Camera Calibration, to appear in *Vision*, MVA/SME's Quarterly on Vision Technology, November.

32. Tsai, R., and Lenz, R., 1987A, Review of the Two-Stage Camera Calibration Technique plus some New Implementation Tips and New Techniques for Center and Scale Calibration, *Second Topical Meeting on Machine Vision*, Optical Society of America, Lake Tahoe, March 18-20.

33. Tsai, R. Y., and Lenz, R., 1987B, A New Technique for Fully Autonomous and Efficient 3D Robotics Hand-Eye Calibration, *4th International Symposium on Robotics Research*, Santa Cruz, CA, August 9-14. Also to appear in IEEE Journal of Robotics and Automation.

34. Tsai, R. Y., and Lenz, R., 1988, Real Time Versatile Robotics Hand/Eye Calibration using 3D Machine Vision, *International Conference on Robotics and Automation*, Philadelphia, PA, April 24-29.

35. Wong, K.W., 1975, Mathematical Formulation and Digital Analysis in Close-Range Photogrammetry, *Photogrammetric Engineering and Remote Sensing*, Vol. 41, No. 11, 1355-1373.

36. Yakimovsky, Y. and Cunningham, R., 1978, A System for Extracting Three-Dimensional Measurements from a Stereo Pair of TV Cameras, *Computer Graphics and Image Processing*, 7, 195-210.

Multisensor Integration and Fusion in Intelligent Systems

REN C. LUO, SENIOR MEMBER, IEEE, AND MICHAEL G. KAY

Abstract—Interest has been growing in the use of multiple sensors to increase the capabilities of intelligent systems. The issues involved in integrating multiple sensors into the operation of a system are presented in the context of the type of information these sensors can uniquely provide. The advantages gained through the synergistic use of multisensory information can be decomposed into a combination of four fundamental aspects: the redundancy, complementarity, timeliness, and cost of the information. The role of multiple sensors in the operation of a particular system can then be defined as the degree to which each of these four aspects is present in the information provided by the sensors. A distinction is made between multisensor integration and the more restricted notion of multisensor fusion to separate the more general issues involved in the integration of multiple sensory devices at the system architecture and control level, from the more specific issues—possibly mathematical or statistical—involved in the actual combination (or fusion) of multisensory information. A survey is provided of the increasing number and variety of approaches to the problem of multisensor integration and fusion that have appeared in the literature in recent years—ranging from general paradigms, frameworks, and methods for integrating and fusing multisensory information, to existing multisensor systems used in different areas of application. General multisensor fusion methods, sensor selection strategies, and world models are surveyed, along with approaches to the integration and fusion of information from combinations of different types of sensors. Short descriptions of the role of multisensor integration and fusion in the operation of a number of existing mobile robots are provided, together with proposed high-level multisensory representations suitable for mobile robot navigation and control. Existing multisensor systems are surveyed in the following areas of application: industrial tasks like material handling, part fabrication (e.g., welding), inspection, and assembly; military command and control for battle management; space; target tracking; inertial navigation; and the remote sensing of coastal waters. A discussion is included of possible problems associated with creating a general methodology for multisensor integration and fusion—focusing on the methods used for modeling error or uncertainty in the integration and fusion process (e.g., the registration problem), the actual sensory information (i.e., the sensor model), and the operation of the overall system (e.g., multisensor calibration).

I. Introduction

IN RECENT YEARS interest has been growing in the synergistic use of multiple sensors to increase the capabilities of intelligent machines and systems. For these systems to use multiple sensors effectively, some method is needed for integrating the information provided by these sensors into the operation of the system. While in many multisensor systems the information from each sensor serves as a separate input to the system, the actual combination or fusion of information prior to its use in the system has been a particularly active area of research. Typical of the applications that can benefit from the use of multiple sensors are industrial tasks like assembly, military command and control for battlefield management, mobile robot navigation, multitarget tracking, and aircraft navigation. Common among all of these applications is the requirement that the system intelligently interact with and operate in an unstructured environment without the complete control of a human operator.

Manuscript received August 10, 1988; revised March 18, 1989. Preliminary versions of portions of this paper were presented at the SPIE Conference on Sensor Fusion, Orlando, FL, April 4–6, 1988, and published in the proceedings thereof, and the IEEE International Workshop on Intelligent Robots, Tokyo, Japan, October 31–November 2, 1988. This work was supported in part by the National Science Foundation under Grant DMC-8716126.

R. C. Luo is with the Robotics and Intelligent Systems Laboratory in the Department of Electrical and Computer Engineering, North Carolina State University, Raleigh, NC 27695-7911.

M. G. Kay is with the Department of Industrial Engineering, North Carolina State University, Raleigh, NC 27695-7906.

IEEE Log Number 8928399.

A. Biological Examples of the Synergistic Integration of Multisensor Information

Two of the major abilities that a human operator brings to the task of controlling a system are the use of a flexible body of knowledge and the ability to integrate synergistically information of different modality obtained through his or her senses. The increasing use of knowledge-based expert systems is an attempt to capture some aspects of this first ability; current research in multisensor integration is an attempt to capture, and possibly extend to additional modalities, aspects of this second ability. Thus a human's or other animal's ability to integrate multisensory information can provide an indication of what is ultimately achievable for intelligent systems (i.e., an existence proof) and insight into possible future research directions.

1) Ventriloquism: A well-known example of human multisensory integration is ventriloquism, in which the voice of the ventriloquist seems to an observer to come from the ventriloquist's dummy. The ability of visual information (the movement of the dummy's lips) to dominate the auditory information coming from the ventriloquist demonstrates the existence of some process of integration whereby information from one modality (audition) is interpreted solely in terms of information from another modality (vision). Howard [1] has reported research that found the discordance between visual and auditory information becomes noticeable only after the source of each has been separated beyond 30° relative to the observer (see Fig. 1). Notwithstanding ventriloquism, the use of information from these two modalities can increase the probability of

Reprinted from *IEEE Transactions on Systems, Man, and Cybernetics*, Vol. 19, No. 5, September/October 1989, pages 901-931. Copyright © 1989 by The Institute of Electrical and Electronics Engineers, Inc. All rights reserved.

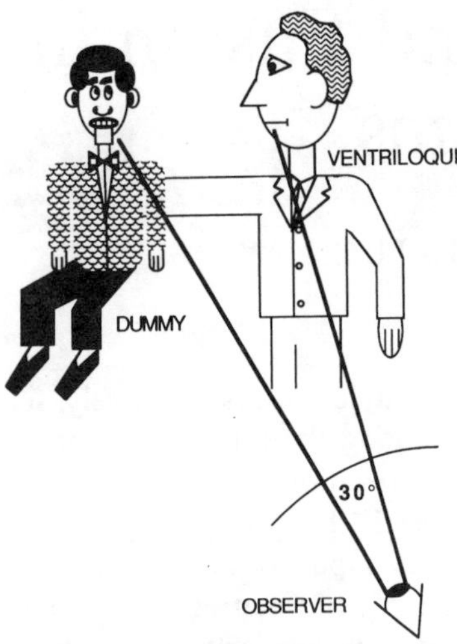

Fig. 1. Ventriloquism demonstrates existence of some process of human multisensory integration through ability of visual information (movement of dummy's lips) to dominate auditory information (from ventriloquist) for up to 30° separation of these information sources relative to observer.

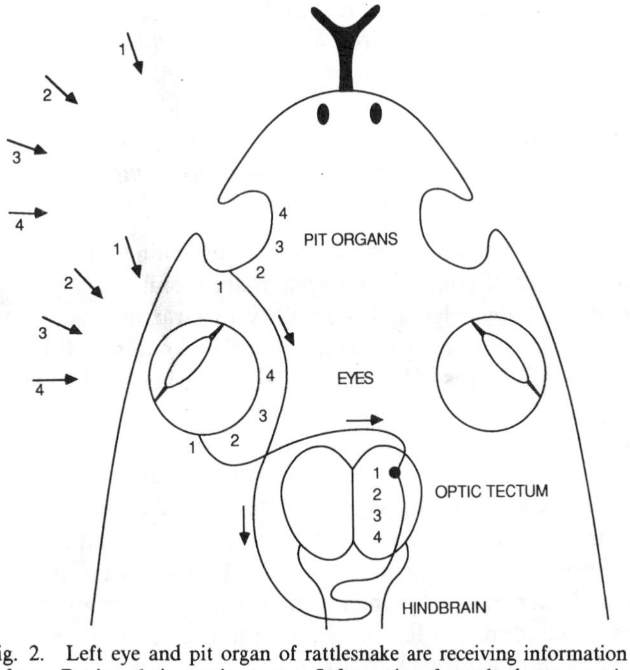

Fig. 2. Left eye and pit organ of rattlesnake are receiving information from Region 1 in environment. Information from both sources is represented on surface of optic tectum in similar spatial orientation. (Adapted from [2, p. 121].)

detecting an event in the environment when compared to the use of either modality alone.

2) Pit Vipers and Rattlesnakes: Although in humans the processes of multisensory integration have not yet been found, research on the less complex nervous systems of the pit viper and the rattlesnake has identified neurons in these snakes' optic tectums (a midbrain structure found in vertebrates) that are responsive to both visual and infrared information [2]. As shown in Fig. 2, both the left eye and pit organ of a rattlesnake are receiving information from Region 1 in the environment. Infrared information from the pit organ, together with visual information from the eye, are represented on the surface of the optic tectum in a similar spatial orientation so that each region of the optic tectum receives information from the same region of the environment. This allows certain "multimodal" neurons to respond to different combinations of visual and infrared information. Certain "or" neurons respond to information from either modality and could be used by the snake to detect the presence of prey in dim lighting conditions, while certain "and" neurons, which only respond to information from both modalities, could be used to recognize the difference between a warm-blooded mouse and a cool-skinned frog. The "and" neurons have been whimsically described as mouse detectors. In evolutionary terms, it seems likely that similar integration processes take place in the tectums of most other vertebrates—although at present only Newman and Hartline's [2] work on pit vipers and rattlesnakes has been reported.

B. Previous Surveys and Reviews

A number of recent papers have surveyed and reviewed different aspects of multisensor integration and fusion. An article on multisensor integration in the *Encyclopedia of Artificial Intelligence* has focused on the issues involved in object recognition [3]. Mitiche and Aggarwal [4] discuss some of the advantages and problems involved with the integration of different image processing sensors, and review recent work in that area. Garvey [5] has surveyed some of the different artificial intelligence approaches to the integration and fusion of information, emphasizing the fundamental role in artificial intelligence of the inference process for combining information. A number of the different knowledge representations, inference methods, and control strategies used in the inference process are discussed in his paper. Mann [6] provides a concise literature review as part of his paper concerning methods for integration and fusion that are based on the maintenance of consistent labels across different sensor domains. Luo and Kay [7], and Blackman [8] have surveyed some of the issues of and different approaches to multisensor integration and fusion, with Blackman providing an especially detailed discussion of the data association problem (Section VI-D). Recent research workshops have focused on the multisensor integration and fusion issues involved in manufacturing automation [9] and spatial reasoning [10].

C. Overview of this Paper

Section II serves both to describe multisensor integration and fusion and to set the stage for the presentation in the four subsequent sections of a broad survey of current approaches to the problem. In each of these subsequent sections, increasingly more specific approaches are surveyed: from general paradigms and methods for integrating and fusing multisensory information, to existing multisensor systems used in different areas of application.

Section II describes the role of multisensor integration and fusion in the operation of an intelligent system. Multisensor integration and the related notion of multisensor fusion are defined and distinguished. A general pattern of multisensor integration and fusion is presented to highlight the distinction between the integration and the fusion of information in the overall operation of a system. The potential advantages in integrating multiple sensors are then discussed in terms of four fundamental aspects of the information provided by the sensors.

Section III presents approaches to different aspects of multisensor integration and fusion that are quite general in terms of their range of applicability. Initially a variety of paradigms, frameworks, and control structures are presented that have been proposed for the overall multisensor integration process. Work is then presented relating to two important integration functions: the preselection and real-time selection of sensors and the use of world models. The section concludes with a survey of general multisensor fusion methods.

Section IV surveys approaches to the integration and fusion of information from combinations of different types of sensors, with special emphasis given to vision and tactile sensor combinations because of the broad range of capabilities that this combination can provide an industrial robot.

Section V details the critical role played by multisensor integration and fusion in enabling mobile robots to operate in uncertain or unknown dynamic environments. A variety of proposed high-level representations for multisensory information are presented that are suitable for mobile robot control and navigation. The section concludes with a discussion of different sensor combinations that have been used in mobile robots, and short descriptions of the role of multisensor integration and fusion in the navigation and control of a number of existing mobile robots.

Section VI surveys a variety of multisensor systems in the following areas of application: industrial tasks like material handling, part fabrication (e.g., welding), inspection, and assembly; military tasks (e.g., command and control for battle management); space; target tracking; inertial navigation; and the remote sensing of coastal waters.

Section VII concludes with a discussion of possible problems and future research directions in the area of multisensor integration and fusion. Discussion of the possible problems centers around the methods used for modeling error or uncertainty in the integration and fusion process, the actual sensory information, and the operation of the overall system.

II. The Role of Multisensor Integration and Fusion in Intelligent Systems

In the operation of an intelligent system, the role of multisensor integration and fusion can best be understood with reference to the type of information that the integrated multiple sensors can uniquely provide the system. The potential advantages gained through the synergistic use of this multisensory information can be decomposed into a combination of four fundamental aspects: the redundancy, complementarity, timeliness, and cost of the information. Prior to discussing these aspects, this section first provides a definition of the distinction between the notions of the integration and the fusion of multisensory information; secondly, a general pattern of multisensor integration and fusion is presented within the context of an overall system architecture to highlight some of the important functions in the integration process.

A. Multisensor Integration versus Fusion

Multisensor integration, as defined in this paper, refers to the synergistic use of the in formation provided by multiple sensory devices to assist in the accomplishment of a task by a system. An additional distinction is made between multisensor integration and the more restricted notion of multisensor fusion. Multisensor fusion, as defined in this paper, refers to any stage in the integration process where there is an actual combination (or fusion) of different sources of sensory information into one representational format. (This definition would also apply to the fusion of information from a single sensory device acquired over an extended time period.) Although the distinction of fusion from integration is not standard in the literature, it serves to separate the more general issues involved in the integration of multiple sensory devices at the system architecture and control level, from the more specific issues involving the actual fusion of sensory information—e.g., in many integrated multisensor systems the information from one sensor may be used to guide the operation of other sensors in the system without ever actually fusing the sensors' information (e.g., Section IV-B).

B. A General Pattern

Fig. 3 is meant to represent a general pattern of multisensor integration and fusion in a system. While the fusion of information takes place at the nodes in the figure, the entire network structure, together with the integration functions, shown as part of the system, are part of the multisensor integration process. In the figure, n sensors are integrated to provide information to the system. The outputs x_1 and x_2 from the first two sensors are fused at the lower left-hand node into a new representation $x_{1,2}$. The output x_3 from the third sensor could then be fused with $x_{1,2}$ at the next node, resulting in the representation $x_{1,2,3}$, which might then be fused at nodes higher in the structure. In a similar manner the output from all n sensors could be integrated into an overall network structure. The dashed lines from the system to each node represent any of the possible signals sent from the integration functions within the system. The three functions shown in the figure are some of the functions typically used as part of the integration process. "Sensor selection" can select the most appropriate group of sensors to use in response to changing conditions, sensory information can be represented within the "world model," and the information from different sensors may need to be "transformed" before it can be fused or represented in the world

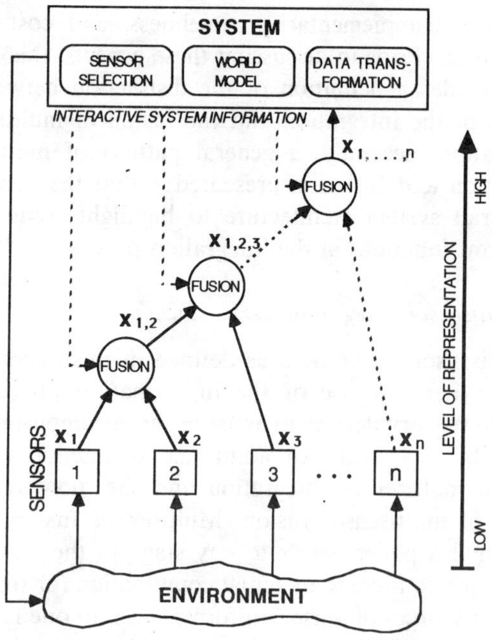

Fig. 3. General pattern of multisensor integration and fusion in system.

model. Shown along the right-side of the figure is a scale indicating the level of representation of the information at the corresponding level in the network structure. The transformation from lower to higher levels of representation as the information moves up through the structure is common in most multisensor integration processes. At the lowest level, raw sensory data are transformed into information in the form of a signal. As a result of a series of fusion steps, the signal may be transformed into progressively more abstract numeric or symbolic representations. This "signals-to-symbols" paradigm is common in computational vision [11].

C. Potential Advantages in Integrating Multiple Sensors

The purpose of external sensors is to provide a system with useful information concerning some features of interest in the system's environment. The potential advantages in integrating and/or fusing information from multiple sensors are that the information can be obtained more accurately, concerning features that are impossible to perceive with individual sensors, in less time, and at a lesser cost. These advantages correspond, respectively, to the notions of the redundancy, complementarity, timeliness, and cost of the information provided the system.

Redundant information is provided from a group of sensors (or a single sensor over time) when each sensor is perceiving, possibly with a different fidelity, the same features in the environment. The integration or fusion of redundant information can reduce overall uncertainty and thus increase the accuracy with which the features are perceived by the system. Multiple sensors providing redundant information can also serve to increase reliability in the case of sensor error or failure.

Complementary information from multiple sensors allows features in the environment to be perceived that are impossible to perceive using just the information from each individual sensor operating separately. If the features to be perceived are considered dimensions in a space of features, then complementary information is provided when each sensor is only able to provide information concerning a subset of features that form a subspace in the feature space, i.e., each sensor can be said to perceive features that are independent of the features perceived by the other sensors; conversely, the dependent features perceived by sensors providing redundant information would form a basis in the feature space.

More timely information, as compared to the speed at which it could be provided by a single sensor, may be provided by multiple sensors due to either the actual speed of operation of each sensor, or the processing parallelism that may be possible to achieve as part of the integration process.

Less costly information, in the context of a system with multiple sensors, is information obtained at a lesser cost when compared to the equivalent information that could be obtained from a single sensor. Unless the information provided by the single sensor is being used for additional functions in the system, the total cost of the single sensor should be compared to the total cost of the integrated multisensor system.

The role of multisensor integration and fusion in the overall operation of a system can be defined as the degree to which each of these four aspects is present in the information provided by the sensors to the system. Redundant information can usually be fused at a lower level of representation compared to complementary information because it can more easily be made commensurate. Complementary information is usually either fused at a symbolic level of representation, or provided directly to different parts of the system without being fused. While in most cases the advantages gained through the use of redundant, complementary, or more timely information in a system can be directly related to possible economic benefits, in one case fused information was used in a distributed network of target tracking sensors just to reduce the bandwidth required for communication between groups of sensors in the network (Section VI-D).

Fig. 4 illustrates the distinction between complementary and redundant information by using the network structure from Fig. 3 to perform, hypothetically, the task of object discrimination. Four objects are shown in Fig. 4(a). They are distinguished by the two independent features, shape and temperature. Sensors 1 and 2 provide redundant information concerning the shape of an object, and Sensor 3 provides information concerning its temperature. Fig. 4(b) and (c) show hypothetical frequency distributions for both square and round objects, representing each sensor's historical (i.e., tested) responses to such objects. The bottom axes of both figures represent the range of possible sensor readings. The output values x_1 and x_2 correspond to some numerical "degree of squareness or roundness" of the object as determined by each sensor, respectively. Because Sensors 1 and 2 are not able to detect the temperature of an object, objects A and C (as well as B and D) cannot be distinguished. The dark portion of the axis in each figure corresponds to the range of output values where there is uncertainty as to the shape of the object being detected. The dashed line in each figure corresponds to the point at which, depending on the output value, objects can be distinguished in terms of a feature. Fig. 4(d) is the frequency distribution resulting from the fusion of x_1 and x_2. Without specifying a particular method of fusion, it is usually true that the distribution corresponding to the fusion of redundant information would have less dispersion than its component distributions. (Under very general assumptions, a plausibility argument can be made that the relative probability of the fusion process not reducing the uncertainty is zero [12].) The uncertainty in Fig. 4(d) is shown as approximately half that of Fig. 4(b) and (c). In Fig. 4(e), complementary information from Sensor 3 concerning the independent feature temperature is fused with the shape information from Sensors 1 and 2 shown in Fig. 4(d). As a result of the fusion of this additional feature, it is now possible to discriminate between all four objects. This increase in discrimination ability is one of the advantages resulting from the fusion of complementary information. As mentioned before, the information resulting from this second fusion could be at a higher representational level (e.g., the result of the first fusion, $x_{1,2}$, may still be a numerical value, while the result of the second $x_{1,2,3}$, could be a symbol representing one of the four possible objects).

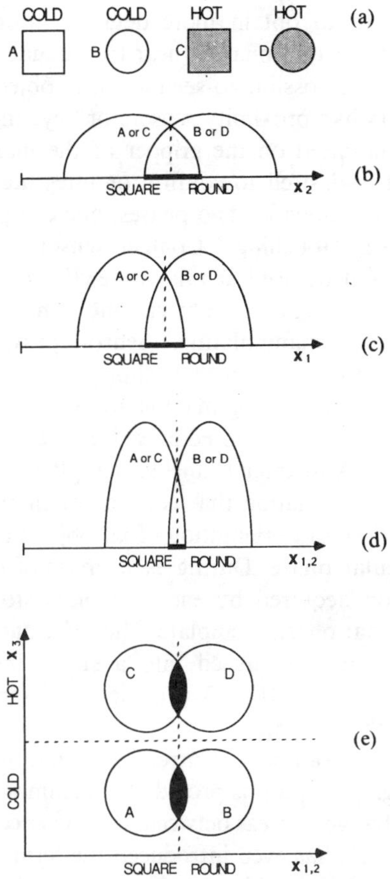

Fig. 4. Discrimination of four different objects using redundant and complementary information from three sensors. (a) Four objects (A, B, C, and D) distinguished by features "shape" (square vs. round) and "temperature" (hot versus cold). (b) Two-dimensional (2-D) distributions from Sensor 1 (shape). (c) Sensor 2 (shape). (d) 2-D distributions resulting from fusion of redundant shape information from Sensors 1 and 2. (e) Three-dimensional (3-D) distributions resulting from fusion of complementary information from Sensors 1 and 2 (shape), and Sensor 3 (temperature).

III. General Approaches to Multisensor Integration and Fusion

This section presents approaches to different aspects of the multisensor integration and fusion problem discussed in the previous section. Although some of the approaches were originally presented in terms of a specific application or combination of sensors, they are distinguished by their applicability to a broad range of systems in a number of possible applications.

A. Paradigms and Frameworks for Integration

1) Hierarchical Phase-Template Paradigm: Luo and Lin [13]–[17] have proposed a general paradigm for multisensor integration in robotic systems based upon four distinct temporal phases in the sensory information acquisition process (see Fig. 5). The four phases, "far away," "near to," "touching," and "manipulation," are distinguished at each phase by the range over which sensing will take place, the subset of sensors typically required, and, most importantly, the type of information desired. During the first phase, "far away," only global information concerning the environment is obtained. Typical information at this stage would be the detection, location, or identity of objects in a scene. The most likely types of sensors to be used during this phase would be noncontact sensors like vision cameras and range finding devices. If the scene is found to be of sufficient interest during the first phase, the manipulator

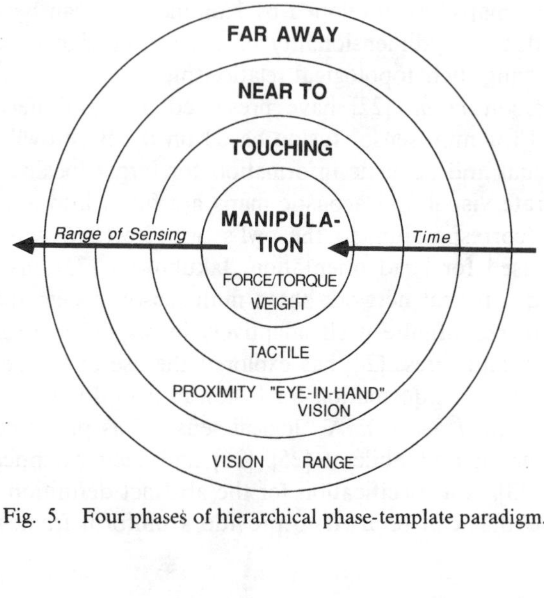

Fig. 5. Four phases of hierarchical phase-template paradigm.

can zoom in to obtain more detailed information. This leads to the second phase, "near to." Usually at this close range it is not possible to see the entire object, so noncontact sensors like proximity sensors or "eye-in-hand" vision systems, mounted on the gripper of the manipulator, are used. If it is desired to confirm or integrate the information from the previous two phases, one can proceed to the third phase, "touching." Contact sensors such as tactile sensors might be used at this phase. Finally if it is necessary to manipulate the object, one can proceed to the fourth phase, "manipulating." Sensors providing information concerning force/torque, slippage, and weight would typically be used during manipulation.

The information acquired at each phase is represented in the form of a distinct framelike template. Each template represents information that is both common to all phases (e.g., position and orientation of an object) and specific to the particular phase. During each phase of operation, the information acquired by each sensor is stored as an instance of that phase's template. The information from each sensor can then be fused into a single instance of the template (Section III-E-3 provides a description of the fusion method used).

2) Neural Networks: Current research in neural networks (e.g., [18]–[20]) is providing a common paradigm for the interchange of ideas between neuroscience and robotics; Pellionisz [21] has even introduced the term "neurobotics" to describe the possible use of brainlike control and representation in robotic systems. Although related to the adaptive learning control structure described in Section III-B-3, neural networks provide a fairly well-established formalism with which to model the multisensor integration process. Neurons can be trained to represent sensory information and, through "associative recall," complex combinations of the neurons can be activated in response to different sensory stimuli. "Simulated annealing" is one of many different techniques that can be used to find a global optimal state in a network based upon the local state of activation of each neuron in the network. Simulated annealing has been used to find optimal global paths for mobile robot navigation (Section V-B-2). "Self-organizing feature maps" as developed by Kohonen [20] can be used to reduce the dimensionality of the sensor signals while preserving their topological relationships.

Pearson *et al.* [22] have presented a neural network model for multisensor fusion based on the barn owl's use of visual and acoustic information for target localization. Separate visual and acoustic maps are fused into a single map (corresponding to the owl's optic tectum) which is then used for head orientation. Jakubowicz [23] has presented a neural network-based multisensor system that is able to reconfigure itself adaptively in response to sensor failure, and Dress [24] has explored the use of frequency-coded sensor information for fusion in neural networks.

3) Logical Sensors: A "logical sensor," as proposed by Henderson and Shilcrat [25], [26] and then extended in [27]–[33], is a specification for the abstract definition of a sensor that can be used to provide a uniform framework for multisensor integration. Through the use of an abstract definition of a sensor, the unnecessary details of the actual physical sensor are separated from their functional use in a system. In a manner similar to how an abstract data type separates the user from unnecessary algorithmic detail, the use of logical sensors can provide any multisensor system with both portability and the ability to adapt to technological changes in a manner transparent to the system.

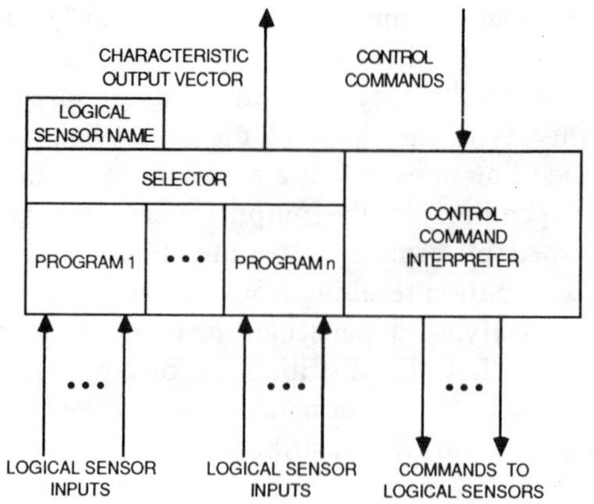

Fig. 6. Basic components of logical sensor. (Adapted from [27, fig. 1].)

Fig. 6 shows the essential elements of a logical sensor. Each logical sensor can serve as an element in a network of logical sensors, which itself can be viewed as a logical sensor. The "logical sensor name" uniquely defines a logical sensor. The "characteristic output vector" describes the data type of the stream of output vectors produced by the logical sensor. The "control commands" input to a logical sensor consist of both commands necessary to control the logical sensor and commands that are just passing through to other sensors lower in the network. The "control command interpreter" processes the incoming commands and sends appropriate commands to logical sensors lower in the network. The "selector" monitors the control commands issued to the logical sensor and the results of the various "program units"—acting as a "microexpert system" which knows the required function of the logical sensor. Each program unit serves to perform any required computation on the inputs to the unit. The logical sensor inputs are the output vectors of logical sensors lower in the network. When the logical sensor is an actual physical sensor, the raw data sensed from the environment can be considered as null inputs.

A hypothetical logical sensor-based range finder is shown in Fig. 7 that incorporates three physical sensors: an ultrasonic range finder and two cameras. Both cameras are used as input to a fast and a slow stereo logical sensor. Each of these logical sensors, which differ in terms of the speed and accuracy of their processing algorithms, are used as input to an overall stereo logical sensor which just serves a switching function based on control commands from the top-level logical sensor. The entire network of logical and physical sensors can provide for a range finder that is both robust in terms of the lighting conditions in which it can operate (i.e., the ultrasonic sensor for poor

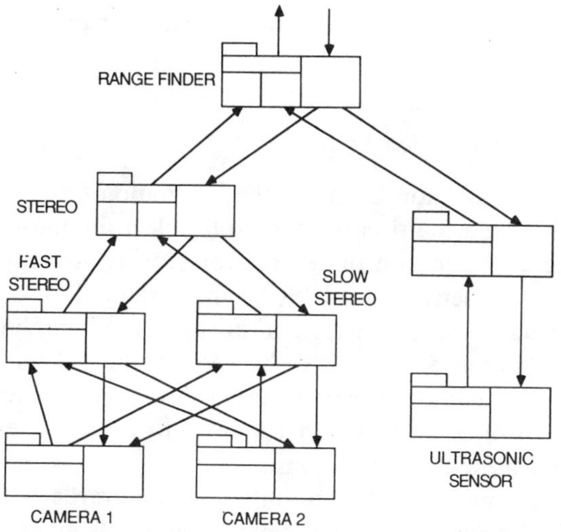

Fig. 7. Logical sensor network for range finder. (Adapted from [25, Fig. 6].)

lighting conditions) and, depending on time constraints, the speed at which it can operate. The range information could possibly be made more accurate if the redundant information from the stereo and ultrasonic sensors is fused at the top-level logical sensor in the network.

4) Object-Oriented Programming: In a similar manner to the logical sensors mentioned earlier, object-oriented programming is a methodology that can be used to develop a uniform framework for implementing multisensor tasks; Henderson and Weitz [29], [31] have, in fact, discussed the development of logical sensor specifications within an object-oriented programming context. In most object-oriented multisensor applications, each sensor is represented as an object. Objects communicate by passing messages that invoke specialized sensor processing procedures ("methods") based on the sensor's attributes and behavior. Each method is transparent to other objects, allowing possibly different physical sensors to be used interchangeably. Rodger and Browse [34] have used object-oriented programming for multisensor object recognition, and Allen [35] has developed an object-oriented framework for multisensor robotic tasks.

B. Control Structures

This section presents different structures that have been used to control the overall integration and fusion process. Control structures based on artificial intelligence (e.g., production systems) have not been included because a thorough discussion of their use in the context of multisensor integration and fusion can be found in [5].

1) The NBS Sensory and Control Hierarchy: The Center for Manufacturing Engineering at the National Bureau of Standards (NBS) is implementing an experimental factory called the Automated Manufacturing Research Facility (AMRF). As part of the AMRF, a multisensor interactive hierarchical robot control system [36]–[40] is being developed based, in part, on the mathematical formalism called the cerebellar model arithmetic computer [41], [42]. As

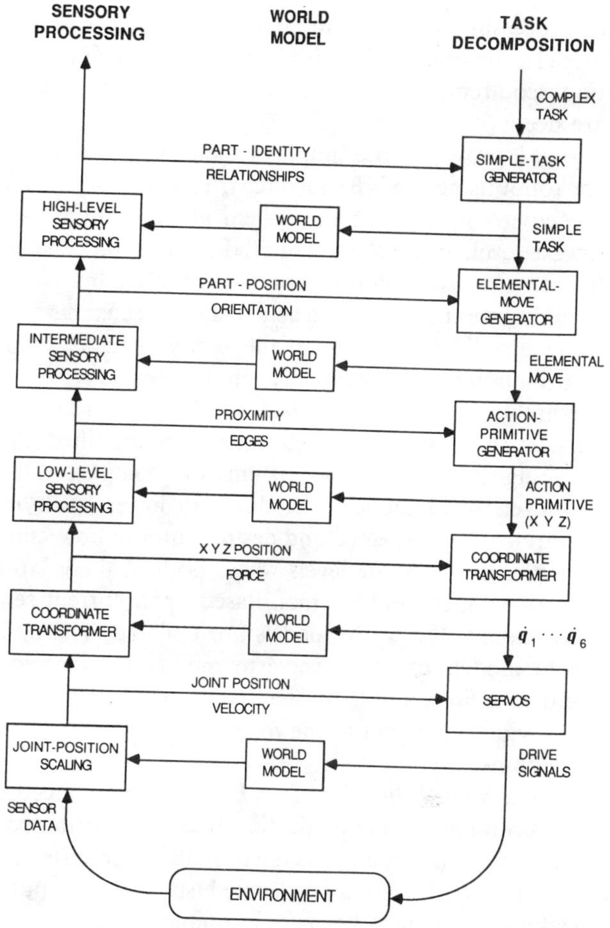

Fig. 8. NBS sensory and control hierarchy used to control multisensor robot. (Adapted from [38, figs. 5.24 and 9.6].)

shown in Fig. 8, the structure of the control system in AMRF consists of an ascending "sensory processing" hierarchy coupled to a descending "task-decomposition" control hierarchy via "world models" at each level. Input to the world model at each level comes from both the task unit at that level, and other unspecified locations in the system. The use of multiple levels is motivated by the observation that the complexity of a control program grows exponentially as the number of sensors and their associated processing increases. By isolating related portions of the required processing at one level, this complexity can be reduced. The large number of low-level processing tasks, which usually have to be done in real time, can be separated from the fewer, more complex, higher level processing tasks so that the required processing time at each level can become nearly equal. Assuming the required communication between processing levels will be much less than the communication within levels, complexity is reduced by requiring only a limited number of communication channels between levels. If the processing at each level can be done in parallel, the addition of more levels will not result in an exponential increase in complexity. The amount of processing at each level is further reduced by the use of *a priori* knowledge from the world model. The world models provide predictions to the sensory system concerning the incoming sensory information so that the amount of processing required can be reduced. The use of a world model promotes modularity because the specific information requirements of the sensory and control hierarchies are decoupled.

Fig. 8 provides an example of the control of a multisensor robot using the NBS hierarchy. Raw sensory data from the environment enter the system at the bottom. At this lowest level, most of the required sensory processing will be continuous monitoring of the robot's joint positions. Any deviation between the actual and expected data is sent as feedback information to the servos, and as summary information to the next level in the sensory processing hierarchy. More complex data, like that from vision sensors, is sent through to higher levels unmodified. At the very highest level in the system, the complex task and top-level world model filter down to lower levels in the hierarchy both expected and desired information values. It is at the intermediate levels where both of these information flows meet and interact. Based upon current sensory information, the world models are updated. The updated world models can then serve to modify the desired task control actions until, at the lowest level, the necessary drive signals are sent to the robot to initiate actions in the environment.

2) Distributed Blackboard: A blackboard architecture allows economical communication between distributed sensory subsystems in an integrated multisensor system. Each subsystem can send time-stamped summary output to a blackboard where it becomes available to any fusion process as well as the integration functions. The time stamp on the output in the blackboard allows for sensor information to be made commensurate before being fused. The blackboard can contain any system information needed by the integration functions. Any number of different fusion methods can be implemented using the output from the blackboard. Harmon *et al.* [43] have used a blackboard architecture to compare different methods of multisensor fusion, and Harmon (Section V-D-3) has used a blackboard architecture for autonomous vehicle control.

3) Adaptive Learning: Miller *et al.* [44]–[46] have applied an adaptive learning approach to the multisensor-based control of robotic manipulators. In experiments using this approach, the performance accuracy was limited by the resolution of the sensor feedback rather than by any limitations in the control structure. Adaptive learning is a method of control in which the system "discovers" the appropriate signals for control based on the output of the sensors. The system is taught a representative sample of correlated control signals and associated sensory outputs over the range of signals and sensory outputs encountered by the system. Based on the associations developed during this teaching phase, it is possible to have the system respond to any combination of sensory outputs with an appropriate control signal. The system requires no *a priori* knowledge of the relationship between the structural kinematics of the robot, or the desired control signals and their associated sensory outputs. It is this feature of the adaptive learning approach that makes it attractive when there are possibly multiple sensors interacting to produce complex output.

C. Sensor Selection Strategies

Sensor selection is one of the integration functions that can enable a multisensor system to select the most appropriate configuration of sensors (or sensing strategy) from among the sensors available to the system. Two different approaches to the selection of the type, number, and configuration of sensors to be used in the system can be distinguished: preselection during design or initialization, and real-time selection in response to changing environmental or system conditions.

1) Preselection: As an initial step towards a general methodology for optimal sensor design, Beni *et al.* [47] have derived a general relationship between the number and operating speed of available sensing elements as a function of their response and processing times. This relationship can be used to determine the optimal arrangement of the sensing elements in a multisensor system. In addition to the actual geometric arrangement of the sensing elements, consideration of the choice between adding sensing elements (static sensing) and moving the elements (dynamic sensing) is used in determining the optimal arrangement.

2) Real-Time Selection: Hutchinson *et al.* [48] have presented an approach to planning sensing strategies for object recognition in a robotic workcell. One sensor is used to form an initial set of object hypotheses and then subsequent sensors are chosen so as to disambiguate maximally the remaining object hypotheses. Grimson [49] has also considered the problem of recognizing objects in the workspace of a robot using minimal sets of sensory information, and Taylor and Taylor [50] have used "dynamic error probability vectors" to select the appropriate sensors necessary for the recovery from errors during an automatic assembly process.

D. World Models

The use of world models enables a multisensor system to both store and reason with previously acquired sensory information. World models are usually defined in terms of a high-level representation. Sensory information can be either added to a predefined model of the world (i.e., the system's environment), or used to create the model dynamically during operation. The majority of the research related to the development of multisensor world models has been within the context of the development of suitable high-level representations for multisensor mobile robot navigation and control. Section V-B describes a number of examples of world models used in mobile robots. Included in this section are two representations that are more general in nature. These representations were developed within the context of a previously mentioned multisensor integration framework (logical sensors) and control structure (the NBS hierarchy).

TABLE I
GENERAL METHODS OF MULTISENSOR FUSION

Method	Operating Environment	Type of Sensory Information	Information Representation	Uncertainty	Measurement Consistency	Fusion Technique
Weighted average	dynamic	redundant	raw sensor readings	—	(thresholding possible)	weighted average
Kalman filter	dynamic	redundant	probability distribution	additive Gaussian noise	(thresholding, calibration)	filtering of system model
Bayesian estimate using consensus sensors	static	redundant	probability distribution	additive Gaussian noise	largest digraph in relation matrix	maximum Bayesian estimate of consensus sensor
Multi-Bayesian	static	redundant	probability distribution	additive Gaussian noise	ε-contamination	maximum Bayesian estimate
Statistical decision theory	static	redundant	probability distribution	additive noise	ε-contamination	robust minimax decision rules
Evidential reasoning	static	redundant and complementary	proposition	level of support versus ignorance	—	logical inference
Fuzzy logic	static	redundant and complementary	proposition	degree of truth	—	logical inference
Production rules	static	redundant and complementary	proposition	confidence factor	—	logical inference

1) The Multisensor Kernel System: Henderson *et al.* [51], [52] have presented the multisensor kernel system as a means of providing a representation for sensor information that is compatible with the specification of logical sensors (Section III-A-3). Object features are extracted from low-level sensory data and organized into a three-dimensional "spatial proximity graph" that makes explicit the neighborhood relations between features. Each feature is defined in terms of a logical sensor and is available to the system as the output of the logical sensor's characteristic vector. Subsequent sensory data can then either be matched in terms of the spatial proximity graph, or a "*k-d* tree" (a binary tree with *k*-dimensional keys that allows the nearest neighbors of one of *k* features to be found) is constructed, using the proximity graph, for faster processing.

2) The NBS Sensory System: Shneier *et al.* [53], [55], and Kent *et al.* [54] have described the kinds of processes involved in the higher levels of the sensory system of the NBS hierarchy (Section III-B-1). World models at each level in the hierarchy are used to create initial expectations about the form of the sensory information available at that level and then to generate predictions for the task control units in the hierarchy so that they do not have to wait for sensory processing to finish. Errors between the sensed information and the world model are used initially to register the model and later to maintain the consistency of the model during operation of the system.

E. General Fusion Methods

This section surveys different methods that have been proposed for general multisensor fusion (discussion of additional fusion methods relating to specific applications is included in Sections IV–VI). Most methods of multisensor fusion make explicit assumptions concerning the nature of the sensory information. The most common assumptions include the use of a measurement model for each sensor that includes a statistically independent additive Gaussian error or noise term (i.e., location data) and an assumption of statistical independence between the error terms for each sensor. Many of the differences in the fusion methods included below center on their particular techniques (e.g., calibration, thresholding) for transforming raw sensory data into a form so that the above assumptions become reasonable and a mathematically tractable fusion method can result. An excellent introduction to the conceptual problems inherent in any fusion method based on these common assumptions has been provided by Richardson and Marsh [12]. Their paper provides a proof that the inclusion of additional redundant sensory information almost always improves the performance of any fusion method based on optimal estimation.

Pau [56], [57] describes a number of statistical pattern-recognition techniques that are appropriate for multisensor fusion. All of these techniques could be used to reduce the error in classifying objects through the use of multiple sensors to provide redundant information concerning features of the objects. To avoid an exponential increase in complexity as sensors are added to a system, a key requirement is that the number of features and levels in the recognition process increase at a slower rate than the number of sensors. To meet this requirement it becomes necessary to improve the overall methods of feature extraction and selection—two major areas of interest in pattern recognition. Thus multisensor fusion becomes a problem within the context of statistical pattern recognition. Pau describes a number of operators and techniques that can fuse the features perceived by the sensors to limit their growth as additional sensors are added [56] and introduces a representation for multisensor fusion that is based on "context truth maintenance" [57].

Table I summarizes for comparison the relevant aspects of each general multisensor fusion method presented in this section. The sequence in which the methods are presented corresponds roughly to the increasingly high levels of representation of the information being fused (see Fig. 3). The representations used extend from low-level probability distributions for statistical inference to high-level logical propositions used in production rules for logical inference. In addition to the level of representation of the multisensory information, distinctions can be made as to whether the method is appropriate when information is assumed to come from *static* or *dynamic* sources in the operating environment, and as to whether the information is redundant or complementary (Section II-C). Included in the table are the means used to represent uncertainty in the measurement and fusion processes, possible methods used to determine the consistency of sensor measurements (e.g., elimination of any spurious sensor measurements), and the actual techniques used for fusion.

1) Weighted Average: One of the simplest and most intuitive general methods of fusion is to take a weighted average of redundant information provided by a group of sensors and use this as the fused value. While this method allows for the real-time processing of dynamic low-level data, in most cases the Kalman filter is preferred because it provides a method that is nearly equal in processing requirements and, in contrast to a weighted average, results in estimates for the fused data that are optimal in a statistical sense. A weighted average has been used for multisensor fusion in the mobile robot HILARE (Section V-D-1), after first thresholding the sensory information to eliminate spurious measurements.

2) Kalman Filter: The Kalman filter (see [66] for a general introduction) is used in a number of multisensor systems when it is necessary to fuse dynamic low-level redundant data in real time. The filter uses the statistical characteristics of the measurement model to determine estimates recursively for the fused data that are optimal in a statistical sense. If the system can be described with a linear model and both the system and sensor error can be modeled as white Gaussian noise, the Kalman filter will provide unique statistically optimal estimates for the fused data. The recursive nature of the filter makes it appropriate for use in systems without large data storage capabilities. Examples of the use of the filter for multisensor fusion include: object recognition using sequences of images (Section IV-A), robot navigation (Section V-D-4), multitarget tracking (Section VI-D-2), inertial navigation (Section VI-E), and remote sensing (Section VI-F). In some of these applications the "U-D (unit upper triangular and diagonal matrix) covariance factorization filter" or the "extended Kalman filter" is used in place of the conventional Kalman filter if, respectively, numerical instability or the assumption of approximate linearity for the system model present potential problems.

3) Bayesian Estimation using Consensus Sensors: Luo and Lin [13]–[17] have developed a method for the fusion of redundant information from multiple sensors that can be used within their hierarchical phase-template paradigm (Section III-A-1). The central idea behind the method is first to eliminate from consideration the sensor information that is likely to be in error and then to use the information from the remaining "consensus sensors" to calculate a fused value.

Fig. 9 shows a functional block diagram of the method. The information from each sensor is represented as a probability density function. Given readings from n sensors in the system, the resulting information is first made commensurate through preprocessing. An n by n distance matrix is created by calculating for each element (i, j) in the matrix the "confidence distance measure" between the information from sensors i and j. A confidence distance measure is defined to be equal to twice the area under the density function of sensor i between the readings from sensor i and sensor j. Use of this measure assumes that the domains of each sensor's density function are commensurate. If the density functions are assumed to be Gaussian, the distance can be computed by use of the error function. The distance matrix determined by the use of this measure will not be symmetric unless the density functions of all the sensors are identical. Threshold values, based on the required sensing accuracy, are then applied to the elements in the matrix. Elements not exceeding their threshold are represented by a one in a binary-valued n by n relation matrix. The largest connected digraph formed from this matrix will determine the group of consensus sensors most likely not to be in error. The optimal fusion of the information is determined by finding the Bayesian estimator that maximizes the likelihood function of the consensus sensors.

4) Multi-Bayesian: Durrant-Whyte [58]–[61] has developed a model of a multisensor system that represents the task environment as a collection of uncertain geometric objects. Each sensor in the system is described by its ability to extract useful static descriptions of these objects. An "ϵ-contaminated" (see paragraph 5, below) Gaussian distribution is used to represent the geometric objects. The sensors in the system are considered as a team of decision-makers. Together, the sensors must determine a team-consensus view of the environment. A multi-Bayesian approach, with each sensor considered a Bayesian estimator, is used to combine the associated probability distributions of each respective object into a joint posterior distribution function. A likelihood function of this joint distribution is then maximized to provide the final fusion of the sensory information. The fused information, together with an *a priori* model of the environment, can then be used to direct the robotic system during the execution of different tasks.

5) Statistical Decision Theory: McKendall and Mintz [62] and Zeytinoglu and Mintz [63], [64] have used statistical decision theory to develop a general two-step method for the fusion of redundant location data from multiple sensors. (Location data refer to sensor measurements that are modeled as additive sensor noise translated by the parameter of interest being sensed.) Sensor noise is modeled as the ϵ-contamination of a variety of possible proba-

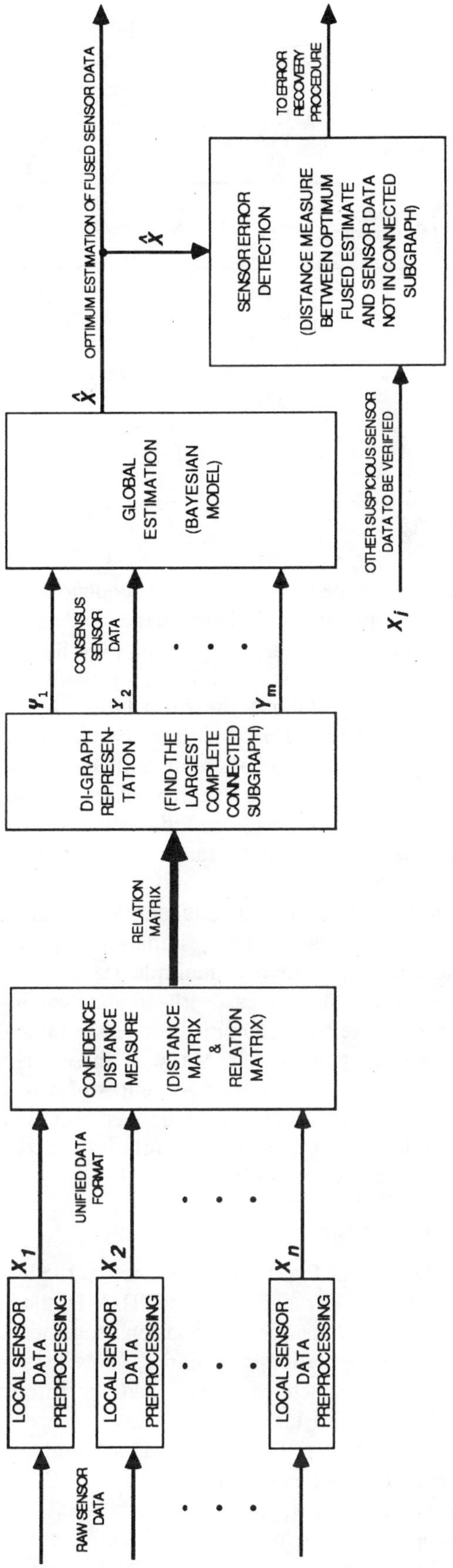

Fig. 9. Functional block diagram of consensus sensor fusion method. (Adapted from [17, fig. 2.)

bility distributions. The use of ϵ-contamination in the sensor model serves to increase the robustness of the decision procedure by removing a certain outlying fraction ϵ of the distribution to account for heavy-tailed deviations from the assumed noise distribution that may have been caused by spurious sensor readings. Initially, the data from different sensors are subject to a robust hypothesis test as to its consistency (see [65] for an introduction to "robust statistics"). Data that passes this preliminary test are then fused using a class of robust minimax decision rules.

6) Shafer–Dempster Evidential Reasoning: Garvey *et al.* [67] introduced the possibility of using Shafer–Dempster evidential reasoning in multisensor fusion. Bogler [68] and Waltz and Buede [69] have explored its possible application in, respectively, multisensor target identification, and military command and control. Shafer–Dempster evidential reasoning [70] is an extension to the Bayesian approach that makes explicit any lack of information concerning a proposition's probability by separating firm support for the proposition from just its plausibility. In the Bayesian approach all propositions (e.g., features in the environment) for which there is no information are assigned an equal *a priori* probability. When additional information from a sensor becomes available and the number of unknown propositions is large relative to the number of known propositions, an intuitively unsatisfying result of the Bayesian approach is that the probabilities of known propositions become unstable. In the Shafer–Dempster approach this is avoided by not assigning unknown propositions an *a priori* probability (unknown propositions are assigned instead to "ignorance"). Ignorance is reduced (i.e., probabilities are assigned to these propositions) only when supporting information becomes available.

7) Fuzzy Logic: Huntsberger and Jayaramamurthy [71] have used fuzzy logic to fuse information for scene analysis and object recognition. Fuzzy logic [72], a type of multiple-valued logic, allows the uncertainty in multisensor fusion to be directly represented in the inference (i.e., fusion) process by allowing each proposition, as well as the actual implication operator, to be assigned a real number from 0.0 to 1.0 to indicate its degree of truth. Consistent logical inference can take place if the uncertainty of the fusion process is modeled in some systematic fashion.

8) Production Rules with Confidence Factors: Kamat [73], Belknap *et al.* [74], and Hanson *et al.* [75] have used production rule-based systems for object recognition using multisensor fusion. Production rules are used to represent symbolically the relation between an object feature and the corresponding sensory information. A confidence factor is associated with each rule to indicate its degree of uncertainty. Fusion takes place when two or more rules, referring to the same object, are combined during logical inference to form one rule. The major problem in using production rule-based methods for fusion is that the confidence factor of each rule is defined in relation to the confidence factors of the other rules in the system, making it difficult

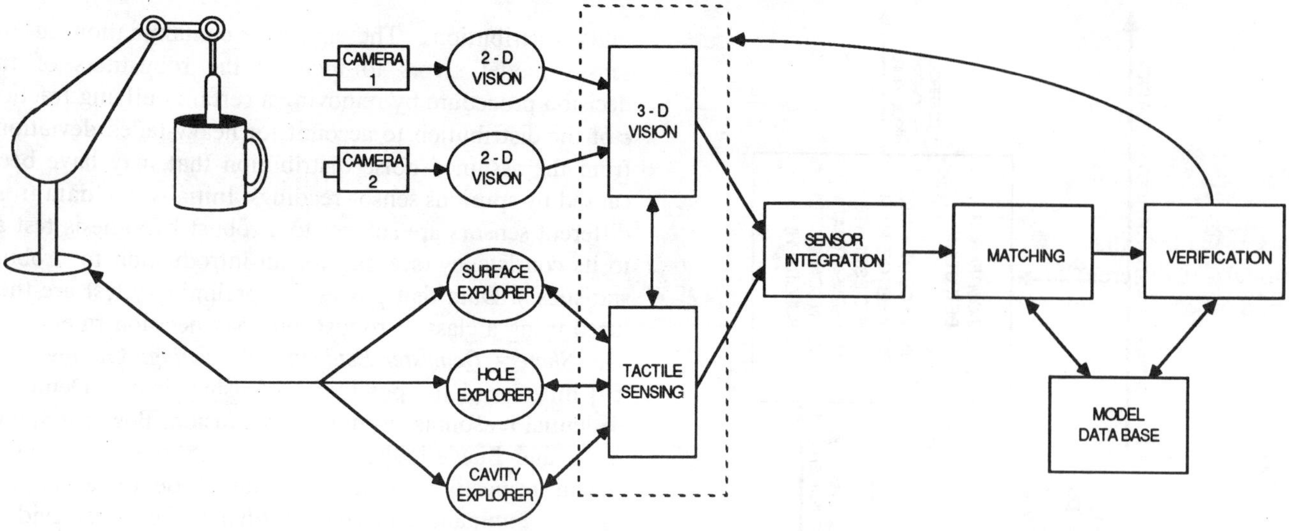

Fig. 10. Overview of Allen's robotic object recognition system. Mug is shown being recognized through vision and active exploratory tactile sensing. (Adapted from [91, Figs. 1.1 and 1.2].)

to alter the system when, for example, new sensors are added that require additional rules.

IV. Integration and Fusion using Specific Sensor Combinations

This section surveys a variety of approaches to the integration and fusion of information from combinations of different types of sensors. An effort is made to present approaches that exploit the specific characteristics of the information provided by each type of sensor and have an area of possible application that is quite general (e.g., object recognition). Sections V and VI present a wide variety of additional sensor combinations that have been used in mobile robot and other applications, respectively.

A. Vision

Visual information is the most powerful single source of sensory information available to a system. Many different types of nonvisual sensors are used in combination with vision sensors to compensate for some of the difficulties encountered in the machine processing of visual information. Tasks such as object recognition can sometimes require the aid of additional types of sensors to approach the capabilities of a human using just visual information. This section will not describe in detail the many integration and fusion techniques that use only vision sensors because there already is an extensive literature, including many reviews and book-length treatments (e.g., [76], [77]) covering the various aspects of computational vision.

Magee and Aggarwal [78] provide a review of recent research efforts aimed at combining intensity and range features derived from visual images to determine the structure of three-dimensional objects. In some of the work reviewed, information from one feature is used to guide the acquisition of information concerning another feature when the second feature requires a much longer processing time (e.g., intensity guided range sensing for object recognition [79] and the determination of motion parameters [80]). Research related to the fusion of sequences of images has used the "optical flow" of the images to determine the motion of objects in the image (see [81] for a recent review), and both a Bayesian [82] and extended Kalman filter [84], [85] to establish the surfaces of three-dimensional objects. Flachs *et al.* [83] have used a "complexity metric" as a mathematical basis for multisensor fusion in vision systems. Much of the research related to the use of multiple visual sensors has used the stereoscopic effect from the sensors to determine range information (see [86] for a recent stereo vision review, and [87] for an approach to binocular fusion that uses simulated annealing). Porrill [88] has used Gauss–Markov estimation together with geometric constraints to fuse multiple stereo views of a wire frame model. In robotics, overhead and "eye-in-hand" vision sensors have been combined for use in three-dimensional object recognition [89]. Many of these techniques, originally developed to fuse both sequences of images from a single vision sensor and images derived simultaneously from multiple sensors, have significantly influenced subsequent work in nonvisual fusion.

B. Vision and Tactile

As can be seen in Fig. 5 with reference to the hierarchical phase-template paradigm (Section III-A-1), the integration of vision and tactile information, together with a robot's own manipulation capabilities, gives that robot a wide range over which to receive sensory information. Combined vision and tactile sensors have been used to a great extent by industrial robots to perform both assembly and inspection tasks (Section VI-A).

1) Allen: Allen [3], [90]–[95] has developed a robotic object-recognition system that uses three-dimensional vision together with active exploratory tactile sensing (see Fig. 10). The system was developed to recognize common kitchen items like mugs, plates, pitchers, and bowls that had no discernible textures and were homogeneous in color. Recognition of objects like these pose a serious problem for many vision-only recognition systems because of the lack of features that can be used for matching and depth analysis. Through the integration of tactile information with vision, Allen's system is able to obtain information concerning any holes, cavities, or curved surfaces that can be used to identify a particular object.

The model data base represents each object in a hierarchical manner that allows the sensory devices to match the models at different levels of detail. The models are independent of viewpoint and scale and contain relational information that can be used to reduce the searching of the data base. Each object is modeled as a collection of components and features: the components correspond to the discernible differences in the surface of the objects (e.g., the body, bottom, and handle of the mug shown in the figure), and the features correspond to the object's holes or cavities. At the lowest level, each surface component is modeled as a grid of bicubic spline curves that form a patch. Both holes and cavities are modeled as right cylinders of constant cross-sectional area—cavities having the additional attribute of depth. The operation of the system can be summarized in a five-step recognition cycle [91], as follows.

Step 1: Initially, two-dimensional vision processing routines are applied to the image to determine bounded regions. The centroid of each region is calculated by using the matched three-dimensional stereo points of its boundary so that it can be used as the starting point for tactile exploration of the region. The depth and surface orientation of each region is determined using binocular stereo. Isolated edge pixels, which could be possible noise points, are excluded from consideration through thresholding. In a system relying only on vision to determine depth and orientation, this elimination of data to reduce error could result in a surface description that is not dense enough for recognition purposes. By allowing tactile sensing to explore further any uncertain regions, the regions that are identified from the remaining data will have a greater accuracy and can be used with higher confidence in later steps of the recognition process.

Step 2: The tactile sensing system explores each region identified by the vision system to determine if it is a surface, hole, or cavity. The tactile sensor used in this system is an octagonal cylinder covered with conducting surfaces mounted perpendicular to the mounting plate of the robot (the tactile sensor in the figure is exploring the cavity formed by the well of the mug). The sensor approaches a region orientated in a direction normal to the centroid of the region until either it contacts a surface, travels beyond a specified threshold used to indicate the presence of a cavity, or, if the sensor is able to travel its full length into the region without contact, the region is assumed to be a hole. If the region is a surface, a surface patch is constructed by integrating vision and touch. If the region is either a hole or cavity, the sensor moves in a sawtooth manner around the region's boundary to determine its shape.

Step 3: For regions that are surfaces, the vision and tactile sensing results of the previous steps, together with additional tactile sensing, are integrated to create three-dimensional surface patches that can be matched with the model data base. Starting from the location of contact with the surface, the tactile sensor uses knot points to determine the directions along which traces of the surface will be made. The points reported along each trace are combined into cubic least squares polynomial curves which can then be used to fill in areas of the surface that still lacked detail after stereo vision processing.

Step 4: The surfaces patches and closed curves (corresponding to holes or cavities) are matched against the model data base to find an object that is consistent with the sensory information. If more than one object is found to be consistent, a probabilistic measure is used to order the objects for verification.

Step 5: Once a consistent object is found the verification procedure is used for further active exploratory sensing to verify components and features of an object's model that have not been sensed. Visually occluded holes and cavities are verified using the tactile sensor. Verification of visually occluded surfaces using only the tactile sensor is difficult because vision is needed both to guide the sensor during traces of the surface and, most importantly, to establish that the region of interest is indeed a smooth surface that can be approximated by a patch. The system provides for robust viewer-independent object recognition because no *a priori* viewpoint or orientation of the object is assumed, e.g., any identifiable part of the model of an object can be used to invoke a search to verify the remaining surfaces or features of the model needed for recognition.

2) Integration using a Decision Tree: Luo and Tsai [96] have developed a system that uses two-dimensional vision, together with two tactile sensing arrays mounted on a gripper of a robot, to recognize objects. Moment invariants of an object's shape are used as features for recognition and to calculate the centroid of a region of the object needed for determining the proper grasping position for the gripper. During the initial learning phase the system creates a decision tree by first presenting all of the objects to be recognized to the vision sensor so that their top-view silhouette boundaries can be determined. If there still exist groups of objects that are indistinguishable in terms of this visual information, tactile information concerning the objects' lateral shape is obtained to make the objects in each group more distinguishable. Different predetermined lateral directions are used until all of the objects can be distinguished. The final result of this clustering process is a hierarchical decision tree with each leaf representing a single discriminable object, each nonterminal node corresponding to a group of objects that are indistinguishable at that level of sensory processing, and each arc associated with the effective lateral direction used by the tactile sensor to distinguish the child from the parent node. The first level below the root node in the tree corresponds to the initial version sensing; levels below the first represent successive stages of tactile sensing. Finally, the recognition phase proceeds by traversing the decision tree in the same direction as the tree was created until the object is able to be discriminated.

3) Object Apprehension: Stansfield [97]–[100] has presented a system which uses vision and tactile sensors for object "apprehension." Apprehension is defined as the determination of the properties of an object and the relationships among these properties without, as in recognition, going on to attach a label to the object as a whole. The system is structured as a modularized hierarchy of knowledge-based experts, each responsible for either the execution of an exploratory procedure or for the further processing of information from other exploratory procedures. An exploratory procedure extracts information concerning a predefined general-purpose primitive (e.g., compliance, elasticity, texture, etc.) or feature (e.g., edges, surface patches, holes, cavities, etc.) that is related to some aspect of the object's form, substance, or function [99]. Each feature is composed of one or more primitives or features. Modules in the lower portion of the hierarchy are responsible for processing the information from each sensor system, while the upper level modules integrate the information coming from the different sensor systems. Many of the higher level modules are able to function with information from a variety of different lower level modules. Spatial polyhedrons have been proposed recently by Stansfield [100] as a generic object representation that can extend the capabilities of the system to include recognition.

C. Vision and Thermal

Nandhakumar and Aggarwal [101]–[104] have presented a technique for the classification of objects in outdoor scenes using thermal and visual sensors. A thermal camera is used to acquire an infrared image of a scene and a vision camera is used to acquire an intensity image. Both cameras are adjusted so that their images are in spatial correspondence. Three features are used in a decision tree to label the objects in a scene. The first, the conductive heat flux of each region in the scene, is determined by integrating complementary information provided by both sensors. Although this feature, characterizing the intrinsic thermal behavior of the imaged object, provides the greatest amount of discriminatory information, two additional features are used to identify the objects unambiguously: the surface reflectance of a region as determined from the visual image and the average region temperature as derived from the thermal image. Production rules are used to implement the decision tree. An object label is assigned in the consequent of each rule and logical combinations of heuristically determined intervals for the value of each feature are used in the antecedent.

D. Range and Tactile

Grimson and Lozano-Pérez [105] describe a technique that uses tactile and range sensors to provide measurements of position and surface normals that can be used to identify and locate objects from among a group of known objects. The objects are modeled as polyhedrons, and constraints are used to keep the number of hypotheses as to an object's identity small. The only assumptions made about the sensors required are that they be able to provide information concerning an object's surface points and surface normals; as a result, the technique should be applicable to a wide variety of different types of sensors besides range and tactile.

E. Laser Radar and Forward Looking Infrared

Roggemann *et al.* [106] and Tong *et al.* [107] have developed a method of fusing the complementary information provided by laser radar and forward looking infrared sensors to segment and enhance features of man-made objects such as tanks and trucks in a cluttered background. Once separated from the background, these features can be used by other methods to identify the object. The laser radar image provides range information for areas within a field of view, and the infrared image can show special features of the area. The gradients of both images are used for both object enhancement and segmentation. Boundaries enclosing areas of small range gradient can be used to identify possible objects because natural backgrounds such as mud, grass, and trees exhibit large range differences from one pixel to another that can be modeled as random noise. The gradient of the infrared image will sharpen the infrared signature and temperature characteristics of objects that are infrared sources and can also serve to identify cold objects in a hot background because only temperature differences are noted. Initially, a binary mask of the laser radar image is created that indicates the location, area, and boundary of possible objects. Assuming that objects exhibit small gradients relative to the background and occupy a small percentage of the overall image, the first derivative of the histogram of the gradient magnitudes in the image is used to estimate a threshold for the mask. The threshold alone is not discriminatory enough to separate every pixel in the image. To distinguish object from background pixels further, both the segmented infrared and segmented range gradient images are first "anded," pixel to pixel, with the mask. The resulting images are then multiplied to produce a final image that shows where the range and infrared gradients match, emphasizing object features and deemphasizing the background.

V. Multisensor-based Mobile Robots

The mobility of robots and other vehicles is required in a variety of applications. In simple well-structured environments, automatic control technology is sufficient to coordinate the use of the required sensor systems (e.g., automatic guided vehicles, missiles, etc.) [108]. When a vehicle must operate in an uncertain or unknown dynamic environment—usually in close to real time—it becomes necessary to consider integrating or fusing the data from a variety of different sensors so that an adequate amount of information from the environment can be quickly per-

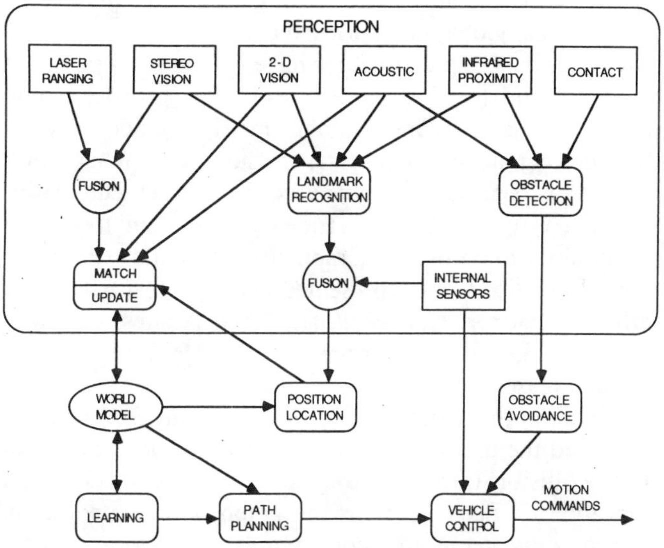

Fig. 11. Role of perception function in hypothetical architecture for mobile robot.

ceived. Because of these factors, mobile robot research has proved to be a major stimulus to the development of concrete approaches to multisensor integration and fusion. Luo and Kay [109] have reviewed the role of multisensor integration and fusion in the operation of mobile robots, and Levi [110] has discussed some of the multisensor fusion techniques appropriate for mobile robot navigation and has reviewed their use in a number of applications. Multiple sensors have been used in mobile robots to enable them to operate in environments ranging from roadways [111], [112] to unstructured indoor environments [113] to unknown natural terrain [114], [115] and to be used for applications including assembly [116] and nuclear power station maintenance [117].

A. The Role of Perception in a Hypothetical Architecture

Fig. 11 illustrates the role of the perception function in a hypothetical architecture for a mobile robot. Perception, together with vehicle control, obstacle avoidance, position location, path planning, and learning, are generic functions necessary for intelligent autonomous mobility [108]. Six different external sensor types are shown in the figure as part of the perception function. Subsets of these sensors are used to perform three tasks that usually comprise the perception function: the matching of sensory data to a world model (or map) representing the environment and then updating the model to reflect the matching results, the recognition of landmarks in the environment for use in determining the location of the robot, and the detection of obstacles so that they can be avoided. The degree of integration and fusion of the sensory data required for each of these tasks can differ. In a simple system, each sensor used for obstacle detection might operate independently of the other sensors if the detection of nearby obstacles is required; in a more complex system, some of the sensor data might be also fused so as to extend the range and accuracy of possible detection [114]. One of the simplest techniques used for position location is trajectory integration, where the location is calculated from the accumulated rotational and translational motion of the vehicle as determined by internal sensors like an odometer. Due to the inherent inaccuracies in any sensor, locational error continues to accumulate as the robot moves. To deduce this cumulative error most mobile robots periodically determine the location of some external landmark. As shown in the figure, the results from landmark recognition are sometimes fused with the location determined by internal sensors after each has been transformed to common coordinates. Any of the techniques of multisensor object recognition presented in this paper may be used first to recognize the landmark. The world model matching and updating task requires that the sensor information and any associated measure of its uncertainty correspond to the representation used in the world model so that integration can take place. Depending on the representational format used in the world model, the information will in most cases have to be made commensurate by applying appropriate space and time transformations. Information from the different sensors might be fused or otherwise transformed before reaching the matching task to reduce the communication bandwidth required or the complexity of the matching process. In the figure, information from the laser ranging and stereo vision sensors are fused before being matched to the world model.

While the hypothetical architecture described represents in broad outline most current approaches to the design of mobile robots, the MIT mobile robot project has adopted a radically different layered approach for autonomous control that they term a "subsumption" architecture [118], [119]. Each layer in this architecture consists of a complete control system similar to that in Fig. 11 for a simple task achieving behavior like avoiding obstacles or wandering. Starting with low-level tasks, new task-achieving behavior can be added incrementally because the layers operate asynchronously communicating over low-bandwidth lines without a central locus of control, central data structure, or global plan.

B. High-Level Representations

Considerable research has been directed at the development of either a single representation or multiple hierarchical levels of representations suitable for use by a mobile robot to perform the reasoning required for its control, path planning, and learning functions. The representations are usually at a high enough level so as not to be sensor specific. As mentioned before, information from different sensors is usually transformed to the common high-level representation and then added to a world model. The function of the world model and the particular form of the high-level representation depend both on the control architecture used in the robot and the complexity of the required reasoning—extremes range from road-following vehicles where sensor information is dynamically processed using feedback loops to produce control commands without ever using a world model, to pure production rule-based representations that assume a static and perfect model of the world that is difficult to modify [120]. In practice, the representations used for robots operating in unknown or unstructured environments allow for their world models to be dynamically modified and updated with uncertain sensor information. Except for explicit learning procedures used in many production rule-based representations, learning takes place implicitly as the world model is updated with new information as the robot traverses the environment.

Included below are some examples of different high-level representations. Many of the papers referenced are distinguished by a discussion of the multisensor integration and fusion issues relevant to their proposed representation. Other representations are discussed as part of the descriptions of different mobile robots found in Section V-D of this section.

1) Spherical Octree: Chen [121]–[124] has proposed a "spherical octree" representation for use in mobile robot navigation. A spherical octree is an 8-ary tree structure that at its first level separates a spherical perspective view of the environment into eight octants corresponding to the children of the root node (the entire spherical environment perceived by the robot). Objects in the environment can be represented in the octree by recursively subdividing octants containing part of the object into eight more octants at the next lowest level and merging octants that are completely contained by the object into one octant at the next highest level, repeating this process to represent the object at increasingly finer resolution. The use of a spherical perspective view eliminates some of the limitations of the typical orthographic and planar perspectives used in optical sensing. It is also appropriate for sensors providing range information because range values can be represented as the radial distances from the sensor to the object. Information from each different sensor is used to reconstruct three-dimensional surfaces using a knowledge base of typical patterns for the sensor. The reconstructed surfaces are then fused together as part of the process of being represented in the octree structure.

2) Occupancy Grids and Neural Nets: Elfes [12]–[127] originally developed a cellular world model representation called the "occupancy grid" for use with a sonar equipped mobile robot. This representation has been extended to allow for the integration of information from many different types of sensors [128], [129]. Bayesian estimation is used to fuse together each sensor's probabilistic estimate as to whether a cell in the grid is occupied by an object. The resulting grid can then be used to determine paths through unoccupied areas of the grid. Jorgensen [130] has proposed dividing the environment into equal-size volumetric cells and associating each with a neuron. In a manner similar to the occupancy grid, the magnitude of each neuron's activation corresponds to the probability that the cell it represents is occupied. The neurons are trained using sensory information from different perspectives. Associative recall (Section III-A-2) can then be used to recognize objects in the environment and simulated annealing can be used to find optimal global paths for navigation.

3) Graphs: Graph structures have been used to represent the local and topological features of the environment to avoid having to define a global metric relation between nonadjacent nodes (or points) in the graph. When landmarks or beacons are not used to correct cumulative position error, a global metric would contain too much uncertainty to be useful. Graph structures allow the topological features to be represented and reasoned with in an efficient manner. Kak *et al.* [131] and Andress and Kak [132] used an attributed graph and Shafer–Dempster evidential reasoning (Section III-E-6) to integrate sensory information for hierarchical spatial reasoning. Brooks [133] proposed the use of a graph to represent regions of potential collision-free motion termed "freeways" and "meadows." Each point in the graph, represented as an "uncertainty manifold," corresponds to the location of a robot in configuration space at which sensory information was acquired; each arc is labeled with a local measurement of the distance traveled between endpoints. The cumulative uncertainty of the robot as it moves from point to point in the graph is taken into account through the cascading of successive uncertainty manifolds.

4) Labeled Regions: Sensory information can be used to segment the environment into regions with properties that are useful for spatial reasoning. The known characteristics of different types of sensory information can be used to label some useful property of each region so that symbolic reasoning can be performed at higher levels in the control structure. Asada [134] proposed a method for building world models that uses the range images from sensors to create height maps of the local environment. Grey levels represent the height of points in the map with respect to an assumed ground plane. The map is then segmented into regions and labeled. Sequences of maps, created as a robot moves, are integrated into a global map by overlaying pairs of height maps and then replacing the labels of corresponding regions in the height maps with a label determined according to a precedence procedure (e.g., if a region is labeled as unexplored in one height map and as an obstacle in another, the global map might label the region as an obstacle). The global map is then used for obstacle detection and path planning. Miller [135] developed a spatial representation that divided a map of an indoor environment into labeled regions. Each label is one of four possible types—each type referring to the number of degrees of freedom in the region (i.e., the two planar dimensions and the robot's orientation) that information from a sensor could be used to eliminate (e.g., the empty space in the center of the room is of type zero, a region near a corner is of type three, etc.). "Voronoi diagrams" are then used to group the regions into areas that correspond to a specific edge (or wall) in the environment.

5) Production Rules: The use of production rules in a control structure allows for a wide range of well-known artificial intelligence methods to be used for path planning and learning purposes. Lawton *et al.* [136] have used

TABLE II
SELECTED EXAMPLES OF MULTISENSOR-BASED MOBILE ROBOTS

Mobile Robot	References	External Sensors	Operating Environment	World Model Representation	Fusion Method
HILARE (1979)	[113], [149]–[154]	vision acoustic laser range finder	unknown man-made	polygon objects in graph of locations	weighted average
Crowley's mobile robot (1984)	[155]–[157]	rotating ultrasonic tactile	known man-made	connected sequences of line segments in two dimensions	best correspondence using integer valued confidence factors
Ground surveillance robot (1984)	[114], [158]	high-resolution grey level vision Low-resolution color vision acoustic laser range finder	unknown natural terrain	triangular segments in blackboard	variety possible (data put in spatial and temporal correspondence)
Stanford mobile robot (1987)	[159]	stereo vision tactile ultrasonic	unknown man-made	hierarchical sensor measurements and symbols	Kalman filter
CMU's ALV's NAVLAB and Terregator (1986)	[111], [160]	color vision sonar laser range finder	unknown roadway	polygon tokens with attribute-value pairs in whiteboard	variety possible (data put in spatial and temporal correspondence)
DARPA ALV (1985)	[112], [115], [161]–[163]	color vision sonar laser range finder	unknown natural terrain	Cartesian elevation maps (CEM's)	average elevation change over small CEM areas

production rules to create schemas to represent both objects in terrain models and certain generic object types. Network hierarchies are created from the schemas that allow inference and matching procedures to take place at multiple levels of abstraction, with each level using an appropriate combination of the available sensors. Isik and Meystel [137] have used fuzzy-valued linguistic variables to represent the attributes of objects as part of a fuzzy logic-based (Section III-E-7) production system for mobile robot control.

C. Sensor Combinations

Due to the advantages and limitations of each type of sensor, most mobile robots use some combination of different sensor types to perform each task of the perception function. Some sensors cannot be used in a particular environment due to their inherent limitations (e.g., acoustic sensors in space), while others are limited due to either technical or economic factors. Obstacle detection with contact sensors necessarily limits the speed of a robot because contact must be made before detection can take place [138]; laser sensors require an intense energy source, and they have a short range and slow scan rate (their use can also cause eye problems [139]); and vision sensors are critically dependent on ambient lighting conditions, and their scene analysis and registration procedures can be complex and time-consuming [138]. Shaky, one of the first autonomous vehicles, used vision together with tactile sensors for obstacle detection [140]. JASON combined acoustic and infrared proximity sensors for obstacle detection and also used these sensors for path planning [141]. The Stanford University Cart used acoustic and infrared sensors together with stereo vision for navigating over a flat terrain while avoiding obstacles [142]. Bixler and Miller [143] used simple low-resolution vision in their autonomous mobile robot to locate the direction of an obstacle, and then used an ultrasonic range finder to determine its depth and shape. Other combinations of sensors used in mobile robot systems have included contact, infrared, and stereo vision [144]; sonar and infrared [145]; contact and acoustic [146]; acoustic and stereo vision [147]; and stereo vision and laser range finding [148].

D. Selected Examples

Short descriptions of a number of different mobile robots are provided next. Emphasis is given to the role of multisensor integration and fusion in their navigation and control. Table II summarizes for comparison the relevant multisensor integration and fusion features of each mobile robot. Included under the name of each mobile robot in the table is the year the initial publication appeared. In cases where there have been major modifications to the mobile robot, the features listed in the table correspond to the most recent published research.

1) Hilare: The mobile robot Hilare combines contact, acoustic, two-dimensional vision, and laser range finding sensors so that it can operate in unknown environments [113], [149]–[154]. Hilare was the first mobile robot to create a world model of an unknown environment using information from multiple sensors [108]. Acoustic and vision sensors are used to create a graph partitioned into a hierarchy of locations. Vision and laser range-finding sensors are then used to develop an approximate three-dimensional representation of different regions in the environment—constraints being used to eliminate extraneous features of the representation. The laser range finder is then used to obtain more accurate range information for each region. To provide a robust and accurate estimation of the robot's position, three independent methods are used: absolute position referencing by use of a beacon, trajectory integration without external reference, and relative position referencing with respect to landmarks in the environment [154]. Each of these methods is used in a complementary fashion to correct or reduce errors and uncertainties in the other methods. The information from a variety of sensors is integrated to provide the position of known objects and places relative to the robot. The shape of each object is represented as a polygon. Depending on the features of an object and its distance from the robot, an appropriate group of redundant sensors is selected to measure the object. The uncertainty of each sensor is modeled as a Gaussian distribution. If the standard deviations of all the sensor's measurements have the same magnitude, a weighted average (Section III-E-1) of their values is used as the fused estimate of a vertex of the object; otherwise, the measurement from the sensor with the smallest standard deviation is used. The estimated vertices of the object can then be matched to known regions of the world model by finding an object in the model that minimizes the weighted sum of the distance between corresponding vertices.

2) Crowley's Mobile Robot: Crowley [155]–[157] describes a mobile robot with a rotating ultrasonic range sensor and a touch sensor capable of autonomous navigation in a known domain. Information from a prelearned global world model is integrated with information from both sensors to maintain dynamically a composite model of the local environment. Obstacles and surfaces are represented as connected sequences of line segments in two dimensions. Confidence as to the actual existence of each line segment is represented by an integer ranging from one (transient) to five (stable and connected). The uncertainty as to the position, orientation, and length of each segment is accounted for by allowing tolerances in the value of these attributes. Information from the global model and the sensors is matched to the composite local model by determining which line segment in the local model has the best correspondence with a given line segment from either the global model or sensors. The best correspondence is found by performing a sequence of tests of increasing computational cost based on the position, orientation, and length of the line segments relative to each other. The results of the matching process are then used to update the composite local model by either adding newly perceived line segments to the model or adjusting the confidence value of the existing segments. The local model can then be used for obstacle detection, local path planning, path execution, and learning.

3) Ground Surveillance Robot: The ground surveillance robot described by Harmon [114], [158] is an autonomous vehicle designed to transit from one known location to another over unknown natural terrain. Vision and acoustic ranging sensors are used for obstacle detection. A laser range finder together with a high-resolution gray-level camera and a low-resolution color camera are used for distant terrain and landmark recognition. The information from the obstacle detection sensors is fused into a single estimate of the position of nearby obstacles by superpositioning distributions that represent each sensor's *a priori* probability of detection. A distributed blackboard is used both to control the various subsystems of the vehicle and as a mechanism through which to integrate and fuse various types of sensor data. As part of the blackboard, a world model is used to organize the data into a class tree with inheritance properties. Each element in the world model has a list of properties to which values are assigned, some values being determined by sensory data. Terrain data are represented as triangular segments with the properties of absolute position, orientation, adjacency, and type of ground cover. To allow for a variety of fusion methods to be used, each element in the world model includes a time stamp and measures of its accuracy and confidence. When two or more sensor values are functionally dependent, changes in one value will propagate throughout the blackboard so that its dependent values reflect the change. When sensor values refer to the same property of an element, either a decision is made as to which of the competing values is to be used (e.g., the most recent value of the most accurate sensor) or the values are fused.

4) Stanford Mobile Robot: The Stanford mobile robot [159] uses tactile, stereo vision, and ultrasonic sensors for navigation in unstructured man-made environments. A hierarchical representation is used in its two-dimensional world model, with features close to the actual sensor measurements at the lowest level and more abstract or symbolic features at the higher levels. The uncertainty as to the location of the robot and the features in the environment is modeled with a Gaussian distribution. A Kalman filter (Section III-E-2) is used to fuse the measurements from a sensor as the robot moves. An example of the application of this method is shown in Fig. 12. Fig. 12(a) shows two points $\boldsymbol{a}$ and $\boldsymbol{b}$ as measured by the stereo vision sensor—first at location $\boldsymbol{p}_1$ and then at location $\boldsymbol{p}_2$. The uncertainty ellipses around the point measurements are elongated toward the sensor because the uncertainty due to distance is much greater than the angular uncertainty when calculated through stereo vision. The uncertainty due to distance is also greater for points further from their location of measurement (i.e., $\boldsymbol{b}$ is more uncertain than $\boldsymbol{a}$). The two measurements of each point (e.g., $\boldsymbol{a}_1$ and $\boldsymbol{a}_2$) are not coincident because of the inherent error in the internal odometer sensor of the robot. In Fig. 12(b), the uncertainty of the points and location $\boldsymbol{p}_2$ are shown with respect to $\boldsymbol{p}_1$. The uncertainty ellipse around $\boldsymbol{p}_2$ is elongated perpendicular to the direction of motion because the angular error of the odometer sensor is greater than its error in determining distance. The uncertainty with respect to $\boldsymbol{p}_1$ of measurements $\boldsymbol{a}_2$ and $\boldsymbol{b}_2$ has increased because their uncertainty with respect to $\boldsymbol{p}_2$ has been compounded with $\boldsymbol{p}_2$ locational uncertainty. In Fig. 12(c), the Kalman filter is used to determine new fused estimates for both points ($\boldsymbol{a}^*$ and $\boldsymbol{b}^*$) that have a reduced uncertainty with

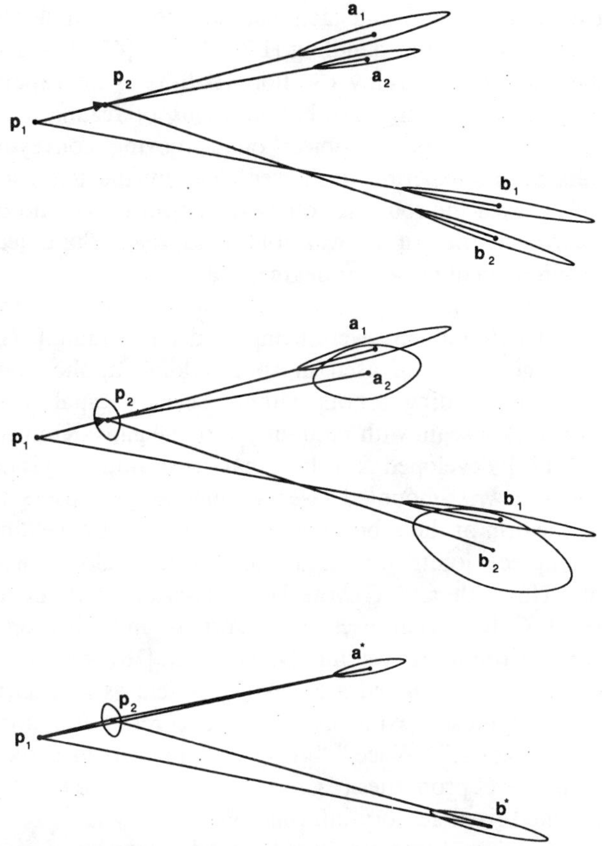

Fig. 12. Reduction in uncertainty as to location of robot (p) and two points (a and b) in environment through use of Kalman filter to fuse measurements as robot moves from p_1 to p_2. (a) Uncertainty before fusion of points a and b as measured from p_1 and p_2. (b) Uncertainty of a, b, and p_2 with respect to p_1. (c) Fused estimates for a, b, and p_2 with respect to p_1. (Adapted from [159, fig. 10].)

respect to both locations p_1 and p_2. The uncertainty of p_2 with respect to p_1 has also been reduced.

5) CMU's Autonomous Land Vehicles: The NAVLAB and Terregator are two vehicles developed at Carnegie–Mellon University's Robotics Institute as part of their research on autonomous land vehicles [111], [160]. Each vehicle is equipped with a color TV camera, laser range finder, and sonar sensors. The sonar sensors are used to detect nearby obstacles. The design of an architecture able to support parallel processing and the development of multisensor integration and fusion techniques have been major goals of the research. The current system consists of several independently running modules that are tied together in what is termed a "whiteboard" control structure, which differs from a blackboard in that each module continues to run while synchronization and data retrieval requests are made. Data in a local world model are represented as tokens with attribute–value pairs. Tokens representing physical objects and geometric locations consist of a two-dimensional polygonal shape, a reference coordinate frame that can be used to transform the location to other frames, and time stamps that record when the token was created and the time at which sensor data were received that led to its creation. When range data, measured by the camera and laser range finder at different times and locations on the vehicle, are to be fused, the coordinate frames of the tokens created by each sensor for these data are first transformed to a common vehicle frame and then transformed forward to the same point in time. The data are now fused, resulting in the creation of a new token representing the fused data.

6) The DARPA Autonomous Land Vehicle: The Autonomous Land Vehicle (ALV) [112], [115], [161]–[163] built by Martin Marietta is part of the Defense Advanced Research Projects Agency's (DARPA) Strategic Computing Program. The ALV is intended to be a testbed designed for demonstrating the state of the art in autonomous vehicle research [163]. A number of companies and universities are currently working on different research aspects of the project. In the initial stages of the project, the ALV was used in road-following applications [112], [161], [164]; in more recent stages, obstacle avoidance [163] and autonomous cross-country navigation [163], [165] capabilities have been demonstrated. Future research is aimed at enhancing the operational speed and robustness of the ALV, and adding capabilities like landmark recognition [163].

In road-following applications [112], [163], the ALV uses sonar to determine its height, tilt, and roll with respect to road surfaces directly beneath it. Complementary information from a laser range scanner and two color video cameras is used for obstacle detection. Color video information is used to locate roads because the laser range information can easily be confused if there is very little difference in depth between the road and surrounding areas. Laser range information is used to obtain accurate descriptions of the geometrical features of obstacles on the road because, unlike the video information, it is not sensitive to poor lighting conditions and shadows. After being transformed to a common world coordinate system, the video information is used both to determine the boundaries of the road for path planning and, after being integrated with similarly transformed laser range information, for obstacle recognition.

In autonomous cross-country navigation applications [115] a hierarchical control system is used to provide the ALV with the flexibility needed for operation over natural terrain. At the lowest level in the hierarchy, "virtual sensors" and "reflexive behaviors" are used as real-time operating primitives for the rest of the control system [165]. Functioning in a similar transparent manner as the logical sensors described in Section III-A-3, virtual sensors combine information from physical sensors with appropriate processing algorithms to provide specific information to associated reflexive behaviors. Combinations of behavior and virtual sensors are used to handle specific subproblems that are part of the overall navigation task. In these applications a laser range-scanner, together with orientation sensors to determine the pitch, roll, and x and y position of the vehicle, were used to provide information needed to create overhead map view representations of the terrain called Cartesian elevation maps (CEM's); other range sensors such as stereo vision could also be used to create CEM's. Smoothing procedures are applied to the CEM's to fill in detail not provided by the sparse laser range information. As the ALV travels over the terrain, CEM's are fused together to provide a means for selecting traversable trajectories for the vehicle.

VI. Applications

This section discusses a variety of intelligent systems in different areas of application to illustrate the role of multisensor integration and fusion in their overall operation. A description of the use of multisensor-based mobile robots in different areas of application can be found in Section V.

A. Industrial

The addition of sensory capabilities to the control of industrial robots can increase their flexibility and allow for their use in the production of products with a low volume or short design life [166]. In many industrial applications, the use of multiple sensors is required to provide the robot with sufficient sensory capabilities. Most of the multisensor integration and fusion techniques discussed in this paper are suitable for industrial applications because the industrial environment is usually well-structured, and descriptions of many of the objects in the environment are available from the data bases of computer-aided design systems.

Nitzan *et al.* [167] divide industrial robot applications into four general areas: material handling, part fabrication (e.g., spot and arc welding, forging, etc.), inspection, and assembly. Material handling is usually the simplest area of application and assembly the most complex.

1) Material Handling: Industrial robots can be used for in-process workpiece handling and the loading and unloading of industrial trucks (e.g., automatic guided vehicles, tow tractors, etc.) and conveyors—the two major material handling equipment types. Much of the research on mobile robots (Section V) can readily be applied to existing industrial trucks to increase their capabilities in areas such as route planning and obstacle avoidance. Choudry *et al.* [168] have developed a simulation system to test designs for the sensory control of an autonomous material handling vehicle. Their sensory control designs take advantage of the relatively well-structured shop floor environment to avoid having to use the more cumbersome hierarchical control structures used in most mobile robots. The hierarchical phase-template paradigm (Section III-A-1) summarizes the integration and fusion issues resulting from the use of a multisensor robot for handling workpieces. Sensory capabilities can enable a robot to grasp workpieces that are randomly oriented in a bin or on a conveyor. Hitachi Ltd. [169] has developed a robot which uses three-dimensional vision and force sensors to pick up randomly positioned connectors and mount them on a printed circuit board. One of the projects of the ESPRIT program (the European Strategic Programme for Research and Development into Information Technology) is developing a system that combines vision and tactile sensors for real-time applications in material handling [170]. Miller [45], [46] has applied adaptive learning (Section III-B-3) to an experiment involving having a robot use sensory feedback to track and intercept an object on a moving conveyor. Results of the experiment showed that, by the tenth attempt to teach the robot to follow the object to the end of the conveyor, the gripper was able to approach the object to within a small (1–4 cm) degree of accuracy.

2) Part Fabrication: As of 1985, almost half of the robots in U.S. industry were being used for welding [171] —the majority being used in spot welding applications because arc welding robots without sensory capabilities cannot track a seam with randomly variable gaps. Kremers *et al.* [172] developed a robot that used both a vision sensor and wrist-mounted laser scanner range sensors to guide the arm of the robot during the one pass arc welding of workpiece joints that had random gaps along their seams. Howarth and Guyote [173] describe work being done at Oxford that uses eddy current and ultrasonic sensors for robot arc welding. Nitzan *et al.* [167] provide a plan for a sensor guided arc welding system as a specific example to illustrate the use of generic robot functions (e.g., "recognize," "place," "grasp," etc.) and their associated high-level properties (e.g., "identity," "location," etc.) determined from sensory information.

3) Inspection: Inspection can be divided into two different types [167]: explicit and implicit. Explicit inspection verifies the integrity of workpieces, as a separate operation, either during or after the manufacturing process. Depending on the nature of the work pieces involved, any of the multisensor integration and fusion techniques described in this paper could be of potential use for explicit inspection. Implicit inspection verifies the integrity of work pieces while handling them during the manufacturing process. Many of the object-recognition approaches that combine vision with tactile sensors (Section IV-B) would be especially appropriate if applied to implicit inspection (as well as assembly) operations because the manipulator would already be in position to grasp and inspect the workpieces. Manufacturing research at Georgia Tech [174] is focused on integrating vision and tactile information for the adaptive control of a robot manipulator that would be useful in just such applications.

4) Assembly: Assembly is the most complex area of industrial robot application because, in addition to operations like insertion that are unique to assembly, different aspects of the other three application areas can be part of the overall assembly process. Smith and Nitzan [175] describe an assembly station consisting of two robots with wrist-mounted vision and force sensors, overhead cameras, and a general-purpose parts feeder. Printer carriages were assembled by first locating components on the feeder using the overhead cameras and then transporting them to the carriage and snapping them in place using the robots. The force and wrist-mounted vision sensors were used to verify that the components were correctly in place. One of the projects of the ESPRIT program is to develop vision and tactile sensors that can be used in an integrated manner for assembly operations [170].

Groen *et al.* [166], [176] describe a multisensor robotic assembly station equipped with vision, ultrasonic, tactile, and force/torque sensors. The assembly process is represented as a sequence of stages that are entered when certain sensor determined conditions are satisfied. A hierarchical control structure, modeled after the NBS control hierarchy (Section III-B-1), is used to enable the entire process to be executed using a set of modular low-level peripheral processes that perform dedicated tasks like sensory processing, robot control, and data communication. The system has been applied to the assembly of three different types of hydraulic lift assemblies for gas water heaters. In operation, vision sensors are used to recognize different parts of the assemblies as they arrive in varying order and at undefined positions. Feedback information from the force/torque sensors and the passive compliance of the robot's gripper are used for bolt insertion operations and to transport and place, with great precision, assembly housings on work spots so that the housing can be used as a reference for the remaining assembly operations. Final inspection is performed with the vision sensors.

Ruokangas *et al.* [177] describe an experimental hierarchically controlled multisensor robot work station developed using Rockwell's Automation Sciences Testbed. Vision, acoustic, and force/torque sensors, all mounted on the end effector of a robot, were used both separately and in different combinations to demonstrate the limitations of the information each sensor was able to provide and the advantages to be gained when two or more sensors are integrated. In one demonstration, distance information from the acoustic sensor was used to position the end effector so that the camera was at the correct focal distance for the visual inspection of workpieces. In another demonstration, the force/torque sensors were used to provide modifications during task execution to the measurement of hole positions that were initially determined by the integrated range information provided by stereo cameras and an acoustic sensor—the acoustic information being used to provide redundancy to the distance information provided by the cameras.

B. Military

As the complexity, speed, and scope of warfare has increased, the military has increasingly turned to automated systems to support many activities traditionally performed manually [178]. The use of large numbers of diverse sensors as part of these systems has resulted in the issues of multisensor integration and fusion assuming critical importance. As an example of the need for highly automated systems, the typical average information requirements for the command and control of tactical air warfare have been estimated to be 25–50 decisions/min based on 50 000–100 000 reports from 156 separate sensor platforms concerning as many as 1000 hostile targets being tracked up to an altitude of 20 km over an area of 800 km^2 [69]. As part of a comprehensive survey of the possible military applications of artificial intelligence technologies, Franklin *et al.* [179] listed multisensor integration and fusion as being of major importance in the areas of general operations, intelligence analysis and situation assessment, force command and control, autonomous vehicles, avionics, and electronic warfare; together with areas of minor importance, multisensor integration and fusion were listed in 20 of the 25 areas of application. Given this wide scope, multisensor integration and fusion will be one of the technologies necessary for the development of the Autonomous Land Vehicle (Section V-D-6), an intelligent Pilot's Associate, and a command and control system for naval battle management—three of the initial projects supported by DARPA's Strategic Computing Program aimed at exploring the potential of AI-based solutions to important military problems [180].

The use of multisensor integration and fusion for object recognition in military applications requires that consideration be given to some additional factors that are not present in nonmilitary applications. Object recognition can be considered as a two-person zero-sum statistical game played against nature by the system performing the recognition [181]. In the terms of game theory, at each move in the game both players select a strategy. Associated with each pair of possible strategies is a payoff. In a zero-sum game, the gain of one player is equal to the loss of the other. The system's strategies are the possible decisions as to the identity of an object; nature's strategies are the *a priori* probabilities corresponding to the occurrence of each of the possible classes to which an object can belong. In most nonmilitary applications, nature can be considered to be indifferent (i.e., the strategies it selects are based upon the *a priori* probability of each object, and they remain constant even though they may not be optimal in the game theoretic sense); in military applications, by contrast, it is possible that the moves of the other player (e.g., the design or operation of the hostile aircraft being recognized) are being selected with the goal of maximizing the potential payoff.

The possibility of a game against a "malevolent nature" (i.e., the enemy) highlights the necessity of being able to make decisions through the use of a variety of different sensors. Any system relying on the information provided by just a single sensor type could be more vulnerable to what might be called "meta-sensor" error. Although the sensor might be functionally providing correct information, it could be spoofed as to the true identity of an object being sensed (e.g., a hostile aircraft using a friendly radar signature). Through the use of redundant information provided by sensors of different types together with the ability to integrate this information intelligently, the system can minimize the effect of the spoofed sensor in the overall determination of the object's identity. Garvey and Fischler [182] discuss the use of multiple sensors for inductive and interpretative perceptual reasoning in hostile environments, Comparato [183] describes the potential role of multisensor fusion in the next generation of tactical com-

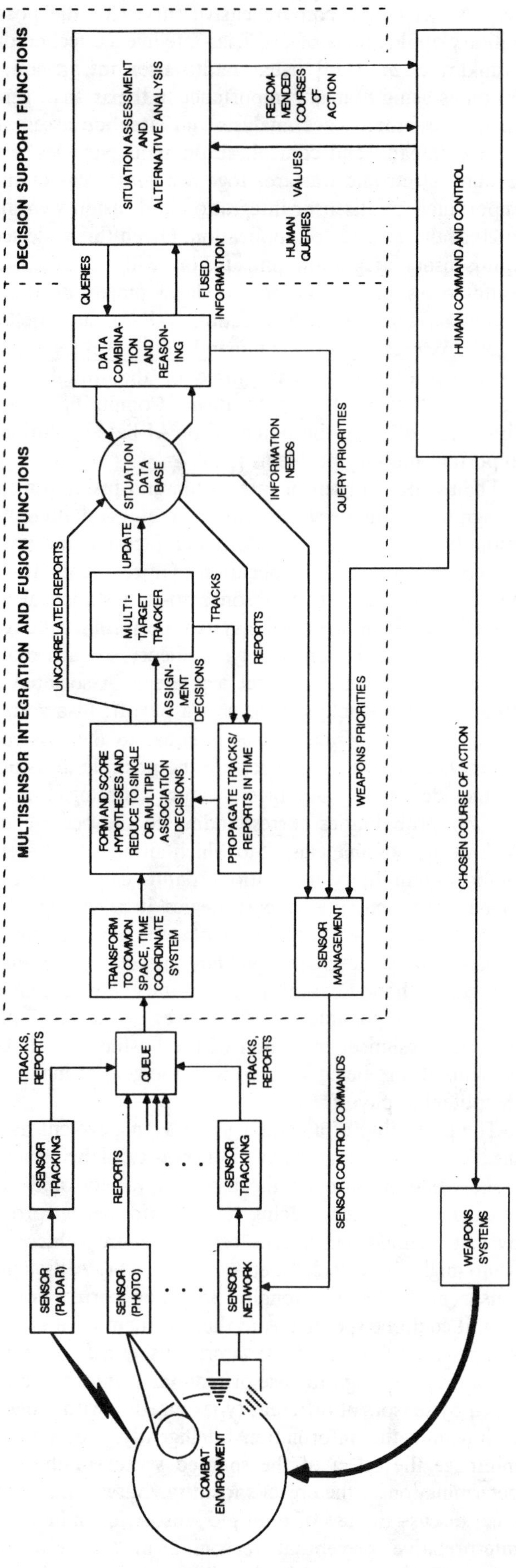

Fig. 13. Generic military command and control architecture. (Adapted from [69, figs. 5 and 6].)

bat platforms, and Mayersak [184] reviews the multisensor fusion aspects of munition (e.g., guided missile) design.

1) A Generic Command and Control Architecture: Waltz and Buede [69] proposed an architecture for a generic command and control system that includes multisensor integration and fusion functions as one of its two major subsystems (see Fig. 13). The operation of the system can be divided into four main steps of a feedback loop. First, a variety of different sensors collect data from the combat environment and then transmit these data on to the multisensor integration and fusion subsystem. The functions comprising this subsystem integrate and fuse the data so that any targets or events can be located and identified. The fused information, representing the possible targets or events that comprise the current situation, is then sent to the decision support subsection where it is used to create, analyze, and rank alternative courses of action. A human commander completes the feedback loop by selecting a course of action which possibly changes the environment. The system initiates operation with a query from the human commander to the decision support subsystem for recommended courses of action. Using certain key parameter values supplied by the human commander and the system's assessment of current situation, the decision support subsystem analyzes alternative courses of action, possibly querying the multisensor integration and fusion subsystem for additional information, to select those actions to recommend to the human commander. The human commander can then either select one of the current recommended actions or query the system for additional information. Although the human commander can always make the final decision, certain routine or time-critical actions can automatically be determined by the system.

The process of data fusion starts with each sensor or network of sensors in the combat environment sending detection reports or target tracks to a queue from which they are made commensurate by being transformed to a common space and time coordinate system. A local target track is maintained by a sensor if the measurement time of the sensor is small relative to the ability of the system to process the data. (In Fig. 13, both the radar and the network of sensors (e.g., ground sensors measuring seismic or acoustic events) may generate a track containing dozens of measurements in the time it takes the photographic sensor to generate a single image.) Because the system is generic, the common spatial reference (e.g., *XYZ*; latitude/longitude, range/azimuth/elevation, etc.) depends on the specific environment, and the methods of transforming the sensed data depend on the differing locations, resolutions, fields of view, and measurement times of the particular sensors used. After the reports or tracks are transformed to a common coordinate system, hypothetical pairwise assignments are made to each existing report from the situation data base and track from the multitarget dynamic tracking filter—each of which has been propagated in time to coincide with the current data. The assignments are then scored using a metric defined over some feature of the data (e.g., the spatial distance between the data and existing reports or tracks). The resulting scores are then used in selecting either a single hypothesis or a small set of hypotheses that are sent to the multitarget tracker. Probabilistic methods are sometimes needed to make the selection because of uncertainties in the sensor measurements or the large number of hypotheses generated when there are many sensors and possible targets. Reports that remain uncorrelated to any existing reports are sent to the situation data base for possible use in the future. When an assignment of a new report is made, the multitarget tracker updates the estimate of the associated target in the data base. The situation data base contains the reports and tracks that correspond to the most likely grouping of the data so far collected. The process of actually combining the collected data to enable the attributes of a target (e.g., its identity, intent, future behavior, etc.) to be inferred is the core function in the multisensor integration and fusion subsystem. While all of the methods of fusion mentioned in this paper can be used in implementing this function, the use of production rules to infer higher-level information (e.g., intent and future behavior) allows for a convenient interface to the decision support subsystem. The arrival of new sensor data causes a forward-chaining process that may result in the antecedent of a production rule being satisfied and the situation data base being updated with the rule's consequent. Reversing the direction of inference, queries from the decision support subsystem as to the support for a hypothetical situation can cause a backward-chaining process that searches to determine if any of the antecedents in the data base support the consequents implied by the query. If the required antecedents to not exist or have too large of an associated uncertainty, the data base can direct the sensor management function to redirect the sensors to search for data to support the required antecedents; for example, if the current possible enemy threat to a specific location X is queried, sensors near to X could be redirected to focus on that location. Priorities can be sent to the sensor management function when multiple queries exist.

2) Analyst: Analyst [185], [186] is a prototype expert system developed for the U.S. Army that generates real-time battlefield tactical situation descriptions using the information provided by multiple sensor sources. (This summary follows that given in [174].) Analyst determines as output a situation map showing the suspected location of enemy units through the use of intelligence input from battlefield sensors, photographs, and intercepted enemy communications and radar transmissions. Each intelligence report input is represented as a frame. The use of frames allows default values to be attached to the slots of the frame in the case where the intelligence report was incomplete. Each frame is then applied simultaneously to production rules contained in the first two of six different knowledge bases. The initial fusion of the sensor information is performed by these rules in the process of inferring a possible battlefield entity (also represented as a frame). Associated with each entity frame is a slot containing a likelihood factor to indicate the strength of the evidence used to infer the presence of the entity. The third knowledge base eliminates possible multiple frames corresponding to the same entity that may have been created from the reports of different sensors. The fourth and fifth knowledge bases refine and reinforce the entity frames through the use of tactical and terrain data together with the presence of other possible related entities in the knowledge bases. The sixth knowledge base serves to remove from consideration those entities that have not been reinforced with additional information for a sufficient length of time. Each knowledge base is applied sequentially to each piece of sensor information as it becomes available, enabling Analyst continuously to provide the most current estimate of the battlefield situation.

C. Space

NASA's permanently manned space station will be the United States' major space program in coming years [187]. Previous NASA programs, including the space shuttle, have used a high degree of participation by both the crew and ground-based personnel to perform the sensing and perception functions required for many tasks. Both to increase productivity and allow tasks beyond the capability of the crew to be performed, the space station will make increasing use of autonomous systems for the servicing, maintenance, and repair of satellites and the assembly of large structures for use as production facilities and commercial laboratories. Probably the most important factor promoting the use of autonomous systems for these applications is the cost of having a human in space. In addition to these economic factors, aspects of the space environment can make the use of multiple sensors an especially important part of these systems. The sensing of objects in space using just optical sensors is difficult because the lack of atmosphere invalidates some of the common assumptions concerning surface reflectance used in many visual recognition methods; also, images in space frequently have deep shadows, missing edges, and intense specular reflections [188].

Shaw *et al.* [188] have presented a system that uses TV images to guide a microwave radar unit in the determination of the shape of objects in space. Microwave radar information serves to complement optical information for a number of important features of objects typically found in space. Scattering cross-sectional data from radar can be used to determine the shape of the metallic surfaces typically found on satellites; optical sensors and even laser range-scanners have difficulty with metallic surfaces because they reflect light in a specular direction. Optical sensors would be more useful in determining the shape of the slightly matte surfaces found on the space shuttle because these surfaces reflect light more equally in all directions. As compared to optical wavelengths, the longer wavelength of radar can be used to penetrate the solar blankets on space objects that sometimes cover important surface details needed by a robot for grasping and may sometimes diffract around objects that are occluded along the optical sensor's line of sight. The optical image is used to provide an initial estimate of the shape of a surface that serves to reduce the ambiguity inherent when interpreting narrow-band radar-scattering cross-sectional data. In operation, the system uses equations defining the orthogonal directions of the polarized radar-scattering cross section to determine surface shapes. Initially, occluding contours are derived from the optical image of the surface by thresholding, and a partial surface shape description is derived from shape-from-shading or stereo. The surface is then either matched to some simple geometric shape or, if no match is found, a grid is constructed over the surface. If a simple match is found, a closed-form expression can be used for the scattering equations; if no match is found, an iterative nonlinear least squares technique is used to approximate the equations.

D. Target Tracking

A variety of different filtering techniques, together with radar, optical, and sonar sensors, have been used for tracking targets (e.g., missiles, planes, submarines, etc.) in the air, space, and under water. Bar-Shalom and Fortmann [189], [190] have surveyed a number of tracking methods that can be used when there are multiple targets in the environment. Recently, researchers have been developing methods of multitarget tracking that integrate and/or fuse the information (or measurements) provided by a number of identical sensors. The key problem in multitarget tracking in general, and in multisensor multitarget tracking in particular, is "data association"—the association of sensor information to a single target [191] (in general multisensor integration and fusion, this problem is termed the registration problem (Section VII-A-1)). For the purposes of this discussion, the different multisensor tracking methods can be distinguished based on the complementary versus redundant nature of the information provided by the sensors (Section II-C).

1) Complementary Tracking Information: The position and velocity of targets can be derived through the use of the complementary information provided by the time of arrival and frequency of signals sent by a small group of sensors. The difference between both the time and frequency of the signals arriving at each sensor can be used to determine the tracks of targets. Arnold *et al.* [192] and Mucci *et al.* [193] have used the Fisher information matrix to evaluate the performance of this approach and have found that a single pair of omnidirectional sensors are sufficient to determine the location of moving targets when they are in the vicinity ("near field") of the sensors; three sensors are required when they are distant or stationary.

2) Redundant Tracking Information: Redundant tracking information can be provided by a network of sensors distributed over a large geographic area to increase the overall reliability and survivability of the tracking system [191]. Two different processing architectures have been developed for the association of the information provided by each sensor in the network. In the first, the measurements of all the sensors are transmitted to a centralized site for processing; in the second, the processing is distributed to local nodes in the network. As compared to the first, the second architecture provides the additional benefits of an increased survivability and the possibility of using a smaller bandwidth for communication within the network.

Chang *et al.* [191] and Chong *et al.* [194] have presented a distributed processing version of the Joint Probabilistic Data Association algorithm, first applied in a centralized processing architecture [195], that reduces the network's bandwidth requirements by fusing, at each local node in the network, measurements from a small number of sensors. The compressed higher level fused information is then propagated to the other local nodes in the network. Each local node can use the fused information it receives from the other nodes to arrive at a solution to the global tracking problem. The algorithm used for fusion was developed by Speyer [196]. It is based on the use of a Kalman filter (Section III-E-2) at each local node and requires the propagation of an additional data-dependent vector beyond the usual filter equations.

In a series of papers Thomopoulos *et al.* [197]–[200] have developed algorithms for fusing target detection information from a distributed network of sensors. In the first paper [197], optimal and suboptimal algorithms are developed for the situation in which a group of parallel sensors transmit their detection decision to a fusion center where, assuming the sensors are conditionally independent, a Neyman–Pearson test maximizes the fused probability of detection for a fixed probability of false alarm. A likelihood-ratio test is used for detection at each sensor. In the second paper [198], a pair of sensors is considered in which one sensor is the primary sensor responsible for the final detection decision and the other is a consulting sensor. Based on an estimated cost associated with any communication between the sensors and the quality of its own raw data, the primary sensor makes a decision as to whether to consult with the other sensor as part of its overall detection decision. In the third paper [199] delays in the network and channel errors are considered, and in the final paper [200] the time origin of the information from each sensor is assumed to be uncertain. An extension to the Kalman filter is developed to account for this uncertainty.

E. Inertial Navigation

The inertial system used in spacecraft and many advanced aircraft for navigation consists of a gyroscope mounted on a platform suspended in a gimbal structure that allows the vehicle to change its angular orientation while maintaining the platform fixed with respect to a reference coordinate frame [66], [201]. The position and velocity of the vehicle can be determined by integrating the signals from accelerometers mounted on the platform. Due to inherent gyro characteristics that cause errors in posi-

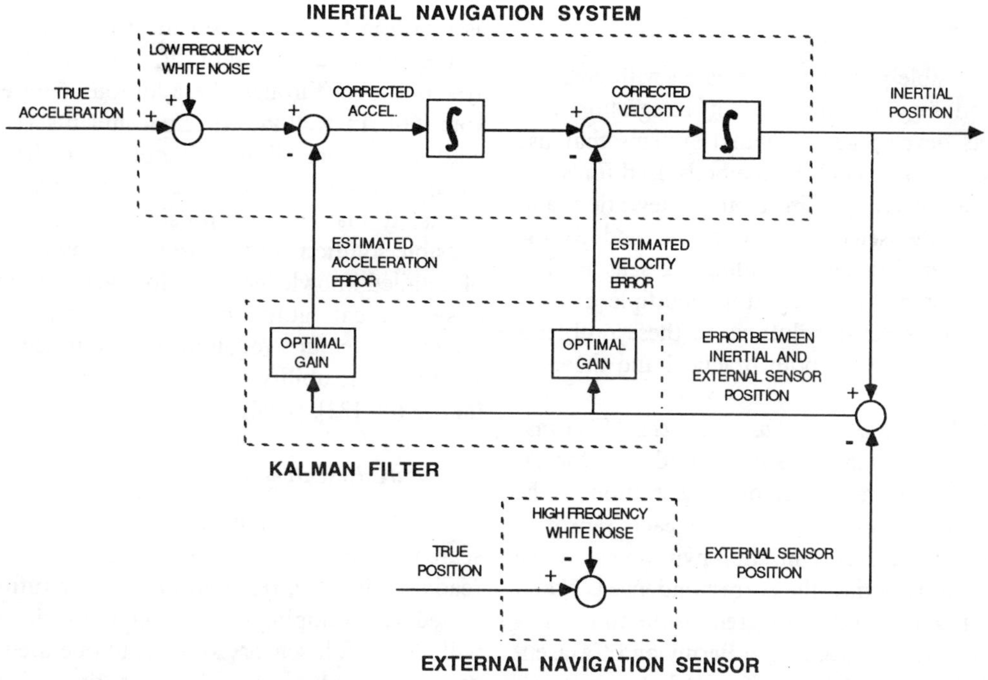

Fig. 14. External sensor-aided indirect feedback inertial navigation system configuration. (Adapted from [66, fig. 6.9(a)].)

tion and velocity to grow unbounded slowly over time (i.e., low-frequency noise), the inertial navigation system requires the aid of other external navigation sensors to bound or dampen these errors. Typical external sensors used to aid the inertial system include on-board or ground-based radar, radio, navigation, satellites, landmarks or star sightings, laser ranging, and altimeters. As opposed to the inertial system which accurately follows the high-frequency motions of the vehicle, each of these external sensors provides information which is good on average (i.e., its error does not increase over time) but is subject to considerable high-frequency noise (i.e., each measurement has considerable error). A Kalman filter (Section III-E-2) can be used to determine the fusion of inertial system and external sensor information that will statistically minimize the error in the estimate of the vehicle's position.

Fig. 14 shows a typical external sensor-aided inertial navigation system configuration. In the figure, the Kalman filter is used in an indirect (error state space) feedback configuration to generate estimates of the errors in the inertial system and then feed these estimates back into the inertial system to correct it. The indirect feedback filter configuration is used in most inertial navigation systems because 1) as opposed to the direct formulation where the vehicle's actual position is estimated, the indirect formulation increases the overall reliability of the navigation system because the inertial system can still operate if the filter should fail due to a temporary computer or external navigation sensor failure; 2) the feedback, as opposed to feedforward, mechanization is used so that inertial errors do not grow unchecked; and 3) the sample rate for this filter is low because only low frequency linear dynamics are modeled.

F. Remote Sensing

In aerial photo mapping over land, known ground points can be used to establish the orientation of photographs; in aerial mapping over water, the orientation must be determined by accurately knowing the position and attitude of the camera at the time the photograph is taken because known ground points are not generally available. Gesing and Reid [202] describe a system that fuses information from multiple navigation sensors to estimate an aircraft's position and attitude accurately enough for use in the mapping of shallow coastal waters. An inertial navigation system is mounted to the top of the aerial survey camera in the aircraft used for mapping. During flight, information is recorded from a number of auxiliary navigation sensors including: a laser bathymeter to measure water depth, a microwave ranging system, barometric and radar altimeters, a radio navigation system, and a Doppler radar. A U-D covariance factorization filter (Section III-E-2) and a modified Bryson–Frazier smoother are then used for post-flight processing to produce time-correlated sensor error estimates that can be subtracted from the recorded inertial system data to yield highly accurate position and attitude information that serves to orient the photographs taken during the flight.

VII. Conclusion

The issues of and approaches to the problem of multisensor integration and fusion presented above demonstrate the wide scope of present research efforts in this area. To conclude this paper, a discussion of the possible problems and future research directions in the area of multisensor integration and fusion is provided.

A. Possible Problems

Many of the possible problems associated with creating a general methodology for multisensor integration and fusion, as well as developing the actual systems that use multiple sensors, center around the methods used for modeling the error or uncertainty in 1) the integration and fusion process, 2) the sensory information, and 3) the operation of the overall system including the sensors. For the potential advantages in integrating multiple sensors (Section II-C) to be realized, solutions to these problems will have to be found that are both practical and theoretically sound.

1) Error in the Integration and Fusion Process: The major problem in integrating and fusing redundant information from multiple sensors is that of "registration"—the determination that the information from each sensor is referring to the same features in the environment. (The registration problem is termed the correspondence [77] and data association [191] problem in stereo vision and multitarget tracking research, respectively.) Barniv and Casasent [203] have used the correlation coefficient between pixels in the grey level of images as a measure of the degree of registration of objects in the images from multiple sensors. Lee and Van Vleet [204] and Holm [205] have studied the registration errors between radar and infrared sensors. Lee and Van Vleet have presented an approach that is able both to estimate and minimize the registration error, and Holm has developed a method that is able to compensate autonomously for registration errors in both the total scene as perceived by each sensor ("macroregistration"), and the individual objects in the scene ("microregistration").

2) Error in Sensory Information: The error in sensory information is usually assumed to be caused by a random noise process that can be adequately modeled as a probability distribution. The noise is usually assumed not to be correlated in space or time (i.e., white), Gaussian, and independent. The major reasons that these assumptions are made are that they enable a variety of fusion techniques to be used that have tractable mathematics and yield useful results in many applications. If the noise is correlated in time (e.g., gyroscope error (Section VI-E-1)) it is still sometimes possible to retain the whiteness assumption through the use of a shaping filter [66]. The Gaussian assumption can only be justified if the noise is caused by a number of small independent sources. In many fusion techniques the consistency of the sensor measurements is increased by first eliminating spurious sensor measurements so that they are not included in the fusion process. Many of the techniques of robust statistics (e.g., ϵ-contamination in Sections III-E-4 and -5) can be used to eliminated spurious measurements. The independence assumption is usually reasonable so long as the noise sources do not originate from within the system (cf. paragraph 3, below).

3) Error in System Operation: When error occurs during operation due to possible coupling effects between components of a system, it may still be possible to make the assumption that the sensor measurements are independent if the error, after calibration, is incorporated into the system model through the addition of an extra state variable [66]. In well-known environments the calibration of multiple sensors will usually not be a difficult problem, but when multisensor systems are used in unknown environments, it may not be possible to calibrate the sensors. Possible solutions to this problem may require the creation of detailed knowledge bases for each type of sensor so that a system can autonomously calibrate itself. One other important feature required of any intelligent multisensor system is the ability to recognize and recover from sensor failure (cf. [23], [206]).

B. Future Research Directions

In addition to multisensor integration and fusion research directed at finding solutions to the problems already mentioned, research in the near future will likely be aimed at developing integration and fusion techniques that will allow multisensory systems to operate in unknown and dynamic environments. As currently envisioned, multisensor integration and fusion techniques will play an important part in the Strategic Defense Initiative in enabling enemy warheads to be distinguished from decoys [207]. Many integration and fusion techniques will likely be implemented on recently developed highly parallel computer architectures (e.g., the Connection Machine [208], etc.) to take full advantage of the parallelism inherent in the techniques. The development of sensor modeling and interface standards would accelerate the design of practical multisensor systems [9]. Continued research in the areas of artificial intelligence and neural networks will continue to provide both theoretical and practical insights. AI-based research may prove especially useful in areas like sensor selection, automatic task error detection and recovery, and the development of high-level representations; research based on neural networks may have a large impact in areas like object recognition through the development of distributed representations suitable for the associative recall of multisensory information, and in the development of robust multisensor systems that are able to self-organize and adapt to changing conditions (e.g., sensor failure).

The development of integrated solid-state chips containing multiple sensors has been the focus of much recent research [209], [210]. As current progress in VLSI technology [211] continues, it is likely that so-called "smart sensors" [212] will be developed that contain many of their low-level signal and fusion processing algorithms in circuits on the chip. In addition to a lower cost, a smart sensor might provide a better signal-to-noise ratio and abilities for self-testing and calibration. Currently, it is common to supply a multisensor system with just enough sensors for it to complete its assigned tasks; the availability of cheap integrated multisensors may enable some recent ideas concerning "highly redundant sensing" [213] to be incorporated into the design of intelligent multisensor systems—in some cases, high redundancy may imply the use of up to ten times the number of minimally necessary sensors to provide the system with a greater flexibility and insensitivity to sensor failure. In the more distant future, the development of micro or "gnat" [214] robots will necessarily entail the advancement of the state of the art in multisensor integration and fusion.

REFERENCES

[1] I. P. Howard, *Human Visual Orientation*. Chichester, UK: Wiley, 1982, ch. 11.

[2] E. A. Newman and P. H. Hartline, "The infrared 'vision' of snakes," *Sci. Amer.*, vol. 246, no. 3, pp. 116–127, Mar. 1982.

[3] R. Bajcsy and P. Allen, "Multisensor integration," in *Encyclopedia of Artificial Intelligence*. New York: Wiley, 1986, pp. 632–638.

[4] A. Mitiche and J. K. Aggarwal, "Multiple sensor integration/fusion through image processing: a review," *Opt. Eng.*, vol. 25, no. 3, pp. 380–386, 1986.

[5] T. D. Garvey, "A survey of AI approaches to the integration of information," in *Proc. SPIE*, vol. 782, *Infrared Sensors and Sensor Fusion*, R. G. Buser and F. B. Warren, Eds., Orlando, FL, May 1987, pp. 68–82.

[6] R. C. Mann, "Multi-sensor integration using concurrent computing," in *Proc. SPIE*, vol. 782, *Infrared Sensors and Sensor Fusion*, R. G. Buser and F. B. Warren, Eds., Orlando, FL, May 1987, pp. 83–90.

[7] R. C. Luo and M. G. Kay, "Multisensor integration and fusion: issues and approaches," in *Proc. SPIE*, vol. 931, *Sensor Fusion*, C. W. Weaver, Ed., Orlando, FL, Apr. 1988, pp. 42–49.

[8] S. S. Blackman, "Theoretical approaches to data association and fusion," in *Proc. SPIE*, vol. 931, *Sensor Fusion*, C. W. Weaver, Ed., Orlando, FL, Apr. 1988, pp. 50–55.

[9] T. C. Henderson, P. K. Allen, A. Mitiche, H. Durrant-Whyte, and W. Snyder, Eds., "Workshop on multisensor integration in manufacturing automation," Dept of Comput. Sci., Univ. of Utah, Snowbird, Tech. Rep. UUCS-87-006, Feb. 1987.

[10] A. Kak and S. Chen, Eds., *Spatial Reasoning and Multi-Sensor Fusion: Proc. 1987 Workshop*. Los Altos, CA: Morgan Kaufmann, 1987.

[11] M. A. Fischler and O. Firschein, *Intelligence: The Eye, the Brain and the Computer*. Reading, MA: Addison-Wesley, 1987, pp. 241–242.

[12] J. M. Richardson and K. A. Marsh, "Fusion of multisensor data," *Int. J. Robot. Res.*, vol. 7, no. 6, pp. 78–96, 1988.

[13] R. C. Luo, M. Lin, and R. S. Scherp, "The issues and approaches of a robot multisensor integration," in *Proc. IEEE Int. Conf. Robotics and Automat.*, Raleigh, NC, Mar. 1987, pp. 1941–1946.

[14] ____, "Dynamic multi-sensor data fusion system for intelligent robots," *IEEE J. Robot. Automat.*, vol. RA-4, no. 4, pp. 386–396, 1988.

[15] ____, "Multi-sensor based intelligent robot system," in *Proc. Int. Conf. Indust. Elect. Contr. and Instrumentation*, Milwaukee, WI, Sept. 1986, pp. 238–243.

[16] R. C. Luo and M. Lin, "Multi-sensor integrated intelligent robot for automated assembly," in *Proc. Workshop Spatial Reasoning and Multi-Sensor Fusion*, St. Charles, IL, Oct. 1987, pp. 351–360.

[17] ____, "Robot multi-sensor fusion and integration: optimum estimation of fused sensor data," in *Proc. IEEE Int. Conf. Robotics and Automat.*, Philadelphia, PA, Apr. 1988, pp. 1076–1081.

[18] G. E. Hinton and J. A. Anderson, Eds., *Parallel Models of Associative Memory*. Hillsdale, NJ: Erlbaum, 1981.

[19] D. E. Rumelhart and J. L. McClelland, Eds., *Parallel Distributed Processing: Explorations into the Microstructure of Cognition*. Cambridge, MA: MIT Press, 1986.

[20] T. Kohonen, *Self-Organization and Associative Memory*. Berlin, Germany: Springer-Verlag, 1984.

[21] A. J. Pellionisz, "Sensorimotor operation: A ground for the co-evolution of brain-theory with neurobotics and neurocomputers," in *Proc. IEEE 1st Int. Conf. Neural Networks*, San Diego, CA, June 1987, pp. IV-593–IV-600.

[22] J. C. Pearson, J. J. Gelfand, W. E. Sullivan, R. M. Peterson, and C. D. Spence, "Neural network approach to sensory fusion," in *Proc. SPIE*, vol. 931, *Sensor Fusion*, C. W. Weaver, Ed., Orlando, FL, Apr. 1988, pp. 103–108.

[23] O. G. Jakubowicz, "Autonomous reconfiguration of sensor systems using neural nets," in *Proc. SPIE*, vol. 931, *Sensor Fusion*, C. W. Weaver, Ed., Orlando, FL, Apr. 1988, pp. 197–203.

[24] W. B. Dress, "Frequency-coded artificial neural networks: An approach to self-organizing systems," in *Proc. IEEE 1st Int. Conf. Neural Networks*, San Diego, CA, June 1987, pp. II-47–II-54.

[25] T. Henderson and E. Shilcrat, "Logical sensor systems," *J. Robot. Syst.*, vol. 1, no. 2, pp. 169–193, 1984.

[26] T. Henderson, E. Shilcrat, and C. Hansen, "A fault tolerant sensor scheme," in *Proc. 7th Int. Conf. Pattern Recognition*, Montreal, PQ, Canada, July 1984, pp. 663–665.

[27] T. Henderson, C. Hansen, and B. Bhanu, "A framework for distributed sensing and control," in *Proc. 9th Int. Joint Conf. Artificial Intell.*, Los Angeles, CA, Aug. 1985, pp. 1106–1109.

[28] ____, "The synthesis of logical sensor specifications," in *Proc. SPIE*, vol. 579, *Intell. Robots and Comput. Vision*, Cambridge, MA, Sept. 1985, pp. 442–445.

[29] ____, "The specification of distributed sensing and control," *J. Robot. Syst.*, vol. 2, no. 4, pp. 387–396, 1985.

[30] T. C. Henderson and E. Weitz, "Multisensor integration in a multiprocessor environment," in *Proc. ASME Int. Comput. in Engr. Conf. and Exhibition*, R. Raghavan and T. J. Cokonis, Eds., New York, NY, Aug. 1987, pp. 311–316.

[31] T. Henderson, E. Weitz, C. Hanson, and A. Mitiche, "Multisensor knowledge systems: interpreting 3-D structure," *Int. J. Robot. Res.*, vol. 7, no. 6, pp. 114–137, 1988.

[32] T. C. Henderson and C. Hanson, "Multisensor knowledge systems," in *Real-Time Object Measurement and Classification*, A. K. Jain, Ed. Berlin, Germany: Springer-Verlag, 1988, pp. 375–390.

[33] R. C. Luo and T. Henderson, "A servo-controlled robot gripper with multiple sensors and its logical specification," *J. Robot. Syst.*, vol. 3, no. 4, pp. 409–420, 1986.

[34] J. C. Rodger and R. A. Browse, "An object-based representation for multisensory robotic perception," in *Proc. Workshop Spatial Reasoning and Multi-Sensor Fusion*, St. Charles, IL, Oct. 1987, pp. 13–20.

[35] P. K. Allen, "A framework for implementing multi-sensor robotic tasks," in *Proc. ASME Int. Computers in Engr. Conf. and Exhibition*, R. Raghavan and T. J. Cokonis, Eds., New York, NY, Aug. 1987, pp. 303–309.

[36] H. G. McCain, "A hierarchically controlled, sensory interactive robot in the automated manufacturing research facility," in *Proc. IEEE Int. Conf. Robotics and Automat.*, St. Louis, MO, Mar. 1985, pp. 931–939.

[37] E. W. Kent and J. S. Albus, "Servoed world models as interfaces between robot control systems and sensory data," *Robotica*, vol. 2, pp. 17–25, 1984.

[38] J. S. Albus, *Brains, Behavior, & Robotics*. Peterborough, NH: Byte Books, 1981.

[39] M. Shneier *et al.*, "Robot sensing for a hierarchical control system," in *Proc. 13th Int. Symp. Industrial Robots and Robots 7*, Chicago, IL, Apr. 1983, pp. 14.50–14.64.

[40] A. J. Barbera, M. L. Fitzgerald, J. S. Albus, and L. S. Haynes, "RCS: The NBS real-time control system," in *Robots 8 Conf. Proc.*, Detroit, MI, June 1984, pp. 19.1–19.33.

[41] J. S. Albus, "Data storage in the Cerebellar Model Articulation Controller (CMAC)," *J. Dynamic Syst., Meas. Contr.*, vol. 97, no. 3, pp. 228–233, 1975.

[42] ____, "A new approach to manipulator control: The cerebellar model articulation controller (CMAC)," *J. Dynamic Syst., Meas. Contr.*, vol. 97, no. 3, pp. 220–227, 1975.

[43] S. Y. Harmon, G. L. Bianchini, and B. E. Pinz, "Sensor data fusion through a distributed blackboard," in *Proc. IEEE Int. Conf. Robotics and Automat.*, San Francisco, CA, Apr. 1986, pp. 1449–1454.

[44] W. T. Miller, III, F. H. Glanz, and L. G. Kraft, III, "Application of a general learning algorithm to the control of robotic manipulators," *Int. J. Robot. Res.*, vol. 6, no. 2, pp. 84–98, 1987.

[45] W. T. Miller, III, "Sensor-based control of robotic manipulators using a general learning algorithm," *IEEE J. Robot. Automat.*, vol. RA-3, no. 2, pp. 157–165, 1987.

[46] ____, "A nonlinear learning controller for robotic manipulators," in *Proc. SPIE*, vol. 579, *Intell. Robots and Comput. Vision*, Cambridge, MA, Sept. 1986, pp. 416–423.

[47] G. Beni, S. Hackwood, L. A. Hornak, and J. L. Jackel, "Dynamic sensing for robots: an analysis and implementation," *Int. J. Robot. Res.*, vol. 2, no. 2, pp. 51–61, 1983.
[48] S. A. Hutchinson, R. L. Cromwell, and A. C. Kak, "Planning sensing strategies in a robot work cell with multisensor capabilities," in *Proc. IEEE Int. Conf. Robotics and Automat.*, San Francisco, CA, Apr. 1986, pp. 1068–1075.
[49] W. E. L. Grimson, "Disambiguating sensory interpretations using minimal sets of sensory data," in *Proc. IEEE Int. Conf. Robotics and Automat.*, Philadelphia, PA, Apr. 1988, pp. 286–292.
[50] G. E. Taylor and P. M. Taylor, "Dynamic error probability vectors: a framework for sensory decisionmaking," in *Proc. IEEE Int. Conf. Robotics and Automat.*, Philadelphia, PA, Apr. 1988, pp. 1096–1100.
[51] T. C. Henderson, W. S. Fai, and C. Hansen, "MKS: A multisensor kernel system," *IEEE Trans. Syst. Man Cybern.*, vol. SMC-14, no. 5, pp. 784–791, 1984.
[52] T. Henderson, B. Bhanu, and C. Hansen, "Distributed control in the multi-sensor kernel system," in *Proc. SPIE*, vol. 521, *Intell. Robots and Comput. Vision*, Cambridge, MA, Nov. 1984, pp. 253–255.
[53] M. O. Shneier, R. Lumia, and E. W. Kent, "Model-based strategies for high-level robot vision," *Comp. Vision Graph. Image Process.*, vol. 33, pp. 293–306, 1986.
[54] E. W. Kent, M. O. Shneier, and T. H. Hong, "Building representations from fusions of multiple views," in *Proc. IEEE Int. Conf. Robotics and Automat.*, San Francisco, CA, Apr. 1986, pp. 1634–1639.
[55] M. O. Shneier, E. W. Kent, and P. Mansbach, "Representing workspace and model knowledge for a robot with mobile sensors," in *Proc. 7th Int. Conf. Pattern Recognition*, Montreal, PQ, Canada, July 1984, pp. 199–202.
[56] L. F. Pau, "Fusion of multisensor data in pattern recognition," in *Pattern Recognition Theory and Applications*, J. Kittler, K. S. Fu, and L. F. Pau, Eds. Reidel, 1982, pp. 189–201.
[57] ——, "Knowledge representation for three-dimensional sensor fusion with context truth maintenance," in *Real-Time Object Measurement and Classification*, A. K. Jain, Ed. Berlin, Germany: Springer-Verlag, 1988, pp. 391–404.
[58] H. F. Durrant-Whyte, "Consistent integration and propagation of disparate sensor observations," *Int. J. Robot. Res.*, vol. 6, no. 3, pp. 3–24, 1987.
[59] ——, "Sensor models and multisensor integration," *Int. J. Robot. Res.*, vol. 7, no. 6, pp. 97–113, 1988.
[60] ——, "Consistent integration and propagation of disparate sensor observations," in *Proc. IEEE Int. Conf. Robotics and Automat.*, San Francisco, CA, Apr. 1986, pp. 1464–1469.
[61] ——, "Sensor models and multi-sensor integration," in *Proc. Workshop Spatial Reasoning and Multi-Sensor Fusion*, St. Charles, IL, Oct. 1987, pp. 303–312.
[62] R. McKendall and M. Mintz, "Robust fusion of location information," in *Proc. IEEE Int. Conf. Robotics and Automat.*, Philadelphia, PA, Apr. 1988, pp. 1239–1244.
[63] M. Zeytinoglu and M. Mintz, "Robust fixed size confidence procedures for a restricted parameter space," *Ann. Statist.*, vol. 16, no. 3, pp. 1241–1253, 1988.
[64] ——, "Optimal fixed size confidence procedures for a restricted parameter space," *Ann. Statist.*, vol. 12, no. 3, pp. 945–957, 1984.
[65] P. J. Huber, *Robust Statistics*. New York: Wiley, 1981.
[66] P. S. Maybeck, *Stochastic Models, Estimation, and Control*, vols. 1 and 2. New York: Academic, 1979 and 1982.
[67] T. D. Garvey, J. D. Lowrance, and M. A. Fischler, "An inference technique for integrating knowledge from disparate sources," in *Proc. 7th Int. Joint Conf. Artificial Intell.*, Vancouver, BC, Canada, Aug. 1981, pp. 319–325.
[68] P. L. Bogler, "Shafer-Dempster reasoning with applications to multisensor target identification systems," *IEEE Trans. Syst. Man Cybern.*, vol. SMC-17, no. 6, pp. 968–977, 1987.
[69] E. L. Waltz and D. M. Buede, "Data fusion and decision support for command and control," *IEEE Trans. Syst. Man Cybern.*, vol. 16, no. 6, pp. 865–879, 1986.
[70] G. Shafer, *A Mathematical Theory of Evidence*. Princeton, NJ: Princeton Univ. Press, 1976.
[71] T. L. Huntsberger and S. N. Jayaramamurthy, "A framework for multi-sensor fusion in the presence of uncertainty," in *Proc. Workshop Spatial Reasoning and Multi-Sensor Fusion*, St. Charles, IL, Oct. 1987, pp. 345–350.
[72] L. A. Zadeh, "Fuzzy sets," *Inform. Contr.*, vol. 8, pp. 338–353, 1965.
[73] S. J. Kamat, "Value function structure for multiple sensor integration," in *Proc. SPIE*, vol. 579, *Intell. Robots and Computer Vision*, Cambridge, MA, Sept. 1985, pp. 432–435.
[74] R. Belknap, E. Riseman, and A. Hanson, "The information fusion problem and rule-based hypotheses applied to complex aggregations of image events," in *Proc. IEEE Conf. Computer Vision and Pattern Recog.*, Miami Beach, FL, June 1986, pp. 227–234.
[75] A. R. Hanson, E. M. Riseman, and T. D. Williams, "Sensor and information fusion from knowledge-based constraints," in *Proc. SPIE*, vol. 931, *Sensor Fusion*, C. W. Weaver, Ed., Orlando, FL, Apr. 1988, pp. 186–196.
[76] D. Ballard and C. Brown, *Computer Vision*. Englewood Cliffs, NJ: Prentice-Hall, 1982.
[77] B. K. P. Horn, *Robot Vision*. Cambridge, MA: MIT Press, 1986.
[78] M. J. Magee and J. K. Aggarwal, "Using multisensory images to derive the structure of three-dimensional objects—A review," *Comp. Vision Graph. Image Process.*, vol. 32, pp. 145–157, 1985.
[79] M. J. Magee, B. A. Boyter, C. Chien, and J. K. Aggarwal, "Experiments in intensity guided range sensing recognition of three-dimensional objects," *IEEE Trans. Pattern Anal. Machine Intell.*, vol. PAMI-7, no. 6, pp. 629–636, 1985.
[80] J. K. Aggarwal and M. J. Magee, "Determining motion parameters using intensity guided range sensing," *Pattern Recognition*, vol. 19, no. 2, pp. 169–180, 1986.
[81] E. Hildreth, "Optical flow," in *Encyclopedia of Artificial Intelligence*. New York: Wiley, 1986, pp. 684–687.
[82] Y. P. Hung, D. B. Cooper, and B. Cernuschi-Frias, "Bayesian estimation of 3-D surfaces from a sequence of images," in *Proc. IEEE Int. Conf. Robotics and Automat.*, Philadelphia, PA, Apr. 1988, pp. 906–911.
[83] G. M. Flachs, J. B. Jordan, and J. J. Carlson, "Information fusion methodology," in *Proc. SPIE*, vol. 931, *Sensor Fusion*, C. W. Weaver, Ed., Orlando, FL, Apr. 1988, pp. 56–63.
[84] N. Ayache and O. D. Faugeras, "Building, registrating, and fusing noisy visual maps," *Int. J. Robot. Res.*, vol. 7, no. 6, pp. 45–65, 1988.
[85] O. D. Faugeras, N. Ayache, and B. Faverjon, "Building visual maps by combining noisy stereo measurements," in *Proc. IEEE Conf. Robotics and Automat.*, San Francisco, CA, Apr. 1986, pp. 1433–1438.
[86] S. T. Barnard and M. A. Fischler, "Stereo vision," in *Encyclopedia of Artificial Intelligence*. New York: Wiley, 1986, pp. 1083–1090.
[87] J. Landa and K. Scheff, "Binocular fusion using simulated annealing," in *Proc. IEEE 1st Int. Conf. Neural Networks*, San Diego, CA, June 1987, pp. IV-327–334.
[88] J. Porrill, "Optimal combination and constraints for geometrical sensor data," *Int. J. Robot. Res.*, vol. 7, no. 6, pp. 66–77, 1988.
[89] R. C. Luo and M. Lin, "3-D object recognition using combined overhead and robot eye-in-hand vision system," in *Proc. SPIE*, vol. 856, *IECON: Indust. Applic. of Robots and Machine Vision*, Cambridge, MA, Nov. 1987, pp. 682–689.
[90] P. K. Allen, "Integrating vision and touch for object recognition tasks," *Int. J. Robot. Res.*, vol. 7, no. 6, pp. 15–33, 1988.
[91] ——, *Robotic Object Recognition using Vision and Touch*. Boston, MA: Kluwer, 1987.
[92] ——, "Sensing and describing 3-D structure," in *Proc. IEEE Int. Conf. Robotics and Automat.*, San Francisco, CA, Apr. 1986, pp. 126–131.
[93] ——, "Surface descriptions from vision and touch," in *Proc. IEEE Int. Conf. Robotics*, 1984, pp. 394–397.
[94] P. Allen and R. Bajcsy, "Integrating sensory data for object recognition tasks," in *Proc. SPIE*, vol. 595, *Computer Vision for Robotics*, Cannes, France, Dec. 1985, pp. 225–232.
[95] ——, "Object recognition using vision and touch," in *Proc. 9th Int. Joint Conf. Artificial Intell.*, Los Angeles, CA, Aug. 1981, pp. 1131–1137.
[96] R. C. Luo and W. Tsai, "Object recognition using tactile array sensors," in *Proc. IEEE Int. Conf. Robotics and Automat.*, San Francisco, CA, Apr. 1986, pp. 1248–1253.
[97] S. A. Stansfield, "A robotic perceptual system utilizing passive vision and active touch," *Int. J. Robot. Res.*, vol. 7, no. 6, pp. 138–161, 1988.
[98] ——, "Visually-aided tactile exploration," in *Proc. IEEE Int. Conf. Robotics and Automation*, Raleigh, NC, Mar. 1987, pp. 1487–1492.

[99] ____, "Primitives, features, and exploratory procedures building a robot tactile perception system," in *Proc. IEEE Int. Conf. Robotics and Automat.*, San Francisco, CA, Apr. 1986, pp. 1274–1279.

[100] ____, "Representing generic objects for exploration and recognition," in *Proc. IEEE Int. Conf. Robotics and Automat.*, Philadelphia, PA, Apr. 1988, pp. 1090–1095.

[101] N. Nandhakumar and J. K. Aggarwal, "Integrated analysis of thermal and visual images for scene interpretation," *IEEE Trans. Pattern Anal. Machine Intell.*, vol. PAMI-10, no. 4, pp. 469–481, 1988.

[102] ____, "Thermal and visual information fusion for outdoor scene perception," in *Proc. IEEE Int. Conf. Robotics and Automat.*, Philadelphia, PA, Apr. 1988, pp. 1306–1308.

[103] ____, "Thermal and visual sensor fusion for outdoor robot vision," in *Proc. ASME Int. Comput. in Engr. Conf. Exhibition*, R. Raghavan and T. J. Cokonis, Eds., New York, NY, Aug. 1987, pp. 301–302.

[104] ____, "Synergetic integration of thermal and visual images for computer vision," in *Proc. SPIE*, vol. 782, *Infrared Sensors and Sensor Fusion*, R. G. Buser and F. B. Warren, Eds., Orlando, FL, May 1987, pp. 28–36.

[105] W. E. L. Grimson and T. Lozano-Pérez, "Model-based recognition and localization from sparse range or tactile data," *Int. J. Robot. Res.*, vol. 3, no. 3, pp. 3–35, 1984.

[106] M. C. Roggemann, J. P. Mills, S. K. Rogers, and M. Kabrisky, "Multisensor information fusion for target detection and classification," in *Proc. SPIE*, vol. 931, *Sensor Fusion*, C. W. Weaver, Ed., Orlando, FL, Apr. 1988, pp. 8–13.

[107] C. W. Tong, S. K. Rogers, J. P. Mills, and M. K. Kabrisky, "Multisensor data fusion of laser radar and forward looking infrared (FLIR) for target segmentation and enhancement," in *Proc. SPIE*, vol. 782, *Infrared Sensors and Sensor Fusion*, R. G. Buser and F. B. Warren, Eds., Orlando, FL, May 1987, pp. 10–19.

[108] S. Y. Harmon, "Autonomous vehicles," in *Encyclopedia of Artificial Intelligence*. New York: Wiley, 1986, pp. 39–45.

[109] R. C. Luo and M. G. Kay, "The role of multisensor integration and fusion in the operation of mobile robots," presented at the IEEE Int. Workshop Intelligent Robots, Tokyo, Japan, Oct. 1988.

[110] P. Levi, "Principles of planning and control concepts for autonomous mobile robots," in *Proc. IEEE Int. Conf. Robotics and Automat.*, Raleigh, NC, Mar. 1987, pp. 874–881.

[111] Y. Goto and A. Stentz, "The CMU system for mobile robot navigation," in *Proc. IEEE Int. Conf. Robotics and Automat.*, Raleigh, NC, Mar. 1987, pp. 99–105.

[112] M. A. Turk, D. G. Morgenthaler, K. D. Gremban, and M. Marra, "Video road-following for the Autonomous Land Vehicle," in *Proc. IEEE Int. Conf. Robotics and Automat.*, Raleigh, NC, Mar. 1987, pp. 273–279.

[113] G. Giralt, R. Chatila, and M. Vaisset, "An integrated navigation and motion control system for autonomous multisensory mobile robots," in *Robotics Research: The First International Symposium*, M. Brady and R. Paul, Eds. Cambridge, MA: MIT Press, 1984, pp. 191–214.

[114] S. Y. Harmon, "The Ground Surveillance Robot (GSR): an autonomous vehicle designed to transit unknown terrain," *IEEE J. Robot. Automat.*, vol. RA-3, no. 3, pp. 266–279, 1987.

[115] M. Daily *et al.*, "Autonomous cross-country navigation with the ALV," in *Proc. IEEE Int. Conf. Robotics and Automat.*, Philadelphia, PA, Apr. 1988, pp. 718–726.

[116] U. Rembold, "The Karlsruhe autonomous mobile assembly robot," in *Proc. IEEE Int. Conf. Robotics and Automat.*, Philadelphia, PA, Apr. 1988, pp. 598–603.

[117] R. C. Mann, W. R. Hamel, and C. R. Weisbin, "The development of an intelligent nuclear maintenance robot," in *Proc. IEEE Int. Conf. Robotics and Automat.*, Philadelphia, PA, Apr. 1988, pp. 621–623.

[118] R. A. Brooks, "A hardware retargetable distributed layered architecture for mobile robot control," in *Proc. IEEE Int. Conf. Robotics and Automat.*, Raleigh, NC, Mar. 1987, pp. 106–110.

[119] ____, "A robust layered control system for a mobile robot," *IEEE J. Robot. Automat.*, vol. 2, no. 1, pp. 14–23, 1986.

[120] J. L. Crowley, "Path planning and obstacle avoidance," in *Encyclopedia of Artificial Intelligence*. New York: Wiley, 1986, pp. 708–715.

[121] S. Chen, "Multisensor fusion and navigation of mobile robots," *Int. J. Intelligent Syst.*, vol. 2, no. 2, pp. 227–251, 1987.

[122] ____, "A geometric approach to multisensor fusion," in *Proc. SPIE*, vol. 931, *Sensor Fusion*, C. W. Weaver, Ed., Orlando, FL, Apr. 1988, pp. 22–25.

[123] ____, "A geometric approach to multisensor fusion and spatial reasoning," in *Proc. Workshop on Spatial Reasoning and Multi-Sensor Fusion*, St. Charles, IL, Oct. 1987, pp. 201–210.

[124] ____, "Adaptive control of multisensor systems," in *Proc. SPIE*, vol. 931, *Sensor Fusion*, C. W. Weaver, Ed., Orlando, FL, Apr. 1988, pp. 98–102.

[125] A. Elfes, "Sonar-based real-world mapping and navigation," *IEEE J. Robot. Automat.*, vol. RA-3, no. 3, pp. 249–265, 1987.

[126] ____, "A sonar-based mapping and navigation system," in *Proc. IEEE Int. Conf. Robotics and Automat.*, San Francisco, CA, Apr. 1986, pp. 1151–1156.

[127] H. P. Moravec, and A. Elfes, "High resolution maps from wide angle sonar," in *Proc. IEEE Int. Conf. Robotics and Automat.*, St. Louis, MO, Mar. 1985, pp. 116–121.

[128] L. Matthies and A. Elfes, "Integration of sonar range data using a grid-based representation," in *Proc. IEEE Int. Conf. Robotics and Automat.*, Philadelphia, PA, Apr. 1988, pp. 727–733.

[129] H. P. Moravec, "Sensor fusion in certainty grids for mobile robots," *AI Mag.*, vol. 9, no. 2, pp. 61–74, 1988.

[130] C. C. Jorgensen, "Neural network representation of sensor graphs in autonomous robot path planning," in *Proc. IEEE 1st Int. Conf. Neural Nets*, San Diego, CA, June 1987, pp. IV-507–IV-515.

[131] A. C. Kak, B. A. Roberts, K. M. Andress, and R. L. Cromwell, "Experiments in the integration of world knowledge with sensory information for mobile robots," in *Proc. IEEE Int. Conf. Robotics and Automat.*, Raleigh, NC, Mar. 1987, pp. 734–740.

[132] K. M. Andress and A. C. Kak, "Evidence accumulation & flow of control in a hierarchical spatial reasoning system," *AI Mag.*, vol. 9, no. 2, pp. 75–94, 1988.

[133] R. A. Brooks, "Visual map making for a mobile robot," in *Proc. IEEE Int. Conf. Robotics and Automat.*, St. Louis, MO, Mar. 1985, pp. 824–829.

[134] M. Asada, "Building a 3-D world model for a mobile robot from sensory data," in *Proc. IEEE Int. Conf. Robotics and Automat.*, Philadelphia, PA, Apr. 1988, pp. 918–923.

[135] D. Miller, "A spatial representation system for mobile robots," in *Proc. IEEE Int. Conf. Robotics and Automat.*, St. Louis, MO, Mar. 1985, pp. 122–127.

[136] D. T. Lawton, T. S. Levitt, C. McConnell, and J. Glicksman, "Terrain models for an autonomous land vehicle," in *Proc. IEEE Int. Conf. Robotics and Automat.*, San Francisco, CA, Apr. 1986, pp. 2043–2051.

[137] C. Isik and A. Meystel, "Decision making at a level of a hierarchical control for unmanned robots," in *Proc. IEEE Int. Conf. Robotics and Automat.*, San Francisco, CA, Apr. 1986, pp. 1772–1778.

[138] S. Harmon, "Robots, mobile," in *Encyclopedia of Artificial Intelligence*. New York: Wiley, 1986, pp. 957–963.

[139] ____, "A report on the NATO workshop on mobile robot implementation," in *Proc. IEEE Int. Conf. Robotics and Automat.*, Philadelphia, PA, Apr. 1988, pp. 604–610.

[140] N. J. Nilsson, "A mobile automation: an application of artificial intelligence techniques," in *Proc. 1st Int. Joint Conf. Artificial Intell.*, Washington, DC, May 1969, pp. 509–520.

[141] M. H. Smith *et al.*, "The design of JASON, a computer controlled mobile robot," in *Proc. Int. Conf. Cybern. Society*, New York, NY, Sept. 1975, pp. 72–75.

[142] H. Moravec, "The Stanford Cart and the CMU Rover," *Proc. IEEE*, vol. 71, no. 7, pp. 872–884, 1983.

[143] J. P. Bixler and D. P. Miller, "A sensory input system for autonomous mobile robots," in *Proc. Workshop Spatial Reasoning and Multi-Sensor Fusion*, St. Charles, IL, Oct. 1987, pp. 211–219.

[144] H. R. Everett, "A second-generation autonomous sentry robot," *Robotics Age*, pp. 29–32, Apr. 1985.

[145] A. M. Flynn, "Combining sonar and infrared sensors for mobile robot navigation," *Int. J. Robot. Res.*, vol. 7, no. 6, pp. 5–14, 1988.

[146] R. Berry, K. Loebbaka, and E. Hall, "Sensors for mobile robots," in *Proc. 3rd Conf. Robot Vision and Sensory Control*, Cambridge, MA, Nov. 1983, pp. 584–588.

[147] R. Wallace *et al.*, "First results in robot road-following," in *Proc. 9th Int. Joint Conf. Artificial Intell.*, Los Angeles, CA, Aug. 1985, pp. 1089–1095.

[148] J. Miller, "A discrete adaptive guidance system for a roving vehicle," in *Proc. IEEE Conf. Decision and Contr.*, New Orleans, LA, Dec. 1977, pp. 566–575.
[149] G. Giralt, R. Sobek, and R. Chatila, "A multi-level planning and navigation system for a mobile robot: A first approach to HILARE," in *Proc. 6th Int. Joint Conf. Artificial Intell.*, Tokyo, Japan, Aug. 1979, pp. 335–337.
[150] M. Ferrer, M. Briot, and J. C. Talou, "Study of a video image treatment system for the mobile robot HILARE," in *Proc. 1st Conf. Robot Vision and Sensory Contrl*, Stratford-upon-Avon, UK, Apr. 1981.
[151] A. R. de Saint Vincent, "A 3D perception system for the mobile robot HILARE," in *Proc. IEEE Int. Conf. Robotics and Automat.*, San Francisco, CA, Apr. 1986, pp. 1105–1111.
[152] M. Briot, J. C. Talou, and G. Bauzil, "The multi-sensors which help a mobile robot find its place," *Sensor Rev.*, vol. 1, no. 1, pp. 15–19, 1981.
[153] G. Bauzil, M. Briot, and P. Ribes, "A navigation sub-system using ultrasonic sensors for the mobile robot HILARE," in *Proc. 1st Conf. Robot Vision and Sensory Contr.*, Stratford-upon-Avon, UK, Apr. 1981.
[154] R. Chatila and J. Laumond, "Position referencing and consistent world modeling for mobile robots," in *Proc. IEEE Int. Conf. Robotics and Automat.*, St. Louis, MO, Mar. 1985, pp. 138–145.
[155] J. L. Crowley, "Navigation for an intelligent mobile robot," *IEEE J. Robot. Automat.*, vol. RA-1, no. 1, pp. 31–41, 1985.
[156] ——, "Dynamic world modeling for an intelligent mobile robot," in *Proc. 7th Int. Conf. Pattern Recognition*, Montreal, PQ, Canada, July 1984, pp. 207–210.
[157] ——, "Dynamic world modeling for an intelligent mobile robot using a rotating ultrasonic ranging device," in *Proc. IEEE Int. Conf. Robotics and Automat.*, St. Louis, MO, Mar. 1985, pp. 128–135.
[158] S. Harmon, "USMC ground surveillance robot: A testbed for autonomous vehicle research," in *Proc. 4th Univ. Alabama Robotics Conf.*, Huntsville, AL, Apr. 1984.
[159] D. J. Kriegman, E. Triendl, and T. O. Binford, "A mobile robot: Sensing, planning and locomotion," in *Proc. IEEE Int. Conf. Robotics and Automat.*, Raleigh, NC, Mar. 1987, pp. 402–408.
[160] S. A. Shafer, A. Stentz, and C. E. Thorpe, "An architecture for sensory fusion in a mobile robot," in *Proc. IEEE Int. Conf. Robotics and Automat.*, San Francisco, CA, Apr. 1986, pp. 2002–2011.
[161] J. M. Lowrie, M. Thomas, K. Gremban, and M. Turk, "The autonomous land vehicle (ALV) preliminary road-following demonstrations," in *Proc. SPIE*, vol. 579, *Intelligent Robots and Computer Vision*, D. P. Casasent, Ed., Sept. 1985, pp. 336–350.
[162] T. A. Linden, J. P. Marsh, and D. L. Dove, "Architecture and early experience with planning for the ALV," in *Proc. IEEE Int. Conf. Robotics and Automat.*, San Francisco, CA, Apr. 1986, pp. 2035–2042.
[163] R. T. Dunlay, "Obstacle avoidance perception processing for the autonomous land vehicle," in *Proc. IEEE Int. Conf. Robotics and Automat.*, Philadelphia, PA, Apr. 1988, pp. 912–917.
[164] A. M. Waxman *et al.*, "A visual navigation system for autonomous land vehicles," *IEEE J. Robot. Automat.*, vol. RA-3, no. 2, pp. 124–141, 1987.
[165] D. W. Payton, "An architecture for reflexive control," in *Proc. IEEE Conf. Robotics and Automat.*, San Francisco, CA, Apr. 1986, pp. 1838–1845.
[166] F. C. A. Groen, E. R. Komen, M. A. C. Vreeburg, and T. P. H. Warmerdam, "Multisensor robot assembly station," *Robotics*, vol. 2, pp. 205–214, 1986.
[167] D. Nitzan, C. Barrouil, P. Cheeseman, and R. Smith, "Use of sensors in robot systems" in *Proc. Int. Conf. Adv. Robotics*, Tokyo, Japan, Sept. 1983, pp. 123–132.
[168] A. Choudry, W. Duinker, L. O. Hertzberger, and F. Tuijnman, "Simulating the sensory-control of an autonomous vehicle," in *Proc. 5th Int. Conf. Robot Vision and Sensory Controls*, Amsterdam, The Netherlands, Oct. 1985, pp. 147–154.
[169] J. Mochizuki, M. Takahashi, and S. Hata, "Unpositioned workpieces handling robot with visual and force sensors," in *Proc. IEEE Int. Conf. Indust. Elect. Contr. Instrumentation*, San Francisco, CA, Nov. 1985, pp. 299–302.
[170] A. Ikonomopoulos, "Trends in computer vision research for advanced industrial automation," in *Advances in Image Processing and Pattern Recognition*, V. Cappellini and R. Marconi, Eds. New York: Elsevier, 1986.
[171] A. L. Porter and F. A. Rossini, "Robotics in the year 2000," *Robotics Today*, pp. 27–28, June 1987.
[172] J. H. Kremers *et al.*, "Development of a machine-vision based robotic arc-welding system," in *Proc. 13th Int. Symp. Industrial Robots and Robots*, Chicago, IL, Apr. 1983, pp. 14.19–14.33.
[173] M. P. Howarth and M. F. Guyote, "Eddy current and ultrasonic sensors for robot arc welding," *Sensor Rev.*, vol. 3, no. 2, pp. 90–93, 1983.
[174] R. J. Didocha, D. W. Lyons, and J. C. Thompson, "Integration of tactile sensors and machine vision for control of robotic manipulators," in *Robots 9 Conf. Proc.*, Detroit, MI, June 1987, pp. 37–71.
[175] R. C. Smith and D. Nitzan, "A modular programmable assembly station," in *Proc. 13th Int. Symp. Industrial Robots and Robots*, Chicago, IL, Apr. 1983, pp. 5.53–5.75.
[176] F. C. A. Groen, E. M. Petriu, M. A. C. Vreeburg, and T. P. H. Warmerdam, "Multi-sensor robot assembly station," in *Proc. 5th Int. Conf. Robot Vision and Sensory Contr.*, Amsterdam, The Netherlands, Oct. 1985, pp. 439–448.
[177] C. C. Ruokangas, M. S. Black, J. F. Martin, and J. S. Schoenwald, "Integration of multiple sensors to provide flexible control strategies," in *Proc. IEEE Int. Conf. Robotics and Automat.*, San Francisco, CA, Apr. 1986, pp. 1947–1953.
[178] C. A. Fowler and R. F. Nesbit, "Tactical C^3, counter C^3 and C^5," *J. Electron. Defense*, Nov. 1980.
[179] J. Franklin, L. Davis, R. Shumaker, and P. Morawski, "Military applications," in *Encyclopedia of Artificial Intelligence*. New York: Wiley, 1986, pp. 604–614.
[180] P. J. Klass, "DARPA envisions new generation of machine intelligence technology," *Aviation Wk. Space Tech.*, vol. 122, no. 16, pp. 46–84, Apr. 22, 1985.
[181] J. T. Tou and R. C. Gonzales, *Pattern Recognition Principles*. Reading, MA: Addison-Wesley, 1974, pp. 111–112.
[182] T. D. Garvey and M. A. Fischler, "Perceptual reasoning in a hostile environment," in *Proc. 1st Annu. Nat. Conf. Artificial Intell.*, Stanford Univ., CA, Aug. 1980, pp. 253–255.
[183] V. G. Comparato, "Fusion—The key to tactical mission success," in *Proc. SPIE*, vol. 931, *Sensor Fusion*, C. W. Weaver, Ed., Orlando, FL, Apr. 1988, pp. 2–7.
[184] J. R. Mayersak, "An alternate view of munition sensor fusion," in *Proc. SPIE*, vol. 931, *Sensor Fusion*, C. W. Weaver, Ed., Orlando, FL, Apr. 1988, pp. 64–73.
[185] R. P. Bonasso, Jr., "ANALYST: An expert system for processing sensor returns," The MITRE Corp., McLean, VA, Tech. Rep. MTP-83W 00002, 1984.
[186] ——, "ANALYST II: A knowledge-based intelligence support system," The MITRE Corp., McLean, VA, Tech. Rep. MTP-84W 00220, 1985.
[187] K. Krishen, R. J. P. de Figueiredo, and O. Graham, "Robotic vision/sensing for space applications," in *Proc. IEEE Int. Conf. Robotics and Automat.*, Raleigh, NC, Mar. 1987, pp. 138–150.
[188] S. W. Shaw, R. J. P. de Figueiredo, and K. Krishen, "Fusion of radar and optical sensors for robotic vision," in *Proc. IEEE Int. Conf. Robotics and Automat.*, Philadelphia, PA, Apr. 1988, pp. 1842–1846.
[189] Y. Bar-Shalom and T. E. Fortmann, *Tracking and Data Association*. Boston, MA: Academic, 1988.
[190] Y. Bar-Shalom, "Tracking methods in a multitarget environment," *IEEE Trans. Automat. Contr.*, vol. AC-23, no. 4, pp. 618–626, 1978.
[191] K. C. Chang, C. Y. Chong, and Y. Bar-Shalom, "Joint probabilistic data association in distributed sensors networks," *IEEE Trans. Automat. Contr.*, vol. AC-31, no. 10, pp. 889–897, 1986.
[192] J. F. Arnold, Y. Bar-Shalom, R. Estrada, and R. A. Mucci, "Target parameter estimation using measurements acquired with a small number of sensors," *IEEE J. Oceanic Eng.*, vol. OE-8, no. 3, pp. 163–171, 1983.
[193] R. Mucci, J. Arnold, and Y. Bar-Shalom, "Track segment association with a distributed field of sensors," *J. Acoust. Soc. Amer.*, vol. 78, no. 4, pp. 1317–1324, 1985.
[194] C. Y. Chong, S. Mori, and K. C. Chang, "Information fusion in distributed sensor networks," in *Proc. American Control Conf.*, Boston, MA, June 1985, pp. 830–835.
[195] T. E. Fortmann, Y. Bar-Shalom, and M. Scheffe, "Sonar tracking of multiple targets using joint probabilistic data association," *IEEE J. Oceanic Eng.*, vol. OE-8, no. 3, pp. 173–184, 1983.

[196] J. L. Speyer, "Computation and transmission requirements for a decentralized Linear-Quadratic-Gaussian control problem," *IEEE Trans. Automat. Contr.*, vol. AC-24, no. 2, pp. 266–269, 1979.
[197] S. C. A. Thomopoulos, D. K. Bougoulias, and L. Zhang, "Optimal and suboptimal distributed decision fusion," in *Proc. SPIE*, vol. 931, *Sensor Fusion*, C. W. Weaver, Ed., Orlando, FL, Apr. 1988, pp. 26–30.
[198] S. C. A. Thomopoulos and N. N. Okello, "Distributed detection with consulting sensors and communication cost," in *Proc. SPIE*, vol. 931, *Sensor Fusion*, C. W. Weaver, Ed., Orlando, FL, Apr. 1988, pp. 31–40.
[199] S. C. A. Thomopoulos and L. Zhang, "Networking delay and channel errors in distributed decision fusion," in *Proc. SPIE*, vol. 931, *Sensor Fusion*, C. W. Weaver, Ed., Orlando, FL, Apr. 1988, pp. 154–160.
[200] ——, "Distributed filtering with random sampling and delay," in *Proc. SPIE*, vol. 931, *Sensor Fusion*, C. W. Weaver, Ed., Orlando, FL, Apr. 1988, pp. 161–167.
[201] K. R. Britting, *Inertial Navigation System Analysis*. New York: Wiley, 1971.
[202] W. S. Gesing and D. B. Reid, "An integrated multisensor aircraft track recovery system for remote sensing," *IEEE Trans. Automat. Contr.*, vol. AC-28, no. 3, pp. 356–363, 1983.
[203] Y. Barniv and D. Casasent, "Multisensor image registration: Experimental verification," in *Proc. SPIE*, vol. 292, *Proc. Images and Data from Optical Sensors*, W. H. Carter, Ed., San Diego, CA, Aug. 1981, pp. 160–171.
[204] R. H. Lee and W. B. Van Vleet, "Registration error analysis between dissimilar sensors," in *Proc. SPIE*, vol. 931, *Sensor Fusion*, C. W. Weaver, Ed., Orlando, FL, Apr. 1988, pp. 109–114.
[205] W. A. Holm, "Air-to-ground dual-mode MMW/IR sensor scene registration," in *Proc. SPIE*, vol. 782, *Infrared Sensors and Sensor Fusion*, R. G. Buser and F. B. Warren, Eds., Orlando, FL, May 1987, pp. 20–27.
[206] T. E. Bullock, S. Sangsuk-iam, R. Pietsch, and E. J. Boudreau, "Sensor fusion applied to system performance under sensor failures," in *Proc. SPIE*, vol. 931, *Sensor Fusion*, C. W. Weaver, Ed., Orlando, FL, Apr. 1988, pp. 131–138.
[207] J. A. Adam, "Star Wars in transition," *IEEE Spectrum*, vol. 26, no. 3, pp. 32–38, Mar. 1989.
[208] W. D. Hillis, *The Connection Machine*. Cambridge, MA: MIT Press, 1985.
[209] K. D. Wise, "Intelligent sensors for semiconductor process automation," in *Proc. Int. Conf. Indust. Elect. Contr. Instrumentation*, Milwaukee, WI, Sept. 1986, pp. 213–217.
[210] R. N. Stauffer, "Integrated sensors extend robotic system intelligence," *Robotics Today*, Aug. 1987.
[211] R. J. Offen, *VLSI Image Processing*. New York: McGraw-Hill, 1985.
[212] S. Middelhoek and A. C. Hoogerwerf, "Smart sensors: When and where?" *Sensors and Actuators*, vol. 8, pp. 39–48, 1985.
[213] NATO Advanced Research Workshop on Highly Redundant Sensing in Robotic Systems, Il Ciocco, Italy, May 1988.
[214] A. M. Flynn and R. A. Brooks, "MIT mobile robots—What's next?" in *Proc. IEEE Int. Conf. Robotics and Automat.*, Philadelphia, PA, Apr. 1988, pp. 611–617.

Ren C. Luo (M'83–SM'88) received the B.S. degree in mechanical engineering and the M.S. degree in electrical engineering both from Feng Chia University, Taiwan, Republic of China, in 1973 and 1975, and the M.S. and Ph.D. degrees in electrical engineering both from Technische Universitat Berlin, Berlin, Germany, in 1980 and 1982, respectively.

From 1974 to 1977 he was employed in industry and served as a Chief Engineer and Research Engineer in Victor, Inc., Taiwan, and Waldrich Siegen GmbH, Seigen, West Germany. From 1977 to 1983 he served on the scientific research staff in the Fraunhofer Institute fur Produktionstechnik and Automatisierung, Division of Industrial Robots and in the Institute fur Mess- und Regelungstechnik in the area of intelligent contact and no-contact sensors in Berlin, West Germany. From 1982 to 1983 he was a Research Project Leader in sensor-based robotics at the NSF-sponsored Robotics Research Center of the University of Rhode Island, Kingston. From 1983 to 1984 he was an Assistant Professor in the Department of Electrical Engineering and Computer Science at the University of Illinois, Chicago. Since September 1984 he has been with the Department of Electrical and Computer Engineering at North Carolina State University, Raleigh, where he is currently an Associate Professor. He is the author and coauthor of a number of technical publications in refereed journals, book chapters, national, and international conference proceedings. These articles deal with a variety of sensor design, sensor data processing, computer vision based 3-D object recognition, and multisensor fusion and integration applied to intelligent robotics systems.

Dr. Luo is a member of Verein Deutsche Ingineure (VDI), West Germany, and SPE and a senior member of RI/SME. He is also the Technical Committee Chairman for transducers and actuators at the IEEE Industrial Electronics Society and Associate Editor of IEEE Transactions on Industrial Electronics. He is listed in *Who's Who in the World* and *Who's Who in Frontiers of Science and Technology*.

Michael G. Kay received the B.A. degree in economics and the M.S. degree in industrial and systems engineering both from the University of Florida, Gainesville, in 1981 and 1984, respectively. He is currently pursuing the Ph.D. degree in industrial engineering at North Carolina State University, Raleigh.

He has worked as a Consultant to municipal utilities for rate making and planning purposes.

Integration of Sonar and Stereo Range Data Using a Grid-Based Representation

Larry Matthies **Alberto Elfes**

Computer Science Department and Robotics Institute
Carnegie-Mellon University
Pittsburgh, PA 15213

Abstract

Multiple range sensors are essential in mobile robot navigation systems. This introduces the problem of integrating noisy range data from multiple sensors and multiple robot positions into a common description of the environment. We propose the use of a cellular representation called the *Occupancy Grid* as a solution to this problem. In this paper, we use Occupancy Grids to combine range information from sonar and one-dimensional stereo into a two-dimensional map of the vicinity of a robot. Each cell in the map contains a probabilistic estimate of whether it is *empty* or *occupied* by an object in the environment. These estimates are obtained from sensor models that describe the uncertainty in the range data. A Bayesian estimation scheme is applied to update the current map using successive range readings from each sensor. The Occupancy Grid representation is simple to manipulate, treats different sensors uniformly, and models uncertainty in the sensor data and in the robot position. It also provides a basis for motion planning and creation of more abstract object descriptions.

1. Introduction

A key issue in autonomous vehicle research is the design of representations and mechanisms for interpreting and integrating range data. Mobile robots travel over extensive areas and must combine sensor views obtained from many different locations into a single, consistent world model. Individual range sensors are subject to limitations that can be overcome only by a multi-sensory approach; therefore, world model representations must accommodate information from different sensors supplying qualitatively different range information. Furthermore, models must cope gracefully with sensor errors and must represent uncertainty arising from sensor noise and from uncertain knowledge of the robot's location over time.

Past mobile robot research has used a variety of world model representations. As in CAD and computer graphics, there is a distinction between surface or boundary-based representations and volumetric or space-filling representations [23]. Some systems recover only sparse sets of 2D or 3D points in space [17, 27]. Others use boundary methods in the form of line segments [6, 7, 11, 13], wire frames [29], or polyhedra [28]. While volumetric descriptions have been used to represent objects in some robot perception applications [2], they have seen little use in mobile robot work to date. One example is the Occupancy Grid approach we have used in previous sonar [9] and stereo [24] work. Related divisions of space into buckets have been used to sort laser scanner data prior to recovering surface descriptions [28].

By fitting data to parameterized models, surface or boundary-based representations impose strong geometric assumptions on the sensor data. This requires segmenting the data into groups and selecting the model most appropriate for each group. Doing this early in the interpretation process is problematic because it commits the system to rigid descriptions based on insufficient evidence. Sensors currently available for unstructured environments are quite noisy; stereo makes matching errors and sonar and laser are coarse and strongly affected by surface characteristics. This makes early geometric interpretation difficult. It also underscores the importance of redundant data for noise reduction and the importance of representing confidence in the raw data and the derived representations.

In previous work, we have addressed these issues by developing a volumetric description called the *Occupancy Grid* for interpreting low level sensor data [9, 10]. Occupancy Grids represent the space around a robot by an array of cells in which each cell carries an estimate of the confidence that it is occupied by objects in the environment. Confidence values vary over a continuous interval, with values at one extreme implying that a cell is empty, values at the other extreme implying that a cell is occupied, and values near the center implying that the state of a cell is unknown. Probabilistic models of sensor data are used to derive stochastic estimates for cell occupancy and to update cells based on successive range measurements. This representation has been used successfully in separate systems for incrementally constructing maps from wide-angle sonar [9] and one-dimensional stereo [24], as well as for motion planning [9, 8, 27], robot position estimation [9, 18] and incorporation of motion uncertainty in the map-building process [10].

In this paper, we emphasize the use of Occupancy Grids for sensor integration. We begin by reviewing the basic representation and the procedures for composing maps from successive sensor readings. We then summarize separate sonar and stereo systems we have built using Occupancy Grids, show maps constructed with each sensor independently, and present results obtained with the sensors combined. The combination successfully exploits complementarity of the sensors, distills valid world models from coarse and noisy data, and copes with conflicting measurements in a consistent manner. The results show the promise of this technique as a low-level mechanism for creating shape representations from noisy sensor data. A great deal of work remains to be done with the representation, for example in formalizing its statistical basis and in extracting higher-level representations from the final map. We touch on these issues at the end of the paper.

2. Using Occupancy Grids and Bayesian Estimation in Sensor Integration

2.1. A Paradigm for Sensor Integration

The traditional paradigm for recovering spatial information from range data emphasizes the extraction of a geometric model directly from the sensor readings. We propose an alternate paradigm that uses Occupancy Grids as an intermediate representation, avoiding early commitment to geometric descriptions. Within this approach, we identify the following components of the data interpretation and integration process [10] (Fig. 1):

- **A Spatial Interpretation Model**, developed for each kind of sensor, translates the sensor data into a statement about areas or volumes of space that are occupied or empty. By incorporating the sensor uncertainty in the form of the conditional probability density function $p($sensor reading $R\,|$state of environment$)$, we arrive at a **Probabilistic Sensor Model.** Using this probabilistic model in the interpretation of each sensor reading produces a grid-based world description called a *sensor-based view*.

This research has been supported by the Office of Naval Research under Contract N00014-81-K-0503. Larry Matthies is supported by a graduate fellowship from FMC Corporation. Alberto Elfes is supported in part by a graduate fellowship from the Conselho Nacional de Desenvolvimento Científico e Tecnológico - CNPq, Brazil, under Grant 200.986-80, and in part by the Mobile Robot Lab, Robotics Institute, Carnegie-Mellon University.

The views and conclusions contained in this document are those of the authors and should not be interpreted as representing the official policies, either expressed or implied, of the funding agencies.

Reprinted from the *Proceedings of the IEEE International Conference on Robotics and Automation*, 1988, pages 727-733.
Copyright © 1988 by The Institute of Electrical and Electronics Engineers, Inc. All rights reserved.

EH0341-8/91/0000/0232$01.00 © 1988 IEEE

- To *compose* views taken from different sensor positions into a unified world model, we must know the coordinate transformation between each position. A **Sensor Position Uncertainty Model** provides a measure of the uncertainty in these transformations. This positional uncertainty is combined with the sensor uncertainty when views are composed.
- A **Sensor Map Updating Model** composes views provided by the same kind of sensor into a single grid representation. If the robot carries several sensors of the same kind, as it does with sonar, combining these in the robot coordinate frame provides a *robot-based view*. Composing robot-based views creates a *global view*. A simple Bayesian updating model is used in both the sonar-based and the stereo-based mapping systems.
- To *integrate* data provided by qualitatively different sensors, which in our case are sonar and stereo, we also need a **Sensor Integration Model**. Expressing data from each sensor in the grid representation allows us to use a Bayesian mechanism for integrating different sensors that is similar to the one used for composing different views. This uniformity between view composition and sensor integration is very convenient.
- A **Decision-Making Model** is needed when cells must be explicitly labelled as OCCUPIED, EMPTY or UNKNOWN. For this, we apply a simple maximum likelihood or thresholding decision rule to the occupancy probability stored in each cell.

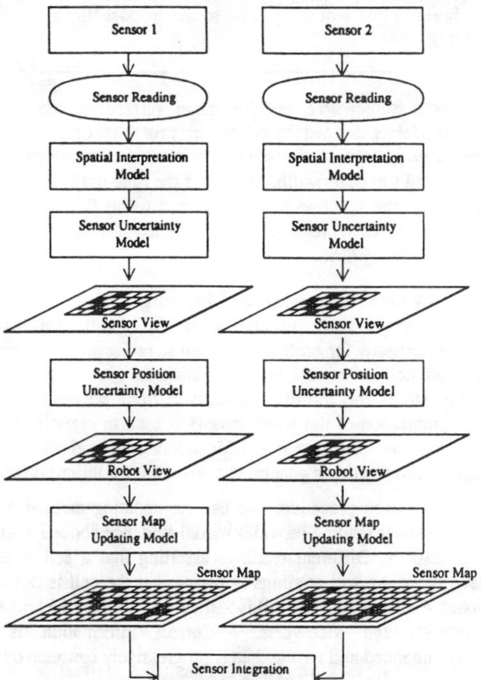

Figure 1: A Framework for Occupancy Grid-Based Sensor Integration.

2.2. Occupancy Grids

Occupancy Grids are 2D or 3D tesselations of space into cells, where each cell contains a probabilistic estimate of its occupancy. Specifically, each cell C of a map has an associated state variable s(C). s(C) is a discrete random variable with two states, *occupied* and *empty*, denoted OCC and EMP. Since the states are exclusive and exhaustive, $P(s(C) = \text{OCC}) + P(s(C) = \text{EMP}) = 1$. In other words, each cell has an associated probability mass function that is estimated by the sensor interpretation and integration process.

Updating the Occupancy Grid — in other words, the mapping process itself — is done with a Bayesian estimation model [10]. A new sensor reading R_{t+1} introduces additional information about the state $s(C)$ of a cell C in the form of the conditional probability density function $p(R_{t+1}|s(C))$. This information is combined with the most recent probability estimate stored in the cell, $P(s(C)|\{R\}_t)$, based on the current set of readings $\{R\}_t=\{R_t, \ldots, R_0\}$, to give a new estimate $P(s(C)|\{R\}_{t+1})$, which is based on all the information available so far. In this way, Occupancy Grids provide a natural way to update the world model by composing information from sensor readings taken from different views.

Because Occupancy Grids supply a common underlying representation for the interpretation of qualitatively different sensors, such as stereo vision and sonar, they also provide a natural framework for sensor integration. Grids can be built separately for each sensor, then merged into an integrated map of the environment, or the probabilistic models that interpret the spatial statements for each sensor can be used directly to update a single map. Since all sensor readings have a common interpretation and make comparable statements in this framework, the sensor integration problem becomes relatively straightforward.

2.3. Developing a Bayesian Model for Updating the Occupancy Grid

Building a map from range data involves reasoning with uncertain pieces of information. There are a number of approaches to inference under uncertainty, including probability theory, Dempster-Shafer theory, and fuzzy sets [30]. We have been influenced by the cogent argument for probability theory presented in [5].

For a given sensor, we derive a Probabilistic Sensor Model in the form of the conditional distribution p(sensor reading R | world is in state S_k). The state S_k of the world is described by the set of states of all cells in the map. In the two-dimensional case, for a map with n^2 cells, each with two possible states, it would be necessary to specify 2^{n^2} conditional probabilities for each reading. To avoid this combinatorial explosion, we assume that the cell states are independent discrete random variables, so that $P(S{=}S_k) = \prod_i P(s(C_i))$, and the state of the map is determined by estimating the state of each cell individually.

For the updating procedure, we start with Bayes' Theorem [1]:

$$P(s_i|e) = \frac{P(e|s_i)\,P(s_i)}{\sum_j P(e|s_j)\,P(s_j)}$$

where s_i is one of n disjoint states being estimated, e is the relevant evidence, $P(s_i)$ is the *a priori* probability of the system being in state i, and $P(e|s_i)$ is the probability that evidence e would be present given that the system is in state s_i. $P(s_i|e)$ is what we need for decision-making, namely the conditional probability that the system is in state i in light of the evidence e.

In our case, the evidence is given by a sensor range reading R and the desired probabilities are $P(s(C) = \text{OCC} \mid R)$ and $P(s(C) = \text{EMP} \mid R)$, which we abbreviate to $P(\text{OCC} \mid R)$ and $P(\text{EMP} \mid R)$. Since $P(\text{OCC} \mid R) = 1 - P(\text{EMP} \mid R)$, we need to specify only one of the probabilities. For all cells in the field of view of the sensor we can express Bayes' Theorem as:

$$P(\text{OCC} \mid R) = \frac{p(R \mid \text{OCC})\,P(\text{OCC})}{p(R \mid \text{OCC})\,P(\text{OCC}) + p(R \mid \text{EMP})\,P(\text{EMP})}$$

and for sequential updating of the map based on multiple readings we write:

$$P(\text{OCC} \mid \{R\}_{k+1}) = \frac{p(R_{k+1} \mid \text{OCC})\,P(\text{OCC} \mid \{R\}_k)}{p(R_{k+1} \mid \text{OCC})\,P(\text{OCC} \mid \{R\}_k) + p(R_{k+1} \mid \text{EMP})\,P(\text{EMP} \mid \{R\}_k)}$$

A similar formula can be derived for the combination of estimates provided by different sensors. Notice that $P(\text{OCC} \mid \{R_k\})$ and $P(\text{EMP} \mid \{R_k\})$, the current prior probabilities that a cell is occupied or empty, are taken from the existing map. The conditional probabilities $p(R \mid \text{OCC})$ and $p(R \mid \text{EMP})$ are determined from the Probabilistic Sensor Model. Formal definitions of these conditionals may not be easy to formulate, since in general the reading R depends on the state of the world as a whole, and cannot be specified independently for each cell. However, we shall see later that approximations perform fairly well.

To initialize a map, we have used non-informative or maximum entropy priors [1], which in this context correspond to assigning equal probability to each state. In other words, the initial map cell prior probabilities are

$P(\text{OCC}) = P(\text{EMP}) = 0.5$, which implies that the state of cell C is UNKNOWN. Different priors can be used if more *a priori* information about the robot's environment is known.

For the experiments discussed later in this paper, we have made the additional simplifying assumption that $p(R \mid \text{EMP}) = 1 - p(R \mid \text{OCC})$. This can be justified in some cases, but is not true in general. However, it provides us with a very simple updating formula. Noting that $P(\text{EMP}) = 1 - P(\text{OCC})$, the formula above becomes:

$$P(\text{OCC} \mid R) = \frac{p(R \mid \text{OCC})\, P(\text{OCC})}{p(R \mid \text{OCC})\, P(\text{OCC}) + (1 - p(R \mid \text{OCC}))\,(1 - P(\text{OCC}))}$$

This formula has several useful properties:

- It is commutative and associative, which means that data in a multisensory system can be incorporated in any order.
- Combining evidence $E = p(R|\text{OCC})$ with the prior probability for UNKNOWN, $P(\text{OCC}) = 0.5$, gives E as result.
- Conflicting measurements cancel: combining *occupied* evidence E^+ with *empty* evidence E^- (of the same strength) produces UNKNOWN.

For decision-making, we use the unconditional *maximum-likelihood* estimate [3]. Other criteria, such as the minimum-cost estimate, can be used. Depending on the application, it may be of interest to define an UNKNOWN band, as opposed to a single threshold value. Note, however, that in many contexts explicitly labelling cells as OCCUPIED or EMPTY may be unnecessary. In motion planning, for example, the cost of a trajectory can be defined by a risk factor directly related to the cell probabilities [9, 10].

3. Sonar-Based Mapping

In this section, we provide a brief overview of the sonar-based mapping system used in our sensor integration experiments. This system was originally developed to provide sonar-based mapping and navigation for an autonomous mobile robot operating in unknown and unstructured environments. A previous version is discussed in [18, 8, 9], where the sonar system, the map building process, the path-planning and navigation mechanisms, the overall architecture of the system and experimental results from indoor and outdoor runs are presented. The system has no *a priori* map of its surroundings. Instead, it interprets the sonar data to build maps of the robot's operating environment. These maps are then used for path-planning, obstacle avoidance and position estimation. Additionally, geometric information can be extracted from the sonar maps to provide a higher-level geometric description of the robot's environment.

3.1. The Sonar Sensor Array

The sonar devices are Polaroid laboratory grade ultrasonic range transducers. Their measuring range spans distances from 0.9 to 35.0 ft [22]. The main lobe of the sensitivity function is contained within a solid angle Ω of 30°, where it falls off to −38 dB. The beamwidth ω at −3 dB is approximately 15°. Range accuracy of the sensors is on the order of 1 %. The associated control circuitry only reports the distance to the first strong reflector. The sonar sensor array consists of a ring of 24 transducers, spaced 15° apart. For indoor experiments, we used the *Neptune* mobile robot [21] (Fig. 2); for outdoor runs, a larger robot vehicle called the *Terregator* [12] was used.

3.2. Interpreting the Sonar Sensor Data

There are a number of difficulties associated with processing sonar data, mainly due to characteristics of the sensor itself. These include low angular resolution and errors due to multiple reflections or specular reflections away from the sensor. This precludes a direct interpretation of the sonar readings and originally led us to consider a probabilistic approach to the recovery of spatial information from sonar range data, using a discrete grid to store the inferences.

A sonar reading R corresponds to a range value r returned by a sensor positioned at (x,y,θ). This reading is interpreted as making an assertion about two volumes in 3D space: one that is *probably empty*, and one that is *somewhere occupied*. To build a two-dimensional map, we only consider the horizontal cross-section of the sonar beam, and define a conditional probability function $P(\text{cell } C \text{ is occupied} \mid \text{sensor reading } R)$ over this domain. Informally, this function measures our confidence that cells inside the cone of the beam are empty, and our uncertainty about the location of the object that

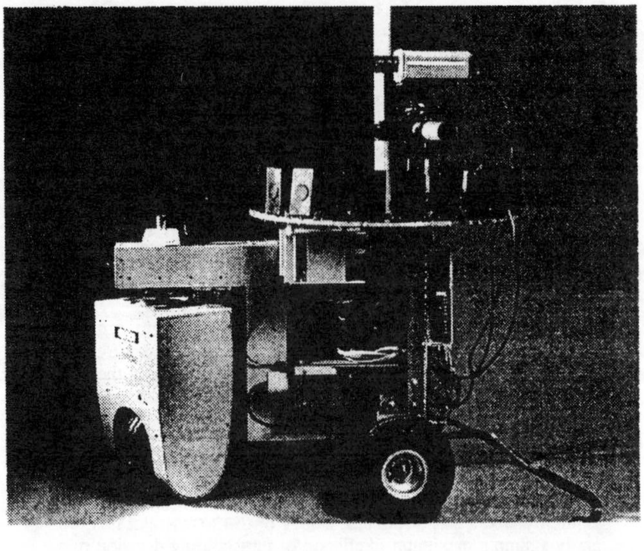

Figure 2: The *Neptune* mobile robot, with a pair of cameras and the sonar ring. For experiments in sensor integration, the cameras were mounted lower, so that their horizon line would be close to the cross-sectional view provided by the sonar ring.

caused the echo, somewhere on the range surface of the cone. This probabilistic model is defined based on the geometry of the beam and the spatial sensitivity pattern of the sonar sensor, and is parameterized by the range distance and the beamwidth. A plot of the function is shown in Fig. 3. For map building, the function can be reversed using Bayes' Theorem; it is then evaluated for each reading and projected on the horizontal two-dimensional Occupancy Grid.

3.3. Occupancy Grid-Based Sonar Mapping

The sonar map is built by computing the conditional sonar sensor probability distributions for each range reading, projecting these probabilities onto the discrete cells of a *view*, and combining the view with the sonar map, which already stores the information derived from previous readings. The position and orientation of the sonar sensors is used to register the view with the map. Typical cell sizes are 0.5 ft for indoor runs and 1.0 ft for outdoor runs, though we have at times generated maps with resolutions down to 0.1 ft.

For the combination procedure, we use the updating method described in section 2. Each sonar reading provides partial evidence about a map cell being OCCUPIED or EMPTY. Different readings asserting that a cell is EMPTY will confirm each other, as will readings implying that the cell is OCCUPIED. On the other hand, evidence that the cell is EMPTY will weaken the probability of it being OCCUPIED and vice-versa. Correct information is therefore incrementally enhanced and wrong data is progressively canceled out.

The resulting sonar maps are very useful for navigation. They are much denser than the ones generated by previous sparse stereo vision programs [17, 27] and computationally about an order of magnitude faster to produce. We have implemented an autonomous navigation system [8, 9] that uses an A*-based path-planner to obtain routes in these maps. Additionally, polygonal object outlines can be extracted from the sonar maps [9].

Fig. 6(a, b) shows experimental results of the sonar mapping procedure. These are discussed in Section 5, together with results from the stereo system and from the integration of both sensors.

4. Stereo-Based Mapping

The stereo system uses two cameras mounted side by side just above the sonar ring (Fig. 2). The cameras are aligned to make their optical axes co-planar, with this plane parallel to the floor of the room. As the robot moves, the portion of the room lying in this plane always projects to approximately the same scanline in the images; we call this scanline the *horizon line*. The stereo system processes a narrow band of scanlines around the horizon line to

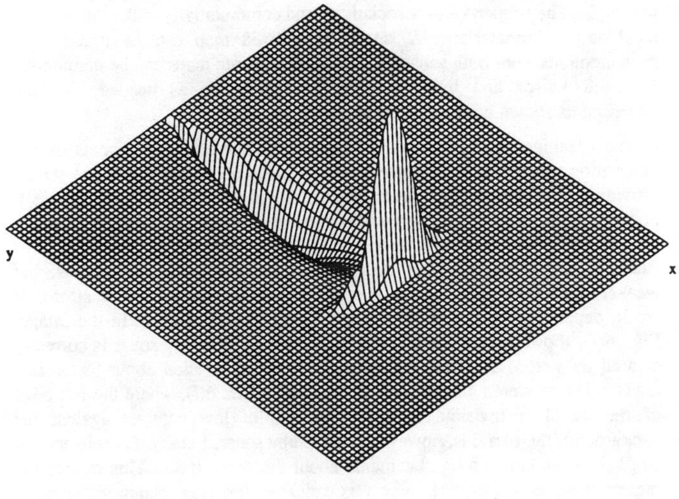

Figure 3: Probabilistic Sonar Sensor Interpretation Model. The probability profile shown corresponds to a reading taken by a sensor positioned at the upper left, pointing to the lower right. The plane shows the UNKNOWN level. Values above the plane represent OCCUPIED probabilities, and values below represent EMPTY probabilities.

produce a 2-D world model commensurate with sonar. Mounting the cameras close to the sonar ring ensures that both sensors detect the same 2-D portion of the world.

Stereo range information is complementary to sonar. Stereo detects and localizes surface boundaries and surface markings, but misses the interior of unmarked surfaces. On the other hand, wide-angle sonar gives better detection of broad surface structure and indicates large *empty* areas between the robot and nearby objects. However, it gives less definition to surface boundaries and fine surface structure. With our current systems, stereo can see through clutter but sonar cannot. Finally, range *uncertainty* differs for the two sensors. Stereo localizes features well in the direction perpendicular to the line of sight and less well along the line of sight, whereas sonar has much more uncertainty perpendicular to the line of sight than along it.

Obtaining Occupancy Grids from stereo images involves three stages of processing: (1) matching features near the horizon line to obtain range data, (2) ascribing an uncertainty model to the range data, and (3) using the range data and the uncertainty model to create an Occupancy Grid. For the first two steps, we match features with a single-scanline version of the Ohta-Kanade dynamic programming stereo algorithm [20], then use approximate models of triangulation error to estimate uncertainty in the resulting range values. We interpret the camera field of view and the range values as a wedge of information analogous to the sonar cones. For each stereo pair, this information is converted to the grid format and merged with the central map as described in section 2. We summarize the main steps of these processes below; more details can be found in [24].

4.1. Obtaining Stereo Correspondence with Dynamic Programming

The stereo matcher proceeds as follows:

- Stereo pairs are digitally warped and resampled to correct alignment errors in the cameras. This guarantees that corresponding epipolar lines lie on corresponding scanlines.
- Near-vertical edges are extracted with a horizontal gradient operator based on the Canny edge detector [4].
- A band of five scanlines centered on the horizon line is extracted from the images, then the stereo matching algorithm is applied to match edge points *independently* on each scanline. The edges matched on all scanlines are combined to create a single row of matched edge points.
- Edges that extend across all five scanlines and that are matched consistently on all scanlines are used to constrain the stereo matcher in the next pair of images. This is done by reducing the range of stereo disparities searched for matches in the dynamic programming algorithm.

The result of this process is a set of edge points on the horizon line where depth is known. Joining these points with line segments creates a depth profile that approximates the surface structure in the scene.

4.2. Ascribing Uncertainty to the Range Profile

We create Occupancy Grids from the continuous range profile output by the stereo matcher. By analogy to sonar, the range profile creates a wedge with the cameras at the apex, the boundaries of the field of view forming the sides of the wedge, and the range profile forming the base (Fig. 4). In general terms, the interior of the wedge is inferred to be empty, an area around the base is inferred to be occupied, and the space outside the wedge is considered to be unknown as far as this stereo pair is concerned.

Aside from correspondence errors, there are two sources of uncertainty in the range profile. One arises from measurement errors in determining edge positions and the other from "inference errors" in assuming that line segments between edges correspond to real surfaces. Eventually, we wish to treat measurement error by determining probability distributions for the position of the range profile [16]. Integrating such distributions over the area of cells in the grid will provide the confidence values needed to build the map. At present we use the following, simpler approach. It has been shown [14] that for a given stereo disparity d, corresponding to a depth Z, an error of δd in disparity yields (approximately) an error of $\delta Z = Z\,\delta d / d$ in the depth estimate. This defines an "uncertainty band" around the depth profile that is wide where the profile is far from the cameras and narrow where the profile is near (Fig. 4). The width of this band defines the probability that cells underlying the band are occupied according to the formula:

$$P(C = \text{OCC} \mid R) = P_{UNK} + \frac{P_{OCC} - P_{UNK}}{N} \times \alpha$$

where $P_{UNK} = 0.5$ is the UNKNOWN probability level, $P_{OCC} \in (0.5, 1.0)$ is a constant indicating the highest occupied probability that will be assigned, and N is the width of the uncertainty band in units of cells, measured along the line of sight from the cameras. α is used to account for inference error; it equals 1.0 at edge points, where depth has actually been measured, and decreases with distance from edge points along line segments in the depth profile.

4.3. Creating an Occupancy Grid

An Occupancy Grid is built from the range profile obtained from each stereo pair. Cells on the interior of the field of view wedge are assigned a fixed empty probability $P_{EMP} \in (0.0, 0.5)$. Within the uncertainty band around the range profile, cells are assigned occupied estimates by scan-converting the range profile and computing probabilities as outlined above. Line segments that are close to parallel to the line of sight are likely to span occluding boundaries and not correspond to actual surfaces; therefore they are treated as unknown. The resulting grid for the current stereo pair is merged with the overall map using the same procedure applied for sonar.

Stereo maps generated by this process (Fig. 6(c, d)) are discussed in the following section.

5. Integration of Sonar and Stereo

In this section, we discuss results obtained with the sonar and stereo systems individually and with both sensors combined. The experimental setup used the *Neptune* robot configured as shown in Fig. 2. The cameras were mounted as close to the sonar ring as possible, so that the sonar and stereo maps would portray the same horizontal slice of the world. The robot was driven on a straight line through the CMU Mobile Robot Lab (Fig. 5), and stopped at 70 positions to gather joint sonar and stereo data. Maps generated at the first and tenth position are shown in Fig. 6.

5.1. Sonar Maps

The sonar maps (Fig. 6(a, b)) show that a single set of sonar readings provides a substantial amount of information about empty spaces, but is weak in recovering object shape. Therefore, immediate motion decisions are possible, but only through the composition of several views is more spatial structure recovered. Additionally, the wide angle of the beam sometimes precludes the detection of free space between close objects, and phantom obstacles can be generated due to multiple reflections. The system operates

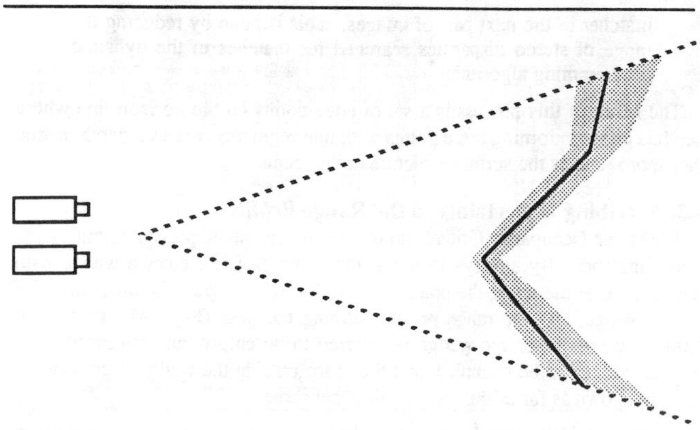

Figure 4: Stereo Depth Profile. The stereo field of view (dotted lines) and range profile (solid line) form a wedge of information analogous to the sonar cone. The shaded region is the uncertainty band caused by measurement error.

best when views are taken from substantially different positions. Over several sets of readings sonar is usually able to recover good descriptions of surfaces, but does not provide adequate definition of sharp spatial features such as corners.

5.2. Stereo Maps

The stereo results are shown in Fig. 6(c, d). Empty space is represented quite well. Nearby occupied areas show up immediately, but distant areas require either several views or a closer approach before cells acquire a high occupied confidence. This reflects the variation in the width of the uncertainty band around the range profile. As in any method, errors are introduced into the map if the stereo algorithm produces correspondence errors. However, these are ameliorated by processing many frames, since the same error is unlikely to be repeated with each frame and the updating mechanism inherently cancels conflicting information. The "weak model" nature of the grid representation side-steps difficult questions faced by representations with stronger geometry; for example, there is no need to segment images into straight lines. This also implies that stereo algorithms based on other matching techniques, such as correlation, can use this representation directly. The grid representation is compatible with sparse or dense stereo data.

Figure 5: The CMU Mobile Robot Lab, shown with objects present at the joint sonar and stereo run.

5.3. Integrated Maps

For sensor integration, we used the same composition formula presented in section 2. The properties of associativity and commutativity make the order of combination immaterial. A single, centralized map can be updated by measurements from both sonar and stereo, or separate maps can be maintained for each sensor and integrated into a single map as needed. For the experiments shown here, the latter approach was used.

The combined maps shown in Fig. 6(e, f) illustrate three facets of the integration process. First, the sensors *complement* each other, with one sensor providing information about areas inaccessible to the other sensor (Fig. 6(e)). Second, the sensors can *correct* each other, when weak false inferences made by one sensor coincide with strong true inferences made by the other. For example, sonar makes strong statements about emptyness of regions, but weaker statements about occupied areas. Stereo statements can be strong or weak, depending on the distance to or distribution of features in the image. Fig. 6(f), shows a case where a region seen as occupied by sonar is correctly cleared by stereo. Similarly, sonar can recover information about featureless areas, whereas stereo cannot. This is the case in Fig. 6(f), where the left edge of the barrel is invisible to stereo because of low contrast against the background; the barrel is, however, detected by sonar. Finally, the sensors can *conflict* by making strong statements about the same space. This moves the region towards UNKNOWN, which is valuable for later planning, since it correctly signals the fact that sensor information does *not* provide an unambiguous interpretation for a given area. Future systems may detect such conflicts and use them to direct the attention of the sensors.

It should be noted that the totally different nature of the raw data provided by both sensors, as well as the qualitatively different information encoded in the range data they provide, precludes any simple analytic or geometric approach to their integration into a coherent world model. Occupancy Grids provide a means to make the data commensurate and comparable.

6. Discussion

We have reviewed the probabilistic formulation of the Occupancy Grid, shown its use in combining range measurements from different views for two separate sensor systems, and demonstrated its utility in integrating range measurements from both sensors into a single unified representation of the space around a robot. The results of the previous section show that the representation exploits the complementary spatial coverage of the sensors and that the map update process extracts a reasonable world description from noisy data.

A key feature of the Occupancy Grid representation is the simplicity and uniformity with which it treats sensor data. Range measurements from each sensor are converted directly to a common representation that makes statements about spatial occupancy. Once in the Occupancy Grid representation, data from different views and different sensors are combined naturally to map out free space and extend occupied regions. Sensors are treated modularly; to incorporate a new sensor, we create a probabilistic model for it that makes statements about occupancy and compartmentalize the sensor in a module that outputs a map to a higher-level integration system.

The uncertainty model embodied in Occupancy Grids, which is essential for the data combination process, has several useful properties. With well designed sensor models, it offers the possibility of taking into account different levels of accuracy for each sensor. It also allows us to account for uncertainty in the robot's position over time. We model this positional uncertainty with multivariate gaussian distributions [25], which form the basis of a convolution-based approach to composing sensor and robot position uncertainty by blurring occupied regions of previous maps before blending in a new view [10].

Several extensions to the work presented here are possible. Although the relatively simple error models and update schemes used in this paper have been fairly successful, we are exploring a more sophisticated statistical formulation of the representation. The extension of the Occupancy Grid to 3-D is natural [10]. One possible application would be in a 3-D sonar system for mobile robots proposed in [19]. In fact, Occupancy Grids have been recently used in the interpretation of sonar data from undersea surveys [26] and are currently being applied to the interpretation of laser range data for land-based navigation [15]. We also anticipate using 3-D maps to represent information from extensions of stereo to full images. Finally, we expect that it will be

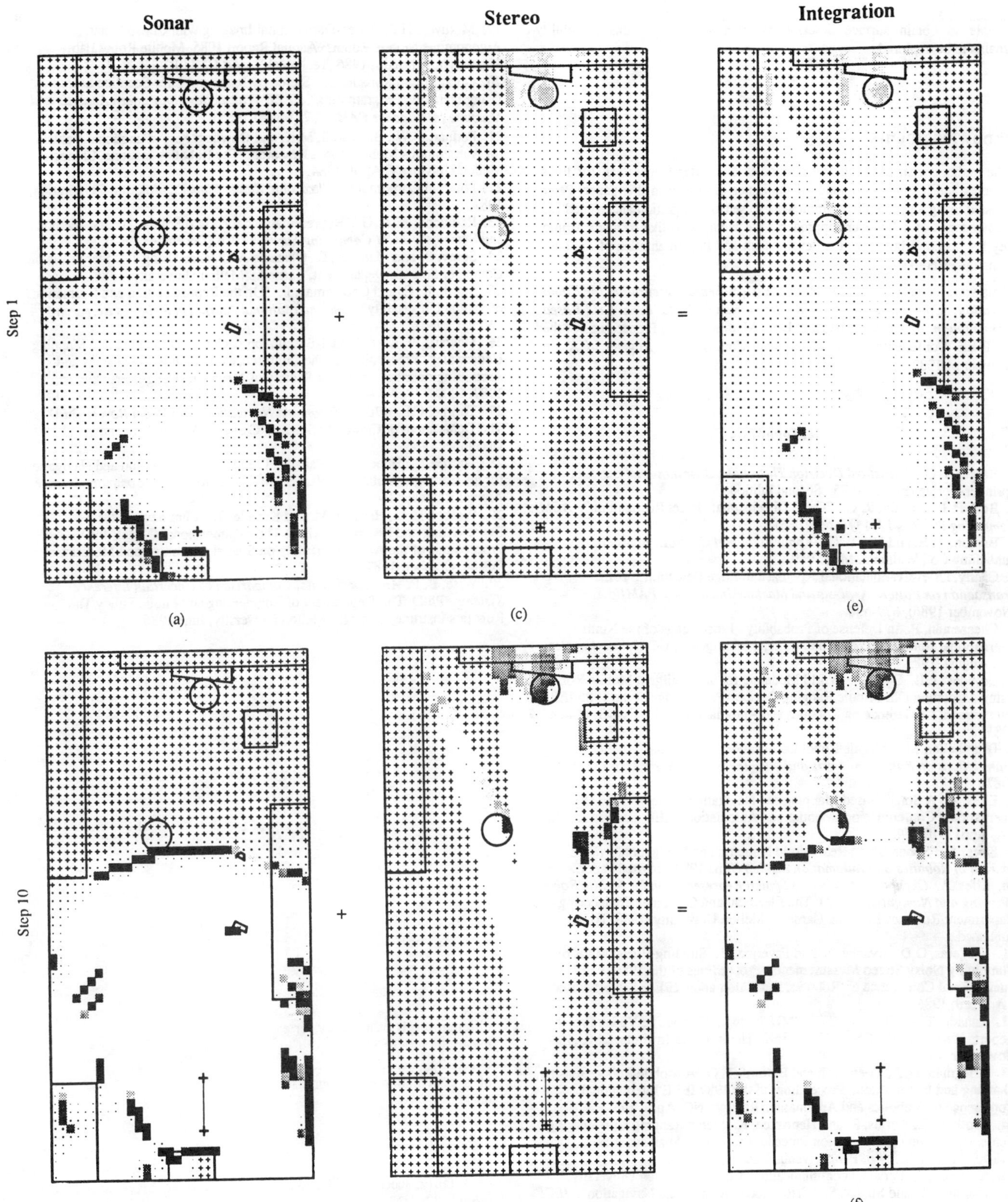

Figure 6: Occupancy Maps Generated by Sonar, Stereo and Sensor Integration. OCCUPIED regions are marked by shaded squares, EMPTY by dots fading to white space, and UNKNOWN by + signs.

possible to obtain surface descriptions from such maps; this possibility remains to be explored.

Acknowledgments

The authors wish to thank Hans Moravec and Peter Cheeseman for their comments and insights into some of the issues discussed in this paper. Peter suggested using a probabilistic framework to sensor integration and argued for the advantages of a Bayesian approach. We would also like to thank Bruno Serey for his contribution to the development and implementation of the one-dimensional stereo system.

This work is an updated version of the paper *Sensor Integration for Robot Navigation: Combining Sonar and Stereo Range Data in a Grid-Based Representation*, by Alberto Elfes and Larry Matthies, in the Proceedings of the 26th IEEE Conference on Decision and Control, Los Angeles, CA, in December of 1987.

References

1. Berger, J. O.. *Statistical Decision Theory and Bayesian Analysis.* Springer-Verlag, Berlin, 1985. Second Edition.

2. Besl, P. J. and Jain, R. C. "Three-Dimensional Object Recognition". *ACM Computing Surveys 17*, 1 (March 1985).

3. Bryson, A. E. and Ho, Y. C.. *Applied Optimal Control.* Blaisdell Publishing Co., Waltham, MA, 1969.

4. Canny, J. "A Computational Approach to Edge Detection". *IEEE Transaction on Pattern Analysis and Machine Intelligence PAMI-8*, 6 (November 1986), 679-698.

5. Cheeseman, P. In Defense of Probability. Proceedings of the Ninth International Joint Conference on Artificial Intelligence, IJCAI, Los Angeles, CA, August, 1985.

6. Crowley, J. L. Dynamic World Modeling for an Intelligent Mobile Robot Using a Rotating Ultra-Sonic Ranging Device. Proceedings of the 1985 IEEE International Conference on Robotics and Automation, St. Louis, MI, March, 1985.

7. Drumheller, M. "Mobile Robot Localization Using Sonar". *IEEE Transactions on Pattern Analysis and Machine Intelligence PAMI-9*, 2 (March 1987).

8. Elfes, A. A Sonar-Based Mapping and Navigation System. 1986 IEEE International Conference on Robotics and Automation, IEEE, San Francisco, CA, April 7-10, 1986.

9. Elfes, A. "Sonar-Based Real-World Mapping and Navigation". *IEEE Journal of Robotics and Automation RA-3*, 3 (June 1987). Invited Paper..

10. Elfes, A. *Occupancy Grids as a Spatial Representation for Mobile Robot Mapping and Navigation.* Ph.D. Th., Electrical and Computer Engineering Department/Robotics Institute, Carnegie-Mellon University, 1988. To be published..

11. Faugeras, O.D., Ayache, N. and Faverjon, B. Building Visual Maps by Combining Noisy Stereo Measurements. Proceedings of the 1986 IEEE International Conference on Robotics and Automation, IEEE, San Francisco, CA, April, 1986.

12. Kanade, T. and Thorpe, C.E. CMU Strategic Computing Vision Project Report: 1984 to 1985. CMU-RI-TR-86-2, The Robotics Institute, CMU, November, 1985.

13. Kriegman, D. J., Triendl, E. and Binford, T. O. A Mobile Robot: Sensing, Planning and Locomotion. Proceedings of the 1987 IEEE International Conference on Robotics and Automation, Raleigh, NC, April, 1987.

14. Krotkov, E., Kories, R. and Henriksen, K. Stereo Ranging with Verging Cameras: A Practical Calibration Procedure and Error Analysis. MS-CIS-86-86, University of Pennsylvania, December, 1986.

15. Kweon, I. S. Personal communication. Carnegie-Mellon University.

16. Matthies, L. and Shafer, S. "Error Modeling in Visual Navigation". *IEEE Journal of Robotics and Automation RA-3*, 3 (June 1987), 239-248.

17. Moravec, H.P. *Obstacle Avoidance and Navigation in the Real World by a Seeing Robot Rover.* Ph.D. Th., Stanford University, September 1980.

18. Moravec, H.P. and Elfes, A. High-Resolution Maps from Wide-Angle Sonar. IEEE International Conference on Robotics and Automation, IEEE, St. Louis, March, 1985. Also published in the 1985 ASME International Conference on Computers in Engineering, Boston, August 1985..

19. Moravec, H.P. Three-Dimensional Imaging with Cheap Sonar. Autonomous Mobile Robots: Annual Report 1985, Mobile Robot Lab, Pittsburgh, February, 1986. Technical Report CMU-RI-TR-86-4.

20. Ohta, Y. and Kanade, T. "Stereo by Intra- and Inter-Scanline Search Using Dynamic Programming". *IEEE Transactions on Pattern Analysis and Machine Intelligence PAMI-7*, 2 (March 1985).

21. Podnar, G.W., Blackwell, M.K. and Dowling, K. A Functional Vehicle for Autonomous Mobile Robot Research. The Robotics Institute, Carnegie-Mellon University, April, 1984.

22. Polaroid Corporation. *Ultrasonic Range Finders.* Polaroid Corporation, 1982.

23. Requicha, A. A. G. "Representations for Rigid Solids: Theory, Methods, and Systems". *ACM Computing Surveys 12*, 4 (December 1980).

24. Serey, B. and Matthies, L. Obstacle Avoidance Using 1-D Stereo Vision. Computer Science Department, Carnegie-Mellon University, 1986.

25. Smith, R. C. and Cheeseman, P. "On the Representation and Estimation of Spatial Uncertainty". *The International Journal of Robotics Research 5*, 4 (Winter 1986).

26. Stewart, W. K. A Model-Based Approach to 3-D Imaging and Mapping Underwater. Proceedings of the Conference on Offshore Mechanics and Arctic Engineering, American Society of Mechanical Engineers, August, 1987.

27. Thorpe, C.E. *FIDO: Vision and Navigation for a Robot Rover.* Ph.D. Th., Department of Computer Science, Carnegie-Mellon University, December 1984.

28. Thorpe, C., Hebert, M., Kanade, T. and Shafer, S. Vision and Navigation for the Carnegie-Mellon NAVLAB. In *Annual Review of Computer Science*, Annual Reviews, Inc., 1987.

29. Walker, E. L., Herman, M. and Kanade, T. A Framework for Representing and Reasoning About Three-Dimensional Objects for Vision. Proceedings of the AAAI Workshop on Spatial Reasoning and Multisensor Fusion, October, 1987.

30. Wise, B. P. *An Experimental Comparison of Uncertain Inference Systems.* Ph.D. Th., Department of Engineering and Public Policy/The Robotics Institute, Carnegie-Mellon University, June 1986.

Chapter 3: Mapping and Real-World Modeling

An autonomous mobile robot must map its environment and recover a sensor-based representation of that environment. Therefore, robot mapping links fundamentally to robotic perception. At the same time, the specific representations and operations performed on robot maps, as well as the kind of information extracted from sensor data, are strongly influenced by the task set for which robots will utilize maps.

Besl and Jain's paper, while not directly tailored to the mobile robotics domain, nevertheless provides an excellent tutorial on different representations for three-dimensional object modeling. Asada presents recent work on a multilevel approach to sensor-based robot mapping that incorporates lower level, sensor-based models and higher level, more abstract representations.

Estimating and updating the uncertainty contained in information encoded in maps is an important issue in spatial models. Smith et al. discuss stochastic frameworks for representing and estimating uncertainty in sensor-derived spatial map features.

Leonard et al. take a new approach to sonar interpretation and dynamic map building, while Ikegami et al. discuss a new representation based on object border distances.

Basye et al. examine topological maps and methods for coping with uncertainty in map-learning strategies. For certain robot tasks, Kuipers and Byun suggest that a map need not encode precise spatial information. They go on to discuss a topological representation.

Three-Dimensional Object Recognition

PAUL J. BESL AND RAMESH C. JAIN

Department of Electrical Engineering and Computer Science, The University of Michigan, Ann Arbor, Michigan 48109-1109

A general-purpose computer vision system must be capable of recognizing three-dimensional (3-D) objects. This paper proposes a precise definition of the 3-D object recognition problem, discusses basic concepts associated with this problem, and reviews the relevant literature. Because range images (or depth maps) are often used as sensor input instead of intensity images, techniques for obtaining, processing, and characterizing range data are also surveyed.

Categories and Subject Descriptors: I.2.10 [**Artificial Intelligence**]: Vision and Scene Understanding—*intensity, color, photometry, and thresholding; modeling and recovery of physical attributes; perceptual reasoning; representations, data structures, and transforms;* I.3.5 [**Computer Graphics**]: Computational Geometry and Object Modeling—*curve, surface, solid, and object representations;* I.4.6 [**Image Processing**]: Segmentation—*edge and feature detection; pixel classification; region growing, partitioning;* I.4.8 [**Image Processing**]: Scene Analysis—*depth cues; range data;* I.5.4 [**Pattern Recognition**]: Applications—*computer vision*

General Terms: Algorithms, Design, Theory

Additional Key Words and Phrases: Range images, depth maps, surface characterization, surface matching, 3-D object recognition, 3-D object reconstruction

INTRODUCTION

One goal of computer vision research is to give computers humanlike visual capabilities so that machines can sense the environment in their field of view, understand what is being sensed, and take appropriate actions as programmed. Vision involves the physical elements of illumination, geometry, reflectivity, and image formation, as well as the intelligence aspects of recognition and understanding. Most computer vision research performed during the past 20 years has concentrated on using digitized gray-scale intensity images as sensor data. Digitized intensity images are arrays of numbers that indicate the *brightness* at points on a regularly spaced grid and contain no explicit information about *depth*. People are able to correctly infer depth relationships between image regions quickly and easily, but automatic inference of such relationships has proved to be quite difficult.

In recent years digitized range data have become available from both active and passive sensors, and the quality of these data has been steadily improving. Range data are often produced in the form of an array of numbers, referred to as a *range image* (*or depth map*), where the numbers quantify the distances from the sensor focal plane to object surfaces within the field of view along rays emanating from points on a regularly spaced grid. Not only are depth relationships between range-image regions

A preliminary version of this paper, "Range Image Understanding," by Paul Besl and Ramesh Jain, appears in the *Proceedings of the Computer Vision and Pattern Recognition Conference* (San Francisco, Calif., June 19–23, 1985), pp. 430–451. ©IEEE 1985. Any portions of this paper that appear in the conference proceedings are reprinted with permission.

"Three Dimensional Object Recognition" by P.J. Besl and R.C. Jain from *Computing Survey*, Vol. 17, No. 1, March 1985, pages 75-145. Copyright © 1985 by The Association for Computing Machinery, Inc., reprinted by permission.

CONTENTS

explicit, the three-dimensional (3-D) *shape* of image regions approximates the 3-D shape of the corresponding object surfaces in the field of view. Since correct depth information depends only on geometry and is independent of illumination and reflectivity, intensity-image problems with shadows and surface markings do not occur. Therefore, the process of recognizing objects by their shape should be less difficult in range images than in intensity images.

Three-dimensional object recognition is closely examined in this paper. An outline of the material we cover follows:

(a) Autonomous single-arbitrary-view 3-D object recognition is defined and treated in mathematical terms. It is intended as a worthwhile goal that computer vision systems might achieve.

(b) The necessary components of a recognition system are discussed from a general, qualitative point of view. Characteristics of an ideal system for solving the object-recognition problem are proposed.

(c) Model representation and rendering are briefly reviewed, followed by image formation and image processing. Because techniques for obtaining and processing range data are fairly new in comparison with the corresponding techniques for intensity-image data, these methods are considered in more detail. The literature on the central topics of surface characterization, object reconstruction, and object recognition is then reviewed.

1. PROBLEM DEFINITION

Three-dimensional object recognition is a rather nebulous term. A brief survey of the literature on this subject demonstrates this point [Brooks 1982; Casasent et al. 1982; Douglass 1981; Fang et al. 1982; Guzman 1968; Ikeuchi 1981; Oshima and Shirai 1983; Sadjadi and Hall 1979; Sato and Honda 1983; Shirai and Suwa 1971; Wallace and Wintz 1980.] Some schemes handle only single presegmented objects, whereas others can interpret multiple-object scenes. However, some of these other schemes are really performing two-dimensional (2-D) processing using 3-D information. There are systems that require objects to be placed on a turntable during the recognition process. A few published methods even require that intermediate data be provided by the person operating the system. Some techniques assume that idealized data will be available from sensors *and* intermediate processors. Others require high-contrast or backlit scenes. Most efforts have limited the class of recognizable objects to polyhedra, spheres, cylinders, cones, generalized cones, or a combination of these. Many papers fail to mention how well the proposed method can recognize objects from a large set of objects (e.g., at least 20 objects). Therefore, we attempt to give a reasonably precise definition of the object-recognition problem. We first discuss human visual capabilities and then describe how these relate to computer vision.

The real world that we see and touch is primarily composed of solid *objects.* When people are given an object they have never seen before, they are typically able to gather information about that object from many different viewpoints. The process of gathering detailed object information and storing that information is referred to as *model formation.* Once we are familiar with many objects, we can normally identify them from an *arbitrary viewpoint* without further investigation.

People are also able to *identify, locate, and qualitatively describe the orientation* of objects in black-and-white photographs. This basic capability is significant to computer vision research because it involves the spatial variation of only a *single* parameter within a framed rectangular region corresponding to a fixed, single view of the real world. Human color vision is more difficult to analyze and is typically treated as a *three*-parameter color variation within a large, almost hemispherical, solid angle that corresponds to a continually changing viewpoint. Because we are interested in an automatic computerized recognition process, sensor input data must be compatible with digital computers. The term *digitized sensor data* is used to refer to an input matrix of numerical values (which can represent intensity, range, or any other scalar parameter) and associated auxiliary information concerning how that matrix of values was obtained.

The above motivates the following definition of the autonomous single-arbitrary-view 3-D object-recognition problem:

(1) *Given* any collection of labeled solid objects, (a) each object may be examined as long as the object is not deformed; (b) labeled models may be created using information from this examination.

(2) *Given* digitized sensor data corresponding to one particular, but arbitrary, field of view of the real world as it existed at the time of data acquisition, *given* any data stored previously during the model formation process, and *given* the list of distinguishable objects, the following issues must be addressed for each object using the capabilities of a single autonomous processing unit:
   - (a) Does the object appear in the digitized sensor data?
   - (b) If so, how many times does it occur?
   - (c) For each occurrence,
     - (i) determine the location in the sensor data,
     - (ii) determine the 3-D location (or translation parameters) of that object with respect to a known coordinate system (if possible with given sensor), and
     - (iii) determine the 3-D orientation (or rotation parameters) with respect to a known coordinate system (if possible with given sensor).

(3) (Optional) If there are regions in the sensor data that do not correspond to any objects in the list, characterize these regions in a way that will make them recognizable should they occur in future images. Presumably, there is an object present that is not known to the system. Ideally, the system should attempt to learn whatever it can about the unknown object from the given view.

We refer to the problem of successfully completing these tasks using real-world sensor data and obeying the given constraints as the 3-D object-recognition problem. This problem is not successfully addressed by the object-recognition systems discussed in the literature; more constrained problems, which are limited to particular surface types or applications, are normally addressed. If the stated 3-D object-recognition problem could be solved successfully by a vision system, that system would be extremely useful in a wide variety of applications, including automatic inspection and assembly, and autonomous vehicle navigation. The problem is stated so that it may be feasible to use computers to solve the problem, and it is also clearly solvable by human beings.

Item (3) above can be interpreted as a partial model-building task to be performed on data that cannot be explained in terms of known objects. It cannot be interpreted as recognition because something is present in the sensor data that the system "knows" nothing about. We are asking a vision system to "learn from experience" in a flexible way.

How do we know whether a given approach solves the problem, and how can we compare different approaches to see whether one is better than another? The performance of object-recognition systems could be measured using the number of errors made by a system in performing the assigned problem tasks on standardized sets of digitized sensor data that challenge the capabilities mentioned in the problem definition. The following list enumerates some of the possible types of errors that are made by such systems.

(1) Miss error: An object's presence is not detected.
(2) False alarm error: The presence of an object is indicated even though there is no evidence of this in the sensor data.
(3) Location error: An object occurrence is correctly identified, but the location of the object is wrong.
(4) Orientation error: The object occurrence and position are determined correctly, but the orientation is wrong.

In the comparison of different object-recognition systems, the term "successful" can be made quantitative by establishing a performance index that quantitatively combines the number, type, and magnitude of the various errors. The information in the literature makes it practically impossible to compare existing object-recognition systems quantitatively because different researchers do not evaluate their systems in any consistent manner. Hence, this literature survey focuses on the basic features of each reviewed approach. Experimental results are mentioned wherever possible. We refrain from subjective comparisons of disparate techniques.

2. MATHEMATICAL PROBLEM FORMULATION

It is often beneficial to define a problem in a stricter mathematical form to eliminate possible problem ambiguities. For example, we have not yet discussed how a system should respond if several distinct objects appear to be identical from a given viewpoint. Therefore, we now redefine *depth-map object recognition* in precise mathematical terms as a *generalized inverse set mapping*. Intensity-image object recognition can then be treated in the same formalism.

First, we consider world-modeling issues. We approximate the world as consisting of N_{tot} objects. The number of distinguishable objects is N_{obj}. Hence, $N_{obj} \leq N_{tot}$. (Two objects are not distinguishable if a person cannot tell them apart using only shape cues.) We refer to the ith distinguishable object as A_i. The number of occurrences, or instances, of that object is denoted as N_i. This means that, in general,

$$N_{tot} = \sum_{i=1}^{N_{obj}} N_i.$$

People can recognize an enormous number of objects depending on personal experience. The number of objects to be recognized by an object-recognition system depends on the application and system training.

It is sometimes difficult to decide what is an object and what is an assembly of objects. To resolve this difficulty, each object could possess its own coordinate system and list of subobjects. We currently consider only simple objects with no subparts and with only one occurrence (or instance) in this discussion. Therefore, $N_{obj} = N_{tot}$. The general case of multiple instances of objects with subparts is not conceptually different; nevertheless, it does present notation problems and important implementation difficulties for higher level recognition processing. We define the origin of the object coordinate system at the center of mass of the object with three orthogonal axes aligned with the principal axes of the object because these parameters can be precisely determined for any given solid object or solid object model.

Each object occupies space, and at most one object can occupy any given point in space. It is necessary to describe the spatial relationships between each object and the rest of the world. One way to describe spatial relationships is through the use of coordinate systems. For reference purposes, we assume the existence of a world coordinate system that is placed at any convenient location. Objects are positioned in space relative to this coordinate system by means of translation and rotation parameters. We refer to the translation parameters of an object as the vector $\boldsymbol{\alpha}$ and to the rotation parameters of an object as the vector $\boldsymbol{\theta}$. The number of parameters for each vector depends on the *dimension* of the depth-map recognition problem. For example, the 2-D problem requires a total of three parameters. For the 3-D case, we write the necessary six parameters as follows:

$$\boldsymbol{\alpha} = (\alpha, \beta, \gamma) \quad \text{and} \quad \boldsymbol{\theta} = (\theta, \phi, \psi).$$

See Figure 1 for the meaning of these parameters. We define our *world model* W as a set of ordered triples (object, translation, rotation):

$$W = \{(A_i, \boldsymbol{\alpha}_i, \boldsymbol{\theta}_i)\}_{i=0}^{N_{obj}}.$$

We consider object A_0 to be the sensor object with position $\boldsymbol{\alpha}_0$ and orientation $\boldsymbol{\theta}_0$. If a time-varying world model is required, all objects and their parameters can be functions of time. For our current purposes of single-view object recognition, we concern ourselves only with static parameter values. We denote the set of all objects, the *object list*, as $L = \{A_i\}$. The set of all translations is denoted R^t, and the set of all rotations is denoted R^r. In the 3-D object-recognition problem, $t = 3$ and $r = 3$. (In 2-D, $t = 2$ and $r = 1$.) R is the set of all real numbers.

A depth sensor (or range finder) obtains a depth-map projection of a scene. We model this projection as a mathematical operator P, which maps elements in the set $\Omega = L \times R^t \times R^r$ into elements in the set of all scalar functions of $t - 1$ variables, which we denote as F:

$$P: \Omega \rightarrow F.$$

These real-valued functions are referred to as depth-map functions. This projection operator might be *orthographic* or *perspective* [Foley and van Dam 1982; Newmann and Sproull 1979]. We write the projection as

$$f(\mathbf{x}) = g_{A,\boldsymbol{\alpha},\boldsymbol{\theta}}(\mathbf{x}) = P(A, \boldsymbol{\alpha}, \boldsymbol{\theta}),$$

where $\mathbf{x}$ is the vector of $t - 1$ spatial variables of the focal plane of the sensor. The spatial parameters of the sensor object (the location $\boldsymbol{\alpha}_0$ and the orientation $\boldsymbol{\theta}_0$) are *implicitly* assumed arguments of the projection operator. This is done to simplify our expressions because we have only one sensor in this formalism. We refer to the depth-map function as f when the identity of the object and its parameters are unknown. The symbol g with subscripts refers to the depth-map function of a known object at a known location and orientation. This notation demonstrates that the set of depth-map functions associated with a single object is an *infinite family of functions*. Two of the rotation parameters in the $\boldsymbol{\theta}$ vector have a particularly profound effect on this family of functions: The 3-D shape of the depth-map function changes as the object rotates. Translation parameters have no effect on shape whatsoever under orthographic projection, and they have negligible effect under the perspective projection unless the sensor is very close to the object of interest (e.g., closer than 10 times the maximum object width).

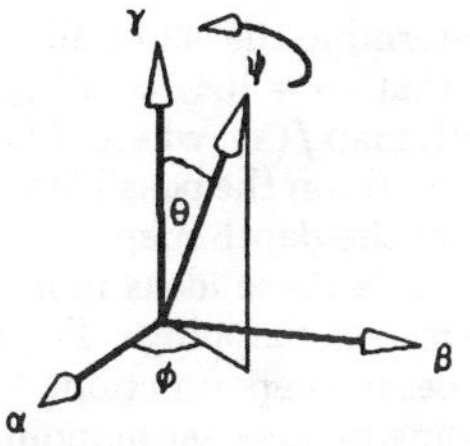

Figure 1. Rigid objects in 3-D space have 6 degrees of freedom. Translation = $\boldsymbol{\alpha}$ = (α, β, γ); rotation $\boldsymbol{\theta}$ = (θ, ϕ, ψ).

Since objects do not occupy all space, we need a convention for the value of the depth-map function for values of the spatial vector $\mathbf{x}$ that do not correspond to object surface points. If the point $(\mathbf{x}, f(\mathbf{x}))$ cannot lie on an object surface, we assign the value of $-\infty$ to $f(\mathbf{x})$. Hence, we write the projection of a set of M objects as

$$f(\mathbf{x}) = \max_{1 \le i \le M} g_{A_i,\boldsymbol{\alpha}_i,\boldsymbol{\theta}_i}(\mathbf{x}).$$

The depth-map object recognition problem is now rephrased as follows: Given a depth-map function $f(\mathbf{x})$, which results from the depth-map projection of a 3-D world scene, determine the sets of possible objects with the corresponding sets of translation and rotation parameters that could be projected to obtain the given depth-map function.

That is, determine the set of all ω_J, subsets of Ω, such that $\omega_J = \{(A_j, \alpha_j, \theta_j)\}_{j \in J}$ projects to the depth map $f(\mathbf{x})$, where J is an index set that depends on the possible valid interpretations of the depth map.

We can write these ideas more precisely using inverse-set mappings. For each *single-object* depth-map function, there is a corresponding inverse-set mapping to yield all single objects that could have created the given object depth-map function. We denote the inverse-set mapping of P as P^{-1}, where

$$P^{-1}(f(\mathbf{x})) = \{(A, \alpha, \theta) \in \Omega \mid P(A, \alpha, \theta) = f(\mathbf{x})\}.$$

An inverse-set mapping takes sets from the power set of the range of the original mapping into sets in the power set of the domain:

$$P^{-1}: 2^F \rightarrow 2^\Omega.$$

For our purposes we restrict the input sets in 2^F to be singletons (i.e., single depth-map functions); therefore, we replace 2^F with F. For multiple-object depth-map functions, we must generalize P^{-1} due to possible combinations of objects, as shown in Figure 2. Hence, given $f(\mathbf{x}) \in F$, we seek a generalized inverse-set mapping $\mathbf{P}^{-1}$ such that

$$\mathbf{P}^{-1}(f(\mathbf{x})) = \{\omega_J \subseteq 2^\Omega \mid \max_{j \in J} P(A_j, \alpha_j, \theta_j) = f(\mathbf{x})\}.$$

Thus, the mapping we seek takes elements of the depth-map function space into the power set of the power set of Ω:

$$\mathbf{P}^{-1}: F \rightarrow 2^{2^\Omega}.$$

The depth-map object-recognition problem can now be stated in terms of the modeled world as follows: Given the world model W with N_{obj} simple objects and given any realizable depth-map function $f(\mathbf{x})$, compute the inverse projection set mapping $\mathbf{P}^{-1}(f(\mathbf{x}))$ to obtain all possible explanations of the function $f(\mathbf{x})$ in terms of the world model. In many cases, there is only one valid scene interpretation. Nonetheless, a general-purpose vision system must generate the list of all valid scene interpretations whenever ambiguous single-view situations are encountered. The problem of determining the next best view from which sensor data should be acquired so as to eliminate ambiguity is a separate issue and has been addressed by Connolly [1985] and Kim et al. [1985].

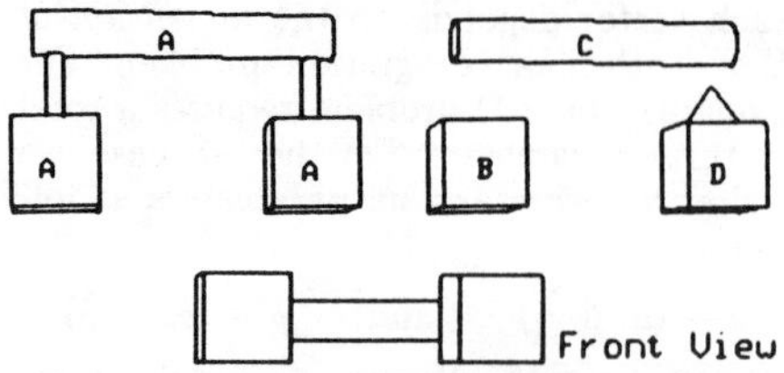

Figure 2. Different valid interpretations of a simple scene. $P^{-1} = \{\{A\}, \{B, C, D\}, \{D, C, B\}\}$.

Since there is no general theory regarding the computation of this mapping, researchers are free to choose their own methods. We propose that the generalized inverse-set mapping can best be computed by recognizing the *individual surface regions* associated with the $g_{A,\alpha,\theta}(\mathbf{x})$ *depth-map surface function families* of the individual objects. That is, we propose that the *object-recognition* problem be solved by decomposing it into a *surface-characterization* problem combined with a *surface-matching* algorithm constrained by known object structures. Although previous work has been done in both of these areas, more research is required to combine these concepts into a working solution.

This formalism is now augmented to state the *object-recognition problem for intensity images*. By adding an illumination–reflectance operator I to the depth-map function f, we obtain an intensity image $\iota(\mathbf{x})$ given by

$$\iota(\mathbf{x}) = I(\max_i P(A_i, \alpha_i, \theta_i)).$$

The intensity-image object-recognition problem is generally much more difficult owing to the additional inversion of the I operator. One needs a priori knowledge of all surface reflectances and all illumination sources of a scene (in addition to object shapes) to be able to invert this I operator. Note that the *max* computation is the multiple-object occlusion operator. To expand our world model for understanding intensity images, it is necessary to add objects that generate light in addition to the single sensor object that receives light. Shape from shading [Ikeuchi and Horn 1981], shape from photometric stereo [Coleman and Jain 1982; Woodham 1981], and shape from texture [Witkin 1981] techniques attempt to uniquely invert the I operator, producing the depth-map function f.

It is interesting to observe that people have the capacity for understanding images even when shading operators other than the illumination–reflectance operator are used to shade visible surfaces. For example, people can correctly interpret photographic negatives or pseudocolored images where the usual color and/or light–dark relationships are completely distorted.

3. RECOGNITION SYSTEM COMPONENTS

The specific tasks to be performed by an object-recognition system are given in the two previous sections. We also suggest that we can measure how well these tasks are performed. Now we must discuss how these tasks can be accomplished.

Recognition implies awareness of something already known. How can one recognize something unless one knows what one is looking for? Even though model formation is not specifically required in the problem definition, it is demanded by the circumstances of the problem. Many different kinds of models, both view independent and view dependent, have been used for modeling real-world objects for recognition purposes. We survey different view-independent object representations because representation is such a critical factor in object-recognition-system design. We do not survey *view-dependent* techniques as a separate topic because these types of representations are clearly not advisable for single-*arbitrary*-view recognition. For now, we assume that the necessity of model formation has been established and that a representation scheme is required to store object model data. That is, a *world model* is a necessary object-recognition-system component.

Once one knows what object to look for, how can one find it in the digitized sensor data? In order to determine how recognition will take place, a method for matching the model data to the sensor data must be considered. A straightforward blind-search approach would entail transforming all possible combinations of all possible known object models in all possible distinguishable orientations and all possible distinguishable locations into the digitized sensor data format and computing a matching error quantity to be minimized. The minimum matching error configuration of object models would correspond to the recognized scene. All the tasks mentioned in the statement of the object-recognition problem would be accomplished, except possibly the characterization of unknown regions not corresponding to known objects. Of course, this would take an enormous amount of processing time, even for the simplest scenes. A better algorithm is needed.

Since object models contain more object information than the sensor data, we are prohibited from transforming sensor data into *complete* model data and matching in the model data format. However, this does not prevent one from matching with *partial* model data. As a result of this problem and the natural desire to reduce the large dimensionality of the input sensor data, it is advantageous to work with an intermediate domain that is computable from both sensor and model data. This domain is referred to as the *symbolic scene description domain.* In the literature review, it is seen that sensor data are processed until they reach the form of symbolic scene description. The model data can also be transformed into an equivalent symbolic scene description. A matching procedure can then be carried out on the quantities in this intermediate domain, which are referred to as *features.* The best matching results occur when the hypothetical object model configuration accurately represents the real-world scene represented in the sensor data. We propose that a matching procedure and intermediate symbolic scene description mechanisms are necessary object-recognition-system components.

Interactions between the individual components of a recognition system are diagrammed in Figure 3. The real world, the

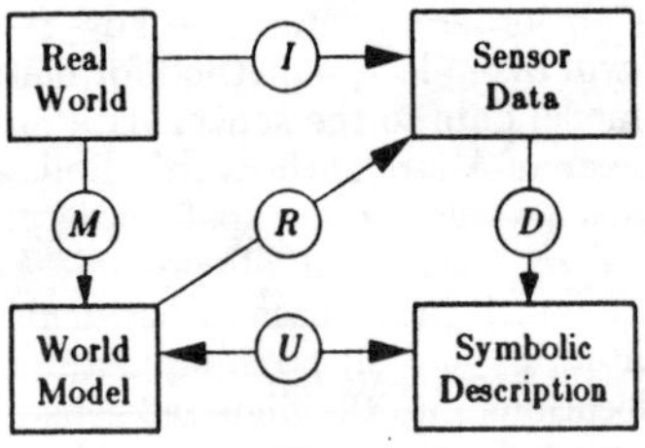

Figure 3. General object-recognition system structure. *I*, image-formation process; *M*, world-modeling process; *D*, description process; *U*, understanding process; *R*, model-rendering process.

digitized sensor data domain, the modeling domain, and the intermediate symbolic scene description are the fundamental domains of the system. Mappings between these domains are listed: The image-formation process (*I*) creates intensity or range data based purely on physical principles. The description process (*D*) acts on the sensor data and extracts relevant application-independent features. This process should be completely data driven and incorporate only the knowledge of the image-formation process and rudimentary facts about the real world. No a priori assumptions about real-world object shapes should be incorporated. The modeling process (*M*) provides object models for real-world objects. Object reconstruction from sensor data is one method for building models automatically. This method is preferred for a fully automated system but is not required by our problem definition, which addresses visual recognition, not modeling per se. Hence, object models could also be constructed manually using geometric solid modeling programs. The understanding, or recognition, process (*U*) involves an algorithm to perform matching between models and data descriptions. This process might include interacting data- and model-driven subprocesses, where segmented sensor data regions seek explanations in terms of models and hypothesized models seek verification from the data. The rendering process (*R*) produces synthetic sensor data from object models. A vision system should be able to communicate its understanding of a scene verbally *and visually* to any person querying the system. Rendering also provides an important feedback link because it allows an autonomous system to check on its own understanding of the sensor data by comparing synthetic images to the sensed images.

It is proposed that any object-recognition system can be discussed within the framework of this system model. This description is in agreement with the ideas brought forth by Brooks [1982], except for the addition of the rendering process.

4. CHARACTERISTICS OF AN IDEAL SYSTEM

What are the characteristics of the ideal system that handles the object-recognition problem as defined? Capabilities that might be realized by object-recognition systems in the near future are summarized below.

(1) The system must be able to handle sensor data from arbitrary viewing directions without giving preference to horizontal, vertical, or other directions. This requires a view-independent modeling scheme that is compatible with recognition processing requirements.

(2) The system must handle arbitrarily complicated real-world objects without giving preference to either curved or planar surfaces.

(3) The system must handle arbitrary combinations of a relatively large number of objects in arbitrary orientations and locations without being sensitive to superfluous occlusions (i.e., if occlusion does not affect human understanding of a scene, then it should not affect the ideal automated object-recognition system either).

(4) The system must be able to handle a certain amount of noise in the sensor data without a significant degradation in system performance.

(5) The system should be able to analyze scenes quickly and correctly.

(6) It should not be difficult for the system to modify the world model data in order to handle new objects and new situations.

(7) It is desirable for the system to be able to express its confidence in its own understanding of the sensor data.

We certainly do not imply that any existing vision systems have all of these capabilities, nor do we imply that the research necessary to build such a system has been completed. Our motivation is to describe a general-purpose object-recognition system in detail before reviewing the methods and systems discussed in the literature.

5. LITERATURE REVIEW

The existing literature and subject matter relevant to 3-D object recognition is now reviewed within the framework established by the previous sections. Individual papers are considered within the context of a general subject area. We consider the following subject areas relevant to the object-recognition problem:

(1) 3-D object-representation schemes,
(2) 3-D surface-representation schemes,
(3) 3-D object- and surface-rendering algorithms,
(4) Intensity- and range image formation,
(5) Intensity- and range image processing,
(6) 3-D surface characterization,
(7) 3-D object-reconstruction algorithms,
(8) 3-D object-recognition systems using intensity images,
(9) 3-D object-recognition systems using range images.

This paper is an expanded version of Besl and Jain [1985c] and is not intended as an overview of computer vision. Gevarter [1983] presents a short, easily understandable introduction to the subject. The book by Ballard and Brown [1982] is a reasonable place to start for the more serious reader who is not already familiar with the field. In addition, several overview papers survey computer vision and treat 3-D issues, but they consider *only intensity images* as input [Bajcsy 1980; Barrow and Tennenbaum 1981; Brady 1981, 1982; Rosenfeld 1984]. Binford [1982] has written a survey of model-based intensity-image analysis systems. Dyer and Chin [1984] have reviewed 2-D and 3-D object-recognition literature, with an emphasis on industrial applications.

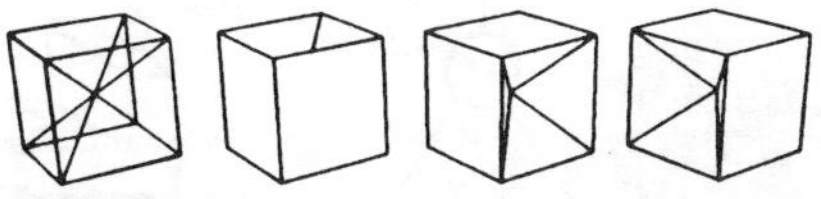

Figure 4. Wire-frame model and three valid object interpretations.

5.1 Object Representation

In order to recognize a particular object, one needs to know how that object may appear in the sensor data context. Computers can "understand" 3-D object structure and appearance through the use of view-independent object models and associated rendering algorithms. How can such a 3-D model be stored in a computer? Different object-representation schemes give distinct answers to this question.

We now review the basic categories of 3-D object representation so that it is easier to understand certain limitations of several object-recognition systems reviewed subsequently. First, we discuss object representations used primarily by systems where the goal is automated rendering of realistic digital images using models (i.e., existing computer *graphics* representations). Then we look at representations used by systems where the goal is automated understanding of digital images using models (i.e., existing computer *vision* representations). Both kinds of systems need the same basic type of geometric object information, but the storage, utilization, and level of detail of that information is quite different.

The object representations commonly used by contemporary computer-aided-design (CAD) geometric solid-object-modeling systems are categorized as one of the following:

(1) *Wire-frame representation.* A wire-frame representation of a 3-D object consists of a 3-D vertex point list and an edge list of vertex pairs, or it can be formatted as such. This representation was quite common owing to its simplicity. Although fast wire-frame displays are still popular, solid modeling has replaced wire-frame modeling because wire frame is an ambiguous representation for determining such quantities as the surface area and volume of an object. See Figure 4 for an example of this wire-

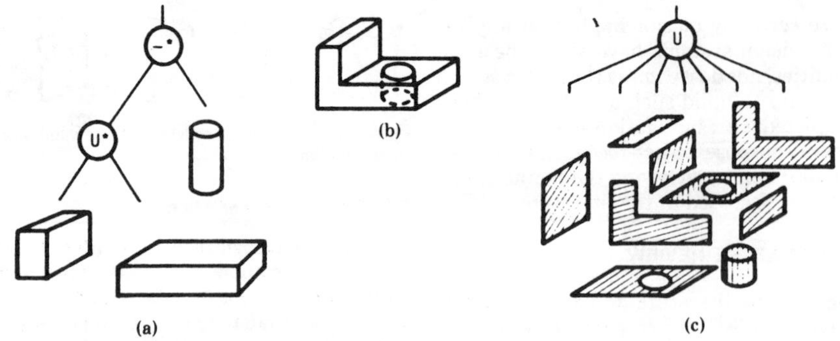

Figure 5. CSG and surface-boundary representations for a solid. (a) A CSG representation; (b) solid; (c) a boundary representation. (From Requicha and Voelcker [1983]; ©IEEE 1983.)

frame ambiguity. Wire-frame models can sometimes be interpreted as different solid objects or as different orientations of the same object.

(2) *Constructive solid geometry representation (CSG)*. The CSG representation of an object is specified in terms of a set of 3-D volumetric primitives (blocks, cylinders, cones, and spheres are typical examples) and a set of Boolean operators: union, intersection, and difference. See Figure 5 for an example of a CSG description of an object. The storage data structure is a binary tree, where the terminal nodes are instances of primitives and the branching nodes represent Boolean set operations and positioning information. CSG trees define object volume and surface area unambiguously and are capable of representing complex objects with a very small amount of data. However, the boundary evaluation algorithms required to obtain usable surface information are very computationally intensive. Also, general sculptured surfaces, such as the human face surface shown in Figure 6, are not easily represented using CSG modelers. A general-purpose modeling system must be able to represent such surfaces.

(3) *Spatial-occupancy representation*. Spatial-occupancy representations use *nonoverlapping* subregions of the 3-D space occupied by an object to define that object. This method unambiguously defines an object's volume. The following single-primitive representations of this type are commonly used:

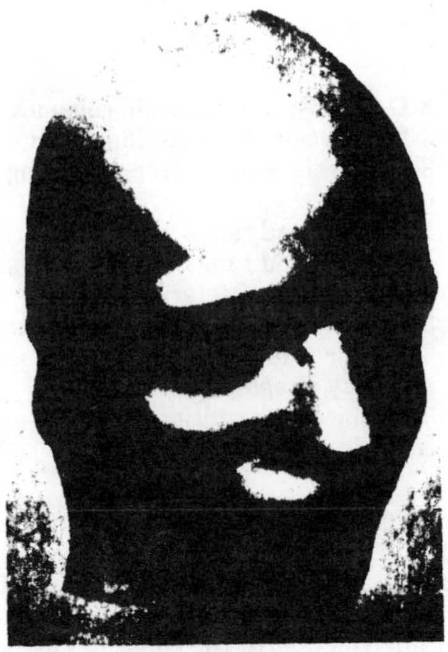

Figure 6. Intensity image of human face surface. (From Tiller [1983]; ©IEEE 1983.)

(a) *Voxel representation*. Voxels are small-volume elements of discretized 3-D space. They are usually fixed-size cubes. Objects are represented by the list of voxels occupied by the object. This representation is very memory intensive, but algorithms using it tend to be very simple.

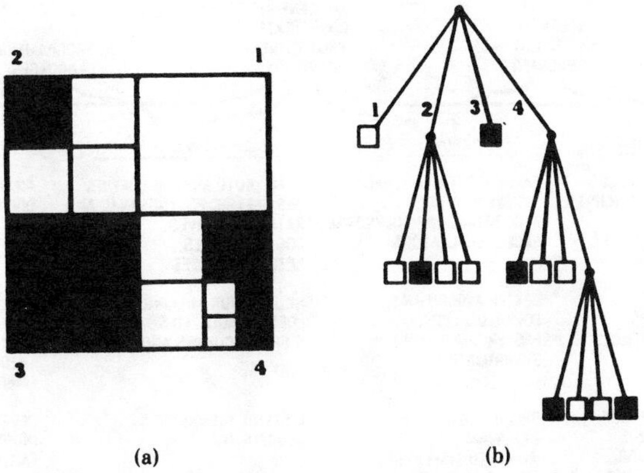

Figure 7. Hierachical approach to spatial occupancy. (a) Image; (b) quadtree. (From Henderson [1983]; ©IEEE 1983.)

(b) *Octree representation* [Meagher 1981, 1982]. An octree is a hierarchical representation of spatial occupancy. Volumes are decomposed into cubes of different sizes, where the cube size depends on the distance from the root node. Each branching node of the tree represents a cube and points to eight other nodes that describe object volume occupancy in the corresponding octant subcubes of the branching node cube. This representation offers the advantages of the voxel description but is more compact. Its compactness requires more complicated algorithms for many computations. The basic idea of octrees is displayed by considering the 2-D analog of octrees (usually referred to as quadtrees), as shown in Figure 7.

(c) *Tetrahedral cell decomposition representation.* Decomposition of 3-D space regions into tetrahedral elements is very similar to the lower dimensional analog of decomposing flat surfaces into triangles. (The tetrahedron is a 3-simplex, whereas the triangle is a 2-simplex.) Tetrahedral decompositions define volume and surface area unambiguously and are useful for mathematical purposes.

(d) *Hyperpatch representation* [Casale and Stanton 1985]. Each volume element is this representation is a hyperpatch: a generalization of bicubic surface patches. A hyperpatch defines volume, surface area, and *internal density variations* of a solid element. It is more general than most solid models, which allow only uniform density within a solid primitive; but a price is paid in memory and algorithm complexity (192 scalars are required for each volume element).

Voxels and octrees are useful for a number of computer graphics applications, whereas tetrahedral and hyperpatch models are useful for finite-element applications. Many other spatial occupancy schemes are possible.

(4) *Surface boundary representation* (*B-Rep*). Surface boundary representations define a solid object by defining the 3-D surfaces that bound the object. Figure 5 shows an example of the boundary representation concept. The simplest boundary representation is the triangle-faced polyhedron, which can be stored as a list of 3-D triangles. Arbitrary surfaces are approximated to any desired degree of accuracy by utilizing many triangles. The human face surface shown in Figure 6 was displayed from a list of triangles and quadrilaterals that approximate a smooth surface (which is stored using a rational B-spline surface representation). A slightly more compact representation allows the replacement of adjacent, connected, coplanar triangles with arbitrary *n*-sided planar pol-

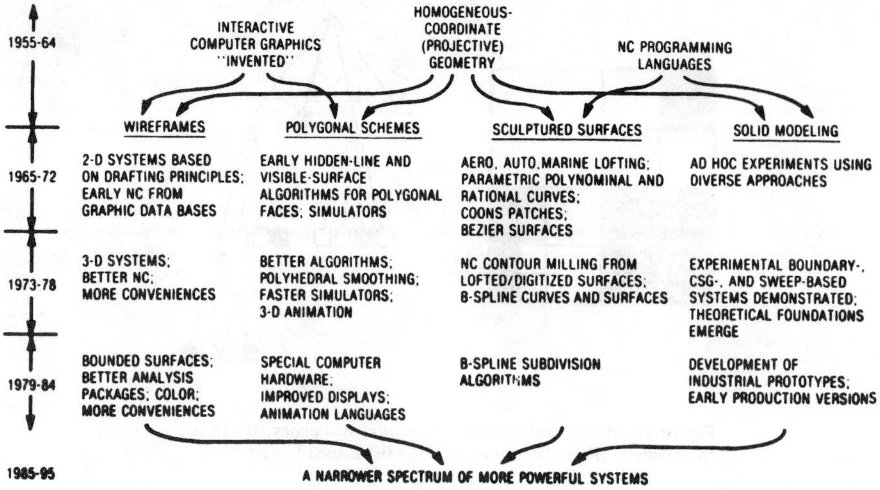

Figure 8. Historical summary of approaches to object representation. (From Requicha and Voelcker [1983]; ©IEEE 1983.)

ygons. This type of representation is popular because model surface area and volume are well defined, and all object operations are carried out using piecewise-planar algorithms. The next step in generality is obtained using quadric-surface boundary representations. More advanced techniques for representing curved surfaces with higher order polynomials or splines are mentioned in the next section. Structural relationships between bounding surfaces may also be included as part of a model. Lin and Fu [1984] have proposed a syntactic model that uses a context-free 3-D plex grammar for this purpose. Many variations are possible, but all boundary representations contain a list of object surfaces.

Any representation that adequately represents a real-world object should be usable by a graphics algorithm to render synthetic sensor images that are similar to real-world sensor data. The above object modeling schemes have been successfully used to provide very realistic renderings of real-world scenes [SIGGRAPH 1984]. On the other hand, any representation chosen for object recognition should also be compatible with matching algorithms. To be suitable for matching, each real-world object shape requires a unique description within the framework of a given representation. Most representations do not guarantee unique numerical descriptions of object shapes. It is often possible to reorder or reorganize points, edges, faces, and/or primitives of a given representation to obtain an identical shape. If a modeling scheme suffers from this problem, model-based matching algorithms for computer vision systems must be made insensitive to this nonuniqueness.

We refer the reader to Badler and Bajcsy [1978], Bajcsy [1979], Brown [1981], Requicha [1980] and Requicha and Voelcker [1982] and [1983] for more details on the relative merits of the above representation schemes. For convenience, we include two self-contained figures from Requicha and Voelcker [1983]: Figure 8 summarizes the history of approaches to 3-D object representation, and Figure 9 is a table of commercially available solid modelers.

Most 3-D object representations mentioned in the computer vision literature are categorized as one of the schemes mentioned above or as one of the following:

(1) *Generalized cone or sweep representation* [Shafer and Kanade 1983; Soroka and Bajcsy 1978]. Generalized cones (or

MODELER	VENDOR/DISTRIBUTOR	CORE SOFTWARE	GENRE
CATIA	IBM	DASSAULT (FRANCE)	B-REP
CATSOFT	CATRONIX		CSG
DDM-SOLIDS	CALMA		B-REP
EUCLID	MATRA DATAVISION/ DEC	CNRS (FRANCE)	B-REP
GEOMOD-II	SDRC/ GENERAL ELECTRIC CAE	SDRC	B-REP
ICEM SOLID MODELLING	CDC	SYNTHAVISION (MAGI)	CSG
ICM GMS	ICM		B-REP
MEDUSA	PRIME	CIS/CV (UK)	B-REP
PADL-1.2	U. ROCHESTER		CSG
PATRAN-G	PDA ENGINEERING		CELL DECOMP
ROMULUS	EVANS & SUTHERLAND	SHAPEDATA (UK)	B-REP
SOLIDESIGN	COMPUTERVISION		B-REP
SOLIDS MODELING-II	APPLICON	SYNTHAVISION (MAGI)	CSG
SYNTHAVISION	MAGI		CSG
TIPS-1	CAM-I	HOKKAIDO U.	CSG
UNIS-CAD	SPERRY UNIVAC	BAUSTEIN GEOMETRIE (T. U. BERLIN)	B-REP
UNISOLIDS	MCAUTO	PADL-2 (U. ROCHESTER)	CSG

Figure 9. Solid modelers available in the United States in May 1983. (From Requicha and Voelcker [1983]; ©IEEE 1983.)

generalized cylinders) are often called "sweep representations" because object shape is represented by a 3-D space curve that acts as the spine or axis of the cone, a 2-D cross-sectional figure, and a sweeping rule that defines how the cross section is to be swept and possibly modified along the space curve. These ideas are shown for a simple cylinder in Figure 10. Generalized cones are well suited to many real-world shapes. However, it is almost impossible to represent certain objects as generalized cones; consider the body of an automobile or the human face in Figure 6. Therefore, this scheme is not general purpose. Many researchers prefer the generalized cone object representation for vision purposes despite its limitations [Agin and Binford 1973; Brooks 1981; Kuan and Drazovich 1984; Nevatia and Binford 1973, 1977; Nishihara 1981].

(2) *Multiple 2-D projection representation.* For some applications it is convenient to store a library of 2-D silhouette projections to represent 3-D objects. For recognition of 3-D objects with a small number of stable orientations on a flat light table, this representation is ideal if silhouettes of objects are different enough. Silhouettes have also been used to recognize aircraft in

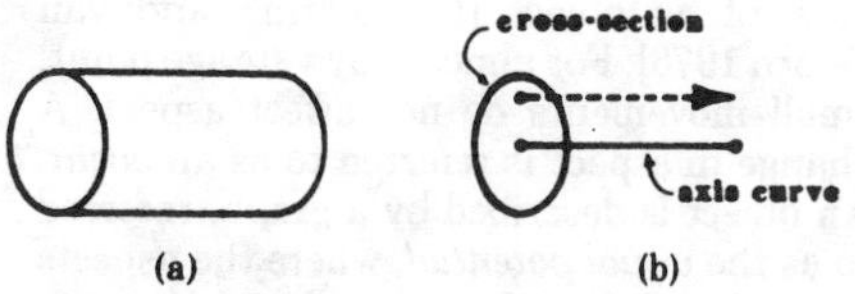

Figure 10. Generalized cylinder concept. (a) Cylinder; (b) sweep circle along axis. (From Henderson [1983]; ©IEEE 1983.)

any orientation against the well-lit sky background [Wallace and Wintz 1980]. It is not a general-purpose technique, however, because it is possible for many different 3-D object shapes to possess the same set of silhouette projections.

A more detailed approach of a similar nature is the *characteristic-views* technique described in Chakravarty [1982] and Chakravarty and Freeman [1982]. All of the infinite 2-D projection views of an object are gouped into a finite number of topological equivalence classes. Different views within an equivalence class are related via linear transformations. This is a general-purpose representation because it specifies the 3-D structure of an object. However, it can require a very large amount of data storage for complex objects. Figure 11

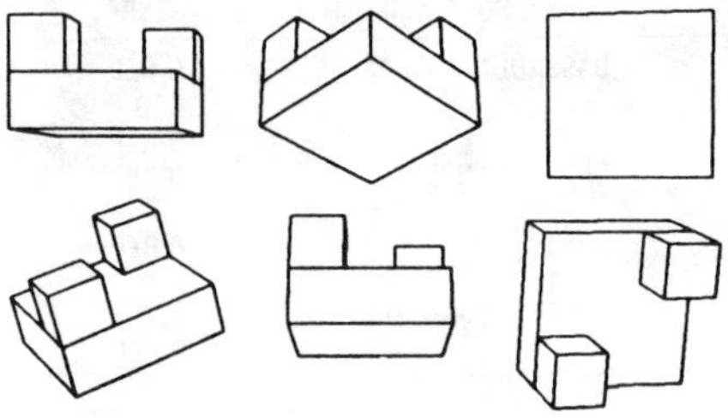

Figure 11. Representative characteristic views for a polyhedron.

shows six representative characteristic views for one nonconvex polyhedron. The characteristic-view idea started with Lavin [1974]. Thorpe and Shafer [1983] have done related work.

Similar ideas have been presented in a different form by Koenderink and van Doorn [1979]. Aspect is defined as the topological structure of singularities in a single view of an object [Koenderink and van Doorn 1976]. For almost any vantage point, small movements do not affect aspect. A change in aspect is referred to as an *event*. An object is described by a graph, referred to as the *visual potential*, where the aspects form the nodes of the graph, and events form the arcs of the graph. Visual object complexity can be measured using the diameter of the visual potential (or aspect graph). Figure 12 shows the aspect graph for a cube; in this case there are three types of aspect: one, two, or three faces are visible. Scott [1984] has studied these ideas with the aim of implementing a graphics system that understands what it is displaying in terms of the projection topology and the visual potential neighborhood of a given view. The property spheres concept [Fekete and Davis 1984] is another 3-D multiple-viewpoint representation.

(3) *Skeleton representation.* Researchers have found it worthwhile to describe shape using (space–curve) skeleton models [Garibotto and Tosini 1982; Udupa and Murthy 1977]. A skeleton can be considered an abstraction of the generalized cone description and consists of only the spines (or axis curves). Interest in skeletal shape description appears to have started with the medial axis (or symmetric axis) transform of Blum [1967]. Nackman [1982] has generalized the symmetric axis transform concept for arbitrary 3-D objects that have *surface skeletons.* Skeleton geometry provides useful abstract information. If a radius function is specified at each point on the skeleton, this representation is capable of general-purpose object description.

(4) *Generalized blob representation.* Generalized blobs have been used by investigators as a 3-D object shape description scheme [Mulgaonkar et al. 1982]. Objects are described by sticks (lines), plates (areas), and blobs (volumes). This abstract representation is not intended for general-purpose shape description.

(5) *Spherical harmonic representation* [*Schudy and Ballard 1978*]. Certain object shapes can be represented by specifying the radius from a point as a function of latitude and longitude angles around that point. This representation is useful only for convex objects and a restricted class of nonconvex objects.

(6) *Overlapping sphere representation.* O'Rourke and Badler [1979] have proposed the use of *overlapping spheres* as a solid object representation. Many spheres are required to yield relatively smooth surfaces. This single-primitive representation is better suited to molecular display algorithms than object modeling. Although it is a general-purpose technique, it is rather awkward for precisely representing most man-made objects.

Several of the object representation methods found in the computer vision literature are not capable of describing arbitrary solid objects. These methods either specialize in a particular class of shapes or ignore geometric details in favor of more abstract information.

The object-recognition problem requires a representation that can model arbitrary solid objects to any desired level of detail *and* can provide abstract shape properties for matching purposes. Perhaps a set of several interrelated representations for the same object is needed to provide multiple levels of geometric information. No matter what representations are used, it will be necessary to evaluate surfaces explicitly in at least one module of a vision system be-

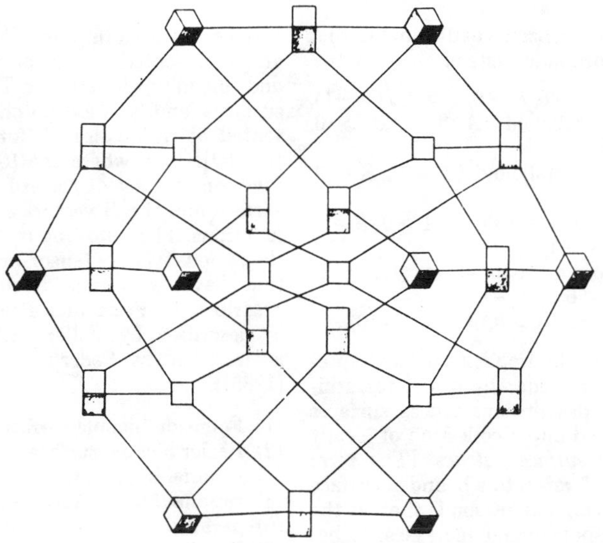

Figure 12. Visual potential of a cube.

cause (1) range images consist of sampled object surfaces, and (2) intensity images are strongly dependent on object surface geometry. We believe that object recognition is dependent on surface perception. When explicit surface information is required, a surface boundary representation is the natural choice to avoid unnecessary computation.

5.2 Surface-Representation Schemes

All surface boundary representations employ lists of surfaces for representing objects. How can these surfaces be represented? We mentioned previously that polyhedral models are represented as a list of planar polygons, so we do not discuss this representation further. How can smooth, curved surfaces be represented?

A general surface in three dimensions is written as

$$S = \{(x, y, z) : F(x, y, z) = 0\}.$$

This is referred to as an *implicit* representation of a surface. If the gradient vector ∇F exists, is continuous, and is nonzero for every point (x, y, z), then S is a smooth surface. The implicit surface representation is useful for *low-order* polynomials of the spatial variables. Planar surfaces are precisely represented with only four coefficients (which describe three degrees of freedom):

$$F_{\text{plane}}(x, y, z) = Ax + By + Cz + D.$$

(A, B, C) specify the direction of the single normal to the surface, whereas D specifies the distance of the plane from the origin of the coordinate system if A, B, C are properly normalized. Quadric surfaces require 10 coefficients in general (which describe 9 degrees of freedom):

$$F_{\text{quadric}}(x, y, z) = Ax^2 + By^2 + Cz^2 + Gxy + Hyz + Izx + Ux + Vy + Wz + D.$$

Only 3 coefficients are needed to describe the *shape* of a quadric surface of a given type, whereas 6 parameters are needed to locate and orient the surface in space. If a quadric surface is properly translated and rotated, at least 6 of the 10 coefficients will be zero. All quadric surfaces can then be classified as one of the 6 following types using the 3 or 4 *nonzero* coefficients in that particular coordinate system:

(1) ellipsoid ($A > 0, B > 0, C > 0, D = -1$),
(2) elliptic paraboloid ($A > 0$, $B > 0$, $W = -1$),
(3) hyperbolic paraboloid ($A > 0$, $B < 0$, $W = -1$),
(4) hyperboloid of one sheet ($A > 0, B > 0, C < 0, D = -1$),
(5) hyperboloid of two sheets ($A > 0$, $B < 0, C < 0, D = -1$),
(6) quadric cone ($A > 0, B > 0, C < 0$).

Unfortunately, the implicit surface representation is not generally useful for arbitrary surface descriptions unless surfaces are decomposed into a collection of locally homogeneous *surface patches*. (The term "surface patch" refers to a bounded surface with finite area.) The reason is that, as the order of the polynomial increases, it becomes more difficult to suppress undesired undulations in polynomial surface functions.

The standard alternative approach is to use an *explicit* parametric surface representation:

$$S = \{(x, y, z): x = h(u, v),\ y = g(u, v),\ z = f(u, v),\ (u, v) \in D \subseteq \mathbf{R}^2\},$$

where f, g, h are smooth scalar functions of two variables. A less general, but still useful, parametric description of surfaces is given via the graph surface (or Monge patch) representation:

$$S = \{(x, y, z): x = u,\ y = v,\ z = f(u, v),\ (u, v) \in D \subseteq \mathbf{R}^2\}.$$

Gray-level surfaces in intensity images and depth surfaces in range images are typically analyzed using this common representation.

Many different types of parametric surface representations are discussed in the computer graphics and CAD literature. The differences between these surface representations are due to different representations for the $f(u, v)$, $g(u, v)$, $h(u, v)$ functions. Parametric polynomial methods are reviewed in Farouki and Hinds [1985]. Coon's patches and tensor-product composite surfaces are commonly used parametric surface representations. Coon's patches are represented using boundary curves and blending functions. Tensor-product surfaces and surface patches are represented as a "quadratic" form $S(u, v) = B(u)[Q]B^T(v)$, where the $[Q]$ matrix is a function of a set of control points for the surface and the B vectors consist of basis functions. The following is a partial list of the various types of tensor-product surfaces that have been considered ((1)–(6) are described in Faux and Pratt [1979], (7) is described by Tiller [1983], and (8) is described by Sederberg and Anderson [1985]):

(1) Ferguson bicubic surface patches,
(2) Bezier bicubic surface patches,
(3) rational biquadratic surface patches,
(4) rational bicubic surface patches,
(5) parametric spline surfaces,
(6) B-spline surfaces,
(7) rational B-spline surfaces,
(8) Steiner surface patches.

The use of homogeneous coordinates allows rational and nonrational tensor-product surfaces to be expressed in the same matrix form. Rational B-spline surfaces are quite general in that they can precisely represent quadric surface primitives, free-form surfaces, and polyhedral objects using one mathematical form. As a result, this representation has been adopted as an Initial Graphics Exchange Specification standard for 3-D surfaces for the CAD/CAM/CAE industry [IGES 1983].

York et al. [1980, 1981] discuss the use of Coon's surface patches with cubic B-spline boundary curves as a surface representation for computer vision. The use of this surface type within the VISIONS system Long Term Memory (LTM) layered network database is outlined. A preliminary example of matching using the shape features of a 3-D circle is presented. Although the scope of these two papers is limited, they are among the few works to discuss CAD surface representations in the computer vision context.

Sometimes it is necessary to represent surfaces over arbitrary domains D. Barnhill [1983] surveys the use of triangular interpolants and distance-weighted interpolants to create smooth surface descriptions from arbitrarily located point data. Smith [1984] discusses a fast surface interpolation algorithm that relies on the fast Fourier transform. Hence, another type of surface representation is one where a set of data points is given, along with an algorithm for computing points on the surface. These surface descriptions are categorized according to whether the smooth surface passes through the given points (interpolation), or whether the surface approximates the given points while minimizing an error criterion (approximation). There are too many different techniques of this sort for us to survey them here. Also, this type of representation is generally not very compact in terms of data storage.

A wide variety of techniques has been developed for representing objects and surfaces for digital computing purposes. Object-recognition systems must use one of these known techniques to describe objects or invent new ones. In order to make informed decisions when constructing an object-recognition system, one needs to be aware of these modeling issues.

5.3 Object- and Surface-Rendering Algorithms

Once an object/surface representation has been selected as a model data storage mechanism for an object-recognition system, it is convenient to have a technique for transforming model data into synthetic sensor data and/or a symbolic scene description. This would allow the introduction of a feedback loop that could be used to evaluate a vision system's understanding automatically. For example, given a plausible model-based interpretation of a scene, the system could generate a synthetic image and process it as if it were real sensor data to obtain a new description. If the new description closely matches the real data description, the hypothesized model is verified; if not, the system must continue to search for a better interpretation. Many existing vision systems are open-loop systems with no feedback. High-level results are not projected back into a low-level form for final error checking. More research is needed in this area.

How is the rendering task accomplished? The computer graphics literature discusses many techniques for generating line drawings and shaded images in color or black and white from geometric models [Foley and Van Dam 1982; Newman and Sproull 1979]. These techniques are divided into display-space algorithms, object-space algorithms, or hybrid algorithms [Sutherland et al. 1974]. Sorting of graphic primitives is generally required and can make these algorithms computationally intensive, but recent advances in graphics hardware and software are alleviating the rendering of speed problems. If model data are adequate for general-purpose vision, relatively realistic intensity images or range images (and their corresponding symbolic scene descriptions) might be generated automatically upon request, within a vision system, by using an appropriate algorithm.

Much research work is specifically interested in range data. The z-buffer (or depth-buffer) algorithm from computer graphics can be used to generate synthetic depth maps from arbitrary polyhedral object models. It is easy to implement such an algorithm in software. In fact, hardware implementations of this algorithm have been commercially available for generating shaded intensity images since 1982. Therefore, it is reasonable to assume that extremely fast rendering algorithms will be available for future object-recognition systems.

5.4 Intensity- and Range-Image Formation

When using sensor data to yield information about the real world, it is important to understand the image-formation process. This process has been studied in detail by both computer vision and computer graphics researchers. Ballard and Brown [1982] provide a thorough treatment of image formation. At each point in an intensity image, the brightness value *encodes information* about surface geometry (shape, orientation, and location), surface reflectance characteristics, surface texture, scene illumination, the distance from the camera to an object surface, the characteristics of the intervening medium, and the camera characteristics (which include spatial resolution, noise parameters, dynamic range, brightness resolution, and lens parameters). Over the years, increased understanding of intensity-image formation [Horn 1977] and of the constraints of the physical world has led to important computer vision research developments, including shape from binocular stereo [Grimson 1980], shape from motion [Jain 1983; Ullman 1978], shape from shading [Ikeuchi and Horn 1981], shape from photometric stereo [Coleman and Jain 1982; Woodham 1981], shape from texture [Witkin 1981], and shape from contours [Kanade 1981]. This group of methods is referred to as the shape-from-(xxx) techniques. These developments are directed toward the goal of correctly inferring the 3-D structure of a scene from brightness values alone. The great difficulty in reaching that goal is certainly related to the large number of factors encoded in each brightness value during the intensity-image-formation process.

Range-image formation is conceptually a simpler process than intensity-image formation. At each pixel in a range image, the depth value *encodes information* about (1) surface geometry and viewing geometry in terms of the distance from the sensor to object surface, and (2) the range-finder characteristics (which include spatial resolution, range resolution, dynamic range, noise parameters, and other range-finder parameters that depend on the type of range finder used). One important difference is that scene illumination and surface reflectance are *not directly* encoded in range values, even though they can definitely affect the accuracy of measured values. Moreover, range finders directly produce the depth (shape) information that the shape-from-(xxx) techniques seek to produce. Although range finders are sometimes regarded as specialized nonvision instruments, since they do not address vision as people experience it, they are currently receiving a great deal of attention and are very useful sensors in many situations. Because range finders are not so common as cameras and video digitization equipment, different techniques for sensing depth are discussed briefly. This review is a condensation of the material found in Ballard and Brown [1982], Jarvis [1983a, 1983b], and Parthasarathy et al. [1982].

Range finders are classified as using either *active* or *passive* methods. Active methods project energy onto a scene to measure range, whereas passive methods do not. Ultrasound and radio-wave techniques are used for range determination, but do not currently possess high enough resolution for most range-imaging purposes.

Lasers are used in pulsed-mode and modulated-mode range sensors. A pulsed-mode time-of-flight laser range finder determines distance by measuring the elapsed time between pulse transmission and signal reception, and therefore requires signal-processing electronics with 70-picosecond time resolution to obtain a depth resolution of 1 centimeter. A laser range finder of this type is discussed in Lewis and Johnston [1977]. A schematic for a range finder of this type is shown in Figure 13.

Amplitude-modulated laser range finders determine distance by measuring the phase difference between the received wave and a reference signal. Svetkoff et al. [1984] discuss a phase-difference laser range finder and the accompanying range ambiguity problem. Range images, from the Environmental Research Institute of Michigan (ERIM) range sensor, of a coffee cup and a computer terminal keyboard are shown in Figure 14. A diagram for the phase-difference type of range finder is shown in Figure 15. For both types of laser range finders, spatial resolution is typically about 128 × 128 pixels, and depth resolution is usually about 1 centimeter for objects in the 1–4-meter range. These instruments tend to be fairly expensive and are often very slow (1 second to several minutes per frame) compared with TV cameras (33 milliseconds per frame). The state of the art in close-range high-resolution laser range finders is perhaps summarized by listing the specifications of the ERIM laser range finder discussed by Svetkoff et al. [1984]:

(1) source: gallium arsenide laser diode, 20 milliwatts;
(2) dynamic range (ambiguity interval): 6 inches–3 feet;

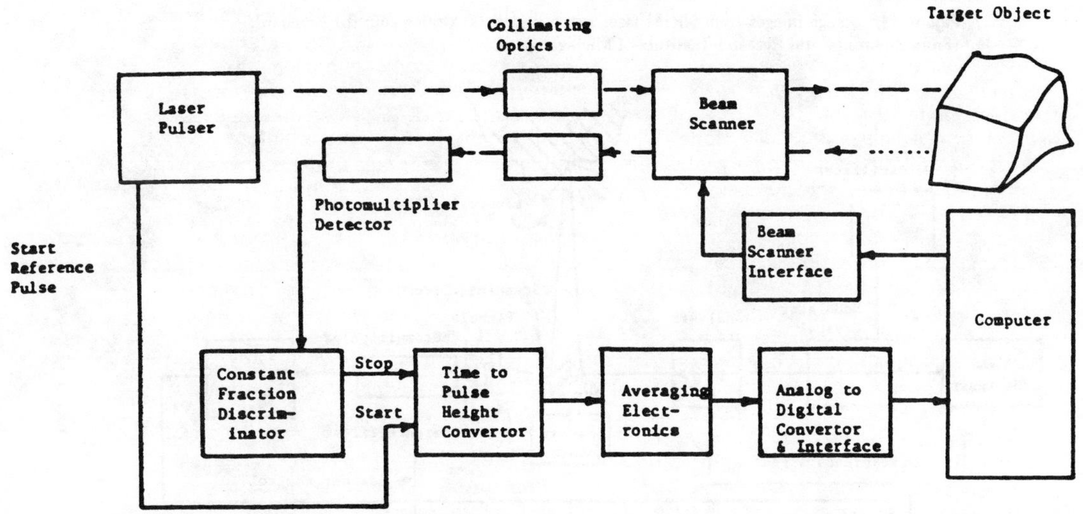

Figure 13. Pulsed-mode time-of-flight laser range-finder schematic. (From Jarvis [1983a]; ©IEEE 1983.)

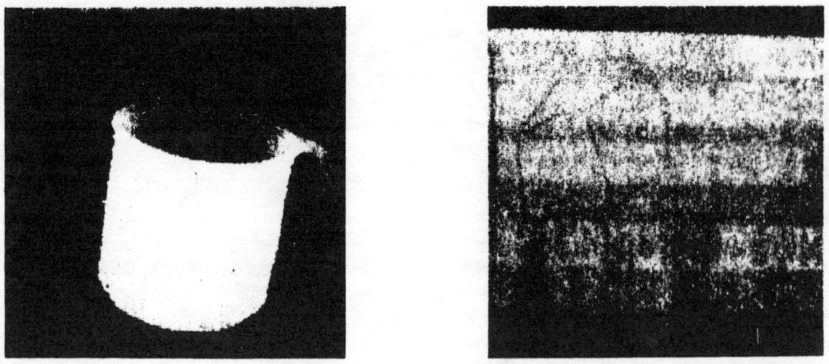

(a) (b)

Figure 14. Range images from ERIM laser range finder. (a) Coffee cup; (b) keyboard. (From Environmental Research Institute of Michigan.)

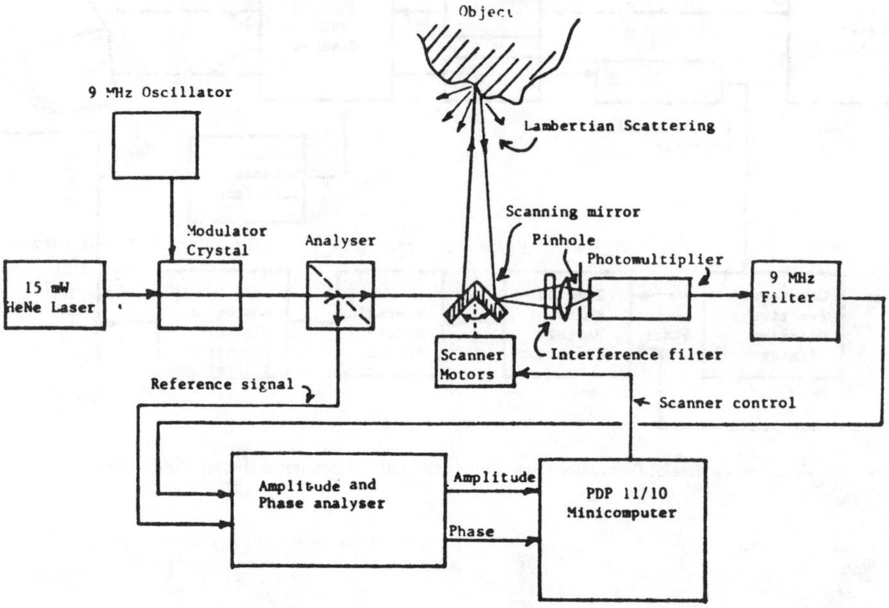

Figure 15. Modulated phase-difference laser range-finder diagram. (From Jarvis [1983a]; ©IEEE 1983.)

(3) range resolution: 0.001–0.1 inch;
(4) field of view: 1.6–35 degrees^2;
(5) frame rate: 1 frame per second typical;
(6) scan control: programmable (capable of 512 × 512).

It is important to note that the accuracy of laser range finders depends on the return signal power, which in turn depends on (1) transmitted power, (2) the distance to the object, and (3) the object's surface reflectance. Laser power must sometimes be so large that human eye damage may result if safety precautions are not exercised. Note that these range finders have problems with shiny surfaces because very little energy is directed back toward the sensor for almost

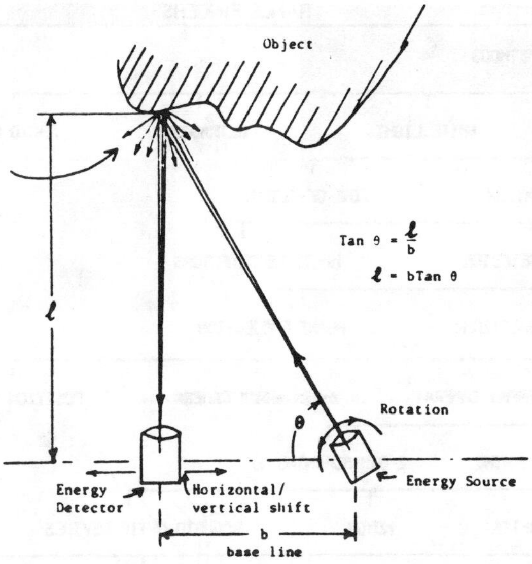

Figure 16. Simple triangulation range-finding geometry. (From Jarvis [1983a]; ©IEEE 1983.)

all reflection angles. Multiple reflections cause difficulties, too.

Lasers are also used in triangulation-based range finders [Pipitone and Marshall 1983], where a spot or line of light is projected onto a scene. Cameras or infrared sensors are used to detect the light, signal- or image-processing techniques are used to determine the position of the spot or pieces of the line, and trigonometry is used to estimate the distance to the detector. See Figure 16 for an example of triangulation range-finder geometry. Depth resolution depends on how well positions, distances, and angles are measured. Triangulation methods always suffer from the "missing parts" (or shadowing) problem owing to the necessary separation of the source and detectors, whereas laser range sensors with coaxial source and detector are not subject to this problem. Figure 17 shows a detailed categorization of the range finder discussed in Parthasarathy et al. [1982]. The outline gives an idea of the variety of laser range finders.

Noncoherent white light is used for range finding in the same way as laser light. A spot, line, or stripe of light, or even an entire grid of light [Hall et al. 1982; Potmesil 1979; Tio et al. 1982] is projected onto a scene so that reflected light is detected by cameras. Image-processing techniques are then needed to isolate the bright pixels in the image, and depth is determined by triangulation. Other patterns and Moire fringe techniques [Idesawa and Yatagai 1980; Lamy et al. 1982] have also been used for range finding.

Light-striping techniques can be improved as follows. Using n binary images of a scene, it is possible to generate the equivalent range information obtained by a light-stripe range finder that has scanned and processed 2^n light-stripe images. For example, Altschuler et al. [1981] discussed an instrument of this type that used noncoded binary patterns at a conference dedicated to 3-D machine perception [SPIE 1981]. Inokuchi et al. [1984] have used Gray-coded bit-mask patterns to limit the position error to ±1 stripe width in the least-significant-bit (LSB) bit-mask image. See Figure 18, which shows the Gray-code patterns and overall system configuration.

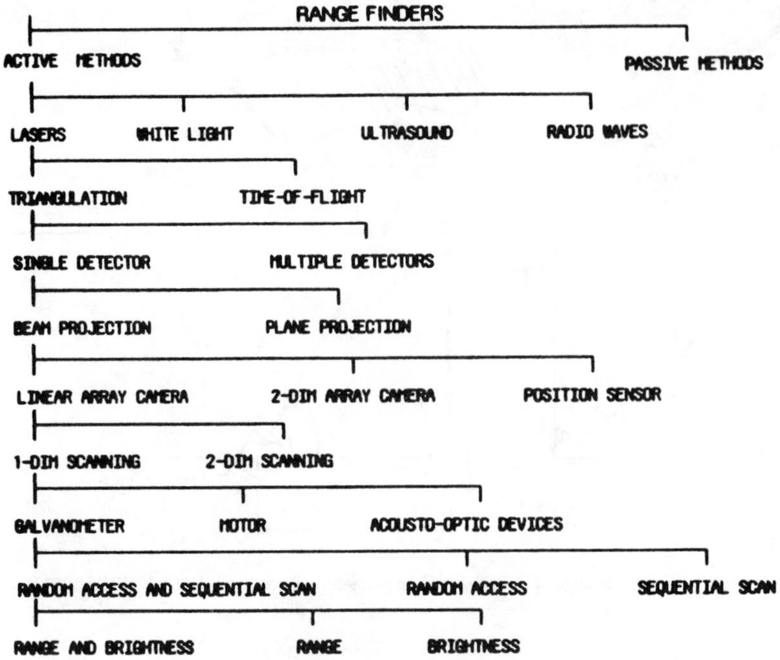

Figure 17. A classification of range finders. (From Parthasarathy et al. [1982].)

Passive range-finding techniques do not project energy onto a scene. Focusing techniques use the shallow depth of field of large aperture lenses for determining the depth to points in a scene. Shape-from-(xxx) methods are other types of passive range-finding techniques. Carrihill and Hummel [1984] have used intensity ratio sensors to determine depth to points in a scene.

The major passive range-finding technique is stereo [Barnard and Fischler 1982; Grimson 1980; O'Brien and Jain 1984; Yakimovsky and Cunningham 1978]. The correspondence problem of matching scene points in images from different views must be solved to obtain good depth values from stereo range finders. Even then, stereo and other passive techniques generally provide depth only at isolated points in the field of view. Surfaces can be interpolated from these points to obtain entire depth maps for a scene [Terzopoulos 1983]. In summary, passive range-finding techniques are ordinarily computationally intensive, but it is expected that these methods will improve with time.

Because of the difficulties involved in passive techniques, range images are obtained from active sensors in practice. As active range-finding techniques are enhanced, high-spatial-resolution (512 × 512 pixels), high-depth-resolution (16 bits) sensors will become available at relatively reasonable costs. High-quality range data are extremely useful for many applications. For example, automatic inspection systems benefit greatly from the capability of sensing 3-D object shape directly. Most current industrial inspection systems are limited to 2-D shape processing.

5.5 Intensity- and Range-Image Processing

Compared with computer vision (or image understanding), intensity-image processing is a relatively mature field. Textbooks on the subject [Gonzalez and Wintz 1978;

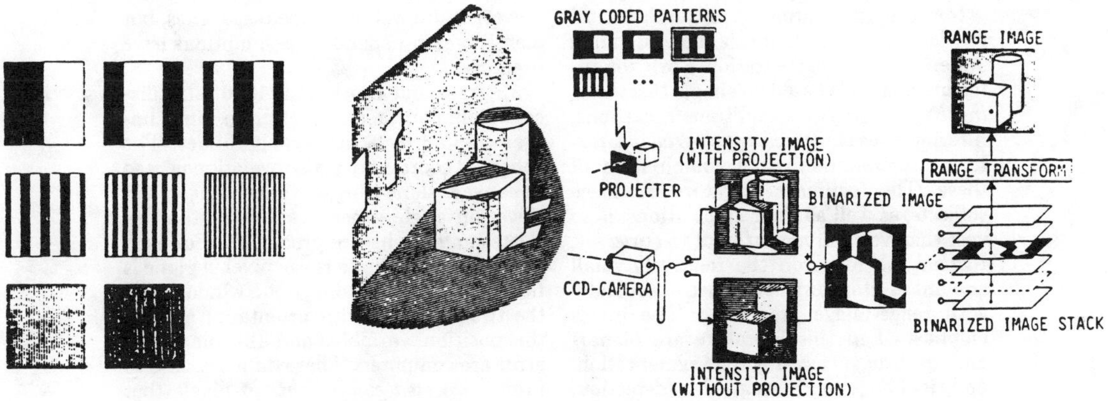

Figure 18. Gray-code patterns and range-finder system configuration. (From Inokuchi et al. [1984]; ©IEEE 1984.)

Figure 19. Different types of edges in range images. (a) Convex roof edge; (b) concave roof edge; (c) concave ramp edge; (d) step edge; (e) convex ramp edge.

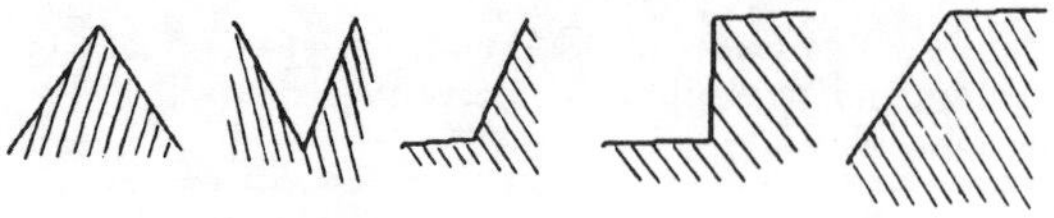

Pratt 1978; Rosenfeld and Kak 1981] discuss the topics of enhancement, restoration, coding, and segmentation of digital intensity images. Since range images have the exact same mathematical form as intensity images, many existing image-processing techniques are directly applicable to range images. We assume below that the reader is familiar with basic image-processing terminology. First occurrences of image-processing-specific terms are denoted in italics in this section. These terms are discussed in the texts listed above and in Ballard and Brown [1982].

Image segmentation is a fundamental low-level process in image-understanding systems. Most segmentation work for intensity images is based on simple *thresholding, correlation, histogram* transformations, filtering, *edge detection, region growing, texture discrimination*, or a combination of these. The technical literature on these subjects, as well as their applications, is so vast that we shall not attempt to survey it. Instead, we focus on the relatively small amount of literature concerned specifically with range-image processing. The issues emphasized in this literature are planar- and quadric-surface-region segmentation, and roof-edge and ramp-edge detection. These issues are seldom pursued in intensity-image processing. Various edge types are shown in Figure 19.

Duda et al. [1979] discuss the use of registered range and reflectance images to find planar surfaces in 3-D scenes. A sequential planar region extraction procedure is utilized on range and reflectance images obtained from a modulated continuous-wave laser range finder [Nitzan et al. 1977]. (Reflectance images are slightly different from intensity images because shadows cannot occur.) A priori scene assumptions concerning man-made horizontal and vertical surfaces motivate the procedure. First, horizontal surface regions of significant size are segmented and removed from the images by means of a filtered range (z) histogram. Second, major vertical surfaces are extracted from the remaining scene data by means of a *Hough transform* method. Third, arbitrary planar surfaces are identified with the help of reflectance histogram data. All planar surfaces are thus segmented and labeled. All unlabeled regions correspond to depth discontinuities, nonplanar regions, or very small planar regions not found in the three main processing steps. The overall technique appears to work quite well on three test scenes, but many domain-dependent assumptions have been used.

Milgram and Bjorklund [1980] also discuss planar surface extraction in range images created by a laser range finder. The spherical coordinate transformation is used to convert slant range, azimuth angle, and elevation angle sensor data into standard Cartesian data before processing. For each Cartesian-coordinate range pixel, a plane is fitted to the surrounding 5×5 window, and the two normal vector orientation angles, the position variable, and the planar fit error are computed. These data are used to form *connected components* of pixels that satisfy planarity constraints. After subsequent region growing is complete, a "sensed plane list" is built. The plane list, the symbolic scene description of the system, is compared with a reference plane list to determine sensor position with respect to a stored scene model. Experimental results are discussed for four real-world range images and two synthetic range images displaying different viewpoints of the same building site. This method appears to be a much better, more straightforward approach to planar surface extraction than that of Duda et al. [1979]. However, no effort was made to handle curved surfaces.

Figure 20 contains a block diagram of the system that was planned for vehicle navigation.

Henderson [1982, 1983] has developed a method for finding planar faces in range data. First, a list of 3-D object points are assumed given by a range finder. To handle multiple depth maps, points are transformed into one object-centered coordinate system using transformation data recorded during range-image formation [Henderson and Bhanu 1982]. These points are stored randomly in a list with no topological connectivity information. The points are then organized into a 3-D binary tree, which can be done in $O(N \log N)$ time (where N is the number of points). Second, each point's neighbors are determined with the aid of the 3-D tree, and the results are stored in a 3-D spatial proximity graph. Third, a spiraling sequential planar-region-growing algorithm, known as the three-point seed method [Henderson and Bhanu 1982], is used to create convex planar faces using the spatial proximity graph as input. The union of these faces form the polyhedral object representation as extracted from the range data. Several processing steps mentioned above are required because of the (x, y, z) list format of the input data. Neighbors are given explicitly in standard range-image formats. This method can be used for either range data segmentation or object reconstruction. It can also work on dense range data or a sparse collection of points. Curved surfaces are approximated by many polygons.

Wong and Hayrapetian [1982] suggest the use of range-image histograms to segment corresponding registered intensity images. All pixels in the intensity image that correspond to pixels in the range image with depth values not in a certain range are set to zero, segmenting all objects in that particular range. This segmentation trick is useful in specific applications, but it hardly begins to take advantage of the explicit range data information. Also, it cannot work on long objects that span the dynamic range of the range image.

Gil et al. [1983] have demonstrated the usefulness of combining intensity and range edges from registered range and intensity images to obtain more reliable edges.

Hebert and Ponce [1982] propose a method of segmenting depth maps into plane, cylindrical, and conical primitives. Surface normals are computed at each depth pixel using the best-fit plane in 3 × 3 windows. These normals are mapped to the *Gaussian sphere* where planar regions become very small clusters, cylinders become unit radius semicircles, and cones become smaller radius semicircles. (This orientation histogram is often referred to as the extended Gaussian image, or EGI.) The Hough transform is used to detect these circles and clusters. Regions are then refined into labeled, connected components. Although somewhat restricted, this technique handles at least certain types of curved surfaces, in addition to handling planes.

Inokuchi et al. [1982] present an edge-region segmentation ring operator for depth maps. The ring operator extracts a one-dimensional (1-D) periodic function of depth values that surround a given pixel. This function is transformed to the frequency domain using an FFT algorithm for either 8 or 16 values. Planar-region, step-edge, convex-roof-edge, and concave-roof-edge pixels are distinguished by examining the zeroth, first, second, and third frequency components of the ring surrounding that pixel. These pixel types are grouped together, and the resulting regions and edges are labeled. Experimental results are shown for one synthetic block world scene range image. The ring operator method appears to compute roof edges fairly well at the boundaries of planar surfaces. But it is not stated how range images with curved surfaces are handled. Two years earlier, Inokuchi and Nevatia [1980] discussed another roof-edge detector that applied a radial line operator at step-edge corners and followed roof edges inward.

Mitiche and Aggarwal [1983] have developed an edge detector that is insensitive to noise owing to use of a probabilistic model that attempts to account for range measurement errors. The computational procedure is as follows: (1) Step edges are extracted first from a depth map. (2) For

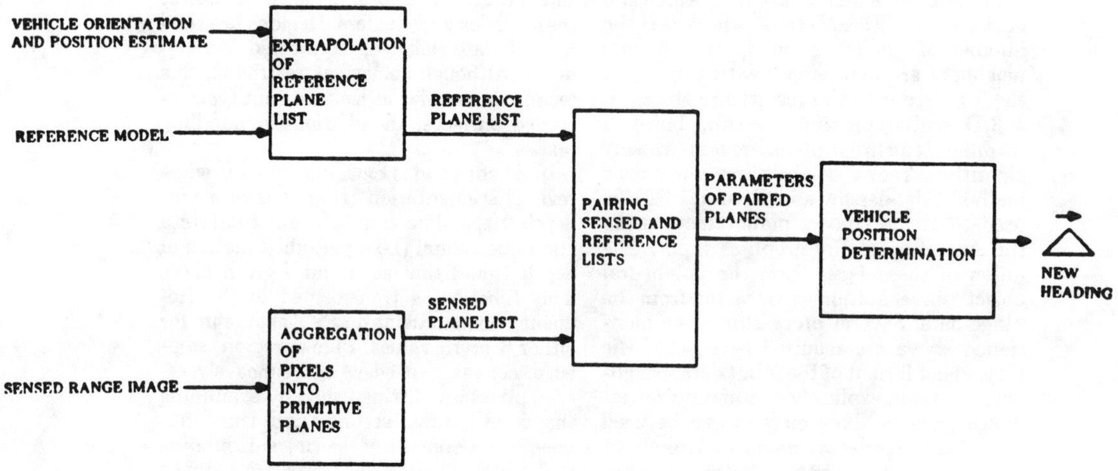

Figure 20. Block diagram of matching system. (From Milgram and Bjorklund [1980]; ©IEEE 1980.)

each direction (usually four) in the image at each pixel, a roof edge is hypothesized. For each hypothetical roof edge, two planes are fitted to the immediate neighborhood of the pixel, and the dihedral angles between these planes are recorded. They call this the "computation of partitions." (3) Pixels are discarded if all dihedral angles are less than a threshold. For every remaining pixel, *a Bayesian likelihood ratio* is computed, and the most likely partition (e[illegible]ection) is chosen. If the angle for th[illegible]n direction is less than another thr[illegible]d, the pixel is also discarded. This is called the "dismissal of flat surfaces." (4) All remaining pixels are passed through a nonmaxima suppression algorithm that theoretically leaves only the desired edge pixels. Experimental results from the paper are shown in Figure 21 for a 64 × 64 depth map of a cube with added noise. The method can handle a large amount of added noise because the system is internally constrained by its model to look for horizontal and vertical edges (a domain-specific constraint).

Lynch [1981] presents a range-image enhancement technique for range data acquired by a 94-gigahertz (3.2-millimeter) radar. In the special case of systems with a shallow (nearly horizontal) line of sight, a strong depth gradient always exists in a range image. This gradient makes it very difficult for people to interpret such a range image using a typical 256-gray-level display device. Two *1-D high-pass* filters (a normalized filter and a differenced filter) are derived, discussed, and applied to an example scene to create a feature image that is more easily interpreted by a human observer than the original range image. Lynch's approach distorts the shape information in range data to achieve high local contrast.

Sugihara [1979] proposes a range-image feature extraction technique for edge junctions similar to the junction features used by Waltz [1972] and others for intensity-image understanding. A junction dictionary of possible 3-D edge junctions is implemented as a directed graph data structure and is useful in aiding 3-D scene understanding. Unlike intensity-image edges, range-image edges are classified as *convex*, *concave*, *obscuring*, or *obscured* without additional higher level information from surrounding image regions; junction knowledge is not necessary for line categorization. This categorization can therefore be used to help predict missing edges. Several junctions are only possible when two or more objects are in a scene; junction information can then be used to segment the range image into different objects. It is noted that junction points are *local curvature maxima* points in the depth-map surface. A system that uses depth discontinuity contours (step edges) and junctions for complete scene segmentation is described. It is limited by the constraint that every vertex is connected to at most three faces.

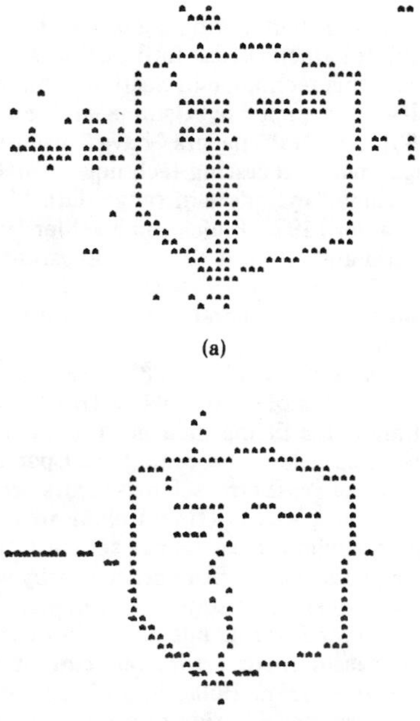

(a)

(b)

Figure 21. Edge maps for depth map of cube with noise. Probabilistic model: (a) edge map before nonmaxima suppression; (b) final edge map. (From Mitiche and Aggarwal [1983]; ©IEEE 1983.)

Other researchers [Popplestone et al. 1975; Shirai and Suwa 1971] have processed range data by analyzing small surface regions and then linking regions with compatible characteristics into surfaces. We discuss this technique in Section 5.9 in our review of a paper by Oshima and Shirai [1983]. Several papers have discussed range-image processing techniques for the detection of cylinders in range data [Agin and Binford 1973; Bolles and Fischler 1981; Nevatia and Binford 1973; Popplestone et al. 1975]. We only review the second paper because of its general-purpose statistical method.

Bolles and Fischler [1981] present the Random Sample Consensus (RANSAC) technique for fitting models to noisy data containing a large percentage (20 percent or more) of gross errors. Gross errors occur, for instance, when fitting a plane to a set of points where most points belong to the plane, but some are from other nearby surfaces. Linear least-square techniques are effective for filtering out normally distributed measurement errors but cannot remove gross errors. Bolles and Fischler propose a two-step filtering process where (1) initial estimates of model parameters are computed to eliminate gross errors, and (2) an improved fit is computed by applying standard smoothing techniques to the prefiltered data. For example, the RANSAC approach to circle fitting is selecting only three points at random, computing a circle, and counting the number of other compatible points in the data that are within the expected measurement error threshold. If there are enough compatible points, then least-squares smoothing is applied to the three initial points and all compatible points. If not, another set of three points is selected and the process is repeated. If the number of trials exceeds a preset threshold, then the process is stopped. The RANSAC technique is applied to finding ellipses, and then cylinders, in light-stripe range data. Although it is slower than standard fitting methods, model parameters are insensitive to gross errors.

5.6 Surface Characterization

The term *surface characteristic* refers to a descriptive feature of a general smooth surface. *Surface characterization* refers to the computational process of partitioning surfaces into regions with equivalent characteristics. Distinctions between range-image processing and surface characterization are made on the basis of generality; general digital surface processing schemes are considered surface characterization. The descriptive quality of the features used to identify surfaces is of critical importance to the performance of object-recognition systems using range data. Surface characterization has also been considered for intensity images [Haralick et al. 1983].

Nackman [1984] discusses surface description using critical-point configuration graphs (CPCGs). All critical points are isolated from each other except in degenerate circumstances. Nondegenerate critical points of surfaces are local maxima, local minima, or saddle points. If critical points of a surface are identified as the nodes of a graph, and the connecting ridge and course lines (the zero crossings of the first partial derivatives) are considered the arcs of a graph, a surface is characterized by a CPCG. Slope districts are the regions bounded by graph cycles. Two important theorems relating to these graphs are discussed: (1) Only eight types of critical points are possible: peaks (local maxima), pits (local minima), and six types of passes (saddle points); (2) only four types of slope districts are possible. All surfaces have a well-defined characterization as the union of slope-district regions, where each region belongs to one of four basic slope-district types. This characterization is a generalization of techniques used to describe 1-D functions $f(x)$. In the 1-D case only two types of nondegenerate critical points exist: local maxima and local minima. Between these critical points are intervals of constant sign of the first derivative. Slope districts are generalizations of these intervals. Nackman also mentions curvature districts determined by the *sign* of the mean and Gaussian curvature of a surface, but does not explore them. This work resulted from research to generalize the symmetric axis transform for 3-D objects [Nackman 1982]. Figure 22 shows equivalent and nonequivalent CPCGs. Figure 23 shows the catalog of the four basic types of slope

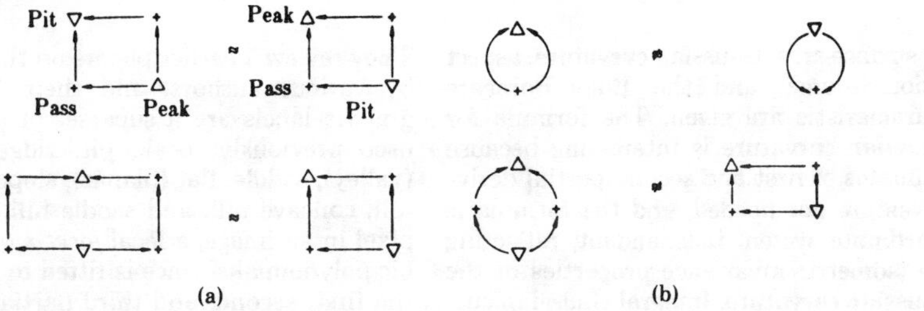

Figure 22. Equivalent and nonequivalent CPCGs. (a) Equivalent CPCGs; (b) nonequivalent CPCGs. (From Nackman [1984]; ©IEEE 1984.)

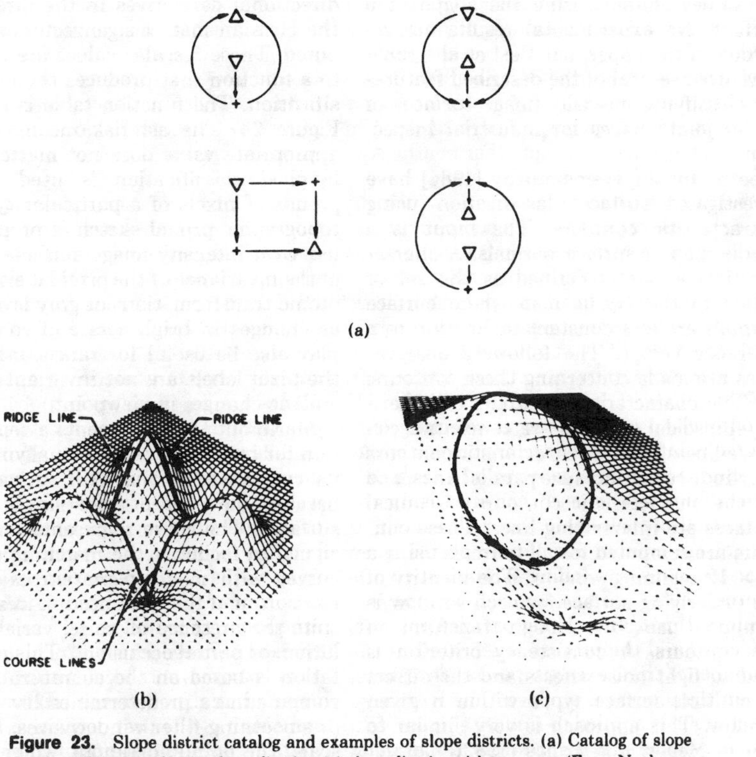

Figure 23. Slope district catalog and examples of slope districts. (a) Catalog of slope districts; (b) example of a slope district; (c) slope district with one pass. (From Nackman [1984]; ©IEEE 1984.)

districts and an example of a slope district with one pass.

Lin and Perry [1982] have investigated surface-shape description using surface triangularization. Differential-geometry-based shape measures are useful if they are reliably computed from sensor data. When a surface is decomposed into a network of triangles, many features are easily computed. Discrete coordinate-free formulas for surface area, Gaussian curvature, aspect ratio, volume, and the Euler–Poincare characteristic are given. The formula for Gaussian curvature is interesting because estimates of first and second partial derivatives are not needed, and the formula is coordinate system independent, reflecting the isometric invariance properties of the Gaussian curvature. Integral Gaussian curvature, integral mean curvature, surface area, volume, surface-area-to-volume ratio, integral curvature to the nth power, and genus (or handle number) are given as scalar values characterizing the shape of a surface. No experimental results are recorded in the paper, but Besl et al. [1985] have used several of the described features for classifying intensity-image surfaces of solder joint images for industrial inspection.

Sethi and Jayaramamurthy [1984] have investigated surface classification using characteristic contours. The input is a needle map of surface normals. A characteristic contour is defined as the set of points in the needle map where surface normals are at a constant inclination to a reference vector. The following observations are made concerning these contours: (1) The characteristic contours of spherical/ellipsoidal surfaces are concentric circles/ellipses, (2) the characteristic contours of cylindrical surfaces are parallel lines, and (3) the characteristic contours of conical surfaces are intersecting lines. These contours are computed for all normals using a 12×12 scanning window. The identity of the underlying surface for each window is computed using the Hough transform on the contours. A consistency criterion is used to fight noise effects and the effects of multiple-surface types within a given window. This approach is very similar to that of Hebert and Ponce [1982]. Correct classification results are discussed for synthetic 40×40 needle maps of adjacent cones and cylinders. It may be difficult to generalize the characteristic contour method to arbitrary surfaces because the characteristic contours of a general surface may not exhibit any usable regularities.

Haralick et al. [1983], Laffey et al. [1982], and Watson et al. [1985] discuss topographic classification of digital surfaces. They review 7 earlier papers on the subject by various authors, and their 10 topographic labels are a superset of all labels used previously: peak, pit, ridge, ravine (valley), saddle, flat (planar), slope, convex hill, concave hill, and saddle hill. At each pixel in an image, a local *facet-model* bicubic polynomial surface is fitted to estimate the first, second, and third partial derivatives of the surface at that pixel. Once the derivatives have been estimated, the magnitude of the gradient vector, the eigenvalues of the 2×2 Hessian matrix, and the directional derivatives in the direction of the Hessian matrix eigenvectors are computed. These 5 scalar values are the input to a function that produces the pixel classification. The function table is shown in Figure 24. The asterisk means that the appropriate value does not matter. Pixel-by-pixel classification is used to form groups of pixels of a particular type. The topographic primal sketch is proposed for use with intensity-image surfaces because of the invariance of the pixel labels to monotonic transformations of gray levels, such as changes in brightness and contrast. It may also be useful for range images, but the pixel labels are not invariant (in general) to changes in viewpoint.

Marimont [1984] presents a representation for image curves and an algorithm for its computation. His work is mentioned because the concepts are also useful for surfaces. The curve representation "is designed to facilitate the matching of image curves with model *plane curves* and the estimation of their orientation in space despite the presence of noise, variable resolution, or partial occlusion." This representation is based on the curvature function computed at a predetermined list of scales or smoothing-filter window sizes. For each scale, the points, or knots, which are the zeros and the extrema of curvature, are stored in a knot list with a tangent direction and a curvature value for each knot. These knots have the following properties:

(1) The zeros of the curvature of a 3-D plane curve almost always project to the zeros of the curvature of the corresponding (projected) 2-D image curve.

Pixel Classification Scheme

$\|\nabla f\|$	λ_1	λ_2	$\nabla f \cdot \omega^{(1)}$	$\nabla f \cdot \omega^{(2)}$	Label
0	-	-	0	0	Peak
0	-	0	0	0	Ridge
0	-	+	0	0	Saddle
0	0	0	0	0	Flat
0	+	-	0	0	Saddle
0	+	0	0	0	Ravine
0	+	+	0	0	Pit
+	-	-	-,+	-,+	Concave Hill
+	-	*	0	*	Ridge
+	*	-	*	0	Ridge
+	-	0	-,+	*	Concave Hill
+	-	+	-,+	-,+	Saddle Hill
+	0	0	*	*	Slope
+	+	-	-,+	-,+	Saddle Hill
+	+	0	-,+	*	Convex Hill
+	+	*	0	*	Ravine
+	*	+	*	0	Ravine
+	+	+	-,+	-,+	Convex Hill

Figure 24. Pixel labels for topographic primal sketch. (From Laffey et al. [1982]; ©IEEE 1983.)

(2) The sign of the curvature value at each point does not change within an entire hemisphere of viewing solid angle. The *pattern-of-curvature sign changes* along a curve are invariant under projection except in the degenerate case when the viewing point lies in the plane of the curve.

(3) Curvature is a local property, which makes it much more suitable than global curve properties for handling occlusion.

(4) Points of maximum curvature of 3-D plane curves project to points that are very close to points of maximum curvature of the projected 2-D image curves. The relationship between these points is stable and predictable, depending on viewpoint. Moreover, the relative invariance of these points increases as the curvature increases. Ideal 3-D corners almost always project to ideal 2-D corners.

The stability of these curvature critical points under orthographic projection is shown in Figure 25. The processing algorithm is outlined as follows: (1) The image curve data are smoothed at multiple scales by Gaussian filters and fitted at each scale with a continuous curve parameterization in the form of composite monotone curvature splines (arcs). (2) Curvature extrema (critical points) are extracted at each scale and stored in a list. (3) Dynamic programming procedures are used to construct a list of critical points that is consistent across the range of scales. (4) The integrated critical-point information is used to match the image curve against the computed critical-point information for the plane curve model. No experimental results for curve matching are quoted in the paper. Related work on hierarchical curve matching has been done by Mackworth and Mokhtarian [1984].

Langridge [1984] reports on an investigation into the problem of detecting and locating discontinuities in the first derivatives of surfaces determined by arbitrarily spaced data. Neighbor computations, smoothing, quadratic variation, and the biharmonic equation are discussed in this paper. The techniques are useful for detecting roof edges in range data.

Medioni and Nevatia [1984] have suggested a *curvature*-based description of 3-D range data. They propose the following features for shape description: (1) zero crossings of the Gaussian curva-

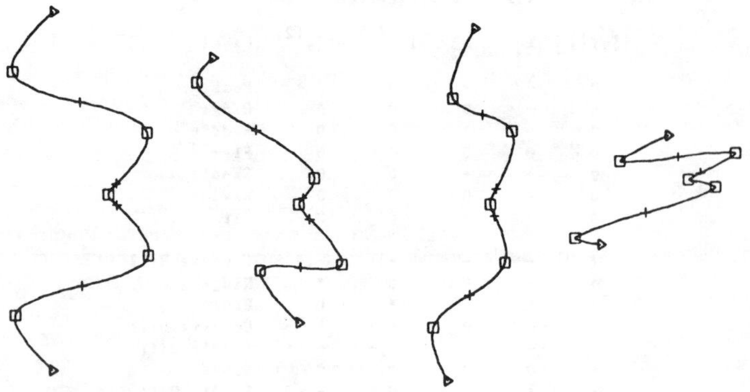

Figure 25. Stability of curvature critical points under rotation: □: max κ; △: min κ, $\kappa \neq 0$; +: $\kappa = 0$. (From Marimont [1984].)

ture, (2) zero crossings of the maximum principal curvature, and (3) maxima of the maximum principal curvature. These features are computed by smoothing a depth map with a large window and by using 1-D windows to compute directional derivatives. This 1-D derivative approach seems to be very sensitive to noise. Experimental results for the vase shape shown in the paper agree with this statement.

Brady et al. [1985] and Ponce and Brady [1985] have also explored the use of differential geometric quantities to describe surfaces, but they concentrate on lines of curvature, asymptotes, bounding contours, surface intersections, planar surface patches, and spherical surface patches. By using surface curves to constrain surface shapes, the curvature primal sketch work of Asada and Brady [1984] can be used for curve-shape description and, thus, surface-shape description. Brady et al. compute principal curvatures and principal directions, and they rely on an ad hoc scheme using a breadth-first search to link principal directions at each point into lines of curvature. They propose a surface primal sketch that combines information on significant surface discontinuities. This approach is domain independent. Experimental results for a light bulb, telephone receiver, coffee mug, and oil bottle look quite good. One problem with this approach is that the Curvature Primal Sketch Algorithm is designed for planar curves but is used to analyze nonplanar curves, such as lines of curvature. Because every surface curve of each surface is processed individually by the Curvature Primal Sketch Algorithm, the method is computationally intensive; 1-hour running times for 128 × 128 range images are mentioned.

Besl and Jain [1984, 1985a, 1985b] have implemented a surface characterization algorithm that computes surface curvature regions (or curvature districts), critical points, step edges, roof edges, and ramp edges as output. Differential geometry states that local surface shape is *uniquely determined* by the first and second fundamental forms of the surface [Lipshutz 1969]. Gaussian and mean curvature combine these first and second fundamental forms intrinsically and extrinsically to obtain scalar surface features that are invariant to *rotations, translations, and changes in parameterization*. Therefore, visible surfaces in depth maps have the same mean and Gaussian curvature from any viewpoint. (The two principal curvatures of a surface are directly computed from Gaussian and mean curvature, and vice versa.) Also, mean curvature uniquely determines graph surfaces if a boundary curve is specified, whereas Gaussian curvature uniquely determines convex regions of a surface. There are eight fundamental viewpoint-independent surface types that are characterized using only the *sign* of the mean curvature (H) and Gaussian curvature (K):

(1) H negative + K positive = peaked surface,
(2) H negative + K zero = ridge surface,
(3) H negative + K negative = saddle ridge surface,
(4) H zero + K zero = plane surface,
(5) H zero + K negative = minimal surface,
(6) H positive + K negative = saddle valley surface,
(7) H positive + K zero = valley surface,
(8) H positive + K positive = cupped surface.

These fundamental surface shapes are shown in Figure 26. Gaussian and mean curvatures are computed directly from a smoothed depth map using window operators that give least-square estimates of first and second partial derivatives [Anderson and Houseman 1942; Beaudet 1978; Bolle and Cooper 1984; Haralick 1984]. Examples of the first-stage output of this surface characterization algorithm are shown in Figure 27 and discussed in detail in Besl and Jain [1985a, 1985b]. The combination of surface curvature images, depth discontinuities, critical points, and other derived images provides rich surface information that is used to identify surfaces. Connected component regions of pixels of a given surface type are easily isolated. Since each surface region is homogeneous, low-order polynomial surfaces can fit these regions quite well; the rms error is typically less than one discrete depth level during this second-stage refinement computation. Hence, higher level accurate surface approximations are obtained without domain-specific assumptions.

5.7 Object-Reconstruction Algorithms

Although our direct concern is object recognition, we are also interested in object reconstruction because it involves many similar ideas. For example, one approach to the object-recognition problem is to reconstruct as much of an object's surface as possible from a single view and then perform a matching of that object's surface with object models in the model domain. In addition, object reconstruction can play a significant role in automating the model-formation process. We survey intensity-image methods first, followed by methods that use depth maps. Object-reconstruction schemes based on the shape-from-(xxx) methods are not considered.

Baumgart [1974] performed perhaps the first intensity-image reconstructions using his GEOMED (GEOMetric EDitor) solid modeling program. Objects are placed on a turntable so that high-contrast video images can be digitized from several known views. Object silhouette contours are obtained by (1) thresholding images, (2) extracting connected component "blobs," (3) converting the blobs to polygons, and (4) "dekinking" the polygons to obtain smooth shapes. The contour shapes are extruded (or swept linearly) to create 3-D polyhedra, where each polyhedron is properly oriented for the corresponding view. The polyhedra are intersected to obtain a polyhedral representation of the object on the turntable. This process is referred to as *object locus solving* using *silhouette cone intersection*. The method is shown in Figure 28 for two silhouette contours. Reconstruction results are shown for two toy horses and a doll; the horses are very well done. Because the doll's hair was black, her head did not show up in one of the silhouette contours (owing to thresholding), resulting in a reconstructed doll model with no head. Baumgart stated in his Ph.D. dissertation that "in computer vision, geometric models provide a goal for descriptive image analysis, an origin for verification image synthesis, and a context for spatial problem solving." His views about the role of modeling and feedback in computer vision are somewhat similar to our ideas expressed above.

Baker [1977] presents a scheme for building object models from many intensity images taken from different known rotated views. A fundamental premise of his work is that "effective vision requires flexible, domain-free, 3-D modeling." His method tracks edge curvature irregularities using correlation techniques over a series of rotated views and creates a wire-mesh exoskeleton (wire-frame) representation. Experimental results are shown in the paper for two complex smooth-surfaced objects.

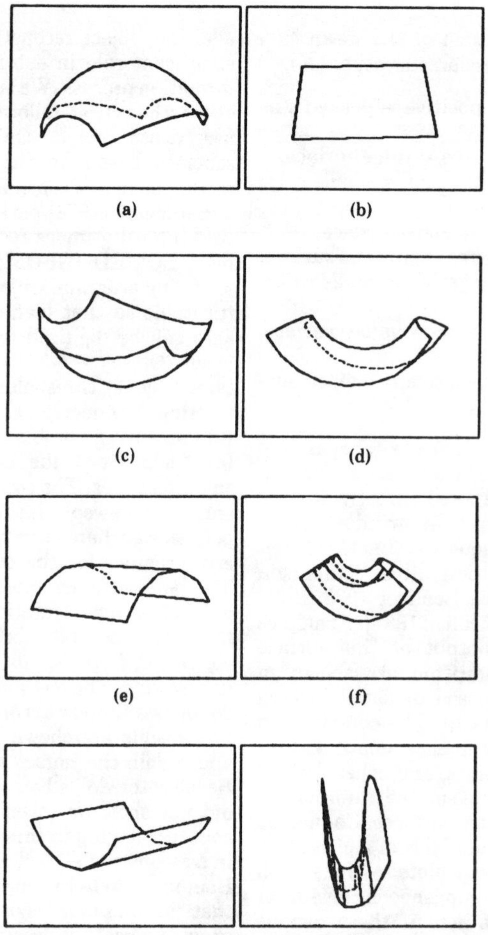

Figure 26. Eight fundamental surface shapes. (a) Peak surface: $H < 0$, $K > 0$; (b) flat surface: $H = 0$, $K = 0$; (c) pit surface: $H > 0$, $K > 0$; (d) minimal surface: $H = 0$, $K < 0$; (e) ridge surface: $H < 0$, $K = 0$; (f) saddle ridge: $H < 0$, $K < 0$; (g) valley surface: $H > 0$, $K = 0$; (h) saddle valley: $H > 0$; $K < 0$.

The results are difficult to visualize owing to the wire-frame line drawings, but they are very detailed. Baker suggests a matching process that uses the maximum-breadth axis of the object and a list of the nth (e.g., $n = 6$) most convex and concave points (points of *high surface curvature*). His algorithm successfully matched two descriptions of the same object that were analyzed in two different orientations.

Bocquet and Tichkiewitch [1982] have an expert system approach to the object-reconstruction problem. Their system accepts input in the form of standard me-

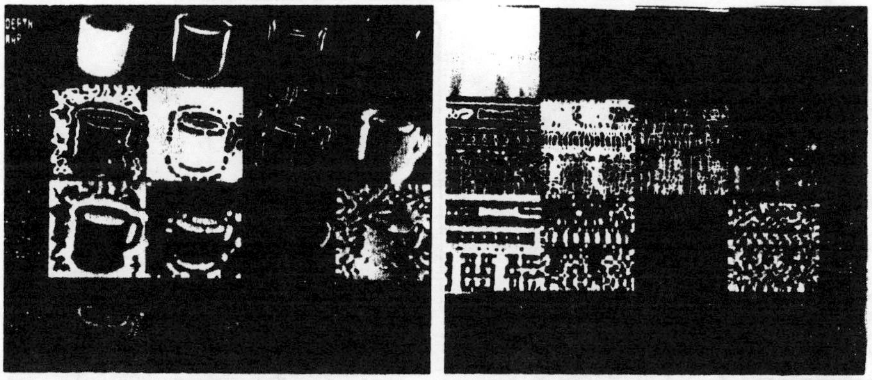

(a) (b)

f_{smooth}	$\sqrt{g}$	Q	ϵ
zeros(H)	zeros(K)	zeros(cosΘ)	cosΘ
sgn(H)	sgn(K)	$\sqrt{H^2 - K}$	Φ_1
$\lvert H \rvert$	$\lvert K \rvert$	$Q \neq 0$ $\nabla f = 0$	$\nabla f = 0$

(c)

Figure 27. Surface characterizations for (a) coffee cup and (b) keyboard. (c) Surface characterization results format: f, surface function; H, mean curvature; K, Gaussian curvature; g, metric determinant; Q, quadratic variation; ϵ, local surface fit error; θ, coordinate angle; Φ_1, principal direction angle.

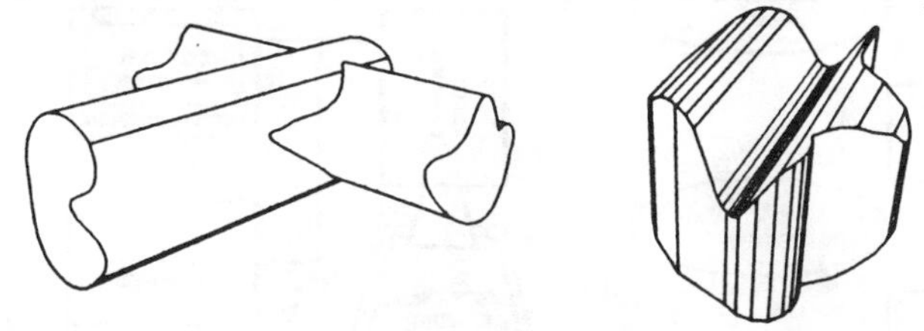

Figure 28. Silhouette intersection process and resulting object.

chanical drawings from three orthogonal views. Drawings are digitized, and line and arc segments are given internal representations. The list of segments is structured into a 2-D relational database that monitors all closed contours. A set of production rules (the knowledge base) are used to infer 3-D surfaces from the 2-D data. Hypothetical surfaces generated from one view are projected into the next view in order to check the corresponding 2-D segments. The system can make adjustments and continue, or it can backtrack. When a surface representation for the object that is compatible with the given views is obtained, the object is drawn from various viewpoints. Figure 29 contains several diagrams that describe the overall structure of the system. When no production rules apply, the system requests a new rule from the operator. If rules have been found for all segments of a contour, but no surface is generated, the system explains the problem and allows the rule base to be modified by the person operating the system. Presumably, a fairly robust system results after many objects

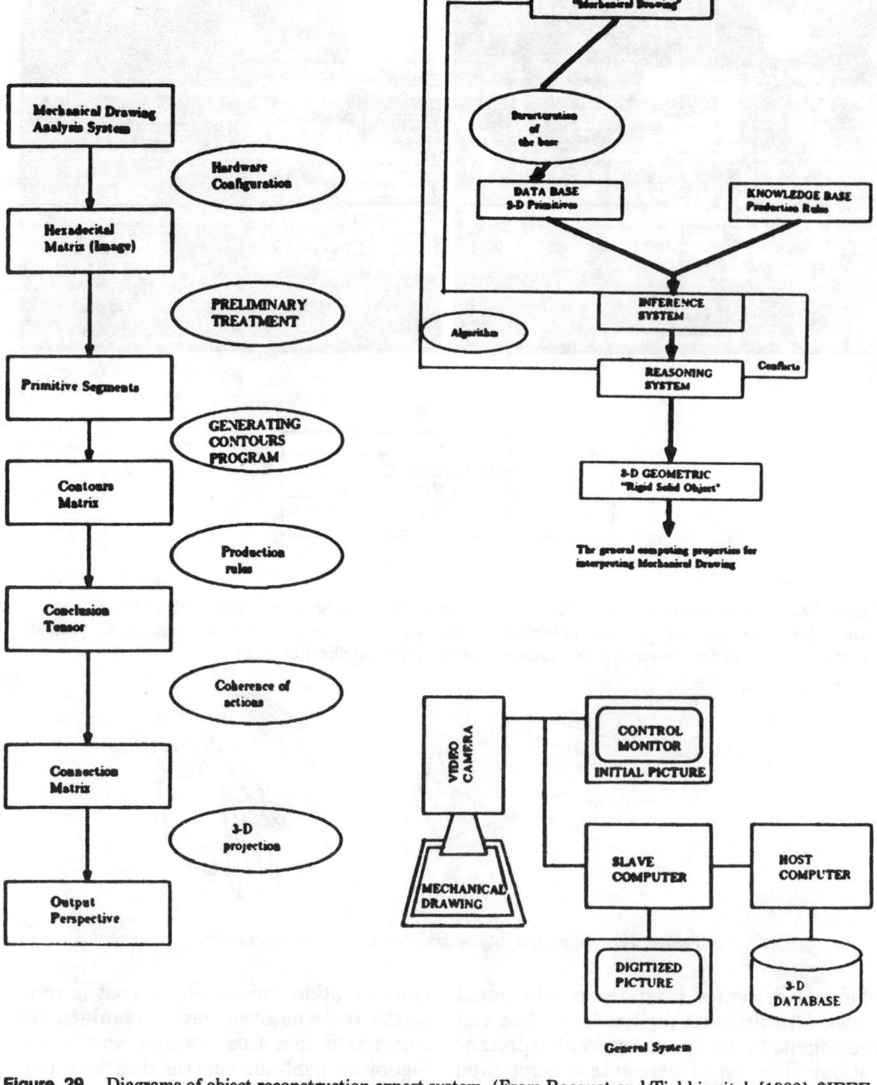

Figure 29. Diagrams of object reconstruction expert system. (From Bocquet and Tichkiewitch [1982]; ©IEEE 1982.)

have been successfully reconstructed. This system could *possibly* be generalized to work from high-quality edge maps rather than from drawings. Reconstruction results are shown for a machined part, along with its input drawings. More object detail is possible from this approach than from those that construct volume descriptions from multiple silhouette boundaries because of the potential ambiguity in silhou-

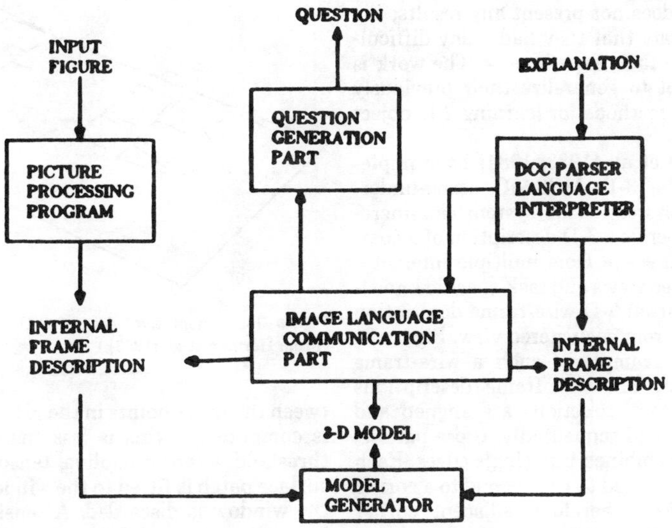

Figure 30. System organization for object reconstruction. (From Abe et al. [1983].)

ettes. Aldefeld [1983] has described a related method for reconstruction of 3-D objects from 2-D drawings.

The Bocquet and Tichkiewitch [1982] paper was preceded by the work of Lafue [1976]. He wrote a program for interpreting orthographic view drawings as 3-D objects. Heuristics were used to resolve ambiguities that occurred during the aggregation of points, edges, and faces into polyhedra. Instead of an expert systems approach, Lafue used a theorem prover to choose the right geometry for each set of local alternatives. His program was also written to prompt the user for more information whenever the system could not proceed on its own.

Shapira and Freeman [1977a] describe a procedure for reconstructing objects bounded by planar or quadric surfaces from a set of photographs of a scene taken from different viewpoints. Line and junction labelings are used to extract model surface descriptions of objects using the "cyclic-order property" described in Shapira and Freeman [1977b]. In this approach, it is necessary that vertices be formed by exactly three edges (three faces). No other object-shape restrictions are used, and the method is designed to handle a limited number of imperfections in the line–junction feature data. Experimental results are shown for one high-contrast scene with several simple objects.

Abe et al. [1983] proposed a system for building 3-D qualitative object models of objects with cylinderlike bodies, given several 2-D intensity-image views and verbal explanations of object structure. Their system mainly consists of a language interpreter, an image-to-language communications subsystem, and a model generator. It also includes an image-processing subsystem and a question generator for interacting with the human operator. Figure 30 shows a block diagram of the system organization. The image inputs for each view and the verbal inputs are processed and stored in separate, internal frame representations. These frame representations are matched for each view to generate a consistent labeling. Different views of an object are constrained by language input and combined using a graph-matching process to create a 3-D model. This paper is preliminary and does not present any results; the authors state that they had many difficulties during their experiments. The work is an attempt to generalize their previously developed methods for learning 2-D object shapes.

Herman et al. [1983, 1984] have implemented the 3-D MOSAIC scene-understanding system. This system can incrementally derive a 3-D description of a complex urban scene from multiple intensity-image stereo views and task-specific knowledge. A partial 3-D wire-frame description is derived from each stereo view. Figure 31 shows an example of such a wire-frame description. The wire-frame descriptions from different directions are aligned and then processed sequentially. Close parallel edges are combined into single edges. Each vertex is assumed to correspond to a corner of an object; therefore, adjacent corner edges correspond to a corner of a planar face. Compatible corners of faces are merged into complete faces. Faces that are probably flat roofs are converted into buildings by adding more faces between the roof and the ground. A complete set of similar rules is applied to the sequence of images. Several rules are shown in Figure 32. Finally, a complete scene description is generated and rendered as a gray-scale image. The technique works well for particular domains in which block shapes predominate (e.g., urban scenes). In this research, all objects are assumed to be block shaped, all surfaces are assumed to be horizontal or vertical, all parallelograms are considered rectangles, unless there is evidence to the contrary, and even the ground plane is assumed known. Unfortunately, there are few general-purpose results available from such an approach.

We now move from intensity-image-based techniques to methods based on range data. Vemuri and Aggarwal [1984] have developed an algorithm for reconstructing 3-D objects using range data from a single view. Their algorithm proceeds as follows: (1) The range image is partitioned into predetermined overlapping $K \times K$ window neighborhoods. The overlap is two pixels. (2) For each neighborhood, the standard deviation of the Euclidean distance between the range points in the $K \times K$ window is computed. If this is less than a preset threshold, a tension-spline, tensor-product surface patch is fitted to the window. If not, the window is discarded. A tension spline is a spline that exhibits properties similar to a mechanical spline subjected to uniform tension. These splines are used as basis functions for surface patches. (3) The principal surface curvatures (minimum and maximum) are computed at each point in the remaining patches. If the magnitude of either curvature value exceeds another preset threshold, the point is labeled as an edge pixel. The depth map is thereby approximated by a set of continuous surface patches, and the edges within those patches are determined. The surface patch model is then passed through a graphics algorithm with a light source model to obtain a shaded image. Experimental results for one synthetic range image and one real range image are displayed in the paper. Figure 33 shows the synthetic range-image results for a car shape. The algorithm's results consist of a shaded image and an edge map. The authors do not point out that shaded images can be obtained directly from the depth map itself, without any intermediate surface fitting. For example, see Figure 34 where the coffee cup and the keyboard range images, shown in Figure 14, have been used to create synthetic intensity images via smoothing and shading algorithms. No actual object reconstruction appears to have been done yet in Vemuri and Aggarwal's work; a set of surface patches has

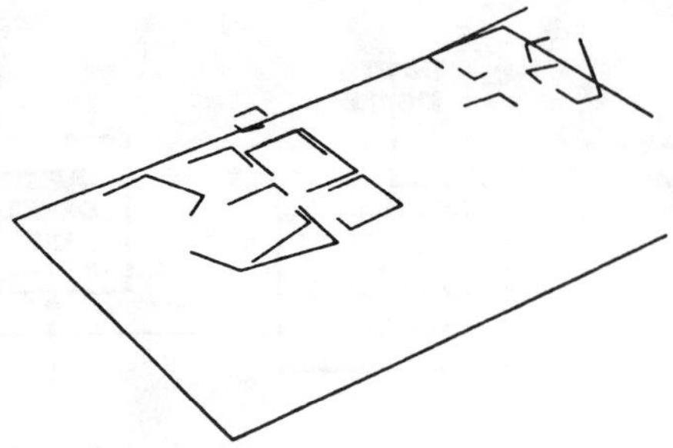

Figure 31. Perspective view of 3-D wire frames. (From Herman et al. [1983].)

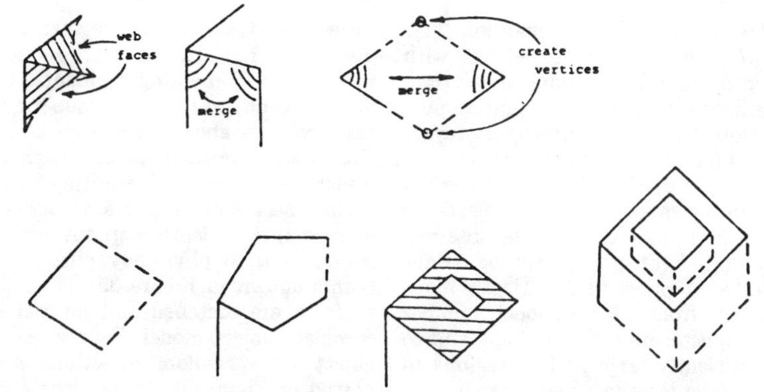

Figure 32. Obtaining surface-based description from wire frames. (From Herman et al. [1983].)

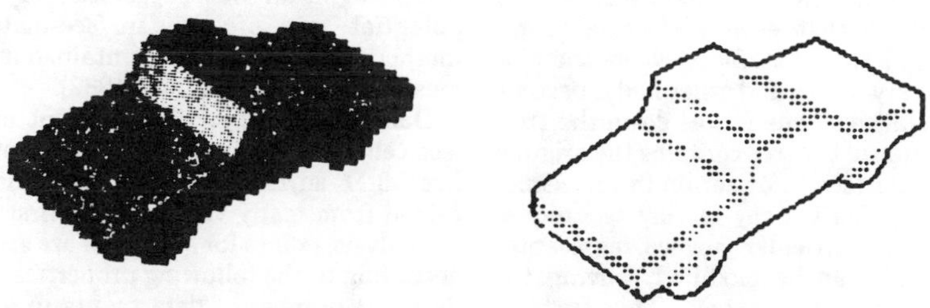

Figure 33. Surface reconstruction: step edges and roof edges. (From Vemuri and Aggarwal [1984]; ©IEEE 1984.)

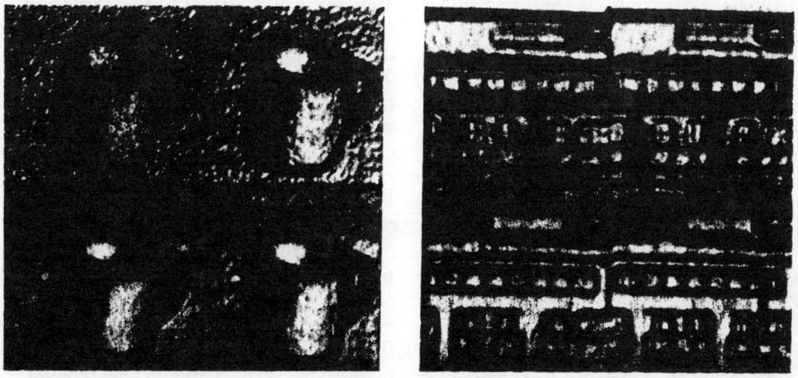

(a) (b)

Figure 34. Intensity images generated from ERIM range images. (a) Coffee cup range image smoothed with size 3, 5, 7, 9 windows; (b) keyboard range image artificially illuminated from different directions.

been fitted to data. They do mention their intention to merge surface patches with similar properties into regions and form a region adjacency graph for recognition purposes. Note the use of *principal surface curvatures* for edge detection.

Potmesil [1979, 1983] describes a method for generating models of solid objects by matching 3-D surface segments obtained using a white-light, grid-projecting, triangulation-based range finder. Depth maps are obtained from a large enough number of views to determine object shape and to allow sufficiently large surface regions to be imaged in at least two views. Range data for object surfaces are fitted with a sheet of parametric bicubic surface patches. Rectangular patches are recursively merged (four at a time) into a quadtree hierarchical structure so that each surface is represented by a tree of surfaces, where the root node is very coarse and other nodes become more detailed as one moves down the tree. The bottom of the tree contains the original patches. Surface information in this structure is queried via ray-casting techniques so that the particular surface representation details can be modified, leaving the rest of the system intact. This surface-model modularization is an important concept for a flexible system. *Surface matching* is defined as "finding a spatial registration of two surface descriptions that maximizes their shape similarities." Given a particular surface to be matched and a set of other potentially matching surfaces, registration transformations are computed for all surfaces to find the one that provides the best surface-segment match. Evaluation points are selected at each level in the quadtree representation, either at surface control points or at points of *maximum curvature.* These evaluation points are used to compute (1) positional differences, (2) orientation differences, and (3) curvature differences for input to the matching algorithm. A heuristic search algorithm is used to control generation of registration transformations. The initial guess corresponds to the alignment of surface normals at the top of the surface tree. When sufficiently close surface-segment matches are found, a merging algorithm generates a new surface for each matched surface segment. A complete object model is created by sequentially matching and merging segments. Experimental results of this technique for a balsa car model are shown in Figures 35 and 36. Thirty-six gray-scale images of grids from 6 views are generated resulting in 18 3-D surface segments. These are matched and merged into 6 depth-map surfaces corresponding to 6 physically different viewpoints shown in Figure 35. Then these 6 surfaces are matched and merged into 1 complete object model. Four views of the object from random directions are displayed in Figure 36. Generalizing Potmesil's approach for object recognition would seem a minor additional step, since surface matching is already implemented, but object recognition was suggested only as a potential use of the surface-matching method. More details are contained in Potmesil's Ph.D. dissertation [1982].

Dane and Bajcsy [1982] present an object-centered 3-D model builder that utilizes 3-D surface point information obtained from many views. In the first stage of analysis, points for each view are grouped according to the following properties of the data: (1) number of data points in a local area, (2) average and deviation of depth values, and (3) average and deviation of X and Y components of the normal vector; and according to the following derived properties of the data: (a) local curvature in an X–Z plane, (b) local curvature in a Y–Z plane, (c) surface orientation discontinuities, and (d) surface depth discontinuities. It is not stated how these derived properties are computed, or how the properties directly affect the grouping process. Points are grouped according to these properties, and a planar or quadric surface primitive is fitted using a least squares technique. Subsequent analysis determines edges and corners, which are stored in an edge-graph data structure. View analysis is followed by view integration, where surface primitives are transformed using the known transformations into a common global coordinate system; identical surfaces are identified; and surface parameters are modified for overall object compatibility. The resulting object description is stand-

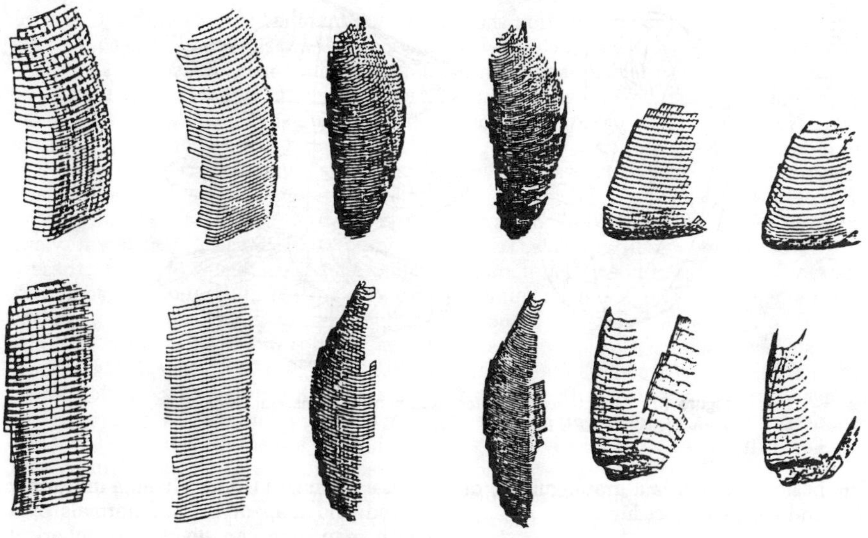

Figure 35. Eighteen surface segments merged into six new segments. (a) Top side; (b) left side; (c) front side; (d) bottom side; (e) right side; (f) back side. (From Potmesil [1983].)

ardized by placing the origin at the center of gravity of the object and aligning the x, y, z directions with the principal axes of the object. This algorithm was tested with nine objects and made only one error. Thirty-six views were used to define the object, although not all points in all views were used in every case. Data acquisition is not described since 3-D data points are assumed to be available from an external source. Dane's Ph.D. dissertation [1982] elaborates on these topics.

We note here that the work of Henderson [1982, 1983] (mentioned in Section 5.5) describes a technique for automatically reconstructing a polyhedral model of an object using points obtained from several views.

Boissonnat [1982], and Boissonnat and Faugeras [1981], and Faugeras et al. [1982] describe an efficient ($O(N \log N)$) way of building a polyhedral approximation of 3-D points obtained from a triangulation-based laser range finder. A 3-D algorithm is presented as a generalization of the triangularization algorithm for a 2-D polygon.

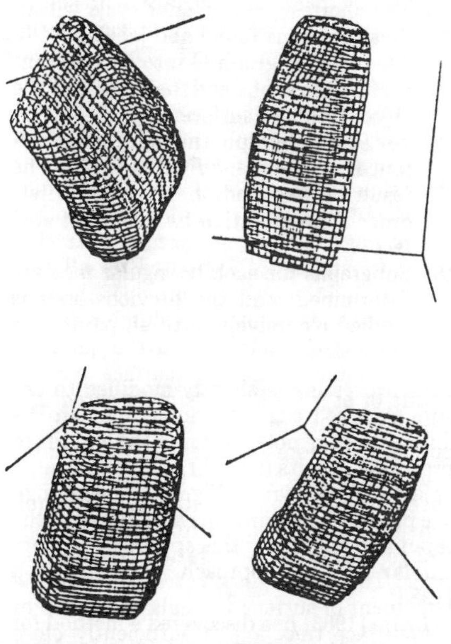

Figure 36. Four views of surface segments matched into one model. (From Potmesil [1983].)

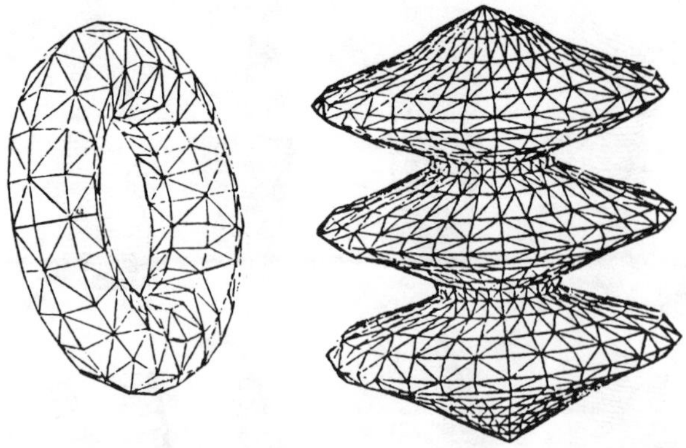

Figure 37. Objects reconstructed from point information. (From Boissonnat [1982]; ©IEEE 1982.)

The basic approach is a graph-guided, divide-and-conquer procedure:

(1) A planar graph $G = (V, A)$ is constructed using the given 3-D points where arcs connect nearest neighbors.
(2) Three nonneighboring points P, Q, R are selected for initialization.
(3) The shortest most planar cycle within these points is found and labeled PQR. This divides graph G into two disconnected subgraphs and the surface of the object into two surfaces.
(4) For each subgraph, the point most distant from the plane PQR is found. The resulting hexahedron is now a first-order approximation to the object surface.
(5) Subgraphs for each triangular face are determined, and the previous step is applied recursively until all points are exhausted.

This processing is slightly modified to ensure that bad edges do not remain in the approximating polyhedron as the algorithm proceeds. Results for two fairly complicated objects are shown in Figure 37. Although the processing algorithms are different, the input and output for this approach are very similar to the approach of Henderson [1983].

Little [1983] has discovered a method for reconstructing convex polyhedral object models from their corresponding extended Gaussian images (EGI). Given a uniformly spaced grid map of surface normals of a depth map, one can divide the spherical solid angle into bins and form an orientation histogram by computing the number of grid points with normal vectors that fall into each bin. This orientation histogram is referred to as a discrete EGI. It can be shown that the continuous analog of this discrete EGI uniquely determines a convex polyhedron via a nonconstructive proof [Minkowski 1897]. Little posed the object reconstruction problem as an iterative constrained-minimization problem and solved it. His results answer a previous open question concerning the inversion of EGIs. This is certainly of mathematical interest regardless of the limited nature of the class of convex polyhedral objects.

Connolly [1985] has developed a technique for building octree object models from range data. He converts a range image into a quadtree format, extends the quadtree into an octree, and merges octree descriptions from different views to create a 3-D object model.

Range data reconstruction methods are more quantitative than the intensity data methods because of the explicit shape information in range images. Higher level spatial reasoning and inference processes must be used in intensity-image reconstruction to compensate for the lack of depth information.

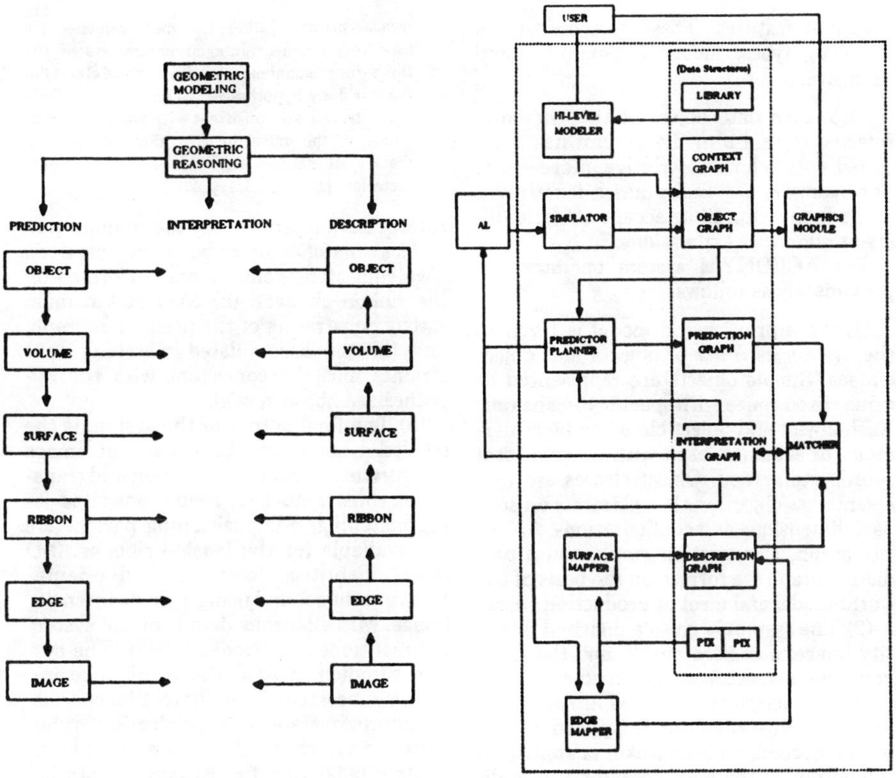

Figure 38. The ACRONYM system. (From Brooks et al. [1979].)

5.8 Object Recognition Using Intensity Images

Brooks [1981, 1983] and Brooks et al. [1979] explain how model-based 3-D interpretations of 2-D images are possible using the rule-based ACRONYM system. ACRONYM is frequently mentioned in the computer vision literature. This is probably because of the flexibility and modularity of its design, its use of view-independent volumetric object models, its domain-independent qualities, and its complex, large-scale nature. Figure 38 shows a block diagram of the ACRONYM system and a diagram of the hierarchical geometric reasoning process. The system is based on the prediction–hypothesis–verification paradigm. The three main data structures of the system are the following:

(1) *Object graph.* The nodes of the object graph are generalized-cone object models. Object-graph arcs correspond to the spatial relationships among the nodes (e.g., relative translations and rotations) and the subpart relations (e.g., is-a-part-of).

(2) *Restriction graph.* Nodes are constraints on the object models of a given object class. The directed arcs of the restriction graph represent subclass inclusions.

(3) *Prediction graph.* Prediction graph nodes are invariant and quasi-invariant observable image features of objects. The arcs specify the image relationships among the

invariant features. These arcs are of the following types: must-be, should-be, and exclusive.

Also, each data object of the system is referred to as a *unit*. Every unit has associated *slots* to hold descriptive information. For example, a cylinder has a length slot and a radius slot. Slots accept numeric *fillers* or *quantifier* expressions.

The ACRONYM system operates approximately as follows:

(1) An a priori world model is given to the system as a set of objects and object classes. Simple objects are represented as generalized cones with specific dimensions. Each object and object class can be a hierarchy of subparts, each with its own local coordinate system. Object classes are represented as objects with constraints on subpart dimensions and configurations. An object graph, a restriction graph, and a prediction graph are formed on the basis of the world model and a set of production rules.

(2) The system is given a digitized intensity image, a camera model, and the three graph data structures created above.

(3) The image is processed in two steps. First, an edge operator is applied to the image. Second, an edge linker is applied to the output of the edge operator and is directed to look for ribbons and ellipses. Ribbons and ellipses are the 2-D image projections of the elongated bodies and the ends of the generalized cone models, respectively. The higher level 3-D geometric reasoning in ACRONYM is based entirely on the 2-D ribbon and ellipse symbolic-scene description.

(4) ACRONYM searches for instances of object models in terms of the ribbons and ellipses. The heart of the system is a nonlinear constraint manipulation system (CMS) that generalizes the linear SUP-INF methods of Presburger arithmetic [Bledsoe 1974; Shostak 1977]. Constraint implications are propagated "downward" during prediction and "upward" during interpretation. The interpretation matching process is described by Brooks as follows:

> Matching does not proceed by comparing image feature measurements with predictions for those measurements. Rather the measurements are used to put constraints on parameters of the three-dimensional models, of which the objects in the world are hypothesized to be instances. Only if constraints are consistent with what is already known of the model in three dimensions, then these local matches are retained for later interpretation. [Brooks 1981, p. 288]

Interpretation proceeds by the combination of local matches of ribbons into clusters. Two consistency checks are performed on the ribbon clusters: (a) each match must satisfy constraints of the prediction graph, and (b) the accumulated matching constraints must be consistent with the hypothesized object model.

(5) The final output of the system is the labeled ribbons of the consistent image interpretation. Since orientation and translation constraints have been propagated using matching, 3-D positioning parameters are available for the labeled ribbons. 3-D object identities, locations, and orientations are thus found using a single intensity image. Miscellaneous details of the system are mentioned in Brooks [1983]. The system is implemented in MACLISP. The predictor subsystem of ACRONYM consists of approximately 280 production rules. During a typical prediction phase, approximately 6000 rule firings occur. Rotation and translation operations are treated as matrix operator strings, where the string length is typically 10 or more.

Despite the detailed 3-D concerns in the ACRONYM design, no 3-D interpretation results have ever been published to our knowledge. Aerial images of jets on runways and jets near airport terminals have been successfully interpreted using ACRONYM. There are other less complicated schemes that could yield similar results on aerial images of this type. Binford once wrote that

> there is no profound reason why ACRONYM could not recognize aircraft in images taken at ground level, although it will probably break when tested on such images because of bugs or missing capabilities that were not exercised previously. [Binford 1982, p. 39]

The theory and the implementation of ACRONYM are two separate issues, but curious readers are left wondering whether this complicated system is as robust as it might seem. Many system difficulties have been blamed on the quality of output from the ribbon-finding mechanism. There are no feedback connections between the final decision-making mechanism and the original data.

ACRONYM's problems provide a reminder for us that any open-loop system is only as robust as its most limited component. Even the best possible geometric reasoning system cannot be successful if its input is consistently unreliable and no feedback paths exist. Feedback could possibly be provided by rendering algorithms that relate object models to sensor data at intermediate levels of interpretation.

There are many other 3-D object-recognition schemes based on intensity images. Mulgaonkar et al. [1982] have devised a scene analysis system that recognized 3-D objects from a single perspective view using geometric and relational reasoning. *Generalized blobs* (sticks, plate, and blobs) are used to represent the 3-D geometry of objects. Object-recognition algorithms work with intensity images that have already been smoothed, thresholded, and segmented to produce a 2-D convex polygon decomposition of image regions. The system can only handle one object per image. It performs direct matching between 3-D images and 3-D objects using constraint propagation and backtracking. All objects are assumed to be in an upright position. The "connect/support" and "triples" relations between 3-D and 2-D primitives play a fundamental role in the recognition process. Thresholds are employed for measures of circularity and relational error. Seventeen out of 22 cases (77 percent) exhibited successful recognition in the experimental results, which used 11 different objects. Camera parameters were estimated from the match to within 10 degress on tilt and 20 degrees on pan. The use of structural relationships is an important feature of this method. Further research is required to generalize these concepts for use with more descriptive object representations.

Fisher [1983] has implemented a data-driven object-recognition program called IMAGINE. Surfaces are used as geometric primitives. There are three major stages in the operation of this program:

(1) Image regions determined by their region boundaries are matched to model object surfaces with the goal of estimating surface orientation parameters. Specific object surfaces are hypothesized.

(2) Hypothesized object surfaces are related to object models constrained by the structural relationships implied by the objects. Specific objects are hypothesized.

(3) Hypothesized objects are verified using consistency checks against constraints due to adjacency and ordering.

The program has four specific goals: (1) to locate instances of 3-D objects in 2-D images, (2) to locate image features corresponding to all features of the model *or* explain why the image features are not present, (3) to verify that all features are consistent with the geometrical and topological predictions of the model, and (4) to extract translation and rotation parameters associated with all objects in the scene. The input to the IMAGINE program is presegmented surface regions. The regions have the property that all boundaries between regions correspond to surface or shape discontinuities. The only information used by IMAGINE is the 2-D boundary shape of the segmented surface regions. The object models of the program are surface boundary models where all surfaces are planar or have only a single axis of curvature. Subcomponent hierarchies for objects determine the joint connections of subparts. Model surface-to-image-region matching is performed using a set of heuristics that generate rotation, slant and tilt, distance, and x–y translation-in-a-plane hypotheses. These heuristics gave reasonable results in 94 out of 100 test cases. Given the hypothesized surfaces and their position and orientation in space, a set of 10 rules is applied to generate object-model hypotheses. Another set of occlusion-handling rules is applied for object verification. Fisher provides his own list of program criticisms, which include the following: (1) The heuristic parameter estimation techniques require mostly planar surfaces. (2) The program's surface modeling does not account for surface shape internal to the region boundary. (3) Surface segmentation is currently done *by hand* with the assumption that adequate techniques will soon be available. (4) Its object models are nongeneric.

Despite these criticisms, the program did achieve its goal of recognizing and locating a PUMA robot and "understanding" its 3-D structure in a test image. Valuable ideas concerning occlusion are presented in the paper.

General convex polyhedra are a special object class. Underwood and Coates [1975] developed a technique that reconstructs object shape from multiple-view intensity images. Edges and planar surface regions extracted from the images are input to the reconstruction algorithm, which constructs internal models. The algorithm, however, is not given any information about viewing parameters for the different vantage points. The internal object model is topological and considers only relationships among surfaces. The models of different views are matched, and a complete topological object model is constructed using a graphical learning tree. These object models can then be used for object recognition. Experimental results for 20 test views matched against an object library of 19 objects yielded an 18 out of 20 success rate. Extensions to more general objects are suggested in the paper. The limitations of this implementation are mainly due to incomplete use of spatial information.

Lee and Fu [1982, 1983] propose a design for a general computer vision system that would be capable of 3-D object recognition using a single image. The system design and system flowchart are shown in Figure 39. Lee and Fu are interested in creating a system that allows for the proper interaction of top-down (model-guided) analysis and bottom-up (data-driven) analysis. The proposed system consists of the following six components:

(1) General-Purpose Primitive and Relation Extractor, which uses no higher level knowledge: Input = Input Image + Requests for More Evidence from Grouping Processor (2); Output = Extracted Primitive and Relation Information for (2).
(2) Primitive and Relation Organizer for Grouping Process: Input = Output from (1) + Requests for Reorganization from the Cognitive Interpreter (4); Output = Organized Set of Primitives and Relations.
(3) Associative Memory Network with Knowledge of World Model Objects: Input = Output from (2); Output = Candidate Object Models Compatible with Input Primitives and Relations.
(4) Cognitive Interpreter: Input = Output from (1) + Replies from (2) + Object Models from (3) + Object Evaluations from Cognitive Description Generator (5); Output = Object Models to be Verified for (5) + Requests for Reorganization for (2) + Final Output Image Description for System User when decision-making processing terminates.
(5) Cognitive Description Generator: Input = Object Models to be Verified from (4) + Results of Verification Search from Special-Purpose Image Processor (6); Output = Request for Finding a Particular Primitive or Relation for (6) + Evaluation of an Object Model for (4).
(6) Special-Purpose Primitive and Relation Finder: Input = Input Image + Requests from (5) Output = Results from Verification Searches.

Note the verification feedback from the original image and the *multilevel* interaction of the different components in this design. The processing consists of three basic processes: object description generation, model retrieval, and model verification. The two papers by Lee and Fu have concentrated on object description generation: Images are converted to the gray-level geographic structure (GLGS) representation. Target regions are selected so that they correspond to the maximum of the "conspicuousness" function. Extracted edges in target regions are classified as one of the following: (1) parallelogram, (2) ellipse, (3) skewed-symmetric arc, or (4) corner. Regularity constraints are applied:

(1) Parallelograms are always projections of rectangles.

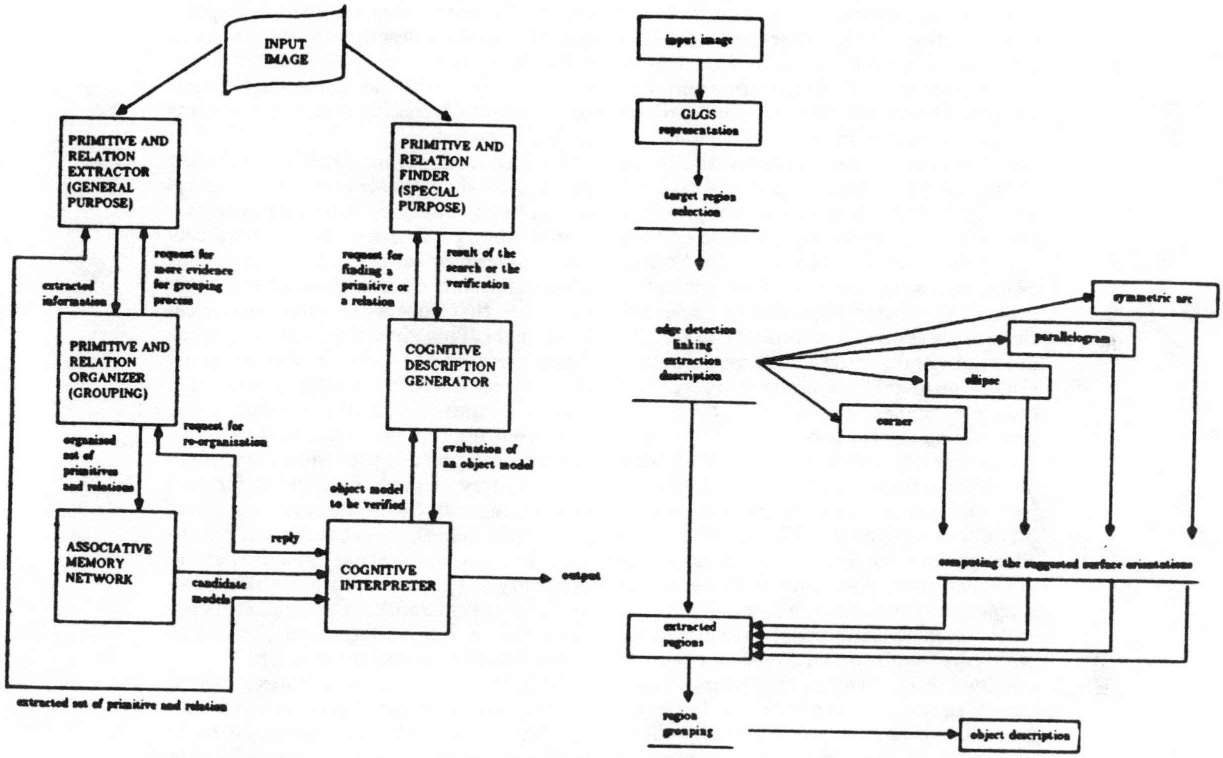

Figure 39. Design of vision system and flowchart. (From Lee and Fu [1982]; ©IEEE 1982.)

(2) Ellipses are always projections of circles.
(3) Skewed-symmetric arcs are always projections of a symmetric planar curve.
(4) Corners are always intersections of orthogonal line segments.

These regularity constraints and the so-called "least-slant-angle" preference rule are used to compute 3-D surface orientations of the selected target regions. Local interpretations of regions are propagated to neighboring regions stored in the edge-region adjacency graph. Constraints and consistency checks interact to yield a rough object description in terms of visible surface orientations. Experimental results are shown for a car and a machine shop tool. The final output is a line drawing, where each surface is drawn with a slant and tilt vector for its normal. This system is limited owing to its use of only four geometric primitive extracted edges and its use of the right-angle assumption throughout. Also, it does not handle curved surfaces in a consistent manner. The preliminary general design principles were followed by a nongeneral implementation.

Chakravarty and Freeman [1982] have developed a technique that uses characteristic views as a basis for intensity-image 3-D object recognition. The set of all possible perspective-projection views of an object is partitioned into a finite set of topological equivalence classes, which are represented by characteristic views. Different views within an equivalence class are obtained from one another using linear transformations. The number of characteristic views is reduced still further by allowing objects to have *only stable orientations* as positioned on a planar surface. Matching is performed using line–junction labeling constraints on detected edges. The method requires silhouette determination to guide the matching process (this is a disadvantage for occlusion handling), and, in addition to identifying objects, it produces position and orientation information as output. A system structure diagram is shown in Figure 40. Chakravarty's Ph.D. dissertation [1982] describes the details.

Some 3-D object-recognition techniques are based purely on object silhouettes and cannot, therefore, distinguish among objects that have the same set of silhouettes. McKee and Aggarwal [1975] have worked on recognizing 3-D curved objects from a partial silhouette description. Three-dimensional object models are not used, however. During the training process, the system learns the global silhouette boundary description for each view of an object and stores the description in an object-view library. The recognition algorithm accepts a partial boundary description and produces a list of all the compatible objects in the library. This work did not include view-independent processing and had problems with noisy edges.

Wallace and Wintz [1980] have used global 2-D shape descriptors to recognize 3-D aircraft shapes by matching against a stored library of shape descriptors. One shape descriptor set is computed and compressed for each discrete viewing angle in a finite set that covers the entire spherical solid angle. This gives the system view independence at the cost of storing many descriptors. Given an arbitrary view of a known aircraft, 2-D shape descriptors are computed for the silhouette and matched against *each* precomputed view description in the library of shape descriptors for *each* possible aircraft. Since the entire outline of an aircraft is available at sufficient resolution in this application, global Fourier boundary shape descriptors are used providing excellent results. Research still continues for a similar technique for partial shape description and recognition.

Global *moment-based* silhouette shape description techniques have also been used [Dudani et al. 1977] and continue to be used [Reeves et al. 1984] for aircraft shape description. Libraries for 3-D recognition are implemented in much the same way as mentioned above. Three-dimensional moment invariants have been discussed by Sadjadi and Hall [1980].

Wang et al. [1984] also match 3-D objects using silhouettes, but their method is somewhat different. For each prototype object, the principal axes, the principal moments, and the Fourier boundary shape descriptors of the three primary silhouettes (silhouettes as viewed from each of the three principal axes) are computed and stored in a

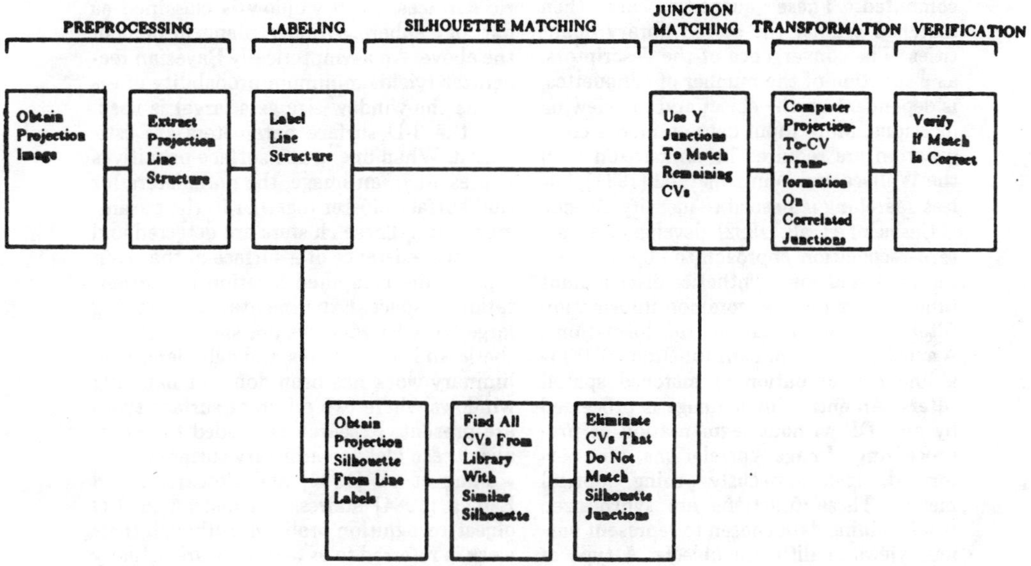

Figure 40. Recognition scheme using characteristic views. (From Chakravarty and Freeman [1982].)

library. They employ 3-D models constructed by the Martin and Aggarwal [1983] approach, but any 3-D model reconstruction method could be used in practice. (We note that the Martin–Aggarwal approach is duplicated using commercial solid modelers, such as SDRC GEOMOD [1982]; object models are constructed by the intersection of extruded silhouette contours, as in Baumgart [1974].) For each unknown object, at least three silhouettes from different views are required for recognition. Silhouette boundaries are combined to produce an object from which the principal moments and Fourier shape descriptors are computed. These quantities are then matched against the stored library quantities. The convergence of the descriptors, as a function of the number of silhouettes, is dependent on the object and the viewing locations. More input data and more computation are required in comparison with the Wallace and Wintz method [1981], but less searching is needed to identify objects.

Casasent et al. [1982] developed a pattern-recognition approach to object recognition based on synthetic discriminant functions, maximum common information filters, and decorrelation transformations. A synthetic discriminant function (SDF) is a linear combination of matched spatial filters. An entire input image is processed by an SDF without segmentation or preprocessing. Image correlations are performed instantaneously using optical means. These functions are synthesized from training data chosen to represent various views of different objects. A type of nongeometric model formation occurs during this training phase. Experimental results are discussed for two different objects. Thirty-six images of each object are obtained (10-degree rotation increments), and 6 of these images (for each object) are used for training. With two SDFs created from 12 images, 60 additional images taken from different views were correctly classified. A two-class mutual-orthogonal-function filter recognized the object and gave the correct orientation for 90 percent of the 72 images used.

Bolle and Cooper [1984], Bolle et al. [1982], and Cernushi-Frias et al. [1982] describe an approach to intensity-image-based object recognition where objects are modeled using composite Lambertian quadric-surface patches. The assumption is that 85 percent of manufactured parts are well represented by combinations of such models as stated in Hakala et al. [1981]. This work is innovative in its use of quadric picture functions. These functions are analytically computed quadric intensity functions that combine Lambertian surface primitives with basic point-source-at-infinity illumination model parameters. Given an intensity image, the image is partitioned into square windows that are fitted to quadric surfaces. Each window is classified as part of a sphere, cylinder, plane, or none of the above. An asymptotically Bayesian recognizer (yields minimum probability of error as the window size gets large) is used, and the 3-D surface parameters are estimated. When one of the surface primitives is present in an image, the parameters for that surface cluster together in the parameter space. These clusters are detected and infer the existence of a surface of the given type at the computed location and orientation in space. Experimental results using large 65×65 windows are shown for synthetic and real spheres and cylinders. Preliminary work has been done for handling windows where two different surface types are present. Research is needed to generalize these ideas to arbitrary surfaces.

Fang et al. [1982] and Stockman and Esteva [1984] address a constrained 3-D object-recognition problem. Although their work is referred to as 3-D, they are actually addressing a 2-D estimation problem using 3-D techniques. The problem is that of determining the (x, y) location and the *single* rotation angle of a 3-D polyhedral object sitting stably on a flat plane. A single intensity image and polyhedral object models are used. Important edges and points are extracted as the primitive features from the input image. Geometric constraints and model matching of grouped primitives determine possible translation and rotation parameters, which are then accumulated in a 3-D histogram. The detection of histogram clusters is used to identify a particular object at a particular

(x, y) location rotated by a particular angle. Perfect experimental recognition results for five toy objects were obtained by Fang et al. [1982]. The transformation clustering technique could possibly be useful for general 3-D object recognition, but it is not shown in this paper. Objects with curved surfaces are not necessarily handled by this method.

In a similar approach, Silberberg et al. [1984] use a generalized Hough transform [illegible] match observed 2-D line segments with [illegible] model line segments and observed 2-D [illegible] junctions to 3-D model vertices. They assume that (1) all objects are polyhedra with single stable positions, (2) the ground plane is known, and (3) the camera position and parameters are known. Experimental results for a synthetic image of a nonconvex polyhedron are discussed.

Tropf and Walter [1983] discuss an augmented transition network (ATN) model for single-image recognition of randomly oriented 3-D solid objects with known geometry. ATN models were developed in the field of natural language understanding and are used in this work to control an analysis-by-synthesis search procedure based on hypothesis generation and verification. The method is explained in the paper with the following example:

(1) Assume that only point primitives (such as edgeless corners) are used, and assume that an object is described using a set of points that are rigidly connected to each other.
(2) A parallel projection of the points is created from an arbitrary view. Pick a point from the projected data and hypothesize that it is a point P1 on a particular known object from the object library.
(3) That object must have a second point P2, which is a distance R from P1. In 3-D, P2 must lie on the surface of a sphere of radius R; in the 2-D projected image, P2 must lie within a circle of radius R. Therefore, choose a second point *within that circle* if one exists. (Otherwise, try another second point of the object. If none of those work, go on to the next object.)
(4) Now it is assumed that the axis P1–P2 in 3-D space is known (to within a near–far ambiguity that is checked later). Next, consider a third point P3 on the object not lying on P1–P2. It must lie on a 3-D circle surrounding the P1–P2 axis and must therefore lie *on a known ellipse* in the projected image.
(5) Now pick a point in the image closest to the ellipse that will fix the object in space. Object verification is the final step in the process.

In some respects, this is similar to the RANSAC approach [Bolles and Fischler 1981] because it uses a few randomly selected data points to estimate model parameters and relies on verification for a better fit. Three-dimensional polyhedral object models and hidden-line algorithms are used by the system. The ATN itself consists of states, arcs, a dictionary, named registers, actions, and conditions. Tropf and Walter claim that the ATN approach differs from block-world-scene approaches in that it is able to cope with heavily distorted data. No experimental results are given, as the system was being implemented at the time the paper was written.

Lin et al. [1984] have specifically addressed the estimation of 3-D object orientation for vision systems *with feedback*. A camera model method and a hierarchical search method are presented for determining the rotation angles necessary to verify object hypotheses via graphic rendering. The two methods are complementary so that if one fails to yield adequate results, the other method is used. Experimental results are discussed for polyhedra, but the methods can be generalized to curved-surface objects.

Douglass [1977, 1981] developed the EYE system for interpreting outdoor scenes using a 3-D model-building approach. He wanted to build a system that did more than label each pixel with an object name; he attempted to use depth cues to create surface descriptions and to use surfaces to create objects. Digitized images are preprocessed and segmented. Image segmentation is performed using a "recognition cone" (or layered network database) and a region-growing algorithm. The placement routine at the heart of the system forms intensity-image segments into 3-D surfaces using a scene model depth map. The depths are iteratively refined by other processes of the system using depth cues. Heuristic visual inference routines interpret perspective, shadows, highlights, occlusions, shading, texture gradients, and monocular motion parallax from multiple images, and adjust depths accordingly. The depth map acts as a blackboard for communication and cooperation among processes. Objects are formed from the scene model after the iterative depth adjustment process terminates. Detailed experimental results were shown for only one house scene.

Goad [1983] presents a technique for object recognition based on special-purpose automatic programming. Individual object descriptions are compiled into programs for which the only task is recognizing one object from any view. Time-consuming shape analysis is performed off line prior to the recognition phase so that actual recognition execution time is minimal (about 1 second). He uses a multiple-view object feature model that incorporates 218 different 3-D views of each object. The features are line segments stored as a pair of endpoints and a 218-bit string. The bit string describes the visibility of the feature in each of 218 discrete views. Edges for objects are ordered by their expected utility for matching purposes. Experimental results are shown for a jumbled pile of key caps for keyboards.

Shneier [1981] proposes a combined multiple-object representation, a "graph of models," where each graph node represents a 3-D surface primitive. The nodes contain a set of properties describing surface shape and a set of pointers to the object names in which the surface is used. The arcs between the nodes describe relationships between surfaces and also contain pointers to the model names where those relationships occur. The integration of multiple objects into a single shared data structure provides a compact representation that is indexed quickly for faster recognition processing.

We conclude this section on a historical note. The pioneering work of Roberts [1965] was a 3-D intensity-image-based object-recognition system. Objects were constrained to be blocks, wedges, prisms, or combinations thereof. Roberts' cross operator was used to detect edges, and collinear segments were merged into lines to produce a line drawing of the scene. Regions were classified as triangles, quadrilaterals, and hexagons. These regions were matched to faces of the prototype objects. Possible object-part model matches were rendered using a hidden-line algorithm to verify the correct object match. Recognized object parts were cut away from the image, and the same process was repeated until all detected edges and vertices were explained. After identifying an object, the system could draw the object from any view to demonstrate its understanding of the object shape. This research was followed by the more advanced work of Guzman [1968], Waltz [1972], and others, which concentrated on line–edge and edge–junction labeling for detecting polygonal regions. These early systems addressed many of the fundamental problems encountered in computer vision, but were limited to processing high-quality images of block world scenes. The algorithms were not robust enough to handle scenes from the real world with noise, curved objects, etc.

5.9 Object Recognition Using Range Images

Nevatia and Binford [1977] is an early paper concerning object recognition in range data. The emphasis in this paper is on the analysis of scenes containing curved objects, which are represented as subpart hierarchies of generalized cones. The recognition processing is summarized as follows:

(1) Range image edges and regions are extracted and organized to create object descriptions that are structured and symbolic.
(2) Key features of these object descriptions are used to retrieve, from an object-model library, a set of models that are similar to the objects in the image.
(3) The image object description is compared to each of the retrieved models, and the best match is chosen.

(4) Verification is performed to check the differences in the best retrieved object model and the image object description. (This step was not implemented.)

Experimental results are presented for a doll, a horse model, a glove, a ring, and a snakelike object. Differently structured objects were easily distinguished, and moderate amounts of occlusion were handled successfully. This work does not appear to have directly evolved into any recent range data systems.

Kuan and Drazovich [1984] have developed a system that attempts to extend the principles of the ACRONYM approach [Brooks 1981] to range imagery. Generalized cylinder object models with model priorities and subpart attachment relations are used to create multilevel coarse-to-fine object descriptions. A model-driven prediction module predicts the following features at different levels to enable coarse-to-fine multilevel interpretation:

(1) *Object level.* Object features include spatial relationships among object components, overall dimensions, *extreme points*, side-view characteristics, and occlusion relationships among object components.

(2) *Cylinder level.* This level is the most important level because cylinders are the basic symbolic entity of the object description. These features include cylinder contour, cylinder position and orientation, parallel edge relationships, edge types, cylinder length, extent of overlap with other cylinders, and overall cylinder visibility and obscuration.

(3) *Surface level.* Surface features include information such as whether the surface is planar or curved, surface edge–boundary information, and spatial surface relationships.

(4) *Edge level.* Edge features include information such as whether the edge is occluding (step), convex, or concave.

The predictions guide the low-level feature-extraction processes, and they also provide mechanisms for feature-to-model matching and interpretation. In contrast to the ACRONYM system, actual measured features are used for matching on the basis of maximizing likelihood rather than creating constraints for later constraint propagation processing. This model-driven approach is limited by its dependence on generalized cylinder-type objects. Experimental results are discussed for one synthetic 64×64 range image of a missile launcher decoy and object models of a missile launcher and a decoy. See the system diagram and object models shown in Figure 41.

Smith and Kanade [1984] discuss a program designed to produce object-centered 3-D object descriptions from depth maps. Conical and cylindrical surfaces are the only shape primitives used. The object descriptions derived from their data-driven approach can be used for matching and object recognition. Coherent relationships between subcylinders of parts aid the extraction of object surfaces. An example of this coherency is the relationship between the handle of a pan and the main body of a pan. Experimental cylinder identification results are displayed for a coffee mug, a shovel, and a pan.

Gennery's [1979] main concern for object recognition was obstacle avoidance for autonomous vehicle navigation. His algorithm is summarized as follows: First, find the ground surface, which is usually big and flat. Second, segment objects above the ground by clustering all range data points that are greater than a threshold distance above the ground. Next, fit ellipsoids to these clusters, and then adjust the clusters according to the ellipsoid fits. He argues that although ellipsoids are very crude object representations, a large scene containing many objects is fairly well described by sets of ellipsoids for the purposes of navigation. Experimental results are shown for a pair of stereo pictures from the Viking Lander, which landed on Mars.

Boyter and Aggarwal [1984] have obtained results from single-view object-recognition experiments using registered range and intensity images. They address our defined recognition problem, except that objects are allowed only 1 degree of rotational freedom. This constraint allows models to be stored in a sensor data format and allows matching to be done using correlation techniques.

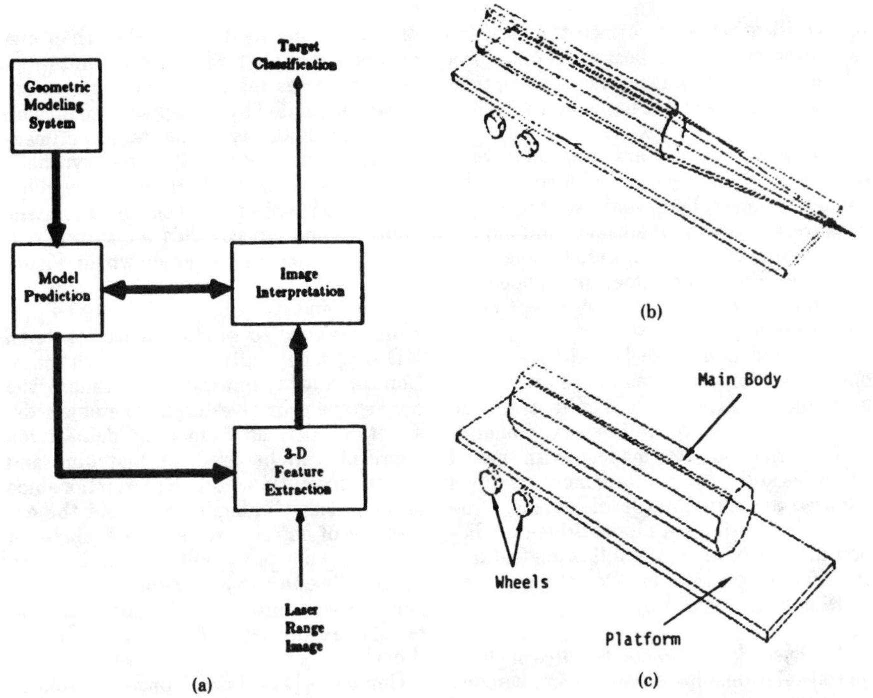

Figure 41. (a) Object-recognition system and two object models: (b) component-level model of missile launcher; (c) component-level model of missile launcher decoy. (From Kuan and Drazovich [1984].)

Bhanu [1982, 1984] presents a 3-D scene analysis system for recognizing 3-D objects in depth maps. The system uses the object representation and surface-extraction method discussed by Henderson [1983]. It constructs object models from physical prototypes using multiple-view depth maps. A complex curved-surface automobile part is discussed. A total of 8334 object surface points were obtained by transforming points from 14 individual views into a common object-centered coordinate system. These surface points are used to fit a convex-face polyhedron using a two step algorithm: (1) The three-point seed algorithm is used to group all points into face regions using convexity and narrowness tests (four threshold values are needed for this); (2) the face regions are then approximated by 3-D planar convex polygons. For the auto part, 85 flat faces are computed to describe its curved surfaces. Object recognition is accomplished after model determination as follows: A depth map from an arbitrary view (same scale as model) is acquired using a range finder. The object points are segmented from the background, and a polygonal-face approximation of the object surface is computed using the same technique mentioned above for object reconstruction. This generates approximately 10–25 faces for unknown views of the auto part. These faces are used to perform object matching using a relaxation-based scheme called *stochastic face labeling*. The face features of area; perimeter; peround; length of maximum, minimum, and mean radius vectors from the face centroid; number of vertices; and angle between maximum and minimum radius vectors are used to compute the initial face-labeling probabilities. (A feature-weighting vector is also used.) In addition,

FACE NUMBER	NEIGHBORS		
1	12	17	0
2	3	13	18
3	2	9	0
4	5	0	0
5	4	9	0
6	15	0	0
7	10	8	13
8	7	10	0
9	3	5	0
10	7	8	21
11	16	21	0
12	1	17	0
13	7	2	18
14	19	0	0
15	6	20	0
16	11	22	21
17	1	12	0
18	2	13	0
19	14	0	0
20	15	0	0
21	10	11	16
22	16	0	0

(a)

FACE NUMBER	NEIGHBORS		
1	6	9	0
2	7	13	0
3	14	8	0
4	12	0	0
5	0	0	0
6	1	10	9
7	2	10	0
8	3	14	0
9	1	6	10
10	6	7	9
11	0	0	0
12	4	0	0
13	2	0	0
14	3	8	0

(b)

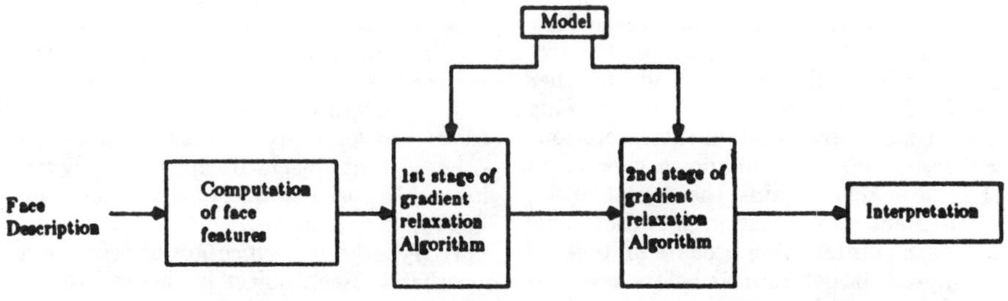

(c)

Figure 42. Face neighbor table for neighbors of a face in (a) 0° and (b) 90° views arranged in descending order by size; (c) block diagram of the 3-D shape-matching algorithm. (From Bhanu [1984]; ©IEEE 1984.)

a face neighbor table is computed where neighbors are ranked according to area. An example of a face neighbor table is shown in Figure 42. The first-stage relaxation iteration involves maximizing the first-stage compatibility measure (defined in terms of a one-largest-area-neighbor compatibility function). Using the face labels from the completion of the first iteration, a second-stage iteration that involves maximizing the second-stage compatibility measure (defined in terms of a two-largest-area-neighbor compatibility function) is performed. Both compatibility functions use the following quantities: the distance between neighboring face centroids, the ratio of the areas of neighboring faces, the difference in face orientations, and the rotation angles for the maximum intersection area of coplanar faces. (These quantities are also weighted.) At the end of the second stage, translation and rotation information concerning the object can be computed. Object recognition is possible by choosing the object in a library of prototype object models that maximizes the compatibility measures. See the block diagram in Figure 42. The method handles arbitrary viewpoints. However, it relies too heavily on the consistency of the output from the face-finding algorithm. Perhaps this is justified, but no evidence is given. All face-adjacency information is not being utilized.

Ballard and Sabbah [1983] have investigated viewer-independent shape recognition by factoring an image object description into an object-centered, view-independent description and a view-dependent view transformation. A decoupling of the three subgroups of scale, orientation, and translation parameters is emphasized. It is assumed that a planar surface-patch (polyhedral) description of an object is available, both as a known prototype model and as sensor data from either processed range data or other sources. They also assume that scale is already known and that the orthographic projection approximation is valid. Their 3-D algorithm consists of two main sequential processing steps: (1) Use the generalized Hough transform (GHT) to compute the three 3-D rotational parameters corresponding to a given view and a given object; (2) determine the two 2-D translational parameters via another GHT. If the correct object is not being matched, only inconsistent interpretations will result. When an unknown view of an object is matched against the correct object, a consistent interpretation is output. This approach is interesting because an object's orientation is determined before its location is computed. It is assumed that there can only be one object per image.

Horaud and Bolles [1984] and Bolles et al. [1983] present the 3DPO system for recognizing and locating 3-D parts in range data. This work extends the 2-D "local-feature-focus" ideas discussed in Bolles and Cain [1982]. Their object-recognition ideas are quite different from those of most other researchers:

(1) Moderately complex parts are preferred instead of polyhedra or quadric-surface models because the abundance of features are helpful for object recognition. The authors point out that most industrial parts are moderately complex; very few ideal spheres, cylinders, and polyhedra are used as industrial parts.

(2) Only a few features should be used for matching. Two or three highly constraining object-specific features are preferred. For example, if a dihedral edge is found in range data, 5 degrees of freedom (2 position and 3 orientation) of that edge are determined, leaving only one unknown (the position along the edge). A preliminary planning system can do as much processing as is required up front to select the best features and the best decision strategy, since this computation is only done once for a given application.

The recognition process is partitioned into five steps:

(1) *Primitive feature detection.* Range edges are detected and linked using two separate techniques, one based on discontinuities and the other based on significant second-derivative zero crossings. Convex, concave, and step edges are distinguished. They argue that edges contain more information than surface patches.

(2) *Feature cluster formation.* Edges are processed to form coplanar edge clusters. Circular arcs are isolated among the coplanar edges.

(3) *Hypothesis generation about possible objects and locations.* The system can hypothesize that objects are appropriately positioned in space so that object edges align with range data edges.

(4) *Hypothesis verification of best object hypotheses.* Each object is checked to determine whether additional object features are found in the image or in the primitive features already extracted.

(5) *Parameter refinement to obtain more precise information.* If additional features are predicted and found, this information is averaged with the existing information to yield more accurate part locations.

The 3DPO system uses an extended CAD model to represent objects. A volume–surface-edge–vertex model is extended to support matching by the addition of redundant pointers and other data structures. (See Figure 43.) Object recognition is considered a two-part process: low-level data-driven analysis followed by high-level model-directed search. The goal of Bolles and coworkers is a flexible system capable of executing quick, customized recognition algorithms for each object. They do not believe "that a single technique will be general

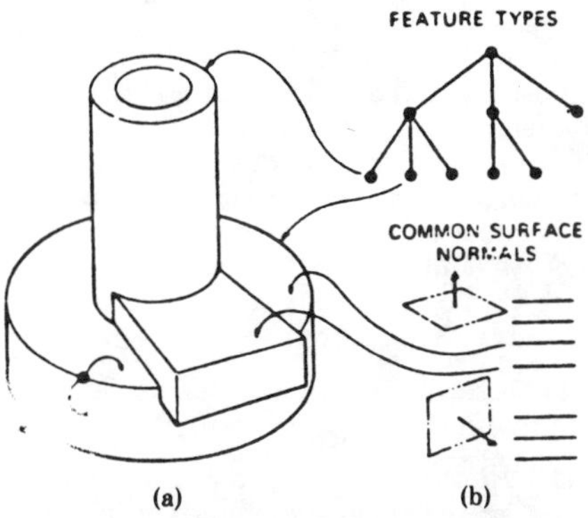

Figure 43. Extended CAD model adds redundancy for matching purposes. (a) Extended CAD model; (b) feature classifications. (From Bolles et al. [1983].)

enough to recognize a large class of objects efficiently." A large collection of experimental results are shown in Horaud and Bolles [1984] for a bin of castings. All the castings in the bin have the same shape. The 3DPO system correctly located six out of seven castings so that the object shape is verified and the 3-D orientation is known. These results rely heavily on circular arcs and straight dihedral edges. Their range images are obtained from a White Scanner 100A, a triangulation-based light-stripe range finder.

Oshima and Shirai [1981, 1983] perform object recognition using 3-D information. Their object-recognition system is based on depth maps obtained from a light-stripe range finder. The range data are processed as follows: (1) Points (range pixels) are grouped into small planar surface elements; (2) surface elements are merged into elementary regions that are classified as either planar or curved; (3) curved elementary regions are merged into consistent global regions that are fitted with quadric surfaces; (4) a scene description is generated from global-region properties and relationships among these regions. This process is indicated in Figure 44. The region properties are based on the best-fit planar region and its boundary, and include the following quantities: perimeter; area; peround; minimum, maximum, and mean region radii about the region centroid; and the standard deviation of the radii of the boundary. The region relationships are characterized by the distance between region centroids, the dihedral angle between best-fit planes, and the type of intersection curve between the regions. There is a learning process that must be executed *for each view* of each object to be recognized. The recognition process compares unknown scene data with learned scene data. Matching is restricted using an algorithmically selected kernel region that has a principal and a subordinate part. The kernel is matched against each learned scene, and each good match is processed further until a consistent scene description is generated (see Figure 44). Two experiments were performed, one using simple objects bounded only by planar or quadric surfaces and the other using machined parts. When an empirically determined set of thresholds for the matching algorithm was used, perfect interpretations resulted. The technique is worthwhile because it can handle many objects at once. However, because of the view-dependent nature of the stored object models (learned scenes), matching will become very slow when many views are allowed. This approach appears to be inadequate for arbitrary-view object recognition.

Sato and Honda [1983] have investigated pseudodistance measures for recognition of objects that are placed on a turntable in a stable vertical orientation. A fixed set of horizontal cross-section boundaries is determined for each recognizable object using a laser projection system and image processor as described by Sato et al. [1982]. Boundary-based Fourier shape descriptors are computed for each horizontal cross section. The object representation consists of N sets of M complex Fourier coefficients, Pseudodistance measures between two object representations are defined for elongatedness, horizontal strain, section shape, torsion, and displacement. Experimental pseudodistance results are shown in the paper for four wooden animal models and a doll in three different positions. Two positions of the doll are shown in Figure 45. Using a weighted sum of pseudodistance measures and, for example, a minimum-distance classifier, unknown curved shapes

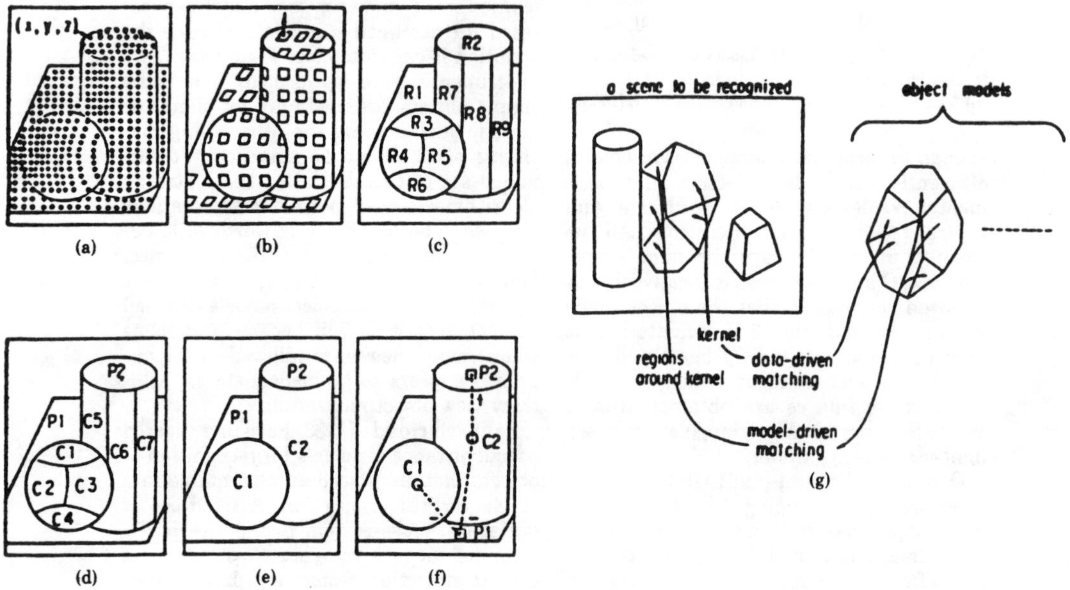

Figure 44. Surface characterization: (a) 3-D coordinates, (b) surface elements, (c) elementary regions, (d) classified regions, (e) global regions, (f) description of the scene; (g) matching process. (From Oshima and Shirai [1983]; ©IEEE 1983.)

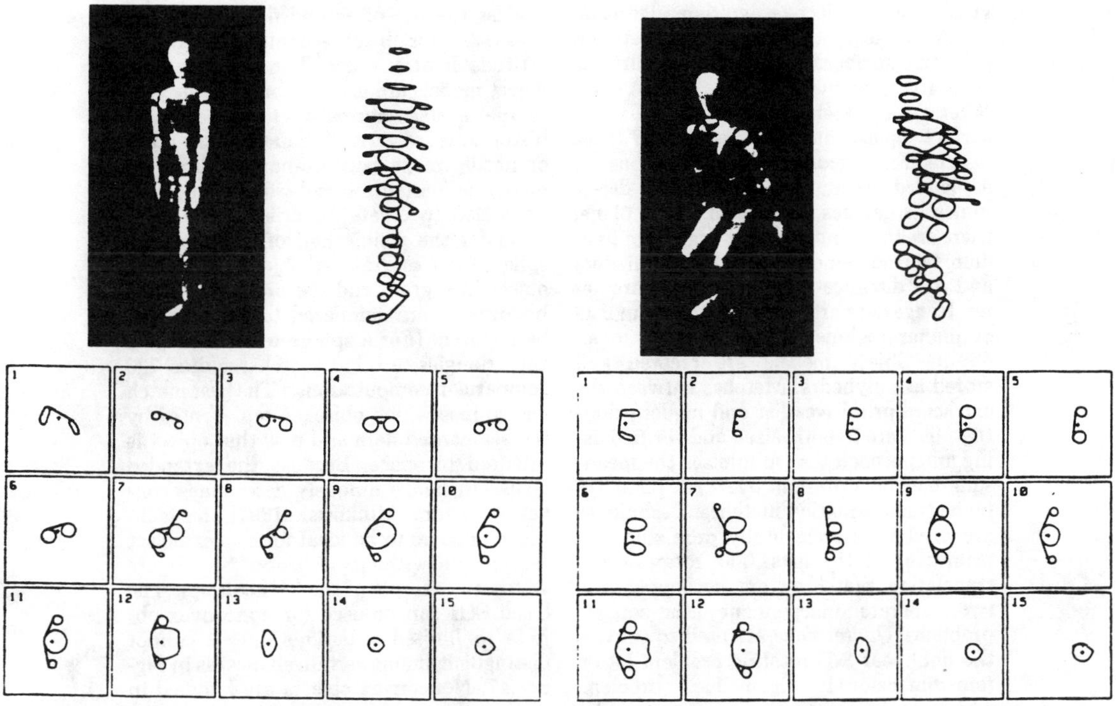

Figure 45. Cross sections of doll used for shape description. (From Sato and Honda [1983]; ©IEEE 1983.)

can be classified. One problem with this method as currently implemented is that disjoint parts of the doll's cross sections had to be linked *manually* to create a simple closed curve usable by the Fourier descriptor algorithm. This system is inadequate for single-arbitrary-view object recognition because of its need to rotate the object.

Faugeras [1984], Faugeras and Hebert [1983], and their group at INRIA have devised a 3-D object-recognition algorithm based on geometrical matching between primitive surfaces. The primitive surfaces currently implemented in the INRIA computer vision system are planes, but quadric-surface algorithms are presented in these papers. Each geometric primitive has an associated parameter vector that determines its degrees of freedom. For a plane, there are three independent degrees of freedom: two independent direction parameters and one distance-from-the-origin parameter. Range data are processed to obtain lists of planar regions that correspond to an object. Object models are created and stored as polyhedra. Matches between the extracted primitives list and model primitives lists are hypothesized and verified using an approach that minimizes the mean-square-error criterion over all plane-to-plane transformation matches. Techniques are used to incrementally drop and add primitives in the lists. The rotation and translation matching are decoupled into two separate independent least-squares problems. *Quaternions* are used to convert the nonlinear 3-D rotation problem into a four-dimensional eigenvalue problem, which can be solved directly. The translation problem permits a standard linear least-squares solution. The rotation and translation matching errors are combined to provide a quantitative measure of the goodness of the match between the data and a hypothetical model; the best match represents the recognized object. Local consistency tests are used to avoid full computation on strongly inconsistent plane lists. These ideas result in a computationally efficient method of identifying objects and determining their translation and rotation parameters. Experimental results are shown in Faugeras [1984] for the same automobile part used by Bhanu [1984] and Henderson [1983]. The precision of rotation angle and translation vector results are stated to be 0.04 radians (2.3 degrees) and 3 millimeters, respectively, where the accuracy of the range data is 1 millimeter. These are probably the best results quoted in the literature.

Horn [1984], Horn and Ikeuchi [1984], Ikeuchi [1981], and Ikeuchi et al. [1983] discuss the use of extended Gaussian images (EGI) for object recognition and object attitude determination. Three-dimensional object models are used to compute the prototype surface-normal-vector orientation histograms for various shapes. Depth maps or needle maps (surface-normal direction maps) computed for real-world scenes are processed to create an orientation histogram for the visible half of the Gaussian sphere for presegmented objects. The scene object histogram and the prototype object histograms are compared to compute the best match. (For a sphere tessellated with 240 triangles, a blind search requires 720 comparison computations.) The best match determines which object is represented by the segmented data and how that object is oriented in space. Because the extended Gaussian image uniquely determines convex polyhedra [Minkowski 1897], this technique appears to be ideal for *convex* object recognition without occlusion. See Figure 46 for an example of the EGI concept. The basic EGI can be used for nonconvex objects in limited situations, but it cannot distinguish among certain shapes, as in Figure 47. Nonconvex objects are handled in general by creating a separate orientation histogram for every view in a discrete set of views and matching against this enlarged data structure. The EGI approach and the INRIA approach are similar in that they both use surface-normal matching procedures, and both do not make explicit use of face-adjacency information.

Besl and Jain [1984] propose an approach to range-image object recognition that combines surfaces, edges, and points that have been computed independently. No predetermined surface shapes are used. This approach is motivated by a theorem from differential geometry stating that the

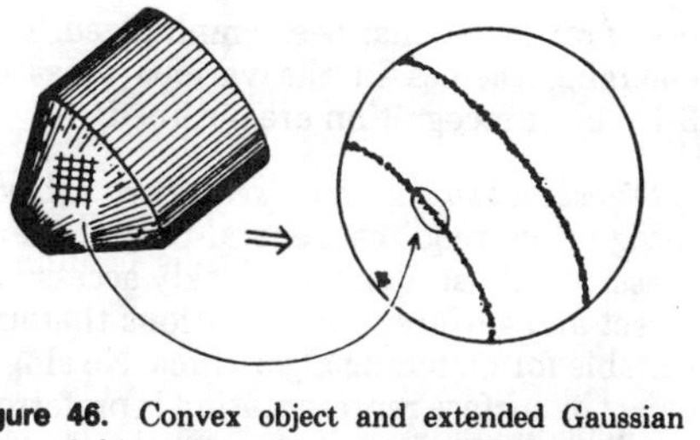

Figure 46. Convex object and extended Gaussian image of object. (From Horn [1984]; ©IEEE 1984.)

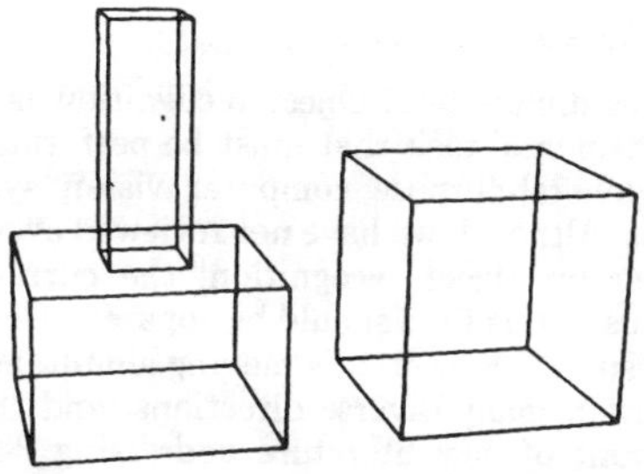

Figure 47. Two objects with same basic EGI representation.

coefficients of the first and second fundamental forms of a smooth surface uniquely characterize the shape of that surface. Gaussian and mean curvatures are very important features because they combine the information of the two fundamental forms and are invariant to rotations and translations and to changes in surface parameterization. These surface curvature characteristics generalize the notion of curvature for plane curves [O'Neill 1966]. Also, the mean curvature and the boundary curve of a surface uniquely determine a graph surface, and Gaussian curvature uniquely determines convex surfaces. The method consists of the following steps:

(1) A smoothed range-image is convolved with window operators to provide least-squares estimates of the first and second partial derivatives of the underlying surface.

(2) These derivative estimates are used to compute mean curvature, Gaussian curvature, critical points, step edges, roof edges, and ramp edges. Roof and ramp edges are detected as local maximum mean-curvature pixels. A critical-point image is generated using the intersection of the zero-crossing images of the two first partial derivatives.

(3) The sign of the surface curvature values is used to place every pixel in one of eight classes: pit, peak, saddle ridge, saddle valley, valley, ridge, minimal, and flat (planar). Each critical point is also categorized using these classes.

(4) Critical points with positive Gaussian curvature are used as starting points for the matching algorithm. The neighborhood of a critical point is extracted to form a view-independent shape descriptor that is used for matching against a library of precomputed matching representations of known objects that enumerates possible types of critical points of each object. The depth map is segmented during the matching process. Edges are used to guide and verify surface-region segmentation.

(5) Possible matches of individual objects are projected back into the depth-map format for verification. Simple image differencing can be used to check compatibility. Best-match depth-map surface regions are extracted when found, and matching continues on the remaining depth-map regions until all objects are explained.

(6) The entire scene model description is then processed by a depth buffer algorithm to create an entire synthetic-scene range image. Occlusion relationships are checked for correct interpretation. The system outputs the final description, which lists each distinguishable object, the number of occurrences of each object, and the location and orientation of each object instance. Depth-map regions that could not be interpreted as an object are segmented and characterized by their curvature properties and stored in the matching library format for future reference.

The proposed approach utilizes the structural scene information available in the eight-level image created by the sign bits of the surface curvature values. Experimental surface characterization results are shown in Figure 27 above and in Besl and Jain [1984, 1985a, 1985b].

6. EMERGING THEMES

Three-dimensional object recognition is a fundamental task that must be performed by general-purpose computer vision systems. Although we have not reviewed every paper on object recognition, the current status of this field should be apparent from our survey. Research is moving simultaneously in many diverse directions, and the amount of new literature addressing 3-D vision issues is rapidly increasing. In the past, many researchers have not presented justification of their methods in logical and/or mathematical terms. The approach has often been "think of something that might work and try it." As this relatively new field matures, investigators must continually learn from the mistakes of previous vision research, not repeat them. Although there are projects that have been successful, general-purpose computer vision does not yet exist and many promising ideas have yet to be realized. Specific long-term goals must be set forth and systematically accomplished.

Range-image understanding is quickly becoming an important and recognized branch of computer vision. High-quality range images contain a wealth of explicit information that is obscured in intensity images. This information is just beginning to be utilized. Vision systems should make maximum use of the information present in the sensor data; they should not rely on a single type of image feature, such as edges or planar regions.

Intensity-image object recognition is an ill-posed problem because light-source parameters and surface reflectance functions are not known. Range-image object recognition is difficult, but it is a well-posed problem when possible object shapes are known. Range-image vision systems can and will surpass the capabilities of intensity-image systems in many environments, perhaps leading to new insights in intensity-image understanding.

We have given a precise definition of object recognition, listed qualitative requirements of recognition systems, reviewed a variety of related topics, and surveyed the literature on these subjects. The use of range data has been emphasized. The emerging themes in the various areas of 3-D object recognition are outlined:

Object and surface representation. Object recognition requires general-purpose, expressive, robust, flexible, quickly accessible object and surface representations that are suitable for matching algorithms. No single object or surface representation is preferred by all computer vision researchers. Generalized cones have received a great deal of attention perhaps owing to their compact representation of appropriate objects. Yet generalized cones lack the generality of surface-boundary representations, which have also been used by computer vision researchers. Range images, for example, provide sampled surface information, not sampled cylinders or cones. Generalized cones are effective for image interpretation only when object surfaces exhibit the appropriate uniformity along a curve. For the recognition problem stated in this paper, a surface-boundary representation appears to be the right choice for detailed general-purpose object models. Of course, it may be necessary to use other, more abstract representations in conjunction with detailed object models to achieve recognition. Some might argue that surface representations require too much data storage, but memory is generally becoming faster and less expensive. Future research and experience will eventually decide the answers to the representation question. One thing is certain, however: A vision system cannot understand what it cannot represent internally.

Object and surface rendering algorithms. The hypothesize-and-verify paradigm is commonly used in image-understanding research. Verification requires a rendering process that puts model data into the sensor data or scene description format. This notion of rendering for verification has been mentioned in past research, but has not been utilized enough. Very fast rendering algorithms are required, but computer graphics researchers will continue to address this display speed issue. Object and surface models must be suitable for both fast matching and fast rendering.

Image formation. Intensity-image formation is a well-understood process. Images with adequate intensity resolution are acquired more quickly than they can be analyzed. Range-image systems need range finders with faster acquisition times and greater depth resolution. Both types of image analysis systems would benefit from sensors with less noise and higher spatial resolution.

Image processing and digital surface characterization. Early (low-level) processing of images should provide a domain-independent rich description for higher level processing. Intensity-image processing techniques are still a topic of active research and debate. Edge detection/linking receives perhaps the most attention because so many intensity-image-understanding approaches are based on edges. Research in range-image processing is heading toward differential geometric features, especially surface curvature features. Critical points, slope districts, curvature districts, and pixel-by-pixel classification are other common ideas. Local features of this sort are not significantly affected by occlusion. Past image-processing techniques tended to impose a high-level model on the sensor data too early in the processing sequence. The last decade in vision research has shown that this approach may not work even in block world scenes. We expect the future to bring faster, more robust, general-purpose data-driven approaches to image processing.

Object reconstruction. The model formation process is critically important to object recognition. Automating the modeling process using object-reconstruction algorithms could allow a vision system to learn object shapes with little human interaction. Extruded-silhouette–volume intersection algorithms allow the construction of object models from sets of intensity images. Expert systems may be able to provide better models than those obtained by straightforward intersection. Sets of range images can provide adequate surface information for constructing reasonably detailed object models. We expect range-image object-reconstruction techniques to improve as surface-characterization and surface-matching algorithms progress. Intensity-image reconstructions should advance as edge detectors and expert reconstruction systems are improved.

Object recognition. Topics for future recognition research were stated earlier as characteristics of an ideal system. We therefore make a few general observations on the survey and the recognition problem. The symbolic-scene description features should be invariant to translations and rotations of object surfaces. Many current systems have addressed this issue. The 3DPO system emphasizes the constraining power of the knowledge of a single dihedral edge shared by planar faces. More systems need to take advantage of these ideas. The hypothesize-and-verify paradigm, constraint propagation, graph data structures, and consistency checking are commonly used tools. It is not difficult to fit surfaces to data, but it is not as easy to determine what image data should be used for fitting. There are definite problems with fitting surfaces to large fixed-position image windows. Multilevel logic and modular communicating structures have been advocated by several authors. An interactive combination of data-driven and model-driven approaches is likely to perform better than systems restricted to one mode of operation. Open-loop systems are only as robust as their most limited component. Feedback via rendering should be able to address this weakness. Visual potentials (aspect graphs) and characteristic views have been mentioned as methods for describing projected 2-D object shape from all views by a finite list. Basic research in surface matching is necessary. Most systems use edge-based techniques, whereas others use region-based techniques for segmentation and recognition. Further research is required for determining methods that use both edges and regions together in the most effective manner.

Miscellaneous remarks. Occlusion is always a problem for vision systems. Can the explicitness of range data be used in such a way that occlusion is not a hindrance to image understanding? Most recognition schemes use linear-time (one-by-one) object-matching algorithms. How can large object libraries be searched fast enough for practical purposes? Recognition systems are bound to make mistakes occasionally. How can systems be made to learn from mistakes when they are corrected? Registered intensity and range images are available for recognition systems. Existing approaches for combining sensor data from different sources are not very sophisticated. What types of sensor integration techniques should be applied to range, intensity, and possibly tactile images to yield the richest descriptions for recognition processes? These questions are just a few of the future research issues in 3-D object recognition.

ACKNOWLEDGMENTS

This work was supported in part by IBM Corporation, Data Systems Division, Kingston, N.Y., under the monitorship of Dr. Jack Contino, and in part by the Air Force Office of Scientific Research under contract F49620-82-C-0089. We thank the reviewers and the technical editor for their helpful comments on improving this paper. We acknowledge the Environmental Research Institute of Michigan (ERIM) for providing us with range images from their laser finder.

REFERENCES

ABE, N., ITHO, F., AND TSUJI, S. 1983. Toward generation of 3-dimensional models of objects using 2-dimensional figures and explanations in language. In *Proceedings of the 8th International Joint Conference on Artificial Intelligence* (Karlsruhe, West Germany, Aug. 8–12). IJCAI, pp. 1113–1115.

AGIN, G. J., AND BINFORD, T. O. 1973. Computer description of curved objects. In *Proceedings of the 3rd International Joint Conference on Artificial Intelligence* (Stanford, Calif., Aug. 20–23). IJCAI, pp. 629–640.

ALDEFELD, B. 1983. Automatic 3D reconstruction from 2D geometric part descriptions. In *Proceedings of the Computer Vision and Pattern Recognition Conference* (Washington, D.C., June 19–23). IEEE, New York, pp. 66–72.

ALTSCHULER, M. D., POSDAMER, J. L., FRIEDER, G., ALTSCHULER, B. R., AND TABOADA, J. 1981. The numerical stereo camera. In *Proceedings of The Society for Photo-Optical Instrumentation Engineers Conference on 3-D Machine Perception*, vol. 283, SPIE, Bellingham, Wash., pp. 15–24.

ANDERSON, R. L., AND HOUSEMAN, E. E. 1942. Tables of Orthogonal Polynomial Values Extended to $N = 104$. Research Bulletin 297, Iowa State College of Agriculture and Mechanic Arts, Ames, Iowa (Apr.).

ASADA, H., AND BRADY, M. 1984. The curvature primal sketch. In *Proceedings of the Workshop on Computer Vision: Representation and Control* (Annapolis, Md., Apr. 30–May 2). IEEE, New York, pp. 8–17.

BADLER, N., AND BAJCSY, R. 1978. Three-dimensional representations for computer graphics and computer vision. *ACM Comput. Graphics 12*, 153–160.

BAJCSY, R. (Ed.) 1979. *Proceedings of the Workshop on Representation of Three-Dimensional Objects* (Univ. of Pennsylvania, Philadelphia, Pa., May 1–2).

BAJCSY, R. 1980. Three-dimensional scene analysis. In *Proceedings of the 5th International Conference on Pattern Recognition* (Miami, Fla., Dec. 1–4). IAPR and IEEE, New York, pp. 1064–1074.

BAKER, H. 1977. Three-dimensional modeling. In *Proceedings of the 5th International Joint Conference on Artificial Intelligence* (Cambridge, Mass., Aug. 22–25). IJCAI, pp. 649–655.

BALLARD, D. H., AND BROWN, C. M. 1982. *Computer Vision.* Prentice-Hall, Englewood Cliffs, N.J.

BALLARD, D. H., AND SABBAH, D. 1983. Viewer independent shape recognition. *IEEE Trans. Pattern Anal. Machine Intell. PAMI-5*, 2 (Mar.), 653–659.

BARNARD, S. T., AND FISCHLER, M. A. 1982. Computational stereo. *ACM Comput. Surv. 14*, 4 (Dec.), 553–572.

BARNHILL, R. E. 1983. A survey of the representation and design of surfaces. *IEEE Comput. Graphics Appl. 3*, 7 (Oct.), 9–16.

BARROW, H. G., AND TENENBAUM, J. M. 1981. Computational vision. In *Proceedings of the IEEE 69*, 5 (May), 572–595.

BAUMGART, B. G. 1974. Geometric modeling for computer vision. Ph.D. dissertation, Computer Science Dept., Stanford Univ., Stanford, Calif.

BEAUDET, P. R. 1978. Rotationally invariant image operators. In *Proceedings of the 4th International Conference Pattern Recognition* (Kyoto, Japan, Nov. 7–10). IAPR and IEEE, New York, pp. 579–583.

BESL, P. J., AND JAIN, R. C. 1984. Surface characterization for three-dimensional object recognition. RSD-TR-20-84, Electrical Engineering and Computer Science Dept., Univ. of Michigan, Ann Arbor, Mich. (Dec.).

BESL, P. J., AND JAIN, R. C. 1985a. Intrinsic and extrinsic surface characteristics. In *Proceedings of the Computer Vision and Pattern Recognition Conference* (San Francisco, Calif., June 9–13). IEEE, New York, pp. 226–233.

BESL, P. J., AND JAIN, R. C. 1985b. Invariant surface characteristics for three-dimensional object recognition in range images. *Comput. Vision, Graphics, Image Processing.* To appear.

BESL, P. J., AND JAIN, R. C. 1985c. Range image understanding. In *Proceedings of the Computer Vision and Pattern Recognition Conference* (San Francisco, Calif., June 9–13). IEEE, New York, pp. 430–451.

BESL, P. J., DELP, E. J., AND JAIN, R. C. 1985. Automatic visual solder joint inspection. *IEEE J. Robotics and Automation 1,* 1 (May), 42–56.

BHANU, B. 1982. Surface representation and shape matching of 3-D objects. In *Proceedings of the Pattern Recognition and Image Processing Conference* (Las Vegas, Nev., June 14–17). IEEE, New York, pp. 349–354.

BHANU, B. 1984. Representation and shape matching of 3-D objects. *IEEE Trans. Pattern Anal. Machine Intell. PAMI-6,* 3 (May), 340–350.

BLEDSOE, W. W. 1974. The sup-inf method in Presburger arithmetic. Dept. of Math and Computer Science Memo ATP-18, Univ. of Texas, Austin, Tex.

BLUM, H. 1967. A transformation for extracting new descriptors of shape. In *Models for the Perception of Speech and Visual Form.* W. Wathen-Dunn, Ed., MIT Press, Cambridge, Mass., pp. 362–380.

BINFORD, T. O. 1982. Survey of model-based image analysis systems. *Int. J. Robotics Res. 1,* 1 (Spring), 18–64.

BOCQUET, J. C., AND TICHKIEWITCH, S. 1982. An 'expert system' for reconstruction of mechanical objects from projections. In *Proceedings of the 6th International Conference on Pattern Recognition* (Munich, West Germany, Oct. 19–22). IAPR and IEEE, New York, pp. 491–496.

BOISSONNAT, J. D. 1982. Representation of object triangulating points in 3-D space. In *Proceedings of the 6th International Conference on Pattern Recognition* (Munich, West Germany, Oct. 19–22). IAPR and IEEE, New York, pp. 830–832.

BOISSONNAT, J. D., AND FAUGERAS, O. D. 1981. Triangulation of 3-D objects. In *Proceedings of the 7th International Joint Conference on Artificial Intelligence* (Vancouver, B.C., Canada, Aug. 24–28). IJCAI, pp. 658–660.

BOLLE, R. M., AND COOPER, D. B. 1984. Bayesian recognition of local 3-D shape by approximating image intensity functions with quadric polynomials. *IEEE Trans. Pattern Anal. Machine Intell. PAMI-6,* 4 (July), 418–429.

BOLLE, R. M., COOPER, D. B., AND CERNUSHI-FRIAS, B. 1982. Three-dimensional surface shape recognition by approximating image intensity functions with quadric polynomials. In *Proceedings of the Pattern Recognition and Image Processing Conference* (Las Vegas, Nevada, June 14–17). IEEE-CS, New York, pp. 611–617.

BOLLES, R. C., AND CAIN, R. A. 1982. Recognizing and locating partially visible objects: the local-feature-focus method. *Int. J. Robotics Res. 1,* 3 (Fall), 57–82.

BOLLES, R. C., AND FISCHLER, M. A. 1981. A RANSAC-based approach to model fitting and its application to finding cylinders in range data. In *Proceedings of the 7th International Joint Conference on Artificial Intelligence* (Vancouver, B.C., Canada, Aug. 24–28). IJCAI, pp. 637–643.

BOLLES, R. C., HORAUD, P., AND HANNAH, M. J. 1983. 3DPO: A three-dimensional part orientation system. In *Proceedings of the 8th International Joint Conference on Artificial Intelligence* (Karlsruhe, West Germany, Aug. 8–12). IJCAI, pp. 1116–1120.

BOYTER, B. A., AND AGGARWAL, J. K. 1984. Recognition with range and intensity data. In *Proceedings of Workshop on Computer Vision: Representation and Control* (Annapolis, Md., Apr. 30–May 2). IEEE, New York, pp. 112–117.

BRADY, M. 1981. Preface—The changing shape of computer vision. *Artificial Intell. 17* (Aug.), 1–15.

BRADY, M. 1982. Computational approaches to image understanding. *ACM Comput. Surv. 14,* 1 (Mar.), 3–71.

BRADY, M., PONCE, J., YUILLE, A., AND ASADA, H. 1985. Describing surfaces. In *Proceedings of the 2nd International Symposium on Robotics Research,* H. Hanafusa and H. Inoue, Eds., MIT Press, Cambridge, Mass.

BROOKS, R. A. 1981. Symbolic reasoning among 3-D models and 2-D images. *Artificial Intell. 17* (Aug.), 285–348.

BROOKS, R. A. 1982. Representing possible realities for vision and manipulation. In *Proceedings of the Pattern Recognition and Image Processing Conference* (Las Vegas, Nevada, June 14–17). IEEE, New York, pp. 587–592.

BROOKS, R. A. 1983. Model-based three-dimensional interpretations of two-dimensional images. *IEEE Trans. Pattern Anal. Machine Intell. PAMI-5,* 2 (Mar.), 140–149.

BROOKS, R. A., GREINER, R., AND BINFORD, T. O. 1979. The ACRONYM model-based vision system. In *Proceedings of the 6th International Joint Conference on Artificial Intelligence* (Tokyo, Japan, Aug. 20–23). IJCAI, pp. 105–113.

BROWN, C. M. 1981. Some mathematical and representational aspects of solid modeling. *IEEE Trans. Pattern Anal. Machine Intell. PAMI-3,* 4 (July), 444–453.

CARRIHILL, B., AND HUMMEL, R. 1984. Experiments with intensity ratio sensors. Tech. Rep., Courant Institute of Mathematical Science, New York Univ., New York, N.Y.; *Comput. Vision, Graphics, Image Processing.* To appear.

CASALE, M. S., AND STANTON, E. L. 1985. An overview of analytic solid modeling. *IEEE Comput. Graphics Appl. 5,* 2 (Feb.), 45–56.

CASASENT, D., VIJAYA-KUMAR, B. V. K., AND SHARMA, V. 1982. Synthetic discriminant functions for three-dimensional object recognition. In

Proceedings of The Society for Photo-Optical Instrumentation Engineers Conference on Robotics and Industrial Inspection, vol. 360 (San Diego, Calif., Aug. 24–27). SPIE, Bellingham, Wash., pp. 136–142.

CERNUSHI-FRIAS, B., COOPER, D. B., AND BOLLE, R. M. 1982. Estimation of location and orientation of 3-D surfaces using a single 2-D image. In *Proceedings of the Pattern Recognition and Image Processing Conference* (Las Vegas, Nevada, June 14–17). IEEE, New York, pp. 605–610.

CHAKRAVARTY, I. 1982. The use of characteristic views as a basis for recognition of three-dimensional objects. IPL-TR-034, Image Processing Lab, Rensselaer Polytechnic Inst., Troy, N.Y. (Oct.).

CHAKRAVARTY, I., AND FREEMAN, H. 1982. Characteristic views as a basis for three-dimensional object recognition. In *Proceedings of The Society for Photo-Optical Instrumentation Engineers Conference on Robot Vision*, vol. 336 (Arlington, Va., May 6–7). SPIE, Bellingham, Wash., pp. 37–45.

COLEMAN, E. N., AND JAIN, R. 1982. Obtaining shape of textured and specular surfaces using four-source photometry. *Comput. Graphics Image Processing 18*, 4 (Apr.), 309–328.

CONNOLLY, C. I. 1984. Cumulative generation of octree models from range data. In *Proceedings of the International Conference on Robotics* (Atlanta, Ga., Mar. 13–15). IEEE, New York, pp. 25–32.

CONNOLLY, C. I. 1985. The determination of next best views. In *Proceedings of the International Conference on Robotics and Automation* (St. Louis, Mo., Mar. 25–28). IEEE, New York, pp. 432–435.

DANE, C. 1982. An object-centered three-dimensional model builder. Ph.D. dissertation, Computer and Information Sciences Dept., Moore School of Electrical Engineering, Univ. of Pennsylvania, Philadelphia, Pa.

DANE, C., AND BAJCSY, R. 1982. An object-centered three-dimensional model builder. In *Proceedings of the 6th International Conference on Pattern Recognition* (Munich, West Germany, Oct. 19–22). IEEE, New York, pp. 348–350.

DOUGLASS, R. M. 1977. Recognition and depth perception of objects in real world scenes. In *Proceedings of the 5th International Joint Conference on Artificial Intelligence* (Cambridge, Mass., Aug. 22–25). IJCAI, p. 657.

DOUGLASS, R. M. 1981. Interpreting 3-D scenes: a model-building approach. *Comput. Graphics Image Processing 17*, 2 (Oct.), 91–113.

DUDA, R. O., NITZAN, D., AND BARRETT, P. 1979. Use of range and reflectance data to find planar surface regions. *IEEE Trans. Pattern Anal. Machine Intell. PAMI-1*, 3 (July), 254–271.

DUDANI, S. A., BREEDING, K. J., AND MCGHEE, R. B. 1977. Aircraft identification by moment invariants. *IEEE Trans. Comput. C-26*, 1 (Jan.), 39–46.

DYER, C. R., AND CHIN, R. T. 1984. Model-based industrial part recognition: systems and algorithms. Computer Sciences Tech. Rep. 538, Univ. of Wisconsin, Madison, Wis. (Mar.).

FANG, T. J., HUANG, Z. H., KANAL, L. N., LAMBIRD, B., LAVINE, D., STOCKMAN, G., AND XIONG, F. L. 1982. Three-dimensional object recognition using a transformation clustering technique. In *Proceedings of the 6th International Conference on Pattern Recognition* (Munich, West Germany, Oct. 19–22). IAPR and IEEE, New York, pp. 678–681.

FAROUKI, R. T., AND HINDS, J. K. 1985. A hierarchy of geometric forms. *IEEE Comput. Graphics Appl. 5*, 5 (May), 51–78.

FAUGERAS, O. D. 1984. New steps toward a flexible 3-D vision system for robotics. In *Proceedings of the 7th International Conference on Pattern Recognition* (Montreal, Canada, July 30–Aug. 2). IEEE, New York, pp. 796–805.

FAUGERAS, O. D., AND HEBERT, M. 1983. A 3-D recognition and positioning algorithm using geometrical matching between primitive surfaces. In *Proceedings of the 7th International Joint Conference on Artificial Intelligence* (Vancouver, B.C., Canada, Aug. 24–28). IJCAI, pp. 996–1002.

FAUGERAS, O. D., HEBERT, M., MUSSI, P., AND BOISSONNAT, J. D. 1982. Polyhedral approximation of 3-D objects without holes. *In Proceedings of the Pattern Recognition and Image Processing Conference* (Las Vegas, Nevada, June 14–17). IEEE, New York, pp. 593–598.

FAUX, I. D., AND PRATT, M. J. 1979. *Computational Geometry for Design and Manufacture*. Ellis Horwood, Chichester, U.K.

FEKETE, G., AND DAVIS, L. 1984. Property spheres: a new representation for 3-D object recognition. In *Proceedings of the Workshop on Computer Vision: Representation and Control* (Annapolis, Md., Apr. 30–May 2). IEEE, New York, pp. 192–201.

FISHER, R. B. 1983. Using surfaces and object models to recognize partially obscured objects. In *Proceedings of the 8th International Joint Conference on Artificial Intelligence* (Karlsruhe, West Germany, Aug. 8–12). IJCAI, pp. 989–995.

FOLEY, J. D., AND VAN DAM, A. 1982. *Fundamentals of Interactive Computer Graphics*. Addison-Wesley, Reading, Mass.

GARIBOTTO, G., AND TOSINI, R. 1982. Description and classification of 3-D objects. In *Proceedings of the 6th International Conference on Pattern Recognition* (Munich, West Germany, Oct. 19–22). IEEE, New York, pp. 833–835.

GENNERY, D. B. 1979. Object detection and measurement using stereo vision. In *Proceedings of the 6th International Joint Conference on Artificial Intelligence* (Tokyo, Japan, Aug. 20–23). IJCAI, pp. 320–327.

GEOMOD User Manual and Reference Manual. 1982. Structural Dynamics Research Corporation (SDRC), Cincinnati, Ohio.

GEVARTER, W. B. 1983. Machine vision: a report on the state of the art. *Comput. Mech. Eng.* (*CIME*), *1*, 4 (Apr.), 25–30.

GIL, B., MITICHE, A., AND AGGARWAL, J. K. 1983. Experiments in combining intensity and range edge maps. *Comput. Vision, Graphics, Image Processing 21*, (Mar.), 395–411.

GOAD, C. 1983. Special purpose automatic programming for 3D model-based vision. In *Proceedings of the Image Understanding Workshop* (Arlington, Va., June 23). DARPA, Science Applications, McLean, Va., pp. 94–104.

GONZALEZ, R. C., AND WINTZ, P. 1978. *Digital Image Processing.* Addison-Wesley, Reading, Mass.

GRIMSON, W. E. L. 1980. A computer implementation of a theory of human stereo vision. MIT Artificial Intelligence Lab Memo 565, Massachusetts Institute of Technology, Cambridge, Mass.

GUZMAN, A. 1968. Computer recognition of three-dimensional objects in a visual scene. MAC-TR-59, Ph.D. dissertation, Project MAC, Massachusetts Institute of Technology, Cambridge, Mass.

HAKALA, D. G., HILLYARD, R. C., AND MALRAISON, P. F. 1981. Natural quadrics in mechanical design. SIGGRAPH '81 Seminar: Solid Modeling, (Dallas Tex., Aug. 3–7).

HALL, E. L., TIO, J. B. K., MCPHERSON, C. A., AND SADJADI, F. A. 1982. Measuring curved surfaces for robot vision. *Computer 15*, 12 (Dec.), 42–54.

HARALICK, R. M. 1984. Digital step edges from zero-crossings of second directional derivatives. *IEEE Trans. Pattern Anal. Machine Intell. PAMI-6*, 1 (Jan.), 58–68.

HARALICK, R. M., LAFFEY, T. J., AND WATSON, L. T. 1983. The topographic primal sketch. *Int. J. Robotics Res. 2*, 1 (Spring), 50–72.

HEBERT, M., AND PONCE, J. 1982. A new method for segmenting 3-D scenes into primitives. In *Proceedings of the 6th International Conference on Pattern Recognition* (Munich, West Germany, Oct. 19–22). IAPR and IEEE, New York, pp. 836–838.

HENDERSON, T. C. 1982. Efficient segmentation method for range data. In *Proceedings of The Society for Photo-Optical Instrumentation Engineers Conference on Robot Vision*, vol. 336 (Arlington, Va., May 6–7). SPJE, Bellingham, Wash., pp. 46–47.

HENDERSON, T. C. 1983. Efficient 3-D object representations for industrial vision systems. *IEEE Trans. Pattern Anal. Machine Intell. PAMI-5*, 6 (Nov.), 609–617.

HENDERSON, T. C., AND BHANU, B. 1982. Three-point seed method for the extraction of planar faces from range data. In *Proceedings of the Workshop on Industrial Applications of Machine Vision* (Research Triangle Park, N.C., May). IEEE, New York, pp. 181–186.

HERMAN, M., AND KANADE, T. 1984. The 3-D MOSAIC scene understanding system. In *Proceedings of the Image Understanding Workshop* (New Orleans, La., Oct. 3–4). DARPA, Science Applications, McLean, Va., pp. 137–148.

HERMAN, M., KANADE, T., AND KUROE, S. 1983. The 3-D MOSAIC scene understanding system. In *Proceedings of the 8th International Joint Conference on Artificial Intelligence* (Karlsruhe, West Germany, Aug. 8–12). IJCAI, pp. 1108–1112.

HORAUD, P., AND BOLLES, R. C. 1984. 3DPO's strategy for matching three-dimensional objects in range data. In *Proceedings of the International Conference on Robotics* (Atlanta, Ga., Mar. 13–15). IEEE, New York, pp. 78–85.

HORN, B. K. P. 1977. Understanding image intensities. *Artificial Intell. 8*, 2 (Apr.), 201–231.

HORN, B. K. P. 1984. Extended Gaussian images. In *Proceedings of the IEEE 72*, 12 (Dec.), pp. 1656–1678.

HORN, B. K. P., AND IKEUCHI, K. 1984. The mechanical manipulation of randomly oriented parts. *Sci. Amer. 251*, 2 (Aug.), 100–111.

IDESAWA, M., AND YATAGAI, T. 1980. 3-D shape input and processing by Moire technique. In *Proceedings of the 5th International Conference on Pattern Recognition* (Miami, Fla., Dec. 1–4). IAPR and IEEE, New York, pp. 1085–1090.

IGES 1983. Initial Graphics Exchange Specification (IGES), Version 2.0 (Document No. PB83-137448, National Technical Information Service (NTIS), 5285 Port Royal Rd., Springfield, Va., 20161.

IKEUCHI, K. 1981. Recognition of 3-D objects using the extended Gaussian image. In *Proceedings of the 7th International Joint Conference on Artificial Intelligence* (Vancouver, B.C., Canada, Aug. 24–28). IJCAI, pp. 595–600.

IKEUCHI, K., AND HORN, B. K. P. 1981. Numerical shape from shading and occluding boundaries. *Artificial Intell. 17* (Aug.), 141–184.

IKEUCHI, K., HORN, B. K. P., NAGATA, S., CALLAHAN, T., AND FEIMGOLD, O. 1983. Picking up an object from a pile of objects. MIT Artificial Intelligence Lab Memo 726. Massachusetts Institute of Technology, Cambridge, Mass.

INOKUCHI, S., AND NEVATIA, R. 1980. Boundary detection in range pictures. In *Proceedings of the 5th International Conference on Pattern Recognition* (Miami, Fla., Dec. 1–4). IAPR and IEEE, New York, pp. 1031–1035.

INOKUCHI, S., NITA, T., MATSUDAY, F., AND SAKURAI, Y. 1982. A three-dimensional edge-region operator for range pictures. In *Proceedings of the 6th International Conference on Pattern Recognition* (Munich, West Germany, Oct. 19–22). IAPR and IEEE, New York, pp. 918–920.

INOKUCHI, S., SATO, K., AND MATSUDA, F. 1984. Range imaging system for 3-D object recognition. In *Proceedings of the 7th International Conference on Pattern Recognition* (Montreal, Canada, July 30–Aug. 2). IAPR and IEEE, New York, pp. 806–808.

JAIN, R. 1983. Dynamic scene analysis. In *Progress in Pattern Recognition*, vol. 2. A. Rosenfeld and L. Kanal, Eds., North-Holland, Amsterdam, The Netherlands.

JARVIS, R. A. 1983a. A perspective on range finding techniques for computer vision. *IEEE Trans. Pattern Anal. Machine Intell. PAMI-5*, 2 (Mar.), 122–139.

JARVIS, R. A. 1983b. A laser time-of-flight range scanner for robotic vision. *IEEE Trans. Pattern Anal. Machine Intell. PAMI-5*, 5 (Sept.), 505–512.

KANADE, T. 1981. Recovery of the three-dimensional shape of an object from a single view. *Artificial Intell. 17* (Aug.), 409–460.

KIM, H. S., JAIN, R. C., AND VOLZ, R. A. 1985. Object recognition using multiple views. In *Proceedings of the International Conference on Robotics and Automation* (St. Louis, Mo., Mar. 25–28). IEEE, New York, pp. 28–33.

KOENDERINK, J. J., AND VAN DOORN, A. J. 1976. The singularities of the visual mapping. *Biol. Cybern. 24*, 1, 51–59.

KOENDERINK, J. J., AND VAN DOORN, A. J. 1979. Internal representation of solid shape with respect to vision. *Biol. Cybern. 32*, 4, 211–216.

KUAN, D. T., AND DRAZOVICH, R. J. 1984. Model-based interpretation of range imagery. In *Proceedings of the National Conference on Artificial Intelligence* (Austin, Tex., Aug. 6–10). AAAI, pp. 210–215.

LAFFEY, T. J., HARALICK, R. M., AND WATSON, L. T. 1982. Topographic classification of digital image intensity surfaces. In *Proceedings of the Workshop on Computer Vision: Representation and Control* (Rindge, N. H., Aug. 23–25). IEEE, New York, pp. 171–177.

LAFUE, G. 1976. Recognition of three-dimensional objects from orthographic views. *ACM Comput. Graphics 10*, 2, 103–108.

LAMY, F., LIEGEOIS, C., AND MEYRUEIS, P. 1982. Three-dimensional automated pattern recognition using the Moire technique. In *Proceedings of The Society for Photo-Optical Instrumentation Engineers Conference on Robotics and Industrial Inspection*, vol. 360, (San Diego, Calif., Aug. 24–27). SPIE, Bellingham, Wash., pp. 345–351.

LANGRIDGE, D. J. 1984. Detection of discontinuities in the first derivatives of surfaces. *Comput. Vision, Graphics, Image Processing 27*, 3 (Sept.), 291–308.

LAVIN, M. A. 1974. An application of line labeling and other scene-analysis techniques to the problem of hidden-line removal. Working Paper 66, MIT Artificial Intelligence Lab, Massachusetts Institute of Technology, Cambridge, Mass. (April).

LEE, H-C., AND FU, K-S. 1982. A computer vision system for generating object descriptions. In *Proceedings of the Pattern Recognition and Image Processing Conference* (Las Vegas, Nevada, June 14–17). IEEE, New York, pp. 466–472.

LEE, H-C., AND FU, K-S. 1983. Generating object descriptions for model retrieval. *IEEE Trans. Pattern Anal. Machine Intell. PAMI-5*, 5 (Sept.), 462–471.

LEWIS, R. A., AND JOHNSTON, A. R. 1977. A scanning laser rangefinder for a robotic vehicle. In *Proceedings of the 5th International Joint Conference on Artificial Intelligence* (Cambridge, Mass., Aug. 22–25). IJCAI, pp. 762–768.

LIN, W. C., AND FU, K. S. 1984. A syntatic approach to 3-D object representation. *IEEE Trans. Pattern Anal. Machine Intell. PAMI-6*, 3 (May), 351–364.

LIN, W. C., FU, K. S., AND SEDERBERG, T. 1984. Estimation of three-dimensional object orientation for computer vision systems with feedback. *J. Robotic Syst. 1*, 1 (Spring), 59–82.

LIN, C., AND PERRY, M. J. 1982. Shape description using surface triangularization. In *Proceedings of the Workshop on Computer Vision: Representation and Control* (Rindge, N. H., Aug. 23–25). IEEE, New York, pp. 38–43.

LIPSCHUTZ, M. M. 1969. *Differential Geometry*. McGraw Hill, New York.

LITTLE, J. J. 1983. An iterative method for reconstructing convex polyhedra from extended Gaussian images. In *Proceedings of the National Conference on Artificial Intelligence* (Washington, D.C., Aug. 22–26). AAAI, pp. 247–250.

LYNCH, D. K. 1981. Range enhancement via one-dimensional spatial filtering. *Comput. Graphics Image Processing 15*, 2 (Feb.), 194–200.

MACKWORTH, A., AND MOKHTARIAN, F. 1984. Scale-based description of planar curves. In *Proceedings of the 5th National Conference of Canadian Society for Computational Studies of Intelligence* (London, Ont., Canada, May). Pp. 114–119.

MARIMONT, D. H. 1984. A representation for image curves. In *Proceedings of the National Conference on Artificial Intelligence* (Austin, Tex., Aug. 6–10). AAAI, pp. 237–242.

MARTIN, W. N., AND AGGARWAL, J. K. 1983. Volumetric descriptions of objects from multiple views. *IEEE Trans. Pattern Anal. Machine Intell. PAMI-5*, 2 (Mar.), 150–158.

MCKEE, J. W., AND AGGARWAL, J. K. 1975. Computer recognition of partial views of three-dimensional curved objects. Computer Science Tech. Rep. 171, Univ. of Texas, Austin, Tex.

MEAGHER, D. J. 1981. Geometric modeling using octree encoding. *Comput. Graphics Image Processing 19*, 2 (June), 129–147.

MEAGHER, D. J. 1982. Efficient synthetic image generation of arbitrary 3-D objects. In *Proceedings of the Pattern Recognition and Image Processing Conference* (Las Vegas, Nevada, June 14–17). IEEE, New York, pp. 473–478.

MEDIONI, G., AND NEVATIA, R. 1984. Description of 3-D surfaces using curvature properties. In *Proceedings of the Image Understanding Workshop* (New Orleans, La., Oct. 3–4). DARPA, Science Applications, McLean, Va., pp. 291–299.

MILGRIM, D. L., AND BJORKLUND, C. M. 1980. Range image processing: planar surface extraction. In *Proceedings of the 5th International Conference on Pattern Recognition* (Miami, Fla., Dec. 1–4). IEEE, New York, pp. 912–919.

MINKOWSKI, H. 1897. Allgemeine lehrsatze uber die konvexen polyeder. Nachrichten von der Koniglichen Gesellschaft der Wissenschaften, Mathematisch-Physikalische Klasse, Gottingen, pp. 198–219.

MITICHE, A., AND AGGARWAL, J. K. 1983. Detection of edges using range information. *IEEE Trans. Pattern Anal. Machine Intell. PAMI-5*, 2 (Mar.), [illegible]178.

[illegible]NKAR, P. G., SHAPIRO, L. G., AND HARALICK, [illegible]M. 1982. Recognizing three-dimensional objects single perspective views using geometric and relational reasoning. In *Proceedings of the Pattern Recognition and Image Processing Conference* (Las Vegas, Nevada, June 14–17). IEEE, New York, pp. 479–484.

NACKMAN, L. R. 1982. Three-dimensional shape description using the symmetric axis transform. Ph.D. dissertation, Computer Science Dept., Univ. of North Carolina, Chapel Hill, N.C.

NACKMAN, L. R. 1984. Two-dimensional critical point configuration graphs. *IEEE Trans. Pattern Anal. Machine Intell. PAMI-6*, 4 (July), 442–449.

NEVATIA, R., AND BINFORD, T. O. 1973. Structured descriptions of complex objects. In *Proceedings of the 3rd International Joint Conference on Artificial Intelligence* (Stanford, Calif., Aug. 20–23). IJCAI, pp. 641–647.

NEVATIA, R., AND BINFORD, T. O. 1977. Description and recognition of curved objects. *Artificial Intell. 8*, 1, 77–98.

NEWMAN, W. M., AND SPROULL, R. F. 1979. *Principles of Interactive Computer Graphics*, 2nd ed. McGraw-Hill, New York.

NISHIHARA, H. K. 1981. Intensity, visible-surface, and volumetric representations. *Artificial Intell. 17* (Aug.), 265–284.

NITZAN, D., BRAIN, A. E., AND DUDA, R. O. 1977. The measurement and use of registered reflectance and range data in scene analysis. In *Proceedings of IEEE 65* (Feb.), 206–220.

O'BRIEN, N., AND JAIN, R. 1984. Axial motion stereo. In *Proceedings of the Workshop on Computer Vision: Representation and Control* (Annapolis, Md., Apr. 30–May 2). IEEE, New York, pp. 88–94.

O'NEILL, B. 1966. *Elementary Differential Geometry.* Academic Press, New York.

O'ROURKE, J., AND BADLER, N. 1979. Decomposition of three-dimensional objects into spheres. *IEEE Trans. Pattern Anal. Machine Intell. PAMI-1*, 3 (July), 295–305.

OSHIMA, M., AND SHIRAI, Y. 1981. Object recognition using three-dimensional information. In *Proceedings of the 7th International Joint Conference on Artificial Intelligence* (Vancouver, B.C., Canada, Aug. 24–28). IJCAI, pp. 601–606.

OSHIMA, M., AND SHIRAI, Y. 1983. Object recognition using three-dimensional information. *IEEE Trans. Pattern Anal. Machine Intell. PAMI-5*, 4 (July), 353–361.

PARTHASARATHY, S., BIRK, J., AND DESSIMOZ, J. 1982. Laser rangefinder for robot control and inspection. In *Proceedings of the Society for Photo-Optical Instrumentation Engineers Conference on Robot Vision*, vol. 336 (Arlington, Va., May 6–7). SPIE, Bellingham, Wash., pp. 2–11.

PIPITONE, F. J., AND MARSHALL, T. G. 1983. A wide-field scanning triangulation rangefinder for machine vision. *Int. J. Robotics Res., 2*, 1 (Spring), 39–49.

PONCE, J., AND BRADY, M. 1985. Toward a surface primal sketch. In *Proceedings of the International Conference on Robotics and Automation* (St. Louis, Mo., Mar. 25–28). IEEE, New York, pp. 420–425.

POPPLESTONE, R. J., BROWN, C. M., AMBLER, A. P., AND CRAWFORD, G. F. 1975. Forming models of plane-and-cylinder faceted bodies from light stripes. In *Proceedings of the 4th International Joint Conference on Artificial Intelligence* (Tbilisi, Georgia, USSR, Sept.). IJCAI, pp. 664–668.

POTMESIL, M. 1979. Generation of 3D surface descriptions from images of pattern-illuminated objects. In *Proceedings of the Pattern Recognition and Image Processing Conference* (Chicago, Ill., Aug. 6–8). IEEE, New York, pp. 553–559.

POTMESIL, M. 1982. Generating three-dimensional surface models of solid objects from multiple projections. IPL-TR-033, Ph.D. dissertation, Image Processing Lab, Rennselaer Polytechnic Institute, Troy, N.Y.

POTMESIL, M. 1983. Generating models of solid objects by matching 3D surface segments. In *Proceedings of the 8th International Joint Conference on Artificial Intelligence* (Karlsruhe, West Germany, Aug. 8–12). IJCAI, pp. 1089–1093.

PRATT, W. K. 1978. *Digital Image Processing.* Wiley Interscience, New York.

REEVES, A. P., PROKOP, R. J. ANDREWS, S. E., AND KUHL, F. P. 1984. Three-dimensional shape analysis using moments and Fourier descriptors. In *Proceedings of the 7th International Conference on Pattern Recognition* (Montreal, Canada, July 30–Aug. 2). IAPR and IEEE, New York, pp. 447–450.

REQUICHA, A. A. G. 1980. Representations for rigid solids: theory, methods, and systems. *ACM Comput. Surv. 12*, 4 (Dec.), 437–464.

REQUICHA, A. A. G., AND VOELCKER, H. B. 1982. Solid modeling: a historical summary and contemporary assessment. *IEEE Comput. Graphics Appl. 2*, 2 (Mar.), 9–24.

REQUICHA, A. A. G., AND VOELCKER, H. B. 1983. Solid modeling: current status and research directions. *IEEE Comput. Graphics Appl. 3*, 7 (Oct.), 25–37.

ROBERTS, L. G. 1965. Machine perception of three-dimensional solids. *Optical and Electro-Optical*

Information Processing. J. T. Tippett et al., Eds., MIT Press, Cambridge, Mass. pp. 159–197.

ROSENFELD, A., AND KAK A. 1981. *Digital Picture Processing,* vols. 1 and 2. Academic Press, New York.

ROSENFELD, A. 1984. Image analysis: Problems, progress, and prospects. *Pattern Recognition 17,* 1 (Jan.), 3–12.

SADJADI, F. A., AND HALL, E. L. 1979. Object recognition by three-dimensional moment invariants. In *Proceedings of the Pattern Recognition and Image Processing Conference* (Chicago, Ill., Aug. 6–8). IEEE, New York, pp. 327–336.

SADJADI, F. A., AND HALL, E. L. 1980. Three-dimensional moment invariants. *IEEE Trans. Pattern Anal. Machine Intell. PAMI-2,* 2 (Mar.), 127–136.

SATO, Y., AND HONDA, I. 1983. Pseudodistance measures for recognition of curved objects. *IEEE Trans. Pattern Anal. Machine Intell. PAMI-5,* 4 (July), 362–373.

SATO, Y., KITAGAWA, H., AND FUJITA, H. 1982. Shape measurement of curved objects using multiple slit ray projections. *IEEE Trans. Pattern Anal. Machine Intell. PAMI-4,* 6 (Nov.), 641–646.

SCHUDY, R. B., AND BALLARD, D. H. 1978. Model-detection of cardiac chambers in ultrasound images. TR-12, Computer Science Dept., Univ. of Rochester, Rochester, N.Y. (Nov.).

SCOTT, R. 1984. Graphics and prediction from models. In *Proceedings of the Image Understanding Workshop* (New Orleans, La., Oct. 3–4). DARPA, Science Applications, McLean, Va., pp. 98–106.

SEDERBERG, T. W., AND ANDERSON, D. C. 1985. Steiner surface patches. *IEEE Comput. Graphics Appl. 5,* 5 (May), 23–36.

SETHI, I. K., AND JAYARAMAMURTHY, S. N. 1984. Surface classification using characteristic contours. In *Proceedings of the 7th International Conference on Pattern Recognition* (Montreal, Canada, July 30–Aug. 2). IAPR and IEEE, New York, pp. 438–440.

SHAFER, S. A., AND KANADE, T. 1983. The theory of straight homogeneous generalized cylinders and taxonomy of generalized cylinders. CMU-CS-83-105. Carnegie-Mellon Univ., Pittsburgh, Pa. (Jan.).

SHAPIRA, R., AND FREEMAN, H. 1977a. Reconstruction of curved surface bodies from a set of imperfect projections. In *Proceedings of the 5th International Joint Conference on Artificial Intelligence* (Cambridge, Mass., Aug. 22–25). IJCAI, pp. 628–634.

SHAPIRA, R., AND FREEMAN, H. 1977b. A cyclic-order property of bodies with three-face vertices. *IEEE Trans. Comput. C-26,* 10 (Oct.), 1035–1039.

SHIRAI, Y., AND SUWA, M. 1971. Recognition of polyhedra with a range finder. In *Proceedings of the 2nd International Joint Conference on Artificial Intelligence* (London, U.K., Aug.). IJCAI, pp. 80–87.

SHNEIER, M. O. 1981. Models and strategies for matching in industrial vision. Computer Science Tech. Rep. TR-1073, Univ. of Maryland, College Park, Md. (July).

SHOSTAK, R. E. 1977. On the sup-inf method for proving Presburger formulas. *J. ACM 24,* 529–543.

SIGGRAPH '84 Conference Proceedings. 1984. *Comput. Graph. (ACM) 18,* 3 (July).

SILBERBERG, T. M., HARWOOD, D., AND DAVIS, L. S. 1984. Object recognition using oriented model points. In *Proceedings of the 1st Artificial Intelligence Applications Conference* (Denver, Colo., Dec. 5–7). AAAI and IEEE, New York, pp. 645–651.

SMITH, G. B. 1984. A fast surface interpolation technique. In *Proceedings of the Image Understanding Workshop* (New Orleans, La., Oct. 3–4). DARPA, Science Applications, McLean, Va., pp. 211–215.

SMITH, D. R., AND KANADE, T. 1984. Autonomous scene description with range imagery. In *Proceedings of the Image Understanding Workshop* (New Orleans, La., Oct. 3–4). DARPA, Science Applications, McLean, Va., pp. 282–290.

SOROKA, B. I., AND BAJCSY, R. K. 1978. A program for describing complex three-dimensional objects using generalized cylinders as primitives. In *Proceedings of the Pattern Recognition and Image Processing Conference* (Chicago, Ill., June). IEEE, New York, pp. 331–339.

SPIE 3-D Machine Perception Conference 1981. The Society for Photo-Optical Instrumentation Engineers, vol. 283, Bellingham, Wash.

STOCKMAN, G., AND ESTEVA, J. C. 1984. Use of geometrical constraints and clustering to determine 3-D object pose. In *Proceedings of the 7th International Conference on Pattern Recognition* (Montreal, Canada, July 30–Aug. 2). IAPR and IEEE, New York, pp. 742–744.

SUGIHARA, K. 1979. Range-data analysis guided by junction dictionary. *Artificial Intell. 12,* 41–69.

SUTHERLAND, I. E., SPROULL, R. F., AND SUMAKER, R. A. 1974. A characterization of ten hidden-surface algorithms. *ACM Comput. Surv. 6,* 1 (Mar.), 293–347.

SVETKOFF, D. J., LEONARD, P. F., SAMPSON, R. E., AND JAIN, R. C. 1984. Techniques for real-time 3D feature extraction using range information. In *Proceedings of The Society for Photo-Optical Instrumentation Engineers Conference on Intelligent Robotics and Computer Vision,* vol. 521 (Cambridge, Mass., Nov. 5–8). SPIE, Bellingham, Wash.

THORPE, C., AND SHAFER, S. 1983. Topological correspondence in line drawings of multiple views of objects. CMU-CS-83-113, Dept. of Computer Science, Carnegie-Mellon Univ., Pittsburgh, Pa. (Mar.).

TILLER, W. 1983. Rational B-splines for curve and surface representation. *IEEE Comput. Graphics Appl. 3,* 6 (Nov.), 61–69.

TIO, J. B. K., MCPHERSON, C. A., AND HALL, E. L. 1982. Curved surface measurement for robot vision. In *Proceedings of the Pattern Recognition and Image Processing Conference* (Las Vegas, Nevada, June 14–17). IEEE-CS, New York, pp. 370–378.

TERZOPOULOS, D. 1983. Multilevel computational processes for visual surface reconstruction. *Comput. Vision, Graphics, Image Processing 24,* 52–96.

TROPF, H., AND WALTER, I. 1983. An ATN model for 3-D recognition of solids in single images. In *Proceedings of the 8th International Joint Conference on Artificial Intelligence* (Karlsruhe, West Germany, Aug. 8–12). IJCAI, pp. 1094–1098.

UDUPA, K. J., AND MURTHY, I. S. N. 1977. New concepts for three-dimensional shape analysis. *IEEE Trans. Comput. C-26,* 10 (Oct.), 1043–1049.

ULLMAN, S. 1979. *The Interpretation of Visual Motion.* MIT Press, Cambridge, Mass.

UNDERWOOD, S. A., AND COATES, C. L., JR. 1975. Visual learning from multiple views. *IEEE Trans. Comput. C-24,* 6 (June), 651–661.

VEMURI, B. C., AND AGGARWAL, J. K. 1984. 3-dimensional reconstruction of objects from range data. In *Proceedings of the 7th International Conference on Pattern Recognition* (Montreal, Canada, July 30–Aug. 2). pp. 752–754.

WALLACE, T. P., AND WINTZ, P. A. 1980. An efficient three-dimensional aircraft recognition algorithm using normalized Fourier descriptors. *Comput. Graphics Image Processing 13,* 96–126.

WALTZ, D. L. 1972. Generating semantic descriptions from drawings of scenes with shadows. AI-TR-271, MIT Artificial Intelligence Lab, Massachusetts Institute of Technology, Cambridge, Mass., (Nov.).

WANG, Y. F., MAGGEE, M. J., AND AGGARWAL, J. K. 1984. Matching three-dimensional objects using silhouettes. *IEEE Trans. Pattern Anal. Machine Intell. PAMI-6,* 4 (July), 513–517.

WATSON, L. T., LAFFEY, T. J., AND HARALICK, R. M. 1985. Topographic classification of digital image intensity surfaces using generalized splines and the discrete cosine transformation. *Comput. Vision, Graphics, Image Processing 29,* 143–167.

WITKIN, A. P. 1981. Recovering surface shape and orientation from texture. *Artificial Intell. 17* (Aug.), 17–45.

WONG, R. Y., AND HAYREPETIAN, K. 1982. Image processing with intensity and range data. In *Proceedings of the Pattern Recognition and Image Processing Conference* (Las Vegas, Nevada, June 14–17). IEEE, New York, pp. 518–520.

WOODHAM, R. J. 1981. Analysing images of curved surfaces. *Artificial Intell. 17,* (Aug.), 117–140.

YAKIMOVSKY, Y., AND CUNNINGHAM, R. 1978. A system for extracting three-dimensional measurements from a stereo pair of TV cameras. *Comput. Graphics Image Processing 7,* 195–210.

YORK, B. W., HANSON, A. R., AND RISEMAN, E. M. 1980. A surface representation for computer vision. In *Proceedings of the 5th International Conference on Pattern Recognition* (Miami, Fla., Dec. 1–4). IAPR and IEEE, New York, pp. 124–129.

YORK, B. W., HANSON, A. R., AND RISEMAN, E. M. 1981. 3D object representation and matching with B-splines and surface patches. In *Proceedings of the 7th International Joint Conference on Artificial Intelligence* (Vancouver, B.C., Canada, Aug. 24–28). IJCAI, pp. 648–651.

Received November 1984; final revision accepted June 1985.

Map Building for a Mobile Robot from Sensory Data

MINORU ASADA, MEMBER, IEEE

Abstract—**The development of an autonomous land vehicle (ALV) is a central problem in artificial intelligence and robotics, and has been extensively studied. To perform visual navigation, a robot must gather information about its environment through external sensors, interpret the output of these sensors, construct a scene map and a plan sufficient for the task at hand, and then monitor and execute the plan. As a first step, real time visual navigation systems for road following were developed in which simple methods for detecting road edges were applied in simple environments. For even slightly more complicated scenes, the difficulty of the problem increases dramatically, therefore a world model such as a map could be very important for successful navigation through such environments. A method for building a three-dimensional (3-D) world model for a mobile robot from sensory data derived from outdoor scenes is presented. The 3-D world model consists of four kinds of maps: a physical sensor map, a virtual sensor map, a local map, and a global map. First, a range image (physical sensor map) is transformed to a height map (virtual sensor map) relative to the mobile robot. Next, the height map is segmented into unexplored, occluded, traversable and obstacle regions from the height information. Moreover, obstacle regions are classified into artificial objects or natural objects according to their geometrical properties such as slope and curvature. A drawback of the height map (recovery of planes vertical to the ground plane) is overcome by using multiple height maps that include the maximum and minimum height for each point on the ground plane. Multiple height maps are useful not only for finding vertical planes but also for mapping obstacle regions into video image for segmentation. Finally, height maps are integrated into a local map by matching geometrical parameters and by updating region labels. The results obtained using landscape models and ALV simulator of the University of Maryland are shown, and constructing a global map with local maps is discussed.**

I. Introduction

THE DEVELOPMENT of an autonomous land vehicle (ALV) is a central problem in artificial intelligence and robotics, and has been extensively studied [1]–[2]. To perform visual navigation, a robot must gather information about its environment through external sensors, interpret the output of these sensors, construct a scene map and a plan sufficient for the task at hand, and then monitor and execute the plan. As a first step, real time visual navigation systems for road following were developed in which simple methods for detecting road edges were applied in simple environments [1], [4], [5], [12]. For even slightly more complicated scenes, the difficulty of the problem increases dramatically, therefore a world model such as a map could be very important for successful navigation through such environments.

Sometimes, accurate, quantitative maps may be available in advance [13], more often, maps are less descriptive and provide only global information as in a conventional geographical map [14]. In other cases, the robot may try to construct the map from sensory date in unknown environments. Hebert and Kanade [15] have analyzed ERIM range images and constructed a surface property map represented in a Cartesian coordinate system viewed from top, which yields surface type of each point and its geometric parameters for segmentation of scene map into traversable and obstacle regions. Tsuji and Zheng [16] discussed the differences between the two-dimensional (2-D) maps in [15] and perspective maps proposed in their stereo-vision-based mobile robot system. Their point is that 2-D maps are easy to understand but do not naturally capture sensor resolution and accuracy. They used perspective maps for navigation in which three-dimensional (3-D) information obtained by stereo vision is represented in the image coordinate system. However, integration of perspective maps obtained at different locations on a single perspective map seems difficult. Having both types of maps in a hierarchical representation and referring to each other when necessary is one solution for the above problem.

Elfes [17] has developed a sonar-based mapping and navigation system which constructs sonar maps of the environments viewed from the top and updates them with recently acquired sonar information. He proposed a hierarchical representation of sonar map which includes three kinds of axes: an abstract axis, a resolution axis, and a geographic axis. In his system, the outputs of sonar sensors are directly mapped to a 2-D map, therefore, the difference between sensor maps and the 2-D maps is implicit, and other sensory data such as video images and range images seem difficult to be represented in this hierarchy.

In this paper, we propose a method for building a hierarchical representation of a 3-D world model for a

Manuscript received April 1, 1989; revised January 15, 1990. This work was supported in part by the Defense Advanced Research Projects Agency and the U.S. Army Engineer Topographic Laboratory under contract DACA76-84-C-0004, and in part by the Grant-In-Aid for Scientific Research from the Ministry of Education, Science, and Culture, Japanese Government. This work was partially presented at the 1988 IEEE International Conference on Robotics and Automation, Philadelphia, PA, 1988.

The author was with Center for Automation Research, University of Maryland, College Park, MD 20742. He is now with the Department of Mechanical Engineering for Computer-Controlled Machinery, Osaka University, Suita, Osaka 565, Japan.

IEEE Log Number 9037596.

Reprinted from *IEEE Transactions on Systems, Man, and Cybernetics*, Vol. 37, No. 6, November/December 1990.
Copyright © 1990 by The Institute of Electrical and Electronics Engineers, Inc. All rights reserved.

mobile robot, making the relation between coordinate systems at different levels explicit. With this model, the mobile robot can derive useful information from a map at adequate level to accomplish various kinds of tasks such as visual navigation, obstacle avoidance, and landmarks and/or objects recognition. The 3-D world model consists of four kinds of maps: a physical sensor map, a virtual sensor map, a local map, and a global map. A physical sensor map usually represents sensory data or analyzed data in the sensor-based coordinate system from which the sensory data is taken (e.g., perspective map in [16], or ERIM range image in [15]).

A virtual sensor map represents the sensory data (in the physical sensor map) in the vehicle-centered Cartesian coordinate system. Any other type of coordinate system such as a cylindrical one can be applicable to the virtual sensor map representation in the context of representing sensory data in the vehicle-centered coordinate system. However, the Cartesian mapping seems more suitable for the virtual sensor map representation because the size of the cell on the map (it corresponds to the resolution of the map) is constant everywhere, therefore, fusing the data at the same point but observed from different view points is much easier than other type of mapping such as a cylindrical mapping or Delaunay triangulation which requires a complex algorithm to access data points and their neighborhoods [18]. As one example of a particular instance of the virtual sensor map, we introduce a height map that represents the height information transformed from a range image in the vehicle-centered Cartesian coordinate system. The 2-D map in [15] and Cartesian elevation map (CEM) in [19] are also categorized into the virtual sensor map representation. Virtual sensor maps are integrated into a local map that is represented in the object-centered coordinate system. The local map has its own reference (object) on which the integration of the virtual sensor maps is based, therefore, a new local map with a new reference is generated when the current reference cannot be observed as the robot moves. Thus, a number of local maps are generated along with the robot navigation. A global map consists of these local maps and the geometrical relationship between them. How to build a global map with relational local maps is proposed by Asada et al. [26]. In this paper, we focus on the building the physical sensor maps, the virtual sensor maps, and the local map from the video and range data, and have not dealt with how to build the global map.

In our system, a range image, one example of the physical sensor map, is transformed to a height map, one example of the virtual sensor map, in the mobile robot centered Cartesian coordinate system. The height is estimated from the assumed ground plane on which the vehicle exists. First, we segment the height map into unexplored, occluded, traversable and obstacle regions from the height information, and then classify obstacle regions into artificial objects or natural objects according to their geometrical properties such as slope and curvature. A drawback of the height map—recovery of planes vertical to the ground plane—is overcome by using a multiple height map that includes the maximum and minimum heights for each point on the ground plane. The multiple height map is useful not only for finding vertical planes but also for mapping obstacle regions into video image (another sensor map) for segmentation. Finally, the system integrates height maps observed at different locations into a local map, matching geometrical parameters of obstacle and traversable regions and updating region labels. We show the results applied to landscape models using ALV simulator of the University of Maryland [4], and discuss about constructing a global map with local maps.

II. System Configuration

A. Physical Simulation System of ALV and Its Environments

Our ALV physical simulation system was developed in our laboratory [4] for providing a low cost experimental environment for navigation (as opposed to an outdoor vehicle [7], [15]). A range finder based on structured light was recently added to this system. Planes of light are projected from a rotating mirror controlled by a stepping motor (see [20] for more detail). More recently, we extended the system in two ways. First, we developed a drive simulator program which controls the speed and steering angle of the vehicle (robot arm) during the motion. The camera height and camera tilt to the ground plane are kept constant during the motion through the position feedback of three leg sensors attached to the camera.

Previously, a wooden terrain board on which a road network is painted was set vertically to increase the flexibility of camera motion simulated by the robot arm. Due to its vertical setting, it was very difficult to put landscape model such as trees, bushes, buildings and other vehicles on the board. Thus, we set the terrain board horizontally so that we can place any landscape model without permanently fixing their positions. Figs. 1(a) and (b) show our experimental setup and a picture of the new simulation board with many landscape models such as trees, bushes, cabins, mail box and cars. The robot arm attached with a TV camera and a light-stripe range scanner is set on the board to input a picture and a range image.

B. Overall of Map Building System

Fig. 2 shows the architecture of our system. In this figure, we omit other modules such as path planner, navigator, pilot and supervisor in [8] in order to concentrate on the map building system.

The 3-D world model for a mobile robot consists of four kinds of maps: a physical sensor map, a virtual sensor map, a local map, and a global map. Elfes [17] proposed multiple axes of representation of a sensor map in his sonar mapping and navigation system (resolution axis, geographical axis and abstraction axis) and adopted three

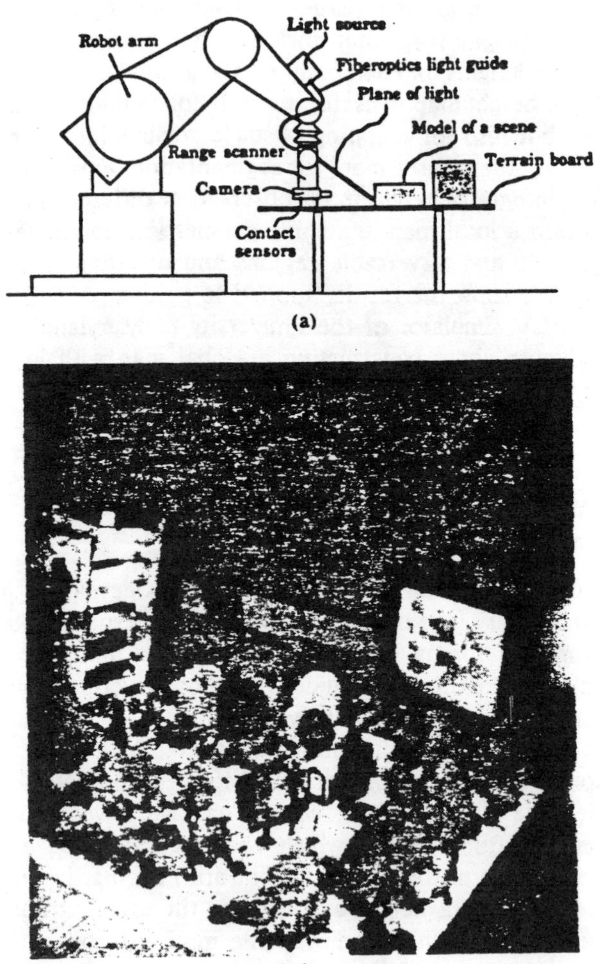

Fig. 1. Simulation board. (a) Experimental set up (from [20]). (b) Photo of new simulation board.

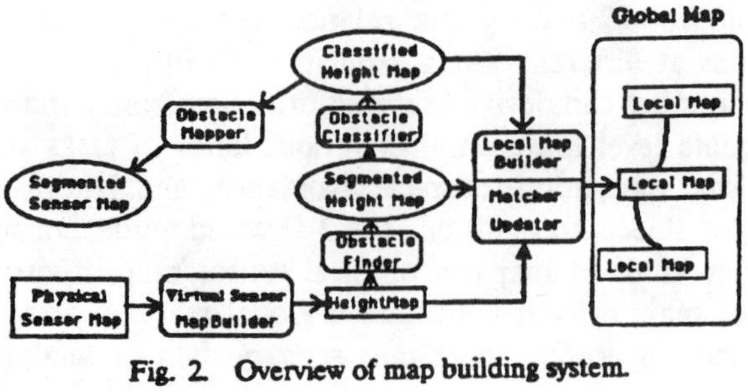

Fig. 2. Overview of map building system.

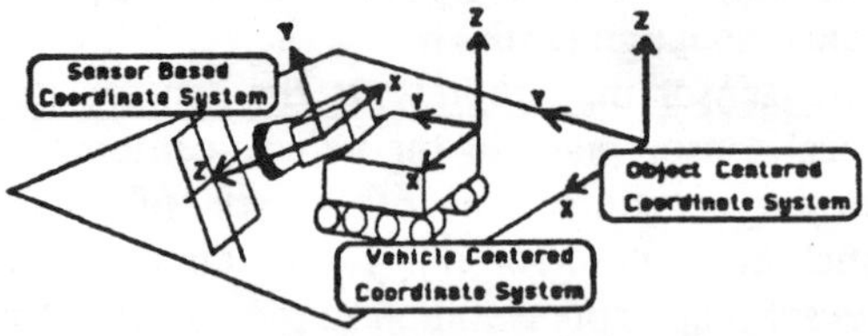

Fig. 3. Geometrical relation of coordinate systems between three maps.

levels (view, local map, global map) in the geographical axis. We extend and generalize these levels so that other sensory data such as range and video data can be represented. Fig. 3 shows the geometrical relation between the three coordinate systems of the physical sensor map (sensor-based), the virtual sensor map (vehicle-centered), and the local map (object-centered). The global map is a set of the local maps and discussed in Section IV. Each sensor has its own coordinate system; for example, an intensity image is represented in the camera-centered coordinate system and a range image in the range-finder-centered coordinate system, both of which are fixed to the robot (vehicle). Here, we assume that the relation between sensor coordinate systems and vehicle coordinate system is known, and that the motion information is available, but not always accurate. A virtual sensor map builder (VSMB) builds the virtual sensor maps from the physical sensor map. Stereo matching, which we do not consider in this paper, is one possible strategy of virtual sensor map building for obtaining the depth map (virtual sensor map). Here, we introduce a height map obtained from the range image as a virtual sensor map. The height map is analyzed by the obstacle finder (OF) and the obstacle classifier (OC) to segment it into unexplored, occluded, traversable and obstacle regions and then to classify obstacle regions into artificial or natural objects. The result of the height map analysis is mapped onto the intensity image by the obstacle mapper (OM) in order to segment the intensity image. The local map builder (LMB) builds a local map, matching and updating virtual sensor maps at different observing stations in the world coordinate system. In the following, each module is described along with some experimental results.

III. Height Map Analysis

A. Virtual Sensor Map Builder (from Range Image to Height Map)

The virtual sensor map builder builds virtual sensor maps from physical sensor maps. Here, we deal with a video image taken by a single camera and a range image obtained from our range finder [20] as physical sensor maps. Even though we use the same camera to take both video and range data, our idea can be applicable to other type of range data such as a range image taken by ERIM range finder [15]. We discuss the differences between two types of the range images in Section IV. Figs. 4 show examples of these physical sensor maps. The input scene includes a straight road, T-type intersection, two cabins, one truck, two cars, a mailbox, a stop sign at the intersection, trees and bushes as shown in Figs. 4(a), (c), and (e). Figs. 4(b), (d), and (f) are the corresponding range image to Fig. 4(a), (c), and (e), respectively. All of the images are 512×512, and both the range and the intensity values are quantized into 256 levels (8 bits). The darker points are closer to the range finder and the brighter points are farther from it. In the white regions, range information is not available due to inadequate reflection or occlusion. Although actual range finders such as the ERIM range scanner [15] measure the radial distances rather than Cartesian coordinates in both axes, the range image ob-

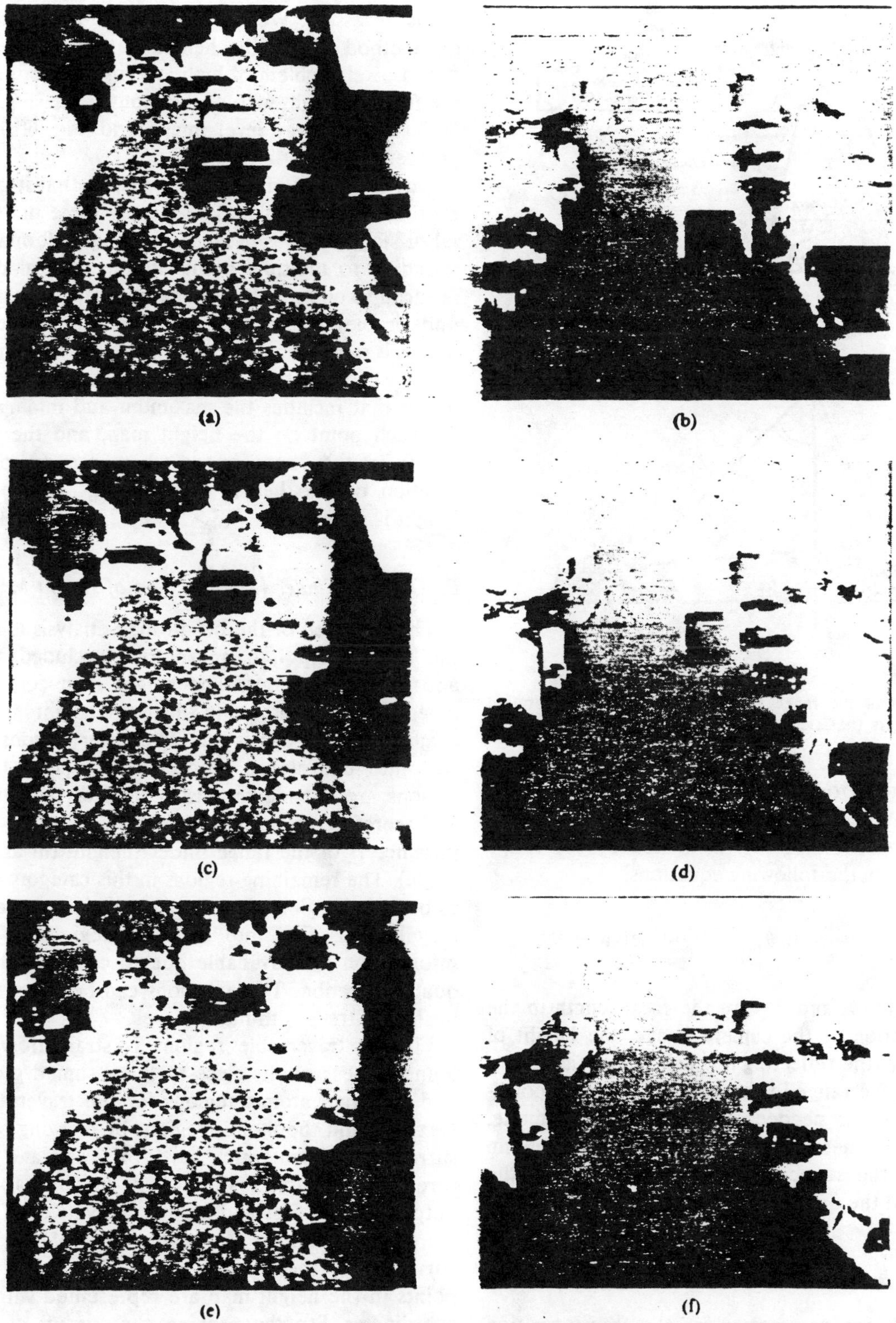

Fig. 4. Sensor maps. (a) First intensity image. (b) First range image. (c) Third intensity images. (d) Third range image. (e) Fifth intensity image. (f) Fifth range image.

tained from our range finder has irregular coordinate aces due to its special ranging geometry [20]; the vertical coordinate is radial (scanning angle of the light plane) but the horizontal coordinate is the same as that of the intensity image because range calculation is based on the triangulation with light planes and a single TV camera. Fig. 5(a) shows calibration parameters required to calculate the range.

The range image is transformed to a height map in the vehicle centered coordinate system based on the known height and tilt of the range finder relative to the vehicle. The position of a point in a given coordinate system can be derived from the obtained range R and the direction to that point (it corresponds to the coordinates (x_r, y_r) on the range image). We use the Cartesian coordinate system O-XYH shown in Fig. 5(b), in which case the XY plane

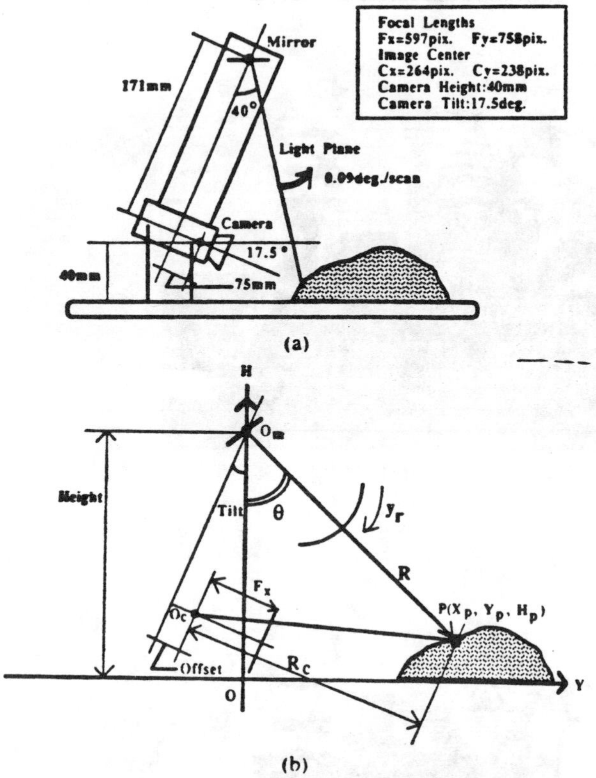

Fig. 5. Calculation of the range and height. (a) Range calibration parameters. (b) Calculation of height from range.

corresponds to the ground plane, and the origin O is just below the range finder (the reflecting mirror). The coordinates of a point $P(X_p, Y_p, H_p)$ estimated by the range finder are given by the following equations:

$$X_p = \frac{R_c x_r}{F_x}, \qquad Y_p = R \sin\theta, \qquad H_p = \text{Height} - R\cos\theta$$

where F_x, Height, θ, and R_c are the focal length in the horizontal direction of the camera image, the height of the range finder (the reflecting mirror), the vertical scanning angle, and the range in the camera-centered coordinate system which is needed to estimate X-coordinate. The last two parameters are derived from the column position y_r in the range image, the tilt angle of the camera Tilt, and the offset parameter Offset by the equations:

$$\theta = a y_r + b, \qquad R_c = R\sin(\theta + \text{Tilt}) - \text{Offset}$$

where a and b are parameters for the transformation from the column position to the vertical scanning angle. An adequate area on the ground plane in front of the vehicle is assigned for a height map and the 3-D coordinates (X_p, Y_p, H_p) are quantized into 8bits numbers. The height map is a 256×256 image, each pixel corresponds to 1 mm^2 on the simulation board. The entire map corresponds to a square of side length 256 mm; the scale of the new simulation board is 87:1 (HO scale). Gray levels encode the height from the assumed ground plane. Since the range is sparse and noisy at far points, smoothing is necessary. We applied an edge-preserving smoothing method [22] to the height map in order to avoid a mixed pixel problem of high and low points. Figs. 6 show the filtered height map of the input scene (Fig. 4(b)); Fig. 6(a) shows a gray level image and Fig. 6(b) shows its perspective view.

One drawback of the height map is that it is unable to represent vertical planes, especially these under horizontal or sloped planes because the range information corresponding to multiple points in the vertical direction is reduced to one point in the height map. This is especially undesirable since the range information on the vertical planes is more accurate than that on the horizontal planes. Thus, we compute a multiple height map for one range image that includes the maximum and minimum heights for each point on the height map, and the number of points in the range image which are mapped to one location on the height map. In Figs. 6, the maximum height is shown.

B. Obstacle Finder (Segmentation of Height Map)

The first step of the height map analysis is to segment the height map into unexplored, occluded, traversable and obstacle regions. The height map consists of two types of regions—these in which the height information is available or these it is not. The latter regions are classified into unexplored or occluded regions. Unexplored regions are outside the visual field of the range finder, and therefore are easily detected by using the calibration parameters of the range finder (height, tilt and scanning angle). The remaining regions in this category are labeled as occluded regions. Some regions which are not occluded may be classified into occluded regions if the height information is unavailable due to causes such as inadequate reflection. These regions can be often seen inside bushes or trees with many leaves.

Finding traversable regions is straightforward. First, identify these points close to the assumed ground plane and construct atomic regions for the traversable region. Next, expand the atomic regions by merging other points surrounding them which have low slope and low curvature. The desirable feature of the height map is that the outputs of the first and second derivatives of the height map correspond to the magnitudes of the slope and curvature of the surface because the locations of the points in the height map are represented with Cartesian coordinates. Strictly speaking, the outputs of the second derivative of height do not directly correspond to the curvature of the surface, but at least, it outputs zeros for the sloped plane such as a roof plane of a house. Figs. 7 show the magnitudes of the first (Sobel) and second derivatives of the height map (Fig. 6(a)) whose mask size is 3 by 3 pixels. The remaining regions are labeled as obstacle regions. Fig. 8 shows the final result of the segmentation of the height map. White, light gray, dark gray and black regions are unexplored, occluded, traversable and obstacle regions, respectively. We can see that

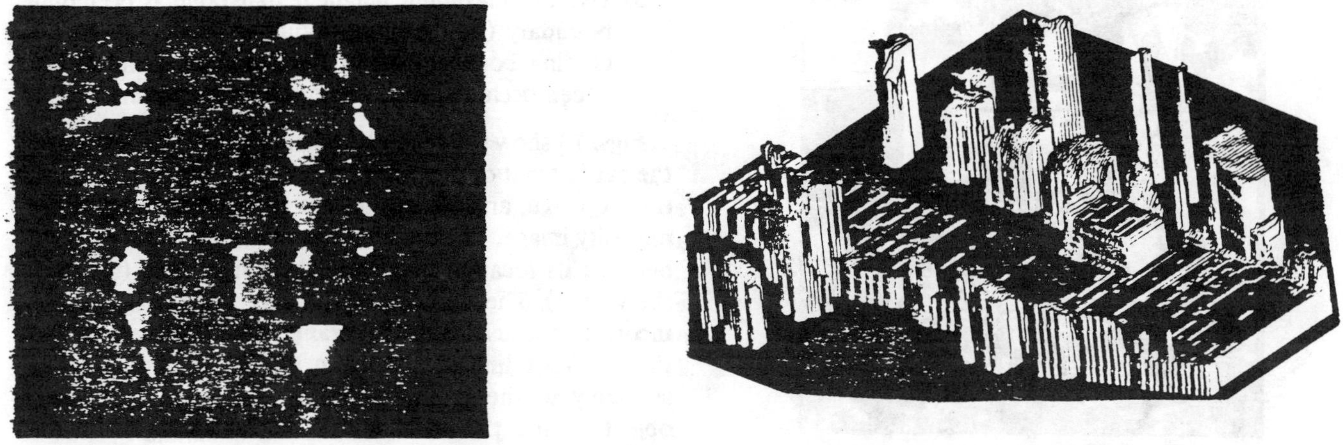

(a) (b)

Fig. 6. Height map. (a) Gray image. (b) Perspective view of (a).

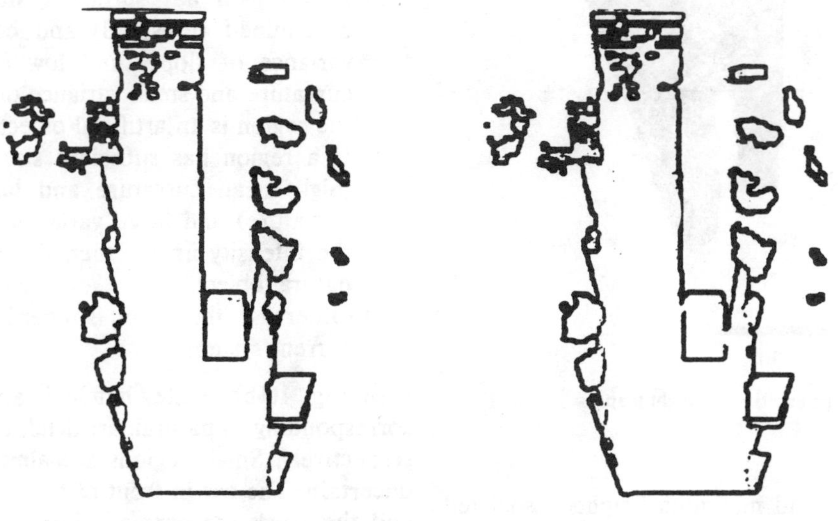

(a) (b)

Fig. 7. (a) Slope map. (b) Curvature map.

the boundary of the obstacle regions has high slope and/or high curvature (see Figs. 7).

The result of segmentation of the height map should be useful for path planning since many path planning algorithm are based on a top view of the configuration of obstacles and free space [23].

C. Obstacle Classifier

The segmented height map constructed by the OF is very useful for navigation tasks such as avoiding obstacles, but does not contain sufficient explicit information enough for higher level tasks such as landmark or object recognition. As a first step in object recognition, we try to classify obstacle regions as artificial objects or parts of natural objects. Many artificial objects such as cabins, cars, mailboxes and road signs shown in Figs. 4 have planar surfaces, which yield constant slope and low curvature in the height map and linear features in the intensity image. On the other hand, natural objects such as trees and bushes have fine structures with convex and concave surfaces, which yield various slopes and/or high curvatures in the height map and therefore large variance of brightness in the intensity image (the reverse is not always true).

Thus, utilization of not only the height map but also the intensity image is useful for obstacle classification. In order to use the brightness information in the intensity image, we map the obstacle regions to the intensity image to segment it. The mapping of obstacles to the intensity image seems at first straightforward based on the geometrical relation between the camera and the range finder. However, it is complicated by the need to correctly choose

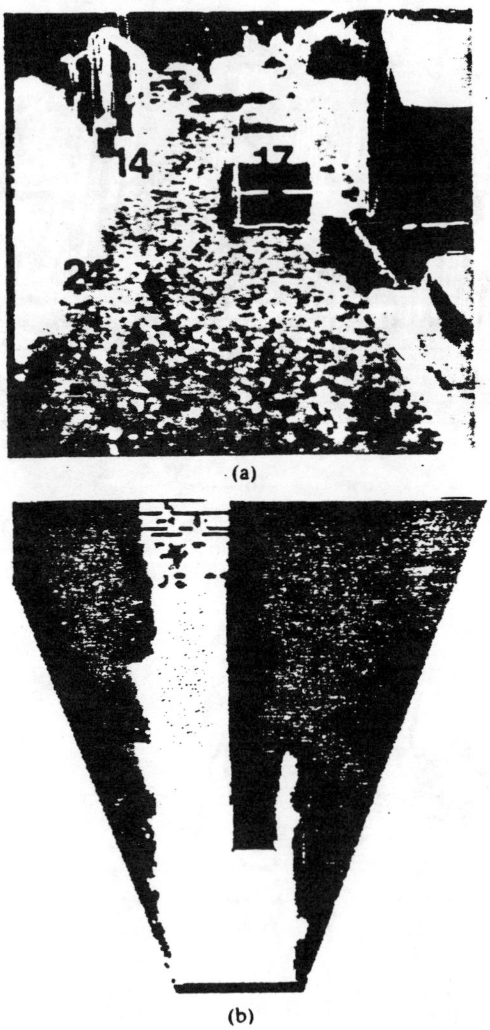

Fig. 8. Segmentation of height map.

between the maximum and minimum heights associated with each point in the height map. Figs. 9 show a perspective view of a cube on a plane and its height map. If we map only the maximum value of the height of the cube, only a top surface of the cube is cut out from its perspective view and visible side surfaces are left undiscovered (see Fig. 9(a)). We should use the minimum height when the object is bounded by traversable regions and use the maximum height when it is occluding other objects behind it (see Fig. 9(c)). Classifying the boundary of obstacle regions in the height map and using the multiple height map, the obstacle mapper maps the obstacles to the intensity image as follows.

1) Classify each boundary point of the obstacle region in the height map according to the geometrical relation between the point, occluded region and traversable region in the segmented height map (Fig. 8):
2) Use the minimum height if that point is adjacent to a traversable region or that point is an occluded boundary (the boundary point is labeled as an occluded boundary when the occluded region is between the boundary point and the range finder).
3) Use the maximum height if that point is occluding boundary (the boundary point is labeled as an occluding boundary when the boundary point is between occluded region and the range finder).

Figs. 10 show the result of this mapping. The truck and the car in front of the cabin on the right side, the mailbox, the stop sign, and the bushes are finely segmented in the intensity image. The car at the intersection is not mapped because its location is outside the height map (far from the viewer). The roof line of the cabin on the left side is incorrect because of the bad range data. The reason that the top roof line of the cabin on the right side drops suddenly to the ground plane is that there is a lower object behind the cabin and the obstacle region in the height map includes both the cabin and the lower object.

The next step is to classify the obstacle using the properties of the height map and the brightness in the intensity image. The obstacle classifier classifies each resegmented region according to the following criteria.

1) If a region has sufficient size (larger than predetermined threshold) and constant slope (small variance of slope) and low curvature (low mean curvature and small variance of the curvature), then the region is an artificial object.
2) If a region has sufficient size and high curvature (high mean curvature and large variance of the curvature) and large variance of the brightness in the intensity image, then the region is a part of a natural object.
3) Otherwise, the region is regarded as uncertain in the current system.

In Fig. 10(b), white, hatched, and black regions are corresponding to natural, artificial, and uncertain objects, respectively. Small regions are almost always labeled as uncertain. The car in front of the cabin on the right side and the truck are correctly interpreted as artificial objects. However, the roofs of two cabins, the mailbox and the stop sign are misinterpreted as natural objects because of the high curvature due to vertical planes and/or insufficient, noisy range data. The region corresponding to the right side cabin includes a roof and the lower object behind it, therefore resegmentation into two regions is necessary [21]. Uncertain regions and some regions with vertical surfaces would require closer examination for correct interpretation.

D. Local Map Builder

During the motion of the vehicle, the system produces a sequence of virtual sensor maps constructed at different observing stations. These virtual sensor maps should be integrated into a local map in the world centered coordinate system. The local map builder consists of two parts; the first part matches a new height map and the current local map to determine the correct motion parameters of the vehicle, and the second one updates the description of region properties on the integrated local map. Matching

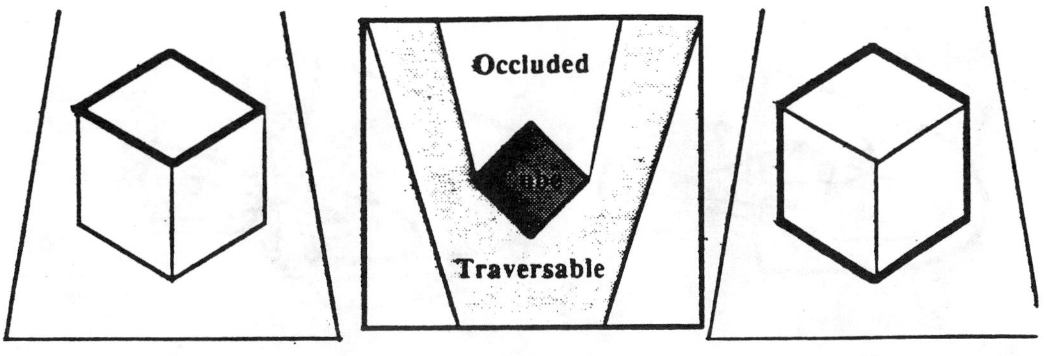

Fig. 9. Cube on a plane and its height map. (a) Only maximum height is used. (b) Height of map (a). (c) Maximum and minimum heights are used.

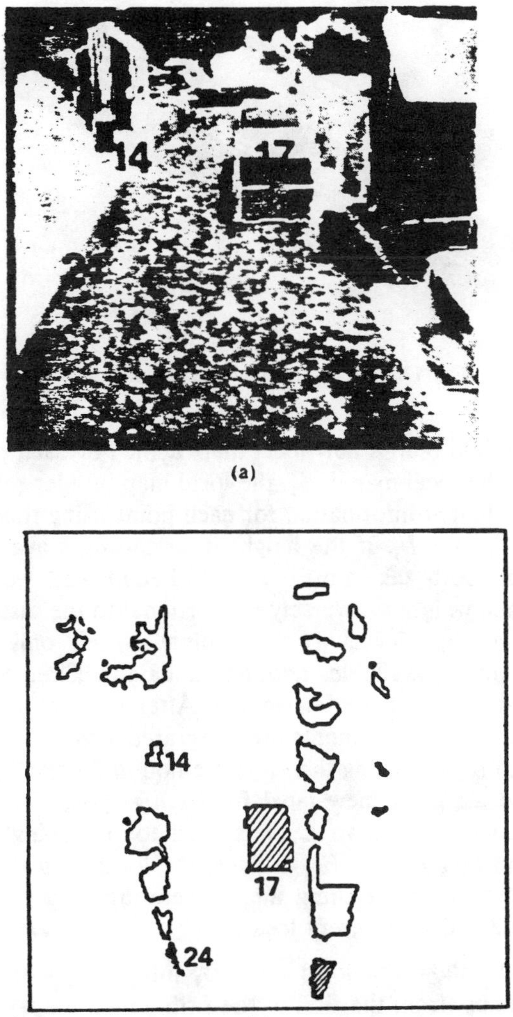

Fig. 10. Mapping obstacle region into the intensity image. (a) Mapped region. (b) Obstacle map.

is performed between the $(i+1)$th height map H_{i+1} and the local map L_i in which the height maps from the first to the ith observing stations are integrated. If $i=1$, then $L_i = H_i$. The point P_l in the local map L_i can be represented as the point P_h in the height map H_{i+1} by the following equation:

$$P_h = R_{lh} P_l + T_{lh}$$

where R_{lh} is a rotation matrix consisting of three rotational components, $\Delta\alpha$, $\Delta\beta$, and $\Delta\theta$ along each axis, and T_{lh} is a translation vector along each axis, that is, $T_{lh}^{-1} = (\Delta X, \Delta Y, \Delta H)$. The following are matching procedures.

1) Match the traversable regions between H_{i+1} and L_i. Since the traversable regions are usually the larger planar regions in the segmented map and rough estimates of the motion parameters (ΔX, ΔY, ΔH, $\Delta\alpha$, $\Delta\beta$, and $\Delta\theta$) are ordinarily available from the internal sensors of the vehicle, this matching is relatively straightforward. The traversable region in H_{i+1} and L_i are on the XY-plane in each coordinate system because we determined the orientation and location of the coordinate system so that the ground plane corresponds to the XY-plane (see Fig. 5(b)). Therefore, matching the traversable region between H_{i+1} and L_i is performed by overlaying the XY-plane in L_i onto the XY-plane in H_{i+1}. By this matching, $\Delta H, \Delta\alpha, \Delta\beta$ are determined (in our experiments, all are zeros), and the remains are ΔX, ΔY, and $\Delta\theta$; the motion parameters on the XY-plane.
2) Determine the remaining motion parameters ΔX, ΔY, and $\Delta\theta$ by matching the obstacle regions between H_{i+1} and L_i. Search for the motion parameters which take the minimum height difference for each obstacle region, starting the rough motion parameters from the vehicle navigation system as an initial value. We evaluate the following criterion function to find the minimum height difference:

$$\Delta H_i^k(\Delta x, \Delta y, \Delta\theta) = \sqrt{\frac{\Sigma\left(H_{i+1} - L_i^k(\Delta x, \Delta y, \Delta\theta)\right)^2}{N_i^k}}$$

where ΔX, ΔY, $\Delta\theta$, L_i^k, and N_i^k are estimated translation, rotation, the height of kth obstacle region in L_i, and the number of height points compared between H_i and L_{i+1} for the kth obstacle region, respectively. If we pick up only obstacle region, the matching could be ambiguous. For example, a small obstacle region with horizontal plane matches at any position of a large region with hori-

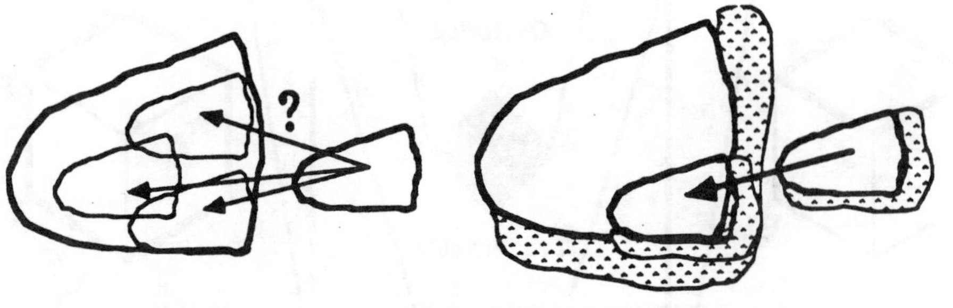

Fig. 11. Expanding obstacle region to avoid the matching ambiguity.

zontal plane of the same height (see Fig. 11(a)). In order to avoid such an ambiguity, we expand obstacle regions so that they include the height information surrounding them (see Fig. 11(b)).

A special care needs to be taken for moving objects because they have different motion parameters from those of the stationary environment. In the current system, we use a heuristic for detecting moving object. The obstacle surrounded with traversable regions is a candidate moving object because the moving objects should be inside the traversable region (except for flying objects). Three regions numbered 14, 17, and 25 in Fig. 10(b) are candidate moving objects while only a region 17 (truck) is actually moving. Although search area for stationary obstacles is ± 2 mm for translation $(\Delta X, \Delta Y)$ and $\pm 2°$ for rotation $\Delta\theta$ with fine resolution of 1 mm and 1°, coarse to fine search algorithm is applied to the candidates for moving objects since moving objects might be outside the search area for the stationary objects. The algorithm is as follows: estimating motion parameters in a large search area with sparse intervals of translation (± 5 mm) and rotation ($\pm 5°$)first, and then, refining it with fine intervals same as for the stationary obstacles. In Fig. 10, motion parameters for almost all regions are correctly obtained, and as a result, only region 17 is interpreted as an object moving farther from the viewer (towards the intersection). The motion parameters ΔX, ΔY, and $\Delta\theta$ for the stationary objects are 0 mm, 15 mm, and 0°, respectively, and those for the moving object are 0 mm, 45 mm, and 0°, respectively. Fig. 12 shows the histogram of the motion parameters when $\Delta\theta = 0°$, in which the inverse number of the minimum height difference for each obstacle region are accumulated into the corresponding motion parameters $(\Delta X, \Delta Y)$. Two peaks correspond to the stationary regions (larger) and the moving region (smaller). Velocities of the moving objects are used to predict their locations and orientations in the next height map H_{i+2}.

3) Integrate the results of the matching into a local map and update region labels. The local map builder overlays all regions in H_{i+1} onto the local map L_i according to their motion parameters. We call the

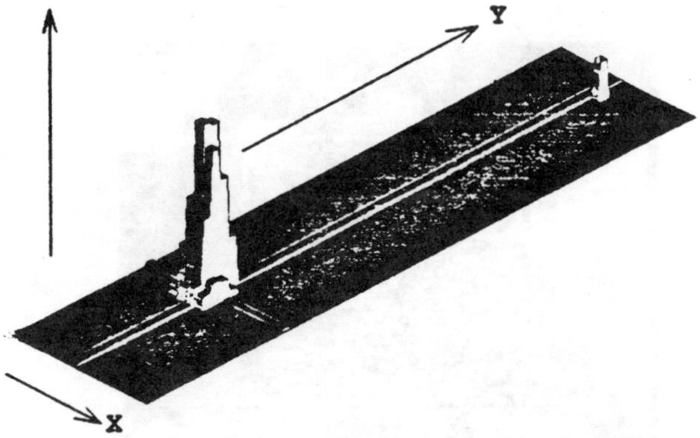

Fig. 12. Histogram of motion parameters.

overlaid map a new local map L_{i+1}. For each point on the local map L_{i+1}, the local map builder refines the height information for each point using those in H_{i+1} and L_i. If the height information is available from both, take a mean of both heights with weights (each weight is inversely proportional to the distance from the viewer to that point). When only one height is available, adopt that height. Else, retain that point left undetermined. After integration, obstacle finder segments the integrated local map L_{i+1} into regions using the same method in Section III-C. and assigns a new label for each region. The segmented local map L_{i+1} is used for matching with next height map H_{i+2}. Thus, the local map is constructed by matching and integrating height maps observed at different locations.

Figs. 13 show the local map L_5 integrated with five height maps (from the first to the fifth height maps). Fig. 13(b) shows the roof of the cabin on the left side in the first height map H_1 (left) and the local map L_5 (right) from the different view angle from Fig. 13(a). In H_1, the roof is worm-eaten and includes noisy heights due to the shallow angle between the light plane from range scanner and the roof plane. While, in L_5, the shape of the roof is clear because the vehicle approached to the cabin, therefore, better height information was obtained. Fig. 13(c) shows the shape difference of moving object (truck) between H_5 (left) and L_5 (right). While it is worm-eaten in H_5 because of long distance from the viewer, the region corresponding to the bed of the truck is clear in L_5 where

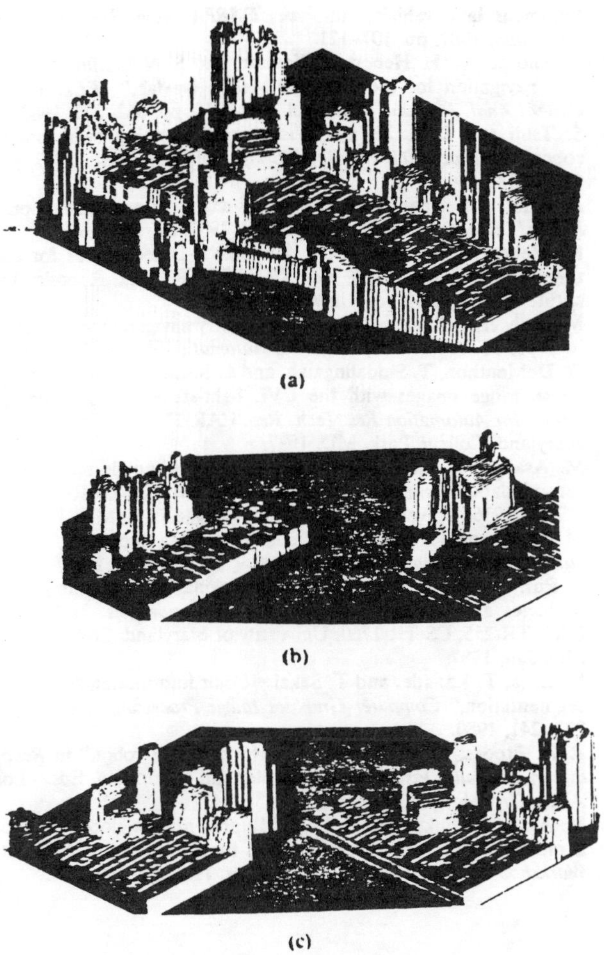

Fig. 13. Local map integration with five height maps. (a) Local map. (b) Roof of the cabin on the left side in the first height map (left) and the local map (right). (c) Shape difference of moving object (truck) between the fifth height man (left) and the local man (right).

the system successfully tracks the moving object and updates its height information.

IV. Conclusion

A map building system for a mobile robot from sensory data has been described. We introduced the virtual sensor map into the 3-D world model construction in order to represent various sensory data such as range and video data and in order to integrate them in a Cartesian coordinate system. We proposed the use of a height map, one example of virtual sensor maps, obtained from the range image for a mobile robot to support various tasks such as path planning and landmark recognition. The height map is easy to recover and calculation of geometrical properties such as slope and curvature is straightforward.

In the experiments, we used the range finder based on the structured lighting method that produces dense range images consisting of 512×512 image points. The largest difference between the range image used and actual one such as an ERIM range image [15] size of which is 256×64 is the resolution of the range image. Since the size of the height map used is smaller than that of the range image, we do not need the sophisticated technique such as the locus algorithm which can produce the dense height map from the sparse range image [18]. In our case, the edge-preserved smoothing on the height map can provide us with the sufficient height information for our task. In spite of the above difference, the proposed idea for the world model representation can be applicable to both types of the range images given the geometrical relation between the range finder, the TV camera, and the vehicle.

The current obstacle classifier uses principally the geometrical properties of height information and makes little use of brightness information in identifying the obstacles as artificial or natural objects. The experimental results include some number of errors. Errors due to vertical planes could be corrected by verifying the existence of the vertical plane in the multiple height maps and by obtaining more correct parameters for them from the range image (physical sensor map). Other errors might be corrected by the use of color images since color information is often useful for segmentation of images of outdoor scene [24].

In our experiments, we have dealt with almost flat terrains. In the case of non-flat terrain, the matching process would be more complicated because the orientation of the *H*-axes in a height map and a local map might be different. One alternative is to adopt the direction of the gravity as the common orientation of the *H*-axes in the different virtual sensor maps utilizing the output from the orientation sensor in the navigation system. This could make the matching process as simple as in the case of flat terrain.

In this paper, we have not dealt with the problem of constructing a global map with a set of lcoal maps. The method of local map builder that numerically integrates virtual sensor maps using height information could not be applicable to the construction of the global map because a traveling of long distance often results in a significant amount of positional error in the global map and because the robot cannot correct this error unless landmarks in the environment are given in advance [25]. Adada, Fukui, and Tsuji [26] developed a method of representing a global map with relational local maps where the rough description of the geometrical relation between adjacent local maps is described. We could apply their method to our scene. However, as the map scale becomes larger, the higher level descriptions are generally required to represent a world model instead of numerical representation of it. Our current system has dealt with only region labels (unexplored, occluded, traversable and obstacle) and sublabels (artificial or natural). In order to accomplish higher tasks such as landmark and/or object recognition, knowledge base of the objects expected in the scene and geometrical modeling for them are needed, and the control structure for them should be exploited. These problems are under investigation.

Acknowledgment

The author wishes to thank Prof. Azriel Rosenfeld and Prof. Larry S. Davis for helpful comments and discussions and Mr. Daniel DeMenthon for providing range images at the University of Maryland, and Prof. Saburo Tsuji and Prof. Yoshiaki Shirai for constructive discussions at Osaka University, Japan. He also wishes to thank his son, Ryu Asada, who helped him assemble and paint the landscape models on the simulation board.

References

[1] A. M. Waxman, J. Le Moigne, and B. Srinivasan, "Visual navigation of roadways," in *Proc. IEEE Int. Conf. Robotics Automation*, 1985, pp. 862–867.

[2] S. Tsuji, Y. Yagi, and M. Asada, "Dynamic scene analysis for a mobile robot in man-made environment," in *Proc. IEEE Int. Conf. Robotics Automation*, 1985, pp. 850–855.

[3] S. Tsuji, J. Y. Zheng, and M. Asada, "Stereo vision of a mobile robot: world constraints for image matching and interpretation," in *Proc. IEEE Int. Conf. Robotics Automat.*, 1986, pp. 1594–1599.

[4] A. M. Waxman, J. LeMoigne, L. S. Davis, E. Liang, and T. Siddalingaiah, "A visual navigation system," in *Proc. IEEE Int. Conf. Robotics Automation*, 1986, pp. 1600–1606.

[5] R. Wallace, K. Matsuzaki, Y. Goto, J. Crisman, J. Webb, and T. Kanade, "Progress in robot road-following," in *Proc. IEEE Int. Conf. Robotics Automat.*, 1986, pp. 1615–1621.

[6] S. A. Shafer, A. Stentz, and C. Thorpe, "An architecture for sensor fusion in a mobile robot," in *Proc. IEEE Int. Conf. Robotics Automat.*, 1986, pp. 2002–2011.

[7] C. Thorpe, S. Shafer, T. Kanade, *et al.*, "Vision and navigation for the Carnegie Mellon Navlab," in *Proc. DARPA Image Understanding Workshop*, 1987, pp. 143–152.

[8] L. S. Davis, D. DeMenthon, R. Gajulapalli, T. R. Kushner, J. Le Moigne, and P. Veatch, "Vision-based navigation: a status report," in *Proc. DARPA Image Understanding Workshop*, 1987, pp. 153–169.

[9] Y. Goto and A. Stentz, "The CMU system for mobile robot navigation," in *Proc. IEEE Int. Conf. Robotics Automat.*, 1987, pp. 99–105.

[10] R. A. Brooks, "A hardware retargetable distributed layered architecture for mobile robot control," in *Proc. IEEE Int. Conf. Robotics Automat.*, 1987, pp. 106–110.

[11] R. S. Wallace, "Robot road following by adaptive color classification and shape trucking," in *Proc. IEEE Int. Conf. Robotics Automat.*, 1987, pp. 258–263.

[12] M. A. Turk, D. G. Morgenthaler, K. D. Gremban, and M. Marra, "Video road-following for the autonomous land vehicle," in *Proc. IEEE Int. Conf. Robotics Automat.*, 1987, pp. 273–280.

[13] S. Tsuji, "Monitoring of a building environment by a mobile robot," in *Proc. 2nd Int. Symp. Robotics Res.*, 1985, pp. 349–365.

[14] D. Lawton, T. S. Levvit, C. C. McConnell, P. C. Nelson, and J. Glicksman, "Environmental modeling and recognition for an autonomous land vehicle," in *Proc. DARPA Image Understanding Workshop*, 1987, pp. 107–121.

[15] C. Thorpe, M. H. Hebert, T. Kanade, and S. A. Shafer, "Vision and navigation for the Carnegie-Mellon Navlab," *IEEE Trans. Pattern Anal. Machine Intell.*, vol. PAMI-10, pp. 362–373, 1988.

[16] S. Tsuji and J. Y. Zheng, "Visual path planning by a mobile robot," in *Proc. 10th Int. Joint Conf. Artificial Intell.*, 1987, pp. 1127–1130.

[17] A. Elfes, "A sonar-based real world mapping and navigation," *IEEE J. Robotics Automat.*, vol. RA-3, 1987, pp. 249–265.

[18] M. Hebert, T. Kanada, and I. Kweon, "3-D techniques for autonomous vehicles," *Tech. Rep.* CMU-RI-TR-88-12, Robotics Institute, Carnegie-Mellon University, 1988.

[19] M. Daily *et al.*, "Autonomous cross-country navigation with ALV," in *Proc. IEEE Int. Conf. Robotics Automat.*, 1988, pp. 718–726.

[20] D. DeMenthon, T. Siddalingaiah, and L. S. Davis, "Production of dense range images with the CVL light-stripe range scanner," *Center for Automation Res. Tech. Rep.* CAR-TR-337, University of Maryland, College Park, MD, 1987.

[21] M. Asada, "Building 3-D world model for a mobile robot from sensory data," *Center for Automation Res. Tech. Rep.* CAR-TR-332, CS-TR-1936, University of Maryland, College Park, MD, 1987.

[22] M. Nagao and T. Matsuyama, "Edge preserving smoothing," *Computer Graphics Image Processing*, vol. 9, pp. 394–407, 1979.

[23] S. Puri and L. S. Davis, "Two dimensional path planning with obstacles and shadows," *Center for Automation Res. Tech. Rep.* CAR-TR-255, CS-TR-1760, University of Maryland, College Park, MD, Jan. 1987.

[24] Y. Ohta, T. Kanade, and T. Sakai, "Color information for region segmentation," *Computer Graphics Image Processing*, vol. 13, pp. 222–241, 1980.

[25] R. A. Brooks, "Visual map making for a mobile robot," in *Readings in Computer Vision*, M. A. Fischler and O. Firscein, Eds. Los Altos, CA: Morgan Kaufman, pp. 438–443.

[26] M. Asada, Y. Fukui, and S. Tsuji, "Representing a global world of a mobile robot with relational local maps," in *Proc. 1988 Int. Workshop Intelligent Robots and Systems*, 1988, pp. 199–204.

Minoru Asada (M'88) was born in Shiga, Japan, on October 1, 1953. He received the B.E., M.Sc., and Ph.D., degrees in control engineering from Osaka University, Osaka, Japan, in 1977, 1979, and 1982, respectively.

From 1982 to 1988, he was an Assistant Professor of Control Engineering, Osaka University, Toyonaka, Osaka, Japan. Since April 1989, he has been an Associate Professor of Mechanical Engineering for Computer-Controlled Machinery, Osaka University, Suita, Osaka, Japan. From August 1986 to October 1987, he was a visiting researcher of Center for Automation Research, University of Maryland, College Park, MD.

Dr. M. Asada is a member of the Institute of Electronics, Information and Computer Engineering (Japan), the Information Processing Society of Japan, and the Robotics Society of Japan. His current research includes Computer Vision, Artificial Intelligence, and Robotics.

A Stochastic Map
For Uncertain Spatial Relationships

Randall Smith[†] Matthew Self[‡] Peter Cheeseman[§]

SRI International
333 Ravenswood Avenue
Menlo Park, California 94025

[‡]Currently at UC Berkeley.

[†]Currently at General Motors Research Laboratories, Warren, Michigan.

[§]Currently at NASA Ames Research Center, Moffett Field, California.

Reprinted from *International Symposium on Robotics Research*, pages 467-474, "A Stochastic Map for Uncertain Spatial Relationships" by R. Smith, M. Self, and P. Cheesman, by permission of The MIT Press, Cambridge, Massachusetts.

In this paper we will describe a representation for spatial relationships which makes explicit their inherent uncertainty. We will show ways to manipulate them to obtain estimates of relationships and associated uncertainties not explicitly given, and show how decisions to sense or act can be made *a priori* based on those estimates. We will show how new constraint information, usually obtained by measurement, can be used to update the world model of relationships consistently, and in some situations, optimally. The framework we describe relies only on well-known state estimation methods.

1 Introduction

Spatial structure is commonly represented at a low level, in both robotics and computer graphics, as local coordinate frames embedded in objects and the transformations among them — primarily, translations and rotations in two or three dimensions. These representations manifest themselves, for example, in transformation diagrams [Paul 1981]. The structural information is *relative* in nature; relations must be chained together to compute those not directly given, as illustrated in Figure 1. In the figure the nominal, initial locations of a beacon and a robot are indicated with coordinate frames, and are defined with respect to a fixed reference frame in the room. The *actual* relationships are $\mathbf{x}_{01}$ and $\mathbf{x}_{02}$, (with the sero subscript dropped for relations defined with respect to the reference frame). After the robot moves, its relation to the beacon is no longer explicitly described.

Generally, nominal information is *all* that is given about the relations. Thus, errors due to measurement, motion (control), or manufacture cause a disparity between the actual spatial structure and the nominal structure we expect. Strategies (for navigation, or automated assembly of industrial parts) that depend on such complex spatial structures, will fail if they cannot accommodate the errors. By utilising knowledge about tolerances and device accuracies, more robust strategies can be devised, as will be subsequently shown.

1.1 Compounding and Merging

The spatial structure shown in Figure 1 represents the *actual* underlying relationships about which we have explicit information. Given a method for combining serial "chains" of given relationships, we can derive the implicit ones. If the explicit relationships are not *known* perfectly, errors will *compound* in a chain of calculations, and be larger than those in any constituent of the chain.

With perfect information, relationship $\mathbf{x}_{21}$ need not be measured — it can be computed through the chain (using $\mathbf{x}_2$ and $\mathbf{x}_1$). However, because of imperfect knowledge, the computed value and the measurement will be different. The difference is resolved by *merging* the pieces of information into a description *at least* as "accurate" as the most accurate piece, no matter how the errors are described. If the merging operation does not do this, there is no point in using it.

The real relationships $\mathbf{x}_1$, $\mathbf{x}_2$, and $\mathbf{x}_{21}$ are mutually constrained, and when information about $\mathbf{x}_{21}$ is introduced, the merging operation should improve the estimates of them all, by amounts proportional to the magnitudes of their initial relative uncertainties. If the merging operation is *consistent*, one updated relation (vector) can be removed from the loop, as the relation can always be recomputed (by compounding the others).

Obviously, a situation may be represented by an arbitrarily complex graph, making the estimation of some relationship, given all the available information, a difficult task.

1.2 Previous Work

Some general methods for incorporating error information in robotics applications([Taylor 1976], [Brooks 1982]) rely on using worst-case bounds on the parameters of individual relationships. However, as worst-case estimates are combined (for example, in the chaining above) the results can become *very conservative*, limiting their use in decision making.

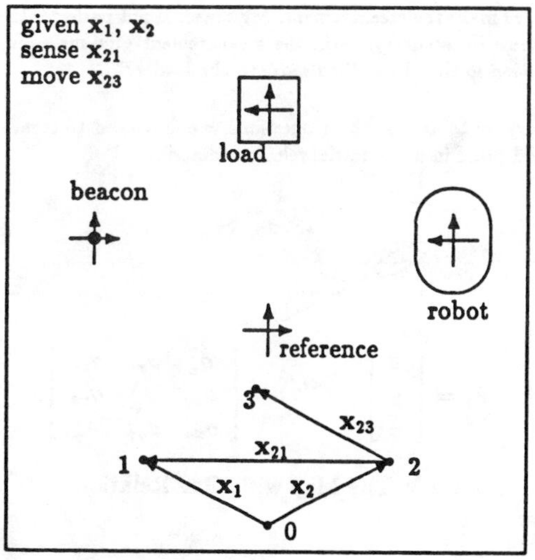

Figure 1: Robot Navigation: Spatial Structure

A probabilistic interpretation of the errors can be employed, given some constraints on their size, and the availability of error models. Smith and Cheeseman[Smith, 1984] described six-degree-of-freedom relationships by their mean vectors and covariance matrices, and produced first-order formulae for compounding them. These formulae were subsequently augmented ([Smith, 1985]) with a merging operation — computation of the conditional mean and covariance — to combine two estimates of the same relation. A similar scalar operation is performed by the HILARE mobile robot [Chatila, 1985].

Durrant-Whyte [Durrant-Whyte 1986] takes an approach to the problem similar to Smith and Cheeseman, but propagates errors differentially rather than using the partial derivative matrices of the transformations. Both are concerned with integrating information consistently across an explicitly represented spatial graph. Others [Faugeras 1986],[Bolle and Cooper, 1986] are exploiting similar ideas for the optimal integration of noisy geometric data in order to estimate global parameters (object localisation).

This paper (amplified in [Smith 1986]) extends our previous work by defining a few simple procedures for representing, manipulating, and making decisions with uncertain spatial information, in the setting of recursive estimation theory.

1.3 The Example

In Figure 1, the initial locations of a beacon and a mobile robot are given with respect to a fixed landmark. Our knowledge of these relations, $\mathbf{x}_1$ and $\mathbf{x}_2$, is imprecise, however. In addition, the location of a loading area (the box) is given very accurately with respect to the landmark. Thus, the vector labeled $\mathbf{x}_3$, has been omitted.

The robot's task is to move to the loading area, so that it's center is within the box. It can then be loaded.

The robot reasons:

"I know where the loading area is, and approximately where I am (in the room). Thus, I know approximately what motion I need to make. Of course, I can't move perfectly, but I have an idea what my accuracy is. *If I move*, will I likely reach the loader (with the required accuracy)? If so, then I will move."

"If not, suppose that I try to sense the beacon. My map shows its approximate location *in the room*; but of course, I don't know exactly where *I am*. Where is it in relation to me? Can I get the beacon in the field of view of my sensor without searching around?"

"*Suppose* I make the measurement. My sensor is not perfect either, but I know its accuracy. Will the measurement give me enough information so that I can then move to the loader?"

Before trying to answer these questions, we first need to create a map, and place in it the initial relations described.

$$\hat{\mathbf{x}} = \hat{\mathbf{x}}_1 = \begin{bmatrix} \hat{x} \\ \hat{y} \\ \hat{\phi} \end{bmatrix}, \quad \mathbf{C}(\mathbf{x}) = \begin{bmatrix} \sigma_x^2 & \sigma_{xy} & \sigma_{x\phi} \\ \sigma_{xy} & \sigma_y^2 & \sigma_{y\phi} \\ \sigma_{x\phi} & \sigma_{y\phi} & \sigma_\phi^2 \end{bmatrix}$$

Figure 2: The Map with One Relation

2 The Stochastic Map

In this paper, uncertain spatial relationships will be tied together in a representation called the *stochastic map*. It contains estimates of the spatial relationships, their uncertainties, and the inter-dependencies of the estimates.

2.1 Representation

A *spatial relationship* will be represented by the vector of its *spatial variables*, $\mathbf{x}$. For example, the position and orientation of a mobile robot can be described by its coordinates, x and y, in a two dimensional cartesian reference frame and by its orientation, ϕ, given as a rotation about the z axis. An *uncertain* spatial relationship, moreover, can be represented by a *probability distribution* over its spatial variables.

The complete probability distribution is generally not available. For example, most measuring devices provide only a nominal value of the measured relationship, and we can estimate the average error from the sensor specifications. However, the full distribution may be unneccesary for making decisions, such as whether the robot will be able to complete a given task (e.g. passing through a doorway). For these reasons, we choose to model an uncertain spatial relationship by estimating the first two moments of its probability distribution—the *mean*, $\hat{\mathbf{x}}$ and the *covariance* (see Figure 2). Figure 2 shows our map with only one object located in it — the beacon. The diagonal elements of the covariance matrix are just the variances of the spatial variables, while the off-diagonal elements are the covariances between the spatial variables. The interpretation of the ellipse in the figure follows in the next section.

Similarly, to model a system of n uncertain spatial relationships, we construct the vector of *all* the spatial variables, called the *system state vector*. As before, we will estimate the mean of the state vector, $\hat{\mathbf{x}}$, and the *system covariance matrix*, $\mathbf{C}(\mathbf{x})$. In Figure 3 the map structure is defined recursively (described below), providing the method for building it by adding one new relation at at time.

The current system state vector is appended with $\mathbf{x}_n$, the vector of spatial variables for a new uncertain relationship being added. Likewise, the current system covariance matrix is augmented with the covariance matrix of the new vector, $\mathbf{C}(\mathbf{x}_n)$, and its cross-covariance with the new vector $\mathbf{C}(\mathbf{x}, \mathbf{x}_n)$, as shown. The cross-covariance matrix is composed of a column of sub-matrices — the cross-covariances of each of the original relations in the state vector with the new one, $\mathbf{C}(\mathbf{x}_i, \mathbf{x}_n)$. These off-diagonal sub-matrices encode the dependencies between the estimates of the different spatial relationships and provide the mechanism for updating all relational estimates that depend on any that are changed.

Thus our "map" consists of the current estimate of the mean of the system state vector, which gives the nominal locations of objects in the map with respect to the world reference frame, and the associated system covariance matrix, which gives the uncertainty of each point in the map and the inter-dependencies of these uncertainties.

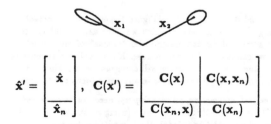

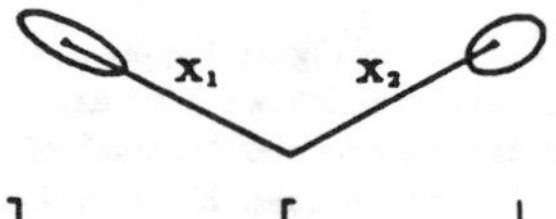

$$\hat{\mathbf{x}}' = \begin{bmatrix} \hat{\mathbf{x}} \\ \hline \hat{\mathbf{x}}_n \end{bmatrix}, \quad \mathbf{C}(\mathbf{x}') = \left[\begin{array}{c|c} \mathbf{C}(\mathbf{x}) & \mathbf{C}(\mathbf{x}, \mathbf{x}_n) \\ \hline \mathbf{C}(\mathbf{x}_n, \mathbf{x}) & \mathbf{C}(\mathbf{x}_n) \end{array}\right]$$

Figure 3: Adding A New Object

The map can now be constructed with the initial estimates of the means and covariances of the relations $\mathbf{x}_1$ and $\mathbf{x}_2$, as shown in Figure 3. If the given estimates are independent of each other, the cross-covariance matrix will be 0.

2.2 Interpretation

For some decisions based on uncertain spatial relationships, we must assume a particular distribution that fits the estimated moments. For example, a robot might need to be able to calculate the probability that a certain object will be in its field of view, or the probability that it will succeed in passing through a doorway.

Given only the mean, $\hat{\mathbf{x}}$, and covariance matrix, $\mathbf{C}(\mathbf{x})$, of a multivariate probability distribution, the principle of maximum entropy indicates that the distribution resulting from assuming the least addtional information is the normal distribution. Furthermore if the relationship is calculated by combining many different pieces of information, the central limit theorem indicates that the resulting distribution will tend to a normal distribution.

We will graph uncertain spatial relationships by plotting contours of constant probability from a normal distribution with the given mean and covariance information. These contours are concentric ellipsoids (ellipses for two dimensions) whose parameters can be calculated from the covariance matrix, $\mathbf{C}(\mathbf{x})$ [Nahi, 1976]. It is important to emphasize that we do not assume that the individual uncertain spatial relationships are described by normal distributions. We estimate the first two central moments of their distributions, and use the normal distribution only when we need to calculate specific probability contours.

In the figures in this paper, a line represents the *actual* relation between two objects (located at the endpoints). The actual object locations are known only by the simulator and displayed for our benefit. The robot's information is shown by the ellipses which are drawn centered on the estimated mean of the relationship and such that they enclose a 99.9% confidence region (about four standard deviations) for the relationships. The mean point itself is not shown.

We have defined our map, and loaded it with the given information. In the next two sections we must learn how to read it, and then change it, before discussing the example.

3 Reading the Map

Having seen how we represent uncertain spatial relationships by estimates of the mean and covariance of the system state vector, we now discuss methods for estimating the first two moments of unknown multivariate probability distributions. See [Papoulis, 1965] for detailed justifications of the following topics.

3.1 Uncertain Relationships

The first two moments computed by the formulae below for non-linear relationships on random variables will be first-order estimates of the true values. To compute the actual values requires knowledge of the *complete* probability density function of the spatial variables, which will not generally be available in our applications. The usual approach is to approximate the non-linear function

$$\mathbf{y} = \mathbf{f}(\mathbf{x})$$

by a Taylor series expansion about the estimated mean, $\hat{\mathbf{x}}$, yielding:

$$\mathbf{y} = \mathbf{f}(\hat{\mathbf{x}}) + \mathbf{F_x}(\mathbf{x} - \hat{\mathbf{x}}) + \cdots,$$

where $\mathbf{F_x}$ is the matrix of partials, or Jacobian, of $\mathbf{f}$ evaluated at $\hat{\mathbf{x}}$:

$$\mathbf{F_x} \triangleq \frac{\partial \mathbf{f}(\mathbf{x})}{\partial \mathbf{x}}(\hat{\mathbf{x}}) \triangleq \begin{bmatrix} \frac{\partial f_1}{\partial x_1} & \frac{\partial f_1}{\partial x_2} & \cdots & \frac{\partial f_1}{\partial x_n} \\ \frac{\partial f_2}{\partial x_1} & \frac{\partial f_2}{\partial x_2} & \cdots & \frac{\partial f_2}{\partial x_n} \\ \vdots & \vdots & \ddots & \vdots \\ \frac{\partial f_r}{\partial x_1} & \frac{\partial f_r}{\partial x_2} & \cdots & \frac{\partial f_r}{\partial x_n} \end{bmatrix}_{\mathbf{x}=\hat{\mathbf{x}}}$$

This terminology is the extension of the f_x terminology from scalar calculus to vectors. The Jacobians are always understood to be evaluated at the estimated mean of the input variables.

Truncating the expansion for $\mathbf{y}$ after the linear term, and taking the expectation produces the linear estimate of the mean of $\mathbf{y}$:

$$\hat{\mathbf{y}} \approx \mathbf{f}(\hat{\mathbf{x}}). \tag{1}$$

Similarly, the first-order estimate of the covariances are:

$$\begin{aligned} \mathbf{C}(\mathbf{y}) &\approx \mathbf{F_x}\mathbf{C}(\mathbf{x})\mathbf{F_x}^T, \\ \mathbf{C}(\mathbf{y},\mathbf{z}) &\approx \mathbf{F_x}\mathbf{C}(\mathbf{x},\mathbf{z}), \\ \mathbf{C}(\mathbf{z},\mathbf{y}) &\approx \mathbf{C}(\mathbf{z},\mathbf{x})\mathbf{F_x}^T. \end{aligned} \tag{2}$$

Of course, if the function $\mathbf{f}$ is *linear*, then $\mathbf{F_x}$ is a constant matrix, and the first two moments of the multivariate distribution of $\mathbf{y}$ are computed exactly, given correct moments for $\mathbf{x}$. Further, if $\mathbf{x}$ follows a normal distribution, then so does $\mathbf{y}$.

In the remainder of this paper we consider only first order estimates, and the symbol "$\approx$" should read as "linear estimate of."

3.2 Coordinate Frame Relationships

We now consider the spatial operations which are necessary to reduce serial chains of coordinate frame relationships between objects to some resultant (implicit) relationship of interest: compounding, and reversal. A useful composition of these operations is also described.

Given two spatial relationships, $\mathbf{x}_2$ and $\mathbf{x}_{23}$, as in Figure 1, *with the second described relative to the first*, we wish to compute the resultant relationship. We denote this binary operation by $\oplus$, and call it compounding.

In another situation, we wish to compute $\mathbf{x}_{21}$. It can be seen that $\mathbf{x}_2$ and $\mathbf{x}_1$ are not in the right form for compounding. We must first invert the sense of the vector $\mathbf{x}_2$ (producing $\mathbf{x}_{20}$). We denote this unary inverse operation $\ominus$, and call it reversal.

The composition of reversal and compounding operations used in computing $\mathbf{x}_{21}$ is very common, as it gives the location of one object coordinate frame *relative* to another, when both are described with a common reference.

These three formulae are:

$$\begin{aligned} \mathbf{x}_{ik} &\triangleq \mathbf{f}(\mathbf{x}_{ij},\mathbf{x}_{jk}) \triangleq \mathbf{x}_{ij} \oplus \mathbf{x}_{jk} \\ \mathbf{x}_{ji} &\triangleq \mathbf{g}(\mathbf{x}_{ij}) \triangleq \ominus\mathbf{x}_{ij} \\ \mathbf{x}_{jk} &\triangleq \mathbf{h}(\mathbf{x}_{ij},\mathbf{x}_{ik}) \triangleq \mathbf{f}(\mathbf{g}(\mathbf{x}_{ij}),\mathbf{x}_{ik}) \triangleq \ominus\mathbf{x}_{ij} \oplus \mathbf{x}_{ik} \end{aligned}$$

Utilizing (1), the first-order estimate of the mean of the compounding operation is:

$$\hat{\mathbf{x}}_{ik} \approx \hat{\mathbf{x}}_{ij} \oplus \hat{\mathbf{x}}_{jk}.$$

Also, from (2), the first-order estimate of the covariance is:

$$\mathbf{C}(\mathbf{x}_{ik}) \approx \mathbf{J}_{\oplus} \begin{bmatrix} \mathbf{C}(\mathbf{x}_{ij}) & \mathbf{C}(\mathbf{x}_{ij},\mathbf{x}_{jk}) \\ \mathbf{C}(\mathbf{x}_{jk},\mathbf{x}_{ij}) & \mathbf{C}(\mathbf{x}_{jk}) \end{bmatrix} \mathbf{J}_{\oplus}^T$$

where the Jacobian of the compounding operation, $\mathbf{J}_\oplus$ is given by:

$$\mathbf{J}_\oplus \triangleq \frac{\partial(\mathbf{x}_{ij} \oplus \mathbf{x}_{jk})}{\partial(\mathbf{x}_{ij}, \mathbf{x}_{jk})} = \frac{\partial \mathbf{x}_{ik}}{\partial(\mathbf{x}_{ij}, \mathbf{x}_{jk})} = \left[\ \mathbf{J}_{1\oplus} \quad \mathbf{J}_{2\oplus}\ \right].$$

The square sub-matrices, $\mathbf{J}_{1\oplus}$ and $\mathbf{J}_{2\oplus}$, are the left and right halves of the compounding Jacobian.

The first two moments of the reversal function can be estimated similarly, utilizing its Jacobian, $\mathbf{J}_\ominus$. The formulae for compounding and reversal, and their Jacobians, are given for three degrees-of-freedom in Appendix A. The six degree-of-freedom formulae are given in [Smith 1986].

The mean of the composite relationship, computed by $\mathbf{h}()$, can be estimated by application of the other operations:

$$\hat{\mathbf{x}}_{jk} = \hat{\mathbf{x}}_{ji} \oplus \hat{\mathbf{x}}_{ik} = \ominus\hat{\mathbf{x}}_{ij} \oplus \hat{\mathbf{x}}_{ik}$$

The Jacobian can be computed by chain rule as:

$$\begin{aligned} {}_\ominus\mathbf{J}_\oplus \triangleq \frac{\partial \mathbf{x}_{jk}}{\partial(\mathbf{x}_{ij}, \mathbf{x}_{ik})} &= \frac{\partial \mathbf{x}_{jk}}{\partial(\mathbf{x}_{ji}, \mathbf{x}_{ik})} \frac{\partial(\mathbf{x}_{ji}, \mathbf{x}_{ik})}{\partial(\mathbf{x}_{ij}, \mathbf{x}_{ik})} \\ &= \mathbf{J}_\oplus \begin{bmatrix} \mathbf{J}_\ominus & \mathbf{0} \\ \mathbf{0} & \mathbf{I} \end{bmatrix} = \left[\ \mathbf{J}_{1\oplus}\mathbf{J}_\ominus \quad \mathbf{J}_{2\oplus}\ \right]. \end{aligned}$$

The chain rule calculation of Jacobians applies to any number of compositions of the basic relations, so that long chains of relationships may be reduced recursively. It may appear that we are calculating first-order estimates of first-order estimates of ..., but actually this recursive procedure produces *precisely* the same result as calculating the first-order estimate of the composite relationship. This is in contrast to min-max methods which make conservative estimates at each step and thus produce *very* conservative estimates of a composite relationship.

3.3 Extracting Relationships

We have now developed enough machinery to describe the procedure for estimating the relationships between objects which are in our map. The map contains, by definition, estimates of the locations of objects with respect to the world frame; these relations can be read of out the estimated system mean vector and covariance matrix directly. Other relationships are implicit, and must be extracted, using methods developed in the previous sections.

For any relationship on the variables in the map we can write:

$$\mathbf{y} = \mathbf{g}(\mathbf{x}).$$

where the function $\mathbf{g}()$ is general (not the function described in the previous section). *Conditioned on all the evidence in the map*, estimates of the mean and covariance of the relationship are given by:

$$\begin{aligned} \hat{\mathbf{y}} &\approx \mathbf{g}(\hat{\mathbf{x}}), \\ \mathbf{C}(\mathbf{y}) &\approx \mathbf{G_x}\mathbf{C}(\mathbf{x})\mathbf{G_x}^T. \end{aligned}$$

4 Changing the Map

Our map represents uncertain spatial relationships among objects referenced to a common world frame. It should change if the underlying world itself changes. It should also change if our knowledge changes (even though the world is static). An example of the former case occurs when the location of an object changes; e.g., a mobile robot moves. An example of the latter case occurs when a constraint is imposed on the locations of objects in the map, for example, by measuring some of them with a sensor.

To change the map, we must change the two components that define it — the (mean) estimate of the system state vector, $\hat{\mathbf{x}}$, and the estimate of the system variance matrix, $\mathbf{C}(\mathbf{x})$.

Figure 4 shows the changes in the system due to moving objects, and the addition of constraints. A similar description appears in Gelb [Gelb 1984] and we adopt the same notation.

We will assume that new constraints are applied at discrete moments, marked by states k. The update of the estimates at state k, based on new information, is considered to be instantaneous. The estimates, at state k, *prior* to the integration of the new information are denoted by $\hat{\mathbf{x}}_k^{(-)}$ and $\mathbf{C}(\mathbf{x}_k^{(-)})$, and *after* the integration by $\hat{\mathbf{x}}_k^{(+)}$ and $\mathbf{C}(\mathbf{x}_k^{(+)})$. At these discrete moments our knowledge is increased, and uncertainty is reduced.

In the interval between states the system may be changing dynamically — for instance, the robot may be moving. When an object moves, we must define a process to extrapolate the estimate of the state vector and uncertainty at state $k-1$, to state k to reflect the changing relationships.

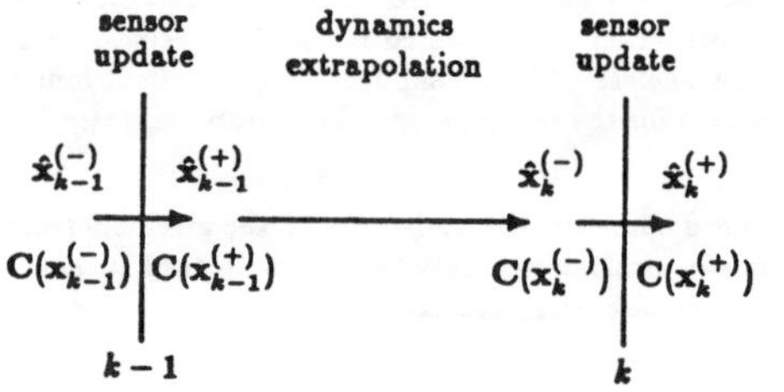

Figure 4: The Changing Map

4.1 Moving Objects

In our example, only the robot moves, so the process model need only describe its motion. A continuous dynamics model can be developed given *a particular robot*, formulated as a function of time (see [Gelb, 1984]). However, if the robot only makes sensor observations at discrete times, then a discrete motion approximation is quite adequate.

Assume the robot is represented by the Rth relationship in the map. When the robot moves, it changes its relationship, $\mathbf{x}_R$, with the world. The robot makes an uncertain relative motion, $\mathbf{y}_R$, to reach a final world location $\mathbf{x}'_R$. Thus,

$$\mathbf{x}'_R = \mathbf{x}_R \oplus \mathbf{y}_R.$$

Only a portion of the map needs to be changed due to the change in the robot's location from state to state — specifically, the Rth element of the estimated mean of the state vector, and the Rth row and column of the estimated variance matrix.

In Figure 5,

$$\hat{\mathbf{x}}'_R \approx \hat{\mathbf{x}}_R \oplus \hat{\mathbf{y}}_R,$$

$$\mathbf{C}(\mathbf{x}'_R) \approx \mathbf{J}_{1\oplus}\mathbf{C}(\mathbf{x}_R)\mathbf{J}_{1\oplus}^T + \mathbf{J}_{2\oplus}\mathbf{C}(\mathbf{y}_R)\mathbf{J}_{2\oplus}^T,$$

$$\mathbf{C}(\mathbf{x}'_R, \mathbf{x}_i) \approx \mathbf{J}_{1\oplus}\mathbf{C}(\mathbf{x}_R, \mathbf{x}_i).$$

For simplicity, the formulae presented assume independence of the errors in the relative motion, $\mathbf{y}_R$, and the current estimated robot location $\mathbf{x}_R$. As in the desciption of Figure 3, $\mathbf{C}(\mathbf{x}, \mathbf{x}'_R)$ is a column of the individual cross-covariance matrices $\mathbf{C}(\mathbf{x}_i, \mathbf{x}'_R)$.

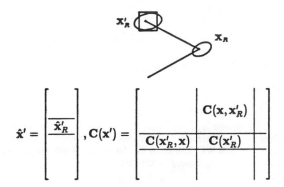

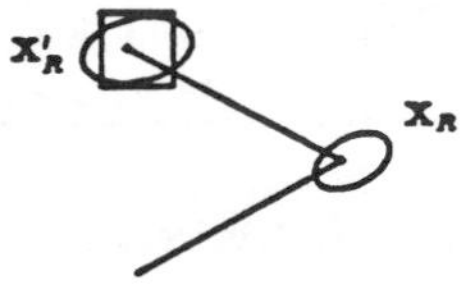

$$\hat{\mathbf{x}}' = \begin{bmatrix} \\ \hline \hat{\mathbf{x}}'_R \\ \hline \\ \end{bmatrix}, \mathbf{C}(\mathbf{x}') = \begin{bmatrix} & \mathbf{C}(\mathbf{x}, \mathbf{x}'_R) & \\ \hline \mathbf{C}(\mathbf{x}'_R, \mathbf{x}) & \mathbf{C}(\mathbf{x}'_R) & \\ \hline & & \end{bmatrix}$$

Figure 5: The Moving Robot

4.2 Adding Constraints

When new information is obtained relating objects *already in the map*, the system state vector and variance matrix do not increase in size; i.e., no new elements are introduced. However, the old elements are *constrained* by the new relation, and their values will be changed.

Constraints can arise in a number of ways:

- A robot measures the relationship of a *known* landmark to itself (i.e., estimates of the world locations of robot and landmark already exist).
- A geometric relationship, such as colinearity, coplanarity, etc., is given for some set of the object location variables.

In the first example the constraint is noisy (because of an imperfect measurement). In the second example, the constraint could be absolute, but could also be given with a tolerance.

There is no mathematical distinction between the two cases; we will describe all constraints as if they came from measurements by *sensors* — real sensors or pseudo-sensors (for geometric constraints), perfect measurement devices or imperfect.

When a constraint is introduced, there are two estimates of the geometric relationship in question — our current best estimate of the relation, which can be extracted from the map, and the new sensor information. The two estimates can be compared (in the same reference frame), and together should allow some improved estimate to be formed (as by averaging, for instance).

For each sensor, we have a *sensor model* that describes how the sensor maps the spatial variables in the state vector into sensor variables. Generally, the measurement, $\mathbf{z}$, is described as a function, $\mathbf{h}$, of the state vector, corrupted by mean-zero, additive noise $\mathbf{v}$. The covariance of the noise, $\mathbf{C}(\mathbf{v})$, is given as part of the model.

$$\mathbf{z} = \mathbf{h}(\mathbf{x}) + \mathbf{v}. \qquad (3)$$

The expected value of the sensor and its covariance are easily estimated as:

$$\hat{\mathbf{z}} \approx \mathbf{h}(\hat{\mathbf{x}}).$$

$$\mathbf{C}(\mathbf{z}) \approx \mathbf{H_x}\mathbf{C}(\mathbf{x})\mathbf{H_x}^T + \mathbf{C}(\mathbf{v}),$$

where:

$$\mathbf{H_x} \triangleq \frac{\partial \mathbf{h}_k(\mathbf{x})}{\partial \mathbf{x}} \left(\hat{\mathbf{x}}_k^{(-)}\right).$$

The formulae describe our best estimate of the sensor's values under the circumstances, and the likely variation. The actual sensor values returned are usually assumed to be conditionally independent of the state, meaning that the noise is assumed to be independent in each measurement, even when measuring the same relation with the same sensor. The actual sensor values, corrupted by the noise, are the second estimate of the relationship.

In Figure 6, an over-constrained system is shown. We have two estimates of the same node, labeled $\mathbf{x}_1$ and $\mathbf{z}$. In our example, $\mathbf{x}_1$ represents the location of a beacon about which we have prior information, and $\mathbf{z}$ represents a second estimate of the beacon location derived from a sensor located on a mobile robot at $\mathbf{x}_2$. We wish to obtain a better estimate of the location of the robot, and perhaps the beacon as well; i.e., more accurate values for the vector $\hat{\mathbf{x}}$. One method is to compute the conditional mean and covariance of $\mathbf{x}$ given $\mathbf{z}$ by the standard statistical formulae:

$$\widehat{\mathbf{x} \mid \mathbf{z}} = \hat{\mathbf{x}} + \mathbf{C}(\mathbf{x}, \mathbf{z})\mathbf{C}(\mathbf{z})^{-1}(\mathbf{z} - \hat{\mathbf{z}})$$

$$\mathbf{C}(\mathbf{x} \mid \mathbf{z}) = \mathbf{C}(\mathbf{x}) - \mathbf{C}(\mathbf{x}, \mathbf{z})\mathbf{C}(\mathbf{z})^{-1}\mathbf{C}(\mathbf{z}, \mathbf{x}).$$

Using the formulae in (2), we can substitute expressions in terms of the sensor function and its Jacobian for $\hat{\mathbf{z}}$, $\mathbf{C}(\mathbf{z})$ and $\mathbf{C}(\mathbf{x}, \mathbf{z})$ to obtain the Kalman Filter equations [Gelb, 1984] given below:

$$\hat{\mathbf{x}}_k^{(+)} = \hat{\mathbf{x}}_k^{(-)} + \mathbf{K}_k\left[\mathbf{z}_k - \mathbf{h}_k(\hat{\mathbf{x}}_k^{(-)})\right],$$

$$\mathbf{C}(\mathbf{x}_k^{(+)}) = \mathbf{C}(\mathbf{x}_k^{(-)}) - \mathbf{K}_k\mathbf{H_x}\mathbf{C}(\mathbf{x}_k^{(-)}),$$

$$\mathbf{K}_k = \mathbf{C}(\mathbf{x}_k^{(-)})\mathbf{H_x}^T\left[\mathbf{H_x}\mathbf{C}(\mathbf{x}_k^{(-)})\mathbf{H_x}^T + \mathbf{C}(\mathbf{v})_k\right]^{-1}.$$

For linear transformations of Gaussian variables, the matrix $\mathbf{H}$ is constant, and the Kalman Filter produces the *optimal minimum-variance Bayesian estimate*, which is equal to the mean of the *a posteriori conditional density function* of $\mathbf{x}$, given the prior statistics of $\mathbf{x}$, and the statistics of the measurement $\mathbf{z}$. Since the transformations are linear, the mean and covariances of $\mathbf{z}$ are exactly determined by (1) and (2). Since the orignal random variables were Gaussian, so is the result. Finally, since a Gaussian distribution is completely defined by its first two moments, the conditional mean and covariance computed define the conditional density.

No non-linear estimator can produce estimates with smaller mean-square errors. For example, if there are *no angular errors* in our coordinate frame relationships, then compounding is linear in the (translational) errors. If only linear constraints are imposed, the map will contain optimal and consistent estimates of the frame relationships.

For linear transformations of non-Gaussian variables, the Kalman Filter is not optimal, but produces the optimal *linear* estimate. The map will again be consistent. A non-linear estimator might be found with better performance, however.

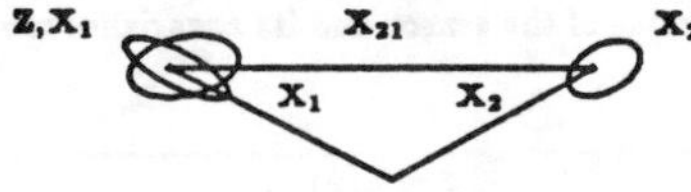

$$\hat{\mathbf{x}}' = \left[\begin{array}{c}\hat{\mathbf{x}}\\ \hline \hat{\mathbf{z}}\end{array}\right], \quad \mathbf{C}(\mathbf{x}') = \left[\begin{array}{c|c}\mathbf{C}(\mathbf{x}) & \mathbf{C}(\mathbf{x},\mathbf{z})\\ \hline \mathbf{C}(\mathbf{z},\mathbf{x}) & \mathbf{C}(\mathbf{z})\end{array}\right]$$

Figure 6: Overconstrained Relationships

For non-linear transformations, Jacobians such as $\mathbf{H}$ will have to be evaluated (they are not constant matrices). The given formulae then represent the Extended Kalman Filter, a sub-optimal non-linear estimator. It is one of the most widely used non-linear estimators because of its similarity to the optimal linear filter, its simplicity of implementation, and its ability to provide accurate estimates in practice.

The error in the estimation due to the non-linearities in $\mathbf{h}$ can be greatly reduced by iteration, using the Iterated Extended Kalman Filter equations [Gelb, 1984]. Such iteration is necessary to maintain consistency in the map when non-linearities become significant. Convergence to the true value of $\mathbf{x}$ cannot be guaranteed, in general, for the Extended Kalman Filter, although as noted, the filter has worked well in practice on a large number of problems, including navigation.

5 The Example

Our example is designed to illustrate a number of uses of the information kept in the Stochastic Map for decision making. An initial implementation of the techniques described in this paper has been performed. The uncertainties represented by ellipses in the illustrations, were originally computed by the system on a set of sample problems with varying error magnitudes. This description, however, will have to remain qualitative until a more extensive investigation can be performed.

5.1 What if I Move?

We combine discrete robot motions by compounding them, as shown in Figure 5. It is assumed that any systematic biases in the robot motion have been removed by calibration. The robot's best estimate of its location is $\hat{\mathbf{x}}_2$, with error covariance $\mathbf{C}(\mathbf{x}_2)$ (given in the map). Since the location of the loading area is known very accurately in room coordinates, the robot can compute the nominal relative motion that it would like to make, $\hat{\mathbf{y}}_{2,load}$. From an internal model of its own accuracy, the robot estimates the covariance of its motion error as $\mathbf{C}(\mathbf{y}_{2,load})$. If there were no errors in the initial estimate of the robot location, and no motion errors incurred in moving, the robot would arrive with its center coincident with the center of the loading area. When the two uncertain relations are compounded, the first-order estimate of the mean of the robot's final location is also the center of the loading area, but the covariance of the error has increased.

In order to compare the likely locations of the robot with the loading zone, we must *now* assume something about the probability distribution of the robot's location. For reasons already discussed, a multi-variate Gaussian distribution which fits the estimated moments is assumed. Given that, we can determine the elliptical region of 2-D space in which the robot should be found, with probability approximately determined by our choice of confidence levels—more than likely corresponding to 4 or 5 standard deviations of the estimated errors, for relative certainty.

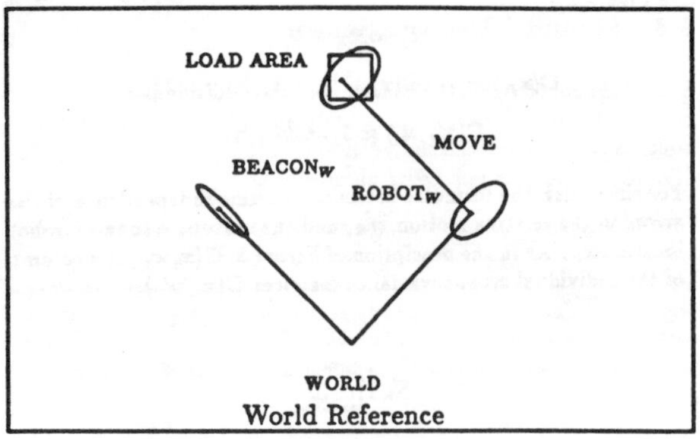

Figure 7: A Direct Move *Might* Fail

All that remains is to determine if the ellipse is completely contained in the desired region; for purposes of illustration, Figure 7 shows that it is not. The robot decides it cannot achieve its goal reliably by moving directly to the load area.

5.2 Where is the Beacon?

Before moving, the robot can attempt to reduce the uncertainty in its initial location by trying to spot the beacon. The relative location of the beacon to the robot is computed by $\ominus\mathbf{x}_2 \oplus \mathbf{x}_1$. The two estimated moments of each relation are pulled from the map, and the moments of the result are estimated, as described in section 3.2.

Given the estimate, an elliptical region in which the beacon should be found with high confidence can be computed as before; but this time the relational estimate, and hence the ellipse are described in robot coordinates. The robot can compare this region with the region swept out by the field of view of its sensor to determine if sensing is feasible (without repositioning the sensor, or worse, turning the robot). The result is illustrated in Figure 8.

The robot determines that the beacon is highly likely to be in its field of view.

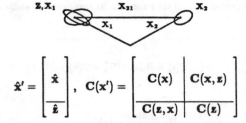

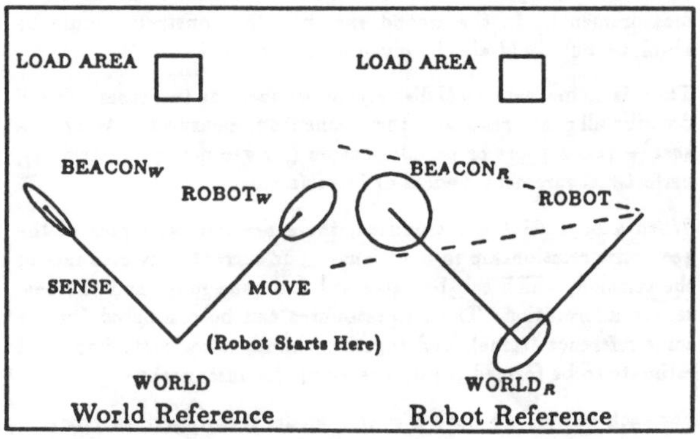

Figure 8: Is the Beacon Viewable?

5.3 Should I Use the Sensor?

Even if the robot sights the beacon, will the additional information help it estimate its location accurately enough so that it can then move to the loader successfully? If not, the robot should pursue a different strategy to reduce its uncertainty.

For simplicity, we assume that the robot's sensor measures the relative location of the beacon in Cartesian coordinates. Thus the sensor function is the functional composition of reversal and compounding, already described. The sensor produces a measurement with additive, mean-zero noise $\mathbf{v}$, whose covariance is given in the sensor model as $C(\mathbf{v})$. Given the information in the map, the conditional mean and covariance of the expected sensor value can be estimated:

$$\mathbf{z} = \mathbf{x}_{21} = \ominus\mathbf{x}_2 \oplus \mathbf{x}_1 + \mathbf{v}.$$

$$\hat{\mathbf{z}} = \hat{\mathbf{x}}_{21} = \ominus\hat{\mathbf{x}}_2 \oplus \hat{\mathbf{x}}_1.$$

$$C(\mathbf{z}) = {}_{\ominus}J_{\oplus} \begin{bmatrix} C(\mathbf{x}_2) & C(\mathbf{x}_2,\mathbf{x}_1) \\ C(\mathbf{x}_1,\mathbf{x}_2) & C(\mathbf{x}_1) \end{bmatrix} {}_{\ominus}J_{\oplus}^T + C(\mathbf{v}).$$

In the Kalman Filter Update equations described in section 4.2, the system covariance matrix can be updated *without an actual sensor measurement having been made*; it depends only on $C(\mathbf{x})$, $C(\mathbf{v})$, and the matrix $\mathbf{H_x}$. In the example, $\ominus J \oplus$ takes the place of $\mathbf{H_x}$, and is evaluated with the current values of $\hat{\mathbf{x}}_2$ and $\hat{\mathbf{x}}_1$, the robot and beacon locations. The updated system covariance matrix can be computed *as if* the sensor were used. The reduction in the robot's locational uncertainty due to applying the sensor can be judged by comparing the old value of $C(\mathbf{x}_2)$ with the updated value. The magnitudes of this "updated" robot covariance estimate, and $C(\mathbf{y}_{2,load})$ (from 5.1), can be used to decide if the robot will be able to reach its goal with the desired tolerance.

In our example, it is determined that the sensor should be useful. Figures 9 shows the result of a simulated measurement, with the location and measurement uncertainties transformed into either map or robot coordinates, respectively. Figure 10 illustrates the improvement in the estimations of the robot and beacon locations following application of the Kalman Filter Update formulae with the given measurement. Finally, Figure 11 shows the result of compounding the uncertain relative motion of the robot with its newly estimated initial location. The robot achieves its goal.

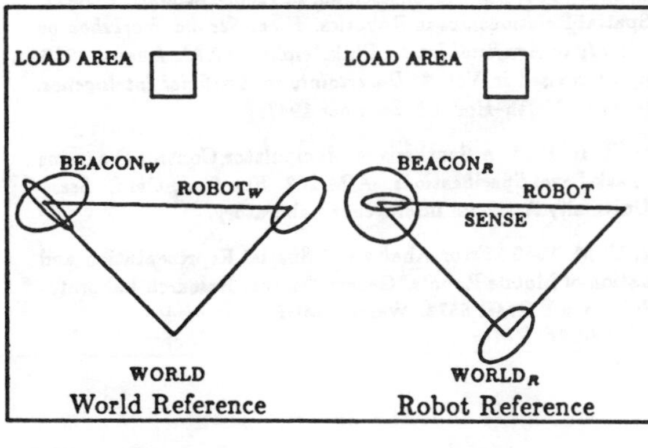

Figure 9: The Robot Senses the Beacon Again

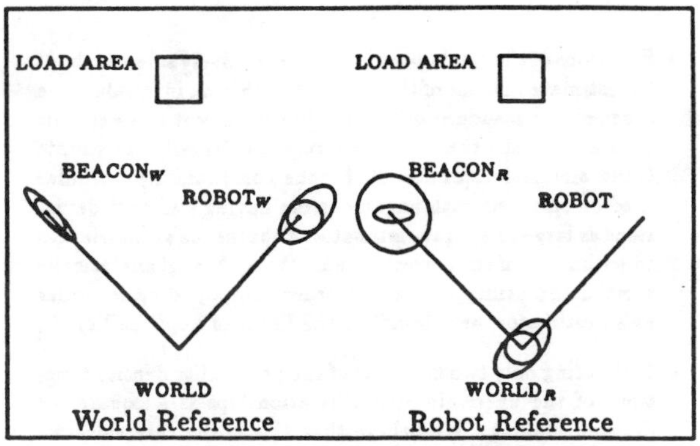

Figure 10: Updated and Original Estimates

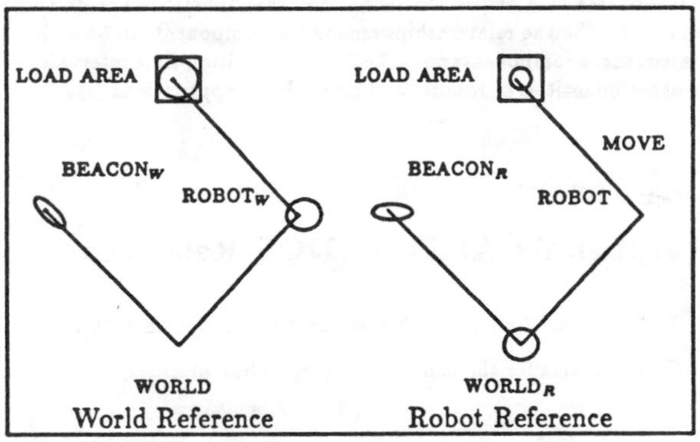

Figure 11: The Robot Moves Successfully

6 Discussion and Conclusions

This paper presents a general method for estimating uncertain relative spatial relationships between reference frames in a network of uncertain spatial relationships. Such networks arise, for example, in industrial robotics and navigation for mobile robots, because the system is given spatial information in the form of sensed relationships, prior constraints, relative motions, and so on. The methods presented in this paper allow the efficient estimation of these uncertain spatial relations and can be used, for example, to compute *in advance* whether a proposed sequence of actions (each with known uncertainty) is likely to fail due to too much accumulated uncertainty; whether a proposed sensor observation will reduce the uncertainty to a tolerable level; whether a sensor result is so unlikely given its expected value and its prior probability of failure that it should be ignored, and so on. This paper applies state estimation theory to the problem of estimating parameters of an entire spatial configuration of objects, with the ability to transform estimates into any frame of interest.

The estimation procedure makes a number of assumptions that are normally met in practice, and can be summarized as follows:

- Functions of the random variables are relatively smooth about the estimated means of the variables within an interval on the order of one standard deviation. In the current context, this generally means that angular errors are "small". In Monte Carlo simulations[Smith, 1985], the compounding formulae used on relations with angular errors having standard deviations as large as 5°, gave estimates of the means and variances to within 1% of the correct values. Wang [Wang] analytically verified the utility of the first-order compounding formulae as an estimator, and described the limits of applicability.

- Estimating only two moments of the probability density functions of the uncertain spatial relationships is adequate for decision making. We believe that this is the case since we will most often model a sensor observation by a mean and variance, and the relationships which result from combining many pieces of information become rapidly Gaussian, and thus are accurately modelled by only two moments.

Although the examples presented in this paper have been solely concerned with *spatial* information, there is nothing in the theory that imposes this restriction. Provided that functions are given which describe the relationships among the components to be estimated, those components could be forces, velocities, time intervals, or other quantities in robotic and non-robotic applications.

Appendix A: Three DOF Relations

Formulae for the full 6DOF case are given in [Smith 1986].

The formulae for the compounding operation are:

$$\mathbf{x}_{ik} \triangleq \mathbf{x}_{ij} \oplus \mathbf{x}_{jk} = \begin{bmatrix} x_{jk}\cos\phi_{ij} - y_{jk}\sin\phi_{ij} + x_{ij} \\ x_{jk}\sin\phi_{ij} + y_{jk}\cos\phi_{ij} + y_{ij} \\ \phi_{ij} + \phi_{jk} \end{bmatrix}.$$

where the Jacobian for the compounding operation, $\mathbf{J}_\oplus$ is:

$$\mathbf{J}_\oplus \triangleq \frac{\partial(\mathbf{x}_{ij} \oplus \mathbf{x}_{jk})}{\partial(\mathbf{x}_{ij}, \mathbf{x}_{jk})} = \frac{\partial \mathbf{x}_{ik}}{\partial(\mathbf{x}_{ij}, \mathbf{x}_{jk})} =$$

$$\begin{bmatrix} 1 & 0 & -(y_{ik} - y_{ij}) & \cos\phi_{ij} & -\sin\phi_{ij} & 0 \\ 0 & 1 & (x_{ik} - x_{ij}) & \sin\phi_{ij} & \cos\phi_{ij} & 0 \\ 0 & 0 & 1 & 0 & 0 & 1 \end{bmatrix}.$$

The formulae for the reverse operation are:

$$\mathbf{x}_{ji} \triangleq \ominus \mathbf{x}_{ij} \triangleq \begin{bmatrix} -x_{ij}\cos\phi_{ij} - y_{ij}\sin\phi_{ij} \\ x_{ij}\sin\phi_{ij} - y_{ij}\cos\phi_{ij} \\ -\phi_{ij} \end{bmatrix}.$$

and the Jacobian for the reversal operation, $\mathbf{J}_\ominus$ is:

$$\mathbf{J}_\ominus \triangleq \frac{\partial \mathbf{x}_{ji}}{\partial \mathbf{x}_{ij}} = \begin{bmatrix} -\cos\phi_{ij} & -\sin\phi_{ij} & y_{ji} \\ \sin\phi_{ij} & -\cos\phi_{ij} & -x_{ji} \\ 0 & 0 & -1 \end{bmatrix}.$$

Acknowledgments

The research reported in this paper was performed at SRI International, and was supported by the National Science Foundation under Grant ECS-8200615, the Air Force Office of Scientific Research under Contract F49620-84-K-0007, and by General Motors Research Laboratories.

References

Bolle R.M., and Cooper, D.B. 1986. On Optimally Combining Pieces of Information, with Application to Estimating 3-D Complex-Object Position from Range Data. *IEEE Trans. Pattern Anal. Machine Intell.*, vol. PAMI-8, pp. 619-638, Sept. 1986.

Brooks, R. A. 1982. Symbolic Error Analysis and Robot Planning. *Int. J. Robotics Res.* 1(4):29-68.

Chatila, R. and Laumond, J-P. 1985. Position Referencing and Consistent World Modeling for Mobile Robots. *Proc. IEEE Int. Conf. Robotics and Automation.* St. Louis: IEEE, pp. 138-145.

Durrant-Whyte, H. F. 1986. Consistent Integration and Propagation of Disparate Sensor Observations. *Proc. IEEE Int. Conf. Robotics and Automation.* San Francisco: IEEE, pp. 1464-1469.

Faugeras, O. D., and Hebert, M. 1986. The Representation, Recognition, and Locating of 3-D Objects. *Int. J. Robotics Res.* 5(3):27-52.

Gelb, A. 1984. *Applied Optimal Estimation.* M.I.T. Press

Nahi, N. E. 1976. *Estimation Theory and Applications.* New York: R.E. Krieger.

Papoulis, A. 1965. *Probability, Random Variables, and Stochastic Processes.* McGraw-Hill.

Paul, R. P. 1981. *Robot Manipulators: Mathematics, Programming and Control.* Cambridge: MIT Press.

Smith, R. C., *et al.* 1984. Test-Bed for Programmable Automation Research. Final Report-Phase 1, SRI International, April 1984.

Smith, R. C., and Cheeseman, P. 1985. On the Representation and Estimation of Spatial Uncertainty. SRI Robotics Lab. Tech. Paper, and *Int. J. Robotics Res.* 5(4): Winter 1987.

Smith, R. C., Self, M., and Cheeseman, P. 1986. Estimating Uncertain Spatial Relationships in Robotics. *Proc. Second Workshop on Uncertainty in Artificial Intell.*, Philadelphia, AAAI, August 1986. To appear revised in Vol. 2, *Uncertainty in Artificial Intelligence.* Amsterdam: North-Holland, Summer 1987.

Taylor, R. H. 1976. A Synthesis of Manipulator Control Programs from Task-Level Specifications. AIM-282. Stanford, Calif.: Stanford University Artificial Intelligence Laboratory.

Wang, C. M. 1986. Error Analysis of Spatial Representation and Estimation of Mobile Robots. General Motors Research Laboratories Publication GMR 5573. Warren, Mich.

Dynamic Map Building for an Autonomous Mobile Robot*

John Leonard
Hugh Durrant-Whyte

Department of Engineering Science
University of Oxford
Parks Road, Oxford OX1 3PJ
ENGLAND

Ingemar J. Cox

NEC Research Institute
4 Independence Way
Princeton, NJ 08540
U.S.A.

Abstract

This paper presents an algorithm for autonomous map building and maintenance for a mobile robot. With each geometric target in the map we associate a *validation measure* to represent our belief in the validity of a target, in addition to the usual covariance matrix to represent spatial uncertainty. At each position update cycle, predicted features are generated for each target in the map and compared to features actually observed . Successful matches to targets with a high validation measure are used for localization. Unpredicted observations are used to initialize target tracks for new environment features, while unobserved predictions result in a target's validation measure being decreased. We describe experimental results obtained with the algorithm that demonstrate successful map-building using real sonar data.

1 Introduction

Navigation is a fundamental requirement of autonomous mobile robots. We understand navigation to be "the science of getting ships, aircraft, or spacecraft from place to place; *esp*: the method of determining position, course, and distance traveled"[1]; we do not interpret it to include motion planning. In this paper we examine the problem of constructing and maintaining a map of an autonomous vehicle's environment for the purpose of navigation. In previous papers[7][11], we examined the problem of mobile robot localization when an accurate map is provided *a priori*. We believe this problem to be well understood[8], particularly in the context of marine and aerospace navigation. However, for many autonomous vehicles, it is desirable to *automatically* build and maintain a map of the robot's environment. Maintenance of the map is, perhaps, the primary motivation because even if a map is provided *a priori*, few robot environments are completely static. Consequently, as the robot's environment alters, so must its model.

In this paper, we describe preliminary work intended to allow an autonomous robot vehicle to construct and maintain a map of its environment. One major hurdle to this goal is that the information received by the vehicle is unreliable because of the sensor's noise and physical operating characteristics. The following comments by Lozano-Pérez describe this problem [8]:

> The robot must be able to determine its relationship to the environment by sensing. There are a wide variety of sensing technologies possible: odometry, ultrasonic, infrared and laser range sensing, and monocular, binocular, and trinocular vision have all been explored. The difficulty is in interpreting this data, that is, deciding what the sensor signals tell us about the external world.
>
> The trend is to attack this problem by so-called sensor fusion, that is, by combining the outputs of multiple feature detectors possibly operating on a variety of sensors or simply multiple observations of the same object. Much of this work has focused on applications of Kalman filtering, which essentially provides a mechanism for weighting the various pieces of data based on estimates of their reliability (covariances when we assume the noise has a Gaussian distribution). This is reasonable as far as it goes. But, no amount of Kalman filtering will restore a missing feature or prevent a totally erroneous feature from having an impact on the estimate.
>
> I believe that part of the problem is that models that characterize a sensor or feature detector by a covariance matrix are too weak; they do not capture the relevant physics. The fact is that in some cases the detector will provide very good estimates in others it will be completely useless data. [18]

We broadly agree with T. Lozano-Pérez's comments. Automated cartography must deal with two distinctly different sensor problems

- the sensor noise and
- validation of sensor measurements.

The first problem is well understood. All sensor measurements have an associated accuracy that is fundamentally noise limited, e.g. thermal or shot noise at the receiver. We can usually handle noise by conventional techniques such as the Kalman filter, the corresponding uncertainty due to noise usually being represented by a covariance matrix.

The second item is often erroneously referred to as noise. However, a sensor's physical and environmental operating conditions are rarely random, but can result in missing, spurious or irrelevant information. For example, spurious information may be due to the specular reflection characteristics if ultrasonic sensing is used; irrelevant information might be caused by a person walking by the vehicle. In this case, a sensor may accurately identify the presence of an object, but the object may be of no relevance to the task at hand - for navigational cartography we are only interested in features that are unchanging.

Covariance matrices can not represent the uncertainty associated with the physical and environmental operating conditions of the sensor and the relevance of a particular sensor reading to a given problem. We are unaware of any general tool for handling such uncertainty. In fact, given that this uncertainty is very problem specific, e.g. sensing a person walking nearby is irrelevant for navigational purposes but critical to obstacle avoidance and motion planning, we think it is unlikely that a general tool can be found.

Clearly there are two uncertainties; (i) associated with the noise in the system and which is amenable to modeling using conventional statistical tools, e.g. covariances, and (ii) associated with the *validity* of a sensor measurement. These two uncertainties are independent of one another; it is entirely reasonable to have a small uncertainty due

*The authors would like to acknowledge the support of NATO collaborative research grant 0204/89. This work is also partially supported in part by ESPRIT 1560 (SKIDS) and SERC-ACME GRE/42419

Reprinted from the *IEEE International Workshop on Intelligent Robots and Systems (IROS 90)*, pages 89-95. Copyright © 1990 by The Institute of Electrical and Electronics Engineers, Inc. All rights reserved.

EH0341-8/91/0000/0331$01.00 © 1990 IEEE

to noise but a high degree of belief uncertainty — this is common for ultrasonic imaging. Thus the main issues facing us in building and maintaining a map of the environment are

1. accurately modeling the sensor noise and

2. accurately modeling the validation/belief uncertainty of a sensor reading.

A third issue is an efficient map representation, but this is not addressed here.

We refer to the task of determining the robot's position as the process of localization. Localization is a top-down, expectation-driven competence; it is a process of "looking" for and tracking *expected* events. In contrast, obstacle avoidance is a bottom-up, data-driven competence; it is a process of detecting and explaining *unexpected* events. An event is "expected" if it can be predicted from either an internal model of the environment or from previously observed events. It is possible to estimate the motion of a vehicle by observing the motion of these expected events [14]. Logically, an "unexpected" event is one that has not been predicted. Such events may arise as the result of spurious sensor readings or as the result of observing an unmodeled object. In a probabilistic sense, all sensor data is either expected or unexpected. Building maps of the environment from sensor information involves maintaining a coherent map of these expected events while explaining and incorporating new, unexpected, events.

Two important observations can be made about about this process:

- Tracking expected events and explaining unexpected events are two complimentary parts of the same problem. When an event or target is observed, predictions based on an environment model or previous observations can be used to suggest possible interpretations or explanations of the data. If a target can be explained through correspondence with prior predictions, then it may be tracked to provide positional information. The rejects from this process are those events that can not be explained. If an observation can not be explained, then a new target can be initiated and subsequently tracked to provide an explanation for the information.

- Map building is a dynamic process that involves providing an interpretation of observed sensor information in terms of physical features in the environment. The prediction of expected events relies on the fact that a correct interpretation has been found for targets that are repeatably observed. Further, tracks that are initiated to explain unexpected events are an attempt to develop such an interpretation over time. Thus the map is built up as new events are observed and explained, and is refined by reobserving events that have been correctly interpreted.

The starting point for formalizing the problem of unifying these different aspects of the navigation problem is to develop a good model of the relation between sensor observations and physical features in the environment. This sensor model is crucial to generating predictions of what can be observed from different sensor locations and correspondingly in providing an interpretation for observed events or targets. Without such a model, expectations and explanations become meaningless. The reader is directed to the appendix for a description of the feature extraction algorithms used in this paper.

Before further developing our algorithm, we briefly summarize previous work in map building. Next, Section 3 summarizes an algorithm for mobile robot localization, which utilizes the competence of tracking geometric beacons which naturally occur in the environment. This navigation algorithm is based around an extended Kalman filter which utilizes matches between observed geometric beacons and an *a priori* map of beacon locations to provide a reliable estimate of vehicle location. This approach is extended in Section 4 for situations in which the map is built up and dynamically maintained, allowing operation in unknown and changing environments. A crucial component of this extension is that we attach a measure of validation/belief uncertainty to each target in the map that is independent of the covariance matrix which represents the target's spatial uncertainty. Experimental results are presented in Section 5.

2 Previous Work in Map Building

Previous automated cartography for autonomous mobile robots has not discriminated between spatial uncertainty due to noise and validation uncertainty due to sensor and environmental conditions. Moreover, much earlier work has attempted to build maps for multiple purposes, e.g. navigation, motion planning, 3-D world modeling. Although a single map is obviously desirable, we believe that multiple maps, each containing only information relevant to a particular task, may be significantly easier to construct.

2.1 Geometric approaches

There are several research groups that have used Kalman filtering techniques to to build models of the environment. Impressive results are reported by Ayache and Faugeras[2] for a mobile vehicle using trinocular stereo to determine depth. Kriegman et al[15] also describe how a map can be built using binocular stereo. Crowley[9] describes a similar approach using ultrasonic sensing. Common to all these approaches is the use of the Kalman filter to model and propagate uncertainty in both the position of the robot and the geometric features using covariance matrices. This is a powerful tool for dealing with noise, but, as discussed earlier, is of little help in modeling the sensor/environment/task operating characteristics. The experimental results presented by these groups are all for static scenes, i.e. the environment is unchanging. It is unlikely that these approaches would be successful in a dynamic environment without extension, perhaps along the lines of the work described here.

2.2 Occupancy Grids

The occupancy grid representation developed by Moravec and Elfes [21, 12] does in fact combine a probability of occupancy (belief) with a spatial uncertainty. Occupancy grids represent space as a 2– or 3-D array of cells, each cell hold an estimate of the confidence that it is occupied. The uncertainty surrounding an objects position is then represented by a spatial distribution of these probabilities within the occupancy grid. The larger the spatial uncertainty, the greater the number of cells occupied by the feature. This representation allows a large neighborhood of cells (high spatial uncertainty) to have a high probability of occupancy (belief).

We feel the occupancy grid map representation is particularly useful for the task of obstacle avoidance[1], since there is an explicit representation of free space. Moreover, the occupancy grid representation provides a powerful tool for dealing with sensory data in which reliable feature extraction is not practical.

Ideally, position estimation should be a process of (i) determining the correspondence between observed sensory features and predicted map features and (ii) then computing an optimal estimate of the vehicle's relative position. The optimality criterion is almost always least mean square error and should weight each measurement by its spatial uncertainty. Grid-based position estimation[21] has been proposed based on the concept of matching occupancy grids via cross correlation. The accuracy of the position estimate must, however, be inferior to the feature based approach because (i) the correlation uses all points in the map, but some points will be spurious (this must reduce accuracy) and (ii) neighborhoods of cells with both a high probability of occupancy and high spatial uncertainty will have a disproportionate affect on the correlation compared with those neighborhoods that have a high probability of occupancy and small spatial uncertainty. Moreover, while the original $O(n^5)$ algorithm was sig-

[1] For example, Borenstein[5] has presented the combination of a grid-based representation with a potential field algorithm to achieve very fast obstacle avoidance.

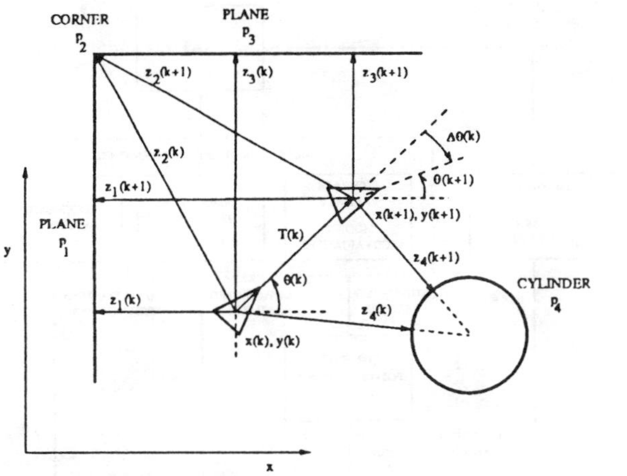

Figure 1: Localization by Tracking Geometric Beacons.

nificantly improved, the cost of matching is still slightly larger than $O(n)$, where n is the number of cells in the map. In contrast, the cost of a feature based approach should only be proportional to the number of features currently present in the map. Under normal conditions, this should be significantly less than the number of grid cells.

We believe the building of a navigational map is a distinctly different process from that used to construct occupancy grids. Map-building for navigation needs to confront the correspondence problem head-on. It is a process of making decisions: is this piece of data good or bad? does this piece of data correspond to this geometric beacon? Consequently, we believe that navigation requires a feature-based approach, in which a precise, concise map, is used to efficiently generate predictions of what the robot should "see" from a given location.

3 Localization by Tracking Geometric Beacons

This section summarizes our approach to the problem of model-based localization. We view model-based navigation as a process of tracking geometric targets. A *geometric beacon* is a special class of target which can be reliably observed in successive sensor measurements (a beacon), and which can be accurately described in terms of a concise geometric parameterization.

With reference to Figure 1, we denote the position and orientation of the vehicle at time step k by the state vector $\mathbf{x}(k) = [x(k), y(k), \theta(k)]^T$ comprising a cartesian location and a heading defined with respect to a global coordinate frame. At initialization the robot starts at a known location, and the robot has an *a priori* map of the locations of geometric beacons $\mathbf{p}_i$. At each time step, observations $\mathbf{z}_j(k)$ of these beacons are taken. Our goal in the cyclic process is to associate measurements $\mathbf{z}_j(k)$ with the correct beacon $\mathbf{p}_i$ to compute an updated estimate of vehicle position.

The Kalman filter relies on two models: a *plant* model and a *measurement model*. The plant model describes how the vehicle's position $\mathbf{x}(k)$ changes with time in response to a control input $\mathbf{u}(k)$ and a noise disturbance $\mathbf{v}(k)$

$$\mathbf{x}(k+1) = \mathbf{F}(\mathbf{x}(k), \mathbf{u}(k)) + \mathbf{v}(k), \quad \mathbf{v}(k) \sim \mathrm{N}(0, \mathbf{Q}(k)) \tag{1}$$

where $\mathbf{F}(\mathbf{x}(k), \mathbf{u}(k))$ is the (non-linear) state transition function. We use the notation $\mathbf{v}(k) \sim \mathrm{N}(0, \mathbf{Q}(k))$ to indicate that this noise source is assumed to be zero mean gaussian with variance $\mathbf{Q}(k)$ [13].

The measurement model expresses a sensor observation in terms of the vehicle position and the geometry of the beacon being observed, and has the form:

$$\mathbf{z}_j(k) = \mathbf{h}_i(\mathbf{p}_i, \mathbf{x}(k)) + \mathbf{w}_j(k), \quad \mathbf{w}(k) \sim \mathrm{N}(0, \mathbf{R}(k)) \tag{2}$$

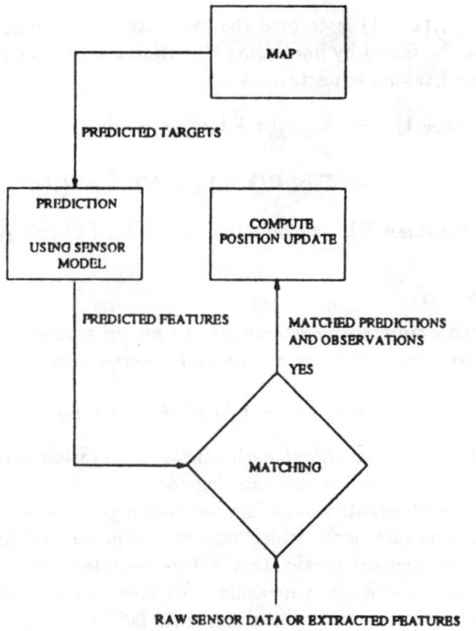

Figure 2: Model-based localization using an *a priori* map.

The observation function $\mathbf{h}_i(\mathbf{p}_i, \mathbf{x}(k))$ expresses an observed measurement $\mathbf{z}_j(k)$ as a function of the vehicle location $\mathbf{x}(k)$ and beacon location $\mathbf{p}_i$. This observation is assumed corrupted by a zero-mean, gaussian noise disturbance $\mathbf{w}_j(k)$ with variance $\mathbf{R}(k)$. The form of the observation function $\mathbf{h}_i(\cdot,\cdot)$ depends on the sensor employed *and* the type of beacon being observed. Functions for a sonar sensor observing corner, plane, and cylinder targets are described in Appendix A.

The goal of the cyclic computation is to produce an estimate of the location of the robot [2] $\hat{\mathbf{x}}(k+1 \mid k+1)$ at time step $k+1$ based on the estimate of the location $\hat{\mathbf{x}}(k \mid k)$ at time step k, the control input $\mathbf{u}(k)$ and the new beacon observations $\mathbf{z}_j(k+1)$. The algorithm employs the following steps: prediction, observation, matching, and estimation. Figure 2 presents an overview of this cyclic process. We state the Kalman Filter equations without derivation. (More detail can be found in [4], [11].)

Prediction

First, using the plant model and a knowledge of the control input $\mathbf{u}(k)$, we predict the robot's new location at time step $k+1$:

$$\hat{\mathbf{x}}(k+1 \mid k) = \mathbf{F}(\hat{\mathbf{x}}(k \mid k), \mathbf{u}(k)) \tag{3}$$

We next compute $\mathbf{P}(k+1 \mid k)$, the variance associated with this prediction:

$$\mathbf{P}(k+1 \mid k) = \nabla\mathbf{F}\ \mathbf{P}(k \mid k)\ \nabla\mathbf{F}^T + \mathbf{Q}(k) \tag{4}$$

where $\nabla\mathbf{F}$ is the Jacobian of $\mathbf{F}(\cdot,\cdot)$ obtained by linearizing about the updated state estimate $\hat{\mathbf{x}}(k \mid k)$. Next, we use this predicted robot location to generate predicted observations of each geometric beacon $\mathbf{p}_i$:

$$\hat{\mathbf{z}}_i(k+1) = \mathbf{h}_i(\mathbf{p}_i, \hat{\mathbf{x}}(k+1 \mid k)), \quad i = 1, \cdots, N_k \tag{5}$$

Observation

The next step is to actually take a number of observations $\mathbf{z}_j(k+1)$ of these different beacons, and compare these with our predicted observations. The difference between a predicted beacon location $\hat{\mathbf{z}}_i(k+1)$ and an observation is written as

$$\begin{aligned} \nu_{ij}(k+1) &= [\mathbf{z}_j(k+1) - \hat{\mathbf{z}}_i(k+1)] \\ &= [\mathbf{z}_j(k+1) - \mathbf{h}_i(\mathbf{p}_i, \hat{\mathbf{x}}(k+1 \mid k))] \end{aligned} \tag{6}$$

[2] The term $\hat{\mathbf{x}}(i \mid j)$ should be read as "the estimate of the vector $\mathbf{x}$ at time step i given all observations up to time step j".

The vector $\nu_{ij}(k+1)$ is termed the innovation. The innovation covariance can be found by linearizing Equation 2 about the prediction, squaring, and taking expectations as

$$\begin{aligned} \mathbf{S}_{ij}(k+1) &\equiv \mathrm{E}\left[\nu_{ij}(k+1)\nu_{ij}^T(k+1)\right] \\ &= \nabla \mathbf{h}_i \; \mathbf{P}(k+1 \mid k) \; \nabla \mathbf{h}_i{}^T + \mathbf{R}_i(k+1) \end{aligned} \tag{7}$$

where the Jacobian $\nabla \mathbf{h}_i$ is evaluated at $\hat{\mathbf{x}}(k+1 \mid k)$ and $\mathbf{p}_i$.

Matching

Around each predicted measurement, we set up a validation gate in which we are prepared to accept beacon observations:

$$\nu_{ij}(k+1) \; \mathbf{S}_{ij}^{-1}(k+1) \; \nu_{ij}^T(k+1) = q_{ij} \tag{8}$$

This equation is used to test each sensor observation $\mathbf{z}_j(k+1)$ for membership in the validation gate for each predicted measurement. When a single observation falls in a validation gate, we get a successful match. Measurements which do not fall in any validation gate are simply ignored for localization. More complex data association scenarios can arise when a measurement falls in two validation regions, or when two or more measurements fall in a single validation region. At this stage, such measurements are simply ignored by the algorithm, as outlier rejection is vital for successful localization.

Estimation

The final step is to use successfully matched predictions and observations to compute $\hat{\mathbf{x}}(k+1 \mid k+1)$, the updated vehicle location estimate. We utilize the standard result [4] that the Kalman gain matrix associated with each successful match can be written as

$$\mathbf{W}_j(k+1) = \mathbf{P}(k+1 \mid k) \; \nabla \mathbf{h}_i^T \; \mathbf{S}_{ij}^{-1}(k+1), \tag{9}$$

Using all successfully matched prediction–observation pairs, we find the optimal (minimum variance) linear estimate for the state as

$$\hat{\mathbf{x}}(k+1 \mid k+1) = \hat{\mathbf{x}}(k+1 \mid k) + \sum_j \mathbf{W}_j(k+1)\nu_j(k+1) \tag{10}$$

with variance

$$\mathbf{P}(k+1 \mid k+1) = \mathbf{P}(k+1 \mid k) - \sum_j \mathbf{W}_j(k+1)\mathbf{S}_j(k+1)\mathbf{W}_j^T(k+1). \tag{11}$$

Equation 10 states that the estimated location is a linear weighted sum of the predicted location, and the difference between expected and matched observed beacon locations. The weighting depends on our relative confidence in prediction $\mathbf{P}(k+1 \mid k)$ and innovations $\mathbf{S}_j(k+1)$.

Note that this algorithm facilitates a *directed* sensing approach, in which we can use our expectations of where useful beacon information can be observed to control the sensors of the mobile robot to only "look" in these locations. By knowing where to "look", the speed and reliability of the sensing process can be greatly enhanced. From time-to-time, the localization process will predict observations which are not supported by measurements; there will also be occasions when sensor measurements are not previously predicted. In these circumstances it is desirable to update our map of the world.

4 Building and Maintaining Localization Maps

The buiding and maintaining of localization maps can be broadly considered a problem in machine learning, an area of research which has received a tremendous amount of interest. The reader is directed to [20, 10] for an overview of the subject. Broadly, learning can be divided into supervised and unsupervised learning. Supervised learning assumes an initial training set of correctly labeled examples is

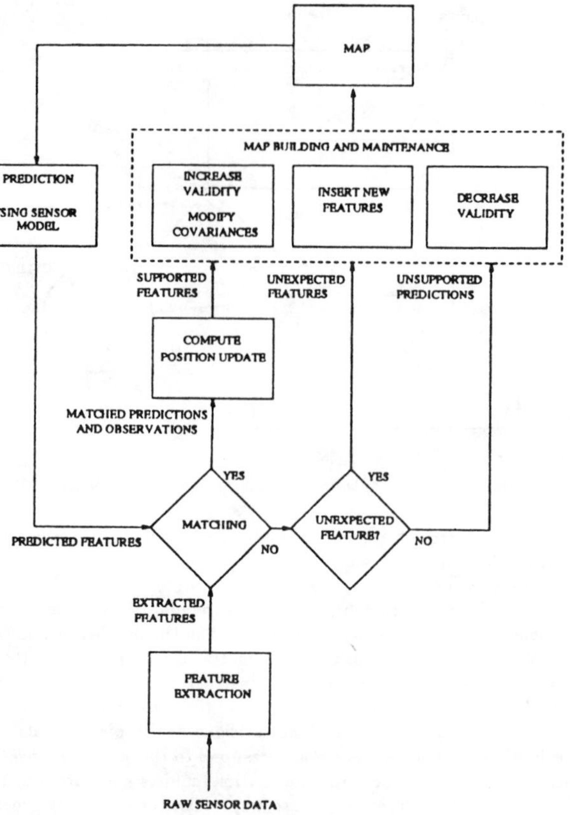

Figure 3: Model-based localization incorporating map building and maintenance.

available. This is not true for the autonomous vehicle which must therefore learn its environment by some unsupervised means. Traditionally, unsupervised learning takes the form of a cluster analysis problem arranging features into clusters or classes based on a numerical measure of feature similarity. In our case, the features are geometric beacons. Clustering of the geometric beacons *in space* is unnecessary since the Kalman filter and validation gate methodologies have already determined which features correspond with each other.

Previous theoretical work on robot learning of an environment includes Lumelsky et al[19] and Rivest and Shapire[23]. The first is concerned with determining a path for the robot, the traversing of which would guarantee that the entire environment is viewed. Navigation is not an issue here, but rather, the development of a sensing strategy for "terrain acquisition". Rivest and Shapire examined the problem of unsupervised learning for deterministic environments. However, dynamic environments are non-deterministic.

We outline an algorithm to perform these learning functions.

4.1 Proposed Algorithm

Figure 3 shows the extension of the localization algorithm presented in section 3 to incorporate map-building and maintenance. Initialization no longer requires an *a priori* map, though initial target locations can be provided. In this way, the algorithm accommodates any level of prior information ranging from none to full. The Kalman filtering steps of **Prediction**, **Observation**, and **Matching** proceed just as before. Matching yields three sets: matched predictions and observations, unobserved predictions, and unpredicted observations. Note in the original algorithm only successful matches are processed, and the rest were considered "rejects."

Estimation is now performed only using matched predictions and observations which correspond to targets in the map which have a high validation measure. In this way, we only use the data we have the most confidence in–data corresponding to beacons–to compute an accurate position estimate. Updates from just a few beacons can quickly provide an accurate, reliable position estimate. This should be contrasted with the position update method proposed for occupancy grids, as discussed above in Section 2.2, which utilizes all the data in the map in a cross-correlation computation to update position.

The primary modification to the algorithm is the addition of a **Map Update** step. Matched features, together with unexpected features and unsupported predictions, are all then sent to an explanation phase which must decide on how best to update the world model. The set of actions consists of inserting additional features, deleting existing features and increasing or decreasing the validity of existing features.

The problem of updating the navigation map can be formulated as: Given sets of predicted **P** and sensed **S** features and the intersection of the two sets $\mathbf{P} \cap \mathbf{S}$, i.e. those features that were both predicted and sensed, how can we build and maintain a model of our environment? There are two error terms; the number of ***unsupported predictions***, $\| \mathbf{P} - \mathbf{P} \cap \mathbf{S} \|$, and the number of ***unexpected measurements***, $\| \mathbf{S} - \mathbf{P} \cap \mathbf{S} \|$. We should forget unsupported predictions and learn unexpected events, provided they persist in time. More precisely, belief in a prediction that is frequently unsupported should be reduced each time until forgotten; belief in an unexpected event should be increased if subsequent predictions of the event are regularly supported by measurements. Intuitively, in a highly dynamic environment in which there are many unmodeled, moving objects there will be many temporarily unsupported predictions and many transient unexpected events. Conversely, in an almost static environment in which there is very little change we will have few unsupported predictions and few unexpected measurements. One might expect that the rate of change of belief with respect to errors will be slower in the former case, faster in the latter case. However, if all errors are due to transient objects moving through the environment, e.g. people, we should update our belief at the same rate, the relative frequency of occurrence of these transient objects is not relevant. If, on the other hand, changes in the (almost) static environment usually represent more permanent change, e.g. the moving furniture, then the rate of changes of belief should be increased.

Precisely what the functional form of this updating should take remains unclear. A probabilistic Bayesian approach does not seem possible since no knowledge is available of the prior probabilities. Instead, we chose to increase our belief in a feature based on the number of times it is supported and decrease our belief based on the number of times it is unsupported. Obviously the asymptotic values for the belief are 0 and 1. Given a model containing a set of features, $\{f_i\}$, then if n_s is the number of supported sightings of f_i and n_u is the number of unsupported sightings of f_i and $p(f_i)$ is the validation probability/confidence of feature f_i then it seems reasonable to assume an update function of the form

$$p(f_i) = 1 - e^{-(n_s/\alpha - n_u/\beta)} \qquad (12)$$

The arbitrary coefficients, α and β are dependent on the sensor and environmental operating conditions. If α and β are equal then we learn at the same rate we forget. However, it may be advantageous to have different rates for increasing and decreasing the belief in a feature. In particular, localization accuracy is unaffected by the number of unsupported predictions, they simply increase the computational requirements of the prediction and verification stages. Figure 4 summarizes the proposed algorithm.

5 Experimental Results

To test our map-building capabilities, a grid of sonar scans were taken from precisely known positions in an uncluttered office scene. Figure 5 shows triangles at each vehicle location for 28 sonar scans, and a hand-measured model of the room in which the scans were taken. The scans were processed off-line in a spiral sequence, starting from

```
for(all predicted and observed points)
{
    if observed = predicted
    {
        then increase confidence in model
    }
    else if predicted is not observed
    {
        then decrease confidence in predicted feature
        (model)
    }
    else if observed is not predicted
    {
        then insert new feature and begin
        increasing confidence
    }
}
```

Figure 4: Map building and maintenance algorithm

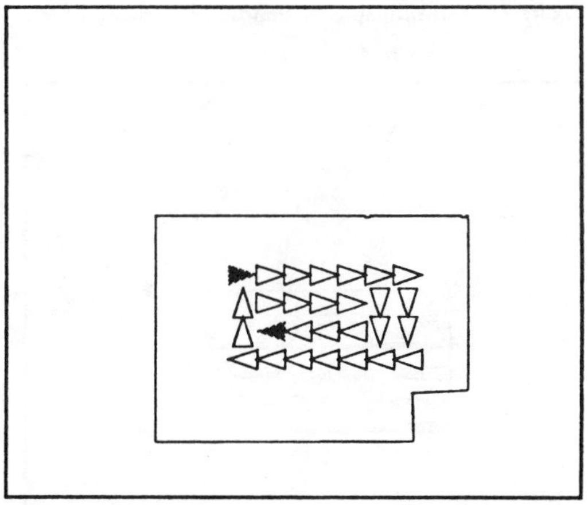

Figure 5: Hand-measured map of the room, with triangles at each of 28 locations where sonar scans were taken. The shaded triangles indicate the start and finish of the run. The room is 3 meters wide, with a closed door in the upper right hand region of the picture.

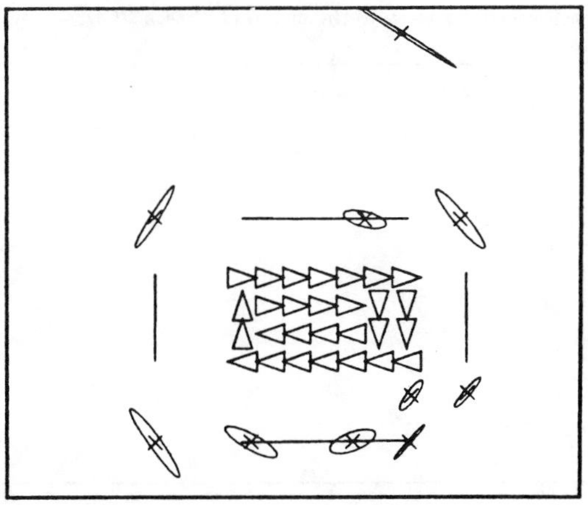

Figure 6: Localization map of the room produced by the algorithm. 8σ (magnified by 64) error ellipses are shown for point (corner) targets.

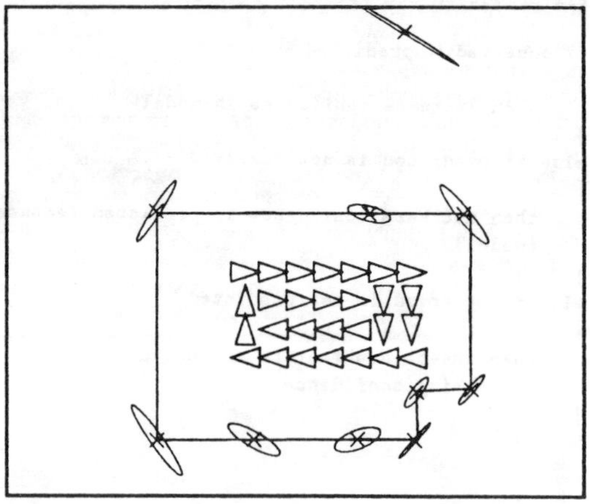

Figure 7: Learned map superimposed over room model.

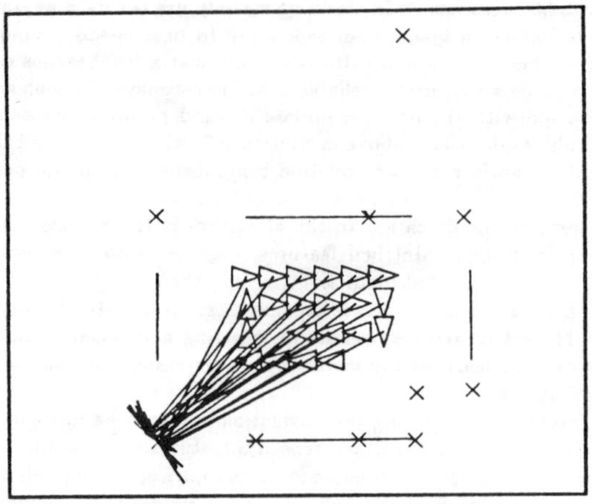

Figure 8: RCD's matched to a typical corner target.

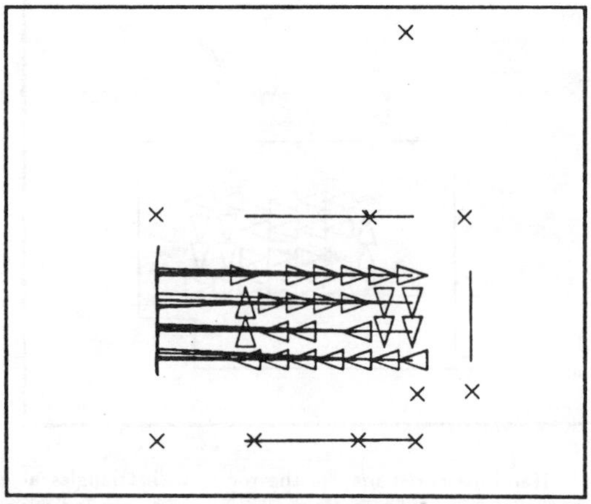

Figure 9: RCD's matched to a plane target. A triangle is shown at each location from which the target was observed. Note the absence of a triangle for scans 2 and 26–these were the scans in which a chair was inserted in the room to block this wall from view. A line is drawn from each vehicle location to the midpoint of each RCD.

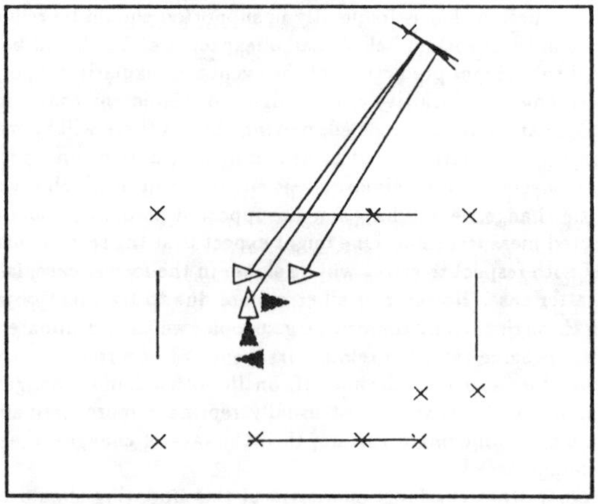

Figure 11: A corner track initiated by multiple reflection RCD's. The open triangles show locations from which the hypothesized target was observed. The shaded triangles show locations "between" the first two sightings from which the hypothesized target was not observed.

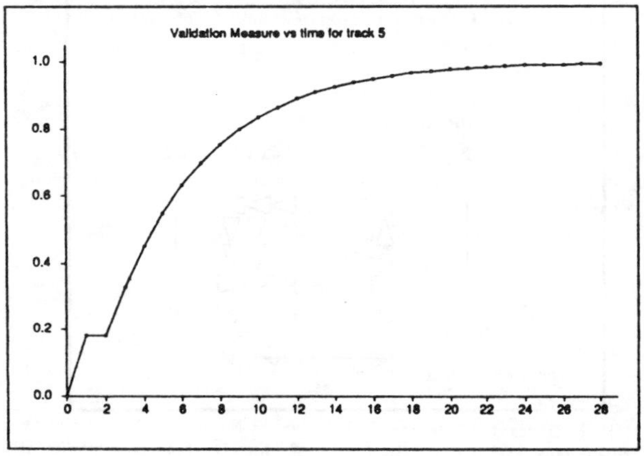

Figure 10: Validation measure vs. time for the line target shown in Figure 9. The validation probability rises exponentially as this target is repeatedly sighted, except for steps 2 and 26 when the environment was modified to occlude this target.

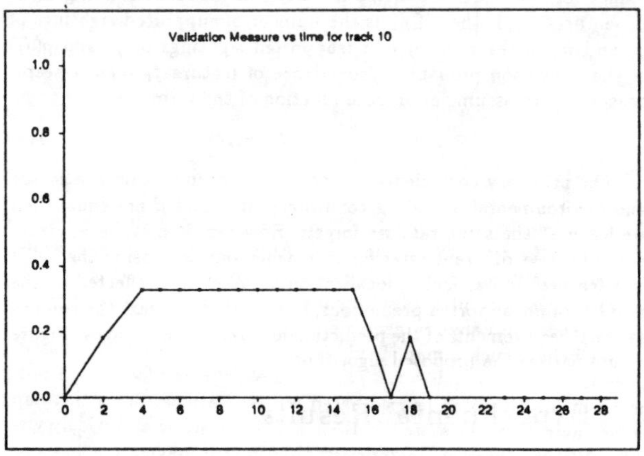

Figure 12: Validation measure vs. time for the target shown in Figure 11. The validation measure is increased for each observation (open triangles in Figure 11) and decreased for each unobserved prediction (shaded triangles in Figure 11).

the upper left of the figure. Scans 2 and 26 were taken with a changed environment, in which a chair was inserted in to the left part of the room, and a door leading into the upper right part of the room was opened, presenting two "transient" targets. For this run, α and β, the learning and forgetting factors for the validation update function (Equation 12), were both set to a value of 5.

A sensor model for the standard Polaroid ultrasonic ranging system [22] which we use in our experiments is presented in Appendix A. We wish to build a map consisting of the locations of walls and corners. The features these targets produce in sonar scans are RCD's. An RCD which cannot be matched to any beacons currently in the map causes a new track to be initiated. Because our sonar model tells us that a single RCD could be caused by either a corner, plane, cylinder, or false multiple reflection, tracks for several hypothesized geometries are initialized for a new, unexpected RCD. Through subsequent matches from different locations the unlikely hypotheses are eliminated to determine the true target geometry.

Figure 6 shows the map of line and corner beacon locations produced by the algorithm. 8σ (magnified by 64) error ellipses are displayed for each corner estimate. Figure 7 shows this map of beacon locations superimposed over the hand-measured model, revealing the close correspondence between the built-up map and the actual geometry of the room. Figures 8 to 12 show RCD's matched to a typical line and point beacons. Figure 9 shows a plot of validation probability vs. time for the target shown in figure 8. We can see that at time step 2 this target is not visible (the chair inserted into the room at step 2 occluded the left wall), and hence the validation measure is not increased at this time step.

With accurate position estimates, two or three RCD's can very accurately determine the spatial location of a target. However, this target could be caused by a multiple reflection. As discussed in Appendix A, the most crucial task in sonar data interpretation is to identify and eliminate these multiple reflections. Figure 13 shows a corner target track initiated by the algorithm for a false multiple reflection. The three open triangles show locations from which a hypothesized corner target was matched. Because of sonar's high range accuracy, this hypothesized target has a low spatial uncertainty (as shown by the magnified error ellipse in Figure 6 for this target.) However, the hypothesized target can be exposed as a false target because it can not be consistently observed from a wide range of vehicle locations. The three shaded triangles in the figure show sensing locations "between" the open triangles from which the hypothesized corner target was not observed. As a result, the validation measure was decreased at each of these time steps in accordance with Equation 12. This can be seen in Figure 14, which shows the validation probability for this target as a function of time.

This track is the only false track with three or more matches produced by the algorithm in the run shown. Thus, we feel that our sonar sensor model coupled with the validation measure presented here provides a way of overcoming false, multiple reflections to build precise maps bottom-up from sonar data.

Conclusion

We are interested in automatically building a map with which a vehicle may navigate, i.e. locate itself. Automatic cartography is important for autonomous mobile robots; costly off-line cartography is unneeded and potential obsolescence of the map is avoided for dynamic (changing) environments.

We believe such a map should be feature rather than grid based, since localization requires the prediction of expected features, a process which is ill-suited to grid-based representations. Since the map is for navigation, the stored features should be static and widely visible. We use these characteristics of geometric navigation beacons to reject those features that, for whatever reason (bad data, moving person), are spurious.

Each feature has an associated spatial uncertainty represented by a covariance matrix and an associated validation/belief uncertainty, represented as a probability. We increase the probability of a feature each time it is both predicted and seen. Features that are predicted but not sensed have their probabilities decreased. Finally, unexpected features, those that are seen but not predicted, are inserted into the map, initially with a low probability. These new features increase in probability if subsequent predictions are supported by measurements.

Our experimental results support this methodology. The results shown in section 5 present a limited form of the algorithm in which a precise knowledge of vehicle position is provided. This is easier than the general map building problem, which calls for the map to be used for localization while it is being constructed. The algorithm presented in section 4 is aimed at this general problem. None-the-less, bypassing the vehicle position update step allows us to demonstrate the map-building process.

As shown by Figure 7, very precise maps can be produced with sonar, provided one has a good sensor model. Our model tells us that sonar measurements take the form of Regions of Constant Depth (RCDs), each RCD corresponding to a different target in the environment. We group RCD's which correspond to the same target together to determine if the target is a corner, plane or higher-order multiple reflection. Observing a corner or plane target from many locations results in a high validation measure, while multiple reflections have a low validation measure, despite having a small amount of geometric uncertainty.

Currently, significant time is spent in the prediction phase, where, for an estimated vehicle position and current map, we must predict which map features should be visible. We are currently determining whether there are known algorithms for efficiently solving this problem. Such an algorithm would presumably define a representation for the map; intuitively, we would expect to do better than the present representation of unordered sets of features.

Acknowledgements

The authors thank Hans P. Moravec and Alberto Elfes for fruitful discussions.

References

[1] *Webster's Ninth New Collegiate Dictionary.* Merriam-Webster, 1985.

[2] N. Ayache and O. D. Faugeras. Maintaining representations of the environment of a mobile robot. *IEEE Trans. on Robotics and Automation*, 804–819, 1989.

[3] D. Ballard and C. Brown. *Computer Vision.* Prentice-Hall, 1982.

[4] Y. Bar-Shalom and T. E. Fortmann. *Tracking and Data Association.* Academic Press, 1988.

[5] J. Borenstein and Y. Koren. Real-time obstacle avoidance for fast mobile robots. *IEEE Trans. on Systems, Man and Cybernetics*, 19:1179–1189, September 1989.

[6] O. Bozma and R. Kuc. Building a sonar map in a specular environment using a single mobile transducer. *Trans. IEEE Pattern Analysis and Machine Intelligence*, 1989. In the Press.

[7] I. J. Cox. Blanche: position estimation for an autonomous robot vehicle. In *IEEE/RSJ Int. Workshop on Intelligent Robots and Systems*, pages 432–439, 1989.

[8] I. J. Cox and G. T. Wilfong, editors. *Autonomous Robot Vehicles.* Springer-Verlag, 1990.

[9] J. L. Crowley. World modeling and position estimation for a mobile robot using ultrasonic ranging. In *Int. Conf. Robotics and Automation*, pages 674–680, 1989.

[10] R. O. Duda and P. E. Hart. *Pattern Classification and Scene Analysis.* Wiley, 1973.

[11] H. F. Durrant-Whyte and J. J. Leonard. Navigation by correlating geometric sensor data. In *IEEE/RSJ Int. Workshop on Intelligent Robots and Systems*, 1989.

[12] A. Elfes. A tesselated probabilistic representation for spatial robot perception and navigation. In *Proc. NASA Conference on Space Telerobotics*, 1989.

[13] A. C. Gelb. *Applied Optimal Estimation.* The MIT Press, 1973.

[14] J. C. T. Hallam. *Intelligent Automatic Interpretation of Active Marine Sonar.* PhD thesis, University of Edinburgh, 1984.

[15] D. J. Kriegman, E. Triendl, and T. O. Binford. Stereo vision and navigation in building mobile robots. *IEEE Trans. on Robotics and Automation*, 5(6):792-803, 1989.

[16] R. Kuc and B. Barshan. Navigating vehicles through an unstructured environment with sonar. In *Proc. IEEE Int. Conf. Robotics and Automation*, pages 1422-1426, May 1989.

[17] R. Kuc and M. W. Siegel. Physically based simulation model for acoustic sensor robot navigation. *IEEE Trans. on Pattern Analysis and Machine Intelligence*, PAMI-9(6):766-778, November 1987.

[18] T. Lozano-Pérez. Foreword. In I. J. Cox and G. T. Wilfong, editors, *Autonomous Robot Vehicles*, Springer-Verlag, 1990.

[19] V. Lumelsky, S. Mukhopadhyay, and K. Sun. Sensor-based terrain acquisition: a "seed spreader" strategy. In *IEEE/RSJ Int. Workshop on Intelligent Robots and Systems*, pages 62-67, 1989.

[20] R. S. Michalski, J. G. Carbonell, and T. M. Mitchell, editors. *Machine Learning, An Artificial Intelligence Approach.* Tioga, 1983.

[21] H. P. Moravec and A. Elfes. High resolution maps from wide angle sonar. In *Int. Conf. Robotics and Automation*, pages 116-121, 1985.

[22] Polaroid Corporation, Commercial Battery Division. Ultrasonic ranging system. 1984.

[23] R. L. Rivest and R. E. Schapire. A new approach to unsupervised learning in deterministic environments. In *FOCS*, 1989.

A Reliable Ultrasonic Feature Extraction Using Regions of Constant Depth

Our model of sonar is inspired by the work of Kuc and Siegel [15]. Their work describes a physically-based model of sonar which considers the responses of corners, walls, and edges in a specular environment. One key conclusion from their work is that corners and walls produce responses that can not be distinguished from a single scan. The responses from these specular targets take the form of a sequence of headings over which the range value measured is *very accurate* (typically within 1 cm.) Figure 13 shows a typical, densely-sampled (612 range readings), sonar scan obtained in an uncluttered office scene. With this high sampling density, one can see that the scan is composed of sequences of headings at which the range value measured is essentially constant. We refer to such sequences as Regions of Constant Depth (RCD's).

Figure 13 shows that in a typical indoor scene, many "false" range readings are produced by the system when the beam is oriented at high angles of incidence to planar targets. At these high angles of incidence, the sound energy emitted from the side-lobes of the beam that strikes the wall perpendicularly is not of sufficient strength to exceed the threshold of the receiving circuit. As a result, the first echo detected by the system is a *multiple reflection* by some other part of the beam reflecting specularly off the wall to some other target and then back. These multiple reflections have been observed by many other researchers, and have commonly been referred to in the literature as "specularities." put a reference in here We feel this term is misleading, because Kuc's model shows that most accurate range measurements produced by planes and corners are in fact due to specular reflections.

To prevent this confusion we define the order of a range measurement as the number of surfaces the sound has reflected from before returning to the transducer. Orienting the transducer perpendicular to a planar surface such as a wall produces a 1st order range reading. Corners produce 2nd order range readings, because the sound has reflected specularly off two surfaces before returning back to the transducer. Multiple reflections will produce 3rd order and higher range readings. A crucial task in interpretation is to eliminate these higher-order reflections which, if naively taken to be the distance to the nearest object, yield false range readings.

Figure 14 shows the result of extracting RCD's visible over 10 degrees or more from the scan in Figure 13 superimposed on a line segment model of the room. Here we can see RCD's corresponding to walls, convex and concave corners, and higher-order targets.

To use 1st and 2nd order RCD's for navigation, we need to define which geometric targets in the environment produce them. 1st order RCD's principally arise from *planes* and *cylinders*. Most 2nd order RCD's arise from *corners*. Kuc and Siegel show that corners and walls will appear the same in a scan from a given location [15]. A cylinder is a 1st-order target that appears similar to a plane or corner from a single location. Higher-order multiple reflections also produce RCD's that if considered in isolation from a single location could be interpreted as any of these targets. Hence, to determine the target geometry which caused an RCD, we must observe the target from more than one location [5]. Figure 15 illustrates this process.

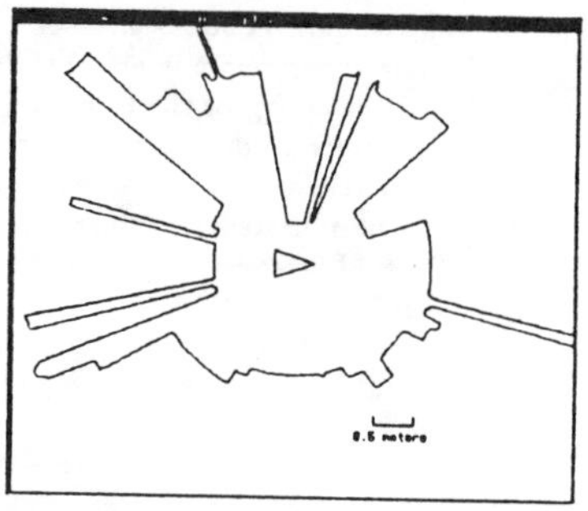

Figure 13: A typical sonar scan.

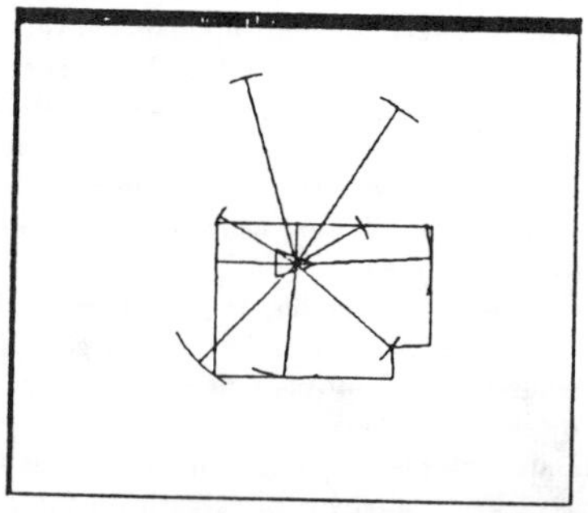

Figure 14: Regions of Constant Depth (RCD's) of width $\geq$ 10 degrees extracted from this sonar scan. A line is drawn from the vehicle position to the mid-point of each RCD. The order of each RCD has been written in by hand. We can see 4 1st-order RCD's from the four walls of the room, 3 2nd-order RCD's from corners, and 2 4th-order RCD's that result from reflections off the top wall into corners on the opposite side of the room. There is a single 0th-order RCD resulting from a diffuse reflection from the edge in the lower right-hand region of the room.

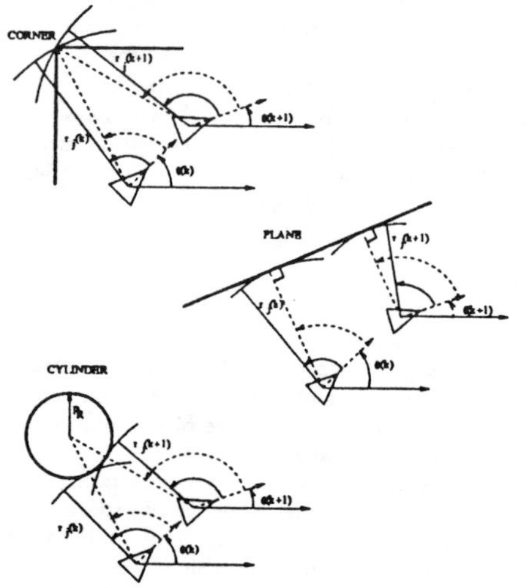

Figure 15: RCD's for corner, wall, and cylinder targets, as observed from two vehicle locations. RCD's which correspond to a plane (or cylinder) will all be tangent to the plane (or cylinder). RCD's which correspond to a corner will all intersect in a point, at the corner.

SENSOR DATA INTEGRATION BASED ON THE BORDER DISTANCE MODEL

Takanori IKEGAMI, Jun-ichi KATO and Shigeo OZONO
Department of Precision Machinery Engineering
The University of Tokyo

Reprinted from the *IEEE/RSJ International Workshop on Intelligent Robots and Systems*, pages 32-39. Copyright © 1989 by The Institute of Electrical and Electronics Engineers, Inc. All rights reserved.

Abstract This paper proposes a new data processing model suitable for an autonomous mobile robot, and shows an application for sensor data integration. First, the Border Distance Model is proposed and the concept of the model is introduced. Next, the representation methods of sensor data and their integration in the model are studied. Finally, a map generation by a mobile robot with range sensors is investigated. In the Border Distance Model, an arbitrary point in the model is represented by a distance value which shows the minimum distance from the point to the border between the occupied area and the free area. This model can deal with various kinds of sensor data or geometric information quantitatively, and can integrate these data without inconsistency.

1. Introduction

Multi-sensor systems are essential in autonomous mobile robots. Therefore, integration of sensor data is one of the important subjects in robot research.[1] We consider the following to be the main purposes of sensor (data) integration :

(1) Elimination of overlapped information

An autonomous robot usually acquires the information about the environment from sensors. Because of the limitation in memory capacity or information processing ability, it is necessary for the robot to extract only effective information from the total amount of sensor data. This sensor integration is considered as the pre-processing stage of information input.

(2) Supplement of information

Since individual sensor information is physically or spatially limited, various types of sensors are necessary for reliable navigation. In this case, each item of sensor data is used supplementally.

(3) Creation of new information

A more advanced function of sensor integration is the creation of new information through the integration of individual sensor data. This level is what we call sensor fusion. Intelligent processing is necessary to realize it.

We investigate levels (1) and (2) only, so our research is not so much sensor fusion as sensor integration.

One of the approaches in sensor integration is based on the probabilistic sensor model.[2] Furthermore, Dempster & Shafer's theory attempted to deal with uncertainty instead of Bayesian theory.[3][4] However, these models need a priori probability and each sensor datum must be independent. Thus, there are several problems in applying these models to actual situations.

In this paper, we propose a new data processing model named the Border Distance Model,[5] and apply the model to sensor data representation and integration. It is unnecessary for the model to assume a priori probability or independence of sensor data. Furthermore, the model can estimate sensor data and environmental knowledge quantitatively.

We explain the concept and application of the Border Distance Model in the following sections.

2. The Concept of the Border Distance Model(B.D. Model)

The Border distance model (B.D. Model) is fundamentally based on grid representation. But the information in a cell is completely different from the other grid-based models. Usually, in grid models, each cell shows to which area it belongs, in an occupied area or free area, and the probabilistic sensor model shows the probability attached to each one of them.

In the B.D. Model, each cell shows the signed minimum distance from a cell to the border of occupied areas and free areas. Positive values designate occupied areas and negative values designate free areas. We call these distances distance values which are represented by $\phi(p)$ for a point p. In other words, the B.D. Model is a model in which geometric information is represented by the distance value.

Fig.1 shows the concept of the B.D. Model. Generally, we define the distance value by the following equation:

$$\phi(p) = \begin{cases} \min\{d(p,BL)\} & \text{if } p \in OA \\ 0 & \text{if } p \in BL \\ -\min\{d(p,BL)\} & \text{if } p \in FA, \end{cases} \qquad (1)$$

where, OA indicates Occupied Area, BL indicate Border Line, FA indicate Free Area, and $d(p,BL)$, $\min\{d(p,BL)\}$ show the distances from p to the border and the minimum of these distances respectively.

In the B.D. Model, for arbitrary points p and q, the following equation is satisfied:

$$|\phi(p) - \phi(q)| \le d(p,q). \qquad (2)$$

Eq.(2) can be rewritten as:

$$\phi(q) - d(p,q) \le \phi(p) \le \phi(q) + d(p,q) \qquad (3)$$

This equation shows the distance values of two arbitrary points are constrained by the distance between them. This constraint condition characterizes the model and is closely concerned with sensor integration.

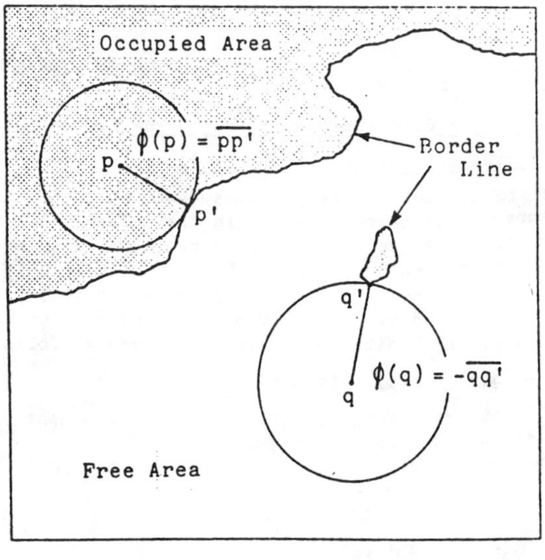

Fig. 1 The concept of the Border Distance Model (B.D. Model)

3. Environmental geometry representation in the B.D. Model

If an environmental geometry is already known, each cell can be transformed into the distance value according to eq.(1). To calculate the distance value, we can use not only Euclidean Distance, but also the other distance functions which satisfy the three axioms of distance. For example, if we use City-Block Distance or Chess-Board Distance, the transformation is easily and quickly accomplished, but the transformed values contain some distance error. In this case, we apply Euclidean Distance for the simulation.

Fig.2 shows the transformed result of the environmental geometry shown in fig.1. This transformation takes 4'03" on EWS HP9000/330(2MIPS) for the environment which is divided into 512 x 512 small cells.

Since the distance value is closely related to the environmental geometry, it is feasible for a robot to use the B.D. Model as an environmental map or a potential function of the environment.

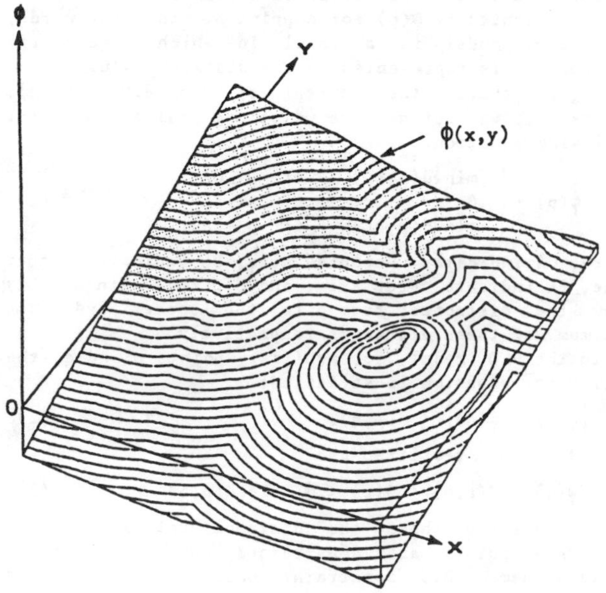

Fig. 2 Environmental geometry represented by the B.D. Model

4. Sensor data representation in the B.D. Model

4.1 General representation of the distance value

As mentioned in the previous section, the known environment can be represented in the B.D. Model. When an environment contains unknown or uncertain areas, the distance value cannot be determined only by eq.(1). For these cases, we shall represent $\phi(p)$ using the maximum and minimum of $\phi(p)$, $\phi max(p)$ and $\phi min(p)$. Here, $\phi(p)$ is limited by these two values as follows:

$$\phi min(p) \leq \phi(p) \leq \phi max(p). \quad (4)$$

In this representation, the known environment is a special case of $\phi max(p) = \phi min(p)$.

Furthermore, eq.(3) can be rewritten as:

$$\phi max(p) \leq \phi max(q) + d(p,q), \qquad \phi min(p) \geq \phi min(q) - d(p,q). \quad (5)$$

The concept of these generalized constraint conditions is illustrated in fig.3.

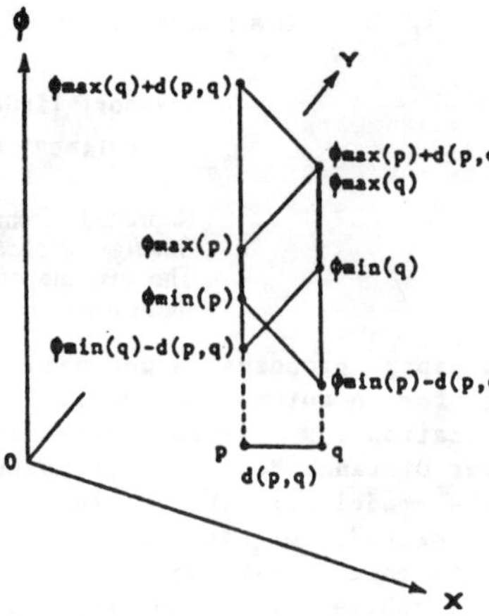

Fig. 3 Constraint conditions in general representation of the distance value.

4.2 Laser ranging data representation[6]

Generally, the following three staged processes are necessary to represent sensor data in the B.D. Model. These processes in the case of laser ranging data are as follows:

(1) Input of the initial condition

First, the initial condition of the distance values is input. If there is no knowledge about or sensor data on the robot's environment, the initial distance values are given by:

$$\phi max(p) = \infty, \quad \phi min(p) = -\infty, \quad \text{for } p \in EV, \quad (6)$$

where EV indicates environment.

(2) Interpretation of the sensor data

Next, the sensor data is interpreted according to the Border Distance Model. Fig.4 shows the sensing of the sample environment by a laser range finder. Generally, the spatial resolution of a laser range finder is so high that the reflecting point is considered as a point and the trajectory of the light beam is a line. Therefore, the following information is considered to be acquired from the sensing data:

inf(a) : there is a border at the point a,

inf(b) : there are free areas on the line $\overline{oa}$.

Let q be an arbitrary point in the model, then inf(a) can be interpreted as the distance value through the use of eq.(1) and expressed as:

$$\phi(q) = 0, \quad \text{if } q \in BL. \quad (7)$$

Similarly, inf(b) can be expressed as:

$$\phi(q) < 0, \quad \text{if } q \in FA. \quad (8)$$

(3) Adaptation of the constraint condition

Finally, the constraint condition of the distance value is adapted. From eq.(3) and (7), the constraint condition for a point p is given by:

$$-d(p,q) \leq \phi(p) \leq d(p,q), \quad \text{if } q \in BL. \quad (9)$$

Similarly, eq.(3) and (8) give another constraint:

$$\phi(p) < d(p,q), \quad \text{if } q \in FA. \quad (10)$$

Let the distance values constrained by eq.(9) or (10) be $\phi^c max(p)$ and $\phi^c min(p)$, then $\phi(p)$ is updated according to the following rules:

$$\text{if } \phi max(p) > \phi^c max(p), \text{ then } \phi max(p) = \phi^c max(p), \qquad \text{if } \phi min(p) < \phi^c min(p), \text{ then } \phi min(p) = \phi^c min(p). \quad (11)$$

After adapting every constraint condition, $\phi(p)$ can be expressed by the following equations:

$$\phi max(p) = \min\left\{\phi^{c} max(p)\right\},$$
$$\phi min(p) = \max\left\{\phi^{c} min(p)\right\}, \qquad (12)$$

where, $\min\left\{\phi^{c}max(p)\right\}$ and $\max\left\{\phi^{c}min(p)\right\}$ show the minimum of $\phi^{c}max(p)$ and maximum of $\phi^{c}min(p)$ respectively. Eq.(11) or (12) is closely concerned with sensor integration which will be discussed in section 5.

Adaptation of the constraint condition takes 9'56" under the same simulation conditions, but the calculation time is almost the same even if the amount or source of sensor data is different.

Fig. 5 shows the laser ranging data acquired by the sensing shown in fig.4. In the B.D. Model, the sensor data is represented in the model by a pair of distance values, φmax and φmin.

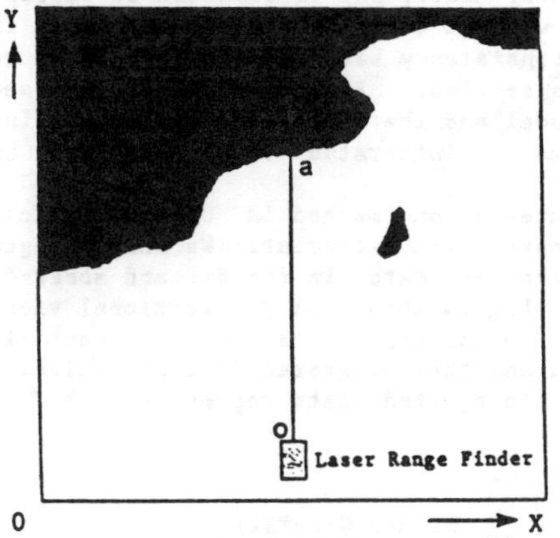

Fig. 4 Environmental sensing by a laser range finder

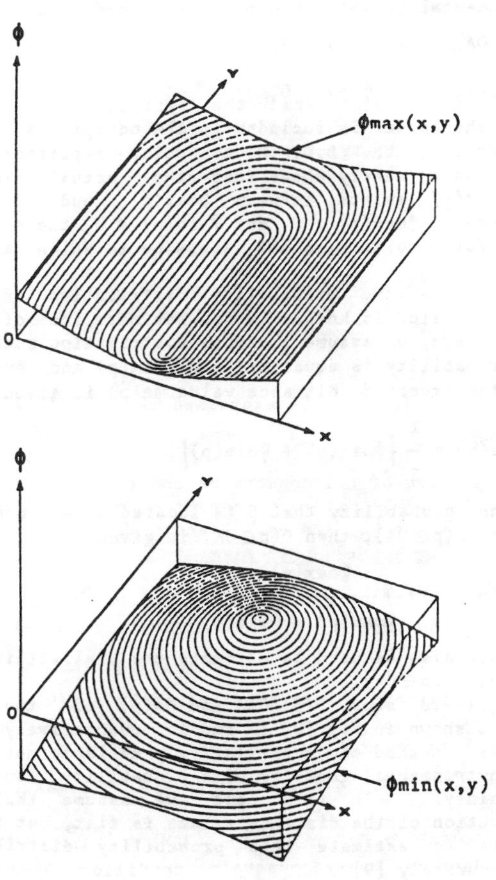

Fig. 5 Laser ranging data represented by the B.D. Model

4.3 Sonar data representation[7][8]

The representation procedure for sonar data is almost the same as that of laser ranging data. However, the procedure is slightly different because of the uncertainty attached to the sensor data.

(1) Input of the initial condition

If the robot does not have any information about its environment, then the initial condition is given by eq.(6). However, the robot has some information about its environment, so the initial condition is equal to the updated data, φmax and φmin.

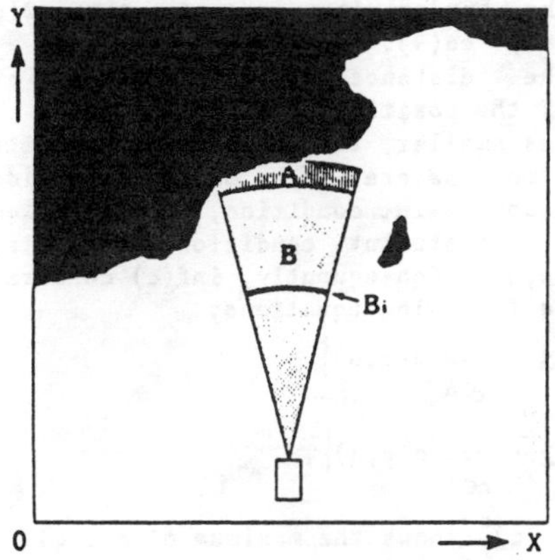

Fig. 6 Environmental sensing by sonar.

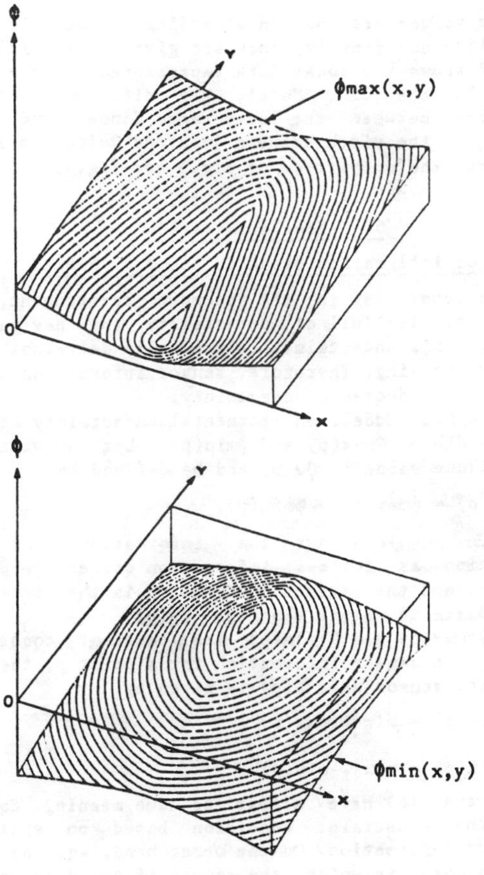

Fig. 7 Sonar data represented by the B.D. Model

(2) Interpretation of the sensor data

Since sonar data contains uncertainty due to multiple reflections or specular reflection, only the first reflected signal is used. Moreover, as shown in fig.6, the sonar data contains errors in position and orientation, so that the obstacle's position cannot be specified correctly.

In this case, we divide the sonar beam into two areas, A and B. Area A shows the maximum area of error for the reflected point, and area B is the rest of the area of the sonar beam. Then we can get at least the following informations from this sonar data:

inf(c) : there is border somewhere in area A,

inf(d) : there is free area somewhere in area Bi,

where, Bi is one of the sliced concentric circle of area B.

(3) Adaptation of the constraint condition

Assume that p is an arbitrary point and q is a point on the border in area A, then $\phi(p)$ is constrained by eq(9). In this case, we cannot calculate the distance d(p,q) because of the uncertainty of the position of q. Eq.(9) shows that as d(p,q) becomes smaller, the constraint condition of $\phi(p)$ becomes more severe. To be on the safe side when adapting the constraint condition, we must select the least severe constraint condition, that is, the maximum of d(p,q). Consequently, inf(c) constrains any point p by the following equations:

$$\phi max(p) \leq \max_{q \in A}\{d(p,q)\},$$
$$\phi min(p) \geq -\max_{q \in A}\{d(p,q)\}, \quad (13)$$

where, $\max\{d(p,q)\}$ shows the maximum of d(p,q).

Similarly, inf.(d) gives the equation below:

$$\phi(p) < \max_{q \in Bi}\{d(p,q)\}. \quad (14)$$

Distance values are updated according to the rule shown in eq.(11), and finally they are given by eq.(12).

Fig.7 shows the sonar data represented in the B.D. Model. In the B.D. Model, the difference of the resolution between the laser range finder and sonar appears in the width between ϕmax and ϕmin, and it is clear from the comparison of fig.5 and fig.7.

5. Sensor integration

5.1 Sensor integration based on the B.D. Model

If a robot is in the unknown environment, the environment is full of uncertainty for navigation. However, this uncertainty is gradually decreases with effective sensing. Therefore, sensor information can be considered to decrease uncertainty.

In the B.D. Model, environmental uncertainty appears in the width of $\phi max(p)$ and $\phi min(p)$. Let the width of the distance value be $\phi w(p)$ and be defined by:

$$\phi w(p) = \phi max(p) - \phi min(p). \quad (15)$$

Here, we consider the information and its integration as follows: information causes $\phi w(p)$ to decrease, and the sensor integration is the selection of the distance values minimize $\phi w(p)$.

If $\phi^i max(p)$ and $\phi^i min(p)$ are constant conditions for the i-th sensor information for a point p, then the integrated sensor data $\phi(p)$ is given by:

$$\phi max(p) = \min\{\phi^i max(p)\},$$
$$\phi min(p) = \max\{\phi^i min(p)\}, \quad (16)$$

Eq.(12) and (16) have almost the same meaning. Eq.(12) shows the constraint condition based on spatially different information. On the other hand, eq.(16) deals with the case in which the source of the data or the time at which it is acquired is different.

In other words, information corresponds to constraint conditions, and integration of the information corresponds to the selection of the strictest constraint conditions.

5.2 Simulation of the sensor integration

Fig.8 shows environment scanning by a laser range finder. In this case, the robot acquires three spatially different sensor data, and each of them is converted into distance values. Fig.9 illustrates the ϕ-x sectional view of the B.D. Model. Integrated distance values are shown in the shaded area. Fig.10 shows the integrated result of this scanning data.

Fig.11 shows three sonar sensings of the same environment. In this case, the sonar beams are partly overlapped, and this causes some inconsistency in integration using a probability model.

In the B.D. Model, any information is susceptible to represent without deficiency or excess and integrate without inconsistency based on the idea as we mentioned in previous section. Fig.12 shows the ϕ-x section of the B.D. Model and the integrated distance values. The three items of integrated sonar data are shown in fig.13.

This integration method is unproblematic even if the data sources are different. We show integration of the laser ranging data in fig.8 and sonar data in fig.11. Fig.14 shows the ϕ-x sectional view of the B.D. Model which represents two different items of sensor data and the integrated distance values. Fig.15 shows the integrated data represented by the B.D. Model.

6. Map Generation

6.1 Estimation of environmental geometry

If the distance value is already known, the environmental geometry can be determined by eq.(1):

$$p \in OA, \quad \text{if } \phi(p) \geq 0,$$
$$p \in FA, \quad \text{if } \phi(p) < 0, \quad (17)$$

where, the border is included in the occupied area.

Generally, the B.D. Model which represents the sensor data merely shows that the actual distance value $\phi(p)$ exists between $\phi max(p)$ and $\phi min(p)$. Therefore, if $\phi max(p) > 0$ and $\phi min(p) < 0$, the distance value for a point p cannot determine to which area p belongs.

However, the probability distribution of the distance value is known, we can estimate the distance value. Here, we assume a simple distribution such that the probability is equal between $\phi max(p)$ and $\phi min(p)$. Then, the expected distance value $\phi e(p)$ is given by:

$$\phi e(p) = \frac{1}{2}\{\phi max(p) + \phi min(p)\}. \quad (18)$$

Let the probability that p is located in an occupied area be $P(p \in OA)$, then $P(p \in OA)$ is given by:

$$P(p \in OA) = \frac{\phi max(p)}{\phi max(p) - \phi min(p)}, \quad (19)$$

$\phi w(p)$ can also be used to estimate the reliability of $\phi e(p)$ or $P(p \in OA)$.

Fig.16-18 show $P(p \in OA)$ and $\phi w(p)$ of the B.D. Model as shown in Fig.10, 13 and 15, respectively.

This method can quantitatively estimate not only the environmental geometry with probability, but its uncertainty. In this case, we assume that the distribution of the distance values is flat, but it is possible to estimate this probability distribution more accurately.[9]

6.2 Map Generation by a mobile robot

In order to verify the effectiveness of this integration method,we simulate the following situation: a mobile robot locomotes in an unknown environment and generates an environment map.

Fig.19 shows the unknown environment and the trajectory of the robot. At the points①,② ,③ and ④, the robot scans the environment using two types of range sensors, a laser range finder and sonar. We assume that the characteristics of range sensors are as follows:

[laser range finder]
scanning intervals is 10°,
scanning angle is 360°,
measuring range is 75.
(for a scale of 512 x 512 environment)

[sonar]
sonars are consist of 8 ringed transducers,
range measurement error of 5 %,
beamwidth is 30°,
measuring range is 50.

Fig.20 and 21 show the sensing areas and orientation of each range sensor. These sensing data are integrated in the B.D. Model and the environmental geometry and its uncertainty is estimated as we mentioned before.

In fig.22A-25A, the areas in black are those areas in which the probability of being occupied is grater than or equal to 50 %. The gray level in fig.22B-25B represents the degree of environmental uncetainty. Fig.22-25 show that the estimated environment gradually closes to the real world and the environmental uncertainty gradually decreases according to the effective sensing. This simulation indicates that the B.D. Model is effective in allowing the robot to recognize its environment.

7. Conclusion

In this paper, we have proposed the Border Distance Model and have applied it to sensor data representation and its integration. The effectiveness of the model is demonstrated by the simulation of map generation.

In the B.D. Model, all geometric information is represented by distance values. Furthermore, the information corresponds to the constraint conditions, and the integration of the information corresponds to the selection of the strictest constraint condition. These ideas enabled us to solve the various kinds of problems related to mobile robot navigation.

The main features of the B.D. model are as follows:
(1) The concept of the model is simple and clear.
(2) Geometric information can be dealt with collectively and quantitatively.
(3) This model can be applied to a complicated environment.

In this paper, we deal with only two types of sensor data, but the B.D. model can represent other geometric information or sensor data. Although it would take too much time to apply the model to the real time processing, and a large amount of memory would be necessary to extend the model to the 3-D world, these are not fundamental problems.

Since the B.D. Model has a high potential for the practical use of mobile robot navigation, we are now investigating B.D. Model-based navigation systems.[10] [11]

Acknowledgment

A part of this research was supported by the TEPCO Research Foundation.

References

[1] S.A.Shafer, A.Stentz and C.E.Thorpe, "An Architecture for Sensor Fusion in a Mobile Robot", Technical Report of SRI International #400, pp.2002-2011, Oct.1986

[2] A.Elfes and L,Matthies, "Sensor Integration for Navigation: Combining Sonar and Stereo Range Data in a Grid-Based Representation", Proc.26th IEEE Conf. on Decision and Control, CA, pp.1802-1807, Dec.1987

[3] T.D.Garvey, J.D.Lowrance and M.A.Fischler, "An Inference Technique for Integrating Knowledge from Disparate Sources", Proc.7th IJCAI, pp.319-325, Aug.1981

[4] M.Tazumi, M.Yatida and S.Tsuji, "Planning of Observation and Motion for a Moblie Robot", IPSJ, Vol.28, No.6, pp.558-566, June 1987 (in Japanese)

[5] T.Ikegami and Y,Yasuda,"Progressive Transmission and Display of Binary Image Using Compound Distance Transformation", IIEEJ Vol.15, No.4, pp.292-301, Oct.1986 (in Japanese)

[6] M.Hebert and T.Kanade, "Outdoor Scene Analysis Using Range Data", Ibid, pp.1426-1432, 1986

[7] A.Elfes, "Sonar-Based Real-World Mapping and Navigation", IEEE J. Robotics and Automation, Vol. RA-3, No.3, pp.249-265, June 1987

[8] M.Drumheller, "Mobile Robot Localization Using Sonar", IEEE Trans. Pattern Anal. Machine Intell., Vol.PAMI-9, No.2, pp.325-332, March 1987

[9] T.Ikegami, J.Kato and S.Ozono, "Estimation of the Environment Geometry Based on Border Distance Model", IEICE Spring National Convention, 6-127, March 1989 (in Japanese)

[10] T.Ikegami, "Navigation system for Autonomous Mobile Robot", Technical Report, Faculty of Engineering The University of Tokyo, pp.5-8, July 1988 (in Japanese)

[11] T.Ikegami, J.Kato and S.Ozono, "Estimation of Position Based on the Border Distance Model", IEICE Autumn National Convention, Sept. 1989 (in Japanese)

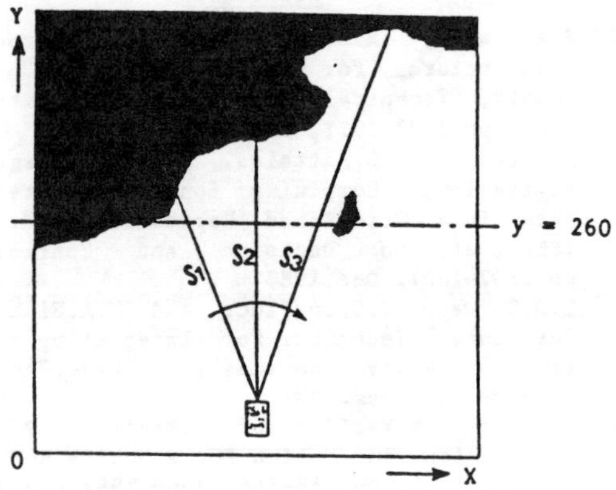

Fig. 8 Environmental scanning by laser range finder.

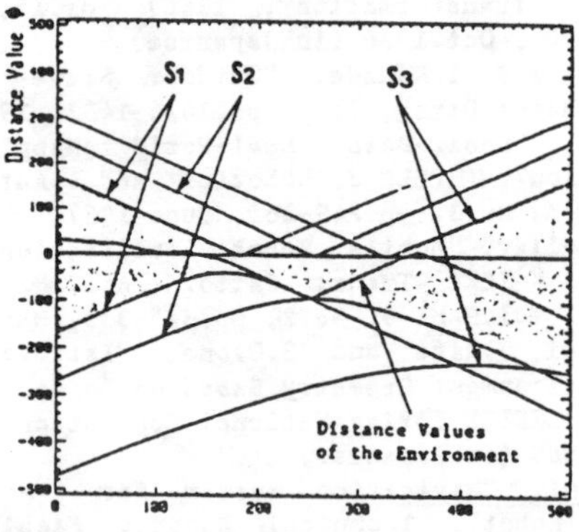

Fig. 9 φ-x sectional view of the B.D. Model (y = 260)
Laser scanning data and integrated distance values.

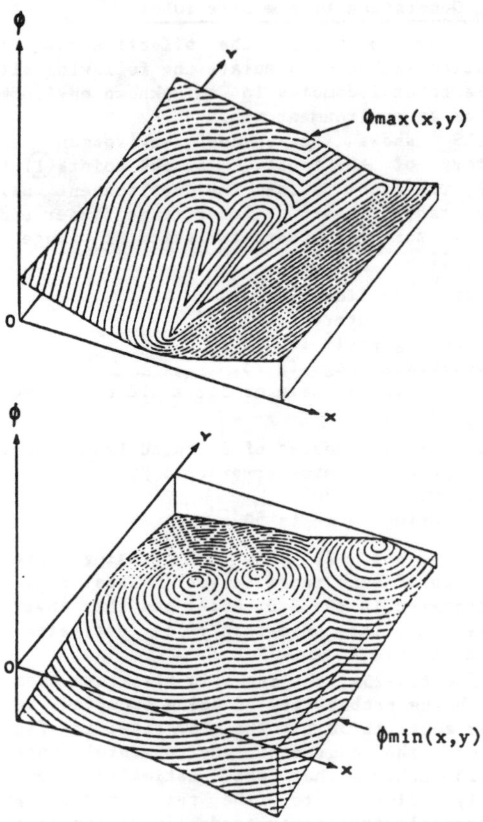

Fig. 10 Integrated result of laser scanning data represented by the B.D. Model

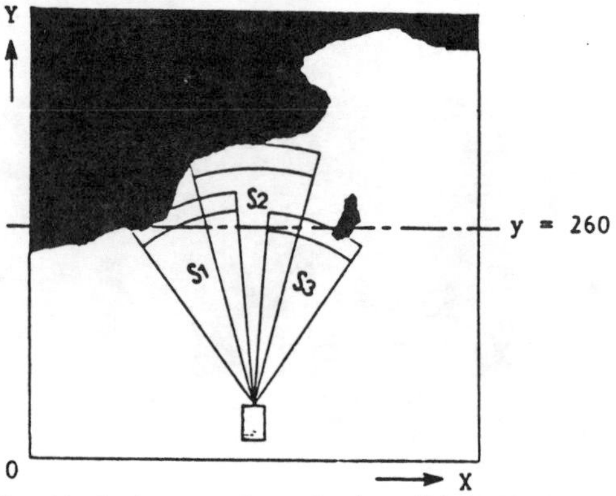

Fig. 11 Environmental sensing by multiple sonar.

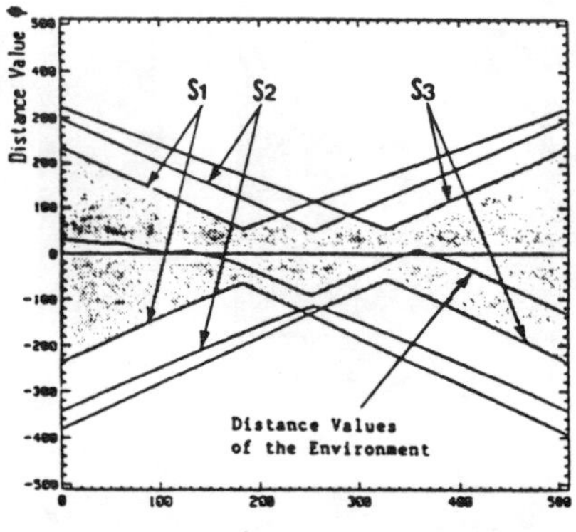

Fig. 12 φ-x sectional view of the B.D. Model (y = 260)
Multiple sonar data and integrated distance values.

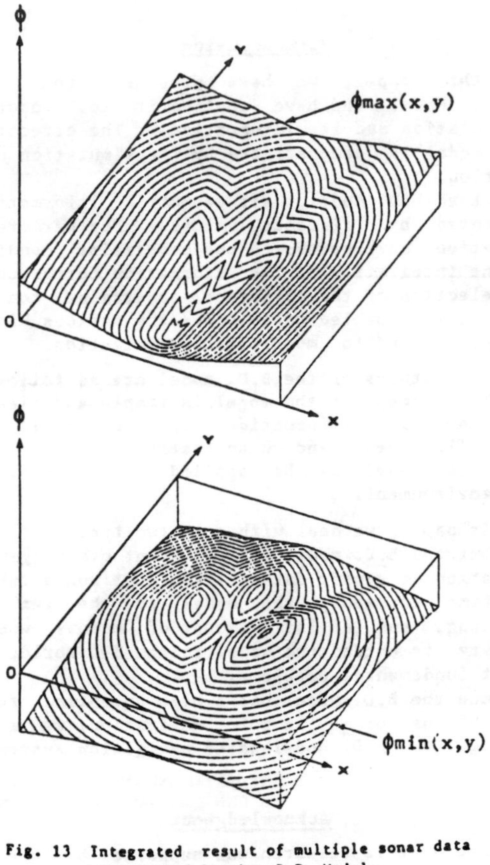

Fig. 13 Integrated result of multiple sonar data represented by the B.D. Model.

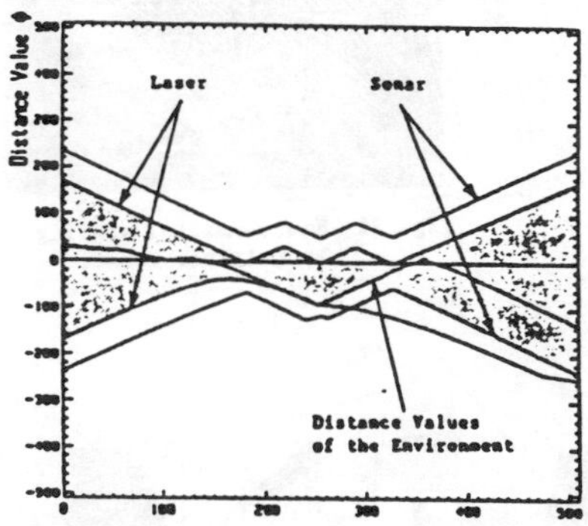

Fig.14 Φ-x section of the B.D. Model (y = 260)
Two kinds of range data and integrated distance values.

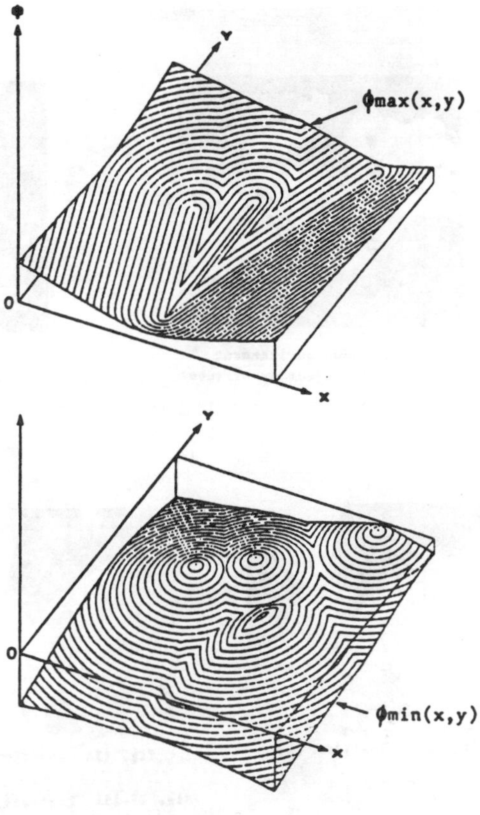

Fig.15 Integrated result of two kinds of range sensor data represented by the B.D. Model.

Fig. 16A Laser

Fig. 17A Sonar

Fig. 18A Laser + Sonar

Fig. 16A-18A Estimated environmental geometry.
(The gray level shows the probability of the occupied Area.)

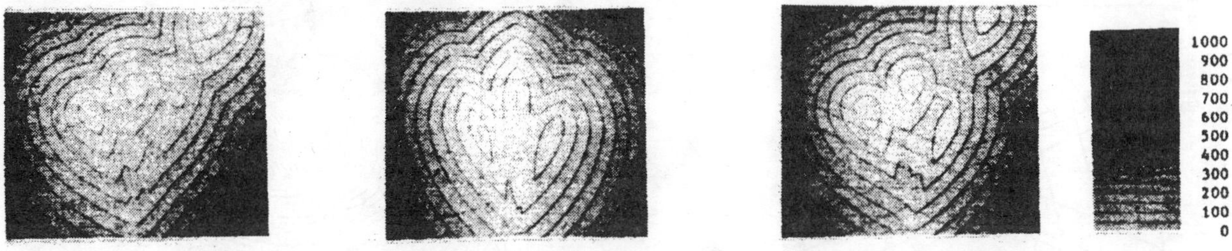

Fig. 16B Laser

Fig. 17B Sonar

Fig. 18B Laser + Sonar

Fig. 16B-18B Uncertainty of environment.
(The gray level shows the environmental uncertainty.)

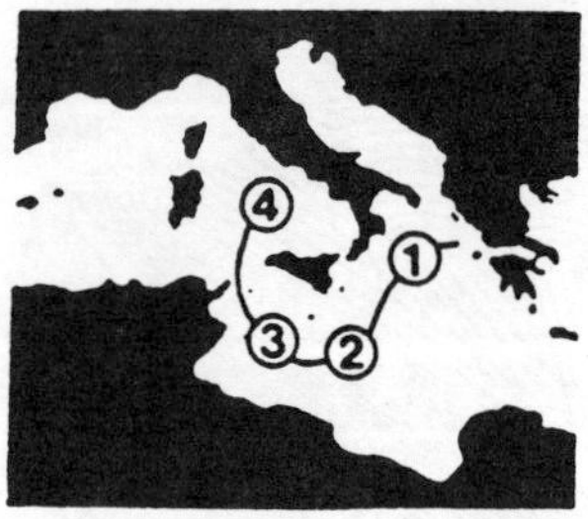

Fig. 19 Unknown environment and trajectory of robot.

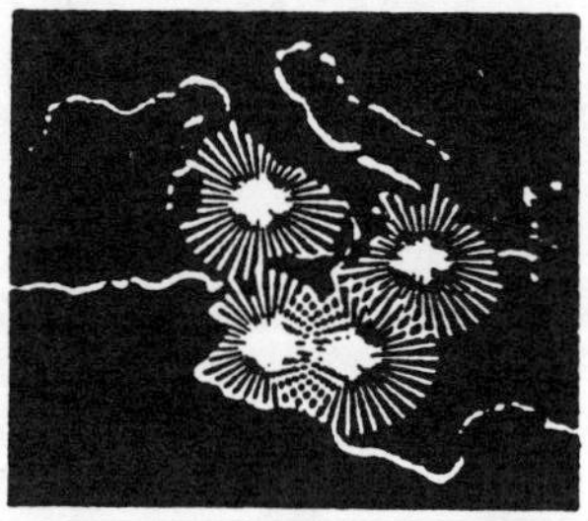

Fig. 20 Environmental scanning by laser range finder

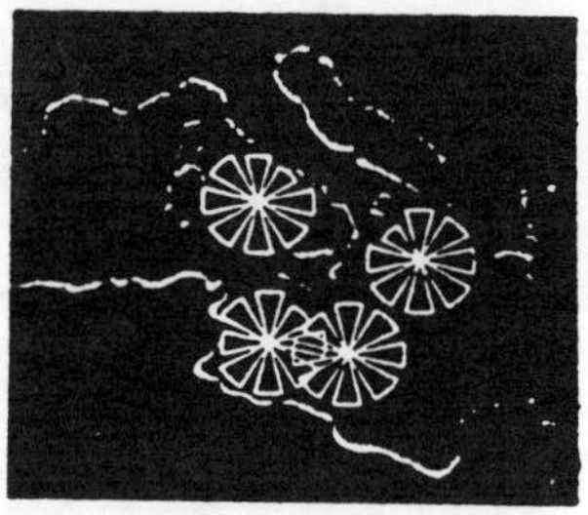

Fig. 21 Environmental sensing by multiple sonar.

Fig. 22A position ①

Fig. 23A position ②

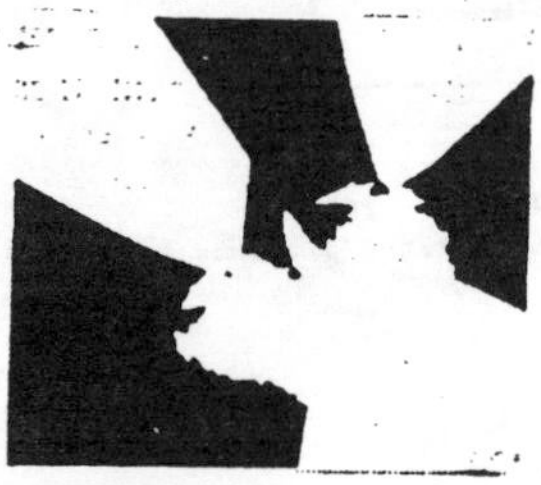

Fig. 24A position ③

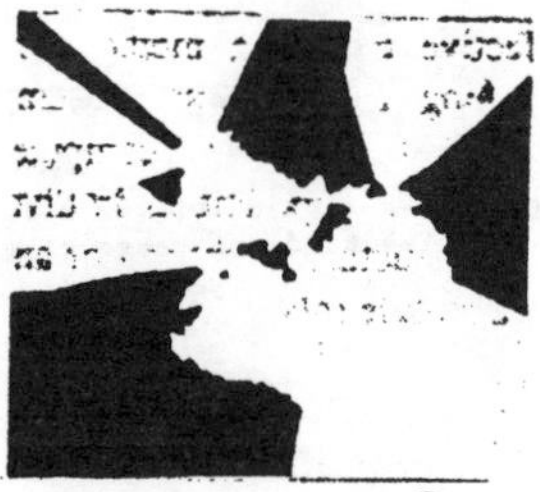

Fig. 25A position ④

Fig. 22-25A Estimated environmental geometry.(The area in black are those areas in which the probability of being occupied is greater than or equal to 50 %.)

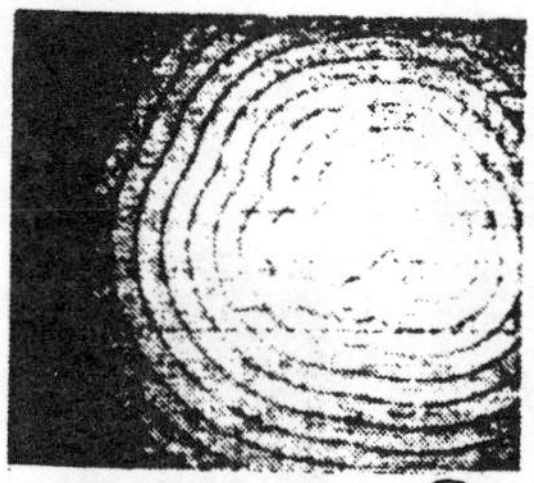

Fig. 22B position ①

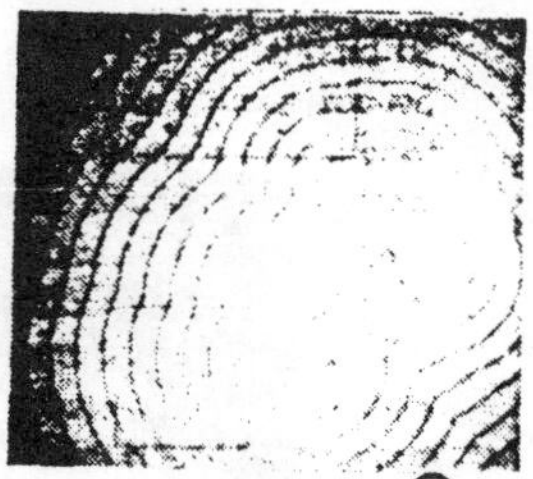

Fig. 23B position ②

Fig. 24B position ③

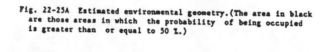

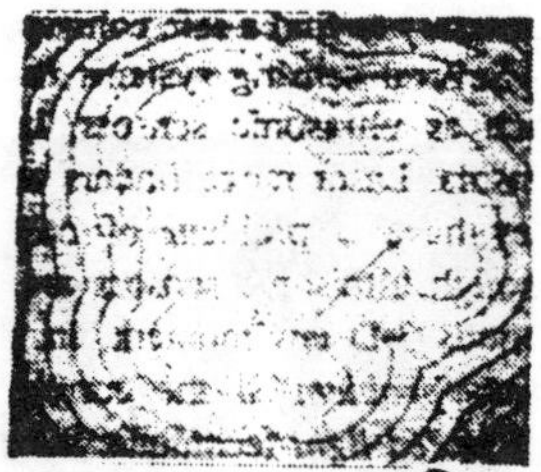

Fig. 25B position ④

Fig 22-25B Environmental uncertainty. (The gray level shows the environmental uncertainty or reliability of estimated environmental geometry.)

Coping With Uncertainty in Map Learning

Kenneth Basye Thomas Dean* Jeffrey Scott Vitter†
Department of Computer Science, Brown University
Box 1910, Providence, RI 02912

"Coping with Uncertainty in Map Learning" by K. Basye, T. Dean, and J.S. Vitter. Reprinted from the *Proceedings of the 11th International Joint Conference on Artificial Intelligence*, 1989. Copyright © 1989 International Joint Conferences on Artificial Intelligence, Inc. Copies of this and other IJCAI Proceedings are available from Morgan Kaufmann Publishers, Inc., 2929 Campus Drive, San Mateo, CA 94403.

Abstract

In many applications in mobile robotics, it is important for a robot to explore its environment in order to construct a representation of space useful for guiding movement. We refer to such a representation as a *map*, and the process of constructing a map from a set of measurements as *map learning*. In this paper, we develop a framework for describing map-learning problems in which the measurements taken by the robot are subject to known errors. We investigate two approaches to learning maps under such conditions: one based on Valiant's *probably approximately correct* learning model, and a second based on Rivest & Sloan's *reliable and probably nearly almost always useful* learning model. Both methods deal with the problem of accumulated error in combining local measurements to make global inferences. In the first approach, the effects of accumulated error are eliminated by the use of reliable and probably useful methods for discerning the local properties of space. In the second, the effects of accumulated error are reduced to acceptable levels by repeated exploration of the area to be learned. Finally, we suggest some insights into why certain existing techniques for map learning perform as well as they do.

1 Introduction

Many of the problems faced by robots navigating in the environment can be facilitated by using expectations in the form of explicit models of objects and the spaces that they occupy. We use the term *map* to refer to any model of large-scale space used for purposes of navigation. *Map learning* involves exploring the environment, making observations, and then using the observations to construct a map. The construction of useful maps is complicated by the fact that observations involving the position, orientation, and identification of spatially remote objects are invariably error prone. In this paper, we explore a number of problems involved in constructing useful maps from measurements taken with sensors subject to known errors.

In previous work [Dean, 1988], we have looked at various optimization problems related to constructing maps (*e.g.*, construct the most accurate map consistent with a set of measurements). Even in cases involving only a single dimension, such optimization problems can turn out to be NP-hard [Yemini, 1979]. In this paper, rather than look at problems that involve doing the best with what you have, we consider problems that involve going out and getting what you need to generate useful representations. In particular, we consider a form of *reliable and probably almost always useful* learning [Rivest and Sloan, 1988] in which the robot gathers information to ensure that it nearly always (with probability $1-\delta$) can provide a guaranteed perfect path from one location to another. A prerequisite to this sort of learning is that the robot, in moving around in its environment, can discern the local properties of space with absolute certainty with high probability having expended an amount of effort polynomial in $\frac{1}{\delta}$ and n, where n is some measure of the size of the environment.

By eliminating local uncertainty, small errors incurred in making local measurements are not allowed to propagate rendering global queries unacceptably inaccurate. In general, local uncertainty accumulates as the product of the distance in generating global estimates. One way to avoid this sort of accumulation is to establish strategies such that the robot can discern properties of its environment with certainty. Most existing map learning schemes exploit this sort of certainty in one way or another (see Section 4). The rehearsal strategies of Kuipers [1988] are one example of how a robot might plan to eliminate uncertainty. Once we have a method for eliminating uncertainty, the problem then reduces to one of planning out and executing the necessary experiments to extract certain information about the environment.

In situations in which it is not possible to eliminate local uncertainty completely, it is still possible to reduce

*This work was supported in part by the National Science Foundation under grant IRI-8612644 and by the Advanced Research Projects Agency of the Department of Defense and was monitored by the Air Force Office of Scientific Research under Contract No. F49620-88-C-0132.

†This work was supported in part by a National Science Foundation Presidential Young Investigator Award CCR-8846714 with matching funds from IBM, and by National Science Foundation research grant CCR-8403613.

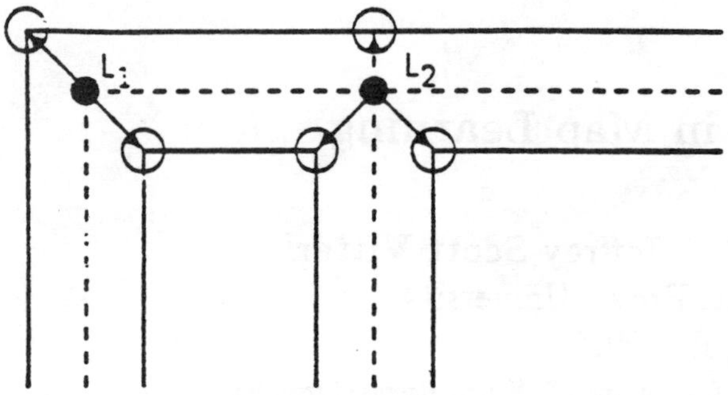

Figure 1: Identifying distinguisned locations

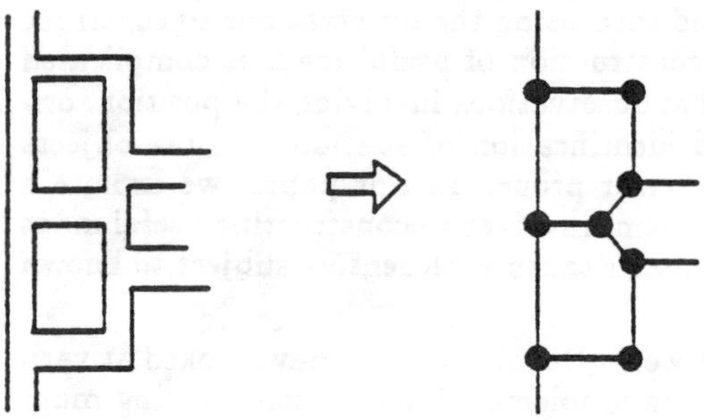

Figure 2: The induced grapn of a building

the effects of accumulated errors to acceptable levels by performing repeated experiments. To support this claim, we describe a map-learning technique based on Valiant's *probably approximately correct* learning model [Valiant, 1984] that, given small $\delta > 0$, constructs a map to answer global queries such that the answer provided in response to any given query is correct with probability $1-\delta$. The techniques presented appiy to a wide range of map-learning problems of which the specific problems addressed in this paper are meant to be merely illustrative.

2 Spatial Representation

We model the world, for the purposes of studying map learning, as a graph with labels on the edges at each vertex. In practice, a graph will be induced from a set of measurements by identifying a set of distinctive locations in the world, and by noting their connectivity. For example, we might model a city by considering intersections of streets to be distinguished locations, and this will induce a grid-like graph. Kuipers [1988] develops a mapping based on locations distinguished by sensed features like those found in buildings (see Figure 1). Figure 2 shows a portion of a building and the graph that might be induced from it. Levitt [1987] develops a mapping based on locations in the world distinguished by the visibility of landmarks at a distance.

In general, different mappings result in graphs with different characteristics, but there are some properties common to most mappings. For example, if the mapping is built for the purpose of navigating on a surface, the graph induced will almost certainly be planar and cyclic. Other properties may include reguiarity or bounded degree. In what follows, we will always assume that the graphs induced are connected and undirected; any other properties will be explicitly noted.

Following [Aleliunas *et al.*, 1979], a graph model consists of a graph, $G = (V, E)$, a set L of labels, and a labeling, $\phi : \{V \times E\} \rightarrow L$, where we may assume that L has a null element $\perp$ which is the label of any pair $(v \in V, e \in E)$ where e is not an edge from v. We will frequently use the word *direction* to refer to an edge and its associated label from a given vertex. With this notation, we can describe a path in the graph as a sequence of labels indicating the edges to be taken at each vertex. We can describe a procedure to follow as a function from $V \rightarrow L$ indicating the preferred direction at each location.

If the graph is a regular tessellation, we may assume that the labeling of the edges at each vertex is consistent, *i.e.*, there is a global scheme for labeling the edges and the labels conform to this scheme at every vertex. For example, in a grid tessellation, it is natural to label the edges at each vertex as North, South, East, and West. In general, we do not require a labeling scheme that is globally consistent. You can think of the labels on edges emanating from a given vertex as local directions. Such local directions might correspond to the robot having a compass that is locally consistent but globally inaccurate, or local directions might correspond to locally distinctive features visible from intersections in learning the map of a city.

In the following, we identify three sources of uncertainty in map learning. First, there may be uncertainty in the movement of the robot. In particular, the robot may occasionally move in an unintended direction. We refer to this as *directional* uncertainty, and we model this type of uncertainty by introducing a probabilistic movement function from $\{V \times L\} \rightarrow V$. The intuition behind this function is that for any location, one may specify a desired edge to traverse, and the function gives the location reached when the move is executed. For example, if G is a grid with the labeling given above, and we associate the vertices of G with points (i, j) in the plane, we might define a movement function as follows:

$$\psi(i,j,l) = \begin{cases} (i, j+1) & 70\% \text{ of the time if } (l = North) \\ (i+1, j) & 10\% \text{ of the time if } (l = North) \\ (i-1, j) & 10\% \text{ of the time if } (l = North) \\ (i, j-1) & 10\% \text{ of the time if } (l = North) \\ \ldots & \end{cases}$$

where the "..." indicate the distribution governing movement in the other three directions. The probabilities associated with each direction sum to 1. If all directions are equally likely regardless of the intended direction, then the movement function is said to be *random*. Throughout this paper, we will assume that movement in the intended direction takes place with probability better than chance.

A second source of uncertainty involves sensors, and in particular recognizing locations that have been seen before. The robot's sensors have some error, and this can cause error in the recognition of places previously visited; the robot might either fail to recognize some previously visited location, or it might err by mistaking some new location for one seen in the past. We refer to this type of uncertainty as *recognition* uncertainty, and model it by partitioning the set of vertices into equivalence classes. We assume that the robot is unable to distinguish between elements of a given class using only its sensors.

A third source of error involves another manifestation of sensor error. In representing the world using a graph, some mapping must be established from a set of distinguished locations in the world to V. Error in the sensors could cause the robot to fail to notice a distinguished location some of the time. For example, a robot taxi might use intersections as distinguished locations, leading to a grid-like graph. But if sensor error causes the robot not to notice that he is passing through an intersection, his map will become flawed. In exploring an office environment, the point in a hallway in front of a door may correspond to a vertex in the induced graph. If the door is closed, there is some chance that the robot will not recognize the vertex in traversing the hall. We model this type of uncertainty by introducing a probabilistic movement function that can skip over vertices. We refer to this type of movement function as *discontinuous* and to the type of uncertainty modeled as *continuity* uncertainty.

Apparently, the three types of uncertainty described above are orthogonal in the sense that none implies or precludes the others. The issues involved in modeling and reasoning about continuity uncertainty are complex and will not be treated further in this paper. In the following, we are concerned with directional and recognition uncertainty.

3 Map Learning

For our purposes, a map is a data structure that facilitates queries concerning connectivity, both local and global. Answers to queries involving global connectivity will generally rely on information concerning local connectivity, and hence we regard the fundamental unit of information to be a connection between two nearby locations (*i.e.*, an edge between two vertices in the induced undirected graph). We say that a graph has been *learned completely* if for every location we know all of its neighbors and the directions in which they lie (*i.e.*, we know every triple of the form (u, l, v) where u and v are vertices and l is the label at u of an edge in G from u to v). We assume that the information used to construct the map will come from exploring the environment, and we identify two different procedures involved in learning maps: *exploration* and *assimilation*. Exploration involves moving about in the world gathering information, and assimilation involves using that information to construct a useful representation of space. Exploration and assimilation are generally handled in parallel, with assimilation performed incrementally as new information becomes available during exploration. In this section, we are concerned with the conditions under which a graph can be completely learned, and how much time will be required for the exploration and assimilation.

3.1 Tessellation Graphs

It's not hard to see that any connected, undirected graph can be completely learned easily if there is no uncertainty; [Kuipers and Byun, 1988] describes a way of doing this by building up an agenda consisting of unexplored paths leading out of locations and then moving about so as to eventually explore all such paths. Nothing about the graph need be known before the exploration begins. Introducing the kinds of uncertainty described in Section 2 complicates things considerably. If, however, the graph has additional structure, then that structure can often be exploited to eliminate uncertainty. In the following, we sketch a proof that it is possible to efficiently learn maps that correspond to regular tessellations with boundaries. It turns out that the exploration component of learning regular tessellations is quite simple; random walks suffice for polynomial-time performance. In the longer version of this paper, we describe an efficient incremental assimilation procedure that is called whenever the robot encounters a location during exploration, and then prove the following[1].

Lemma 1 *The assimilation algorithm provided will learn a finite tessellation completely if the exploration tour traverses every edge in the graph. The overall cost of assimilation is $O(m)$ where m is the length of the tour.*

We now have to ensure that during exploration the robot traverses each edge in the graph at least once with high probability. The following two lemmas establish that, for any connected, regular, undirected graph G and any $\delta > 0$, a random walk of length polynomial in $\frac{1}{\delta}$ and the size of G is sufficient for traversing every edge in G with probability $1 - \delta$.

Lemma 2 *For any $d > 1$, there exists a polynomial $p(d, \frac{1}{\delta})$ of order $O(d \log \frac{d}{\delta})$ such that with probability $1-\delta$, p visits to a vertex of order d result in traversing all edges out of the vertex at least once.*

Lemma 3 *For any connected, regular, undirected graph $G = (V, E)$ with order d, any $\delta > 0$, and any $m \geq 1$, there exists a polynomial $p(|E|, m, \frac{1}{\delta})$ such that with probability $1 - \delta$, a random tour on G of length p visits every vertex in V at least m times.*

In most cases, we can do better than random exploration. If the robot moves in the direction it is pointing with probability better than chance, then the robot can traverse every edge in the graph with high probability in time linear in the size of the graph. Using the above three lemmas it is easy to prove the following.

Theorem 1 *Any finite regular tessellation $G = (V, E)$ can be reliably, probably almost always usefully learned.*

The lemmas and form of the proof described above provide a framework for proving that other kinds of graphs can be reliably probably almost always usefully learned in a polynomial number of steps. In general, all

[1] To meet the submission length requirements, all proofs have been omitted. The longer version of the paper, including all proofs[Basye *et al.*, 1989], is available upon request.

we require is that a polynomial number of visits to every vertex provides enough information to learn the graph. Perhaps, the most important lesson to extract from this exercise is that the effects of multiplicative error in learning maps of large-scale space can be eliminated if there is a reliable method for eliminating local uncertainty that works with high probability. The above approach to map learning was inspired by Rivest's model of learning [Rivest and Sloan, 1988], in which complex problems are broken down into simple subproblems that can be learned independently. In order to learn a useful representation of the global structure of its environment, it is sufficient that a robot have reliable and usually effective methods for sensing the local structure of its environment and a method for composing the local structure to generate an accurate global structure. The sensing methods need not always provide useful answers; they need only guarantee that the answer returned is not wrong. The problem then becomes largely one of determining a sequence of sensing and movement tasks that will provide useful answers with high probability. There are situations, however, in which reliable sensing methods are not available, and it is still possible to learn useful maps of large-scale space.

3.2 General Graphs

The next problem we look at involves both recognition and directional uncertainty with general undirected graphs. We show that a form of Valiant's probably approximately correct learning is possible when applied to learning maps. In this section, we consider the case in which movement in the intended direction takes place with probability better than chance, and that, upon entering a vertex, the robot knows with certainty the local name of the edge upon which it entered. We call the latter requirement *reverse movement certainty*. Results for related models are summarized in the next section.

At any point in time, the robot is facing in a direction defined by the label of a particular edge/vertex pair—the vertex being the location of the robot and the edge being one of the edges emanating from that vertex. We assume that the robot can turn to face in the direction of any of the edges emanating from the robot's location. We also assume that upon entering a vertex the robot can determine with certainty the direction in which it entered. Directional uncertainty arises when the robot attempts to move in the direction it is pointing. Let $\gamma > 0.5$ be the probability that the robot moves in the direction it is currently pointing. More than 50% of the time, the robot ends up at the other end of the edge defining its current direction, but some percentage of the time it ends up at the other end of some other edge emanating from its starting vertex. While the robot won't know that it has ended up at some unintended location, it will know the direction to follow in trying to return to its previous location.

To model recognition uncertainty, we assume that the vertices V are partitioned into two sets, the distinguishable vertices D and the indistinguishable vertices I. We are able to distinguish only vertices in D. We refer to the vertices in D as *landmarks* and to the graph as a *landmark graph*. We define the *landmark distribution parameter*, r, to be the maximum distance from any vertex in I to its nearest landmark (if $r = 0$, then I is empty and all vertices are landmarks). We say that a procedure learns the *local connectivity within radius r* of some $v \in D$ if it can provide the shortest path between v and any other vertex in D within a radius r of v. We say that a procedure learns the *global connectivity of a graph G within a constant factor* if, for any two vertices u and v in D, it can provide a path between u and v whose length is within a constant factor of the length of the shortest path between u and v in G.

We begin by showing that the multiplicative error incurred in trying to answer global path queries can be kept low if the local error can be kept low, that the transition from a local uncertainty measure to a global uncertainty measure does not increase the complexity by more than a polynomial factor, and that it is possible to build a procedure that directs exploration and map building so as to answer global path queries that are accurate and within a small constant factor of optimal with high probability.

Lemma 4 *Let G be a landmark graph with distribution parameter r, and let c be some integer > 2. Given a procedure that, for any $\delta_l > 0$, learns the local connectivity within cr of any landmark in G in time polynomial in $\frac{1}{\delta_l}$ with probability $1-\delta_l$, there is a procedure that learns the global connectivity of G with probability $1-\delta_g$ for any $\delta_g > 0$ in time polynomial in $\frac{1}{\delta_g}$ and the size of the graph. Any global path returned as a result will be at most $\frac{c}{c-2}$ times the length of the optimal path.*

The procedure presented in the proof of Lemma 4 searches outward from a vertex $v \in D$ to a distance cr, and then uses the edges found while entering vertices on the outward path to attempt to return to v. The directions used on the way out form an expectation for the labels observed on the way back. When these expectations are not met, the traversal is said to have failed, and the procedure tries again. The procedure keeps track of the edge/vertex labels associated with vertices visited during exploration in order to ensure that it explores all paths of length cr or less emanating from each vertex in D with high probability.

There is a possibility that some combination of movement errors could result in false positive or false negative tests. But we show by exploiting reverse certainty that we can statistically distinguish between the true and false test results. By attempting enough traversals, the procedure can ensure with high probability that the most frequently occurring sets of directions corresponding to perceived traversals actually correspond to paths in G. What is required, then, is for the learning procedure to do enough exploration to identify all paths of length cr or less in G with high probability.

Lemma 5 *There exists a procedure that, for any $\delta_l > 0$, learns the local connectivity within cr of a vertex in any landmark graph with probability $1-\delta_l$ in time polynomial in $\frac{1}{\delta_l}$, $\frac{1}{1-2\gamma}$ and the size of G, and exponential in r.*

Theorem 2 *It is possible to learn the global connectivity of any landmark graph with probability $1-\delta$ in time polynomial in $\frac{1}{\delta}$, $\frac{1}{1-2\gamma}$, and the size of G, and exponential in r.*

Theorem 2 is a simple consequence of Lemma 4 and 5. It has an immediate application to the problem of learning the global connectivity of a graph where all the vertices are landmarks. In this case, the parameter $r = 0$, and we need only explore paths of length 1 in order to establish the global connectivity of the graph. This process works even if there is no reverse certainty.

Corollary 1 *It is possible to learn the connectivity of a graph G with only distinguishable locations with probability $1-\delta$ in time polynomial in $\frac{1}{\delta}$, $\frac{1}{1-2\gamma}$, and the size of G, even if there is reverse uncertainty.*

Given the notion of global connectivity defined above, no attempt is made to *completely learn* the graph (*i.e.*, to recover the structure of the entire graph). It is assumed that the indistinguishable vertices are of interest only in so far as they provide directions necessary to traverse a direct path between two landmarks. But it is easy to imagine situations where the indistinguishable vertices and the paths between them are of interest. For instance, the indistinguishable vertices might be partitioned further into equivalence classes so that one could uniquely designate a vertex by specifying its equivalence class and some radius from a particular global landmark (*e.g.*, the bookstore just across the street from the Chrysler building).

We shall apply our above approach and try to completely learn the graph by first completely learning local neighborhoods of each landmark. Let us define $G_d(v)$ to be the subgraph of G consisting of all vertices and edges within radius d of v.

Lemma 6 *Let G be a landmark graph with distribution parameter r. Given a procedure that, for any $\delta_l > 0$, completely learns G_{2r-1} in time polynomial in $\frac{1}{\delta_l}$ with probability $1-\delta_l$, there is a procedure that completely learns G with probability $1-\delta_g$ for any $\delta_g > 0$ in time polynomial in $\frac{1}{\delta_g}$ and the size of G.*

The above algorithms used for determining the local connectivity of landmarks can be thought of as building a search tree emanating from each landmark, in which each indistinguishable node in G may correspond to several nodes in the search tree. In order to completely learn $G_{2r+1}(v)$ (as opposed to just learning the local connectivity), we must avoid this redundant representation.

It turns out that we can extend our methods to completely learn $G_{2r+1}(v)$. The algorithm builds G_{2r+1} via an incremental breadth-first search in which each vertex encountered is tested via repeated walks from v to determine with high probability if it has already been added to G_{2r+1}. The proof requires a careful examination of the probabilities of true and false test results.

Lemma 7 *There exists a procedure that, for any $\delta_l > 0$, completely learns $G_{2r-1}(v)$ for any landmark v in a landmark graph with probability $1-\delta_l$ in time polynomial in $\frac{1}{\delta_l}$, $\frac{1}{1-2\gamma}$ and the size of G_{2r+1}, and exponential in r.*

Theorem 3 *It is possible to completely learn any landmark graph with probability $1-\delta$ in time polynomial in $\frac{1}{\delta}$, $\frac{1}{1-2\gamma}$, and the size of G, and exponential in r.*

3.3 Related Models

We can get the same results as in the last section if we allow movement uncertainty in the reverse direction, but demand forward movement certainty. The algorithms are similar, the justifications different. In this case, the graph can be reliably navigated by the same agent that did the map learning.

We are also investigating ways to remove the requirement of either reverse certainty or forward certainty. Reverse certainty is used in the last section to help distinguish probabilistically between true and false results in our testing procedures. We can show, for example, that if $r(1-\gamma)$ is bounded by a small constant, then efficient map learning is possible without either the reverse certainty or forward certainty requirement. Another way around this restriction is to allow the exploring agent to drop pebbles or beacons to remember where it has been.

4 Related Work

There have been many approaches to dealing with uncertainty in spatial reasoning [Brooks, 1984, Davis, 1986, Durrant-Whyte, 1988, Kuipers, 1978, Lozano-Perez, 1983, McDermott and Davis, 1982, Moravec and Elfes, 1985, Smith and Cheeseman, 1986], but most of these methods suffer from the effects of multiplicative error in estimating relative position and orientation. This paper is concerned with eliminating the effects of multiplicative error by either eliminating local uncertainty altogether or by taking enough measurements to ensure that such effects are reduced to tolerable levels. In this section, we consider two related approaches.

Kuipers defines the notion of "place" in terms of a set of related visual events [Kuipers, 1978]. This notion provides a basis for inducing graphs from measurements. In Kuipers' framework [1988], locations are arranged in an unrestricted planar graph. There is recognition uncertainty, but there is no directional uncertainty (if a robot tries to traverse a particular hall, then it will actually traverse that hall; it may not be able to measure exactly how long the hall is, but it will not mistakenly move down the wrong hall). Kuipers goes to some length to deal with recognition uncertainty. To ensure correctness, he has to assume that there is some reference location that is distinguishable from all other locations. Since there is no directional uncertainty, any two locations can be distinguished by traversing paths to the reference location. Given a procedure that is guaranteed to uniquely identify a location if it succeeds, and succeeds with high probability, we can show that a Kuipers-style map can be reliably probably almost always usefully learned using an analysis similar to that of Section 3. In fact, we do not require that the edges emanating from each vertex be labeled, just that they are cyclically ordered.

Dudek *et al* [1988] consider the problem of learning a graph in which all vertices are indistinguishable and upon entering a vertex the robot can leave by any arc indexed from the one it entered on. The robot can always retrace its steps if it remembers the directions it took at each point during exploration. The authors show that the problem is unsolvable in general, but that by providing the robot with a number of distinct markers ($k \geq 1$) the robot can learn the graph in time polynomial in the graph's size. In order to place a marker on a particular vertex, the robot must visit that vertex; in order to recover the marker at later time, the robot must return to the vertex. A vertex with a marker on it acts as a temporary landmark. No assumption is made regarding the planarity of the graph. The problem with a single marker that can be placed once but not recovered is also unsolvable, but, if you allow a compass in addition, the problem can be solved in polynomial time.

Levitt *et al* [1987] describe an approach to spatial reasoning that avoids multiplicative error by introducing local coordinate systems based on landmarks. Landmarks correspond to environmental features that can be acquired and, more importantly, reacquired in exploring the environment. Given that landmarks can be uniquely identified, one can induce a graph whose vertices correspond to regions of space defined by the landmarks visible in that region. The resulting problem involves neither recognition nor movement uncertainty. Our results in Section 3 bear directly on any extension of Levitt's work that involves either recognition or movement uncertainty.

5 Conclusion

This paper examines the role of uncertainty in map learning. We assume an environmental model that provides for a finite set of distinctive locations that can be reliably detected and repeatedly found. Under this assumption, the problem of map learning reduces to one of extracting the structure of a graph through a process of exploration in which only small parts of the structure can be sensed at a time and sensing is subject to error. We are particularly interested in showing that cumulative errors in reasoning about the global properties of the environment based on local measurements can be reduced to acceptable levels using a polynomial (in the size of the graph) amount of exploration. The results in this paper shed light on several existing approaches to map learning by showing how they might be extended to handle various types of uncertainty. Our basic framework is general enough to be applied to a wide variety of map learning problems. We have identified one particular source of uncertainty, namely continuity uncertainty (see Section 2), that we believe of particular interest in learning maps of buildings and other environments possessing an easily discernable structure.

References

[Aleliunas *et al.*, 1979] Romas Aleliunas, M. Karp, Richard, Richard J. Lipton, Laszlo Lovasz, and Charles Rackoff. Random walks, universal traversal sequences, and the complexity of maze problems. In *Proceedings of the 20th Symposium on the Foundations of Computer Science*, pages 218–223, 1979.

[Basye *et al.*, 1989] Kenneth Basye, Thomas Dean, and Jeffrey Scott Vitter. Coping with uncertainty in map learning. Technical Report CS-89-27, Brown University Department of Computer Science, 1989.

[Brooks, 1984] Rodney A. Brooks. Aspects of mobile robot visual map making. In H. Hanafusa and H. Inoue, editors, *Second International Symposium on Robotics Research*. pages 325–331, Cambridge, Massachusetts, 1984. MIT Press.

[Davis, 1986] Ernest Davis. *Representing and Acquiring Geographic Knowledge*. Morgan-Kaufman, Los Altos, California, 1986.

[Dean, 1988] Thomas Dean. On the complexity of integrating spatial measurements. In *Proceedings of the SPIE Conference on Advances in Intelligent Robotic Systems*. SPIE, 1988.

[Dudek *et al.*, 1988] Gregory Dudek, Michael Jenkins, Evangelos Milios, and David Wilkes. Robotic exploration as graph construction. Technical Report RBCV-TR-88-23, University of Toronto, 1988.

[Durrant-Whyte, 1988] Hugh F. Durrant-Whyte. *Integration, Coordination and Control of Multi-Sensor Robot Systems*. Kluwer Academic Publishers, 1988.

[Kuipers and Byun, 1988] Benjamin J. Kuipers and Yung-Tai Byun. A robust qualitative method for robot spatial reasoning. In *Proceedings AAAI-88*. pages 774–779. AAAI, 1988.

[Kuipers, 1978] Benjamin Kuipers. Modeling spatial knowledge. *Cognitive Science*, 2:129–153, 1978.

[Levitt *et al.*, 1987] Tod S. Levitt, Daryl T. Lawton, David M. Chelberg, and Philip C. Nelson. Qualitative landmark-based path planning and following. In *Proceedings AAAI-87*, pages 689–694. AAAI, 1987.

[Lozano-Perez, 1983] Tomas Lozano-Perez. Spatial planning: A configuration space approach. *IEEE Transactions on Computers*, 32:108–120, 1983.

[McDermott and Davis, 1982] Drew V. McDermott and Ernest Davis. Planning routes through uncertain territory. *Artificial Intelligence*, 22:107–156, 1982.

[Moravec and Elfes, 1985] H. P. Moravec and A. Elfes. High resolution maps from wide angle sonar. In *IEEE International Conference on Robotics and Automation*, pages 138–145, March 1985.

[Rivest and Sloan, 1988] Ronald L. Rivest and Robert Sloan. Learning complicated concepts reliably and usefully. In *Proceedings AAAI-88*, pages 635–640. AAAI, 1988.

[Smith and Cheeseman, 1986] Randall Smith and Peter Cheeseman. On the representation and estimation of spatial uncertainty. *The International Journal of Robotics Research*, 5:56–68, 1986.

[Valiant, 1984] L. G. Valiant. A theory of the learnable. *Communications of the ACM*, 27:1134–1142, 1984.

[Yemini, 1979] Yechiam Yemini. Some theoretical aspects of position-location problems. In *Proceedings of the 20th Symposium on the Foundations of Computer Science*, pages 1–7, 1979.

A Robust, Qualitative Approach to a Spatial Learning Mobile Robot

Benjamin J. Kuipers and Yung-Tai Byun

University of Texas at Austin, Department of Computer Sciences
Austin, TX 78712

"A Robust, Qualitative Approach to a Spatial Learning Mobile Robot" by B.J. Kuipers and Y.-T. Byun. Reprinted from *SPIE Vol. 1003 Sensor Fusion: Spatial Reasoning and Scene Interpretation*, 1988. Copyright © 1988 by the International Society for Optical Engineering (SPIE), reprinted with permission.

ABSTRACT

In traditional approaches to spatial learning, mobile robots in unstructured unknown environments try to build metrically accurate maps in an absolute coordinate system, and therefore have to cope with device errors. We use a qualitative method which can be robust in the face of various possible errors in the real world. Our method uses a multi-layered map which consists of *procedural knowledge* for movement, *topological model* for the structure of the environment, and *metrical information* for geometrical accuracy. The topological model consists of *distinctive places* and *local travel edges* linking nearby distinctive places. A distinctive place is defined as the local maximum of some measure of distinctiveness appropriate to its immediate neighborhood, and is found by a *hill-climbing search*. Local travel edges are defined in terms of *local control strategies* required for travel. The identification of distinctive places and travel edges and the use of local control strategies make it possible to largely eliminate cumulative error, resulting in an accurate topological map even in the face of sensory or motor errors. How to find distinctive places and follow edges is the procedural knowledge in the map, and the distinctive places and the travel edges have metrical descriptions for local geometry on the top of the topological map. The metrical descriptions are integrated gradually for global geometry by using *local coordinate frames* and a *best-fit method* to connect them. With a working simulation in which a robot, NX, with range sensors explores a variety of 2-D environments, we show its successful results in the face of random and simple systematic device errors.

1. INTRODUCTION

In unfamiliar large-scale environments, where the whole structure of the environments is not available at an instant, humans explore the environments and build mental models, cognitive maps, from exploration. While or after building maps, humans succesfully do route-planning, navigation, and place-finding. Our goal is to build a mobile robot which can do what humans do in spatial learning and reasoning. Researchers have studied several spatial representation methods and exploration strategies to learn the structure of environments, the *robot exploration and map-learning problem*. We review that work briefly and introduce the background of our qualitative method to the robot exploration and map-learning problem by discussing how humans perform well at spatial learning and spatial problem-solving in spite of sensory and processing limitations and partial knowledge [Kuipers, 1979, 1983].

1.1. Related Research

Traditional spatial representation methods for known environments and approaches to the robot exploration and map-learning problem in unknown environments are based on the accumulation of accurate geometrical descriptions of the environment. These methods include Configuration Space [Lozano-Perez, 1981], Generalized Cones [Brooks, 1982], the Voronoi Diagram [Rosenberg and Rowat, 1981; Miller, 1985; Meng 1987; Iyengar *et al.*, 1985; Weisbin 1987], the Grid Model [Moravec and Elfes, 1985; Elfes, 1986; Moravec, 1988], the Segment Model [Crowley, 1985; Turchan and Wong, 1985], the Vertex Model [Koch *et al.*, 1985], the Convex Polygon Model [Laumond, 1983; Giralt, 1983; Chatila and Laumond, 1985], the Graph Model [Iyengar *et al.*, 1985; Rao *et al.*, 1986; Weisbin, 1987; Oomen *et al.*, 1987; Turchan and Wong, 1985], and the Polygonal Region Model [Miller, 1985]. Some of these and [Kadonoff *et al.*, 1986; Kuan *et al.*, 1985] use several methods together.

Because of low mechanical accuracy and sensory errors, it is very hard in large-scale space to get accurate metrical information [Brooks, 1985; Chatila and Laumond, 1985]. Some of the traditional methods perform reasonably well where environments are small enough to get most important features from a single position. But the problem is more difficult in large-scale space, as discussed by Brooks [1985], Kuipers and Byun [1987], and Levitt *et al.* [1987]. Recent work taking a more qualitative approach [Kuipers and Byun, 1987, 1988; Kuipers and Levitt, 1988; Levitt *et al.*, 1987] shows promise of overcoming the fragility of purely metrical methods.

1.2. Cognitive Maps

Many scientists [Lynch, 1960; Piaget and Inhelder, 1967; Siegel and White, 1975] have observed that a cognitive map is organized into successive layers, and suggested that the basic element of a useful and powerful description of the environment in large-scale space is a topological description. The layered model consists of the identification and recognition of landmarks and places from local sensory information, procedural knowledge of routes from one place to another, a topological model of connectivity, order, and containment, and metrical information of shapes, distance, direction, orientation, and local and global coordinate systems. It appears that the layered structure of the cognitive map is responsible for humans' robust performance in large-scale space. Our approach attempts to apply these methods to the problem of robot exploration and map-learning.

1.3. A Topological Model

The central description of environments in our qualitative approach is a topological model as in the TOUR model [Kuipers, 1978]. The model consists of a set of nodes and arcs, where nodes represent distinctively recognizable places in the environment, and arcs represent travel edges connecting them. The nodes and arcs are defined procedurally in terms of the sensorimotor capabilities of the robot. Metrical information is added on top of the topological model.

A place corresponding to a node must be locally distinctive within its immediate neighborhood by one geometric criterion or another. We introduce locally meaningful *distinctiveness* measures defined on a subset of sensory features, by which some distinctive features can be maximized at a distinctive place. We define the *signature* of a distinctive place to be the subset of features, the distinctiveness measures, and the feature values, which are maximized at the place. A hill-climbing search is used to identify and recognize a distinctive place when a robot is in its neighborhood. When a robot is exploring, both the signature and the local maximum must be found for a distinctive place. When returning to a known distinctive place, a robot is guided by the known signature. Travel edges corresponding to arcs are defined by local control strategies which describe how a robot can follow the link connecting two distinctive places. This local control strategy depends on the local environment. For example, in one environment, following the midline of a corridor may be reasonable; in another environment, maintaining a certain distance from a single boundary on one side is appropriate.

1.4. Others w.r.t. the Topological Model

Several researchers use various types of graph model or topological model to show connectivity. Laumond [1983] and Chatila and Laumond [1985] build a topological model from the geometric model and then derive a semantic model from the topological model. But this is only to represent the information in the map at higher abstract levels. There is no use of the topological model to cope with metrical inaccuracy. Turchan and Wong [1985] use the attributed graph to represent the world, in which line segments and their attributes become vertices and relations between adjacent vertices are used for arcs. This graph becomes complete by integrating local sensory information from several different locations. It is proposed as a way of finding a correct segment model for a large-scale environment with an error-free assumption, but it is vulnerable to errors. With the same assumption, Oommen *et al.* [1987] use the visibility graph where vertices represent noticeable or actually visited meaningful points in the environment and, arcs show the connectivity of vertices from travel. The purpose of using graphs in most works is to get the connectivity and for use in route-finding and guidance of further exploration. Our purposes of using the topological model are to build an accurate metrical map from the local metrical information of each node and edge in the model, as well as to get the connectivity, order, and regions.

In the layered control system proposed by Brooks [1986], our qualitative exploration approach and building of a map correspond to level 2, Explore, and level 3, Building maps. Level 0, Avoid Objects, corresponds to the Obstacle Avoider in the work of Kadonoff *et al.* [1986], and it is implemented in our approach as a method for taking an immediate reaction to too close objects while performing a local control strategy or a hill-climbing search.

Levitt *et al.* [1987] propose qualitative methods of place definition and navigation based on visual landmark recognition. Like us, they argue the weakness of traditional navigation techniques and show the possibility of navigation and guidance using a coordinate-free model of visual landmark memory, without an accurate map or metrical information. Their place definition is based on regions, whose boundaries are defined by line-segments connecting remote landmarks, whereas our place definition is based on the distinctive place and its neighborhood. Their methods are most appropriate in environments where remote point-like landmarks are easily observable.

Little of the literature discussing movement control strategy explicitly relates it to the topological model. Most researchers have used simply a goal-directed movement control strategy: repeating until getting to a goal place (X, Y), "Try to go straight to the place and if there is an obstacle in the direction, move around until there is a possibly straight way to the goal place". Kadonoff *et al.* [1986] use several local navigation strategies to avoid unexpected obstacles along a path without knowledge of the robot's position in a known world. Several different sensors are used to perform the local navigation strategies which are Obstacle Avoider, Path Follower, Beacon Tracker, Wall Follower, Aisle Centerer, and Vector Summer. One of them is dynamically chosen at any time by an arbitor using a production system. But local navigation strategy information is neither used in describing the world nor saved for later use. It must be computed and decided every time.

2. BUILDING THE TOPOLOGICAL MODEL

The basic structure of a map in our approach is the topological model in which nodes are distinctive places and arcs are travel edges. We discuss in detail how to define distinctive places and travel edges, and their procedural and metrical descriptions with respect to the topological description. We also discuss a basic control strategy to build the topological model.

2.1. Distinctive Places

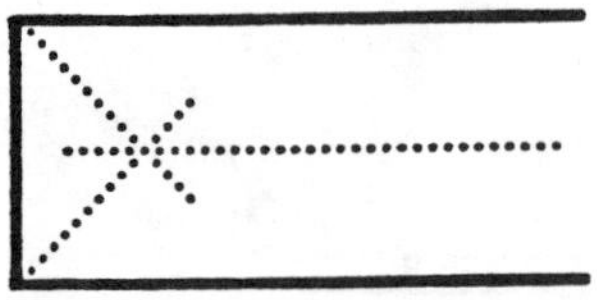

Figure 1: Distinctive Points and Most Distinctive Points

We need to look for distinctive places (DPs) in order to have the nodes of the network-structured topological model. If we consider the geometry of a simple 2-D local neighborhood in the figure, we can argue that the dotted lines define a set of places that are qualitatively distinctive for one reason or another. There is clearly a place which is the most distinctive compared to its surroundings. Our approach attempts to find a suitable criterion for defining a maximally distinctive place in any given neighborhood.

It is required to determine which sensory characteristics provide the distinguishing features by which a place becomes locally distinctive in order to formulate locally meaningful *distinctiveness* measures. We hypothesize that any reasonably rich sensory system will have distinctiveness measures that can be defined in terms of low level sensory input. For a given sensory-motor system, we can specify what features can be maximized by corresponding distinctiveness measures and how to move to a DP to maximize feature values. The given knowledge is specific to the sensory-motor system of the robot. But the notion of distinctive places is still general.

Once the robot recognizes that it is in the neighborhood of a DP, it applies a hill-climbing search to move to the DP where some distinctiveness measures have local maximum values. We will see examples of several DPs with a specific instance of a robot in Section 4 and 5. Note that it is not necessary for a place to be globally distinctive; it is only necessary to be distinguished from other points in its immediate neighborhood.

When connecting edges from or to a DP are found, the DP is described topologically in the model in terms of connecting edges and adjacent DPs. Besides this connectivity information, metrical information from sensory devices is essential to describe a DP. We have three description levels for a DP and summarize the levels of description of DPs in the following.

- Procedural knowledge for a DP: Ability to recognize the neighborhood, knowledge of what features can be maximized in the neighborhood, and ability to perform the hill-climbing search to get to the DP. Learned in the exploration stage and used in the navigation stage.

- Topological descriptions of a DP: A node in the topological model, connected to edges and other DPs. Added to the topological model when it is found and possibly updated during the process of constructing the model.

- Metrical information about a DP: Local geometry like directions to false and true open space, shape of near objects, distances and directions to objects, etc. Continuously accumulated in the exploration and navigation stage and averaged to minimize metrical error.

2.2. Travel Edges

Travel edges connecting two DPs are defined in terms of local control strategies (LCS). Once a DP has been identified, a robot moves to another place by choosing an appropriate control strategy. While following an edge with a chosen strategy, the robot continues to analyze its sensory input for evidence of new distinctive features. Once the next distinctive place has been identified and defined, the arc connecting the two DPs is defined procedurally in terms of the LCS required to follow it.

The edges followed during exploration are defined by some distinctiveness criterion that is sufficient to specify a one-dimensional set of points. Therefore, following our current set of control strategies, the robot will follow the midline of a corridor, or walk along the edge of a large space, but will not venture into the interior of a large space, where the points have no qualitatively distinctive characteristics.

The knowledge for selecting and performing the proper LCS is dependent on the robot's sensory motor system. Besides the connectivity information, metrical information is accumulated to describe geomtric features of an edge. We need to emphasize that this metrical information for edges and DPs is completely local. Examples are given in Section 4. We summarize three level descriptions as follow.

- Procedural knowledge: Ability to choose and perform a proper LCS and knowledge of which control strategy defines the edge. Learned in the exploration stage and used in the navigation stage.

- In the Topological model: An edge with direction, connected to two end-places. Added to the topological model when the second end-place is found.

- Metrical information: Curvature, distance, change of orientation, lateral width while traveling, etc. Continuously accumulated in the exploration and navigation stage and averaged to minimize metrical error.

2.3. A Control Strategy

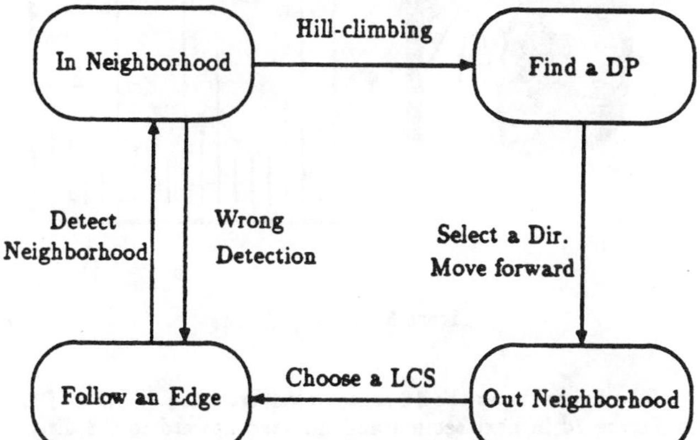

Figure 2: A State-Event Diagram

Figure 2 shows a state-event diagram for the basic exploration strategy to build the topological model. Since the topological model is the basis of our model of the environment, the robot is supposed to be either at a DP or on an edge in the topological model. For example, let's start in the figure from a state when a robot is in the neighborhood of a DP. Once a robot recognizes that it is in the neighborhood of a known place, it knows how to do the hill-climbing search to maximize the distinctive features associated with that place. The hill-climbing search is based on continuous sensory feedback movement control.

Once it finds a DP, the robot may have several directions which it should explore to finish the exploration. By selecting one of directions and moving toward that direction, it leaves the neighborhood of the DP after a while and chooses a proper LCS depending on sensory readings. Then it follows an edge and checks continuously the possibility of neighborhood of a DP.

When it detects the neighborhood of a DP, it goes to the first state. But sometimes it may detect a false neighborhood after getting into the first state and doing the hill-climbing search. If it can choose a proper LCS again in this situation, it goes back to the previous state and continues its exploration. In pathological cases, if it cannot select a LCS, it can wander around until the available sensory information becomes sufficiently clear to support the exploration task. This is similar to the relation between "Explore" and "Wander" processes in Brooks' [1986] subsumption architecture.

2.4. Movement with Errors

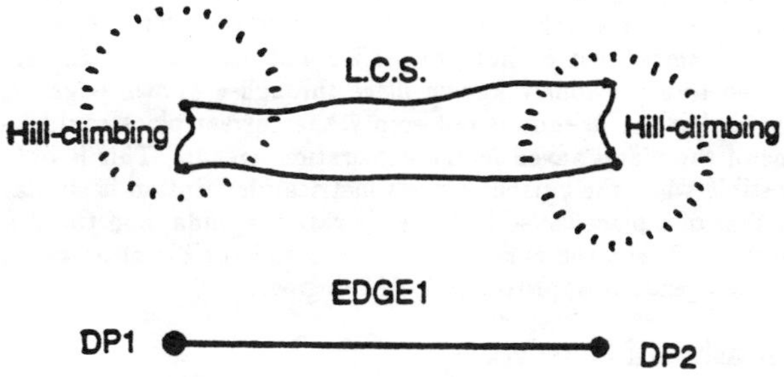

Figure 3: Movement with Errors

Once a robot recognizes that it is in the neighborhood of a DP, it performs the hill-climbing search. Although there are sensory and movement errors, continuous sensory feedback to movement will not fail to bring it very near to a distinctive place. But it is not necessary for a robot to be located at the same (X, Y) coordinates in an absolute coordinate frame.

Then the robot is following an edge from one node to another. It starts from the DP and once a proper LCS is available, it follows the LCS associated with the arc and ends up somewhere in the neighborhood of the second place. Then the hill-climbing algorithm brings it to the DP corresponding to the second node. When the robot moves through a known area, it uses knowledge from previous exploration. Because of device errors, we can not guarantee that the robot can reach exactly the same (X, Y) position. But our method guarantees that a robot gets to the same DP with the same movement control strategies in the same topological model. Since our strategy builds a consistent topological model for an environment in spite of device errors, our method eliminates cumulative error. This is why our model can be robust in face of several types of real errors.

3. MORE EXPLORATION STRATEGIES

We discuss several additional exploration strategies which are significantly different from traditional approaches: our qualitative + quantitative approach vs. traditional quantitative approaches, namely (X, Y) approaches.

3.1. Position Referencing Problem

While a robot explores a given environment, it needs to know its current position in the map. This is the most important task in the robot exploration and map-learning problem. In traditional approaches, the current position is represented by (X, Y) in a global coordinate frame. As discussed in Section 1, it is not easy to get correct coordinates.

In our method, the current position is described at two levels: topological and metrical. At the topological level, the current position is a DP or an edge. At the metrical level, if the robot is at a DP, its current orientation is given in degrees. If it is on an edge, the current position may be described in terms of the place it is coming from, the distance it has travelled, and its current orientation.

3.2. Exploration Agenda

While the robot explores, it uses an *exploration agenda* to keep the information about *where* and *in which direction* it should explore further to complete its exploration. If (Place1 Direction1) is in the exploration agenda, it means that a robot has previously visited Place1 and left it in some direction(s) other than Direction1. Therefore, in order to delete (Place1 Direction1) from the exploration agenda, a robot should either visit Place1 later and leave Place1 in the direction Direction1, or return to Place1 from the opposite direction to Direction1.

When a robot gets to a place in the exploration stage, the exploration agenda can be either empty or not empty. If the exploration agenda is empty, it means that there is no known place with directions which require further exploration. Therefore the current place must be new, unless a robot has intentionally returned to a previously known place through a known edge. If the exploration agenda is not empty, the current place could be one of the places saved in the exploration agenda. This is only possible when the current place's metrical description is similar to that of a place saved in the exploration agenda, and the difference between the current orientation and the direction saved on the agenda is approximately 180 degrees.

3.3. Rehearsal Procedure

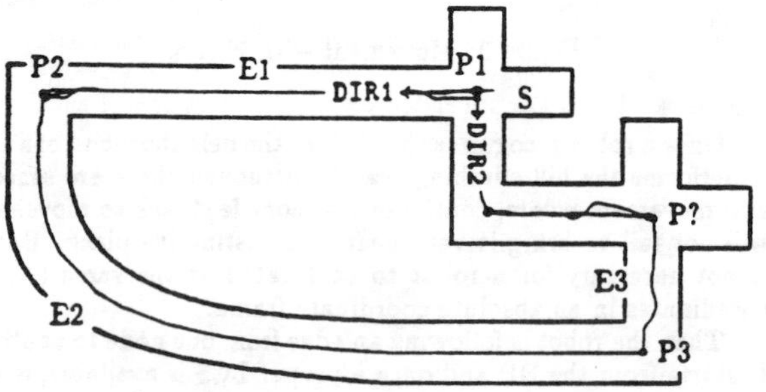

Figure 4: An Environment Requiring the Rehearsal Procedure

When a robot reaches a place during its exploration, the identification of the place is the most important task. If a place has been visited before and a robot comes back to that place, the robot should recognize it. A new place must be recognized as new, even if it is very similar to one of the previously visited places. Our matching process is done using global topological constraints as well as local metrical comparison.

The current and stored place descriptions are compared metrically, allowing a certain amount of looseness of match to provide robustness in the face of small variations in sensory input. But mismatching is possible. If there is any possibility, the topological matching process is initiated. From the topological model and procedural knowledge of edges and nearby DPs, the *rehearsal procedure* [Kuipers 1985] is activated to test the hypothesis that the current place is equal to a previously known place. A robot constructs routes between the known place and adjacent DPs. It then tries to follow the routes and return to the current place. If the routes performed as predicted, then the current place matches the previously known one, and a robot has identified the current place. If not, then the current place must be a new place with the same sensory description as the old one.

Figure 4 shows an environment in which this procedural is absolutely necessary. Let's start a robot from S. It finds a DP *P1* and follows an edge *E1*. When it leaves *P1*, it keeps (*P1 Dir2*) in the exploration agenda. Then it finds *P2*, follows E2, and finds *P3*. And then it follows *E3* and gets to a place *P?* where the local sensory information is very similar to that of *P1*. *P?* may be *P1* or a different DP. It sets up a hypothesis: If *P?* is *P1* and I follow *E1*, then I will get to *P2*. Then it tests this hypothesis by trying to visit *P2*. But all of sudden it get to a place (*P5*) in which the local sensory information is quite different from that at *P2*. Therefore it decides that the hypothesis is wrong and *P?* is *P4*.

For any fixed search radius of this topological match, it is possible to construct an environment that will yield a false positive match. However, if there is a reference place that is somehow marked so as to be globally unique (e.g., "home"), false positives can be eliminated.

4. A Robot Instance: NX

We hypothesize that our approach is supported by any sensorimotor system that provides sufficiently rich sensory input, and takes sufficiently small steps through the environment. For simplicity and concreteness, we currently define a specific instance of a robot NX.

4.1 Egocentric Sensory Motor System

NX has **sixteen sonar-type distance sensors** covering 360 degrees with equal angle difference between adjacent sensors, **two tractor-type chains** for movement, and an **absolute compass** for global orientation. Thus the input to NX is a vector of time-varying, real-valued functions $[S_1(t), S_2(t), ..., S_{16}(t), Compass(t)]$. Although we use NX to test our qualitative method, our approach does not depend critically on the choice of sensors and movement actuators.

We have developed a simulation system NX-SIM. Our simulator is implemented on the Symbolics 3600 and is written in Common Lisp. We discuss our simulator NX-SIM and its system architecture. Readers should refer to [Kuipers and Byun, 1988] to see a copy of a simulation window.

4.2. How to Handle Errors

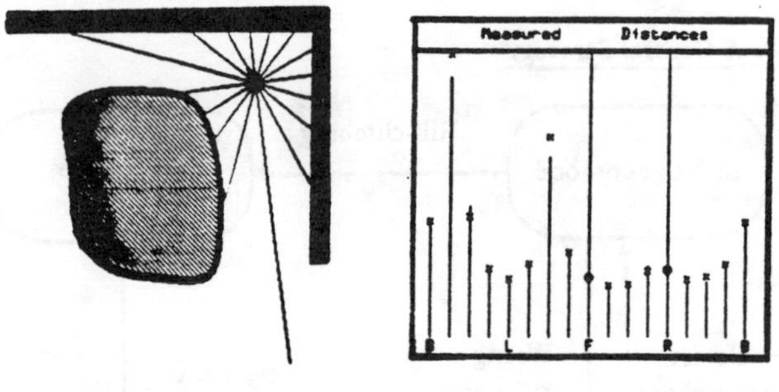

a b

Figure 5: Sensory Errors

The figure shows the 16 sensor readings when NX is at P11 in Figure 7d in next section and directed upward to the direction between a upper wall and E8. F, L, R and B in the figure stand for Front, Left, Right, and Back, respectively. The measured distance is the length of each line in each direction, and the distance from the bottom to an "x" or "o" for each sensor is the true distance (perceived only by the researchers). For an "x" case, it shows a random error and for an "o" case, it shows a systematic error. This error simulation is based on Hickling and Marin [1986], Walter [1987], Flynn [1985], and Drumheller [1985].

The first step to handle errors is a smoothing operation. Basically it tries to smooth sensory information so that it can detect a hyperbolic shape (e.g., one made by six sensors centered near L in Figure 5b). By this operation we can smooth out random errors and also ignore the first false open space reading shown above F in the figure. But the second false open space reading shown above R in the figure still remains and NX considers there is a free space to explore between two objects on the right side.

The second operation is a being-patient operation. Rather than making a decision from one snap-shot, it accumulates sensory information over several small steps, and describes near surroundings by analyzing the accumulated sensory information.

This might help to describe near surroundings correctly. In some cases, the second false OPEN-SPACE in the figure can eliminated by this operation.

The third operation is a hypothesis-testing operation. Once it describes near surroundings, it knows where objects are and where OPEN-SPACEs are. This first belief becomes a hypothesis at this operation stage. It tries to check the hypothesis by visiting nearby and decides the truth. Readers can see the trace of this operation in Figure 7d. By this operation, the second false OPEN-SPACE is completely eliminated from the description of the current surroundings.

We also simulate a five percent random error of movement control in Figure 7d. As mentioned earlier, this can result in incorrect metrical information about edges. But all metrical information is local until it is propagated into a global metrical map. This does not affect the first two levels of description of the model and the error can be compensated by the accumulation of information from several traversals. A possible systematic error may be also corrected by the similar method.

4.3. Distinctive Places and Travel Edges

With NX, we need to formulate methods to implement what we have mentioned in earlier sections. What distinctiveness measures, how to detect the event DETECT-NEIGHBORHOOD, how to implement the hill-climbing search, and what information is recorded for a DP are all issues. We discuss these briefly here by showing examples. All examples are from Figure 7d.

The individual distinctiveness measures are an open-ended, domain- and sensor-specific set of measures. For our current robot, the measures we can define include the following.

- Extent of distance differences to near objects.
- Extent and quality of symmetry across the center of the robot or a line.
- Temporal discontinuity in one or more sensors, given a small step.
- Number of directions of reasonable motion into open spaces around the robot.
- Temporal change in number of directions of motion provided by the distinct open spaces, with a small step.
- The point along a path that minimizes or maximizes lateral distance readings.

A set of production rules is used to decide whether NX is in the neighborhood of a DP and what distinctive features can be maximized in that neighborhood. Each rule consists of assumptions and a decision for the distinctive features. Here is an example:

```
(defrule dpr10
 (if (>= (no-of-walls) 3)
     (not (all-objects-far-away))
     (not (there-is-wide-open-space))
     (not (and (there-are-two-walls-near-to)
               (two-walls-facing-to-each-other)))
 (then (am-I-in-neighborhood-dp is 'dp-symm-equal)))
```

Once NX knows what distinctive features can be maximized locally in the neighborhood of a DP, NX performs a hill-climbing search around the neighborhood looking for the point of maximum distinctiveness (e.g., minimizing differences of distances to near objects, if dpr10 is true). When a DP is identified, it is added to the topological model with its distinctiveness measures, connectivity to edges, and metrical information.

```
PLACE
 Name = P8
 Procedural: SYMM-EQUAL
 Topological: E9, E8
 Metrical:
   More-explore-dirs:
     True :(357deg.(E9) 269deg.(E8))
     False:(110deg.)
   Near-obstacles:
     ((172deg. 52units) (82 34) (306 44))
   Raw-sensory-information:
     (52units 69 108 205 118 44 65 216
      195 68 41 34 37 48 59 51 194deg.)
```

The current local control strategies for edges are:

- Follow-Midline
- Walk-along-Object-Right
- Walk-along-Object-Left
- Blind-Step

A set of production rules selects a proper LCS depending on the current sensory information. Examples of a rule and an edge are given below. Units are not listed except the first one of each line. DIR+ in the example means the movement from P1 to P8, whereas DIR- means the movement from P8 to P1. Refering to [Kuipers and Byun, 1988] is recommended for more details.

```
(defrule r42
 (if (>= (no-of-walls) 1)
     (not (all-objects-far-away))
     (there-is-an-object-on-left)
     (or (there-is-wide-open-space-on-right)
         (there-is-wide-open-space-in-front)))
 (then (proper-local-c-st is
              'move-along-object-on-left)))
```

```
EDGE
 Name = E8
 Procedural: PASS-ON-THE-MIDLINE
 Topological: From P1 To P8
 Metrical:
   Leaving-orientation: 97deg.
   Arriving-orientation: 85deg.
   Travel-history
     ((DIR- (14units 6units) (38 42) (15 15))
      (DIR+ (8units 12units) (31 29) (12 8) (8 12)
            (11 9) (9 11) (3 7)))
   Distance:(65units 85)
   Lateral-width:
     ((DIR- (76units INC 46) (77 ALMOST-STD 19))
      (DIR+ (85units DEC 55) (76 ALMOST-STD 30)))
   Minimum-width: 73
   Delta-orientation: ((DIR- 16) (DIR+ -21))
```

4.4 System Architecture

The overall system architecture of our simulation system NX-SIM is presented in Figure 6. However we need to emphasize again that our exploration strategy does not depend on the types of sensors and movement actuators at all. The modules shown in the bottom half of Figure 6 depend on the kinds of devices and those modules shown in the top half do not.

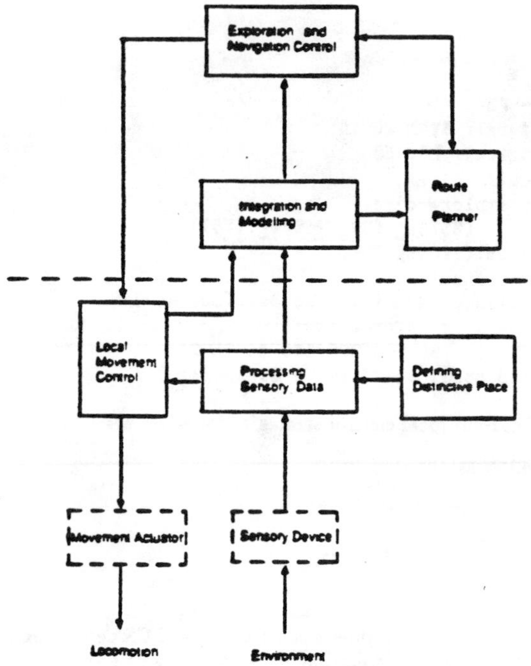

Figure 6: Overall System Architecture

5. EXPLORATION RESULTS

We explain how NX explores and builds a map in detail with the environment shown in Figure 7d. Thick black rectangles are considered surfaces which can cause systematic errors by specular reflection [Drumheller, 1985]. In order to show the effects of errors, we also show the graphic exploration results for three different random error rates: the error-free case in Figure 7a, five percent error in Figure 7b, and ten percent error in Figure 7c. The starting place is marked S in each figure, P*i* means Place-*i* and E*i* means Edge-*i*. We will trace NX's movement in Figure 7d. NX constructs the correct map successfully in all cases, but careful examination of figures Figure 7a-d will reveal subtle differences. Figure 8 shows the topological model with procedural descriptions and Figure 9 shows the metrical map of the environment in Figure 7d.

In Figure 7d, NX starts its exploration from S between P1 and P8 and directed to P1. It chooses the Pass-on-the-midline LCS and moves to P1. In the neighborhood of P1, it recognizes that there is a wide open-space in front and there will be a change of the angle made by directions to two near objects while maintaining the same LCS. When NX is in the area between S and P1, the angle is 180 degrees (In terms of number of sensors, the angle is N/2 where N is the number of sensors). As NX moves into the wide open-space near to P1, the angle becomes less than N/2 after being constant over a period. As external observers, we notice that there is an important change of a feature of sensory information. We call it Temporal-discontinuity in a sense that there is a big change of one sensor reading when NX moves back and forth near P1. However it is more robustly implemented by detecting the change of angles made by directions to near objects. No connectivity information is stared for P1 at this time. The metrical information is recorded as shown graphically in Figure 9. We see two "O"s with a small dot inside around P1. They indicate the distances and directions to the nearest objects, and several small dots show the rough shapes of nearby objects. There are three directions NX can choose from P1. If there is no particular reason to choose an indicated direction, it chooses the direction which requires the least rotation. When NX finds P1, the rotation angle to the direction toward P2 happens to be less than that toward P7 or P8. Therefore it rotates to the direction toward P2 and saves two other directions on the exploration agenda.

When NX leaves P1, it chooses Move-along-object-on-left, since it has selected a direction for travel and there is a wide open-space on the right side. However the story is different sometimes. When NX leaves the same place in Figure 7c, it decides that maintaining the same distance to the objects on the right side and an object on the left side is an appropriate local control strategy. You can see a line stretching to the direction between E2 and E6 in Figure 7c. But it soon realizes that Move-along-object-on-left or Move-along-object-on-right are more appropriate. Because it prefers a smaller rotation angle, it chooses Move-along-object-on-left. The choice of an incorrect LCS is the result of random sensory errors. However, the exploration process recovers, and is successful in all four cases.

While it moves, NX continously checks for the possibility of getting into a neighborhood of a DP. It finds P2 where it locally maximizes the value of distinctiveness measure Symmetry-Equal-Distances to near objects. Two tasks are being performed at this point. The first task concerns the edge, E1, and the second task which is the same as what has been done for P1, concerns P2.

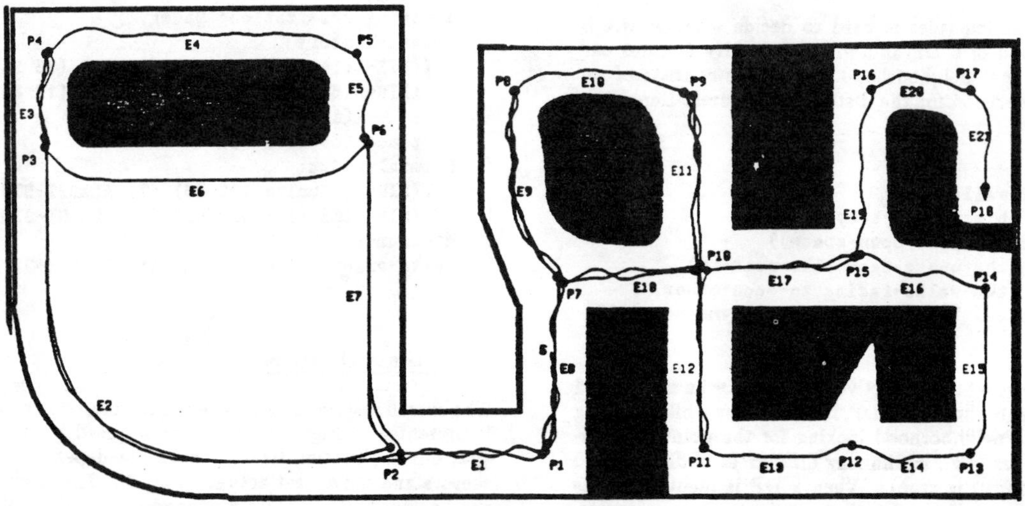

a. No Sensor Error.

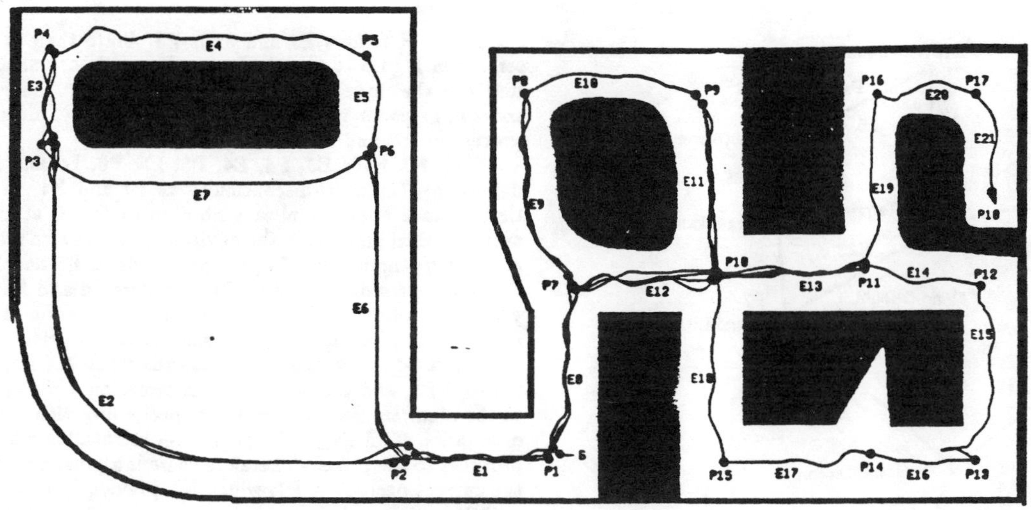

b. Five Percent Random Sensor Error.

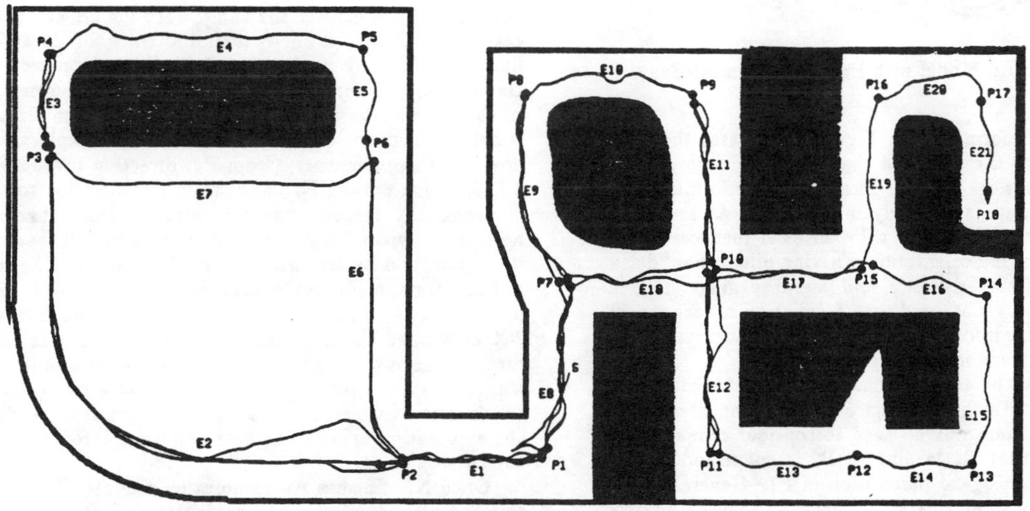

c. Ten Percent Random Sensor Error.

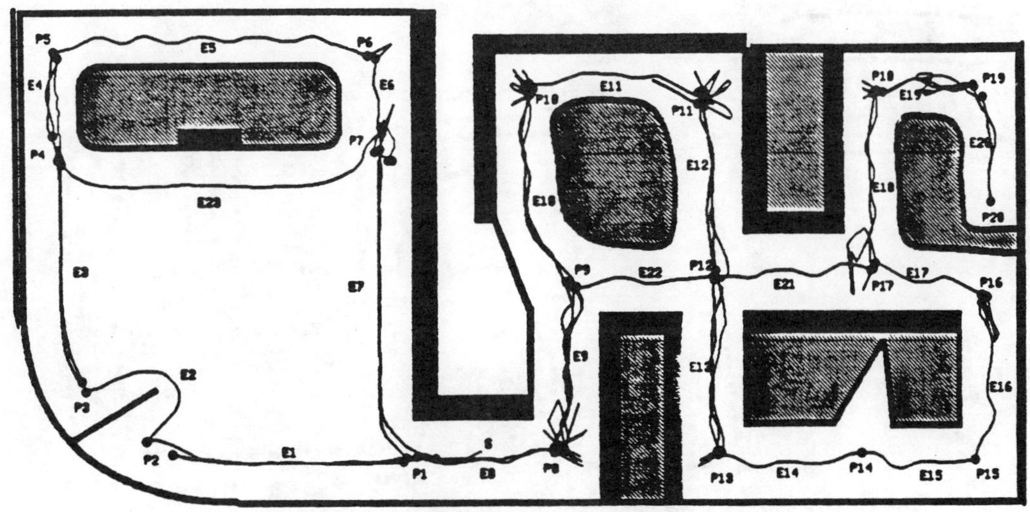

d. Systematic Error and Ten Percent Random Sensory Error.

Figure 7: Exploration Results with Errors

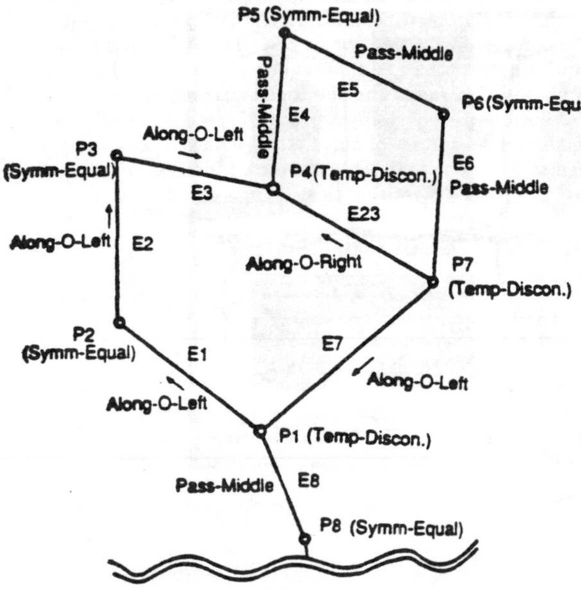

Figure 8: Topological Model with Procedural Knowledge

The procedural information in Figure 8 indicates the control strategy used for the edge and the topological information says that E1 connects P1 to P2. The metrical information is also saved and is shown graphically in Figure 9. As shown in the previous section, NX records a lot of local metrical information, including leaving orientation, arriving orientation, delta orientation, travel history, distance, and lateral readings. This is sufficiently rich that a generalized cone description of each travel edge is derivable. This information becomes more accurate when more traversals are made for the edge.

As well as saving the edge information and the definition of P2, it needs to update the topological information about P1. When it found P1, there was no need to consider connectivity. But now it knows that E1 leading to P2 is connected to P1, so it updates the topological description of P1. Generally, NX updates the topological information of places whenever it finds a new edge.

Then NX follows E2 and finds P3. Since there is one more feature in Figure 7d than in figures Figure 7a-c, a small straight line from the curved wall in the lower left corner, there are differences in maps for two different environments. NX does the same process for E2 and P3 as before.

Then NX finds E3, P4, E4, P5, E5, P6, E6 and P7. P7 is defined by Temporal-discontinuity as P1 and P4. Notice here that a place does not always need to be found at exactly the same physical location in the environment. We also see the trace of the hypothesis test of open-space around P6 and P7. From P7, it explores downward to P1 in Figure 7d and finds a place which could be P1. The local sensory information at the place is very similar to that of P1. Besides that, the orientation information saved in the exploration agenda when NX found P1 and followed E1, and the orientation information while approaching to the place are matching in the opposite direction (i.e., upward orientation from P1 = 180 degrees + orientation when NX approaches to the place). Therefore there is a great possibility that the current position is P1, which NX previously visited.

NX performs the rehearsal procedure with the following reasoning. If the current position is really P1, then NX knows from the topological information that it can reach P2 with the information of E1. It actually follows E1 and visits P2. Notice here that NX makes neither the same trace for E1 nor the same location for P2. But it shows a similar trace for E1 and a nearby finding of P2. By performing the rehearsal procedure, it concludes that the current position is P1. The information saved in the exploration agenda for the direction from P1 toward P7 is deleted from the exploration agenda. The exploration agenda now has three elements, (from P4, direction toward P7), (from P7, direction toward P4), and (from P1, direction toward P8).

Then NX removes the last element from the exploration agenda and visits P8 through E8. It explores all areas and finish its exploration by traversing E23. We see that NX had a more difficult time in Figure 7d than figures Figure 7a-c. Difficulties occurred when NX traversed between P18 and P19 and when NX performed the hill-climbing search for P17. In all four figures, NX shows a slightly different exploration order. Since we consider random sensor error, the order of exploration is nondeterministic. NX continues its exploration until there is nothing in the exploration agenda and no more unexplored directions from the current place.

Once NX finishes its exploration completely, it repeatedly selects a place randomly and navigates to the place. While NX is navigating, it accumulates more metrical information to increase

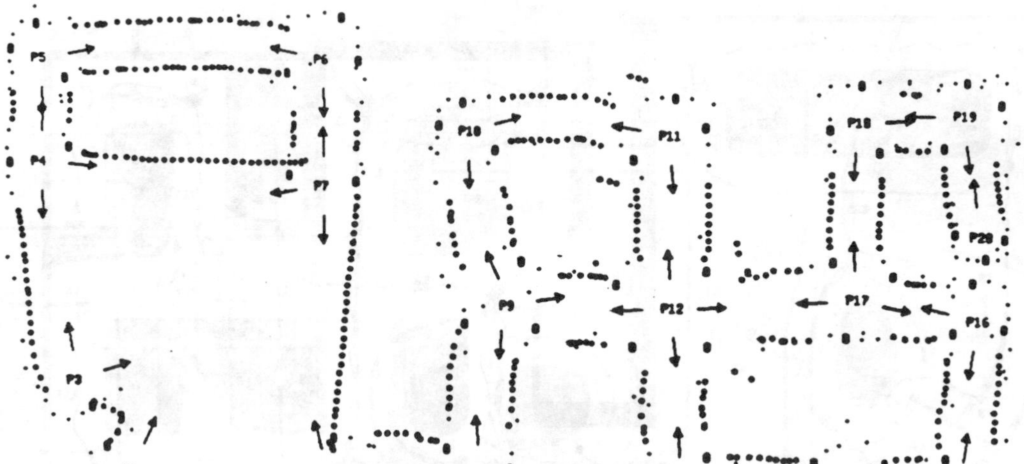

Figure 9: Metrical Map

the metrical accuracy of edges and places. NX also shows a robust self-orientation ability when it is dropped anywhere in an already-explored environment without any information about the current place. We present more exploration results of various environments in Figure 10.

6. SUMMARY

We believe that once there are errors in the real world, it is almost impossible in large-scale space for a mechanical robot to build a globally metrically accurate map through exploration in unknown unstructured environments. There are several important features in our method. Instead of using a global absolute coordinate frame, we use a successive layered model in which the topological model provides the basic structure. Procedural knowledge is used to find nodes and arcs for the topological model, and each component accumulates local metrical information. With our robot exploration and map-learning strategy, we solve the cumulative metrical error problem of traditional approaches.

When the robot is following a known edge from one node to another, it starts by using the hill-climbing algorithm to locate itself at the DP corresponding to the first node. It then follows the LCS associated with the arc and ends up somewhere in the neighborhood of the second place. Then the hill-climbing algorithm brings it to the DP corresponding to the second node, eliminating cumulative error.

Our matching process is also important. We use a loose matching process for metrical information and the rehearsal procedure for a topological matching process. The procedural knowledge for distinctive places and travel edges, the metrical matching process with looseness, and the topological matching process make our approach robust in the face of metrical errors. We have demonstrated success of our method with NX, and we plan to extend this work so that NX explores a wider variety of environments including dynamic worlds.

7. ACKNOWLEDGEMENTS

Support for this research is provided by NASA, under grant number NAG9-200.

8. REFERENCES

[R.A.Brooks, 1982] Solving the Find-Edge problem by Good Representation of Free Space, *Proc. AAAI-82*, pp 381-387

[R.A.Brooks, 1985] Visual Map Making for a Mobile Robot, *IEEE Proc. International Conference on Robotics and Automation*, pp 824-829

[R.A.Brooks, 1986] A Robust Layered Control system for a Mobile Robot. *IEEE Journal of Robotics and Automation* VOL RA-2 No.1, pp 14-23

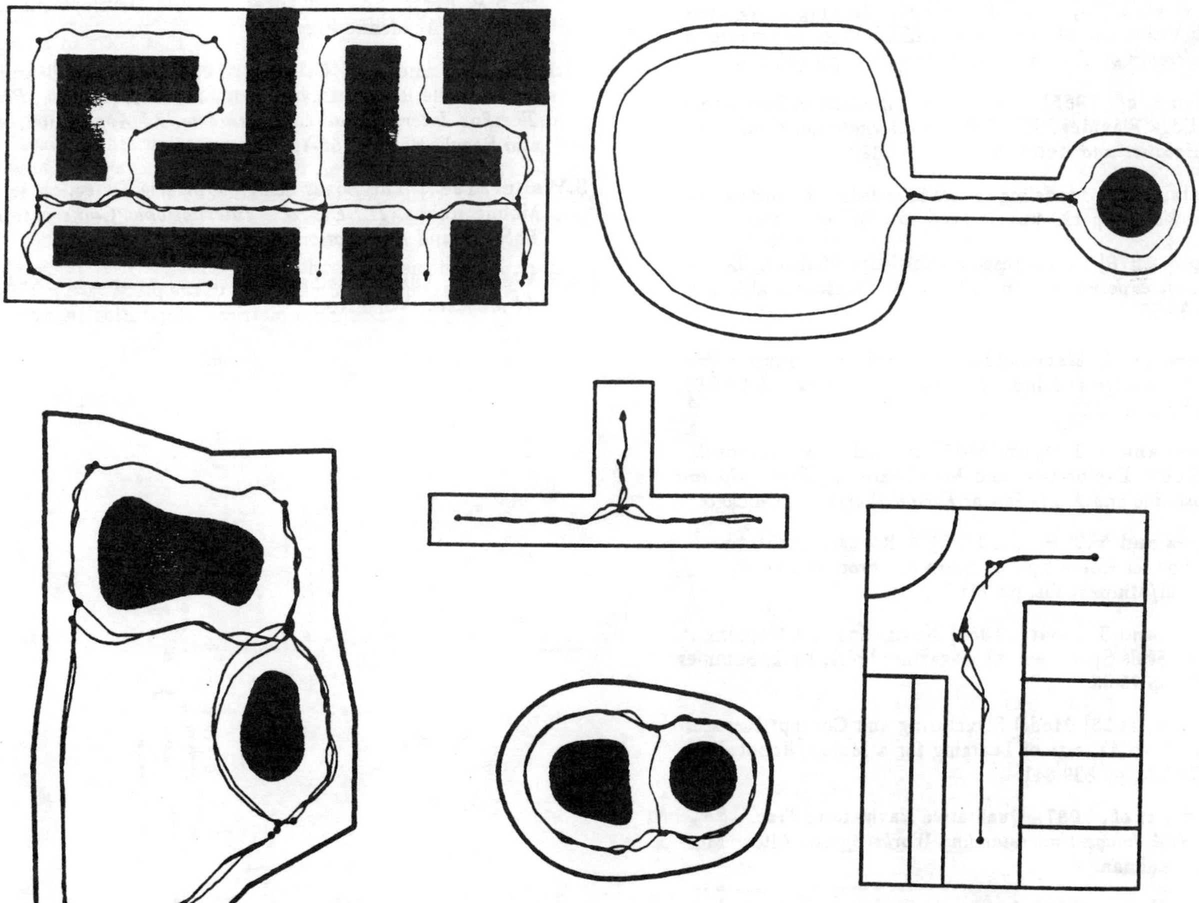

Figure 10: More Exploration Results

[R.Chatila and J.Laumond, 1985] Position Referencing and Consistent World Modeling for Mobile Robots, *IEEE Proc. International Conference on Robotics and Automation*, pp 138-170

[J.L.Crowley, 1985] Navigation for an Intelligent Mobile Robot, *IEEE Journal of Robotics and Automation*, Vol RA-1, No.1, pp 31-41

[M.Drumheller, 1985] Mobile Robot Localization Using Sonar, MIT AI-Lab TR AI-M-826

[A.Elfes, 1986] A Sonar-Based Mapping and Navigation System, *IEEE Proc. International conference of Robotics and Automation*. pp 1151-1156

[A.M.Flynn, 1985] Redundant Sensors for Mobile Robot Navigation, TR 859, MIT A.I. Lab.

[G.Giralt, 1983] *Mobile Robots, Robotics and Artificial Intelligence*, Edited by M.Brady L.A.Gerhardt and H.F.Dividson, NATO ASI series, pp 375-393

[R.Hickling and S.P.Marin, 1986] The Use of Ultrasonics for Gauging and Proximity Sensing in Air, *Journal of the Acoustical Soc. of America*, Vol 79, pp 1151-1159

[S.S.Iyengar *et al.*, 1985] Learned Navigation Edges for a Robot in unexplored Terrain. *IEEE Proc. the 2nd Conference on AI Application*, pp 148-155

[M.B.Kadonoff *et al.*, 1986] Arbitration of Multiple Control Strategies for Mobile Robots, *Proc. SPIE. Mobile Robots*. Cambridge MA.

[E.Koch *et al.*, 1985] Simulation of Path Planning for a system with Vision and Map Updating, *IEEE Proc. International Conference on Robotics and Automation*, pp 146-160

[D.T.Kuan *et al.*, 1985] Natural Decompostion of Free Space for Edge Planning. *IEEE Proc. International Conference on Robotics and Automation*. pp 178-183

[B.Kuipers, 1978] Modeling spatial knowledge. *Cognitive Science*. Vol 2. pp 129-153

[B.Kuipers, 1979] Commonsense knowledge of space: Learning from experience. *Proc. IJCAI-79*. Stanford. California, pp 499-501

[B.Kuipers, 1983] Modeling human knowledge of routes: Partial knowledge and individual variation. *Proc. AAAI-83*, Washington,D.C.

[B.Kuipers and Y.T.Byun, 1987] A Qualitative Approach to Robot Exploration and Map-Learning, *Proc. Spatial Reasoning and Multi-Sensor Fusion Workshop*, Chicago.

[B.Kuipers and Y.T.Byun, 1988] A Robust, Qualitative Method for Robot Spatial Learning, *Proc. AAAI-88*, St.Paul/Minneapolis, pp 774-779

[B.Kuipers and T.Levitt, 1988] Navigation and Mapping in Large-Scale Space, em AI magazine, Vol.9, No.2, Summer 1988, pp 25-43

[J.Laumond, 1983] Model Structuring and Concept Recognition: Two Aspects of Learning for a Mobile Robot *Proc. IJCAI-83*, pp 839-841

[T.S.Levitt *et al.*, 1987] Qualitative Navigation, *Proc. DARPA Image Understanding Workshop* Los Altos: Morgan Kaufmann

[T.Lozano-Perez, 1981] Automatic Planning of Manipulator Transfer Movements, *IEEE Trans. on Systems Man and Cybernetics*, pp 781-798

[K.Lynch, 1960] The Image of the City, Cambridge: MIT Press

[D.Miller, 1985] A Spatial Representation System for Mobile Robots, *IEEE Proc. International Conference on Robotics and Automation*, pp 122-128

[A.C.C.Meng, 1987] Free Space Modeling and Geometric Motion Planing Under Location Uncertainty, *Proc. Spatial Reasoning and Multi-Sensor Fusion Workshop*, Chicago.

[H.P.Moravec, 1988] Sensor Fusion in Certainty Grids for Mobile Robots, em AI magazine, Vol.9, No.2, Summer 1988, pp 61-74

[H.Moravec and A.Elfes, 1985] High Resolution Maps from Wide Angle Sonar, *IEEE Robotics and Automation*. pp 116-121

[B.J.Oommen *et al.*, 1987] Robot Navigation in Unknown Terrains Using Learned Visibility Graphs. Part 1: The Disjoint Convex Obstacle Case, *IEEE Journal of Robotics and Automation*, Vol.RA-3. No.6, December 1987, pp 672-681

[J.Piaget and B.Inhelder, 1967] The child's conception of space, New York: Norton

[N.S.V.Rao *et al.*, 1986] Robot Navigation in an Unexplored Terrain, *Journal of Robotic Systems*, Vol.3 No.4.

[R.S.Rosenberg and P.F.Rowat, 1981] Spatial Problems for a simulated Robot. *Proc. IJCAI-81*, pp 758-765

[A.W.Siegel and S.H. White, 1975] The Development of Spatial Representations of Large-Scale Environments. In H.W.Reese, Ed., Advances in Child Development and Behavior; Academic Press

[M.P.Turchan and A.K.C.Wong, 1985] Low Level Learning for a Mobile Robot: Environment Model Aquisition, *Proc. The 2nd International Conference on AI Application*, Miami Beach, FL. pp 156-161

[S.Walter, 1987] The Sonar Ring: Obstacle Detection for a Mobile Robot, *IEEE Proc. International Conference on Robotics and Automation*, pp 1574-1579

[C.R.Weisbin, 1987] Intelligent-Machine Research at CESAR, *AI Magazine* Vol.8 No.1

Chapter 4: Path Planning and Navigation

Recently, interest has grown in the design of algorithms for robotics applications such as motion planning, position estimation, and terrain acquisition. In particular, algorithms for robotic motion planning have provided a vast research area. Typically, this area deals with algorithms that plan paths through which robotic systems must navigate – either mobile robots or manipulators, from a source configuration to a destination configuration, subject to certain motion constraints. This chapter focuses on an algorithmic framework that provides the basis for robotic navigational algorithms in known and unknown terrains. Algorithms presented by various researchers try to solve sensor-based algorithms for mobile robotic navigation.

Schwartz and Sharir survey recent developments in motion planning and related geometric algorithms, focusing on exact algorithmic solutions to the motion-planning problem. Methods surveyed include the "projection" method, the "retraction" method, optimal motion planning, adaptive motion planning, and motion planning in the presence of moving obstacles. Also discussed are various results in computational geometry relevant to motion planning. Hart et al. discuss the A* algorithm, which has been used extensively in graph search problems.

Lozano-Pérez and Wesley describe a collision-avoidance algorithm, based on visibility graphs, that plans collision-free paths for a polyhedral object moving among other known polyhedral objects. The mechanism magnifies stationary objects while shrinking the moving object to a point, and finds trajectories by searching a network that indicates – for each vertex in the transformed obstacles – other vertices that can be reached safely. In his second paper in this chapter, Lozano-Pérez introduces the configuration space approach to path planning.

Mitchell examines the finding of optimal paths from one point to another, given a terrain "map" to work from, and discusses several models of shortest-path problems relevant to route planning for autonomous vehicles – including obstacle avoidance in the plane, shortest path on a surface, and weighted regions.

Khatib bases a real-time obstacle avoidance approach for manipulators and mobile robots on artificial potential field concepts. He presents an operational space formulation of manipulator control that provides the basis for this approach. Freund and Hoyer present a general approach for solving the path-finding problem, including collision avoidance in multirobotic systems, based on a systematical design method that uses a structure of the overall system (including the dynamics of all robots involved).

Krogh and Feng discuss a feedback algorithm for selecting subgoals that guide an autonomous mobile robot, using only local information about the locations of visible objects. The algorithm generates a sequence of visible, safe subgoals that are perused by a dynamic steering-control algorithm. Iyengar et al. present a method of robotic navigation that requires no prelearned model; the authors use a terrain model consisting of a spatial graph and a Voronoi diagram. As more information is gained, the obstacle's bounding polygons shrink, and polygons representing free space grow. In unknown terrain, the terrain model is not available for path-planning purposes. The moving body is equipped with a sensor system that obtains information about terrain in the immediate vicinity. Compared with results related to navigation in known terrains, relatively few results are reported that deal with navigation in unknown terrains. Navigation in unknown terrain is sensor based and differs significantly from navigation in known terrain. Doshi et al. discuss multilevel path planning.

Oommen et al. present an algorithm for robotic navigation in unexplored terrain arbitrarily populated with disjoint convex polygonal obstacles. Terrain information is integrated into an incrementally constructed visibility graph. Rao and Iyengar discuss several interesting algorithms on the terrain-acquisition problem; their chosen algorithm learns the environment, and planned paths improve as knowledge accumulates. However, paths occur due to navigational strategy.

Lumelsky et al. consider path planning for an automaton equipped only with tactile sensors moving in a two-dimensional space populated with unknown obstacles; the authors develop a worst-case lower bound on the pattern length generated by any algorithm operating within the accepted model's framework, and present an algorithm that tests for target reachability and guarantees reaching the target (if it can be reached).

A Survey of Motion Planning and Related Geometric Algorithms

J.T. Schwartz and M. Sharir
Courant Institute of Mathematical Sciences, New York University, New York, NY 10003, U.S.A.; and School of Mathematical Sciences, Tel-Aviv University, Ramat-Aviv, Tel-Aviv 69978, Israel

1. Introduction

This paper surveys recent developments in motion planning and related geometric algorithms, a theoretical research area that has grown rapidly in response to increasing industrial demand for automatic manufacturing systems which use robotic manipulators and sensory feedback devices, and, more significantly, in anticipation of a future generation of substantially more autonomous and intelligent robots. These future robots are expected to possess advanced capabilities of sensing, planning, and control, enabling them to gather knowledge about their environment, construct a symbolic world model of the environment, and use this model in planning and carrying out tasks set to them in high-level style by an application programmer.

Current research in theoretical robotics therefore aims to identify these basic capabilities that an autonomous intelligent robot system will need to advance understanding of the mathematical and algorithmic principles fundamental to these capabilities.

Among these capabilities, planning involves the use of an environment model to carry out significant parts of a robot's activities automatically. The aim is to allow the robot's user to specify a desired activity in very high-level general terms, and then have the system fill in the missing low-level details. For example, the user might specify the end product of some assembly process, and ask the system to construct a sequence of assembly substeps; or, at a less demanding level, to plan collision-free motions which pick up individual subparts of an object to be assembled, transport them to their assembly position, and insert them into their proper places.

Techniques for the automatic planning of robot motions have advanced substantially during the last several years. This area has shown itself to have significant mathematical content; tools drawn from classical geometry, topology, algebraic geometry, algebra, and combinatorics have all been used in it. This work relates closely to work in computational geometry, an area which has also progressed very rapidly during the last few years.

This survey concentrates on exact algorithmic solutions to the motion planning problem, and does not address other heuristic or approximating approaches which have also been recently developed, some of which may have significant practical advantages. These include works by Lozano-Pérez, Brooks, Mason and Taylor (cf. [4–6, 30, 31]) of the MIT school, to which researchers at many other places are beginning to contribute.

"A Survey of Motion Planning and Related Geometric Algorithms" by J.T. Schwartz and M. Sharir. Reprinted from *Artificial Intelligence Journal*, Vol. 37, 1988, pages 157-169.
Copyright © 1988 Elsevier Science Publishers B.V., reprinted with permission.

2. Statement of the Problem

In its simplest form, the motion planning problem can be defined as follows. Let B be a robot system consisting of a collection of rigid subparts (some of which may be attached to each other at certain joints while others may move independently) having a total of k degrees of freedom, and suppose that B is free to move in a two- or three-dimensional space V amidst a collection of obstacles whose geometry is known to the robot system. The *motion planning problem* for B is: Given an initial position Z_1 and a desired final position Z_2 of B, determine whether there exists a continuous obstacle-avoiding motion of B from Z_1 to Z_2, and if so plan such a motion.

This problem has been studied in many recent papers (cf. [4–6, 14, 15, 17, 19, 21–24, 26–28, 30–32, 37, 39–41, 45, 49–52, 60, 63, 68, 70]). It is equivalent to the problem of calculating the path-connected components of the (k-dimensional) space FP of all *free positions* of B (i.e. the set of positions of B in which B does not contact any obstacle), and is therefore a problem in "computational topology". In general FP is a high-dimensional space with irregular boundaries, and is thus hard to calculate efficiently.

This standard motion planning problem can be extended and generalized in many possible ways. For example, if the geometry of the environment is not fully known to the robot system, one must employ "exploratory" approach in which plan generation is tightly updated to gathering of data on the environment and to dynamic updating of a world model. Another interesting extension of the motion planning problem is to the case in which the environment contains objects moving in some known and predictable manner. Although this problem has been little studied, the few results obtained so far seem to indicate that it is inherently harder than the static problem.

All the problem variants mentioned so far aim to determine whether a collision-free path exists between two specified system positions, and, if so, to produce such a path. A further issue is to produce a path which satisfies some criterion of optimality. For example, if a mobile robot is approximated as a single moving point, one might want to find the shortest Euclidean path between the initial and final system positions. In more complex situations the notion of optimal motion is less clearly defined, and has as yet been little studied.

Studies of the motion planning problem tend to make heavy use of many algorithmic techniques in computational geometry. Various motion-planning-related problems in computational geometry will also be reviewed.

3. Motion Planning in Static and Known Environments

As above, let B be a moving robot system, k be its number of degrees of freedom, V denote the two- or three-dimensional space in which B is free to move, and FP denote the space of free positions of B, as defined above. The space FP is determined by the collection of algebraic inequalities which express the fact that at position Z the system B avoids collision with any of the obstacles present in its workspace. We will denote by n the number of inequalities needed to define FP, and call it the "geometric (or combinatorial) complexity" of the given instance of the motion planning problem. As noted, we make the reasonable assumption that the parameters describing the degrees of freedom of B can be chosen in such a way that each of these inequalities is algebraic. Indeed, the group of motions (involving various combinations of translations and rotations) available to a given robot can ordinarily be given algebraic representation, and the system B and its environment V can typically be modeled as objects bounded by a collection of algebraic surfaces (e.g., polyhedral, quadratic, or spline-based).

3.1. The general motion planning problem

Assuming then that *FP* is an algebraic or semi-algebraic set in E^k, Schwartz and Sharir [50] show that the motion planning problem can be solved in time polynomial in the number *n* of algebraic constraints defining *FP* and in their maximal degree, but double exponential in *k*. The general procedure described uses a decomposition technique due to Collins [9] and originally applied to Tarski's theory of real closed fields. Though hopelessly inefficient in practical terms, this result nevertheless serves to calibrate the computational complexity of the motion planning problem.

3.2. Lower bounds

The result just cited suggests that motion planning becomes harder rapidly as the number *k* of degrees of freedom increases; this conjecture has in fact been proved for various model "robot" systems. Specifically, Reif [45] proved that motion planning is PSPACE-hard for a certain 3-D system involving arbitrarily many links and moving through a complex system of narrow tunnels. Since then PSPACE-hardness has been established for simpler moving systems, including 2-D systems of mechanical linkages (Hopcroft, Joseph, and Whitesides [15]), a system of 2-D independent rectangular blocks sliding inside a rectangular box (Hopcroft, Schwartz and Sharir [17]), and a single 2-D arm with many links moving through a 2-D polygonal space (Joseph and Plantinga [21]). Several weaker results establishing NP-hardness for still simpler systems have also been obtained.

3.3. The "projection method"

In spite of these negative worst-case results, algorithms of varying levels of efficiency for planning the motions of various single robot systems have been developed. These involve several general approaches to the design of motion planning algorithms. The first such approach, known as the *projection method*, uses ideas similar to those appearing in the Collins decomposition procedure described above. One fixes some of the problem's degrees of freedom (for the sake of exposition, suppose just one parameter *y* is fixed, and let $\bar{x}$ be the remaining parameters); then one solves the resulting restricted $(k-1)$-dimensional motion planning problem. This subproblem solution must be such as to yield a discrete combinatorial representation of the restricted free configuration space (essentially, a cross-section of the entire space *FP*) that changes only at a finite collection of "critical" values of the final parameter *y*. These critical values of *y* are then calculated; they partition the entire space *FP* into connected cells, and by calculating relationships of adjacency between these cells one can describe the connectivity of *FP* by a discrete *connectivity graph CG*. This graph has the aforesaid cells as vertices, and has edges which represent relationships of cell adjacency in *FP*. The connected components of *FP* correspond in a one-to-one manner to the connected components of *CG*, reducing the problem to a discrete path searching problem in *CG*.

This relatively straightforward technique was applied in a series of papers by Schwartz and Sharir on the "piano movers" problem, to yield polynomial-time motion planning algorithms for various specific systems, including a rigid polygonal object moving in 2-D polygonal space [49], two or three independent discs moving in coordinated fashion in 2-D polygonal space [51], certain types of multi-arm linkages moving in 2-D polygonal space [60], and a rod moving in 3-D polyhedral space [52]. These initial solutions were coarse and not very efficient; subsequent refinements have improved substantially their efficiency. For example, Leven and Sharir [26] obtained an $O(n^2 \log n)$ algorithm for the case of a line segment (a "rod") moving in 2-D polygonal space (improving the $O(n^5)$ algorithm of [49]); the Leven–Sharir result was later shown to be nearly optimal.

3.4. The "retraction method" and other approaches to the motion planning problem

Several other important algorithmic motion planning techniques were developed subsequent to the series of papers just reported. The so-called *retraction method* proceeds by retracting the configuration space *FP* onto a lower-dimensional (usually a one-dimensional) subspace *N*, so that two system positions in *FP* lie in the same connected component of *FP* if and only if their retractions to *N* lie in the same connected component of *N*. This reduces the dimension of the problem, and if *N* is one-dimensional the problem becomes one of searching a graph.

O'Dunlaing and Yap [39] introduced this retraction technique in the simple case of a disc moving in 2-D polygonal space. Here the subspace *N* can be taken to be the Voronoi diagram associated with the set of given polygonal obstacles. Their technique yields an $O(n \log n)$ motion planning algorithm. After this first paper, O'Dunlaing, Sharir and Yap [40, 41] generalized the retraction approach to the case of a rod moving in 2-D polygonal space by defining a variant Voronoi diagram in the 3-D configuration space *FP* of the rod, and by retracting onto this diagram. This achieves $O(n^2 \log n \log^* n)$ performance in this case (a substantial improvement on the naive projection technique first applied to this case, but nevertheless a result shortly afterward superceded by Leven and Sharir [26]).

A similar retraction approach was used by Leven and Sharir [27] to obtain an $O(n \log n)$ algorithm for planning the purely translational motion of a simple convex object amidst polygonal barriers. This last result uses another generalized variant of Voronoi diagram. (A somewhat simpler $O(n \log^2 n)$ algorithm, based on a general technique introduced by Lozano-Pérez and Wesley [32], was previously obtained by Kedem and Sharir [23] (cf. also [22]); this last result exploits an interesting topological property of intersecting planar Jordan curves.)

Recently Sifrony and Sharir [63] devised another retraction-based algorithm for the motion of a rod in 2-D polygonal space. The retraction used maps the rod's free configuration space *FP* onto a network containing all edges of the boundary of *FP*, plus some additional arcs which connect particular vertices of *FP*. They obtain an $O(n^2 \log n)$ algorithm which has the advantage that it runs much faster than that of Leven and Sharir [26] if the obstacles do not lie close to one another.

Hybrid techniques are also appropriate for certain cases of motion planning. For example, in an analysis of the motion planning problem for a convex polygonal object moving in 2-D polygonal space, Kedem and Sharir [24] obtained an $O(n^2\beta(n)\log n)$ motion planning algorithm (where $\beta(n)$ is a very slowly growing function of n), using a hybrid approach which involves projection of *FP* onto a 2-D space in which the orientation θ of the object is fixed, followed by retraction of the 2-D space roughly onto its boundary. This result makes use of a combinatorial result of Leven and Sharir [28].

4. Variants of the Motion Planning Problem

4.1. Optimal motion planning

The only optimal motion planning which has been studied extensively thus far is that in which the moving system is represented as a single point, in which case one aims to calculate the shortest Euclidean path connecting initial and final system positions, given that specified obstacles must be avoided. Most existing work on this problem assumes that the obstacles are either polygonal (in 2-space) or polyhedral (in 3-space).

The 2-D case is considerably simpler than the 3-D case. When the free space V in 2-D is bounded by n straight edges, it is easy to calculate the desired shortest path in time $O(n^2 \log n)$. This is done by constructing a *visibility graph* VG whose edges connect all pairs of boundary corners of V which are visible from each other through V, and then by searching for a shortest path through VG (see [60] for a sketch of this idea). This procedure was improved to $O(n^2)$ by Asano et al. [1], by Welzl [69], and by Reif and Storer [47], using a cleverer method for constructing VG. Their quadratic-time bound has been improved in certain special cases. However, it is not known whether shortest paths for a general polygonal space V can be calculated in subquadratic time. Among the special cases allowing more efficient treatment the most important is that of calculating shortest paths inside a simple polygon P. Lee and Preparata [25] gave a linear-time algorithm for this case, assuming that a triangulation of P is given in advance. (As a matter of fact, a recent algorithm of Tarjan and van Wyk shows that triangulation in $O(n \log \log n)$ time is possible.) The Preparata–Lee result was recently extended by Guibas et al. [12], who gave a linear-time algorithm which calculates all shortest paths from a fixed source point to all vertices of P.

Other results on 2-D shortest paths include an $O(n \log n)$ algorithm for *rectilinear* shortest paths avoiding n rectilinear disjoint rectangles [48]; an $O(n^2 \log n)$ algorithm for Euclidean shortest motion of a circular disc in 2-D polygonal space [7]; algorithms for cases in which the obstacles consist of a small number of disjoint convex regions [47]; algorithms for the "weighted region" case (in which the plane is partitioned into polygonal regions and the path has a different multiplicative cost weight when it passes through each of these regions) [35]; and some other special cases.

The 3-D polyhedral case is substantially more difficult. To date, only exponential-time algorithms for the general polyhedral case have been developed [47, 62], and it is not yet known whether the problem is really intractable. However, more efficient algorithms exist in certain special cases. The simplest case is that in which we must calculate the shortest path between two points lying on the surface of a convex polyhedron; it was shown that this can be done in $O(n^2 \log n)$ time (see [30, 62]). Generalizations of this result include algorithms for shortest paths along a (not necessarily convex) polyhedral surface [36], algorithms for shortest paths in 3-space which must avoid a fixed number of convex polyhedra [3, 57], and an approximating pseudo-polynomial scheme for the general case [43].

4.2. Adaptive and exploratory motion planning

If the environment is not known to the robot system a priori, but the system is equipped with sensory devices, motion planning assumes a more "exploratory" character. If only tactile (or proximity) sensing is available, then a plausible strategy might be to move along a straight line (in physical or configuration space) directly to the target position, and when an obstacle is reached, to follow its boundary until the original straight line of motion is reached again [33]. If vision is also available, then other possibilities need to be considered, e.g. the system could obtain partial information about its environment by viewing it from the present position, and then "explore" it to gain progressively more information until the desired motion can be fully planned. However, problems of this sort have hardly begun to be investigated.

Even when the environment is fully known to the system, other interesting issues arise if the environment is changing. For example, when some of the objects in the robot's environment are picked up by the robot and moved to a different position, one wants fast techniques for incremental updating of the environment model and the data structures used for motion planning. Moreover, whenever the robot grasps an object to move it, robot plus grasped object become a new moving system and may require a different motion planning algorithm, but one whose relationship to motions of the robot alone needs to be investigated. Adaptive motion planning problems of this kind have hardly been studied as yet.

4.3. Motion planning in the presence of moving obstacles

Interesting generalizations of the motion planning problem arise when some of the obstacles in the robot's environment are assumed to be moving along known trajectories. In this case the robot's goal will be to "dodge" the moving obstacles while moving to its target position. In this "dynamic" motion planning problem, it is reasonable to assume some limit on the robot's velocity and/or acceleration. Two initial studies of this problem by Reif and Sharir [46] and by Sutner and Maass [65] indicate that the problem of avoiding moving obstacles is substantially harder than the corresponding static problem. By using time-related configuration changes to encode Turing machine states, they show that the problem is PSPACE-hard even for systems with a small and fixed number of degrees of freedom. However, polynomial-time algorithms are available in a few particularly simply special cases.

5. Results in Computational Geometry Relevant to Motion Planning

The various studies of motion planning described above make extensive use of efficient algorithms for the geometric subproblems which they involve, for which reason motion planning has encouraged research in computational geometry. Problems in computational geometry whose solutions apply to robotic motion planning are described in the following subsections.

5.1. Intersection detection

The problem here is to detect intersections and to compute shortest distances, e.g. between moving subparts of a robot system and stationary or moving obstacles. Simplifications which have been studied include that in which all objects involved are circular discs (in the 2-D case) or spheres (in the 3-D case). In a study of the 2-D case of this problem, Sharir [55] developed a generalization of Voronoi diagrams for a set of (possibly intersecting) circles, and used this diagram to detect intersections and computing shortest distances between discs in time $O(n \log^2 n)$ (an alternative approach to this appears in [20]). Hopcroft, Schwartz and Sharir [16] present an algorithm for detecting intersections among n 3-D spheres which also runs in time $O(n \log^2 n)$. However, this algorithm does not adapt in any obvious way to allow proximity calculation or other significant problem variants.

Other intersection detection algorithms appearing in the computational geometry literature involve rectilinear objects and use multi-dimensional searching techniques for achieving high efficiency (see [34] for a survey of these techniques).

5.2. Generalized Voronoi diagrams

The notion of Voronoi diagram has proven to be a useful tool in the solution of many motion planning problems. We have also mentioned the use of various variants of Voronoi diagram in the retraction-based algorithms for planning the motion of a disc [39], or of a rod [40, 41], or the translational motion of a convex object [27], and in the intersection detection algorithm for discs mentioned above [55]. The papers just cited, and some related works [29, 71] describe the analysis of these diagrams and the design of efficient algorithms for their calculations.

5.3. Davenport–Schinzel sequences

Davenport–Schinzel sequences are combinatorial sequences of n symbols which do not contain certain forbidden subsequences of alternating symbols. Sequences of this sort appear in studies of efficient techniques for calculating the lower envelope of a set of n continuous functions, if it is assumed that the graphs of any two functions in the set can intersect in some fixed number of points at most. These sequences, whose study was initiated in [10, 11], have proved to be powerful tools for analysis (and design) of a variety of geometric algorithms, many of which are useful for motion planning.

More specifically, an (n, s) Davenport–Schinzel sequence is defined to be a sequence U composed of n symbols, such that (i) nor two adjacent elements of U are equal, and (ii) there do not exist $s+2$ indices $i_1 < i_2 < \cdots < i_{s+2}$ such that $u_{i_1} = u_{i_3} = u_{i_5} = \cdots = a$, $u_{i_2} = u_{i_4} = u_{i_6} = \cdots = b$, with $a \neq b$. Let $\lambda_s(n)$ denote the maximal length of an (n, s) Davenport–Schinzel sequence. Early study by Szemeredi [66] of the maximum possible length of such sequences shows that $\lambda_s(n) \leq C_s n \log^* n$, where C_s is a constant depending on s. Improving on this result, Hart and Sharir [13] proved that $\lambda_3(n) = \Theta(n\alpha(n))$ where $\alpha(n)$ is the very slowly growing inverse of the Ackermann function. In [56, 59] Sharir established the bounds

$$\lambda_s(n) = O(n\alpha(n)^{O(\alpha(n)^{s-3})})$$

and

$$\lambda_s(n) = \Omega(n\alpha^{\lfloor (s-1)/2 \rfloor}(n))$$

for $s > 3$. These results show that, in practical terms, $\lambda_s(n)$ is an almost linear function of n (for any fixed s).

Recently, numerous applications of these sequences to motion planning have been found. These include:

(i) an upper bound of $O(kn\lambda_6(kn))$ on the number of simultaneous triple contacts of a convex k-gon translating and rotating in 2-D polygonal space bounded by n edges [28]; an extension of this result was used to produce an $O(kn\lambda_6(kn)\log kn)$ motion planning algorithm for a moving convex k-gon in such a 2-D space [24];

(ii) an $O(mn\alpha(mn)\log m \log n)$ algorithm for separating two interlocking simple polygons by a sequence of translations [44], where it assumed that the polygons have m and n sides respectively;

(iii) an $O(n^2\lambda_{10}(n)\log n)$ algorithm for finding the shortest Euclidean path between two points in 3-space avoiding the interior of two disjoint convex polyhedra having n faces altogether [3].

Other applications are found in [2, 8, 13, 41, 61].

5.4. Topological results related to motion planning

Motion planning is equivalent to the topological problem of calculating the connected components of semi-algebraic varieties in E^k (namely free configuration spaces of robot systems). It is therefore of interest to study topological properties of such varieties which have close relationships to motion planning. A result of this kind by Hopcroft and Wilfong [19], which applies techniques drawn from homology theory, shows that when an object A moves in the presence of just a single (connected) planar obstacle B, then if collision-free motion of A is possible between two positions in which it makes contact with B, then A can move between these two positions so that it always stays in contact with B.

REFERENCES

1. Asano, T., Asano, T., Guibas, L., Hershberger, J. and Imai, H., Visibility polygon search and Euclidean shortest paths, in: *Proceedings 26th Symposium on Foundations of Computer Science* (1985) 155–164.
2. Atallah, M., Dynamic computational geometry, in: *Proceedings 24th Symposium on Foundations of Computer Science* (1983) 92–99.
3. Baltsan, A. and Sharir, M., On shortest paths between two convex polyhedra, *J. ACM* **35** (1988) 267–287.
4. Brooks, R.A., Solving the find-path problem by good representation of free space, *IEEE Trans. Syst. Man Cybern.* **13** (1983) 190–197.

5. Brooks, R.A., Planning collision-free motions for pick-and-place operations, *Int. J. Rob. Res.* **2** (4) (1983) 19–40.
6. Brooks, R.A. and Lozano-Pérez, T., A subdivision algorithm in configuration space for findpath with rotation, AI Memo 684, MIT, Cambridge, MA (1982).
7. Chew, L.P., Planning the shortest path for a disc in $O(n^2 \log n)$ time, in: *Proceedings ACM Symposium on Computational Geometry* (1985) 214–223.
8. Cole, R. and Sharir, M., Visibility problems for polyhedral terrains, *J. Symbolic Comput.* (to appear).
9. Collins, G.E., Quantifier elimination for real closed fields by cylindrical algebraic decomposition, in: *Proceedings Second GI Conference on Automata Theory and Formal Languages*, Lecture Notes in Computer Science **33** (Springer, New York, 1975) 134–183.
10. Davenport, H., A combinatorial problem connected with differential equations, II, *Acta Arithmetica* **17** (1971) 363–372.
11. Davenport, H. and Schinzel, A., A combinatorial problem connected with differential equations, *Am. J. Math.* **87** (1965) 684–694.
12. Guibas, L., Hershberger, J., Leven, D., Sharir, M. and Tarjan, R.E., Linear time algorithms for shortest path and visibility problems inside triangulated simple polygons, *Algorithmica* **2** (1987) 209–233.
13. Hart, S. and Sharir, M., Nonlinearity of Davenport–Schinzel sequences and of generalized path compression schemes, *Combinatorica* **6** (1986) 151–177.
14. Hopcroft, J.E., Joseph, D.A. and Whitesides, S.H., On the movement of robot arms in 2-dimensional bounded regions, *SIAM J. Comput.* **14** (1985) 315–333.
15. Hopcroft, J.E., Joseph, D.A. and Whitesides, S.H., Movement problems for 2-dimensional linkages, *SIAM J. Comput.* **13** (1984) 610–629.
16. Hopcroft, J.E., Schwartz, J.T. and Sharir, M., On the complexity of motion planning for multiple independent objects; PSPACE hardness of the 'warehouseman's problem', *Int. J. Rob. Res.* **3** (4) (1984) 76–88.
17. Hopcroft, J.E., Schwartz, J.T. and Sharir, M., Efficient detection of intersections among spheres, *Rob. Res.* **2** (4) (1983) 77–80.
18. Hopcroft, J.E., Schwartz, J.T. and Sharir, M. (Eds.), *Planning, Geometry and Complexity of Robot Motion* (Ablex, Norwood, NJ, 1987).
19. Hopcroft, J.E. and Wilfong, G., On the motion of objects in contact, Tech. Rept. 84-602, Computer Science Department, Cornell University, Ithaca, NY (1984).
20. Imai, H., Iri, M. and Murota, K., Voronoi diagram in the Laguerre geometry and its applications, Tech. Rept. RMI 83-02, Department of Mathematical Engineering and Instrumentation Physics, University of Tokyo (1983).
21. Joseph, D.A. and Plantinga, W.H., On the complexity of reachability and motion planning questions, in: *Proceedings ACM Symposium on Computational Geometry* (1985) 62–66.
22. Kedem, K., Livne, R., Pach, J. and Sharir, M., On the union of Jordan regions and collision-free translational motion amidst polygonal obstacles, *Discrete Comput. Geometry* **1** (1986) 59–71.
23. Kedem, K. and Sharir, M., An efficient algorithm for planning collision-free translational motion of a convex polygonal object in 2-dimensional space amidst polygonal obstacles, in: *Proceedings ACM Symposium on Computational Geometry* (1985) 75–80.
24. Kedem, K. and Sharir, M., An efficient motion-planning algorithm for a convex polygonal object in two-dimensional polygonal space, *Discrete Comput. Geometry* (to appear).
25. Lee, D.T. and Preparata, F.P., Euclidean shortest paths in the presence of rectilinear barriers, *Networks* **14** (1984) 393–410.
26. Leven, D. and Sharir, M., An efficient and simple motion planning algorithm for a ladder moving in two-dimensional space amidst polygonal barriers, *J. Algorithms* **8** (1987) 192–215.
27. Leven, D. and Sharir, M., Planning a purely translational motion for a convex object in two-dimensional space using generalized Voronoi diagrams, *Discrete Comput. Geometry* **2** (1987) 9–31.
28. Leven, D. and Sharir, M., On the number of critical free contacts of a convex polygonal object moving in 2-D polygonal space, *Discrete Comput. Geometry* **2** (1987) 255–270.
29. Leven, D. and Sharir, M., Intersection and proximity problems and Voronoi diagrams, in: Schwartz, J. and Yap, C. (Eds.), *Advances in Robotics* **I** (Erlbaum, Hillsdale, NJ, 1987) 187–228.
30. Lozano-Pérez, T., Spatial planning: A configuration space approach, *IEEE Trans. Comput.* **32** (2) (1983) 108–119.
31. Lozano-Pérez, T., A simple motion planning algorithm for general robot manipulators, Tech. Rept., AI Lab, MIT, Cambridge, MA (1986).
32. Lozano-Pérez, T. and Wesley, M., An algorithm for planning collision-free paths among polyhedral obstacles, *Comm. ACM* **22** (1979) 560–570.
33. Lumelsky, V.J. and Stepanov, A., Path planning strategies for a traveling automaton in an environment with uncertainty, Tech. Rept. 8504, Center for Systems Science, Yale University, New Haven, CT (1985).
34. Mehlhorn, K., *Data Structures and Algorithms, III: Multidimensional Searching and Computational Geometry* (Springer, New York, 1984).
35. Mitchell, J., Mount, D. and Papadimitriou, C., The discrete geodesic problem, Tech. Rept., Department of Operations Research, Stanford University, Stanford, CA (1985).

36. Mitchell, J. and Papadimitriou, C., The weighted region problem, Tech. Rept., Department of Operations Research, Stanford University, Stanford, CA (1985).
37. Moravec, H.P., Robot rover visual navigation, Ph.D. Dissertation, Stanford University, Stanford, CA (1981).
38. Mount, D.M., On finding shortest paths on convex polyhedra, Tech. Rept., Computer Science Department, University of Maryland, College Park, MD (1984).
39. O'Dunlaing, C. and Yap, C., A 'retraction' method for planning the motion of a disc, *J. Algorithms* **6** (1985) 104–111.
40. O'Dunlaing, C., Sharir, M. and Yap, C., Generalized Voronoi diagrams for a ladder, I: Topological analysis, *Comm. Pure Appl. Math.* **39** (1986) 423–483.
41. O'Dunlaing, C., Sharir, M. and Yap, C., Generalized Voronoi diagrams for a ladder, II: Efficient construction of the diagram, *Algorithmica* **2** (1987) 27–59.
42. O'Rourke, J., Lower bounds on moving a ladder, Tech. Rept. 85/20, Department of EECS, Johns Hopkins University, Baltimore, MD (1985).
43. Papadimitriou, C., An algorithm for shortest path motion in three dimensions, *Inf. Proc. Lett.* **20** (1985) 259–263.
44. Pollack, R., Sharir, M. and Sifrony, S., Separating two simple polygons by a sequence of translations, *Discrete Comput. Geometry* **3** (1988) 123–136.
45. Reif, J.H., Complexity of the mover's problem and generalizations, in: *Proceedings 20th IEEE Symposium on Foundations of Computer Science* (1979) 421–427.
46. Reif, J.H. and Sharir, M., Motion planning in the presence of moving obstacles, Tech. Rept. 39/85, The Eskenasy Institute of Computer Science, Tel-Aviv University (1985).
47. Reif, J.H. and Storer, J.A., Shortest paths in Eucidean space with polyhedral obstacles, Tech. Rept. CS-85-121, Computer Science Department, Brandeis University, Waltham, MA (1985).
48. de Rezende, P.J., Lee, D.T. and Wu, Y.F., Rectilinear shortest paths with retangular barriers, in: *Proceedings ACM Symposium on Computational Geomedtry* (1985) 204–213.
49. Schwartz, J.T. and Sharir, M., On the piano movers' problem, I: The case of a two-dimensional rigid polygonal body moving amidst polygonal barriers, *Comm. Pure Appl. Math.* **36** (1983) 345–398.
50. Schwartz, J.T. and Sharir, M., On the piano movers' problem, II: General techniques for computing topological properties of real algebraic manifolds, *Adv. Appl. Math.* **4** (1983) 298–351.
51. Schwartz, J.T. and Sharir, M., On the piano movers' problem, III: Coordinating the motion of several independent bodies: The special case of circular bodies moving amidst polygonal barriers, *Rob. Res.* **2** (3) (1983) 46–75.
52. Schwartz, J.T. and Sharir, M., On the piano movers' problem, V: The case of a rod moving in three-dimensional space amidst polyhedral obstacles, *Comm. Pure Appl. Math.* **37** (1984) 815–848.
53. Schwartz, J.T. and Sharir, M., Efficient motion planning algorithms in environments of bounded local complexity, Tech. Rept. 164, Computer Science Department, Courant Institute, New York (1985).
54. Schwartz, J.T. and Sharir, M., Mathematical problems and training in robotics, *Notices Am. Math. Soc.* **30** (1983) 478–481.
55. Sharir, M., Intersection and closest-pair problems for a set of planar discs, *SIAM J. Comput.* **14** (1985) 448–468.
56. Sharir, M., Almost linear upper bounds on the length of general Davenport-Schinzel sequences, *Combinatorica* **7** (1987) 131–143.
57. Sharir, M., On shortest paths amidst convex polyhedra, *SIAM J. Comput.* **16** (1987) 561–572.
58. Sharir, M., On the two-dimensional Davenport Schinzel problem, Tech. Rept. 193, Computer Science Department, Courant Institute, New York (1985).
59. Sharir, M., Improved lower bounds on the length of Davenport Schinzel sequences, *Combinatorica* **8** (1988) 117–124.
60. Sharir, M. and Ariel-Sheffi, E., On the piano movers' problem, IV: Various decomposable two-dimensional motion planning problems, *Comm. Pure Appl. Math.* **37** (1984) 479–493.
61. Sharir, M. and Livne, R., On minima of functions, intersection patterns of curves, and Davenport-Schinzel sequences, in: *Proceedings 26th Symposium Foundations of Computer Science* (1985) 312–320.
62. Sharir, M. and Schorr, A., On shortest paths in polyhedral spaces, *SIAM J. Comput.* **15** (1986) 193–215.
63. Sifrony, S. and Sharir, M., An efficient motion planning algorithm for a rod moving in two-dimensional polygonal space, *Algorithmica* **2** (1987) 367–402.
64. Spirakis, P. and Yap, C.K., Strong NP-hardness of moving many discs, *Inf. Proc. Lett.* **19** (1984) 55–59.
65. Sutner, K. and Maass, W., Motion planning among time dependent obstacles, Manuscript (1985).

66. Szemeredi, E., On a problem by Davenport and Schinzel, *Acta Arithmetica* **25** (1974) 213–224.
67. Tarjan, R.E. and van Wyk, C., An O(n log log n) time algorithm for triangulating simple polygons, *SIAM J. Comput.* **17** (1988).
68. Udupa, S., Collision detection and avoidance in computer controlled manipulators, Ph.D. Dissertation, California Institute of Technology, Pasadena, CA (1977).
69. Welzl, E., Constructing the visibility graph for n line segments in O(n^2) time, *Inf. Proc. Lett.* **20** (1985) 167–172.
70. Yap, C.K., Coordinating the motion of several discs, Tech. Rept. 105, Computer Science Department, Courant Institute, New York (1984).
71. Yap, C.K., An O(n log n) algorithm for the Voronoi diagram of a set of simple curve segments, Tech. Rept. 161, Computer Science Department, Courant Institute, New York (1985).

A Formal Basis for the Heuristic Determination of Minimum Cost Paths

PETER E. HART, MEMBER, IEEE, NILS J. NILSSON, MEMBER, IEEE, AND BERTRAM RAPHAEL

Abstract—Although the problem of determining the minimum cost path through a graph arises naturally in a number of interesting applications, there has been no underlying theory to guide the development of efficient search procedures. Moreover, there is no adequate conceptual framework within which the various ad hoc search strategies proposed to date can be compared. This paper describes how heuristic information from the problem domain can be incorporated into a formal mathematical theory of graph searching and demonstrates an optimality property of a class of search strategies.

I. Introduction

A. *The Problem of Finding Paths Through Graphs*

MANY PROBLEMS of engineering and scientific importance can be related to the general problem of finding a path through a graph. Examples of such problems include routing of telephone traffic, navigation through a maze, layout of printed circuit boards, and mechanical theorem-proving and problem-solving. These problems have usually been approached in one of two ways, which we shall call the *mathematical approach* and the *heuristic approach.*

1) The mathematical approach typically deals with the properties of abstract graphs and with algorithms that prescribe an orderly examination of nodes of a graph to establish a minimum cost path. For example, Pollock and Wiebenson[1] review several algorithms which are guaranteed to find such a path for any graph. Busacker and Saaty[2] also discuss several algorithms, one of which uses the concept of dynamic programming.[3] The mathematical approach is generally more concerned with the ultimate achievement of solutions than it is with the computational feasibility of the algorithms developed.

2) The heuristic approach typically uses special knowledge about the domain of the problem being represented by a graph to improve the computational efficiency of solutions to particular graph-searching problems. For example, Gelernter's[4] program used Euclidean diagrams to direct the search for geometric proofs. Samuel[5] and others have used ad hoc characteristics of particular games to reduce

Manuscript received November 24, 1967.

The authors are with the Artificial Intelligence Group of the Applied Physics Laboratory, Stanford Research Institute, Menlo Park, Calif.

Reprinted from *IEEE Transactions on Systems, Man, and Cybernetics*, Vol. SSC-4, No. 2, July 1968, pages 100-107.
Copyright © 1968 by The Institute of Electrical and Electronics Engineers, Inc. All rights reserved.

the "look-ahead" effort in searching game trees. Procedures developed via the heuristic approach generally have not been able to guarantee that minimum cost solution paths will always be found.

This paper draws together the above two approaches by describing how information from a problem domain can be incorporated in a formal mathematical approach to a graph analysis problem. It also presents a general algorithm which prescribes how to use such information to find a minimum cost path through a graph. Finally, it proves, under mild assumptions, that this algorithm is optimal in the sense that it examines the smallest number of nodes necessary to guarantee a minimum cost solution.

The following is a typical illustration of the sort of problem to which our results are applicable. Imagine a set of cities with roads connecting certain pairs of them. Suppose we desire a technique for discovering a sequence of cities on the shortest route from a specified start to a specified goal city. Our algorithm prescribes how to use special knowledge—e.g., the knowledge that the shortest road route between any pair of cities cannot be less than the airline distance between them—in order to reduce the total number of cities that need to be considered.

First, we must make some preliminary statements and definitions about graphs and search algorithms.

B. Some Definitions About Graphs

A *graph* G is defined to be a set $\{n_i\}$ of elements called nodes and a set $\{e_{ij}\}$ of directed line segments called arcs. If e_{pq} is an element of the set $\{e_{ij}\}$, then we say that there is an *arc* from node n_p to node n_q and that n_q is a *successor* of n_p. We shall be concerned here with graphs whose arcs have *costs* associated with them. We shall represent the cost of arc e_{ij} by c_{ij}. (An arc from n_i to n_j does not imply the existence of an arc from n_j to n_i. If both arcs exist, in general $c_{ij} \neq c_{ji}$.) We shall consider only those graphs G for which there exists $\delta > 0$ such that the cost of every arc of G is greater than or equal to δ. Such graphs shall be called *δ graphs*.

In many problems of interest the graph is not specified explicitly as a set of nodes and arcs, but rather is specified implicitly by means of a set of source nodes $S \subset \{n_i\}$ and a successor operator Γ, defined on $\{n_i\}$, whose value for each n_i is a set of pairs $\{(n_j, c_{ij})\}$. In other words, applying Γ to node n_i yields all the successors n_j of n_i and the costs c_{ij} associated with the arcs from n_i to the various n_j. Application of Γ to the source nodes, to their successors, and so forth as long as new nodes can be generated results in an explicit specification of the graph thus defined. We shall assume throughout this paper that a graph G is always given in implicit form.

The subgraph G_n from any node n in $\{n_i\}$ is the graph defined implicitly by the single source node n and some Γ defined on $\{n_i\}$. We shall say that each node in G_n is *accessible* from n.

A *path* from n_1 to n_k is an ordered set of nodes $(n_1, n_2, \ldots, n_k)$ with each n_{i+1} a successor of n_i. There exists a path from n_i to n_j if and only if n_j is accessible from n_i. Every path has a cost which is obtained by adding the individual costs of each arc, $c_{i,i+1}$, in the path. An *optimal path* from n_i to n_j is a path having the smallest cost over the set of all paths from n_i to n_j. We shall represent this cost by $h(n_i, n_j)$.

This paper will be concerned with the subgraph G_s from some single specified ***start node*** s. We define a nonempty set T of nodes in G_s as the ***goal nodes***.[1] For any node n in G_s, an element $t \in T$ is a *preferred* goal node of n if and only if the cost of an optimal path from n to t does not exceed the cost of any other path from n to any member of T. For simplicity, we shall represent the unique cost of an optimal path from n to a preferred goal node of n by the symbol $h(n)$; i.e., $h(n) = \min_{t \in T} h(n,t)$.

C. Algorithms for Finding Minimum Cost Paths

We are interested in algorithms that search G_s to find an optimal path from s to a preferred goal node of s. What we mean by searching a graph and finding an optimal path is made clear by describing in general how such algorithms proceed. Starting with the node s, they generate some part of the subgraph G_s by repetitive application of the successor operator Γ. During the course of the algorithm, if Γ is applied to a node, we say that the algorithm has *expanded* that node.

We can keep track of the minimum cost path from s to each node encountered as follows. Each time a node is expanded, we store with each successor node n both the cost of getting to n by the lowest cost path found thus far, and a pointer to the predecessor of n along that path. Eventually the algorithm terminates at some goal node t, and no more nodes are expanded. We can then reconstruct a minimum cost path from s to t known at the time of termination simply by chaining back from t to s through the pointers.

We call an algorithm *admissible* if it is guaranteed to find an optimal path from s to a preferred goal node of s for any δ graph. Various admissible algorithms may differ both in the order in which they expand the nodes of G_s and in the number of nodes expanded. In the next section, we shall propose a way of ordering node expansion and show that the resulting algorithm is admissible. Then, in a following section, we shall show, under a mild assumption, that this algorithm uses information from the problem represented by the graph in an optimal way. That is, it expands the smallest number of nodes necessary to guarantee finding an optimal path.

II. An Admissible Searching Algorithm

A. Description of the Algorithm

In order to expand the fewest possible nodes in searching for an optimal path, a search algorithm must constantly make as informed a decision as possible about which node to expand next. If it expands nodes which obviously cannot be on an optimal path, it is wasting effort. On the other hand, if it continues to ignore nodes that might be on an

[1] We exclude the trivial case of $s \in T$.

optimal path, it will sometimes fail to find such a path and thus not be admissible. An efficient algorithm obviously needs some way to evaluate available nodes to determine which one should be expanded next. Suppose some *evaluation function* $\hat{f}(n)$ could be calculated for any node n. We shall suggest a specific function below, but first we shall describe how a search algorithm would use such a function.

Let our evaluation function $\hat{f}(n)$ be defined in such a way that the available node having the smallest value of $\hat{f}$ is the node that should be expanded next. Then we can define a search algorithm as follows.

Search Algorithm A:*

1) Mark s "open" and calculate $\hat{f}(s)$.

2) Select the open node n whose value of $\hat{f}$ is smallest. Resolve ties arbitrarily, but always in favor of any node $n \epsilon T$.

3) If $n \epsilon T$, mark n "closed" and terminate the algorithm.

4) Otherwise, mark n closed and apply the successor operator Γ to n. Calculate $\hat{f}$ for each successor of n and mark as open each successor not already marked closed. Remark as open any closed node n_i which is a successor of n and for which $\hat{f}(n_i)$ is smaller now than it was when n_i was marked closed. Go to Step 2.

We shall next show that for a suitable choice of the evaluation function $\hat{f}$, the algorithm A^* is guaranteed to find an optimal path to a preferred goal node of s and thus is admissible.

B. The Evaluation Function

For any subgraph G_s and any goal set T, let $f(n)$ be the actual cost of an optimal path *constrained to go through n*, from s to a preferred goal node of n.

Note that $f(s) = h(s)$ is the cost of an unconstrained optimal path from s to a preferred goal node of s. In fact, $f(n) = f(s)$ for every node n on an optimal path, and $f(n) > f(s)$ for every node n not on an optimal path. Thus, although $f(n)$ is not known a priori (in fact, determination of the true value of $f(n)$ may be the main problem of interest), it seems reasonable to use an estimate of $f(n)$ as the evaluation function $\hat{f}(n)$. In the remainder of this paper, we shall exhibit some properties of the search algorithm A^* when the cost $f(n)$ of an optimal path through node n is estimated by an appropriate evaluation function $\hat{f}(n)$.

We can write $f(n)$ as the sum of two parts:

$$f(n) = g(n) + h(n) \tag{1}$$

where $g(n)$ is the actual cost of an optimal path from s to n, and $h(n)$ is the actual cost of an optimal path from n to a preferred goal node of n.

Now, if we had estimates of g and h, we could add them to form an estimate of f. Let $\hat{g}(n)$ be an estimate of $g(n)$. An obvious choice for $\hat{g}(n)$ is the cost of the path from s to n having the smallest cost so far found by the algorithm. Notice that this implies $\hat{g}(n) \geq g(n)$.

A simple example will illustrate that this estimate is easy to calculate as the algorithm proceeds. Consider the

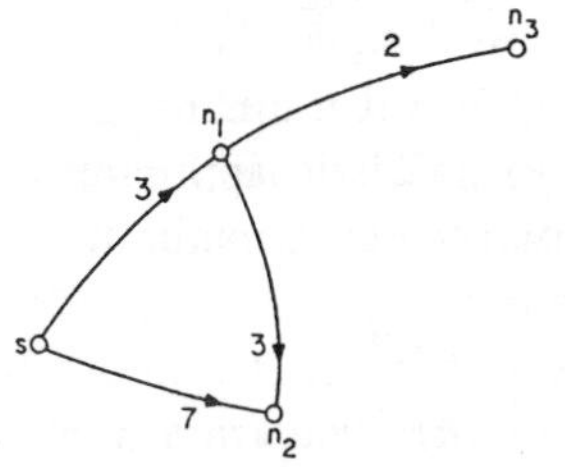

Fig. 1.

subgraph shown in Fig. 1. It consists of a start node s and three other nodes, n_1, n_2, and n_3. The arcs are shown with arrowheads and costs. Let us trace how algorithm A^* proceeded in generating this subgraph. Starting with s, we obtain successors n_1 and n_2. The estimates $\hat{g}(n_1)$ and $\hat{g}(n_2)$ are then 3 and 7, respectively. Suppose A^* expands n_1 next with successors n_2 and n_3. At this stage $\hat{g}(n_3) = 3 + 2 = 5$, and $\hat{g}(n_2)$ is lowered (because a less costly path to it has been found) to $3 + 3 = 6$. The value of $\hat{g}(n_1)$ remains equal to 3.

Next we must have an estimate $\hat{h}(n)$ of $h(n)$. Here we rely on information from the problem domain. Many problems that can be represented as a problem of finding a minimum cost path through a graph contain some "physical" information that can be used to form the estimate $\hat{h}$. In our example of cities connected by roads, $\hat{h}(n)$ might be the airline distance between city n and the goal city. This distance is the shortest possible length of any road connecting city n with the goal city; thus it is a lower bound on $h(n)$. We shall have more to say later about using information from the problem domain to form an estimate $\hat{h}$, but first we can prove that if $\hat{h}$ is any lower bound of h, then the algorithm A^* is admissible.

C. The Admissibility of A*

We shall take as our evaluation function to be used in A^*

$$\hat{f}(n) = \hat{g}(n) + \hat{h}(n) \tag{2}$$

where $\hat{g}(n)$ is the cost of the path from s to n with minimum cost so far found by A^*, and $\hat{h}(n)$ is any estimate of the cost of an optimal path from n to a preferred goal node of n. We first prove a lemma.

Lemma 1

For any nonclosed node n and for any optimal path P from s to n, there exists an open node n' on P with $\hat{g}(n') = g(n')$.

Proof: Let $P = (s = n_0, n_1, n_2, \cdots, n_k = n)$. If s is open (that is, A^* has not completed even one iteration), let $n' = s$, and the lemma is trivially true since $\hat{g}(s) = g(s) = 0$. Suppose s is closed. Let Δ be the set of all closed nodes n_i in P for which $\hat{g}(n_i) = g(n_i)$. Δ is not empty, since by assumption $s \epsilon \Delta$. Let n^* be the element of Δ with highest index. Clearly, $n^* \neq n$, as n is nonclosed. Let n' be the successor of n^* on P. (Possibly $n' = n$.) Now $\hat{g}(n') \leq \hat{g}(n^*) + c_{n^*,n'}$ by definition of $\hat{g}$; $\hat{g}(n^*) = g(n^*)$ because n^* is in Δ, and $g(n') = g(n^*) + c_{n^*,n'}$ because P is an optimal path. Therefore, $\hat{g}(n') \leq g(n')$. But in general, $\hat{g}(n') \geq g(n')$, since the lowest cost $\hat{g}(n')$ from s to n' discovered at any time is certainly not lower than the optimal cost $g(n')$. Thus $\hat{g}(n') = g(n')$, and moreover, n' must be open by the definition of Δ.

Corollary

Suppose $\hat{h}(n) \leq h(n)$ for all n, and suppose A^* has not terminated. Then, for any optimal path P from s to any preferred goal node of s, there exists an open node n' on P with $\hat{f}(n') \leq f(s)$.

Proof: By the lemma, there exists an open node n' on P with $\hat{g}(n') = g(n')$, so by definition of $\hat{f}$

$$\begin{aligned}\hat{f}(n') &= \hat{g}(n') + \hat{h}(n') \\ &= g(n') + \hat{h}(n') \\ &\leq g(n') + h(n') = f(n').\end{aligned}$$

But P is an optimal path, so $f(n') = f(s)$ for all $n' \epsilon P$, which completes the proof. We can now prove our first theorem.

Theorem 1

If $\hat{h}(n) \leq h(n)$ for all n, then A^* is admissible.

Proof: We prove this theorem by assuming the contrary, namely that A^* does not terminate by finding an optimal path to a preferred goal node of s. There are three cases to consider: either the algorithm terminates at a nongoal node, fails to terminate at all, or terminates at a goal node without achieving minimum cost.

Case 1

Termination is at a nongoal node. This case contradicts the termination condition (Step 3) of the algorithm, so it may be eliminated immediately.

Case 2

There is no termination. Let t be a preferred goal node of s, accessible from the start in a finite number of steps, with associated minimum cost $f(s)$. Since the cost on any arc is at least δ, then for any node n further than $M = f(s)/\delta$ steps from s, we have $\hat{f}(n) \geq \hat{g}(n) \geq g(n) > M\delta = f(s)$. Clearly, no node n further than M steps from s is ever expanded, for by the corollary to Lemma 1, there will be some open node n' on an optimal path such that $\hat{f}(n') \leq f(s) < \hat{f}(n)$, so, by Step 2, A^* will select n' instead of n. Failure of A^* to terminate could then only be caused by continued reopening of nodes within M steps of s. Let $\chi(M)$ be the set of nodes accessible within M steps from s, and let $\nu(M)$ be the number of nodes in $\chi(M)$. Now, any node n in $\chi(M)$ can be reopened at most a finite number of times, say $\tilde{\rho}(n, M)$, since there are only a finite number of paths from s to n passing only through nodes within M steps of s. Let

$$\rho(M) = \max_{n \epsilon \chi(M)} \tilde{\rho}(n, M),$$

the maximum number of times any one node can be reopened. Hence, after at most $\nu(M)\rho(M)$ expansions, all nodes in $\chi(M)$ must be forever closed. Since no nodes outside $\chi(M)$ can be expanded, A^* must terminate.

Case 3

Termination is at a goal node without achieving minimum cost. Suppose A^* terminates at some goal node t with $\hat{f}(t) = \hat{g}(t) > f(s)$. But by the corollary to Lemma 1, there existed just before termination an open node n' on an optimal path with $\hat{f}(n') \leq f(s) < \hat{f}(t)$. Thus at this stage, n' would have been selected for expansion rather than t, contradicting the assumption that A^* terminated.

The proof of Theorem 1 is now complete. In the next section, we shall show that for a certain choice of the function $\hat{h}(n)$, A^* is not only admissible but optimal, in the sense that no other admissible algorithm expands fewer nodes.

III. On the Optimality of A^*

A. Limitation of Subgraphs by Information from the Problem

In the preceding section, we proved that if $\hat{h}(n)$ is any lower bound on $h(n)$, then A^* is admissible. One such lower bound is $\hat{h}(n) = 0$ for all n. Such an estimate amounts to assuming that any open node n might be arbitrarily close to a preferred goal node of n. Then the set $\{G_n\}$ is unconstrained; anything is possible at node n, and, in particular, if $\hat{g}$ is a minimum at node n, then node n must be expanded by every admissible algorithm.

Often, however, we have information from the problem that constrains the set $\{G_n\}$ of possible subgraphs at each node. In our example with cities connected by roads, no subgraph G_n is possible for which $h(n)$ is less than the airline distance between city n and a preferred goal city of n. In general, if the set of possible subgraphs is constrained, one can find a higher lower bound of $h(n)$ than one can for the unconstrained situation. If this higher lower bound is used for $\hat{h}(n)$, then A^* is still admissible, but, as will become obvious later, A^* will generally expand fewer nodes. We shall assume in this section that at each node n, certain information is available from the physical situation on which we can base a computation to limit the set $\{G_n\}$ of possible subgraphs.

Suppose we denote the set of all subgraphs from node n by the symbol $\{G_{n,\omega}\}$ where ω indexes each subgraph, and ω is in some index set Ω_n. Now, we presume that certain information is available from the problem domain about the state that node n represents; this information limits the set of subgraphs from node n to the set $\{G_{n,\theta}\}$, where θ is in some smaller index set $\Theta_n \subset \Omega_n$.

For each $G_{n,\theta}$ in $\{G_{n,\theta}\}$ there corresponds a cost $h_\theta(n)$ of the optimum path from n to a preferred goal node of n. We shall now take as our estimate $\hat{h}(n)$, the greatest lower bound for $h_\theta(n)$. That is,

$$\hat{h}(n) = \inf_{\theta \epsilon \Theta_n} h_\theta(n). \tag{3}$$

We assume the infimum is achieved for some $\theta \epsilon \Theta_n$.

In actual problems one probably never has an explicit representation for $\{G_{n,\theta}\}$, but instead one selects a procedure for computing $\hat{h}(n)$, known from information about the problem domain, to be a lower bound on $h(n)$. This selection itself induces the set $\{G_{n,\theta}\}$ by (3). Nevertheless, it is convenient to proceed with our formal discussion as if $\{G_{n,\theta}\}$ were available and as if $\hat{h}(n)$ were explicitly calculated from (3). For the rest of this paper, we assume that the algorithm A^* uses (3) as the definition of $\hat{h}$.

B. A Consistency Assumption

When a real problem is modeled by a graph, each node of the graph corresponds to some state in the problem domain. Our general knowledge about the structure of the problem domain, together with the specific state represented by a node n, determines how the set Ω_n is reduced to the set Θ_n. However, we shall make one assumption about the uniformity of the manner in which knowledge of the problem domain is used to impose this reduction. This assumption may be stated formally as follows. For any nodes m and n,

$$h(m,n) + \inf_{\theta \epsilon \Theta_n} h_\theta(n) \geq \inf_{\theta \epsilon \Theta_m} h_\theta(m). \tag{4}$$

Using the definition of $\hat{h}$ given in (3), we can restate (4) as a kind of triangle inequality:

$$h(m, n) + \hat{h}(n) \geq \hat{h}(m). \tag{5}$$

The assumption expressed by (4) [and therefore (5)] amounts to a type of consistency assumption on the estimate $\hat{h}(n)$ over the nodes. It means that any estimate $\hat{h}(n)$ calculated from data available in the "physical" situation represented by node n alone would not be improved by using corresponding data from the situations represented by the other nodes. Let us see what this assumption means in the case of our example of cities and roads. Suppose we decide, in this example, to use as an estimate $\hat{h}(n)$, the airline distance from city n to its closest goal city. As we have stated previously, such an estimate is certainly a lower bound on $h(n)$. It induces at each node n a set $\{G_{n,\theta}\}$ of possible subgraphs from n by (3). If we let $d(m, n)$ be the airline distance between the two cities corresponding to nodes n and m, we have $h(m, n) \geq d(m, n)$ and, therefore, by the triangle inequality for Euclidean distance

$$h(m, n) + \hat{h}(n) \geq d(m, n) + \hat{h}(n) \geq \hat{h}(m),$$

which shows that this $\hat{h}$ satisfies the assumption of (5).

Now let us consider for a moment the following $\hat{h}$ for the roads-and-cities problem. Suppose the nodes of the graph are numbered sequentially in the order in which they are discovered. Let $\hat{h}$ for cities represented by nodes with odd-numbered indexes be the airline distance to a preferred goal city of these nodes, and let $\hat{h} = 1$ for nodes with even-numbered indexes. For the graph of Fig. 2, $\hat{f}(s) = \hat{h}(n_1) = 8$. Nodes n_2 and n_3 are the successors of n_1 along arcs with costs as indicated. By the above rule for computing $\hat{h}$, $\hat{h}(n_2) = 1$ while $\hat{h}(n_3) = 5$. Then $\hat{f}(n_2) = 6 + 1 = 7$, while $\hat{f}(n_3) = 3 + 5 = 8$, and algorithm A^* would erroneously

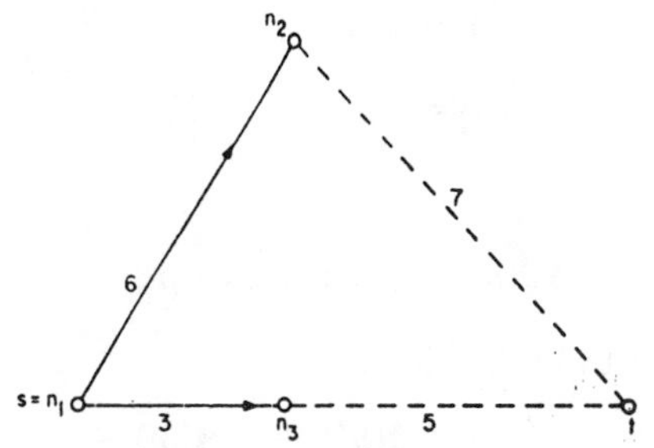

Fig. 2.

choose to expand node n_2 next. This error occurs because the estimates $\hat{h}(s) = 8$ and $\hat{h}(n_2) = 1$ are inconsistent in view of the fact that n_2 is only six units away from s. The information that there cannot exist a path from s to a goal with total cost less than eight was somehow available for the computation of $\hat{h}(s)$, and then ignored during the computation of $\hat{h}(n_2)$. The result is that (5) was violated, i.e.,

$$h(s, n_2) + \hat{h}(n_2) = 6 + 1 < 8 = \hat{h}(s).$$

For the rest of this paper, we shall assume that the family $\{\Theta_n\}$ of index sets satisfies (4) or, equivalently, the procedures for computing the estimates $\hat{h}$ always lead to values that satisfy (5). We shall call this assumption the *consistency assumption.* Note that the estimate $\hat{h}(n) = 0$ for all n trivially satisfies the consistency assumption. Intuitively, the consistency assumption will generally be satisfied by a computation rule for $\hat{h}$ that uniformly uses measurable parameters of the problem state at all nodes; it will generally be violated if the computation rule depends upon any parameter that varies between nodes independently of the problem state (such as a parity count or a random variable), or if the computations at some nodes are more elaborate than at others.

C. Proof of the Optimality of A^*

The next lemma makes the important observation about the operation of A^* that, under the consistency assumption, if node n is closed, then $\hat{g}(n) = g(n)$. This fact is important for two reasons. First, it is used in the proof of the theorem about the optimality of A^* to follow, and second, it states that A^* need never reopen a closed node. That is, if A^* expands a node, then the optimal path to that node has already been found. Thus, in Step 4 of the algorithm A^*, the provision for reopening a closed node is vacuous and may be eliminated.

Lemma 2

Suppose the consistency assumption is satisfied, and suppose that node n is closed by A^*. Then $\hat{g}(n) = g(n)$.

Proof: Consider the subgraph G_s just before closing n, and suppose the contrary, i.e., suppose $\hat{g}(n) > g(n)$. Now there exists some optimal path P from s to n. Since $\hat{g}(n) > g(n)$, A^* did not find P. By Lemma 1, there exists an open node n' on P with $\hat{g}(n') = g(n')$. If $n' = n$, we have proved the lemma. Otherwise,

$$g(n) = g(n') + h(n',n)$$

$$= \hat{g}(n') + h(n',n).$$

Thus,

$$\hat{g}(n) > \hat{g}(n') + h(n',n).$$

Adding $\hat{h}(n)$ to both sides yields

$$\hat{g}(n) + \hat{h}(n) > \hat{g}(n') + h(n',n) + \hat{h}(n).$$

We can apply (5) to the right-hand side of the above inequality to yield

$$\hat{g}(n) + \hat{h}(n) > \hat{g}(n') + \hat{h}(n')$$

or

$$\hat{f}(n) > \hat{f}(n'),$$

contradicting the fact that A^* selected n for expansion when n' was available and thus proving the lemma.

The next lemma states that $\hat{f}$ is monotonically nondecreasing on the sequence of nodes closed by A^*.

Lemma 3

Let $(n_1, n_2, \cdots, n_t)$ be the sequence of nodes closed by A^*. Then, if the consistency assumption is satisfied, $p \leq q$ implies $\hat{f}(n_p) \leq \hat{f}(n_q)$.

Proof: Let n be the next node closed by A^* after closing m. Suppose first that the optimum path to n does not go through m. Then n was available at the time m was selected, and the lemma is trivially true. Then suppose that the optimum path to n does, in fact, go through m. Then $g(n) = g(m) + h(m, n)$. Since, by Lemma 2, we have $\hat{g}(n) = g(n)$ and $\hat{g}(m) = g(m)$,

$$\hat{f}(n) = \hat{g}(n) + \hat{h}(n)$$

$$= g(n) + \hat{h}(n)$$

$$= g(m) + h(m, n) + \hat{h}(n)$$

$$\geq g(m) + \hat{h}(m)$$

$$= \hat{g}(m) + \hat{h}(m)$$

where the inequality follows by application of (5). Thus we have

$$\hat{f}(n) \geq \hat{f}(m).$$

Since this fact is true for any pair of nodes n_k and n_{k+1} in the sequence, the proof is complete.

Corollary

Under the premises of the lemma, if n is closed then $\hat{f}(n) \leq f(s)$.

Proof: Let t be the goal node found by A^*. Then $\hat{f}(n) \leq \hat{f}(t) = f(t) = f(s)$.

We can now prove a theorem about the optimality of A^* as compared with any other admissible algorithm A that uses no more information about the problem than does A^*. Let Θ_n^A be the index set used by algorithm A at node n. Then, if $\Theta_n^{A^*} \subset \Theta_n^A$ for all nodes n in G_s, we shall say that algorithm A is *no more informed* than algorithm A^*.

The next theorem states that if an admissible algorithm A is no more informed than A^*, then any node expanded by A^* must also be expanded by A. We prove this theorem for the special case for which ties never occur in the value of $\hat{f}$ used by A^*. Later we shall generalize the theorem to cover the case where ties can occur, but the proof of the no-ties theorem is so transparent that we include it for clarity.

Theorem 2

Let A be any admissible algorithm no more informed than A^*. Let G_s be any δ graph such that $n \neq m$ implies $\hat{f}(n) \neq \hat{f}(m)$, and let the consistency assumption be satisfied by the $\hat{h}$ used in A^*. Then if node n was expanded by A^*, it was also expanded by A.

Proof: Suppose the contrary. Then there exists some node n expanded by A^* but not by A. Let t^* and t be the preferred goal nodes of s found by A^* and A, respectively. Since A^* and A are both admissible,

$$\hat{f}(t^*) = \hat{g}(t^*) + \hat{h}(t^*) = g(t^*) + 0 = f(t^*) = f(t) = f(s).$$

Since A^* must have expanded n before closing t^*, by Lemma 3 we have

$$\hat{f}(n) < \hat{f}(t^*) = f(t).$$

(Strict inequality occurs because no ties are allowed.)

There exists some graph $G_{n,\theta}$, $\theta \in \Theta_n$, for which $\hat{h}(n) = h(n)$ by the definition of $\hat{h}$. Now by Lemma 2, $\hat{g}(n) = g(n)$. Then on the graph $G_{n,\theta}$, $\hat{f}(n) = f(n)$. Since A is no more informed than A^*, A could not rule out the existence of $G_{n,\theta}$; but A did not expand n before termination and is, therefore, not admissible, contrary to our assumption and completing the proof.

Upon defining $N(A,G_s)$ to be the total number of nodes in G_s expanded by the algorithm A, the following simple corollary is immediate.

Corollary

Under the premises of Theorem 2,

$$N(A^*, G_s) \leq N(A, G_s)$$

with equality if and only if A expands the identical set of nodes as A^*.

In this sense, we claim that A^* is an optimal algorithm. Compared with other no more informed admissible algorithms, it expands the fewest possible nodes necessary to guarantee finding an optimal path.

In case of ties, that is if there exist two or more open nodes $n_1, \cdots, n_k$ with $\hat{f}(n_1) = \cdots = \hat{f}(n_k) < \hat{f}(n)$ for every other open node n, A^* arbitrarily chooses one of the n_i. Consider the set $\mathfrak{A}^*$ of all algorithms that act identically to A^* if there are no ties, but whose members resolve ties differently. An algorithm is a member of $\mathfrak{A}^*$ if it is simply the original A^* with any arbitrary tie-breaking rule.

The next theorem extends Theorem 2 to situations where ties may occur. It states that for any admissible algorithm A, one can always find a member A^* of $\mathcal{A}^*$ such that each node expanded by A^* is also expanded by A.

Theorem 3

Let A be any admissible algorithm no more informed than the algorithms in $\mathcal{A}^*$, and suppose the consistency assumption is satisfied by the $\hat{h}$ used in the algorithms in $\mathcal{A}^*$. Then for any δ graph G_s there exists an $A^* \epsilon \mathcal{A}^*$ such that every node expanded by A^* is also expanded by A.

Proof: Let G_s be any δ graph and A_1^* be any algorithm in $\mathcal{A}^*$. If every node of G_s that A_1^* expands is also expanded by A, let A_1^* be the A^* of the theorem. Otherwise, we will show how to construct the A^* of the theorem by changing the tie-breaking rule of A_1^*. Let L be the set of nodes expanded by A, and let $P = (s, n_1, n_2, \cdots, n_k, t)$ be the optimal path found by A.

Expand nodes as prescribed by A_1^* as long as all nodes selected for expansion are elements of L. Let n be the first node selected for expansion by A_1^* which is not in L. Now $\hat{f}(n) \leq f(s)$ by the corollary to Lemma 3. Since $\hat{f}(n) < f(s) = f(t)$ would imply that A is inadmissible (by the argument of Theorem 2), we may conclude that $\hat{f}(n) = f(s)$. At the time A_1^* selected n, goal node t was not closed (or A_1^* would have been terminated). Then by the corollary to Lemma 1, there is an open node n' on P such that $\hat{f}(n') \leq f(s) = \hat{f}(n)$. But since n was selected for expansion by A_1^* instead of n', $\hat{f}(n) \leq \hat{f}(n')$. Hence $\hat{f}(n) \leq \hat{f}(n') \leq \hat{f}(n)$, so $\hat{f}(n) = \hat{f}(n')$. Let A_2^* be identical to A_1^* except that the tie-breaking rule is modified just enough to choose n' instead of n. By repeating the above argument, we obtain for some i an $A_i^* \epsilon \mathcal{A}^*$ that expands only nodes that are also expanded by A, completing the proof of the theorem.

Corollary 1

Suppose the premises of the theorem are satisfied. Then for any δ graph G_s there exists an $A^* \epsilon \mathcal{A}^*$ such that $N(A^*, G_s) \leq N(A, G_s)$, with equality if and only if A expands the identical set of nodes as A^*.

Since we cannot select the most fortuitous tie-breaking rule ahead of time for each graph, it is of interest to ask how all members of $\mathcal{A}^*$ compare against any admissible algorithm A in the number of nodes expanded. Let us define a *critical* tie between n and n' as one for which $\hat{f}(n) = \hat{f}(n') = f(s)$. Then we have the following as a second corollary to Theorem 3.

Corollary 2

Suppose the premises of the theorem are satisfied. Let $R(A^*, G_s)$ be the number of critical ties which occurred in the course of applying A^* to G_s. Then for any δ graph G_s and *any* $A^* \epsilon \mathcal{A}^*$,

$$N(A^*, G_s) \leq N(A, G_s) + R(A^*, G_s).$$

Proof: For any noncritical tie, all alternative nodes must be expanded by A as well as by A^* or A would not be admissible. Therefore, we need merely observe that each node expanded by A^* but not by A must correspond to a different critical tie in which A^*'s tie-breaking rule made the inappropriate choice.

Of course, one must remember that when A does expand fewer nodes than some particular A^* in $\mathcal{A}^*$, it is only because A was in some sense "lucky" for the graph being searched, and that there exists a graph consistent with the information available to A and A^* for which A^* would not search more nodes than A.

Note that, although one cannot keep a running estimate of R while the algorithm proceeds because one does not know the value of $f(s)$, this value is established as soon as the algorithm terminates, and R can then be easily computed. In most practical situations, R is not likely to be large because critical ties are likely to occur only very close to termination of the algorithm, when $\hat{h}$ can become a perfect estimator of h.

IV. Discussion and Conclusions

A. Comparisons Between A^* and Other Search Techniques

Earlier we mentioned that the estimate $\hat{h}(n) \equiv 0$ for all n trivially satisfies the consistency assumption. In this case, $\hat{f}(n) = \hat{g}(n)$, the lowest cost so far discovered to node n. Such an estimate is appropriate when no information at all is available from the problem domain. In this case, an admissible algorithm cannot rule out the possibility that the goal might be as close as δ to that node with minimum $g(n)$. Pollack and Wiebenson[1] discuss an algorithm, proposed to them by Minty in a private communication, that is essentially identical to our A^* using $\hat{f}(n) = \hat{g}(n)$.

Many algorithms, such as Moore's "Algorithm D"[6] and Busacker and Saaty's implementation of dynamic programming, keep track of $\hat{g}(n)$ but do not use it to order the expansion of nodes. The nodes are expanded in a "breadth-first" order, meaning that all nodes one step away from the start are expanded first, then all nodes two steps away, etc. Such methods must allow for changes in the value of $\hat{g}(n)$ as a node previously expanded is later reached again by a less costly route.

It might be argued that the algorithms of Moore, Busacker and Saaty, and other equivalent algorithms (sometimes known as "water flow" or "amoeba" algorithms) are advantageous because they first encounter the goal by a path with a minimum number of steps. This argument merely reflects an imprecise formulation of the problem, since it implies that the number of steps, and not the cost of each step, is the quantity to be minimized. Indeed, if we set $c_{ij} = 1$ for all arcs, this class of algorithms is identical to A^* with $\hat{h} = 0$. We emphasize that, as is always the case when a mathematical model is used to represent a real problem, the first responsibility of the investigator is to ensure that the model is an adequate representation of the problem for his purposes.

It is beyond the scope of the discussion to consider how to define a successor operator Γ or assign costs c_{ij} so that the resulting graph realistically reflects the nature of a specific problem domain.[2]

B. The Heuristic Power of the Estimate $\hat{h}$

The algorithm A^* is actually a family of algorithms; the choice of a particular function $\hat{h}$ selects a particular algorithm from the family. The function $\hat{h}$ can be used to tailor A^* for particular applications.

As was discussed above, the choice $\hat{h} = 0$ corresponds to the case of knowing, or at least of using, absolutely no information from the problem domain. In our example of cities connected by roads, this would correspond to assuming a priori that roads could travel through "hyperspace," i.e., that any city may be an arbitrarily small road distance from any other city regardless of their geographic coordinates.

Since we are, in fact, "more informed" about the nature of Euclidean space, we might increase $\hat{h}(n)$ from 0 to $\sqrt{x^2 + y^2}$ (where x and y are the magnitudes of the differences in the x, y coordinates of the city represented by node n and its closest goal city). The algorithm would then still find the shortest path, but would do so by expanding, typically, considerably fewer nodes. In fact, A^* expands no more nodes[3] than any admissible algorithm that uses no more information from the problem domain; viz., the information that a road between two cities might be as short as the airline distance between them.

Of course, the discussion thus far has not considered the cost of computing $\hat{h}$ each time a node is generated on the graph. It could be that the computational effort required to compute $\sqrt{x^2 + y^2}$ is significant when compared to the effort involved in expanding a few extra nodes; the optimal procedure in the sense of minimum number of nodes expanded might not be optimal in the sense of minimum total resources expended. In this case one might, for example, choose $\hat{h}(n) = (x + y)/2$. Since $(x + y)/2 < \sqrt{x^2 + y^2}$, the algorithm is still admissible. Since we are not using "all" our knowledge of the problem domain, a few extra nodes may be expanded, but total computational effort may be reduced; again, each "extra" node must also be expanded by other admissible algorithms that limit themselves to the "knowledge" that the distance between two cities may be as small as $(x + y)/2$.

Now suppose we would like to reduce our computational effort still further, and would be satisfied with a solution path whose cost is not necessarily minimal. Then we could choose an $\hat{h}$ somewhat larger than the one defined by (3). The algorithm would no longer be admissible, but it might be more desirable, from a heuristic point of view, than any admissible algorithm. In our roads-and-cities example, we might let $\hat{h} = x + y$. Since road distance is usually substantially greater than airline distance, this $\hat{h}$ will usually, but not always, result in an optimal solution path. Often, but not always, fewer nodes will be expanded and less arithmetic effort required than if we used $\hat{h}(n) = \sqrt{x^2 + y^2}$.

Thus we see that the formulation presented uses one function, $\hat{h}$, to embody in a formal theory all knowledge available from the problem domain. The selection of $\hat{h}$, therefore, permits one to choose a desirable compromise between admissibility, heuristic effectiveness, and computational efficiency.

[2] We believe that appropriate choices for Γ and c_{ij} will permit many of the problem domains in the heuristic programming literature[7] to be mapped into graphs of the type treated in this paper. This could lead to a clearer understanding of the effects of "heuristics" that use information from the problem domain.

[3] Except for possible critical ties, as discussed in Corollary 2 of Theorem 3.

References

[1] M. Pollack and W. Wiebenson, "Solutions of the shortest-route problem—a review," *Operations Res.*, vol. 8, March–April 1960.

[2] R. Busacker and T. Saaty, *Finite Graphs and Networks: An Introduction with Applications.* New York: McGraw-Hill, 1965, ch. 3.

[3] R. Bellman and S. Dreyfus, *Applied Dynamic Programming.* Princeton, N. J.: Princeton University Press, 1962.

[4] H. Gelernter, "Realization of a geometry-theorem proving machine," in *Computers and Thought.* New York: McGraw-Hill, 1963.

[5] A. Samuel, "Some studies in machine learning using the game of checkers," in *Computers and Thought.* New York: McGraw-Hill, 1963.

[6] E. Moore, "The shortest path through a maze," *Proc. Internat'l Symp. on Theory of Switching* (April 2–5, 1957), pt. 2. Also, *The Annals of the Computation Laboratory of Harvard University*, vol. 30. Cambridge, Mass.: Harvard University Press, 1959.

[7] E. Feigenbaum and J. Feldman, *Computers and Thought.* New York: McGraw-Hill, 1963.

Scientific Applications

F.N. Fritsch
Editor

An Algorithm for Planning Collision-Free Paths Among Polyhedral Obstacles

Tomás Lozano-Pérez and Michael A. Wesley
IBM Thomas J. Watson Research Center

This paper describes a collision avoidance algorithm for planning a safe path for a polyhedral object moving among known polyhedral objects. The algorithm transforms the obstacles so that they represent the locus of forbidden positions for an arbitrary reference point on the moving object. A trajectory of this reference point which avoids all forbidden regions is free of collisions. Trajectories are found by searching a network which indicates, for each vertex in the transformed obstacles, which other vertices can be reached safely.

Key Words and Phrases: path finding, collision-free paths, polyhedral objects, polyhedral obstacles, graph searching, growing objects

CR Categories: 3.15, 3.64, 3.66, 8.1

"An Algorithm for Planning Collision-Free Paths Among Polyhedral Obstacles" by T. Lozano-Pérez and M.A. Wesley from *Communications of the ACM*, Vol. 22, No. 10, 1979, pages 560-570. Copyright © 1979 by the Association for Computing Machinery, Inc., reprinted by permission.

Permission to copy without fee all or part of this material is granted provided that the copies are not made or distributed for direct commercial advantage, the ACM copyright notice and the title of the publication and its date appear, and notice is given that copying is by permission of the Association for Computing Machinery. To copy otherwise, or to republish, requires a fee and/or specific permission.

Authors' present addresses: T. Lozano-Pérez, Artificial Intelligence Laboratory, Massachusetts Institute of Technology, Cambridge, MA 02139; M.A. Wesley, IBM Thomas J. Watson Research Center, P.O. Box 218, Yorktown Heights, NY 10598.

1. Introduction

The problem of avoiding collisions when operating on computer models of physical objects is central to model-based manipulation systems. This paper describes an algorithm for planning safe, that is collision-free, paths for a polyhedral object among similarly described obstacles.[1] The algorithm is required to:

(1) find safe paths that might involve going near obstacles, and
(2) guarantee that these paths are short relative to a prespecified distance metric.

The simplest collision avoidance algorithms fall into the generate and test paradigm. A simple path from start to goal, usually a straight line, is hypothesized and then the path is tested for potential collisions. If collisions are detected, a new path is proposed, possibly using information about the detected collision to help hypothesize the new path. This is repeated until no collisions are detected along the path. Roughly, the three steps in this type of algorithm are:

(1) calculate the volume swept out by the moving object along the proposed path,
(2) determine the overlap between the swept volume and the obstacles, and
(3) propose a new path.

The second step, determining the overlap between the swept volume and the obstacles, is also known as an intersection or interference calculation [2, 3]. Current computer modeling techniques employ large numbers of simple surfaces to model accurately even the most common objects. It can be quite difficult to determine whether two such models overlap. This general method, which we will call the *swept volume* method, has a more fundamental drawback. The problem is in the relationship between the second and third steps. Each proposed path provides only local information about potential collisions, for example, the shape of the intersections of the volumes involved, or the identity of the obstacle giving rise to the collision. This information suggests local path changes but is not sufficient to determine when a radically different path would be better. This lack of a global view can result in an expensive search of the space of possible paths with a very large upper bound on the worst case length of the path.

A radical alternative to the swept volume method is to compute explicitly the constraints on the position of the moving object relative to the obstacles. The desired trajectory is the shortest path which satisfies all the position constraints. If the objects are modeled as collections of convex polyhedra, the position constraints can be stated in terms of the position of the vertices of the mov-

[1] We will henceforth use the term "polyhedron" for closed figures bounded by "flats" in two or three dimensions.

ing object relative to the planes of the obstacle surfaces. The trajectory problem can then be posed as an optimization problem as in Ignat'yev [5]. The difficulty with this formulation is that these position constraints, although linear, do not all apply simultaneously. It is not necessary for each point on the moving object to be outside *all* the planes of the obstacles; it is sufficient for each point to be outside *at least one* of the planes of each obstacle. This property makes traditional linear optimization methods inapplicable.

The algorithm presented in this paper is closely related to the optimization approach. The constraints on the position of an arbitrary reference point on the moving object are computed. Polyhedral obstacles in two or three dimensions give rise to sets of polyhedral *forbidden regions*; that is, regions corresponding to positions of the reference point where collisions would occur. This transformation reduces the problem of finding a safe path for the polyhedron to the simpler problem of finding a safe path for a point. This last task is accomplished by finding a path through a graph connecting vertices of the forbidden regions.

The technique of computing the position constraints on an object as constraints on a reference point is extremely powerful and has been applied independently to different problems. It has been used by Udupa [9] for planning safe paths for computer-controlled manipulators, by Lozano-Perez [6] for identifying feasible grasp points on an object, and by Adamowicz and Albano [1] for two-dimensional template layout.

Udupa uses a simple "growing" transformation on obstacles to compute approximations to the forbidden regions for the three-dimensional reference point of a three degree of freedom subset of a manipulator. The system maintains a variable resolution description of the legal positions of the reference point (the *free space*). Safe paths for the subset manipulator are found by recursively introducing intermediate goals into a straight line path until the complete path is in free space. This method has two drawbacks:

(1) Because the complete manipulator has more than three degrees of freedom, the three-dimensional forbidden regions cannot model all the constraints on the manipulator. When a trajectory fails, Udupa's system makes a correction using manipulator-dependent heuristics. The use of heuristics tends to limit the performance of the algorithm in cluttered spaces.

(2) The recursive path finder uses only local information to determine a safe path and therefore suffers from some of the same drawbacks as the swept volume method.

The algorithm presented in this paper uses a more accurate growing operation to compute the forbidden regions in both two and three dimensions. It introduces a graph searching technique for path finding which produces optimum two-dimensional paths when only translations are involved. This technique is then generalized to deal with three-dimensional obstacles and extended to deal uniformly with more than three degrees of freedom. The resulting algorithm no longer guarantees optimum paths. This algorithm has been used to plan safe trajectories for a seven degree of freedom manipulator. These trajectories have been successfully executed.

A detailed survey of previous work in collision avoidance, specifically in connection with computer-controlled manipulators, can be found in Udupa [9].

The nature of the models used for the obstacles affects the details of any collision avoidance algorithm. For concreteness, the detailed discussions and examples in this paper assume that all objects are modeled as sets of, possibly overlapping, convex polyhedra. Any object can be modeled to any desired degree of accuracy in this fashion. A method for finding collision-free paths for a single convex polyhedron among sets of convex polyhedra can be simply extended to plan safe paths for a complex moving object among complex obstacles. The extension involves finding the constraints due to each of the convex components of the moving object relative to each of the components of all obstacles. The constraints for the composite moving object are the union of the constraints on its components.

The collision avoidance algorithm is defined for three dimensions. However, the presentation is easier to follow in two dimensions; for clarity the next sections first develop the complete algorithm for the two-dimensional case and then consider the extension to three dimensions. Section 2 presents a simple form of the algorithm for the case of a polygonal object translating in the plane among polygonal obstacles. Section 3 considers the effect of allowing the moving object to rotate as well as translate. Section 4 deals with more complex moving objects with more degrees of freedom. Section 5 discusses generalization to three dimensions. Discussion of the two steps of the algorithm that are directly affected by the choice of modeling methodology is relegated to the appendices. These steps will be functionally described in the body of the paper.

2. Collision Avoidance on the Plane

Consider the problem, shown in Figure 1, of moving a point object A from position S to position G while avoiding the obstacles (shown shaded); the shortest collision-free path from S to G is also shown. The important property of this path is that it is composed of straight lines joining the origin to the destination via a possibly empty sequence of vertices of obstacles. In the case of motion in the plane with arbitrary polygonal objects, the shortest collision-free path connecting any two accessible points always has this property.

The undirected graph VG(N, L) is defined: The node

Fig. 1.

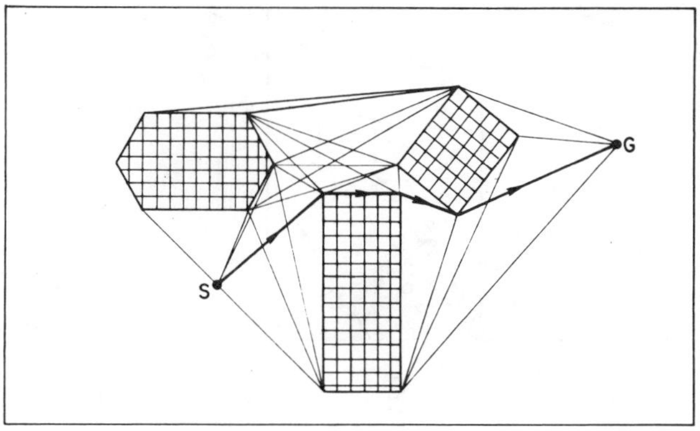

Fig. 2.

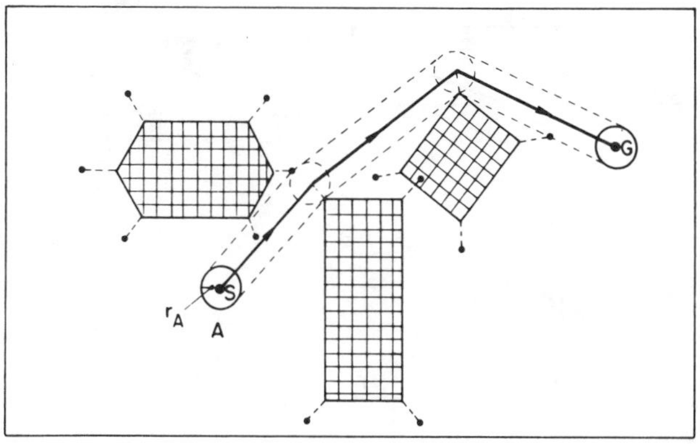

Fig. 3(a).

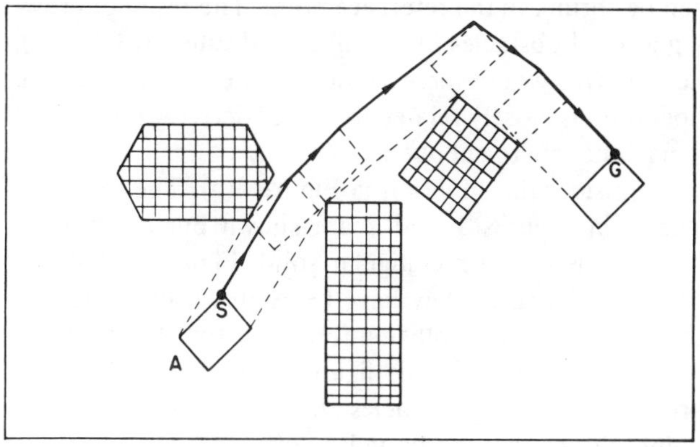

Fig. 3(b).

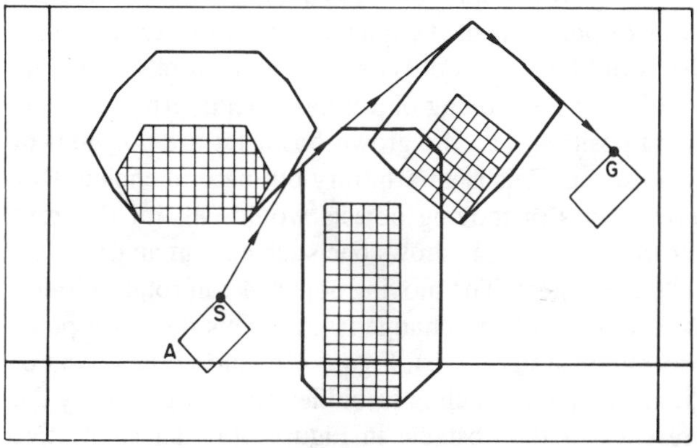

set N is $V \cup \{S, G\}$ where V is the set of all vertices of obstacles and the Link set L is the set of all links (n_i, n_j) such that a straight line connecting the ith element of N to the jth does not overlap any obstacle. The graph VG(N, L) is called the *visibility graph* (VGRAPH) of N since connected vertices in the graph can see each other. The VGRAPH is shown in Figure 1. The shortest collision-free path from S to G on the plane is the shortest path in the VGRAPH from the node corresponding to S to that corresponding to G when the euclidean metric is used on the links. We will call this method for finding collision-free paths for a point by finding the shortest path in a visibility graph the VGRAPH algorithm. This method was used for navigating SHAKEY [8], an early robot vehicle, and is also described in some detail in Ignat'yev [5].

The simplicity of the VGRAPH algorithm stems from the fact that the moving object A is a point. This is a good approximation for moving objects which are small in relation to the obstacles, but causes problems otherwise. Ignat'yev [5, p. 241] puts it as follows:

> The robot begins to move from the point y_0 (S in our example) along the direction to the x_{19} (a vertex). Here he must consider his dimensions in order not to run into the obstacles and walls.

This paper shows how a more general form of the collision avoidance problem can be reduced to the VGRAPH problem. In other words, it concerns how the robot "must consider his dimensions."

A simple generalization of the problem in Figure 1 is to make the moving object A a circle with nonnegligible radius r_A. The VGRAPH algorithm can be adapted to this situation by moving the vertices away from the obstacles so that they are at least r_A away from all the sides (Figure 2). Moving A so that its center point moves through the new displaced vertices will still produce a minimum distance, collision-free path. Notice, however, that the path found is different from that in Figure 1. This technique of displacing the vertices was also used in SHAKEY [8].

The VGRAPH algorithm requires that the moving object be a point; the obstacles then represent the forbidden regions for the position of that point. If the moving object is not a point, a new set of obstacles must be computed which are the forbidden regions of some reference point on the moving object. These new obstacles must describe the locus of positions of this reference point which would cause a collision with any of the original obstacles. The displaced vertices of Figure 2 are, in fact, approximations to the vertices of these new obstacles when the reference point is the center of A.

The operation of computing a new obstacle O' from an original obstacle O and a moving object A will be called *growing O by A*. This name reflects the fact that the obstacles are being grown so that the moving object can be shrunk to the reference point. The result of growing a set of obstacles by A will be indicated by GOS(A), i.e., the *Grown Obstacle Set* of A. Note that the growing operation is closely related to that of deriving the path of a machine tool to cut out a part.

Consider the situation in Figure 3(a). The same obstacles in Figures 1 and 2 are shown but the moving object A is now a rectangular solid. Figure 3(b) shows the obstacles after they have been grown by A. It also shows the shortest collision-free path for A's reference point from S to G. This figure demonstrates how the process of growing obstacles allows representing A as a point. Notice that the boundary of the obstacle space is treated as an obstacle and is also grown, thus avoiding paths which involve moving outside the space.

The growing operation was defined as computing the locus of positions of the moving object's reference point that would cause a collision with a given obstacle. The position of the moving object has been interpreted as its (x, y) position, i.e., the grown obstacles are polygons in (x, y) space. This is an arbitrary but natural choice. Different types of moving objects would call for different choices. Figure 4(a) shows one such case in an (x, y) coordinate system. The moving object A can rotate about a fixed point and can change length. This defines a polar coordinate system (r, α). Figure 4(b) shows the region of the (r, α) space which is forbidden to the tip of A by the presence of the obstacle in Figure 4(a); an alternative way of representing this region is shown in (x, y) coordinates in Figure 4(c). The choice of representation depends on:

(1) the ease of computing the forbidden regions, i.e., growing the obstacles, versus
(2) the ease of building the VGRAPH from the grown obstacles.

The use of polyhedra as the basic unit of shape description influences our choice of obstacle representation. Polyhedra (polygons when on the plane) have boundaries which are linear equations in the coordinate variables. This property makes them computationally attractive. In this section we have represented objects as polygons in a planar cartesian coordinate system. The natural choice is to express the grown obstacles in the same space, thus making the growing operation a mapping from polyhedra to polyhedra. Notice that in Figure 4(b) the object O was interpreted as a polygon in (x, y) space and the resulting grown obstacle O' in (r, α) is not a polygon in that space.

Another factor in the choice of obstacle representation is the shape of the path between two nodes in a VGRAPH. A link connecting two nodes in the VGRAPH implies that the path between the corresponding locations does not overlap any of the obstacles. Paths have so far been shown as straight lines in cartesian space; since the grown obstacles were in this coordinate space, the use of straight lines simplifies the detection of overlap. Of course, paths could be more complicated curves which are best expressed in different coordinate systems. For example, the object in Figure 4 might move in straight lines in the (r, α) system. In that case it might be more efficient to use the polar form of the grown obstacles in detecting overlap.

The choice of representation for the grown obstacles depends on the geometric details of the application domain. The choice should be made so as to simplify the overall computation. For the sake of simplicity the next section will continue to assume that the grown obstacles are polygons in (x, y) space.

Fig. 4(a).

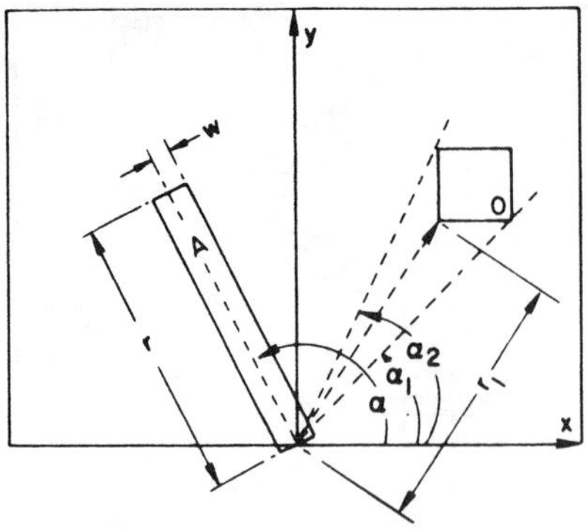

Fig. 4(b).

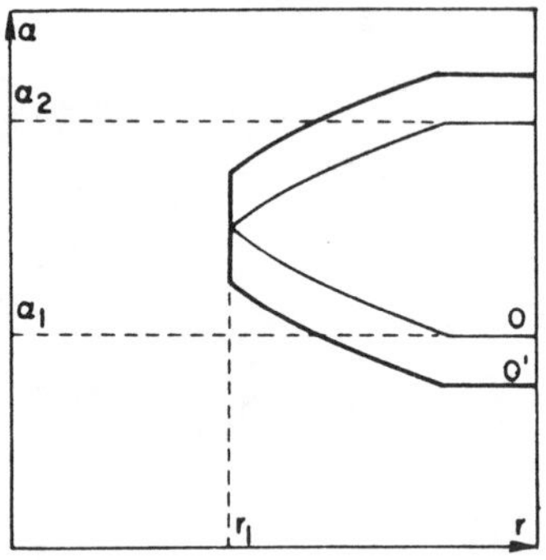

Fig. 4(c).

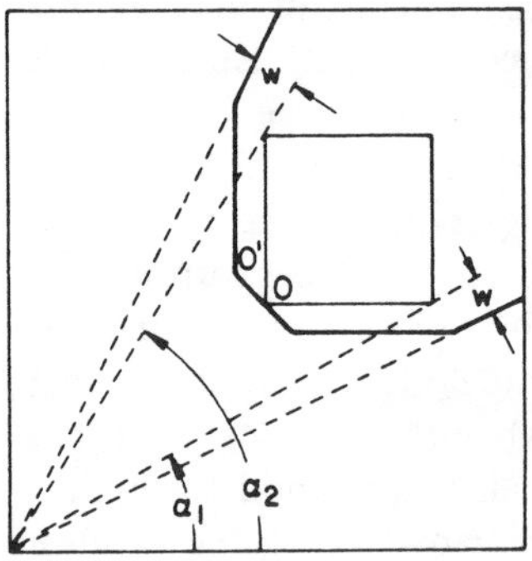

Fig. 5.

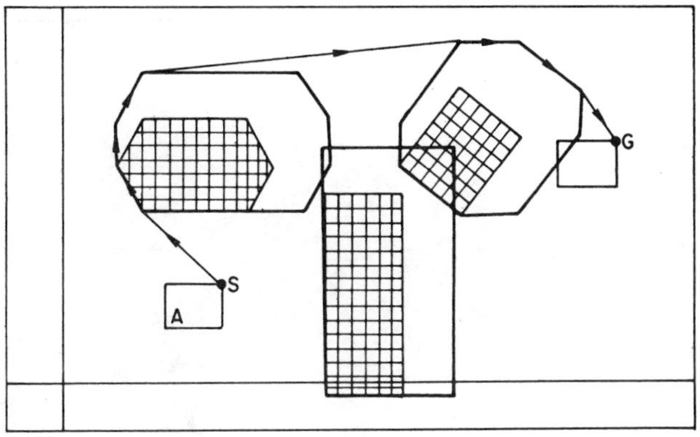

Fig. 6.

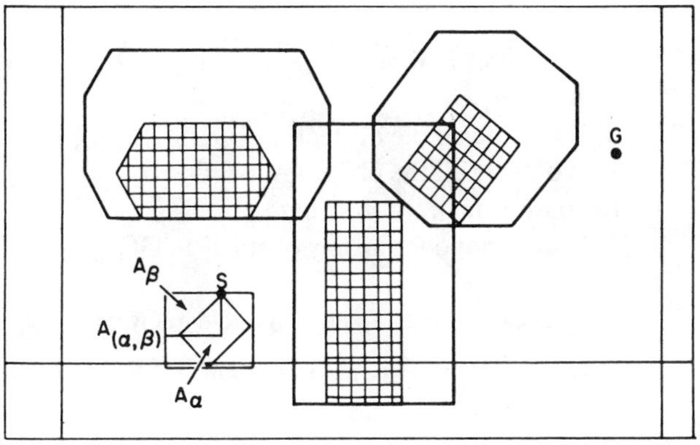

3. The Effect of Rotation

It is important to notice that the growing operation as shown in Figures 2 and 3 is sensitive to the orientation of A. This was not apparent in Figure 2 because the moving object was a circle. The orientation dependence follows from the fact that a grown obstacle is defined as the forbidden region for a reference point. The position of a point on the plane can encode only two degrees of freedom, whereas differentiating the legality of two positions of A with different orientations requires at least three degrees of freedom. Figure 5 shows that a different orientation of A from that in Figure 3 will produce different grown obstacles and a different path. To make the orientation explicit, we will denote the result of growing all the obstacles with a moving object A, whose orientation parameter is the angle α, GOS(A_α). The set of vertices of these grown obstacles will be called V_α.

To summarize, any position of A at orientation α for which A's reference point is outside all the elements of the grown obstacle set is free of collisions. The sides of each obstacle in GOS(A_α) are computed by tracing the path of A's reference point around each of the original objects while keeping A in contact with the obstacle. Before two objects collide they must first touch; therefore any position of the reference point that would cause a collision must be inside the obstacle, and any position outside must be safe. Clearly this condition presupposes that the orientation of A does not change.

Consider the problem of moving object A from position S with orientation α to G with a different orientation β. A safe trajectory cannot be found by simply computing a path that is free of collisions in GOS(A_α) and GOS(A_β) since, in changing the orientation from α to β, A must pass through the whole range of intermediate orientations. One way to find a path requires knowing what positions on the plane will allow the desired rotation to take place. The algorithm can then plan a path from the start to one of these positions, rotate to the desired orientation, and move in that orientation to the goal.

For a position to allow a change in orientation there must be no overlap between the rotating object in any of its intermediate orientations and any of the obstacles. Figure 6 shows the area that A traverses in going from orientation α to β; this area may be approximated by another polygon $A_{[\alpha,\beta]}$ shown rectangular for simplicity. This new object, called an *envelope*, can be used to grow a new obstacle set GOS($A_{[\alpha,\beta]}$), also shown in Figure 6, which represents the forbidden regions for the reference point of A in any of the orientations within the interval $[\alpha,\beta]$. We will refer to this as a *transition obstacle set*. By analogy to the vertex set V_α, the set $V_{[\alpha,\beta]}$ represents the set of vertices of obstacles in the transition obstacle set. In general we can associate with all the elements of a vertex set an orientation interval (possibly singular) as well as a position.

The problem in Figure 6 can now be solved by:

(1) finding a path starting with orientation α at S which avoids the obstacles in GOS(A_α) and which ends at a point clear of the obstacles in GOS($A_{[\alpha,\beta]}$),
(2) rotating to orientation β, and
(3) finding a path to G avoiding the obstacles in GOS(A_β).

This can be stated as a VGRAPH problem of finding the shortest path from S to G in a visibility graph defined as follows:

$$VG_{\alpha,\beta}(N_{\alpha,\beta}, L_{\alpha,\beta})$$

where

$$N_{\alpha,\beta} = V_{[\alpha,\beta]} \cup V_{[\alpha,\alpha]} \cup V_{[\beta,\beta]}$$
$$V_{[\alpha,\alpha]} = V_\alpha \cup \{S\}$$
$$V_{[\beta,\beta]} = V_\beta \cup \{G\}$$
$V_{[\alpha,\beta]}$ defined as above

and

$$L_{\alpha,\beta} = \{(n_i, n_j)\}$$

$n_i \in V_{[a,b]}$ and $n_j \in V_{[c,d]}$ where a, b, c, d are either α or β

such that the following *visibility conditions* hold on the link:

(1) the orientation intervals $[a, b]$ and $[c, d]$ must not be disjoint,
(2) n_i is outside all the obstacles in GOS($A_{[a, b]}$),
(3) n_j is outside all the obstacles in GOS($A_{[c, d]}$),
(4) the path from n_i to n_j either:
  (a) does not overlap any obstacle in GOS($A_{[a, b]}$), or
  (b) does not overlap any obstacle in GOS($A_{[c, d]}$).

A solution path in $VG_{\alpha, \beta}$ is a sequence of nodes starting at S and ending at G:

$$S, n_1, n_2, \ldots, n_k, G$$

in which adjacent nodes are connected by a link in $L_{\alpha, \beta}$. *Each* $n_j \in V_{[a, b]}$ is defined such that if n_j is outside all obstacles in GOS($A_{[a, b]}$), then the reference point of the moving object A can be at position n_j in any orientation within the interval $[a, b]$ without danger of collisions. Following the link from n_j to n_{j+1} means that the reference point of A must make the corresponding translation. Also, if n_j and n_{j+1} belong to different vertex sets, $V_{[a, b]}$ and $V_{[c, d]}$ respectively, then a change of orientation may also be required. The conditions on $L_{\alpha, \beta}$ require that the orientation intervals corresponding to the endpoints of a link must not be disjoint. This means that there is some orientation x, such that if $a \leq b$ and $c \leq d$ then $\max(a, c) \leq x \leq \min(b, d)$, for which A can safely be at either node of the link. Moving along the link requires first rotating to the orientation x and then translating from the first node to the second. Since the translation happens in an orientation compatible with both nodes of the link, the visibility conditions on the link require only checking for overlap with the obstacles in the obstacle set of *either* one of the nodes. Alternatively, if the path from n_j to n_{j+1} is outside all obstacles in both GOS($A_{[a, b]}$) and GOS($A_{[c, d]}$), then the rotation may take place in conjunction with the translation along the link.

The use of transition sets, e.g., GOS($A_{[\alpha, \beta]}$), has two important drawbacks. The shortest solution path in $VG_{\alpha, \beta}$ is no longer guaranteed to be an optimum solution to the original problem, and failure to find a solution path in the VGRAPH does not necessarily mean that no safe trajectory exists. The reasons are twofold. The first and most basic is that paths found in this VGRAPH will change the orientation of the moving object only at locations where the full rotation can be performed. If the optimum path involves traversing a narrow passage where the orientation of A must be within a small subrange of the orientations between α and β, then this path could not be a solution path in this version of the VGRAPH algorithm. Secondly, even if the first problem were avoided, the current formulation considers orientations only in the range $[\alpha, \beta]$; it could not negotiate a passage where the moving object could only fit at an orientation outside the specified range. The latter problem can be solved simply by expanding the orientation

Fig. 7.

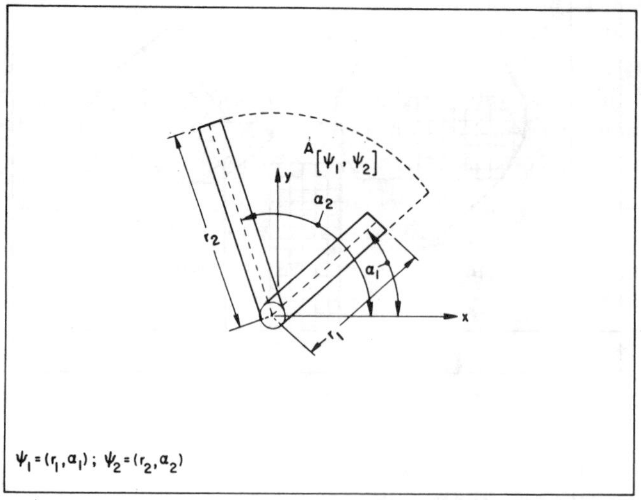

interval, but only at the expense of making the former problem worse.

The two problems mentioned above can be alleviated by replacing the single transition obstacle set, GOS ($A[_{\alpha, \beta}]$), by the union of several other obstacle sets, each generated with a smaller orientation interval for the moving object. In this fashion the range of legal orientations can also be extended beyond the interval $[\alpha, \beta]$. As the number of transition obstacle sets increases, the VGRAPH becomes a better match to the original problem. Unfortunately, the computational burden also increases rapidly. Each new obstacle set requires growing all the obstacles with the moving object in a new configuration, though the growing operation can be speeded up by using approximations, as will be shown later. Also, the added vertices from the extra obstacle sets make searching the visibility graph much more time consuming. Alternatively, it may be possible to derive automatically transition sets to handle narrow passages specifically, and combine these with wider-range transition sets.

4. More Degrees of Freedom

Transition obstacle sets can be used whenever the moving object has more degrees of freedom than can be represented by a point in the obstacle coordinate space. The only requirement is that it be possible to compute an envelope $A_{[x, y]}$ which is an object of the same type as A, e.g., a polygon, such that any point inside an A_z, $x \leq z \leq y$, is also inside $A_{[x, y]}$. This object then can be used in the growing operation to generate a transition obstacle set. There are no other restrictions on the nature of the parameter range $[x, y]$; in particular, it need not be an orientation range and both x and y may also be vectors. A point outside all of the obstacles in GOS($A_{[x, y]}$) indicates a position where each of A_z's *configuration parameters*, z_i, can safely take on values such that $x_i \leq z_i \leq y_i$.

Figure 7 repeats the example of Figure 4 except that now the moving object can translate in x and y as well as rotate and change its length. The choice of a coordinate system for the grown obstacles will also determine which of the coordinate variables is to be used for the configuration parameters. For example, if the grown obstacles are represented as polygons in (x, y), then (r, α) are configuration parameters and vice-versa.

Configuration parameters can also be used to deal with a moving object whose shape can change due to changes in the relative positions of its components. The object shown in Figure 8 is composed of two rectangles that are free to rotate about a common point. The shape of this object relative to a stationary obstacle can be described by:

(1) the shape of its components,
(2) their relative displacements,
(3) the two angles θ and ρ indicated in Figure 8.

In this example only the angles can change during a motion; therefore the obstacle set for this moving object must be parameterized by the value of both θ and ρ. Generally the configuration parameters describe not only the global orientation or position of the object but also the relative positions of its components.

In general, objects need not be grown in the full dimensional configuration space; instead, repeated use is made of operations on lower dimensional, partitioned, configuration spaces which allows the growing operation to work in a convenient subspace of the full configuration space. The VGRAPH algorithm described in Sections 2 and 3 remains unchanged except that the scalar parameters and intervals are replaced by vector parameters and intervals.

5. Collision Avoidance in Three Dimensions

The VGRAPH algorithm has so far been presented as an algorithm for collision avoidance on the plane. This section examines how three-dimensional obstacles affect the algorithm. This generalization does not affect the statement of the algorithm but does affect the details of the obstacle growing and graph searching. These subjects are discussed in the appendices.

The generalization to three dimensions has an unfortunate side effect. The shortest path around a polyhedral obstacle does not in general traverse only vertices of the polyhedron (Figure 9). That is, the shortest path in a VGRAPH whose node set contains only vertices of the grown obstacles is not guaranteed to be the shortest collision-free path. In general, the shortest path will involve going via points on edges of the obstacles. Our approach is to introduce additional vertices along edges of the grown obstacles so that no edge is longer than a prespecified maximum length. This method generally results in a good approximation to the optimum path.

The use of three-dimensional obstacles also has a

Fig. 8.

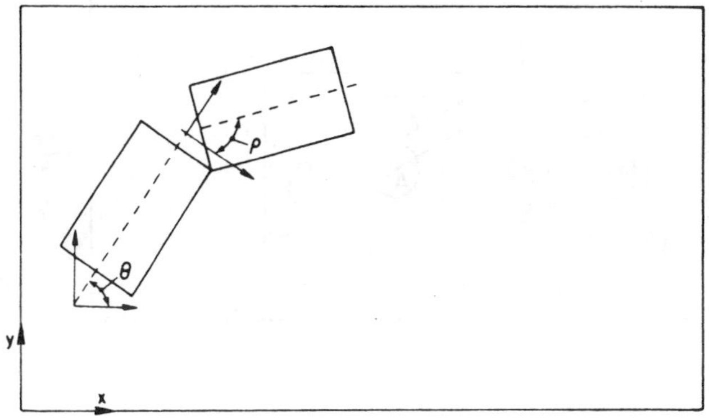

Fig. 9.

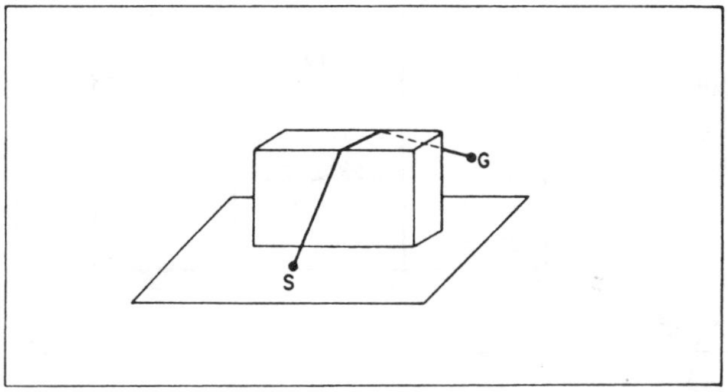

significant effect on the execution time of the algorithm. The three-dimensional growing operation is much more time consuming than the corresponding operation in two dimensions. Grown obstacles in three dimensions are generally much more complex than the underlying objects (Appendix 1). The larger vertex sets also increase the time necessary to search the visibility graph. These effects make the use of approximations necessary for practical applications.

A great saving can be realized by using the detailed growing operation sparingly. Many application domains have the property that the moving object need only be close to obstacles at a small number of points along the path. These *care points* usually include the start and goal of the path. Elsewhere the requirements on the path are less strict; in fact, it is often undesirable to move close to the obstacles when away from the care points. This property can easily be exploited in the VGRAPH algorithm; instead of executing the detailed growing operation on each of the known obstacles, it need only be executed on those obstacles close to the care points. Away from the care points drastic approximations can safely be used. Complex objects, built up from many polyhedra, can be approximated by a single enclosing polyhedron. The moving object can be similarly approximated so as to further simplify the process. In addition, a very simple form of the growing operation (Appendix 1) can be used

Fig. 10.

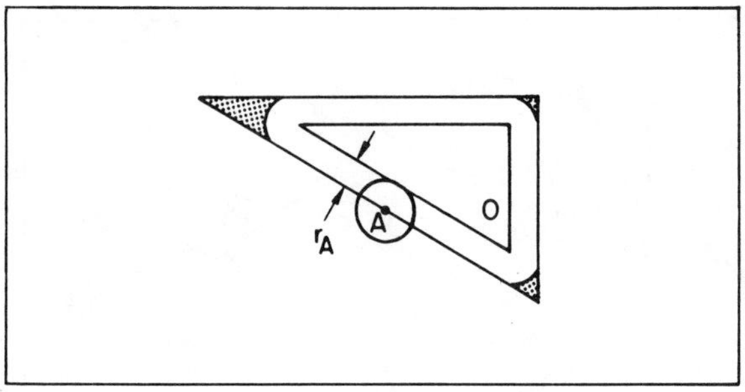

Fig. 11.

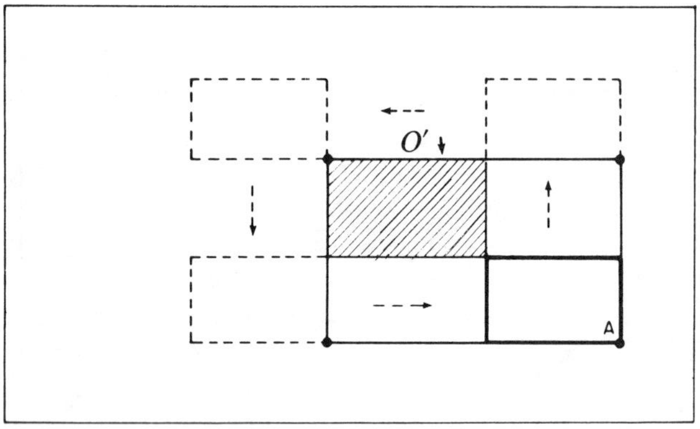

which, at the expense of accuracy, is faster and results in simpler objects.

The key to using this approximation technique is an effective way of determining which objects are close to the care points. Clearly a care point is close to an object if it is inside or close to one of the sides of the grown obstacle resulting from it. This means that the moving object when located at the care point is either inside or close to a side of the object. Approximating both the moving and the stationary objects will cause the care point to be inside the grown obstacle. This condition can be used as a criterion for careful growing. When the moving object is large relative to the obstacles, approximating it as a single object results in detailed growing of too many obstacles. The larger the moving object, the worse a simple approximation is likely to be. In particular, some part of the moving object, relatively far from the care point, will cause the grown obstacle to include the care point. The solution is to have a hierarchic decomposition of the moving object; that is, if the test fails for the roughest description, then use a slightly better approximation. In this way the source of potential collision can be better isolated. The other components of the moving object which are not involved need not be considered carefully. Udupa [9] proposed a similar variable level of detail approximation scheme.

Another way to increase the efficiency of the algorithm is to use heuristics in the graph search operation. This is discussed briefly in Appendix 2 which deals with searching the VGRAPH.

6. Summary and Discussion

This paper has shown how the simple visibility graph algorithm used for navigation of SHAKEY [8] can be extended to more general collision avoidance problems. The mechanism necessary to achieve this involves growing the obstacles and shrinking the moving object to a point. This approach has the desirable property of providing two subproblems, growing the obstacles and searching a visibility graph, which can be pursued independently. A description of our current approach to these problems is included in the appendices.

The most important remaining problem with the VGRAPH algorithm is the quantization of configuration parameters into intervals. Paths that require almost continuous changes of orientation as well as position require small quantization intervals, resulting in many transition obstacle sets, and are therefore expensive to compute.

The VGRAPH algorithm as described in this paper has been implemented in PL/1 on an IBM 370/168. It has been used to plan collision-free trajectories for a seven degree of freedom computer-controlled manipulator [10]; these trajectories have been successfully executed in the laboratory.

Appendix 1: Growing the Obstacles

This appendix describes the component of the VGRAPH algorithm that computes from an obstacle description the shape of the forbidden regions for the position of the moving object's reference point. This is called growing the obstacle. The ideas will be developed in two dimensions and then extended to their three-dimensional counterparts. In the initial two-dimensional case the degrees of freedom used for *growing* will be the x and y position of the moving object.

Consider growing a polygonal obstacle by a circular solid as in Figure 10. The simplest growing algorithm moves each of the sides of the original obstacle by a constant amount r_A and then intersects the lines to obtain the vertices of the grown polygon. The drawbacks of this algorithm are twofold:

(1) It works well only for moving objects that are nearly circular.

(2) It generates wasted space near pointed corners, as seen by the dark shaded regions in Figure 10. This problem can be alleviated by clipping the corners of the grown polygon.

This was the form of the growing algorithm used by Udupa [9].

Fig. 12(a).

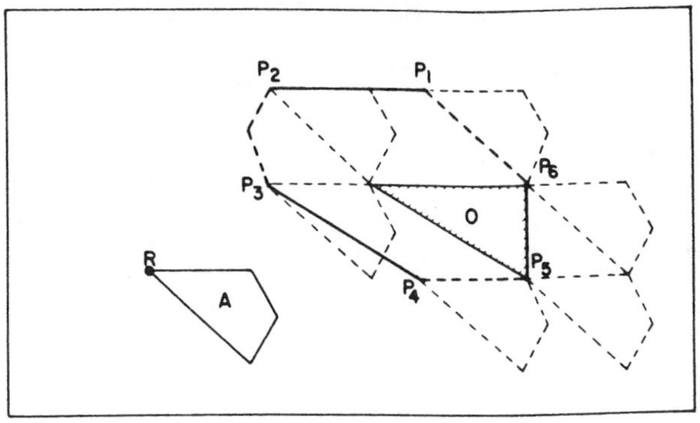

Fig. 12(b).

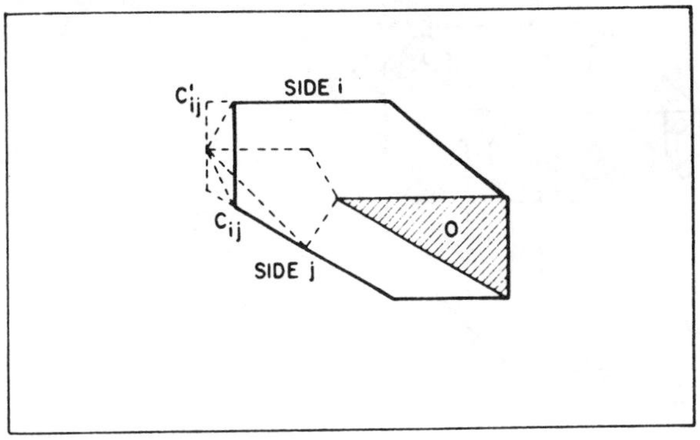

A simple variation of this procedure will solve problem 1 above. Figure 11 shows a convex polygon O and a moving object A; both are rectangular and both are aligned with the global coordinate axes. R, the reference point of A, coincides with one of the vertices of A. The boundary of the forbidden region for the position of A is the locus of positions of R for which A is in contact with O. This locus defines another convex polygon O' shown in Figure 11. Clearly any point inside this polygon implies a collision between A and O. This grown polygon has side $nside_i$ corresponding to each side $side_i$ of the original obstacle. The distance from $nside_i$ to $side_i$ is the perpendicular distance of R from $side_i$ minus the perpendicular distance to the point where A would first contact $side_i$. The distance from $side_i$ to this contact point on A is the minimum perpendicular distance of all of the vertices of A from $side_i$. Once the sides are displaced by this amount, the lines can be intersected to generate the grown polygon.

This method only makes use of the distance from the reference point to the contact point for a side. In polygons with interior angles of less than a right angle the method described above produces wasted space at the vertices. This waste can be reduced by simply cutting the corner at a conservative distance. A more accurate growing procedure can be obtained by determining the actual locus of motion of R along each $side_i$ as the contact point slides along the side. Figure 12(a) shows the line segments traced out by this procedure (bold lines). Notice that the line segments do not intersect and that the endpoints of these line segments correspond to the position of R when the contact point of A with O is at a vertex of O. These positions will be referred to as *maximal locus points*.

To complete the figure, notice that the locus of R as A moves from its contact point with $side_i$ to its contact point with the adjacent $side_j$ traces successive edges of A between the two contact points on A. In the course of connecting all the maximal locus points, all the edges of A are traced out in reverse order (heavy dashed lines). A simple algorithm for growing convex polygons exists which is based on merging a list of displaced edges of O with a reverse order list of displaced edges of A. To simplify the geometry of the grown object, the locus of R between successive contact points can be conservatively estimated by a straight line c'_{ij} which is parallel to the line c_{ij} connecting the maximal locus points but displaced to the position of the point on the actual locus furthest from the line c_{ij} as shown in Figure 12(b).

The *approximate* method for growing a convex polyapproach is to grow each face of the polyhedron independently and then introduce new faces to complete the grown polyhedron. The steps in the process of growing a rectangular solid O with a rotated rectangular solid A are shown in Figure 13. The locus of R as the contact point of A moves along each edge of $face_i$ is called the *maximal locus edge*, Figure 13(a). Such edges define potential new edges for the faces of the grown polyhedron. Each edge of O generates two adjacent maximal loci. These edges have to be connected in a manner analogous to the way in which maximal locus points are connected in a grown polygon, Figure 12(b). The edge has to be displaced to compensate for points on A which are closer to O than the plane defined by the two adjacent maximal locus edges and passing through the corresponding contact points. Faces are introduced to connect each pair of edges of the grown faces arising from a common edge, Figure 13(b). These new faces introduce new edges, each of which connects two points on the grown faces arising from a common vertex of A. All the edges corresponding to a single vertex also define a new set of faces, Figure 13(c). The total number of faces in a polyhedron grown from an object O in this fashion is equal to the sum of the numbers of faces, edges, and vertices of O.

The operation of growing a polyhedron is related to an operation known as *mixing* polyhedra [7]. A mixed polyhedron is the set of points which can be expressed as a linear combination of points from the two starting polyhedra. A polyhedron isomorphic to a grown obstacle can be obtained by mixing the underlying obstacle with a negative image of the moving object, as in convolution.

Fig. 13(a).

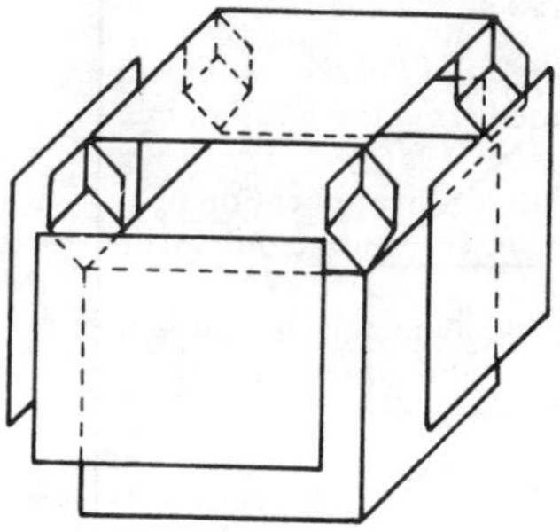

Fig. 13(b).

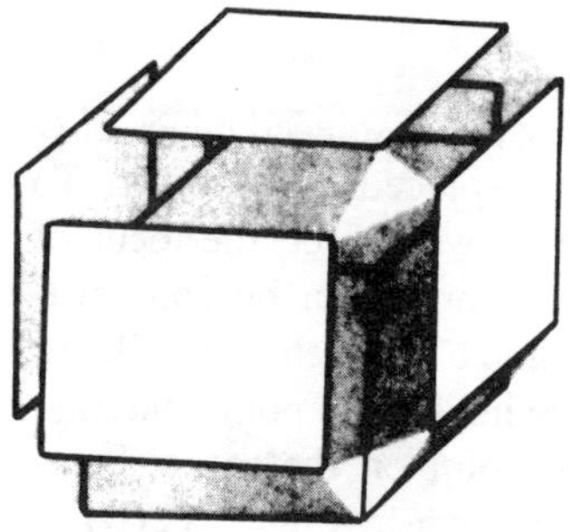

Fig. 13(c).

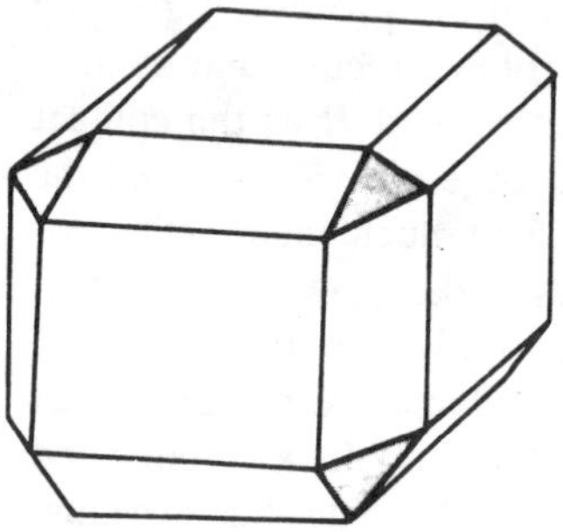

Appendix 2: Finding a Path

A generalized *visibility graph* VG(N, L) contains a node set N and a link set L_{ij} of links between node pairs (n_i, n_j) for which a visibility function $l(n_i, n_j)$ is true. A node is a representation of a region in an n-dimensional parameter space; for each node the associated region is represented by two n-dimensional parameter vectors ϕ_1 and ϕ_2. Individual elements of the difference vector $\delta\phi = \phi_1 - \phi_2$ will be zero when the corresponding parameter has a fixed value, or nonzero when the parameter has a range of values.

The path finding problem is defined as: Given a node set N with an associated parameter vector set, a start node n_s, and a goal node n_g, find a sequence of nodes from n_s to n_g, by way of an ordered set of intermediate nodes $n_1, \ldots. n_k$, which may be null, such that the visibility function for each node pair in sequence:

$$L_1 = l(n_s, n_1)$$
$$L_2 = l(n_1, n_2)$$
$$\ldots$$
$$L_{k+1} = l(n_k, n_g)$$

is true, and that a cost function

$$C = \Sigma c(n_i, n_j)$$

where $c(n_i, n_j) \geq 0$ is minimized.

A direct approach to finding an optimum path is to enumerate all possible paths and choose one for which C is minimum. For node sets whose cardinality is of practical interest (e.g., > 50) the computational load of the direct approach is prohibitive, and more efficient heuristic based search methods may be used.

The $A*$ algorithm of Hart et al. [4] allows use of efficient heuristic information. For each node, an estimate *hhat* is made of the cost h to travel from the node to the goal. Initially, n_s is placed on a list of candidate nodes for examination (the OPEN list). At each step of the algorithm, the node with minimum total path cost estimate (i.e., actual cost of reaching the node along the trial path plus *hhat*) is moved onto a CLOSE list and its minimum cost estimate visible successor nodes are placed on the OPEN list.

Hart et al. have shown that the $A*$ algorithm finds an optimum path when *hhat* is a lower bound estimate of the true cost h. When the estimator for *hhat* is zero and some $c(n_i, n_j) \neg \neq 0 \; n_i, n_j \in N$, the estimate gives no heuristic information to assist in the choice of a path; as $hhat \rightarrow h$, the heuristic information increases and the average number of unsuccessful trial paths is reduced. In the context of this paper, the lower bound requirement for *hhat*, i.e., $hhat \leq h$, may be met by assuming $l(n_i, n_g)$ to be true and computing $hhat_i = c(n_i, n_g)$.

The cost function $c(n_i, n_j)$ may be tailored to suit the requirements of a particular problem environment, for example:

- –distance to be traveled in a subspace of the parameter space;
- –functions of distances in parameter space, for example, time to complete a change based on allowable rate of change of parameters, with the option of selecting the limiting (i.e., slowest)dimension;
- –special costs may be assigned to particular node sequence pairs to allow, for example, costs to be assigned depending on whether the pair allows the motion to proceed without a speed change, as opposed to pairs requiring a change of speed or an intermediate halt.

The form of the visibility function $l(n_i, n_j)$ depends on the semantics of the parameters. Two mutually exclusive classes of parameters are considered as described in Section 3:

—those that may vary continuously, that is, those embodied in the growing operation;
—those that may occupy only discrete ranges, that is, those that are represented by transition obstacle sets.

Note that this distinction between continuous and discrete parameters is an artifact introduced to simplify the handling of spaces of high dimensionality (i.e., $n > 3$) and that the partitioning of the parameter set is not unique: In general, either linear or rotary motions may be represented by continuous or discrete ranges. In the case of discrete range parameters, specific values must be chosen to enable *l* and *hhat* to be evaluated. In all cases of parameter change, a path function defines the motion effect of the change, as either linear or nonlinear motions in parameter space.

In the formalization for path planning described in the body of this paper, an obstacle *A* is grown under some parameter dependent transformation to produce GOS($A_{\phi 1, \phi 2}$) where ($\phi 1$, $\phi 2$) represents a range of parameters. Three parameters represent continuous motion, and the rest represent discrete motions. The visibility function is line of sight in three-dimensional orthogonal cartesian space, with the provision that visibility is possible only when the discrete parameter ranges at the start and end of the path segment overlap, and values assigned to these parameters are in the overlap region.

In many practical situations, the computational cost of evaluating *l* is very much greater than that of evaluating *hhat* for candidate successor nodes. In such cases it is computationally efficient to select a candidate successor node in terms of minimum *hhat* before computing *l*.

Acknowledgments. We would like to thank P. Will for providing the support, guidance, and encouragement for this work. D. Grossman and L. Lieberman contributed useful discussions and criticism as well as programming advice and assistance. Conversations with R. Taylor triggered the development of the multiple obstacle set approach.

Received June 1978; revised August 1979

References

1. Adamowicz, M., and Albano, A. Nesting two-dimensional shapes in rectangular modules. *Comptr. Aided Design 8,* 1 (Jan. 1976), 27-33.
2. Boyse, J.W. Interference detection among solids and surfaces. *Comm. ACM. 22,* 1 (Jan. 1979), 3-9.
3. Braid, I.C. *Designing with Volumes.* Cantab Press, Cambridge, England, 1973.
4. Hart, P., Nilsson, N.J., and Raphael, B. A formal basis for the heuristic determination of minimum cost paths. *IEEE Trans. Syst. Sci. Cybernetics SSC-4,* 2 (July 1968), 100-107.
5. Ignat'yev, M.B., Kulakov, F.M., and Pokrovskiy, A.M. Robot manipulator control algorithms. Rep. No. JPRS 59717, NTIS, Springfield, Va., Aug. 1973.
6. Lozano-Pérez, T. The design of a mechanical assembly system. Rep. No. AI-TR-397, Artif. Intell. Lab., MIT, Cambridge, Mass., Dec. 1976
7. Lyusternik, L.A. *Convex Figures and Polyhedra.* Dover Publications, N.Y., 1963. (Translated from the Russian by T.J. Smith; original copyright Moscow, 1956.)
8. Nilsson, N.J. A mobile automaton: An application of artificial intelligence techniques. Proc. Int. Joint Conf. Artif. Intell., 1969, pp. 509-520.
9. Udupa, S. Collision detection and avoidance in computer controlled manipulators. Ph.D. Th., Calif. Inst. of Technology, Pasadena, Calif., 1977.
10. Will, P.M., and Grossman, D.D. An experimental system for computer controlled mechanical assembly. *IEEE Trans. Comptrs.* (1975), 879-888.

Figure 1. The visibility graph and the shortest collision-free path from **S** to **G**.

Figure 2. The shortest path from **S** to **G** for small circular object **A**. The vertices of the obstacles have been displaced by the radius of **A**.

Figure 3a. The obstacles after growing them by **A**. The shortest path of **A**'s reference point from from **S** to **G** is shown. Each segment of the path is a link in the visibility graph.

Figure 3b. The volume of **A** as it moves along the path shown in Figure 3a.

Figure 4a. A moving object **A** and an obstacle **O** in a polar coordinate system (r,α).

Figure 4b. Representation of **O** and the grown obstacle **O′** in the polar coordinate system. This indicates the ranges of coordinates forbidden to the tip of **A**. Note that neither **O** nor **O′** are polygons.

Figure 4c. Representation of **O** and**O′** in cartesian coordinates (**x,y**). Note that both **O** and **O′** are polygons.

Figure 5. Shortest path from **S** to **G** for **A**. Comparison with Figure 3 shows the effect that rotation of **A** has on the grown obstacles. Note that the shortest path has changed.

Figure 6. The obstacles grown with the rectangular envelope $A_{[\alpha,\beta]}$. Each grown obstacle represents a forbidden region for **A**'s reference point in any orientation between α and β. More accurate envelopes are clearly possible.

Figure 7. An envelope for the moving object of Figure 4 with the addition of **x** and **y** degrees of freedom. This envelope shows the choice of **x** and **y** as the coordinate system for the obstacles and **r** and α as configuration parameters. The area $A_{[\Psi_1,\Psi_2]}$ enclosed by the dotted line is the envelope of the moving object over the range of **r** and α.

Figure 8. Configuration parameters θ and ρ describe the internal geometry of the moving object.

Figure 9. In general, the shortest path amongst three-dimensional objects does not necessarily traverse vertices of the obstacles.

Figure 10. Growing a polyhedral object by a circular solid of radius r_A. Note the wasted space (shown shaded) generated by growing the obstacle by simply moving the sides and re-intersecting.

Figure 11. Growing a rectangular object **O** with a rectangular moving object **A**. The outline of the grown obstacle **O′** is the locus of the reference point of **A** as it slides along the outline of **O**.

Figure 12a. An exact growing algorithm. The lines in **bold** are the locus of **R** as **A** slides along each edge of **O**. The P_i are called *maximal locus points*. The heavy dashed lines trace the edges of **A** in reverse order.

Figure 12b. The line c'_{ij} indicates a conservative estimate of the shape of **A** to simplify the resulting grown obstacle.

Figure 13a. The surfaces generated by a moving object sliding along the surfaces of the obstacle. Each new edge is a *maximal locus edge*.

Figure 13b. Connecting the maximal locus edges with rectangular faces. This is done with an approximation similar to that shown in Figure 12b.

Figure 13c. Adding triangular patches over the vertices to complete the figure. This is also an approximation.

Spatial Planning: A Configuration Space Approach

TOMÁS LOZANO-PÉREZ

***Abstract*—This paper presents algorithms for computing constraints on the position of an object due to the presence of other objects. This problem arises in applications that require choosing how to arrange or how to move objects without collisions. The approach presented here is based on characterizing the position and orientation of an object as a single point in a configuration space, in which each coordinate represents a degree of freedom in the position or orientation of the object. The configurations forbidden to this object, due to the presence of other objects, can then be characterized as regions in the configuration space, called *configuration space obstacles*. The paper presents algorithms for computing these configuration space obstacles when the objects are polygons or polyhedra.**

***Index Terms*—Computational geometry, obstacle avoidance, robotics.**

I. Introduction

INCREASINGLY, computer applications deal with models of two- and three-dimensional objects. Partly because of this, there has been rapid growth of interest in efficient algorithms for geometric problems. For example, research has focused on algorithms for 1) computing convex hulls [16], [32], 2) intersecting convex polygons and polyhedra [6], [27], [36], [38], 3) intersecting half-spaces [11], [33] 4) decomposing polygons [35], and 5) closest-point problems [37].[1] Another class of geometric problems involves placing an object among other objects or moving it without colliding with nearby objects. We call this class of problems: *spatial planning* problems. The following are representative applications where spatial planning plays an important role:

1) the layout of templates on a piece of stock [1]–[3], [13] so as to minimize the area of stock required:

2) machining a part using a numerically controlled machine tool [50], which requires plotting the path of one or more cutting surfaces so as to produce the desired part;

3) the layout of an IC chip [48] to minimize area, subject to geometric design constraints;

4) automatic assembly using an industrial robot [22], [23], [43], which requires grasping objects, moving them without collisions, and ultimately bringing them into contact.

One common spatial planning problem is to determine where an object A can be placed, inside some specified region R, so that it does not collide with any of the objects B_j already

Manuscript received August 15, 1980; revised June 29, 1981 and June 15, 1982. This work was supported in part by the Office of Naval Research under Contract N00014-81-K-0334 and in part by the Defense Advanced Research Projects Agency under Office of Naval Research Contracts N00014-80-C-0505, N00014-82-K-0494.

The author is with the Artificial Intelligence Laboratory, Massachusetts Institute of Technology, Cambridge, MA 02139.

[1] The references cited here are representative of the current literature; they are by no means a complete survey.

Reprinted from *IEEE Transactions on Computers*, Vol. C-32, No. 2, February, 1983. Copyright © 1983 by The Institute of Electrical and Electronics Engineers, Inc. All rights reserved.

placed there. We call this the **Findspace** problem. Finding where to place another suitcase in the trunk of a car is an example of Findspace, where the new suitcase is A, the previous suitcases are the B_j, and the inside of the trunk is R. A related problem is to determine how to move A from one location to another without causing collisions with the B_j. We call this the **Findpath** problem. For example, moving the suitcase mentioned above from its initial position outside the trunk to the desired position in the trunk, requires computing a path for the suitcase (and the mover's arms) that avoids the rest of the car. These two geometric problems, Findspace and Findpath, are the subject of this paper. Previous work on Findspace and Findpath is surveyed in Section VIII.

Findspace and Findpath can be defined more formally as follows.

Definition: Let R be an object that completely contains k_B other, possibly intersecting, objects B_j.

1) *Findspace*—Find a position for A, inside R, such that for all B_j, $A \cap B_j = \emptyset$. This is called a *safe position.*

2) *Findpath*—Find a path for A from position s to position g such that A is always in R and all positions of A on the path are safe. This is called a *safe path.*

Throughout this paper, the objects R and B_j are fixed convex polyhedra (or polygons). We take A to be the set union of k_A (possibly intersecting) convex polyhedra (or polygons) A_i. For example, A may be a convex decomposition of a nonconvex polyhedron [35]. Fig. 1 illustrates the definitions of Findspace and Findpath for convex polygons.

The algorithm presented here for the Findspace and Findpath problems has two main steps: 1) building a data structure that captures the geometric constraints and 2) searching the data structure to find the solution. In this paper we focus on algorithms for constructing the appropriate data structure. In this sense, the approach is similar to many geometric search algorithms, for example, the Voronoi polygon approach to closest-point problems [37]. In the Findspace and Findpath algorithms described here, we build geometric objects, called *configuration space obstacles*, that represent all the positions of the object A that cause collisions with the B_j. Given these objects, Findspace and Findpath correspond to the simpler problems of finding a single point (a position of A) or a path (a sequence of positions of A), outside of the configuration space obstacles. The advantage of this formulation is that the intersection of a point relative to a set of objects is easier to deal with than the intersection of objects among themselves.

Representing the positions of rigid objects requires specifying all their degrees of freedom, both translations and rotations. We will use the notion of configuration to unify our treatment of degrees of freedom. The *configuration* of a polyhedron is a set of independent parameters that characterize the position of every point[2] in the object. The configuration of

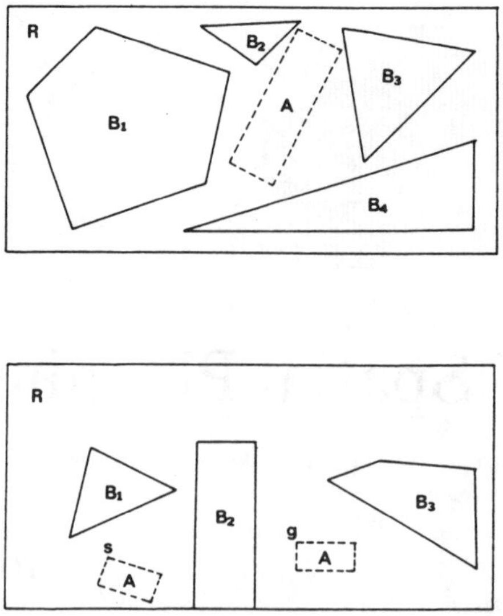

Fig. 1. R, B_j, and A for Findspace and Findpath problems in two dimensions. (a) The Findspace problem is to find a configuration for A where A does not intersect any of the B_j. (b) The Findpath problem is to find a path for A from s to g that avoids collisions with the B_j.

a polyhedron is defined relative to an initial configuration. In this initial configuration, by convention, a fixed vertex of the polyhedron coincides with the origin of the global coordinate frame. For a polyhedron A, this vertex is called the *reference vertex of A*, or rv_A.

The number of parameters required to specify the configuration of a k-dimensional polyhedron, A, relative to its initial configuration, is d, where $d = k + \binom{k}{2}$ [7, p. 10]; k parameters are required to specify the position of rv_A in $\mathcal{R}^k$ and $\binom{k}{2}$ are required to specify the orientation[3] of A. Thus, the configuration of A can be regarded as a point $x \in \mathcal{R}^d$; this d-dimensional space of configurations of A is denoted $Cspace_A$. A in configuration x is $(A)_x$; A in its initial configuration is $(A)_0$. When an object's configuration is fixed, e.g., the B_j mentioned earlier, we leave it unspecified.

If A is a polygon in $\mathcal{R}^2$, the configuration of A is specified by (x, y, θ), where (x, y) is the position of rv_A and θ is the rotation of A, about rv_A, relative to $(A)_0$. That is, for polygons in $\mathcal{R}^2$, $k = 2$, configurations are elements of $\mathcal{R}^3$, $d = 2 + 1$. If the orientation of A is fixed, (x, y) alone is sufficient to specify the polygons configuration; therefore, $Cspace_A$ is simply the (x, y) plane. If A is a polyhedron in $\mathcal{R}^3$, $k = 3$, the configurations of A are elements of $\mathcal{R}^6$, $d = 3 + 3$. That is, three translations and three rotations are needed to specify the position and orientation of a rigid three-dimensional object [7].

Not all possible configurations in $Cspace_A$ represent legal configurations of A; in particular, configurations of A where $A \cap B_j \neq \emptyset$ are illegal because they would cause collisions.

[2] In what follows, all geometric entities—points, lines, edges, planes, faces, and objects—will be treated as (infinite) sets of points. All of these entities will be in some $\mathcal{R}^n$, an n-dimensional real Euclidean space. a, b, x, and y shall denote points of $\mathcal{R}^n$, as well as the corresponding vectors. A, B, and C shall denote sets of points in $\mathcal{R}^n$, while I and K shall denote sets of integers. γ, θ, and β, shall denote reals, while i, j, k, l, m, n shall be used for integers. The coordinate representation of a point $c \in \mathcal{R}^n$, shall be $c = (\gamma_i) = (\gamma_1, \cdots, \gamma_n)$.

[3] The relative rotation of one coordinate system relative to another can be specified in terms of $\binom{k}{2}$ angles usually referred to as Euler angles [7]. These angles indicate the magnitude of three successive rotations about specified axes. Many conventions for the choice of axes exist, any of which is suitable for our purposes.

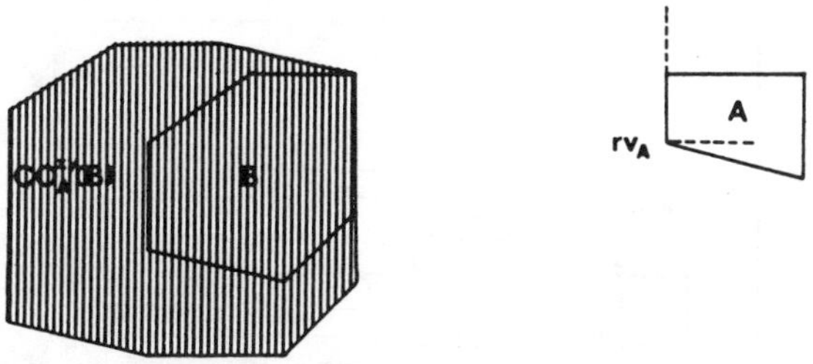

Fig. 2. The $Cspace_A$ obstacle due to B, for fixed orientation of A.

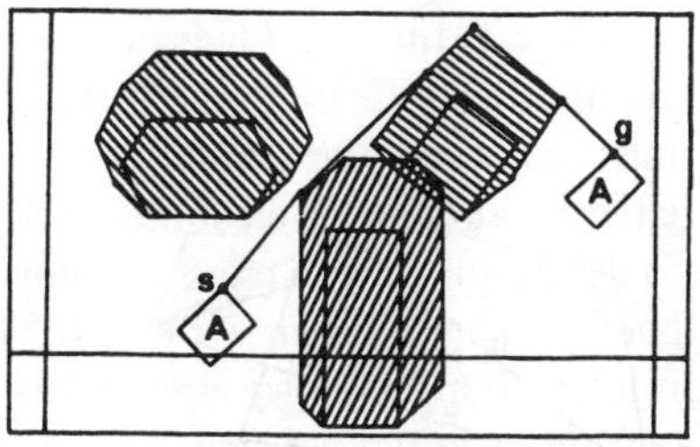

Fig. 3. The Findpath problem and its formulation using the $CO_A^{xy}(B_j)$. The shortest collision-free paths connect the origin and the destination via the vertices of the $CO_A^{xy}(B_j)$ polygons.

These illegal configurations are the result of a mapping of the B_j into $Cspace_A$. This mapping exploits two fundamental properties of objects: 1) their *rigidity*, which allows their configurations to be characterized by a few parameters and 2) their *solidity*, which requires that a point not be inside more than one object.

Definition: The $Cspace_A$ obstacle due to B, denoted $CO_A(B)$, is defined as follows:

$$CO_A(B) \equiv \{x \in Cspace_A \mid (A)_x \cap B \neq \varnothing\}.$$

Thus, if $x \in CO_A(B)$ then $(A)_x$ intersects B, therefore x is not safe. Conversely, any configuration $x \notin CO_A(B_j)$ (for all objects B_j) is safe. If A is a convex polygon with fixed orientation, the presence of another convex polygon B constrains the configuration of A, in this case simply the position of rv_A, to be outside of $CO_A(B)$, a larger convex polygon, shown as the shaded region in Fig. 2. The choice of a different vertex as rv_A would result in translating $CO_A(B)$ relative to B in the figure.

Just as $CO_A(B)$ defines those configurations for which A intersects B, $CI_A(B)$ defines those configurations for which A is completely inside B.

Definition: The $Cspace_A$ *interior* of B, denoted $CI_A(B)$, is defined as follows:

$$CI_A(B) \equiv \{x \in Cspace_A \mid (A)_x \subseteq B\}.$$

Clearly, $CI_A(B) \subseteq CO_A(B)$. Moreover, it is easy to see that for A to be inside B, it must be outside of B's complement. Therefore, letting $-X$ represent the complement of the set X, $CI_A(B) = -CO_A(-B)$.

A superscript to $CO_A(B)$ and $CI_A(B)$ will be used to indicate the coordinates of the configurations in the sets, e.g., $CO_A^{xy}(B)$ and $CO_A^{xy\theta}(B)$ denote sets of (x, y) and (x, y, θ) values, respectively. When no superscript is used, as in $CO_A(B)$, we mean sets of configurations in the complete $Cspace_A$ for a polyhedron of A's dimension, e.g., $\mathcal{R}^6$ for a three-dimensional polyhedron.

Using the definitions of *Cspace* obstacle and *Cspace* interior, Findspace and Findpath can be expressed as equivalent problems that involve placing one point, the configuration of A, relative to the $Cspace_A$ objects $CO_A(B_j)$ and $CI_A(R)$. In general, these problems are equivalent to finding either a single configuration of A or a connected sequence of configurations of A (a path), outside all of the $CO_A(B_j)$, but inside $CI_A(R)$

If A and all of the B_j are polygons and if the orientation of A is fixed, then the $CO_A^{xy}(B_j)$ are also polygons. In that case, the shortest[4] safe paths for A are piecewise linear paths connecting the start and the goal configurations via the vertices of the $CO_A^{xy}(B_j)$ polygons; see Fig. 3. Therefore, Findpath can be formulated as a graph search problem. The graph is formed by connecting all pairs of $CO_A^{xy}(B_j)$ vertices (and the start and goal) that can "see" each other, i.e., can be connected by a straight line that does not intersect any of the obstacles. The shortest path from the start to the goal in this *visibility graph* (**Vgraph**) is the shortest safe path for A among the B_j [24]. This algorithm solves two-dimensional Findpath problems when the orientation of A is fixed, but the paths it finds are very susceptible to inaccuracies in the object model. These paths touch the $Cspace_A$ obstacles; therefore, if the model were exact, an object moving along this type of path would just touch the obstacles. Unfortunately, an inaccurate model or a slight error in the motion may result in a collision. Furthermore, the Vgraph algorithm does not find optimal paths among three-dimensional obstacles [24]. Alternative techniques for pathfinding are treated in [23].

Here is a brief summary of the rest of the paper. Section II presents algorithms for computing $CO_A^{xy}(B)$. Section III characterizes $CO_A^{xy\theta}(B)$, the $Cspace_A$ obstacle for polygons that are allowed to rotate. Section IV describes an algorithm for computing $CO_A^{xyz}(B)$, the $Cspace_A$ obstacle for polyhedra with fixed orientation. Section V characterizes $CO_A(B)$, the $Cspace_A$ obstacle for polyhedra that are allowed to rotate. Section VI deals with slice projection, an approximation technique for higher dimensional $Cspace_A$ obstacles, for example, those obtained when a polyhedron is allowed to rotate. Section VII discusses the extensions to the Findspace and Findpath algorithms needed to plan the motions of industrial robots. Section VIII discusses related work in spatial planning.

II. Computing $CO_A^{xy}(B)$

The crucial step in the *Cspace* approach to Findspace and Findpath is computing the $Cspace_A$ obstacles for the B_j. Thus far, we have only provided an implicit definition of $CO_A(B)$; we now provide, in Theorem 1, a characterization of $CO_A^{xy}(B)$ and $CI_A^{xy}(B)$ in terms of set sums that will lead us to an efficient algorithm for computing $Cspace_A$ obstacles.

Set sum, *set difference*, and *set negation* are defined on sets of points, eqivalently vectors, in $\mathcal{R}^n$ as follows:

[4] This assumes Euclidean distance as a metric. For the optimality conditions using a rectilinear (Manhattan) metric, see [19].

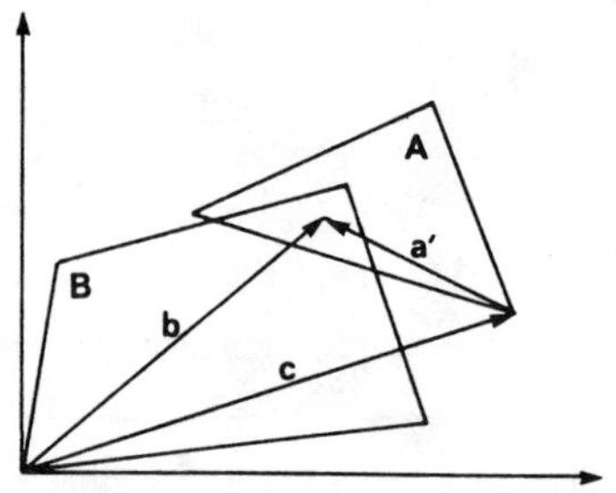

Fig. 4. Illustration of Theorem 1. Any location of rv_A, denoted c, for which A and B have a point in common (expressible as b and a'), can be expressed as $c = b - a'$. Therefore, $CO_A^{xy}(B) = B \ominus (A)_0$.

$$A \oplus B = \{a + b \mid a \in A, b \in B\}$$

$$A \ominus B = \{a - b \mid a \in A, b \in B\}$$

$$\ominus A = \{-a \mid a \in A\}.$$

If a set A consists of a single point a, then $a \oplus B = \{a\} \oplus B = A \oplus B$. Also, $A \ominus B = A \oplus (\ominus B)$. Note that, typically $A \oplus A \neq \{2a \mid a \in A\}$ and $A \ominus A \neq \emptyset$, although $A \oplus B = B \oplus A$.

We can characterize the $Cspace_A$ obstacle for objects with fixed orientation as a set difference of the objects' point sets:

Theorem 1: For A and B, sets in $\mathcal{R}^2$, $CO_A^{xy}(B) = B \ominus (A)_0$.

Proof: If c is an (x, y) configuration of A, then $(A)_c = c \oplus (A)_0$. Therefore, if $a \in (A)_c$, then there exists $a' \in (A)_0$ such that $a = c + a'$; see Fig. 4. Thus, if $b \in B \cap (A)_c$, and therefore $b \in (A)_c$, then, for some $a' \in (A)_0$, $b = c + a'$ and therefore $c = b - a'$. Clearly, the converse is also true. ■

This theorem extends naturally to higher dimensions, e.g., $CO_A^{xyz}(B)$, as long as the orientation of A is fixed.

If A and B are convex, then $A \oplus B$ and $A \ominus B$ are also convex [16, p. 9]; therefore $CO_A^{xy}(B)$ is convex. In fact, if A and B are convex polygons, then $CO_A^{xy}(B)$ is also a convex polygon. We can now use well-known properties of convex polygons and the set sum operation [5] to state an $O(n)$ algorithm for computing $CO_A^{xy}(B)$ when A and B are convex n-gons.

Let $\pi(A, u)$ denote the *supporting line* [5] of A with outward normal u. All of A is in one of the closed half-spaces bounded by $\pi(A, u)$ and u points away from the interior of A. $\pi(A, u) \cap A$ is the set of boundary points of A on the supporting line.

Lemma 1 [5]*:* If A and B are convex sets and u is an arbitrary unit normal, then

$$\pi(A \oplus B, u) \cap (A \oplus B) = (\pi(A, u) \cap A) \oplus (\pi(B, u) \cap B). \quad (1)$$

Fig. 5 illustrates this lemma.

Lemma 2[5]*:*

a) Let $s(a_1, a_2)$ be a line segment and b a point, then $s(a_1, a_2) \oplus b = s(a_1 + b, a_2 + b)$ is a line segment parallel to $s(a_1, a_2)$ and of equal length. See Fig. 6(a).

b) Let $s(a_1, a_2)$ and $s(b_1, b_2)$ be parallel line segments such that $(a_2 - a_1) = k(b_2 - b_1)$ for $k > 0$. Then $s(a_1, a_2) \oplus s(b_1, b_2) = s(a_1 + b_1, a_2 + b_2)$ and the length of the sum is the sum of the lengths of the summands. See Fig. 6(b).

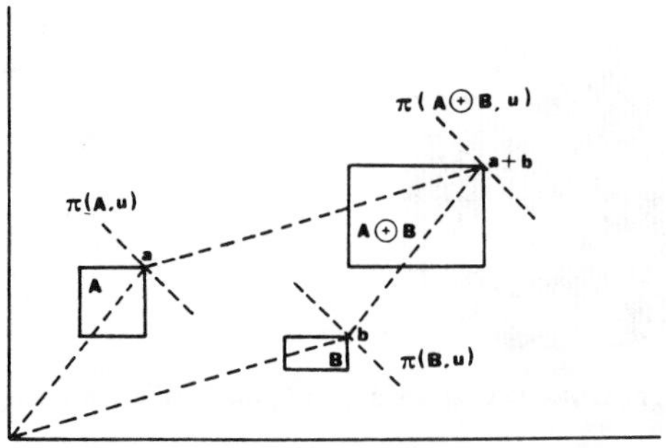

Fig. 5. Illustration for Lemma 1.

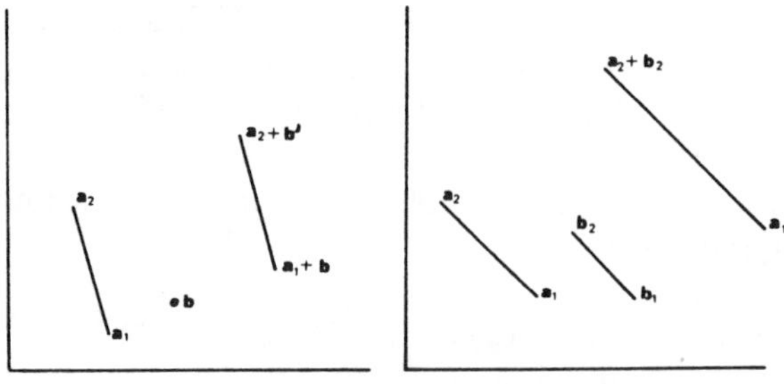

Fig. 6. Illustration for Lemma 2.

Theorem 2: For A a convex n-gon and B a convex m-gon, $CO_A^{xy}(B)$ can be computed in time $O(n + m)$.

Proof: For a polygon P, assume the jth edge, $s(v_j, v_{j+1})$, makes the angle θ_j with the x-axis. If $\theta(u)$ is the angle u makes with the x axis, then

$$\pi(P, \mathrm{u}) \cap P = \begin{cases} v_j, & \text{if } \theta_{j-1} < \theta(u) < \theta_j \\ s(v_j, v_{j+1}), & \text{if } \theta(u) = \theta_j \\ v_{j+1}, & \text{if } \theta_j < \theta(u) < \theta_{j+1}. \end{cases}$$

We now apply Lemmas 1 and 2. Depending on the angle $\theta(u)$, each term on the right-hand side of (1) is either a line segment (edge) or a single point (vertex). It follows from Lemma 2 that the term on the left of (1) is one of:

a) a new vertex, when two vertices are combined;

b) a displaced edge, when an edge and a vertex are combined (Lemma 2a);

c) an edge, corresponding to a pair of displaced end-to-end edges, when two edges are combined (Lemma 2b).

As u rotates counterclockwise, the boundary of $A \oplus B$ is formed by joining a succession of these elements. Note that, because of the convexity of A and B, each edge is encountered exactly once [25, p. 13].

Polygons are stored as lists of vertices in the same order as they are encountered by the counterclockwise sweep of u. This is equivalent to a total order on the edges, based on the angle that the edge makes with the x axis. These lists for A and B can be merged into a single total order on the angle in linear time, as they are traversed. At each step, we construct a new vertex (edges need not be represented explicitly) by the method indicated in the lemmas. The time for constructing the new vertices is bounded by a constant, since it involves at most two

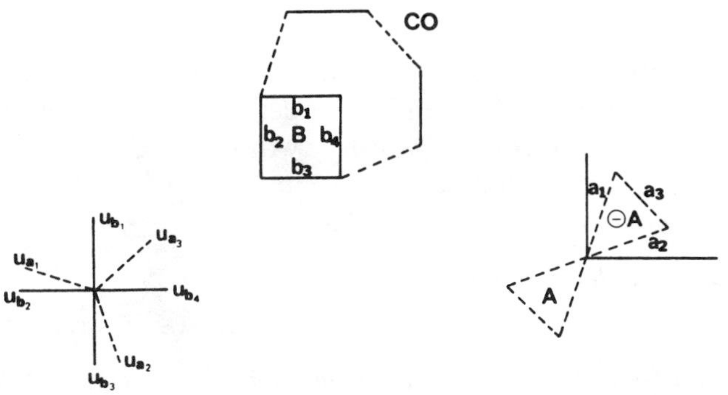

Fig. 7. The edges of $B \ominus (A)_0$, when A and B are convex polygons, are found by merging the edge lists of B and $\ominus(A)_0$, ordered on the angle their normals make with the positive x axis.

vector additions. Thus $A \oplus B$ can be computed in linear time during a scan of the vertices of A and B; see Fig. 7. An implementation of this algorithm is included in Appendix I. Thus, $B \ominus (A)_0$ can be computed in linear time by first converting each vertex a_i to $rv_A - a_i$; see Fig. 7. ■

This algorithm is similar to Shamos' diameter algorithm using antipodal pairs [36], but instead of dealing with two supporting lines on one polygon, it deals with two polygons and one supporting line on each. An algorithm, essentially identical to the one in Theorem 2, has been recently described in [39]. The proof Theorem 2, however, will be used in subsequent sections to derive characterizations and algorithms for other *Cspace* entities.

When A or B are nonconvex polygons, $CO_A^{xy}(B)$ can be computed by an extension of the algorithm above. The extension relies on decomposing the boundaries of the polygons into a sequence of polygonal arcs whose internal angles, i.e., the angle facing the inside of the polygon, are each less than π. The algorithm of Theorem 2 can then be applied to pairs of arcs; the result is a polygon whose boundary, in general, intersects itself. The algorithm requires, in the worst case, $O(n \times m)$ steps.

An alternative method of computing $CO_A^{xy}(B)$ for nonconvex A and B can be used when convex decompositions of A and B are available, e.g., the objects may have been designed by set operations on convex primitives. If A is represented as the union of k_A objects A_i, and B is the union of k_B objects B_j, then Theorem 3 follows directly from the definition of $CO_A(B)$.

Theorem 3: If $A = \bigcup_{i=1}^{k_A} A_i$ and $B = \bigcup_{j=1}^{k_B} B_j$:

$$CO_A(B) = \bigcup_{i=1}^{k_A} \bigcup_{j=1}^{k_B} CO_{A_i}(B_j).$$

This theorem simply says that the set of configurations that cause a collision between A and B are those for which any part of A intersects any part of B.

III. Characterizing $CO_A^{xy\theta}(B)$

We have so far restricted our attention to cases where A remains in a fixed orientation. In these cases, all the geometric constraints for spatial planning are embodied in $CO_A^{xy}(B)$. However, $CO_A^{xy}(B)$ is only the cross section, for fixed θ, of the three-dimensional full configuration space obstacle for polygons, $CO_A^{xy\theta}(B)$. In this section, we consider $CO_A^{xy\theta}(B)$, when A and B are convex, by examining changes in its cross section as θ changes.

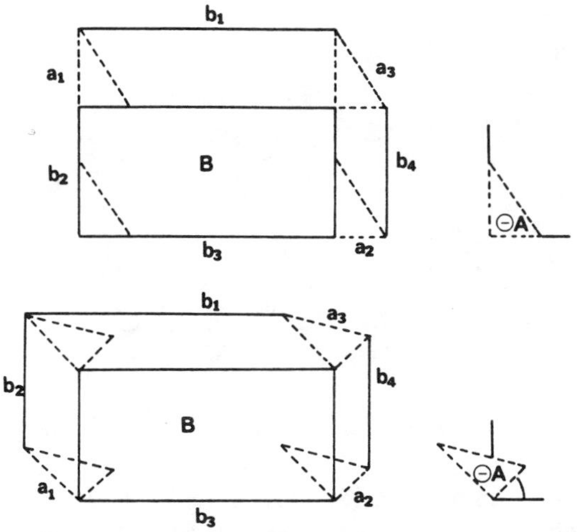

Fig. 8. When A (and $\ominus(A)_0$) rotates by θ, the e_i^a rotate around b_j and the e_j^b are displaced. When an e_i^a is aligned with an e_j^b for some θ, any additional rotation of A will interchange the order in which they are encountered during the counterclockwise scan of Theorem 2. For example, in the top figure a_1 appears before b_2 on the boundary of $CO_A^{xy}(B)$ and is (nearly) aligned with b_2; in the bottom figure, after additional rotation of A, a_1 appears after b_2 on the boundary of $CO_A^{xy}(B)$. Therefore, the top figure illustrates A in the orientation $\theta_{i,j}^*$.

For fixed θ, we know from Theorem 2 that the edges of $CO_A^{xy}(B)$ are either displaced edges of A or displaced edges of B. Therefore, for a small rotation of A, we expect that those edges of $CO_A^{xy}(B)$ corresponding to the edges of A will rotate, but the edges corresponding to edges in B will not. As A rotates, however, the rate of displacement of these edges changes discontinuously when edges of A and B become parallel, as illustrated below.

Let $vert(B)$ denote the set of vertices of a polygon B, b_j be the position vector of the jth member of $vert(B)$, and $a_i(\theta)$ be the position of the ith member of $vert(\ominus(A)_0)$, which depends on θ. Assume that A and B have no parallel edges. For fixed θ, the proof of Theorem 2 shows that each edge of $CO_A^{xy}(B)$ can be expressed as one of

$$e_i^a = b_j \oplus s(a_i(\theta), a_{i+1}(\theta)) \tag{2a}$$

$$e_j^b = a_i(\theta) \oplus s(b_j, b_{j+1}). \tag{2b}$$

The order in which the a_i and b_j are encountered in the counterclockwise scan described in Theorem 2 determines the (i, j) pairings of vertices and edges. For example, in (2b), $a_i(\theta)$ is the vertex of $\ominus(A)_0$ that is on the supporting line of $\ominus(A)_0$ which is parallel to $s(b_j, b_{j+1})$, i.e., if u_j is the normal to $s(b_j, b_{j+1})$, then $a_i = \pi(\ominus(A)_0, u_j) \cap \ominus(A)_0$.

Equation (2) shows that, for a given pairing of edges and vertices, the e_i^a rotate around b_j, while the e_j^b are simply displaced by the vector $a_i(\theta)$; see Fig. 8. The discontinuous changes occur at values of θ, denoted $\theta_{i,j}^*$, where the ith edge of A becomes parallel to the jth edge of B. For values of θ just greater than these $\theta_{i,j}^*$, some pair of edges has a different order in the scan of Theorem 2 from what they had when θ was just less than $\theta_{i,j}^*$; see Fig. 8. Therefore, at each $\theta_{i,j}^*$, the pairings between edges and vertices change. For A a convex n-gon and

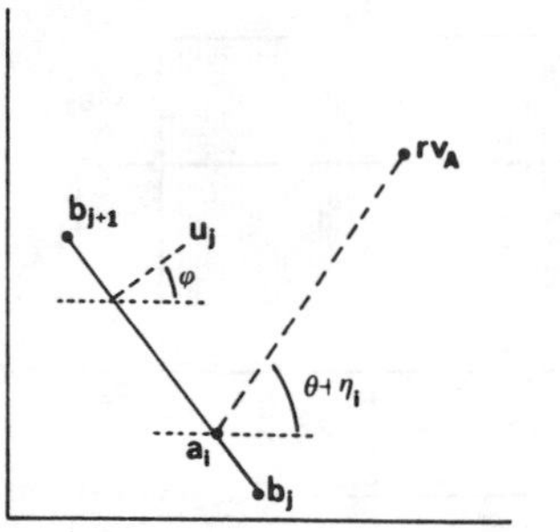

Fig. 9. Illustration of terms used in (3).

B a convex m-gon, there are $O(n \times m)$ such $\theta^*_{i,j}$ in $CO_A^{xy\theta}(B)$.

Between discontinuities, the lines defined by e_j^b edges have a simple dependence on θ. The edge $s(b_j, b_{j+1})$ is on a line whose vector equation[5] is: $\langle u_j, x\rangle = \langle u_j, b_j\rangle$ where u_j is the constant unit normal to $s(b_j, b_{j+1})$. Let $a_i(\theta)$ make the angle $\theta + \eta_i$ with the x axis, with η_i constant, and u_j make the angle ϕ_j with the x axis. Then, if $\|a_i\|$ represents the vector magnitude of a_i, the equation for the line including e_j^b is

$$\begin{aligned}\langle u_j, x\rangle &= \langle u_j, a_i(\theta) + b_j\rangle \\ &= \|a_i\| \cos(\theta + \eta_i - \phi_j) + \langle b_j, u_j\rangle. \qquad (3)\end{aligned}$$

The terms are illustrated in Fig. 9. This equation holds between discontinuities.

The equation for the e_i^a edges is not as simple, because the orientation of the edge changes with θ, i.e., the edge is a cross section of a curved surface in (x, y, θ) space. Let $v_i(\theta)$ be the normal vector to $s(a_i(\theta), a_{i+1}(\theta))$; then the vector equation of this curved surface is

$$\langle v_i(\theta), x\rangle = \langle v_i(\theta), a_i(\theta) + b_j\rangle. \qquad (4)$$

This equation also holds only between discontinuities, i.e., for each pairing of vertices and edges.

Equations (3) and (4) define the shape of $CO_A^{xy\theta}(B)$. Since the resulting object is not polyhedral, however, it cannot be used for the Vgraph algorithm. Section VI discusses a technique for constructing lower dimension polyhedral approximations of $Cspace_A$ obstacles and an extended Vgraph algorithm to use them (see also [23]). These equations for $CO_A^{xy\theta}(B)$ are the basis for the Findpath algorithm described in [10].

IV. Algorithm for $CO_A^{xyz}(B)$

Theorem 1 applies also to $CO_A^{xyz}(B)$, but the algorithm of Theorem 2 cannot be extended to polyhedra, since there is no similar total ordering of the faces of a polyhedron. However, Theorem 4 below follows easily from Theorem 1 and provides a way to compute $CO_A^{xyz}(B)$ for convex polyhedra A and B. The method of Theorem 4 also applies to polygons, but is much less efficient than the linear algorithm of Theorem 2. Theorem 4 provides the basis for approximating $CO_A^{xyz}(B)$ when A and B are nonconvex, simply by replacing A or B by their convex hulls.

Let $conv(A)$ denote the *convex hull* of a polygon A, i.e., the smallest convex polygon enclosing A. We know that $conv(A)$, for a nonempty set $A \subseteq \mathcal{R}^d$, is $\{\sum_{i=1}^n \gamma_i x_i \mid x_i \in A, \gamma_i \geq 0, \sum_{i=1}^n \gamma_i = 1\}$, for some n [16, p. 15]. This definition says that every point in the convex hull of A can be written as a convex linear combination of points in A.

Theorem 4: For polyhedra A and B,

$$\begin{aligned}conv(A \oplus B) &= conv(A) \oplus conv(B) \\ &= conv(vert(A) \oplus vert(B)).\end{aligned}$$

Proof: First show that $conv(A \oplus B) = conv(A) \oplus conv(B)$.

($\supseteq$): By the definition of $\oplus$, if $x \in conv(A) \oplus conv(B)$, then there exist $a \in conv(A)$ and $b \in conv(B)$ such that $x = a + b$. The definition of convex hull states that any $a \in conv(A)$ can be expressed as a convex linear combination of points in A; likewise for any $b \in conv(B)$. Therefore, there exist $a_i \in A$, $b_i \in B$, $\sum_i \gamma_i = 1$, $\gamma_i \geq 0$, $\sum_j \beta_j = 1$, and $\beta_j \geq 0$ such that $a = \sum_i \gamma_i a_i$ and $b = \sum_j \beta_j b_j$ and thus

$$\begin{aligned}x = a + b &= \left(\sum_i \gamma_i a_i\right) + b = \sum_i \gamma_i (a_i + b) \\ &= \sum_i \gamma_i \left(a_i + \sum_j \beta_j b_j\right) \\ &= \sum_i \gamma_i \left(\sum_j \beta_j a_i + \sum_j \beta_j b_j\right) \\ &= \sum_i \gamma_i \sum_j \beta_j (a_i + b_j) = \sum_i \sum_j \gamma_i \beta_j (a_i + b_j).\end{aligned}$$

But, since $\sum_i \sum_j \gamma_i \beta_j = 1$ and $\gamma_i \beta_i \geq 0$, x is a convex linear combination of points in $A \oplus B$ and therefore belongs to its convex hull. Therefore $conv(A) \oplus conv(B) \subseteq conv(A \oplus B)$.

($\subseteq$): If $x \in conv(A \oplus B)$, then there exist $\gamma_i \geq 0$ with $\sum_i \gamma_i = 1$, $a_i \in A$, and $b_i \in B$, such that

$$x = \sum_i \gamma_i (a_i + b_i) = \sum_i \gamma_i a_i + \sum_i \gamma_i b_i.$$

Therefore, $x \in conv(A) \oplus conv(B)$.

This establishes that $conv(A \oplus B) = conv(A) \oplus conv(B)$. Replacing A by $vert(A)$ and B by $vert(B)$, and using the fact that $conv(A) = conv(vert(A))$ [16], shows that $conv(vert(A) \oplus vert(B)) = conv(A) \oplus conv(B)$. ■

Corollary: For convex polyhedra A and B, $CO_A^{xyz}(B) = conv(vert(B) \ominus vert((A)_0))$.

Proof: $A \oplus B$ is convex, when A and B are both convex; thus, $A \oplus B = conv(A \oplus B)$. By Theorem 4, $A \oplus B = conv(vert(A) \oplus vert(B))$. Using Theorem 1 establishes the corollary. ■

Several algorithms exist for finding the convex hull of a finite set of points on the plane, e.g., [15] and [32]. The latter [32] also describes an efficient algorithm for points in $\mathcal{R}^3$. These algorithms are known to run in worst case time $O(v \log v)$, where v is the size of the input set. Therefore, Theorem 4 leads immediately to an algorithm for computing $CO_A^{xyz}(B)$ and an upper bound on the computational complexity of the problem or convex polyhedra.

Theorem 5: For convex polyhedra $A, B \subseteq \mathcal{R}^3$, each with

[5] The *scalar (dot) product* of vectors a and b will be denoted $\langle a, b\rangle$.

$O(n)$ vertices, $CO_A^{xyz}(B)$ can be computed in time $O(n^2 \log n)$.

Proof: The set $vert(B) \ominus vert((A)_0)$ is of size $O(n^2)$. Applying an $O(v \log v)$ convex hull algorithm to this set gives an $O(n^2 \log n)$ algorithm for computing $CO_A^{xyz}(B)$. ■

The Vgraph algorithm discussed in Section I can be extended directly to deal with three-dimensional $Cspace_A$ obstacles, $CO_A^{xyz}(B_j)$. However, the paths found are not, in general, optimal paths [24]. Furthermore, with three-dimensional obstacles, the Vgraph algorithm is not even guaranteed to find a solution when one exists. This happens when the vertices of the $CO_A^{xyz}(B_j)$ are inaccessible, because they are outside of $CI_A^{xyz}(R)$. In that case, there may exist collision-free paths (via edges of the $CO_A^{xyz}(B_j)$), but the Vgraph algorithm will not find them. An alternative suboptimal, but complete, path searching strategy is described in [23]. A path searching algorithm based on mathematical optimization of a path along a fixed set of edges is described in [12].

V. Characterizing $CO_A(B)$

The surfaces of $CO_A(B)$, when A is three-dimensional and allowed to rotate, can be characterized in the same manner as the surfaces of $CO_A^{xy\theta}(B)$ were characterized in Section III. There are three types of surfaces that need to be considered, rather than two types as in two-dimensional objects. Let $f_i(\Theta)$ be the ith face of the convex polyhedron A, with Θ being the vector of three Euler angles indicating the orientation of A relative to its initial orientation. Similarly, let g_j be the jth face of the convex polyhedron B. As before, we let $a_i(\Theta)$ and b_j denote vertices of A and B, respectively. Each face of $CO_A(B)$ can be expressed as one of

$$f_i^a = b_j \oplus f_i(\Theta) \tag{5a}$$

$$f_j^b = a_i(\Theta) \oplus g_j \tag{5b}$$

$$f_{i,j}^{a \times b} = s(a_i(\Theta), a_{i+1}(\Theta)) \oplus s(b_j, b_{j+1}). \tag{5c}$$

The faces defined by (5a) and (5b) are parallel to the faces of A and B, respectively. Each face defined by (5c) is a parallelogram, with edges parallel to the edges of A and B that give rise to the face. The vector equation for each type of surface follows the pattern of (3) and (4) above:

$$\langle N, x \rangle = \langle N, a_i(\Theta) + b_j \rangle \tag{6}$$

where N is 1) the normal to $f_i(\Theta)$ for (5a) faces, 2) the normal to g_j for (5b) faces, or 3) the cross product of the vectors along $s(a_i(\Theta), a_{i+1}(\Theta))$ and $s(b_j, b_{j+1})$ for (5c) faces. As above, this characterization only holds between discontinuities.

VI. Approximating High-Dimension *Cspace* Obstacles

We have seen that when A is a three-dimensional solid which is allowed to rotate, $CO_A(B)$ is a complicated curved object in a six-dimensional $Cspace_A$. An alternative to computing these objects directly is to use a sequence of low-dimensional projections of the high-dimensional $Cspace_A$ obstacles. For example, a three-dimensional (x, y, θ) $Cspace_A$ obstacle can be simply approximated by its projection on the (x, y) plane, and any path of A that avoided the projection would be safe for all orientations of A. On the other hand, there may be no paths that completely avoid the projection. A better approach is to divide the complete range of θ values into k smaller ranges and, for each of these ranges, find the section of the (x, y, θ) obstacle in that range of θ. These are called θ-slices of the obstacle. The projection of these slices serves as an approximation of the obstacle. Paths that avoid individual slices are safe for orientations of A in the θ-range defining the slice.

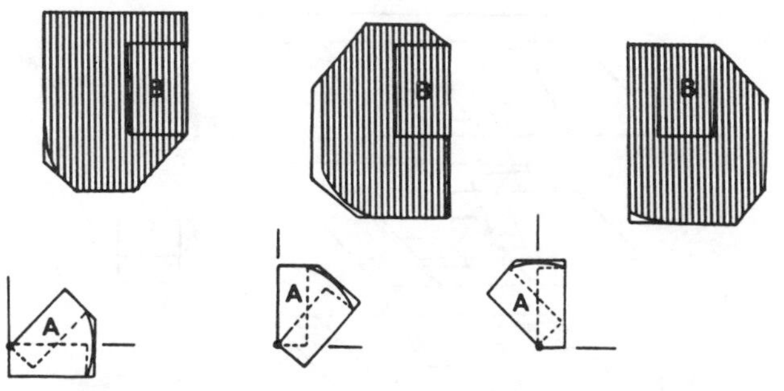

Fig. 10. Slice projections of $Cspace_A$ obstacles computed using the (x, y)-area swept out by A over a range of θ values. Each of the shaded obstacles is the (x, y)-projection of a θ-slice of $CO_A(B)$. The figure also shows a polygonal approximation to the slice projection and the polygonal approximation to the swept volume from which it derives.

The shaded areas in Fig. 10 are the (x, y) projection of θ-slices of $CO_A(B)$ when A and B are rectangles. These slices represent configurations where A overlaps B for some orientation of A in the specified range of θ. We will show that these slice projections are the $Cspace_A$ obstacles of the area swept out by A over the range of orientations of the slice. The swept area under rotation of a polygon is not pologonal. To use the $CO_A^{xy}(B)$ algorithm developed earlier, we approximate the swept area as the union of polygons [23]. This polygonal approximations leads to a polygonal approximation for the projected slices, as shown in Fig. 10. Similar considerations apply to polyhedra.

The crucial properties of slice projection are: 1) a solution to a Findspace or Findpath problem in any of the slices is a solution to the original problem, although not all actual solutions can be found in the slices; and 2) the degree of approximation can be controlled by choosing the range of parameters of the slice, in particular the approximation need not be uniform across the range of parameters.

The Vgraph algorithm for Findpath, has been extended [24], by means of slice projection, to find paths when A and all the B_j are three-dimensional polyhedra that are allowed to rotate. Fig. 11 illustrates the basic idea of this algorithm. An alternative path-searching technique, also using slice projection is described in [23]. Because the slice projections are approximations to the $Cspace_A$ obstacles, neither of these algorithms is guaranteed to find solutions to Findpath problems. Paths found by Findpath algorithms that use slice projections are composed of sequences of translations interspersed with rotations, but where the rotations happen in quantized increments corresponding to the ranges of orientations that define the slices. Not all paths can be expressed in this fashion. For

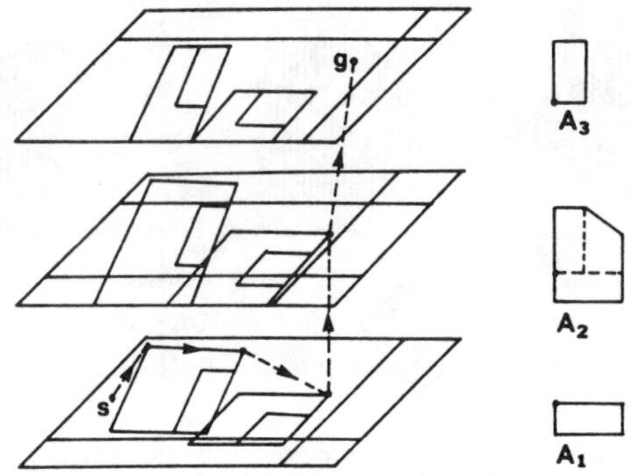

Fig. 11. An illustration of the Findpath algorithm using slice projection described by Lozano-Pérez and Wesley in [24]. A number of slice projections of the *Cspace* obstacles are constructed for different ranges of orientations of A. The problem of planning safe paths in the high-dimensional $Cspace_A$ is decomposed into 1) planning safe paths via $CO_A^{xyz}(B_j)$ vertices within each slice projection and 2) moving between slices, at configurations that are safe in both slices. A_1 represents A in its initial configuration, A_3 represents A in its final configuration and A_2 is a simple polyhedral approximation to the swept volume of A between its initial and final orientation.

example, the classic problem of moving a rectangular sofa through a rectangular bend in a hallway that just fits the sofa requires continuous rotation during translation. However, a large class of useful problems can be solved using slice projection.

In the rest of the section we show how slice projections of $CO_A(B)$ may be computed using the $CO_A^{xy}(B)$ and $CO_A^{xyz}(B)$ algorithms of Section IV. The idea is simply that if a collision would occur for A in some orientation, it would also occur for a swept volume of A that includes A in that orientation.

More formally, a j-slice of an object $C \in \mathcal{R}^n$ is defined to be $\{(\beta_1, \cdots, \beta_n) \in C \mid \gamma_j \le \beta_j \le \gamma_j'\}$, where γ_j and γ_j' are the lower and upper bounds of the slice, respectively. Then, if $I = \{1, \cdots, n\}$ and $K \subseteq I$, then a K-slice is the intersection of the j-slices for $j \in K$. Note that a K-slice of C is an object of the same dimension as C. Slices can then be *projected* onto those coordinates in I not in K, i.e., $I - K$, to obtain objects of lower dimension. A j-cross section is a j-slice whose lower and upper bounds are equal, e.g., $CO_A^{xy}(B)$, for some orientation of A, is the projection on the (x, y) plane of a θ-cross-section of $CO_A^{xy\theta}(B)$.

Slice projections are related to cross-section projections by the *swept volume* of an object. Intuitively, the swept volume of A is all the space that A covers when moving within a range of configurations. In particular, given two configurations for A, called c and c', then the union of $(A)_a$ for all $c \le a \le c'$ is the swept volume of A over the configuration range $[c, c']$. Generally, c and c' differ only on some subset, K, of the configuration coordinates. For example, if c and c' are of the form $(\beta_1, \beta_2, \beta_3)$ and $K = \{3\}$, then the swept volume of A over the range $[c, c']_K$ refers to the union of A over a set of configurations differing only on β_3. The swept volume of A over a configuration range is denoted $A[c, c']_K$.

If $A[c, c']_K$ overlaps some object B then, for some configuration a in that range, $(A)_a$ overlaps B. The converse is also true. $CO_{A[c,c']_K}(B)$ is the set of $I - K$ projections of those configurations of A within $[c, c']_K$ for which A overlaps B.

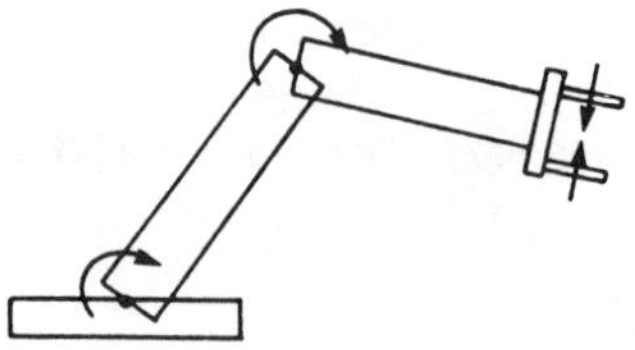

Fig. 12. Linked polyhedra can be used to model the gross geometry of industrial robot manipulators.

Equivalently, $CO_{A[c,c']_K}(B)$ is the $I - K$ projection of the $[c, c']_K$ slice of $CO_A(B)$. If the configurations of the swept volume are one of (x, y), (x, y, z), or (x, y, θ) then the algorithms of the previous sections can be used to compute $CO_{A[c,c']_K}(B)$ and thereby compute the required slice projections.[6]

A formal statement and proof of this result is included in Appendix II as Theorem 6. This theorem is of practical importance since it provides the mechanism underlying the Findspace and Findpath implementations described in [23] and [24].

VII. Automatic Planning of Industrial Robot Motions

One application of the algorithms for Findspace and Findpath developed above is in the automatic planning of industrial robot motions [23], [43]. However, some extensions of the results for polyhedra are needed. In this section, we briefly discuss these extensions.

Industrial robots are open kinematic chains in which adjacent links are connected by prismatic or rotary joints, each with one degree of freedom [29]. We model them by *linked polyhedra*, kinematic chains with polyhedral links, each of which has either a translational or rotational degree of freedom relative to the previous joint in the chain; see Fig. 12. The relative position and orientation of adjacent links, A_i and A_{i+1}, is determined by the ith *joint parameter* [29], [7]. The set of joint parameters of a linked polyhedron completely specifies the position and orientation of all the links. This type of model is clearly an approximation to the actual geometry; in particular, the shape of the joints is not represented and some values of the joint parameters may cause overlap of adjacent links.

The natural $Cspace_A$ for a linked polyhedron is that defined by the set of joint parameters. A point in this space determines the shape of the linked polyhedron and the configuration of each of its links. Unfortunately, the presence of rotary joints prevents the use of the $CO_A^{xyz}(B)$ algorithm of Section IV to plan the motions of linked polyhedra. However, there is an increasingly popular class of industrial robots, known as *Cartesian* robots, where the translational degrees of freedom of the robot are separate from the rotational. With this class of robots, we can use the $CO_A^{xyz}(B)$ algorithm and slice projection approach to plan collision-free paths and to plan how to grasp objects [23]. Actually doing this requires constructing the swept volume, over the rotational parameters, of the linked

[6] Of course, this requires computing a convex polyhedral approximation to the swept volume of A. Simple approximations are not difficult to compute [23], but this is an area where better algorithms are required. Nevertheless, the swept volume computation is a three-dimensional operation which can be defined and executed without recourse to six-dimensional constructs.

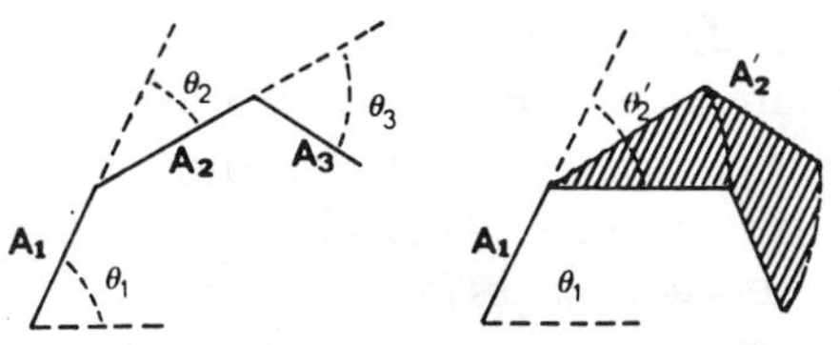

Fig. 13. Changes in the second joint angle from θ_2 to θ_2' causes changes in the configurations of both link A_2 and link A_3.

polyhedron modeling the robot. The resulting swept volume can be viewed as a polyhedron with only translational degrees of freedom, for which the $CO_A^{xyz}(B)$ algorithm is applicable.

Using swept volumes of linked polyhedra for slice projection requires taking into consideration the interdependence of the joint parameters. Note that for a linked polyhedron, the position of link j typically depends on the positions of links $k < j$, which are closer to the base than link j. Let $K = \{j\}$, $c = (\theta_i)$, $c' = (\theta_i')$, and $[c, c']_K$ define a range of configurations differing on the jth $Cspace_A$ parameter. Since joint j varies over a range of values, links $l \geq j$ will move over a range of positions which depend on the values of c and c', as shown in Fig. 13. The union of each of the link volumes over its specified range of positions is the swept volume of the linked polyhedron. The swept volume of links j through n can be taken as defining a new jth link. The first $j - 1$ links and this new jth link define a new manipulator whose configuration can be described by the first $j - 1$ joint parameters. On the other hand, the shape of the new link j depends not only on the K-parameters of c and c', i.e., θ_j and θ_j', but also on θ_l for $l > j$. This implicit dependence on parameters of c and c' that are not in K is undesirable, since it means that the *shape* of the new jth link will vary as the swept volume is displaced, i.e., the $(I - K)$-parameters are changed. If we let $K = \{j, \cdots, n\}$, then the shape of the swept volume depends only on the K-parameters of c and c', while its configuration is determined by the $(I - K)$-parameters. A swept volume that satisfies this property is called *displaceable*. This property plays a crucial role in proving the fact, mentioned in Section VI, that slice projections of a $Cspace_A$ obstacle can be computed as the $Cspace_A$ obstacles of the swept volume of A (see Theorem 6 in Appendix II).

In summary, with the extensions discussed in this section, the spatial planning algorithms developed for the case of rigid polyhedra can serve as the basis for planning the motions of industrial robots.

VIII. Related Work in Spatial Planning

The definition of the Findspace problem used here is based on that in [49]. One previous approach to this problem is described in [30]; it is an application of the Warnock algorithm [47] for hidden line elimination. The idea is to recursively subdivide the workspace until an area "large enough" for the object is found. This approach has several drawbacks: 1) any nonoverlapping subdivision strategy will break up potentially useful areas and 2) the implementation of the predicate "large enough" is not specified (in general, the $CI_A(B)$ computation is required to implement this predicate). However, once the $Cspace_A$ obstacles have been computed, the Warnock search provides a good way of solving Findspace; since we only need space for a point, any free area is "large enough" [10].

The work by Udupa, reported in [45], [46], was the first to approach Findpath by explicitly using transformed obstacles and a space where the moving object is a point. Udupa used only rough approximations to the actual *Cspace* obstacles and had no direct method for representing constraints on more than three degrees of freedom. A survey of previous heuristic approaches to the Findpath problem for manipulators, for example, [20], [31], has been given in [46]. An early paper on Shakey [28] describes a technique for Findpath using a simple object transformation that defines safe points for a circular approximation to the mobile robot and uses a graph search formulation of the problem. More recent papers on navigation of mobile robots are also relevant to two-dimensional Findpath [14], [26] [44]. An early paper [18] reports on a program for planning the path of a two-dimensional sofa through a corridor. This program does a brute-force graph search through a quantized *Cspace*.

The *Cspace* approach to Findspace and Findpath described here is an extension of that reported in [24]. In that paper, an approximate algorithm for $CO_A^{xyz}(B)$ is described and the Vgraph algorithm for high-dimensional Findpath is first represented. An application of the Findpath and Findspace approach described in the current paper to automatic planning of manipulator motions is described in [23]. Alternative approaches to path searching in the presence of obstacles are described in [19], [12], [23]. The visibility computation needed in Vgraph is treated, in the context of hidden-line elimination, in [4], [51].

The basic idea of representing position constraints as geometric figures, e.g., $CO_A^{xy}(B)$, has been used (independently) in [1]–[3], where an algorithm to compute $CO_A^{xy}(B)$ for nonconvex polygons is used in a technique for two-dimensional layout. The template packing approach described in [13] uses a related computation based on a chain-code description of figure boundaries. Algorithms for packing of parallelopipeds, in the presence of forbidden volumes, using a construct equivalent to the $CO_A^{xyz}(B)$, but defined as "the hodograph of the close positioning function" are reported in [42]. The only use of this construct in the paper is for computing $CO_A^{xyz}(B)$ for aligned rectangular prisms.

An extension of the approach in [24] to the general Findpath problem is proposed in [34]. The proposal is based on the use of an exact representation of the high-dimensional *Cspace* obstacles. The basic approach is to define the general configuration constraints as a set of multinomials in the position parameters of A. However, the proposal still requires elaboration. It defines the configuration space constraints in terms of the relationships of vertices of one object to the faces of the other. This is adequate for polygons, but the equations in the paper only express the constraints necessary for vertices of A to be outside of B, i.e., they are of the form of (3). They do not account for the positions of A where vertices of B are in contact with A [see (4)]. The new equations will have terms of the form $x \cos \theta$ and $y \cos \theta$. Furthermore, the approach of defining the configuration constraints by examining the interaction of vertices and faces does not generalize to three-dimensional polyhedra. It is not enough to consider the interaction of vertices and faces; the interaction of edges and faces must also be taken into account (see Section V and [8]).

Two recent papers describe solutions for the Findpath problem with rotations in two [40] and three dimensions [41]. In [41], the *Cspace* surfaces are represented as algebraic manifolds in a 12-dimensional space; in this way the surfaces can be described as polynomials, allowing the use of some powerful mathematical machinery. The resulting algorithm has (large) polynomial time complexity, for fixed dimensionality of the *Cspace*.

A *Cspace* algorithm is described in [10] for solving Findpath, allowing rotations of the moving objects. The algorithm is based on recursively subdividing *Cspace* until a path of cells completely outside of the obstacles is found.

An alternative approach to two-dimensional Findpath with rotations is described in [9]. The algorithm is based on representing the empty space outside the objects B_j explicitly as *generalized cones.* Motions of A are restricted to be along the spines of the cones. The algorithm bounds the moving object by a convex polygon and characterizes the legal rotations of the bounding polygon along each spine.

APPENDIX I

ALGORITHM FOR $CO_A^{xy}(B)$

This Appendix shows an algorithm for computing $A \oplus B$, called SET-SUM(A, B, C), when A and B are convex polygons. Section II used this operation to compute $CO_A^{xy}(B)$.

Each polygon is described in terms of its vertices and the angles that the edges make with the positive x axis. The edges and vertices are ordered in counterclockwise order, i.e., by increasing angle. The implementation assumes that a POLYGON record is available with the following components:

1) size—number of edges in the polygon.

2) vert [1:size + 1, 1:2]—an array of vectors representing the coordinates of a vertex. The ith edge, $i = 1, \cdots$, size, has the endpoints vert[i, k] and vert[i + 1, k], for $k \in \{1, 2\}$. Note that vert[size + 1, k] = vert[1, k].

3) angle [0:size]—the edge normal's angle (in the range $[0, 2\pi]$) with the x axis, monotonically increasing. For convenience angle [0] = 2π − angle [size].

References to the components of a polygon, a, are written as one of a.size, a.vert, and a.angle.

The algorithm implements the angle scan in the proof of Theorem 2; in particular, the edges of the input polygons, a and b, are examined in order of angle. The algorithm determines the position of the vertices of c. It is clear that vertices can occur only at angles where there is either a vertex of a, or a vertex of b, or both. From Lemmas 1 and 2, it is easy to see that the position of the vertex of c is the sum of the positions of the corresponding vertices of a and b. The algorithm starts the scan at the angle determined by the first edge of b, the first loop in the program below serves to find the edges of a that straddle that angle. From there, the algorithm increments the edge index into a or b depending on which makes the smaller angle increment. In general, the algorithm requires incrementing the angle beyond 2π so as to consider all the edges of a before the edge found by the first loop of the program. Since the edges are stored with angles between 0 and 2π, an offset variable is used to add 2π to the angle when the wraparound on polygon a is detected.

```
PROCEDURE setsum (a, b, c);
        POLYGON a, b, c;
        BEGIN INTEGER ea, eb, vc, i;
                REAL ang, offset;
                COMMENT Initialize an index into a, one into b, and one into c.
                    The value of offset will be either 0 or 2*pi, and it is used
                    to handle angle wraparound as described above;
                ea := 1;
                eb := 1;
                vc := 1;
                offset := 0;
                COMMENT Find adjacent edges in a whose angles straddle the angle
                    of the first edge of b;
                WHILE (a.angle[ea] <= b.angle[1] OR a.angle[ea - 1] >= b.angle[1])
                    DO ea := ea + 1;
                FOR i := 1 STEP 1 UNTIL 2 DO
                     c. vert[1, i] := a.vert[ea, i] + b.vert[1, i];
                COMMENT This loop implements the scan of Theorem 2 in the body of the
                    paper. The result of the loop is to fill the vertex array of c;
                WHILE (eb <= b.size) DO
                    BEGIN
                        vc := vc + 1;
                        ang := offset + a.angle[ea];
                        IF (ang <= b.angle[eb])
```

```
            THEN IF (ea >= a.size)
                     THEN BEGIN offset := 2*pi; ea := 1 END
                     ELSE ea := ea + 1;
            IF (ang >= b.angle[eb])
            THEN eb := eb + 1;
            FOR i := 1 STEP 1 UNTIL 2 DO
                  c.vert[vc, i] := a.vert[ea, i] + b.vert[eb, i]
         END;
      c.size := vc;
      FOR i := 1 STEP 1 UNTIL 2 DO
         c.vert[vc + 1, i] := c.vert[1, i];
   END
```

Appendix II

Proof of Theorem 6

Assume that $Cspace_A \subseteq \mathcal{R}^d$, let $I = \{1, 2, \cdots, d\}$ and $K \subseteq I$. Let I, K and $I - K$ denote sets of indexes for the coordinates of $a \in Cspace_A$. Define the following vectors, all in $Cspace_A$: $b = (\beta_i), c = (\gamma_i)$ and $c' = (\gamma_i')$ for $i \in I$. Then,

$$\Phi_K(c, c') \equiv \left\{ b \in \mathcal{R}^d \mid \bigwedge_{k \in K} \gamma_k \leq \beta_k \leq \gamma_k' \right\}$$

$$\Phi_K(c) \equiv \Phi_K(c, c)$$

$$\Theta_K(c, c') \equiv \Phi_K(c, c') \cap \Phi_{I-K}(c, c)$$

These definitions are illustrated in Fig. 14.

The *projection operator*, denoted $P_K[\cdot]$: $\mathcal{R}^d \to \mathcal{R}^{|K|}$ is defined, for vectors and sets of vectors, by

$$P_K[b] = (\beta_k) \qquad k \in K$$

$$P_K[B] = \{P_K[b] \mid b \in B\}$$

Superscripts on vectors indicate projection, e.g., $b^K = P_K[b]$. In addition, the vector in $\mathcal{R}^{|I|}$ composed from one vector in $\mathcal{R}^{|K|}$ and one in $\mathcal{R}^{|I-K|}$ is denoted $(a^{I-K}: b^K)$, where $P_{I-K}[(a^{I-K}: b^K)] = a^{I-K}$ and $P_K[(a^{I-K}: b^K)] = b^K$.

In this notation, precise definitions for the notions of *cross section projection* and *slice projection* can be provided. The cross section projection of a $Cspace_A$ obstacle is written as follows:

$$P_{I-K}[CO_A(B) \cap \Phi_K(c)].$$

The slice projection, is similar to the cross section projection,

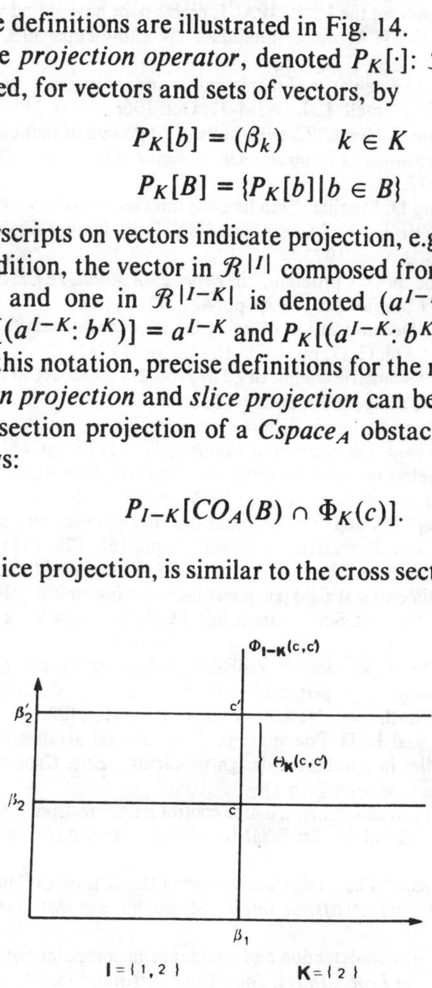

Fig. 14. Illustration of the definition of $\Phi_K(c,c')$ and $\Theta_K(c,c')$.

but carried out for all configurations between two cross sections:

$$P_{I-K}[CO_A(B) \cap \Phi_K(c, c')].$$

The K-parameters of the two configurations, c and c', define the bounds of the slice. Similarly, the swept volume can be defined in this notation.

Definition: The *swept volume* of A over the configuration range $[c, c']_K$ is

$$(A[c, c']_K)_c \equiv \bigcup_{a \in \Theta_K(c,c')} (A)_a.$$

The requirement discussed in Section VII that the swept volume of A be *displaceable* is embodied in the following condition:

$$\forall a: \bigcup_{x \in \Theta_K(c,c')} (A)_{(a^{I-K}:x^K)} = (A[c, c']_K)_{(a^{I-K}:c^K)} \qquad (7)$$

Note that the $I - K$ parameters may be changed, as in (7), but not those parameters in K. Therefore, $(A[c, c']_K)_a$ is defined only if $a \in \Phi_K(c)$.

Theorem 6: If (7) holds, then

$$P_{I-K}[CO_A(B) \cap \Phi_K(c, c')] = P_{I-K}[CO_{A[c,c']_K}(B) \cap \Phi_K(c)]$$

Proof of Theorem: Assume that the configuration a is in the slice projection of $CO_A(B)$, that is,

$$a \in P_{I-K}[CO_A(B) \cap \Phi_K(c, c')].$$

This assumption and the definition of the projection operator allows us to deduce that some configuration in $Cspace_A$, whose $I - K$-projection is a, is in $CO_A(B)$:

$$\Leftrightarrow \exists x_1 \in \Phi_K(c, c'):((a^{I-K}:x_1^K) \in CO_A(B)).$$

In fact, since we are only interested in the K-parameters of x_1 and $\Theta_K(c, c') \subseteq \Phi_K(c, c')$, we can assume without loss of generality that x_1 is in the smaller set, i.e.,

$$\Leftrightarrow \exists x_1 \in \Theta_K(c, c'):((a^{I-K}:x_1^K) \in CO_A(B)).$$

Simply using the definition of $CO_A(B)$, it follows that

$$\Leftrightarrow (A)_{(a^{I-K}:x_1^K)} \cap B \neq \emptyset,$$

but if A in this configuration intersects B, then any set including A in that configuration will also intersect B. In particular,

$$\Leftrightarrow \bigcup_{x \in \Theta_K(c,c')} (A)_{(a^{I-K}:x^K)} \cap B \neq \emptyset.$$

We are assuming that swept volumes are displaceable, i.e., that (7) holds. Therefore, using (7), we get

$$\Leftrightarrow (A[c, c']_K)_{(a^{I-K}:c^K)} \cap B \neq \emptyset$$

Hence, by the definitions of $CO_A(B)$ and $\Phi_K(c)$,

$$\Leftrightarrow (a^{I-K}:c^K) \in CO_{A[c,c']}(B) \text{ and } (a^{I-K}:c^K) \in \Phi_K(c)$$

Applying the definition of the projection operator completes the proof:

$$\Leftrightarrow a \in P_{I-K}[CO_{A[c,c']_K}(B) \cap \Phi_K(c)]. \quad \blacksquare$$

Acknowledgment

The author would like to thank M. Brady, R. Brooks, J. Hollerbach, B. Horn, T. Johnson, M. Mason, P. Winston, B. Woodham, and the referees for their suggestions on the content and presentation of this paper. J. Hollerbach suggested the proof of Theorem 6 in the current version of the paper, which is much simpler than the original proof.

References

[1] M. Adamowicz, "The optimum two-dimensional allocation of irregular, multiply-connected shapes with linear, logical and geometric constraints," Ph.D. dissertation, Dep. Elec. Eng., New York Univ., 1970.

[2] M. Adamowicz and A. Albano, "Nesting two-dimensional shapes in rectangular modules," *Computer Aided Design*, vol. 8, pp. 27–32, Jan. 1976.

[3] A. Albano and G. Sapuppo, "Optimal allocation of two-dimensional irregular shapes using heuristic search methods," *IEEE Trans. Syst., Man., Cybern.*, vol. SMC-10, pp. 242–248, May 1980.

[4] D. Avis and G. T. Toussaint, "An optimal algorithm for determining the visibility of a polygon from an edge," Sch. Comput. Sci., McGill Univ., Rep. SOCS-80.2, Feb. 1980.

[5] R. V. Benson, *Euclidean Geometry and Convexity*. New York: McGraw-Hill, 1966.

[6] J. I. Bentley and T. Ottman, "Algorithms for reporting and counting geometric intersections," *IEEE Trans. Comput.*, vol. C-28, Sept. 1979.

[7] O. Bottema and B. Roth, *Theoretical Kinematics*. Amsterdam: North-Holland, 1979.

[8] J. W. Boyse, "Interference detection among solids and surfaces," *Commun. Ass. Comput. Mach.*, vol. ACM 22, pp. 3–9, Jan. 1979.

[9] R. A. Brooks, "Solving the Find-Path problem by representing free space as generalized cones," M.I.T. Artificial Intell. Lab., Rep. AIM-674, May 1982.

[10] R. A. Brooks and T. Lozano-Pérez, "A subdivision algorithm in configuration space for Findpath with rotation," M.I.T. Artificial Intell. Lab., Rep. AIM-684, Dec. 1982.

[11] K. Q. Brown, "Fast intersection of half spaces," Dep. Comput. Sci., Carnegie-Mellon Univ., Rep. CS-78-129, June 1978.

[12] C. E. Campbell and J. Y. S. Luh, "A preliminary study on path planning of collision avoidance for mechanical manipulators," Sch. Elec. Eng., Purdue Univ., Rep. TR-EE 80-48, Dec. 1980.

[13] H. Freeman, "On the packing of arbitrary-shaped templates," in *Proc. 2nd USA-Japan Comput. Conf.*, 1975, pp. 102–107.

[14] G. Giralt, R. Sobek, and R. Chatila, "A multilevel planning and navigation system for a mobile robot," in *Proc. 6th Int. Joint Conf. Artificial Intell.*, Tokyo, Japan, Aug. 1979, pp. 335–338.

[15] R. L. Graham, "An efficent algorithm for determining the convex hull of a finite planar set," *Inform. Processing Lett.*, vol. 1, pp. 132–133, 1972.

[16] B. Grunbaum, *Convex Polytopes*. New York: Wiley-Interscience, 1967.

[17] L. J. Guibas and F. F. Yao, "On translating a set of rectangles," in *Proc. 12th Annu. ACM Symp. Theory of Computing*, Los Angeles, CA, Apr. 1980, pp. 154–160.

[18] W. E. Howden, "The sofa problem," *Comput. J.*, vol. 11, pp. 299–301, Nov. 1968.

[19] R. C. Larson and V. O. K. Li, "Finding minimum rectilinear distance paths in the presence of obstacles," *Networks*, vol. 11, pp. 285–304, 1981.

[20] R. A. Lewis, "Autonomous manipulation on a robot: Summary of manipulator software functions," Jet Propulsion Lab., California Inst. Technol., TM 33-679, Mar. 1974.

[21] L. Lieberman and M. A. Wesley, "AUTOPASS: An automatic programming system for computer controlled assembly," *IBM J. Res. Develop.*, vol. 21, July 1977.

[22] T. Lozano-Pérez, "The design of a mechanical assembly system," M.I.T. Artificial Intell. Lab., Rep. TR-397, Dec. 1976.

[23] ——, "Automatic planning of manipulator transfer movements," *IEEE Trans. Syst., Man., Cybern.*, vol. SMC-11, pp. 681–698, Oct. 1981.

[24] T. Lozano-Pérez and M. A. Wesley, "An algorithm for planning collision-free paths among polyhedral obstacles," *Commun. Ass. Comput. Mach.*, vol. ACM 22, pp. 560–570, Oct. 1979.

[25] L. Lyusternik, *Convex Figures and Polyhedra*. Dover, 1963, original copyright, Moscow, 1956.

[26] H. P. Moravec, "Visual mapping by a robot rover," in *Proc. 6th Int. Joint Conf. Artificial Intell.*, Tokyo, Japan, Aug. 1979.

[27] D. Mueller and F. Preparata, "Finding the intersection of two convex polyhedra," Coordinated Sci. Lab., Univ. Illinois, Urbana, Rep. R-793, Oct. 1977.

[28] N. Nilsson, "A mobile automaton: An application of artificial intelligence techniques," in *Proc. 2nd Int. Joint Conf. Artificial Intell.*, 1969, pp. 509–520.

[29] R. P. Paul, "Manipulator Cartesian path control," *IEEE Trans. Syst., Man., Cybern.*, vol. SMC-9, pp. 702–711, Nov. 1979.

[30] G. Pfister, "On solving the FINDSPACE problem, or how to find where things aren't," M.I.T. Artificial Intell. Lab., Working Paper 113, Mar. 1973.

[31] D. L. Pieper, "The kinematics of manipulators under computer control," Stanford Artificial Intell. Lab., AIM-72, Oct. 1968.

[32] F. Preparata and S. Hong, "Convex hulls of finite sets of point in two and three dimensions," *Commun. Ass. Comput. Mach.*, vol. 20, pp. 87–93, Feb. 1977.

[33] F. Preparata and D. Mueller, "Finding the intersection of a set of half spaces in time $O(n \log n)$," Coordinated Sci. Lab., Univ. Illinois, Urbana, Rep. R-803, Dec. 1977.

[34] J. Reif, "On the movers problem," in *Proc. 20th Annu. IEEE Symp. Foundation of Comput. Sci.*, 1979, pp. 421–427.

[35] B. Schachter, "Decomposition of polygons into convex sets," *IEEE Trans. Comput.*,vol. C-27, pp. 1078–1082, Nov. 1978.

[36] M. I. Shamos, "Geometric complexity," in *Proc. 7th Annu. ACM Symp. Theory of Computing*, 1975, pp. 224–233.

[37] M. I. Shamos and D. Hoey, "Closest-point problems," in *Proc. 16th Annu. IEEE Symp. Foundation of Comput. Sci.*, 1975, pp. 151–161.

[38] ——, "Geometric intersection problems," in *Proc. 17th Annu. IEEE Symp. Foundation of Comput. Sci.*, 1976, pp. 208–215.

[39] J. T. Schwartz, "Finding the minimum distance between two convex polygons," *Inform. Processing Lett.*, vol. 13, pp. 168–170, 1981.

[40] J. T. Schwartz and M. Sharir, "On the piano movers' problem I. The case of a two-dimensional rigid polygonal body moving amidst polygonal barriers," Dep. Comput. Sci., Courant Inst. Math. Sci., New York Univ., Rep. 39, 1981.

[41] ——, "On the piano movers' problem II. General techniques for computing topological properties of real manifolds," Dep. Comput. Sci., Courant Inst. Math. Sci., New York Univ., Rep. 41, 1982.

[42] Y. G. Stoyan and L. D. Ponomarenko, "A rational arrangement of geometric bodies in automated design problems," *Eng. Cybern.*, vol. 16, Jan. 1978.

[43] R. Taylor, "A synthesis of manipulator control programs from task-level specifications," Stanford Artificial Intell. Lab., Rep. AIM-282, July 1976.

[44] A. M. Thompson, "The navigation system of the JPL robot," in *Proc. 5th Int. Joint Conf. Artificial Intell.*, Massachusetts Inst. Technol., 1977.

[45] S. Udupa, "Collision detection and avoidance in computer controlled manipulators," in *Proc. 5th Int. Joint Conf. Artificial Intell.*, Massachusetts Inst. Technol., 1977.

[46] ——, "Collision detection and avoidance in computer controlled manipulators," Ph.D. dissertation, Dep. Elec. Eng., California Inst. Technol., 1977.

[47] J. E. Warnock, "A hidden line algorithm for halftone picture representation," Dep. Comput. Sci., Univ. Utah, Rep. TR4-5, May 1968.

[48] J. Williams, "STICKS—A new approach to LSI design," S.M. thesis, Dep. Elec. Eng., Massachusetts Inst. Technol., 1977.

[49] T. Winograd, *Understanding Natural Language.* New York: Aademic, 1972.

[50] T. C. Woo, "Progress in shape modelling," *IEEE Computer*, vol. 10, Dec. 1977.

[51] F. F. Yao, "On the priority approach to hidden-surface algorithms," in *Proc. 21st Symp. Foundation of Comput. Sci.*, 1980, pp. 301–307.

Tomás Lozano-Pérez was born in Guantanamo, Cuba, on August 21, 1952. He received the S.B. degree in 1973, the S.M. degree in 1977, and the Ph.D. degree in 1980, all in computer science from the Massachussetts Institute of Technology, Cambridge.

In 1981 he joined the Department of Electrical Engineering and Computer Science, Massachusetts Institute of Technology, where he is currently an Assistant Professor. He has been associated with the M.I.T. Artificial Intelligence Laboratory since 1973. His research interests include robotics, geometric modeling, computational geometry, computer vision, and artificial intelligence.

An Algorithmic Approach to Some Problems in Terrain Navigation

Joseph S. B. Mitchell
Operations Research
Cornell University
Ithaca, NY, 14853
email: jsbm@cs.cornell.edu

Abstract

Recent advances in the field of computational geometry have provided efficient algorithms for a variety of shortest path problems. Many problems in the field of terrain navigation can be cast as optimal path problems in a precise geometric model. With such a model one can develop and analyze algorithms for the solution of the original problem and can gain insights into how to design more efficient heuristics to deal with more complex problems. We examine the path planning problem in which we are given a "map" of a region of terrain and we are expected to find optimal paths from one point to another. This, for example, is a task which must be done repeatedly for the guidance of an autonomous vehicle. We examine how to formulate some path planning problems precisely, and we report algorithms to solve certain special cases.

1 Introduction

One of the basic tasks required of an autonomous vehicle is to move from one location to another. Usually, the robot has some knowledge (which we refer to as a *map*) of the terrain on which it is navigating, although it may initially have none. It should try to exploit as much of this knowledge as possible in order to make the best decisions about its path. Two types of navigation problems immediately arise: How does the robot *locate* its position in the map, and how does one use the map and the current location of the robot to *plan* a route to the goal? In this paper, we concentrate on the latter type of problem. We use the word "terrain" to refer to both indoor and outdoor environments for a vehicle.

The problem of location involves using sensory data to compute the coordinates of the robot. This can be done only to within some error bounds; however, we will assume that the errors involved are small enough in comparison with the scale of the planning problem that we can assume the exact position of the robot is known at each point in time and that the robot can be controlled exactly to follow a specified trajectory. The issue of location uncertainty is very important, and it is an area of continuing research.

In planning paths through the map, we typically wish to minimize some objective function which specifies the *cost* of motion along the path (the *path length*). Depending on how the terrain is modeled and what the optimization criteria are, this results in various versions of the *shortest path problem.* The shortest path problem for a mobile robot was one of the early problems addressed by researchers in artificial intelligence (see Nilsson [65] and Moravec [64]).

In this paper we discuss some of the issues that go into selecting a representation of a map, and we report some of the recent results from the field of *computational geometry* which are of use in solving terrain navigation problems. Computational geometry and algorithmic motion planning have recently erupted as important areas of research in computer science. We certainly cannot do justice here to the field of motion planning, so we refer the interested reader to the superb new books of Schwartz and Yap [84] and Schwartz et al. [83], which collect together many of the landmark papers of this exciting field. We also refer the reader to the award-winning thesis of Canny [10]. Much of our discussion uses concepts from the field of computational geometry. We refer the interested reader to three recent books on the subject: Edelsbrunner [22], Mehlhorn [47], and Preparata and Shamos [71].

While many of the problems of terrain navigation have been addressed before from a heuristic or approximation point of view, we feel that it is important to

"An Algorithmic Approach to Some Problems in Terrain Navigation" by J.S.B. Mitchell. Reprinted from *Artificial Intelligence*, 37, 1988, pages 171-201. Copyright © 1988 Elsevier Science Publishers B.V., reprinted with permission.

formalize some of the models and to discuss the computational complexity of exact algorithms. We have found that an understanding of the underlying geometry can lead to new heuristics and faster algorithms for a variety of problems.

How do we go about representing varied terrain? What representations yield particularly natural or efficient algorithms for path planning? What do shortest paths "look like"; that is, can we understand enough about the local optimality criteria to cut down the search for optimal paths?

We begin addressing some of these issues in the following sections, where we describe the obstacle avoidance problem, the *Discrete Geodesic Problem*, the *Weighted Region Problem*, and some special cases of these problems that arise naturally in finding shortest paths through terrain. Throughout our discussions, we concentrate on the fundamental algorithmic issues of computing paths, so we are interested in worst-case assymptotic running times. Where appropriate, we will allude to the practicality of the algorithms and approaches we suggest.

2 Obstacle Avoidance in the Plane

The simplest surface to model is a plane, and the simplest nontrivial terrain map on a plane is a binary partitioning of the plane into regions that are obstacles and regions that are traversable by the vehicle. Usually, the robot can be modeled as a point, which is a reasonable assumption if it is small in comparision with the dimensions of the terrain map. Also, by *configuration space* techniques, many problems of moving a non-point robot among obstacles can be reduced to the case of moving a point (possibly in higher dimensions), by "growing" the obstacles in an appropriate way, as in Lozano-Pérez and Wesley [44]. The problem of planning a shortest Euclidean path from one point to another is then the usual *obstacle-avoidance shortest path problem*. This problem has been studied by many researchers over the last twenty years. A survey of many algorithms is given in Mitchell [52,56,57], Alt and Welzl [3].

Some recent work of Papadimitriou and Silverberg [70] and O'Rourke [67] has started to address the tough problem of *shortest* paths for a non-point noncircular body in the plane. While this can be mapped into a shortest path problem in three-dimensional configuration space, calculating shortest paths in three dimensions has been shown to be quite difficult, as we will mention later.

There are two standard approaches to modeling the terrain for the obstacle avoidance problem. The first subdivides the plane into small regular pieces ("pixels") and labels each piece as obstacle or nonobstacle. If all of the pieces are squares of exactly the same size, then we get the standard grid tesselation, and we can represent the map as a binary array. A measure of the size of the problem instance is given by n, the total number of pixels. Of course, n depends on the resolution of the grid used to approximate the set of obstacles. Picking a very small pixel (very high resolution) allows a better model of the obstacle space, but it costs us significantly in problem size. If using a regular grid, we may be forced to have pixels smaller than a certain size in order to assure that skinny passageways between obstacles or tiny little obstacles appear in the map. Other subdivision representations exist which are hierarchical in nature, such as the *quadtree* (see Samet and Webber [82] or Kambhampati and Davis [34]). These schemes have the advantage that they conglomerate groups of contiguous small squares into single larger squares.

Another standard approach to modeling terrain which is obstacle or non-obstacle is to give a description of a polygonal space in terms of its boundary representation. Obstacles are given as a list of k simple polygons, each represented by a doubly-linked list of vertices (each vertex just being a pair of coordinates, either integer or real). Usually, the obstacles are assumed to be disjoint. Let n be the total number of vertices of all polygonal obstacles. We refer to n as the "complexity of the scene", as it is proportional to the number of bytes of storage needed to represent the map. To be more precise, though, the complexity of the scene should be the pair (n, k), since some of the complexities of algorithms depend not only on n, but also on k. (Of course, k is always less than or equal to n, but it may be significantly less than n, in which case it is important to know precisely how the running times depend on k.) The objective of the algorithmic study of the shortest path problem is to determine an algorithm whose worst-case running time is a low-degree polynomial in n and k. Most work in computational geometry focuses on *worst-case* running times; however, it is an important issue (not addressed here) to consider the *average-case* running times of shortest path algorithms.

An example of a set of obstacles is given in Figure 1(a)-(c). In Figure 1(a), we show the representation of the obstacles as a set of $k = 2$ polygons with $n = 8$ vertices and the shortest path from start to goal. Then, in Figure 1(b) we show the digitized rep-

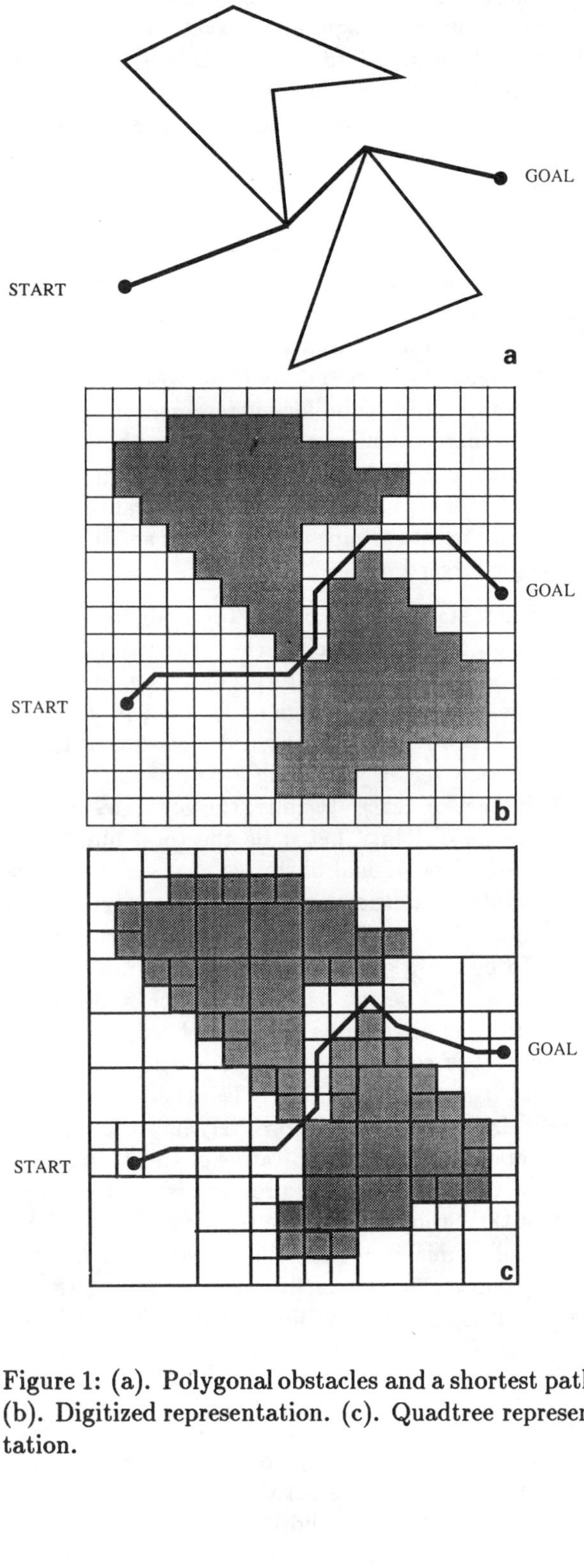

Figure 1: (a). Polygonal obstacles and a shortest path. (b). Digitized representation. (c). Quadtree representation.

resentation on a grid of 16-by-16 pixels ($n = 256$). In Figure 1(c) we show the same set of obstacles represented as a quadtree, where the free space is also subdivided into a quadtree representation. The total size of the representation in this case is $n = 124$.

There are other methods of modeling the terrain in the obstacle-avoidance problem. The obstacles may be given as a list of circles, ellipses, or other nonpolygonal objects. The shortest path problem among circular or elliptical obstacles has been addressed by Baker [6], Chew [14], and Mitchell [52].

Souvaine [87] introduced the notion of *splinegons* as a means of describing simple closed *curved* figures, and she describes many generalizations of standard computational geometry algorithms to the case of splinegons. A different approach to modeling obstacle maps is given by Brooks [8] and Brooks and Lozano-Pérez [9], who give a decomposition of free space into generalized cylinders. Their (heuristic) method is similar to the retraction method of motion planning (see Schwartz and Yap [84]).

Shortest paths according to the Euclidean metric will be "taut-string" paths: imagine threading a string from START to GOAL among the obstacles, then pulling the string until it is taut. The question really comes down to how the string should be threaded. One cannot afford to try all of the exponentially many possible threadings.

The algorithm one uses to find shortest paths among obstacles depends on the representation used in the map. For the case of a binary grid representation, the search for shortest paths is easy and straightforward. Shortest paths are found by restricting the robot's motion to connect centers of adjacent free squares. Then, we can consider the graph (so-called *grid graph*) whose nodes are the free space squares and the edges connect adjacent free squares. The degree of a node is thus at most 8, so the graph has at most n nodes and no more than $8n$ edges. Lengths are assigned to edges in the obvious way: diagonal edges are of length $\sqrt{2}$ times the size of a square, and horizontal and vertical edges are of length equal to the size of a square. We can then search the graph for shortest paths by applying the A^* algorithm (Nilsson [65]) or the Dijkstra [21] algorithm (which is "uninformed" A^*). The worst-case running time of Dijkstra's algorithm (and hence of A^*) is $O(|E| + |V| \log |V|)$ for a graph with $|E|$ edges and $|V|$ nodes (Fredman and Tarjan [24]); thus, the worst-case running time of the shortest path algorithm is $O(n \log n)$ for a grid with n pixels. Note that for A^* there is an obvious admissible heuristic function, $h(v)$, for node v, namely, the straight-line Euclidean distance from v to the goal.

Figure 1(b) shows the path from START to GOAL which results from finding the shortest path through the grid graph. Note that the path is not shortest according to the Euclidean metric. This is a common problem with the grid graph approach – it introduces a "digitization bias", a product of requiring paths to stay on the grid graph. Within the same time bounds ($O(n \log n)$), an algorithm of Keirsey and Mitchell [36] can be run which attempts to correct for this digitization bias imposed by requiring paths to stay on the grid graph. See also Mitchell et al. [63] and Thorpe [90,91]. These grid-based planners do not, however, solve the exact Euclidean shortest path problem.

For maps given in the form of a list of polygonal obstacles, there are two basic approaches to finding shortest paths. A survey of these methods is given in Mitchell [52,57]; see also the survey article of Alt and Welzl [3].

The first approach is to build a *visibility graph* of the obstacle space, whose nodes are the vertices of the obstacles and whose edges connect pairs of vertices for which the segment between them does not cross the interior of any obstacle. The resulting graph can be searched using Dijkstra's algorithm or A^*, where the heuristic function may again be taken to be the straight-line distance from a point to the goal. The time required for searching the graph is $O(K+n \log n)$, where K is the number of edges in the visibility graph. The result of the search is actually more than just the shortest path from START to GOAL; we actually get a *shortest path tree* on the set of obstacle vertices which gives a shortest path from the START (or the GOAL) to every other obstacle vertex.

The bottleneck in the worst-case complexity for the visibility graph approach is in the construction of the graph. We need to determine for every pair of vertices whether or not the vertices are visibile to each other. There are $O(n^2)$ such pairs. A trivial $O(n^3)$ algorithm simply checks every pair of vertices against every obstacle edge. With a little more care, $O(n^2 \log n)$ suffices to compute the visibility graph (Lee [41], Sharir and Schorr [85], or Mitchell [52]). The basic idea is to sort (angularly) the set of all vertices about each vertex and then to do an angular sweep about each vertex, keeping track of the closest obstacle boundary. By mapping the vertices to their dual lines and constructing the line arrangement, Lee and Ching [42] showed that the n sorts can be done in total time $O(n^2)$ rather than $O(n^2 \log n)$. This fact allowed Welzl [92] and Asano et al. [4] to achieve an $O(n^2)$ algorithm for constructing visibility graphs. This running time is not likely to be improved for the worst case since there are visibility graphs of quadratic size.

However, a recent development in the analysis of visibility graphs is the algorithm of Ghosh and Mount [26] which computes the visibility graph of a set of n disjoint line segments in time $O(K + n \log n)$, where K is the number of edges in the visibility graph. This *output-sensitive* algorithm takes advantage of the fact that K is not always of size $\Theta(n^2)$, but rather can be as small as $\Theta(n)$. So, when the visibility graph is sparse (say, $O(n \log n)$ edges), the algorithm is very fast, building the visibility graph in time $O(n \log n)$. Thus, when $K = O(n \log n)$, Dijkstra's algorithm (or A^*) requires worst-case complexity $O(K + n \log n) = O(n \log n)$, so the shortest path can be found in time $O(n \log n)$.

A second technique for finding shortest paths is that of building a *shortest path map*. A shortest path map is a subdivision of the plane into regions each of which is the locus of all goal points whose shortest paths from the START have the same topology (i.e., pass through the same sequence of obstacle vertices). With such a subdivision, the shortest path from START to GOAL is obtained simply by locating the GOAL point in the subdivision (which takes time $O(\log n)$) and then backtracing an optimal path. See Lee and Preparata [43] or Mitchell [52] for precise definitions and examples. Reif and Storer [75] have given an algorithm to compute the shortest path map in time $O(kn + n \log n)$, where k is the number of obstacles. Note that this is an improvement over $O(n^2)$ when the number of obstacles is small in comparison with the number of obstacle vertices. Their algorithm simulates the propagation of a "wavefront" from the START, in much the same way that expansion operates in the Dijkstra or A^* algorithms. Because of its similarity to the discrete Dijkstra algorithm, this technique has come to be called the "continuous Dijkstra" paradigm.

The continuous Dijkstra paradigm, in the form of a plane sweep algorithm, has lead to $O(n \log n)$ algorithms in a few special cases (Lee and Preparata [43], Mitchell [52], Sharir and Schorr [85]). Basically, if the obstacles are known to have disjoint projections onto some line (e.g., if the obstacles are vertical line segments), then the shortest path will always be monotone with respect to that line (except, perhaps, at the ends of the path). This monotonicity allows one to use plane sweep algorithms for the iterative construction of the shortest path map.

In the simple case of finding shortest paths required to stay inside a single simple polygonal room with n

sides (without other obstacles), the shortest path can be found in linear ($O(n)$) time once a triangulation of the polygon is known (Lee and Preparata [43], Guibas et al. [28]). With the recent breakthrough result of Chazelle [14], triangulation of a simple polygon can be done in optimal time $O(n)$. Furthermore, a triangulated simple polygon can be processed in linear time so that in time $O(\log n)$ one can find the length of the shortest path from any given START to any given GOAL (Guibas and Hershberger [29]), after which, in time proportional to the number of bends in the optimal path, one can output the actual path from START to GOAL.

Lower bounds are known for the shortest obstacle-free path problem. In the worst case, computing the shortest path among k obstacle with a total of n vertices requires $\Omega(n+k\log k)$ time. One can use a shortest path algorithm to find convex hulls, which in turn can be used to sort integers, for which the lower bound is well known.

The big open problem in this field is to determine whether or not the shortest path among n disjoint line segments can be computed in optimal time $\Theta(n\log n)$, and more generally if shortest paths among k simple polygons can be found in time $\Theta(n + k\log k)$. We should note that for some metrics other than Euclidean, such as the L_1 ("Manhattan") metric and "fixed orientation" metrics (Widmayer et al. [93]), optimal or near-optimal shortest path algorithms are known (see Clarkson et al. [18], de Rezende et al. [20], and Mitchell [53,58]).

3 Shortest Paths on a Surface

A natural generalization of the obstacle-avoidance problem in the plane is to consider the case of finding shortest paths for a point which is constrained to move on a nonplanar (possibly nonconvex) surface. To see that this is a generalization of the above problem, consider the surface which results from making each obstacle into a very tall cylinder whose base is in the plane of motion. Then, one can show that the shortest path on such a (nonconvex) surface will stay in the plane and follow the shortest obstacle-avoidance path. Note that the problem we wish to solve is a calculus of variations problem: Find the shortest *geodesic* path between two points on a surface. Since the surface is modeled in a discrete manner, we refer to this problem as the *Discrete Geodesic Problem.*

Here, we assume that our vehicle is not able to fly, but is able to travel over all types of terrain with uniform efficiency. We can go up hills, down hills, and

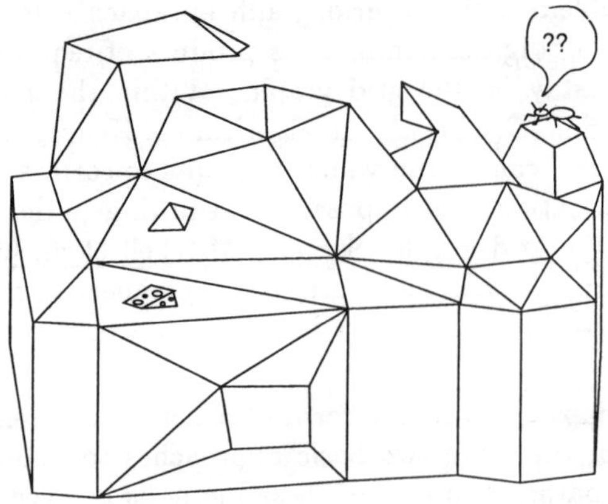

Figure 2: The problem of navigating on a surface

even perhaps on the underside of an overhang. (Our tires are "magnetic" and the ground has a very high iron content.) Such is the problem faced by the crawling insect in Figure 2. He desires to get to the cheese in the shortest possible path, but he does not have wings to be able to fly there. Along what route should he crawl?

First, let us mention something about the representation of the problem. What is the map in this case? It is simply some representation of a surface in three dimensions. Formally, the surface might be represented as a polyhedral surface, giving faces, edges, and vertices, along with all of the usual adjacency relationships that allow one to give the faces on each side of any edge and to "walk around" a vertex while enumerating adjacent faces. (The winged-edge data structure of Baumgart [7] or the similar quad-edge data structure of Guibas and Stolfi [30] would work nicely here.) Another possible surface representation that is often used in practice is that of contour lines. We might be given a set of iso-elevation curves, as is usually the case with geological maps of terrain. A third common representation is that of an *elevation array.* In this case, we are simply given a two-dimensional array of numbers which represent the altitude at each grid point. Digital terrain data bases of the form compiled by the Defense Mapping Agency fall into this category. (Typically, pixels are of size 5, 12.5, or 100 meters.) These are just some of the methods of specifying a discrete approximation of a surface.

Note that the last two representations require that the surface be the image of a single-valued function of the plane (i.e., there are no "overhangs"), an assumption that generally holds for most outdoor terrain.

The case of a surface represented with an elevation array can be solved easily if one makes the approximation that paths are constrained to follow the corresponding grid graph. We just connect each pixel to each of its eight neighbors, assigning a length measured by the three-dimensional Euclidean distance, and then search the graph for a shortest path using Dijkstra or an A^* algorithm. This requires $O(n \log n)$ time in the worst case, where n is the number of array elements (pixels).

Research in the field of algorithmic motion planning has favored the representation of a surface in terms of a polyhedron. The general problem of finding shortest paths for a point in three dimensions amidst polyhedral obstacles has been of interest for some time (Sharir and Schorr [85], Reif and Storer [75], Franklin and Akman [23], Akman [1,2], Papadimitriou [69], Clarkson [17]). The general problem seems very hard to solve, as Canny and Reif [11] have shown the problem to be NP-hard, and the best known exact algorithm is the singly-exponential algorithm of Reif and Storer [75]. Papadimitriou [69] and Clarkson [17] have solved the problem approximately with fully polynomial approximation schemes (meaning that the running time is polynomial in n, the number of bits, K, to represent the problem instance, and $1/\epsilon$). The complexity of their algorithms is $O(n^3 K^2/\epsilon)$. The special case, though, of our bug crawling on the surface of a single polyhedron is interesting because it admits a polynomial-time algorithm which is exact.

Franklin and Akman [23] and Akman [1,2] have given exponential (worst-case) algorithms for the shortest path problem on a surface, and have shown that their algorithms perform reasonably well in practice. The first polynomial-time algorithm was an $O(n^3 \log n)$ algorithm of Sharir and Schorr [85] which worked only for the case of convex polyhedra (where n is the number of edges of the polyhedron).

Most interesting terrain is not convex, though, since it will usually have many mountains and valleys. An algorithm which works for non-convex surfaces was devised by O'Rourke et al. [68], and its complexity is $O(n^5)$. Both of these running times were improved by the algorithm of Mitchell et al. [60], which runs in time $O(n^2 \log n)$. The algorithm actually constructs, for a fixed start point, a subdivision (a *Shortest Path Map*) of the surface such that after the $O(n^2 \log n)$ preprocessing required to build the subdivision, the distance to any particular destination can be found by locating the goal in the subdivision (a task which is well-known to take $O(\log n)$ time; see, for example, Kirkpatrick [37]). Listing of a path can be done in time proportional to the number of turns in the optimal path. (The algorithm of Mitchell et al. [60] has now been improved to yield a time bound of $O(n^2)$ and a space bound of $\Theta(n)$; see Chen and Han [13].)

Let us make a few simple observations that were critical to all of the algorithms mentioned above. We can state a few known facts about how optimal paths must behave:

1). Optimal paths are simple (not self-intersecting), and they can pass through any one face at most once.

2). Optimal paths "unfold" into straight lines. This is an observation that any high school geometry student has made if he has tried to solve the problem of finding the shortest path between two points on the surface of a box: Simply "cut" the box and flatten it in such a way that the start and goal can be connected by a straight line segment that stays on the flattened cardboard. One such unfolding is guaranteed to produce a shortest path. An optimal path which passes through an edge of the polyhedron must obey the *local optimality criterion* that it unfolds to give a straight line. More generally, for any type of surface there is a similar statement of the principle of optimality, and this produces a characterization of the *geodesics* of that type of surface. (Recall that the geodesics of a sphere are arcs of great circles; the geodesics of a polyhedral surface are characterized by the above unfolding property.) Of course, the question that must be answered by an algorithm is which of the (possibly exponentially many) unfoldings yields a shortest path?

3). On a convex surface, optimal paths will not go through vertices (except perhaps at their endpoints), while on a non-convex surface, an optimal path may pass through a vertex only if the angle it makes in so doing is greater than π (see Mitchell et al. [60] for a more precise statement).

Obviously, any algorithm which attempts to solve a shortest path problem should exploit as much of the structure of the paths as possible. It is also sometimes possible to exploit special structure when designing good heuristics. For example, to solve a shortest path problem on a given surface, we could use the simple heuristic which connects the start with the goal along the path "as the crow flies" (that is, the intersection of the surface with the vertical plane through the start and goal), and then iteratively improves the path by applying the local optimality criterion of (1) above. This will result in a *locally* optimal path which will frequently be very good in practice (while it is certainly not guaranteed to be *globally* optimal). Perhaps a planner which combines some "intelligence" with this heuristic would produce reasonably good paths. For example, one might have knowledge that there is only one feasible way through a mountain range, say through a specific gorge, and we would then want to route our initial guess through this passage, so that the locally optimal path produced will be reasonable. Further experimental and theoretical research is needed on these problems.

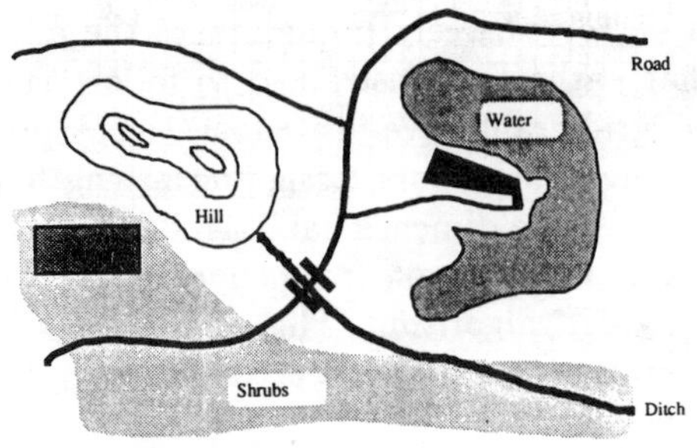

Figure 3: Map for terrain navigation

4 Weighted Regions

We now consider a more general terrain navigation problem. Assume that our map represents a set of *regions* in the plane, each of which has an associated *weight*, or cost, α. The weight of a region specifies the "cost per unit distance" of a vehicle (considered to be a point) traveling in that region. Our objective is then to find a path from START to GOAL in the plane which minimizes total cost according to this weighted Euclidean metric. For example, the weights may represent the reciprocal of the maximum speed in each region, in which case minimizing the weighted Euclidean length of a path is simply minimizing the length of time it takes the vehicle to execute the path.

This problem has been termed the *Weighted Region Problem.* Note that it is a generalization of the obstacle-avoidance problem in the plane, for the obstacle-avoidance problemis simply the weighted region problem in which the "weights" are either 1 or $+\infty$ depending on whether a region is "free space" or obstacle, respectively. Thus, an important theoretical reason for considering the weighted region problem is that it is a natural generalization of the well-studied obstacle-avoidance problem.

An example of a typical map is illustrated in Figure 3. Basically, the ground surface is subdivided into uniform regions, with a label (or some traversability index) attached to each region which gives information about how fast one can move in that region (or how costly it is to do so). Presumably, the robot can move through different types of terrain at different speeds. Speeds may also depend on other factors such as time of day, precipitation, or the location of other vehicles; however, we assume that a given problem instance has fixed weights. In military applications, there may be regions which correspond to high threat risk, perhaps because the enemy has good visibility of you when you are in these regions. Costs can be assigned to traveling in these risky regions as well.

We will discuss two basic types of map representations: regular tesselations (e.g., grids of pixels, or quadtrees), and straight-line planar subdivisions.

Representing terrain in the form of a regular grid of pixels is natural and simple. Figure 4 shows the map of Figure 3 in digitized form, on an grid of size 26-by-40 (1040 pixels). Frequently, terrain data is given in the form of a set of arrays, with each array giving information about some aspect of the terrain (e.g., ground cover, land usage, hydrography data, man-made features, and traversability indices). This, for example, is the form in which the Defense Mapping Agency terrain data is supplied. Pixels are usually squares 5, 12.5, or 100 meters on a side. From this type of data, we can specify the map by taking the superposition of all the data arrays to form one composite map (a "terrain array"), whose regions of equal attributes will have descriptions such as "brush-covered, drainage plain", or "forested land, sandy soil, small boulders". Each composite region may be assigned a weight representing the cost of motion in that type of terrain. Certain types of terrain features have priority over others. For example, if the "roads array" tells us there is a road surface occupying pixel (i, j), while the ground-cover array describes the pixel as "grassland", then the composite region attribute at (i, j) should say that there is a road there.

If the terrain map is given as a grid of weighted pixels, then a straightforward solution to the weighted

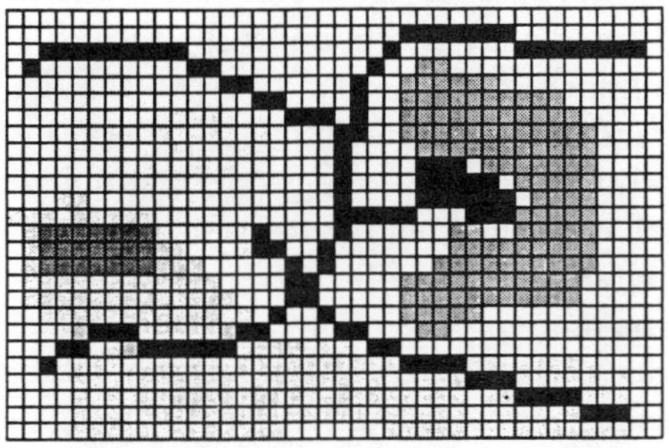

Figure 4: Digitized terrain map

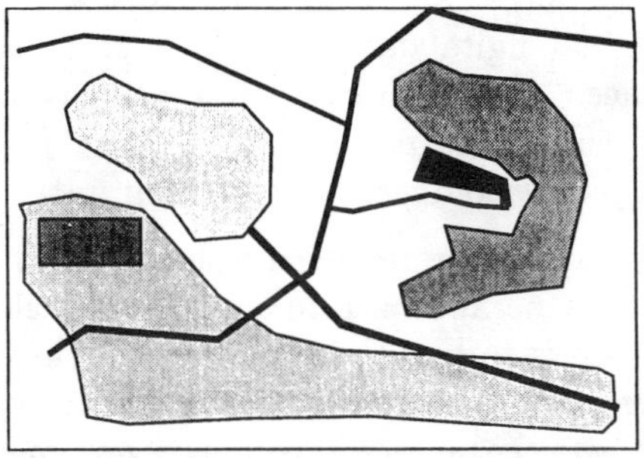

Figure 5: Polygonal subdivision map

region problem is to search the corresponding grid graph for shortest paths (Jones [32], Keirsey and Mitchell [36], Quek et al. [72]). This is simply a generalization of the approach described earlier for the case of digitized obstacle maps. The grid graph will have either 4- or 8-connectivity depending on whether or not we consider diagonal neighbors of pixels. Costs may be assigned in the natural way to arcs connecting adjacent pixels of the grid, taking into account the weights of the two pixels by, say, taking the cost of the arc between them to be their average weight. Then, Dijkstra's algorithm or A^* may be used to compute minimum-cost paths in the grid graph. For a grid with n pixels, this results in an algorithm that runs in worst-case time $O(n \log n)$. This approach is particularly appealing in cases in which the data is given to us in the form of feature arrays.

One problem with the grid graph approach is that it may require extremely fine grids to capture the content of a relatively simple piece of terrain.

Frequently, there may be just a couple of large uniform regions, perhaps with a road running through them, and the pixel representation of the map requires, say, a 512-by-512 array. The shortest path may indeed be obvious (such as a simple straight line segment), but the algorithm must still search and expand thousands of pixels during its execution (although hierarchical algorithms can avoid much of the expansion). Another problem with the grid graph approach is that it creates a *digitization bias* because of the metrication error imposed by confining movements to 4 or 8 orientations. See Keirsey and Mitchell [36] for a more detailed discussion of digitization bias and the various possible remedies for it. See Quek et al. [72] for a technique that uses grid graphs of multiple resolutions to design a hierarchical algorithm.

If, instead of being a regular tesselation, the map is modeled as a straight-line planar subdivision then the regions are just simple polygons.

Figure 5 shows the map of Figure 3 drawn as a straight-line planar subdivision with 85 edges. An appropriate data structure would be either a winged-edge (Baumgart [7]) or a quad-edge (Guibas and Stolfi [30]) data structure. Roads and other linear features (such as ditches, fences, streams, etc.) can be modeled as very skinny polygonal regions or approximated as sets of line segments. In the planar subdivision, then, we allow edges to be assigned a weight which may be different from the weights of the faces on either side of it. If an edge is a line segment of a road, then, presumably, its weight will be less than the weights on either side of it (assuming that it is cheaper to travel on roads than off roads). If an edge is a line segment of a ditch or a fence, then it may be assigned a fixed cost of crossing it. The fixed cost would be $+\infty$ if the edge cannot be crossed.

The representation of regions as polygonal patches and of roads as linear features is being used in the ongoing work on the DARPA Autonomous Land Vehicle project. One advantage of encoding the map information originally in the form of polygonal patches is that it can save greatly on the storage costs, while increasing the resolution of the data. The centerline of a road need not be specified only to the resolution of a pixel (say 12.5 meters), but can be specified to any desired degree of accuracy and can be specified to a higher resolution than other less critical features of the terrain.

In order to put the terrain array into the representation of a polygonal subdivision, we must fit polygonal boundaries about the sets of pixels with uniform attributes. At the highest resolution, each pixel can be considered to be a polygonal patch, namely, a square; however, this extreme will usually result in many more regions than desired, and it does not reflect the fact that the original digital data was only approximate in the first place. Usually, we will also want to represent roads, fences, and power lines as linear features by fitting a piecewise-linear path to them.

Assume now that we are given a planar subdivision with weights. One approach to solving the weighted region problem is to build a "region graph" in which nodes correspond to regions and arcs correspond to boundaries between adjacent regions. Assume that the subdivision is fine enough that all regions are convex. (We could, for example, make the subdivision a triangulation by triangulating each of the simple polygonal regions; however, we may want to subdivide in such a way that each convex region is as close to being circular as possible.) Then we can think of placing a node at the "center" (e.g., center of mass) of each region. Two nodes are joined by an edge if the corresponding regions are adjacent. We then assign costs to arcs according to the weighted distance between adjacent nodes. (An alternative graph can be constructed by placing nodes at the midpoints of region edges and linking two nodes if the corresponding edges share a common region.) Note that a grid graph is a special case of a region graph in which all regions are equal-sized squares. Searching this graph for shortest paths yields a "region path" from the START to the GOAL, giving a sequence of regions through which a "good" path should pass. We could then do some post-processing (e.g., using Snell's Law of refraction from optics, a local optimality criterion that we will mention below) to make the path locally optimal at region boundaries. The problem with this approach is that it can produce paths that are far from being optimal, for the optimal path need not have any relationship to the shortest region-path. In practice, however, it promises to be a very useful technique, particularly in cases in which the data is not very accurate or the assignment of weights is somewhat subjective. (Can one really say that grassland deserves a weight of 5 while brushland deserves a weight of 15? It seems likely that weights will be highly subjective.)

While the region graph approach does not guarantee that the resulting path is close to being optimal, another approach, given by Mitchell and Papadimitriou [62], yields a polynomial-time algorithm for computing ϵ-optimal paths. (A path is said to be *ϵ-optimal* if its length is guaranteed to be within a factor of $1+\epsilon$ of the length of an optimal path.) The complexity of the algorithm is $O(ES)$, where E is the number of "events" in the algorithm, and S is the time it takes to run a numerical search procedure. It is shown that the worst-case complexity of E is $\Theta(n^4)$, and that S is bounded by $O(n^4L)$, where L is the number of bits needed to represent the problem data and ϵ. The algorithm is a generalization of the algorithm for the discrete geodesic problem given by Mitchell et al. [60], and it too uses the continuous Dijkstra paradigm.

Let us be more precise about the problem solved in [62]. We are given a (straight-line) planar subdivision, $\mathcal{S}$, specified by a set of faces (regions), edges, and vertices, with each edge occurring in two faces and two faces intersecting either at a common edge, a vertex, or not at all. We consider faces to be *closed* polygons (they include their boundaries) and edges to be *closed* line segments (they include their endpoints, which are vertices). We are also given an initial point, START. Without loss of generality, we assume that all faces are triangles and that s and t are vertices of the triangulation. Assume that $\mathcal{S}$ has n edges (and hence $O(n)$ triangular faces and $O(n)$ vertices). Our complexity measures will be written in terms of n.

Each face f has associated with it a weight $\alpha_f \in [0,+\infty]$ which specifies the cost per unit distance of traveling on face f. Similarly, each edge e has associated with it a weight $\alpha_e \in [0,+\infty]$. The weighted distance between any two points x and y on edge e is simply the product $\alpha_e|xy|$, where $|xy|$ is the usual Euclidean distance between x and y. Likewise, the weighted distance between any two points x and y on face f (but not both on the same edge of f) is the product $\alpha_f|xy|$. The weighted length of a path through the subdivision is then the sum of the weighted lengths of its subpaths through each face.

We are asked to find the minimal length (in the weighted sense) path from START to some goal point. The algorithm of [62] actually solves the *query* form of the weighted region problem: Build a structure which allows one to compute a shortest path (in the weighted Euclidean metric) from START to *any* query point t. It turns out that this single-source version of the problem is no harder (in worst-case assymptotic time) than the problem with a given fixed goal point.

As with the discrete geodesic problem, the algorithm for the weighted region problem exploits certain facts about the local behavior of optimal paths:

1). Shortest paths are piecewise-linear, bending only at points where they cross an edge or pass through a vertex. This follows from the fact that each

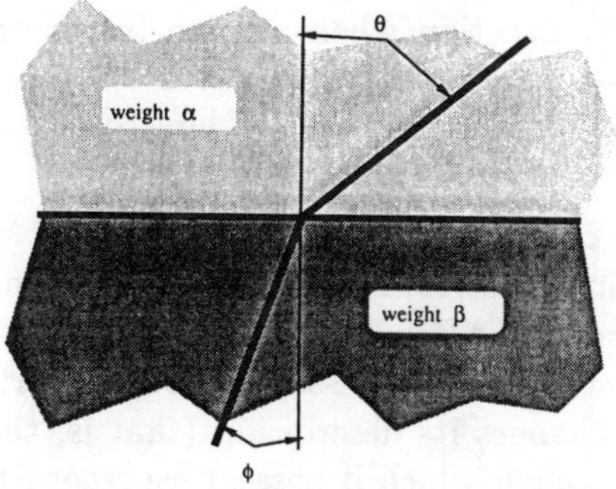

Figure 6: Light ray crossing a boundary

region is assumed to have a *uniform* weight. (The case of *non*uniform weighted regions gives rise to other types of curves and is an interesting area for further research.)

2). When an optimal path passes through an edge between regions, it does so while obeying Snell's Law of Refraction at that edge. This is the local optimality criterion which is analogous to the fact that shortest paths on the surface of a polyhedron unfold to be straight lines. It is a simple property of locally-optimal paths through weighted regions which has been observed by many researchers (Lyusternik [46], Mitchell and Papadimitriou [61, 62], Richbourg [76], Richbourg et al. [77,78], Rowe and Richbourg [80], Smith [86]).

Snell's Law of Refraction The path of a light ray passing through a boundary e between regions f and f' with indices of refraction α and β obeys the relationship that $\alpha \sin\theta = \beta \sin\phi$, where θ and ϕ are the angles of incidence and refraction (respectively).

The *angle of incidence*, θ, is defined to be the counterclockwise angle between the incoming ray and the normal to the region boundary. The *angle of refraction*, ϕ, is defined as the angle between the outgoing ray and the normal. Refer to Figure 6.

The fact that light obeys Snell's Law comes from the fact that light seeks the path of minimum time (this is *Fermat's Principle*). The index of refraction for a region is proportional to the speed at which light can travel through that region. Hence, the shortest paths in our weighted region problem must also obey Snell's Law. This can be shown formally (Mitchell and Papadimitriou [62], Mitchell [52]).

3). The angle of incidence of an optimal path at a boundary that it crosses will not be greater (in absolute value) than the *critical angle* defined at that boundary. This fact comes from examining the relationship specified in Snell's Law and making sure it is well-defined.

Without loss of generality, assume that $\alpha_f > \alpha_{f'} > 0$. Then, we must assure that

$$-1 \le \sin\alpha_{f'} = \frac{\alpha}{\alpha_{f'}} \sin\theta \le 1.$$

The angle, $\theta_c(e) = \theta_c(f, f')$, at which

$$\frac{\alpha_f}{\alpha_{f'}} \sin[\theta_c(f, f')] = 1$$

is called the *critical angle* defined by edge e. If $\alpha_f > 0$ and $\alpha_{f'} \ge 0$, then the critical angle is defined and is given by $\theta_c(f, f') = \sin^{-1}(\alpha_{f'}/\alpha_f)$, *provided* that $\alpha_f \ge \alpha_{f'}$. A ray of light which strikes e at the critical angle will (theoretically) travel along e rather than enter into the interior of f'. We can think of it as if the light ray wishes to be "just inside" the "cheap" region on the other side of e.

4). Optimal paths will not reflect from a boundary, except when they are "critically reflected" as shown in Figure 7: A path is incident (from a face f of weight β) on edge e (of weight $\gamma < \min\{\alpha, \beta\}$) at the critical angle $\phi = \theta_c$ at some point $y \in int(e)$, it then travels along edge e for some positive distance, and then exits edge e *back into face* f at some point $y' \in int(e)$, leaving the edge at an angle $\phi = \theta_c$. We then say that the path is *critically reflected* by edge e and that segment $\overline{yy'}$ is a *critical segment* of p along e. The phenomenon of critical reflection is not just an unlikely event of degeneracy. It is commonly the case that the shortest path between two points consists of a single critical reflection along an edge.

5). If $\alpha_e > \min\{\alpha_f, \alpha_{f'}\}$, then an optimal path will never travel along edge e, since it would be better to travel "just inside" either region f or region f' (whichever is cheaper).

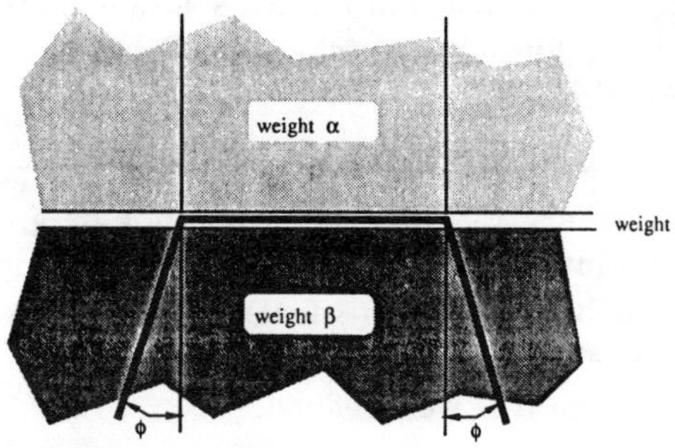

Figure 7: An optimal path which is critically reflected

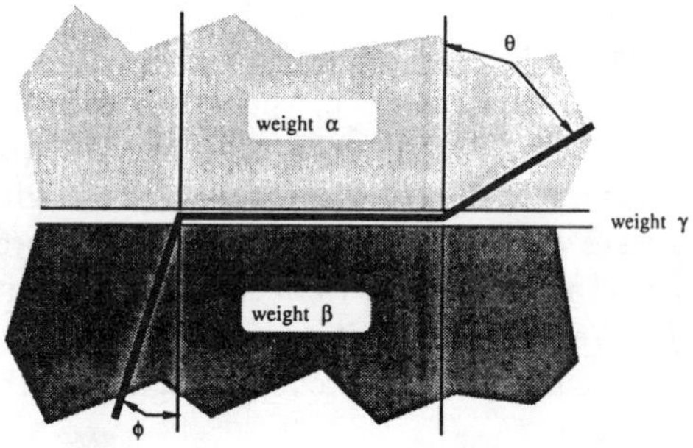

Figure 8: Shortest path "hitching a ride" on a road

6). An optimal path can cross a "road" (that is, an edge whose cost is less than that of the regions on either side of it) in one of two ways: either by obeying Snell's Law while passing through the boundary, or by "hitching a ride" along the edge, as illustrated in Figure 8. The path hits edge e at the *incoming* critical angle $\phi = \theta_c(f,e) = \sin^{-1}(\alpha_e/\alpha_f)$, then travels along edge e (of weight γ) for some distance, then leaves edge e into f' at the *outgoing* critical angle $\theta = \theta_c(f',e) = \sin^{-1}(\alpha_e/\alpha_{f'})$.

7). On an optimal path, between any critical point of exit and the next critical point of entry, there must be a vertex of $\mathcal{S}$. This fact turns out to be very important in the proofs of the complexity of the algorithm of Mitchell and Papadimitriou [62].

The local optimality criteria outlined above are sufficiently strong that they uniquely specify a locally optimal path which passes through a given sequence of edges. (This follows from the global convexity of the objective function when written in terms of the coordinates at which the path crosses each edge.) The problem remains, however, of how to select the proper sequence edges for the optimal path.

The continuous Dijkstra algorithm of Mitchell and Papadimitriou [62] simulates propagation of a wavefront from START. Each time the wavefront first hits a region boundary, passes through a vertex, or otherwise changes its description (that is, the list of regions through which it passes) we record the effect of the event in an appropriate data structure. When the wave has propagated throughout the surface, we have completed the construction of a shortest path map. For the application to the weighted region problem, it can be shown that the worst-case number of *event* points is $E = \Theta(n^4)$, thereby establishing a polynomial time bound.

Processing each event involves calling certain functions. One such function locates a point in an "access channel" between representative optimal paths that have already been established. This is easy and can be done by binary search in $O(\log n)$ time. However, we must also do $O(n^4)$ operations of the form "Find the refraction path from a given point r that passes through a given sequence of edges (while obeying Snell's Law) to hit a given point x". This is not an easy problem to solve exactly. It is handled by a numerical routine that does binary search to find a path whose length is at most $(1+\epsilon)$ times the length of the optimal path, where $\epsilon > 0$ is an arbitrary error tolerance value that is given in the problem specification. The search requires time which is polynomial in n and L, the number of bits necessary to represent the problem instance (namely, $\log(nNW/\epsilon w)$, where N is the largest integer value of any vertex coordinate, W is the largest finite weight, and w is the smallest nonzero weight). A crude calculation shows that the time to run the numerical search is bounded by $O(n^4 L)$.

It would be great if we could calculate exactly the refraction path from point r to point x, but this problem seems to be hard. The complication is that when we write down the equations that represent the local optimality criterion (Snell's Law) at each edge, and then try to solve for the points where the path crosses edges, we get k quartic equations in k unknowns (where k is the length of the edge sequence through which the refraction path is known to pass). Elimination among these equations yields a very high degree polynomial (of degree exponential in k). Alternatively, we could apply the cylindrical decomposition technique of Collins [19] to arrive at a doubly exponential time procedure to determine whether or not a given rational path length is achievable. This would also provide us with a technique of comparing the lengths of paths through two different edge sequences, and thus would solve the entire problem precisely in doubly exponential time (outputting the sequence of edges and vertices along the optimal path). The same technique led Sharir and Schorr [85] to a doubly exponential time solution to the three-dimensional shortest path problem.

In practice, a very straightforward numerical approach to solving the search problem can be used. For example, one technique begins with a path from r to x that connects the edges in the sequence along their midpoints. We then iteratively shorten the path by applying the local optimality criterion to adjacent segments of the path. We simply pick an edge at which Snell's Law is violated, and then let the crossing point "slide" along the edge until Snell's Law is obeyed. (This is a coordinate descent method for searching for the minimum of the convex function which describes the path length.) Each iteration can be computed in constant time and results in a strict decrease in the weighted path length. Hence, this procedure will converge to the locally optimal path from r to x, which is also the retraction path from r to x. (This type of numerical procedure is also applicable to the three-dimensional shortest path problem, in which the local optimality criterion is that paths unfold to be straight.) The problem with this approach is that we do not have good bounds on the rate of convergence or on the number of iterations necessary to guarantee that the solution is close (say, within ϵ%) to optimal.

However, the convergence rate has been observed to be "very fast" practice. In particular, in some recent implementation work of Karel Zikan [94], he has observed that a version of the above-mentioned coordinate descent method tends to take only about 5-10 iterations to get within 1 percent of optimality for problems with up to about 20 edges. By employing certain "overrelaxation" methods, he is able to improve the convergence rate, cutting approximately in half the number of iterations needed.

5 Special Cases of the Weighted Region Problem

There are a few special cases of the weighted region problem which admit alternative polynomial-time algorithms and are of interest in terrain navigation. We have already mentioned the special case in which all weights are 1 or $+\infty$. This leads to the usual obstacle-avoidance problem in the plane.

Consider now the special case in which all the weights on faces and edges are 0, 1, or $+\infty$. Infinite-cost regions are obstacles, cost one regions are simply "free space" (through which we can move at some fixed speed), and cost zero regions are "freebies" (through which motion costs nothing; we can move at infinite speed). If there are no zero-cost regions, then the problem is simply the usual obstacle avoidance shortest path problem in the plane. When there are zero-cost regions, they behave like "islands" between which one may want to "hop" to get from the source to the destination.

As a motivation for this special case, let us first note that it may be a reasonable approximation to the case in which the terrain is divided into three categories: obstacles (or very costly regions), very low cost regions, and regions whose costs are in between, but similar. Such would be the case for a runner (who is not a very good swimmer) trying to find a route from one point (say, on a flat island) to another point (say, in the water). There are many islands around. Some are flat and can be crossed easily on foot. Others have huge cliffs surrounding them which make them impassable. Swimming through the water is possible, but the runner would much prefer to be running across a flat island. So how can he get from one point to another in the shortest time? His path will "hop" from one flat island to another, swimming in between, and avoiding the mountainous islands.

An application of this problem which has arisen in practice is the *maximum concealment* problem. There are regions of the plane which are "visible" to an enemy threat, and these are hopefully to be avoided. There are also obstacles, which *must* be avoided. Then there are regions (e.g., between mountain ranges, or behind rocks) which are hidden from the enemy's view, and are hence extremely cheap in comparison with the visible regions. See Figure 9 for an example in which there is a single enemy observer whose view is obstructed by tall obstacles, but who can see over the bodies of water. The tall obstacles cast "shadows", which are regions of essentially free travel. The problem is to find an obstacle-avoiding path between two

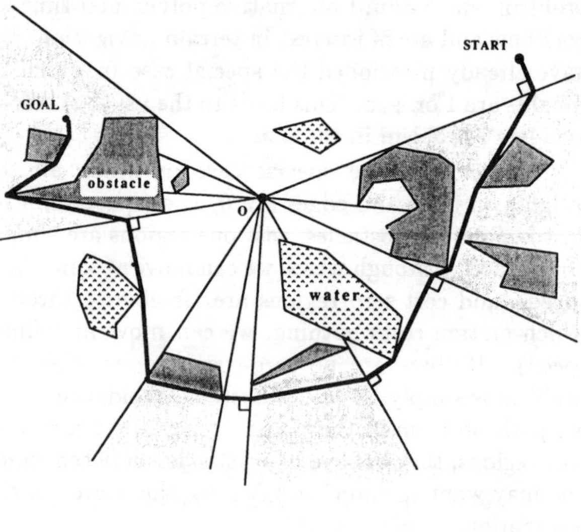

Figure 9: Maximal concealment from a single enemy observer

points which maximizes the concealment (minimizing the length of time in the exposed regions). In the context of the weighted region problem, the model is that water and mountain ranges are infinite-weight regions, shadowed regions behind mountains are zero-cost regions (for purposes of concealment), and all other terrain has weight 1. (It makes sense that the cost of being seen by a threat should depend on the distance from the threat. Models that make this assumption will be discussed in a forthcoming paper, Mitchell [59], where polynomial-time algorithms are given for some special cases.)

Other applications also use the $\{0, 1, +\infty\}$ special case of the weighted region problem. It can be used to find the shortest path from one *region* to another *region* in the presence of obstacles (by letting the source and destination regions be zero-weight regions), as was independently considered by Asano et al. [5]. It can also be used to find *lexicographically* shortest paths through weighted regions: minimize the length of path in the most expensive region, then, subject to this being minimal, minimize the length in the next most costly region, etc. Lexicographically shortest paths may be good initial guesses for heuristic algorithms that try to find shortest paths through weighted regions. Also, there is an intimate connection between shortest path problems and *maximum flow* problems, and the algorithms for maximum flow need to find shortest paths through $\{0, 1, +\infty\}$ weighted regions. These and other applications are discussed in Gewali et al. [25] and Mitchell [54,55].

An interesting application discussed in Gewali et al. [25] is that of the "least-risk watchman route problem". The watchman problem asks one to find the shortest path that a watchman would have to make from his current location in order to be able to "see" all of a given space (e.g., a polygonal room, or a polygonal room with "holes"). Chin and Ntafos [15,16] have shown that the optimum watchman route problem is NP-complete in a polygonal room with holes, but that there is a polynomial-time algorithm for polygonal rooms without holes. The *least-risk watchman route problem* asks one to find the path of minimum exposure to threats which sees the entire space. Then, for rectilinear polygons with n sides, Gewali et al. [25] give an $O(k^2n^3)$ algorithm for the case of k threats.

The algorithm described by Mitchell and Papadimitriou [62] for the weighted region problem does in fact solve the $\{0, 1, +\infty\}$ special case in polynomial time. But the special case has structure which allows an alternative approach which is faster and easier to implement. The approach in Mitchell [54] and Gewali et al. [25] to this case is to build a special kind of "extended visibility graph", VG^*, after shrinking the zero-cost regions to single nodes. The graph VG^* can be constructed in polynomial time (in fact, quadratic time), and it can then be searched for shortest paths in the usual way using Dijkstra or A^*. The result is an exact polynomial-time algorithm. The basic idea is to exploit the local optimality criteria, which allow the problem to be discretized by limiting our search to the graph VG^*. We now sketch these three simple criteria.

Criterion 1. The behavior of optimal paths with respect to the obstacles is the same as it was in the obstacle-avoidance problem (see Mitchell [52]).

In other words, the path behaves around obstacles as if it were a "taut string", being locally tangent to obstacles and forming a convex angle when it comes in contact with an obstacle vertex.

Criterion 2. An optimal path will either enter a zero-cost region at a vertex or at a normal to the interior of an edge bounding the region. If it enters at a vertex, then it must make an angle greater than $\pi/2$ with the edges incident to the vertex.

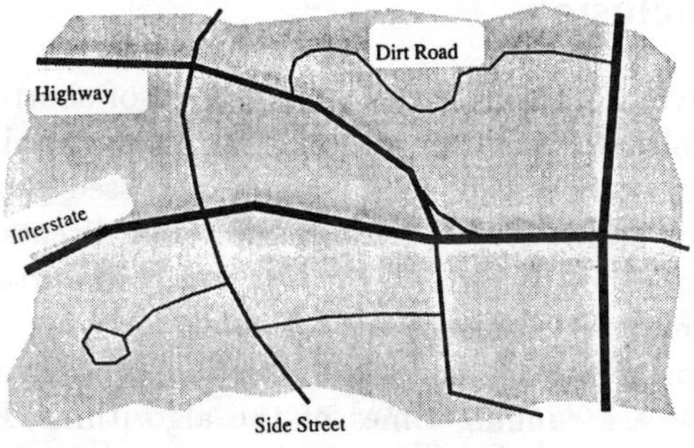

Figure 10: A map consisting of just roads

Thus, the extended visibility graph has edges of two types: those that join two vertices (of obstacles or zero-cost regions), and those that join a vertex (of either an obstacle or a zero-cost region) to an edge of a zero-cost region (in such a way as to be perpendicular to the edge). This immediately implies a quadratic bound on the size of the graph. The exact details of the construction are given in either of the above-mentioned papers ([54] or [25]). The result is an $O(n^2)$ algorithm for shortest paths.

Another special case of interest in terrain navigation is that in which all of the terrain has the same (finite) weight, but there exists a (piecewise-linear) network of roads or other linear features (of possibly different sizes, which admit different speeds of traversal). This is the case illustrated in Figure 10, and is simply the weighted region problem in which all face weights, α_f, are identical, and edge weights, α_e, are arbitrary. Then, we apply the following local optimality criterion:

Criterion 3. Shortest paths will enter a road segment either at one of its endpoints, or by hitting at the critical angle of incidence defined by the ratio of the edge's weight to that of the surrounding terrain.

This observation again leads to a discrete graph (a *critical graph*) which can be searched for shortest paths. Mitchell [54] and Gewali et al. [25] show how to construct this critical graph in time $O(n^2)$ and search it in time $O(n^2 \log n)$, where n is the number of road edges. See also Rowe [79].

A generalization of this technique is possible by putting together all of the above-mentioned local optimality criteria. Assume now that we have a weighted region problem in which the weights on faces are in the set $\{0, 1, +\infty\}$, and the weights on edges are arbitrary ($\in [0, +\infty]$). Additionally, there can be a fixed cost, ξ_e, associated with crossing an edge e. The model is that there are obstacles, zero-cost regions, and background regions (all of the same weight). Additionally, there are roads, ditches, fences, etc. of arbitrary weights. Then, the algorithm given in Mitchell [54] or Gewali et al. [25] finds shortest paths in time $O(n^2 \log n)$, again by building a critical graph in time $O(n^2)$.

6 Generalizations and Extensions

Several generalizations to the discrete geodesic and weighted region problems are possible. Some are solved by minor changes to the algorithms of Mitchell et al. [60] and Mitchell and Papadimitriou [62], while others are open research problems.

1). First, one can generalize to the case of *multiple source points*. This means that instead of building a shortest path map for one point, START, we consider several given point sites (called "source points"), and we compute a subdivision of the plane or the surface which tells us for each point t which source point is closest to it and how to backtrace a shortest path to the closest source point. The resulting structure is then a classic *Voronoi diagram* according to the metric of interest (either shortest Euclidean paths on a surface or weighted Euclidean lengths through a weighted map).

As a further generalization along these lines, we could allow the source and goal to be *regions* rather than single points. This generality is actually already incorporated in the model of Mitchell and Papadimitriou [62] by allowing there to be zero-cost regions. One simply makes the starting region have zero weight, and places the starting point anywhere inside; a similar accomodation is made for the goal region. The applicability of this problem is apparent: Goals are frequently specified in the form "Get me to the town square," or "Get me to Highway 95." Any point within the square or on the highway will suffice as a goal.

2). We can generalize the cost structure from being that of a *uniform* cost per distance in a region to that of allowing a cost function (such as a linear function) in each region. The function should be specified by some fixed number of parameters, and should allow computation of geodesics. In the uniform cost case, the geodesics within a given region are simply straight lines. In more general cases, geodesics would have to be computed by techniques of calculus of variations (Lyusternik [46]). The algorithm must then be modified to apply the local optimality criterion (Snell's law) to curved arcs at the boundaries of regions. If an optimal path passes through the point y interior to edge e, then the tangent lines at y to the geodesic curves in f and in f' must meet at y such that they obey Snell's law. Using this local optimality criterion, then, it should be possible to run an analogous algorithm to that of Mitchell and Papadimitriou [62], this time piecing together curves from region to region. The details of this generalization need to be examined further.

3). We could allow boundaries between regions to be curved (e.g., splinegons; see Souvaine [87]). Then, for the weighted region problem, Snell's law at a point y interior to a boundary curve must be applied as if the boundary at y is the straight line tangent to the boundary at y. Again, more research is needed on this problem.

4). The weighted region problem can be generalized to the case of a weighted polyhedral surface. Then the local optimality criterion becomes Snell's law applied to the "unfolded problem" at each boundary. That is, for an edge $e = f \cap f'$, we must "unfold" f' about e so that f and f' are made coplanar, and then we can apply Snell's law to points interior to e in this rotated coordinate frame.

5). We could allow the traversal of a region to be directionally dependent. This comment applies both to the (unweighted) discrete geodesic problem and to the weighted region problem. For example, it makes sense that it should be more costly to go "up" a hill than to go down it. We are currently investigating this case.

6). An important open problem is that of incorporating other types of vehicle constraints into the shortest path problem. For example, we may want a minimum time path for a vehicle which has a bounded amount of power (implying bounded acceleration). Or it may be that the vehicle will tip over if it is on too steep of a slope. Or the vehicle may have fuel consumption constraints. Solving the exact shortest path problem in the presence of other such constraints is an important area for continuing research.

7 Conclusion

We have examined some of the models of shortest path problems which are relevant to route planning for an autonomous vehicle. We have discussed problems which admit exact solutions, and have discussed the currently best known algorithms. Many important issues remain to be addressed.

Obviously, there remains the question of reducing the worst-case running times of the algorithms discussed. It is a challenging problem to determine the *expected* running times of shortest path algorithms.

Problems involving the design of a good terrain map need to be studied more. We have discussed regular tesselations and polygonal subdivisions as representations for map data, and we have mentioned some of the benefits of each. What is probably needed is some kind of hybrid representation which can take advantage of the strengths of each, while also providing a hierarchical representation of the terrain.

If we were to compose a "wish list" of features for a map representation, we would most certainly want to include the ability to handle the following.

Discovery of new information. Ideally, as the vehicle moves about through terrain, it will update its map with new information it "learns". It may, for example, discover that road construction is causing congestion along some stretch of highway or that a certain bridge it thought was there has been closed. It is therefore important to have a representation of a map which allows updating and annotations.

This also brings up the issue of interfacing the map-based path planner with some *sensor*-based planner that accumulates local information during the execution of a plan and must then modify the path accordingly. We have not addressed here the issue of planning routes in unknown or uncertain terrains, but we can refer the reader to a few of the myriad of papers on the subject: Goldstein et al. [27], Iyengar et al. [31], Jorgensen et al. [33], Kauffman [35], Kuan [38], Kuan et al. [39], Kuan et al. [40], Lumelsky and Stepanov [45], Meng [48,49], Metea and Tsai [50], Mitchell [51], Mitchell et al. [63], Moravec [64], Oommen et al. [66], Rao et al. [73,74], Rueb and Wong [81], Stern [88], Sutherland [89], and Thorpe [91].

Temporal nature of information. Annotations might carry with them some estimated *duration* which bounds the interval of time for which the information is valid. For example, road construction is usually known to affect traffic flow only during certain times of the day, and it remains in effect only until it achieves completion.

Of course, uncertainty assures that the planner will never be able to guarantee that the routes it selects in advance are optimal. All types of map information are subject to some uncertainty. For example, roads can be closed or become impassable, open terrain can become flooded or reforested, boundaries of lakes and rivers certainly change with the seasons and with the rainfall pattern. A truly "intelligent" planner should have information about the kinds of circumstances which potentially affect different types of terrain and be able to reason about it. If it has knowledge, for example, that there was recently a greater than average rainfall, it should adjust the probabilities it places on the traversability of a low-water bridge.

Many open questions remain. How can these issues be addressed within the framework of exact or approximate algorithms? What are the appropriate mathematical models of the path-planning process? How can our knowledge of algorithmic path planners be translated into useful heuristics for complex terrain problems?

Acknowledgements

Part of this research was conducted while J. Mitchell was supported by a Howard Hughes Doctoral Fellowship at Stanford University and affiliated with the Hughes Artificial Intelligence Center in Calabasas, California. This research was also partially supported by grants from the Hughes Research Laboratories and the National Science Foundation (IRI-8710858 and ECSE-8857642).

References

[1] V. Akman, "Shortest Paths Avoiding Polyhedral Obstacles In 3-Dimensional Euclidean Space", PhD Thesis, Computer and Systems Engineering, Rensselaer Polytechnic Institute, June, 1985.

[2] V. Akman, "Unobstructed Shortest Paths in Polyhedral Environments", *Lecture Notes in Computer Science*, Vol. 251, Edited by G. Goos and J. Hartmanis, Springer-Verlag, 1987.

[3] H. Alt and E. Welzl, "Visibility Graphs and Obstacle-Avoiding Shortest Paths", *Zeitschrift für Operations Research*, **32** (1988), pp. 145-164.

[4] T. Asano, T. Asano, L. Guibas, J. Hershberger, and H. Imai, "Visibility of Disjoint Polygons", *Algorithmica*, **1** (1986), pp. 49-63.

[5] T. Asano, T. Asano, and H. Imai, "Shortest Paths Between Two Simple Polygons", *Information Processing Letters*, **24** (1987), pp. 285-288.

[6] B. Baker, "Shortest Paths With Unit Clearance Among Polygonal Obstacles", *SIAM Conference on Geometric Modeling and Robotics*, Albany, NY, July 15-19, 1985.

[7] B.G. Baumgart, "A Polyhedron Representation for Computer Vision", *AFIPS Conference Proceedings*, Vol. 44, 1975 National Computer Conference, pp. 589-596.

[8] R.A. Brooks, "Solving the Find-Path Problem by Good Representation of Free Space", *IEEE Transactions on Systems, Man, and Cybernetics*, **13** (1983), pp. 190-197.

[9] R.A. Brooks and T. Lozano-Pérez, "A Subdivision Algorithm in the Configuration Space for Find Path with Rotation", *IEEE Transactions on Systems, Man, and Cybernetics*, **15** (1985), pp. 224-233.

[10] J.F. Canny, "The Complexity of Robot Motion Planning", Ph.D. Thesis, Department of Electrical Engineering and Computer Science, Massachusetts Institute of Technology, May 1987.

[11] J. Canny and J. Reif, "New Lower Bound Techniques for Robot Motion Planning Problems", *Proc. 28th Annual IEEE Foundations of Computer Science*, pp. 49-60, 1987.

[12] B. Chazelle, "Triangulating a Simple Polygon in Linear Time", Technical Report CS-TR-264-90, Princeton Univ., May 1990.

[13] J. Chen and Y. Han, "Shortest Paths on a Polyhedron", *Proc. 6th Annual ACM Symposium on Computational Geometry*, Berkeley, CA, June 6-8, 1990, pp. 360-369.

[14] L.P. Chew, "Planning the Shortest Path for a Disk in $O(n^2 \log n)$ Time", *Proc. First Annual ACM Conference on Computational Geometry*, Baltimore, MD, pp. 214-220, 1985.

[15] W.P. Chin and S. Ntafos, "Optimum Watchman Routes", *Information Processing Letters*, **28** (1988), pp. 39-44.

[16] W.P. Chin and S. Ntafos, "Watchman Routes in Simple Polygons", Technical Report, Computer Science, University of Texas at Dallas, 1987. To appear: *Discrete and Computational Geometry.*

[17] K. Clarkson, "Approximation Algorithms for Shortest Path Motion Planning", *Proc. 19th Annual ACM Symposium on Theory of Computing*, New York City, May 25-27, 1987, pp. 56-65.

[18] K. Clarkson, S. Kapoor, and P. Vaidya, "Rectilinear Shortest Paths through Polygonal Obstacles in $O(n \log^2 n)$ Time", *Proc. Third Annual ACM Conference on Computational Geometry*, Waterloo, Ontario, 1987, pp. 251-257. To appear: *Algorithmica.*

[19] G.E. Collins, "Quantifier Elimination for Real Closed Fields by Cylindric Algebraic Decomposition", *Proc. Second GI Conference on Automata Theory and Formal Languages*, in *Lecture Notes in Computer Science*, Vol. 35, Springer-Verlag, pp. 134-183, 1975.

[20] P.J. de Rezende, D.T. Lee, and Y.F. Wu, "Rectilinear Shortest Paths With Rectangular Barriers", *First ACM Conference on Computational Geometry*, June 1985, pp. 204-213.

[21] E.W. Dijkstra "A Note On Two Problems in Connexion With Graphs", *Numerische Mathematik*, **1** (1959), pp. 269-271.

[22] H. Edelsbrunner, *Algorithms in Combinatorial Geometry*, Springer-Verlag, Heidelberg, Germany, 1987.

[23] W. R. Franklin and V. Akman, "Shortest Paths Between Source and Goal Points Located On/Around a Convex Polyhedron", *22nd Allerton Conference on Communication, Control and Computing*, pp. 103-112, 1984.

[24] M. Fredman and R. Tarjan, "Fibonacci Heaps and Their Uses in Improved Network Optimization Algorithms", *Proc. 25th Annual IEEE Symposium on Foundations of Computer Science*, pp. 338-346, 1984.

[25] L. Gewali, A. Meng, J.S.B. Mitchell, and S. Ntafos, "Path Planning in $0/1/\infty$ Weighted Regions With Applications", *ORSA Journal on Computing*, **2** (1990), pp. 253-272.

[26] S.K. Ghosh and D.M. Mount, "An Output Sensitive Algorithm for Computing Visibility Graphs", *Proc. 28th Annual IEEE Symposium on Foundations of Computer Science*, 1985.

[27] M. Goldstein, F.G. Pin, G. de Saussure, and C.R. Weisbin, "3-D World Modeling With Updating Capability Based on Combinatorial Geometry", Technical Report CESAR-87/02, Oak Ridge National Laboratory, 1987.

[28] L.J. Guibas, J. Hershberger, D. Leven, M. Sharir, and R. Tarjan, "Linear Time Algorithms for Visibility and Shortest Path Problems Inside Triangulated Simple Polygons", *Algorithmica*, **2** (1987), pp. 209-233.

[29] L.J. Guibas and J. Hershberger, "Optimal Shortest Path Queries in a Simple Polygon", *Proc. Third Annual ACM Conference on Computational Geometry*, Waterloo, Ontario, 1987, pp. 50-63.

[30] L.J. Guibas and J. Stolfi, "Primitives for the Manipulation of General Subdivisions and the Computation of Voronoi Diagrams", *ACM Trans. Graphics*, **4** (1985), pp. 74-123.

[31] S.S. Iyengar, C.C. Jorgensen, S.V.N. Rao, and Weisbin, "Robot Navigation Algorithms Using Learned Spatial Graphs", *Robotica*, **4** (1986), pp. 93-100.

[32] S.T. Jones, "Solving Problems Involving Variable Terrain. Part 1: A General Algorithm", *Byte*, **5** (1980).

[33] C. Jorgensen, W. Hamel, and C. Weisbin, "Autonomous Robot Navigation", *BYTE*, **11** (1986), pp. 223-235.

[34] S. Kambhampati and L.S. Davis, "Multiresolution Path Planning for Mobile Robots", *IEEE Journal of Robotics and Automation*, **2** (1986), pp. 135-145.

[35] S. Kauffman, "An Algorithmic Approach to Intelligent Robot Mobility", *Robotics Age*, **5** (1983), pp. 38-47.

[36] D.M. Keirsey and J.S.B. Mitchell, "Planning Strategic Paths Through Variable Terrain Data", *Proc. SPIE Applications of Artificial Intelligence*, Vol. 485, pp. 172-179, 1984.

[37] D.G. Kirkpatrick, "Optimal Search in Planar Subdivisions", *SIAM Journal on Computing*, **12** (1983), pp. 28-35.

[38] D.T. Kuan, "Terrain Map Knowledge Representation for Spatial Planning", *Proc. The First Conference of Artificial Intelligence Applications*, IEEE Computer Society, Denver, CO, December 1984, pp. 578-584.

[39] D.T. Kuan, R.A. Brooks, J.C. Zamiska, and M. Das, "Automatic Path Planning for a Mobile Robot Using a Mixed Representation of Free Space", *Proc. The First Conference of Artificial Intelligence Applications*, IEEE Computer Society, Denver, CO, December 1984, pp. 70-74.

[40] D.T. Kuan, J.C. Zamiska, and R.A. Brooks, "Natural Decomposition of Free Space for Path Planning", *Proc. IEEE International Conference on Robotics and Automation*, IEEE Computer Society, St. Louis, MO, March 1985, pp. 168-173.

[41] D.T. Lee, "Proximity and Reachability in the Plane", Ph.D. Thesis, Technical Report R-831, Coordinated Science Laboratory, University of Illinois, 1978.

[42] D.T. Lee and Y.T. Ching, "The Power of Geometric Duality Revisited", *Information Processing Letters*, **21** (1985), pp. 117-122.

[43] D.T. Lee and F.P. Preparata, "Euclidean Shortest Paths in the Presence of Rectilinear Boundaries", *Networks*, **14** (1984), pp. 393-410.

[44] T. Lozano-Perez and M.A. Wesley, "An Algorithm for Planning Collision-Free Paths Among Polyhedral Obstacles", *Communications of the ACM*, **22** (1979), pp. 560-570.

[45] V.J. Lumelsky and A.A. Stepanov, "Path-Planning Strategies for a Point Mobile Automaton Moving Amidst Unknown Obstacles of Arbitrary Shape", *Algorithmica*, **2** (1987), pp. 403-430.

[46] L.A. Lyusternik, *Shortest Paths: Variational Problems*, Macmillan Company, New York, 1964.

[47] K. Mehlhorn, *Multi-dimensional Searching and Computational Geometry*, EATCS Monographs on Theoretical Computer Science, Springer-Verlag, Berlin, Heidelberg, New York, Tokyo, 1984.

[48] A. Meng, "Free-Space Modeling and Geometric Motion Planning Under Unexpected Obstacles", Technical Report, Texas Instruments Artificial Intelligence Lab, 1987.

[49] A. Meng, "Free-Space Modeling and Path Planning Under Uncertainty for Autonomous Air Robots", Technical Report, Texas Instruments Artificial Intelligence Lab, 1987.

[50] M.B. Metea and J.J.-P. Tsai, "Route Planning for Intelligent Autonomous Land Vehicles Using Hierarchical Terrain Representation", *Proc. IEEE International Conference on Robotics and Automation*, Raleigh, NC, March/April 1987.

[51] J.S.B. Mitchell, "An Autonomous Vehicle Navigation Algorithm", *Proc. SPIE Applications of Artificial Intelligence*, Vol. 485, pp. 153-158, 1984.

[52] J.S.B. Mitchell, "Planning Shortest Paths", PhD Thesis, Department of Operations Research, Stanford University, August, 1986.

[53] J.S.B. Mitchell, "Shortest Rectilinear Paths Among Obstacles", Technical Report No. 739, School of Operations Research and Industrial Engineering, Cornell University, April, 1987. To appear, *Algorithmica*, titled "L_1 Shortest Paths Among Polygonal Obstacles in the Plane".

[54] J.S.B. Mitchell, "Shortest Paths Among Obstacles, Zero-Cost Regions, and Roads", Technical Report No. 764, School of Operations Research and Industrial Engineering, Cornell University, 1987.

[55] J.S.B. Mitchell, "On Maximum Flows in Polyhedral Domains", *Journal of Computer and System Sciences*, **40** (1990), pp. 88-123.

[56] J.S.B. Mitchell, "A New Algorithm for Shortest Paths Among Obstacles in the Plane", *Annals of Mathematics and Artificial Intelligence*, **3** (1991), pp. 83-106.

[57] J.S.B. Mitchell, "Algorithmic Approaches to Optimal Route Planning", Technical Report 937, School of Operations Research and Industrial Engineering, Cornell University, October, 1990. *Proc. SPIE Conference on Mobile Robots*, November 4–9, Boston, MA, 1990.

[58] J.S.B. Mitchell, "An Optimal Algorithm for Shortest Rectilinear Paths Among Obstacles", *First Canadian Conference on Computational Geometry*, Montreal, Canada, August, 1989. Full paper is in preparation.

[59] J.S.B. Mitchell, "On the Maximum Concealment Problem", Manuscript in preparation, Department of Operations Research, Cornell University.

[60] J.S.B. Mitchell, D.M. Mount, and C.H. Papadimitriou, "The Discrete Geodesic Problem", *SIAM Journal on Computing*, **16** (1987), pp. 647-668.

[61] J.S.B. Mitchell and C.H. Papadimitriou, "Planning Shortest Paths", *SIAM Conference on Geometric Modeling and Robotics*, Albany, NY, July 15-19, 1985.

[62] J.S.B. Mitchell and C.H. Papadimitriou, "The Weighted Region Problem: Finding Shortest Paths Through a Weighted Planar Subdivision", *Journal of the ACM*, **38** (1991), pp. 18-73.

[63] J.S.B. Mitchell, D.W. Payton, and D.M. Keirsey, "Planning and Reasoning for Autonomous Vehicle Control", *International Journal of Intelligent Systems*, **II** (1987), pp. 129-198.

[64] H.P. Moravec, "Obstacle Avoidance and Navigation in the Real World by a Seeing Robot Rover", Carnegie-Mellon Robotics Institute, Pittsburgh, PA, Technical Report CMU-RI-TR-3, September, 1980.

[65] N.J. Nilsson, *Principles of Artificial Intelligence*, Tioga Publishing Co., Palo Alto, CA, 1980.

[66] B.J. Oommen, S.S. Iyenger, S.V. Rao, and R.L. Kashyap, "Robot Navigation in Unknown Terrains Using Learned Visibility Graphs. Part I: The Disjoint Convex Obstacle Case", Technical Report SCS-TR-86, School of Computer Science, Carleton University, February 1986.

[67] J. O'Rourke, "Finding a Shortest Ladder Path: A Special Case", Technical Report, Institute for Mathematics and its Applications, Preprint Series No. 353, University of Minnesota, October 1987.

[68] J. O'Rourke, S. Suri, and H. Booth, "Shortest Paths on Polyhedral Surfaces", *Proc. 2nd Symposium on Theoretical Aspects of Computer Science*, pp. 243-254, 1984.

[69] C.H. Papadimitriou "An Algorithm for Shortest-Path Motion in Three Dimensions", *Information Processing Letters*, **20** (1985), pp. 259-263.

[70] C.H. Papadimitriou and E.B. Silverberg, "Optimal Piecewise Linear Motion of an Object Among Obstacles", *Algorithmica*, **2** (1987), pp. 523-539.

[71] F.P. Preparata and M.I. Shamos, *Computational Geometry*, Springer-Verlag, New York, Berlin, Heidelberg, Tokyo, 1985.

[72] F.K.H. Quek, R.F. Franklin, and F. Pont, "A Decision System for Autonomous Robot Navigation Over Rough Terrain", *Proc. SPIE Applications of Artificial Intelligence*, Boston, 1985.

[73] N.S.V. Rao, S.S. Iyengar, C.C. Jorgensen, and C.R. Weisbin, "Robot Navigation in an Unexplored Terrain", *Journal of Robotic Systems*, **3** (1986), pp. 389-407.

[74] N.S.V. Rao, S.S. Iyengar, C.C. Jorgensen, and C.R. Weisbin, "On Terrain Acquisition by a Finite-Sized Mobile Robot in Plane", *Proc. IEEE International Conference on Robotics and Automation*, Raleigh, NC, March/April 1987, pp. 1314-1319.

[75] J.H. Reif and J.A. Storer, "Shortest Paths in Euclidean Space with Polyhedral Obstacles", Technical Report CS-85-121, Computer Science Department, Brandeis University, April, 1985.

[76] R.F. Richbourg, "Solving a Class of Spatial Reasoning Problems: Minimal-Cost Path Planning in the Cartesian Plane", Ph.D. Thesis, Naval Postgraduate School, Monterey, CA, June, 1987.

[77] R.F. Richbourg, N.C. Rowe, and M.J. Zyda, "Exploiting Capability Constraints To Solve Global, Two-Dimensional Path Planning Problems", Technical Report NPS-86-006, Department of Computer Science, Naval Postgraduate School, 1986.

[78] R.F. Richbourg, N.C. Rowe, M.J. Zyda, and R. McGhee, "Solving Global Two-Dimensional Routing Problems Using Snell's Law and A^* Search", *Proc. IEEE International Conference on Robotics and Automation*, Raleigh, NC, March/April 1987, pp. 1631-1636.

[79] N.C. Rowe, "Roads, Rivers, and Obstacles: Optimal Two-Dimensional Path Planning around Linear Features for a Mobile Agent", *International Journal of Robotics Research*, **9** (1990), pp. 67-73.

[80] N.C. Rowe and R.F. Richbourg, "An Efficient Snell's Law Method for Optimal-Path Planning across Multiple Two-Dimensional, Irregular, Homogeneous-Cost Regions", *International Journal of Robotics Research*, **9** (1990), pp. 48-66.

[81] K.D. Rueb and A.K.C. Wong, "Structuring Free Space as a Hypergraph for Roving Robot Path Planning and Navigation", *IEEE Transactions on Pattern Analysis and Machine Intelligence*, **PAMI-9** (1987), pp. 263-273.

[82] H. Samet and R.E. Webber, "Storing a Collection of Polygons Using Quadtrees", *ACM Transactions on Graphics*, **4** (1985), pp. 182-222.

[83] J.T. Schwartz, M. Sharir, and J. Hopcropt, Ed., *Planning, Geometry, and Complexity of Robot Motion*, Ablex Series in Artificial Intelligence, Ablex Publishing Corporation, Norwood, NJ, 1987.

[84] J.T. Schwartz and C. Yap, Ed., *Algorithmic and Geometric Aspects of Robotics, Vol. 1*, Lawrence Erlbaum Associates, Hillsdale, NJ, 1987.

[85] M. Sharir and A. Schorr, "On Shortest Paths in Polyhedral Spaces", *SIAM Journal of Computing*, **15** (1986), pp. 193-215.

[86] T. Smith, Private communication, Department of Computer Science, University of California, Santa Barbara, 1986.

[87] D. Souvaine, "Computational Geometry in a Curved World", Ph.D. Thesis, Technical Report CS-TR-094-87, Department of Computer Science, Princeton University, October, 1986.

[88] H.I. Stern, "A Routing Algorithm for a Mobile Search Robot", in *Impacts of Microcomputers on Operations Research*, Gass, Greenburg, Hoffman, and Langley, Eds., Elsevier Science Publishing Co., 1986, pp. 73-87.

[89] I.E. Sutherland, "A Method for Solving Arbitrary Wall Mazes by Computer", *IEEE Transactions on Computers*, **C-18** (1969), pp. 1092-1097.

[90] C.E. Thorpe, "Path Relaxation: Path Planning for a Mobile Robot", *Proc. AAAI National Conference on Artificial Intelligence*, Austin, TX, August 6-10, 1984, pp. 318-321.

[91] C.E. Thorpe, "FIDO: Vision and Navigation for a Robot Rover", Ph.D. Thesis, Technical Report CMU-CS-84-168, Department of Computer Science, Carnegie-Mellon University, 1984.

[92] E. Welzl, "Constructing the Visibility Graph for n Line Segments in $O(n^2)$ Time", *Information Processing Letters*, **20** (1985), pp. 167-171.

[93] P. Widmayer, Y.F. Wu, and C.K. Wong, "On Some Distance Problems in Fixed Orientations", *SIAM Journal on Computing*, **16** (1987), pp. 728-746.

[94] K. Zikan, Private communication, Department of Operations Research, Stanford University, Stanford, CA, 1986.

Oussama Khatib
Artificial Intelligence Laboratory
Stanford University
Stanford, California 94305

Real-Time Obstacle Avoidance for Manipulators and Mobile Robots

Abstract

This paper presents a unique real-time obstacle avoidance approach for manipulators and mobile robots based on the artificial potential field concept. Collision avoidance, traditionally considered a high level planning problem, can be effectively distributed between different levels of control, allowing real-time robot operations in a complex environment. This method has been extended to moving obstacles by using a time-varying artificial potential field. We have applied this obstacle avoidance scheme to robot arm mechanisms and have used a new approach to the general problem of real-time manipulator control. We reformulated the manipulator control problem as direct control of manipulator motion in operational space—the space in which the task is originally described—rather than as control of the task's corresponding joint space motion obtained only after geometric and kinematic transformation. Outside the obstacles' regions of influence, we caused the end effector to move in a straight line with an upper speed limit. The artificial potential field approach has been extended to collision avoidance for all manipulator links. In addition, a joint space artificial potential field is used to satisfy the manipulator internal joint constraints. This method has been implemented in the COSMOS system for a PUMA 560 robot. Real-time collision avoidance demonstrations on moving obstacles have been performed by using visual sensing.

1. Introduction

In previous research, robot collision avoidance has been a component of higher levels of control in hierarchical robot control systems. Collision avoidance has been treated as a planning problem, and research in this area has focused on the development of collision-free path planning algorithms (Lozano-Perez 1980; Moravec 1980; Chatila 1981; Brooks 1983). These algorithms aim at providing the low level control with a path that will enable the robot to accomplish its assigned task free from any risk of collision.

From this perspective, the function of low level control is limited to the execution of elementary operations for which the paths have been precisely specified. The robot's interaction with its environment is then paced by the time cycle of high level control, which is generally several orders of magnitude slower than the response time of a typical robot. This places limits on the robot's real-time capabilities for precise, fast, and highly interactive operations in a cluttered and evolving environment. We will show that it is possible to extend greatly the function of low level control and to carry out more complex operations by coupling environment sensing feedback with the lowest level of control.

Increasing the capability of low level control has been the impetus for the work on real-time obstacle avoidance that we discuss here. Collision avoidance at the low level of control is not intended to replace high level functions or to solve planning problems. The purpose is to make better use of low level control capabilities in performing real-time operations. At this low level of control, the degree or *level of competence* (Brooks 1984) will remain less than that of higher level control.

The *operational space formulation* is the basis for the application of the potential field approach to robot manipulators. This formulation has its roots in the work on end-effector motion control and obstacle avoidance (Khatib and Le Maitre 1978) that has been implemented for an MA23 manipulator at the Laboratoire d'Automatique de Montpellier in 1978. The operational space approach has been formalized by constructing its basic tool, the equations of motion in the operational space of the manipulator end effector.

Reprinted from *The International Journal of Robotics Research*, Vol. 5, No. 1, Spring 1986, pages 90-98, "Real-Time Obstacle Avoidance for Manipulators and Mobile Robots" by O. Khatib, by permission of The MIT Press, Cambridge, Massachusetts.

Details of this work have been published elsewhere (Khatib 1980; Khatib 1983; Khatib 1985). In this paper, we review the fundamentals of the operational space formulation and the artificial potential field concept. We present the integration of this collision avoidance approach into the operational space control system and its real-time implementation. The extension of this work to link collision avoidance is also developed.

2. Operational Space Formulation

An *operational coordinate system* is a set x of m_0 *independent* parameters describing the manipulator end-effector position and orientation in a frame of reference R_0. For a nonredundant manipulator, these parameters form a set of configuration parameters in a domain of the operational space and constitute a system of generalized coordinates. The kinetic energy of the holonomic articulated mechanism is a quadratic form of the generalized velocities,

$$T(\mathbf{x}, \dot{\mathbf{x}}) = \tfrac{1}{2}\dot{\mathbf{x}}^T \Lambda(\mathbf{x})\dot{\mathbf{x}}, \tag{1}$$

where $\Lambda(\mathbf{x})$ designates the symmetric matrix of the quadratic form, *i.e.*, the kinetic energy matrix. Using the Lagrangian formalism, the end-effector equations of motion are given by

$$\frac{d}{dt}\left(\frac{\partial L}{\partial \dot{\mathbf{x}}}\right) - \frac{\partial L}{\partial \mathbf{x}} = \mathbf{F}, \tag{2}$$

where the Lagrangian $L(\mathbf{x}, \dot{\mathbf{x}})$ is

$$L(\mathbf{x}, \dot{\mathbf{x}}) = T(\mathbf{x}, \dot{\mathbf{x}}) - U(\mathbf{x}), \tag{3}$$

and $U(\mathbf{x})$ represents the potential energy of the gravity. The symbol F is the operational force vector. These equations can be developed (Khatib 1980; Khatib 1983) and written in the form,

$$\Lambda(\mathbf{x})\ddot{\mathbf{x}} + \mu(\mathbf{x}, \dot{\mathbf{x}}) + \mathbf{p}(\mathbf{x}) = \mathbf{F}, \tag{4}$$

where $\mu(\mathbf{x}, \dot{\mathbf{x}})$ represents the centrifugal and Coriolis forces and $\mathbf{p}(\mathbf{x})$ the gravity forces.

The control of manipulators in operational space is based on the selection of F as a command vector. In order to produce this command, specific forces Γ must be applied with joint-based actuators. The relationship between F and the joint forces Γ is given by

$$\Gamma = J^T(\mathbf{q})\,\mathbf{F}, \tag{5}$$

where q is the vector of the n joint coordinates and $J(\mathbf{q})$ the Jacobian matrix.

The decoupling of the end-effector motion in operational space is achieved by using the following structure of control,

$$\mathbf{F} = \Lambda(\mathbf{x})\mathbf{F}^* + \mu(\mathbf{x}, \dot{\mathbf{x}}) + \mathbf{p}(\mathbf{x}), \tag{6}$$

where $\mathbf{F}^*$ represents the command vector of the decoupled end effector that becomes equivalent to a *single unit mass*.

The extension of the operational space approach to redundant manipulator systems is discussed in Khatib (1980); Khatib (1983). The integration of active force control for assembly operations is presented in Khatib (1985).

3. The Artificial Potential Field Approach

We present this method in the context of manipulator collision avoidance. Its application to mobile robots is straightforward. The philosophy of the artificial potential field approach can be schematically described as follows:

The manipulator moves in a field of forces. The position to be reached is an attractive pole for the end effector and obstacles are repulsive surfaces for the manipulator parts.

We first consider the collision avoidance problem of a manipulator end effector with a single obstacle O. If $\mathbf{x}_d$ designates the goal position, the control of the manipulator end effector with respect to the obstacle O can be achieved by subjecting it to the artificial potential field,

$$U_{art}(\mathbf{x}) = U_{\mathbf{x}_d}(\mathbf{x}) + U_O(\mathbf{x}). \tag{7}$$

This leads to the following expression of the potential

energy in the Lagrangian (Eq. 3),

$$U(\mathbf{x}) = U_{art}(\mathbf{x}) + U_g(\mathbf{x}), \quad (8)$$

where $U_g(\mathbf{x})$ represents the gravity potential energy. Using Lagrange's (Eq. 2) and taking into account the end-effector dynamic decoupling (Eq. 6), the command vector $\mathbf{F}^*_{art}$ of the decoupled end effector that corresponds to applying the artificial potential field U_{art} (Eq. 7) can be written as

$$\mathbf{F}^*_{art} = \mathbf{F}^*_{\mathbf{x}_d} + \mathbf{F}^*_O, \quad (9)$$

with

$$\begin{aligned} \mathbf{F}^*_{\mathbf{x}_d} &= -\mathrm{grad}[U_{\mathbf{x}_d}(\mathbf{x})], \\ \mathbf{F}^*_O &= -\mathrm{grad}[U_O(\mathbf{x})]. \end{aligned} \quad (10)$$

The symbol $\mathbf{F}^*_{\mathbf{x}_d}$ is an attractive force allowing the point $\mathbf{x}$ of the end effector to reach the goal position $\mathbf{x}_d$, and $\mathbf{F}^*_O$ represents a *Force Inducing an Artificial Repulsion from the Surface* (FIRAS, from the French) of the obstacle created by the potential field $U_O(\mathbf{x})$. The symbol $\mathbf{F}^*_{\mathbf{x}_d}$ corresponds to the proportional term, *i.e.* $-k_p(\mathbf{x} - \mathbf{x}_d)$, in a conventional PD servo where k_p is the position gain. The attractive potential field $U_{\mathbf{x}_d}(\mathbf{x})$ is simply

$$U_{\mathbf{x}_d}(\mathbf{x}) = \tfrac{1}{2} k_p(\mathbf{x} - \mathbf{x}_d)^2. \quad (11)$$

$U_O(x)$ is selected such that the artificial potential field $U_{art}(x)$ is a positive continuous and differentiable function which attains its zero minimum when $\mathbf{x} = \mathbf{x}_d$. The articulated mechanical system subjected to $U_{art}(\mathbf{x})$ is stable. Asymptotic stabilization of the system is achieved by adding dissipative forces proportional to $\dot{\mathbf{x}}$. Let k_v be the velocity gain; the forces contributing to the end-effector motion and stabilization are of the form,

$$\mathbf{F}^*_m = -k_p(\mathbf{x} - \mathbf{x}_d) - k_v\dot{\mathbf{x}}. \quad (12)$$

This command vector is inadequate to control the manipulator for tasks that involve large end-effector motion toward a goal position without path specification. For such a task, it is better for the end effector to move in a straight line with an upper speed limit.

Rewriting Eq. (12) leads to the following expression, which can be interpreted as specifying a desired velocity vector in a pure velocity servo-control,

$$\dot{\mathbf{x}}_d = \frac{k_p}{k_v}(\mathbf{x}_d - \mathbf{x}). \quad (13)$$

Let V_{max} designate the assigned speed limit. The limitation of the end-effector velocity magnitude can then be obtained (Khatib, Llibre, and Mampey 1978),

$$\mathbf{F}^*_m = -k_v(\dot{\mathbf{x}} - \nu\dot{\mathbf{x}}_d), \quad (14)$$

where

$$\nu = \min\left(1, \frac{V_{max}}{\sqrt{\dot{\mathbf{x}}_d^T \dot{\mathbf{x}}_d}}\right). \quad (15)$$

With this scheme, the velocity vector $\dot{\mathbf{x}}$ is controlled to be pointed toward the goal position while its magnitude is limited to V_{max}. The end effector will then travel at that speed in a straight line, except during the acceleration and deceleration segments or when it is inside the repulsive potential field regions of influence.

4. FIRAS Function

The artificial potential field $U_O(\mathbf{x})$ should be designed to meet the manipulator stability condition and to create at each point on the obstacle's surface a potential barrier which becomes negligible beyond that surface. Specifically, $U_O(\mathbf{x})$ should be a nonnegative continuous and differentiable function whose value tends to infinity as the end effector approaches the obstacle's surface. In order to avoid undesirable perturbing forces beyond the obstacle's vicinity, the influence of this potential field must be limited to a given region surrounding the obstacle.

Using analytic equations $f(\mathbf{x}) = 0$ for obstacle description, the first artificial potential field function used (Khatib and Le Maitre 1978) was based on the values of the function $f(\mathbf{x})$,

$$U_O(\mathbf{x}) = \begin{cases} \dfrac{1}{2}\eta\left(\dfrac{1}{f(\mathbf{x})} - \dfrac{1}{f(\mathbf{x}_0)}\right)^2 & \text{if } f(\mathbf{x}) \leq f(\mathbf{x}_0), \\ 0 & \text{if } f(\mathbf{x}) > f(\mathbf{x}_0). \end{cases} \quad (16)$$

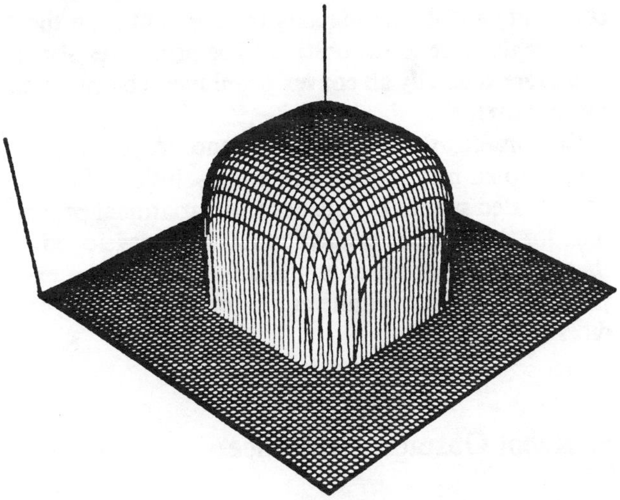

Fig. 1. An n-ellipsoid with $n = 4$.

The region of influence of this potential field is bounded by the surfaces $f(\mathrm{x}) = 0$ and $f(\mathrm{x}) = f(\mathrm{x}_0)$, where x_0 is a given point in the vicinity of the obstacle and η a constant gain. This potential function can be obtained very simply in real time since it does not require any distance calculations. However, this potential is difficult to use for asymmetric obstacles where the separation between an obstacle's surface and equipotential surfaces can vary widely.

Using the shortest distance to an obstacle O, we have proposed (Khatib 1980) the following artificial potential field;

$$U_O(\mathrm{x}) = \begin{cases} \frac{1}{2}\eta\left(\frac{1}{\rho} - \frac{1}{\rho_0}\right)^2 & \text{if } \rho \le \rho_0, \\ 0 & \text{if } \rho > \rho_0, \end{cases} \tag{17}$$

where ρ_0 represents the limit distance of the potential field influence and ρ the shortest distance to the obstacle O. The selection of the distance ρ_0 will depend on the end effector operating speed V_{max} and on its deceleration ability. End-effector acceleration characteristics are discussed in Khatib and Burdick (1985).

Any point of the robot can be subjected to the artificial potential field. The control of a *Point Subjected to the Potential* (PSP) with respect to an obstacle O is achieved using the FIRAS function,

$$\mathrm{F}^*_{(O,psp)} = \begin{cases} \eta\left(\frac{1}{\rho} - \frac{1}{\rho_0}\right)\frac{1}{\rho^2}\frac{\partial\rho}{\partial \mathrm{x}} & \text{if } \rho \le \rho_0, \\ 0 & \text{if } \rho > \rho_0, \end{cases} \tag{18}$$

where $\frac{\partial\rho}{\partial \mathrm{x}}$ denotes the partial derivative vector of the distance from the PSP to the obstacle,

$$\frac{\partial\rho}{\partial \mathrm{x}} = \left[\frac{\partial\rho}{\partial x}\ \frac{\partial\rho}{\partial y}\ \frac{\partial\rho}{\partial z}\right]^T. \tag{19}$$

The joint forces corresponding to $\mathrm{F}^*_{(O,psp)}$ are obtained using the Jacobian matrix associated with this PSP. Observing Eqs. (6) and (9), these forces are given by

$$\Gamma_{(O,psp)} = J^T_{psp}(\mathrm{q})\Lambda(\mathrm{x})\mathrm{F}^*_{(O,psp)}. \tag{20}$$

5. Obstacle Geometric Modelling

Obstacles are described by the composition of *primitives.* A typical geometric model base includes primitives such as a point, line, plane, ellipsoid, parallelepiped, cone, and cylinder. The first artificial potential field (Eq. 16) requires analytic equations for the description of obstacles. We have developed analytic equations representing envelopes which best approximate the shapes of primitives such as a *parallelepiped, finite cylinder,* and *cone.*

The surface, termed an *n-ellipsoid,* is represented by the equation,

$$\left(\frac{x}{a}\right)^{2n} + \left(\frac{y}{b}\right)^{2n} + \left(\frac{z}{c}\right)^{2n} = 1, \tag{21}$$

and tends to a parallelepiped of dimensions (2a, 2b, 2c) as n tends to infinity. A good approximation is obtained with $n = 4$, as shown in Fig. 1. A cylinder of elliptical cross section ($2a$, $2b$) and of length $2c$ can be approximated by the so-called *n-cylinder* equation,

$$\left(\frac{x}{a}\right)^{2} + \left(\frac{y}{b}\right)^{2} + \left(\frac{z}{c}\right)^{2n} = 1. \tag{22}$$

The analytic description of primitives is not necessary for the artificial potential field (Eq. 17) since the continuity and differentiability requirement is on the shortest distance to the obstacle. The primitives above, and more generally all convex primitives, comply with this requirement.

Determining the orthogonal distance to an n-ellipsoid or to an n-cylinder requires the solution of a complicated system of equations. A variational procedure for the distance evaluation has been developed that avoids this costly computation. The distance expressions for other primitives are presented in Appendices I through III.

Fig. 2. Displacement of a 4 dof manipulator inside an enclosure.

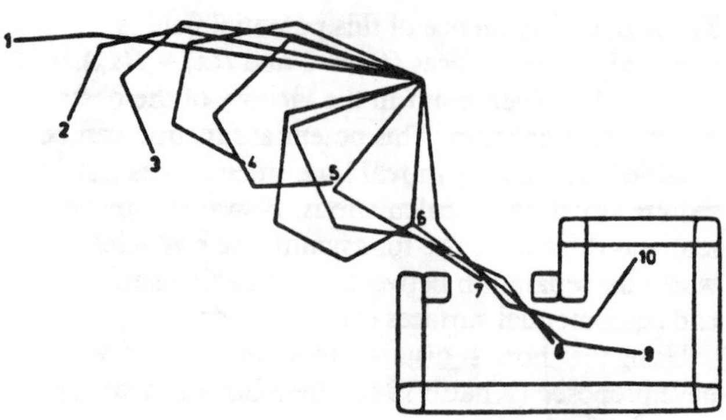

6. Robot Obstacle Avoidance

An obstacle O_i is described by a set of primitives $\{P_p\}$. The superposition property (additivity) of potential fields enables the control of a given point of the manipulator with respect to this obstacle by using the sum of the relevant gradients,

$$\mathbf{F}^*_{O_i,psp} = \sum_p \mathbf{F}^*_{(P_p,psp)}. \quad (23)$$

Control of this point for several obstacles is obtained using

$$\mathbf{F}^*_{psp} = \sum_i \mathbf{F}^*_{(O_i,psp)}. \quad (24)$$

It is also feasible to have different points on the manipulator controlled with respect to different obstacles. The resulting joint force vector is given by

$$\Gamma_{obstacles} = \sum_j J^T_{psp_j}(\mathbf{q})\Lambda(\mathbf{x})\mathbf{F}^*_{psp_j}. \quad (25)$$

Specifying an adequate number of PSPs enables the protection of all of the manipulator's parts. An example of a dynamic simulation for a redundant 4 *dof* manipulator operating in the plane is shown in the display of Fig. 2.

The artificial potential field approach can be extended to *moving obstacles* since stability of the mechanism persists with a continuously time-varying potential field.

The manipulator obstacle avoidance problem has been formulated in terms of *collision avoidance of links* rather than points. Link collision avoidance is achieved by continuously controlling the link's closest point to the obstacle. At most, *n* PSPs then have to be considered. Additional links can be artificially introduced or the length of the last link can be extended to account for the manipulator tool or load. In an articulated chain, a link can be represented as the line segment defined by the Cartesian positions of its two neighboring joints. In a frame of reference *R*, a point $m(x, y, z)$ of the link bounded by $m_1(x_1, y_1, z_1)$ and $m_2(x_2, y_2, z_2)$ is described by the parametric equations,

$$\begin{aligned} x &= x_1 + \lambda(x_2 - x_1), \\ y &= y_1 + \lambda(y_2 - y_1), \\ z &= z_1 + \lambda(z_2 - z_1). \end{aligned} \quad (26)$$

The problem of obtaining the link's shortest distance to a parallelepiped can be reduced to that of finding the link's closest point to a vertex, edge, or face. The analytic expressions of the link's closest point, the distance, and its partial derivatives are given in Appendix I. In Appendices II and III these expressions are given for a cylinder and a cone, respectively.

7. Joint Limit Avoidance

The potential field approach can be used to satisfy the manipulator internal joint constraints. Let $\underline{q}_i$ and $\overline{q}_i$ be respectively the minimal and maximal bounds of the ith joint coordinate q_i. These boundaries can contain q_i by creating barriers of potential at each of the hyperplanes $(q_i = \underline{q}_i)$ and $(q_i = \overline{q}_i)$. The corresponding joint forces are

$$\gamma_{\underline{q}_i} = \begin{cases} \eta\left(\frac{1}{\underline{\rho}_i} - \frac{1}{\underline{\rho}_{i(0)}}\right)\frac{1}{\underline{\rho}_i^2} & \text{if } \underline{\rho}_i \leq \underline{\rho}_{i(0)}, \\ 0 & \text{if } \underline{\rho}_i > \underline{\rho}_{i(0)}, \end{cases} \tag{27}$$

and

$$\gamma_{\bar{q}_i} = \begin{cases} -\eta\left(\frac{1}{\bar{\rho}_i} - \frac{1}{\bar{\rho}_{i(0)}}\right)\frac{1}{\bar{\rho}_i^2} & \text{if } \bar{\rho}_i \leq \bar{\rho}_{i(0)}, \\ 0 & \text{if } \bar{\rho}_i > \bar{\rho}_{i(0)}, \end{cases} \tag{28}$$

where $\underline{\rho}_{i(0)}$ and $\bar{\rho}_{i(0)}$ represent the distance limit of the potential field influence. The distances $\underline{\rho}_i$ and $\bar{\rho}_i$ are defined by

$$\begin{aligned} \underline{\rho}_i &= q_i - \underline{q}_i, \\ \bar{\rho}_i &= \bar{q}_i - q_i. \end{aligned} \tag{29}$$

8. Level of Competence

The potential field concept is indeed an attractive approach to the collision avoidance problem and much research has recently been focused on its applications to robot control (Kuntze and Schill 1982; Hogan 1984; Krogh 1984). However, the complexity of tasks that can be implemented with this approach is limited. In a cluttered environment, local minima can occur in the resultant potential field. This can lead to a stable positioning of the robot before reaching its goal. While local procedures can be designed to exit from such configurations, limitations for complex tasks will remain. This is because the approach has a *local* perspective of the robot environment.

Nevertheless, the resulting potential field does provide the global information necessary and a collision-free path, if attainable, can be found by linking the absolute minima of the potential. Linking these minima requires a computationally expensive exploration of the potential field. This goes beyond the real-time control we are concerned with here but can be considered as an integrated part of higher level control. Work on high level collision-free path planning based on the potential field concept has been investigated in Buckley (1985).

9. Real-Time Implementation

Finally, the global control system integrating the potential field concept with the operational space approach has the following structure:

$$\Gamma = \Gamma_{motion} + \Gamma_{obstacles} + \Gamma_{joint-limit}, \tag{30}$$

where Γ_{motion} can be developed (Khatib 1983) in the form,

$$\begin{aligned} \Gamma_{motion} = {} & J^T(q)\Lambda(q)F_m^* + \tilde{B}(q)[\dot{q}\dot{q}] \\ & + \tilde{C}(q)[\dot{q}^2] + g(q), \end{aligned} \tag{31}$$

where $\tilde{B}(q)$ and $\tilde{C}(q)$ are the $n \times n(n-1)/2$ and $n \times n$ matrices of the joint forces under the mapping into joint space of the end-effector Coriolis and centrifugal forces. The symbol g(q) is the gravity force vector, and the symbolic notations $[\dot{q}\dot{q}]$ and $[\dot{q}^2]$ are for the $n(n-1)/2 \times 1$ and $n \times 1$ column matrices,

$$\begin{aligned} [\dot{q}\dot{q}] &= [\dot{q}_1\dot{q}_2\ \dot{q}_1\dot{q}_3\ \ldots\ \dot{q}_{n-1}\dot{q}_n]^T, \\ [\dot{q}^2] &= [\dot{q}_1^2\ \dot{q}_2^2\ \ldots\ \dot{q}_n^2]^T. \end{aligned} \tag{32}$$

In this control structure, dynamic decoupling of the end effector is obtained using the end-effector dynamic parameters (EEDP) $\Lambda(q)$, $\tilde{B}(q)$, $\tilde{C}(q)$ and g(q), which are configuration dependent. In real time, these parameters can be computed at a lower rate than that of the servo control. In addition, the integration of an operational position and velocity estimator allows a reduction in the rate of end-effector position computation, which involves evaluations of the manipulator geometric model. This leads to a two-level control system architecture (Khatib 1985):

- A low rate *parameter evaluation level* that updates the end-effector dynamic coefficients, the Jacobian matrix, and the geometric model.
- A high rate *servo control level* that computes the command vector using the estimator and the updated dynamic coefficients.

The control system architecture is shown in Fig. 3 where *np* represents the number of PSPs. The Jacobian matrices $J_{psp_i}^T(q)$ have common factors with the end-effector Jacobian matrix $J^T(q)$. Thus, their evaluation does not require significant additional computation.

Fig. 3. Control system architecture.

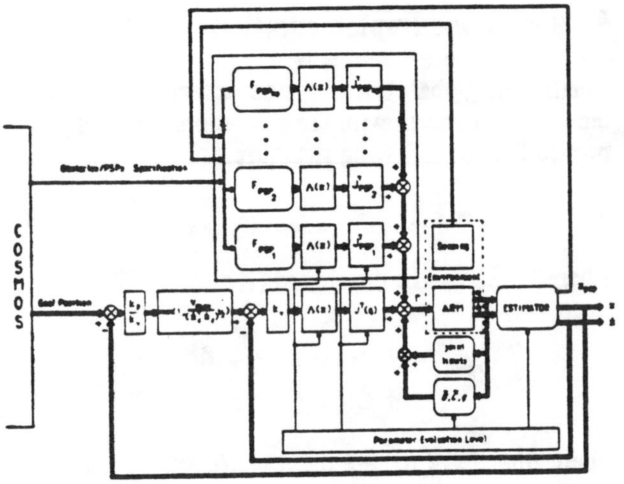

10. Applications

This approach has been implemented in an experimental manipulator programming system *Control in Operational Space of a Manipulator-with-Obstacles System* (COSMOS). Demonstration of real-time collision avoidance with links and moving obstacles (Khatib et al. 1984) have been performed using a PUMA 560 and a Machine Intelligence Corporation vision module.

We have also demonstrated real-time end-effector motion and active force control operations with the COSMOS system using wrist and finger sensing. These include contact, slide, insertion, and compliance operations (Khatib, Burdick, and Armstrong 1985).

In the current multiprocessor implementation (PDP 11/45 and PDP 11/60), the rate of the servo control level is 225 Hz while the coefficient evaluation level runs at 100 Hz.

11. Summary and Discussion

We have described the formulation and implementation of a real-time obstacle avoidance approach based on the artificial potential field concept. Collision avoidance, generally treated as high level planning, has been demonstrated to be an effective component of low level real-time control in this approach. Further, we have briefly presented our operational space formulation of manipulator control that provides the basis for this obstacle avoidance approach, and have described the two-level architecture designed to increase the real-time performance of the control system.

The integration of this low level control approach with a high level planning system seems to be one of the more promising solutions to the obstacle avoidance problem in robot control. With this approach, the problem may be treated in two stages:

- at high level control, generating a global strategy for the manipulator's path in terms of intermediate goals (rather than finding an accurate collision-free path);
- at the low level, producing the appropriate commands to attain each of these goals, taking into account the detailed geometry and motion of manipulator and obstacle, and making use of real-time obstacle sensing (low level vision and proximity sensors).

By extending low level control capabilities and reducing the high level path planning burden, the integration of this collision avoidance approach into a multi-level robot control structure will improve the real-time performance of the overall robot control system. Potential applications of this control approach include moving obstacle avoidance, collision avoidance in grasping operations, and obstacle avoidance problems involving multimanipulators or multifingered hands.

Appendix I: Link Distance to a Parallelepiped

The axes of the frame of reference R are chosen to be the parallelepiped axes of symmetry. The link's length is l and the dot product is $(\cdot)$.

Distance to a Vertex

The closest point m of the line (Eq. 26) to the vertex v is such that

$$\lambda = \frac{(\mathbf{vm}_1) \cdot (\mathbf{m}_1\mathbf{m}_2)}{l^2}. \qquad \text{(A1-1)}$$

The link's closest point m is identical to m_1 if $\lambda \leq 0$; it is identical to m_2 if $\lambda \geq 1$ and it is given by Eq. (26) otherwise. The shortest distance is therefore,

$$\rho = \begin{cases} [\rho_1^2 - \lambda^2 l^2]^{1/2}, & \text{if } 0 \leq \lambda \leq 1, \\ \rho_1, & \text{if } \lambda < 0, \\ \rho_2 & \text{if } \lambda > 1, \end{cases} \qquad \text{(A1-2)}$$

where ρ_1 and ρ_2 are the distance to the vertex from m_1 and m_2, respectively. The distance partial derivatives are

$$\frac{\partial \rho}{\partial x} = \left[\frac{x}{\rho}\,\frac{y}{\rho}\,\frac{z}{\rho}\right]^T. \qquad \text{(A1-3)}$$

Distance to an Edge

By a projection in the plane perpendicular to the considered edge (xoy, yoz, or zox), this problem can be reduced to that of finding the distance to a vertex in the plane. This leads to expressions similar to those of (A1-1)–(A1-3) with a zero partial derivative of the distance *w.r.t.* the axis parallel to the edge.

Distance to a Face

In this case, the distance can be directly obtained by comparing the absolute values of the coordinates of m_1 and m_2 along the axis perpendicular to the face. The partial derivative vector is identical to the unit normal vector of this face.

Appendix II: Link Distance to a Cylinder

The frame of reference R is chosen such that its z-axis is the cylinder axis of symmetry and its origin is the cylinder center of mass. The cylinder radius and height are designated by r and h, respectively.

Distance to the Circular Surface

The closest point of the link (Eq. 27) to the circular surface of the cylinder can be deduced from the distance to a vertex considered in the xoy plane and by allowing for the radius r.

Distance to the Circular Edges

The closest distance to the cylinder circular edge can be obtained from that of the circular surface by taking into account the relative z-coordinate of m to the circular edge, *i.e.*, $(z + h/2)$ for the base and $(z - h/2)$ for the top. The distance partial derivative vector results from the torus equation,

$$[x^2 + y^2 + (z \pm h/2)^2 - r^2 - \rho^2]^2 = 4r^2[\rho^2 - (z \pm h/2)^2]. \qquad \text{(A2-1)}$$

This vector is

$$\frac{\partial \rho}{\partial x} = \left[\zeta\frac{x}{\rho}\,\zeta\frac{y}{\rho}\,\frac{z \pm h/2}{\rho}\right]^T, \qquad \text{(A2-2)}$$

with

$$\zeta = \frac{x^2 + y^2 + (z \pm h/2)^2 - r^2 - \rho^2}{x^2 + y^2 + (z \pm h/2)^2 + r^2 - \rho^2}. \qquad \text{(A2-3)}$$

The distance to the planar surfaces is straightforward and can be simply obtained as in Appendix I.

Appendix III: Link Distance to a Cone

In this case, the frame of reference R is chosen such that its z-axis is the cone axis of symmetry and its origin is the center of the cone circular base. The cone base radius, height and half angle are represented respectively by r, h, and β.

Distance to the Cone-Shaped Surface

The problem of locating $m(x, y, z)$ is identical to that for the cylinder case. The distance can be written as

$$\rho = z \sin(\beta) + (\sqrt{x^2 + y^2} - r)\cos(\beta). \qquad \text{(A3-1)}$$

The partial derivatives come from the equation,

$$x^2 + y^2 = r_z^2, \qquad \text{(A3-2)}$$

where

$$r_z = \tan(\beta)[h + \rho \sin(\beta) - z]. \qquad \text{(A3-3)}$$

They are

$$\frac{\partial \rho}{\partial x} = \left[\frac{x}{r_z \tan(\beta)} \; \frac{y}{r_z \tan(\beta)} \; \frac{1}{\sin(\beta)} \right]^T. \qquad \text{(A3-4)}$$

The problem of the distance to the cone circular edge is identical to that of the cylinder circular edge in Appendix II. The distance to the cone vertex is solved as in Appendix I.

Acknowledgments

Tom Binford and Bernie Roth have encouraged and given support for the continuation of this research at Stanford University. I thank also Harlyn Baker, Peter Blicher, and Jeff Kerr for their help in the preparation of the original manuscript.

References

Brooks, R. 1983. Solving the find-path problem by good representation of free space. *IEEE Sys., Man Cyber.* SMC-13:190–197.

Brooks, R. 1984 Aug. 20–23, Kyoto, Japan. Aspects of mobile robot visual map making. *2nd Int. Symp. Robotics Res.*

Buckley, C. 1985. *The application of continuum methods to path planning.* Ph.D. Thesis (in progress). Stanford University, Department of Mechanical Engineering.

Chatila, R. 1981. *Système de navigation pour un robot mobile autonome: modélisation et processus décisionnels.* Thèse de Docteur-Ingénieur. Université Paul Sabatier. Toulouse, France.

Hogan, N. 1984 June 6–8, San Diego, California. Impedance control: an approach to manipulation. 1984 *Amer. Control Conf.*

Khatib, O., Llibre, M. and Mampey, R. 1978. Fonction decision-commande d'un robot manipulateur. *Rapport No. 2/7156.* DERA/CERT. Toulouse, France.

Khatib, O. and Le Maitre, J. F. 1978 September 12–15, Udine, Italy. Dynamic control of manipulators operating in a complex environment. *Proc. 3rd CISM-IFToMM Symp. Theory Practice Robots Manipulators,* 267-282. Elsevier. 1979.

Khatib, O. 1980. *Commande dynamique dans l'espace opérationnel des robots manipulateurs en présence d'obstacles.* These de Docteur-Ingénieur. École Nationale Supérieure de l'Aéronautique et de l'Espace (ENSAE). Toulouse, France.

Khatib, O. 1983 December 15–20, New Delhi. Dynamic control of manipulators in operational space. *6th CISM-IFToMM Congress Theory Machines Mechanisms,* 1128–1131.

Khatib, O., et al. 1984 June, *Robotics in three acts* (Film). Stanford University. Artificial Intelligence Laboratory.

Khatib, O. 1985 September 11–13, Tokyo. The operational space formulation in robot manipulators control. *15th Int. Symp. Indust. Robots.*

Khatib, O. and Burdick, J. 1985 November, Miami, Florida. Dynamic optimization in manipulator design: the operational space formulation. *ASME Winter Annual Meeting.*

Khatib, O., Burdick, J., and Armstrong, B. 1985. *Robotics in three acts–Part II* (Film). Stanford University, Artificial Intelligence Laboratory.

Krogh, B. 1984 August, Bethlehem, Pennsylvania. A generalized potential field approach to obstacle avoidance control. *SME Conf. Proc. Robotics Research: The Next Five Years and Beyond.*

Kuntze, H. B., and Schill, W. 1982 June 9–11, Paris. Methods for collision avoidance in computer controlled industrial robots. *12th ISIR.*

Lozano-Perez, T. 1980. Spatial planning: a configuration space approach. *AI Memo 605.* Cambridge, Mass. MIT Artificial Intelligence Laboratory.

Moravec, H. P. 1980. *Obstacle avoidance and navigation in the real world by a seeing robot rover.* Ph.D. Thesis. Stanford University, Artificial Intelligence Laboratory.

E. Freund
H. Hoyer
Institut für Roboterforschung (IRF)
University of Dortmund
4600 Dortmund, West Germany

Real-Time Pathfinding in Multirobot Systems Including Obstacle Avoidance

Abstract

The paper describes an approach to the solution of the findpath problem, including obstacle avoidance in multirobot systems. The design of multirobot systems requires an overall approach where not only parts of the structure, including the hierarchy are considered, but a complete concept, including the dynamics of the robots, has to be developed. In this paper the structure of these systems is based on the nonlinear control approach. The method for real-time pathfinding itself uses a systematic design procedure for multirobot systems, which includes the hierarchical coordinator. This hierarchical coordinator is designed for real-time collision avoidance, where the collision avoidance strategy is based on an analytically described avoidance trajectory that serves for collision detection as well as avoidance. The efficiency of the new approach for real-time pathfinding is demonstrated by several cases of practical interest, such as collision avoidance between three robots, interaction of three stationary robots and a stationary obstacle, as well as interaction of mobile robots and moving obstacles, including obstacles of variable size and different shapes.

1. Introduction

With increased automation and application of CAD/CAM methods in connection with automatically generated robot programs (e.g., in car manufacturing), an overall consideration of multirobot systems becomes a basic requirement. This seems to be the only possibility to reduce the programming effort and the fault rate and to provide flexibility in manufacturing.

In this context the findpath problem gains an increasing importance in view of advanced assembly automation and partial realizations of unstaffed factories. In Brady et al. (1982) the findpath problem was defined as the problem of planning free motions, given models of the robots and of any objects in the workspace, which is fundamentally the geometric problem of finding a path for a moving solid among other solid obstacles. Approaching these problems, a main subject is a coordinated operation of the robots involved, including automatic on-line collision avoidance between all robots as well as between robots and obstacles. The solution of these problems, however, requires an overall approach based on the hierarchical structure of the system containing robots and obstacles (multirobot system). In these systems not only parts of the structure including the hierarchy should be considered, but a complete concept including the dynamics of the robots involved has to be developed. The performances of the basic control circuits are quite essential, because these circuits have to react fast and accurately to the superimposed input signals of the hierarchy and must allow precise changes of their control dynamics by command of the hierarchy as well.

In spite of the importance for future developments relatively few results are known in this field. Considering the collision avoidance problem as one of the important aspects, it can be distinguished between simple realizations and complex theoretical approaches. In simple realizations the common working space is completely blocked for the other robot arm if the one robot arm enters (Dunne 1979, Tyridal 1980). These methods lack flexibility, of course, since they are designed for a specific working sequence and therefore are not suitable for a broader range of applications. The more complex theoretical approaches, however, apply optimization and search methods for the plan-

Reprinted from *The International Journal of Robotics Research*, Vol. 7, No. 1, February 1988, pages 42-70, "Real-Time Pathfinding in Multirobot Systems Including Obstacle Avoidance" by E. Freund and H. Hoyer, by permission of The MIT Press, Cambridge, Massachusetts.

ning of a collision free path (Lozano-Pérez and Wesley 1979, Lozano-Pérez 1986), so due to computational efforts the practical use for industrial applications is limited. In Erdmann and Lozano-Pérez (1986) for each moving object a configuration space-time is constructed that represents the time-varying constraints on the moving object. This space-time is searched for a collision-free path. Another approach (Brooks 1983, 1985) uses elongated regions of free space, so-called freeways, for straight-line motions of the moving object. A potential field concept is used in Krogh (1984), Hogan (1984), and Khatib (1986). In Freund (1983) the hierarchical coordinator is designed for collision avoidance of two robots and is based on an approach using various hierarchical strategies in connection with extremely time-efficient evaluation of decision tables (Freund and Hoyer 1983, 1984). The method allows a simultaneous movement of both robots through the collision space and is based on a suitable description of this actual possible space. This collision avoidance strategy approach is very effective for two robots; however, it becomes more complicated for more than two robots.

In this paper a general approach is given for the solution of the findpath problem including collision avoidance in multirobot systems. This is based on a systematic design method for multirobot systems (Freund 1983, 1984), which uses a hierarchical structure of the overall system including the dynamics of all robots involved. The design procedure permits a change of the basic control dynamics, the introduction of useful couplings as well as the change of the input gains of the robots by means of the hierarchical coordinator. Hereby the control of the variables of motion of the robots is based on the nonlinear control and decoupling method (Freund 1976, Freund and Syrbe 1976). This approach leads to a dynamic overall behavior of the robot that is characterized by complete decoupling of all variables of motion and arbitrary pole assignment where a suitable partition of the dynamic equations of the robot provides directly applicable, explicit control laws for each drive (Freund 1982).

The hierarchical coordinator is structured in such a way that various strategies for the coordination of the robots can be realized. The collision avoidance strategies that matter in the pathfinding problem are realized via a specific matrix by the hierarchical coordinator on the basis of avoidance trajectories (Hoyer 1984, 1985) that serve for collision detection as well as avoidance.

It is shown that this new systematic approach solves the findpath problem in several cases of practical interest without changes of strategies and algorithms. The method is applicable to systems with stationary and mobile robots, stationary and moving obstacles, and obstacles of variable size; it is suitable for off-line as well as on-line pathfinding. For clarity, we consider systems of three robots only in this paper. The resulting equations, however, demonstrate the usefulness of the design method for multirobot systems and provide a systematic approach to the solution of the findpath problem when more robots are involved.

2. Basic Nonlinear Control of Robots

For an efficient overall behavior of a multirobot system the basic controls of the robots are required to be accurate and fast and to meet real-time constraints in order to insure practical applicability. For very accurate path control the invariance of the dynamics of all axes is necessary to avoid deviation of the desired path under all working conditions. Arbitrary pole placement is another condition for the design of an efficient robot control.

For the consideration of the findpath problem including collision avoidance for several robots, the nonlinear control approach (Freund 1973, 1976, Freund and Syrbe 1976) has to be treated first. The resulting nonlinear control laws are then substituted in the general multirobot system, which is derived in Section 3. For this purpose one of the direct design methods (Freund 1982) is most suitable where the basic equations are briefly described in the following.

In a system of r robots, the kth robot ($k = 1, 2, \ldots, r$) can be characterized by the state-space description

$$\dot{\mathbf{x}}_k(t) = \mathbf{A}_k(\mathbf{x}_k) + \mathbf{B}_k(\mathbf{x}_k)\mathbf{u}_k(t), \tag{1}$$

$$\mathbf{y}_k(t) = \mathbf{C}_k(\mathbf{x}_k), \qquad k = 1, 2, \ldots, r, \tag{2}$$

with $\mathbf{u}_k(t)$ and $\mathbf{y}_k(t)$ as the m_k-dimensional input and output vectors, respectively. If the variables of motion of the kth robot are denoted by $q_{k,i}(t)$ and the corresponding velocities by $\dot{q}_{k,i}(t)$, the state variables of (1), (2) are chosen as

$$\begin{aligned} x_{k,(2i-1)}(t) &= q_{k,i}(t) \\ x_{k,2i}(t) &= \dot{q}_{k,i}(t), \qquad i = 1, 2, \ldots, m_k. \end{aligned} \tag{3}$$

Then, the state vector in the state space description (1), (2) has dimension $2m_k$, where m_k is the number of axes or degrees of freedom of the kth robot. The input vector $\mathbf{u}_k(t)$ consists of external forces and torques that drive the corresponding variable of motion $q_{k,i}(t)$. $\mathbf{y}_k(t)$ contains the outputs of the robot that are the variables of motion or combinations of them, depending on the chosen coordinate system. For simplification it is assumed in the application considered there that the output variables of the robot are directly equal to certain state variables; that is,

$$y_{k,i}(t) = x_{k,(2i-1)}(t) = q_{k,i}(t), \qquad i = 1, 2, \ldots, m_k. \tag{4}$$

On the basis of the state vector $\mathbf{x}_k(t)$ as given in (4) in connection with (3), $\mathbf{A}_k(\mathbf{x}_k)$ and $\mathbf{B}_k(\mathbf{x}_k)$ in (1) have a compatible order and the following form, which is specific for robots and manipulators:

$$\mathbf{A}_k(\mathbf{x}_k) = \begin{bmatrix} x_{k2} \\ f_{k1}(\mathbf{x}_k) \\ \hline \cdot \\ \cdot \\ \cdot \\ \hline x_{k,2m_k} \\ f_{k,m_k}(\mathbf{x}_k) \end{bmatrix} \tag{5}$$

$$\mathbf{B}_k(\mathbf{x}_k) = \begin{bmatrix} 0 \\ \mathbf{B}_{k1}(\mathbf{x}_k) \\ \hline \cdot \\ \cdot \\ \cdot \\ \hline 0 \\ \mathbf{B}_{k,m_k}(\mathbf{x}_k) \end{bmatrix} \tag{6}$$

Each of the subsections in $\mathbf{A}_k$ and $\mathbf{B}_k$ corresponds to one variable of motion, where the total number is m_k.

For the control of the robotic system as given by (1) and (2) or (4) in connection with (5) and (6) a feedback of the form

$$\mathbf{u}_k(t) = \mathbf{F}_k(\mathbf{x}_k) + \mathbf{G}_k(\mathbf{x}_k)\mathbf{w}_k(t) \tag{7}$$

is applied to this system, where $\mathbf{w}_k(t)$ is the new m_k-dimensional reference input vector and $\mathbf{F}_k(\mathbf{x}_k)$ and $\mathbf{G}_k(\mathbf{x}_k)$ are of compatible order. From these results the closed-loop system

$$\dot{\mathbf{x}}_k(t) = [\mathbf{A}_k(\mathbf{x}_k) + \mathbf{B}_k(\mathbf{x}_k)\mathbf{F}_k(\mathbf{x}_k)] + \mathbf{B}_k(\mathbf{x}_k)\mathbf{G}_k(\mathbf{x}_k)\mathbf{w}_k(t) \tag{8}$$

with the output equation (2) or (4), respectively. Applying the nonlinear control and decoupling method to the closed-loop system (8), we want to find $\mathbf{F}_k(\mathbf{x}_k)$ and $\mathbf{G}_k(\mathbf{x}_k)$ such that system (8) is decoupled from the inputs to the outputs and has arbitrarily designed poles. For the state-space representation (1), (2) or (4) with $\mathbf{A}_k(\mathbf{x}_k)$, $\mathbf{B}_k(\mathbf{x}_k)$ from (5) and (6), these matrices are (Freund 1976, 1982, Freund and Syrbe 1976)

$$\mathbf{F}_k(\mathbf{x}_k) = -\mathbf{D}_k^{*-1}(\mathbf{x}_k)\,[\mathbf{C}_k^*(\mathbf{x}_k) + \mathbf{M}_k^*(\mathbf{x}_k)], \tag{9}$$

$$\mathbf{G}_k(\mathbf{x}_k) = \mathbf{D}_k^{*-1}(\mathbf{x}_k)\Lambda_k, \tag{10}$$

where in (9) the term $-\mathbf{D}_k^{*-1}(\mathbf{x}_k)\mathbf{C}_k^*(\mathbf{x}_k)$ represents the part of the feedback that yields decoupling, and $-\mathbf{D}_k^{*-1}(\mathbf{x}_k)\mathbf{M}_k^*(\mathbf{x}_k)$ performs the control part with arbitrary pole placement. Λ_k characterizes the desired input gains.

Substituting (9) and (10) in (8) with (4) gives the following overall behavior for all input/output pairs $y_{k,i}(t)$, $w_{k,i}(t)$:

$$\ddot{y}_{k,i}(t) + \alpha_{k,i}^1 \dot{y}_{k,i}(t) + \alpha_{k,i}^0 y_{k,i}(t) = \lambda_{k,i} w_{k,i}(t) \quad \text{for } i = 1, 2, \ldots, m_k. \tag{11}$$

In (11) $w_{k1}(t)$ to $w_{k,m_k}(t)$ are the reference inputs of the closed-loop system (8) which are the components of the reference input vector $\mathbf{w}_k(t)$. Equation (11) means that all input/output pairs $y_{k,i}(t)$, $w_{k,i}(t)$ for $i = 1, 2, \ldots, m_k$ are completely decoupled from each other and have a dynamic that can be chosen arbitrarily via the coefficients $\alpha^1_{k,i}$ and $\alpha^0_{k,i}$ (with $\lambda_{k,i}$ as input gain). The matrix $\mathbf{M}^*_k(\mathbf{x}_k)$ in (9) contains the $\alpha^1_{k,i}$ and $\alpha^0_{k,i}$, the matrix Λ_k in (10) contains the $\lambda_{k,i}$. By the nonlinear control approach it is, therefore, possible to gain a decoupled overall behavior of the form (11) for each link of the robots by feeding back link positions and velocities (Eq. (7)).

3. Multirobot System

Based on nonlinear robot control as briefly described in the foregoing paragraph, the structure of a multirobot system including the control is considered (Freund 1983, 1984). The multirobot system consists of r robots described by Eqs. (1) and (4). Controlling these robots by corresponding nonlinear feedbacks of the type (7) leads to a robot system that consists of r equations of the form (8). As the robots are involved on coordinated operation with a common working space, the reference inputs $\mathbf{w}_1(t), \ldots, \mathbf{w}_r(t)$ are not independent of each other anymore but have to be coordinated by hierarchical control.

This hierarchical coordinator can be set up in the general form

$$\begin{bmatrix} \mathbf{w}_1(t) \\ \cdot \\ \cdot \\ \cdot \\ \mathbf{w}_r(t) \end{bmatrix} = \mathbf{H}(\mathbf{x}_1, \ldots, \mathbf{x}_r; \mathbf{v}_1, \ldots, \mathbf{v}_r), \quad (12)$$

where $\mathbf{v}_1(t), \ldots, \mathbf{v}_r(t)$ are the new external input vectors of dimension $m_1, \ldots, m_r$, respectively, referring to the robots $1, \ldots, r$. The hierarchical coordinator is provided with the actual information about the positions of all variables of motion of the robots as well as about the corresponding velocities by the state vectors $\mathbf{x}_1(t), \ldots, \mathbf{x}_r(t)$, which are also included in Eq. (12). Applying (12) to the system of r robots in the form of (8) leads to the general form of the hierarchical multirobot system:

$$\begin{bmatrix} \dot{\mathbf{x}}_1(t) \\ \cdot \\ \cdot \\ \cdot \\ \dot{\mathbf{x}}_r(t) \end{bmatrix} = \begin{bmatrix} \mathbf{A}_1(\mathbf{x}_1) + \mathbf{B}_1(\mathbf{x}_1)\mathbf{F}_1(\mathbf{x}_1) \\ \cdot \\ \cdot \\ \cdot \\ \mathbf{A}_r(\mathbf{x}_r) + \mathbf{B}_r(\mathbf{x}_r)\mathbf{F}_r(\mathbf{x}_r) \end{bmatrix} + \begin{bmatrix} \mathbf{B}_1(\mathbf{x}_1)\mathbf{G}_1(\mathbf{x}_1) & \cdots & 0 \\ \vdots & \ddots & \vdots \\ 0 & \cdots & \mathbf{B}_r(\mathbf{x}_r)\mathbf{G}_r(\mathbf{x}_r) \end{bmatrix} \mathbf{H}(\mathbf{x}_1, \ldots, \mathbf{x}_r; \mathbf{v}_1, \ldots, \mathbf{v}_r), \quad (13)$$

where the output equations are given by (4). The structure of the multirobot system is demonstrated in Fig. 1. This hierarchical overall system (14) describes in general form the coordinated operation of r robots, including the hierarchical coordinator and the dynamics of the robots, where the inputs of the system are the external input vectors $\mathbf{v}_1(t), \ldots, \mathbf{v}_r(t)$ and outputs are the output vectors $\mathbf{y}_1(t), \ldots, \mathbf{y}_r(t)$ of the robots.

In application of the nonlinear control approach as described in the foregoing paragraph, the feedback matrices in (9) and (10) are substituted into the overall system (14). This yields in consideration of (11) the following explicit form of the hierarchical multirobot system:

$$\begin{bmatrix} \dot{\mathbf{x}}_1(t) \\ \cdot \\ \cdot \\ \cdot \\ \dot{\mathbf{x}}_r(t) \end{bmatrix} = \begin{bmatrix} \mathbf{A}^*_1 & \cdots & 0 \\ \vdots & \ddots & \vdots \\ 0 & \cdots & \mathbf{A}^*_r \end{bmatrix} \begin{bmatrix} \mathbf{x}_1(t) \\ \cdot \\ \cdot \\ \cdot \\ \mathbf{x}_r(t) \end{bmatrix} + \mathbf{B}^*\mathbf{H}(\mathbf{x}_1, \ldots, \mathbf{x}_r; \mathbf{v}_1, \ldots, \mathbf{x}_r) \quad (14)$$

with

$$y_{k,i}(t) = x_{k,(2i-1)}(t) \quad \text{for } k = 1, 2, \ldots, r, \quad i = 1, 2, \ldots, m_k,$$

where the matrices in (14) have the following form:

$$\mathbf{A}^*_k = \mathrm{diag}\{\mathbf{A}^*_{k,i}\} \quad (15)$$

Fig. 1. Structure of the multirobot system.

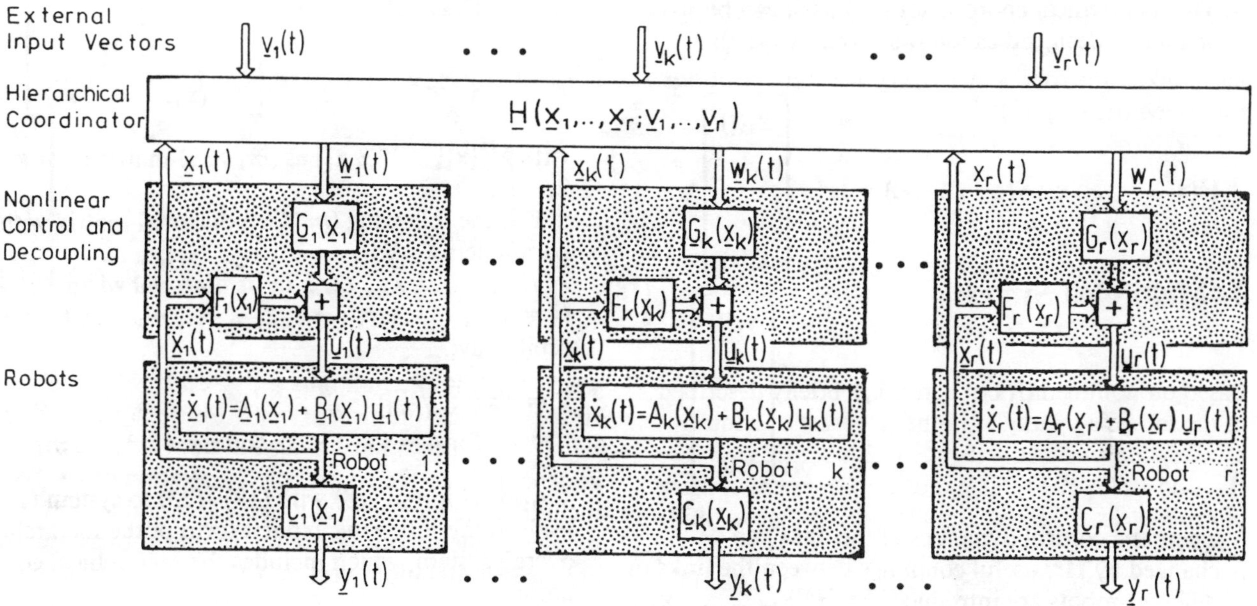

with

$$\mathbf{A}^*_{k,i} = \begin{bmatrix} 0 & 1 \\ -\alpha^0_{k,i} & -\alpha^1_{k,i} \end{bmatrix} \tag{16}$$

for $k = 1, 2, \ldots, r$ and $i = 1, 2, \ldots, m_k$.

The $(2m \times m)$-matrix $\mathbf{B}^*$ is

$$\mathbf{B}^* = [\mathbf{b}^*_{11}, \ldots, \mathbf{b}^*_{1,m_1}; \ldots; \mathbf{b}^*_{r1}, \ldots, \mathbf{b}^*_{r,m_r}] \tag{17}$$

with $m = m_1 + \cdots + m_k$ and

$$\mathbf{b}^*_{k,i} = \begin{bmatrix} 0 \\ \cdot \\ \cdot \\ \cdot \\ 0 \\ \lambda_{k,i} \\ 0 \\ \cdot \\ \cdot \\ \cdot \\ 0 \end{bmatrix} \begin{matrix} \\ \\ \\ \\ \\ \leftarrow \{2(m_1 + \cdots + m_{k-1} + i)\}\text{th position} \\ \\ \\ \\ \\ \\ \end{matrix}$$

$$k = 1, 2, \ldots, r, \quad i = 1, 2, \ldots, m_k.$$

The structure of the hierarchical overall system (14) in Fig. 1, which is based on nonlinear control, shows that this system will in general not be decoupled anymore. This means that the external input $v_{k,i}(t)$ does not only affect the corresponding output $y_{k,i}(t)$ for $k = 1, 2, \ldots, r$ and $i = 1, 2, \ldots, m_k$, where $v_{k,i}(t)$ are the components of the external input vector $\mathbf{v}_k(t)$. This is due to the fact that the robots cooperate or interfere in coordinated operation so that these movements are in general not independent anymore.

In system (14), the dynamics of all variables of motion as represented by $\mathbf{A}^*_{k,i}$ seem to be independent, combined with arbitrary pole placement. This is true if the hierarchical coordinator is of the type $\mathbf{H}(\mathbf{v}_1, \ldots, \mathbf{v}_r)$—that is, if only the external inputs are coordinated with no comparison to the actual positions relying on the feedback systems of the basic controls. Including the state variables in the design of the hierarchical coordinator (i.e., $\mathbf{H}(\mathbf{x}_1, \ldots, \mathbf{x}_r; \mathbf{v}_1, \ldots, \mathbf{v}_r)$) implies that the hierarchical overall system has a second feedback system (Fig. 1). The design of this hierarchical coordinator is the subject of the following section.

4. Design of the Hierarchical Coordinator

The hierarchical coordinator (12) itself can be structured and designed as follows (Freund 1984):

$$\mathbf{H}(\mathbf{x}_1, \ldots, \mathbf{x}_r; \mathbf{v}_1, \ldots, \mathbf{v}_r) = \mathbf{H}^a \cdot \begin{bmatrix} \mathbf{x}_1(t) \\ \cdot \\ \cdot \\ \cdot \\ \mathbf{x}_r(t) \end{bmatrix} + \mathbf{H}^b(\mathbf{x}_1, \ldots, \mathbf{x}_r) + \mathbf{E} \cdot \begin{bmatrix} \mathbf{v}_1(t) \\ \cdot \\ \cdot \\ \cdot \\ \mathbf{v}_r(t) \end{bmatrix}, \tag{18}$$

where the control dynamics of the links can be changed by $\mathbf{H}^a$, useful couplings between the links of different robots are introduced by $\mathbf{H}^b(\mathbf{x}_1, \ldots, \mathbf{x}_r)$, and the input gains of the external inputs $\mathbf{v}_1(t), \ldots, \mathbf{v}_r(t)$ are chosen by $\mathbf{E}$. The $(m \times 2m)$-matrix $\mathbf{H}^a$ is set up as follows:

$$\mathbf{H}^a = \operatorname{diag}\{\mathbf{H}_k^a\} \tag{19}$$

with

$$\mathbf{H}_k^a = \operatorname{diag}\{\mathbf{H}_{k,i}^a\} \quad \text{for } k = 1, 2, \ldots, r, \quad i = 1, \ldots, m_k \tag{20}$$

and

$$\mathbf{H}_{k,i}^a = \begin{bmatrix} \dfrac{\bar{\alpha}_{k,i}^0}{\lambda_{k,i}} & \dfrac{\bar{\alpha}_{k,i}^1}{\lambda_{k,i}} \end{bmatrix} \quad \text{for } k = 1, 2, \ldots, r \text{ and } i = 1, 2, \ldots, m_k \tag{21}$$

where the parameters $\bar{\alpha}_{k,i}^0$ and $\bar{\alpha}_{k,i}^1$ can be chosen arbitrarily.

The $(m \times 1)$-matrix $\mathbf{H}^b$ has, in general, the nonlinear form

$$\mathbf{H}^b(\mathbf{x}_1, \ldots, \mathbf{x}_r) = \begin{bmatrix} \mathbf{H}_1^b(\mathbf{x}_1, \ldots, \mathbf{x}_r) \\ \mathbf{H}_2^b(\mathbf{x}_1, \ldots, \mathbf{x}_r) \\ \cdot \\ \cdot \\ \cdot \\ \mathbf{H}_r^b(\mathbf{x}_1, \ldots, \mathbf{x}_r) \end{bmatrix} \tag{22}$$

with $\mathbf{H}_k^b(\mathbf{x}_1, \ldots, \mathbf{x}_r)$ as $(m_k \times 1)$-matrices for $k = 1, 2, \ldots, r$.

$\mathbf{E}$ is the following $(m \times m)$-diagonal matrix with the parameters $\bar{\lambda}_{11}/\lambda_{11}, \ldots, \bar{\lambda}_{1,m_1}/\lambda_{1,m_1}, \ldots, \bar{\lambda}_{r_1}/\lambda_{r_1}, \ldots, \bar{\lambda}_{r,m_r}/\lambda_{r,m_r}$ in its diagonal where the $\bar{\lambda}_{k,i}$ $(k = 1, 2, \ldots, r;\ i = 1, 2, \ldots, m_k)$ can be chosen arbitrarily:

$$\mathbf{E} = \operatorname{diag}\{\bar{\lambda}_{k,i}/\lambda_{k,i}\} \quad \text{for } k = 1, \ldots, r \text{ and } i = 1, \ldots, m_k. \tag{23}$$

Application of (18) with (19)–(23) to system (14) leads to the following explicit form of the hierarchical overall system, which includes the hierarchical coordinator:

$$\begin{bmatrix} \mathbf{x}_1(t) \\ \cdot \\ \cdot \\ \cdot \\ \mathbf{x}_r(t) \end{bmatrix} = \begin{bmatrix} \mathbf{A}_1^{**} & \cdot & \cdot & \cdot & 0 \\ \cdot & \cdot & & & \cdot \\ \cdot & & \cdot & & \cdot \\ \cdot & & & \cdot & \cdot \\ 0 & \cdot & \cdot & \cdot & \mathbf{A}_r^{**} \end{bmatrix} \begin{bmatrix} \mathbf{x}_1(t) \\ \cdot \\ \cdot \\ \cdot \\ \mathbf{x}_r(t) \end{bmatrix} + \begin{bmatrix} \mathbf{H}_1^b(\mathbf{x}_1, \ldots, \mathbf{x}_r) \\ \cdot \\ \cdot \\ \cdot \\ \mathbf{H}_r^b(\mathbf{x}_1, \ldots, \mathbf{x}_r) \end{bmatrix} + \mathbf{B}^{**} \begin{bmatrix} \mathbf{v}_1(t) \\ \cdot \\ \cdot \\ \cdot \\ \mathbf{v}_r(t) \end{bmatrix} \tag{24}$$

with

$$y_{k,i}(t) = x_{k,(2i-1)}(t) \quad \text{for } k = 1, 2, \ldots, r, \quad i = 1, 2, \ldots, m_k.$$

The matrices $\mathbf{A}_k^{**}$ in (24) correspond to the matrices $\mathbf{A}_k^{*}$ in (14) and have a similar structure, where instead of $\alpha_{k,i}^0, \alpha_{k,i}^1$

$$\tilde{\alpha}_{k,i}^0 = \alpha_{k,i}^0 - \bar{\alpha}_{k,i}^0, \qquad \tilde{\alpha}_{k,i}^1 = \alpha_{k,i}^1 - \bar{\alpha}_{k,i}^1 \tag{25}$$

has to be used.

The $(2m \times m)$-matrix $\mathbf{B}^{**}$ is

$$\mathbf{B}^{**} = \mathbf{B}^{*}\mathbf{E} \tag{26}$$

and has, due to the proposed diagonal form of $\mathbf{E}$ in (23), the same structure as $\mathbf{B}^{*}$ in (17), with $\bar{\lambda}_{k,i}$ replacing $\lambda_{k,i}$.

A nondiagonal form of $\mathbf{E}$ results in linear combinations of the external inputs driving the dynamics of the links. This case is also of interest for certain practical applications, but since it is based on the same principle, it is not considered here.

The hierarchical overall system (24) shows as a new result that by the hierarchical coordinator (18) the overall behavior of system (14) consisting of r robots can be changed. The dynamics of each link as presented by (15) and (16) with the characteristic coefficients $\alpha^0_{k,i}$ and $\alpha^1_{k,i}$ can be changed via $\mathbf{H}^a_{k,i}$ in (21) to the dynamics with the characteristic coefficients $\bar{\alpha}^0_{k,i}$ and $\bar{\alpha}^1_{k,i}$ (Eq. (25)). This has the advantage that by approaching a target point or by executing a certain working procedure the dynamical behavior of each link can be altered appropriately by the hierarchical coordinator. By the $(m_k \times 1)$-matrices $\mathbf{H}^b_k$ $(k = 1, 2, \ldots, r)$ in the hierarchical overall system useful couplings between the links of different robots can be introduced which are in general nonlinear for different coordinates systems of the robots and depend on the kind of application of the multirobot system. It can be seen from $\mathbf{B}^{**}$ (Eq. (26)) that the input gains of the hierarchical overall system (24) are $\bar{\lambda}_{k,i}$, which can be arbitrarily assigned by the matrix $\mathbf{E}$ in (23).

For the realization of the collision avoidance between robots as well as between robots and obstacles, $\mathbf{H}^b(\mathbf{x}_1, \ldots \mathbf{x}_r)$ in (18) is the essential part of the hierarchical coordinator in the multirobot system. The procedure itself requires special strategies for collision detection as well as collision avoidance. In the approach considered here, the strategies are based on analytically described avoidance trajectories (Hoyer 1984, 1985).

5. The Collision Avoidance Trajectory

For derivation of the collision avoidance trajectory, the kth and the jth robot with a common working space are considered out of the r robots of the multirobot system, which means $k \neq j$. It is assumed without loss of generality that the main axes of the robots have a cylindrical configuration of the joints (i.e., the kth robot has a translational joint $r_k(t)$ and a rotational joint $\varphi_k(t)$ in the (x, y)-plane of the world coordinate system and another translational joint $z_k(t)$ vertical to it). The jth robot has the joints $r_j(t)$, $\varphi_j(t)$, and $z_j(t)$, respectively. With these variables the state vectors $\mathbf{x}_k(t)$, $\mathbf{x}_j(t)$ and the corresponding external input vectors $\mathbf{v}_k(t)$ and $\mathbf{v}_j(t)$ of the kth and the jth robot have the form

$$\mathbf{x}_i = \begin{bmatrix} x_{i1} \\ x_{i2} \\ x_{i3} \\ x_{i4} \\ x_{i5} \\ x_{i6} \end{bmatrix} = \begin{bmatrix} r_i \\ \dot{r}_i \\ \varphi_i \\ \dot{\varphi}_i \\ z_i \\ \dot{z}_i \end{bmatrix}, \qquad \mathbf{v}_i = \begin{bmatrix} v_{i1} \\ v_{i2} \\ v_{i3} \end{bmatrix} = \begin{bmatrix} v_{ir} \\ v_{i\varphi} \\ v_{iz} \end{bmatrix}, \qquad i = j,k. \tag{27}$$

The cylindrical coordinate system is chosen as the basic coordinate system for $\mathbf{H}(\mathbf{x}_1, \ldots, \mathbf{x}_r; \mathbf{v}_1, \ldots, \mathbf{v}_j)$, because most robot coordinates can be easily transformed into this system. These transformations allow a general applicability without changes of the algorithms for different types of robots.

The derivation of the collision avoidance trajectories is based on the description of a fictitious permanent colliding robot. Relating the position of this fictitious robot to the estimated position of the robot, which has to avoid a collision, two parameters p and q can be introduced. These so-called collision parameters are the basis for the calculation of the collision avoidance trajectories, where the parameter p is related to translational movement and the parameter q to the rotational movement of the arms. In most practical cases the vertical movement is restricted by such things as conveyor belts and assembly stands. Therefore, the vertical movement is not considered in the avoidance strategies, but it can be involved without principal difficulties.

For the derivation of the collision avoidance trajectory the configuration of the two robots in Fig. 2 is used, where the length a_{kj} is the shortest distance between the origins of the robots, which together with

Fig. 2. Geometrical configuration of the robots k *and* j.

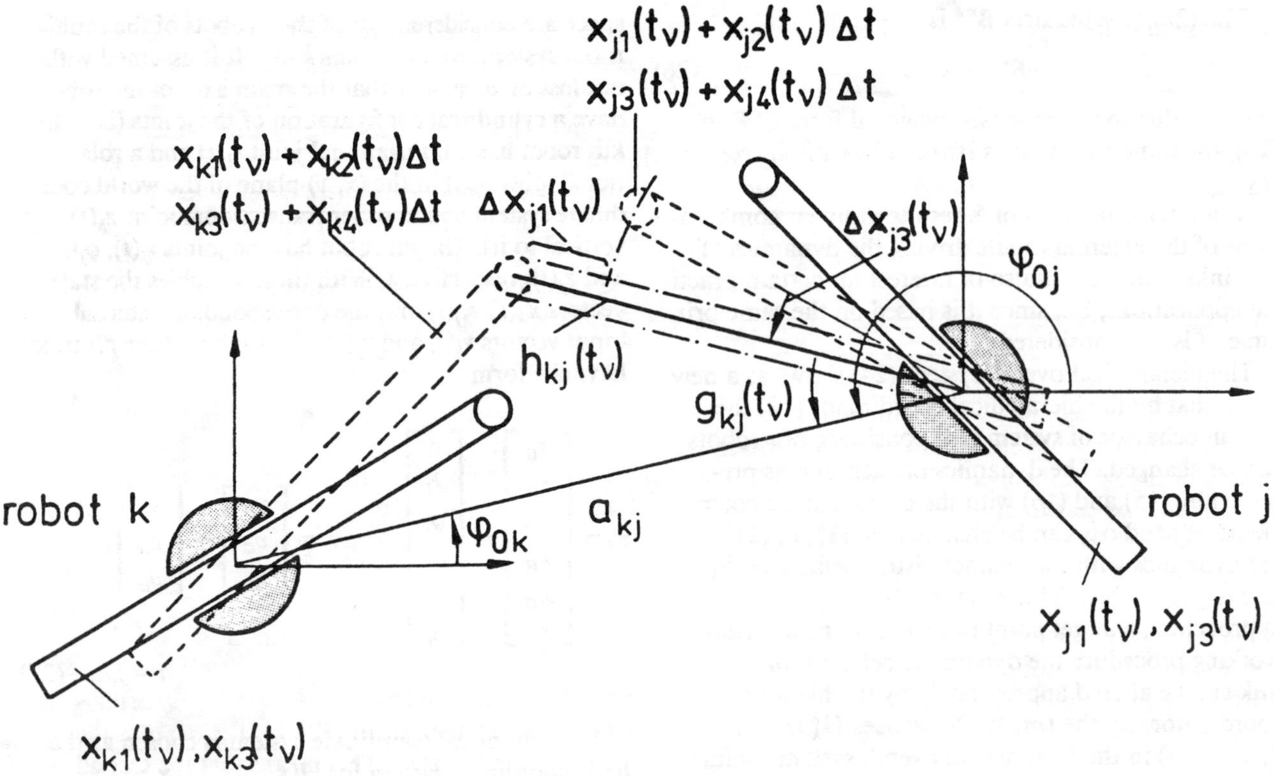

the angles φ_{0k} and φ_{0j} denote the relative positions. The right-of-way precedence is given to robot k. This means that robot k can follow its desired path described by $\mathbf{v}_k(t)$, while robot j has to avoid a collision.

The actual positions of the robots k and j at the time t_v are described by $x_{k1}(t_v)$, $x_{k3}(t_v)$ and $x_{j1}(t_v)$, $x_{j3}(t_v)$. Denoting the sampling interval of the robot control by Δt, the predicted positions of both robots for $t_{v+1} = t_v + \Delta t$ are $x_{k1}(t_v) + x_{k2}(t_v)\Delta t$; $x_{k3}(t_v) + x_{k4}(t_v)\Delta t$ and $x_{j1}(t_v) + x_{j2}(t_v)\Delta t$; $x_{j3}(t_v) + x_{j4}(t_v)\Delta t$, respectively. These predicted positions are shown in Fig. 2 by broken lines.

To detect a danger of collision, the motion of the robot with right-of-way precedence has to be brought in relation to the other robot. This is done in the approach considered here by describing a fictitious robot. The handpoint of this fictitious robot is permanently colliding with the handpoint of the robot with right-of-way precedence, shown in Fig. 2 by semidotted lines. The position of the permanent colliding robot is described by a length $h_{kj}(t_v)$ and an angle $g_{kj}(t_v)$ (Hoyer 1984).

From Fig. 2 follows for the length

$$h_{kj}(t_v) = \sqrt{a_{kj}^2 + [x_{k1}(t_v) + x_{k2}(t_v)\Delta t]^2 - 2a_{kj}[x_{k1}(t_v) + x_{k2}(t_v)\Delta t] \cos (x_{k3}(t_v) + x_{k4}(t_v)\Delta t - \varphi_{0k})}, \quad (28)$$

and for the angle

$$g_{kj}(t_v) = \arctan \frac{[x_{k1}(t_v) + x_{k2}(t_v)\Delta t] \sin (x_{k3}(t_v) + x_{k4}(t_v)\Delta t - \varphi_{0k})}{a_{kj} - [x_{k1}(t_v) + x_{k2}(t_v)\Delta t] \cos (x_{k3}(t_v) + x_{k4}(t_v)\Delta t - \varphi_{0k})}. \quad (29)$$

Relating now the position of the permanent colliding robot (28), (29) to the estimated position of robot j, the so-called collision parameters $p_{kj}(t_v)$ and $q_{kj}(t_v)$ can be introduced, which are a criterion for an actual danger of collision. Setting the difference $\Delta x_{j1}(t_v)$ between the predicted length of robot j and the length of the permanent colliding robot defined by

$$x_{j1}(t_v) = [x_{j1}(t_v) + x_{j2}(t_v)\Delta t] - h_{kj}(t_v) \quad (30)$$

in relation to the translational movement $\bar{x}_{j2}\,\Delta t$, the parameter $p_{kj}(t_v)$ can be introduced in the form of

$$p_{kj}(t_v) = \Delta x_{j1}(t_v)/\bar{x}_{j2}\,\Delta t. \quad (31)$$

The translational movement $\bar{x}_{j2}\,\Delta t$ results from a nominal velocity $\bar{x}_{j2}$ of the axis, which is specific to the type of robot used. The parameter $p_{kj}(t_v)$ therefore gives an indication how many time intervals Δt are required for the arm movement of the robot j to get the length of the permanent colliding robot. For $p_{kj}(t_v) > 0$, the arm has to draw back in order to avoid a collision; for $p_{kj}(t_v) < 0$ the arm can still go forward; for a collision of the handpoints of both robots, it is $p_{kj}(t_v) = 0$.

In the same way, the parameter $q_{kj}(t_v)$ which corresponds to the rotational movement of the robots can be introduced as

$$q_{kj}(t_v) = \frac{|\Delta x_{j3}(t_v)|}{\hat{x}_{j4}(t_v)\Delta t} \quad (32)$$

with

$$\Delta x_{j3}(t_v) = [\varphi_{0j} - g_{kj}(t_v)] - [x_{j3}(t_v) + x_{j4}(t_v)\Delta t] \quad (33)$$

as the difference between the angle of the permanent colliding robot and the predicted angle of the robot j and $\hat{x}_{j4}(t_v)\Delta t$ as a rotational movement. The angular velocity $\hat{x}_{j4}(t_v)$ is given by

$$\hat{x}_{j4}(t_v) = \begin{cases} \bar{x}_{j4} & \text{for } \left| x_{j4}(t_v) - \dfrac{g_{kj}(t_{v-1}) - g_{kj}(t_v)}{\Delta t} \right| \leq \bar{x}_{j4}, \\ \left| x_{j4}(t_v) - \dfrac{g_{kj}(t_{v-1}) - g_{kj}(t_v)}{\Delta t} \right| & \text{otherwise,} \end{cases} \quad (34)$$

and consists of the nominal angular velocity $\bar{x}_{j4}$ (specific to the type of robot used) as well as of the actual velocity $x_{j4}(t_v)$ of the robot j and via $g_{kj}(t_v)$ of the movement of the robot k. Based on this information, $q_{kj}(t_v)$ indicates similar to $p_{kj}(t_v)$ the number of time intervals Δt required for robot j to get the angular position of the permanent colliding robot.

The variables $h_{kj}(t_v)$, $g_{kj}(t_v)$ in (28), (29) and $p_{kj}(t_v)$, $q_{kj}(t_v)$ in (31), (32) are the basic parameters for the description of the collision avoidance trajectory consisting of the components $f_{j1}(t_v)$, $f_{j2}(t_v)$, where $f_{j1}(t_v)$ belongs to the translational movement and $f_{j2}(t_v)$ belongs to the rotational movement of the robot j.

Considering the translational movement, the corresponding value $f_{j1}(t_v)$ of the trajectory can be derived as

$$f_{j1}(t_v) = [h_{kj}(t_v) - s_{j1}(t_v)] + g_{kj}(t_v)[\bar{x}_{j2}\,\Delta t + h_{kj}(t_v) - h_{kj}(t_{v-1})]. \quad (35)$$

$f_{j1}(t_v)$ represents for the translational motion of the robot j the actual allowed range of movement in order to avoid a collision with robot k. It consists of two parts: The first term on the right side of Eq. (35) describes the maximal length $h_{kj}(t_v) - s_{j1}(t_v)$ of the arm of robot j for the safe passing with respect to robot k (which has the right of way), where $s_{j1}(t_v)$ is a safety factor. When the arm has this length in the passing position (Fig. 2), then from (32), (33) $q_{kj}(t_v)$ is at the same time equal to zero. This means that the second term on the right side of (35) is equal to zero while passing. The performance of the algorithm is improved by the second term on the right of (35) because it takes via $\bar{x}_{j2}\,\Delta t$ the possible translational movement of the arm in a time interval Δt with a nominal velocity $\bar{x}_{j2}$ of the arm into account as well as the difference between the length $h_{kj}(t_v)$ of the permanent colliding robot at the time t_v and the length $h_{kj}(t_{v-1})$ of this system at the time $t_{v-1} = t_v - \Delta t$. As these terms are weighted by the parameter $q_{kj}(t_v)$ in (32), this results in an extension of the allowed translational range of the arm as well as in faster reaction proportional to the distance to the position of the permanent colliding robot.

In a similar way the rotational movement of robot j is guided by the corresponding value $f_{j2}(t_v)$ in the form

$$f_{j2}(t_v) = [\varphi_{0j} - g_{kj}(t_v) + \gamma s_{j2}(t_v)] + \gamma p_{kj}(t_v)\hat{x}_{j4}(t_v)\Delta t. \quad (36)$$

In Eq. (36) $s_{j2}(t_v)$ is a safety factor while the predicted position of the arm of robot j in relation to the angle $(\varphi_{0j} - g_{kj}(t_v))$—that is, the position of the permanent colliding robot—is regarded by the parameter γ.

Figure 3 shows an implicit description of the colli-

Fig. 3. The collision avoidance trajectory.

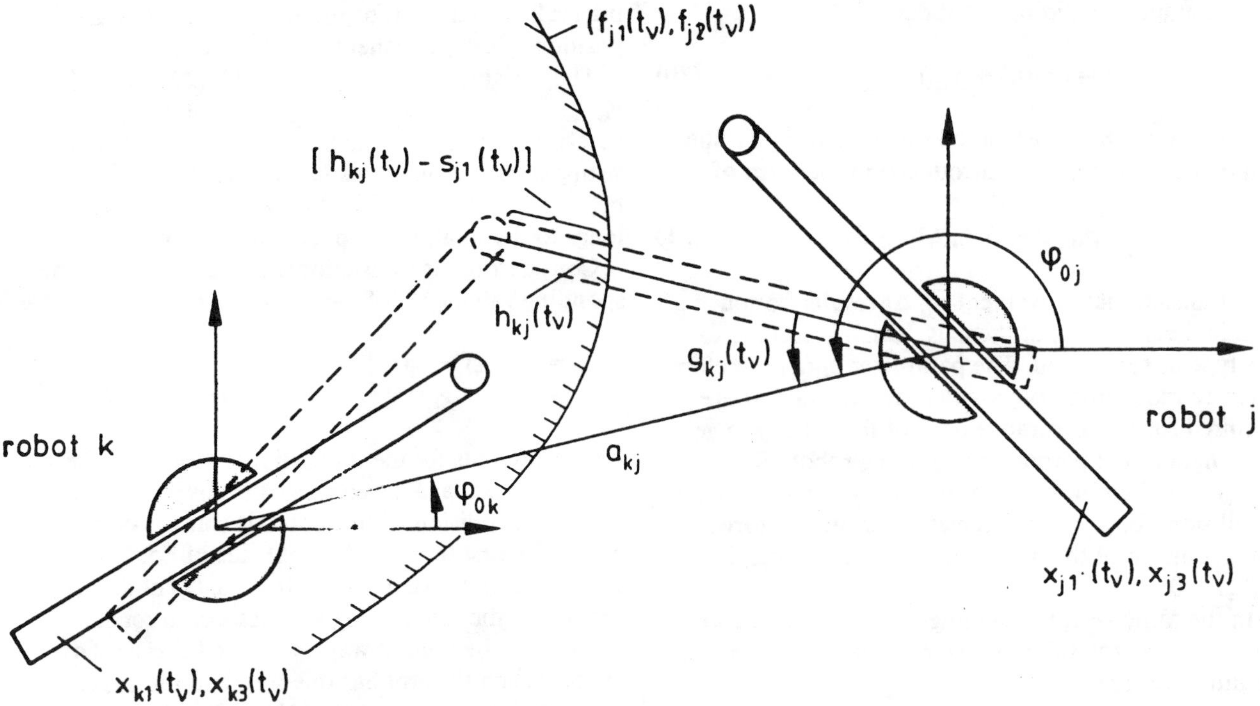

sion avoidance trajectory consisting of the components $f_{j1}(t_v)$ for the translational movement and $f_{j2}(t_v)$ for the rotational movement of the arm of the robot *j*. This trajectory determines the allowed range of movement for the robot *j* guaranteeing collision avoidance. It follows from (35) and (36) that this trajectory depends on the actual movement of the robots and is permanently updated.

In current research work a distinction is made between methods for collision detection and methods for collision avoidance. In the approach described before, this is not necessary, since the components f_{j1} and f_{j2} serve for collision detection as well as avoidance because an arm movement beyond f_{j1}, f_{j2} means collision detection which leads to a substitution of the external input v_{j1} by f_{j1} and/or v_{j2} by f_{j2} (Hoyer 1984).

6. Collision Avoidance Strategy

Using the collision avoidance trajectory (35), (36) in the foregoing section, the hierarchical coordinator can be designed for collision avoidance (Freund and Hoyer 1985). As a basis for the solution of the collision avoidance problem, the hierarchical coordinator in (18) is used in a form extended by a dependency of $\mathbf{H}^b$ on the reference inputs. For the *j*th robot the coordinator therefore has the form

$$\mathbf{H}_j = \mathbf{H}_j^a(\mathbf{H}_j^b) \cdot \begin{bmatrix} x_{j1}(t) \\ \vdots \\ x_{j6}(t) \end{bmatrix} + \underbrace{\begin{bmatrix} H_{j1}^b(\mathbf{x}_1, \ldots, \mathbf{x}_r; \mathbf{v}_j) \\ H_{j2}^b(\mathbf{x}_1, \ldots, \mathbf{x}_r; \mathbf{v}_j) \\ H_{j3}^b(\mathbf{x}_1, \ldots, \mathbf{x}_r; \mathbf{v}_j) \end{bmatrix}}_{\mathbf{H}_j^b} + \mathbf{E}_j(\mathbf{H}_j^b) \cdot \begin{bmatrix} v_{j1}(t) \\ v_{j2}(t) \\ v_{j3}(t) \end{bmatrix}, \quad j = 1, 2, \ldots, r, \tag{37}$$

where the vector $\mathbf{H}_j^b$, which represents useful couplings between the robots, has a key position for collision avoidance. This is due to the fact that the avoidance trajectory (35), (36) sets the motions of robots *k* and *j* into relation in order to detect and avoid collision.

This leads to the following designs of $\mathbf{H}_j^b$. For the first element of $\mathbf{H}_j^b$ related to the translational movement follows

$$H^b_{j1} = \begin{cases} 0 & \text{for } v_{j1}(t_{\nu+1}) \leq f_{j1}(t_\nu), \\ f_{j1}(t_\nu) & \text{for } v_{j1}(t_{\nu+1}) > f_{j1}(t_\nu). \end{cases} \quad (38)$$

This structure of H^b_{j1} means that the trajectory $f_{j1}(t_\nu)$ functions as a switching criterion. If the original reference input $v_{j1}(t_{\nu+1})$ is smaller than or equal to the corresponding component of the trajectory for collision avoidance $f_{j1}(t_\nu)$ in (35), then H^b_{j1} is set equal to zero; that is, the original reference input $v_{j1}(t_{\nu+1})$ is applied. Otherwise it is replaced by $f_{j1}(t_\nu)$.

For the rotational movement follows with q_0 as a safety factor in the same way

$$H^b_{j2} = \begin{cases} 0 & \text{for } p_{kj}(t_\nu) \leq q_{kj}(t_\nu) + q_0 \text{ or} \\ & \text{for } p_{kj}(t_\nu) > q_{kj}(t_\nu) + q_0 \\ & \quad \text{with } \gamma = -1 \text{ and } v_{j2}(t_{\nu+1}) \leq f_{j2}(t_\nu) \\ & \text{or with } \gamma = 1 \text{ and } v_{j2}(t_{\nu+1}) \geq f_{j2}(t_\nu); \\ f_{j2}(t_\nu) & \text{for } p_{kj}(t_\nu) > q_{kj}(t_\nu) + q_0 \\ & \quad \text{with } \gamma = -1 \text{ and } v_{j2}(t_{\nu+1}) > f_{j2}(t_\nu) \\ & \text{or with } \gamma = 1 \text{ and } v_{j2}(t_{\nu+1}) < f_{j2}(t_\nu). \end{cases} \quad (39)$$

As the vertical movements of the robots are not involved in collision avoidance, the third element H^b_{j3} of the vector $\mathbf{H}^b_j$ is given by $H^b_{j3} = 0$. For the substitution of the external input $v_{j,i}(t_\nu)$ by the corresponding value of the collision avoidance trajectory $f_{j,i}(t_\nu)$ in case of danger of collision, the element $e_{j,i}$ $(i = 1, 2, 3)$ of the diagonal matrix $\mathbf{E}_j$ in (37) is switched between 0 and 1, depending on the corresponding element $H^b_{j,i}$ $(i = 1, 2, 3)$ of the vector $\mathbf{H}^b_j$. The factor 1 follows from the choice of the input gain $\bar{\lambda}_{j,i}$ equal to $\lambda_{j,i}$. From this results

$$e_{j,i}(H^b_{j,i}) = \frac{\bar{\lambda}_{j,i}}{\lambda_{j,i}} = \begin{cases} 0 & \text{for } H^b_{j,i} \neq 0, \\ 1 & \text{for } H^b_{j,i} = 0 \quad (i = 1, 2, 3). \end{cases} \quad (40)$$

This is due to the fact that the original reference inputs are applied to the basic robot control whenever the level of collision avoidance does not intervene.

The structure of the hierarchical coordinator in (37) offers the possibility of a variation of the dynamics of the robots via $\mathbf{H}^a$. This is used to change the dynamics of the basic robot control circuits in order to increase the fastness of reaction in case of danger of collision. Afterwards the control system can return to the original dynamics which guarantee accurate path control. Therefore, only the dynamics of those variables of motion are changed via $\mathbf{H}^a$ which are involved in the collision avoidance strategies.

The change of the dynamics is realized on the basis of Eqs. (19)–(21) for $\mathbf{H}^a$, where for the jth robot $\mathbf{H}^a_{j,i}$ has the form

$$\mathbf{H}^a_{j,i} = \left[\frac{\bar{\alpha}^0_{j,i}}{\lambda_{j,i}} \;\; \frac{\bar{\alpha}^1_{j,i}}{\lambda_{j,i}} \right], \quad j = 1, 2, \ldots, r, \quad i = 1, 2, 3. \quad (41)$$

The parameters $\bar{\alpha}^0_{j,i}$, $\bar{\alpha}^1_{j,i}$ can be chosen arbitrarily. One possibility is an aperiodic behavior of the axes of motion together with a dependency on the parameter $q_{kj}(t_\nu)$ in (32), which is a criterion for the actual danger of collision. This leads to

$$\bar{\alpha}^0_{j,i}(H^b_{j,i}) = \begin{cases} 0 & \text{for } H^b_{j,i} = 0, \\ \dfrac{\bar{\alpha}^0_{j,i\max} - \alpha^0_{j,i}}{\mu q_{kj}(t_\nu) + 1} & \text{for } H^b_{j,i} \neq 0, \end{cases} \quad (42)$$

and

$$\bar{\alpha}^1_{j,i}(H^b_{j,i}) = 2\sqrt{\bar{\alpha}^0_{j,i}(H^b_{j,i})}, \quad j = 1, 2, \ldots, r, \quad i = 1, 2, 3. \quad (43)$$

In (42) μ is a constant coefficient and $\bar{\alpha}^0_{j,i\max}$ denotes the maximal value of $\bar{\alpha}^0_{j,i}$ with respect to technical limitations. This equation shows that the original dynamics of the robot j remain unchanged for $H^b_{j,i} = 0$. That means there is no danger of collision, and the original inputs $v_{j,i}$ are applied to the basic robot controls. Otherwise the dynamics will be changed via $q_{kj}(t_\nu)$, depending on the actual danger of collision. This shows that similar to $\mathbf{E}(\mathbf{H}^b)$ the matrix $\mathbf{H}^a$ depends on $\mathbf{H}^b$ as well, which is characteristic for the hierarchical coordinator (37) designed for collision avoidance.

It can be seen by the algorithms described that the collision detection as well as the collision avoidance is based on the state variables and not on the inputs of the robot with right-of-way precedence. This has the advantage that the collision avoidance strategy is based on the actual movement of the robots and still func-

Fig. 5. Structure of the hierarchical coordinator for collision avoidance.

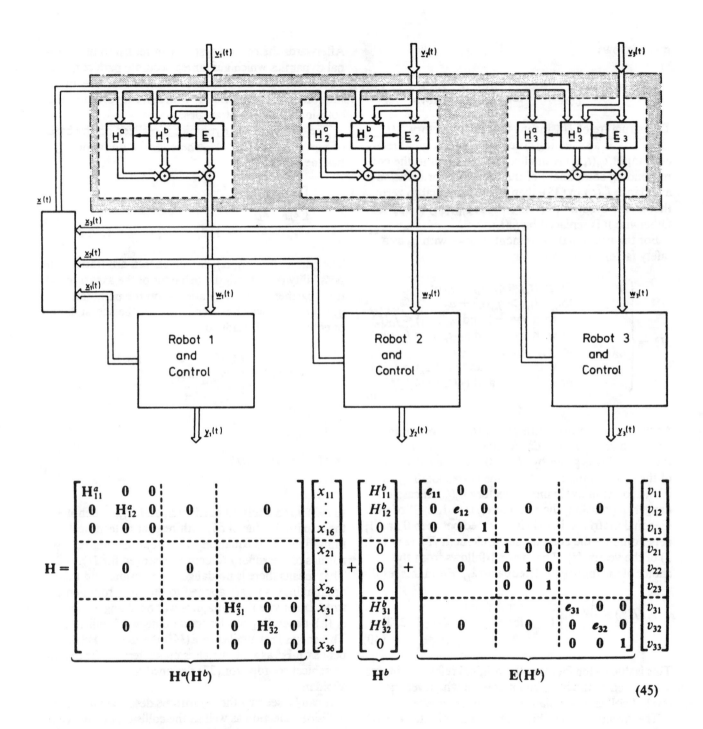

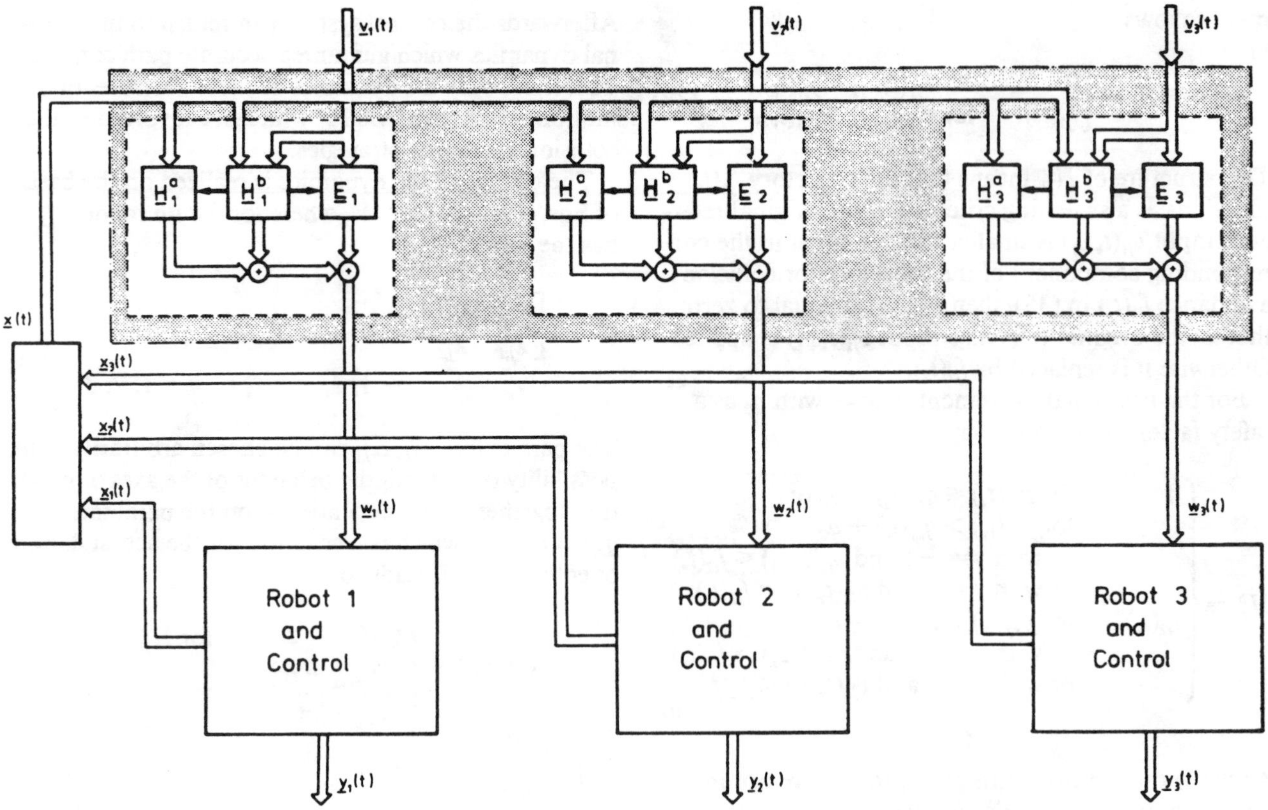

$$
\mathbf{H} = \underbrace{\left[\begin{array}{ccc:c:ccc}
\mathbf{H}^a_{11} & 0 & 0 & & & & \\
0 & \mathbf{H}^a_{12} & 0 & 0 & & 0 & \\
0 & 0 & 0 & & & & \\
\hdashline
& 0 & & 0 & & 0 & \\
\hdashline
& & & & \mathbf{H}^a_{31} & 0 & 0 \\
& 0 & & 0 & 0 & \mathbf{H}^a_{32} & 0 \\
& & & & 0 & 0 & 0
\end{array}\right]}_{\mathbf{H}^a(\mathbf{H}^b)}
\left[\begin{array}{c} x_{11} \\ \vdots \\ x_{16} \\ \hdashline x_{21} \\ \vdots \\ x_{26} \\ \hdashline x_{31} \\ \vdots \\ x_{36} \end{array}\right]
+ \underbrace{\left[\begin{array}{c} H^b_{11} \\ H^b_{12} \\ 0 \\ \hdashline 0 \\ 0 \\ 0 \\ \hdashline H^b_{31} \\ H^b_{32} \\ 0 \end{array}\right]}_{\mathbf{H}^b}
+ \underbrace{\left[\begin{array}{ccc:ccc:ccc}
e_{11} & 0 & 0 & & & & & & \\
0 & e_{12} & 0 & & 0 & & & 0 & \\
0 & 0 & 1 & & & & & & \\
\hdashline
& & & 1 & 0 & 0 & & & \\
& 0 & & 0 & 1 & 0 & & 0 & \\
& & & 0 & 0 & 1 & & & \\
\hdashline
& & & & & & e_{31} & 0 & 0 \\
& 0 & & & 0 & & 0 & e_{32} & 0 \\
& & & & & & 0 & 0 & 1
\end{array}\right]}_{\mathbf{E}(\mathbf{H}^b)}
\left[\begin{array}{c} v_{11} \\ v_{12} \\ v_{13} \\ \hdashline v_{21} \\ v_{22} \\ v_{23} \\ \hdashline v_{31} \\ v_{32} \\ v_{33} \end{array}\right]
\tag{45}
$$

The structure of Eq. (45) and of Fig. 5 holds for many robots as well because the application to three robots (as demonstrated by the derivation) does not restrict the generality of this approach.

Fig. 6. Collision of three robots.

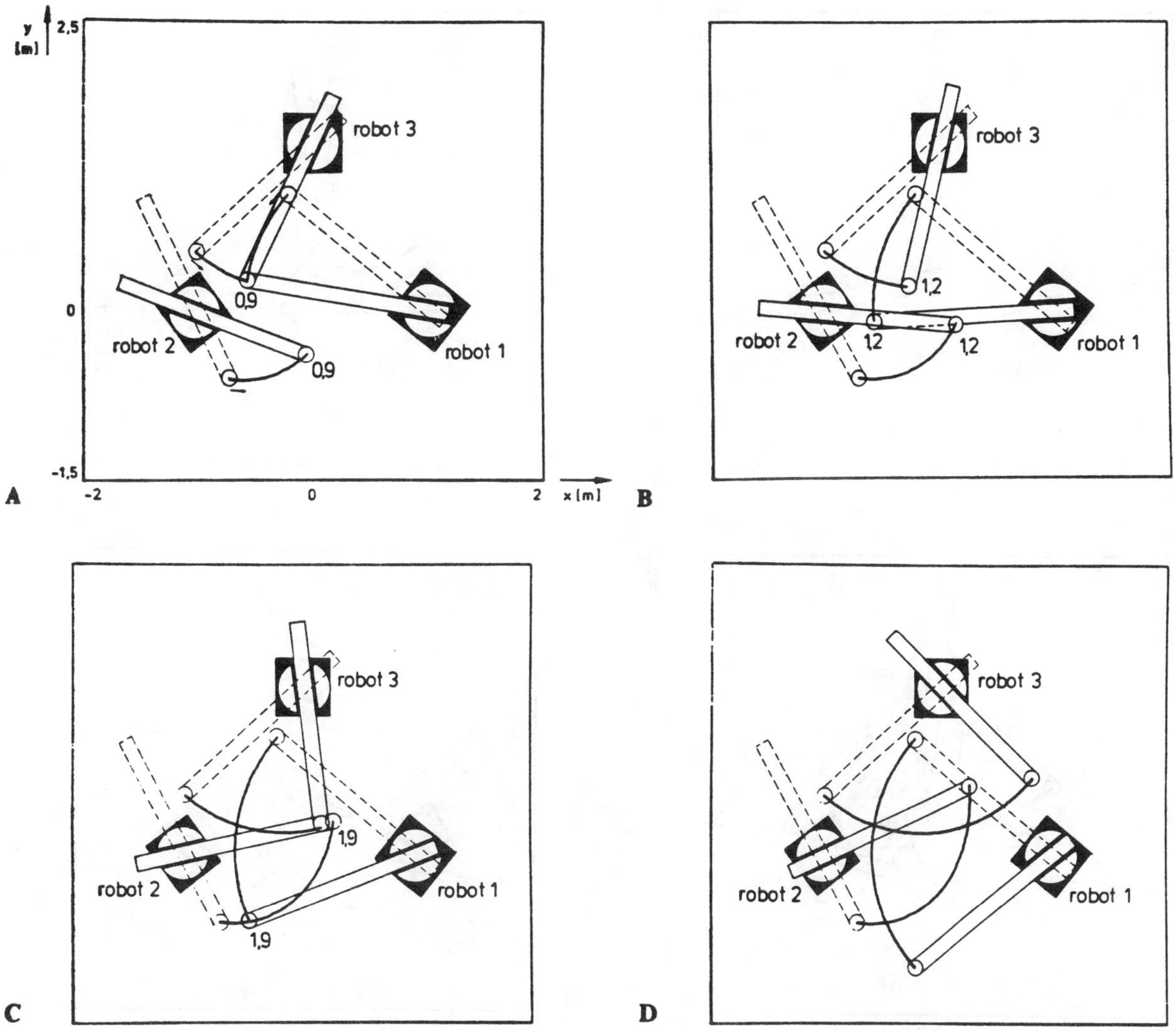

8. Collision Avoidance between Robots

The performance of the hierarchical coordinator as presented in Eq. (45) is demonstrated by simulations in Figs. 6 and 7, which are based on the configuration of the three robots in Fig. 4. For the simulation itself the block-oriented digital simulation language SIMAIT was used, which was developed at the Institut für Roboterforschung (IRF). In these simulations the highest right-of-way precedence is given to robot 2 and the lowest to robot 3. This means that robot 2 can follow its desired path, while robot 1 has to avoid a collision with robot 2 and at the same time robot 3 has to avoid a collision with robot 1 and robot 2.

Figure 6 shows the trajectories of the three robots if no collision avoidance strategy is provided. Robot 1 collides with robot 3 after 0.9 time units (Fig. 6A) and with robot 2 after 1.2 time units (Fig. 6B) while robot 2 collides with robot 3 after 1.9 time units (Fig. 6C). The desired paths of the robots are given in Fig. 6D.

Applying now the described collision avoidance strategies for the robots, the resulting collision-free trajectories provided by on-line pathfinding are given in

Fig. 7. Automatic pathfinding of three robots.

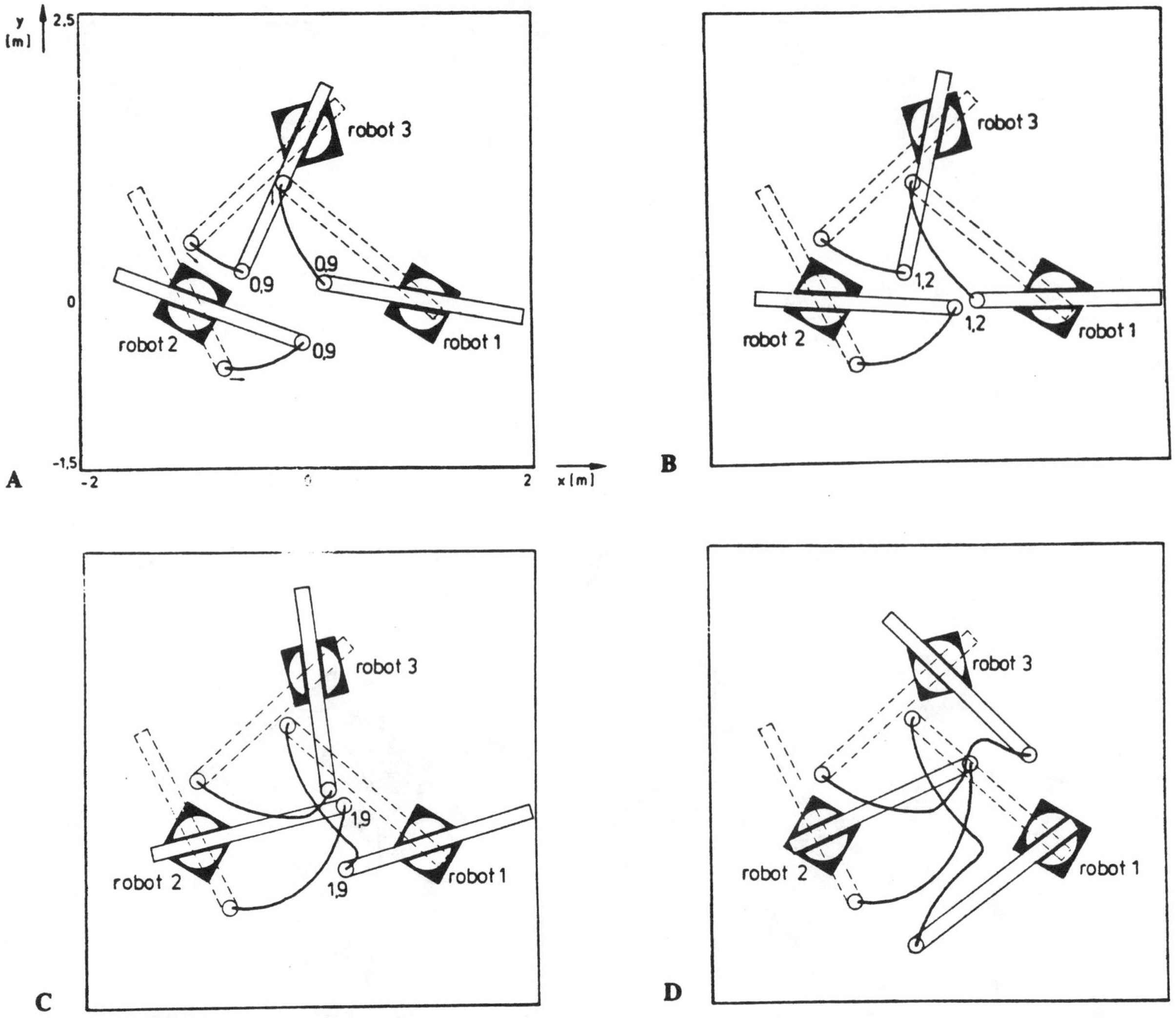

Fig. 7. The deviations from the original paths can be extended by the safety factors in Eqs. (35), (36) but are, of course, mainly a function of the positions and the speeds of the robots. A comparison with Fig. 6 demonstrates the usefulness of the automatic on-line pathfinding algorithms and shows that robots 1 and 3 reach their desired final points on the modified paths.

Based on these simulations the required amount of time for real-time processing of the collision detection and avoidance was estimated. It turned out that based on Eq. (45) the computation of the detection and avoidance strategies will require approximately 5.5 ms for a system of two robots and 20 ms for three robots (like Fig. 4) on a single-microprocessor system of type TI 9995. This computational time can be further reduced by parallel processing which can be realized according to the structure of the level of collision avoidance in Eq. (45) and Fig. 5.

9. Strategies for Robots and Obstacles

In current research work a distinction is made between methods for collision avoidance for robots and

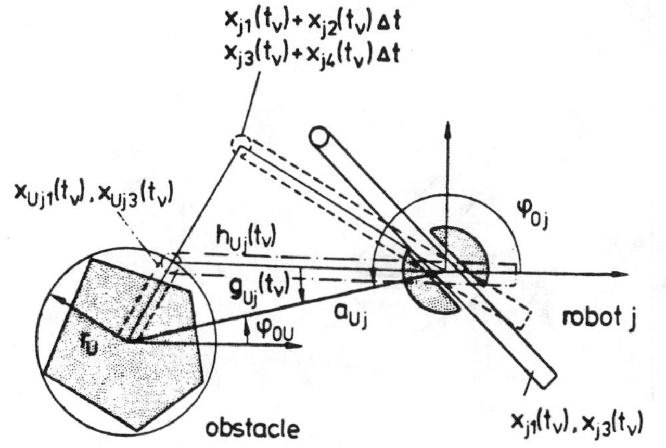

Fig. 8. Geometrical configuration of the robot j and an obstacle.

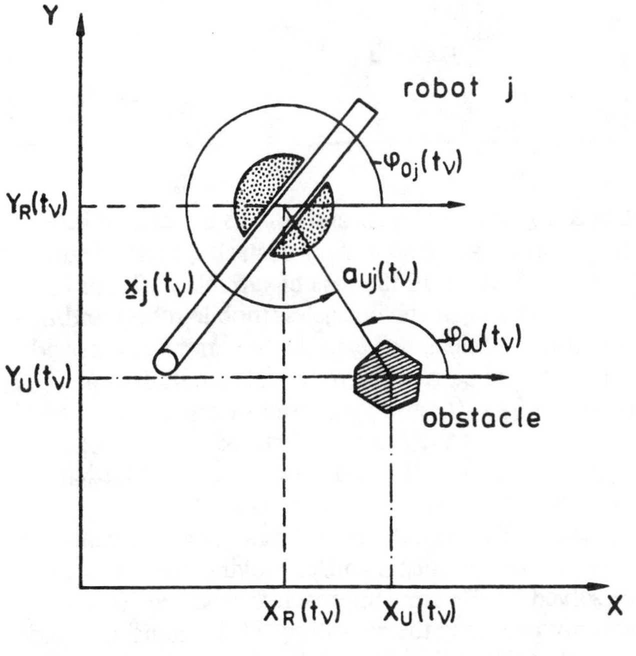

Fig. 9. Configuration of a mobile robot and a moving obstacle.

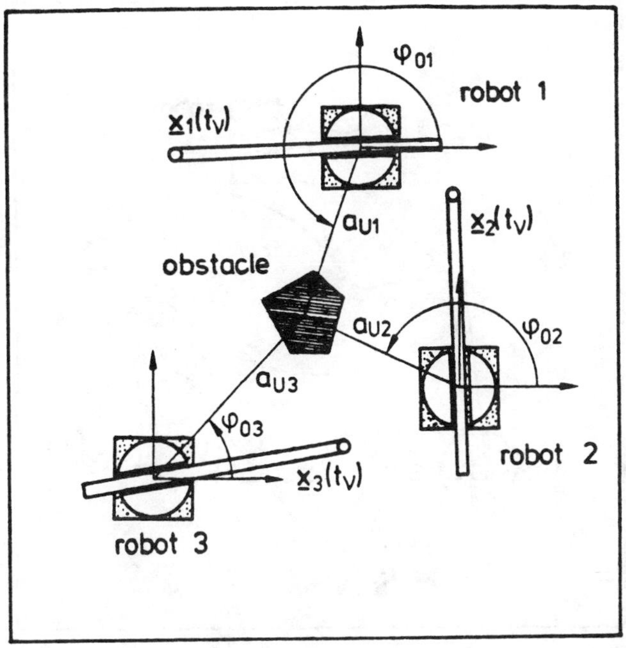

Fig. 10. Arrangement of three robots and a polygonal obstacle.

methods for obstacle avoidance. However, this is not necessary, since an obstacle can be considered as a special type of robot with right-of-way precedence, such that the basic pathfinding strategy is directly applicable (Hoyer 1985, Freund and Hoyer 1986a,b). This is realized by the description of an obstacle in form of a circle around the obstacle with the radius r_U, which can be measured by an optical sensor. The relative position of the center of this circle to robot j is denoted by the length a_{Uj} and the angles φ_{0U} and φ_{0j} as shown in Fig. 8. Based on the predicted position of the robot j, given by $x_{j1}(t_v) + x_{j2}(t_v)\Delta t$, $x_{j3}(t_v) + x_{j4}(t_v)\Delta t$, a fictitious robot is constructed with the positions $x_{Uj1}(t_v)$ and $x_{Uj3}(t_v)$ in Fig. 8, where

$$x_{Uj1}(t_v) = r_U \tag{46}$$

is the constant length of the arm and

$$x_{Uj3}(t_v) = \varphi_{0U} + \arctan \frac{[x_{j1}(t_v) + x_{j2}(t_v)\Delta t] \sin(\varphi_{0j} - x_{j3}(t_v) - x_{j4}(t_v)\Delta t)}{a_{Uj} - [x_{j1}(t_v) + x_{j2}(t_v)\Delta t] \cos(\varphi_{0j} - x_{j3}(t_v) - x_{j4}(t_v)\Delta t)} \tag{47}$$

is the angle belonging to the rotational movement. This means that the fictitious robot moves in dependence on the predicted position of the robot j. Its origin is the center of the circle while the handpoint moves on the periphery of the circle. Using this description of the fictitious robot, the problem of collision avoidance between robots and obstacles can be treated by the same approach as derived in the foregoing sections. This means that based on the permanent colliding robot with positions h_{Uj} and g_{Uj} (Fig. 8) the

Fig. 11. Collision of the three robots and the polygonal obstacle.

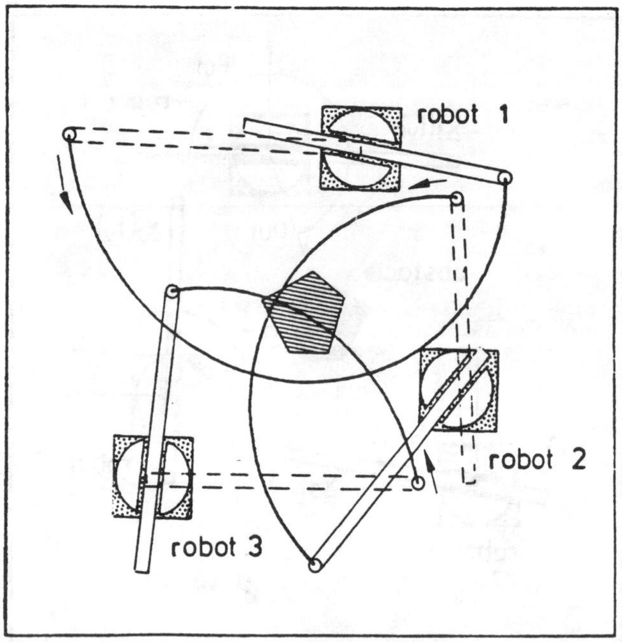

Fig. 12. Automatic pathfinding for the three robots and the polygonal obstacle.

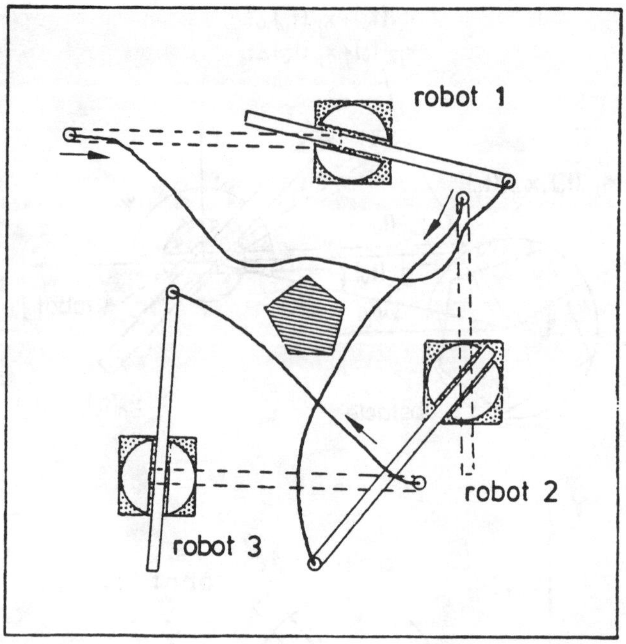

collision avoidance strategy derived in Section 6 can be applied without any changes of the algorithms (Hoyer 1984, Freund and Hoyer 1986a). This procedure is used in Case 1 in the following section (Figs. 10–12).

Obstacles of variable size or of different geometrical forms can be included in the findpath algorithm by changing the radius of the circle around the obstacle with respect to the actual size. In this case the constant radius r_U is replaced by

$$x_{Uj1}(t_v) = r_U(t_v). \quad (48)$$

This means that the structure and parameters of the findpath algorithm itself remain unchanged even for different shapes of the obstacles (Freund and Hoyer 1986b). This approach is applied in Cases 1 and 3 in the following section.

The method presented is a general approach using state feedback, which allows an extension of the solution to the findpath problem to mobile robots and moving obstacles as well (Freund and Hoyer 1986a). In Fig. 9 the positions of robot *j* and an obstacle are shown in relation to the origin of the world coordinate system (denoted by *X* and *Y*). The relative position of the robot *j* to the center of the obstacle is given by length a_{Uj} and angles φ_{0j} and φ_{0U}. In the case of stationary robots and stationary obstacles, these parameters are constant. In the case of variable positions of the robots and/or the obstacles (mobile robots and/or moving obstacles), however, these parameters depend on the actual path of the robot coordinate system (denoted by $x_R(t_v)$ and $y_R(t_v)$) and/or on the actual path of the center of the obstacle (denoted by $x_U(t_v)$ and $y_U(t_v)$). Using now the variables $a_{Uj}(t_v)$, $\varphi_{0j}(t_v)$, $\varphi_{0U}(t_v)$ instead of the corresponding constant parameters a_{Uj}, φ_{0j}, φ_{0U} in Eq. (47), the problem of collision avoidance between mobile robots and/or moving obstacles can be solved by the same approach as described before without changing the algorithms of the collision avoidance strategy. This approach is used in Cases 2 and 4 in the following section.

10. Applications

In application of the new approach for the solution of the findpath problem in multirobot systems four cases

Fig. 13. Exhibit of the Institut für Roboterforschung at the Hannover Fair in 1987.

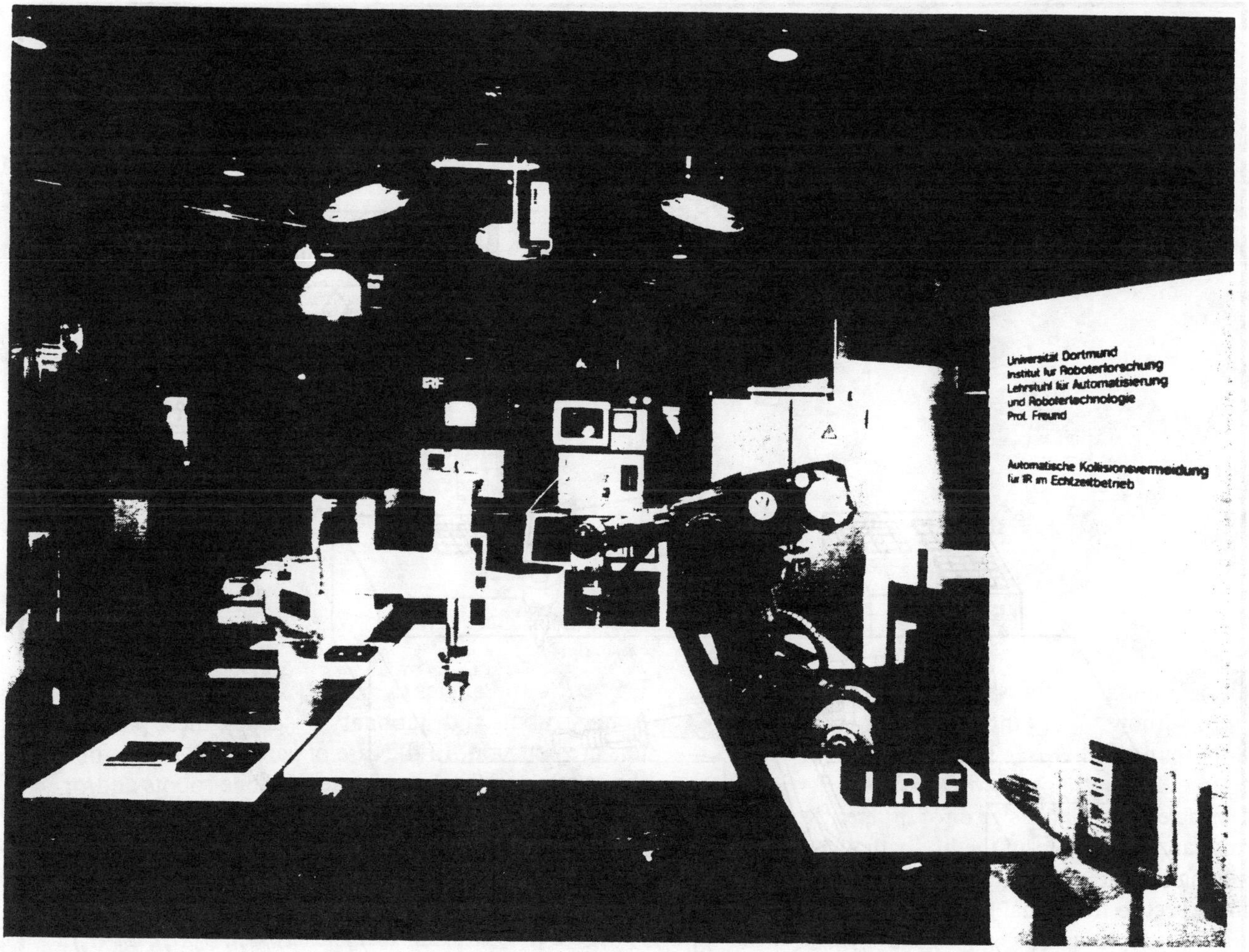

of practical importance are considered in this section. All cases are realized by the same strategy and algorithms for collision detection and collision avoidance in (37)–(43), where the restriction to a smaller number of robots and obstacles is chosen for clarity. As the algorithms used are very simple and effective, all computations of the collision-free paths can be done on-line. To show the efficiency of this new approach, simulations were made for these cases where the results are presented on a graphic display and documented by hard copies. These simulations include the dynamics of the joints of all robots involved.

The distinction between the following four cases is given by *stationary* or *mobile* robots, *stationary* or *moving* obstacles as well as *constant* or *variable* size of the obstacles (Freund and Hoyer 1986a).

Case 1

a. *Moving* robot arms ← (modified by collision avoidance strategy)
b. *Stationary* positions of the robots
c. *Stationary* positions of the obstacles
d. *Constant* size of the obstacles

Fig. 14. Structure of the system.

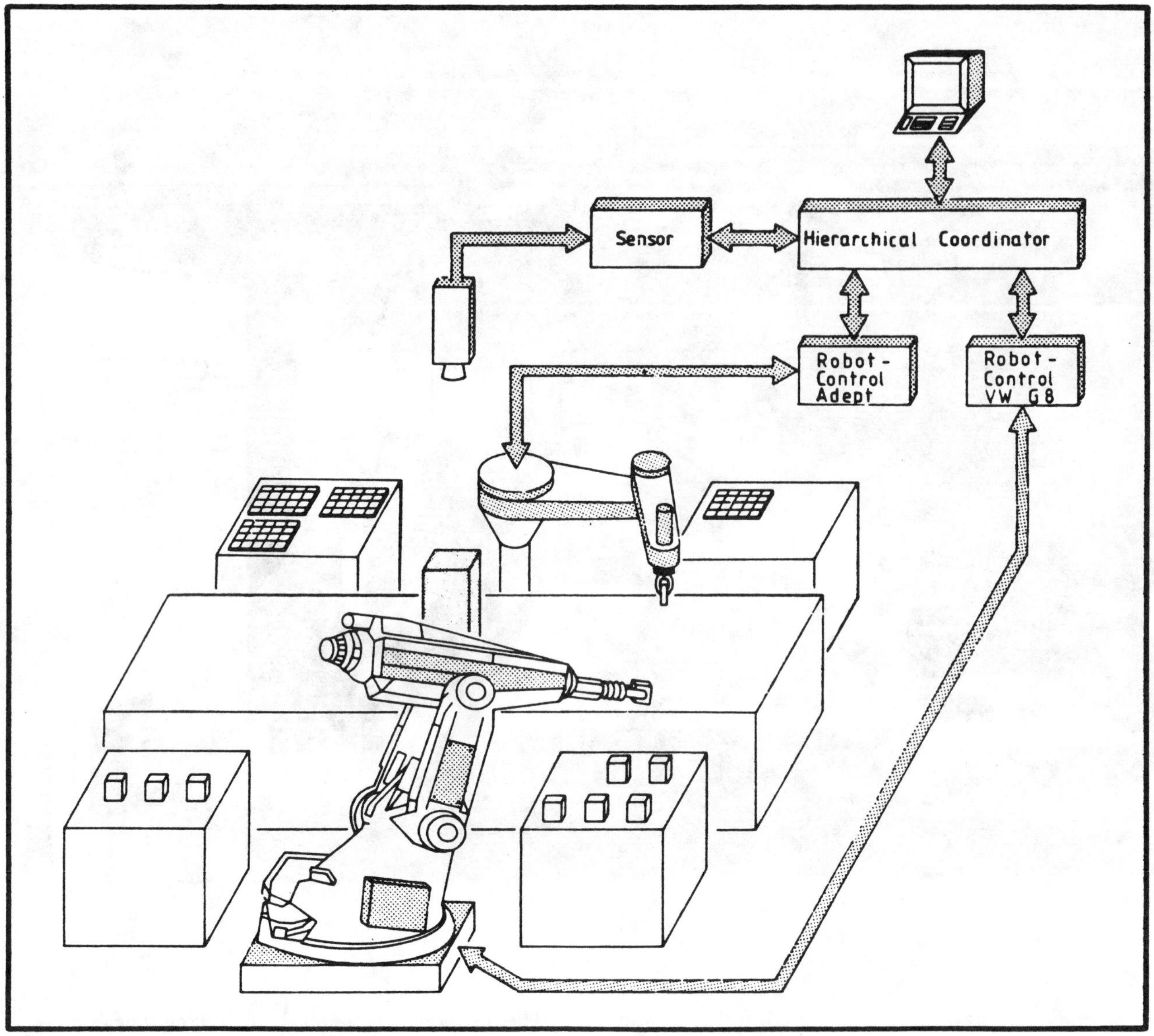

Case 2

a. *Stationary* robot arms
b. *Variable* positions of the robots (mobile robots) ← (modified by collision avoidance strategy)
c. *Variable* positions of the obstacles (moving obstacles)
d. *Constant* size of the obstacles

Case 3

a. *Moving* robot arms ← (modified by collision avoidance strategy)
b. *Variable* positions of the robots (mobile robots)
c. *Variable* positions of the obstacles (moving obstacles)
d. *Variable* size of the obstacles

Fig. 15. Arrangement of three robots and a rectangular obstacle.

Fig. 16. Collision of the three robots and the rectangular obstacle.

Fig. 17. Automatic pathfinding for the three robots and the rectangular obstacle.

Fig. 18. Arrangement of the three robots and an obstacle.

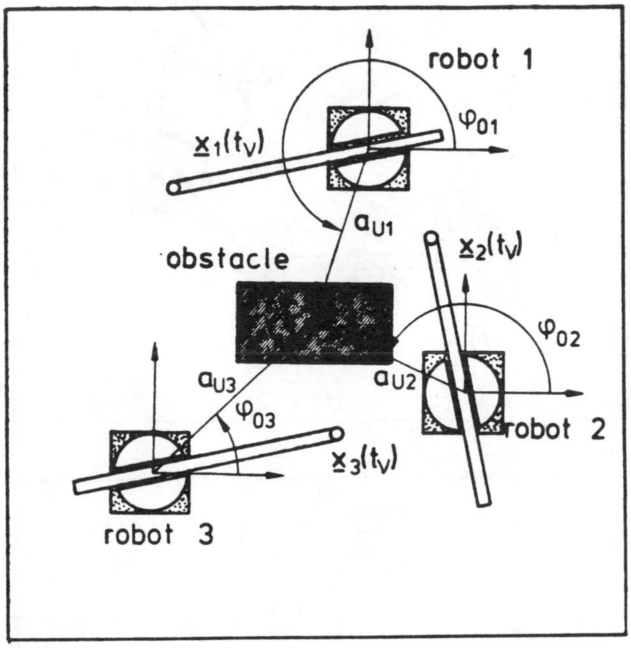

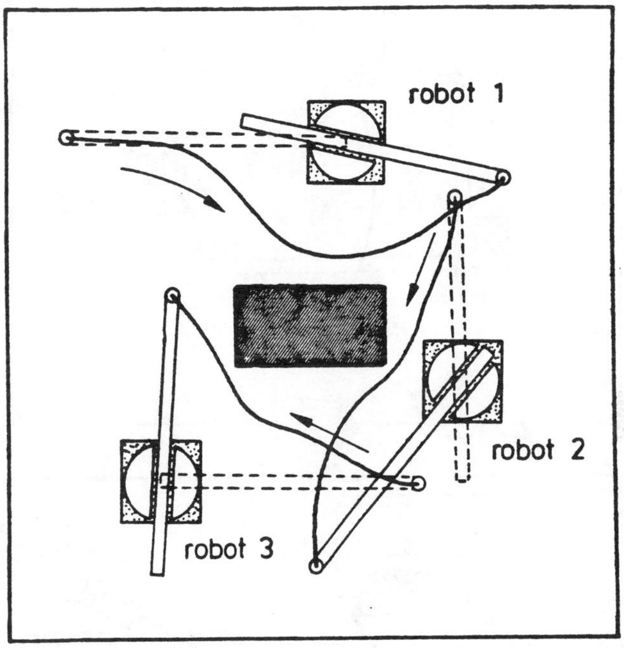

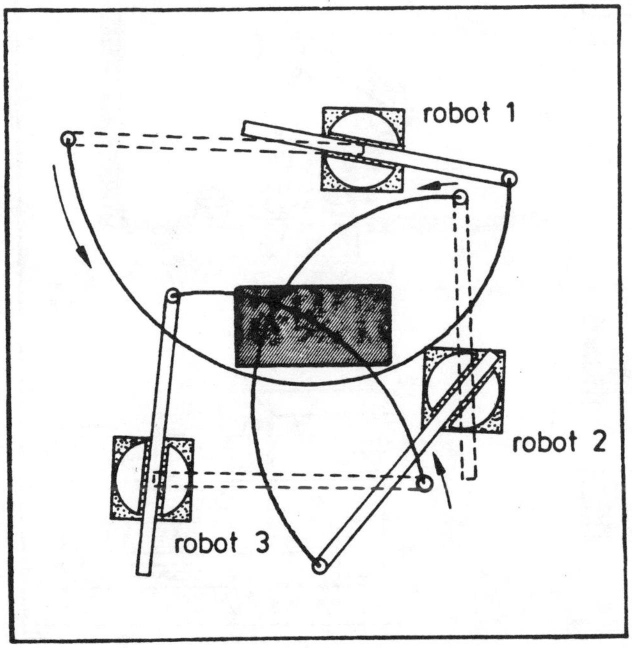

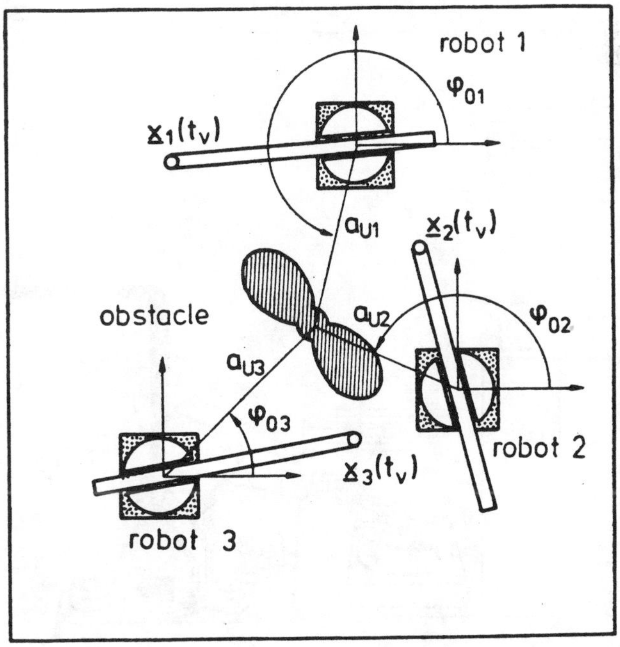

Fig. 19. Collision of the three robots and the obstacle.

Fig. 20. Automatic pathfinding for the three robots and the obstacle.

Fig. 21. Mobile robots and a moving obstacle in an unstaffed factory.

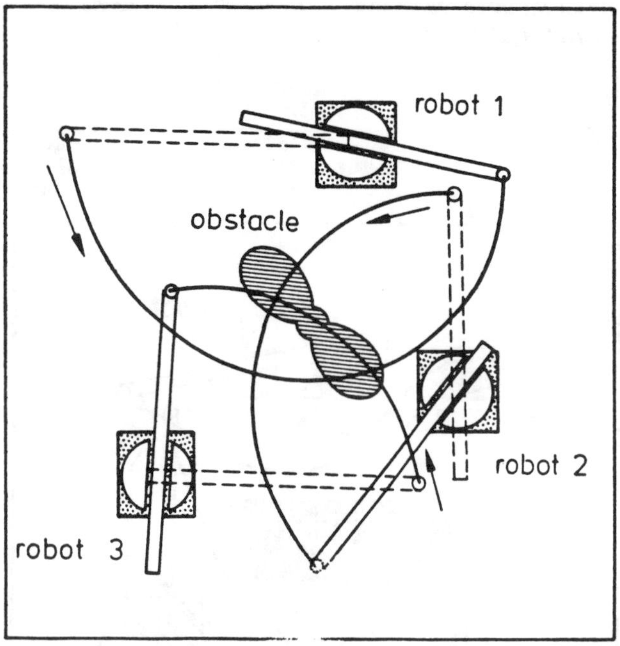

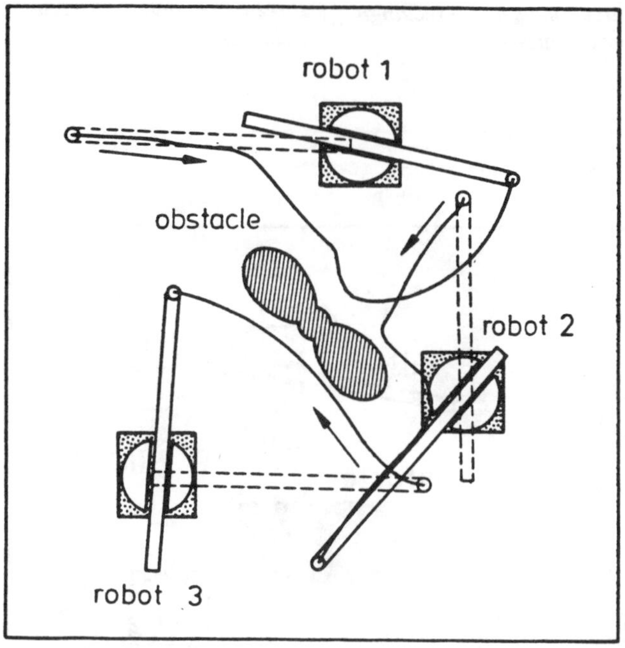

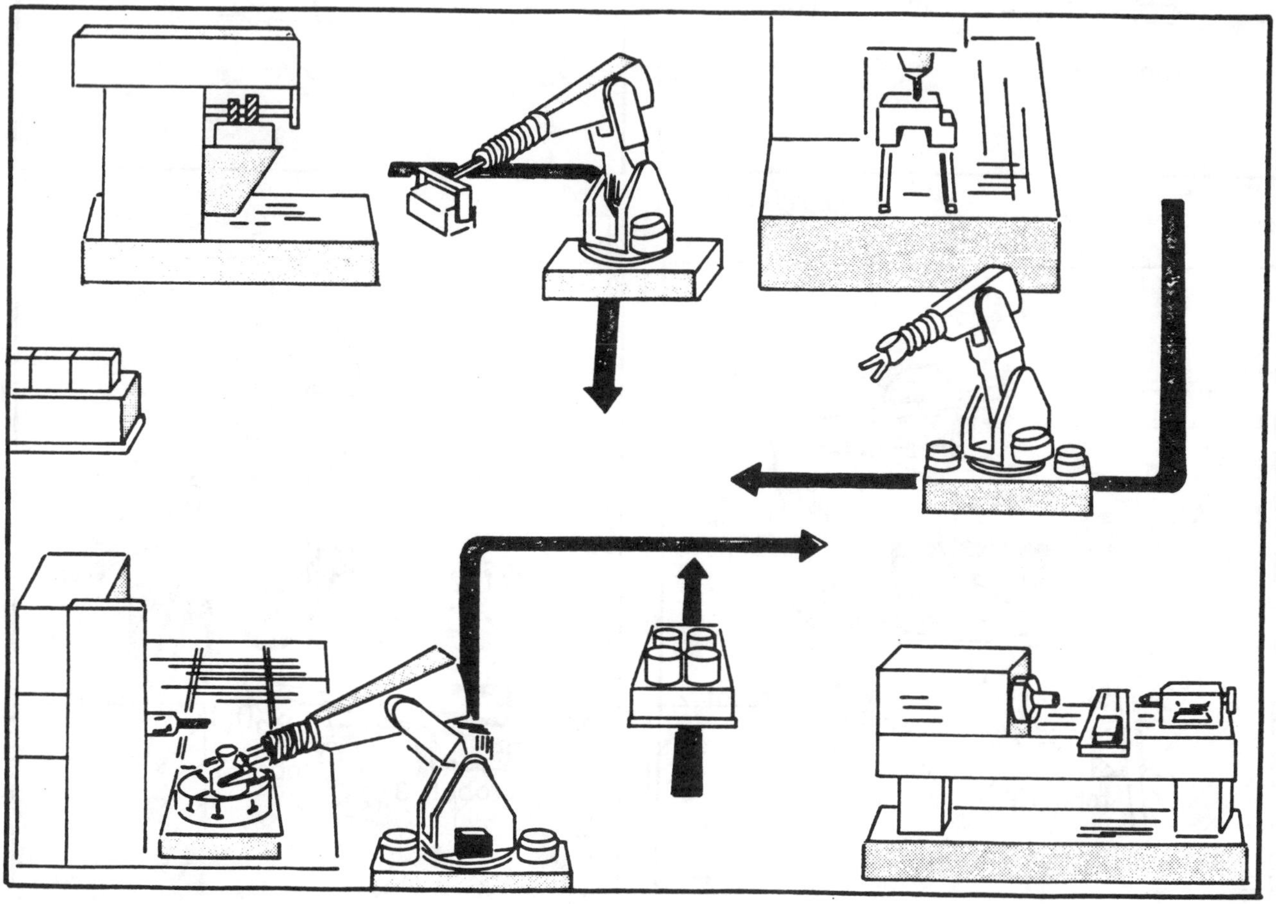

Fig. 22. Collision of three mobile robots with the moving obstacle.

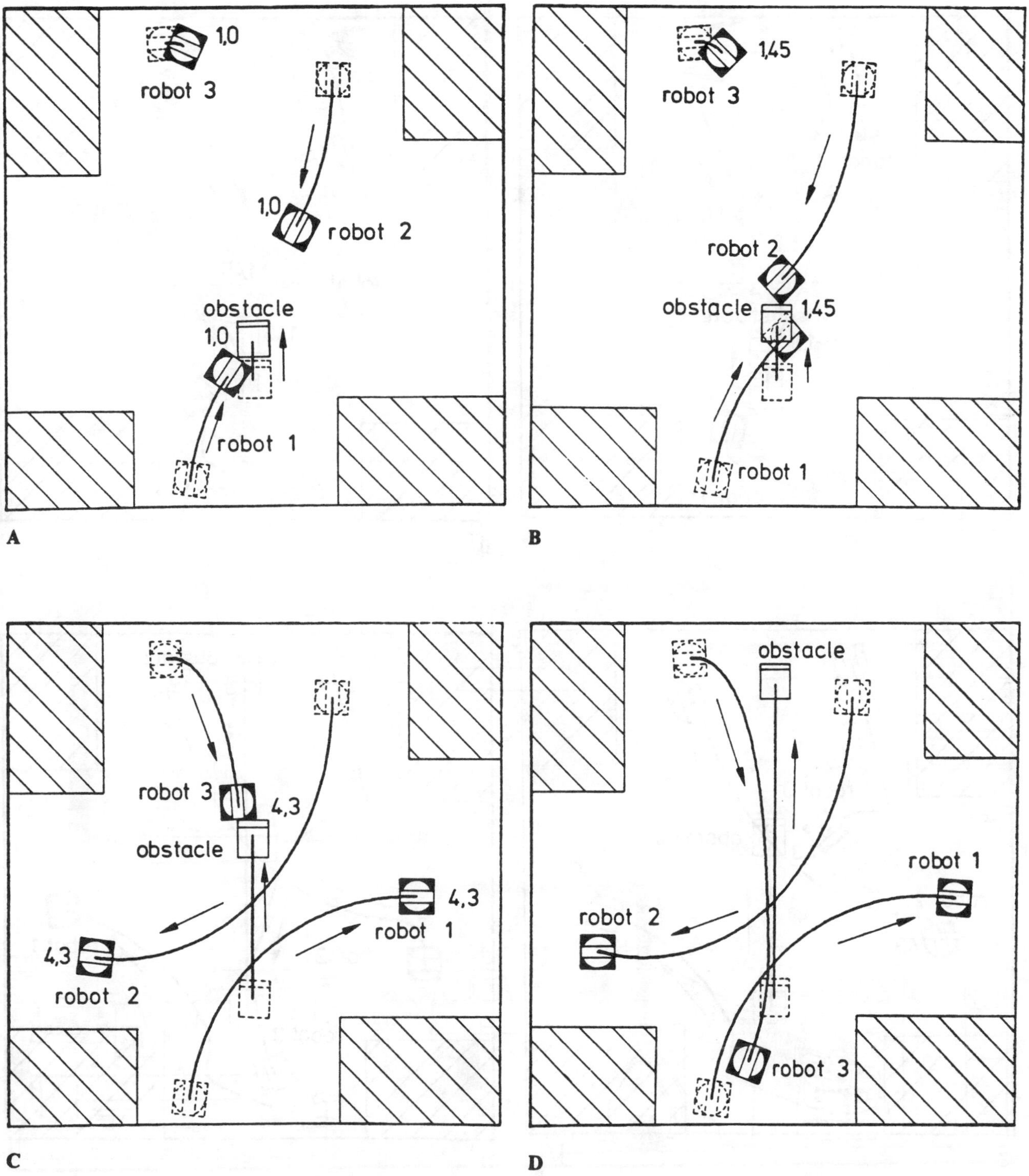

Fig. 23. Automatic pathfinding for three mobile robots and the moving obstacle.

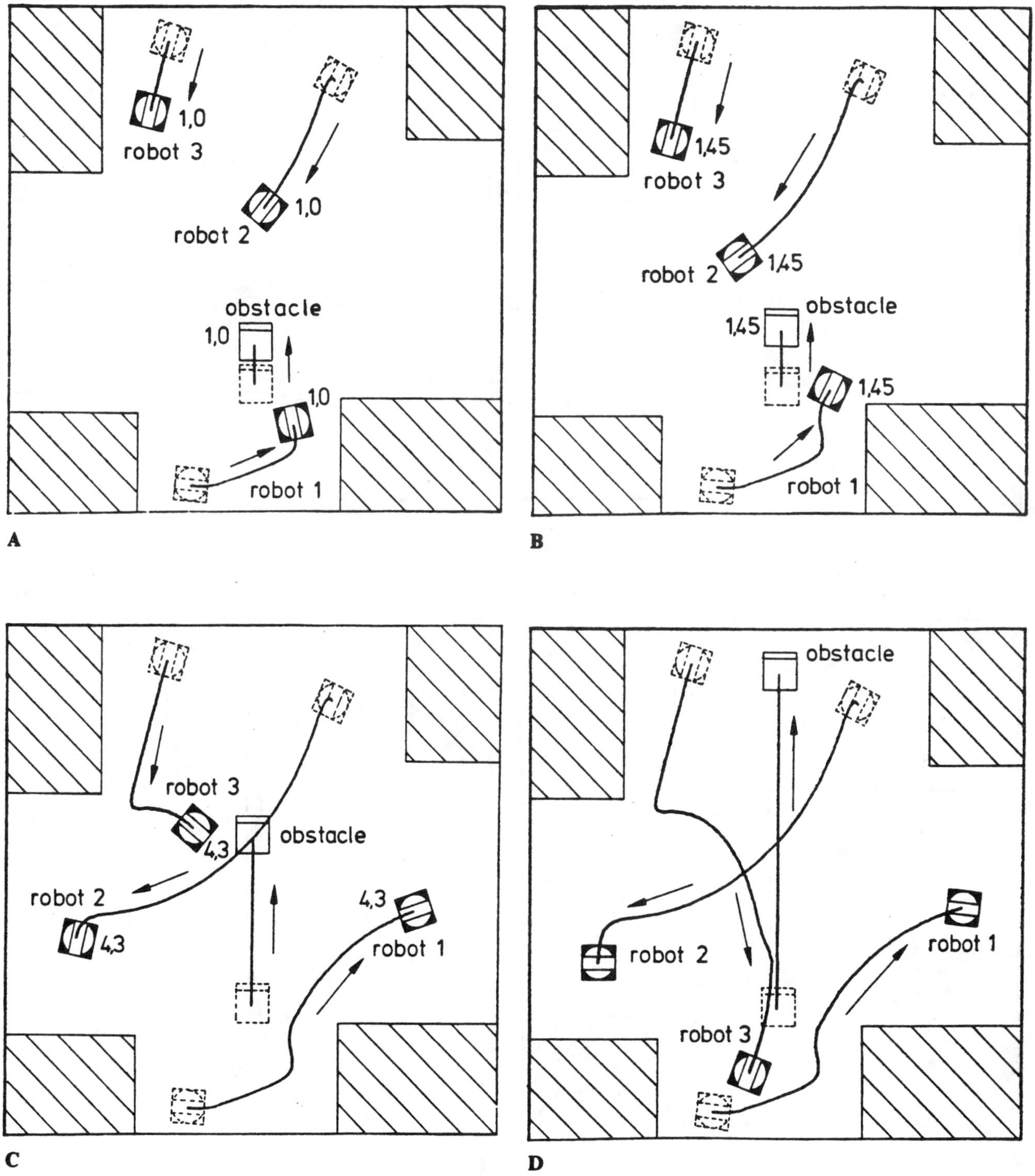

Fig. 24. Collision of a mobile robot with a moving obstacle of variable size.

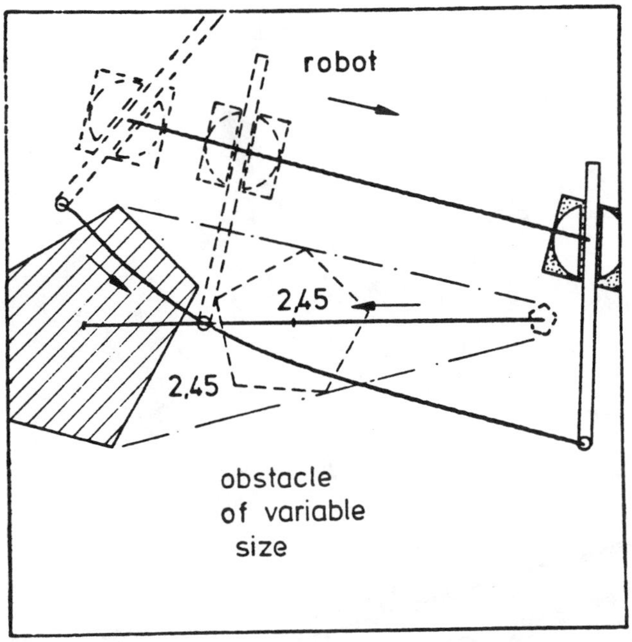

Fig. 25. Automatic pathfinding for a mobile robot and the moving obstacle of variable size.

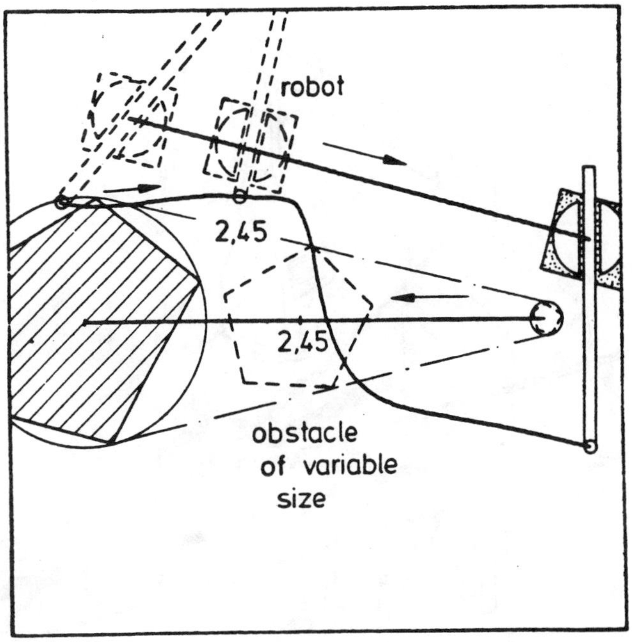

Case 4

a. *Moving* robot arms ← (modified by collision avoidance strategy)
b. *Variable* positions of the robots (mobile robots)
c. *Variable* positions of the obstacles (moving obstacles)
d. *Constant* size of the obstacles.

Case 1 is described by Figs. 10, 15, and 18 and shows three stationary robots with a common working space containing a stationary obstacle. To demonstrate the efficiency of the algorithms, the shape of the obstacle is varied while the constellation of the robots is kept the same. For the solution of this problem, the hierarchical coordinator (37) with (38)–(43) is used for each robot involved, where the algorithms remain completely unchanged for all applications. The obstacle is described by Eqs. (46)–(48), respectively.

In Fig. 10 the obstacle has the form of a polygon. The description of this obstacle can be realized in form of a circle with the constant radius r_U around the polygon as it is shown in Fig. 8. In Fig. 11 the desired paths of the three robots and the collisions with the obstacle are given. Figure 12 demonstrates the application of the findpath method via the strategy of the hierarchical coordinator. A comparison with Fig. 11 shows that all robots reach their final points.

The solution of this problem was realized by the IRF at the Hannover Fair in 1987 (Fig. 13). The structure of this system containing two robots with different kinematics (Adept-One and Volkswagen G8) is shown in Fig. 14. The visitors were allowed to put an obstacle at an arbitrary position in the common working space. The position and the size of the obstacle were recognized by an optical sensor system and the information was given to the hierarchical coordinator. Based on these sensor dates and the actual information about the positions and speeds of the robot, a danger of collision was detected and the collisions of both robots were avoided with the on-line strategies described in the paragraphs before. The hierarchical coordinator was realized on a Motorola 68020 microprocessor system.

Figures 15 and 18 show the arrangement of the three stationary robots and a stationary obstacle like in Fig. 10 where contrary to Fig. 10 the obstacle has the shape of a rectangle in Fig. 15 and a propellerlike form in Fig. 18. In these cases, a description of the obstacle by a circle reduces the allowed range of movement for the robots unnecessarily. Using now Eq. (48)

Fig. 26. Two mobile robots working on a conveyor belt containing an obstacle.

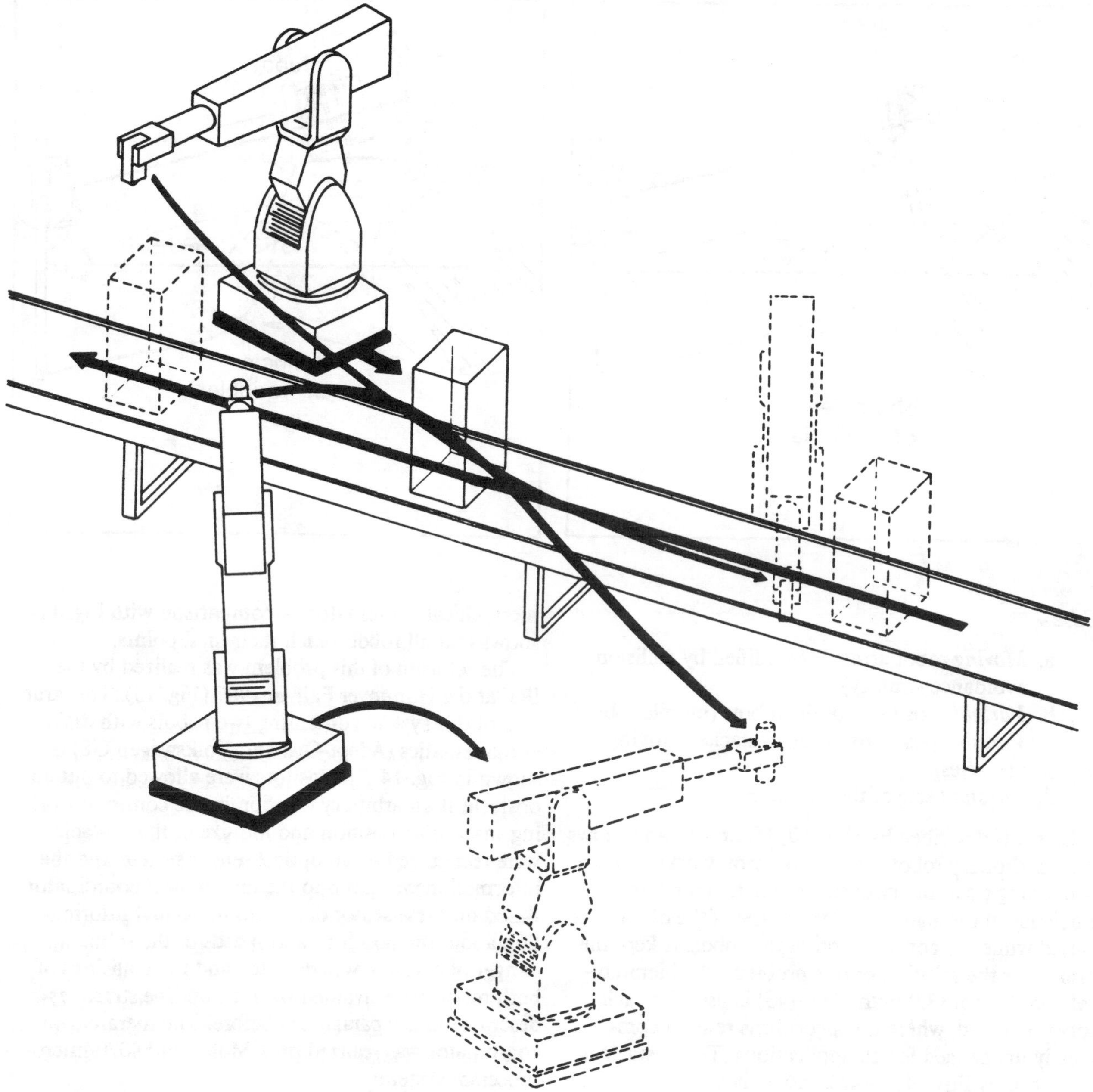

instead of the constant radius r_U, the obstacle can be modeled more effectively.

Figure 16 shows the desired paths and the collisions of the three robots with the rectangular obstacle for the case that no collision avoidance strategy is provided. The results of the automatic on-line pathfind method are demonstrated in Fig. 17, where no collision occurs. A comparison with Fig. 16 shows that even on collision-free paths all robots reach the desired target points. The same results are achieved for the

Fig. 27. Collision of the two mobile robots with the moving obstacle.

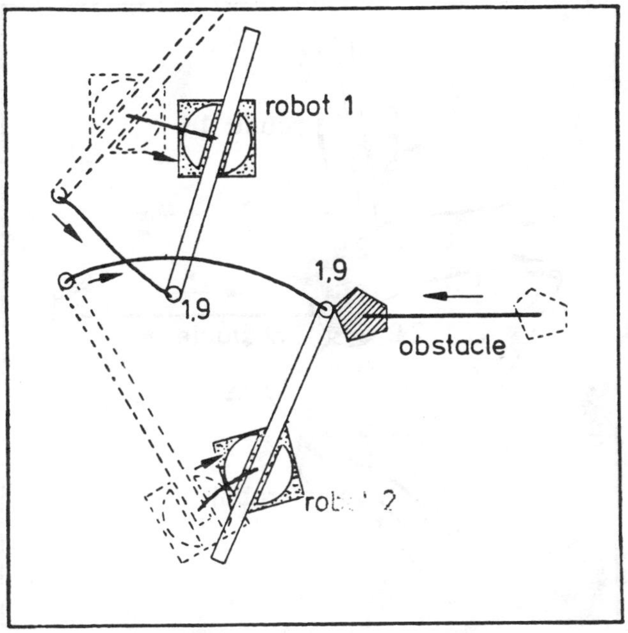

A

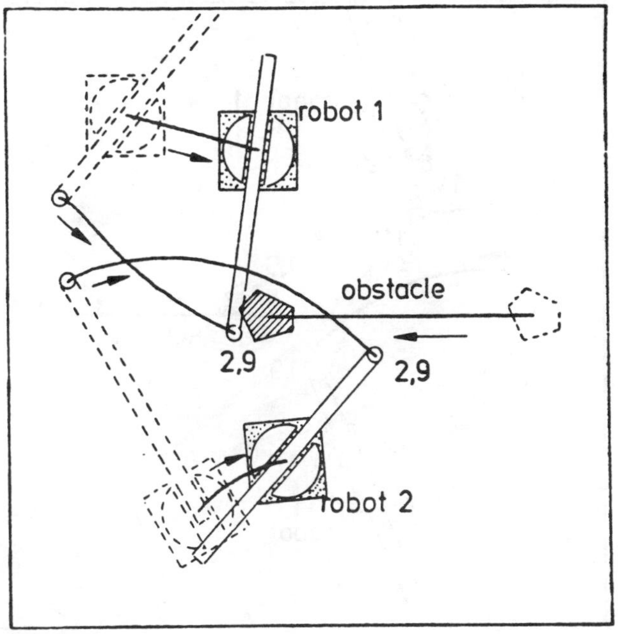

B

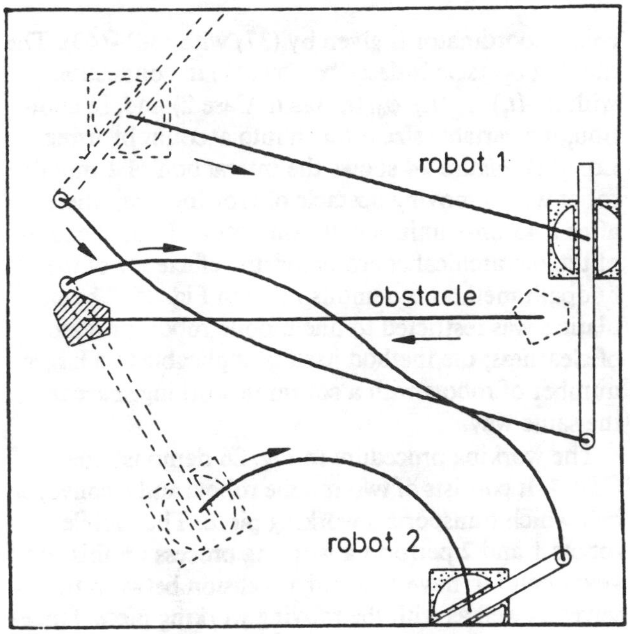

C

configuration of the three robots with the propellerlike obstacle (Fig. 18), which are demonstrated in Figs. 19 and 20, respectively.

The pathfinding in unstaffed factories (Fig. 21) is subject of Case 2 which is demonstrated by three mobile robots and a moving obstacle on colliding trajectories. For the solution of this case, the hierarchical coordinator (37) with (38)–(43) is applied for the robots, while the moving obstacle is described by (46), (47) with the variables $a_{Uj}(t_v)$, $\varphi_{0j}(t_v)$, and $\varphi_{0U}(t_v)$.

Figure 22D shows the desired paths of the robots and the obstacle. These paths, however, are not possible because robots 1, 2, and 3 collide with the moving obstacle after 1.0 time units (Fig. 22A), after 1.45 time units (Fig. 22B), and after 4.3 time units (Fig. 22C), respectively. Figures 23A–D prove that the pathfinding method applied avoids these collisions. A comparison with Figs. 22A–D demonstrates the usefulness of the automatic on-line findpath algorithms and shows that the robots reach their desired final points on the modified collision-free paths.

Case 3 demonstrates the efficiency of the collision avoidance trajectories for moving obstacles of variable size (contrary to the obstacle in Case 1 which is stationary and of constant size). In this case the hierar-

Fig. 28. Automatic pathfinding for the two mobile robots and the moving obstacle.

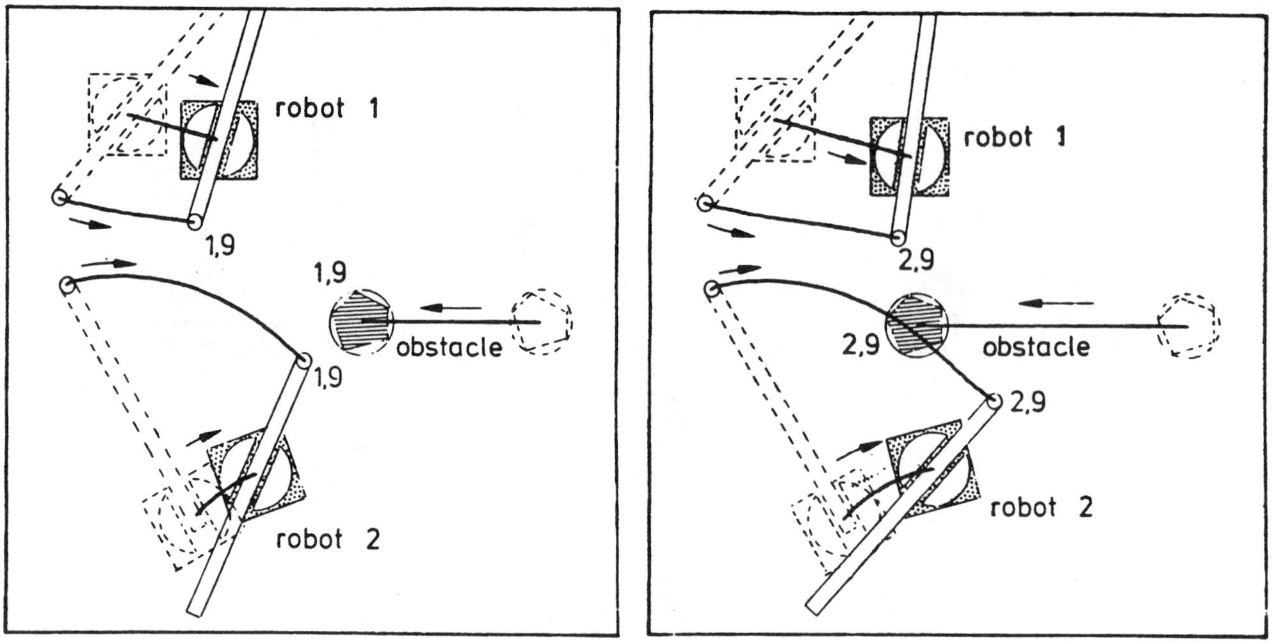

A

B

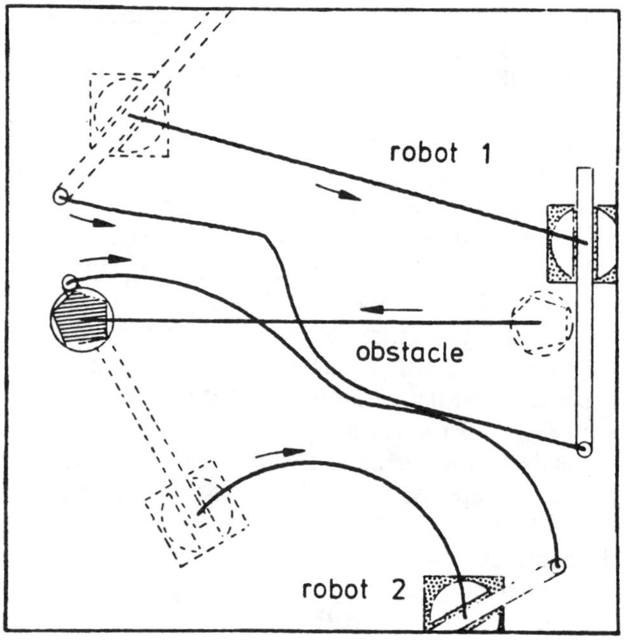

C

chical coordinator is given by (37) with (38)–(43). The moving obstacle is described by (47) in connection with $a_{Uj}(t_v)$, $\varphi_{0j}(t_v)$, $\varphi_{0U}(t_v)$ (as in Case 2) and, in addition, the variable size is taken into account by using Eq. (48). Figure 24 shows the interaction of a mobile robot with a moving obstacle of growing size, where after 2.45 time units a collision occurs. In application of the hierarchical coordinator, the efficiency of the findpath method is demonstrated in Fig. 25. The simulation was restricted to one mobile robot for reasons of clearness; the method itself is applicable to a larger number of robots with a common working space in the same way.

The working procedure in Fig. 26 demonstrates Case 4. It consists of two mobile robots and a conveyor belt which transports a working piece. The mobile robots 1 and 2 perform a working process on this conveyor belt but have to avoid a collision between themselves as well as with the moving working piece. Since this working piece does not belong to the working process of the robots in the example considered, it is treated as an obstacle. Therefore both robots have to modify the desired paths of their arms with respect to this moving obstacle. In this case, the hierarchical coordinator (37) in connection with (38)–(43) for the mobile robots and the description of the moving obstacle in (46), (47) with $a_{Uj}(t_v)$, $\varphi_{0j}(t_v)$, and $\varphi_{0U}(t_v)$ gives the solution of the findpath problem. Figure 27C shows the desired working procedure of Fig. 26 in principle, where the first collision occurs after 1.9 time units (Fig. 27A) and the second collision after 2.9 time units (Fig. 27B).

The results of the automatic on-line findpath method are demonstrated in Figs. 28A–C, where no collision occurs. Figures 28A and B show the situation in comparison to Fig. 27A after 1.9 time units and to Fig. 27B after 2.9 time units, respectively. The modified paths of the robot arms are given completely in Fig. 28C. A comparison to the desired paths in Fig. 27C shows that in spite of the collision avoidance both robots reach the desired target points.

The resulting algorithms for the simulations of Cases 1–4 were considered with respect to real-time processing on a single-microprocessor system TI 9995, too. It turned out that based on the hierarchical coordinator (37) the computation of the collision detection and avoidance strategies including the description of a stationary obstacle in (46)–(48) will require about 9 ms in Case 1 for each robot involved. In the Cases 2–4 the calculation of $a_{Uj}(t_v)$, $\varphi_{0j}(t_v)$, $\varphi_{0U}(t_v)$ has to be considered in addition to Case 1. In these cases the required computational time is about 10 ms for each robot involved. These computation times, however, can be further reduced by application of multiprocessor systems according to the subdivision of the structure of the hierarchical coordinator in (18) and (37). The estimated computational times show (as well as the results of the simulations) that the new approach on the solution of the findpath problem presented in the paper is well suited for on-line processing, where it is applicable to multirobot systems with stationary and mobile robots, stationary and moving obstacles as well as obstacles of variable size and different shapes.

Acknowledgment

This research is supported by a grant of the Minister für Wissenschaft und Forschung des Landes Nordrhein-Westfalen, West Germany.

References

Brady, M., et al. 1982. *Robot motion: planning and control.* Cambridge, Mass.: MIT Press.

Brooks, R. A. 1983. Planning collision free motions for pick and place operations. *Int. J. Robotics Res.* 2(4):19–44.

Brooks, R. A. 1985. Aspect of mobile robot visual map making. In *Robotics research: the Second International Symposium.* Cambridge, Mass.: MIT Press, pp. 369–375.

Dunne, M. 1979. An advanced assembly robot. In *Industrial robots*, Vol. 2. Society of Manufacturing Engineers, Michigan, pp. 249–262.

Erdmann, M., and Lozano-Pérez, T. 1986 (San Francisco, California). On multiple moving objects. *Proc. IEEE Conf. on Robotics & Automation,* pp. 1419–1424.

Freund, E. 1973. Decoupling and pole assignment in nonlinear systems. *Electron. Lett.* 9(16):373–374.

Freund, E. 1976. Verfahren und Anordnung zur Regelung von Manipulatoren und industriellen Robotern. Deutsches Patentamt, *Auslegeschrift 25 56 433.* Method and arrangement for the control of manipulators and industrial robots. U.S. Patent 4218172; Patent Japan Sho-60-10 876.

Freund, E. 1982. Fast nonlinear control with arbitrary poleplacement for industrial robots and manipulators. *Int. J. Robotics Res.* 1(1):65–78.

Freund E. 1983. (Atlanta, Georgia). On the design of multirobot systems. *Proc. IEEE Int. Conf. on Robotics,* pp. 477–490.

Freund, E. 1984. Hierarchical nonlinear control for robots. In *Robotics research: the First International Symposium.* Cambridge, Mass.: MIT Press, pp. 817–840.

Freund, E., and Hoyer, H. 1983 (Tokyo, Japan). Hierarchical control of guided collision avoidance for robots in automatic assembly. *Proc. Int. Conf. on Assembly Automation,* pp. 91–102.

Freund, E., and Hoyer, H. 1984. Collision avoidance for industrial robots with arbitrary motion. *Int. J. Robotic Syst.* 1(4):317–329.

Freund, E., and Hoyer, H. 1985. Collision avoidance in multi-robot systems. In *Robotics research: the Second International Symposium.* Cambridge, Mass.: MIT Press, pp. 135–146.

Freund, E., and Hoyer, H. 1986a. On the on-line solution of the findpath problem in multi-robot systems. In *Robotics research: the Third International Symposium.* Cambridge, Mass.: MIT Press, pp. 253–262.

Freund, E., and Hoyer, H. 1986b (San Francisco, California). Pathfinding in multi-robot systems: solution and applications. *Proc. IEEE Conf. on Robotics & Automation,* pp. 103–111.

Freund, E., and Syrbe, M. 1976. Control of industrial robots by means of microprocessors. *Lecture Notes in Control and Information Sciences.* New York: Springer-Verlag, pp. 167–185.

Hogan, N. 1984 (San Diego, California). Impedance control: an approach to manipulation. *Proc. American Control Conf.*

Hoyer, H. 1984. Verfahren zur automatischen Kollisionsvermeidung von Robotern im koordinierten Betrieb (Automatic collision avoidance for robots in coordinated operation). Ph.D. dissertation, FernUniversität Hagen, Electrical Engineering Department.

Hoyer, H. 1985 (Barcelona, Spain). On-line collision avoidance for industrial robots. *Proc. 1st IFAC Symp. Robot Control,* pp. 477–485.

Khatib, O. 1986. Real-time obstacle avoidance for manipulators and mobile robots. *Int. J. Robotics Res.* 5(1):90–98.

Krogh, B. 1984 (Bethlehem, Pennsylvania). A generalized potential field approach to obstacle avoidance control. *SME Conf. Proc. Robotics Research: The Next Five Years and Beyond.*

Lozano-Pérez, T., and Wesley, M. 1979. An algorithm for planning collision free paths among polyhedral obstacles. *Comm. ACM* 22(10):560–570.

Lozano-Pérez, T. (1986). Motion planning for simple robot manipulators. In *Robotics research: the Third International Symposium.* Cambridge, Mass.: MIT Press, pp. 133–140.

Tyridal, P. 1980 (Milan, Italy). New ideas in multi-task real-time control systems for industrial robots. *Proc. 10th Int. Symp. Industrial Robots,* pp. 659–670.

Dynamic Generation of Subgoals for Autonomous Mobile Robots Using Local Feedback Information

BRUCE H. KROGH, MEMBER, IEEE, AND DAI FENG, STUDENT MEMBER, IEEE

Reprinted from *IEEE Transactions on Automatic Control*, Vol. 34, No. 5, May 1989, pages 483-493. Copyright © 1989 by The Institute of Electrical and Electronics Engineers, Inc. All rights reserved.

Abstract—**An algorithm is presented for using local feedback information to generate subgoals for driving an autonomous mobile robot (AMR) along a collision-free trajectory to a goal. The *subgoals selection algorithm* (SSA) updates subgoal positions while the AMR is moving so that continuous motion is achieved without stopping to replan a path when new sensor data become available. Assuming a finite number of polygonal obstacles (i.e., the internal representation of the local environment is in terms of a 2-D map with linear obstacle boundaries) and a dynamic steering control algorithm (SCA) capable of driving the AMR to *safe* subgoals, it is shown that the feedback algorithm for subgoal selection will direct the AMR along a collision-free trajectory to the final goal in finite time. Properties of the algorithm are illustrated by simulation examples.**

I. INTRODUCTION

A MAJOR objective of current robotics research is to develop autonomous mobile robots (AMR's) which can navigate in unstructured environments [1]–[4]. In this paper we present a feedback algorithm for selecting subgoals to steer an AMR to a goal along a collision-free path. The subgoals are chosen while the robot is moving using feedback information about the locally visible environment. The objective of the proposed subgoal selection algorithm is to close to the higher level feedback loop in real time so that collisions with unanticipated obstacles can be avoided without stopping to replan the path to the goal.

There has been a considerable amount of research on path planning for AMR's. Many of the proposed algorithms require a map of the entire region to be navigated and give no consideration to kinematic and dynamic constraints [5]–[9]. Application of these algorithms for autonomous navigation requires a "stop-look-and move" approach to incorporating new sensor information [2]. In such a scheme, the computer uses the most recent map of the environment to plan a collision-free path to the goal while the AMR is stopped. The AMR moves a short distance along this path and then stops to update the map. While the AMR is stopped, the computer modifies the original path, taking into account new information from the sensors. The robot then moves a short distance along the updated path and the cycle is repeated until the robot reaches the goal. This is essentially an open-loop feedback approach to the problem of autonomous navigation.

The algorithm proposed in the present paper offers an alternative to stop-look-and-move schemes for autonomous navigation in unstructured environments. The subgoal selection algorithm (SSA) uses the most recent sensor data to generate subgoals while the AMR is moving. The current subgoal is pursued by a steering control algorithm (SCA) which takes into account the kinematic and dynamic operating constraints for the AMR. The SCA is designed to drive the AMR toward the current subgoal along a collision-free path in the local obstacle-free space. The subgoal is updated whenever the next subgoal generated by the SSA can be safely pursued by the SCA (the concept of safe subgoals is defined in Section II).

The proposed algorithm is in the spirit of so-called *reflexive* control schemes for autonomous navigation in that the subgoals and resulting trajectory are generated "on the fly" as the local environment becomes visible. Most of the research in reflexive control for AMR's has focused on hardware and software architectures for sensor fusion and intelligent navigation using *a priori* information [10]–[13]. Although the integration of *a priori* information and higher level planning into an overall navigation system is essential for practical applications, the purpose of the present paper is to focus on the navigation problem in domains where no *a priori* information is available and to demonstrate rigorously that a collision-free, finite-time trajectory to the goal can be generated using only local feedback information.

Among the many papers which have appeared in the area of path planning, our problem formulation and basic objective are most similar to the work of Lumelsky [14], [15]. Lumelsky considers the navigation problem for a *point mobile automation* using only *tactile* sensing, that is, the sensors only indicate whether the system (AMR) is in contact with an obstacle. The most significant ways in which our problem formulation differs from Lumelsky's work are as follows: 1) we assume that proximity sensors provide a map of the locally visible obstacles (within, perhaps, a finite range); 2) the system dynamics are explicitly taken into account by the SCA and the notion of safe subgoals; and 3) obstacles are modeled as convex polygons. The latter assumption makes our algorithm less general than Lumelsky's with respect to the model of the environment. However, our assumptions about the AMR sensing capabilities and dynamics lead to trajectories which are more acceptable for actual AMR's, such as the autonomous land vehicle at Carnegie-Mellon University [16].

The basic approach of the SSA is as follows. If an obstacle lies between the AMR and the final goal, subgoals are chosen near the obstacles vertices as they become visible, guiding the AMR around the obstacle until it does not block the line-of-sight to the final goal. Further subgoals are generated when there is another obstacle between the AMR and the goal, or when a previously generated subgoal is not visible due to another obstacle. As subgoals are generated they are stored until they can be safely pursued by the SCA. The algorithm also saves information about the extreme edges of obstacles which have been circumvented to avoid cycling indefinitely around the final goal.

Two aspects of our approach contribute to the complexity of the convergence proof for the SSA. First, we make no assumptions about the particular path followed by the AMR. Thus, no assumptions can be made concerning the vantage points from which the local obstacle maps are generated. Second, the subgoals generated by the SSA are not necessarily reached by the AMR because the subgoal being pursued can be updated while the AMR is moving. Thus, although the subgoals are associated with obstacle vertices, the convergence proof cannot rely exclusively upon the topological properties of the obstacle surfaces. With respect to the general problem of autonomous navigation, the

Manuscript received July 2, 1987; revised June 7, 1988. Paper recommended by Past Associate Editor, S. S. Sastry. This work was supported in part by the National Science Foundation under Grants ECS-8404607 and DMC-8451493, by the Ben Franklin Partnership Program, Commonwealth of Pennsylvania, and by the Robotics Institute, Carnegie-Mellon University.

The authors are with the Department of Electrical and Computer Engineering, Carnegie-Mellon University, Pittsburgh, PA 15213.

IEEE Log Number 8926481.

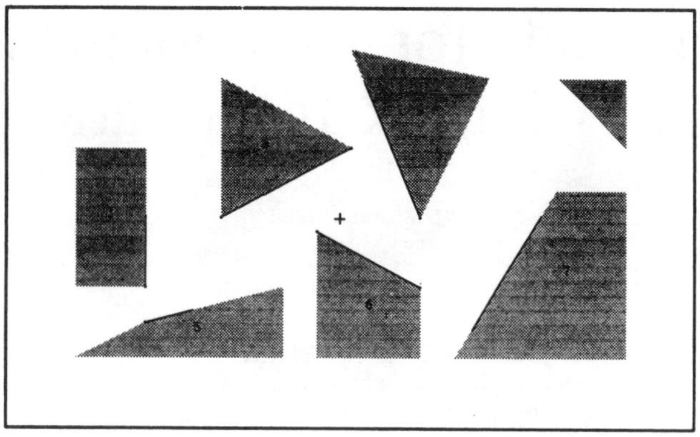

Fig. 1. Illustration of local obstacle map (bold lines) from the position $p_k(+)$.

present work addresses the problem of local guidance through a field of obstacles. Our objective is to explore the interface between navigation decisions and dynamic steering of an AMR. Consequently, we restrict our model of the environment to include only convex polygonal obstacles. Extending our algorithm to handle more complex environments is a topic for future research.

In the following section we formulate the subgoal selection problem and introduce terminology, notation, and assumptions used in the remainder of the paper. In Section III we present the SSA and describe each step in the algorithm. Section IV contains the analytical results which establish the convergence properties of the algorithm. The performance of the algorithm is demonstrated with simulations in Section V using a point mass model of the AMR and a previously proposed SCA based on artificial potential fields [17], [18]. In the concluding section we discuss directions for future research. As a convenience to the reader, the Appendix provides a summary of the notation, parameters, and functions defined in the text.

II. Problem Formulation

We model an AMR as a point on a plane which is to be steered along a continuous path from an initial position $\boldsymbol{p}_0$ to $\boldsymbol{p}_g$, the *final goal.* The space to be navigated contains a finite number of obstacles which are modeled as convex polygons in the plane. We assume that the minimum distance between points on any two obstacles is greater than a given parameter κ, which is greater than the physical width of the AMR. This assures that there exists a feasible path for the AMR from $\boldsymbol{p}_0$ to $\boldsymbol{p}_g$. In this paper we deal only with a point model of the AMR. The finite dimensions of the AMR are easily taken care of by appropriately "growing" the obstacles in the local maps to account for the AMR dimensions.

We assume that on-board sensors provide a local map of the environment which is *visible* from the AMR at discrete points in time. Letting $\boldsymbol{p}_k$ denote the AMR position at the kth iteration of the SSA, the faces of obstacles which can be detected from the position $\boldsymbol{p}_k$ are modeled by line segments in a two-dimensional *local map*. At the kth execution of the SSA, the local map consists of a set $\boldsymbol{O}(\boldsymbol{p}_k)$ of line segments representing obstacle faces which are visible from $\boldsymbol{p}_k$. A local map is illustrated in Fig. 1 where the bold lines indicate the faces of obstacles which are visible from the AMR position $\boldsymbol{p}_k$. We note that since the AMR is not necessarily stopped, the position vector $\boldsymbol{p}_k$ and the local map reflect the situation at the time the sensor data are taken. The processing of the sensor data will take a finite amount of time, which means the state of the AMR at the point when the SSA is executed is different from the time at which the local map was produced. As we shall show below, this time delay is accounted for by the evaluation of the safeness of subgoals.

In general, an AMR is a complex dynamic system. We denote its dynamic state including position at time t by $\boldsymbol{x}(t)$ and assume that there is a dynamic SCA which can drive the AMR to specified points in the plane provided certain conditions are satisfied. The minimum requirement is that SCA be capable of driving the AMR along a finite-time, collision-free trajectory from rest at an initial point to stop at any point visible from the initial point. If this is all the SCA is capable of doing, our subgoal algorithm will generate a sequence of points at which the AMR will stop en route to the goal.

Our objective, however, is to generate subgoals for more sophisticated SCA's which can dynamically steer the system to visible subgoals without necessarily starting at rest. The purpose of the SSA is to provide the SCA with a temporary direction to pursue when the final goal $\boldsymbol{p}_g$ is not visible. Continuous motion is achieved by reevaluating and updating the subgoal while the AMR is moving. Each subgoal is pursued by the SCA until a new subgoal is provided by the SSA.

One condition which must be satisfied before a new subgoal can be introduced is directly related to the capabilities of the SCA, as characterized by the following definition.

Definition: A candidate subgoal $\boldsymbol{g}^*$ is said to be *safe* at time t with respect to a local map $\boldsymbol{O}(\boldsymbol{p}_k)$ and the current dynamic state $\boldsymbol{x}(t)$ if the SCA can bring the AMR to rest at $\boldsymbol{g}^*$ along a collision-free trajectory in the obstacle-free space visible in $\boldsymbol{O}(\boldsymbol{p}_k)$.

The safeness of a subgoal is evaluated with respect to the AMR state at the current time t, which is different from the time at which the sensor data for local map $\boldsymbol{O}(\boldsymbol{p}_k)$ were taken. This reflects the fact that the time required to generate the local map and the subgoal is not negligible.

We assume there exists an algorithm for evaluating whether a candidate subgoal is safe. Any SCA can be used which satisfies these assumptions. For example, in Section V we use an SCA developed by Feng based on potential fields in the local obstacle-free space [18]. Other possible SCA's can be found in the literature on so-called *avoidance control* [19]–[22].

We can now state the problem solved by the subgoal selection algorithm as follows. Given an AMR modeled as a point in the plane and a finite number of convex obstacles, use the local map of visible obstacles at discrete points in time to generate safe subgoals to be pursued by an SCA such that AMR arrives at the final goal in finite time.

III. Subgoal Selection Algorithm (SSA)

In this section we describe and illustrate the steps in the SSA. The purpose of the subgoal is to provide the steering algorithm a temporary direction to pursue when the final goal is not in sight. The subgoal is chosen to direct the AMR towards the final goal while avoiding collisions with the obstacles. To achieve it, we determine the subgoals based on the data (local map) collected by the sensors on board the AMR. As new sensor data become available, the SSA determines when to update the subgoal.

To present the algorithm in a clear, concise manner, we use functions and data structures in a pseudoprogramming language. The following notation is used: small bold-faced letters represent points in the plane, large bold-faced letters represent other data structures, and small capitals are used for function names with brackets $\langle , \rangle$ delimiting function arguments. Before presenting the algorithm we define some of the variables, data structures, and functions.

A straight line (unbounded) containing two points $\boldsymbol{a}$ and $\boldsymbol{b}$ is denoted by $\boldsymbol{ab}$. The open or closed termination of a line segment is indicated by a parenthesis or bracket, respectively. For example, $(\boldsymbol{ab}]$ denotes the line segment from $\boldsymbol{a}$ to $\boldsymbol{b}$ which contains $\boldsymbol{b}$ but does not contain $\boldsymbol{a}$, and $[\boldsymbol{ab}$ denotes the ray beginning at and including $\boldsymbol{a}$ passing through $\boldsymbol{b}$. The function INT $\langle \boldsymbol{ab}, \boldsymbol{cd} \rangle$ equals the point at which the two lines $\boldsymbol{ab}$ and $\boldsymbol{cd}$ intersect. If the arguments of INT $\langle .,. \rangle$ do not intersect, the function returns the symbol NIL, which is used throughout the algorithm definition to indicate empty sets or vacuous conditions.

At the kth execution of the SSA, the local map consists of a set $O(p_k)$ of line segments representing obstacle faces which are visible from p_k. Connected lines in $O(p_k)$ are faces of the same obstacle. Given two points a and b, the function OBS $\langle a, b\rangle$ is defined as the line segment (obstacle face) in $O(a)$ which intersects the line segment (ab), along with all other faces in $O(a)$ connected to the intersected face. When point b is visible from point a, OBS $\langle A, B\rangle$ equals NIL.

Connected line segments in $O(p_k)$ are faces of a single obstacle, where the points of connection are vertices of the obstacle. Each set of connected line segments will have two *extreme points*; that is, two ends not connected to any other line segments in $O(p_k)$. We refer to these extreme points as *edges* and denote the set of edges at iteration k by $E(p_k)$. Note that if some faces of obstacle A are only partially visible from p_k, due to the presence of another obstacle, B, then at least one extreme point on obstacle A is not actually a vertex. To distinguish edges which are actually vertices of the obstacle models from extreme points arising from obstructions, we let $E'(p_k)$ denote the set of edges which are actually vertices on the visible obstacles in $O(p_k)$. Given two points a and b, if OBS $\langle a, b\rangle \neq$ NIL, then the two extreme points on the connected lines in OBS $\langle a, b\rangle$ correspond to two edges in $E(p_k)$, but not necessarily in $E'(p_k)$. From the perspective of p_k, we distinguish the two edges on OBS $\langle p_k, b\rangle$ as "right edge" and "left edge," with the obvious meaning. Note that a particular physical obstacle can lead to more than one set of connected faces in $O(p_k)$ when it is partially obstructed from view by other obstacles, but it can lead to no more than two visible edges in $E'(p_k)$.

Subgoals are always associated with obstacle edges or vertices in a local map. A subgoal g and its associated obstacle edge (or vertex) e make up the fundamental data structure referred to as a subgoal–edge pair, denoted by (g, e), the associated subgoal is defined as a point g a distance $\epsilon \geq 0$ from e. The precise locations of subgoals g with respect to associate edges e are given below in the descriptions of the functions NEXT_VERTEX, CHOOSE_EDGE, OTHER_EDGE, EXTEND_FACE, OTHER_VERTEX, and SUBGOAL. We choose $\epsilon \leq \kappa/2$ to assure that the subgoal for an edge is not on another obstacle (which must be at least a distance κ from e). The purpose of setting the subgoal a certain distance from the edge is to guide the AMR around a vertex to a vantage point from which it is guaranteed that further subgoals will be generated.

Subgoal–edge pairs generated in the SSA are stored on a dynamic stack S which is operated on using the standard functions POP $\langle S\rangle$ and PUSH $\langle (g, e), S\rangle$. In the remainder of the paper (g, e) denotes the subgoal–edge pair currently being pursued by the SCA, and (g^T, e^T) denotes the top of the stack S, that is, (g^T, e^T) = POP $\langle S\rangle$. The function HEIGHT $\langle S\rangle$ equals the number of subgoal–edge pairs in the stack S.

$C(a_1, \cdots, a_n)$ denotes the open convex hull of a set of n points in the plane $\{a_1, \cdots, a_n\}$. L is a particular subset of subgoals generated by the SSA. The meaning of the points in L is described below in the discussion of Step 4 of the SSA.

The SSA is given in Fig. 2. In the remainder of this section we discuss each step in turn, providing further definitions of variables, data structures, and functions as needed.

The algorithm is initialized in Step 1 when $k = 0$ and the AMR is at rest at p_0. If there are no obstacles between the AMR and the final goal p_g, the subgoal g is set equal to p_g and the SSA terminates. It is assumed that the SCA can drive the AMR from rest to any visible point. Thus, p_g is a safe subgoal at this stage and no other subgoals are needed. When p_g is not visible initially, it is pushed on the stack (with a default value of p_g for the edge), L and h (defined below) are initialized, and the algorithm is continued at Step 5. It will be shown in the following section that in this case more subgoals will be put on the stack in Step 6 and a safe subgoal–edge pair will be generated.

The SSA terminates at Step 2 when the final goal is already being pursued (no further subgoals are needed), or when the ray from the goal on the top of the stack g^T through p_k does not intersect the ray from the current edge e through the goal g. As illustrated in Fig. 3, this second condition assures that a new subgoal is sought only after the AMR has gone beyond the current edge e so that visibility of the subgoal g^T on the top of the stack is not blocked by the obstacle face that produced e.

```
step 1: If k=0 then
            If (OBS<p0,pg> = NIL) then
                (g,e) ← (pg,pg)
                exit
            else
                PUSH<(pg,pg),S>
                L ← NIL
                h ← p0
                (g', e') ← CHOOSE_EDGE<OBS<p0, pg>>
                goto step 5
step 2: If ( g=pg or INT<[gT pk, (eg> = NIL) then
            exit
step 3: SAVE<S,(g,e),[ab],L,h>
        If (HEIGHT<S> ≥ 3) then
            goto step 6
step 4: If (HEIGHT<S> = 1) then
            If (OBS<pk,pg> = NIL) then
                goto step 7
            If (O(pk) ∩ C(pk, g, e) ≠NIL) then
                goto step 8
            (g', e') ← NEXT_VERTEX < O(pk), (g, e)>
            If (INT < (ge'), (epg) > ≠ NIL or e'=e) then
                If (OBS<pk,pg> ∩ C(g,e,pg)=NIL) then
                    goto step 8
                else
                    (g', e') ← CHOOSE_EDGE<OBS<pk, pg>>
                    L ← L U {h}
                    h ← g
            If (INT<(hg'],(pg x>≠NIL and INT<(hg'],(pg x)>=NIL for some x∈L) then
                (g',e') ← OTHER_EDGE<O(pk),(g',e')>
        else
            If (O(pk) ∩ C(pk,gT,eT) ≠ NIL) then
                    goto step 6
            (g', e') ← EXTEND_FACE < O(pk), (gT, eT)>
            If (INT<(hg'],(pg x>≠NIL and INT<(hg'],(pg x)>=NIL for some x∈L) then
                (g',e') ← OTHER_VERTEX<O(pk),(g',e')>
            POP<S>
step 5: [ab] ← [pk e']
        (g, e) ← (pk, pk)
        PUSH < (g', e'), S >
step 6: while (O(pk) ∩ C(pk, gT, eT) ≠ NIL or OBS<pk,gT> ≠ NIL)
            (g',e') ← CREATEG<pk,(gT,eT),(g,e),[ab]>
            If ((g',e') = NIL) then
                goto step 8
            else
                PUSH<(g',e'),S>
step 7: (g,e) ← POP<S>
        If (SAFE<g,x(t), O(pk)> then
            exit
step 8: RESTORE<S,(g,e),[ab],L,h>
        end
```

Fig. 2. Subgoal selection algorithm (SSA).

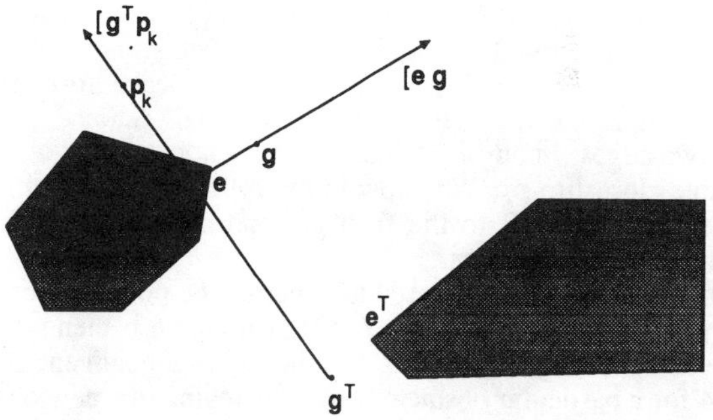

Fig. 3. Condition tested in Step 2: p_k must be past current edge e before the visibility of subgoal g^T is evaluated.

In Step 3 the function SAVE stores the current *context* which consists of the stack S, the current subgoal–edge pair (g, e), line segment $[ab]$, a set of previous subgoals L, and vector h ($[ab]$, L, and h are defined below). After the context is saved in Step 3, the algorithm jumps to Step 6 if there are three or more subgoal–edge pairs in the stack S. If it is determined in Steps 4–7 that a new subgoal should not be selected on the current iteration, the context stored in Step 3 is retrieved by the function RESTORE in Step 8.

Step 4 is executed for two mutually exclusive cases, namely,

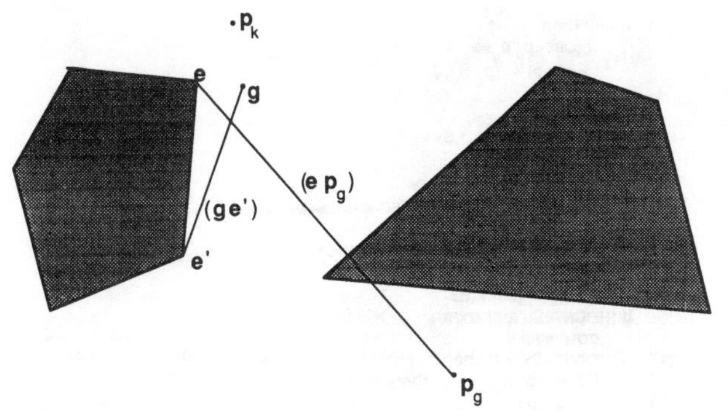

Fig. 4. Condition tested in Step 4: if vertex e' is not a face blocking the view of p_g, a new obstacle is used to generate a subgoal.

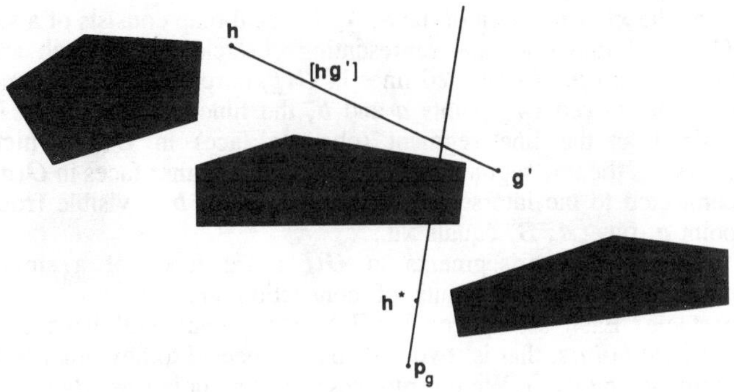

Fig. 5. Example of candidate subgoal g' taking the line $[hg']$ "behind" a previous subgoal $h*$ in L in Step 4.

when only the final goal p_g remains in the stack (HEIGHT $\langle S \rangle = 1$) or when the stack contains one subgoal in addition to the final goal (HEIGHT $\langle S \rangle = 2$). In the first case, if the final goal is visible from p_k, the algorithm jumps to Step 7 to evaluate whether the final goal is safe. If the final goal is not visible, the next condition in Step 4 assures that the line segment (e, g) is visible. If it is not visible, the SSA jumps to Step 8 to restore the original context. Otherwise, Step 4 continues and the function NEXT_VERTEX generates a subgoal–edge pair (g', e') where e' is the next visible vertex on the obstacle containing the current edge e in the direction of g. The associated subgoal g' is on the extension of the face containing e', a distance ϵ from e'. If the current edge e is an extreme point in $O(p_k)$, then NEXT_VERTEX returns $(g', e') = (g, e)$, the current subgoal–edge pair.

The next **if** condition in Step 4 determines whether the edge e' is on a face between the current edge and the goal. This condition is illustrated in Fig. 4. If INT $\langle (ge'), (ep_g) \rangle \neq$ NIL, or $e' = e$, the current edge is the last one that must be circumvented on that obstacle. In this case, before choosing an edge on another obstacle, a test is performed to see if the obstacle between p_k and p_g intersects the region $C(g, e, p_g)$. The purpose of this test is to assure that the next obstacle lies between the current subgoal and p_g. If OBS $\langle p_k, p_g \rangle \cap C(g, e, p_g) =$ NIL, the SSA jumps to Step 8 and terminates. If OBS $\langle p_k, p_g \rangle \cap C(g, e, p_g) \neq$ NIL, a subgoal–edge pair is generated by the function CHOOSE_EDGE from one of the two edges for OBS $\langle P_K, p_g \rangle$. Since only local information is being used in the SSA, there is no "optimal" choice between these two edges. In our implementation of CHOOSE_EDGE we select the edge closest to p_g. We show in the following section that the SSA steers the AMR to the final goal no matter which edge is selected by CHOOSE_EDGE.

When a new obstacle is used to generate the potential subgoal–edge pair, h is added to the set L. The variable h is then set equal to the current subgoal g. Thus, h is the final subgoal generated in Step 4 for a particular obstacle before switching to a new obstacle and L is the set of previous values of h. The set of subgoals L serves as a "memory" of where the AMR has been so that it does not cycle around the final goal indefinitely. This is assured by the next condition tested in Step 4 which, in words, checks to see if the subgoal g' will take the AMR "behind" a previous subgoal in L. This condition is illustrated in Fig. 5. If it occurs, the function OTHER_EDGE generates a subgoal–edge pair from the other edge on the obstacle containing e'. This concludes Step 4 when HEIGHT $\langle S \rangle = 1$.

In the second case of Step 4 (when HEIGHT $\langle S \rangle = 2$), if the line segment $(g^T e^T)$ is not completely visible from p_k [which is the case when there are obstacles in $C(p_k, g^T, e^T)$], the SSA jumps to Step 6. Otherwise, function EXTEND_FACE chooses e' as the most extreme visible point on the face collinear with $(g^T e^T)$ in the direction of g^T from e^T. The subgoal g' is generated as the extension of this face a distance ϵ from e'. Before going to Step 5 to push this new subgoal–edge pair on the stack, the same test is applied as described above for case one of Step 4 to make sure the subgoal generated by EXTEND_FACE is not taking the AMR behind a previous subgoal in L. If it is, the other vertex of the face is chosen by OTHER_VERTEX to generate the subgoal–edge pair (g', e'). This subgoal–edge pair will replace the subgoal–edge pair on top of the stack (g^T, e^T) which is "popped" and discarded.

```
CREATEG ⟨p_k, (g^T, e^T), (g, e), [ab], E'(p_k)⟩
    A ← E'(p_k) ∩ {C(e, g, e^T, g^T)U[g, g^T]}
    if (A = NIL) then
        (g', e') ← NIL
    else
        e' ← MIN_DIST ⟨A, [ab]⟩
        g' ← SUBGOAL ⟨e', [ab]⟩
    return (g', e')
```

Fig. 6. Function CREATEG.

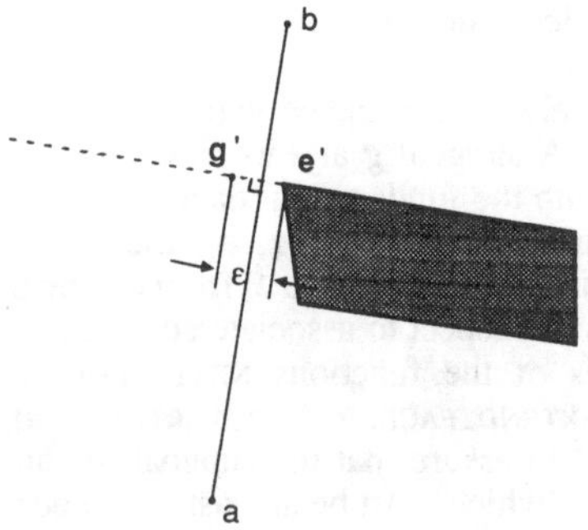

Fig. 7. Location of subgoal g' generated by function SUBGOAL for edge e'.

Step 5 updates the appropriate variables with the subgoal–edge pair (g', e') generated in Step 4. The line $[ab]$ is set equal to line segment $[p_k e']$, which is guaranteed not to intersect the interior of any obstacles. The current edge e is set equal to p_k for use in the function CREATEG in Step 6. The subgoal–edge pair (g', e') is then pushed on the stack. We shall see that the line segment $[ab]$ provides the "memory" required to assure that if any other subgoal–edge pairs are pushed on top of (g', e'), they will guide the AMR to a point from which the subgoal g' is visible.

In Step 6, the function CREATEG generates subgoal–edge pairs when the stack already contains a subgoal–edge pair generated previously in Step 4. The function CREATEG is defined in Fig. 6. An edge selected by CREATEG must satisfy conditions related to its proximity to the current edge e, the edge e^T on top of the stack, and the line $[ab]$. If no such edges exists, NIL is returned. The function CREATEG chooses the edge e' in $E'(p_k)$ (the set of visible edges) which is closest to the line $[ab]$ while being in the region $C(g, e, g^T, e)$. The significance of these conditions is explained in the following section where we prove the trajectory converges to the final goal in finite time.

As illustrated in Fig. 7, function SUBGOAL generates a subgoal g' a distance ϵ from e' on a line perpendicular to, and in the

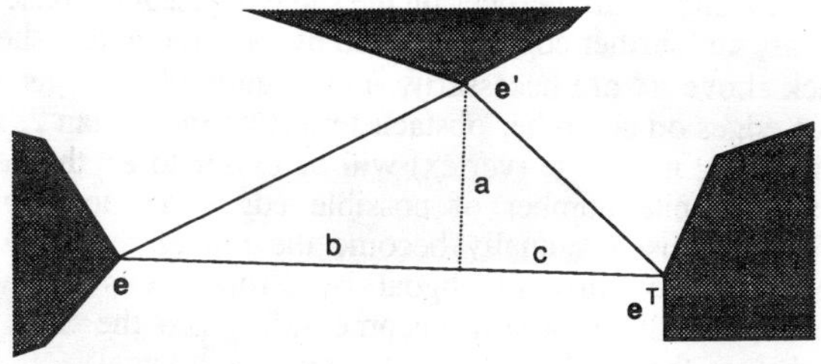

Fig. 8. Triangle with vertices e', e, e^T and line segments a, b, c, used in proof of Lemma 1.

direction of, the line segment $[ab]$. Subgoal–edge pairs are generated by CREATEG and pushed on the stack until the line segment $[g^Te^T]$ is visible. If CREATEG returns NIL, the algorithm jumps to Step 8 and the original context is restored. In Step 7 the top of the stack is popped and the (visible) subgoal is evaluated by the function SAFE which returns the logical condition TRUE if the SCA can drive the AMR to g along a collision-free path given the current system state $x(t)$ and the local map $O(p_k)$. If g is a safe subgoal, the algorithm terminates and this new subgoal is pursued by the SCA. If g is not safe, Step 8 is executed and the original context is restored.

IV. Proof of Convergence

In this section we prove that the SSA generates a sequence of subgoals which guide the AMR to the final goal p_g in finite time, that is, we show that p_g eventually becomes a safe subgoal for the SCA under the assumptions delineated in Section II. We prove this by establishing certain properties of the sequence of subgoals which are generated by consecutive iterations of the SSA. In particular, we show that the height of the stack S is finite bounded, that any goal generated by CREATEG eventually becomes a safe subgoal, and that the height of the stack is always reduced to one in a finite number of iterations. Finally, we demonstrate that the SSA can generate only a finite number of subgoal–edge pairs in Step 4 when the stack height is one before the final goal is visible and safe.

We begin with three lemmas on the properties of subgoal–edge pairs generated by the function CREATEG in Fig. 6. The first lemma establishes a bound on the distances between an edge generated by CREATEG and the two edges passed to the function as arguments. Lemmas 2 and 3 are existence results which state that under certain conditions a new subgoal–edge pair will always be generated by CREATEG.

Lemma 1: Suppose that at iteration k of the SSA, (g', e'), is a subgoal–edge pair generated in Step 6 by CREATEG for given values of p_k, (g^T, e^T), (g, e), $[ab]$, and $E'(p_k)$. If e', e, and e^T are on three different obstacles, then

$$\|e'-e\|^2 \le \|e-e^T\|^2 - 8\epsilon^2, \text{ and } \|e'-e^T\|^2 \le \|e-e^T\|^2 - 8\epsilon^2.$$

Proof: The first condition in the definition of CREATEG guarantees that the distance from e' to the line $[ee^T]$ is less than ϵ, since g and g^T are a distance ϵ from e and e^T, respectively. Also, the pairwise distances between e', e, and e^T are all more than $\kappa \ge 2\epsilon$ since they are on different obstacles. The geometry of this situation is illustrated by the triangle with vertices e', e, and e^T in Fig. 8. From the figure we have

$$b^2 = \|e'-e\|^2 - a^2 \ge 3\epsilon^2 \text{ and } c^2 = \|e'-e^T\|^2 - a^2 \ge 3\epsilon^2.$$

Since $\|e - e^T\|^2 = (b + c)^2$, we have from the above inequalities for b and c

$$\|e-e^T\|^2 \ge b^2 + c^2 + 6\epsilon^2. \tag{1}$$

Substituting the above equation for b^2 and inequality for c^2 in inequality (1) and noting $a \le \epsilon$, we have

$$\|e-e^T\|^2 \ge \|e'-e\|^2 + 8\epsilon^2.$$

Alternatively, substituting the above inequality for b^2 and equation for c^2 in inequality (1) gives

$$\|e-e^T\|^2 \ge \|e'-e^T\|^2 + 8\epsilon^2.$$

These last two inequalities prove the lemma. △

The bounds in Lemma 1 will be used to prove that only a finite number of edges can be generated in one iteration of the SSA, or added to the stack in multiple iterations of the SSA. In the following lemma we show that if the AMR has arrived at the current goal at some iteration m, that is, $p_m = g$, and CREATEG is executed, then it is guaranteed to generate a subgoal–edge pair.

Lemma 2: Suppose the current subgoal–edge pair (g, e) at iteration m of the SSA was generated by CREATEG at some previous iteration $k < m$. If $O(g) \cap C(g, g^T, e^T) \ne$ NIL, then

$$\text{CREATEG}\langle g, (g^T, e^T), (g, e), [ab], E'(g)\rangle \ne \text{NIL}.$$

Proof: It must be shown that under the hypotheses there is always an edge e' in $E'(g)$ which is in the region $C(e, g, e^T, g^T)$, since this will give $A \ne$ NIL in CREATEG. Since $O(g) \cap C(g, g^T, e^T) \ne$ NIL at least one obstacle intersects the region $R = C(e, g, e^T, g^T)$. Suppose there are no visible edges [elements of $E'(g)$] in R. This would imply that any line segment which intersects both $[eg]$ and $[e^Tg^T]$ necessarily intersects an obstacle in R. We now show that $[ab]$ interests both $[eg]$ and $[e^Tg^T]$, which is a contradiction since $[ab]$ does not intersect any obstacles except possibly at b.

Consider the first time CREATEG is executed in Step 6 following the definition of $[ab]$ in Step 5. In this case $a = p_k = g$ and $b = e^T$ and the assertion is trivially true, that is, $[ab]$ intersects $[eg]$ and $[e^Tg^T]$. The other case is that (e, g) and possibly (e^T, g^T) were generated by CREATEG. If, however, CREATEG generates the subgoal–edge pair (g', e') from two subgoal–edge pairs (e, g) and (e^T, g^T) and $[ab]$, where $[ab]$ intersects $[eg]$ and $[e^Tg^T]$, then $[ab]$ must intersect $[g'e']$. This follows from the fact that e' is within ϵ of $[ab]$ and $g' =$ SUBGOAL $\langle e', [ab]\rangle$, which puts g' a distance ϵ from e' on a line perpendicular to, and in the direction of, $[ab]$. Thus, it can be shown by induction that $[ab]$ intersects $[eg]$ and $[e^Tg]$ in all cases and the lemma is proved. △

Lemma 2 applies to the first time CREATEG is invoked in Step 6 if the AMR has arrived at the current goal. In the following lemma, we show that if CREATEG is executed multiple times in Step 6 when the AMR is at g, a new subgoal–edge pair is always generated.

Lemma 3: Suppose for two subgoal–edge pairs (g, e), (g^T, e^T) there exists a subgoal–edge pair (g_1, e_1) such that $(g_1, e_1) =$ CREATEG $\langle g, (g^T, e^T), (g, e), [ab], E'(g)\rangle$. If $O(g) \cap C(g, g_1, e_1) \ne$ NIL, then

$$\text{CREATEG}\langle g, (g^1, e^1), (g, e), [ab], E'(g)\rangle \ne \text{NIL}.$$

Proof: As shown in the proof of Lemma 2, for any subgoal–edge pair (g_1, e_1) generated by CREATEG in Step 6 we have INT $\langle [ab], [g_1e_1]\rangle \ne$ NIL. Therefore, since $O(g) \cap C(g, g_1, e_1) \ne$ NIL, there must be a visible edge in $C(e, g, e_1, g_1)$ by the proof of Lemma 2, *mutatis mutandis.* △

We now use these results to prove the following three propositions which characterize the sequences of subgoals generated by the SSA when there are more than two subgoal–edge pairs on the stack. First, we show in the following proposition that the SSA never gets "stuck" in an infinite loop.

Proposition 1: On any given iteration, the SSA always terminates in a finite number of steps.

Proof: Since Step 6 is the only point at which there exists the possibility of an infinite loop, it is sufficient to show that whenever Step 6 is executed, the function CREATEG eventually returns NIL or it returns a subgoal–edge pair (g', e') for which $O(p_k) \cap C(p_k, g', e') =$ NIL and OBS $\langle p_k, g^T\rangle =$ NIL. In fact, we shall show that on a given execution of Step 6, no more than $1 + \|e - e_0\|^2/8\epsilon^2$ subgoal–edge pairs can be pushed on the stack S, where e is the current edge and e_0 is the edge on top of S when Step 6 is initiated.

Let e_j be the edge returned on the jth call to CREATEG in a given execution of Step 6. From Lemma 1 we have for $j > 1$

$$\|e_j - e\|^2 \geq \|e - e_{j-1}\|^2 - 8\epsilon^2$$

where e_{j-1} is the edge on top of S when e_j is generated. Therefore, if e_j returned by CREATEG, it satisfies $\|e_j - e\|^2 \leq \|e - e_0\|^2 - (j-1)8\epsilon^2$ which implies $j \leq 1 + \|e - e_0\|^2/8\epsilon^2$, proving the lemma. △

The next proposition states that when there are more than two subgoal–edge pairs on the stack, the SSA always generates a new safe subgoal to pursue after a finite number of iterations. In other words, the current subgoal is eventually replaced with a new subgoal to pursue.

Proposition 2: If (g, e) is the current subgoal–edge pair and HEIGHT $\langle S\rangle \geq 3$ following iteration k_1 of the SSA, then for some $k > k_1$, the SSA will terminate with a new subgoal–edge pair (g^*, e^*) where g^* is safe with respect to the state x_k and local map $O(p_k)$.

Proof: If HEIGHT$\langle S\rangle \geq 3$, the SSA terminates either at Step 2, Step 7, or Step 8. If it terminates at Step 7, a new visible, safe subgoal has been found. It must be shown that on consecutive iterations the SSA cannot terminate indefinitely at Steps 2 or 8 while the SCA continues to pursue the same subgoal–edge pair (g, e). In the absence of a new subgoal from the SSA, we have assumed that the SCA will bring the AMR to rest at the current subgoal g in finite time. Moreover, we have assumed that when the AMR is stopped, the SCA can drive the AMR safely to any visible point. Thus, it suffices to show that if at some iteration $k > k_1$, the AMR is at rest with $p_k = g$, then the SSA cannot terminate at Step 2 to Step 8.

Under these assumptions neither condition in Step 2 can be satisfied. The first condition $g = p_g$ occurs only when p_g has become a visible, safe subgoal, in which case the stack is empty, but we have assumed HEIGHT $(S) \geq 3$. The second condition for terminating at Step 2 cannot be satisfied since $p_k = g$ implies INT $\langle [g^T p_k, (eg\rangle \neq$ NIL.

We now consider the ways in which the SSA can terminate at Step 8. First note that if HEIGHT $(S) \geq 3$, Step 4 is not executed, which means that the SSA can terminate at Step 8 only if CREATEG returns NIL in Step 6, or a visible subgoal generated by CREATEG in Step 6 is not safe. Since we have assumed that the AMR is at rest at $p_k = g$, any visible subgoal will be safe. Hence, it suffices to show that CREATEG will generate a visible subgoal in Step 6 in this case. From Lemmas 1 and 2 it follows by induction that CREATEG cannot return NIL when $p_k = g$. Moreover, by Proposition 1, CREATEG eventually returns a subgoal–edge pair (g^*, e^*) for which $O(g) \cap C(g, g^*, e^*) =$ NIL and OBS $\langle p_k, g^T\rangle =$ NIL. Since the subgoal g^* is visible, and hence safe, the proposition is proved. △

The implication of Proposition 2 is that no subgoal g generated by CREATEG will be persued indefinitely by the SCA since, in the worst case, the SCA eventually brings the AMR to rest at g from which it is guaranteed that a new safe subgoal will be generated. We now show in the next proposition that any subgoal generated by CREATEG and pushed on the stack will eventually become a visible, safe subgoal.

Proposition 3: If (g^*, e^*) is a subgoal–edge pair generated by CREATEG and pushed on the stack S during iteration k of the SSA, then at some iteration $m \geq k$, g^* will become the current subgoal.

Proof: Let e be the current edge at iteration k when (g^*, e^*) was pushed on the stack. Noting that all edges generated by CREATEG are obstacle vertices, we consider the following cases.

Case 1: e and e^* are vertices on the same physical obstacle O_A. In this case any further edges generated by CREATEG and pushed on the stack above g^* are necessarily on the same obstacle as e and e^* since edges on any other obstacles must be more than 2ϵ from O_A. Since any new edge (vertex) will be closer to e^* than e and there are a finite number of possible edges (vertices) on an obstacle, e^* must eventually become the top of the stack and visible from the current subgoal by Propositions 1 and 2. Moreover, it will eventually become safe since the SCA ultimately brings the AMR to rest at the current subgoal.

Case 2: e and e^* are on the different physical obstacles O_A and O_B. If any more subgoals are pushed on the stack they are either on O_A, O_B, or on yet another obstacle. If an edge e_B is generated on O_B and pushed on the stack, we have by the same argument as in Case 1 that there can be no edges between e^* and e_B on the stack which are not on O_B. Therefore, if e_B ever becomes the current edge, e^* will eventually become the current edge by Case 1. If, on the other hand, an edge e_A on O_A is generated and pushed on the stack, any edges pushed on top of e_A must also be on O_A, and by Case 1, e_A will eventually become the current edge. Moreover, only a finite number of edges on O_A can be pursued before the top of the stack is on another obstacle.

Finally, if an edge e' is not on O_A or O_B, then by Lemma 1 we have $\|e' - e\|^2 \leq \|e - e^*\|^2 - 8\epsilon^2$, that is, when the edges are on different obstacles the distance between the current edge and the new edge is always less than the distance between the current edge and the top of the stack by a finite amount. Therefore, only a finite number of different obstacles can lead to edges which are pushed on top of e^*. Since there are a finite number of obstacles, g^* eventually becomes visible and safe. △

In summary, the above propositions demonstrate that any subgoals generated by CREATEG which get put on the stack are eventually removed from the stack and used as the current subgoal for the SCA. This implies that if the height of the stack ever exceeds two, it will always return to that height. We now deal with the subgoals generated in Step 4 which is executed when only the stack has less than three subgoal–edge pairs.

We now consider the subgoal–edge pairs generated in Step 4 when HEIGHT $\langle S\rangle = 1$ or 2. Step 4 generates subgoal–edge pairs for visible obstacles which are between the AMR and the final goal. As described in the previous section, when only (p_g, p_g) is in the stack and there is an obstacle O between p_k and p_g, Step 4 is executed and a subgoal–edge pair is generated from O and pushed on the stack. In future iterations, whenever the stack height is 2, new subgoal–edge pairs are generated on the same face of the obstacle O until a real vertex is used as the edge. Subsequent vertices on the obstacle are used as edges until the final goal becomes visible and safe, or until subgoals can be generated for another obstacle which lies between the AMR and p_g. To analyze the sequence of subgoals generated in Step 4 we make the following definition.

Definition: A vertex v on an obstacle O is said to be an *extreme vertex* (with respect to p_g) if the ray $[p_g v$ does not intersect the interior of O and if a face F of O is contained in $[p_g v$, then v is the vertex of F which is the greatest distance from p_g.

Clearly every obstacle has exactly two extreme vertices. The objective of Step 4 is to generate a sequence of subgoal–edge pairs on an obstacle until one of the extreme vertices is used as an edge. At that point either p_g is visible or another obstacle is used to generate subgoal–edge pairs in Step 4. A subgoal for an extreme vertex on an obstacle, referred to as an *extreme subgoal,* is saved as the variable h in Step 4, and then placed in the set L when an extreme vertex of the next obstacle has been reached.

When a candidate subgoal g' is generated in Step 4, it is tested to assure the line from h (the last extreme subgoal) to g' does not go behind one of the previous extreme subgoals in L, as described in the previous section. The following lemma implies that if the line segment $[hg']$ does go behind a subgoal in L, then the sequence of subgoals which will be generated in the other direction on the obstacle (starting with the subgoal generated by

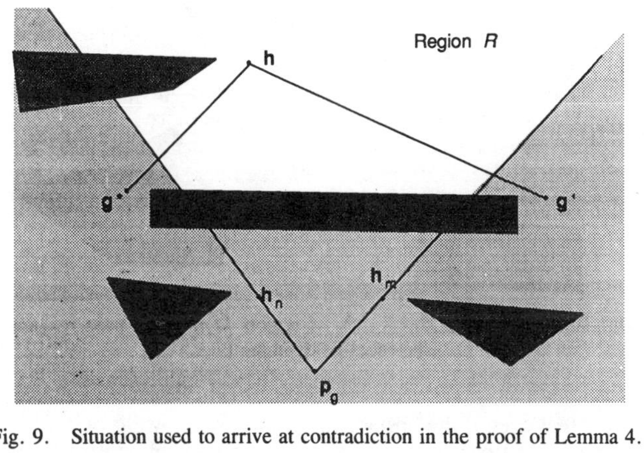

Fig. 9. Situation used to arrive at contradiction in the proof of Lemma 4.

OTHER_EDGE or OTHER_VERTEX) will not lead the AMR behind any other subgoals in L.

Lemma 4: Suppose Step 4 is executed during an iteration of the SSA and (g', e') is a subgoal–edge pair generated for an obstacle O by the function NEXT_VERTEX, CHOOSE_EDGE, or EXTEND_FACE. Let g^* be the extreme vertex of O in the opposite direction of e' as g'. If the line segment $[hg']$ goes behind one of the subgoals in L, then the line segment $[hg^*]$ will not go behind any subgoals in L.

Proof: Let $L = \{h_1, \cdots, h_j\}$, where j is the number of extreme subgoals in L and the index indicates the order in which the extreme subgoals have been added to L. For notational convenience we assign the index $j + 1$ to the most recent extreme subgoal h, that is, $h_{j+1} = h$. Note that for any $h_k \in L$ the line segment $[h_k h_{k-1}]$ cannot go behind any of the subgoals $h_1, \cdots, h_{k-2}$.

We now prove the lemma by contradiction. Suppose at some iteration of the SSA a candidate subgoal g' is generated in Step 4 for which the line segment $[hg']$ goes behind $h_m \in L$ and the line segment $[hg^*]$ goes behind $h_n \in L$, where g^* is the extreme subgoal on the obstacle as defined in the statement of the lemma. This situation is illustrated in Fig. 9. As illustrated in the figure, the most recent extreme subgoal h_{j+1} is the region R behind the obstacle O (with respect to p_g) and between the rays $[p_g h_m$ and $[p_g h_n$. Without loss of generality we assume $m < n$. Since $h_{j+1} \in R$ and $h_n \notin R$, there is an index i where $n < i \leq j + 1$ such that $\{h_i, \cdots, h_{j+1}\} \subset R$ and $h_{i-1} \notin R$. By the property of the extreme subgoals in L, the line segment $[h_{i-1}h_i]$ cannot pass behind h_m or h_n. Therefore, the line segment $[h_{j-1}h_i]$ must go through the obstacle O. However, since h_i must be an extreme subgoal generated in Step 4 for an obstacle encountered while pursuing the subgoal h_{i+1}, this situation contradicts the assumption that the obstacles are convex, which proves the lemma. △

The implication of Lemma 4 is that the direction in which consecutive vertices of a given obstacle are selected as edges in Step 4 can change at most once. Thus, an extreme vertex will always be reached in a finite number of steps. This result is stated as the following proposition.

Proposition 5: If at some iteration k_0 of the SSA, h is the most recent extreme subgoal saved in Step 4 and the subgoal–edge pair (g_0, e_0) is generated from obstacle O in Step 4 and pushed on the stack in Step 5, then at some iteration $k^* \geq k_0$, the SSA will terminate with $(g, e) = (g^*, e^*)$ where g^* is a visible, safe subgoal, and with $g^* = p_g$, the final goal, or e^* is an extreme edge of the obstacle O.

Proof: First note that whenever Step 5 is executed, (p_g, p_g) is the only subgoal–edge pair in the stack. Thus, (g_0, e_0) is the second entry in the stack. Also, the line segment $[gh_0]$ does not go behind any of the extreme subgoals in the set L. This follows from the condition tested in Step 4 for subgoals generated by either NEXT_VERTEX, CHOOSE_EDGE, or EXTEND_FACE. If the subgoal initially generated by one of these functions went behind a subgoal in L, the function OTHER_EDGE or OTHER_VERTEX would generate the subgoal g_0 which, by Lemma 4, cannot take the line segment $[hg_0]$ behind a subgoal in L.

Any subgoals pushed on top of g_0 in the stack are necessarily generated by CREATEG in Step 6. By Proposition 3, any such subgoals eventually become visible and safe. Therefore, in a finite number of iterations the SSA terminates with HEIGHT $\langle S \rangle = 2$ and (g_0, e_0) on top of the stack. On the next iteration, if the entire line segment $(g_0 e_0)$ is not visible, Step 6 is executed and more subgoal–edge pairs are pushed on the stack. By the same argument as used in the proof of Proposition 3, this can occur for only a finite number of iterations before (g_0, e_0) is the top of the stack and the entire line segment $(g_0 e_0)$ is visible. Suppose this occurs at iteration $k_1 \geq k_0$. At this point Step 4 is executed and (g_0, e_0) is replaced by a new subgoal–edge pair (g_1, e_1) where e_1 is one of the two visible extreme points of the face containing e_0 (generated by either EXTEND_FACE or OTHER_VERTEX). Again, by Lemma 4, the line segment $[hg_1]$ does not go behind any subgoal in L.

Applying the above argument again, (g_1, e_1) becomes the top of the stack in a finite number of iterations and is replaced by a new subgoal–edge pair (g_2, e_2) with e_2 on the same obstacle face as e_0. This cycle continues until at some iteration k_{n1} a subgoal–edge pair (g_{n1}, e_{n1}) is generated in Step 4 where e_{n1} is an actual vertex of the obstacle face containing e_0 and the line segment $[hg_{n1}]$ does not go behind any subgoals in L. This follows from Lemma 4 since at least one of the vertices of the obstacle face cannot take the line from h behind any subgoals in L. By the same arguments as in the proof of Proposition 3, g_{n1} eventually becomes a visible, safe subgoal.

If e_{n1} is an extreme edge of O, the theorem is proved. Otherwise, in the next iteration Step 4 is executed with HEIGHT $\langle S \rangle = 1$ at which point there are several possibilities. If the final goal is visible and safe, the SSA returns $g = g_g$ and the theorem is proved. If p_g is visible but not a safe subgoal, a new subgoal is not generated and the former context is restored. If p_g remains visible on future iterations, then it eventually becomes a safe subgoal since the SCA will stop the AMR at the current subgoals, again proving the proposition.

If p_g is not visible when Step 4 is executed with HEIGHT $\langle S \rangle = 1$, the SSA first checks to see if the region $C(p_k, g, e)$ is free of obstacles. If it is not, the SSA terminates and the former context is restored. This condition must eventually be satisfied, however, since the SCA drive the AMR to g. Therefore, if p_g does not become a safe subgoal while pursuing g_{n1}, the function NEXT_VERTEX is executed in Step 4 after a finite number of iterations and a subgoal–edge pair (g', e') is generated. Since we have assumed that e_{n1} is not an extreme edge, the edge e' is either the same as e_{n1}, or e' is on a different face on O. In the former case, p_k is not a position from which the next face on O is visible and the SSA terminates after restoring the context. This can only happen a finite number of times before the SCA drives the AMR to a position from which the next face is visible. In the latter case, e' is retained if $[hg']$ does not go behind a subgoal in L, otherwise an alternative subgoal–edge pair is generated in the other direction by OTHER_EDGE which is guaranteed not to go behind a subgoal in L by Lemma 4. The subgoal–edge pair generated in Step 4 is pushed on the stack in Step 5.

By the discussion above, after a finite number of iterations there is some iteration $k_{n2} \geq k_{n1}$ such that a subgoal–edge pair (g_{n2}, e_{n2}) is generated in Step 4 with e_{n2} the next vertex on the same obstacle face as e_{n1}. This cycle will continue, generating a sequence of edges $e_{n1}, e_{n2}, \cdots$ which are consecutive vertices of obstacle O. By Lemma 4 and the assumption that O has a finite number of vertices, this sequence must eventually terminate with a subgoal–edge pair (g^*, e^*) where e^* is an extreme edge of O. Moreover, there will be an iteration k^* for which g^* is returned by the SSA as a visible, safe subgoal and HEIGHT $\langle S \rangle = 1$. △

The previous proposition implies that the height of the stack will always be reduced to one in a finite number of iterations. The following final lemma states that the SSA returns a valid subgoal on the first iteration when $k = 0$.

Lemma 5: When $k = 0$ with the AMR stopped at p_0, the SSA terminates with a safe, visible subgoal.

Proof: If p_g is visible from p_0, the subgoal $g = p_g$ returned by the SSA in Step 1 is safe by the assumption that the SCA can steer the AMR to any visible point when it is initially at rest. If p_g is not visible, (p_g, p_g) is pushed on the stack and a subgoal–edge pair (g', e') is generated by CHOOSE_EDGE from the obstacle between p_0 and p_g. The SSA skips to Step 5 where the line segment $[ab]$ is set equal to $[p_0e']$, the current subgoal–edge pair (g, e) is set equal to (p_0, p_0), and (g', e') is pushed on the stack (which now has height 2). Step 6 is then executed. By Lemmas 2 and 3, CREATEG will not return NIL because $p_0 = g$. Moreover, by Proposition 1, Step 6 will terminate in a finite number of steps at which point a visible subgoal will be on top of the stack and Step 7 will be executed. The visible subgoal on top of the stack is necessarily safe since the AMR is at rest, which proves the lemma. △

We now use the previous results to prove the fundamental property of the SSA which is stated as the following theorem.

Theorem: Given an AMR modeled as a point in two-dimensions, initially at rest at position p_0, a finite number of polygonal convex obstacles with a finite number of vertices separated by a distance of at least 2ϵ, and an SCA (steering control algorithm) capable of driving the AMR from rest to stop at any point visible from the initial point. If the final goal position p_g is not contained in an obstacle, then the SSA (subgoal selection algorithm) will always return a visible, safe subgoal and will return p_g as a visible safe subgoal after a finite number of iterations.

Proof: By Lemma 5 the SSA returns a visible, safe subgoal when $k = 0$. If $g = p_g$, then the theorem is proved. If $g \neq p_g$, the stack will contain at least one subgoal–edge pair, namely, (p_g, p_g). By Proposition 1 the SSA always terminates in subsequent iterations with a visible, safe subgoal. If after any iteration there is more than one subgoal–edge pair in the stack, Proposition 5 states that after a finite number of iterations either p_g will be returned as a visible, safe subgoal, or HEIGHT $\langle S \rangle = 1$ and the visible, safe subgoal will be an extreme subgoal on an obstacle. If the former case occurs, the theorem is proved. The latter case can occur only a finite number of times since there are a finite number of extreme subgoals and, by Lemma 4, no extreme subgoal can become the current subgoal when HEIGHT $\langle S \rangle = 1$ more than once. Therefore, after a finite number of iterations, an extreme subgoal will become the current subgoal g with HEIGHT $\langle S \rangle = 1$ and, since the SCA will eventually bring the AMR to rest at g, p_g will become a visible, safe subgoal. △

V. Simulation Examples

To illustrate the properties of the SSA we have implemented a simulation program using a simple point mass model of an AMR in a two-dimensional planar region containing polygonal obstacles. The dynamics of the point mass are given by

$$\ddot{p} = \dot{v} = u \tag{2}$$

where $p(t)$ and $v(t)$ are the AMR position and velocity, respectively, and $u(t)$, the acceleration vector, is the control input for the dynamic steering control algorithm (SCA). We assume the magnitude of the acceleration is constrained as

$$\|u(t)\| \leq \alpha. \tag{3}$$

The control problem is to transfer the point mass with dynamics (2) along a collision-free trajectory in the plane from an initial position $p(0) = p_0$ to a final goal position $p(t_f) = p_g$ with $v(0) = v(t_f) = O$, subject to the control constraint (3). The final time t_f is unspecified.

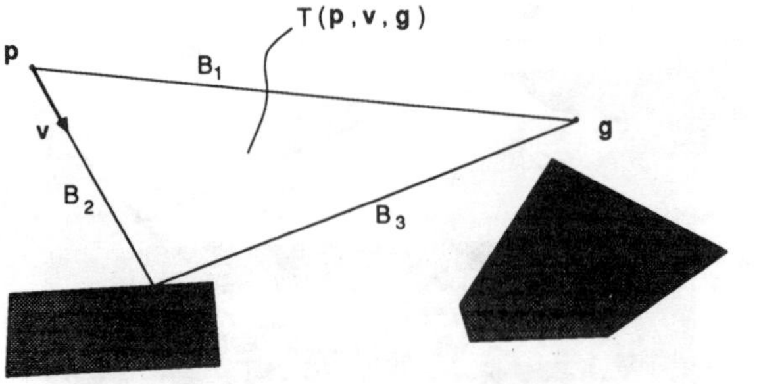

Fig. 10. Boundaries B_1, B_2, B_3 of region $T(p, v, g)$ used to generate potential fields in the PFA.

Note that even if the obstacle locations are known *a priori*, this problem would be difficult to solve using optimal control theory because the obstacles impose complex state-variable constraints. Obtaining a feedback solution using only the information about obstacles visible from $p(t)$ would be even more difficult. Our approach is to use the SSA to generate subgoals for an SCA which steers the point mass with the local obstacle-free space. The particular SCA used in the simulation examples is a dynamic feedback scheme based on artificial potential fields which "attract" the point mass to the subgoal while "repelling" it from a set of boundaries defining an obstacle-free region in the most recent local map. Since our purpose here is to illustrate the properties of the SSA, we only briefly describe the basic elements of the SCA which we shall refer to as the potential field algorithm (PFA). General properties of the potential field approach are discussed by Krogh in [17], and details of the PFA used in the simulation examples are given in Feng's Master's Thesis [18]. The objective of the PFA to choose the acceleration at each instant based on the available local information so as to guarantee the resulting trajectory is acceptable (but not necessarily "optimal").

For a given position p, velocity v, and subgoal g, the PFA computes the acceleration vector u to maximize the projection of u onto the gradient of a potential function $P(p, v, g)$ given by

$$P(p, v, g) = P_a(p, v, g) + P_r(p, v, g) \tag{4}$$

where $P_a(p, v, g)$ is computed to attract the point mass to the subgoal g and $P_r(p, v, g)$ is computed to repulse the point mass from the obstacles. The dependence of P_a and P_r on the system velocity as well as position takes into account the dynamic aspects of the trajectory and control constraints.

$P_r(p, v, g)$ is the sum of a set of repulsive potential functions $P_r(p, v, B_i)$ for three linear boundaries B_1, B_2, B_3 which define a triangular obstacle-free region $T(p, v, g)$ in the local map $O(p)$. As illustrated in Fig. 10, $B_1 = [pg]$, B_2 extends from p in the direction of v, and B_3 extends from g to $B2$ so as to maximize the area of $T(p, b, g)$ while keeping the angle at g less than or equal to $\pi/2$. For this SCA we define a subgoal to be *safe* when there is sufficient deceleration capability in the direction of v to bring the point mass to a stop before crossing boundary B_3.

The repulsive potential field $P_r(p, v, B_i)$ is defined as $(\tau_M(B_i) - \tau_m(B_i))^{-1}$, where $\tau_m(B_i)$ is the minimum time in which the velocity toward boundary B_i can be brought to zero using maximum deceleration, and $\tau_M(B_i)$ is the maximum time in which the velocity toward the obstacle can be brought to zero under constant deceleration without crossing boundary B_i. This gives

$$P_r(p, v, B_i) = v_i(v_i^2 - 2\alpha p_i)^{-1}$$

where $-v_i$ is the speed at which the AMR is moving toward B_i and p_i is the distance to B_i. The essential property of $P_r(p, v, B_i)$ is that it grows to infinity as the capability to avoid crossing the boundary goes to zero.

The goal attraction potential function $P_a(p, v, g)$ is computed so that the gradient results in an acceleration which will eventually

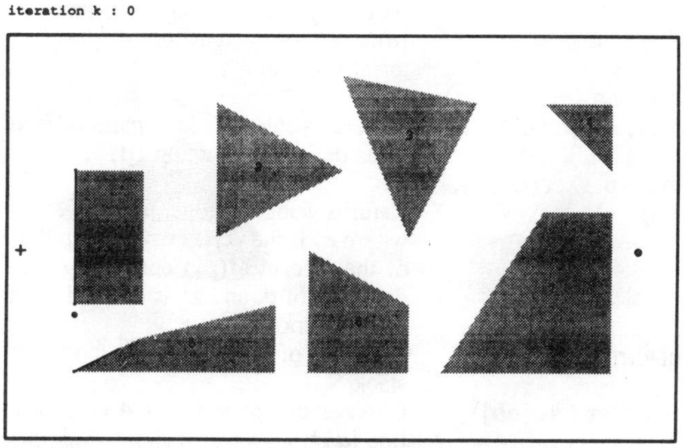

Fig. 11. Initial position (+) and local map for simulation example 1.

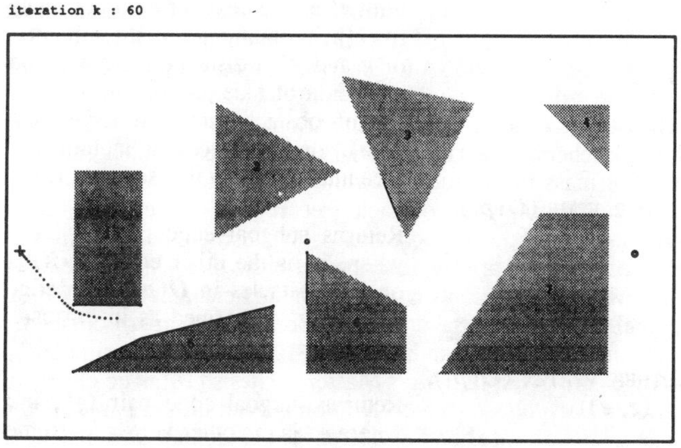

Fig. 12. Trajectory after $k = 60$ iterations for simulation example 1.

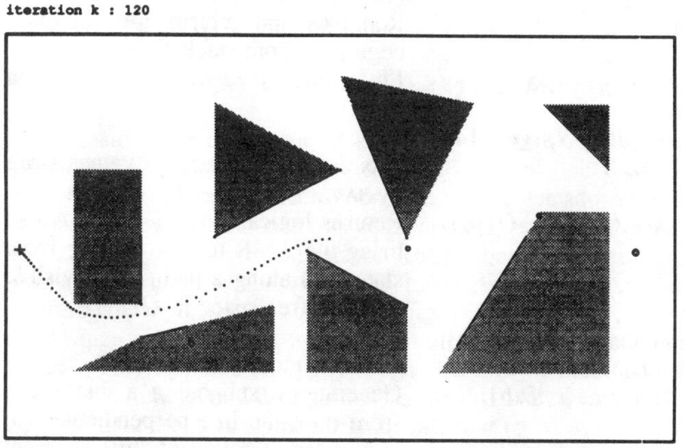

Fig. 13. Trajectory after $k = 120$ iterations for simulation example 1.

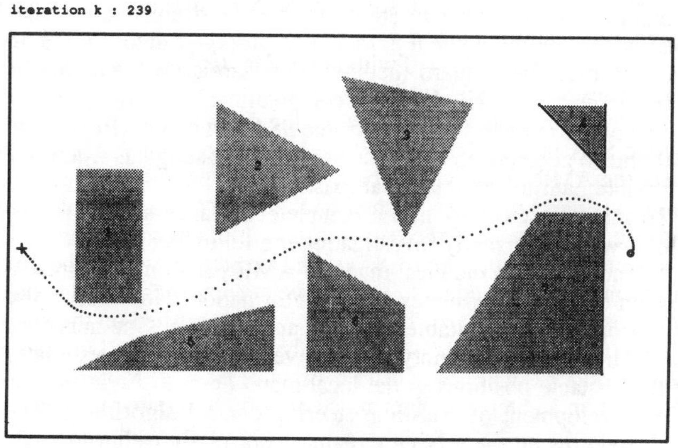

Fig. 14. Complete trajectory (242 iterations) for simulation example 1.

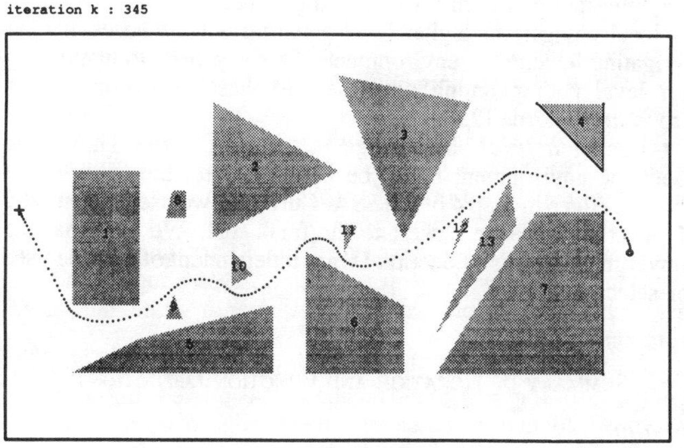

Fig. 15. Complete trajectory (345 iterations) for simulation example 2.

bring the system to rest at p_g. In the simulation examples we use a heuristic proposed by Krogh to compute orthogonal components of the goal attracting potential gradient based on the time required to reach the goal using the maximum acceleration available in each orthogonal direction in the absence of obstacles [17].

We implement the PFA as a sampled-data control scheme in which the acceleration u at each sampling instant is chosen to maximize the rate of decrease in the function $P(p, v, g)$, subject to the constraints 3. Additional constraints are imposed on u to guarantee the subgoal will be safe at the next control instant. In the simulations presented below the sample frequency for the PFA is the same as the frequency for the SSA iterations.

Figs. 11–14 show the evolution of the trajectory generated by the SSA and SCA for a particular configuration of obstacles. Points along the trajectory indicate the AMR position at constant time intervals. In each frame, bold lines indicate the obstacle faces which are visible from the current position, + indicates the initial position p_0, and small circles indicate the locations of the current subgoals in the stack. For example, in Fig. 13, there is one intermediate subgoal in addition to the final goal which is not visible from the current position. Note that the resulting trajectory provides smooth, continuous motion and the system is brought to rest at the final goal without overshoot.

The second case shown in Fig. 15 demonstrates that the SSA guides the point mass to the goal even in a very cluttered environment with virtually the same amount of computational effort as required for the previous case. The significance of these simulation results is that smooth, acceptable dynamic trajectories are being generated using feedback information without path planning.

VI. Conclusions

In this paper we have presented a feedback algorithm for selecting subgoals for guiding an autonomous mobile robot (AMR) using only local information about the locations of visible obstacles. We model the AMR as a point in a planar environment with convex, polygonal obstacles. The subgoal selection algorithm (SSA) generates a sequence of visible, safe subgoals which are pursued by a dynamic steering control algorithm (SCA). We have shown analytically that the subgoals generated by the SSA guide the AMR to the goal along a finite-time trajectory. Properties and performance of the algorithm were demonstrated by simulation examples using a point-mass model of an AMR with limited acceleration and an SCA based on artificial potential fields.

From the perspective of control theory, the fundamental contribution of this paper is a feedback algorithm for steering a dynamic system subject to state-variable constraints (the obstacles) which are unknown *a priori*. The proposed algorithm is an alternative to the standard methods of dynamic programming or optimal control which become computationally intractable for even trivial examples. In contrast, the SSA can be implemented in real time to generate an acceptable sequence of subgoals whenever a feasible path to the final goal exists.

To integrate the SSA into a complete guidance system for an AMR, we are currently investigating the following issues.

• Uncertainty in the local map and AMR position. The present version of the SSA assumes perfect information. However, noise in sensor data is inevitable in AMR applications. One approach for dealing with uncertainty is to use very conservative estimates of the obstacle positions in the local map.

• Development of real-time steering control algorithms. The SCA requires an underlying dynamic control algorithm to drive the AMR to the subgoals. Such algorithms must be designed for the specific kinematic and dynamic configuration of the ARM.

• Integration of high-level planning. Since the SSA is intended for local navigation, higher level planning will be necessary for navigating in complex environments. One approach to integrating high-level path planning with the SSA has been proposed by Krogh and Thorpe [23].

• Use of *a priori* knowledge in the SSA. Prior knowledge about the environment might be used to direct the selection of obstacle edges in Step 4 of the SSA. Currently we use the heuristic of selecting the edge closest to the final goal. We note that the convergence results of Section IV are independent of the rule used for selecting edges.

Appendix

Summary of Notation and Function Definitions

Notation

NIL	Empty set or vacuous result.
ϵ	Distance between a subgoal g and its associated edge e.
ab	Straight line (unbounded) through points a and b.
$[ab$	Ray from point a through point b containing point a.
$(ab]$	Line segment from point a to point b containing b but not a.
κ	Minimum distance between obstacles.
(g, e)	Subgoal-edge pair currently being pursued.
(g^T, e^T)	Subgoal-edge pair on top of stack S.
h	Most recent extreme subgoal generated in Step 4.
p_0	Initial position.
p_g	Final goal position.
p_k	Position at beginning of the kth execution of the SSA.
x_k	Dynamic state of the AMR at beginning of the kth iteration.
$C(a_1 \cdots, a_n)$	Open convex hull of points $a_1, \cdots, a_n$.
$E(a)$	Set of obstacle edges in $O(a)$, that is, extreme points of the connected obstacle faces in $O(a)$.
$E'(a)$	Set of visible edges in $E(a)$.
L	Set of extreme subgoals generated in Step 4 of SSA.
$O(a)$	Local map: set of obstacle faces (line segments) visible from point a.
S	Stack of subgoal-edge pairs maintained by SSA.

Functions

CHOOSE_EDGE $\langle$OBS $\langle p_k, p_g\rangle\rangle$	Returns subgoal-edge pair (g', e') with e' one of the two edges in $E(p_k)$ on OBS $\langle p_k, p_g\rangle$ and g' a distance E from e' on extension of the fact terminating at e'.
CREATEG $\langle p_k, (g^T, e^T), (g, e), [ab], E'(p_k)\rangle$	Generates subgoal-edge pairs in Step 6, as defined in Section III.
EXTEND_FACE$\langle O(p_k), (g, e)\rangle$	Returns subgoal-edge pair (g', e') where e' is the vertex (extreme point) of the face in $O(p_k)$ containing e in direction of g, and g' is defined as in CHOOSE_EDGE.
HEIGHT $\langle S\rangle$	Number of subgoal-edge pairs in stack S.
MIN_DIST $\langle A, [ab]\rangle$	Chooses one point in set A closest to line $[ab]$.
NEXT_VERTEX $\langle O(p_k), (g, e)\rangle$	Returns subgoal-edge pair (g', e') with e' as the next vertex on obstacle in $O(p_k)$ containing e in the direction of g and g' a distance ϵ from e' on extension of face $[ee']$.
OBS $\langle a, b\rangle$	Set of obstacle faces in local map $O(a)$ connected to and including a face intersected by line segment (ab).
OTHER_EDGE $\langle O(p_k), (g, e)\rangle$	Returns subgoal-edge pair (g', e') where e' is the other edge in $E(p_k)$ on the obstacles in $O(p_k)$ with edge e', and g' is defined as in CHOOSE_EDGE.
OTHER_VERTEX $\langle O(p_k), (g, e)\rangle$	Returns subgoal-edge pair (g', e') where e' is the other vertex (extreme point) for the face in $O(p)$ containing e, and g' is defined as in CHOOSE_EDGE.
POP $\langle S\rangle$	Removes and returns top subgoal-edge pair from stack S.
PUSH $\langle (g, e), S\rangle$	Puts subgoal-edge pair (g, e) on stack S.
RESTORE $\langle S, (ge), [ab], L, h\rangle$	Sets arguments equal to values store by SAVE.
SAFE $\langle g, x(t), O(p_k)\rangle$	Returns logical TRUE if the SCA can bring the AMR to a strop at g from state $x(t)$ along a path in the visible obstacle-free space in $O(p_k)$.
SAVE $\langle S, (g, e), [ab], L, h\rangle$	Stores context of SSA.
SUBGOAL $\langle e, [ab]\rangle$	Generates a subgoal g a distance ϵ from e on the line perpendicular to, and in the direction of, line segment $[ab]$.

References

[1] G. Giralt, R. Chatila, and M. Vaisset, "An integrated and motion control system for autonomous multisensory mobile robots," presented at the 1st Int. Symp. Robotic Research, Bretton Woods, NH, Sept. 1983.

[2] H. P. Moravec, "The Stanford cart and the CMU rover," *Proc. IEEE*, vol. 71, no. 7, July 1983.

[3] A. M. Parodi, J. J. Nitao, and L. S. McTamaney, "An intelligent system for an autonomous vehicle," in *Proc. 1986 IEEE Int. Conf. Robotics Automat.*, San Francisco, CA, Apr. 1986, pp. 1657–1663.

[4] R. Wallace *et al.*, "Progress in robot road-following," *IEEE Int. Conf. Robotics Automat.*, San Francisco, CA, Apr. 1986, pp. 1615–1621.

[5] S. K. Kambhampati and L. S. Davis, "Multiresolution path planning for mobile robots," *IEEE J. Robotics Automat.*, vol. RA-2, pp. 135-145, Sept. 1986.

[6] C. E. Thorpe, "Path relaxation: Path planning for a mobile robot," in *Proc. AAAI Conf.*, Aug. 1984.

[7] E. Koch *et al.*, "Simulation of path planning for a system with vision and map updating," in *Proc. 1985 IEEE Int. Conf. on Robotics Automat.*, St. Louis, MO, Mar. 1985.
[8] C. K. Yap, *Algorithmic and Geometric Aspects of Robotics.* Hillsdale, NJ: Lawrence Erlbaum Associates, Advances in Robotics, vol. 1, ch. Algorithmic Motion Planning, 1987.
[9] B. J. Oommen *et al.*, "Robot navigation in unknown terrains using learned visibility graphs. Part 1: The disjoint convex obstacle case," *IEEE J. Robotics Automat.*, vol. RA-3, pp. 672–681, Dec. 1987.
[10] D. W. Payton, "An architecture for reflexive autonomous vehicle control," in *Proc. 1986 IEEE Int. Conf. Robotics Automat.*, San Francisco, CA, Apr. 1986, pp. 1838–1845.
[11] J. J. Nitao and A. M. Parodi, "A real-time reflexive pilot for an autonomous land vehicle," *IEEE Contr. Syst. Mag.*, vol. 6, pp. 14–23, Feb. 1986.
[12] R. A. Brooks, "A robust layered control system for a mobile robot," *IEEE J. Robotics Automat.*, vol. RA-2, pp. 14–23, Mar. 1986.
[13] A. M. Waxman *et al.*, "A visual navigation system for autonomous land vehicles," *IEEE J. Robotics Automat.*, vol. RA-3, pp. 124–141, Apr. 1987.
[14] V. J. Lumelsky and A. A. Stepanov, "Dynamic path planning for a mobile automaton with limited information on the environment," *IEEE Trans Automat. Contr.*, vol. AC-31, pp. 1058–1063, Nov. 1986.
[15] V. J. Lumelsky and A. A. Stepanov, "Path planning strategies for a point mobile automaton moving amidst unknown obstacles of arbitrary shape," *Algorithmica,* 1987.
[16] T. Kanade, C. Thorpe, and W. Whittaker, "Autonomous land vehicles project at CMU," in *Proc. 1986 ACM Comput. Conf.*, Cincinnati, OH, Feb. 1986.
[17] B. H. Krogh, "A generalized potential field approach to obstacle avoidance control," in *Proc. Robotics Int. Robotics Research Conf.*, Bethlehem, PA, Aug. 1984.
[18] D. Feng, "Dynamic steering control," Master's thesis, Dep. Elec. Comput. Eng., Carnegie-Mellon Univ., Pittsburgh, PA, Mar. 1986.
[19] R. Aggarwal and G. Leitmann, "Avoidance control," *ASME J. Dynam. Syst. Meas. Contr.*, pp. 152–154, June 1972.
[20] B. R. Barmish, W. E. Schmitendorf, and G. Leitmann, "A note on avoidance control," *ASME J. Dynam. Syst., Meas. Contr.*, pp. 69–70, Mar. 1981.
[21] W. E. Schmitendorf, B. R. Barmish, and B. S. Elenbogen, "Guaranteed avoidance control and holding control," *ASME J. Dynam. Syst., Meas. Contr.*, pp. 166–172, June 1982.
[22] G. Leitmann, "Guaranteed avoidance strategies," *J. Optimiz. Theory Appl.*, vol. 32, Dec. 1980.
[23] B. H. Krogh and C. E. Thorpe, "Integrated path planning and dynamic steering control for autonomous vehicles," in *Proc. IEEE Int. Conf. Robot. Automat.*, San Francisco, CA, Apr. 1986, pp. 1664–1669.

Bruce H. Krogh (S'82–M'82) received the B.S. degree in mathematics and physics from Wheaton College, Wheaton, IL, in 1975, and the M.S. and Ph.D. degrees in electrical engineering from the University of Illinois, Urbana, in 1978 and 1983, respectively.

He is currently an Associate Professor in the Department of Electrical and Computer Engineering, Carnegie-Mellon University, Pittsburgh, PA, where he conducts research in feedback algorithms for supervisory control of robotic systems and synthesis of control logic for discrete-event systems.

Dr. Krogh is an Associate Editor of the IEEE TRANSACTIONS ON AUTOMATIC CONTROL and received the Presidential Young Investigator Award from the National Science Foundation in 1985.

Dai Feng (S'83) was born in Beijing, China, in 1961. He received the B.S. degree in electrical engineering from Grove City College, Grove City, PA, in 1984 and the M.S. degree in electrical engineering from Carnegie-Mellon University, Pittsburgh, PA, in 1986.

He is currently working towards the Ph.D. degree at Carnegie-Mellon University. Since 1984 he has been a Research Assistant in the Department of Electrical and Computer Engineering and the Robotics Institute, Carnegie-Mellon University.

Satisficing Feedback Strategies for Local Navigation of Autonomous Mobile Robots

DAI FENG, MEMBER, IEEE, AND BRUCE H. KROGH, MEMBER, IEEE

Reprinted from *IEEE Transactions on Systems, Man, and Cybernetics*, Vol. 20, No. 6, November/December 1990.
Copyright © 1990 by The Institute of Electrical and Electronics Engineers, Inc. All rights reserved.

***Abstract*—A general approach to the local navigation problem for autonomous mobile robots (AMRs) and its application to omnidirectional and conventionally steered wheelbases are presented. The problem of driving an AMR to a goal in an unknown environment is formulated as a dynamic feedback control problem in which local feedback information is used to make steering decisions while the AMR is moving. To obtain a computationally tractable algorithm, we propose a class of satisficing feedback strategies that generate reasonable, collision-free trajectories to the goal using simplified representations of the AMR dynamics and constraints. Realizations of the feedback strategy are presented and illustrated by simulation under the assumptions of perfect feedback information and zero servo error. Straight forward extensions of the approach to handle uncertainties in real systems are briefly described.**

I. Introduction

THIS PAPER concerns the problem of guiding an autonomous mobile robot (AMR) in an unknown environment. In contrast to a "stop-look-and-move" approach to local navigation, we formulate a dynamic feedback control problem in which the navigator makes steering decisions on the fly while the AMR is moving. Because navigation decisions must be made in real time, optimal control methodologies are computationally prohibitive for this problem. We propose feedback strategies in which approximations for the complex nonlinear system dynamics and operating constraints (including obstacles) admit a computationally feasible method for generating reasonable, collision-free trajectories to the goal.

Various forms of the obstacle avoidance problem (also called the path-finding problem) have been studied extensively by other researchers. The assumption in much of the previous work is that complete information about the environment is known *a priori*. The most notable works in this area have been reported by Lozano-Perez [1], [2], Reif [3], Schwartz [4], [5] and O'Dunlaing *et al* [6]. The objective is to move a finite-size object (a robot) from a starting position to a goal position among obstacles, typically modeled as polygons. In these works, consideration of the kinematic and dynamic constraints is absent, which severely limits the application of their results to mobile robots.

Trajectory-generation algorithms based on global optimization incorporating kinematic and dynamic constraints have been proposed by Gilbert [7], Shin [8], and Kim and Shin [9]. Due to the intensive computations required for global optimization, this approach is not suitable for real-time navigation of mobile robots. Another approach to dynamic obstacle avoidance is to introduce artificial potential fields that "attract" the robot to the goal and "repulse" the robot from obstacles. This approach is represented by the work of Khatib [10], Hogan [11], Krogh [12], [13], Newman and Hogan [14], and Volpe and Khosla [15]. Although the potential field approach is attractive for steering omnidirectional wheelbases, it is not clear how the approach can be extended when there are nonholonomic constraints on the system, as is the case for the conventionally steered mobile robot discussed in Section V.

Lumelsky approached the navigation problem with the assumption that only local information about the environment is available [16], [17]. Kinematic and dynamic constraints are not taken into account by Lumelsky's algorithms. The navigation strategies presented in the present paper are similar in spirit to Lumelsky's approach in that we use only information about the locally visible environment. However, our method takes into account the dynamic constraints and capabilities of the system. Because of the conservative nature of our approach and the lack of any global optimization criterion, we call our algorithms satisficing feedback strategies.[1]

We envision the navigation algorithms developed in this paper be used in an AMR planning/navigation system with a structure shown in Fig. 1. Such systems consist of four major components: planner, map generator, navigator, and the servocontroller. The *servocontroller* is

Manuscript received February xx, 1989; revised October 16, 1989. This work was supported in part by the National Science Foundation under Grants ECS-8404607 and DMC-8451493, and in part by the Robotics Institute, Carnegie Mellon University, Pittsburgh, PA.
The authors are with Department of Electrical and Computer Engineering and the Robotics Institute, Carnegie Mellon University, Pittsburgh, PA 15213.
IEEE Log Number 9037599.

[1]"to satisfice: to decide on and pursue a course of action that will satisfy the minimum requirements necessary to achieve a particular goal." (Supplement to the Oxford English Dictionary, 1982)

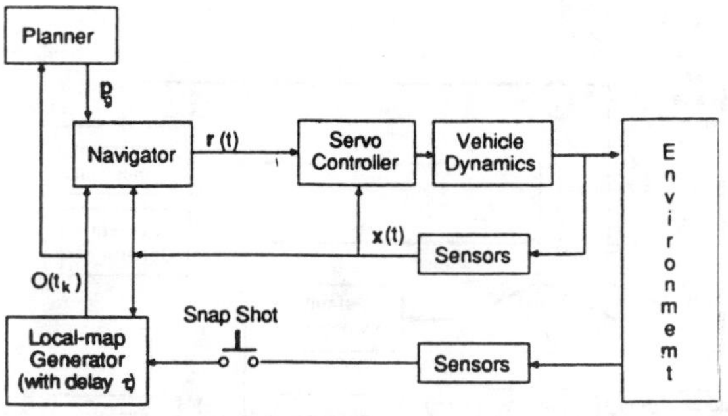

Fig. 1. Structure of the control/navigation system for an AMR.

designed to track reference trajectories with a known accuracy, provided the reference trajectories satisfy the kinematic and dynamic constraints of the AMR [18]--[21]. Under such conditions, the servocontroller, the AMR, and its actuator dynamics can be modeled by a simple kinematic model, which we call the *reference model*.

The *planner* determines the ultimate goal, $\boldsymbol{p}_g$, for the AMR. The *map generator* uses information from on-board sensors (such as cameras, sonar rings, or infrared scanners) to compute a local map of the obstacles in the local environment. The local maps represent "snap shots" of the environment taken at discrete points in time. We denote the time at which the kth snap shot is taken by t_k, and denote the local map at t_k by $O(t_k)$. Because of the finite time required to process the sensor data, there is a time delay of τ associated with the map generator. The strategies proposed in this paper are realistic for implementation with currently available sensors and map-building algorithms, such as those proposed by Elfes [22] and Cox *et al.* [23].

The feedback strategies developed in this paper address the problem of the *navigator*. Given an initial position $\boldsymbol{p}_o$, the navigator must generate a reference trajectory $\boldsymbol{r}(t)$ for the servocontroller that will guide the AMR to the goal $\boldsymbol{p}_g$ while avoiding collisions with obstacles. The navigator uses two sources of feedback information: local maps of environment $O(t_k)$ from the map generator, and estimates of the AMR state $\boldsymbol{x}(t)$ (including its position and velocity) obtained from wheel encoders and other system sensors. At any instant, the obstacles represent future constraints on the reference trajectory. The basic idea of our approach is to map these future constraints into the current navigation decision space so that it is guaranteed the AMR will be in a *safe state* when the next navigation decision is made. The precise definition of safe states depends on the kinematic and dynamic characteristics of the AMR.

We decompose the navigation strategy into two parts: *subgoal selection* and *steering decision*. The subgoal selection algorithm (SSA), described in detail in [24], generates temporary subgoals to be pursued when the final goal is not in sight. This paper focuses on the steering decision algorithm (SDA), which chooses steering vectors to drive the AMR toward the current subgoal. We develop the SDA for two AMRs: an omnidirectional AMR (OAMR), a prototype of which is operated in our laboratory [25], and a conventionally steered AMR (CAMR) with front wheel steering and rear wheel drive, such as the CMU NAVLAB [26].

The remainder of the paper is organized as follows. In Section II we describe our basic approach to the local navigation problem and its decomposition into the problems of subgoal selection steering decisions. The general structure of the SDA as it applies to any AMR configuration is presented in Section III. We then describe realizations of the SDA for the OAMR (Section IV) and CAMR (Section V). Section VI illustrates the combined performance of the SSA and SDA for local navigation through complex environments. Simulation experience demonstrates that the computational requirements for the algorithms are realistic for real-time implementation. Throughout the presentation we assume perfect sensor information and servoing. In the concluding section we discuss how the satisficing framework can be easily extended to accommodate sensing and servo errors inherent in real systems. Details of these extensions are described elsewhere [27], [28].

II. Decomposition of the Local Navigation Problem

Referring to the general AMR control system illustrated in Fig. 1, the local navigation problem can be stated as follows: Using local maps $O(t_k)$ and state estimates $\boldsymbol{x}(t)$ from on board sensors, generate a finite-time reference trajectory $\boldsymbol{r}(t)$ to drive an AMR along a collision-free path from an initial position $\boldsymbol{p}_o$ to a goal position $\boldsymbol{p}_g$.

In this paper we propose a solution to the local navigation problem under the following assumptions:

1) The AMR has a servocontrol system that is capable of following a given class of reference trajectories;
2) The AMR state is known, including its position, orientation and velocity in the global coordinate frame;
3) Visible portions of obstacles in the local map $O(t_k)$ are represented accurately by linear boundaries of convex polygons grown to account for the finite size of the AMR; and
4) Obstacles in the local map do not intersect.

The first three assumptions characterize a system without sensing or servo errors. We make these assumptions in this paper to simplify the presentation of the basic satisficing strategies; they are not fundamental limitations of our approach. The fourth assumption assures that local decisions will never lead to dead ends. Since our interest is in dynamic navigation for local obstacle avoidance, this assumption holds in our simulation examples so that the final goal is actually reached. To guarantee the final goal will be reached in more general cases, it would be necessary to augment our algorithm with higher-level rules to redirect the search for a feasible path when a dead end is

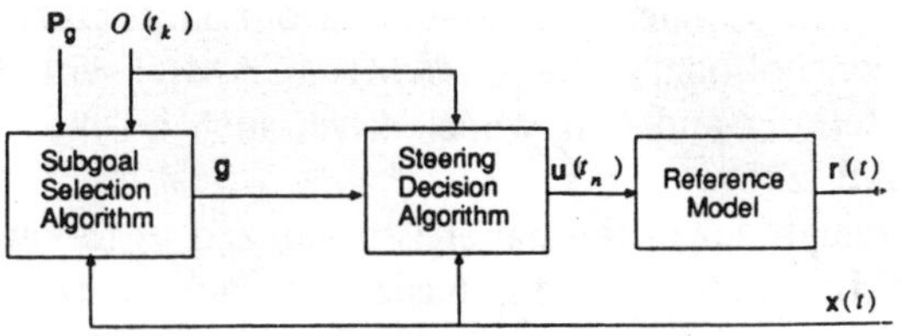

Fig. 2. Internal structure of the navigator.

encountered. The strategies developed in this paper guarantee the AMR will not collide with obstacles even when assumption 4 does not hold.

When a new local map of the environment is available, two decisions must be made. First, because the goal p_g may be obscured by the obstacles, the navigator must select a temporary subgoal for the AMR to pursue. Second, it must generate a reference trajectory to the chosen subgoal. The first decision is influenced largely by the consideration of the final goal position p_g and the obstacle locations, while the second decision is influenced by the constraints imposed on the reference trajectory by the locally visible obstacles and the AMR operating limits. This leads to a natural decomposition of the navigation problem into two sub-problems: *subgoal selection* and *steering decision*.

The subgoal selection process provides a temporary direction to pursue while the final goal is not in sight. Subgoals maybe updated each time a new local map becomes available. This task is accomplished by the SSA. Given a subgoal from the SSA, the (SDA) drives the AMR to the subgoal along a feasible trajectory, taking into account the system and obstacle constraints. The subgoal can be updated by the SSA while the AMR is moving. Thus the AMR does not necessarily reach a given subgoal, or stop before reaching the final goal.

The internal structure of the navigator is shown in Fig. 2. We use index k to denote the kth execution of the SSA at time t_k. At each time t_k when a new local map of the obstacles is available, the SSA attempts to update the subgoal. This new subgoal must be validated to ensure its safeness. A subgoal is safe if the SDA can generate a reference trajectory bringing the AMR to rest at the subgoal without colliding with any obstacles in the local map. Once a new subgoal is validated as being safe, it becomes the current subgoal, denoted by $\boldsymbol{g}$, which is used by the SDA along with map $O(t_k)$ from which the subgoal is generated. Index n denotes the nth execution of the SDA at time t_n. The task of the SDA is to generate a sequence of *steering vectors* $\boldsymbol{u}(t_n)$, which is applied to the reference model to produce the desired reference trajectory $\boldsymbol{r}(t)$. In general, the SDA is executed much more frequently then the SDA, that is, $t_{n+1} - t_n \ll t_{k+1} - t_k$.

The SSA, which is independent of the particular AMR dynamics, has been described in detail elsewhere [24]. In the remainder of the paper we focus on the SDA, the details of which depend on the specific AMR dynamics. In Section III we present the general structure of the SDA, followed by specific realizations of the algorithm for the OAMR and CAMR in Sections IV and V, respectively.

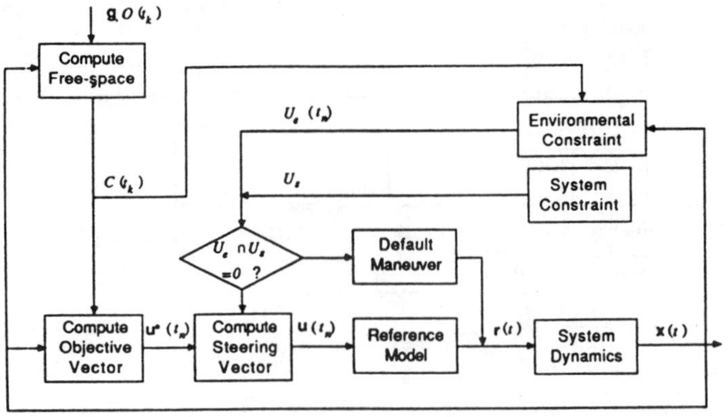

Fig. 3. Feedback structure of the steering control algorithm.

III. The Steering Decision Algorithm

The overall structure of the SDA is illustrated in Fig. 3. As described in the previous section, the SDA generates the reference trajectory $\boldsymbol{r}(t)$ by selecting a steering vector $\boldsymbol{u}(t_n)$ at discrete increments of time. The steering vector is an input to the reference model for an AMR, which is integrated to obtain $\boldsymbol{r}(t)$.

To explain the feedback structure of the SDA in more detail, we start with the notion of the local *free-space*, $C(t_n)$. The free-space is defined to reduce the complexity of the representation of obstacles in the local map. Although the obstacles are already represented in a relatively simple form as polygons, the arbitrary orientation and location of their boundaries lead to computational difficulties when trying to implement a general obstacle avoidance algorithm in real time. We show in the following sections that by considering the dynamic characteristics of the AMR, it is possible to simplify the representation of the obstacle boundaries without sacrificing performance.

We imposed the obstacle constraints by transforming them into constraints in the steering vector space at each decision instant t_n. Given a free-space $C(t_n)$, we find limits on the steering vector that guarantee the reference trajectories will be confined to the free-space during the interval (t_n, t_{n+1}). Moreover, we guarantee collisions can be avoided beyond time t_{n+1} using the notions of *default maneuvers* and *safe states*. For a given reference state $\boldsymbol{r}_o, \dot{\boldsymbol{r}}_o$ for the AMR and free-space $C(t_o)$ with $\boldsymbol{r}_o \in C(t_o)$, a default maneuver is a predefined, unique reference trajectory that will bring the AMR to rest at some future time. The default maneuver can be thought of as an emergency operating mode in which the AMR is deterministically brought to a stop so as to avoid leaving the free-space $C(t_o)$. It is predetermined (as a function of $\boldsymbol{r}_o$ and $\dot{\boldsymbol{r}}_o$) so that it can be invoked and executed without delay. It is also predetermined to be executable by the AMR, satisfying the braking and steering limitations of the system. The known response of the system under the default maneuver determines the safety of the AMR at a given state. A reference state $r_o, \dot{r}_o$ is said to be a *safe state* with respect to a free-space $C(t_o)$ if the entire trajectory of the default maneuver from that state remains within $C(t_o)$.

Using the concept of safe states, the transformation of free space boundaries into steering vector constraints is explained as follows. Given a free space $C(t_n)$ at time t_n, a constraint on $u(t_n)$ is computed for each boundary of $C(t_n)$. Each of these constraints are chosen so that for any $u(t_n)$ satisfying the constraint it is guaranteed that: 1) the trajectory $r(t)$ does not cross the boundary of $C(t_n)$ for $t_n < t \leqslant t_{n+1}$; and 2) the reference state $r(t_{n+1}), \dot{r}(t_{n+1})$ is safe with respect to the boundary of $C(t_n)$. Thus, the resulting constraints on $u(t_n)$, which form the *environment constraint set* $U_e(t_n)$, guarantee that at the next decision time t_{n+1}, at least the default maneuver will keep the AMR within the known free-space $C(t_n)$. In general, the exact constraints imposed on $u(t_n)$ by the boundaries of $C(t_n)$ are nonlinear and difficult to express analytically. As illustrated in the following sections, simple linear approximations to the environmental constraint set can be used in applications so that real-time performance can be achieved.

The system constraint set U_s represents the physical operating limits on the AMR which impose constraints on the current steering vector. The *objective vector* $u^*(t_n)$ is determined from the current state as a desired direction and magnitude for the steering vector $u(t_n)$ to drive the AMR to the subgoal g. In the following sections we describe heuristics for determining $u^*(t_n)$. We note that our simulation results have demonstrated that the feedback structure of the SDA makes the exact value of the objective vector less critical than one might expect. Given the constraint sets $U_e(t_n)$ and U_s, and the objective vector $u^*(t_n)$, we devise simple rules to choose $u(t_n) \in U_e(t_n) \cap U_s$ to approximate the vector $u^*(t_n)$ without solving a complete optimization problem. Again, our simulation results indicate that finding an optimal approximation to $u^*(t_n)$ is not necessary provided $u(t_n)$ is chosen at frequent time intervals.

IV. An SDA for Omnidirectional Mobile Robots

In this section we present a realization of the SDA for an omnidirectional autonomous mobile robot (OAMR). An OAMR is characterized by three degrees of freedom in the plane, in contrast to the two degrees of freedom of normal wheelbases. Starting from rest, an OAMR is capable of translating along any direction in the plane. In addition, it can also rotate. Examples of AMRs constructed with omnidirectional characteristics include the Unimation robot [29], Uranus [30], and the Ilonator Cart [25].

A. OAMR Reference Model

For the purpose of reference trajectory generation, we consider the OAMR as a point in the global coordinate frame. When navigating through obstacle environments, the finite size of the OAMR can be compensated for by means of expanding the obstacle regions in the local map. Because of the omnidirectional property, we consider only the position of the OAMR when generating the reference trajectory. Thus, the reference state variables are the position vector $r = p$, and velocity vector $\dot{r} = v$, of the robot in R^2. The steering vector is $u = a$, the acceleration of the point p, also in R^2. Thus, we can model the reference trajectory dynamics by two-dimensional double integrator dynamics:

$$\begin{pmatrix} \dot{p} \\ \dot{v} \end{pmatrix} = \begin{pmatrix} 0 & I \\ 0 & 0 \end{pmatrix} \begin{pmatrix} p \\ v \end{pmatrix} + \begin{pmatrix} 0 \\ a \end{pmatrix}. \tag{1}$$

Integrating (1), we obtain the reference trajectory for the time interval $t \in [t_n, t_{n+1}]$ as

$$p(t) = p(t_n) + (t - t_n)v(t_n) + 0.5(t - t_n)^2 a(t_n)$$
$$v(t) = v(t_n) + (t - t_n)a(t_n). \tag{2}$$

We represent the OAMR's operating limits by magnitude constraints on the acceleration and velocity vectors given by: $\|a\| \leqslant a_{\max}$ and $\|v\| \leqslant v_{\max}$. These limits constitute the system constraint set U_s for the OAMR.

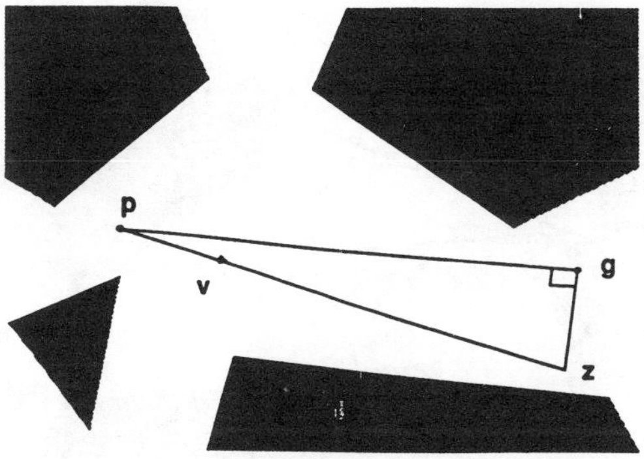

Fig. 4. Construction of free-space for OAMR, case 1.

B. OAMR Free-space

Before defining the free-space for the OAMR, we first introduce some notation. A straight, unbounded line containing two points a and b is denoted by ab. The open or closed termination of a line segments is indicated by a parenthesis or bracket, respectively. For example, $(ab]$ denotes the line segment from a to b that contains b but not a, and $[ab$ denotes the ray beginning at and including a passing through b.

The OAMR free-space is defined in association with the current subgoal g, the current reference state, and the current local map of the environment $O(t_k)$. Consider the situation when a subgoal g has just become the currently pursued subgoal. Suppose the OAMR is currently at reference state p, v and beginning to pursue the subgoal g as shown in Fig. 4. The local map is indicated in the figure by the bold lines on the obstacles, which are the obstacle faces visible from the AMR position p. The subgoal selection algorithm guarantees that g is a safe and visible subgoal from p. We choose one of the free-

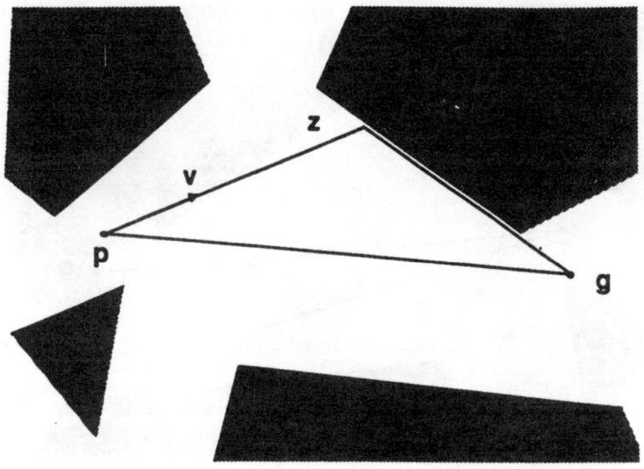

Fig. 5. Construction of free-space for OAMR, case 2.

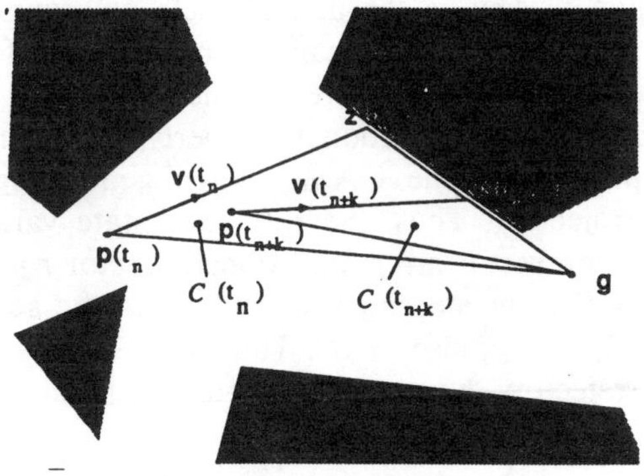

Fig. 6. Construction of free-space for OAMR after the initial free-space.

space boundaries to be the line segment $[pg]$. The second boundary is determined by the ray emitting from p in the direction of the velocity vector v. These boundary choices are motivated by the following observation. With reference to Fig. 4, when the velocity vector v is pointing into the half-plane to the right of the line pg, the trajectory toward the subgoal should stay in the right half-plane determined by pg, otherwise the resulting path would be longer than necessary.

The last boundary is defined by a line segment gz where the point z is chosen such that gz and gp are perpendicular, and zp is colinear with v. Choosing the boundary gz to be perpendicular to gp constrains the trajectory from overshooting the goal. For the case shown in Fig. 4, the free space is defined as the convex hull of three points p, g, and z denoted by $C(p,g,z)$. For the case shown in Fig. 5, the previous definition would result in a free-space that intersects obstacle regions. In this case, to exclude the obstacle region from $C(p,g,z)$, we rotate $[gz$ around g towards p until $C(p,g,z) \cap O(t_k) = \emptyset$. A free-space that has its vertex z to the right(left) of the subgoal g as viewed from the point p is called a right(left)-hand free-space, as is the case in Figs. 4 and 5.

Once an initial free-space is defined for a subgoal, the subsequent free-spaces for the subgoal are defined with respect to the first free-space as the OAMR moves toward the subgoal. Consider the situation shown in Fig. 6. The first free-space is defined for the subgoal g at t_n as $C(p(t_n), g, z(t_n))$. For the OAMR state $p(t_n), v(t_n)$, the free-space for the subgoal g is defined as $C(p(t_n), g, z(t_n))$ where the point $z(t_n)$ is the intersection between the extension of the velocity vector $v(t_n)$ from $p(t_n)$ and the boundary $gz(t_n)$. Thus, the sequence of free-spaces for a subgoal all have a single common boundary $[gz(t_n)]$ defined when the first free-space for the subgoal is constructed. We simplify the notation for the free-space for a particular subgoal at time t_n to $C(t_n)$. The free-space can be represented analytically as a set of linear inequalities of the form

$$C(t_n) = \{q | [n_i(t_n)]^T q \leqslant d_i(t_n) \quad i = 1,2,3\}.$$

where the vectors $n_i(t_n)$ are normal to the boundaries $[gz], [zp(t_n)], [gp(t_n)]$, respectively, and the constants $d_i(t_n)$ determine the boundary lines in the plane.

C. OAMR Default Maneuver and Safe States

The default maneuver for the OAMR is defined by applying the maximum constant deceleration opposing the current velocity of the OAMR until the OAMR comes to a stop. This maneuver results in a straight line trajectory. For an OAMR with velocity $v(t_o)$, the time required to bring the OAMR to a stop using the default maneuver is

$$t^* = \frac{\|v(t_o)\|}{a_{\max}}, \tag{3}$$

and the steering vector for the OAMR default maneuver is given by

$$a(t) = -\frac{a_{\max} v(t_o)}{\|v(t_o)\|} \qquad \text{for} \quad t_o \leqslant t \leqslant t^*.$$

We denote the position of the OAMR when it comes to rest by $p_f^*(p(t_o), v(t_o))$, which is given by

$$p_f^*(p(t_o), v(t_o)) = p(t_o) + \frac{\|v(t_o)\|}{a_{\max}} v(t_o). \tag{4}$$

D. OAMR Environment Constraints

We now present a method for mapping the constraints on the state of the OAMR due to the free-space boundaries into constraints on the steering vector to produce the environment constraint set $U_e(t_n)$. Assuming the OAMR is in a safe state with respect to free-space $C(t_n)$, we want to constrain the steering vector $a(t_n)$ so that the state $p(t_{n+1}), v(t_{n+1})$ is also safe with respect to $C(t_n)$. In other words, the steering vector $a(t_n)$ is constrained so that a default maneuver trajectory starting from $p(t_{n+1}), v(t_{n+1})$ would remain within the boundaries of $C(t_n)$.

Using the definition of the safe state for $p(t_{n+1})$ and $v(t_{n+1})$, the free-space constraint can be expressed as

$$n(t_n)_i^T p_f^*(p(t_{n+1}), v(t_{n+1})) \leqslant d_i(t_n), \tag{5}$$

where $i = 1$, 2, and 3 for the first, second, and third boundaries of the free-space $C(t_n)$ respectively. Using (2) and (4), and substituting for $p_f^*(p(t_{n+1}), v(t_{n+1}))$ in inequality (5) we obtain the following constraints on $a(t_n)$:

$$n_i(t_n)^T\left[p(t_n)+\Delta v(t_n)+\frac{1}{2}a(t_n)\Delta^2 + \frac{\|v(t_n)+\Delta a(t_n)\|}{a_{\max}}(v(t_n)+\Delta a(t_n))\right]\leqslant d_i(t_n) \quad (6)$$

for $i=1$, 2, and 3, where $\Delta = t_{n+1}-t_n$. These constraints on $a(t_n)$ guarantee the trajectory remains in the free space for $t\in[t_n,t_{n+1}]$ and that the OAMR is in a safe state at $t=t_{n+1}$.

Although the constraints on the state of the OAMR have been mapped into the steering vector space, the resulting constraints (6) are nonlinear. These constraints are too complicated for application in a real-time feedback scheme. In [27] it is shown that the constraints (6) can be approximated by linear constraints of the form

$$c_i(t_n)^T a(t_n)\leqslant \bar{a}_i(t_n) \quad (7)$$

where $c_i(t_n)=n_i(t_n)$, for $i=1$ and 2, $c_3(t_n)=g-p(t_n)+0.5\Delta v(t_n)$, and

$$\bar{a}_1(t_n)=\frac{-(2+\Delta a_{\max})-\sqrt{\Delta^2 a_{\max}+\Delta \nu_1(t_n)a_{\max}-8(\Delta\nu_1(t_n)-2d_1(t_n))a_{\max}}}{2\Delta}$$

where $\nu_1(t_n)$ is the projection of the OAMR velocity in the direction of n_1 and d_1 is the distance from the OAMR to boundary 1 of $C(t_n)$, $\bar{a}_2(t_n)=0$, and

$$\bar{a}_3(t_n)=-\frac{(g-p(t_n))^T v(t_n)}{\Delta\|g-p(t_n)+0.5\Delta v(t_n)\|}. \quad (8)$$

E. Objective Vector and Steering Vector Selection

The acceleration vector $a(t_n)$ for generating the reference trajectory over the time interval $[t_n,t_{n+1}]$ is computed by choosing an approximation to an objective vector $a^*(t_n)$ subject to the constraint $a(t_n)\in U_e\cap U_s$. The objective vector is defined to produce a "good" reference trajectory to the subgoal in the sense that it produces a sequence of steering vectors leading to a direct and smooth trajectory to the goal. There are many possible ways to define and compute an acceptable objective vector. If achieving the goal in minimum time is of high priority, it can be defined as the time optimal acceleration vector using the method developed in [31]. For the simulation examples presented in Section VI of this paper, we used the generalized potential fields to produce the objective vector [12].

The final stage of the SDA is to select a steering vector to approximated the objective vector subject to the constraints $U_e(t_n)$ and U_s. The task is to select a steering vector $a(t_n)$ that best approximates the objective vector $a^*(t_n)$. This problem can be formulated as an optimization problem by defining an objective function and solved using standard numerical procedures. Alternatively, the special structure of the constraints can be exploited to allow a reasonable steering vector to be selected with minimal computational burden using heuristic procedures [25]. Our simple heuristic is illustrated in Fig. 7. The steering vector for the OAMR is chosen to be in the direction of the objective vector, truncated by the constraints imposed by $U_e(t_n)$ and U_s. The computation time for this heuristic is negligible, and experience with this and other heuristics has shown that the trajectory is not particularly sensitive to the method used to choose the approximation to $a^*(t_n)$

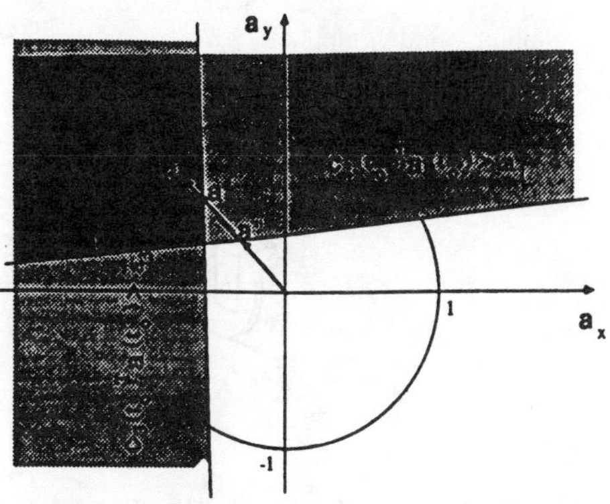

Fig. 7. Selecting the steering vector.

V. An SDA for Conventionally-Steered Mobile Robots

In this section, we present a realization of the SDA for a conventionally-steered autonomous mobile robot (CAMR). The CAMR differs from the OAMR in that its steering mechanism imposes constraints on the path it can follow. Steering is achieved by turning the wheel axes. Examples of CAMRs include Neptune [32], Hero-1 [33], and Blanche [23]. A typical configuration is the tricycle design shown in Fig. 8(a). For simplicity, such CAMRs can be modeled by the "bicycle" model depicted in Fig. 9(b) with the rear wheels collapsed into a single wheel [34]. In this model, we assume the front wheel is the steering wheel and the rear wheel the driving wheel. The wheel base, denoted by l, is the distance between the axes of the front and rear wheels. The local coordinate frame for the CAMR is assigned at a distance of l_r from the rear wheel axle on the center line of the CAMR body. We refer to the velocity of the local coordinate frame on the CAMR as the CAMR velocity, denoted by v, and the speed of CAMR at the rear wheel as the rear wheel speed, denoted by ν_r. In contrast to the OAMR, the orientation of the CAMR, denoted by $\theta(t)$, is critical in determining trajectory constraints for the CAMR.

A. CAMR Reference Model

For the purpose of reference trajectory generation, we consider the CAMR as a point concentrated at the origin of the robot coordinate system. As with the OAMR, when

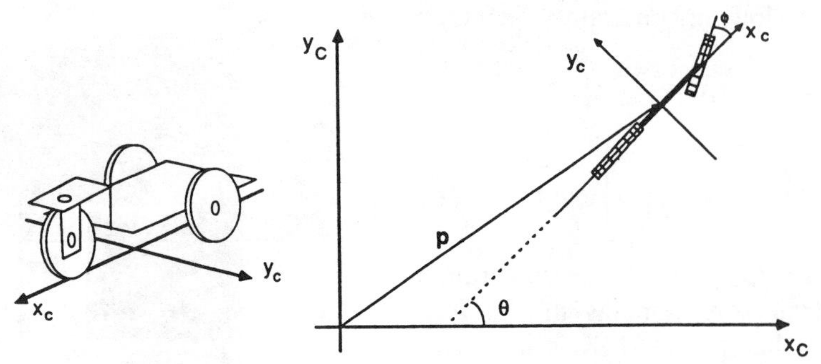

Fig. 8. (a) Typical configuration of conventionally steered mobile robot. (b) Its bicycle model and the coordinate system assignment.

navigating through obstacle environments, the finite size of the OAMR can be compensated for by means of expanding the obstacle regions in the local map. For the ease of generating a reference trajectory to be tracked by the CAMR servocontroller, we choose the components of the reference signal $r(t)$ to be the steering angle of the CAMR front wheel $\delta(t)$ and the speed of the rear wheel $\nu_r(t)$, and define the steering vector as $u(t)=[\dot{\nu}_r(t), \dot{\delta}(t)]^T$. We assume throughout this section that the CAMR is operated conservatively so that no slip occurs at the wheel contact points.

Five state variables are required to specify a complete state of the CAMR. (In some papers, the CAMR is said to have two and a half of degrees of freedom.) For our reference model, we have selected the angle of the steering wheel δ and the speed of the rear wheel ν_r as state variables. For the ease of considering the obstacles when generating a reference trajectory, we choose the other state variables to be the position vector in the global frame, $p(t)$, and the orientation of the CAMR with respect to the global coordinate frame, $\theta(t)$. We note that for purpose of generating the reference trajectory for the servocontroller, only δ and ν_r are required as the output of the reference model. The state variables δ, ν, p, θ and the steering vector u are related by the following differential equations [34]:

$$\begin{aligned} \dot{\nu}_r &= u_1 \\ \dot{\delta} &= u_2 \\ \dot{p}_x &= \nu_r(\cos\theta - \sin\theta\tan\delta) \\ \dot{p}_y &= \nu_r(\sin\theta + \cos\theta\tan\delta) \\ \dot{\theta} &= \frac{1}{l_r}(\dot{p}_y\cos\theta - \dot{p}_x\sin\theta). \end{aligned} \tag{9}$$

Another variable used in the SDA is the instantaneous curvature of the CAMR trajectory when $\dot{\delta}(t)=0$. This curvature is a function of the steering angle, denoted by $K(\delta(t))$, given by

$$K(\delta(t)) = \frac{SIGN(\delta(t))\tan(\delta(t))}{\sqrt{l^2 + l_r^2\tan^2(\delta(t))}}.$$

The curvature $K(\delta(t))$ is a signed value where a positive curvature corresponds to a left hand turning motion in the global coordinate frame.

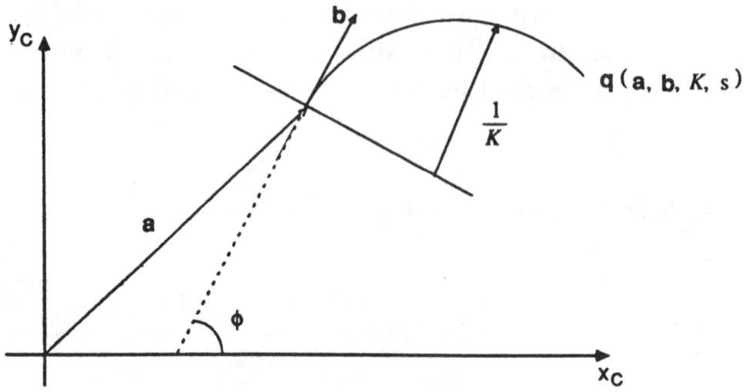

Fig. 9. Definition of notations for arc segment.

Limits on the steering and propulsion systems for the CAMR impose constraints on the steering vector u. We represent the motor limits and the effects of the dynamics and kinematics by constraints on the components of the steering vector u,

$$|u_1| \leqslant \dot{\nu}_{\max} \quad \text{and} \quad |u_2| \leqslant \dot{\delta}_{\max}. \tag{10}$$

The constraint $\dot{\nu}_{\max}$ indicates the maximum rate of change that can be effected on the rear wheel speed given the limit on the torque that can be generated by the driving motor, and $\dot{\delta}_{\max}$ represents the maximum rate of change that can be effected on the steering angle given the operating limits of the steering system. The reference states ν_r and δ are constrained as

$$\begin{aligned} |\nu_r| &\leqslant \nu_{\max} \\ |\delta| &\leqslant \delta_{\max}. \end{aligned} \tag{11}$$

The constraint on the rear wheel speed represents the limitations of the propulsion system while the constraint on δ is a result of the steering angle limits as dictated by the configuration of the CAMR. Constraints (10) and (11) define the system constraint set U_s for the CAMR.

B. CAMR Free-Space

To define the free-space for the CAMR, we introduce the following additional notation. The angle between two line segments $[ab]$ and $[bc]$ is denoted by $\angle abc$. The notation $\angle v$ denotes the orientation of the vector v in

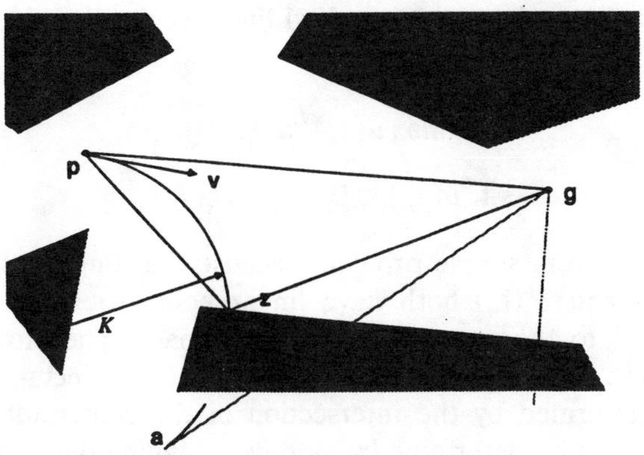

Fig. 10. Construction of free-space for CAMR; case 2, currently turning away from the subgoal q.

the global coordinate frame. A point $\boldsymbol{q}$ on an arc is denoted by $\boldsymbol{q}(\boldsymbol{a},\boldsymbol{b},K,s)$ which indicates the point is a curvilnear distance s from point $\boldsymbol{a}$ on the arc of curvature K that is tangent to the vector $\boldsymbol{b}$ at position $\boldsymbol{a}$. This definition is illustrated in Fig. 9. The curvature K and curvilinear length s are signed quantities. A positive K indicates that the center of the arc is to the left of the vector $\boldsymbol{b}$, and a positive s indicates the arc is a result of clockwise rotation around the center. An arc segment from point $\boldsymbol{a}$ to $\boldsymbol{q}$ is denoted by $[\boldsymbol{a}{}^\wedge\boldsymbol{q}(\boldsymbol{a},\boldsymbol{b},K,s)]$. Fig. 9 shows the curve $[\boldsymbol{a}{}^\wedge\boldsymbol{q}(\boldsymbol{a},\boldsymbol{b},K,s)]$ with $s>0$ and $K<0$. Using the above definitions, an analytical expression can easily be obtained for the point $\boldsymbol{q}(\boldsymbol{a},\boldsymbol{b},K,s)$.

For the CAMR we fix the free-space for each subgoal once it is defined. We denote the free-space associated with subgoal $\boldsymbol{g}$ by $C_{\boldsymbol{g}}$, and use the parameters $\boldsymbol{n}_i^g, d_i^g$ to describe the free-space $C_{\boldsymbol{g}}$ as the set of points

$$C_g=\left\{q \mid [n_i^g]^T q \leqslant d_i^g \quad i=1,2,3\right\}.$$

To define $C_{\boldsymbol{g}}$ we consider two cases for the situation when a subgoal $\boldsymbol{g}$ that has just become the currently pursued subgoal and the CAMR has velocity $\boldsymbol{v}$ at position $\boldsymbol{p}$ and orientation θ with reference state ν_r, δ. When the CAMR is turning toward the subgoal, the definition of free-space for the CAMR is identical to the free-space definition for the OAMR. The second case, when CAMR is turning away from the subgoal, as illustrated in Fig. 10, requires a different definition for the free space. The arc $[\boldsymbol{p}{}^\wedge\boldsymbol{q}(\boldsymbol{p},\boldsymbol{v},K(\delta),s)]$ is the trajectory of the CAMR position when the steering angle and speed of the CAMR are held constant. The free-space is again a triangular convex hull $C(\boldsymbol{p},\boldsymbol{g},\boldsymbol{z})$. The point $\boldsymbol{z}$ is first chosen on the arc $[\boldsymbol{p}{}^\wedge\boldsymbol{q}(\boldsymbol{p},\boldsymbol{v},K(\delta),s)]$ so that $[\boldsymbol{gz}$ is tangent to the arc. If the angle $\angle \boldsymbol{pgz}$ exceeds 90 degrees, the point is moved along the arc toward the point $\boldsymbol{p}$ until $\angle \boldsymbol{pgz}=90°$. If the resulting convex set does not intersect any obstacle region, it becomes the free-space. In the situation shown in Fig. 10, however, this is not the case. This requires the point $\boldsymbol{z}$ be further moved toward $\boldsymbol{p}$ until the convex set $C(\boldsymbol{p},\boldsymbol{g},\boldsymbol{z})$ no longer intersects any obstacle regions in the local map.

C. CAMR Default Maneuver and Safe States

The default maneuver for the CAMR is defined by holding the steering angle constant and applying the maximum constant deceleration at the rear wheel of the CAMR until it comes to a stop. The steering vector for the default maneuver is therefore $\boldsymbol{u}=[0,-\dot{\nu}_{\max}]^T$. This maneuver results in a CAMR trajectory of constant curvature, denoted by

$$p(t_o){}^\wedge p^*(p(t_o),v(t_o),K(\delta(t_o)),s) \tag{12}$$

where $\boldsymbol{p}(t_o)$, $\delta(t_o)$, and $\boldsymbol{v}(t_o)$ are the position, steering angle and velocity of the CAMR at the time t_o when the default maneuver is initiated and $\boldsymbol{p}^*(\boldsymbol{p}(t_o),\boldsymbol{v}(t_o), K(\delta(t_o)),s)$ is a point on the default maneuver trajectory at a curvilinear length of s from $\boldsymbol{p}(t_o)$. For a CAMR with initial velocity $\boldsymbol{v}(t_o)$, the curvilinear distance required to bring the CAMR to a stop using the default maneuver is [27]

$$s^*(\|v(t_o)\|,K(\delta(t_o)))=\frac{\|v(t_o)\|^2}{2a'_{\max}(K(\delta(t_o)))} \tag{13}$$

where

$$a'_{\max}(K(\delta(t_o)))=\frac{\tan(\delta(t_o))}{lK(\delta(t_o))}\dot{\nu}_{\max}. \tag{14}$$

D. CAMR Environment Constraints

Assuming the CAMR state at t_n is safe with respect to free-space $C_{\boldsymbol{g}}$, we want to constrain $\boldsymbol{u}(t_n)$ so that the CAMR state remains safe with respect to $C_{\boldsymbol{g}}$ at t_{n+1}. In other words, the steering vector $\boldsymbol{u}(t_n)$ must be constrained so that the default maneuver trajectory starting from $\boldsymbol{p}(t_{n+1})$, $\theta(t_{n+1})$, and $\boldsymbol{v}(t_{n+1})$ remains within $C_{\boldsymbol{g}}$. Using the definition of the safe state for the CAMR at time t_{n+1}, the free-space constraint can be expressed as

$$[n_i^g]^T p^*(s,p(t_{n+1}),v(t_{n+1}))\leqslant d_i^g. \tag{15}$$

where $i=1$, 2, and 3 for the first, second, and third boundaries of the free-space $C_{\boldsymbol{g}}$, respectively. To obtain a linear approximation to the environmental constraint set $U_e(t_n)$ characterized by the inequalities (15), we consider the effect of the free-space boundaries separately for each steering vector component $\dot{\delta}$ and $\dot{\nu}_r$. This is motivated by the fact that the constraints on the components of the steering vector (the steering angle rate and the rear wheel acceleration limits) are independent.

We first consider the problem of mapping the free-space boundaries into constraints on the steering rate $u_2=\dot{\delta}$. Given the CAMR position $\boldsymbol{p}(t_n)$, the *zero-input trajectory* resulting from setting the steering vector to $\boldsymbol{u}(t_n)=[0,0]$ follows the arc $\boldsymbol{p}(t_n){}^\wedge\boldsymbol{b}(\boldsymbol{p}(t_n),\boldsymbol{v}(t_n),K(\delta(t_n)),s)$. From the same position, we can also find the steering angle δ_d, which if held constant, would generate a position trajectory that passes through the subgoal position $\boldsymbol{g}$. We call this the desired trajectory that follows the arc $\boldsymbol{p}(t_n)$ ${}^\wedge\boldsymbol{a}(\boldsymbol{p}(t_n),\boldsymbol{v}(t_n),K_d,s)$, where $K_d=K(\delta_d)$. We can guarantee the trajectory of the CAMR from the time t_n to t_{n+1} followed by the default trajectory starting at time t_{n+1}

will lie between the zero-input and desired trajectories by imposing the following constraints on the steering vector component $u_2(t_n)$:

$$\underline{u}_2 \equiv \frac{\min(\delta(t_n), \delta_d) - \delta(t_n)}{\Delta} \leqslant u_2(t_n) \leqslant \frac{\max(\delta(t_n), \delta_d) - \delta(t_n)}{\Delta} \equiv \bar{u}_2. \quad (16)$$

By constraining the steering vector component $u_2(t_n)$ as in (16), we only needed to consider portions of the free-space boundaries between the zero-input and desired trajectories in order to guarantee that the CAMR state is safe at t_{n+1}.

Turning to the constraints on $u_1 = \dot{v}_r$, the rear wheel acceleration, we note that for each u_2 within the constraint range (16) the resulting trajectory of the CAMR from the time t_n to t_{n+1} and the default trajectory thereafter will intersect the linear boundary at an unique point. Thus, the problem becomes one of constraining the rear wheel acceleration $u_1(t_n)$ so that the resulting trajectory from t_n to t_{n+1} and the default trajectory thereafter is bounded by a known linear boundary of the free-space. In [27], an explicit expression is derived for a limit $\bar{u}_1(t_n)$ such that $u_1(t_n) \leqslant \bar{u}(t_n)$ guarantees the CAMR is in a safe state at time t_{n+1}, assuming $u_2(t_n)$ satisfies (16).

Thus, the obstacle constraints represented by the free-space boundaries can be transformed into three linear constraints on the steering vector $\boldsymbol{u}$ to obtain an environmental constraint set $U_e(t_n)$ given by

$$U_e(t_n) = \{\boldsymbol{u} | u_1(t_n) \leqslant \bar{u}_1(t_n);\ \underline{u}_2(t_n) \leqslant u_2(t_n) \leqslant \bar{u}_2(t_n)\}.$$

We note that there is no lower bound on $u_1(t_n)$, reflecting our assumption that the CAMR does not move in reverse.

E. Objective Vector and Steering Vector Selection

To define an objective steering vector that would guide the CAMR toward the subgoal when no obstacles are present, we use the following heuristic: Turn the steering wheel toward the subgoal until the orientation of the CAMR is such that it is pointing directly at the subgoal. This heuristic is motivated by the bang-bang form of the optimal two-dimensional (2-D) trajectory when curvature constraints are imposed [35]. Thus, we define the steering rate component of the objective vector, $u_2^*(t_n)$, to be equal to the signed steering rate limits with damping coefficients and a dead-zone to avoid oscillations [27].

Once the steering angle change rate is determined for the objective steering vector, the rear wheel acceleration is computed using a heuristic bang-bang acceleration control along the curvilinear path. In this case, the curvilinear length from the current position to the subgoal is estimated. Given the speed of the CAMR, we can use the time-optimal switching curve to determine if the maximum acceleration or deceleration should be applied.

Given the components of then objective vector $\boldsymbol{u}^*(t_n)$ as defined previously, and the environmental and system constraint sets $U_e(t_n)$ and $U_s(t_n)$, the steering vector is selected by solving the standard linear optimization problem:

$$\max \boldsymbol{u}(t_n)^T \boldsymbol{u}^*(t_n)$$

$$\text{s.t. } \boldsymbol{u}(t_n) \in U_e(t_n) \cap U_s(t_n).$$

This is a very simple problem because the constraint sets $U_e(t_n)$ and $U_s(t_n)$ both have linear boundaries that are parallel to the axes of the decision space. Therefore, a solution will always be on a vertex of the rectangular region formed by the intersection of the constraint sets and it can be determine by merely examining the signs of the components of the objective vector.

VI. Simulation Results

In this section we present results of simulation implementations of the SSA and SDA for both the OAMR and the CAMR. In the simulation experiments, the obstacle environment is a rectangular region scanned to generate a simulated local map before each execution of the SSA. The simulated local map contains line segments representing the portions of the obstacle faces which are visible from the AMR. An obstacle growing algorithm is also executed to account for the AMR size. In the figures illustrating the simulation results, the obstacle regions are shown by shaded areas, the local map for the current AMR position is shown by the solid lines on the boundaries of the obstacles, and the grown map by dotted lines. Also shown is the current free-space. The initial position of the AMR is indicated by a cross.

We used the following parameters for the OAMR system constraints: $a_{\max} = 10$ m/s^2, $v_{\max} = 80$ m/s. These limits are based on the experimental data from the Ilonator Cart [25]. For the CAMR, we chose the following limits to facilitate comparisons between the CAMR and OAMR trajectories: $\dot{v}_{\max} = 1.0$ ft/s^2, $\hat{\delta}_{\max} = 0.35°$/s, $v_{\max} = 8.0$ ft/s, $\delta_{\max} = 28.6°$.

All simulations were run on a DEC Microvax II computer. For the OAMR, the SSA was performed at 1-s intervals (simulation time) while the SDA was performed at 0.5-s intervals. For the CAMR, the SSA was also performed at 1-s intervals, but the SDA was performed at 0.1-s intervals. On the average, it takes 50.31 ms to execute the SDA for the CAMR at each iteration. The SDA implemented in the simulation has not been optimized for speed. For the OAMR, it takes only 5 ms to execute the SDA on the average. The SSA takes an average of about 11 ms per iteration for both the CAMR and the OAMR. These numbers suggest the feasibility of the real-time implementation of the SSA and SDA to achieve the sampling rates used in the simulations. The length of the SSA computations depends linearly on the number of obstacles in the local map. In actual applications, the sensors are limited in their ranges (both direction and distance). These limits will constrain the number of obstacles in each frame of the local map, which will in turn bound the maximum computation time.

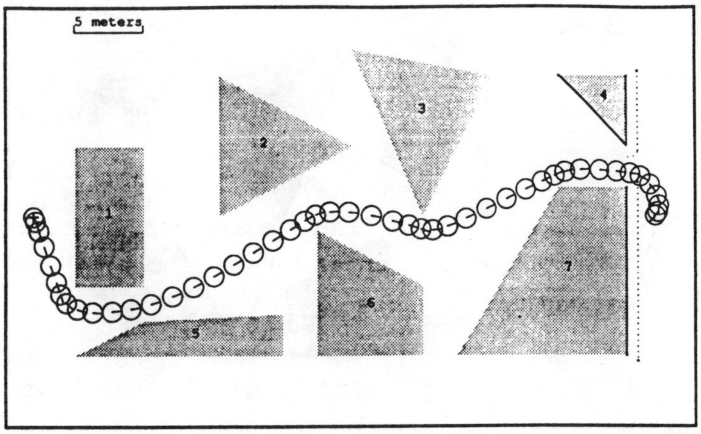

Fig. 11. Trajectory of the OAMR. Total trajectory time is 93 s; OAMR displayed at 2-s intervals.

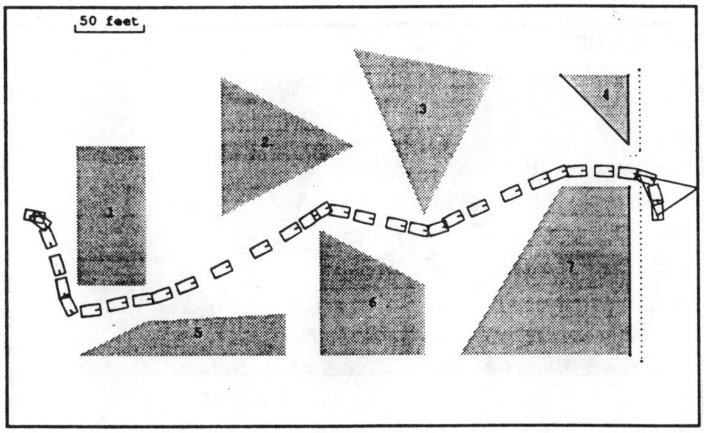

Fig. 12. Trajectory of CAMR. Total trajectory time is 94.2 s; CAMR displayed at 2 s intervals.

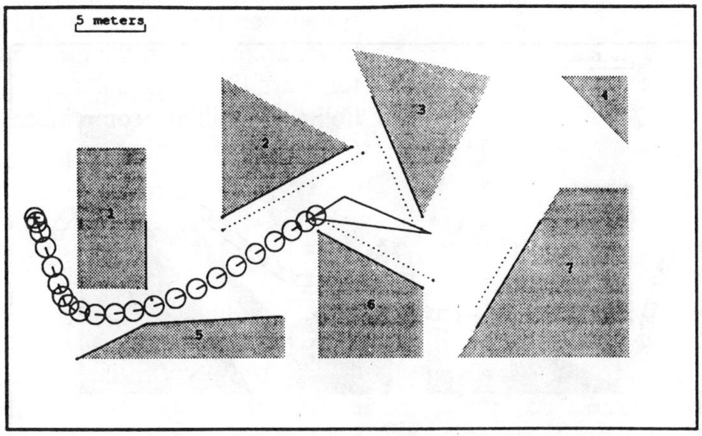

Fig. 13. Trajectory of OAMR after 39.5 s.

Figs. 11 and 12 show the trajectories in the same environment traversed by the OAMR and the CAMR respectively. The OAMR trajectory is very smooth going around the corners of the obstacles. In comparison, the CAMR makes sharper turns around the corners. We note the difference in the trajectories as the AMRs move from the starting position, with an initial orientation of -0.2 rad. The OAMR takes a straight path toward the corner of the obstacle 1, because of its omnidirectional mobility. On the other hand, the CAMR must follow a curved path because its trajectory is dependent on its initial orientation and limited by its maximum steering angle. We note also the slight oscillation in the CAMR trajectory after it turns around the first corner of the obstacle 1. This is caused by the limit on the rate by which the steering angle can be changed.

Figs. 13–15 show intermediate positions of the OAMR and CAMR along their trajectories. These figures illustrate the current local map, grown local map, and the free-space definitions.

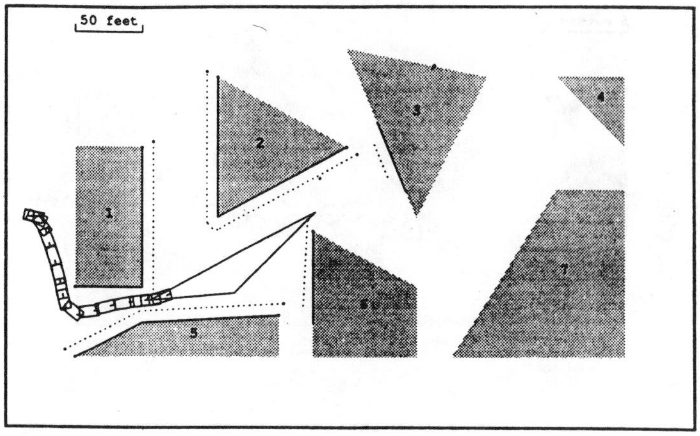

Fig. 14. Trajectory of CAMR after 30.1 s.

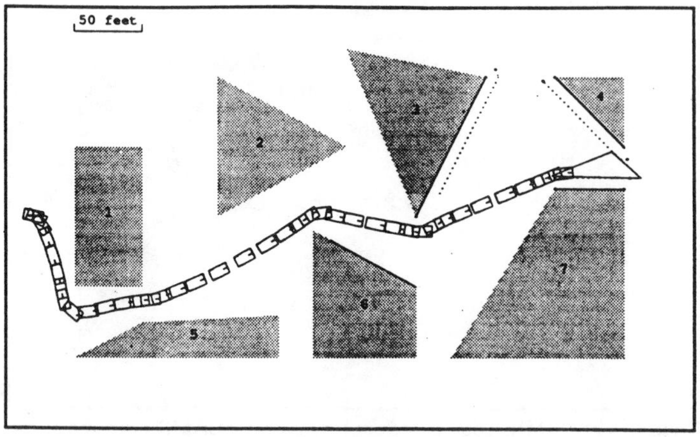

Fig. 15. Trajectory of CAMR after 78.1 s.

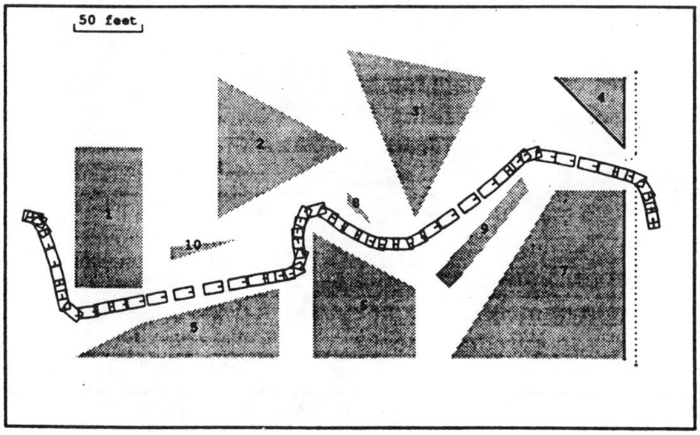

Fig. 16. CAMR trajectory in more cluttered environment. Total trajectory time is 104.7 s. CAMR displayed at 2-s intervals.

Fig. 16 shows the CAMR navigating in a more cluttered environment. We notice that the change in the shape of obstacle 5 caused a major alteration in the trajectory which appears rather unreasonable. Instead of going directly to the top corner of the obstacle 6, the CAMR takes a large detour. This can be explained by the navigation being based solely on the local map. After circumventing the second corner of the obstacle 1, the local map shows only the portion of the obstacle 6 that is not blocked by obstacle 5. Based on this information, the CAMR is steered toward the right edge of the obstacle 6 because that edge is closer to the final goal in the local map. When the CAMR eventually passes the corner of the obstacle 5 where the complete face of the obstacle 6 is visible, it appears that the top corner of the obstacle 6 will be a shorter way to get to the final goal. This results in

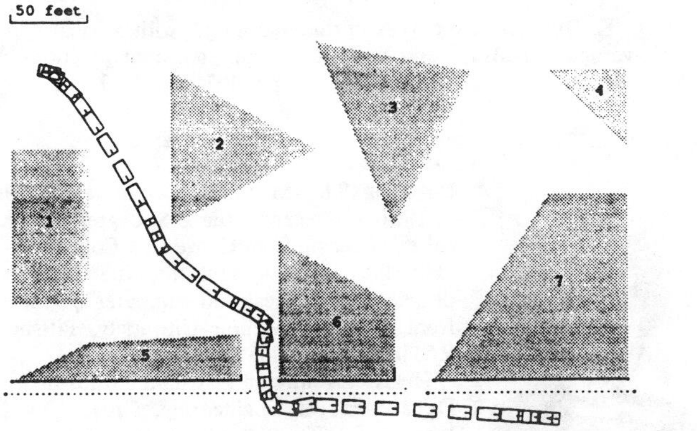

Fig. 17. CAMR trajectory between alternate set of initial and goal positions. Total trajectory time is 72.9 s. CAMR displayed at 2-s intervals.

the sharp turn in the trajectory. Fig. 17 shows the CAMR trajectory starting from a different initial position to a different subgoal.

VII. Conclusion

In this paper, we present a feedback approach to the problem of AMR navigation in obstacle strewn environments. The proposed satisficing feedback strategies generate feasible trajectories to the goal position using only local information. We demonstrate how constraints on the AMR due to the obstacles and the operating limits can be accommodated for two classes of AMRs, namely, omnidirectional and conventionally-steered AMRs. Implementations of the navigation algorithms based on this approach for both types of the AMRs show the effectiveness of the approach under ideal conditions where the sensor, servo, and state estimation errors are neglected.

The satisficing feedback strategies can be easily modified to accommodate sensor and servo errors arising in real systems. In particular, sensor inaccuracies will lead to errors in the representations of the obstacles in the local maps. Servo errors will also introduce inaccuracies in the estimate of the AMR position. Assuming a bound on these errors, possibly as a function of the distances to the obstacle, obstacles can be grown appropriately to account for the noisy data. Sensor noise will also introduce inconsistencies in the local maps generated at different times. To make the feedback algorithms robust with respect to these inaccuracies, the subgoal selection algorithm can be modified to use only the most recent map for generating subgoals. Delays in the map generation and limited sensor range can also be accommodated. Details of these enhancements and results from simulation experiments including sensor and servo errors are described in [27] and [28]. The net effect of these modifications to the basic feedback strategy is that the AMR is driven more conservatively.

The CAMR version of the SDA was recently implemented on the CMU NAVLAB, an autonomous land vehicle with a laser ranger as the principal source of environmental data. The implemented version of the algorithm includes features to handle uncertainties caused by the sensor range limit, sensor noise and servo errors. In experimental runs, the algorithm was able to drive the NAVLAB around obstacles placed in its path to the goal. Results of these experiments will be reported in a future paper.

References

[1] T. Lozano-Perez and M. A. Wesley, "An algorithm for planning collision-free paths among polyhedral obstacles," *Comm. ACM*, vol. 22, no. 10, pp. 560–570, Oct. 1979.

[2] T. Lozano-Perez, "A simple motion-planning algorithm for general robot manipulators," *IEEE J. Robotics Automat.*, vol. RA-3, no. 3, pp. 224–238, 1987.

[3] J. H. Reif, "Complexity of the mover's problem and generalizations," in *Proc. IEEE Symp. Found. Comput. Sci.*, 1979, pp. 421–427.

[4] J. T. Schwartz and M. Sharir, "On the piano movers' problem: I. The special case of a rigid polygonal body moving amidst polygonal barriers," *Advances in Appl. Math.*, vol. XXXVI, pp. 345–398, 1983.

[5] ____, "On the piano movers' problem: II. General techniques for computing topological properties of real algebraic manifolds," *Advances in Appl. Math.*, vol. XXXVI, pp. 345–398, 1983.

[6] C. O'Dunlaing, M. Sharir, and C. K. Yap, "Retraction: A new approach to motion planning," *15th FOCS*, pp. 207–220, 1983.

[7] E. G. Gilbert and D. W. Johnson, "Distance functions and their application to robot path planning in the presence of obstacles," *IEEE J. Robotics Automat.*, vol. RA-1, no. 1, pp. 21–30, Mar. 1985.

[8] K. G. Shin and N. D. McKay, "Minimum-time control of robotics manipulators with geometric path constrains," *IEEE Trans. Automat. Contr.*, vol. AC-30, no. 6, pp. 531–541, June 1985.

[9] B. K. Kim and K. G. Shin, "Minimum-time path planning for robot arms and their dynamics," *IEEE Trans. Syst. Man Cybern.*, vol. SMC-15, no. 2, pp. 213–223, Mar./Apr. 1985.

[10] O. Khatib, "Real-time obstacle avoidance for manipulators and mobile robots," *The Int. J. Robotics Res.*, vol. 5, no. 1, 1986.

[11] N. Hogan, "Impedance control: An approach to manipulation," *J. Dyn. Syst. Measurement Contr.*, vol. 107, pp. 1–24, Mar. 1985.

[12] B. H. Krogh, "Feedback obstacle avoidance control," *21st Allerton Conf. Commun. Contr. Computing*, Urbana, IL, pp. 325–334, Oct. 1983.

[13] ____, "A generalized potential field approach to obstacle avoidance control," in *Proc. Robotics Int. Robotics Res. Conf.*, Bethlehem, PA, Aug. 1984.

[14] W. S. Newman and N. Hogan, "High speed robot control and obstacle avoidance using dynamic potential functions," in *Proc. Robotics Automat.*, IEEE, Raleigh, NC, 1987, pp. 14–22.

[15] R. Volpe and P. Khosla, "Artificial potentials with elliptical isopotential contours for obstacle avoidance," in *Proc. 26th Conf. Decision Contr.*, IEEE, Los Angeles, CA, Dec. 1987.

[16] V. J. Lumelsky and A. A. Stepanov, "Dynamic path planning for a mobile automaton with limited information on the environment," *IEEE Trans. Automat. Contr.*, vol. AC-31, no. 11, pp. 1058–1063, Nov. 1986.

[17] ____, "Path planning strategies for a point mobile automaton moving amidst unknown obstacles of arbitrary shape," *Algorithmica*, 1987.

[18] D. J. Daniel, "Analysis, design, and implementation of microprocessor control for a mobile platform," Master's thesis, Carnegie Mellon Univ., Pittsburgh, PA, Aug. 1984.

[19] T. Hongo *et al.*, "An automatic guidance system of a self-controlled vehicle: The command system and the control algorithm," in *IEEE Proc. IECON*, San Francisco, CA, Nov. 1985, pp. 18–22.

[20] W. L. Nelson and I. J. Cox, "Local path control for an autonomous vehicle," Tech. Rep. AT&T Bell Laboratories, 1988.

[21] P. F. Muir and C. P. Neuman, "Kinematic modeling of wheeled mobile robots," Tech. Rep. CMU-RI-TR-86-12, Carnegie Mellon Univ., Pittsburgh, PA, 1986.

[22] A. Elfes, "A sonar-based mapping and navigation system," in *Proc. IEEE Int. Conf. Robotics Automat.*, San Francisco, CA, Apr. 1986, pp. 1151–1156.
[23] I. J. Cox, "An autonomous robot vehicle for structured environments," Tech. Rep. 311401-2599, AT&T, Sept. 1987.
[24] B. H. Krogh and D. Feng, "Dynamic generation of subgoals for autonomous mobile robots using local feedback information," *IEEE Trans. Automat. Contr.*, May 1989.
[25] D. Feng *et al.*, "The servo-control system for an omnidirectional mobile robot," in *Proc. IEEE Int. Conf. Robotics Automat.*, 1989.
[26] R. Wallace *et al.*, "Progress in robot road-following," in *Proc. IEEE Int. Cont. Robotics Automat.*, San Francisco, CA, Apr. 1986, pp. 1615–1621.
[27] D. Feng, "Satisficing feedback strategy for local navigation of autonomous mobile robot," PhD dissertation, Carnegie Mellon Univ., Pittsburgh, PA, May 1989.
[28] D. Feng and B. Krogh, "A robust satisficing strategy for autonomous navigation," in *Proc. Fourth IEEE Int. Symp. Intelligent Contr.*, Albany, NY, Sept. 1989.
[29] B. Carlisle, *Developments in Robotics*. Kempston, England: IFS Publishing, 1983, ch. Omni-Directional Mobile Robot.
[30] H. P. Moravec, "Autonomous mobile robots annual report-1985," Robotics Institute Tech. Rep. CMU-RI-MRL 86-1, Carnegie Mellon Univ., Jan. 1986.
[31] D. Feng, "Dynamic steering control," Master's thesis, Dept. Elec. Comput. Eng., Carnegie Mellon Univ., Pittsburgh, PA, Mar. 1986.
[32] G. Podnar, K. Dowling and M. Blackwell, "A functional vehicle for autonomous mobile robot research," Tech. Rep. CMU-RI-TR-84-28, Carnegie Mellon Univ., Pittsburgh, PA, Apr. 1984.
[33] C. Helmers, "Ein Heldenleben, (Or, A Hero's Life, With Apologies to R. Strauss)," *Robotics Age*, vol. 5, no. 2, pp. 7–16, Mar./Apr. 1987.
[34] T. J. Graettinger and B. H. Krogh, "Evaluation and time-scaling of trajectories for wheeled mobile robots," *ASME J. Dynamic Syst. Measurement Contr.*, pp. 222–231, June 1989.
[35] L. E. Dubins, "On curves of minimal length with a constraint on average curvature, and with initial and terminal positions and targents," *Amer. J. Math.*, vol. 79, pp. 497–516, 1957.

Dai Feng (S'82–M'89) was born in Beijing, China in 1961. He received the B.S. degree in electrical engineering from Grove City College, Grove City, PA, in 1984, and the M.S. and Ph.D. degrees in electrical and computer engineering from Carnegie Mellon University, Pittsburgh, PA, in 1986 and 1989, respectively.

He is currently an Assistant Professor in the Department of Engineering, Grove City College, Grove City, PA. His research interests are in the area of control and navigation of autonomous vehicles.

Bruce H. Krogh (S'82–M'82) received the B.S. degree in mathematics and physics from Wheaton College, Wheaton, IL, in 1975, and the M.S. and Ph.D. degrees in electrical engineering from the University of Illinois, Urbana, in 1978 and 1983, respectively.

He is currently an Associate Professor in the Department of Electrical and Computer Engineering at Carnegie Mellon University Pittsburgh where he conducts research in supervisory control, discrete event systems, and Petri nets.

Dr. Krogh was an associate editor for the IEEE Transactions on Automatic Control and received the Presidential Young Investigator Award from the National Science Foundation in 1985.

Region based route planning: multi-abstraction route planning based on intermediate level vision processing

Rajkumar S. Doshi, Raymond Lam, James E. White

Jet Propulsion Laboratory, California Institute Of Technology
4800 Oak Grove Drive, Pasadena, CA 91109

"Region-Based Route Planning: Multi-Abstraction Route Planning Based on Intermediate Level Vision Processing" by R.S. Doshi, R. Lam, and J.E. White. Reprinted from *SPIE Vol. 1003 Sensor Fusion: Spatial Reasoning and Scene Interpretation*, 1988. Copyright © 1988 by the International Society for Optical Engineering (SPIE), reprinted with permission.

ABSTRACT

The Region Based Route Planner performs intermediate-level and high-level processing on vision data to organize the image into more meaningful higher-level topological representations. A variety of representations are employed at appropriate stages in the route planning process. A variety of abstractions are used for the purposes of problem reduction and application of multiple criteria at different phases during the navigation planning process.

The Region Based Route Planner operates in terrain scenarios where some or most of the terrain is occluded. The Region Based Route Planner operates without any priori maps. The route planner uses two dimensional representations and utilizes gradient and roughness information. The implementation described here is being tested on the JPL Robotic Vehicle.

The Region Based Route Planner operates in two phases. In the first phase, the terrain map is segmented to derive global information about various features in it. The next phase is the actual route planning phase. The route is planned with increasing amounts of detail by successive refinement. This phase has three abstractions. In the first abstraction, the planner analyses high level information and so a coarse, region-to-region plan is produced. The second abstraction produces a list of pairs of entry and exit waypoints for only these selected regions. In the last abstraction, for every pair of these waypoints, a local route planner is invoked. This planner finds a detailed point-to-point path by searching only within the boundaries of these relatively small regions.

1. RESEARCH ON SPATIAL REPRESENTATIONS

A principal objective in autonomous mobility using sensors is to analyze and understand the environment in terms of freedom in mobility. The outside world has to be cast into a representation. The choice of the representation is critical, because some representations allow certain kinds of reasoning more conveniently than other representations. There does not exist a consensus on what is the "right" representation. The representation of the environment depends on various factors, including detail provided in a priori maps, whether mobility is in two, two and a half, or three dimensions; type of surface to be traversed; types of sensors; range of sensors; geometry of obstacles to be avoided; static domains; number of interacting robots; demands on real-time performance; and non-availability of a priori maps. The importance of such factors in an application influences the choice of the route planning algorithm to be employed.

In general, there are two popular ways to represent space: representing the free space explicitly and representing the obstacle space explicitly. Within the free space approach, there are two methods: Cell Decomposition and Retraction [Yap 87]. The obstacle space representations include the Visibility Graph.

Cell decomposition methods include the Quad Tree [Samet 83] and Oct Tree Methods.The advantage of the grid or quad tree (hierarchical grid) is the simplicity of the shape of the cells.

Retraction methods [O'Dunlaing *et al.* 83] map the free space into a representation of lower dimensions. The complement of the obstacle space can be represented by Generalized Cylinders or Voronoi Diagrams. Generalized Ribbons are used by the Freeway Method [Brooks 83] to represent free space between obstacles in the form of a Freeway Net Graph.

The visibility graph is a representation for the obstacle space. The path has to be planned with reference to the vertices of the polygonal obstacles. A graph of only those vertices that are visible to one another is called the visibility graph. Visibility graphs are also conceptually simple and can be combined with other representations [Gewali *et al.* 88, and Meng 88].

As discussed above, spatial representations extend from sensor-level to syntax-level to semantic-level. Each representation offers unique advantages and disadvantages. In practice, representations range from grids (where the format is generally similar to the inputs received from sensors) to multi-resolution and multi-representation topological maps (where a considerable amount of processing has to be done to the input representation to derive the new representation). The spectrum of representations includes grids [Moravac 87, Slack *et al.* 87, Thorpe 85], visibility graphs [Lozano-Perez *et al.* 79, Rao *et al.* 1987], voronoi diagrams [Iyengar *et al.* 85, O'Dunlaing *et al.* 1982], hierarchical representations [Samet 83, Slack 88], hierarchical and heterogeneous representations [Giralt *et al.* 84, Kambhampati *et al.* 86, Kuipers *et al.* 88, Laumond 83].

It must be noted that "multiple resolution representations" refer to hierarchical single representations such as the quad tree or oct tree. In contrast, "multiple representations" refer to a collection of (possibly hierarchical) diverse representations such as geometric, topological, and semantic data structures. Multiple representations offer more advantages than single representations (which includes multiple resolution representations) if the goals of the systems include several of the following research areas: achieving high-level goals, path planning, map integration, map learning, landmark tracking, generating expectations, monitoring expectations, diagnosing anomalies, and recovering from failures. For many applications this seems to be a current trend [Chatila *et al.* 85, Elfes 86, Moravac 88], although complete implementations do not yet exist.

In our route planning application, a grid is used as the lowest-level description of space. The higher-level representations include regions, a graph of region interconnections, and certain geometric properties of regions (to be defined later). The regions are constructed by segmentation which will be described in the next section.

The voronoi diagram could have been used to construct regions. But as one of the basic inputs to constructing the voronoi diagram, this method needs a set of points. In the route planning domain, these points would relate to the known obstacles. However, in our domain, a hill may be an obstacle if a vehicle is below it, but may not an obstacle if a vehicle is on top of the hill. Also, obstacles in our domain have irregular shapes and are not point-size. It must be noted that a new method for constructing voronoi regions from obstacles of arbitrary shapes [Meng 88] and navigation algorithms [Gewali *et al.* 88] deserve further research and attention.

The quad tree is another method to subdivide a 2D map. Additionally, quad trees are appropriate for representing low-level sensory data. In our research we use the quad tree for maintaining consistency in the integrated map [Doshi *et al.* 1987] and as the lowest-level representation of space. The quad tree imposes an explicit rigid symmetry on the obstacles and is appropriate when the environment can be modeled by well-defined tesselations. In our application, the terrain is not so well-defined. Hence, the quad tree approach does not apply well to topographies where the regions (objects) do not conform to regular, straight line symmetries.

2. REGION SEGMENTATION

2.1 Fundamental Approaches

Understanding a computer image would require dissecting, and then interpreting the anatomy of the image. These two steps are the intermediate-level processing that is concerned with region segmentation, and the high-level processing involving interpreting, reasoning, or building a model with the dissected image.

An image can be looked upon as a two dimensional (2D) brightness function measured in gray scale. Segmentation is the process of partitioning this function into meaningful coherent areas [Bajcsy *et al.* 1986]. There are two approaches to computer image segmentation: segmentation by edge detection and by region growing.

2.1.1 Edge Detection technique. In this technique, boundary points (and hence boundaries) are detected by sharp changes in a 2D brightness function. The technique assumes that pronounced changes in brightness imply the demarcation of boundaries of objects in the image. This method forms boundaries by using the property of difference or discontinuity. Boundaries are detected by linear or non-linear (window) methods.

2.1.2 Region Growing technique. Region growing is a procedure that groups points in an image into larger regions. This method forms regions by using the property of similarities. The simplest of these approaches is pixel aggregation, where we start with a set of seed points and from these grow regions by appending to each seed point those neighboring pixels that have similar properties (e.g., gradient, height, gray level, texture, color). There exist three prominent ways to form regions [Riseman *et al.* 77]:

- Local spatial examination: Local areas are merged by recursively comparing syntactic features. Our approach is based on an extensive modification of the method in [Brice *et al.* 1970].
- Global examination of feature: The image is pre-processed to produce histograms that are then used to recursively split the image into homogeneous regions.
- Combining semantic and syntactic methods: Syntactic clustering based on histograms is then refined by semantic knowledge.

2.2 Basic Concepts

2.2.1 Definitions and Notations. The convention adopted is similar to [Gonzalez *et al.* 87]. Let f(x, y) denote an image. A location (x, y) in the image will be referred to as a pixel In the following sections, we will refer to an image by the term grid and pixels will be referred to as either quads or cells. Lower case letters such as p and q will be used to refer to particular pixels in f(x, y). Every pixel in an image is said to have a value. For example, in binary images, a pixel p can have a value of either zero or one. Uppercase letters such as S will be used to denote sets of pixels in f(x, y), i.e., S is a cluster of pixels of an image. Clusters will also be referred to as segments or regions.

2.2.2 Neighbors, Connectivity and Adjacency of Pixels. A pixel p at coordinates (x, y) has has two horizontal and two vertical neighbors, a unit distance away from p. The coordinates of these neighbors are given by:

$$(x + 1, y), (x - 1, y), (x, y + 1), (x, y - 1).$$

Let N4(p) denote the four non-diagonal neighbors of p. A subset of N4(p) may lie outside of f(x, y) if p is on the perimeter of f(x, y). Also, p has these four diagonal neighbors denoted by ND(p):

$$(x + 1, y + 1), (x + 1, y - 1), (x - 1, y + 1), (x-1, y-1)$$

The union of diagonal and non-diagonal neighbors gives the eight neighbors of p denoted by N8(p). Members of N8(p) that lie inside f(x, y) are legal neighbors of p. In the remainder of this paper we will concern ourselves with only the legal neighbors of pixels. Hence, let N4(p), ND(p) and N8(p) refer to the legal non-diagonal neighbors, legal diagonal neighbors and legal eight-neighbors of p. A pixel p is 4-connected to q, if $q \in N4(p)$. A pixel p is 8-connected to q, if $q \in N8(p)$. The terms, 8-connected, connected and adjacent will be used interchangeably when referring to pixels.

2.2.3 Similarity between pixels. Let P be a set of Pi, where i = 1, 2, ... n, such that each partition Pi defines a disjoint set of values which together cover the entire range of values in an image. Each Pi defines a similarity criterion. The partitions (along with other criteria to be defined) are used to decide if a pixel is a member of a cluster of other pixels. Pixel p is similar to q, if p and q both have values that belong an identical partition Pi. p and q are dissimilar if they belong to non-identical partitions.

For example, in the case of binary images, P = {P1, P2}, where P1 = {0} and P2 = {1}. Pixels having a value in the range of P1 are said to be black and those in the range of P2 are said to be white. A black pixel is not similar to a white pixel. Two white pixels or two black pixels are similar to each other.

2.2.4 Paths. A path exists between p and q if there is a sequence of one or more pixels p, r, s, ..., t, q such that every consecutive pair (such as p and r, or r and s) is connected. The pixels p & q will be referred to as the end-points of the path p, r, s, ..., t, q. The pixels r, s, ..., t are called the intermediate points of the path. If all the end-points and intermediate points are similar, then we will refer to the path as a legal-path. If all the intermediate points are dissimilar to p, the path will be termed as a dissimilar-path. If some of the intermediate points are similar and the rest are dissimilar, the path will be referred to as a mixed-path.In a path, all pixels except possibly the first and the last are distinct.

2.2.5 Boundary Cell. Let p belong to a region Ri. A pixel p is called a boundary cell if at least one of the members of N8(p) does not belong to Ri.

2.3 Basic Definitions for Regions

2.3.1 Definition of a region. A region Ri (i = 1, 2, ..., n) is a cluster of one or more pixels such that there is a legal-path from one pixel in the region to every other pixel in the same region. Each pixel in this cluster is said to belong to the region Ri.

2.3.2 Boundary of a region. A boundary Bi of a region Ri is the set of all boundary cells p, where $p \in Ri$.

2.3.3 Pixels inside a region. A pixel p is inside a region Ri, if $p \in Ri$ and $p \notin Bi$.

2.3.4 Outer Boundary of a region. The outer boundary Oi of a region Ri is the set of boundary cells p in Ri such that there exists a dissimilar-path from p to some point on the border of the image. Any region Ri will always have an outer boundary Oi.

2.3.5 Engulfing of regions. A region Ri engulfs another region Rj (i≠j) if for every pixel q in Oj, members of N8(q) that do not belong to Rj, must all belong to Ri.

2.3.6 Inner Boundary of a region. The inner boundary Ii of a region Ri is the set of boundary cells p in Ri such that there does not exist a dissimilar-path from p to some point on the border of the image, but there does exist a dissimilar-path from p to some point on the border of the image.

The following properties hold: Ii and Oi cannot have common members. In fact, even if a region Ri engulfs Rj (i≠j), Ii may not exist (as in the case where Ri is only one pixel wide and Ri engulfs Rj). Hence, Bi = (Ii U Oi) = Oi if Ri is only one pixel wide and Ri engulfs Rj.

2.3.7 Adjacency and border-contacts of regions. A region Ri is adjacent to (connected to) region Rj if ∃ a p that is adjacent to q, such that $p \in Bi$, and $q \in Bj$ ($i \neq j$). The pair of pixels (p, q) is termed a border-contact between Ri and Rj.

2.3.8 Directional relationships between regions. A pixel p at (x1, y1) is north of pixel q at (x2, y2) if:

- p and q are adjacent
- p and q belong to f(x, y)
- x1 = x2
- y1 = y2 + 1

Then, a region Ri is north of region Rj ($i \neq j$) if p is north of q where $p \in Bi$ and $q \in Bj$. (Other directions are defined similarly.)

2.4 Basic Axioms for Region Segmentation

The axioms can be stated in many ways [Duda *et al.* 73, and Stanton 69]. The formulation here has been adopted from [Gonzalez *et al.* 87]. Let R represent the entire image region. Let Ri (where i = 1, 2, ..., n) represent a unique subregion in R. Let P(Ri) represent the application of a boolean predicate to properties of a region Ri. The process of region segmentation would partition R into n subregions, R1, R2, . . . , Rn, such that:

(Axiom 1) The segmentation must be complete:

$$\bigcup_{i=1}^{n} Ri = R, \quad \text{and } R = f(x, y)$$

(Axiom 2) Every pixel in the regions must be connected: Ri is a connected region, i = 1, 2, . . . , n

(Axiom 3) Every region is disjoint from other regions:

$$Ri \cap Rj = \varphi \text{ for all i and j, } i \neq j$$

(Axiom 4) There must exist a predicate P such that all pixels in a unique region Ri satisfy P and all pixels in Rj, where $i \neq j$, cannot satisfy P:

$$P(Ri) = \text{TRUE for } i = 1, 2, \ldots, n, \text{ and}$$
$$P(Ri \cup Rj) = \text{FALSE for } i \neq j$$

3. THE REGION BASED ROUTE PLANNER

This route planner embodies several important themes. These themes are significant in the context of our work. Their significance would vary with the application. We will examine these themes, the route planner, and then its merits and limitations.

3.1 Theme: Organization of information and utilizing global information

Analysis and understanding requires organization. Partitioning an image leads to transforming the image into meaningful parts. Region segmentation is one way to produce newer representations of the information implicit but inaccessible prior to processing. Higher layers of newer representations facilitate higher forms of behavior and reasoning. Every unique class of representations have a unique utility. Also, it is not mandatory to use every representation simultaneously. Different representations should be used at different stages in the route planning process.

Another emphasis is to use as much information about the terrain as is available. Even more important is to collect all this information before any route planning begins. The idea is not simply to optimize the path length or energy expended. The central idea is to incorporate global constraints into the earliest possible stages of the path planning (Figure 1). Examples of such constraints include ensuring concealment and completely avoiding high cost areas.

Depending on the size and nature of one's application, utilizing global constraints could lead to intelligent behavior because the blind or automatic backtracking behavior of local route planners is avoided.

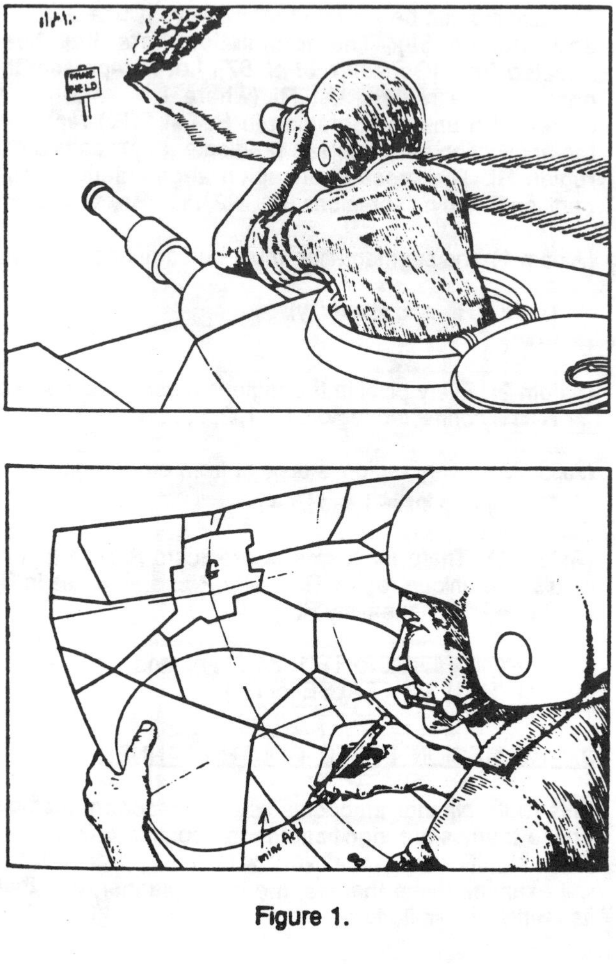

Figure 1.

3.2 Theme: Abstractions and problem reduction allow applicability of multiple criteria

The idea of abstractions has existed for centuries. The process of creating abstractions is an important part of problem solving. One trick to solving problems is to reduce the original larger problem into sets of smaller and simpler problems using abstractions [Polya 73]. In many cases, problem reduction is just another way of restating the problem [Wickelgren 74]. The trick to finding a solution to most problems is largely the trick of turning vague generalities into precise abstractions [Zimmer 85]. In route planning, abstraction involves building the "walls", "rooms," and "corridors" for traversal.

Route planning in complex terrains involves using multiple and possibly complex criteria. All the detail and criteria do not apply to the problem solving right away. It is natural therefore to solve the problem avoiding irrelevant details when necessary.

The idea is to create a coarse route plan as the first step. This skeleton plan only identifies general regions of traversal. A skeleton path plan could be used to effectively guide a more detailed path planning process. Since each region has to be traversed, the problem is then reduced to finding the route for every (smaller) region. Global constraints are used to influence the intermediate route planners to search for a cohesive, connected exact route plan.

Collectively, the reduction in avoidable and needless backtracking, and a coarse guidance of a less detailed top-level path plan would lead to large savings in search time. Abstractions provide a clean, efficient, powerful, and practical approach to problem solving, reasoning, and debugging. Abstractions allow application of different criteria at different layers (Figure 2).

3.3 Two phases: Levels of processing and levels of reasoning

The Region Based Route Planner operates in 2 phases (Figure 3). In the first phase, an input array is converted into regions (predicates). This first pre-processing phase is called the Region Segmentation and Analysis Phase. The inputs to this phase are:

- A start point S
- An goal point G
- The set of partitions P (as defined in sections 2), and
- A 2D array which represents a cost composite terrain matrix [Cameron *et al.* 87 A].

The second phase is the actual route planning phase and is called the Global Route Planning Phase. During this phase, abstractions are used to reduce the problem of traversing a large terrain into a problem of traversing through an ordered sequence of relatively very small regions. The outputs of this phase are:

- A set of disjoint unique regions
- A directional adjacency graph representing the topographical layout of these regions
- A point-to-point path from the start point S to the goal point G.

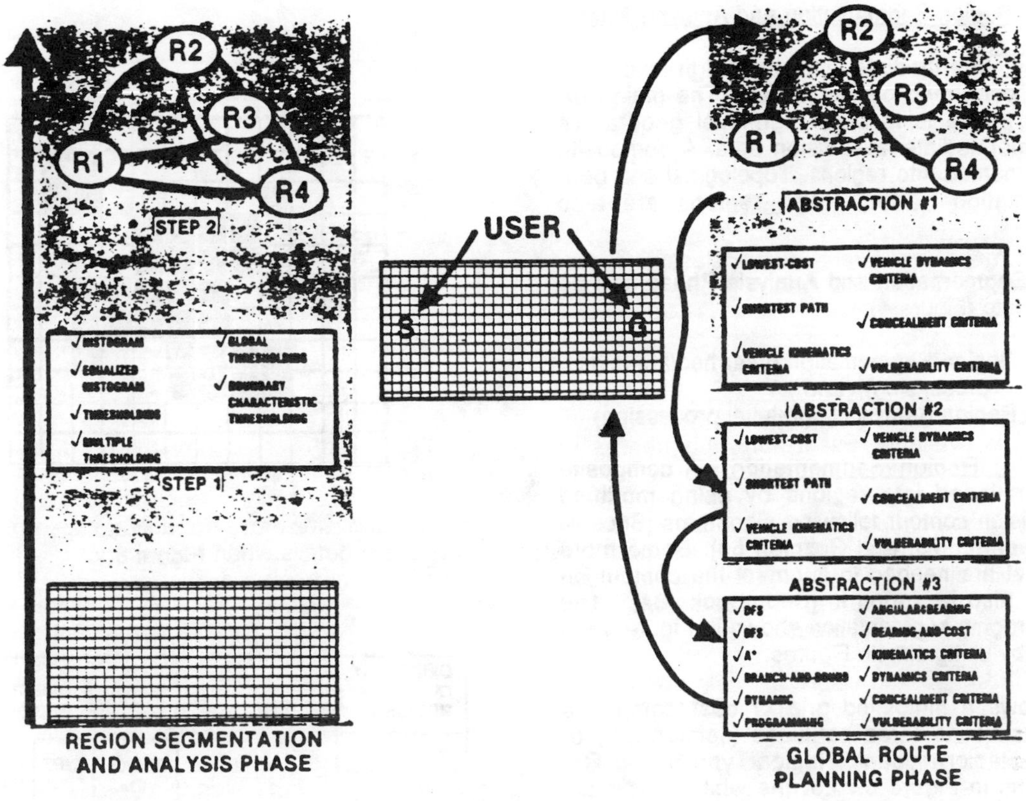

Figure 2.

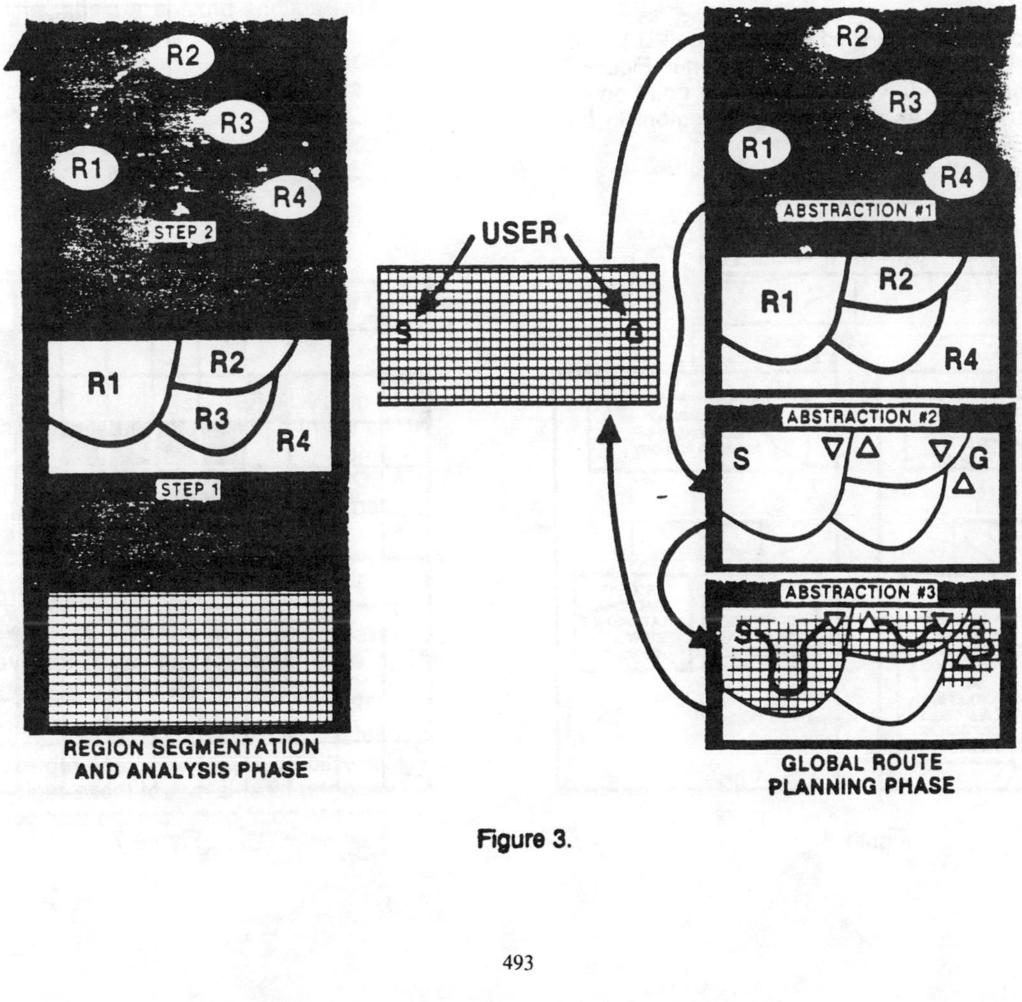

Figure 3.

3.4 Phase 1: Region Segmentation and Analysis Phase

In this phase, intermediate-level and high-level processing are performed on image data. The main purpose of this phase is to build high-level geographic maps from intermediate-level vision data. A composite matrix is segmented into regions. Topological and geometric information about these regions are also computed.

The Region Segmentation and Analysis Phase can be broken down into (Figure 4):

- STEP 1: Region Segmentation (intermediate-level processing), and
- STEP 2: Region Analysis (high-level processing)

3.4.1 STEP 1: Region Segmentation. A composite matrix is partitioned into regions by using modified versions of vision contour following algorithms [Brice *et al.* 70, Duda *et al.* 73, and Guzman 68]. Some more algorithms that are needed to augment the contour algorithms can also be found in [Sedgewick 84]. The algorithm for region segmentation shown in Figure 4 will be illustrated by using simple Figures.

Figure 5 shows a simplified grid or cost composite matrix. Let us define three partitions (section 2.2) or region types Region-Type-A, Region-Type-B and Region-Type-C as in Figure 6. Let the white cells have values belonging to partition Region-Type-A. The black cells have values belonging to partition Region-Type-B. Note that in our example, we do not have any cells belonging to Region-Type-C. The initial region directory data structures are also shown in Figure 6.

The search for an unprocessed boundary cell U begins at cell (1, 1), the initial seed point, as shown in Figure 7. An unprocessed boundary cell is always a point on the outer boundary of an as yet undetected region. In this

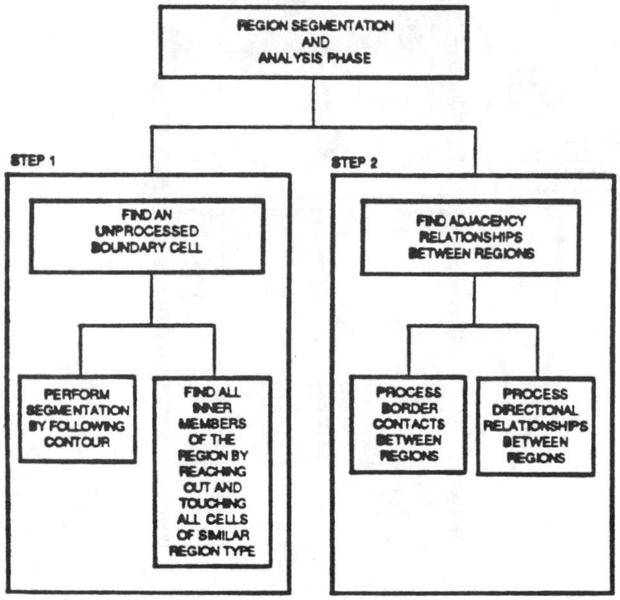

Figure 4.

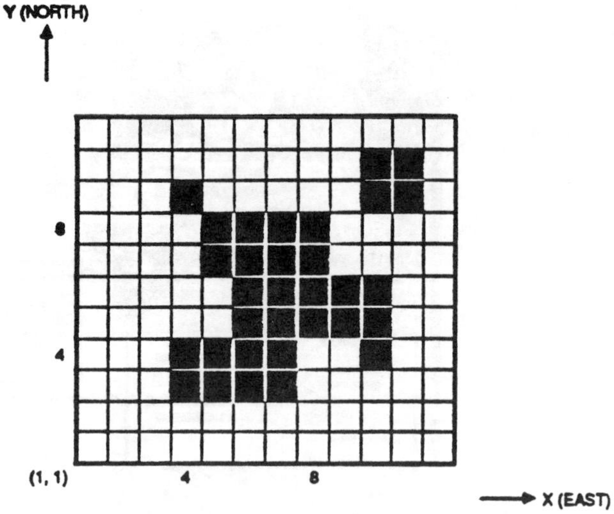

Figure 5.

DIRECTORY OF REGIONS
NIL

DIRECTORY OF REGION TYPES			
REGION IDs	TYPE	MIN OF RANGE	MAX OF RANGE
NIL	REGION-TYPE-A	0	0.3279999
NIL	REGION-TYPE-B	0.328	0.4859999
NIL	REGION-TYPE-C	0.486	1.000

Figure 6.

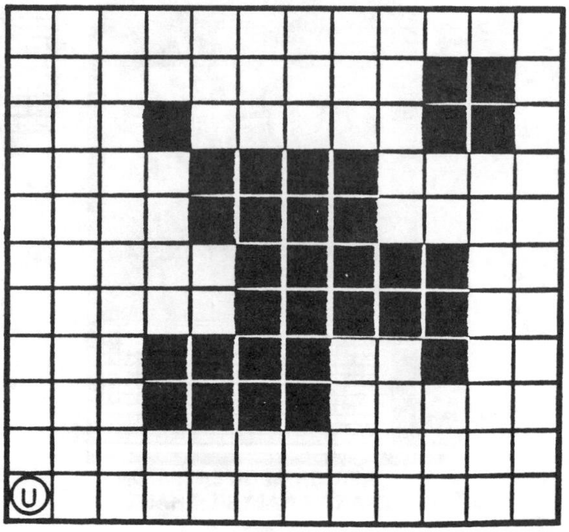

Figure 7.

case, cell (1, 1) has never been processed before. The contour following begins immediately after an unprocessed boundary cell is found. In Figure 8 the symbol C represents the contour (edge) following process. The cell C at (1, 1) is of Region-Type-A. The search for the next contour cell proceeds by getting the set N8(C) (as defined in section 2). Cells in N8(C) that are not of Region-Type-A are removed from consideration. Among the remaining legal neighbors of cell (1, 1), the search for the next edge cell proceeds in the following order: East, North-East, North, North-West, West, South-West, South and South-East. The contour following process begins from the first cell in this list of ordered neighbors that meets these criteria. In our example, this next edge cell is cell (2, 1) as shown in Figure 9. The contour following finds all the cells belonging to the outer boundary of a region. For our discussion, let O1 denote the set of pixels p that form this contour. In our example, O1 = ((1,1), (1,2), ... (1,12), ... (2,11), ...(11,12), (11,11), ... (11,1), ... (2,1)). The end of the contour following process is shown in Figure 10. The next activity is to mark all those pixels q that are in the set N8(p) where p ∈ O1 and have the same region type (in our current example it is Region-Type-A). This activity, often called region painting, marks the inner non-boundary pixels of the region. Figure 11, Figure 12 and Figure 13 show the intermediate steps during region painting. Figure 14 shows the last pixel involved in region painting.

Contour following and region painting result in the identification of regions. Figure 15 shows that the marked region has been identified as REGION-1. Figure 16 shows the updated region directory data structures.

Region Segmentation is not complete until all regions have been identified. The search for an unprocessed boundary cell is re-initiated (Figure 17). Figure 18 shows a cell adjacent to the next unprocessed boundary cell.

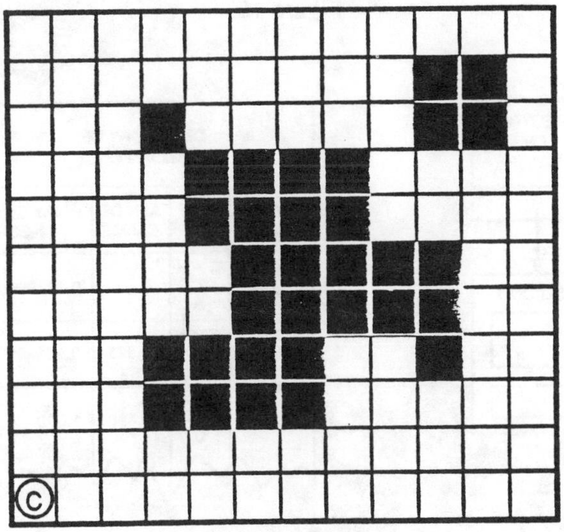

Figure 8.

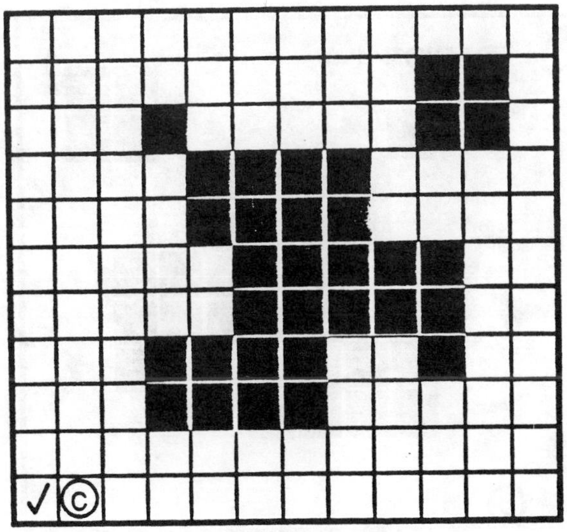

Figure 9.

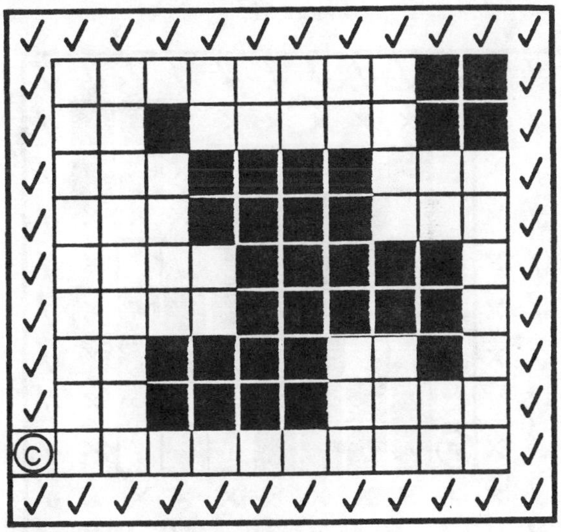

Figure 10.

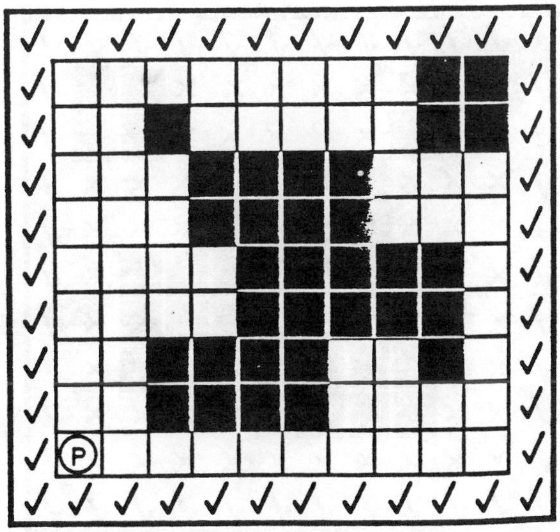

Figure 11.

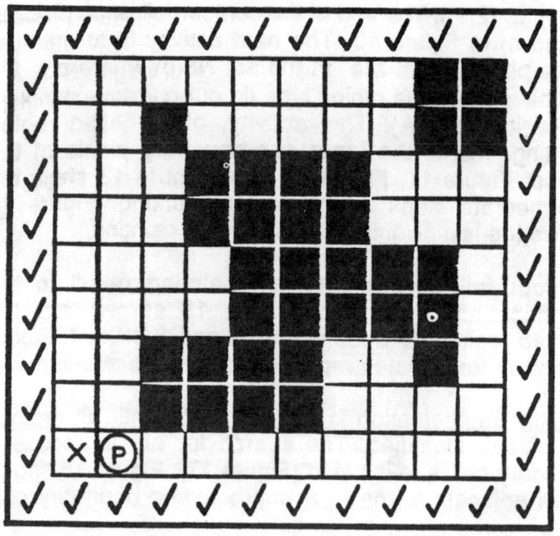

Figure 12.

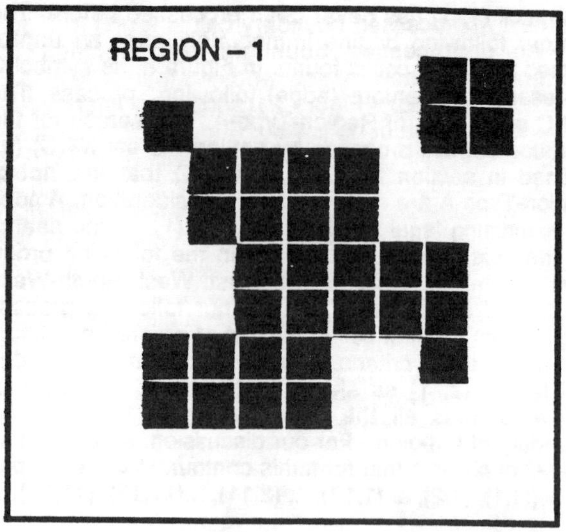

Figure 15.

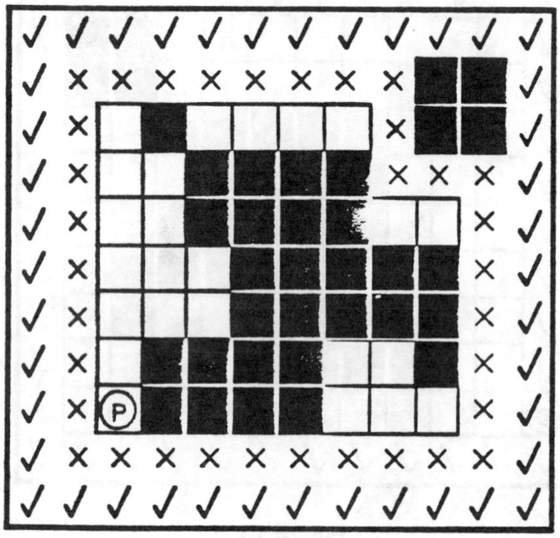

Figure 13.

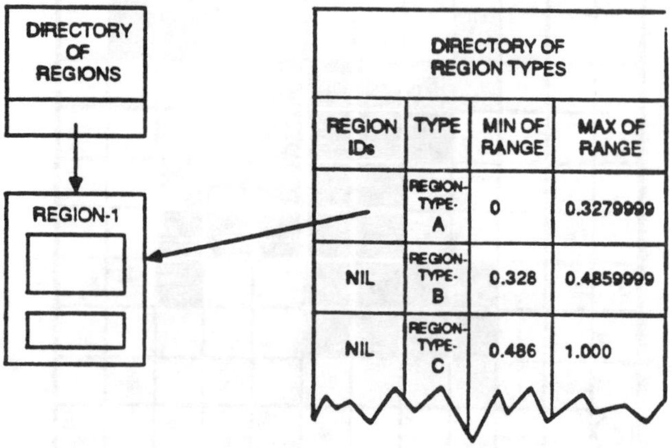

REGION IDs	TYPE	MIN OF RANGE	MAX OF RANGE
	REGION-TYPE-A	0	0.3279999
NIL	REGION-TYPE-B	0.328	0.4859999
NIL	REGION-TYPE-C	0.486	1.000

Figure 16.

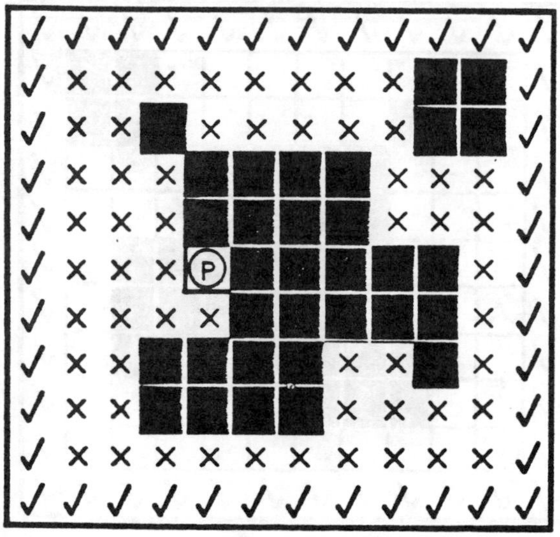

Figure 14.

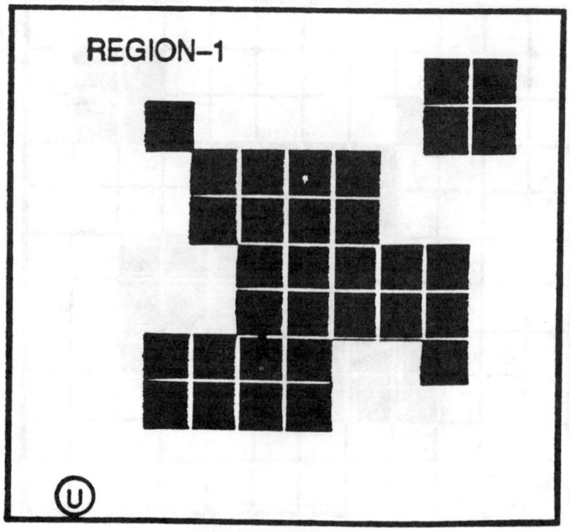

Figure 17.

The next unprocessed boundary cell is found in Figure 19. An unprocessed boundary cell signifies the existence of a new region and contour following is initiated (Figure 20). Figure 21 shows an intermediate step during contour following. Figure 22 shows the identification of REGION-2 which is of REGION-TYPE-B. The updated region directories are shown in Figure 23. Similarly, Figure 24 shows the identification of REGION-3 (of REGION-TYPE-B) and Figure 25 shows the final update of region directory data structures.

<u>3.4.2 STEP 2: Region Analysis.</u> The purpose of region analysis is to compute topological information about the terrain. The following kinds of high-level information (Figure 26) are deduced from the cost composite matrix:

- A graph of directional relationships of every region with respect to its adjacent regions.
- The inner boundary of regions (where applicable)
- The border contacts for every pair of adjacent regions.

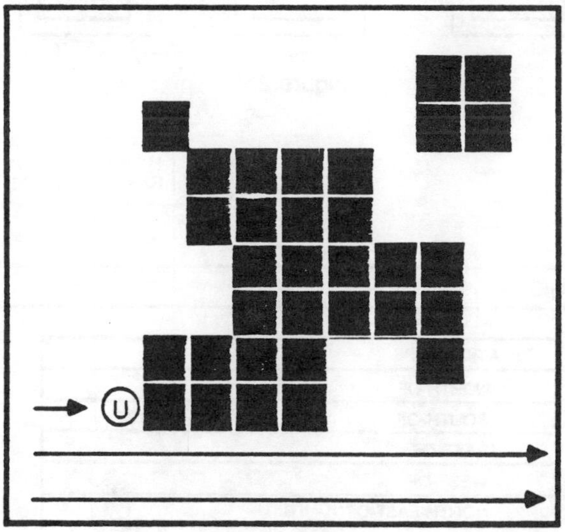

Figure 18.

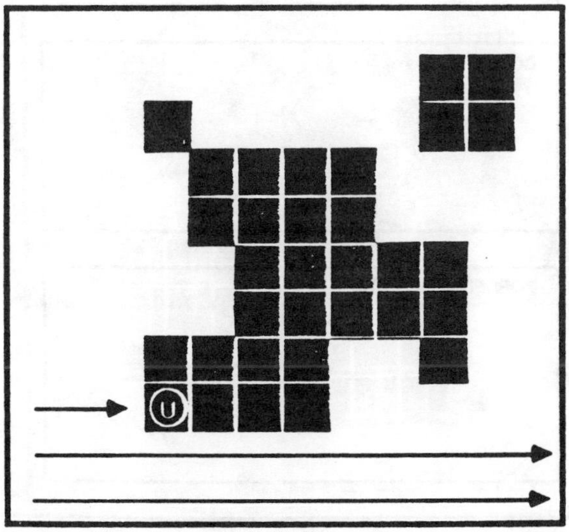

Figure 19.

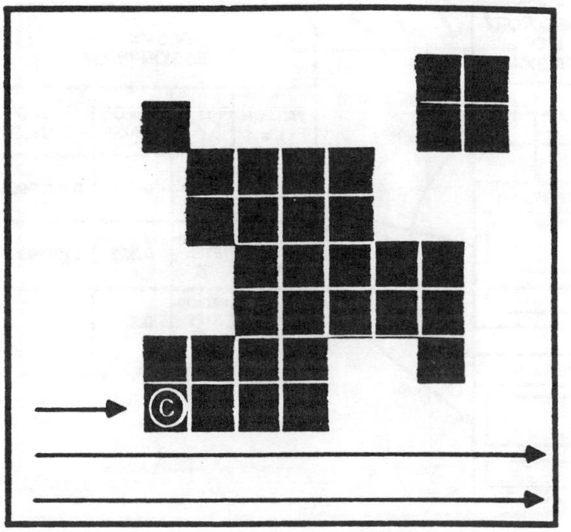

Figure 20.

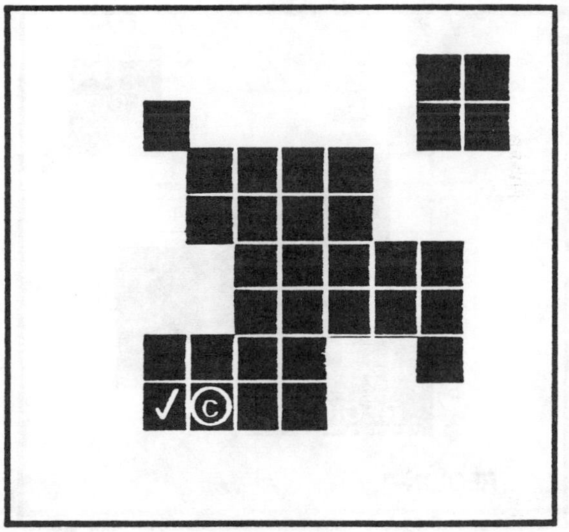

Figure 21.

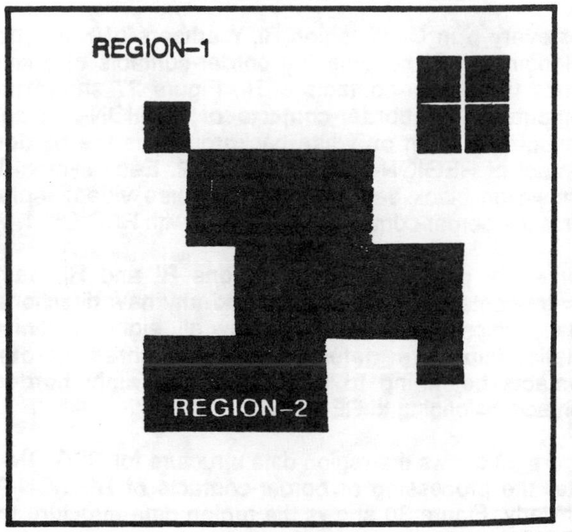

Figure 22.

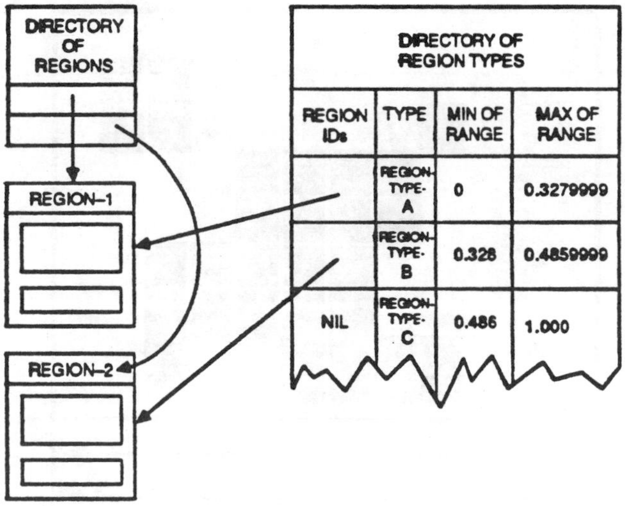

Figure 23.

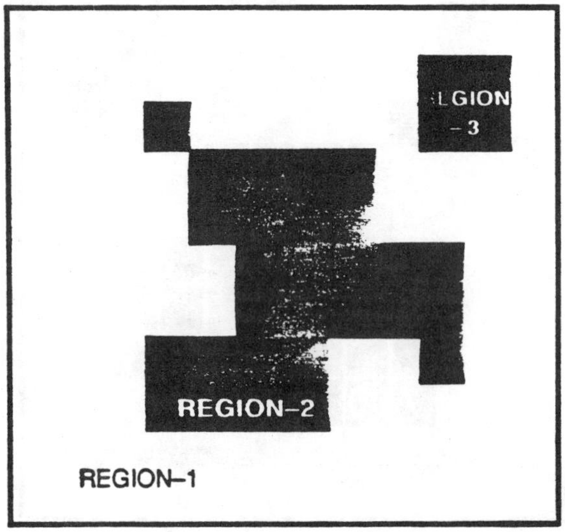

Figure 24.

For every p in Oi of region Ri, members of N8(p) not belonging to Ri become the border-contacts of p and hence the border-contacts of Ri. Figure 27 shows the computation of border-contacts of REGION-2. Each symbol B (shown on white background) is the border-contact of REGION-1 with REGION-2. Each symbol B (shown on black background as reverse video) represents the border-contact of REGION-2 with REGION-1.

For every pair of adjacent regions Ri and Rj, each border-contact is processed to find any new directional relationships. Figure 28 shows how all eight directional relationships are detected between three border-contacts belonging to REGION-2 and eight border-contacts belonging to REGION-1.

Figure 29 shows the region data structure for REGION-1 after the processing of border-contacts of REGION-2. Similarly, Figure 30 shows the region data structure for REGION-3. Finally, Figure 31 shows the outer boundary and inner boundary of REGION-1.

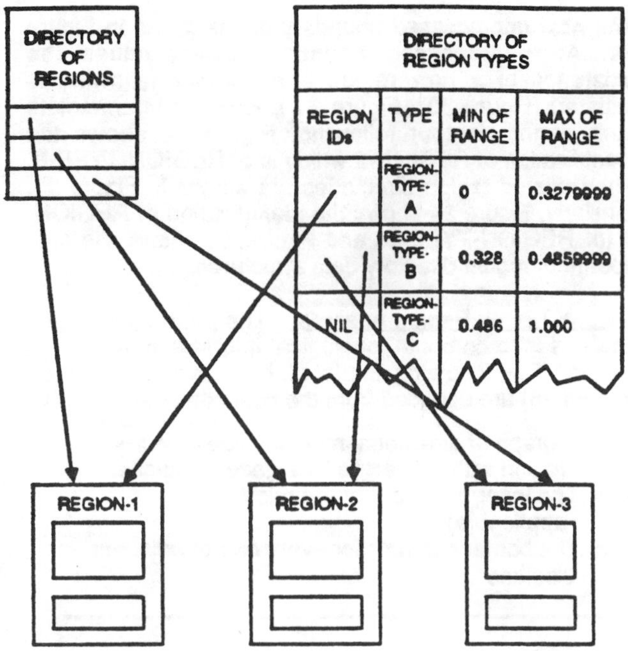

Figure 25.

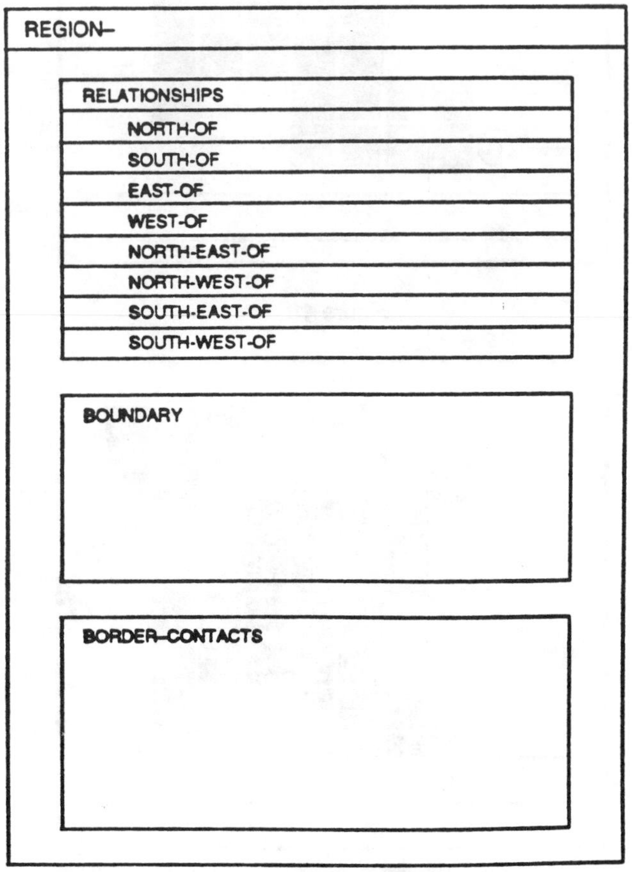

Figure 26.

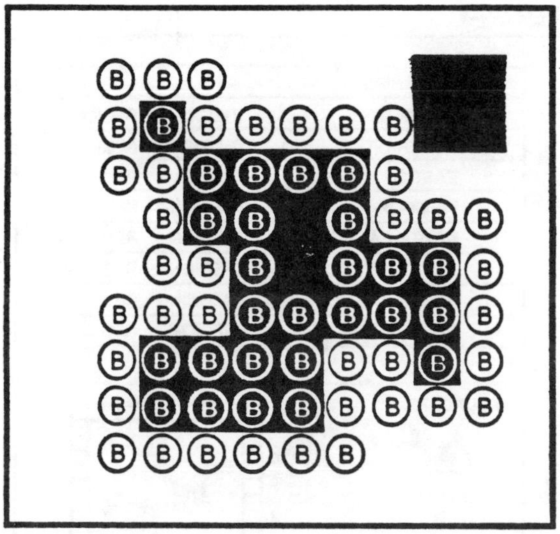

Figure 27.

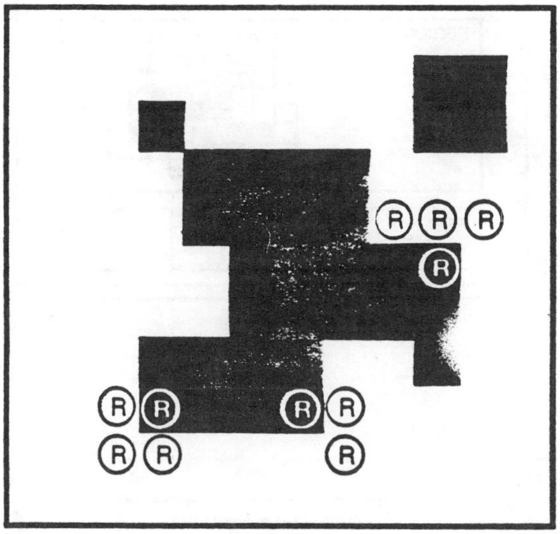

Figure 28.

3.5 Phase 2: Global Route Planning Phase

The second phase is the actual route planning phase and is called the Global Route Planning Phase. Figure 32 illustrates this phase. Figure 2 has shown how abstractions and problem reduction allow application of multiple criteria. Each abstraction contributes to route planning. In each abstraction different constraints for reasoning and searching can be applied and detected. The Figure 3 shows how this phase uses the data computed in the Region Segmentation and Analysis phase.

The main ideas are that since route planning involves many activities of different grain sizes, a variety of representations could be used. Each representation allows reasoning about some but not all of these activities.

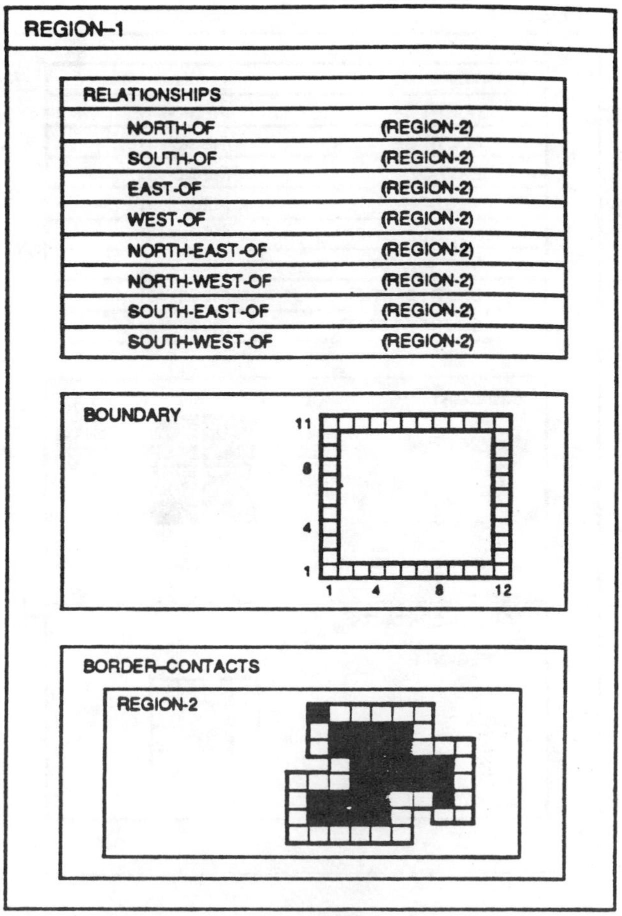

Figure 29.

3.5.1 The Stepping Stone Planner. The previous phase allows application and identification of constraints to be used in determining a coarse route. This planner uses as much information as is available to specify a region-to-region path. A coarse route is vague but good enough for the purpose of determining a framework of constraints for the lower-level route planners to work with. In effect, each route planner uses guidelines given by the user or the route planner at an abstraction above it, to further refine the path in some small measure. This new refinement then becomes the input to the next route planner and so on.

In this first abstraction, the granularity of reasoning is very high i.e., symbolic data which are gross abstractions of detailed lower-level data, are analyzed. The search for a stepping stone direction for the route is the first abstraction in the route planning process. Predefined criteria are applied to the region adjacency graph computed in the previous phase. Although the region-to-region path is coarse and inexact, a large number of regions are eliminated from consideration by lower-level route planners. Elimination of regions implies reduction in search space for the lower-level route planners.

3.5.2 The Waypoint Planner. The Waypoint Planner is the second abstraction in route planning. The purpose of the second step is to use simple geometry and cost

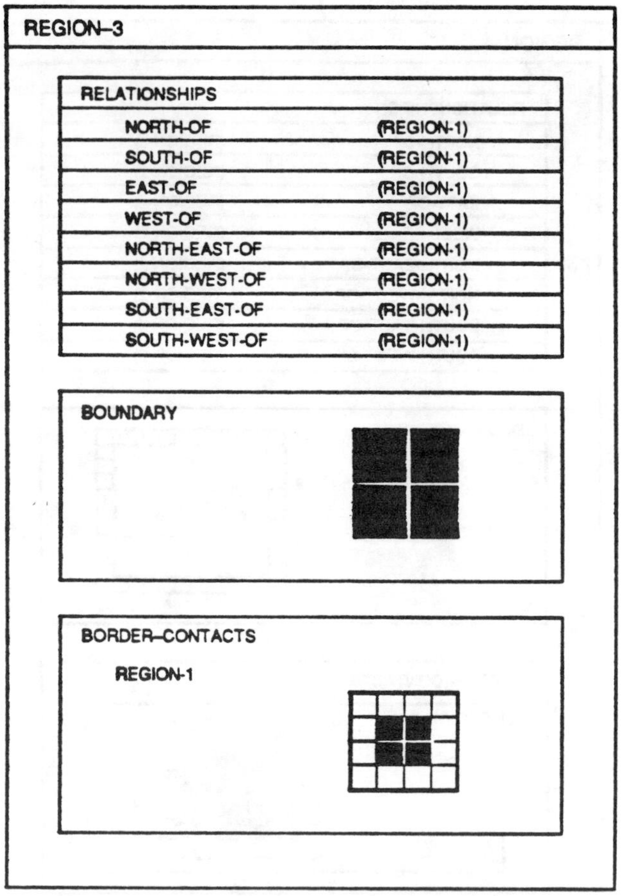

Figure 30.

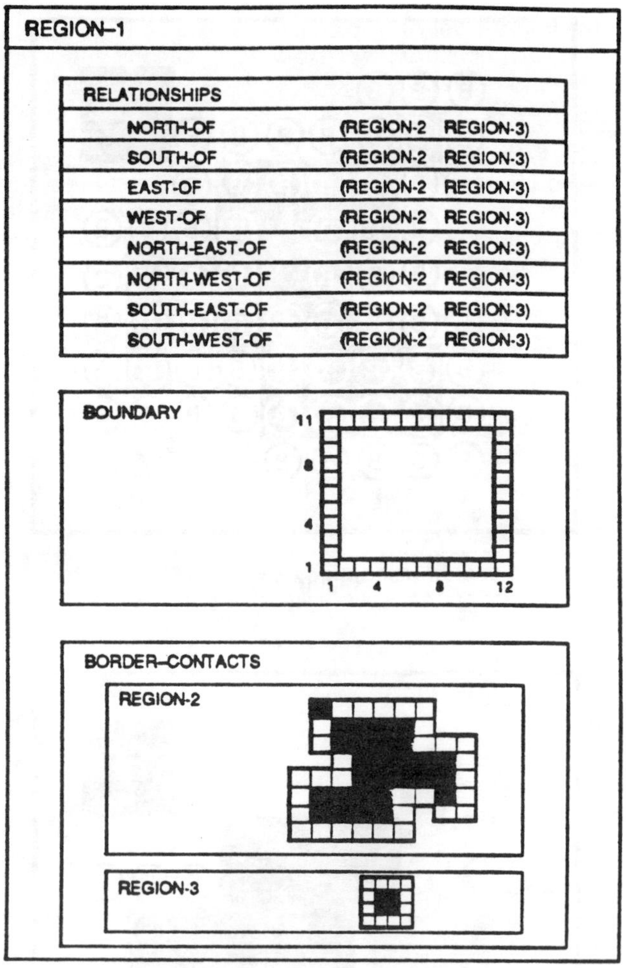

Figure 31.

criteria to refine the coarse stepping stone route. The grain size of reasoning is at the level of individual regions and their properties.

The Waypoint Planner uses user-defined criteria to impose further constraints on lower-level route planners. It searches through the region-to-region path in cycles for regions in descending order of cost. In each cycle, for each region, it searches through their border contacts on both sides of this region, to find a pair of border contacts that minimizes travel through that region. In effect, the purpose of the second abstraction is to compute entry and exit waypoints for each of these region stepping stones.

Hence , this is a problem reduction or subdivision step. For example, lets suppose that the start position in a Start-Region is S and the goal position in a Goal-Region is G. The Waypoint Planner reduces the problem of finding a path in the regions Start-Region, Region-1, ..., Region-N, Goal-Region, into a problem of traveling from S to exit-waypoint-of-Start-Region, and then from entry-waypoint-of-Region-1 to exit-waypoint-of-Region-1, and then from entry-waypoint-of-Region-2, ... to exit-waypoint-of-Region-N, and then from entry-waypoint-of-Goal-Region to G.

3.5.3 Local Route Planner. The purpose of the last abstraction is to find an exact pixel to pixel path connecting all the waypoints. In our implementation this is

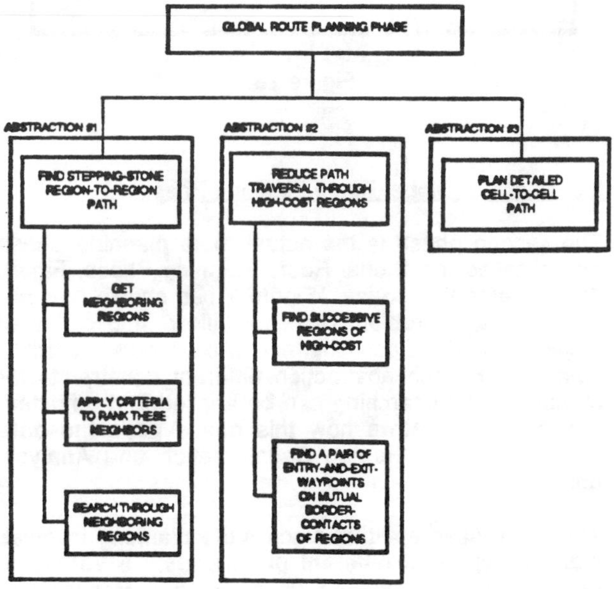

Figure 32.

the lowest-level route planner. The search is done on a grid and therefore the grain size of reasoning is at the level of individual pixels and its immediate eight neighbors.

Any local (blind, exhaustive, heuristic) planner can be used i.e. where the attention is focused on individual pixels and the scope is confined to the immediate eight neighboring pixels. However, this local planner must be modified so that it accepts the following inputs:

- A pair of waypoints for a region to be traversed,
- A corresponding boundary specification of this region, and
- A specification of the border contacts between this region and the next region that has to be traversed.

The Local Route Planner must search strictly in the given boundary of the given region. This constraint would have been determined during the coarse, region-to-region search using global information. It is precisely this constraint that will prevent the Local Route Planner from wandering aimlessly or exhaustively all over the grid. The search space can be reduced by tens of thousands of pixels.

4. RESULTS

4.1 Simulation Results.

For simulation, the partitions (section 2) were based on elevation. Elevation by itself is not a good measure of cost. For actual outdoor tests, we have used gradients in x- and y- direction and a probability of roughness as a measure of cost. We have not incorporated the kinematics of the vehicle in the simulation either. In the simulation we will refer to the partitions as costs.

The simulation used an elevation map provided by the Defense Mapping Agency. The elevation in this map ranged from 1040.0 feet to 1150.0 feet. The grid or cost composite matrix normalized these elevations to lie in the range 0.0 to 1.0. The range was partitioned into eight cost partitions. The partitions, their associated color, their minimum value and their maximum value are as follows (in ascending order of costs):

- Cost-1: red, 0.0 - 0.16375
- Cost-2: white, 0.16376 - 0.032750998
- Cost-3: green, 0.032750999 - 0.049126
- Cost-4: light brown, 0.049126 - 0.06550
- Cost-5: dark green, 0.06551 - 0.081875995
- Cost-6: orange, 0.081875995 - 0.09825099
- Cost-7: blue, 0.09825099 - 0.11462598
- Cost-8: gray, 0.11462598 - 1.0

The input terrain matrix ranges in size from 10,000 pixels to 400,000 pixels. For 196,000 pixels, the Region Segmentation and Analysis phase of the route planner takes about 12 minutes (with 10 meg RAM and 100K virtual address space on Symbolics version 7.1). The Global Route Planning phase takes less than 6 seconds (this includes graphics I/O on a high-resolution color monitor).

Pictures A through F in Figure 33 demonstrate this route planner being applied to the JPL Arroyo terrain. Picture A shows the elevation terrain data base. Elevations are given in varying shades of green. In B, each unique color represents different cost criteria. In C, D and E, only the boundaries of the regions are shown for clarity. Picture D shows the minimal number of high-cost regions that must be traversed. For every displayed region it also shows its entry waypoint (smaller circle) and its exit waypoint (larger circle). Picture E shows the exact point-to-point path connecting the start point (shown by a green "+" sign), the intermediate waypoints (the small and large circles) and the goal point G (shown as a red "+" sign).

The local route planner used in the simulation has been developed by Jonathan Cameron and the author. This local route planner ranks the neighboring pixels according to their angular bearing to the next waypoint and also according to their cost values.

5. OTHER INFORMATION

A user's manual [Cameron *et al.* 87 A] is available. The advantages and disadvantages of this route planner can be only demonstrated adequately through color copies of color photographs. If interested, you can request copies from the authors.

6. CONCLUSIONS

This route planner has provided many advantages. The combination of incorporating constraints first and using problem reduction methods at several abstractions has produced savings in search space and time. It is a testbed for the application of many diverse heuristics for the segmentation and selection of regions. This route planner provides many advantages which include naturalness, interruptibility and extensibility.

This route planner has many disadvantages also. For robust and more intelligent route planning, the complete system must employ a variety of route planners in different circumstances. The Region Based Route Planner is not applicable to all domains. In the simulation described above, the route planner compromised path length to avoid high cost regions. Hence the paths were very long in many test cases.

Also, since constraints are posted at higher abstractions and cannot be modified at lower abstractions, the route planner plans paths that may not be executable by the rover (due to kinematic constraints). Many times the Local Planner cannot find legal-paths through a given region because the Waypoint Planner chose bad exit waypoints. However, this is not a problem inherent with the framework of the planner; it is a problem with the implementation we have described here. Theoretically, the problem could be solved by implementing backtracking mechanisms between every pair of abstractions in route planning.

7. ACKNOWLEDGEMENTS

The authors wish to thank Jonathan Cameron, Brian Cooper, Don Gennery, Ken Holmes, Mark James, Teri Lawton, Keith Lay, Andrew Mishkin, Carl Ruoff, and Brian Wilcox for support and dedication.

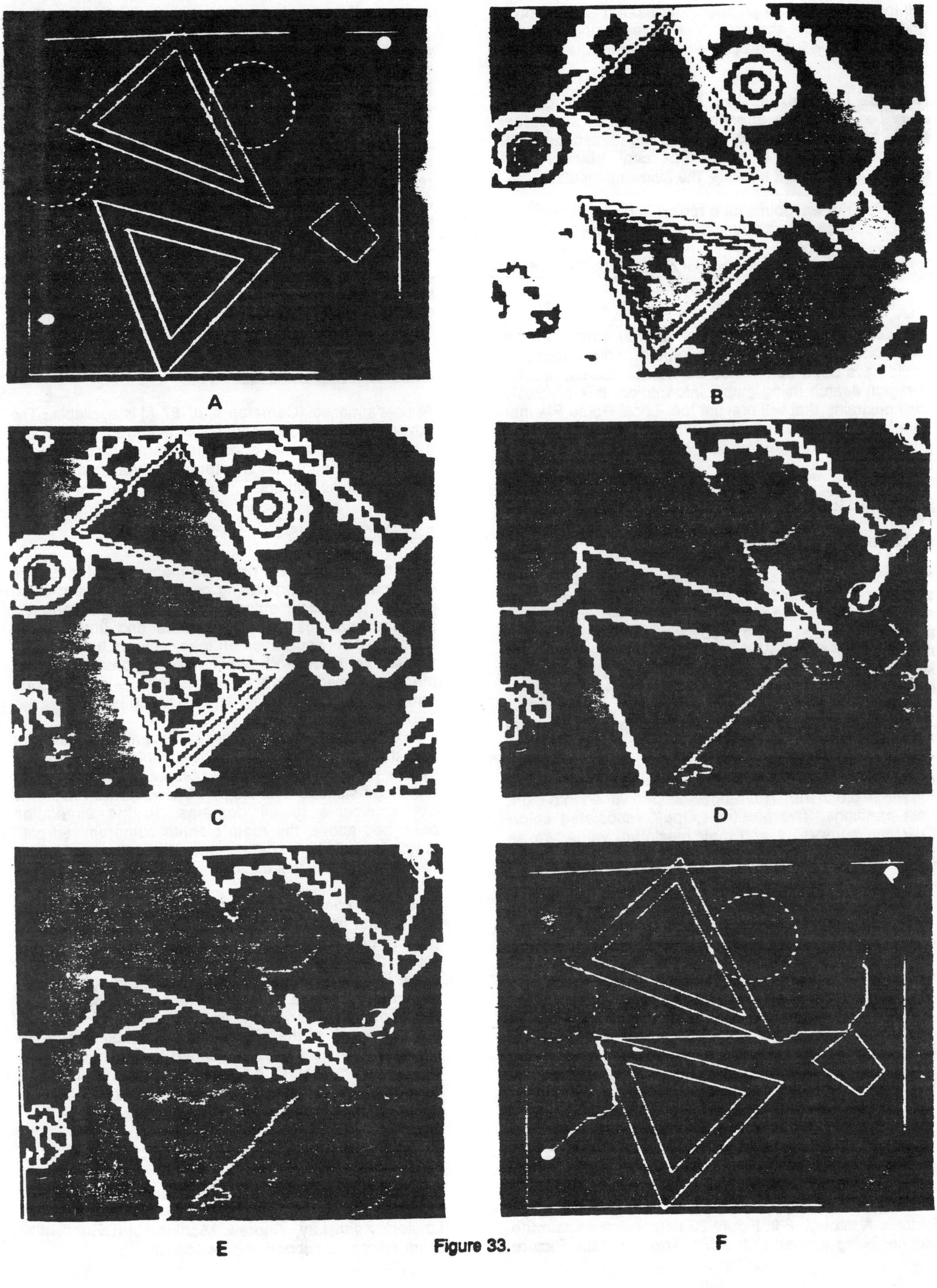

Figure 33.

The authors also wish to thank Harry Joseph Porta and Marc Slack for helping correct technical errors in this paper, and David Atkinson, C. A. Beswick and David Miller for helpful suggestions.

The research work leading to this paper was performed with funding from the JPL's Director's Discretionary Fund (DDF), 1986 and 1987. The work on Region Based Route Planning was also conducted under contract with the U.S. Army Engineer Topographic Laboratories (ETL) under the technical management and cognizance of Matthew Swanson. We are grateful to both, JPL's DDF and ETL for their financial support and assistance.

The research described in this publication was carried out by the Jet Propulsion Laboratory, California Institute of Technology, and was sponsored by ETL, DDF and the National Aeronautics and Space Administration.

8. REFERENCES

Ruzena Bajcsy, Max Mintz, Erica Liebman, *A Common Framework for Edge Detection and Region Growing*, University of Pennsylvania, Department of Computer and Information Science, MS-CIS-86-13, February 1986.

Claude R. Brice, Claude L. Fennema, "Scene Analysis Using Regions," *AI Journal*, 1970, pp. 205 - 226.

Rodney A. Brooks, "Solving the Find-Path Problem by Good Representation of Free Space," *IEEE Transactions on Systems, Man and Cybernetics*, SMC-13, Number 3, 1983.

Jonathan M. Cameron, Rajkumar S. Doshi, Kenneth G. Holmes, "Route Planner Development Workstation," *Volume 3 A - Operational Extensions*, JPL internal document , JPL D-2733 Volume 3 A, May 1987.

R. Chatila, Jean-Paul Laumond, "Position Referencing and Consistent World Modeling for Mobile Robots," *Proceedings of the IEEE International Conference on Robotics and Automation*, St. Louis, Missouri, March 1985.

James P. Crowley, "Dynamic World Modelling for an Intelligent Mobile Robot Using a Rotating Ultra-Sonic Ranging Device," *Proceedings of the IEEE International Conference on Robotics and Automation*, St. Louis, Missouri, March 1985.

Rajkumar S. Doshi, James White, Raymond Lam, *Reasoning with Inaccurate Spatial Knowledge, Proceedings of SPIE Robotics*, 1987, Cambridge, MA.

Richard Duda, Peter E. Hart, *Pattern Classification And Scene Analysis*, John Wiley & Sons, 1973, pp. 284 - 293.

Alberto Elfes, "A Distributed Control Architecture for an Autonomous Mobile Robot," *Artificial Intelligence*, Volume 1, No. 2, 1986.

Laxmi Gewali, Alex C.-C. Meng, Joseph S. B. Mitchell, Simeon Ntafos, "Path Planning in 0/1/∞ weighted Regions with Applications," *Proceedings of the Fourth ACM Symposium on Computational Geometry*, 1988.

Georges Giralt, Raja Chatila, Marc Vaisset, "An Integrated Navigation and Motion Control System for Autonomous Multisensory Robots," *Robotics Research*, edited by M. Brady and R. Paul, MIT Press, 1984.

Rafael C. Gonzalez, Paul Wintz, *Digital Image Processing*, Addison Wesley, Second Edition, 1987.

Adolfo Guzman, *Decomposition of a Visual Scene into Three-dimensional Bodies*, Fall Joint Computer Conference, 1968.

S. S. Iyengar, C. C. Jorgensen, S. V. N. Rao, C. R. Weisbin, *Robot Navigation Algorithms Using Learned Spatial Graphs*, Engineering Physics and Mathematics Division, Oak Ridge National Laboratory, technical memorandum, ONRL/TM-9782, December 1985.

Subbarao Kambhampati, Larry S. Davis, "Multiresolution Path Planning for Mobile Robots," *IEEE Journal of Robotics and Automation*, Volume RA-2, No. 3, September 1986.

Benjamin Kuipers, Tod Levitt, "Navigation and Mapping in Large-Scale Space," *AI Magazine*, Special Issue on Spatial Reasoning, Volume 9, Number 2, Summer 1988.

Jean-Paul Laumond, "Model Structures and Concept Recognition: Two Aspects of Learning for a Mobile Robot," *Proceedings of the Eight International Joint Conference on Artificial Intelligence*, Karlsruhe, Germany, August 1983.

Tomas Lozano-Perez, M. A. Wesley, "An Algorithm for Planning Collision-free Paths Among Polyhedral Obstacles," *Communications of the ACM*, Volume 22, October 1979.

Alex C.-C. Meng, "AMPES: Adaptive Mission Planning Expert System for Air Mission Tasks," *Proceedings of the IEEE National Aerospace and Electronics Conference (NAECON)*, 1988.

Hans P. Moravac, *Certainty Grids for Mobile Robots, Proceedings of the Workshop on Space Telerobotics*, Jet Propulsion Lab, Pasadena, California, January 1987.

Hans P. Moravac, "Sensor Fusion in Certainty Grids for Mobile Robots," *AI Magazine*, Special Issue on Spatial Reasoning, Volume 9, Number 2, Summer 1988.

Andrew Mishkin, Brian Wilcox, Rajkumar Doshi, David Atkinson, Donald Gennery, Carl Ruoff, "Intelligent Control Research For Space Robots and Autonomous Planetary Rovers," *Annual Report of Investigations Carried Out Under The Director's Discretionary Fund*, JPL internal document D-4176, June 30, 1988.

C. O'Dunlaing, Micha Sharir, Chee-Keng Yap, "Retraction: A New Approach to Motion Planning," *Proceedings of the 15th ACM Symposium on Theory of Computing*, Boston, 1983.

George Polya, *How To Solve It*, Princeton University Press, second edition, 1973.

N. S. V. Rao, S. S. Iyengar, C. R. Weisbin, *On Autonomous Terrain Model Acquisition by a Mobile Robot, Proceedings of the Workshop on Space Telerobotics*, Jet Propulsion Lab, Pasadena, California, January 1987.

Edward M. Riseman, Michael A. Arbib, "Computational Techniques in the Visual Segmentation of Static Scenes," *Computer Graphics and Image Processing*, 6, 1977.

Hanan Samet, *The Quadtree and Related Data Structures*, University of Maryland, Center for Automation Research, Technical Report #23, November 1983.

Robert Sedgewick, *Algorithms*, Addison-Wesley Publishing Company, 1984.

Marc G. Slack, D. P. Miller, *Route Planning in a Four-Dimensional Environment, Proceedings of the Workshop on Space Telerobotics*, Jet Propulsion Lab, Pasadena, California, January 1987.

Marc G. Slack, "Planning Paths Through a Spatial Hierarchy: Eliminating Stair-Stepping Effects," *Proceedings of SPIE Robotics*, 1988, Cambridge, MA.

R. B. Stanton, "Plane Regions: A Study in Graphical Communication," in *Picture Language Machines*, edited by S. Kaneff, Academic Press, 1970.

Charles E. Thorpe, *FIDO: Vision and Navigation for a Mobile Robot*, Ph. D. Thesis, Computer Science Department, Carnegie-Mellon University, September 1985.

Wayne A. Wickelgren, *How to Solve Problems*, Freeman Press, 1974.

Chee-Keng Yap, "Algorithmic Motion Planning," in *Algorithmic and Geometric Aspects of Robotics*, edited by J. T. Schwartz and C. K. Yap, Lawrence Erlbaum Associates, 1987.

J. A. Zimmer, *Abstraction for Programmers*, McGraw Hill Book Company, 1985.

Robot navigation algorithms using learned spatial graphs†

S.S. Iyengar,‡ C.C. Jorgensen,§ S.V.N. Rao,‡ C.R. Weisbin§

(Received: April 28, 1981)

SUMMARY

Finding optimal paths for robot navigation in known terrain has been studied for some time but, in many important situations, a robot would be required to navigate in completely new or partially explored terrain. We propose a method of robot navigation which requires no pre-learned model, makes maximal use of available information, records and synthesizes information from multiple journeys, and contains concepts of learning that allow for continuous transition from local to global path optimality. The model of the terrain consists of a spatial graph and a Voronoi diagram. Using acquired sensor data, polygonal boundaries containing perceived obstacles shrink to approximate the actual obstacles surfaces, free space for transit is correspondingly enlarged, and additional nodes and edges are recorded based on path intersections and stop points. Navigation planning is gradually accelerated with experience since improved global map information minimizes the need for further sensor data acquisition. Our method currently assumes obstacle locations are unchanging, navigation can be successfully conducted using two-dimensional projections, and sensor information is precise.

1. INTRODUCTION

Robotics has become an actively pursued research area of computer science and has proven to be full of a variety of issues ranging from abstract mathematical to highly pragmatic problems. In many industrial applications, which are repetitive and tedious (e.g., normal maintenance or inspection), it would be desirable to utilize mobile robots. Other tasks requiring rapid response in emergency situations are also appropriate for intelligent machines; this is particularly true in hazardous environments. Some of the more active robotics research areas today include knowledge representation, task planning, multi-sensor interpretation, dynamics and control, advanced computer architectures, algorithms for concurrent computation, and coordinated manipulation and navigation.

A robot may be characterized as a autonomous machine capable of decision making and action. To perform complex tasks which cannot be fully programmed *a priori*, effect sensing becomes crucial for monitoring both the robot's environment, as well as the status of its own internal system. There have been several efforts to design an automated mobile robot. Examples are SHAKEY,[1,2] the JPL robot,[3] HILARE,[4,5] the Stanford Cart,[6] the CMU Terregator and Neptune robots,[7] Yamabico,[8] and HERMIES.[9]

Navigation planning is one of the vital aspects of any mobile robot. One approach toward navigation, called the find-path problem, addresses itself to determining a collision free path for a robot moving through a terrain cluttered with obstacles whose positions are known. This problem is well understood and solved in many cases.[10–18] The techniques for navigation described in these papers assume that a complete global model of our obstacle laden environment is known. Most of the techniques above model the obstacles and the free space of a robotic environment as mathematical and geometric entities. When a robot must navigate in an unexplored environment, the algorithms are not directly applicable.

Navigation in the more general case calls for the collision-free movement of a mobile robot in entirely or partially unexplored terrain. The problem of planning optimal or near optimal paths that avoid collisions with obstacles in such an environment is a challenging task. Contrary to the known environment case, there has not been as much work reported in the literature about navigation problems in unexplored terrain. This can be attributed to the inherent ambiguity of the problem due to the lack of global information about the obstacles. Early attempts to navigate in unexplored terrain were based solely on image understanding.[3,6] More recently, Crowley[19] and Parodi[20] have suggested hierarchical approaches with global and local models updated based on sensor feedback. Chattergy[21] describes some novel heuristic strategies to aid the navigation of a robot in an unexplored terrain. This paper builds upon many of these ideas but specifically aims toward a method for which no pre-learned model is required, information from multiple journeys is explicitly synthesized, all information is used to maximum extent, and a global path optimization is achieved in a continuous transition from local path optimization as more information is acquired.

In this paper, we assume that the robot (HERMIES)[9] begins his task in a completely unexplored terrain of finite dimensions. HERMIES has to complete a number of different traversals (e.g., carrying objects from place

† Research sponsored by Office of Basic Energy Science, U.S. Department of Energy under contract number DEAC05-840R21400 with Martin Marietta Energy Systems, Inc.

‡ Dept. of Computer Science, Louisiana State Univ., Baton Rouge, LA 70803 (USA).

§ Engineering Physics and Mathematics Division, Oak Ridge National Laboratory, Oak Ridge, Tennessee 37831 (USA).

"Robot Navigation Algorithms Using Learned Spatial Graphs" by S.S. Iyengar, C.C. Jorgensen, S.V.N. Rao, and C.R. Weisbin. Reprinted from *Robotica*, Vol. 4, Part 2, April-June 1986, pages 93-100. Copyright © 1986 Cambridge University Press, reprinted with permission.

to place) and the goal of this paper is to provide the method by which he can navigate more efficiently with each successive trip, based upon experience acquired to date. The terrain can be randomly populated with obstacles, but the world is assumed to be static. The robot, assumed to be a point in a two-dimensional plane, can recognize line-of-sight distances to objects and detect their edges without imprecision.

The terrain is modelled using an attribute graph, called a *spatial graph,* and a *Voronoi diagram.* Initially, both are empty, and they are updated as more and more paths are traversed. Each path of navigation is composed of a sequence of stop points, where the robot stops to take sensor readings, or to access the terrain model to compute the next stop point. The robot travels in straight lines in between two successive stop points. Initially, obstacle avoidance techniques use local optimization for the navigation of the robot. By local optimization we mean optimal path selection based only on sensor information at the time of the decision.

Traversal of paths includes sensor exploration of the regions in which the robot navigates. Information gained while on new paths is consolidated into the existing graph structures. In planning any path, the content of the current graphs is made use of to the maximum possible extent, and local optimization occurs in the regions where no model is available. Initially, since no graph is available, the paths are only locally optimal. As more and more paths are traversed, the graphs become more complete ('learned') and gradually improve from local optimality to global optimality.

2. OPTIMIZATION OF LOCAL OBJECT AVOIDANCE

When a robot navigates in new terrain without *a priori* information, its path of navigation is completely decided by the sensor readings and presumed goal destination. The localized nature of the sensors makes a true globally optimal path determination impossible in a terrain with arbitrary distribution of obstacles. Thus, a local optimization scheme must be used to determine the path of navigation in the immediate proximity of obstacles.

We consider the obstacle that is nearest to the source point S in the direction of the robot's goal destination. The sensor readings obtained allow for determination of the distance from the source to the edges of the obstacle, and also the corresponding edge angles relative to the line between the robot center and the goal. In Figure 1,

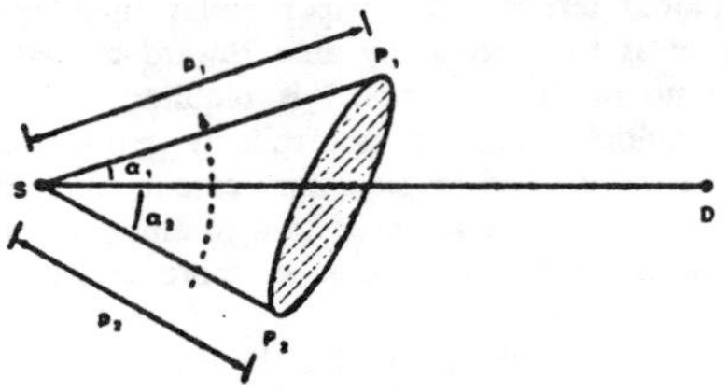

Fig. 1. The sensor readings include α_1, α_2, p_1, and p_2. S - Source point, D - Destination point, P_1, P_2 - The edge points.

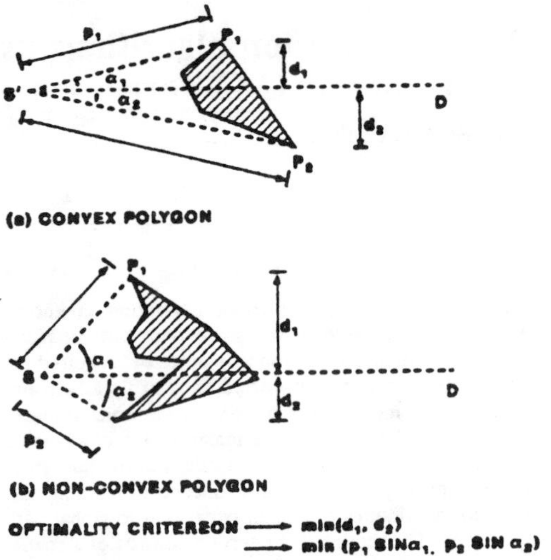

Fig. 2. First case for local optimization.

the angles α_1 and α_2, and the distances p_1 and p_2 are obtained from sensor readings. Our local optimization approach considers two cases. Figure 2 depicts the first case for which no part of the obstacle extends beyond the source point in the direction opposite to the direction of the next robot destination. The local optimization criterion is to minimize the distance traversed in the direction perpendicular to the line joining the source point S, to the destination point D i.e. the locally optimal path is given by the condition $\min(d_1, d_2)$ or $\min(p_1 \sin\alpha_1, p_2 \sin\alpha_2)$. This method may not yield a globally optimal path as shown in Figure 3. The path

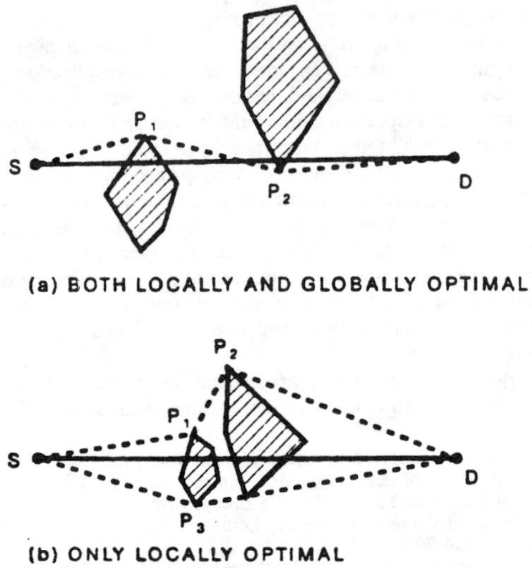

Fig. 3. Local optimality does not mean global optimality.

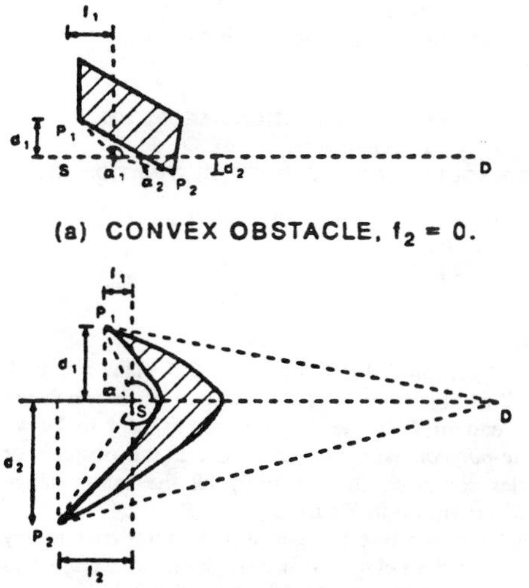

Fig. 4. Second case for local optimization.

SP_1P_2D will be followed according to the local optimality criterion, but the path SP_3D will be globally optimal. The second case of local optimality involves the obstacles that extend beyond the source in the direction opposite to the direction of the destination point as shown in Figure 4. In this case the distance traversed (f_1, f_2) in the direction opposite to that of destination point D also has to be minimized. Referring to Figure 4, the criterion for local optimization is given by $\min(\sqrt{d_1^2+f_1^2}, \sqrt{d_2^2+f_2^2})$. Again, it is to be noted that this method may not give rise to globally optimal solution. The local optimization algorithm is as follows:

LOCAL OPTIMIZATION ALGORITHM

```
ALGORITHM NAVIGATE-LOCAL (S, D);
S THE SOURCE POINT. D IS THE DESTINATION POINT
BEGIN
1. IF D IS DIRECTLY REACHABLE
2. THEN GO STRAIGHT
3. ELSE
   BEGIN
4.   SCAN THE TERRAIN AROUND THE
     DIRECTION OF SD:
5.   P* <- OPTIMUM (P1, P2);
6.   GO STRAIGHT TO P*
7.   IF P* ≠ D
8.   THEN NAVIGATE-LOCAL (P*, D);
   END;
END;
```

3. TERRAIN MODEL

Figure 5 shows an illustrative rectangular terrain populated with four obstacles. Four paths are traversed using local optimization. The paths start at S_1, S_2, S_3, and S_4 and end at D_1, D_2, D_3, and D_4, respectively. The terrain in which the robot navigates is represented by both a *spatial graph*, and a *Voronoi diagram* (Figures 6 and 7, respectively).

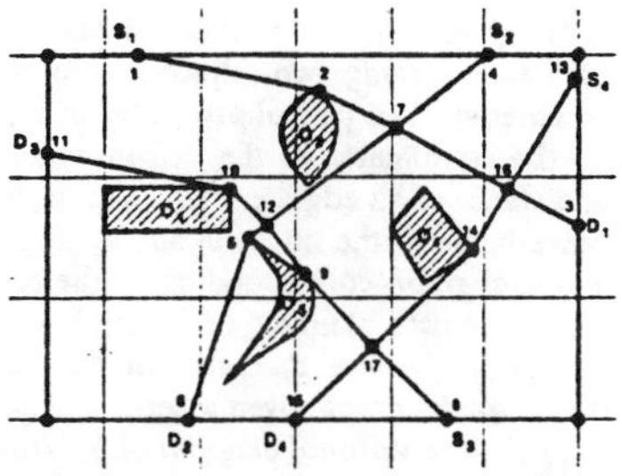

Fig. 5. The terrain. Four paths from S_1, S_2, S_3 and S_4 to D_1, D_2, D_3 and D_4 respectively using local optimisation. All paths are consolidated by finding intersection points such as 7, 16, 12, and 17.

A *spatial graph* G is defined as the ordered triple (V, E, ψ), where V is the set of nodes, E is the set of edges, and ψ is an *attribute mapping* that defines a pair of attributes (e.g., coordinate locations) for each vertex. For an edge $e = (v_i, v_j) \subset E$, we say that v_i and v_j are *connected* to each other. We also have a *distance* $d(e)$ defined for each edge $e = (v_i, v_j) \subset E$, $\psi(v_i) = (i_1, i_2)$, $\psi(v_j) = (j_1, j_2)$, as

$$d(e) = [(i_1 - j_1)^2 + (i_2 - j_2)^2]^{1/2}. \quad (1)$$

Initially a uniform grid is superimposed on the terrain of navigation. The granularity or grid size is chosen to be smaller than the expected size of the smallest obstacle of interest. The grid cells are numbered in the usual manner using x and y coordinate systems. Any path of navigation on the grid consists of straight lines and stop points.

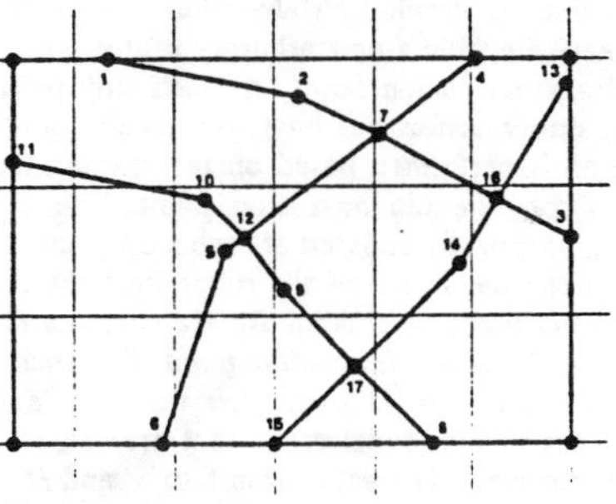

Fig. 6. The spatial graph.

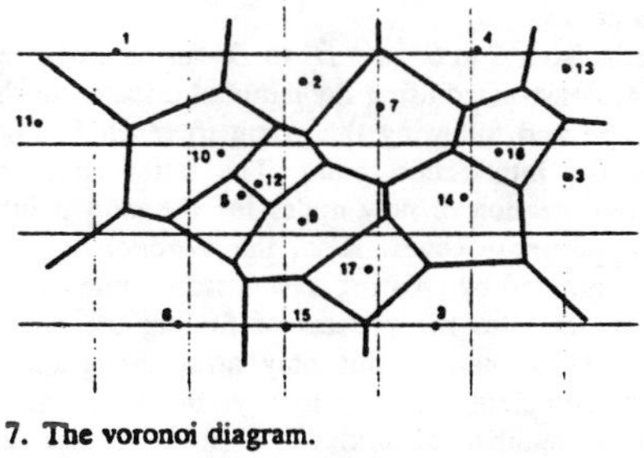

Fig. 7. The voronoi diagram.

Each stop point corresponds to a node of the spatial graph, and each path joining two adjacent stop points corresponds to an edge. The pair of attributes of a node corresponds to the coordinates of the cell in which the node lies. The distance of an edge, $e = (v_i, v_j) \subset E$, is the euclidian distance between the nodes v_i and v_j. Figure 6 illustrates the *spatial graph* corresponding to the terrain and local optimization path planning of Figure 5.

We next obtain a *Voronoi diagram* for the set of vertices, V, of the spatial graph given a set S of n points of $\{p_1, p_2, \ldots, p_n\}$. The *Voronoi diagram* of S, Vor (S), partitions the plane into n equivalence classes, each of which corresponds to a point. Specifically, the equivalence class corresponding to point p_i is the *Voronoi polygon* $VP(p_i)$, defined[22] such that any point x in $VP(p_i)$ is closer to p_i than to any other point in S. Figure 7 illustrates the *Voronoi diagram* corresponding to the spatial graph of Figure 6.

Initially, when HERMIES is first placed in a new terrain, the spatial graph is empty or null and the Voronoi diagram contains no points. the new paths are integrated into the terrain models when they are traversed. The spatial graph is updated for every new path as follows: (i) create new nodes corresponding to new stop points, (ii) create new edges corresponding to the paths in between two adjacent stop points, (iii) create new intersection nodes, corresponding to the intersection points of new edges with the existing edges. When this process is complete, the *Voronoi diagram* is updated accordingly.

4. PATH PLANNING AND LEARNING

In this section we develop an algorithm that plans safe paths to navigate from a new arbitrary source point to a new arbitrary destination point. At each stop point on the path, either sensor readings are taken, or graph computation is performed based on the existing terrain models to compute the next stop point. The terrain model is appropriately updated at each stop point.

Consider the navigation of the robot from the source point S to the destination point D. We compute *virtual source* S' and *virtual destination* point D'. such that $S \in VP(S')$ and $D \in VP(D')$. In other words, S' and D' are the nodes of the spatial graph that are nearest to S and D, respectively. The paths from S to S' and D' to D are traversed according to the local optimization described in Section 2. The path $S'D'$ is planned using the spatial graph model and sensor readings, as will be described below.

The paths from S to S' and D' to D can be navigated directly, or constructed using the minimal distance to the spatial graph and following the graph to reach S' (and D') from the intersection point. The latter approach involves the creation of new nodes for the stop points, and the appropriate edges. Also, the *Voronoi diagram* should be updated by creating new Voronoi regions for the new nodes. But the process of finding the virtual points should be carried out only after the graph is reasonably complete. That is to say, initially, until a considerable number of nodes are inserted into the spatial graph, all the navigation should be determined using sensor based algorithms. The basic algorithm is as follows:

COMPLETE NAVIGATION ALGORITHM

ALGORITHM NAVIGATE (S, D);
S IS THE SOURCE POINT, D IS THE DESTINATION POINT.
BEGIN
1. FIND S' AND D' SUCH THAT $S \in VP(S')$,
   AND $D \in VP(D')$;
2. NAVIGATE-LOCAL (S, S');
3. NAVIGATE-GLOBAL (S', D');
4. NAVIGATE-LOCAL (D', D);

END.

The algorithm **NAVIGATE** (S', D') plans the path $S'D'$. This algorithm tests the polygon P, in which the source end of $S'D'$ lies. A polygon is said to be an *obstacle-polygon* with respect to S', if the obstacle or obstacles contained in P entirely fill the sensor range from S', as shown in Figure 8.

A polygon is a free polygon if it does not contain any obstacles. If the polygon P is unexplored with respect to S', then the algorithm **EXPLORE** (P, S') is involved. Sensor readings from S' distinguish two types of regions – visible and invisible – as shown in Fig. 9. The *invisible* regions are the regions of the polygon that are not reachable by the sensor when the obstacles contained in the region are absent. The regions that are not invisible are called the *visible regions*. Based on the sensor readings, the polygon P can be partitioned into regions as shown in Figure 10.

A region could be an unexplored polygon, a free-polygon, or an obstacle-polygon. The invisible

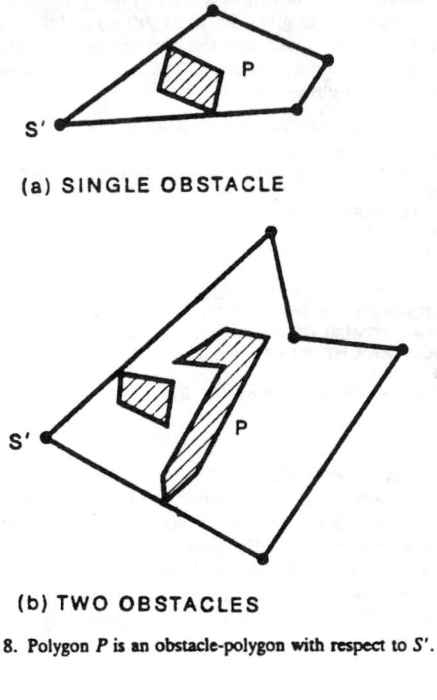

Fig. 8. Polygon P is an obstacle-polygon with respect to S'.

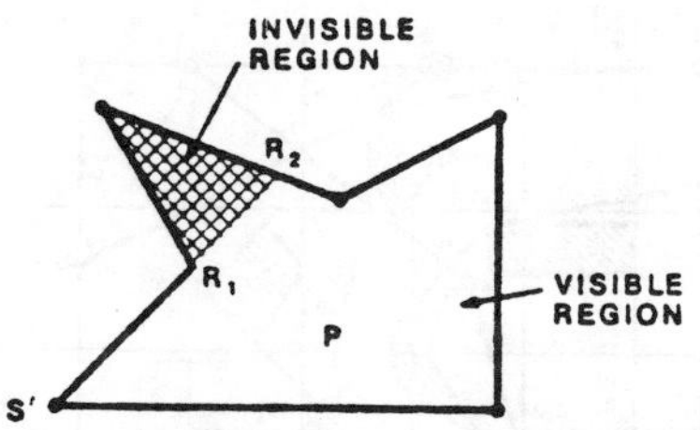

Fig. 9. Visible and invisible regions with respect to S'. Visible region: The region reachable by a sweeping sensor, when all obstacles are removed. Invisible region: the region not reachable by sweeping sensor, when all obstacles are removed.

regions are declared as unexplored with respect to the vertices on the line that limits the range of sensor from S'. The visible region is partitioned into obstacle-polygons and free-polygons. In Figure 10 the region R_1 is unexplored with respect to the vertices P_1 and P_2. The regions R_2 and R_4 are free-polygons, and the region R_3 is an obstacle polygon with respect to S'. It is to be noted that, in general, a polygon can be an obstacle-polygon with respect to the other vertices. But, a polygon is a free-polygon with respect to all the vertices of the polygon.

The algorithm **CONSOLIDATE** checks for any adjacent free-regions from a convex region. If they form a convex region then it combines them and forms a single free-polygon. The consolidation algorithm is described as follows:

CONSOLIDATION ALGORITHM

```
ALGORITHM CONSOLIDATE (P, S);
P IS AN EXPLORED POLYGON WITH RESPECT TO
   VERTEX S
BEGIN
1. FOR EACH FREE-POLYGON P1, BELONGING TO THE
   PARTITION OF P  DO
   BEGIN
2.  FIND ALL ADJACENT FREE-POLYGONS OF P1;
3.  FIND THE MAXIMAL SUBSET OF THEM THAT
    FORMS A CONVEX POLYGON AND COMBINE THEM
    INTO A SINGLE POLYGON;
   END:
END:
```

The complete navigation algorithm for $S'D'$ is described in the Pascal-like syntax. The overall effect of this navigation algorithm is summarized as follows:

1. In general, all free-polygons are convex and these polygons increase in size as learning proceeds.

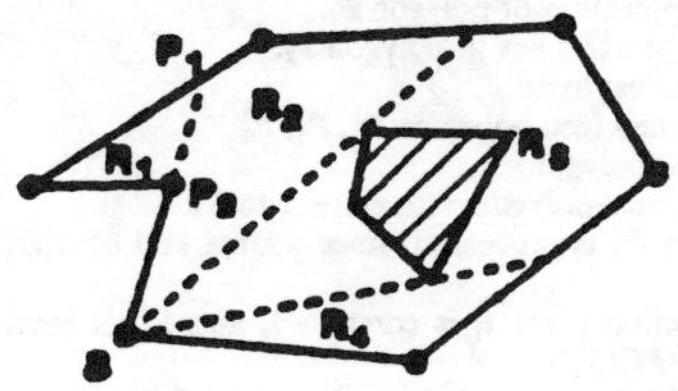

Fig. 10. Partitioning a polygon based on sensor readings.

2. Initially, all the obstancles are bounded by larger polygons, and as learning proceeds the bounding polygons are reduced in size to enclose the obstacles more closely.

3. If the path of navigation runs through all free-polygons, then the complete path from S' to D' can be directly computed.

4. If the path contains unexplored polygons, then the robot halts at the appropriate stop point to explore the regions, and then the next stop point is computed only after the information about the currently explored region is incorporated into the terrain model.

5. Learning is incorporated along with path planning.

6. The paths are locally optimal initially, and they gradually become globally optimal as learning proceeds.

```
ALGORITHM NAVIGATE-GLOBAL (S', D');
S' AND D' ARE THE SOURCE AND DESTINATION
POINTS, RESPECTIVELY,
ON THE SPATIAL GRAPH
S'D' STANDS FOR THE STRAIGHT LINE JOINING S' AND
D'
BEGIN
1. FIND THE POLYGON P, THAT CONTAINS SOURCE
   END OF S'D';
2. IF (P IS AN OBSTACLE-POLYGON)
3. THEN
   BEGIN
4.  FIND THE NEAREST INTERSECTION POINT s, OF
    S'D' AND P;
5.  FIND S*, SUCH THAT s ∈ VP(S*);
6.  MOVE TO S* ALONG EDGES OF P;
7.  NAVIGATE-GLOBAL (S*, D');
   END
8. ELSE IF (P IS A FREE-POLYGON)
9. THEN
   BEGIN
10. FIND THE INTERSECTION POINT s, OF S'D' AND P;
11. GO DIRECTLY TO s;
12. NAVIGATE-GLOBAL (s, D');
   END
13. ELSE IF (P IS UNEXPLORED WITH RESPECT TO S')
14. THEN
   BEGIN
15. EXPLORE (P, S');
16. CONSOLIDATE (P, S');
17. NAVIGATE (S', D');
   END;
END;
```

In the above algorithm we assumed the robot to be a point. However, the same can be applied to any finite

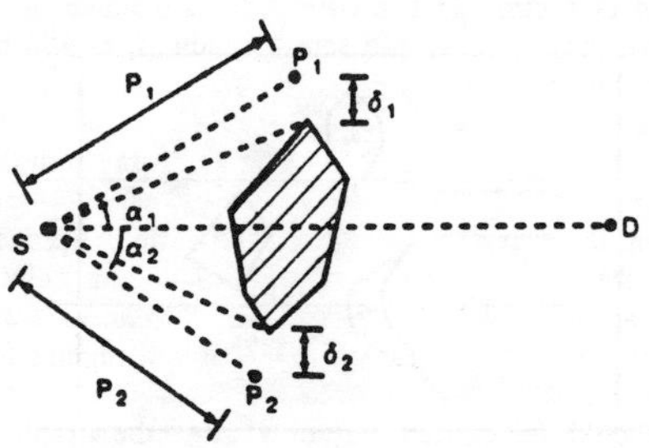

Fig. 11. Modifications for a finite sized robot: δ_1 and δ_2 account for the finite robot dimensions.

sized robot by allowing suitable leeway in computing $\dot{\alpha}_1$, α_2, P_1 and P_2 from the sensor readings as shown in Figure 11. However, a more generalized problem would be to consider the exact shape of the robot and plan the motion that involves both translation and rotation. Other natural extensions of the problem include the use of more than one sensor, and also by talking into account the errors in distance measurement.

5. ILLUSTRATIVE EXAMPLE

In this section we illustrate our technique by tracing the algorithm of the previous section using a sample terrain. Figure 12 shows an unexplored terrain that contains four obstacles O_1, O_2, O_3, and O_4. Initially, four paths are traversed using local optimization from the source points S_1, S_2, S_3, and S_4 to D_1, D_2, D_3 and D_4, respectively. These paths are shown in Figure 5, and the corresponding spatial graph and Voronoi diagram are shown in Figure 6 and Figure 7, respectively. Now consider applying the method of this paper to determining a path from S_5 to D_5. First, the virtual-source S', and virtual-destination D', are found as the nearest graph vertices corresponding to S_5 and D_5, respectively, as in Figure 13.

The path from S_5 to S_5^1 is traversed according to the local optimization method. The polygon P_2 contains the source end of the line $S_5^1D_5^1$. The polygon P_2 is unexplored and hence algorithm EXPLORE (P_2, S_5^1) is invoked. The region P_2 is scanned using the sensor, and the polygon P_2 is partitioned into the regions P_2^1, P_2^2 and P_2^3 as in Figure 14.

The regions P_2^1 and P_2^3 are free-polygons, and the region P_2^2 is an obstacle-polygon with respect to the vertex S'. At this point, the source end of $S_5^1D_5^1$ is contained in the polygon P_2^2. The intersection point I_1, of $S_5^1D_5^1$ with the farther edge of P_2^2 is computed, and its nearest vertex S_5^{11} of the spatial graph is found. Then, the nearest path to S_5^{11} via the edges of the polygon P_2^2 is computed by finding the corresponding euclidian distance. The robot navigates along the edges of the polygon P_2^2 to reach S_5^{11}. Next the path is planned from $S_5^{11}D_5^1$. The polygon P_3 contains the source end of $S_5^{11}D_5^1$, and is unexplored. Based on sensor readings, the polygon P_3 is partitioned into the regions P_3^1, P_3^2 and $P_3^3 \cdot P_3^1$ and P_3^3 are free-polygons and P_3^2 is an obstacle-polygon. At this stage, P_3^1 contains the source

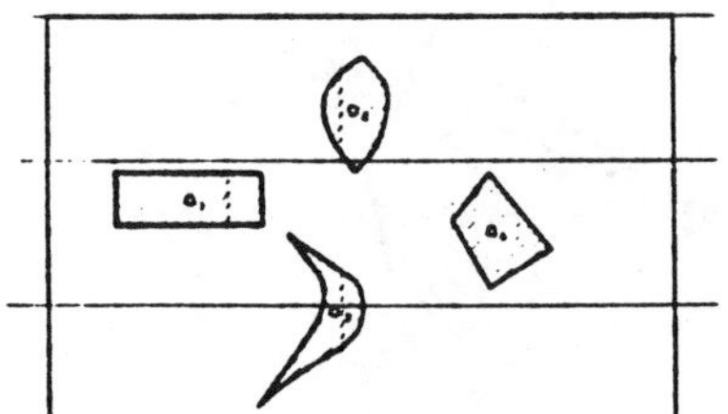

Fig. 12. Unexplored terrain.

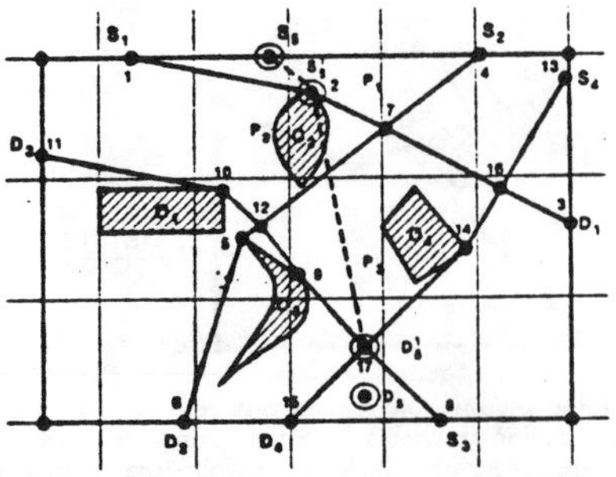

Fig. 13. S_5 source point D_5 destination point. Consider navigation from S_5 to D_5. S_5^1 is the virtual source $S_5 \in VPC\ (S_5^1)$. D_5^1 is the virtual destination $D_5 \in VPC\ (D_5^1)$. Path from S_5 to S_5^1 is according to local optimization.

end of $S_5^{11}D_5^1$. The intersection point of $S_5^{11}D_5^1$ with P_3^1 is D_5^1. The path the $S_5^{11}D_5^1$ is directly traversed as in Figure 15. No update of the model is carried out since $S_5^{11}D_5^1$ is entirely contained in a free-polygon. The navigation from $D_5^1D_5$ is based on local optimization. The final spatial graph of the terrain is given in Figure 16. Note that the obstacles O_2 and O_4 are bounded by smaller polygons that those shown in Figure 5. Also, the polygons P_2^1, P_2^3, P_3^1 and P_3^3 are declared to be free-polygons. Regions P_2^3 and P_3^3 are combined to form a single free-polygon. Clearly, the information about the obstacles and free space of *Figure* 16 is more consolidated and available for utilization than that in Figure 5. Consider another navigation path from S_6 to D_6. The result of this traversal is shown in Figure 17. Now the regions P_4, P_6 and P_9 are declared to be free-polygons. The objects O_3 and O_4 are bounded by much smaller polygons than the ones in Figure 16. Thus, the example illustrates the shrinking of the bounding

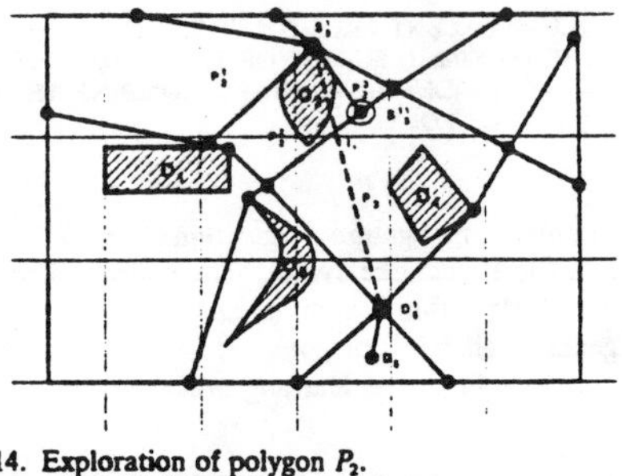

Fig. 14. Exploration of polygon P_2.
Source end of $S_5^1D_5^1$ lies in polygon P_2.
Polygon P_2 is explored.
P_2 is partitioned into polygons $P_2^1, P_2^2, \cdot P_2^3$.
P_2^1, P_2^3 – Free-polygons.
P_2^2 is an obstacle-polygon with respect to S_5^1.
The polygon P_2^2 is processed, since source end of $S_5^1D_5^1$ lies in P_2^2.
The intersection point I_1 is computed, and S_5^{11} is found, such that $I_1 \in VPC(S_5^{11})$.
Path S_5^1 to S_5^{11} is traversed along the minimal length path along the edges of P_2^2.

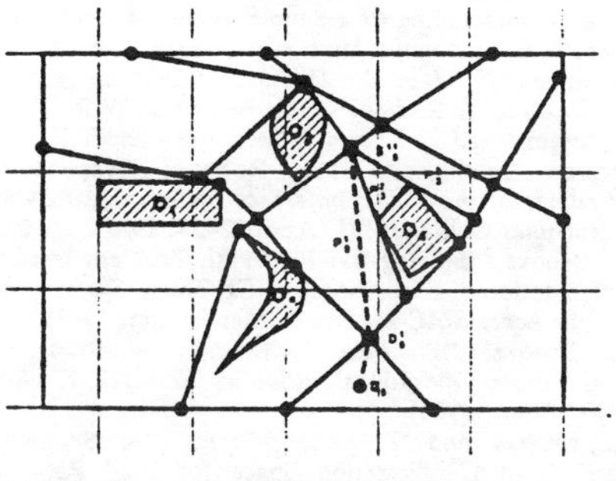

Fig. 15. Exploration of Polygon P_3.

polygons of the obstacles and widening of the free polygons, as learning proceeds. Again, as more paths are traversed, more and more polygons are explored and the spatial graph becomes consolidated.

6. CONCLUSIONS

In this paper, we describe a method that enables a mobile robot to navigate in an unexplored terrain and learn more about the terrain as it navigates paths. Our method requires no prelearned model, makes maximal use of available information, records and synthesizes information from multiple journeys, and contains concepts of learning that allow for continuous transition from local to global optimality. The model of the terrain consists of a *spatial graph* and a *Voronoi diagram*. As more information is consolidated into the terrain model, the bounding polygons of the obstacles fit more closely and the polygons representing free space grow larger. In this way, the robot learns and applies the results of dynamically acquired sensor information to improve performance and relax navigational ambiguity on a continual basis up to the point where the environment is fully described, i.e. all obstacle-polygons are tightly bounded.

This paper has introduced the concept of learning in the domain of robot navigation and movement, namely path traversal and planning through a two-dimensional Cartesian environment. The utilization of concepts of *spatial graphs* has much broader implications however.

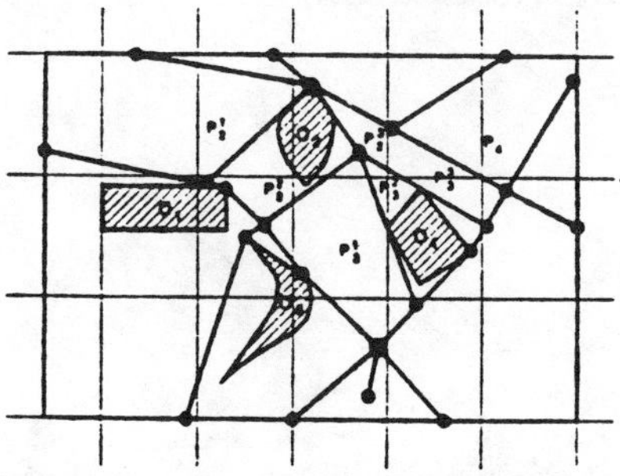

Fig. 16. Terrain model after the path from S_5 to D_5 is consolidated.

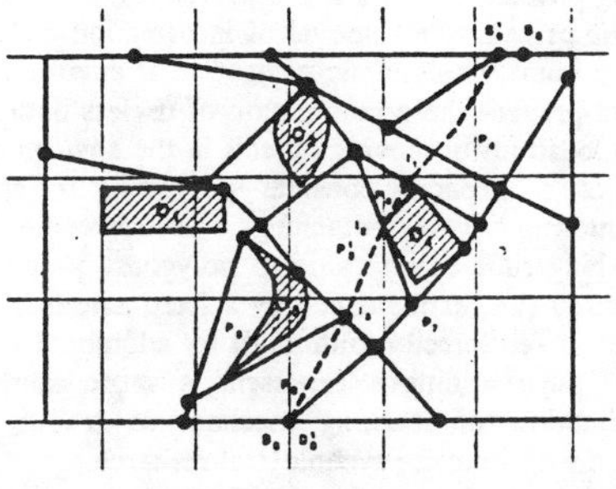

Fig. 17. Terrain model after the path from S_6 to D_6 is consolidated. Traversal of yet another path from S_6 to D_6. 1. The obstacles O_4 and O_3 are bounded by smaller polygons. 2. Polygons P_4, P_6, P_9 are declared free-polygons. 3. Polygons P_7 and P_8 are declared obstacle polygons with respect to l_1 and l_3 respectively. 4. Path is globally optimal from S_6^1 to D_6^1.

For example, the rates at which sensor data updates are applied to the spatial graph directly affect the potential of the robot to navigate in a changing environment. Voronoi regions under a learning navigation paradigm can expand or shrink as a result of changing environmental conditions. The present spatial graph reflects only decisions arrived at from analysis of sensor data, but the method also permits the fusion of multiple sensor sources such as simultaneous use of line-of-sight (visual) and sonar to compose a simple graph space.

Similarly, there is no reason to confine the dimensionality of the graph to an N of two. For example, by extending the two-dimensional polygons to three-dimensional volumes, traversal in three dimensions and the learning of three-dimensional spaces become possible. A typical extension could be three-dimensional path planning of a robot end effector during grasping behavior scenarios. Further, the *spatial graph* nodes do not have to represent a single value. They can, for example, be pointers to complex data structures which contain a variety of relational data about a robot environment. In this way during path planning a spatial graph can serve as a context sensitive procedure for data base searching by limiting the potential set of world data to local Voronoi regions and their associated data sets. In this manner, decisionmaking can be aided through "context focusing" which makes use of the spatial localization of the robot. Details of data structure and complexity analysis of the proposed algorithms are covered in a different paper.

7. FUTURE DIRECTIONS

Research is currently underway to extend the completeness of learning concepts to HERMIES navigation. In reality, true learning involves the utilization of more extensive sets of information such as those contained in complex data structures. Typical data include time tags, inter-object relations, tentative object classifications or labels. At present, we are extending learning to demonstrate performance on the HERMIES-II robot at CESAR by incorporating consolidation, abstraction, and forgetting processes. The latter deserves some comment. Forgetting or selective removal of information becomes more important for dynamic navigation if environments change to prevent the accumulation of useless data such as graph locations of moving objects in the environment. We propose to explicitly consider "forgetting" of spatial graph information by attaching a reinforcement or extinguishing time-based value to polygons. Values are decremented (i.e, extinguished by a fixed amount) uless a polygon is reinforced (confirmed) by additional sensor contacts. Such additions represent a more complete implementation of learning mechanisms traditionally associated with human psychological research.

Acknowledgments

The authors gratefully acknowledge comments by the ORNL reviewers. Special thanks to G. deSaussure for improving the presentation of the paper. We sincerely appreciate the support of O. Manley of the Office of Basic Energy Sciences and the encouragement of Alex Zucker and F. C. Maienschein. This manuscript was expertly prepared by A. B. Weil, E. S. Howe, and C. Zeigler.

References

1. P. Hart et al., "A Formal Basis for the Heuristic Determination of Minimum Cost Paths" *IEEE Trans. Sys. Sci. Cyber*, **SSC-4(2)**, 100–107 (1968).
2. N. Nilsson, "Mobile Automation: An Application of Artificial Intelligence Techniques" *Proceedings of First International Joint Conference on Artificial Intelligence* 509–520 (May, 1969).
3. A.M. Thompson, "The Navigation System of the JPL Robot" *Proceedings of the Fifth International Joint Conference on Artificial Intelligence* Cambridge, MA, 749–757 (August 22–25, 1977).
4. G. Giralt et al., "A Multilevel Planning and Navigation System for a Mobile Robot" *Proceedings of the Sixth International Joint Conference on Artificial Intelligence* Tokyo, Japan 335–338 (August 20–23, 1979).
5. G. Giralt et al., "An Integrated Navigation and Motion Control System for Autonomous Multisensory Mobile Robots" *Robotics Research* 191–214 (1984).
6. H. P. Moravec, "Obstacle Avoidance and Navigation in the Real World by a Seeing Robot Rover" Carnegie-Mellon Robotics Institute, Pittsburgh, PA. *Technical Report* CMU-RI-TR- (September 3, 1980).
7. R. Wallace et al., "First Results in Robot Road Following" *Proceedings of the Ninth International Joint Conference on Artificial Intelligence* Los Angeles, CA 1089–1095 (August 18–23, 1985).
8. Y. Kanamaya et al., "A Mobile Robot with Sonic Sensors and Its Understanding of a Simple World" *Proceedings of Seventh International Joint Conference on Artificial Intelligence* (1981) (see also IECON '84 1984, p. 303).
9. C.R. Weisbin, J. Barhen, G. de Saussure, W.R. Hamel, C.C. Jorgensen, J.L. Lucius, E.M. Oblow and T.E. Swift, "Machine Intelligence for Robotics Applications" *Proceedings of the 1985 Conference on Intelligent Systems and Machines* Oakland, MI (April 22–24, 1985).
10. R.A. Brooks, "Solving the Find-Path Problem by Good Representation of Free-Space" *IEEE Trans. Systems, Man and Cybernetics* **SMC-13**, No. 3 (March/April 1983).
11. R.A. Brooks, "Planning Collision-Free Motions for Pick-and-Place Operations" *Robotics Research* **2**, No. 4, 19–44 (Winter, 1983).
12. R.A. Brooks and T. Lozano-Perez, "A Subdivision Algorithm in Configuration Space for Find Path with Rotation" *IEEE Trans. Systems, Man and Cybernetics* **SMC-15**, No. 2, 224–233 (March/April, 1985).
13. T. Lozano-Perez, "Automatic Planning of Manipulator Transfer Movements" *IEEE Trans. Systems. Man, and Cybernetics* **SMC-11**, 681–689 (October, 1981).
14. T. Lozano-Perez, "Spatial Planning: A Configuration Space Approach" *IEEE Trans. Computers* **C-32**, 108–120 (February, 1983).
15. T. Lozano-Perez, and M.A. Wesley, "An Algorithm for Planning Collision-Free Paths Among Polyhedral Obstacles" *Commun. ACM* **22**, No. 10, 560–570 (October, 1979).
16. J.E. Hopcroft, J.T. Schwartz and M. Sharir, "On the Complexity of Motion Planning for Multiple Independent Objects; PSPACE Hardness of the Warehouseman's Problem" *Robotics Research* 3(4), 76–88, (1984).
17. M. Sharir and E. Ariel-Sheffi, "On the Piano Movers' Problem: IV. Various Decomposable Two-Dimensional Motion Planning Problems" *Comm. Pure and Applied Mathematics* **37**, 479–493 (1984).
18. J.T. Schwartz and M. Sharir, "On the Piano Movers' Problem: V. The Case of a Rod Moving in Three-Dimensional Space Amidst Polyhedral Obstacles" *Comm. Pure and Applied Mathematics* **37**, 815–848 (1984).
19. J. Crowley, "Navigation for an Intelligent Mobile Robot" *IEEE Journal of Robotics and Automation* **RA-1**, No. 1, 31, ff (March, 1985).
20. A. Parodi, "Multi-Goal Real-time Global path Planning for an Autonomous Land Vehicle Using a High-Speed Graph Search Processor" *IEEE International Conference on Robotics and Automation*, St. Louis, Missouri, 161–167 (1985).
21. A. Chattergy, "Some Heuristics for the Navigation of a Robot" *Robotics Research* **4**, No. 1, Spring, 59–66 (Spring, 1985).
22. D.T. Lee and F.P. Preparta, "Computational Geometry—A Survey" *IEEE Transactions on Computers* **C-33**, No. 12, 1072–1101 (December, 1984).

Robot Navigation in Unknown Terrains Using Learned Visibility Graphs. Part I: The Disjoint Convex Obstacle Case

B. JOHN OOMMEN, MEMBER, IEEE, S. SITHARAMA IYENGAR, NAGESWARA S. V. RAO, AND R. L. KASHYAP, FELLOW, IEEE

Abstract—**The problem of navigating an autonomous mobile robot through unexplored terrain of obstacles is discussed. The case when the obstacles are "known" has been extensively studied in literature. *Completely* unexplored obstacle terrain is considered. In this case, the process of navigation involves both learning the information about the obstacle terrain and path planning. An algorithm is presented to navigate a robot in an unexplored terrain that is arbitrarily populated with disjoint convex polygonal obstacles in the plane. The navigation process is constituted by a number of traversals; each traversal is from an arbitrary source point to an arbitrary destination point. The proposed algorithm is proven to yield a convergent solution to each path of traversal. Initially, the terrain is explored using a rather primitive sensor, and the paths of traversal made may be suboptimal. The *visibility graph* that models the obstacle terrain is incrementally constructed by integrating the information about the paths traversed so far. At any stage of learning, the partially learned terrain model is represented as a *learned visibility graph*, and it is updated after each traversal. It is proven that the learned visibility graph converges to the visibility graph with probability one when the source and destination points are chosen randomly. Ultimately, the availability of the complete visibility graph enables the robot to plan globally optimal paths and also obviates the further usage of sensors.**

I. Introduction

Robotics is one of the most important and challenging areas of computer science. Robots have been increasingly applied in carrying out tedious and monotonous tasks, such as normal maintenance, inspection, etc., in industries. In hazardous environments, such as nuclear power plants, underwater, etc., robots are employed to carry out humanlike operations. However, as a scientific discipline, the area of robotics is fascinating from the directions of challenge, application, and results. Perhaps the most interesting aspect of robotics is the gamut of underlying problems that spans from extremely abstract mathematical problems to highly pragmatic ones.

There are many existing robots capable of carrying out intelligent and autonomous operations. Examples are SHAKEY [18], the JPL robot [21], HILARE [8], the Stanford Cart [17], the CMU terrigator and Neptune robots [24], HERMIES [25], etc. There are many facets to a completely autonomous robot; among them some of the actively pursued fields are knowledge representation, task planning, sensor interpretation, terrain model acquisition, dynamics and control, specialized computer architectures, algorithms for concurrent computations, path planning and navigation, and coordinated manipulation.

Path planning and navigation is one of the most important aspects of autonomous roving vehicles. The *find-path* problem deals with navigating a robot through a completely known terrain of obstacles. This problem is extensively studied and solved by many researchers—Brooks and Lozano-Perez [3], Gouzenes [9], Lozano-Perez [14], Lozano-Perez and Wesley [15], and Oommen and Reichstein [19] are some of the most important contributors. Whitesides [26] is an excellent reference for various strategies used to solve the find-path problem. Another problem deals with navigating a robot through an unknown or partially explored obstacle terrain. Unlike the find-path problem, this problem has not been subjected to a rigorous mathematical treatment, and this could be attributed, at least partially, to the inherent nature of this problem. However, this problem is also researched by many scientists—Brooks [2], Chatila [4], Chattergy [5], Crowley [6], Giralt *et al.* [8], Iyengar *et al.* [10], [11], Laumond [12], Lumelsky and Stepanov [16], Rao *et al.* [20], Turchen and Wong [22], and Udupa [23] present many important results. As pointed out in the literature, the navigation through unknown terrain involves activities such as model acquisition and learning, sensing, etc., which are absent in the find-path problem.

In this correspondence we deal with the problem of navigation through an unexplored terrain. A rather elementary method involves sensing the obstacles and avoiding them in a localized manner. In more sophisticated methods the terrain is explored as the robot navigates. Iyengar *et al.* [10], [11] propose a technique that "learns" the terrain model as the robot navigates. Initially, the robot uses the sensor information to avoid obstacles, and the terrain model is incrementally learned by integrating the information extracted from the earlier traversals. In this method the partially built model is used to the maximum extent in path planning, and the regions where no model is available are explored using sensors. Another important aspect is to bound the obstacles using simple polygons. The free space is spanned by convex polygons. These constituent polygons are updated as the learning proceeds; as a result, the polygons that bound the obstacle shrink in size, and the polygons that span the free space grow in size. Such a strategy provides a way to approximate arbitrary-shaped obstacles by polygons and is also benefited by the available computational geometry and other related algorithms found in [1], [7], [13], [26]. However, there are limitations on the technique of Iyengar *et al.* [10], [11]. The proposed algorithm does not yield a convergent solution in all cases.

We propose a technique for navigation in an unexplored terrain when the terrain is populated with disjoint convex polygonal obstacles. In precise terms, the method proposed here is proven for convergence in terms of planning paths and also in acquiring the entire terrain model through learning. As an endeavor to include learning in the navigation process and to formalize the scheme, we now view the problem in a completely new framework. We assume that the robot begins the navigation in a completely unexplored terrain of finite dimensions. The terrain is populated with stationary obstacles. However, as opposed to the work done earlier, we shall not crystallize our terrain in terms of a Voronoi diagram. Rather, we shall compute and maintain a graph termed as the *Learned Visibility Graph* (LVG). To obtain the LVG, the robot initially navigates through the obstacles using a local navigation technique. This technique, which is a "hill climbing technique," is shown to converge in a slightly restricted workspace. In the process of local navigation, the robot manipulates the LVG. It is shown that the LVG

Manuscript received March 19, 1986. This work was supported in part by the National Sciences and Engineering Council of Canada, and by the National Science Foundation under Grant CDR-85-00022. This correspondence was presented in part at the 1986 National Conference on Artificial Intelligence, Philadelphia, PA.

B. J. Oommen is with the School of Computer Science, Carleton University, Ottawa, ON K1S 5B6, Canada.

S. S. Iyengar and N. S. V. Rao are with the Department of Computer Science, Louisiana State University, Baton Rouge, LA 70803.

R. L. Kashyap is with the Department of Electrical Engineering, Purdue University, West Lafayette, IN 47907.

IEEE Log Number 8716968.

Reprinted from *IEEE Journal of Robotics and Automation*, Vol. RA-3, No. 6, December 1987, pages 672-681. Copyright © 1987 by The Institute of Electrical and Electronics Engineers, Inc. All rights reserved.

ultimately converges to the actual visibility graph (VG) of the obstacle terrain with probability one. The use of the LVG in global navigation and its acquisition during the local navigation phase is the essential difference between our technique and the techniques used by other researchers [2], [4]-[6], [12], [16].

The organization of this correspondence is as follows: Section II introduces the definitions and notations used subsequently. The local navigation technique that incorporates learning and path planning is presented in Section III. The convergence of the proposed algorithm is proven. In Section IV, the power of local navigation algorithm is enhanced by incorporating backtracking. As a result, the interior restriction on the obstacle terrain is relaxed. The modified procedure is also proven for correctness. In Section V, a global navigation strategy that makes use of the existing terrain model to the maximum extent is presented. The important result, that the learning eventually becomes complete, is presented in Section VI. The execution of the navigation algorithms on a sample obstacle terrain is presented in Section VII.

II. Notations and Definitions

The robot is initially placed in a completely unexplored terrain, and it is required to undertake a number of traversals; each traversal is from an arbitrary source point S to an arbitrary destination point D. The robot is treated as a point in a plane that is arbitrarily populated with stationary disjoint and convex polygonal obstacles. Let $W = \{w_1, w_2, \cdots, w_k\}$ be the set of obstacles in the terrain R, where w_i is a convex polygonal obstacle. Furthermore, the obstacles' polygons are mutually nonintersecting and nontouching. Let V be the union of the vertices of all the obstacles and Z be the set of all edges of the obstacle polygons.

Of paramount importance to this entire problem is a graph termed as the *visibility graph*. The VG is a pair (V, E) where the following hold.

1) V is the set of vertices of the obstacles.

2) E is the set of edges of the graph. A line joining the vertices v_i and v_j forms an edge $(v_i, v_j) \in E$ if and only if it is an edge of an obstacle or it is not intercepted by any other obstacle. Formally, if $L(v_i, v_j)$ is the set of points on the line joining v_i and v_j, then $(v_i, v_j) \in E$ iff a) $L(v_i, v_j) \in Z$ or b) $L(v_i, v_j) \cap Z = \phi$.

Visibility graphs have been extensively studied in the computational geometry literature and are used in motion planning by Lozano-Perez [15] and many other researchers (see the survey paper of Whitesides [26]). However, in this context it is important to note that the VG is initially unknown to the robot inasmuch as the obstacles and their locations are unknown. Although the VG is completely unknown initially, it is learned during the initial stages of the navigation process. The partially learned VG is augmented after each traversal by integrating the information extracted from the local navigation.

The process of *learning* is completed when the entire VG of the obstacle terrain is completely built. Before the robot attains this state, the VG is only partially built. The robot graduates through various intermediate stages of learning during which the VG is incrementally constructed. These intermediate stages of learning are captured in terms of the *learned visibility graph* which is defined as follows: LVG = (V^*, E^*), where $V^* \subseteq V$ and $E^* \subseteq E$. The LVG is initially empty and is incrementally built. Ultimately, the LVG converges to the exact VG.

We assume throughout this correspondence that the robot is equipped with a sensor capable of measuring the distance to an obstacle in any specified direction. The availability of the present-day range sensors justifies this assumption. Also, we assume that the robot is equipped with sensors which enable the navigation along the edges of the obstacles. A sensor system constituted by a set of primitive proximity sensors can impart such an ability to the robot. Hence the robot can navigate arbitrarily close to the obstacle edges. These sensors are assumed to be error-free.

The interior of any polygon ξ is denoted by INT (ξ). The straight line from the point P to the point Q is denoted by $\overrightarrow{PQ}$. Further, η_{PQ} denotes the unit vector along the straight line $\overrightarrow{PQ}$. We assume throughout that the robot is operating in the plane. Thus when we use the word "terrain," we use it in a more restricted sense than it is customarily used in the literature. Undoubtedly, navigating in a three-dimensional (3-D) terrain is a far more difficult problem, and we do not claim that the technique we propose is applicable to it. Indeed, even the concept of visibility graphs is not all too meaningful in the latter problem because whereas paths along the edges of polyhedra may not exist, paths along faces of the polyhedra may [26].

III. Local Navigation and Learning

When the robot navigates in a completely unexplored terrain, its path of navigation is completely decided by the sensor readings. The obstacles in the proximity of the source point are scanned, and a suitable path of navigation is chosen. This localized nature of the local navigation makes a globally optimal path unattainable in a terrain with an arbitrary distribution of obstacles. However, local navigation is essential during the initial stages of the navigation. The information acquired during the local navigation is integrated into the partially built terrain model. No local navigation is resorted to in the regions where the existing terrain model is sufficient for planning globally optimal paths.

In this section we propose a local navigation technique that enables the robot to detect and avoid obstacles along the path from an arbitrary source point S to an arbitrary destination point D. The robot is equipped with a primitive motion command MOVE(S, A, λ), where

a) S is the source point, namely, the place where the robot is currently located;
b) A is the destination point which may or may not be specified;
c) λ is the direction of motion, which is always specified.

If A is specified, then the robot moves from S to A in a straight-line path. In this case, the direction of motion λ is the vector η_{SA}, the unit vector is the direction of $\overrightarrow{SA}$. If A is not specified, then the robot moves along the direction λ as follows. If the motion is alongside an edge of an obstacle, then the robot moves to the end point of the edge along the direction λ. This end point is returned to the calling procedure as point A as in Fig. 1(a). If motion is not alongside an edge of an obstacle, then the robot traverses along the direction λ until it reaches a point on the edge of an obstacle as shown in Fig. 1(b). This point is returned as the point A to the calling procedure.

In the remainder of this section we describe the local navigation algorithm. For the treatment in this section we assume that the obstacles do not touch or intersect the boundaries of the terrain R. In other words, the obstacles are properly contained in the terrain R. This is formally represented as

$$\bigcup_{i=1}^{k} \text{INT}\,(w_i) \subseteq \text{INT}\,(R). \quad (1)$$

As a consequence of this assumption, a path always exists from a source point S to a destination point D. However, this restriction is removed in the next section.

We present the procedure NAVIGATE-LOCAL that uses a *hill-climbing* technique to plan and execute a path from an arbitrary source point S to an arbitrary destination point D. The outline of this procedure is as follows. The robot moves along $\overrightarrow{SD}$ until it gets to the nearest obstacle. It then circumnavigates this obstacle using a local navigation strategy. The technique is then recursively applied to reach D from the intermediate point. Further, apart from path planning, the procedure also incorporates the learning phase of acquiring the VG.

We now concentrate on local navigation strategy. The robot moves along the direction η_{SD} till it encounters an obstacle at a point A which is on the obstacle edge joining the two vertices, say, A_1 and A_2. At this point the robot has two possible directions of motion: along $\overrightarrow{AA_1}$ or $\overrightarrow{AA_2}$ as shown in Fig. 2. We define a local optimization

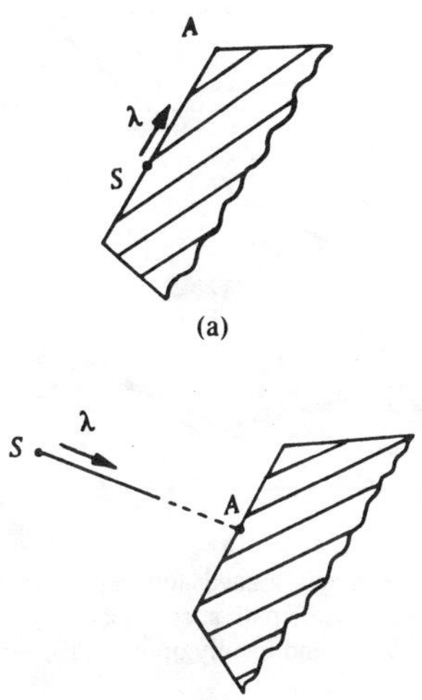

Fig. 1. Value returned by operation MOVE(S, A, λ), when A is not specified. (a) Motion along edge of obstacle. (b) Motion till obstacle is encountered.

criterion function J as follows:

$$J = \eta_{SD} \cdot \lambda \tag{2}$$

where λ is a unit vector along the direction of motion.

Let λ_1 and λ_2 be the unit vectors along $\overrightarrow{AA_1}$ and $\overrightarrow{AA_2}$, respectively. Let $\lambda^* \in \{\lambda_1, \lambda_2\}$ maximize the function J given in (2). The robot then undertakes an exploratory traversal along the direction $-\lambda^*$ until it reaches the corresponding vertex called the *exploratory* vertex. At this exploratory point the terrain is explored using the procedure UPDATE-VGRAPH. Then the robot retraces along the locally optimal direction λ^* until it reaches the other vertex S^*, whence it again calls UPDATE-VGRAPH. The procedure NAVIGATE-LOCAL is recursively applied to navigate from S^* to D.

The procedure UPDATE-VGRAPH implements the learning component of the robot navigation. Whenever the robot reaches a new vertex v_i, this vertex is added to the LVG. From this vertex the robot beams its sensor in the direction of all the *existing* vertices of the LVG. The edge (v_i, v) is added to the edge set E^*, corresponding to each vertex $v \in V^*$ visible from v_i. The algorithm is formally presented as follows:

```
procedure UPDATE-VGRAPH(v);
input: the vertex v which is newly encountered.
output: the updated LVG = (V*, E*).
        Initially, the LVG is set to (φ,φ).
comment: DIST(v1,v2) indicates the Euclidian distance
        between vertices v1 and v2, if they are visible
        to each other.
        This is the auxiliary information stored along with the LVG.
begin
    V* = V* ∪ {v};
    for all v1 ∈ V* − {v} do
        if (v1 is visible from v) then
            DIST(v1,v) = |v1v|;
            E* = E* ∪ {(v1,v)};
        else
            DIST(v1,v) = ∞;
        endif
    endfor;
end.
```

The procedure NAVIGATE-LOCAL uses the motion primitive motion command MOVE and the procedure UPDATE-VGRAPH during execution. This procedure is formally described as follows:

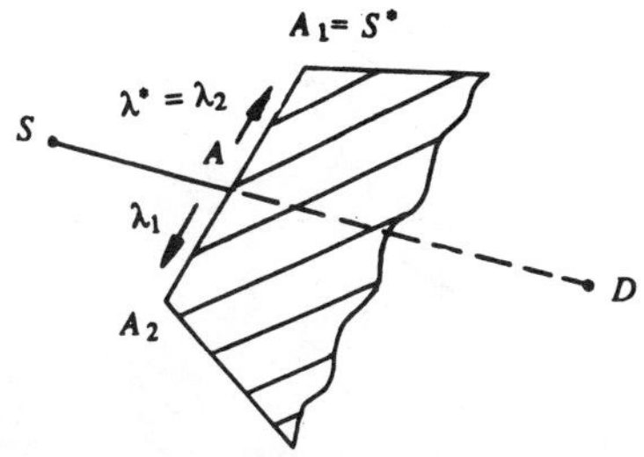

Fig. 2. Robot reached point on obstacle.

```
procedure NAVIGATE-LOCAL(S,D);
Input: The source point S and the destination point D.
Output: A sequence of elementary MOVE commands.
begin
1.  if (D is visible from S) then
2.      MOVE(S,D,ηSD)
3.  else
4.      if (S is on an obstacle and the obstacle obstructs its view) then
5.          compute {λ1,λ2}, the two possible directions of motion;
6.          λ* = direction maximizing λi·ηSD;
7.          if (S is a vertex) then
8.              if (S ≠ V*) then UPDATE-VGRAPH(S);
9.              MOVE(S,S*,λ*);
10.         else
11.             MOVE(S,S1,−λ*); {make exploratory trip to S1}
12.             if (S1 ≠ V*) then UPDATE-VGRAPH(S1);
13.             MOVE(S1,S*,λ*); {retrace steps to S*}
14.             if(S* ≠ V*) then UPDATE-VGRAPH(S*);
            endif;
15.         NAVIGATE-LOCAL(S*,D);
16.     else {move to next obstacle}
17.         MOVE(S,S*,ηSD); {move to next obstacle along ηSD}
18.         NAVIGATE-LOCAL(S*,D);
        endif;
    endif;
end.
```

We shall now prove that the procedure NAVIGATE-LOCAL converges. In a subsequent section we shall show that the LVG updated using UPDATE-VGRAPH ultimately converges to the exact VG.

Theorem 1: For a noninterlocking workspace, the procedure NAVIGATE-LOCAL always finds a path from S to D in finite time.[1]

Proof: There is always a path from S to D as per the assumption in (1). Hence it suffices to prove that the recursion is correctly applied. We shall prove that this is indeed the case and also that the MOVE operations minimize the projected distance along η_{SD}. Then the theorem follows from the fact that total the number of vertices of all the obstacles is finite.

Case I—Terminating Step: If D is visible from S, the procedure terminates as per line 2 in NAVIGATE-LOCAL. In this case, the projected distance of the path traversed by the robot is reduced from $|SD|$ to zero in one step.

Case II—Recursive Steps: This step consists of three mutually exclusive and collectively exhaustive cases. In each case we shall show that each execution of MOVE(S, S^*, λ^*) forces the following *strict* inequality:

$$|SD| > \overrightarrow{S^*D} \cdot \eta_{SD}. \tag{3}$$

Case IIa: The point D is not visible from S, and S is not on the boundary of the obstructing obstacle. Fig. 3(a) depicts this scenario. The lines 17 and 18 of NAVIGATE-LOCAL give the corresponding actions.

[1] Please see the Appendix for the reason why the workspace should be restricted for the current version of NAVIGATE-LOCAL. The details of a workspace being noninterlocking are included in the Appendix.

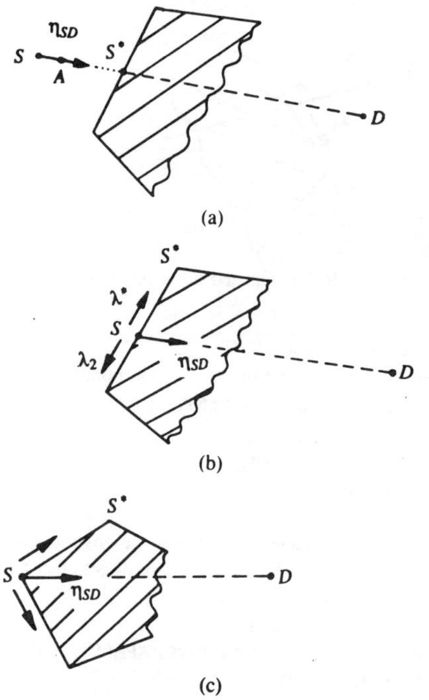

Fig. 3. Robot at S is obstructed by obstacle. (a) S is not on edge of obstacle. (b) S is on edge of obstacle but not at vertex. (c) S is at vertex of obstacle.

In this case, the motion is along η_{SD} and every point A along this vector satisfies the relation $|\overrightarrow{AD}| = \overrightarrow{AD} \cdot \eta_{SD}$. Thus the motion along η_{SD} to A gives the following equality: $|\overrightarrow{SD}| = |\overrightarrow{SA}| + \overrightarrow{AD} \cdot \eta_{SD}$, whence

$$|\overrightarrow{SD}| > \overrightarrow{AD} \cdot \eta_{SD}. \tag{4}$$

The equation is particularly true for $A = S^*$, and hence the result.

Case IIb: The point D is not visible from S, and S lies on the edge of the obstructing obstacle.

Case 1): S does not correspond to a vertex of the LVG as shown in Fig. 3(b). If $\lambda_i \cdot \eta_{SD} = 0$, then the edge is orthogonal to η_{SD}. In this case, either direction does not decrease the projected distance. Note that this situation can occur at most once for an obstacle encountered during the navigation and hence cannot persist. However, after the robot completes the MOVE corresponding to this step, the robot is located at a vertex. Since this is covered in case 2, we shall only consider the case when $\lambda_i \cdot \eta_{SD} \neq 0$. Let λ^* be the direction in which $\lambda_i \cdot \eta_{SD} > 0$. The situation is shown in Fig. 3(b). Hence the included angle S^*SD is less than $\Pi/2$, and $|\overrightarrow{SD}| - |\overrightarrow{S^*D} \cdot \eta_{SD}| = |\overrightarrow{SS^*}| \cos(S^*SD) > 0$. Hence the execution of MOVE(S, S^*, λ^*) ensures the inequality given in (3).

In this case the robot temporarily diverges from the locally optimal path. This trip being purely exploratory does not contribute to the navigation. Note that this trip takes finite time.

Case 2): The point S is located at a vertex of an obstacle. Fig. 3(c) depicts the situation. In this case the edges of the convex polygon meet at S. Let the direction λ^* be the direction that maximizes $\lambda_i \cdot \eta_{SD}$. Because the obstacle is convex, the angle between the edges at S is less than Π. Thus the angle S^*SD is less than $\Pi/2$. Using the arguments of case 1, we conclude that the execution of the MOVE operation satisfies $|\overrightarrow{SD}| - |\overrightarrow{S^*D} \cdot \eta_{SD}| > 0$, and hence the theorem.

Observe that if we had only *one* polygonal obstacle, we could have gone around the obstacle in a systematic way, i.e., either in a clockwise or an anticlockwise direction, until we reached a point from which D is visible. However, the problem becomes more difficult when more than one obstacle exists. In this case, the motion must be made in such a way that a criterion function is minimized. We have chosen to minimize the projected distance along SD by maximizing the function J in (2). This method may not give rise to a

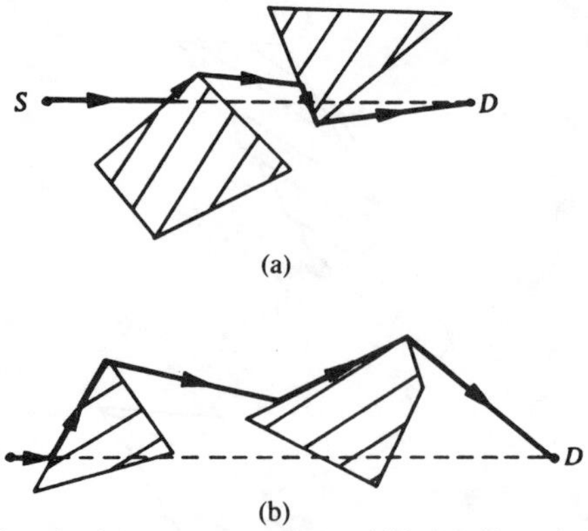

Fig. 4. Local navigation strategy need not yield globally optimal solution. Dark lines with arrows indicate path according to local navigation strategy. (a) Solution is both globally and locally optimal. (b) Solution is only locally optimal.

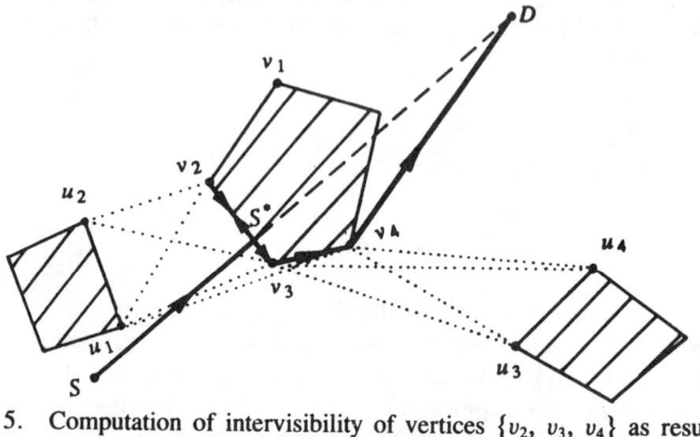

Fig. 5. Computation of intervisibility of vertices $\{v_2, v_3, v_4\}$ as result of local navigation. Dotted lines indicate visibility between vertices.

globally optimal path as shown in Fig. 4. Such counter examples exist for any localized navigation scheme for the want of global information about the obstacles. The modification of NAVIGATE-LOCAL for interlocking workspaces is shown in the Appendix.

It is easy to conceive of a scheme in which the sensor readings can give all the visible edges and vertices of the obstacles. In such a case, there may be a shorter path for navigation. However, we choose to go along the path dictated by NAVIGATE-LOCAL so that the LVG can be updated in the process of navigation while the projected distance along η_{SD} is minimized. The procedure UPDATE-VGRAPH makes sure that edges to all the visible vertices of LVG are added to E^* when a new vertex is added to V^*. Fig. 5 shows the salient features of the approach. The vertices u_2, u_2, u_3, u_4, v_1 are presently existing in LVG, and the edges (u_1, u_2), (u_3, u_4) are also present. A globally optimal path is Sv_4D. However, we choose the path $SS^*v_2S^*v_3v_4D$—which is only suboptimal. The exploratory traversal to v_2 yields the visibility information about the vertices v_1, v_3, v_4, which is obtained from the sensor information. It is conceivable that the procedure NAVIGATE-LOCAL can be modified to avoid exploratory trips along the explored edges of the obstacles. However, we regard this issue as rather straightforword and prefer not to elaborate on it.

IV. Limitations of Local Navigation and a Solution

The procedure NAVIGATE-LOCAL introduced in the previous section always yields a path in a noninterlocking workspace if one exists and if the obstacles do not touch the terrain boundaries. These preconditions are implicitly satisfied as a consequence of the assumption in (1). The relaxation of this assumption results in two conditions in which the procedure NAVIGATE-LOCAL is not guaranteed to halt.

a) There is no path existing between the source point S to the destination point D. In this case, a single obstacle blocks all the paths from S to D. Fig. 6 shows some such cases. Note that when the robot starts moving around the obstacle, its way is blocked in both possible directions.

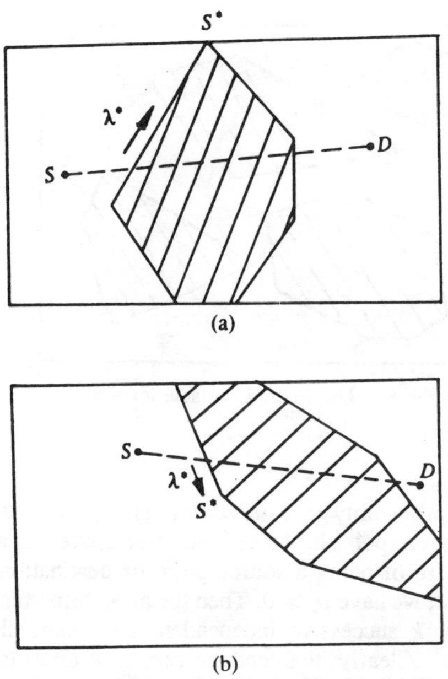

Fig. 6. No path from S to D. (a) Case 1. (b) Case 2.

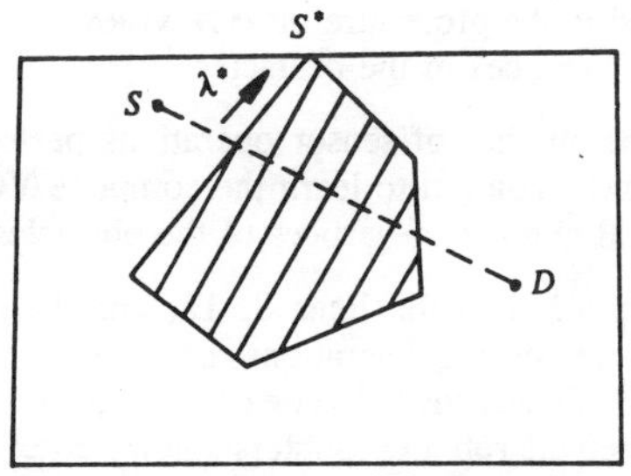

Fig. 7. Dead corner S^* formed by obstacle and terrain boundary.

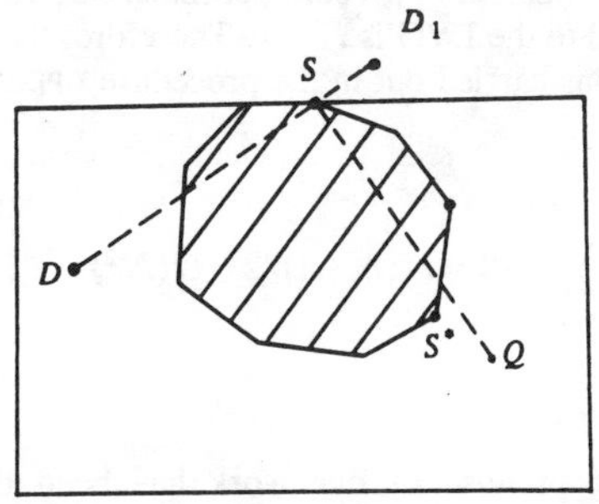

Fig. 8. Proof of convergence of procedure BACKTRACK.

b) The angle between the obstacle edge and the terrain boundary is less than $\Pi/2$. In this case we assume that a path exists between S and D, or, stated equivalently, no obstacle blocks all the paths between S and D. In such a case the robot may be forced to move to the dead corner formed by the obstacle and terrain boundary. At this point the robot has no further defined moves. The robot starting at S gets into the dead corner at S^*. This situation is depicted in Fig. 7.

In this section, we relax the condition in (1) and enhance the capability of NAVIGATE-LOCAL by imparting to it the ability to *backtrack*. The robot backtracks (by invoking procedure BACKTRACK) whenever it reaches a point from which no further moves are possible (see Fig. 8). This procedure intelligently guides the robot in the process of retracing steps. That is, the robot backtracks along the edges of the obstructing obstacle till an edge (S, S_1), that makes an angle less than $\Pi/2$ with η_{SD} is encountered. The fact that such an edge exists is guaranteed because of the convexity of the obstacles. The search for this edge is performed by the while loop of lines 3–6 of procedure BACKTRACK. As a result, the robot moves to a point from which the NAVIGATE-LOCAL can take over. If for the same obstacle the robot has to backtrack twice, then there is no path between S and D. In other words, if a path from S to D exists, then the robot needs to backtrack at most once along the edges of any obstacle. These aspects are further discussed subsequently in this section. The following is the BACKTRACK algorithm:

procedure BACKTRACK(S,D,S^*);
Input: The point D is the destination point.
S is a dead corner, i.e., a vertex of an obstacle and is also on the boundary of the terrain. The terrain is noninterlocking.
Output: A sequence of MOVEs from S in such a way that if a path exists, then it can be determined using NAVIGATE-LOCAL. The location S^* is returned to the calling procedure.

begin
1. $S_1 = S$;
2. λ^* = only permitted direction of motion on the obstacle;
3. **while** ($\overrightarrow{SD} \cdot \lambda^* < 0$) **do**
4. MOVE(S_1, S^*, λ^*);
5. $S_1 = S^*$;
6. λ^* = only permitted direction of motion on the obstacle;
endwhile;
end.

The convergence of the procedure BACKTRACK is proved in the following theorem.

Theorem 2: The procedure BACKTRACK leads to a solution to the navigation problem in a noninterlocking workspace, if one exists.

Proof: The crux of the theorem is to prove that the procedure BACKTRACK terminates in all cases. In other words, an edge exists that makes an angle less than $\Pi/2$ with η_{SD}. Fig. 8 shows the scenario. Consider the line SQ, a normal to SD at S. Because the obstacle is convex, the normal line SD at S must intersect the obstacle again, and at this point the corresponding edge makes angle less than $\Pi/2$ with $\overrightarrow{SD}$. Thus the required vertex is found just after this edge because after this edge the first vertex indeed has a smaller value for the projected distance along η_{SD}. Hence by moving along the boundary in this direction, the procedure BACKTRACK will take the robot to a place from which NAVIGATE-LOCAL can be applied.

We note that if a path exists, the further execution of NAVIGATE-LOCAL will not lead to a dead end formed by the same obstacle. That is because if the procedure BACKTRACK leads the robot to another dead end on the same obstacle, clearly, the robot cannot navigate across the obstacle. Hence no path exists between S and D.

Let the procedure NAVIGATE-LOCAL with the enhanced capability to backtrack be called procedure NAVIGATE-LOCAL-WITH-BACKTRACK. This procedure utilizes NAVIGATE-LOCAL to navigate till the robot encounters a dead end. At this point the procedure BACKTRACK is invoked, after which the NAVIGATE-LOCAL is used. The navigation is stopped if no path exists between S and D. The correctness of the proof of procedure NAVIGATE-LOCAL-WITH-BACKTRACK easily follows from the arguments of this section. Similarly, the formal statement of procedure NAVIGATE-LOCAL-WITH-BACKTRACK easily follows from those of NAVIGATE-LOCAL and BACKTRACK and is omitted for the sake of brevity.

V. Global Navigation

The procedures described in the preceding sections enable a robot to navigate in an unexplored terrain. Such a navigation involves the usage of sensor equipment and traversing the exploratory trips. The navigation paths are not necessarily globally optimal from the path planning point of view. However, the extra work carried out in the form of learning is inevitable because of the lack of information about the obstacles. Furthermore, the LVG is gradually built as a result of learning.

In the regions where the visibility graph is available, the optimal path can be found by computing the shortest path from the source point to the destination point on the graph. The computation can be carried out in quadratic time in the number of nodes of the graph by using the Dijkstra's algorithm [1]. Such a trip can be obtained by using only computations on the LVG and not involving any sensor operations.

We shall now propose a technique that utilizes the available LVG to the maximum extent in planning navigation paths. In the regions where no LVG is available, the procedure NAVIGATE-LOCAL is used for navigation. In these regions the LVG is updated for future navigation. The outline of the global navigation strategy as follows:

procedure NAVIGATE-GLOBAL(S, D);
begin
1. Compute-Best-Vertices(S^*, D^*);
2. NAVIGATE-LOCAL-WITH-BACKTRACK(S, S^*);
3. Move-On-LVG(S^*, D^*);
4. NAVIGATE-LOCAL-WITH-BACKTRACK(D^*, D);

end.

Given S and D, two nodes S^* and D^* on the existing LVG are computed. The robot navigates from S to S^* using local navigation. Then the navigation from S^* and D^* is along the optimal path computed using the LVG. Again, from D^* to D the local navigation is used. Computation of S^* and D^*, corresponding to line 1 of NAVIGATE-GLOBAL, can be carried out using various criteria. We suggest three such possible criteria as follows.

Criterion A: S^* and D^* are the nodes of the LVG closest to S and D. The computation of these nodes involves $O(|V^*|)$ distance computations.

Criterion B: S^* is a vertex such that it is the closest to the line $\overrightarrow{SD}$. D^* is similarly computed. Again, the complexity of this computation is $O(|V^*|)$.

Criterion C: S^* is a vertex which minimizes the angle S^*SD. Again, the complexity of this computation is $O(|V^*|)$.

The closeness of the paths planned by NAVIGATE-GLOBAL to the globally optimal path depends on the degree to which the LVG is built. The paths tend to be globally optimal as the LVG converges to the VG. We shall now prove that the LVG indeed converges to VG after a sufficient number of invocations of NAVIGATE-LOCAL.

VI. Complete Learning

Learning is an integral part of NAVIGATE-LOCAL, primarily because the robot is initially placed in a completely unexplored obstacle terrain, and the LVG is incrementally constructed as the robot navigates. The central goal of the learning is to eventually construct the VG of the entire obstacle terrain. Once the VG is completely constructed, the globally optimal path from S to D can be computed *before* the robot sets into motion as in [15]. Furthermore, the availability of the complete VG obviates the further usage of sensors. Hence the focus of our navigation scheme is to continually augment the LVG with the information extracted from sensor readings with the aim of ultimately obtaining the complete VG. In this section we prove that the learning incorporated in our technique is *complete*, i.e., the LVG ultimately converges to VG with probability one if the source and destination points are randomly selected in the free space.

Theorem 3: If no point in the free space has a zero probability measure of being a source or destination point or a point on a path of traversal, then the LVG converges to the VG with a probability one.

Proof: As per the procedure UPDATE-VGRAPH, when a new vertex is included in V^*, all the edges corresponding to the visible nodes of the present LVG are added to E^*. Hence it suffices to prove that every vertex of the VG is eventually added to the LVG. Equivalently, it is sufficient to prove that every edge of the obstacle is eventually explored by the robot.

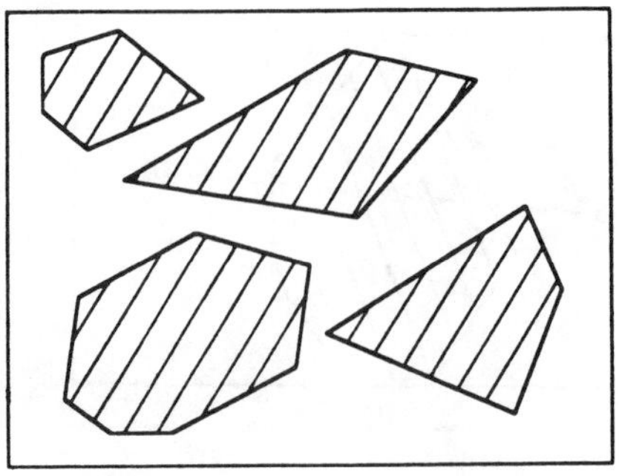

Fig. 9. Unexplored obstacle terrain.

Let p_i be the probability that an edge e_i is explored during any traversal. Since every point in the compact free space has a nonzero probability measure of being a source point or destination point or intermediate point, we have $p_i > 0$. Then the probability that e_i is *not* encountered after k successive independent and randomly chosen paths is $(1 - p_i)^k$. Clearly, this tends to zero as k tends to infinity. Hence the theorem is proved.

We conclude this section with an interesting result that for the complete convergence of the LVG to the VG, the number of sensing operations involved in the procedure UPDATE-VGRAPH is quadratic in the total number of vertices of the obstacles.

Theorem 4: The number of sensor operations performed within the procedure UPDATE-VGRAPH to learn the complete VG is $O(N^2)$, where N is the total number of vertices of the obstacles.

Proof: A explained in the lines 8, 12, and 14 of procedure NAVIGATE-LOCAL, no sensing operations are carried out when the robot encounters an already visited vertex. The sensor operations are performed only when the robot encounters a new vertex. Suppose the LVG presently has $i - 1$ vertices $\{v_1, v_2, \cdots, v_{i-1}\}$ when a new vertex v_i is encountered. At this time, the robot beams its sensors in the direction of $v_j \in \{v_1, v_2, \cdots, v_{i-1}\}$ to determine if v_j is visible from v_i. Hence the number of sensor operations carried out when the ith vertex is added to the LVG is $i - 1$. Therefore, the total number of sensor operations carried out in the procedure UPDATE-VGRAPH is given by

$$\sum_{i=1}^{N} i-1 = N(N-1)/2 = O(N^2),$$

hence the theorem.

The underlying premise of our work has been that we have assumed that the sensors used and the navigation technology used are error free. This, of course, is a serious limitation. The question of operating in an environment prone to errors (with these errors described either by a bound or by a probability distribution) is a problem that is far more complex. An initial (but noteworthy) move in this direction of solving the problem has been made by Brooks [2]. We are currently investigating the formalization of the convergence properties of a path-planning algorithm in the midst of uncertainties using the principle of adaptive learning. In the next section we present a practical example for the technique described in this paper.

VII. An Illustrative Example

In this section we describe an illustrative example of our scheme for a rectangular obstacle terrain shown in Fig. 9. Initially, the terrain is unexplored and the LVG is empty. A sequence of five paths is undertaken in succession by the robot. In other words, the robot moves first to 2 from 1, then to 3 from 2, etc., until it reaches 6. Figs.

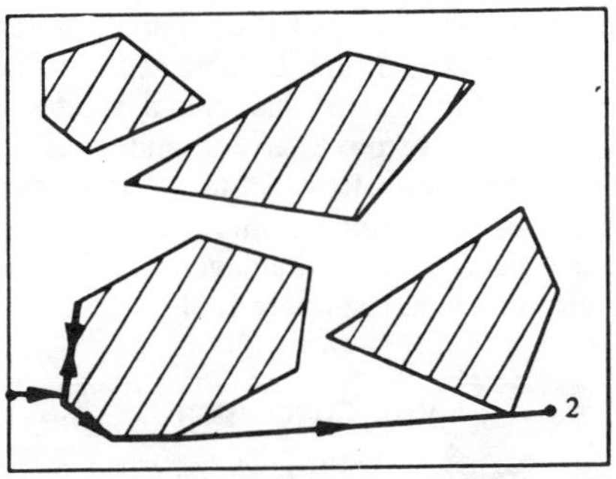

(a)

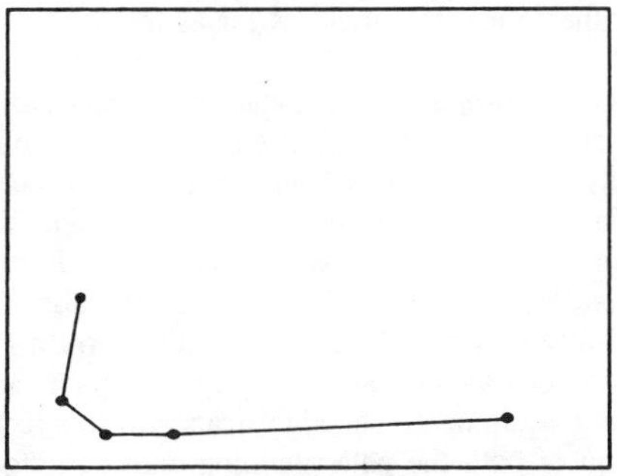

(b)

Fig. 10. NAVIGATE-LOCAL from 1 to 2. (a) Obstacle terrain. (b) Present LVG, portion of VG.

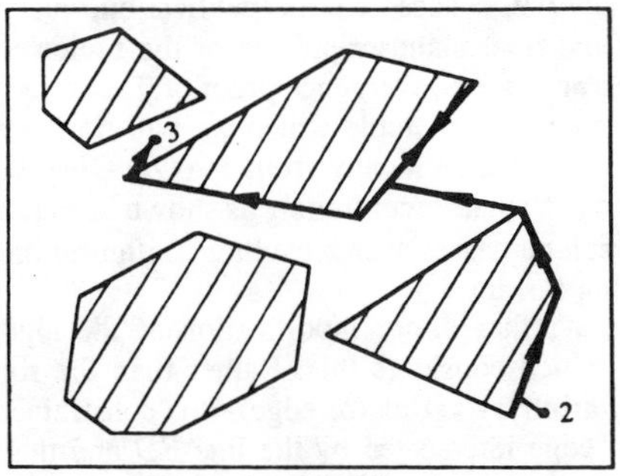

(a)

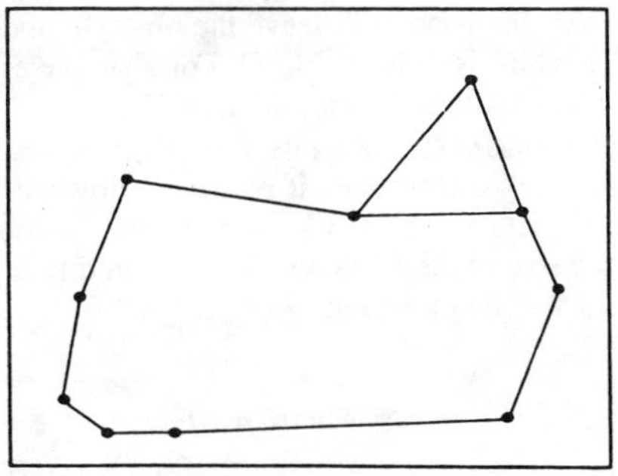

(b)

Fig. 11. NAVIGATE-LOCAL from 2 to 3. (a) Obstacle terrain. (b) Present LVG.

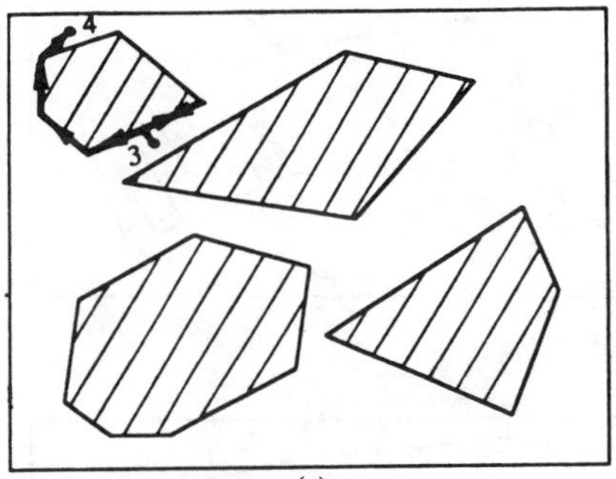

(a)

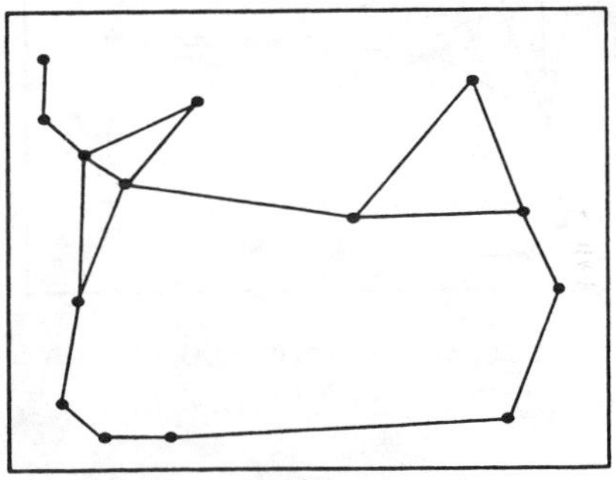

(b)

Fig. 12. NAVIGATE-LOCAL from 3 to 4. (a) Obstacle terrain. (b) Present LVG.

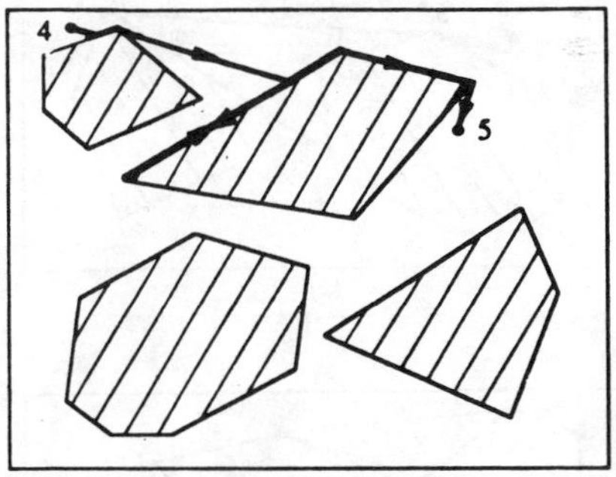

(a)

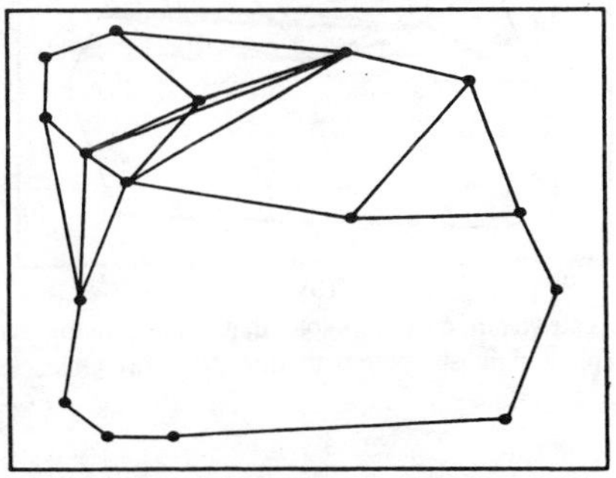

(b)

Fig. 13. NAVIGATE-LOCAL from 4 to 5. (a) Obstacle terrain. (b) Present LVG.

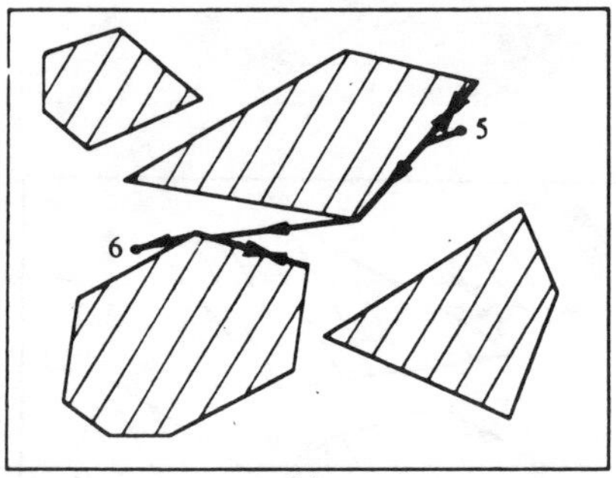

(a)

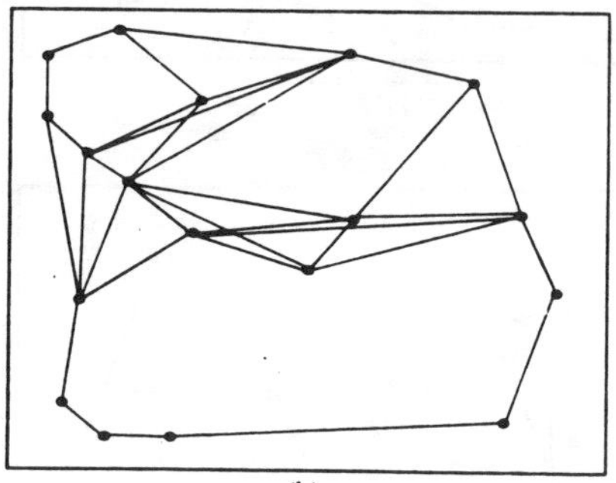

(b)

Fig. 14. NAVIGATE-LOCAL from 5 to 6. (a) Obstacle terrain. (b) Present LVG.

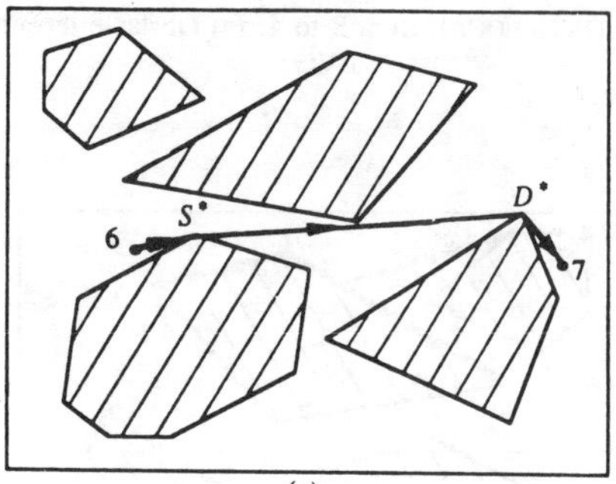

(a)

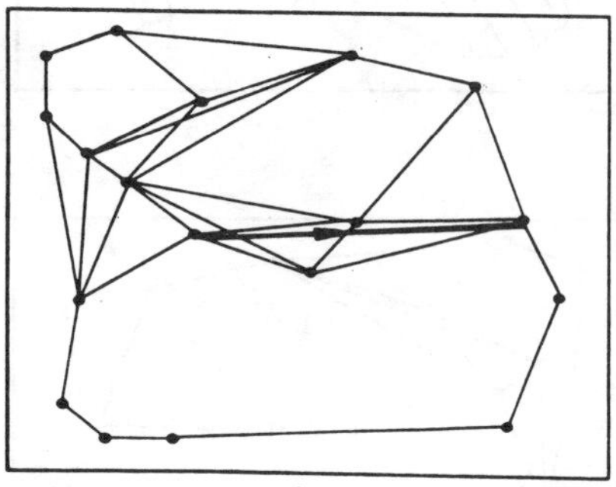

(b)

Fig. 15. Navigate from 6 to 7. Note that path actually computed uses visibility graph and is shortest path on LVG. (a) Obstacle terrain. (b) Present LVG.

10–14 illustrate the various paths traversed and the corresponding LVG's.

Initially, during the motion from 1 to 2, the robot learns four edges of the VG shown in Fig. 10(b). In the next traversal, seven more edges of the VG are learned. A curve showing the number of edges learned as a function of the number of traversals is given in Fig. 16. Note that as many as 31 out of a total of 39 edges of VG are learned in five traversals.

Suppose that at this point the global navigation strategy is invoked to navigate to 7 from 6. The S^* and D^* obtained by using Criterion A of Section V are shown in Fig. 15. The robot navigates locally from S to S^*, then along the LVG from S^* to D^*, and, finally, locally from D^* to D. Note that the path from S^* to D^* does not involve any sensor operations but only quadratic time computation on the LVG to find the shortest path. Actual simulation results obtained using random paths are presented elsewhere [27].

VIII. Conclusion

The terrain model acquisition and path-planning problems are very important aspects of an autonomous robot navigating in an unexplored terrain. In the literature this problem has not been subjected to a rigorous mathematical treatment as far as the model acquisition is concerned.

In this paper, we propose a technique that enables an autonomous robot to navigate in a totally unexplored terrain. The robot builds the terrain model as it navigates and stores the processed sensor information in terms of a learned visibility graph. The proposed technique is proven to obtain a path if one exists. Furthermore, the terrain is guaranteed to become *completely learned* when the complete visibility graph of the entire obstacle terrain is built. After this stage the robot traverses along the optimal paths and no longer needs the sensor equipment. The significance of this technique is the characterization of both the path planning and learning in a precise mathematical framework. The convergence of the path planning and the learning processes is proven.

Appendix

After the paper was accepted for publication, just prior to the publication of the final manuscript, one of the reviewers noted that there was an error in the convergence proof of Theorem I. He did this by presenting a counter example which we will now present.

The robot is to navigate locally from S to D. Observe that in this case the robot can "cycle" indefinitely as shown in Fig. 17. We refer to a terrain which possesses such a cycling configuration of obstacles as a interlocking terrain.

We propose a rather minor modification of the algorithm NAVIGATE_LOCAL which considers this. Rather than the robot leave an obstacle at any arbitrary vertex (or edge), it is constrained to leave an obstacle on an edge intersected by the line SD and that *only* at the point where the edge intersects the line SD. Clearly, since the obstacles are nonintersecting and convex, there will be exactly two such eligible edges. Since the projected distance along the line SD is always minimized, the robot will leave the obstacle under consideration at the edge which is "closer" to D. For a simple example, Fig. 18 shows the edges traversed using the modification.

Observe that NAVIGATE_LOCAL is now slightly less optimal in terms of the number of traversals it requires. However, it must be noted that every extra traversal yields more information about the VG, and thus the size of the LVG will increase in this case. Thus the learning process will be catalyzed.

Acknowledgment

We would like to thank Prof. T. Lazano-Perez of the Massachusetts Institute of Technology for carefully reading the manuscript and for his helpful and critical comments. He also suggested various

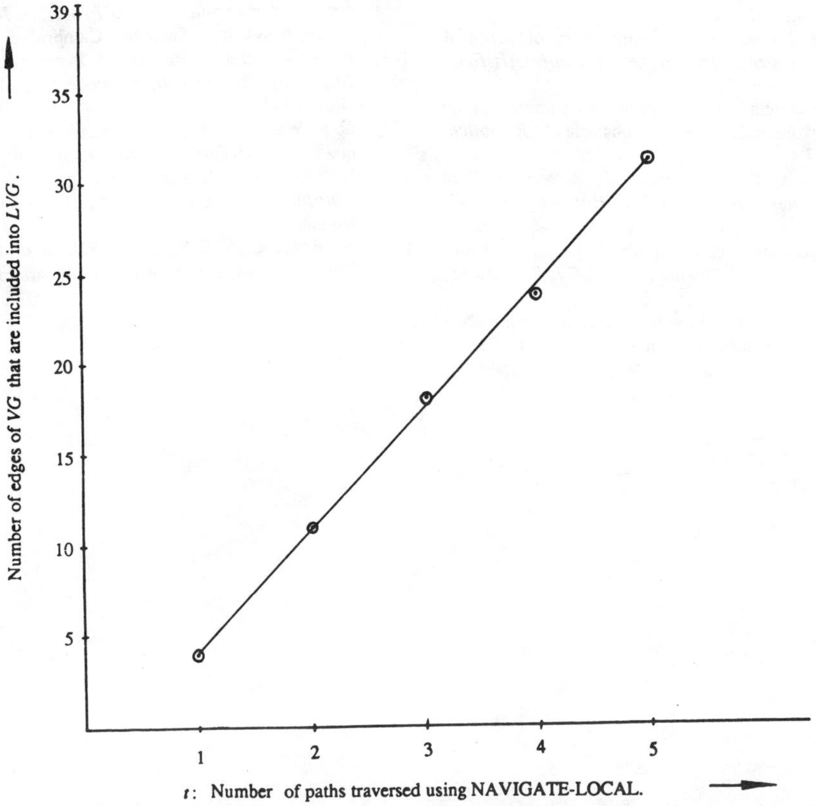

Fig. 16. Graph showing number of edges in LVG as function of number of paths traversed using NAVIGATE-LOCAL. Out of total of 39, robot learns 31 edges in five traversals.

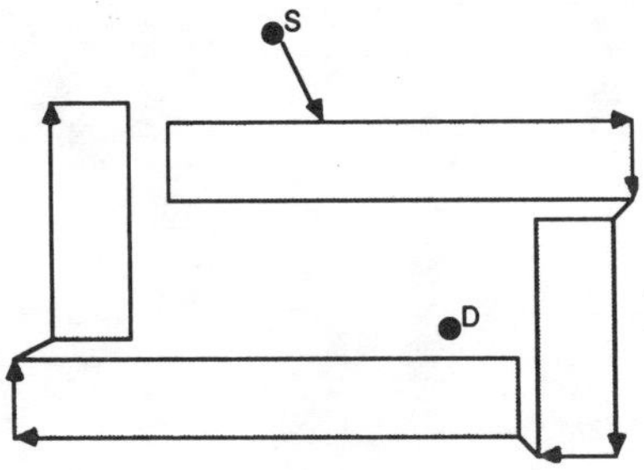

Fig. 17.

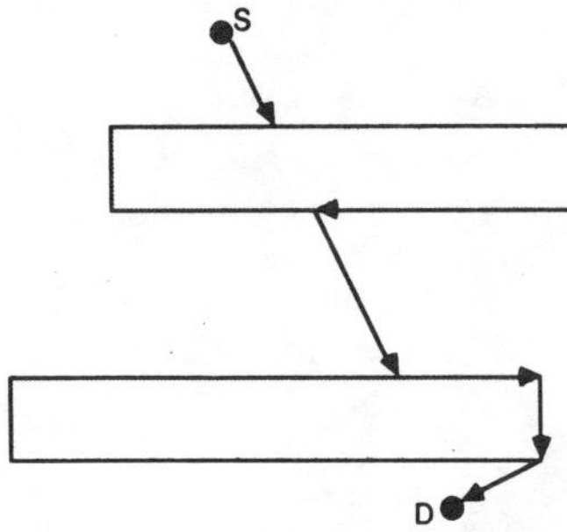

Fig. 18.

avenues for further work. We are also grateful to the anonymous referees who gave us both critical and extremely encouraging feedback, which later led to the Appendix.

References

[1] A. Aho, J. Hopcroft, and J. Ullman, *The Design and Analysis of Computer Algorithms*. Reading, MA: Addison-Wesley, 1974.

[2] R. A. Brooks, "Visual map making for a mobile robot," in *Proc. 1985 IEEE Int. Conf. Robotics and Automation*, 1985, pp. 824–829.

[3] R. A. Books and T. Lozano-Perez, "A subdivision algorithm in configuration space for path with rotation," *IEEE Trans. Syst., Man, Cybern.*, vol. SMC-15, no. 2, pp. 224–233, Mar./Apr. 1985.

[4] R. Chatila, "Path planning and environment learning in a mobile robot system," in *Proc. European Conf. Artificial Intelligence*, Torsey, France, 1982.

[5] R. Chattergy, "Some heuristics for the navigation of a robot," *Int. J. Robotics Res.*, vol. 4, no. 1, pp. 59–66, Spring 1985.

[6] J. L. Crowley, "Navigation of an intelligent mobile robot," *IEEE J. Robotics Automat.*, vol. RA-1, no. 2, pp. 31–41, Mar. 1985.

[7] N. Deo, *Graph Theory with Applications to Engineering and Computer Science*. Englewood Cliffs, NJ: Prentice-Hall, 1974.

[8] G. Giralt, R. Sobek, and R. Chatila, "A multilevel planning and navigation system for a mobile robot," in *Proc. 6th Int. Joint Conf. Artificial Intelligence*, Aug. 1979, Tokyo, Japan, pp. 335–338.

[9] L. Gouzenes, "Strategies for solving collison-free trajectories problems for mobile and manipulator robots," *Int. J. Robotics Res.*, vol. 3, no. 4, pp. 51–65, Winter 1984.

[10] S. S. Iyengar, C. C. Jorgensen, S. V. N. Rao, and C. R. Weisbin, "Robot navigation algorithms using learned spatial graphs," *Robotica*, vol. 4, pp. 93–100, Jan. 1986.

[11] ——, "Learned navigation paths for a robot in unexplored terrain," in *Proc. 2nd Conf. Artificial Intelligence Applications and Engineering of Knowledge Based Systems*, Miami Beach, FL, Dec. 1985.

[12] J. Laumond, "Model structuring and concept recognition: Two aspects of learning for a mobile robot," in *Proc. 8th Conf. Artificial Intelligence*, Aug. 1983, Karlsruhe, W. Germany, p. 839.

[13] D. T. Lee and F. P. Preparata, "Computational geometry—A survey," *IEEE Trans. Comput.*, vol. C-33, pp. 1072–1101, Dec. 1984.

[14] T. Lozano-Perez, "Spatial planning: A configuration space approach," *IEEE Trans. Comput.*, vol. C-32, pp. 108–120, Feb. 1983.

[15] T. Lozano-Perez and M. A. Wesley, "An algorithm for planning collision-free paths among polyhedral obstacles," *Commun. ACM*, vol. 22, pp. 560–570, Oct. 1979.

[16] V. J. Lumelsky and A. A. Stepanov, "Effect of uncertainty on continuous path planning for an autonomous vehicle," in *Proc. 23rd IEEE Conf. Decision and Control*, 1984, pp. 1616–1621.

[17] H. P. Moravec, "The CMU rover," in *Proc. Nat. Conf. Artificial Intelligence*, Aug. 1982, pp. 377-380.

[18] N. J. Nilsson, "Mobile automation: An application of artificial intelligence techniques," in *Proc. 1st Int. Joint Conf. Artificial Intelligence*, May 1969, pp. 509-520.

[19] J. B. Oommen and I. Reichstein, "On the problem of translating an elliptic object through a workspace of elliptic obstacles," *Robotica*, vol. 5, pp. 187-196, 1987.

[20] N. S. V. Rao, S. S. Iyengar, C. C. Jorgensen, and C. R. Weisbin, "On the robot navigation in an unexplored terrain," *J. Robotic Syst.*, vol. 3, pp. 389-407, 1986.

[21] A. M. Thompson, "The navigation system of the JPL robot," in *Proc. 5th Int. Joint Conf. Artificial Intelligence*, Aug. 1977, Cambridge, MA, pp. 749-757.

[22] M. P. Turchen and A. K. C. Wong, "Low level learning for a mobile robot: Environmental model acquisition," in *Proc. 2nd Int. Conf. Artificial Intelligence and Its Applications*, Dec. 1985, pp. 156-161.

[23] S. M. Udupa, "Collision detection and avoidance in computer controlled manipulators," in *Proc. 5th Int. Conf. Artificial Intelligence*, Mass. Inst. Technol., Cambridge, Aug. 1977, pp. 737-748.

[24] R. Wallace *et al.*, "First results in robot road following," in *Proc. 9th Int. Conf. Artificial Intelligence*, Aug. 1985, Los Angeles, CA, pp. 1089-1095.

[25] C. R. Weisbin *et al.*, "Machine intelligence for robotics applications," in *Proc. 1985 Conf. Intelligent Systems and Machines*, Apr. 1985.

[26] S. Whitesides, "Computational geometry and motion planning," in *Computational Geometry*, G. Toussaint, Ed. New York: North Holland, 1985.

[27] N. Andrade, M.C.S. thesis, School of Computer Science, Carlton Univ., Ottawa, ON, Canada, in preparation.

Autonomous Robot Navigation in Unknown Terrains: Incidental Learning and Environmental Exploration

NAGESWARA S. V. RAO AND S. S. IYENGAR

Abstract —**The navigation of autonomous mobile machines, which are referred to as robots, through *unknown* terrains, i.e., terrains whose models are not *a priori* known is considered. We deal with point-sized robots in two- and three-dimensional terrains and circular robots in two-dimensional terrains. The two-dimensional (three-dimensional) terrains are finite-sized and populated by an unknown, but, finite, number of simple polygonal (polyhedral) obstacles. The robot is equipped with a sensor system that detects all vertices and edges that are visible from its present location. In this context, the work deals with two basic navigational problems. In the visit problem, the robot is required to visit a sequence of destination points, in a specified order, using the sensor system. In the terrain model acquisition problem, the robot is required to acquire the complete model of the terrain by exploring the terrain with the sensor. A framework that yields solutions to both the visit problem and the terrain model acquisition problem using a single approach is presented. The approach consists of incrementally constructing, in an algorithmic manner, an appropriate geometric graph structure (1-skeleton), called the navigational course. A point robot employs the restricted visibility graph and the visibility graph as the navigational course in two- and three-dimensional cases respectively. A circular robot employs the modified visibility graph. The algorithms to solve the visit problem and the terrain model acquisition problem based on the abovementioned structures are presented and analyzed.**

I. Introduction

A vital component of unmanned machines or rovers is the navigation system that enables these machines to autonomously navigate to the required destinations. The machines with autonomous navigation capability can be employed in various applications such as autonomous land navigation, unmanned extraterrestrial and underwater exploration, maintenance and repairs in nuclear power plants, operation in chemical and toxic industries, unmanned vigilance and security systems, etc. Such systems must be capable of navigating in known environments, i.e., the environments whose precise models are known, as well as in unknown environments, i.e., environments whose models are not known.

The problem of planning collision-free paths for moving a body through a known terrain has been extensively studied under the popular generic name of the piano movers problem. There has been a surge of research activities in this area due to the important contributions of Lozano-Perez and Wesley [6], O'Dunlaing and Yap [9], Reif [14], and Schwartz and Sharir [15]. Yap [17] and Sharir [16] present excellent surveys of various formulations of this problem and their solutions.

The problem of robot navigation through an unknown terrain, has been studied by several researchers although not to the extent of its counterpart in known terrains. Lumelsky and Stepanov [8] present two algorithms for a point robot to move from a source point to a destination point using touch sensing. In his survey paper, Lumelsky [7] presents a comprehensive discussion on several algorithmic and complexity issues dealing with a point robot in unknown terrains. For the terrains populated by convex polygonal obstacles, Oommen *et al.* [10] develop algorithms for a point robot to navigate to a destination point, and at the same time "learn" about the parts of terrain that are encountered on the way to the destination. Here the robot uses a combination of touch sensing and distance probing. In this treatment, several interesting obstacle configurations such as the mazes, traps etc., are not dealt with. The above problems can be grouped under a broad title, the visit problem, wherein a robot is required to visit a sequence of destination points through an unknown terrain. Another problem, called the terrain model acquisition problem is discussed by Rao *et al.* [13]. Here, a point robot is required to acquire the complete model of the terrain.

The visit problem and the terrain model acquisition problem have been solved independent of each other. Here we present a framework to solve these two problems using a *single approach* that implements a graph search on an incrementally-constructed graph called the *navigation course*. A general outline of this approach has been presented by Rao [11]. Here we present the visibility graph methods to implement this approach, by presenting the technical issues, such as proofs of the properties of various navigational courses, extension to circular robots, lower bounds on sensor operations, etc. We deal with point robots and circular robots. The method of Oommen *et al.* [10] uses the visibility graph (in plane) in their algorithms. We extend their work to terrains with nonconvex obstacles which include mazes and traps. In two-dimensional (2-D) terrains, we show that a subgraph of visibility graph, *the restricted visibility graph*, with only the convex obstacle vertices as nodes, suffices to solve these two problems. This results in a reduction in the number of sensor operations and the storage, if the terrain consists of non-convex corners. Also, we establish the lower bounds on the worst-cast number of scan operations performed by these algorithms.

Motivation for Navigational Problems

The visit problem and the terrain model acquisition problem have been motivated by a specific practical application involving the development of an autonomous rescue robot. However, our treatment is more general than this specific application. In this application, the robot is required to carry out rescue operations in nuclear power plants in the event of radiation leakages, and other events that prevent human operation. A solution to the visit problem enables the robot to carry out a set of operations

Manuscript received April 1, 1989; revised January 4, 1990.

N. S. V. Rao is with the Department of Computer Science, Old Dominion University, Norfolk, VA 23529-0162. Preliminary results on the part of this paper that deals with the visit problem have been presented at the 1988 IEEE Int. Conf. Robotics and Automation, Philadelphia, PA, April 25–29, 1988.

S. S. Iyengar is with the Department of Computer Science, Louisiana State University, Baton Rouge, LA 70803.

IEEE Log Number 9037604.

Reprinted from *IEEE Transactions on Systems, Man, and Cybernetics*, Vol. 20, No. 6, November/December 1990.
Copyright © 1990 by The Institute of Electrical and Electronics Engineers, Inc. All rights reserved.

in different locations in unfamiliar environments. Since the motion planning in this case is essentially sensor-based, the robot may be required to perform a number of expensive sensor operations. Furthermore, the robot could temporarily navigate into local detours because of the partial nature of the information returned by the sensors. By incorporating the incidental learning feature, we reduce the expected number of sensor operations, and the expected number of detours, as the robot visits newer locations. Further, if the complete terrain model is available, the robot can avoid 1) local detours, 2) sensor operations. These two important points motivate the terrain model acquisition problem. In general, a dedicated rescue robot typically idles in between two rescue operations, and the rescue operations could be fairly infrequent. Thus there are definite advantages if the robot is employed in the terrain model acquisition process during this period. Our methodology provides a basic algorithmic framework that aids the design of a navigational system for the abovementioned rescue robot.

The organization of the paper is as follows: Preliminaries are presented in Section II. In Section III, we define the restricted visibility graph and the modified visibility graph, and prove some of their properties. rIn Section IV, we present solutions for the terrain model acquisition and the visit problems. We compare our method with the other methods in Section V.

II. Preliminaries

We consider a point robot R in two- and three-dimensional (3-D) terrains. Here, the location of R is also called the *position* of R. Additionally, in 2-D terrains, we consider a circular body R of radius δ, ($\delta \geqslant 0$). The location of the center of R is called the *position* of R. The R houses a computational device with storage capability. The point robot R is capable of moving along a straight-line path in two- and three-dimensional terrains. Additionally, the circular R is capable of rotating around its center and also around a point on the circumference. R takes a finite amount of time to move through a finite amount of distance. Further, R is equipped with an algorithm B that plans a collision-free path (for R) through a known terrain. For example, in two dimensions, we can use the $O(N \log N)$ algorithm of O'Dunlaing and Yap [9] or Leven and Sharir [5] or Bhattacharya and Zorbas [1] to plan a path from a source location to a destination location for a circular robot, where N is the total number of obstacles corners. In three dimensions, R can use the algorithms of Reif [14]. For a circular robot, we can also use the algorithm of Chew [2] or Hershberger and Guibas [3], if shortest paths are required.

We consider a finite-sized *terrain* populated by a finite set $O = \{O_1, O_2, \cdots O_n\}$. Each O_i is called an obstacle; O_i is a simple polygon in the two-dimensional case and a polyhedron in the three dimensional case. In either case, O_i has a finite number of vertices. The terrain is completely unknown to R, i.e., the number of obstacles, and also the number and locations of vertices of each obstacle are unknown to R. The free-space is given by $\Omega = \cap_{i=1}^{n} O_i^C$, where O_i^C is the complement of O_i. The closure of the free-space is denoted by $\overline{\Omega}$. Let N denote the total number of vertices of all obstacles. A vertex v, of a polygon, is called *convex* if the angle included by the obstacle edges that are incident at v is less than Π. The vertex v, of an obstacle polygon, is called *nonconvex* if it is not convex. For two-dimensional terrains, let C be the number of nonconvex vertices.

We imagine a logical point of reference x on R for the sensor. A point $y \in \overline{\Omega}$ is said to be visible to R if the straight line joining x and y is entirely contained in $\overline{\Omega}$. R is equipped with a sensor that detects the maximal set of points on the obstacle boundaries that are visible from x. Such an operation is termed as the scan operation. We assume that a scan operation is error-free.

Two Navigational Problem

Initially, R is located at the position d_0 without intersecting any obstacle and at a finite distance from an obstacle. In the terrain model acquisition problem, R is required to acquire the model of the terrain to a degree such that it can navigate to any reachable destination location by planning a path using the known terrain algorithm B alone. In the case of a point robot this is tantamount to acquiring the entire model of the terrain. For a circular robot, an appropriate subset of the terrain boundary is to be identified depending on the radius of R and the initial location d_0. Note that after the terrain model is completely acquired, no sensor operations are needed for navigational purposes. Second, in the *visit problem*, R is required to visit the positions $d_1, d_2, \cdots, d_M$ in the specified order if there exists a path through these positions. If no such path exists, then R must report this fact in a finite amount of time.

Basic Algorithm

Here, R performs a "graph exploration type" of navigation using a combinatorial graph called the *navigation course*, $\xi(O)$, of the terrain O. A detailed treatment on this basic algorithm can be found in [11]. $\xi(O)$ is a 1-skeleton embedded in $\overline{\Omega}$. The nodes (edges) of $\xi(O)$ are called ξ-nodes (ξ-edges). In this paper, each ξ-node corresponds to an obstacle vertex, and specifies position for R such that it is entirely contained in $\overline{\Omega}$. For a point robot, a ξ-edge (v_1, v_2) specifies a line segment $v_1 + t(v_2 - v_1)$, $0 \leqslant t \leqslant 1$, that is entirely contained in $\overline{\Omega}$. Thus the edge provides a collision-free path to move from v_1 to v_2. For a circular robot, the edge (v_1, v_2) specifies a collision-free path, of finite length, from v_1 to v_2 for R. The $\xi(O)$ is initially unknown and it is incrementally constructed using the data obtained through the sensor operations. The navigational course $\xi(O)$ has to satisfy a set of properties, in order to yield correct solutions to the visit problem and the terrain model acquisition problem. The property of *local-constructibility*, means that the adjacency list of a ξ-vertex v can be computed from the information obtained by a scan operation performed from v. The *finiteness* property requires that $\xi(O)$ has a finite number of vertices. Also the graph *connectivity* property requires that any two ξ-vertices be connected by a path of ξ-edges. Now the fact that any graph exploration algorithm visits all the nodes of a finite connected graph in a finite amount of time, translates to the following observation:

Observation 1: If $\xi(O)$ satisfies the properties of finiteness, connectivity and local-constructibility, then, R, executing graph search algorithm, visits all vertices of $\xi(O)$ in a finite amount of time.

III. Navigational Courses

First, we present a $\xi(O)$ for a point robot, and in this case $\xi(O)$ is the visibility graph of O for three-dimensional terrains. For two-dimensional terrains, we consider the restricted visibility graph. For circular robots ($\delta > 0$) we present a $\xi(O)$ based on the restricted visibility graph. In each of the cases we show that the proposed structure satisfies the properties of finiteness, connectivity, and local-constructibility. We also consider an additional property, namely *terrain-visibility*, which means that every point in the required subset of $\overline{\Omega}$, is visible from some ξ-vertex.

A. Point Robot

For a point robot we consider finite-sized 3-D terrains populated by polyhedral obstacles, i.e., O_i is a finite-sized polyhedron

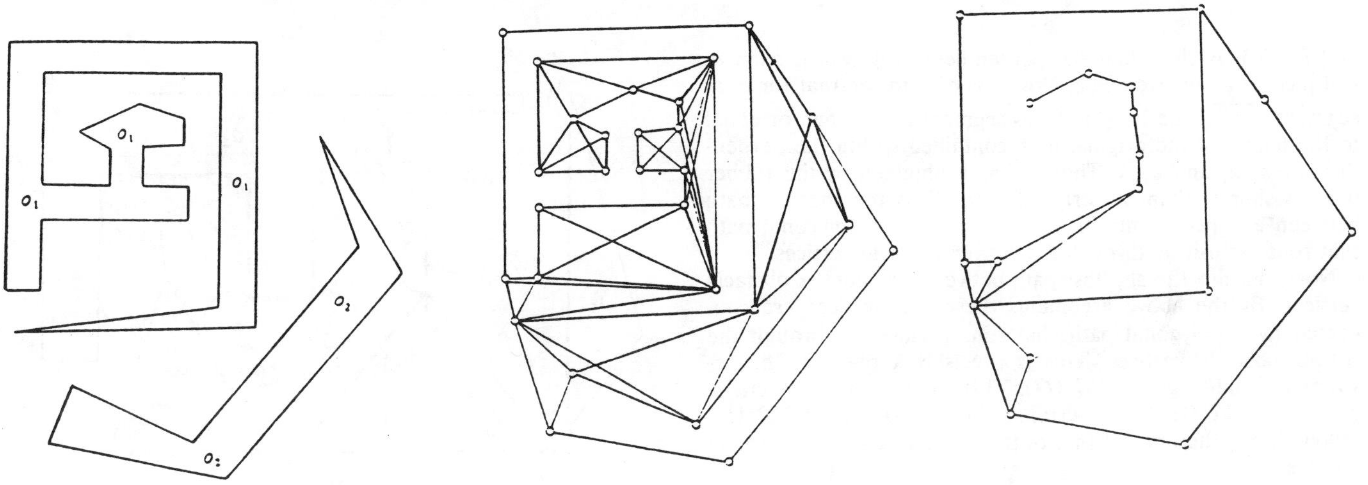

(a) (b) (c)

Fig. 1. Navigation courses based on visibility graphs. (a) Obstacle terrain $O = \{O_1, O_2\}$. (b) $VG(O)$. (c) $VG^*(O)$.

with a finite number of vertices. The visibility graph, $VG(O) = (V, E)$, of a terrain populated by the obstacle set O is defined as follows [6]: 1) V is the union of vertices of all obstacle polyhedra, 2) A line joining the vertices v_i and v_j forms an edge $(v_i, v_j) \in E$ if and only if it is either an obstacle edge or it is not intersected by any obstacle. See Fig. 1 for an example. The visibility graph is connected [13]. To show the terrain-visibility property, consider a point $x \in \bar{\Omega}$. Consider an infinitesimally small polyhedron P at x which can be imagined as a point at x. Consider the visibility graph of $O \cup \{P\}$. This graph is connected. Let v be a vertex of some $O_i \in O$ to which a vertex of P is connected. Then x is visible from v. We shall now summarize the properties of $VG(O)$:

Properties 1: The visibility graph $VG(O)$ satisfies the properties of finiteness, connectivity, local-constructibility and terrain-visibility.

We define the *restricted visibility graph* $VG^*(O) = (V, E)$ of a 2-D terrain O as follows: 1) V is the union of all convex vertices of obstacle polygons, 2) A line joining the vertices v_i and v_j forms an edge $(v_i, v_j) \in E$ if and only if it is either an obstacle edge of it is not intersected by any obstacle polygon. See Fig. 1 for an example. The $VG^*(O)$ is a subgraph of $VG(O)$, and it coincides with $VG(O)$ if every $O_i \in O$ is a convex polygon. The number of nodes of $VG^*(O)$ is $N - C$, where C is the number of non-convex obstacle vertices. In a general case where O contains non-convex vertices, the $VG^*(O)$ has a lesser number of nodes than $VG(O)$. We now have the following properties of $VG^*(O)$.

Lemma 1: The restricted visibility graph $VG^*(O)$ satisfies the properties of connectivity and terrain-visibility.

Proof: The key observation is that the shortest path between any two points in free-space is a polygonal path that runs through the obstacle vertices. Additionally we can show that such a path passes through convex obstacle vertices only. We can show this as follows: Let us say that the shortest path passes through a non-convex vertex v. Let v_1 and v_2 be the obstacles vertices adjacent to v on a shortest path i.e., the shortest path passes along the edges (v_1, v) and (v, v_2). Imagine a rubber band stretched (in the free-space) along the vertices v_1, v and v_2, and then released. The action of the rubber band can be visualized

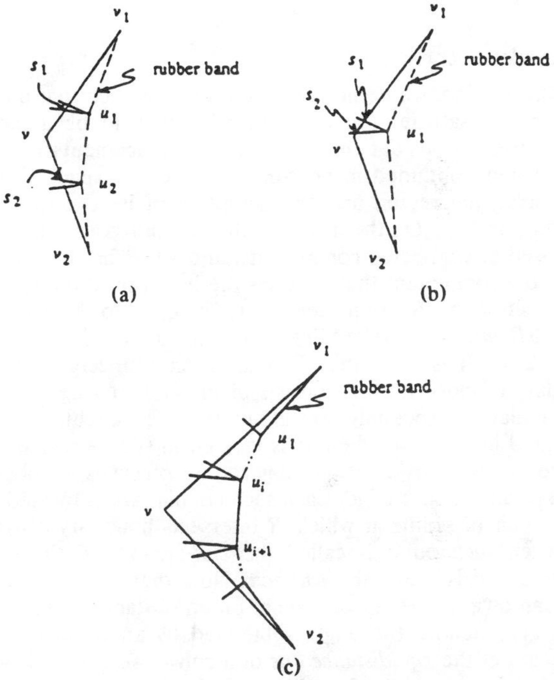

Fig. 2. Shortest path runs through the convex vertices only.

as follows: Imagine a long line segment (a ray) extending from v_1 through v. Rotate this ray around v_1 into the concavity until it encounters v_2 or a convex vertex, say u_1. Now rotate the ray around u_1 in a similar fashion. Note that each such rotation brings the line closer to v_2, and there can be only a finite number of rotations. Thus the rubber band will touch the convex vertices, say u_i, $i = 1, 2, \cdots k$, contained in the triangle formed by v_1, v and v_2 (see Fig. 2). It is clear from Fig. 2(a) and Fig. 2(b) that for cases $k = 1, 2$ the path followed by rubber band is shorter that the original path. For $k = 1$, draw perpendiculars at u_1 to segments $\overline{v_1 u_1}$ and $\overline{u_1 v_2}$. Here length of $\overline{v_1 u_1}$ ($\overline{u_1 v_2}$) is less than that of $\overline{v_1 s_1}$ ($\overline{s_2 v_2}$). Thus the path v_1, u_1, v_2 is shorter. If $k = 2$, the key idea is to note that the length of the original path contained in between the end perpendiculars of $\overline{u_1 u_2}$ is greater than or equal to the length of $\overline{u_1 u_2}$. Thus the path v_1, u_1, u_2, v_2 is shorter than v_1, v, v_2. For $k \geqslant 3$ we use the same argument. Draw perpendiculars at the end of each line segment joining u_i

and u_{i+1}. It is clear that the perpendiculars drawn at each u_i will include a positive angle. Now it is easy to see that for each segment $\overline{u_i u_{i+1}}$, the length of this segment is less than or equal to the length of the original path contained within the perpendiculars at u_i and u_{i+1}. Thus the path obtained by the rubber band is shorter than the original path. Thus the shortest path between any two points in the free-space is a polygonal path that runs exclusively through the convex obstacle vertices.

Now consider the shortest path between any convex obstacle vertices. By the above arguments these two vertices are connected by a polygonal path that runs exclusively through the convex obstacle vertices. This is precisely a path on the restricted visibility graph $VG^*(O)$. This proves the connectivity property of $VG^*(O)$. The terrain-visibility property of $VG^*(O)$ follows along the lines of that of the visibility graph. Hence the Lemma.

In summary we have the following properties.

Properties 2: The restricted visibility graph $VG^*(O)$ satisfies the properties of finiteness, connectivity, terrain-visibility and local-constructibility.

B. Circular Robot

In this section, we define a family of graphs such that each of its members satisfies the required properties to be a navigational course. Consider the set FP of free-placements in which R is entirely contained in Ω. Note that the free-space Ω is an open polygonal region and the boundary of its closure is the boundary of $\bigcup_{i=1}^{n} O_i$, the union of obstacle polygons. The *FP* is composed of connected components, and let Ψ be the maximal connected component that contains the initial position x_0 of R. Any position of R connected to x_0 belongs to Ψ. Consider $\Gamma = \Psi \oplus R$, where $\oplus$ is the Minkowski sum, i.e., $\Gamma = \{x + y | x \in \Psi$ and $y \in R\}$. It is clear that Γ is an open connected set. The boundary of closure of Γ consists of intervals of edges of O_i's and circular arcs (possibly zero in number). The circular arcs are generated in the case when R is located in such a way that its closure intersects two distinct objects; an object is an obstacle vertex or an obstacle edge. Each such circular arc is formed by a unique pair of points at which $\bar{R}$ intersects boundary of obstacles; each such point is called the *end-vertex* and the corresponding pair is called the *end-pair*. Note that an end-vertex is either an obstacle vertex or a point on an obstacle edge.

Let $\theta(v)$ denote the angle subtended by an obstacle at its vertex v. Let the *equidistance line* of a convex vertex v, denoted by $EL(v)$, be a portion of the bisector of $\theta(v)$ that extends from v to the outwards of the obstacle. Now we have the property that any obstacle vertex contained in $\bar{\Gamma}$ is a convex vertex. These convex vertices can belong to one of the two categories. First category consists of all the convex vertices that form an end-pair. And second category consists of all *free* vertices which are convex vertices contained in $\bar{\Gamma}$ and do not form an end-pair. Note that by definition, we can place R so that it touches a free vertex v and we can rotate it around v. Let $\overline{v_1 v}$ and $\overline{v v_2}$ be the segments of obstacle edges contained in Γ. We can slide R along $v_1 v$ to v (at least through infinitesimally small distance) and rotate it around v and then slide it along the edge to $\overline{v v_2}$. Then during the rotation the center of R meets $EL(v)$ at one position. This shows that all points on $EL(v)$ within a distance of 2δ (from a free vertex v) are in $\bar{\Omega}$.

Let V be the union of the free vertices contained in $\bar{\Gamma}$. Consider a function f: $V \rightarrow \bigcup_{v \in V} EL(v)$ called the *sensing function*. This function assigns a unique point on $EL(v)$ for each $v \in V$, i.e. $f(v) \in EL(v)$. The *modified visibility graph* (MVG) of the obstacle terrain O with respect to a sensing fûnction f, denoted by $VG_f(O)$, is a graph (V, E) such that there exists an edge $(v, w) \in E$ if and only if the line joining w and $f(v)$ lies entirely in $\bar{\Gamma}$, and does not cross the boundary of $\bar{\Gamma}$. For a given obstacle terrain O, there exists a *family* of modified visibility graphs, denoted by $\{VG_f(O)\}$ corresponding to all possible fs. See Fig. 3 for an example of $VG_f(O)$. We have the following lemma:

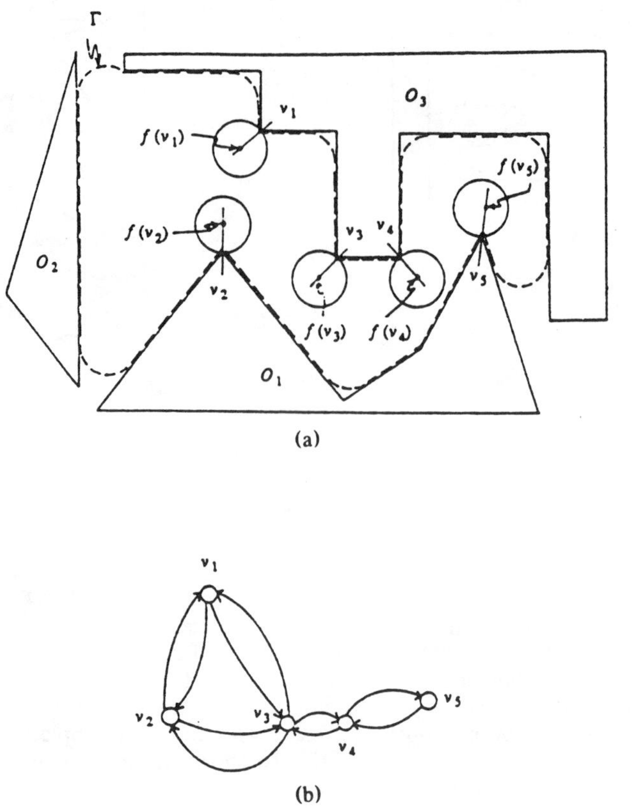

Fig. 3. Example of $VG_f(O)$. (a) Definition of f and the positions of R at the free vertices. (b) The graph $VG_f(O)$ for f and O of (a).

Lemma 2: The modified visibility graph $VG_f(O) \in \{VG_f(O)\}$ satisfies the connectivity, and terrain-visibility properties for all f such that $\|v - f(v)\| < 2\delta$, for all $v \in V$.

Proof: We first discuss the connectivity property. Consider 1) two free vertices $v_1, v_2 \in V$. Consider a shortest path from v_1 to v_2 that runs through $\bar{\Gamma}$ such that the path does not cross the boundary of $\bar{\Gamma}$. Such a path exists because Γ is a connected set. This path runs through only convex vertices of Γ. Using the arguments similar to those in the proof of Lemma 1 (using rubber band) we can show that the path runs through only the free vertices of Γ. Here the convex vertices that form an end-pair can be essentially treated as concave corners, and it the shortest path can be shown not to pass through them. Consider an edge (v_1, v_2) of such shortest path. Now consider a rubber band stretched from v_1 to v_2. Then move the v_1 end of the rubber band along $EL(v_1)$ to $f(v_1)$. In this state the rubber band might touch some other free vertices. Let the rubber band run through the free vertices $u_1, u_2, \cdots, u_r$. Here u_1 is visible from $f(v_1)$. Hence (v_1, u_1) is an edge of $VG_f(O)$. Apply the same technique from each of u_i's. It is clear that there is a path from v_1 to v_2. Thus the $VG_f(O)$ is connected.

Now consider the terrain-visibility property. Consider $x \in \Omega$. Now consider a shortest path from x to a free vertex such that the path lies entirely in Γ as described above. Move on this path from x to the first free vertex u. Then imagine a rubber band stretched from x to u, and move its u end along $EL(u)$ to $f(u)$. If the line from x to $f(u)$ is not intercepted by any obstacle then we are done. Otherwise move from x along the stretched rubber band to the first free vertex, and apply the same procedure. The repeated application of the procedure results in free vertex u_1 such that x is visible from $f(u_1)$. Hence the Lemma.

It is clear that $VG_f(O)$ has at most $N-C$ vertices and $O(N^2)$ edges. Note that all free vertices that are visible from $f(v)$ can be obtained from the information from a single scan. Thus $VG_f(O)$ satisfies the local-constructibility property. We summarize all these properties as follows.

Properties 3: The graph $VG_f(O)$ for an f that satisfies the condition stated in Lemma 2, satisfies the properties of finiteness, connectivity, terrain-visibility and local-constructibility.

Chew [2] proposed the *path graph* that is an extension of the visibility graph. This path graph is used to plan an optimal path between two points through a two-dimensional terrain, and this graph has $O(N^2)$ vertices and $O(N^4)$ edges. This path graph can be used as a $\xi(O)$ for a circular robot. The modified visibility graph contains at most N vertices, which is important because the required number of scan operations in the solution to terrain model acquisition problem and the visit problem (in a worst-case) is equal to the number of vertices of $\xi(O)$.

IV. Navigation Algorithms

A. Circular Robot

A vertex of $VG_f(O)$ is a convex obstacle vertex v contained in $\bar{\Gamma}$ such that it does not form an end-pair. Consequently, we can place R such that it touches v since $v \in \bar{\Gamma}$. Since v does not form an end-pair, we can rotate R around v such that its center moves along a circular arc of radius δ. This arc extends between the perpendiculars to the obstacle edges incident on v. The Minkowski sum of R and this arc is free of obstacles. A vertex v of $VG_f(O)$ defines a position for R as follows. It is clear that R can be located such that its center lies on $EL(v)$ at a distance of exactly δ from v. Then $f(v)$ precisely defines the 'logical' position of sensor corresponding to vertex v. First, R locates its center on $EL(v)$ at a distance of δ from v. Then R rotates around its center until the reference point of the sensor lies on $EL(v)$. The R can rotate either clockwise or anti-clockwise to achieve this and in either case the logical position of the sensor corresponding to $f(v)$ that satisfies the condition in Lemma 2, i.e., $\|v - f(v)\| < 2\delta$. Thus a vertex v of $VG_f(O)$ specifies a position for R and for the sensor. Further, we use the depth-first graph search for R, which chooses a ξ-vertex that is closest to the present location of R. Subsequently, we establish the following aspects: (a) the information stored along the edges of $VG_f(O)$ suffices for the intermediate navigation that is required to move R from one vertex to the other, and (b) the appropriate vertices and adjacency lists of $VG_f(O)$ can be correctly computed from the scan information.

1) Navigation Along Edges: Consider the navigation of R from u to v, $u, v \in V$. Now $EL(u)$ is known, and $EL(v)$ may or may not be known. When v is detected, the portions (that are close to v) of the edges that are incident at v will be visible in a scan operation. If both the edges incident on v are visible during an earlier scan operation then $EL(v)$ can be computed. Note that at least an infinitesimally small portion of one of the edges incident at v will be seen in the scan operation in which v is detected. If the $EL(v)$ is known then, the navigation from u to v is carried out as follows: The subset of Ψ that corresponds to the free-space visible from $f(u)$ is computed (as subsequently described). Let v_1 (u_1) be a point on $EL(v)$ $(EL(u))$ at a distance δ from v (u). Consider $\overline{u_1v_1}$ the line joining u_1 and v_1. This line intersects the boundary of the computed part of Ψ zero or more times. R moves along the $\overline{u_1v_1}$ in the portions that lie in the free-space, and follows the computed boundary of free-space in the other portions of $\overline{u_1v_1}$. There are only a finite number of detours during which R follows the boundary of free-space, and each detour specifies only a finite number of translational and rotational motions for R. If $EL(v)$ is not known, then R moves from u_1 to a point v_2 at a distance δ from v and lies on the perpendicular to the known edge of v. The motion of R from u_1 to v_2 can be handled similar to the above case. From v_2, R rotates around v until it can not rotate further. Then $EL(v)$ is computed and R rotates back it its position on $EL(v)$. The path corresponding to the navigation along an edge of $VG_f(O)$ is computed the first time R moves along this edge. This path is stored and used in subsequent traversals along this edge. In summary we have the following lemma.

Lemma 3: R can compute a path of finite number of translations and rotations to navigate along an edge of $VG_f(O)$.

2) Processing Scan Information: The scan information is to be processed so that the portion of $VG_f(O)$ corresponding to the "seen part" is constructed. We can use a variation of the algorithm of [5] to compute this part. More specifically, we compute the vertices of local non-convexity corresponding to the Minkowski sum of the disc corresponding to R and the visibility polygon returned by the sensor. Here the Minkowski sum is bounded by line segments and the circular arcs. The vertices corresponding to the arcs that do not correspond to points of non-convexity are the nodes of $VG_f(O)$. Conservatively, the complexity of this operation is $O(N \log N)$.

B. Terrain Model Acquisition

The algorithm *ACQUIRE*, for terrain model acquisition, is a direct implementation of a graph search algorithm. From the observation 1, R will visit all the ξ-vertices in a finite amount of time. And by the terrain visibility property $\xi(O)$, R would have seen the required portions of the free-space, after visiting all ξ-vertices. For completeness we state the following theorem which can be proved along the lines of [13].

Theorem 1: The algorithm *ACQUIRE* solves the terrain model acquisition problem in a finite amount of time such that for a point robot and a circular robot, the number of scan operations performed is $N-C$ in 3-D terrains. For a point robot the number of scan operations is N in 3-D terrains. The complexities of various tasks carried out by *ACQUIRE* are as follows: 1) the storage complexity is $O(N^2)$, 2) the cost of construction of $\xi(O)$ is $O(N^2 \log N)$ and $O(N^2)$ for circular robot and point robot respectively, 3) the total cost of path planning is $O(N^3)$.

C. Visit Problem

The algorithm *LNAV*, the navigates R from d_i to d_{i+1}, is obtained by simulating the graph search algorithm. Initially a scan operation is performed from d_i and if d_{i+1} is found reachable, then R moves to d_{i+1}. If d_{i+1} is not found reachable then R computes a ξ-vertex v_0 and moves to v_0. From v_0, the graph search algorithm *NAV* is invoked. Let R be located at v. After a scan is performed from v, R checks if d_{i+1} is reachable. If d_{i+1} is reachable, then R moves to d_{i+1} and terminates *NAV*. If not, R continues to execute *NAV* until the d_{i+1} is found reachable or until completion. The following theorem can be established by specializing the result in [11].

Theorem 2: Algorithm *LNAV* navigates R from d_i to d_{i+1} in a finite amount of time if the latter is reachable. If d_{i+1} is not reachable then R declares so in a finite amount of time. In executing the algorithm *LNAV* by a point R, the number of scan operations is at most $N-C$ and N respectively for two-

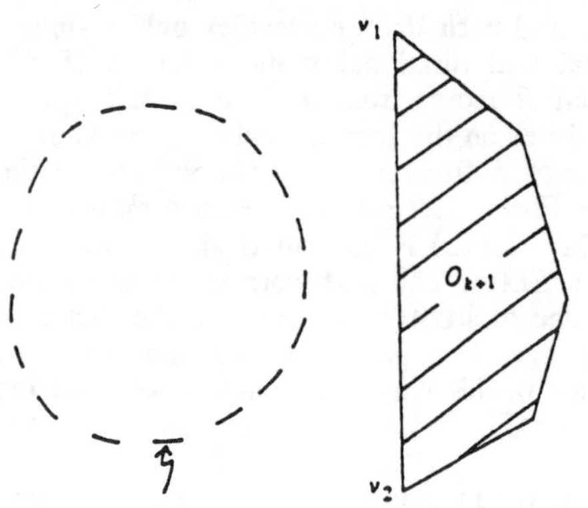

(a)

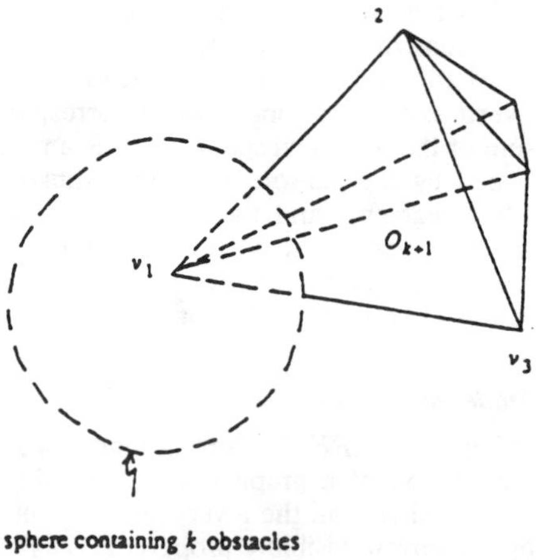

(b)

Fig. 4. Addition of O_{k+1} in the proof of Theorem 5. (a) Two-dimensional case. (b) Three-dimensional case.

and three-dimensional terrains. For a circular R, the number of scan operations is at most $N-C$ in two-dimensional terrains.

The computational complexity of *LNAV* is similar to that of *ACQUIRE*. We obtain the algorithm *GNAV* by extending *LNAV* as follows: We store the adjacency lists computed by R during earlier traversals. Further we store S, which is the set of all vertices that have been detected but not visited yet. Consider the navigation from d_i to d_{i+1}. Then *GNAV* computes a ξ-vertex that is reachable from d_i and moves to this vertex. Then R computes a ξ-vertex d^* that is closest to d_{i+1} according to some criterion such as distance. Then R moves along a path on $\xi(O)$ to d^*. From d^*, R uses *LNAV* to navigate to d_{i+1}. It is direct to see that *GNAV* correctly solves the visit problem. Moreover, R checks the set S after every scan operation. After S becomes empty, R switches-off its sensor and navigates using the algorithm B alone. At this stage R has acquired the terrain model that is sufficient to navigate to any reachable point [11]. Thus we have the following theorem.

Theorem 3: The terrain model will be completely built by R in at most $N+M-C$ and $N+M$ scans respectively for 2-D and 3-D terrains for a point R, then the execution of each traversal involves no scan operations. For a circular R the performance is same as that of a point robot in 2-D terrains.

Here the process by which R acquires the terrain js *incidental*, i.e., the present model of the terrain depends on the previous traversals. Let p_v, $v \in V$, $(\xi(O)=(V,E))$ be the probability that R visits v during a traversal. In this case, we can show that for $M \geq 1/\min_{v \in V}\{p_v\}$, the expected number of scan operations performed by *GNAV* is strictly less than the expected number of scan operations performed by *LNAV* [11]. Now consider that R has successfully navigated to d_i and it is now required to navigate to d_{i+1}. Let s_L and s_G be the random variables that denote the number of scan operations performed by R in cases of using *LNAV* and *GNAV* respectively in navigating from d_i to d_{i+1}. Let $E[x]$ denote the expected value of the random variable x. It is direct to see the following results: 1)$E[s_G]/E[s_L] < 1$, for $p_v < 1$, for $v \in V$, 2) $E[s_G] \to 0$, for $i \to \infty$.

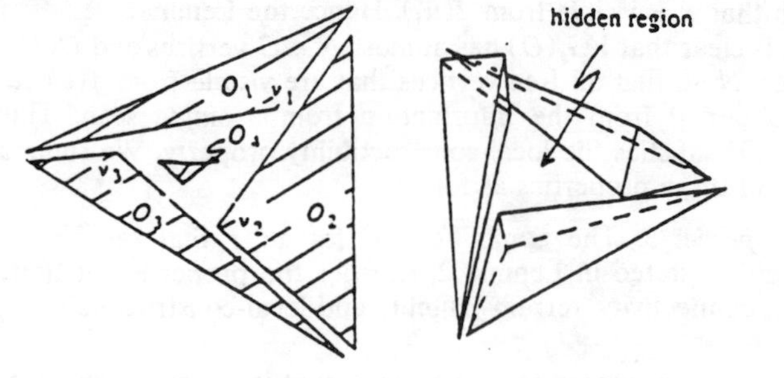

(a) (b)

Fig. 5. Vertices can not be randomly skipped. (a) Two-dimensional case. (b) Three-dimensional case.

E. Lower Bound on Number of Scan Operations

We discuss the case of a point robot in two and three-dimensional terrains. We obtain a lower bound on the number of scan operations that are occasionally necessary. These algorithms are required to ensure that at the time of termination every point in the free space is "sensed." Consider a terrain of one obstacle. Now during execution of the algorithm, no more than *one* (*two*) vertices per obstacle can be left unexplored in a 2-D (3-D) terrains. If R starts at a vertex it detects one new vertex with one scan operation (except when the first vertex is explored) as the robot moves along the circumference of the obstacle. In other words at no point of time the terrain acquisition could be declared complete if there are two unexplored vertices say v_1 and v_2. This is because the robot does not, in general, know what lies on the hinder (unexplored) side of the line joining v_1 and v_2. For the three-dimensional terrain, if three vertices say v_1, v_2 and v_3 are left unexplored then the information on the hinder side of the plane formed by the vertices v_1, v_2 and v_3 is not known in general.

Theorem 4: In the execution of *ACQUIRE* or *LNAV*, for a given positive integer n there exists a terrain $\{O_1, O_2, \cdots, O_n\}$ of polygonal and polyhedral obstacles such that the necessary number of scan operations is $N-n$ and $N-2n$ for two and three dimensional terrains respectively.

Proof: We use induction on the number of obstacles in the terrain. For $n=1$ the claim is true as explained above. Now, assume that the claim is true for $n=k$. Let the set of obstacles in this case be $\{O_1, O_2, \cdots, O_k\}$. Now construct a terrain of $k+1$ obstacles as follows: In two dimensions add a big polygon O_{k+1} outside the circle inscribing the terrain of k obstacles (that satisfies the induction hypothesis) as shown in Fig. 4(a). The $k+1$th polygon has a long edge joining v_1 and v_2 that obscures the remaining edges of the polygon from the scan operations carried out in the terrain of k obstacle. Thus the scan operations needed during the exploration of the $k+1$th obstacle is

TABLE I
COMPUTATIONAL COMPLEXITY

Quantity for Comparison	Restricted Visibility Graph	Modified Visibility Graph	Retraction Method
Storage	$O(N^2)$	$O(N^2)$	$O(N)$
Construction	$O(N^2)$	$O(N^2 \log N)$	$O(N^2 \log N)$
Path planning	$O(N^3)$	$O(N^3)$	$O(N^2\sqrt{\log N})$
Overall time complexity	$O(N^3)$	$O(N^3)$	$O(N^2 \log N)$

$N(O_{k+1})-1$, where $N(O_{k+1})$ is the number of vertices of O_{k+1}. For three-dimensional terrains the obstacle O_{k+1} is such that the plane formed by three vertices v_1, v_2 and v_3 obscures the rest of the obstacle from the scan operations performed in the terrain of k obstacles as in Fig. 4(b). The O_{k+1} lies outside the sphere the encloses the terrain of k obstacles. Thus the necessary number of scan operations to acquire O_{k+1} is $N(O_{k+1})-2$. Thus the theorem follows by induction.

In the above theorem we have seen that no more than one (two) vertices per obstacle can be left unexplored in two (three) dimensional terrain. The natural question is to ask if we can always skip one (two) vertices per obstacle for two (three) dimensional terrains. The answer is no if the vertices are to be arbitrarily skipped. This is illustrated in Fig. 5(a). In two dimensions, if the robot skips the vertices v_1, v_2 and v_3 then the obstacle O_4 will not be detected. Fig. 5(b) shows a 3-D example.

V. Comparison of Performance

We use the worst-case execution of the algorithm *LNAV* or equivalently an invocation of algorithm *ACQUIRE* as a basis for comparison. We consider two-dimensional terrains. We compare the visibility graph methods with the retraction methods of [12] (based on the Voronoi diagram of O). R using the visibility graph methods, may be required to navigate along the boundaries of the obstacles. The paths based on the retraction method always keep R as far away from the obstacle boundaries as possible. In general, the paths generated by the retraction methods tend to be longer than those generated by the visibility graph methods. Using the visibility graph methods, a point robot always navigates along line segments. A circular robot using the visibility graph method will be required to rotate around a vertex. Whereas a point robot or a circular robot will be required to navigate along line segments and second order curves (the parabolic Voronoi edges) in the retraction method. In our methods, for point circular robots, the number of scan operations is at most $N-C$. In the retraction method, the upper bound on the number of scan operations is $4N-n-C-2$.

A summary of the computational complexities is presented in Table I. Consider point robots. It is clear that the adjacency list of the restricted visibility graph can be directly obtained from the scan information. Thus the construction cost for this case is $O(N^2)$ as opposed to the construction cost of $O(N^2 \log N)$ of the retraction based method. Similarly the retraction method has a better complexity for the path planning operations. In terms of the total computational complexity the retraction method has better complexity of $O(N^2 \log N)$ compared to $O(N^3)$ of the visibility graph method. Note that the overall time complexity of the visibility graph based method is dominated by the path planning part whereas that of the retraction method is dominated by the construction cost. Further more the storage complexity in case of retraction methods is $O(N)$ as opposed to $O(N^2)$ of the visibility graph method. For circular robots, the situation remains more or less the same, except that the construction cost of the modified visibility graph is $O(N^2 \log N)$. Thus, in both the cases, the retraction method has better overall time complexity compared to that of visibility graph method.

VI. Conclusion

We presented a framework that solves both the visit problem and the terrain model acquisition problem using a single approach of implementing a graph search on an incrementally constructed geometric structure called the navigational course. A point robot employs the restricted visibility graph and the visibility graph in two and three dimensions respectively. The restricted visibility graph extends the existing solution of [10] to non-convex obstacles for the visit problem. Further, it is better in terms of the bound on the number of scan operations if the terrain contains non-convex corners. A circular robot employs a modified visibility graph in two dimensions. We analyze the algorithms that solve both the visit problem and the terrain model acquisition problem. The proposed framework could be extended to consider more detailed models for the mobile robots in terms of geometric shape, and motion primitives. It would also be interesting to see if there exist general principles to design navigational courses in more detailed cases.

References

[1] B. K. Bhattacharya, and J. Zorbas, "Solving the two-dimensional find-path problem using a line-triangle representation of the robot," *J. Algorithms*, vol. 9, pp. 449–469, 1988.

[2] L. P. Chew, (1985), "Planning shortest path for a disc in $O(n^2 \log n)$ time, in *Proc. Symp. on Computational Geometry*, pp. 214–220, 1985.

[3] J. Hershberger and L. J. Guibas, (1988), "An $O(n^2)$ shortest path algorithm for a nonrotating convex body," *J. Algorithms*, vol. 9, pp. 18–46, 1988.

[4] S. S. Iyengar, C. C. Jorgensen, S. V. N. Rao, and C. R. Weisbin, "Robot navigation algorithms using learned spatial graphs," *Robotica*, vol. 4, pp. 93–100, Jan. 1986.

[5] D. Leven and M. Sharir, "Planning a purely translational motion for a convex object in two-dimensional space using generalized Voronoi diagrams," *Discrete and Computational Geometry*, vol. 2, pp. 9–31, 1987.

[6] T. Lozano-Perez, and M. Wesley, "An algorithm for planning collision-free paths among polyhedral obstacles," *Commun. ACM*, vol. 22, pp. 560–570, 1979.

[7] V. J. Lumelsky, "Algorithmic and complexity issues of robot motion in an uncertain environment," *J. Complexity*, vol. 3, pp. 146–182, 1987.

[8] V. J. Lumelsky, and A. A. Stepanov, "Path-planning strategies for a point mobile automaton moving amidst unknown obstacles of arbitrary shape," *Algorithmica*, vol. 2, pp. 403–430, 1987.

[9] C. O'Dunlaing, and C. K. Yap, "A retraction method for planning the motion of a disc," *J. Algorithms*, vol. 6, pp. 104–111, 1985.

[10] J. B. Oommen, S. S. Iyengar, S. V. N. Rao, and R. L. Kashyap, Robot navigation in unknown terrains using visibility graphs. Part I: The disjoint convex obstacle case, *IEEE J. Robotics and Automation*, vol. RA-12, pp. 672–681, 1987.

[11] N. S. V. Rao, "An algorithmic framework for learned robot navigation in unknown terrains, *IEEE Computer*, pp. 37–43, July 1989.

[12] N. S. V. Rao, N. Stoltzfus and S. S. Iyengar, "A 'retraction' method for terrain model acquisition," *Proc. 1988 IEEE Int. Conf. Robotics and Automation*, pp. 1224–1229, 1988.

[13] N. S. V. Rao, S. S. Iyengar, J. B. Oommen and R. L. Kashyap, Terrain acquisition by a point robot amidst polyhedral obstacles, *IEEE J. Robotics and Automation*, vol. 4, no. 4, pp. 450–451, 1988.

[14] J. Reif, "Complexity of mover's problems and generalizations," *Proc. 20th Ann. Symp. Found. Comput. Sci.*, pp. 421–427, 1979.

[15] J. T. Schwartz and M. Sharir, "On the piano-movers problem: I. The case of two dimensional rigid body moving amidst polygonal barriers," *Comm. Pure Appl. Math.*, vol. 36, pp. 345–398, 1983.

[16] M. Sharir, "Algorithmic motion planning in robotics," *IEEE Computer*, pp. 9–20, Mar. 1989.

[17] C. K. Yap, "Algorithmic motion planning," in *Advances in Robotics: Vol. I: Algorithmic and Geometric Aspects of Robotics*, Ed. J. T. Schwartz and C. K. Yap. Hillsdale, NJ: Lawrence Erlbaum Associated Pub., 1987, pp. 95–144.

Dynamic Path Planning in Sensor-Based Terrain Acquisition

VLADIMIR J. LUMELSKY, SENIOR MEMBER, IEEE, SNEHASIS MUKHOPADHYAY, AND KANG SUN, STUDENT MEMBER, IEEE

Abstract—**The problem of terrain acquisition presents a special case of robot motion planning. In it, a robot that operates in an unfamiliar scene populated with a finite number of objects (obstacles) of unknown shapes and dimensions is asked to cover the scene and/or build its complete map using some sort of sensory feedback (e.g., vision) and generating as short a path during the operation as possible. Algorithms considered thus far in literature make strong assumptions about the obstacles—for example, that they are polygonal—and measure the algorithm performance in terms of the number of constraints describing obstacles, such as the number of obstacle vertices. In this paper, the terrain acquisition problem is formulated as that of continuous motion planning, and no constraints are imposed on obstacle geometry. Two algorithms are described for acquiring planar terrains with obstacles of arbitrary shape. Estimates of the algorithm performance are derived as upper bounds on the lengths of generated paths.**

I. INTRODUCTION

THE PROBLEM of terrain acquisition is an aspect of the more general problem of robot motion planning (robot navigation). It is stated here as follows. A point robot is placed in an unexplored planar terrain (scene) that is randomly populated with a finite number of obstacles (objects, landmarks) of arbitrary sizes and shapes. Depending on the underlying model, the terrain can be finite or infinite. The robot is required to autonomously navigate in the terrain with the purpose of building the complete terrain model (a map), i.e., the boundaries of all the obstacles.[1] One can envision a need for such learning automatic systems in various applications (e.g., in hazardous environments, such as mapping the extent of damage after an explosion at a nuclear power station, in military and space applications, for lawn mowing, etc.).

To gather new information, the robot uses a vision sensor. The sensor is capable of identifying coordinates of all visible points within a limited radius (radius of vision) around the robot. For example, the sensor can be a stereo vision system or a laser range finder. The robot is also capable of plotting identified obstacle boundaries on the map and making use of the current map whenever it is required. During the course of terrain acquisition, the robot is expected to cover as small a distance as possible.

Manuscript received February 28, 1989; revised April 18, 1990. This work was supported by the National Science Foundation under Grant DMC-8712357 and Grant IRI-8805943.

The authors are with the Department of Electrical Engineering, Yale University, New Haven, CT 06520.

IEEE Log Number 9037979.

[1] As used here, the term *obstacle* is free of the connotation of something undesired. It is borrowed from the literature on robot motion planning, where it is used widely. Although in the latter area, obstacles are something to avoid, in terrain acquisition "obstacles" may be something to be attracted to.

Reprinted from *IEEE Transactions on Robotics and Automation*, Vol. 6, No. 4, August 1990. Copyright © 1990 by The Institute of Electrical and Electronics Engineers, Inc. All rights reserved.

A closer look at the terrain acquisition problem reveals two distinct task models, task a) and task b), both of which produce terrain maps. In task a), the map is the sole goal of the acquisition operation—consider, for example, mapping a nuclear power plant site after an explosion. In principle, if obstacles are visible from one another, the robot does not need to physically visit every point of the terrain—it can, for example, accomplish the task by walking from one obstacle to the other.

In task b), the robot is required to physically pass through or near every reachable area of the (finite) terrain—consider, for example, an automatic vacuum cleaner or lawn mower. Here, the map of the terrain can appear as a side effect of the main operation.

Note that these two tasks reflect two different meanings of what can be called "complete exploration" of a terrain. Accordingly, the models of the environment assumed in each task can differ. In this paper, two such models, one for each task, will be considered:

Model (a): Here, the terrain can be finite or infinite, and obstacles are mutually visible, that is, any pair of obstacles are connected through a sequence of obstacles that are visible from one another.

Model (b): Here, the terrain must be finite, and obstacles need not be visible from each other.

The main results presented in this paper are as follows: 1) The requirement commonly used in literature—that obstacles must be algebraic manifolds—is removed, and the problem of terrain acquisition is presented as one of continuous path planning amongst obstacles of arbitrary shapes; 2) two algorithms, one for each of the **Models (a)** and **(b)**, are formulated; 3) the algorithms' performance is analyzed, and the upper bounds on the lengths of the paths that they generate are derived as functions of the obstacle perimeters.

The terrain acquisition problem is of recent interest. Works by Brooks [1], Turchen and Wong [2] and Oommen *et al.* [3] are among the first in this area. Typically, the robot is assumed to be a point that operates in an environment with polygonal obstacles. In [3], the robot is capable of two elementary operations: scanning, to identify all visible obstacle vertices from its current location, and moving, to move the robot along a straight line. Under the proposed strategy, the robot moves from vertex to vertex. To acquire an environment with n obstacle vertices, the algorithm requires n scanning operations and, at most, $2(n-1)$ motions between the vertices. A later modification of this procedure by Rao *et al.* [4] extends the approach to a finite-sized circular mobile robot in the plane.

The assumption in the work above that obstacles in the terrain are polygons (or, in general, sets constrained by algebraic manifolds) is quite important; in its essence, it follows the "piano movers" approach to motion planning, which was developed in general form in [5]. Although it gives rise to an interesting problem of computational geometry and computational complexity, the underlying model of this approach is only one of those that appear in the context of robotics. First, real obstacles do come in arbitrary shapes. It is known that given the accuracy of approximation, the problem of approximating nonlinear surfaces with linear constraints is at least of exponential complexity [6]. On the other hand, one may find this approximation problem rather artificial—for example, walking along an unknown smooth curve may be even easier for a robot than turning at polygon corners.

Second, the performance of planning algorithms is measured in the "piano movers" approach in terms of the total number of obstacle vertices. This is quite natural for computational geometry, but in robotics, one may prefer to measure the cost of planning simply by the length of paths generated by the robot. This way, walking, for example, around a circular obstacle or around its approximating polygon with a very large number of edges would present tasks of the same complexity. Realizing such a performance criterion requires a continuous, as opposed to discrete, formulation of the motion planning problem.

This work has been motivated by the search for a methodology that would, first, put no constraints on the shape of obstacles in the environment and, second, allow measuring algorithm performance directly in terms of the length of generated paths. Consequently, the model we assume differs from that in the work mentioned above. We draw extensively on the algorithmic machinery and tools for convergence analysis developed in [7]. Our robot has a continuous and costless vision in a limited area around its current location as well as sufficient memory to store the current map of the terrain (this can be thought of as a pencil-and-paper drawing in the hands of a cartographer). In practice, this would simply mean that the vision operation is done in parallel with and not slower than the motion planning operation and that the robot has enough memory to store the acquired map.

Given incompleteness of input information, optimality of generated paths is ruled out in any formulation of the terrain acquisition problem [1]–[4] including ours. Our emphasis is on *convergence* and *completeness*, as follows: Given the uncertainties assumed in our model—unknown obstacles of arbitrary shape and local input information—are procedures for terrain acquisition with guaranteed convergence feasible? If so, what would be the upper bounds on the performance of such procedures? The performance of an algorithm is assessed in terms of the length of generated paths as a function of obstacle perimeters and distances between obstacles in the environment. One can envision that in practice, such algorithms would be used as global procedures, complemented wherever possible by local heuristics.

The first algorithm presented below (the Sightseer Strategy) applies to **Model (a)** and is reminiscent of a sightseer who walks from one landmark to the other and then circles each landmark to see it all. The robot moves from its starting position to the nearest obstacle and circumnavigates it while acquiring more knowledge about the visible parts of the environment. This knowledge is then used to identify the nearest partially unknown obstacle and move to it. The process is then repeated till all the obstacles in the terrain are visited and fully known. The strategy converges, and its upper bound on the length of generated paths is shown to be linear in the perimeters of obstacles and distance between obstacles in the scene.

The second procedure (Seed Spreader Strategy) applies to the **Model (b)** and is reminiscent of a landowner who, in order to plant grass in a lawn, divides the terrain into a suitable number of rectangular strips and then covers the strips with seeds sequentially, one by one. While moving along the boundary of a strip, the robot is acquiring (that is, putting on the map) all the information that appears. If, however, it is becoming clear that some obstacle cannot be acquired fully in this fashion, the robot diverges from the current strip boundary to the obstacle, acquires it (and other obstacles if needed), returns to the strip boundary, and repeats the process. Although the upper bound of this procedure turns out to be quadratic, it will be shown that in some general cases, it is as good or better than the first algorithm.

II. Model and Terminology

A. Robot

The robot is a point. It can thus pass through an opening of any size between two obstacles. The robot is capable of two types of motions: *linearnavigate*, which means moving towards a target point along a straight line, and *circumnavigate*, which means moving along an obstacle boundary. The robot always knows coordinates of its current location. The robot is equipped with a *vision sensor*, which allows it to identify coordinates of all visible points on obstacle boundaries within a circle of radius R_v, which is known as the *radius of vision* and is centered at the current location. (For example, the sensor can be a stereo vision system or a sensor range finder). Once a piece of an obstacle boundary is identified, the robot can *register* (i.e., plot) it on its incremental *map*. It can then make use of this information thereafter. An obstacle is said to have been *acquired* if its boundary is completely known and connected. The term *scanner* refers to the robot's sensing/vision system, and scanning refers to the act of seeing. The suffix "navigate" (as in the term "circumnavigate") implies simultaneous actions of moving, scanning, and registering. As for its computational capabilities, the robot can compute the (shortest) Euclidean distances between two points, between a point and an obstacle, and between two obstacles.

B. Environment

The terrain to be explored is a planar area. The robot starts at some *starting point* S in the terrain. The terrain is populated with a finite but unknown number of obstacles of arbitrary shapes. Obstacles have no holes in them. (This assumption is needed simply for completeness; a robot living in a 2D space will not be able to see a hole inside a 2D obstacle). The boundary of an obstacle is a simple closed curve. Let d

be the closest distance (or distance, for short) between two obstacles, and let d_{max} be the maximum such distance. Since no information is available about obstacles, d_{max} becomes a characteristic of the terrain. If, for example, the terrain is a rectangle, then d_{max} is the length of the corresponding diagonal.

The following assumptions produce two somewhat different models of the environment. In **Model (a)**, the terrain can be finite or infinite. At least one obstacle is visible from the robot starting position, and all the obstacles are mutually visible from each other, that is, for any pair of obstacles X and Y, there is a sequence of obstacles visible from one another that leads from X to Y. Clearly, mutual visibility is a function of the obstacle distribution in the terrain and of the radius of vision R_v. In **Model (b)**, the terrain must be finite, and obstacles need not be visible from each other.

C. Performance Criterion

Let n be the number of obstacles that lie, wholly or partially, in the terrain, let p_i be the (unknown) perimeter (i.e., the length of the boundary) of the ith obstacle, let p_{max} be the maximum such perimeter, and let P be the length of the path that the robot traverses in the course of terrain acquisition. The performance of an algorithm is assessed in terms of the upper bound on the length P of the generated path as a function of obstacle perimeters and distances between obstacles. Any computations due to vision, incremental map construction, planning the shortest path between obstacles, etc., are assumed to be of no cost.

Definition 1: A *local direction* is the direction of passing around an obstacle once the robot comes in contact with it. It can be either left or right. A local direction is chosen at the time an obstacle is encountered (contacted) and can be reversed when needed.

Definition 2: A *hit point* H is defined on an obstacle when, while linearnavigating towards an intermediate target, the robot contacts the obstacle at H. A *leave point* L is defined on an obstacle when the robot leaves the obstacle at L in order to continue its linear navigation towards the intermediate target.

Definition 3: An obstacle becomes a *visited* obstacle if it has been explicitly circumnavigated by the robot; otherwise, it is an *unvisited* obstacle. Hence, any visited obstacle must have been acquired and is completely known. Note that the opposite is not necessarily true; the robot might have completed an obstacle boundary while observing it "from afar" (e.g., while visiting other obstacles). Thus, an unvisited obstacle may be known to the robot either partially or completely. A partially known obstacle is also called *incomplete*; it is easily recognized from the fact that its boundary in the map presents a simple open curve; the boundary of a completely known (acquired) obstacle is a simple closed curve.

III. The Sightseer Strategy

A. General Idea

This algorithm applies to **Model (a)**. Consider the following hypothetical strategy. While standing at the start position, the robot scans for and registers all the visible obstacle boundaries. If nothing can be seen, the task is accomplished. Otherwise, the robot linearnavigates toward the nearest obstacle and then circumnavigates it completely while updating the map. Then, the robot marks the obstacle as visited, finds from the map the nearest unvisited obstacle, and linearnavigates to it. The procedure repeats until no unvisited obstacles remain. If a new obstacle is encountered along the way toward the next targeted obstacle, the robot first circumnavigates the newly found obstacle and then chooses a new unvisited obstacle, which becomes the new target. On the other hand, if the robot encounters a visited obstacle on its way, it simply passes it around using a local direction that leads to a shorter path and resumes its linear-navigation motion; since the geometry of the obstacle is already known, choosing the shorter path presents no difficulties.

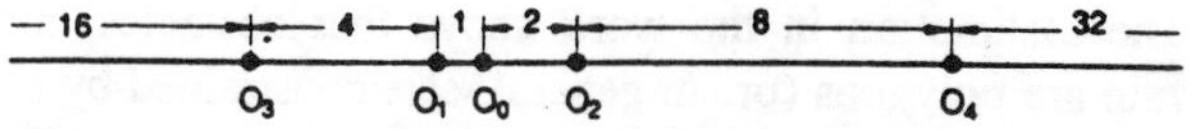

Fig. 1. Case where the procedure sketched in Section III-B would result in performance quadratic in obstacle perimeters and distances between obstacles.

In each act of linear navigation, one unvisited obstacle is targeted for acquisition, and at least one unvisited obstacle is actually acquired. Since the terrain has a finite number of finite-size obstacles, and since no visited obstacle can become a target of linear-navigation, the above procedure guarantees convergence. However, as the following example demonstrates, the upper bound on the path length performance of this simple strategy is not very good. Consider a one-dimensional case with n obstacles shown in Fig. 1. For simplicity, each obstacle, $O_0, O_1, \cdots, O_{n-1}$, is presented as a single point. Note that the obstacles with odd subscripts are placed to the left of O_0, and those with even subscripts are placed to the right of O_0. Let O_1 be $2^0 = 1$ unit away from O_0, let O_2 be $2^1 = 2$ units away from O_0, let O_3 be $2^2 = 4$ units away from O_1, let O_4 be $2^3 = 8$ units away from O_2, etc., and finally, let O_{n-1} be 2^{n-2} units away from O_{n-3}. Assume $R_v > 2^{n-2}$ so that each pair of adjacent obstacles are visible from each other.

Starting from O_0, the robot can see both O_1 and O_2; it goes to O_1 since it is closer. At O_1, it sees a new obstacle O_3 but since O_2 is now the nearest unvisited obstacle, it goes to O_2 first. Continuing with the above procedure, all the obstacles will be eventually visited in the numerical order of their subscripts. Tracing the path, one will notice that O_0 has been passed $(n-1)$ times, O_1 is passed $(n-2)$ times, etc. Clearly, the worst-case performance of this algorithm is quadratic in the number of obstacles, and hence, it is quadratic in the perimeters of obstacles and distances between obstacles.

It will be shown later that the following change in the procedure improves its performance from quadratic to linear. After the robot circumnavigates an obstacle and marks it as visited, the robot shall choose among all the visible obstacles (as opposed to all the unvisited obstacles) the nearest unvisited obstacle and explore it next; if no such obstacle exists, it simply backtracks to the latest visited obstacle and from there looks for the nearest unvisited visible obstacle. Indeed, since the next obstacle to explore is visible from the current obstacle, the robot can always find a path to it without encountering any

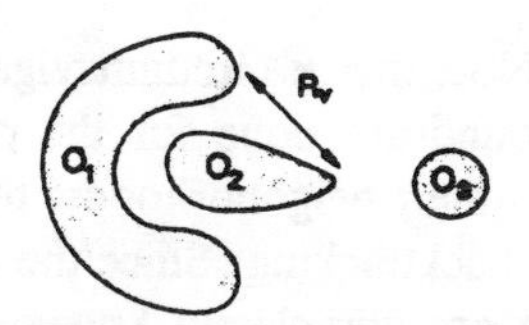

Fig. 2. Obstacle O_3 can only be seen from O_2. Although O_2 is completely known after exploring O_1, the robot still has to visit O_2 in order not to miss other obstacles, such as O_3.

other obstacles. The only time it will visit a previously visited obstacle is during the backtracking. The process is repeated until no unvisited obstacles remain. The elementary operation of exploring an individual obstacle follows the approach described in [7].

It may happen that none of the currently identified unvisited obstacles can be seen from the robot's current position. Using the map, the robot can choose to visit one of them (e.g., the nearest). What path should the robot use to reach the obstacle? Again, this is done by backtracking via previously visited obstacles. Backtracking guarantees (see Section III-C) that the generated path presents a spanning tree, which simplifies the algorithm.[2] In addition, the straight line path segments between obstacles are bounded by the radius of vision R_v.

Note that if an obstacle has not been visited but is completely known, it still cannot be marked as visited. Every unvisited obstacle must be explicitly visited because there might be some other obstacles that are visible only from it. This is shown in an example in Fig. 2; in the process of circumnavigating obstacle O_1, the robot will completely acquire obstacle O_2. Nevertheless, obstacle O_2 has to be explicitly visited since, otherwise, given the radius of vision R_v, the robot will never discover obstacle O_3.

B. Algorithm

This section contains the following notations:

curr-obstacle	obstacle at whose boundary the robot is positioned currently
next-obstacle	obstacle to be visited next
prev-obstacle	obstacle that has led to *curr-obstacle*.

In the procedure, the planning decisions, such as what obstacle to visit next, are made when the robot is either at the start point S or at an obstacle boundary. To keep track of obstacles for backtracking, a linear *stack* is maintained. The stack contains labels of all identified visited or unvisited obstacles. The operation *a = pop(stack)* removes the top item on the stack and assigns it to the item *a*. The operation *push(a, stack)* places item *a* on top of the stack. The procedure operates as follows.

Step 0 (Initialization).

Let *curr-obstacle* = S, and mark it visited. Register all the visible obstacle boundaries in the map and mark them unvisited. Go to *Step 1*.

Step 1 (Determine the next obstacle to visit).

If no unvisited obstacles remain in the map, the map is complete; the procedure terminates. If there are no unvisited obstacles visible from *curr-obstacle*, go to *Step 2*. Otherwise, let *next-obstacle* = nearest unvisited visible obstacle, and go to *Step 3*.

Step 2 (Backtracking).

If *curr-obstacle* = S, the map is complete; the procedure terminates. Otherwise, let *prev-obstacle = pop(stack)*. Backtrack from *curr-obstacle* to *prev-obstacle*. Let *curr-obstacle = prev-obstacle*. Go to *Step 1*.

Step 3 (Linear navigation).

Do *push(curr-obstacle, stack)*. Choose a straight line collision-free path between *curr-obstacle* and *next-obstacle*, and follow it until *next-obstacle* is reached. Let *curr-obstacle = next-obstacle*. Go to *Step 4*.

Step 4 (Circumnavigation).

Circumnavigate *curr-obstacle*; mark it visited. Go to *Step 1*.

An example of the algorithm performance is shown in Fig. 3.

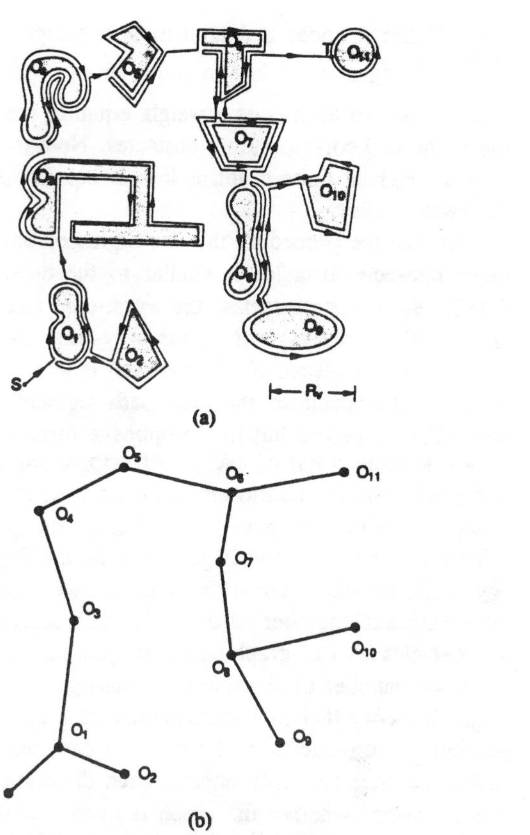

Fig. 3. Example of performance of the Sightseer Strategy. (a) The operation of terrain acquisition starts at S and terminates at T; each obstacle is explicitly visited and circumnavigated; (b) the corresponding spanning tree; obstacles are numbered in the order at which they are visited.

C. Performance Analysis

Given a scene with n obstacles, the corresponding visibility graph (VG) is constructed as follows. Every obstacle forms a node in VG; point S is also treated as an obstacle and thus produces a separate node. An edge is drawn between two nodes if and only if the entities that the nodes represent are visible from each other. Given the assumption of mutual visibility of point S and obstacles, VG is a connected graph. Assuming for simplicity that point S lies on the boundary of an obsta-

[2] One can see that an alternative option—going straight towards the chosen unvisited obstacle—is problematic; it can produce loops in the generated paths as well as straight line path segments whose length exceeds R_v.

cle, VG has n nodes and is at most a complete graph with $\frac{n(n-1)}{2}$ edges and at least a graph with $(n-1)$ edges (as in Fig. 1). Assign each edge a weight equal to the distance between the two corresponding obstacles. Note that any weight in this weighted finite graph is less or equal to R_v, which is the vision radius.

Note that the procedure that the Sightseer Strategy uses to move between obstacles is similar to the depth-first search (DFS) [8]. For n obstacles, the worst-case size of the stack is $(n-1)$. In the algorithm, the only time the robot walks to a previously visited obstacle (node) is when it traces back to a visited obstacle by the same path segment (edge) that it had followed before but in the opposite direction. Since the reason and the result of every such trip is acquiring another unvisited obstacle, no closed loops are possible in the path. Mapped into VG, the path produces a spanning tree.

To evaluate the total path generated by the Sightseer Strategy in the course of terrain acquisition, we need to estimate the (maximum) number of times the path segments between the obstacles (that is, graph edges) are passed, and we must estimate the number of obstacle circumnavigations. It is known in graph theory that in a finite-connected graph, it is always possible to construct a cyclic directed path passing through each edge once and only once in each direction [9]. This is true no matter whether the graph is known or unknown beforehand. One method to construct such a path is known as Tarry's rule [9]. The rule is as follows. Mark each edge along the path, including the direction of passage. When some node g is arrived at for the first time, this entering edge should be marked in a special way. When a node g is reached, always proceed next via an edge (g, r) that has not been traversed before, but if it has, only traverse it in the reversed direction. However, the entering edge should be used only as a last resort when there are no other edges available.

The subgraph generated by removing all edges of VG never traversed by the Sightseer Strategy is a spanning tree (see Fig. 3(b)). Applying Tarry's rule to the subgraph, one can see that DFS follows Tarry's rule exactly. Since we are dealing with a tree structure, each and every edge will be marked at some moment as an entering edge; then, each of them must be backtracked again when there are no more untraversed edges incident to it. As a result, the performance index of Tarry's rule [9] holds—each edge is traversed exactly twice, once in each direction. Since the Sightseer Strategy stops when no unvisited obstacles remain, each edge of the spanning tree will be traversed at least once and at most twice. This proves the following lemma:

Lemma 1: Under the Sightseer Strategy, a path segment between two obstacles can be traversed at most twice.

In a tree structure, each node, except for the root, is connected to its parent node by one edge; a tree structure with n nodes consists of $(n-1)$ edges. Thus, traversing k edges leads to k nodes. Hence, the following statement.

Lemma 2: Following the Sightseer Strategy, the robot circumnavigates n obstacles at most $2(n-1)$ times.

Although each path segment between obstacles is traversed no more than two times, an obstacle can be visited many times. **Lemma 2** provides an upper bound on the total number of obstacle visits. Note that n circumnavigations that contribute to this upper bound are done for the purpose of acquiring the obstacles and their neighborhoods; the remaining $(n-2)$ visits are due to backtracking. Since the obstacles met during the backtracking are completely known, the robot does not need to circumnavigate them completely; instead, the robot can maneuver around the obstacle via a shorter path, which is never more than half of the obstacle perimeter. Hence, the portion of the total path length due to moving around obstacles is bounded by $1.5\sum_{i=1}^{n} p_i + (n-2)\frac{p_{\max}}{2}$. On the other hand, since each path segment between obstacles is bounded by min $(R_v, d_{\max})$, the portion of the path length due to traveling between obstacles is bounded by $2(n-1)\cdot\min(R_v, d_{\max})$. Summing it up, obtain the following upper bound, see the following theorem.

Theorem 1: The total path P traversed by the robot in the process of terrain acquisition under the Sightseer Strategy algorithm is linear in the perimeters of the obstacles in the scene and is bounded by

$$P = 2(n-1)\cdot\min(R_v, d_{\max}) + 1.5\sum_{i=1}^{n} p_i + p_{\max}\cdot\frac{n-2}{2}. \tag{1}$$

Assuming for specificity that the robot's vision is not sufficient to observe the whole terrain, min $(R_v, d_{\max}) = R_v$, this produces

$$P = 2R_v(n-1) + 1.5\sum_{i=1}^{n} p_i + p_{\max}\cdot\frac{n-2}{2}. \tag{2}$$

Note that choosing the nearest visible obstacle in *Step 1* of the algorithm, although reasonable in actual implementations, does not effect the worst-case performance. As an illustration, observe that in the example in Fig. 3(a), the algorithm performance is well above the worst-case bound; only four obstacles, O_1, O_6, O_7, and O_8, are visited more than once during backtracking and thus appear in the third term in (2). Fig. 3(b) presents the corresponding spanning tree. As another example, consider the case shown in Fig. 1. The order in which the obstacles will be visited under the Sightseer Strategy algorithm is $O_0, O_1, O_3, O_5, \cdots, O_2, O_4, \cdots$, that is, obstacle O_0 and each of the obstacles with odd subscripts (except the last one) is visited twice, whereas each obstacle with an even subscript is visited only once.

IV. The Seed Spreader Strategy

A. General Idea

This algorithm applies to the **Model (b)**. Assume that the terrain is rectangular, with dimensions A by B. If the terrain has many obstacles in it, then moving from obstacle to obstacle and circumnavigating every obstacle, as in the Sightseer Strategy algorithm, may produce very long paths. On the other hand, if the obstacles are "nicely distributed" and are of "nice geometry," then encircling a group of obstacles—for example, by a rectangular path—may make them completely known without actually visiting and circumnavigating each of them. We divide the terrain into a number of strips of equal width by a set of lines parallel, say, to the side A of the terrain. It is hoped that most of the obstacles in a strip will be acquired without actually visiting them in the course of the movement along the strip boundaries. If, however, it becomes apparent that an obstacle cannot be acquired from the strip boundaries, the obstacle is explicitly visited via a divergence route.

To choose the strip width, observe that the strip cannot be wider than $2R_v$; otherwise, some obstacles may be missed. Within this limit, a strip can be as narrow as one wishes. In general, very narrow strips would result as a plus in few diversions from the strip boundaries but as a minus in many obstacle crossings and hence in many obstacle circumnavigations; as another minus, there would be too many strip boundaries. On the other hand, very wide strips would result in fewer strip boundaries but in many diversions, each of which ends, again, in an obstacle circumnavigation. This suggests that there might be an optimal value for the width strip. Unfortunately, finding this value does not seem to be feasible given that it depends on the unknown and difficult-to-generalize distributions of obstacle shapes and positions.

We make the strip width equal to the radius of vision R_v. One advantage of this choice is that in many cases, the robot will be able to acquire obstacles in a strip by simply moving along the strip boundaries. Exceptions will be cases with mutual obstacle occlusions and cases where parts of obstacle boundaries are not visible from the circumscribing convex hulls. If $R_v > B$, then there is only one strip, which is of width B.

By proper alignment of the axes of reference, we define the directions E, N, W, and S (for east, north, west, and south). Define the main line (M-line) as a piecewise path that consists of south-north (SN) and east-west (EW) straight line strip boundaries, as is shown in Fig. 4. Consider the M-line to be a one-dimensional space in which a metric is defined such that the distance between two points x and y, $\tilde{d}(x, y)$, is measured along M-line. If points x and y are on two different strip boundaries, with k turning points $t_i, t_{i+2}, \cdots, t_{i+k-1}$ along the M-line between them, then $\tilde{d}(x, y) = d(x, t_i) + d(t_i, t_{i+1}) + \cdots + d(t_{i+k-1}, y)$, where d is the regular Euclidean distance.

Assume for simplicity that point S is at the SW corner of the terrain. Let S be one end of the M-line, and let T be its other end (Fig. 4). Assume that S is always obstacle-free. At the beginning, the map will contain the border of the terrain and M-line; the latter is marked to distinguish it from the terrain border and obstacle boundaries. In the algorithm, the robot registers all the information that comes its way from those (one to three) strips that it can observe simultaneously. Its diversions though are organized to complete the acquisition of one strip before switching to the next strip. If no obstacles are present in the scene, the robot will learn this fact after having completed the path along the M-line from S to T. If, however, the robot encounters an unknown or an incomplete obstacle that crosses the M-line, it will circumnavigate it and then resume its motion along the M-line. If the robot encounters a completely known obstacle, it simply passes it around following a shorter path along its boundary. If the robot concludes that a given incomplete obstacle cannot be acquired directly from the M-line, it diverges to it, acquires the obstacle by circumnavigating it, returns to the M-line and resumes its motion along the M-line.

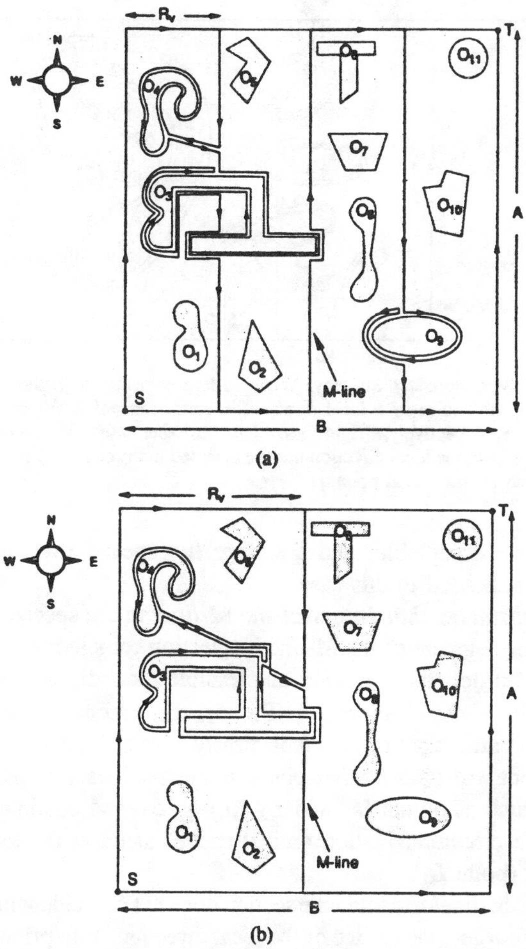

Fig. 4. Example of performance of the Seed Spreader Strategy. The terrain is the same as in Fig. 3; the performance for two values of radius of vision R_v is shown. Out of 11 obstacles, (a) only three obstacles need be explicitly circumnavigated, and (b) only two obstacles need be explicitly circumnavigated.

To define the diversion strategy, the acquisition strategy, and the local direction for handling individual obstacles, we divide obstacles into two types, depending on their locations relative to the M-line and the terrain border: 1) obstacles that lie fully inside the strips and 2) obstacles that intersect the M-line. Consider both types separately.

1) Obstacles that Lie Fully Inside the Strips: If such an obstacle cannot be acquired from the M-line, a diversion is undertaken (see obstacle O_4 in Fig. 4(a)). The diversion starts only at the "very last moment" when, according to the available information, the robot cannot expect to see the obstacle again by simply following the M-line. More specifically, the robot diverges to an obstacle only while moving along an eastern SN-oriented strip boundary, when it is about to lose sight of the obstacle. This corresponds to one of two cases: 1) when the visible obstacle boundary in question is about to go out of the robot's vision range and 2) when the boundary is occluded by another obstacle or by another part of the same obstacle. This implies that an obstacle sometimes may be acquired by diversion even though it intersects the M-line. Out of two possible local directions (right or left), any one

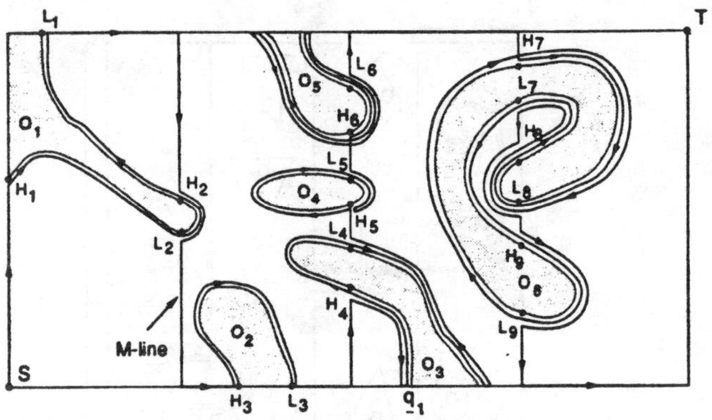

Fig. 5. Seed Spreader Strategy. When *M*-line coincides with the terrain border, the choice of a local direction is unique; at points H_1 and H_3, these are, respectively, "right" and "left." In other cases, the choice may be uncertain; the local direction may be reversed in the course of acquiring an obstacle, such as at point q_1.

is equally acceptable, and therefore, no special rule for its choice is needed in this case.

2) Obstacles that Intersect the M-line. In the special case when the segment of the M-line in question coincides with the terrain border, there is only one feasible local direction. For example, at the hit point H_1 (Fig. 5), the direction is clearly "right," and at point H_3, it is clearly "left." In such cases, the robot will ignore other obstacle intersections with the M-line—such as points H_2 and L_2 on Fig. 5—and continue the obstacle circumnavigation until it arrives again at the terrain border (point L_1).

If the M-line/obstacle intersection does not coincide with the terrain border, the choice of the local direction is, in principle, not important—say, it is "right." In the course of maneuvering around obstacles, the robot may switch its local direction. Switching is caused either by intersections with the terrain border (as at point q_1 on Fig. 5) or by a decision to shorten the path along an already known obstacle boundary (as at point H_5). In the process, some path segments may be traversed more than once (see Section IV-C).

Intersections of the M-line with an obstacle boundary form two or more crossing points, such as those with obstacle O_6 (Fig. 5). Unless the M-line ends "inside" the obstacle, the number of crossing points is even. Denote $\{c_i\}$, $i = 1, 2, \cdots, m$, the set of crossing points numbered in the order of the intersections between the M-line and the obstacle boundary. Once the robot encounters an obstacle and defines on it the hit point H, it has to also define on it the corresponding leave point L. Both H and L are crossing points.

In general, the obstacle shape may be such that the M-line crosses it a number of times; correspondingly, a number of pairs of crossing points (H_i, L_i) will be formed. Such obstacles are called *complex obstacles* (see obstacle O_6 on Fig. 5), and the corresponding segments of the M-line "inside" the obstacle are called *enclosed segments* (segments L_7H_8, L_8H_9 on Fig. 5). Since some unknown obstacle may happen to be visible only from the M-line, all the enclosed M-line segments of a complex obstacle have to be traversed. Furthermore, to maintain algorithm consistency, each enclosed M-line segment has to be traversed in the direction toward T. Note that the crossing points on a complex obstacle form nodes of a graph, which is known as a local graph and is limited to this obstacle only. The graph edges are segments of the obstacle boundaries and enclosed M-line segments; each of its nodes has three adjacent edges. The term local leave point refers to a leave point incident to an enclosed M-line segment of a local graph, and global leave point refers to the leave point corresponding to the final exit of the local graph. For example, the local graph of obstacle O_6 (Fig. 5) has six vertices and eight edges; vertex L_9 is the global leave point.

The necessity to explicitly visit every enclosed segment of the M-line suggests a need for a full exploration of the local graph, with a possibility of worsening the algorithm performance due to the visitation of some segments of obstacle boundaries many times. However, since Tarry's rule guarantees that any known or unknown graph can be fully explored with no more than two traverses of every edge (one in each direction (Section III-C)), this rule is used in the algorithm to limit the path length along obstacle boundaries (see Section IV-C) while guaranteeing that each enclosed M-line segment is traversed at least once in the direction towards T.

It may happen that while exploring a local graph under Tarry's rule and moving for the first time along an enclosed M-line segment of a complex obstacle X, the robot will encounter still another obstacle Y. If Y is a complex obstacle, it has an associated local graph that must be connected with the local graph X by virtue of its sitting on at least one edge of X. Then, both local graphs will be considered parts of a bigger graph, and its processing will proceed as before, and it will still guarantee that each edge of the bigger graph is traversed at most twice.

B. Algorithm

This section contains the following notations:

curr-obstacle is the obstacle that the robot is currently acquiring; *next-obstacle* is an incomplete obstacle that the robot is about to lose sight of while moving along the eastern SN boundary of a strip; *curr-position* is the robot's current position; S is the start position; T is the ending position; D is a point of diversion; Q is the point on the boundary of *next-obstacle* that is closest to D. A flag F is used to remember the fact that the robot is currently processing a complex obstacle.

Step 0 (Initialization)

Register the terrain border in the map. Divide the terrain into strips of width min (R_v, B). Draw the M-line in the map beginning at S; the other endpoint of the M-line is T. Set F = false. Go to *Step 1*.

Step 1 (Navigating along the M-line)

Navigate the M-line towards T until one of the following occurs:

a) T is reached; the procedure terminates.
b) A new obstacle is encountered. Let H = *curr-position*. Go to *Step 2*.
c) A previously visited noncomplex obstacle is encountered. Using a local direction that produces a shorter path, pass around the obstacle until the M-line is met again; iterate *Step 1*.
d) A previously visited complex obstacle is encountered. If all the enclosed M-line segments have been visited,

then pass around the obstacle using a local direction that produces a shorter path until the M-line is again met at the global leave point; set F = false; iterate *Step 1*. Otherwise, move along the segment of obstacle boundary to the next local leave point L chosen according to Tarry's rule; set F = true; iterate *Step 1*.

e) An incomplete obstacle to the west of the current SN oriented strip boundary is about to go out of robot's vision range. Let D = *curr-position*, let *next-obstacle* = the incomplete obstacle, and let Q = a point on *next-obstacle* closest to D. Go to *Step 3*.

Step 2 (Circumnavigating an obstacle that crosses the M-line)

Circumnavigate *curr-obstacle*; then

a) If *curr-obstacle* is not complex, proceed to the corresponding L point via the shorter path along the obstacle boundary. Go to *Step 1*.

b) If *curr-obstacle* is a complex obstacle, set F = true, and proceed to the first local leave point L using Tarry's rule. Go to *Step 1*.

Step 3 (Diversion)

Linearnavigate toward Q until one of the following occurs:

a) Q is reached. Circumnavigate *next-obstacle* and return to Q. Choose a shorter path to D and return to D. Go to *Step 1*.

b) A previously acquired obstacle is encountered. Using a local direction that produces a shorter path, pass around the obstacle until the line DQ is met again. Iterate *Step 3*.

c) An unknown or incomplete obstacle is encountered. Define a hit point H. Circumnavigate the obstacle. Choose $L = c$ such that $\tilde{d}(c, Q) = \min(\tilde{d}(c_i, Q))$, where $\{c_i\}$ are the crossing points between the straight line DQ and the obstacle boundary. Choose the shorter path from H to L along the obstacle boundary; walk to L. Iterate *Step 3*.

An example of the algorithm performance is shown in Figs. 4 and 5.

C. Performance Analysis

We would like to produce an upper bound on the length of paths generated by the Seed Spreader Strategy in the course of terrain acquisition. Since the path consists of segments of the M-line, of straight-line diversions to obstacles, and of segments of obstacle boundaries due to maneuvering around obstacles, the path length is a function of the M-line length, of the number of visits to obstacles by way of diversion, and of obstacle perimeters.

Let $w = \min(R_v, B)$ denote the width of a strip, and let $r = \text{ceil}(B/R_v)$ denote the number of strips, where ceil(x) returns the smallest integer larger or equal to x. If no obstacles are present in the scene, or if all the obstacles can be fully acquired without deviating from the M-line, the total length of the generated path will be equal to the length of the M-line (see Fig. 4). Otherwise, to acquire some or all the obstacles in the scene, the robot will do obstacle circumnavigation by way of diversions or otherwise, and the length of the generated path will be longer. We will first estimate the number of visits to obstacles by way of diversion. However, for a simple obstacle that does intersect the terrain borders, its circumnavigation/acquisition can produce a longer path segment but no more than $2p_i$ (see obstacle O_5 on Fig. 5); walking back to L then adds at most another p_i. The total cost of acquiring the obstacle is thus at most $3p_i$.

Each obstacle can call for at most one diversion. If before the diversion the robot is at some obstacle, the corresponding leave point L is chosen in *Step 3* of the algorithm as the crossing point closest to the encounter point Q. Since Q lies on some other obstacle, once the robot leaves the current obstacle at L, it will not encounter it again between points L and Q. This means that each obstacle can be visited at most once during a diversion. In a scene with n obstacles, at most n diversions are needed.

Let n_j be the number of obstacles totally inside strip j, $j = 1, 2, \cdots, r$; let n_0 be the total number of obstacles crossing the M-line; let n_{0_j} be the number of obstacles crossing the boundary of the strip j but not crossing the boundary of any other strips to the west of strip j. Then, $n_0 = \sum_{j=0}^{r} n_{0_j}$. Note that n_j, $j = 0, 1, \cdots, r$ accounts for all the obstacles in the terrain, and therefore, $n = \sum_{j=0}^{r} n_j$. Consider the acquisition process in the first strip. Since in an unfortunate case each of the n_{0_1} obstacles will be spotted and acquired by way of diversion, before the robot reaches it via the M-line, then at most $(n_{0_1} + n_1)$ diversions are needed to acquire the first strip. Number these obstacles in the order of their being visited. On the diversion to acquire the ith obstacle of the first strip, in the worst case, all the $(i-1)$ already-acquired obstacles are visited. Hence, for the first strip, the total number of obstacle visits by way of diversion is bounded by

$$(1 + 2 + \cdots + (n_{0_1} + n_1 - 1) + n_{0_1} + n_1) = \frac{(n_{0_1} + n_1)\cdot(n_{0_1} + n_1 + 1)}{2}. \quad (3)$$

This also means that the ith obstacle will be visited $(n_{0_1} + n_1 - i)$ times during the diversions required for acquiring the remaining $(n_{0_1} + n_1 - i)$ obstacles of the first strip.

Now, consider the acquisition process in the jth strip, $j = 1, 2, 3, \cdots, r$. Again, number the obstacles in strip j in the order of their being visited. Then, the ith obstacle in the strip j will be visited at most $(n_j + n_{0_j} - i)$ times; in addition, since each of the $\sum_{k=0}^{j-1} n_{0_k}$ obstacles can also intersect the strip boundary, it can be visited at most $(n_j + n_{0_j})$ times. Therefore, the total number of obstacle visits for strip j is bounded by

$$\frac{(n_j + n_{0_j})\cdot(n_j + n_{0_j} + 1)}{2} + (n_j + n_{0_j})\cdot\sum_{k=0}^{j-1} n_{0_k} \quad (4)$$

Adding together the number of visits for all the strips, obtain the total worst-case number of obstacle visits n_v:

$$n_v = \sum_{j=1}^{r}\left[\frac{(n_j + n_{0_j})\cdot(n_j + n_{0_j} + 1)}{2} + (n_j + n_{0_j})\cdot\sum_{k=0}^{j-1} n_{0_k}\right]$$

$$= \sum_{j=1}^{r} \left[\frac{n_j \cdot (n_j + 1)}{2} + \frac{n_{0_j} \cdot (n_{0_j} + 1)}{2} \right.$$

$$\left. + (n_j + n_{0_j}) \cdot \sum_{k=0}^{j} n_{0_k} - n_{0_j}^2 \right]. \quad (5)$$

Since

$$\sum_{j=1}^{r} \frac{n_{0_j} \cdot (n_{0_j} + 1)}{2} \leq \frac{n_0 \cdot (n_0 + 1)}{2} \quad (6)$$

and

$$\sum_{j=1}^{r} (n_j + n_{0_j}) \cdot \sum_{k=0}^{j} n_{0_k} - n_{0_j}^2 \leq \sum_{j=1}^{r} n_j \cdot n_0 = n \cdot n_0 \quad (7)$$

the following upper bound appears:

Lemma 3: The total number n_v of obstacle visits required for acquiring the whole terrain using the Seed Spreader Strategy is bounded by

$$n_v \leq n \cdot n_0 + \sum_{j=0}^{r} \frac{n_j \cdot (n_j + 1)}{2}. \quad (8)$$

We turn now to the evaluation of the length of generated paths. Clearly, whether or not the whole operation can be accomplished without leaving the M-line, the length of the M-line $(A + r(w + A))$ will appear in our estimate. A more difficult case involving the M-line is that of a complex obstacle. Since the corresponding local graph is processed in the algorithm via Tarry's rule (see above), each graph edge, including each M-line segment, may be traversed at most twice. This means that the upper bound on the path length should include a component $2(A + r(w + A))$.

Another set of straight-line path components relates to the diversions to obstacles. A "round trip" estimate for each such diversion gives at most $2w$ or, for all obstacles, at most $2nw$.

Consider now the length of path segments due to the circumnavigation of obstacles. Let p_i be the perimeter of the ith obstacle, and let $p_{\max} = \max\{p_i\}$. Passing around a known obstacle produces a path component of length at most p_i. Processing unknown obstacles is more difficult. In the case of a simple obstacle that does not cross terrain borders, its circumnavigation/acquisition produces a path segment of length p_i, and then walking back to the point of departure L along the obstacle boundary can add at most $p_i/2$. For the same case, passing around a known obstacle requires a path of length at most $p_i/2$. Altogether, this amounts to $2p_i$. If, however, such a simple obstacle does intersect the terrain borders, its circumnavigation/acquisition can produce a longer path segment but no more than $2p_i$ (see obstacle O_5 on Fig. 5); walking back to L then adds at most another p_i. The total cost of acquiring the obstacle is thus at most $3p_i$.

Still another case relates to the acquisition of a complex obstacle. After the obstacle is first encountered along the M-line, its first circumnavigation produces at most a path p_i. At this point, information is sufficient to generate the corresponding local graph. As mentioned above, traversing all enclosed M-line segments related to this graph may add maneuvering around some parts of the obstacle, which adds at most $2p_i$. Altogether, this produces the length $3p_i$. The worst situation appears when the last two cases are combined (a complex obstacle crosses terrain borders). This may add at most p_i resulting in the worst case cost of $4p_i$ needed for acquisition of one obstacle.

Among the obstacle visits accounted for in (8), $n = \sum_{j=0}^{r} n_j$ visits are for the purpose of acquiring obstacles. The corresponding circumnavigation of obstacles will produce a path whose length is bounded by $\sum_{i=1}^{n} 4p_i$. The remaining visits accounted for in (8) are the result of already-acquired obstacles being passed by during diversions to unknown obstacles. The length of the path segment related to each such passaround visit is bounded by $2p_{\max}$, where the coefficient 2 accounts for the "round trip" during the diversion. Summing together the terms above, we arrive at the following statement:

Theorem 2: The total path P traversed by a robot under the Seed Spreader Strategy in an environment with n obstacles is bounded by

$$P = 2(A + r(w + A)) + 2nw + 4\sum_{i=1}^{n} p_i$$

$$+2nn_0 p_{\max} + p_{\max} \sum_{j=0}^{r} n_j(n_j - 1). \quad (9)$$

Assuming, for specificity, that $w = \min(R_v, B) = R_v$ and substituting $r = \frac{B}{R_v}$ in (9)

$$P = 2(A + B + rA) + 2nR_v + 4\sum_{i=1}^{n} p_i$$

$$+2nn_0 p_{\max} + p_{\max} \sum_{j=0}^{r} n_j(n_j - 1). \quad (10)$$

is produced. For easier interpretation, (10) can be presented as a sum of five terms

$$P = a_1 + a_2 + a_3 + a_4 + a_5 \quad (11)$$

where each term accounts for the following path segments: $a_1 = 2(A + B + rA)$ accounts for path segments due to motion along the M-line; $a_2 = 2nR_v$ accounts for straight-line path segments due to diversions from the M-line to obstacles (see, e.g., diversion to O_4 on Fig. 4(a)); $a_3 = 4\sum_{i=1}^{n} p_i$ accounts for first-time circumnavigation of obstacles (see, e.g., obstacle O_6 on Fig. 5); $a_4 = 2nn_0 p_{\max}$ accounts for passing around known obstacles that cross the M-line during diversions to incomplete obstacles; $a_5 = p_{\max} \sum_{j=0}^{r} n_j(n_j - 1)$ accounts for passing around known obstacles that lie fully inside the strips during diversions to incomplete obstacles—a case very similar to that of passing around obstacle O_3 on Fig. 4b on the way to the M-line after acquisition of obstacle O_4. Fig. 4(a) and b demonstrate a rather complex interplay between the components a_j in (11).

As the term a_5 indicates, the upper bound on the performance of this algorithm is quadratic in the number of obstacles and their perimeters. This is clearly worse than the upper bound of the Sightseer Strategy. In practice, however, the Seed Spreader Strategy can perform better than the Sightseer Strategy and vice versa, depending on the particular environment and the radius of vision or/and strip width (compare Figs. 3 and 4). In some general cases considered below, the upper bound of the Seed Spreader Strategy can be improved considerably so that it becomes quite competitive. Assume for simplicity that $w = R_v$.

First, observe that obstacle convexity helps rather insignificantly. If all the obstacles in the scene are convex, a diversion is needed only when one obstacle is occluded by another. Otherwise, any obstacle can be acquired from the M-line. Thus, in every diversion to an occluded obstacle, the robot circumnavigates and acquires at least one unvisited occluding obstacle. This means that the number of diversions is at most $\frac{1}{2}n$, which reduces the number of visits to each obstacle by half. Unfortunately, this does not affect the quadratic term in (10). Consider, however, the following cases, which take place either by chance or because of the availability of some additional information and subsequent guided choice of the M-line or/and the borders of the terrain.

Case 1—Acquisition from the M-line: Assume that no obstacles cross strip boundaries and all the obstacles can be acquired from the M-line without explicitly visiting them by way of diversion. (Imagine, for example, that obstacles O_3 and O_4 did not exist in Fig. 4(b)). Then, all terms in (10) and (11) except a_1 vanish, and the generated path becomes a constant that is twice the length of the M-line:

$$P = 2(A + B + rA). \qquad (12)$$

(One can see from Figs. 4 and 5 that the coefficient "2" will appear only in the most unusual cases). In other words, if one has reason to believe that all the obstacles in the scene can be acquired directly from the M-line without resorting to diversions, the Seed Spreader Strategy may be preferable to the Sightseer Strategy.

Case 2—Diversion Paths Do Not Cross Known Obstacles: The only quadratic term in the upper bound (10) and (11) a_5 appears because of the worst-case assumption that on the diversion to an unknown obstacle, the robot will encounter and will be forced to pass around all already-known obstacles. Such would be rare and rather artificial cases. In addition, paths of diversions might be planned more carefully to avoid encounters with known obstacles. Assuming that the diversion paths do not cross already-known obstacles, the terms a_4 and a_5 in (10) and (11) vanish, and the upper bound becomes linear in the obstacle perimeters (it gives, for example, a more realistic estimate for the length of generated path in Figs. 4 and 5):

$$P = 2(A + B + rA) + 2nR_v + 4\sum_{i=1}^{n} p_i. \qquad (13)$$

V. Conclusion

The problem of terrain acquisition has been formulated here as that of continuous path planning in an uncertain environment. Within this context, two somewhat different models have been considered: a) When the terrain is finite or infinite and all the obstacles are mutually visible; b) when the terrain is finite and the obstacles are not necessarily visible from each other. Two terrain acquisition algorithms have been described (one for each of the models), and their convergence properties and performance were analyzed.

Both algorithms, ad hoc as they might look at first glance, guarantee convergence and take full advantage of the input information available within the context of the corresponding models. Given the general character of these models, the obtained performance—linear lower bound for one algorithm and quadratic for the other algorithm—looks surprisingly good.

In general, the two algorithms address different problems and thus do not replace each other. However, for a wide range of situations, both algorithms might be applicable, and this raises a question—which one should be used? One answer, in terms of the upper bounds on the length of generated paths, gives a clear bias: For one algorithm, the Sightseer Strategy, the bound is linear, and for the second algorithm, the Seed Spreader Strategy, it is quadratic in the perimeters of the obstacles in the scene. In practice, however, these estimates may be of limited help unless additional information suggests that one is dealing with a worst-case situation or with one of the special cases considered above. If, for example, obstacle circumnavigation presents a significant part of the path, the Seed Spreader Strategy may successfully compete with the Sightseer Strategy. This effect can be seen in Figs. 3 and 4, where under the Sightseer Strategy the robot circumnavigates, as it must, all 11 (convex and nonconvex, polygonal and nonpolygonal) obstacles, whereas under the Seed Spreader Strategy, only three or even one obstacle need be circumnavigated.

From the implementation standpoint, the main question is: how realistic is the underlying model, and how computationally intensive are the described techniques? Compared with the commonly used formulation of the problem of terrain acquisition, the distinct feature of our model is that obstacles (objects, landmarks) in the robot environment do not have to be described analytically and thus can be of arbitrary shapes. We hope this removes an important stumbling block in the development of robot systems capable of dealing with uncertainties of real world applications. In addition, the presented performance criterion—the length of paths generated by the robot during its operation—seems to be more natural and realistic than the more common time-complexity measures that estimate algorithm complexity, e.g., as a function of the number of obstacle vertices.

The assumption that robot sensors provide coordinate information within a limited radius is directly tied to the existing technology of range finding and is thus quite realistic. The assumption of a point robot, though important from the theoretical standpoint, does not seem to present a serious limitation in real-world problems; consider, for example, a typical lawn mowing operation. The capability to "plot" the incoming data in the map is realizable via digitization schemes akin to those used in digital image processing.

A more stringent assumption, of course, is that the coordinates of the robot and obstacles are measured precisely—a constraint that is quite typical in the literature on motion planning algorithms. From the standpoint of path planning, the assumption can be softened somewhat, e.g., coordinate information is not crucial for following an obstacle boundary. As for map construction, recall that regular cartography is also impossible without precise registration of the cartographer. In other words, the question is not to what extent terrain acquisition algorithms can tolerate uncertainty in robot registration but rather by what means the uncertainty in registration can be made tolerable in a specific application. Although it is important, this issue goes beyond our topic; for possible approaches, refer, e.g., to [12] and [13].

Of all the operations that need to be done in real time, two might be suspected of being computationally intensive: scanning visible obstacle boundaries in the field of vision and computation of the shortest distances between obstacles. Both are unavoidable and not specific to the considered methods; both were shown to allow fast efficient computation [10], [11].

This work raises a variety of questions that require further research. These questions include the following:

- Two distinct "application" models have been presented here—one does not require complete covering of the terrain (it produces the Sightseer algorithm), whereas the other requires terrain coverage (it produces the Seed Spreader algorithm). Are there other distinct "application" models that are not reducible to ours?
- What are the lower bounds on the terrain acquisition problem for each "application" model?
- Can the presented upper bounds be improved?
- How can finite dimensions of the robot be incorporated into the algorithms? Although from a practical standpoint, this seems to be a secondary issue in the context of terrain acquisition, it is certainly an interesting algorithmic question. A related issue is the algorithmic effects of nonholonomic constraints on the robot motion.

References

[1] R. Brooks, "Visual map making for a mobile robot," in *Proc. 1985 IEEE Int. Conf. Robotics Automat.* (St. Louis), Mar. 1985.

[2] M. Turchen and A. Wong, "Low level learning for a mobile robot: Environmental model acquisition," in *Proc. 2nd Int. Conf. AI Appl.*, Dec. 1985.

[3] J. Oommen, S. Iyenger, N. Rao, and R. Kashyap, "Robot navigation in unknown terrains using learned visibility graph. Part I: The disjoint convex obstacle case," *IEEE J. Robotics Automat.*, vol. RA-3, Dec. 1987.

[4] N. Rao, S. Iyenger, C. Jorgensen, and C. Weisbin, "On terrain acquisition by a finite-sized mobile robot in plane," in *Proc. 1987 IEEE Int. Conf. Robotics Automat.* (Raleigh), Apr. 1987.

[5] J. T. Schwartz and M. Sharir, "On the piano movers' problem II: General techniques for computing topological properties of real algebraic manifolds," *Advances Appl. Math.*, pp. 298–351, Apr. 1983.

[6] J. Reif, "A survey on advances in the theory of computational robotics," in *Adaptive and Learning Systems* (K. Narendra, Ed.). New York: Plenum, 1986.

[7] V. Lumelsky and A. A. Stepanov, "Path planning strategies for a point mobile automaton moving amidst unknown obstacles of arbitrary shape," *Algorithmica*, vol. 2, pp. 403–430, 1987.

[8] A. Aho, J. Hopcroft, and J. Ullman, *The Design and Analysis of Computer Algorithms*. Reading, MA: Addison-Wesley, 1974.

[9] O. Ore, "Path problems," in *Theory of Graphs*, American Mathematical Society, Colloquium Publications, 1962, vol. XXXVIII, ch. 3.

[10] V. Lumelsky and T. Skewis, "A paradigm for incorporating vision in the robot navigation function," in *Proc. 1988 IEEE Int. Conf. Robotics Automat.* (Philadelphia), Apr. 1988.

[11] E. Gilbert, D. Johnston, and S. Keerthi, "A fast procedure for computing the distance between complex objects in three-dimensional space," *IEEE J. Robotics Automat.*, vol. 4, pp. 193–203, 1988.

[12] H. Moravec, "The Stanford cart and the CMU rover," *Proc. IEEE*, vol. 71, July 1983.

[13] A. Elfes, "Sonar-based real-world mapping and navigation," *IEEE J. Robotics Automat.*, vol. RA-3, pp. 249–265, June 1987.

Vladimir J. Lumelsky (SM'83) received the Ph.D. degree in applied mathematics from the Institute of Control Sciences (ICS), U.S.S.R. National Academy of Sciences, Moscow, in 1970.

From 1967 to 1975, he held the academic positions of Junior Researcher and Senior Research Fellow at ICS conducting research in pattern recognition, cluster analysis, factor analysis, and control systems. Concurrently, from 1970 to 1975, he served as Adjunct Professor at Moscow Institute of Radioelectronics and Automation. From 1976 to 1980, he was with Ford Motor Company Scientific Laboratories, Dearborn, MI, doing research in robotics, image processing, and industrial automation. From 1980 to 1985, he was on the research staff at General Electric Research Center, Schenectady, NY, doing research in robotics, pattern recognition, system engineering, and control theory. Since 1985, he has been on the faculty of the Department of Electrical Engineering, Yale University, New Haven, CT. His research interests are in robotics, image processing, pattern recognition, and control theory.

Dr. Lumelsky is a member of ACM and Robotics International of SME.

Snehasis Mukhopadhyay was born near Calcutta in India in 1964. He received the B.E. degree in electronics and telecommunications engineering from Jadavpur University, Calcutta, India, in 1985 and the M.E. degree in systems science and automation from the Indian Institute of Science, Bangalore, India in 1987.

Currently, he is a graduate student pursuing the Ph.D. degree in electrical engineering at Yale University, New Haven, CT. His research interests are in learning systems including neural networks and learning automata, possible applications of learning systems in control problems, learning for path planning in robotics, and application of learning algorithms in computer vision.

Kang Sun (S'85) was born in the People's Republic of China in 1961. He received the B.S. degree in computer science from Shanghai Jiao Tong University in 1982, the M.S. degree in computer science from Polytechnic Institute of New York, Brooklyn, in 1985, and the M.S. and M.Phil. degrees in electrical engineering from Yale University New Haven, CT, in 1987. He is currently completing the Ph.D. degree in electrical engineering at Yale University.

His research interests include robot motion planning, pattern recognition, algorithm design and analysis.

Mr. Sun is a student member of ACM, AMS, and SIAM.

About the authors

S. Sitharama Iyengar – chairman of the Computer Science Department and professor of computer science at Louisiana State University – has directed LSU's Robotics Research Laboratory since its inception in 1986. He has been actively involved with research in high-performance algorithms and data structures since receiving his PhD in 1974, and has directed over 10 PhD dissertations at LSU. He has served as principal investigator on research projects supported by the Office of Naval Research, the National Aeronautics and Space Administration, the National Science Foundation/Laser Program, the California Institute of Technology's Jet Propulsion Laboratory, the Department of Navy-NORDA, the Department of Energy (through Oak Ridge National Laboratory, Tennessee), the LEQFS-Board of Regents, and Apple Computers.

In addition to this two-volume tutorial, he has edited two other books and over 150 publications – including 85 archival journal papers in areas of high-performance parallel and distributed algorithms and data structure for image processing and pattern recognition, autonomous navigation, and distributed sensor networks. Iyengar was a visiting professor (fellow) at JPL, the Oak Ridge National Laboratory, and the Indian Institute of Science. He is also an Association for Computing Machinery national lecturer, a series editor for *Neuro Computing of Complex Systems*, and area editor for the *Journal of Computer Science and Information.* He has served as guest editor for the *IEEE Transactions on Software Engineering* (1988); *Computer* magazine (1989); the *IEEE Transactions On System, Man, and Cybernetics;* the *IEEE Transactions on Knowledge and Data Engineering*; and the *Journal of Computers and Electrical Engineering.* He was awarded the Phi Delta Kappa Research Award of Distinction at LSU in 1989, won the Best Teacher Award in 1978, and received the Williams Evans Fellowship from the University of Otago, New Zealand, in 1991.

Alberto Elfes – a robotics scientist at the Department of Computer Sciences, IBM T.J. Watson Research Center, Yorktown Heights, New York – is one of the principal investigators at that institution's Autonomous Robotics Laboratory. He obtained his E.Eng. degree in electronics engineering and the M.Sc. in computer science from the Instituto Tecnólogico de Aeronáutica, Brazil. From 1976 to 1982 he was on the faculty of ITA's Computer Science Department, where he performed research and directed dissertations in computer graphics, image processing for remote sensing applications, pattern recognition and computer-based medical diagnosis, and where he was awarded CNPq (National Council for Scientific and Technological Development) and FAPESP (Foundation for Research Support of the State of São Paulo) research awards and fellowships. He obtained his PhD in electrical and computer engineering at Carnegie Mellon University in 1989. His doctoral research was performed at the Mobile Robot Lab, CMU, where he concentrated on distributed architectures for mobile robot control, and stochastic approaches to robotic perception and navigation. From 1987 to 1989 he was a researcher with the Engineering Design Research Center and the Robotics Institute, CMU, where he worked on integrated environments for computer-aided design and on large-scale architectures for autonomous mobile robots. Since 1989 he has been with IBM's Autonomous Robotics Laboratory, where he is conducting research on the dynamic planning and control of robotic perception, and on real-time architectures for robotic control. He has published over 35 research papers in refereed journals, conference proceedings, and books in the areas of computer-based medical diagnosis, pattern recognition, computer-aided design, autonomous mobile robots, and computer vision.

IEEE Computer Society Press

Press Activities Board

Vice President: Barry W. Johnson, University of Virginia
James H. Aylor, University of Virginia
James Farrell, III, VLSI Technology Inc.
Michael Mulder, IBM Research Division
Guylaine Pollock, Sandia National Laboratories
Murali Varanasi, University of South Florida
Rao Vemuri, University of California, Davis
Ben Wah, University of Illinois
Staff Representative: True Seaborn, Publisher

Editorial Board

Editor-in-Chief: Rao Vemuri, University of California, Davis
Joydeep Ghosh, University of Texas, Austin
Uma G. Gupta, University of Central Florida
A.R. Hurson, Pennsylvania State University
Krishna Kavi, University of Texas, Arlington
Frederick E. Petry, Tulane University
Dhiraj K. Pradhan, University of Massachusetts
Charles Richter, MCC
Sol Shatz, University of Illinois, Chicago
Ajit Singh, Siemens Corporate Research
Pradip K. Srimani, Colorado State University
Murali R. Varanasi, University of South Florida
Staff Representative: Henry Ayling, Editorial Director

Press Staff

T. Michael Elliott, Executive Director
True Seaborn, Publisher

Henry Ayling, Editorial Director
Catherine Harris, Production Editor
Anne MacCallum, Production Editor
Lisa O'Conner, Production Editor
Robert Werner, Production Editor
Penny Storms, Editorial Production Assistant
Edna Straub, Editorial Production Assistant

Douglas Combs, Assistant Publisher
Thomas Fink, Advertising/Promotions Manager
Frieda Koester, Marketing/Customer Service Manager
Susan Roarke, Customer Service/Order Processing Supervisor
Becky Jacobs, Marketing/Customer Service Admin. Asst.
Beverly Anthony, Order Processor

Offices of the IEEE Computer Society

Headquarters Office
1730 Massachusetts Avenue, N.W.
Washington, DC 20036-1903
Phone: (202) 371-0101 — Fax: (202) 728-9614

Publications Office
P.O. Box 3014
10662 Los Vaqueros Circle
Los Alamitos, CA 90720-1264
Membership and General Information: (714) 821-8380
Publication Orders: (800) 272-6657 — Fax: (714) 821-4010

European Office
13, avenue de l'Aquilon
B-1200 Brussels, BELGIUM
Phone: 32-2-770-21-98 — Fax: 32-3-770-85-05

Asian Office
Ooshima Building
2-19-1 Minami-Aoyama, Minato-ku
Tokyo 107, JAPAN
Phone: 81-3-408-3118 — Fax: 81-3-408-3553

IEEE Computer Society

IEEE Computer Society Press Publications

Monographs: A monograph is an authored book consisting of 100-percent original material.
Tutorials: A tutorial is a collection of original materials prepared by the editors, and reprints of the best articles published in a subject area. Tutorials must contain at least five percent of original material (although we recommend 15 to 20 percent of original material).
Reprint collections: A reprint collection contains reprints (divided into sections) with a preface, table of contents, and section introductions discussing the reprints and why they were selected. Collections contain less than five percent of original material.
Technology series: Each technology series is a brief reprint collection — approximately 126-136 pages and containing 12 to 13 papers, each paper focusing on a subset of a specific discipline, such as networks, architecture, software, or robotics.
Submission of proposals: For guidelines on preparing CS Press books, write the Editorial Director, IEEE Computer Society Press, PO Box 3014, 10662 Los Vaqueros Circle, Los Alamitos, CA 90720-1264, or telephone (714) 821-8380.

Purpose

The IEEE Computer Society advances the theory and practice of computer science and engineering, promotes the exchange of technical information among 100,000 members worldwide, and provides a wide range of services to members and nonmembers.

Membership

All members receive the acclaimed monthly magazine *Computer*, discounts, and opportunities to serve (all activities are led by volunteer members). Membership is open to all IEEE members, affiliate society members, and others seriously interested in the computer field.

Publications and Activities

Computer **magazine:** An authoritative, easy-to-read magazine containing tutorials and in-depth articles on topics across the computer field, plus news, conference reports, book reviews, calendars, calls for papers, interviews, and new products.

Periodicals: The society publishes six magazines and five research transactions. For more details, refer to our membership application or request information as noted above.

Conference proceedings, tutorial texts, and standards documents: The IEEE Computer Society Press publishes more than 100 titles every year.

Standards working groups: Over 100 of these groups produce IEEE standards used throughout the industrial world.

Technical committees: Over 30 TCs publish newsletters, provide interaction with peers in specialty areas, and directly influence standards, conferences, and education.

Conferences/Education: The society holds about 100 conferences each year and sponsors many educational activities, including computing science accreditation.

Chapters: Regular and student chapters worldwide provide the opportunity to interact with colleagues, hear technical experts, and serve the local professional community.

Other IEEE Computer Society Press Titles

MONOGRAPHS

Analyzing Computer Architectures
Written by Jerome C. Huck and Michael J. Flynn
(ISBN 0-8186-8857-2); 206 pages

Branch Strategy Taxonomy and Performance Models
Written by Harvey G. Cragon
(ISBN 0-8186-9111-5); 150 pages

Desktop Publishing for the Writer: Designing, Writing, and Developing
Written by Richard Ziegfeld and John Tarp
(ISBN 0-8186-8840-8); 380 pages

Digital Image Warping
Written by George Wolberg
(ISBN 0-8186-8944-7); 340 pages

Integrating Design and Test — CAE Tools for ATE Programming
Written by Kenneth P. Parker
(ISBN 0-8186-8788-6); 160 pages

JSP and JSD — The Jackson Approach to Software Development (Second Edition)
Written by John R. Cameron
(ISBN 0-8186-8858-0); 560 pages

National Computer Policies
Written by Ben G. Matley and Thomas A. McDannold
(ISBN 0-8186-8784-3); 192 pages

Physical Level Interfaces and Protocols
Written by Uyless Black
(ISBN 0-8186-8824-2); 240 pages

Protecting Your Proprietary Rights in Computer and High-Technology Industries
Written by Tobey B. Marzouk, Esq.
(ISBN 0-8186-8754-1); 224 pages

X.25 and Related Protocols
Written by Uyless Black
(ISBN 0-8186-8976-5); 304 pages

TUTORIALS

Advanced Computer Architecture
Edited by Dharma P. Agrawal
(ISBN 0-8186-0667-3); 400 pages

Advances in Distributed System Reliability
Edited by Suresh Rai and Dharma P. Agrawal
(ISBN 0-8186-8907-2); 352 pages

Autonomous Mobile Robots: Perception, Mapping and Navigation — Volume 1
Edited by S. S. Iyengar and A. Elfes
(ISBN 0-8186-9116-6); 425 pages

Autonomous Mobile Robots: Planning, Control, and Architecture — Volume 2
Edited by S. S. Iyengar and A. Elfes
(ISBN 0-8186-9018-6); 425 pages

Broadband Switching: Architectures, Protocols, Design, and Analysis
Edited by C. Dhas, V. K. Konangi, and M. Sreetharan
(ISBN 0-8186-8926-9); 528 pages

Computer and Network Security
Edited by M. D. Abrams and H. J. Podell
(ISBN 0-8186-0756-4); 448 pages

Computer Architecture
Edited by D. D. Gajski, V. M. Milutinovic, H. J. Siegel, and B. P. Furht
(ISBN 0-8186-0704-1); 602 pages

Computer Arithmetic I
Edited by Earl E. Swartzlander, Jr.
(ISBN 0-8186-8931-5); 398 pages

Computer Arithmetic II
Edited by Earl E. Swartzlander, Jr.
(ISBN 0-8186-8945-5); 412 pages

Computer Communications: Architectures, Protocols, and Standards (Second Edition)
Edited by William Stallings
(ISBN 0-8186-0790-4); 448 pages

Computer Graphics Hardware: Image Generation and Display
Edited by H. K. Reghbati and A. Y. C. Lee
(ISBN 0-8186-0753-X); 384 pages

Computer Graphics: Image Synthesis
Edited by Kenneth Joy, Nelson Max, Charles Grant, and Lansing Hatfield
(ISBN 0-8186-8854-8); 380 pages

Computer Vision: Principles
Edited by Rangachar Kasturi and Ramesh Jain
(ISBN 0-8186-9102-6); 700 pages

Computer Vision: Advances and Applications
Edited by Rangachar Kasturi and Ramesh Jain
(ISBN 0-8186-9103-4); 720 pages

Digital Image Processing (Second Edition)
Edited by Rama Chellappa
(ISBN 0-8186-2362-4); 400 pages

Digital Private Branch Exchanges (PBXs)
Edited by Edwin Coover
(ISBN 0-8186-0829-3); 394 pages

Distributed Computing Network Reliability
Edited by Suresh Rai and Dharma P. Agrawal
(ISBN 0-8186-8908-0); 357 pages

Distributed–Software Engineering
Edited by Sol Shatz and Jia-Ping Wang
(ISBN 0-8186-8856-4); 294 pages

Domain Analysis and Software Systems Modeling
Edited by Ruben-Prieto Diaz and Guillermo Arango
(ISBN 0-8186-8996-X); 312 pages

DSP–Based Testing of Analog and Mixed Signal Circuits
Edited by Matthew Mahoney
(ISBN 0-8186-0785-8); 272 pages

Fault Tolerant Computing
Edited by V. Nelson and B. Carroll
(ISBN 0-8186-8677-4); 428 pages

Formal Verification of Hardware Design
Edited by Michael Yoeli
(ISBN 0-8186-9017-8); 340 pages

Hard Real-Time Systems
Edited by J. A. Stankovic and K. Ramamritham
(ISBN 0-8186-0819-6); 624 pages

Integrated Services Digital Networks (ISDN) (Second Edition)
Edited by William Stallings
(ISBN 0-8186-0823-4); 406 pages

For further information call 1-800-CS-BOOKS or write:

IEEE Computer Society Press, 10662 Los Vaqueros Circle, PO Box 3014,
Los Alamitos, California 90720-1264, USA

IEEE Computer Society, 13, avenue de l'Aquilon,
B-1200 Brussels, BELGIUM

IEEE Computer Society, Ooshima Building, 2-19-1 Minami-Aoyama,
Minato-ku, Tokyo 107, JAPAN

Local Network Technology (Third Edition)
Edited by William Stallings
(ISBN 0-8186-0825-0); 512 pages

Microprogramming and Firmware Engineering
Edited by V. M. Milutinovic
(ISBN 0-8186-0839-0); 416 pages

Modeling and Control of Automated Manufacturing Systems
Edited by Alan A. Desrochers
(ISBN 0-8186-8916-1); 384 pages

Nearest Neighbor Pattern Classification Techniques
Edited by Belur V. Dasarathy
(ISBN 0-8186-8930-7); 464 pages

New Paradigms for Software Development
Edited by William Agresti
(ISBN 0-8186-0707-6); 304 pages

Object-Oriented Computing, Volume 1: Concepts
Edited by Gerald E. Petersen
(ISBN 0-8186-0821-8); 214 pages

Object-Oriented Computing, Volume 2: Implementations
Edited by Gerald E. Petersen
(ISBN 0-8186-0822-6); 324 pages

Parallel Architectures for Database Systems
Edited by A. R. Hurson, L. L. Miller, and S. H. Pakzad
(ISBN 0-8186-8838-6); 478 pages

Reduced Instruction Set Computers (RISC) (Second Edition)
Edited by William Stallings
(ISBN 0-8186-8943-9); 448 pages

Software Engineering Project Management
Edited by Richard H. Thayer
(ISBN 0-8186-0751-3); 512 pages

Software Maintenance and Computers
Edited by David H. Longstreet
(ISBN 0-8186-8898-X); 304 pages

Software Management (Third Edition)
Edited by Donald J. Reifer
(ISBN 0-8186-0678-9); 526 pages

Software Quality Assurance: A Practical Approach
Edited by T.S. Chow
(ISBN 0-8186-0569-3); 506 pages

Software Reuse — Emerging Technology
Edited by Will Tracz
(ISBN 0-8186-0846-3); 400 pages

Software Risk Management
Edited by Barry W. Boehm
(ISBN 0-8186-8906-4); 508 pages

Standards, Guidelines and Examples on System and Software Requirements Engineering
Edited by Merlin Dorfman and Richard H. Thayer
(ISBN 0-8186-8922-6); 626 pages

System and Software Requirements Engineering
Edited by Richard H. Thayer and Merlin Dorfman
(ISBN 0-8186-8921-8); 740 pages

Test Access Port and Boundary-Scan Architecture
Edited by Colin M. Maunder and Rodham E. Tulloss
(ISBN 0-8186-9070-4); 400 pages

Test Generation for VLSI Chips
Edited by V. D. Agrawal and S. C. Seth
(ISBN 0-8186-8786-X); 416 pages

Visual Programming Environments: Paradigms and Systems
Edited by Ephraim Glinert
(ISBN 0-8186-8973-0); 680 pages

Visual Programming Environments: Applications and Issues
Edited by Ephraim Glinert
(ISBN 0-8186-8974-9); 704 pages

Visualization in Scientific Computing
Edited by G. M. Nielson, B. Shriver, and L. Rosenblum
(ISBN 0-8186-8979-X); 304 pages

VLSI Testing and Validation Techniques
Edited by H. K. Reghbati
(ISBN 0-8186-0668-1); 616 pages

Volume Visualization
Edited by Arie Kaufman
(ISBN 0-8186-9020-8); 494 pages

REPRINT COLLECTIONS

Distributed Computing Systems: Concepts and Structures
Edited by A. L. Ananda and B. Srinivasan
(ISBN 0-8186-8975-0); 416 pages

Expert Systems: A Software Methodology for Modern Applications
Edited by Peter G. Raeth
(ISBN 0-8186-8904-8); 476 pages

Milestones in Software Evolution
Edited by Paul W. Oman and Ted G. Lewis
(ISBN 0-8186-9033-X); 332 pages

Object-Oriented Databases
Edited by Ez Nahouraii and Fred Petry
(ISBN 0-8186-8929-3); 256 pages

Validating and Verifying Knowledge-Based Systems
Edited by Uma G. Gupta
(ISBN 0-8186-8995-1); 400 pages

ARTIFICIAL NEURAL NETWORKS TECHNOLOGY SERIES

Artificial Neural Networks — Concept Learning
Edited by Joachim Diederich
(ISBN 0-8186-2015-3); 160 pages

Artificial Neural Networks — Electronic Implementation
Edited by Nelson Morgan
(ISBN 0-8186-2029-3); 144 pages

Artificial Neural Networks — Theoretical Concepts
Edited by V. Vemuri
(ISBN 0-8186-0855-2); 160 pages

SOFTWARE TECHNOLOGY SERIES

Computer-Aided Software Engineering (CASE)
Edited by E. J. Chikofsky
(ISBN 0-8186-1917-1); 110 pages

Software Reliability Models: Theoretical Development, Evaluation, and Applications
Edited by Yashwant K. Malaiya and Pradip K. Srimani
(ISBN 0-8186-2110-9); 136 pages

MATHEMATICS TECHNOLOGY SERIES

Computer Algorithms
Edited by Jun-ichi Aoe
(ISBN 0-8186-2123-0); 154 pages

Multiple-Valued Logic in VLSI Design
Edited by Jon T. Butler
(ISBN 0-8186-2127-3); 128 pages

COMMUNICATIONS TECHNOLOGY SERIES

Multicast Communication in Distributed Systems
Edited by Mustaque Ahamad
(ISBN 0-8186-1970-8); 110 pages

ROBOTICS TECHNOLOGY SERIES

Multirobot Systems
Edited by Rajiv Mehrotra and Murali R. Varanasi
(ISBN 0-8186-1977-5); 122 pages